P9-CKX-613

Beginning Algebra

Beginning Algebra

Sixth Edition

Elayn Martin-Gay

University of New Orleans

PEARSON

Boston Columbus Indianapolis New York San Francisco Upper Saddle River
Amsterdam Cape Town Dubai London Madrid Milan Munich Paris Montréal Toronto
Delhi Mexico City São Paulo Sydney Hong Kong Seoul Singapore Taipei Tokyo

Editorial Director, Mathematics: *Christine Hoag*
Acquisitions Editor: *Mary Beckwith*
Executive Project Manager: *Kari Heen*
Project Editor: *Christine Whitlock*
Editorial Assistant: *Matthew Summers*
Editor-in-Chief–Development: *Carol Trueheart*
Development Editor: *Dawn Nuttall*
Senior Managing Editor: *Karen Wernholm*
Production Project Manager: *Patty Bergin*
Senior Design Specialist: *Heather Scott*
Associate Director of Design, USHE North and West: *Andrea Nix*
Digital Assets Manager: *Marianne Groth*
Supplements Production Project Manager: *Katherine Roz*
Executive Manager, Course Production: *Peter Silvia*
Media Producer: *Audra Walsh*
Executive Marketing Manager: *Michelle Renda*
Marketing Assistant: *Susan Mai*
Senior Author Support/Technology Specialist: *Joe Vetere*
Senior Media Buyer: *Ginny Michaud*
Permissions Project Supervisor: *Michael Joyce*
Procurement Specialist: *Linda Cox*
Production Management, Interior Design, Composition, and Answer Art: *Integra*
Text Art: *Scientific Illustrators*
Cover Design and Image: *Tamara Newnam*

Library of Congress Cataloging-in-Publication Data

Martin-Gay, K. Elayn
Beginning algebra / Elayn Martin-Gay.—6th ed.
p. cm.
ISBN-13: 978-0-321-78491-9
ISBN-10: 0-321-78491-X
1. Algebra—Textbooks. I. Title.
QA152.3.M353 2013
512.9—dc23
2011013310

2 3 4 5 6 7 8 9 10—CRK—15 14 13 12

www.pearsonhighered.com

ISBN-10: 0-321-78491-X
ISBN-13: 978-0-321-78491-9

This book is dedicated to students everywhere—and we should all be students. After all, is there anyone among us who really knows too much? Take that hint and continue to learn something new every day of your life.

Best of wishes from a fellow student: Elayn Martin-Gay

Contents

CHAPTER 4 SOLVING SYSTEMS OF LINEAR EQUATIONS AND INEQUALITIES 254

CHAPTER 5 EXPONENTS AND POLYNOMIALS 306

CHAPTER 6 FACTORING POLYNOMIALS 366

CHAPTER 7 RATIONAL EXPRESSIONS 432

Student Resources 628

These resources, located in the back of the text, give you a variety of tools conveniently located in one place to help you succeed in math.

A New Tool to Help You Succeed

Introducing Martin-Gay's New Student Organizer

The new **Student Organizer** guides you through three important parts of studying effectively–note-taking, practice, and homework.

It is designed to help you organize your learning materials and develop the study habits you need to be successful. The Student Organizer includes:

- How to prepare for class
- Space to take class-notes
- Step-by-step worked examples
- Your Turn exercises (modeled after the examples)
- Answers to the Your Turn exercises as well as worked-out solutions via references to the Martin-Gay text and videos
- Helpful hints and directions for completing homework assignments

A flexible design allows instructors to assign any or all parts of the Student Organizer.

The Student Organizer is available in a loose-leaf, notebook-ready format. It is also available for download in MyMathLab.

For more information, please go to

www.pearsonhighered.com/martingay

www.mypearsonstore.com
(search Martin-Gay, Beginning Algebra, Sixth Edition)
your Martin-Gay MyMathLab® course

Martin-Gay Video Resources to Help You Succeed

Interactive DVD Lecture Series

Active Learning at Your Pace

Designed for use on your computer or DVD player, these interactive videos include a 15–20 minute lecture for every section in the text as well as Concept Checks, Study Skills Builders, and a Practice Final Exam.

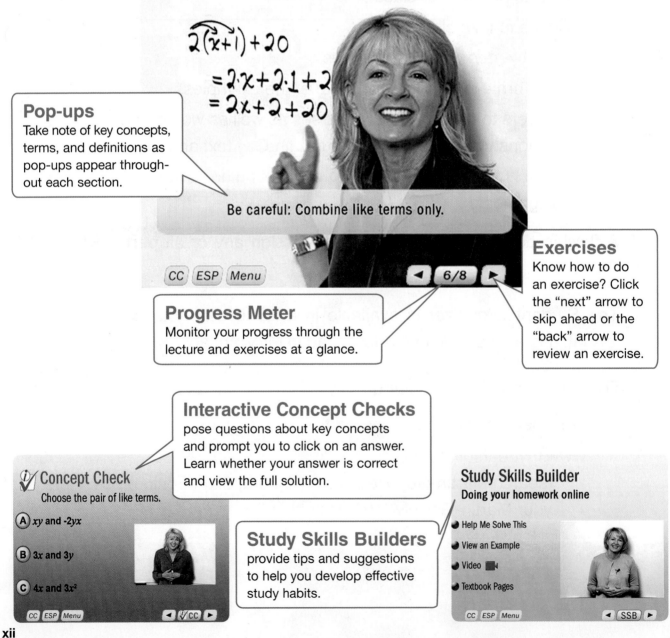

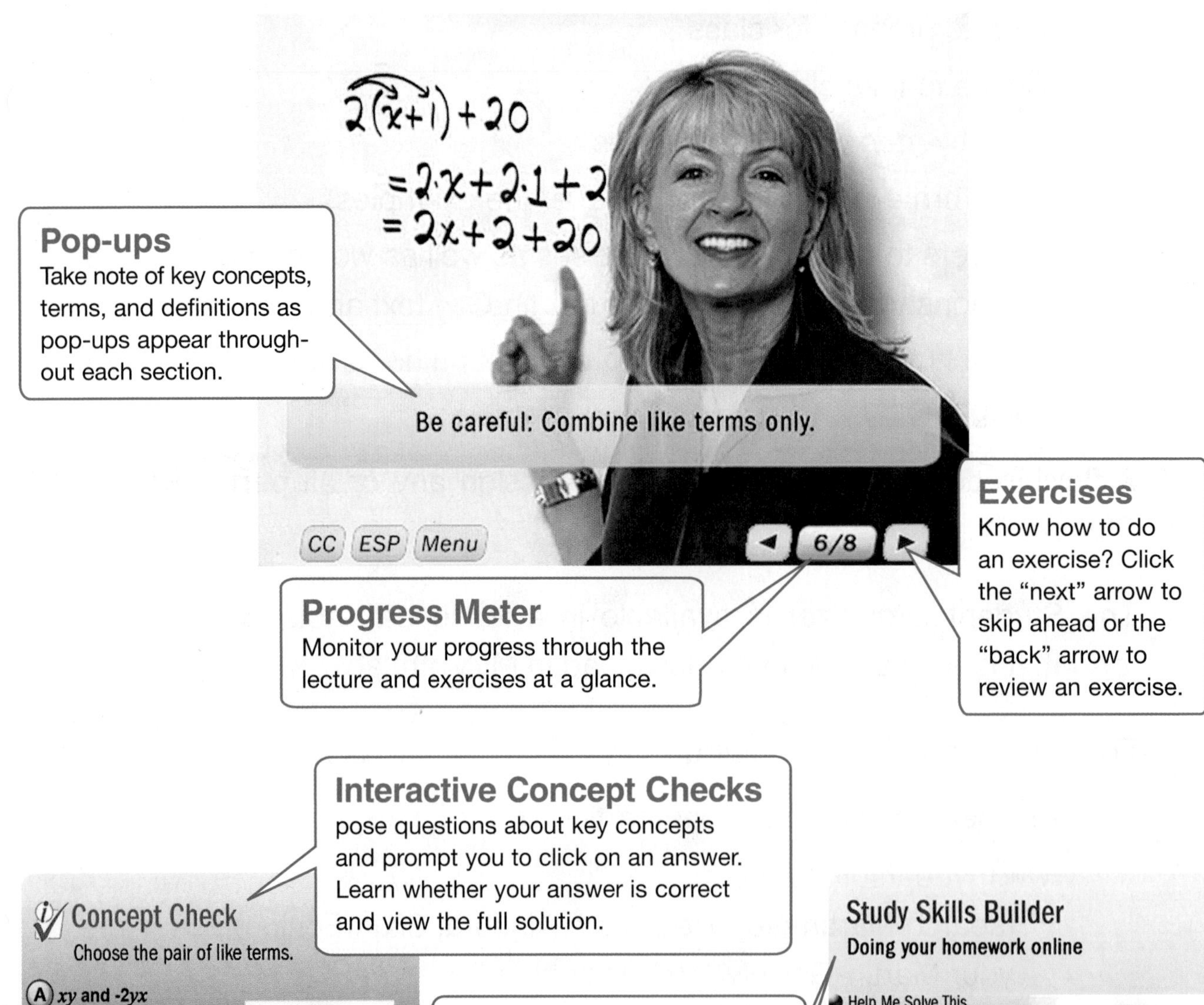

$2(x+1)+20$
$=2\cdot x+2\cdot 1+2$
$=2x+2+20$
Pop-ups
Take note of key concepts, terms, and definitions as pop-ups appear throughout each section.
Be careful: Combine like terms only.
CC ESP Menu
6/8
Exercises
Know how to do an exercise? Click the "next" arrow to skip ahead or the "back" arrow to review an exercise.
Progress Meter
Monitor your progress through the lecture and exercises at a glance.
Interactive Concept Checks
pose questions about key concepts and prompt you to click on an answer. Learn whether your answer is correct and view the full solution.
Concept Check
Choose the pair of like terms.
A xy and $-2yx$
B $3x$ and $3y$
C $4x$ and $3x^2$
CC ESP Menu
CC
Study Skills Builders
provide tips and suggestions to help you develop effective study habits.
Study Skills Builder
Doing your homework online
Help Me Solve This
View an Example
Video
Textbook Pages
CC ESP Menu
SSB

Chapter Test Prep Videos

Step-by-step solutions on video for all chapter tests exercises from the text. Available via:

- Interactive DVD Lecture Series
- MyMathLab®
- YouTube

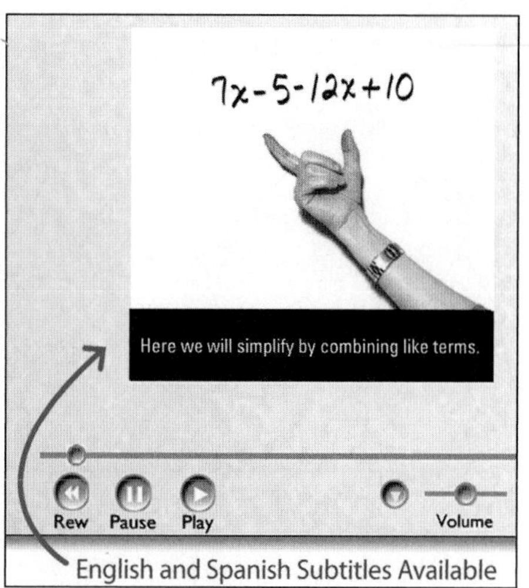

AlgebraPrep Apps for the iPhone™ and iPod Touch®

Your 24/7 Algebra Tutor–Anytime, Anywhere!

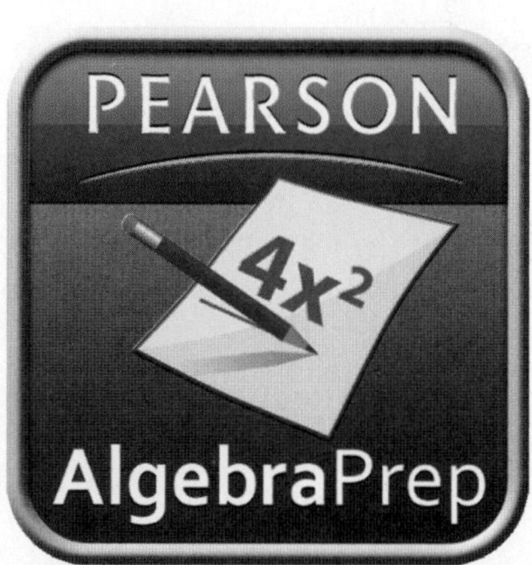

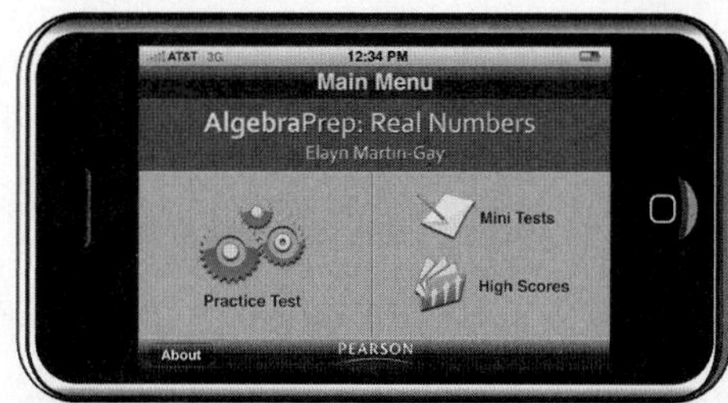

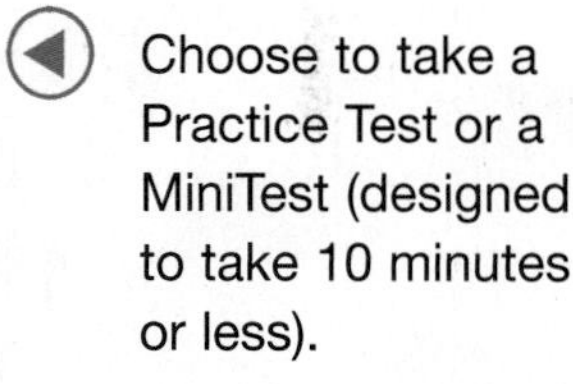

Choose to take a Practice Test or a MiniTest (designed to take 10 minutes or less).

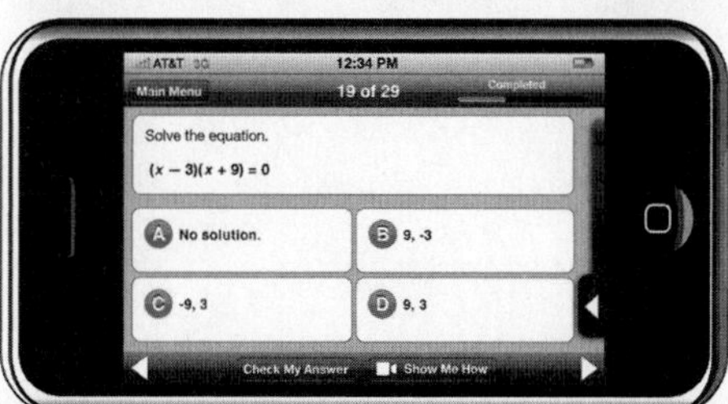

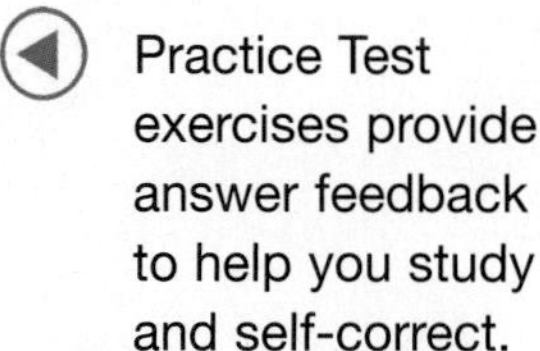

Practice Test exercises provide answer feedback to help you study and self-correct.

Step-by-step video solutions give you the guidance of an expert tutor whenever you need help.

Preface

***Beginning Algebra,* Sixth Edition** was written to provide a solid foundation in algebra for students who might not have previous experience in algebra. Specific care was taken to make sure students have the most up-to-date relevant text preparation for their next mathematics course or for nonmathematical courses that require an understanding of algebraic fundamentals. I have tried to achieve this by writing a user-friendly text that is keyed to objectives and contains many worked-out examples. As suggested by AMATYC and the NCTM Standards (plus Addenda), real-life and real-data applications, data interpretation, conceptual understanding, problem solving, writing, cooperative learning, appropriate use of technology, mental mathematics, number sense, estimation, critical thinking, and geometric concepts are emphasized and integrated throughout the book.

The many factors that contributed to the success of the previous editions have been retained. In preparing the Sixth Edition, I considered comments and suggestions of colleagues, students, and many users of the prior edition throughout the country.

What's New in the Sixth Edition?

- **The Martin-Gay Program** has been revised and enhanced with a new design in the text and MyMathlab to actively encourage students to use the text, video program, and Student Organizer as an integrated learning system.
- The **Student Organizer** is designed by me to help students develop the study habits they need to be successful. This Organizer guides students through the three main components of studying effectively—note-taking, practice, and homework—and helps them develop the habits that will enable them to succeed in future courses. The Student Organizer can be packaged with the text in loose-leaf, notebook-ready format and is also available for download in MyMathLab.
- **New Vocabulary, Readiness & Video Check** questions have been added to the start of every section exercise set. These exercises quickly check a student's understanding of new vocabulary words. The **readiness** exercises center on a student's understanding of a concept that is necessary in order to continue with the exercise set. **New video check questions for the Martin-Gay Interactive Lecture videos** are now included in every section for each learning objective. **These exercises are all available for assignment in MyMathLab** and are a great way to assess whether students have viewed and understood the key concepts presented in the videos.
- **Interactive DVD Lecture Series,** featuring your text author (Elayn Martin-Gay), provides students with active learning at their own pace. The videos offer the following resources and more:

  A complete lecture for each section of the text highlights key examples and exercises from the text. New "pop-ups" reinforce key terms, definitions, and concepts.

  An interface with menu navigation features allows students to quickly find and focus on the examples and exercises they need to review.

  Interactive Concept Check exercises measure students' understanding of key concepts and common trouble spots.

  The Interactive DVD Lecture Series also includes the following resources for test prep:

  The Practice Final Exam helps students prepare for an end-of-course final. Students can watch full video solutions to each exercise.

The Chapter Test Prep Videos help students during their most teachable moment–when they are preparing for a test. This innovation provides step-by-step solutions for the exercises found in each Chapter Test. The videos are captioned in English and Spanish. For the Sixth Edition, the chapter test prep videos are also available on YouTube™.

- **The Martin-Gay MyMathLab course** has been updated and revised providing more exercise coverage, including assignable video check questions, and an expanded video program. There are section lecture videos for every section, students can also access at the specific objective level, and there are an increased number of watch clips at the exercise level to help students while doing homework in MathXL. Suggested homework assignments have been premade for assignment at the instructor's discretion.
- **New MyMathLab Ready to Go courses** (access code required) provide students with all the same great MyMathLab features that you're used to but make it easier for instructors to get started. Each course includes preassigned homeworks and quizzes to make creating your course even simpler. Ask your Pearson representative about the details for this particular course or to see a copy of this course.
- **New Student Resources** section located in the back of the text, gives students a variety of tools that are conveniently located in one place to help them achieve success in math.
  - **Study Skills Builders** give students tips and suggestions on successful study habits and help them take responsibility for their learning. Assignable exercises check students' progress in improving their skills.
  - The **Bigger Picture—Study Guide Outline** covers key concepts of the course—simplifying expressions and solving equations and inequalities—to help students transition from thinking section-by-section to thinking about how the material they are learning fits into mathematics as a whole. This outline provides a model for students on how to organize and develop their own study guide.
  - The **Practice Final Exam** helps students prepare for the end-of-the-course exam. Students can also watch the step-by-step solutions to all the Practice Final Exam exercises on the new Interactive DVD Lecture Series and in MyMathLab.
  - **Answers to Selected Exercises** allows students to check their answers for all odd-numbered section exercises.
- **New guided application exercises** appear in many sections throughout the text, beginning with Section 2.5. These applications prompt students on how to set up the problem and get started with the solution process. These guided exercises will help students prepare to solve application exercises on their own.
- **Enhanced emphasis on Study Skills** helps students develop good study habits and makes it more convenient for instructors to incorporate or assign study skills in their courses. The following changes have been made in the Sixth Edition:

  Section 1.1, Tips for Success in Mathematics, has been updated to include helpful hints for doing homework online in MyMathLab. Exercises pertaining to doing homework online in MyMathLab are now included in the exercise set for 1.1.

  The Study Skills Builders, formerly located at the end of select exercise sets, are now included in the new **Student Resources** section at the back of the book and organized by topic for ease of assignment. This section now also includes new Study Skills Builders on doing homework online in MyMathLab.
- All exercise sets have been reviewed and updated to ensure that even- and odd-numbered exercises are paired.

Key Pedagogical Features

The following key features have been retained and/or updated for the Sixth Edition of the text:

Problem Solving Process This is formally introduced in Chapter 2 with a four-step process that is integrated throughout the text. The four steps are **Understand, Translate, Solve,** and **Interpret.** The repeated use of these steps in a variety of examples shows their wide applicability. Reinforcing the steps can increase students' comfort level and confidence in tackling problems.

Exercise Sets Revised and Updated The exercise sets have been carefully examined and extensively revised. Special focus was placed on making sure that even- and odd-numbered exercises are paired.

Examples Detailed step-by-step examples were added, deleted, replaced, or updated as needed. Many of these reflect real life. Additional instructional support is provided in the annotated examples.

Practice Exercises Throughout the text, each worked-out example has a parallel Practice Exercise. These invite students to be actively involved in the learning process. Students should try each Practice Exercise after finishing the corresponding example. Learning by doing will help students grasp ideas before moving on to other concepts. Answers to the Practice Exercises are provided in the back of the text.

Helpful Hints Helpful Hints contain practical advice on applying mathematical concepts. Strategically placed where students are most likely to need immediate reinforcement, Helpful Hints help students avoid common trouble areas and mistakes.

Concept Checks This feature allows students to gauge their grasp of an idea as it is being presented in the text. Concept Checks stress conceptual understanding at the point-of-use and help suppress misconceived notions before they start. Answers appear at the bottom of the page. Exercises related to Concept Checks are included in the exercise sets.

Mixed Practice Exercises Found in the section exercise sets, these require students to determine the problem type and strategy needed to solve it just as they would need to do on a test.

Integrated Reviews A unique, mid-chapter exercise set that helps students assimilate new skills and concepts that they have learned separately over several sections. These reviews provide yet another opportunity for students to work with "mixed" exercises as they master the topics.

Vocabulary Check Provides an opportunity for students to become more familiar with the use of mathematical terms as they strengthen their verbal skills. These appear at the end of each chapter before the Chapter Highlights. Vocabulary and Readiness exercises provide practice at the section level.

Chapter Highlights Found at the end of every chapter, these contain key definitions and concepts with examples to help students understand and retain what they have learned and help them organize their notes and study for tests.

Chapter Review The end of every chapter contains a comprehensive review of topics introduced in the chapter. The Chapter Review offers exercises keyed to every section in the chapter, as well as Mixed Review exercises that are not keyed to sections.

Chapter Test and Chapter Test Prep Video The Chapter Test is structured to include those problems that involve common student errors. The **Chapter Test Prep Videos** give students instant author access to a step-by-step video solution of each exercise in the Chapter Test.

Cumulative Review Follows every chapter in the text (except Chapter 1). Each odd-numbered exercise contained in the Cumulative Review is an earlier worked example in the text that is referenced in the back of the book along with the answer.

Writing Exercises ╲ These exercises occur in almost every exercise set and require students to provide a written response to explain concepts or justify their thinking.

Applications Real-world and real-data applications have been thoroughly updated and many new applications are included. These exercises occur in almost every exercise set and show the relevance of mathematics and help students gradually, and continuously develop their problem solving skills.

Review Exercises These exercises occur in each exercise set (except in Chapter 1) and are keyed to earlier sections. They review concepts learned earlier in the text that will be needed in the next section or chapter.

Exercise Set Resource Icons Located at the opening of each exercise set, these icons remind students of the resources available for extra practice and support:

See Student Resource descriptions page xix for details on the individual resources available.

Exercise Icons These icons facilitate the assignment of specialized exercises and let students know what resources can support them.

- ▶ Video icon: exercise worked on the Interactive DVD Lecture Series and in MyMathLab.
- △ Triangle icon: identifies exercises involving geometric concepts.
- ╲ Pencil icon: indicates a written response is needed.
- ▦ Calculator icon: optional exercises intended to be solved using a scientific or graphing calculator.

Optional: Calculator Exploration Boxes and Calculator Exercises The optional Calculator Explorations provide key strokes and exercises at appropriate points to give an opportunity for students to become familiar with these tools. Section exercises that are best completed by using a calculator are identified by ▦ for ease of assignment.

Student and Instructor Resources

STUDENT RESOURCES

<table>
<tr>
<td>Student Organizer
Guides students through the 3 main components of studying effectively–notetaking, practice, and homework.
The organizer includes before-class preparation exercises, notetaking pages in a 2-column format for use in class, and examples paired with exercises for practice for each section. It is 3-hole punched. Also available in MyMathLab.</td>
<td>Student Solutions Manual
Provides complete worked-out solutions to<ul><li>the odd numbered section exercises; all Practice Exercises; all exercises in the Integrated Reviews, Chapter Reviews, Chapter Tests, and Cumulative Reviews</li></ul></td>
</tr>
<tr>
<td>Interactive DVD Lecture Series
Provides students with active learning at their pace. The videos offer:<ul><li>A complete lecture for each text section. The new interface allows easy navigation to examples and exercises students need to review</li><li>Interactive Concept Check exercises</li><li>Study Skills Builders</li><li>Practice Final Exam</li><li>Chapter Test Prep Videos</li></ul></td>
<td>Chapter Test Prep Videos<ul><li>Step-by-step solutions to every exercise in each Chapter Practice Test.</li><li>Available in MyMathLab® and on YouTube, and in the Interactive DVD Lecture Series.</li></ul></td>
</tr>
</table>

INSTRUCTOR RESOURCES

<table>
<tr>
<td>Annotated Instructor's Edition
Contains all the content found in the student edition, plus the following:<ul><li>Answers to exercises on the same text page</li><li>Answers to graphing exercises and video exercises</li><li>Teaching Tips throughout the text placed at key points.</li></ul></td>
<td>Instructor's Resource Manual with Tests and Mini-Lectures<ul><li>Mini lectures for each text section</li><li>Additional Practice worksheets for each section</li><li>Several forms of test per chapter–free response and multiple choice</li><li>Answers to all items</li></ul>Instructor's Solutions Manual
TestGen® (Available for download from the IRC)</td>
</tr>
<tr>
<td></td>
<td>Online Resources
MyMathLab® (access code required)

MathXL® (access code required)</td>
</tr>
</table>

Acknowledgments

There are many people who helped me develop this text, and I will attempt to thank some of them here. Cindy Trimble was *invaluable* for contributing to the overall accuracy of the text. Lisa Collette and Suellen Robinson were *invaluable* for their many suggestions and contributions during the development and writing of this Sixth Edition. Allison Campbell of Integra-Chicago provided guidance throughout the production process.

A very special thank you goes to my editor, Mary Beckwith, for being there 24/7/365, as my students say. Last, my thanks to the staff at Pearson for all their support: Patty Bergin, Heather Scott, Michelle Renda, Chris Hoag, and Greg Tobin.

I would like to thank the following reviewers for their input and suggestions:

Sheila Anderson, *Housatonic Community College*
Tom Blackburn, *Northeastern Illinois University*
Gail Burkett, *Palm Beach State College*
James Butterbach, *Joliet Junior College*
Laura Dyer, *Southwestern Illinois College*
Sharon Edgemon, *Bakersfield College*
Hope Essien, *Olive-Harvey College*
Randa Kress, *Idaho State University*
Ted Lai, *Hudson Community College*
Nicole Lang, *North Hennepin Community College*
Lee LaRue, *Paris Junior College*
Jeri Lee, *Des Moines Area Community College*
Jean McArthur, *Joliet Junior College*
Michael Montano, *Riverside Community College*
Lisa J. Music, *Big Sandy Community and Technical College*
Linda Padilla, *Joliet Junior College*
Scott Perkins, *Lake Sumter Community College*
Marilyn Platt, *Gaston College*
Sandy Spears, *Jefferson Community College*
Ping Charlene Tintera, *Texas A & M University*
Jane Wampler, *Housatonic Community College*
Peter Zimmer, *West Chester University*

I would also like to thank the following dedicated group of instructors who participated in our focus groups, Martin-Gay Summits, and our design review for the series. Their feedback and insights have helped to strengthen this edition of the text. These instructors include:

Billie Anderson, *Tyler Junior College*
Cedric Atkins, *Mott Community College*
Andrea Barnett, *Tri-County Technical College*
Lois Beardon, *Schoolcraft College*
Michelle Beerman, *Pasco Hernando Community College*
Laurel Berry, *Bryant & Stratton College*
John Beyers, *University of Maryland*
Bob Brown, *Community College of Baltimore County–Essex*
Lisa Brown, *Community College of Baltimore County–Essex*
NeKeith Brown, *Richland College*
Sue Brown, *Guilford Technical Community College*
Gail Burkett, *Palm Beach State College*
Cheryl Cantwell, *Seminole Community College*
Jackie Cohen, *Augusta State College*
Julie Dewan, *Mohawk Valley Community College*
Janice Ervin, *Central Piedmont Community College*
Richard Fielding, *Southwestern College*
Cindy Gaddis, *Tyler Junior College*

Nita Graham, *St. Louis Community College*
Pauline Hall, *Iowa State College*
Elizabeth Hamman, *Cypress College*
Pat Hussey, *Triton College*
Dorothy Johnson, *Lorain County Community College*
Sonya Johnson, *Central Piedmont Community College*
Irene Jones, *Fullerton College*
Paul Jones, *University of Cincinnati*
Kathy Kopelousos, *Lewis and Clark Community College*
Nancy Lange, *Inver Hills Community College*
Judy Langer, *Westchester Community College*
Lisa Lindloff, *McLinnan Community College*
Sandy Lofstock, *St. Petersburg College*
Kathy Lovelle, *Westchester Community College*
Jamie Malek, *Florida State College*
Jean McArthur, *Joliet Junior College*
Kevin McCandless, *Evergreen Valley College*
Daniel Miller, *Niagara County Community College*
Marica Molle, *Metropolitan Community College*
Carol Murphy, *San Diego Miramar College*
Greg Nguyen, *Fullerton College*
Eric Ollila, *Jackson Community College*
Linda Padilla, *Joliet Junior College*
Davidson Pierre, *State College of Florida*
Marilyn Platt, *Gaston College*
Susan Poss, *Spartanburg Community College*
Natalie Rivera, *Estrella Mountain Community College*
Judy Roane, *Pearl River Community College*
Claudinna Rowley, *Montgomery Community College, Rockville*
Ena Salter, *Manatee Community College*
Carole Shapero, *Oakton Community College*
Janet Sibol, *Hillsborough Community College*
Anne Smallen, *Mohawk Valley Community College*
Barbara Stoner, *Reading Area Community College*
Jennifer Strehler, *Oakton Community College*
Ellen Stutes, *Louisiana State University Eunice*
Tanomo Taguchi, *Fullerton College*
MaryAnn Tuerk, *Elsin Community College*
Gwen Turbeville, *J. Sargeant Reynolds Community College*
Walter Wang, *Baruch College*
Leigh Ann Wheeler, *Greenville Technical Community College*
Valerie Wright, *Central Piedmont Community College*

A special thank you to those students who participated in our design review: Katherine Browne, Mike Bulfin, Nancy Canipe, Ashley Carpenter, Jeff Chojnachi, Roxanne Davis, Mike Dieter, Amy Dombrowski, Kay Herring, Todd Jaycox, Kaleena Levan, Matt Montgomery, Tony Plese, Abigail Polkinghorn, Harley Price, Eli Robinson, Avery Rosen, Robyn Schott, Cynthia Thomas, and Sherry Ward.

Elayn Martin-Gay

About the Author

Elayn Martin-Gay has taught mathematics at the University of New Orleans for more than 25 years. Her numerous teaching awards include the local University Alumni Association's Award for Excellence in Teaching, and Outstanding Developmental Educator at University of New Orleans, presented by the Louisiana Association of Developmental Educators.

Prior to writing textbooks, Elayn Martin-Gay developed an acclaimed series of lecture videos to support developmental mathematics students in their quest for success. These highly successful videos originally served as the foundation material for her texts. Today, the videos are specific to each book in the Martin-Gay series. The author has also created Chapter Test Prep Videos to help students during their most "teachable moment"—as they prepare for a test, along with Instructor-to-Instructor videos that provide teaching tips, hints, and suggestions for each developmental mathematics course, including basic mathematics, prealgebra, beginning algebra, and intermediate algebra. Her most recent innovations are the Algebra Prep Apps for the iPhone and iPod Touch. These Apps embrace the different learning styles, schedules, and paces of students and provide them with quality math tutoring.

Elayn is the author of 12 published textbooks as well as multimedia interactive mathematics, all specializing in developmental mathematics courses. She has participated as an author across the broadest range of educational materials: textbooks, videos, tutorial software, and courseware. This offers an opportunity of various combinations for an integrated teaching and learning package offering great consistency for the student.

Applications Index

CHAPTER 1

Review of Real Numbers

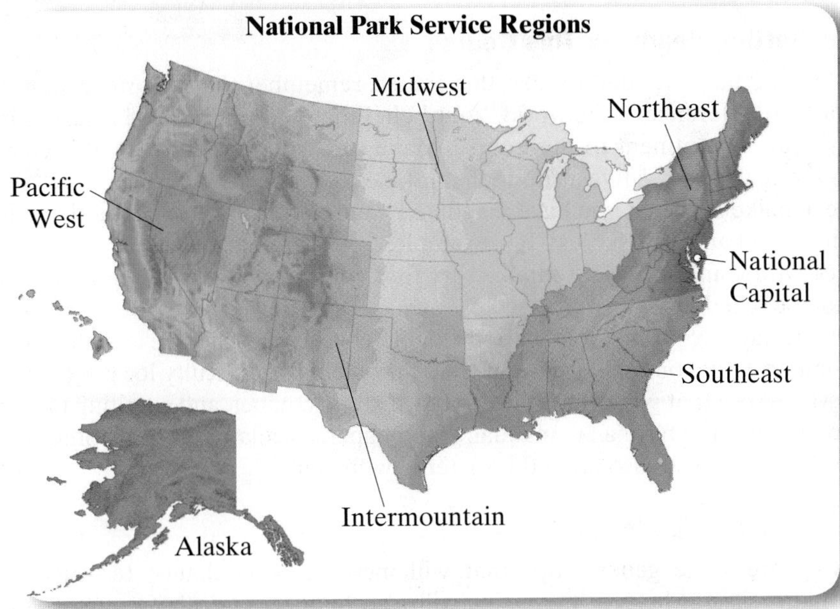

In this chapter, we review the basic symbols and words—the language—of arithmetic and introduce using variables in place of numbers. This is our starting place in the study of algebra.

The National Park Service (NPS) is a federal agency that manages all national parks and many other historic monuments and properties. The NPS has 21,989 employees and had a total of 281 million recreational visitors in 2010 alone. The map above shows the different geographic regions of the NPS, and the bar graph below shows the number of recreational visits per year shown.

In Section 1.2, Exercises 71 through 76, we shall study this bar graph further.

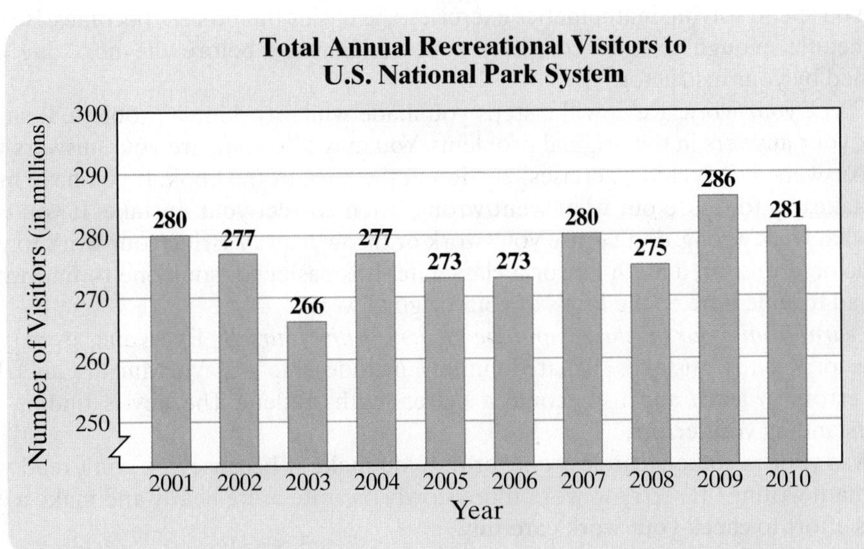

 Tips for Success in Mathematics 

OBJECTIVES

1 Get Ready for This Course.
2 Understand Some General Tips for Success.
3 Understand How to Use This Text.
4 Get Help as Soon as You Need It.
5 Learn How to Prepare for and Take an Exam.
6 Develop Good Time Management.

Before reading this section, remember that your instructor is your best source of information. Please see your instructor for any additional help or information.

OBJECTIVE

1 Getting Ready for This Course 

Now that you have decided to take this course, remember that a *positive attitude* will make all the difference in the world. Your belief that you can succeed is just as important as your commitment to this course. Make sure you are ready for this course by having the time and positive attitude that it takes to succeed.

Next, make sure that you have scheduled your math course at a time that will give you the best chance for success. For example, if you are also working, you may want to check with your employer to make sure that your work hours will not conflict with your course schedule.

On the day of your first class period, double-check your schedule and allow yourself extra time to arrive on time in case of traffic problems or difficulty locating your classroom. Make sure that you bring at least your textbook, paper, and a writing instrument. Are you required to have a lab manual, graph paper, calculator, or some other supplies besides this text? If so, also bring this material with you.

OBJECTIVE

2 General Tips for Success 

Following are some general tips that will increase your chance for success in a mathematics class. Many of these tips will also help you in other courses you may be taking.

Exchange names and phone numbers or email addresses with at least one other person in class. This contact person can be a great help if you miss an assignment or want to discuss math concepts or exercises that you find difficult.

Choose to attend all class periods. If possible, sit near the front of the classroom. This way, you will see and hear the presentation better. It may also be easier for you to participate in classroom activities.

> **Helpful Hint**
> **MyMathLab® and MathXL®** If you are doing your homework online, you can work and rework those exercises that you struggle with until you master them. Try working through all the assigned exercises twice before the due date.

Do your homework. You've probably heard the phrase "practice makes perfect" in relation to music and sports. It also applies to mathematics. You will find that the more time you spend solving mathematics exercises, the easier the process becomes. Be sure to schedule enough time to complete your assignments before the next due date assigned by your instructor.

> **Helpful Hint**
> **MyMathLab® and MathXL®** If you are completing your homework online, it's important to work each exercise on paper before submitting the answer. That way, you can check your work and follow your steps to find and correct any mistakes.

Check your work. Review the steps you made while working a problem. Learn to check your answers in the original problems. You may also compare your answers with the "Answers to Selected Exercises" section in the back of the book. If you have made a mistake, try to figure out what went wrong. Then correct your mistake. If you can't find what went wrong, don't erase your work or throw it away. Bring your work to your instructor, a tutor in a math lab, or a classmate. It is easier for someone to find where you had trouble if he or she looks at your original work.

Learn from your mistakes and be patient with yourself. Everyone, even your instructor, makes mistakes. (That definitely includes me—Elayn Martin-Gay.) Use your errors to learn and to become a better math student. The key is finding and understanding your errors.

Was your mistake a careless one, or did you make it because you can't read your own math writing? If so, try to work more slowly or write more neatly and make a conscious effort to check your work carefully.

Did you make a mistake because you don't understand a concept? Take the time to review the concept or ask questions to understand it better.

Did you skip too many steps? Skipping steps or trying to do too many steps mentally may lead to preventable mistakes.

Know how to get help if you need it. It's all right to ask for help. In fact, it's a good idea to ask for help whenever there is something that you don't understand. Make sure you know when your instructor has office hours and how to find his or her office. Find out whether math tutoring services are available on your campus. Check on the hours, location, and requirements of the tutoring service.

> **Helpful Hint**
> **MyMathLab®** and **MathXL®** When assignments are turned in online, keep a hard copy of your complete written work. You will need to refer to your written work to be able to ask questions and to study for tests later.

Organize your class materials, including homework assignments, graded quizzes and tests, and notes from your class or lab. All of these items will make valuable references throughout your course and when studying for upcoming tests and the final exam. Make sure that you can locate these materials when you need them.

Read your textbook before class. Reading a mathematics textbook is unlike reading a novel or a newspaper. Your pace will be much slower. It is helpful to have paper and a pencil with you when you read. Try to work out examples on your own as you encounter them in your text. You should also write down any questions that you want to ask in class. When you read a mathematics textbook, sometimes some of the information in a section will be unclear. But after you hear a lecture or watch a lecture video on that section, you will understand it much more easily than if you had not read your text beforehand.

Don't be afraid to ask questions. You are not the only person in class with questions. Other students are normally grateful that someone has spoken up.

> **Helpful Hint**
> **MyMathLab®** and **MathXL®** Be aware of assignments and due dates set by your instructor. Don't wait until the last minute to submit work online. Allow 6–8 hours before the deadline in case you have technology trouble.

Turn in assignments on time. This way, you can be sure that you will not lose points for being late. Show every step of a problem and be neat and organized. Also be sure that you understand which problems are assigned for homework. If allowed, you can always double-check the assignment with another student in your class.

OBJECTIVE

3 Using This Text 

Many helpful resources are available to you. It is important to become familiar with and use these resources. They should increase your chances for success in this course.

- *Practice Exercises*. Each example in every section has a parallel Practice exercise. As you read a section, try each Practice exercise after you've finished the corresponding example. This learn-by-doing approach will help you grasp ideas before you move on to other concepts. Answers are at the back of the text.
- *Chapter Test Prep Videos*. These videos provide solutions to all of the Chapter Test exercises worked out by the author. This supplement is very helpful before a test or exam.
- *Interactive DVD Lecture Series*. Exercises marked with a ▶ are fully worked out by the author on the DVDs. The lecture series provides approximately 20 minutes of instruction per section.

> **Helpful Hint**
> **MyMathLab®** In MyMathLab, you have access to the following video resources:
> - Lecture Videos for each section
> - Chapter Test Prep Videos
>
> Use these videos provided by the author to prepare for class, review, and study for tests.

- *Symbols at the Beginning of an Exercise Set*. If you need help with a particular section, the symbols listed at the beginning of each exercise set will remind you of the numerous supplements available.
- *Objectives*. The main section of exercises in each exercise set is referenced by an objective, such as 1 or 2, and by an example(s). There is also often a section of exercises entitled "Mixed Practice," which is referenced by two or more objectives or sections. These are mixed exercises written to prepare you for your next exam. Use all of this referencing if you have trouble completing an assignment from the exercise set.
- *Icons (Symbols)*. Make sure that you understand the meaning of the icons that are beside many exercises. ▶ tells you that the corresponding exercise may be viewed on the video segment that corresponds to that section. ✎ tells you that this exercise is a writing exercise in which you should answer in complete sentences. △ tells you that the exercise involves geometry.
- *Integrated Reviews*. Found in the middle of each chapter, these reviews offer you a chance to practice—in one place—the many concepts that you have learned separately over several sections.
- *End-of-Chapter Opportunities*. There are many opportunities at the end of each chapter to help you understand the concepts of the chapter.

  Vocabulary Checks contain key vocabulary terms introduced in the chapter.

  Chapter Highlights contain chapter summaries and examples.

  Chapter Reviews contain review problems. The first part is organized section by section and the second part contains a set of mixed exercises.

Chapter Tests are sample tests to help you prepare for an exam. The Chapter Test Prep Videos, found in this text, contain all the Chapter Test exercises worked by the author.

Cumulative Reviews consist of material from the beginning of the book to the end of that particular chapter.

- *Student Resources in Your Textbook*. You will find a **Student Resources** section at the back of this textbook. It contains the following to help you study and prepare for tests:

  Study Skill Builders contain study skills advice. To increase your chance for success in the course, read these study tips and answer the questions.

  Bigger Picture—Study Guide Outline provides you with a study guide outline of the course, with examples.

  Practice Final provides you with a Practice Final Exam to help you prepare for a final. The video solutions to each question are provided in the Interactive DVD Lecture Series and within MyMathLab®.

- *Resources to Check Your Work*. The **Answers to Selected Exercises** section provides answers to all odd-numbered section exercises and all integrated review and chapter test exercises.

> **Helpful Hint**
> **MyMathLab®** and **MathXL®**
> - Use the **Help Me Solve This** button to get step-by-step help for the exercise you are working. You will need to work an additional exercise of the same type before you can get credit for having worked it correctly.
> - Use the **Video** button to view a video clip of the author working a similar exercise.

OBJECTIVE

4 Getting Help 

If you have trouble completing assignments or understanding the mathematics, get help as soon as you need it! This tip is presented as an objective on its own because it is so important. In mathematics, usually the material presented in one section builds on your understanding of the previous section. This means that if you don't understand the concepts covered during a class period, there is a good chance that you will not understand the concepts covered during the next class period. If this happens to you, get help as soon as you can.

Where can you get help? Many suggestions have been made in this section on where to get help, and now it is up to you to get it. Try your instructor, a tutoring center, or a math lab, or you may want to form a study group with fellow classmates. If you do decide to see your instructor or go to a tutoring center, make sure that you have a neat notebook and are ready with your questions.

OBJECTIVE

5 Preparing for and Taking an Exam 

> **Helpful Hint**
> **MyMathLab®** and **MathXL®**
> Review your written work for previous assignments. Then, go back and rework previous assignments. Open a previous assignment and click **Similar Exercise** to generate new exercises. Rework the exercises until you fully understand them and can work them without help features.

Make sure that you allow yourself plenty of time to prepare for a test. If you think that you are a little "math anxious," it may be that you are not preparing for a test in a way that will ensure success. The way that you prepare for a test in mathematics is important. To prepare for a test:

1. Review your previous homework assignments.
2. Review any notes from class and section-level quizzes you have taken. (If this is a final exam, also review chapter tests you have taken.)
3. Review concepts and definitions by reading the Chapter Highlights at the end of each chapter.
4. Practice working out exercises by completing the Chapter Review found at the end of each chapter. (If this is a final exam, go through a Cumulative Review. There is one at the end of each chapter except Chapter 1. Choose the review found at the end of the latest chapter that you have covered in your course.) *Don't stop here!*
5. It is important to place yourself in conditions similar to test conditions to find out how you will perform. In other words, as soon as you feel that you know the material, get a few blank sheets of paper and take a sample test. A Chapter Test is available at the end of each chapter, or you can work selected problems from the Chapter

Review. Your instructor may also provide you with a review sheet. During this sample test, do not use your notes or your textbook. Then check your sample test. If you are not satisfied with the results, study the areas that you are weak in and try again.

6. On the day of the test, allow yourself plenty of time to arrive where you will be taking your exam.

When taking your test:

1. Read the directions on the test carefully.
2. Read each problem carefully as you take the test. Make sure that you answer the question asked.
3. Watch your time and pace yourself so that you can attempt each problem on your test.
4. If you have time, check your work and answers.
5. Do not turn your test in early. If you have extra time, spend it double-checking your work.

OBJECTIVE

6 Managing Your Time 

As a college student, you know the demands that classes, homework, work, and family place on your time. Some days, you probably wonder how you'll ever get everything done. One key to managing your time is developing a schedule. Here are some hints for making a schedule:

1. Make a list of all your weekly commitments for the term. Include classes, work, regular meetings, extracurricular activities, etc. You may also find it helpful to list such things as laundry, regular workouts, grocery shopping, etc.
2. Next, estimate the time needed for each item on the list. Also make a note of how often you will need to do each item. Don't forget to include time estimates for the reading, studying, and homework you do outside of your classes. You may want to ask your instructor for help estimating the time needed.
3. In the exercise set that follows, you are asked to block out a typical week on the schedule grid given. Start with items with fixed time slots like classes and work.
4. Next, include the items on your list with flexible time slots. Think carefully about how best to schedule items such as study time.
5. Don't fill up every time slot on the schedule. Remember that you need to allow time for eating, sleeping, and relaxing! You should also allow a little extra time in case some items take longer than planned.
6. If you find that your weekly schedule is too full for you to handle, you may need to make some changes in your workload, classload, or other areas of your life. You may want to talk to your advisor, manager or supervisor at work, or someone in your college's academic counseling center for help with such decisions.

1.1 Exercise Set

MyMathLab®  

1. What is your instructor's name?
2. What are your instructor's office location and office hours?
3. What is the best way to contact your instructor?
4. Do you have the name and contact information of at least one other student in class?
5. Will your instructor allow you to use a calculator in this class?
6. Why is it important that you write step-by-step solutions to homework exercises and keep a hard copy of all work submitted?
7. Is a tutoring service available on campus? If so, what are its hours? What services are available?
8. Have you attempted this course before? If so, write down ways that you might improve your chances of success during this second attempt.
9. List some steps that you can take if you begin having trouble understanding the material or completing an assignment. If you are completing your homework in MyMathLab® and MathXL®, list the resources you can use for help.

10. How many hours of studying does your instructor advise for each hour of instruction?
11. What does the ╲ icon in this text mean?
12. What does the ▶ icon in this text mean?
13. What does the △ icon in this text mean?
14. What are Practice exercises?
15. When might be the best time to work a Practice exercise?
16. Where are the answers to Practice exercises?
17. What answers are contained in this text and where are they?
18. What and where are the study skills builders?
19. What and where are Integrated Reviews?
20. How many times is it suggested that you work through the homework exercises in MathXL® before the submission deadline?
21. How far in advance of the assigned due date is it suggested that homework be submitted online? Why?
22. Chapter Highlights are found at the end of each chapter. Find the Chapter 1 Highlights and explain how you might use it and how it might be helpful.
23. Chapter Reviews are found at the end of each chapter. Find the Chapter 1 Review and explain how you might use it and how it might be useful.
24. Chapter Tests are at the end of each chapter. Find the Chapter 1 Test and explain how you might use it and how it might be helpful when preparing for an exam on Chapter 1. Include how the Chapter Test Prep Videos may help. If you are working in MyMathLab® and MathXL®, how can you use previous homework assignments to study?
25. Read or reread objective 6 and fill out the schedule grid below.

	Monday	*Tuesday*	*Wednesday*	*Thursday*	*Friday*	*Saturday*	*Sunday*
1:00 a.m.							
2:00 a.m.							
3:00 a.m.							
4:00 a.m.							
5:00 a.m.							
6:00 a.m.							
7:00 a.m.							
8:00 a.m.							
9:00 a.m.							
10:00 a.m.							
11:00 a.m.							
Noon							
1:00 p.m.							
2:00 p.m.							
3:00 p.m.							
4:00 p.m.							
5:00 p.m.							
6:00 p.m.							
7:00 p.m.							
8:00 p.m.							
9:00 p.m.							
10:00 p.m.							
11:00 p.m.							
Midnight							

1.2 Symbols and Sets of Numbers 

OBJECTIVES

1. Use a Number Line to Order Numbers.
2. Translate Sentences into Mathematical Statements.
3. Identify Natural Numbers, Whole Numbers, Integers, Rational Numbers, Irrational Numbers, and Real Numbers.
4. Find the Absolute Value of a Real Number.

OBJECTIVE

1 Using a Number Line to Order Numbers 

We begin with a review of the set of natural numbers and the set of whole numbers and how we use symbols to compare these numbers. A **set** is a collection of objects, each of which is called a **member** or **element** of the set. A pair of brace symbols { } encloses the list of elements and is translated as "the set of" or "the set containing."

Natural Numbers

The set of **natural numbers** is $\{1, 2, 3, 4, 5, 6, \ldots\}$.

Whole Numbers

The set of **whole numbers** is $\{0, 1, 2, 3, 4, \ldots\}$.

The three dots (an ellipsis) means that the list continues in the same manner indefinitely.

These numbers can be pictured on a **number line.** We will use number lines often to help us visualize distance and relationships between numbers.

To draw a number line, first draw a line. Choose a point on the line and label it 0. To the right of 0, label any other point 1. Being careful to use the same distance as from 0 to 1, mark off equally spaced distances. Label these points 2, 3, 4, 5, and so on. Since the whole numbers continue indefinitely, it is not possible to show every whole number on this number line. The arrow at the right end of the line indicates that the pattern continues indefinitely.

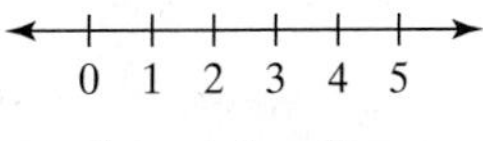

A number line

Picturing whole numbers on a number line helps us see the order of the numbers. Symbols can be used to describe concisely in writing the order that we see.

The **equal symbol** $=$ means "is equal to."

The symbol $\neq$ means "is not equal to."

These symbols may be used to form a **mathematical statement.** The statement might be true or it might be false. The two statements below are both true.

$2 = 2$ states that "two is equal to two."

$2 \neq 6$ states that "two is not equal to six."

If two numbers are not equal, one number is larger than the other.
The symbol $>$ means "is greater than."
The symbol $<$ means "is less than." For example,

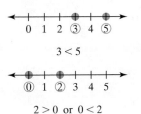

$3 < 5$ states that "three is less than five."

$2 > 0$ states that "two is greater than zero."

0 1 2 ③ 4 ⑤

$3 < 5$

⓪ 1 ② 3 4 5

$2 > 0$ or $0 < 2$

On a number line, we see that a number **to the right of** another number is **larger.** Similarly, a number **to the left of** another number is smaller. For example, 3 is to the left of 5 on a number line, which means that 3 is less than 5, or $3 < 5$. Similarly, 2 is to the right of 0 on a number line, which means 2 is greater than 0, or $2 > 0$. Since 0 is to the left of 2, we can also say that 0 is less than 2, or $0 < 2$.

The symbols $\neq$, $<$, and $>$ are called **inequality symbols.**

Helpful Hint

Notice that $2 > 0$ has exactly the same meaning as $0 < 2$. Switching the order of the numbers and reversing the direction of the inequality symbol does not change the meaning of the statement.

$3 < 5$ has the same meaning as $5 > 3$.

Also notice that, when the statement is true, the inequality arrow points to the smaller number.

EXAMPLE 1 Insert $<$, $>$, or $=$ in the space between each pair of numbers to make each statement true

a. 2 3 **b.** 7 4 **c.** 72 27

Solution

a. $2 < 3$ since 2 is to the left of 3 on the number line.

b. $7 > 4$ since 7 is to the right of 4 on the number line.

c. $72 > 27$ since 72 is to the right of 27 on the number line. □

PRACTICE

1 Insert $<$, $>$, or $=$ in the space between each pair of numbers to make each statement true.

a. 5 8 **b.** 6 4 **c.** 16 82

Two other symbols are used to compare numbers.
The symbol $\leq$ means "is less than or equal to."
The symbol $\geq$ means "is greater than or equal to." For example,

$7 \leq 10$ states that "seven is less than or equal to ten."

This statement is true since $7 < 10$ is true. If either $7 < 10$ or $7 = 10$ is true, then $7 \leq 10$ is true.

$3 \geq 3$ states that "three is greater than or equal to three."

This statement is true since $3 = 3$ is true. If either $3 > 3$ or $3 = 3$ is true, then $3 \geq 3$ is true.

The statement $6 \geq 10$ is false since neither $6 > 10$ nor $6 = 10$ is true. The symbols $\leq$ and $\geq$ are also called **inequality symbols.**

EXAMPLE 2 Tell whether each statement is true or false.

a. $8 \geq 8$ **b.** $8 \leq 8$ **c.** $23 \leq 0$ **d.** $23 \geq 0$

Solution

a. True. Since $8 = 8$ is true, then $8 \geq 8$ is true.

b. True. Since $8 = 8$ is true, then $8 \leq 8$ is true.

c. False. Since neither $23 < 0$ nor $23 = 0$ is true, then $23 \leq 0$ is false.

d. True. Since $23 > 0$ is true, then $23 \geq 0$ is true. □

PRACTICE

2 Tell whether each statement is true or false.

a. $9 \geq 3$ **b.** $3 \geq 8$ **c.** $25 \leq 25$ **d.** $4 \leq 14$

OBJECTIVE

2 Translating Sentences 

Now, let's use the symbols discussed to translate sentences into mathematical statements.

EXAMPLE 3 Translate each sentence into a mathematical statement.

a. Nine is less than or equal to eleven.

b. Eight is greater than one.

c. Three is not equal to four.

Solution

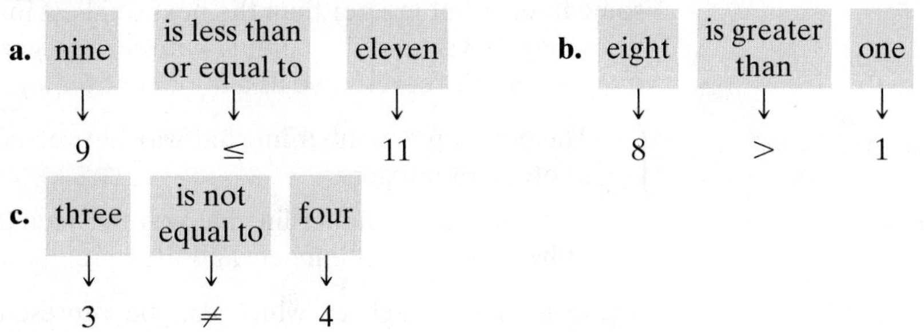

PRACTICE

3 Translate each sentence into a mathematical statement.

a. Three is less than eight.

b. Fifteen is greater than or equal to nine.

c. Six is not equal to seven.

OBJECTIVE

3 Identifying Common Sets of Numbers 

Whole numbers are not sufficient to describe many situations in the real world. For example, quantities less than zero must sometimes be represented, such as temperatures less than 0 degrees.

Numbers Less Than Zero on a Number Line

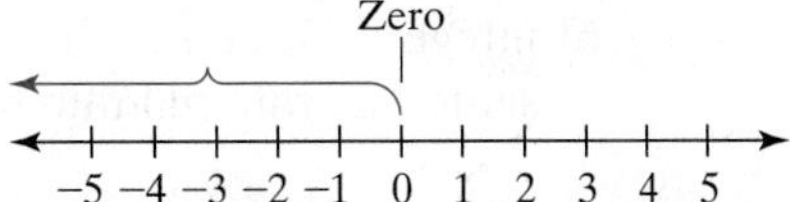

Numbers less than 0 are to the left of 0 and are labeled −1, −2, −3, and so on. A − sign, such as the one in −1, tells us that the number is to the left of 0 on a number line. In words, −1 is read "negative one." A + sign or no sign tells us that a number lies to the right of 0 on the number line. For example, 3 and +3 both mean positive three.

The numbers we have pictured are called the set of **integers.** Integers to the left of 0 are called **negative integers;** integers to the right of 0 are called **positive integers.** The integer **0 is neither positive nor negative.**

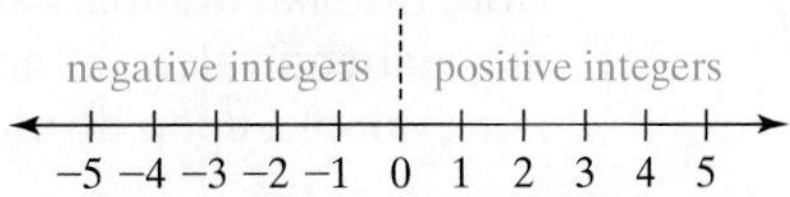

Integers

The set of **integers** is {. . . , −3, −2, −1, 0, 1, 2, 3, . . .}.

The ellipses (three dots) to the left and to the right indicate that the positive integers and the negative integers continue indefinitely.

EXAMPLE 4 Use an integer to express the number in the following. "Pole of Inaccessibility, Antarctica, is the coldest location in the world, with an average annual temperature of 72 degrees below zero." (*Source: The Guinness Book of Records*)

Solution The integer −72 represents 72 degrees below zero.

PRACTICE

4 Use an integer to express the number in the following. The elevation of Laguna Salada in Mexico is 10 meters below sea level. (*Source: The World Almanac*)

A problem with integers in real-life settings arises when quantities are smaller than some integer but greater than the next smallest integer. On a number line, these quantities may be visualized by points between integers. Some of these quantities between integers can be represented as a quotient of integers. For example,

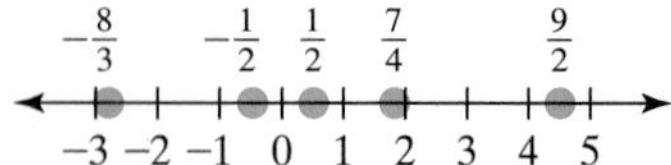

The point on a number line halfway between 0 and 1 can be represented by $\frac{1}{2}$, a quotient of integers.

The point on a number line halfway between 0 and -1 can be represented by $-\frac{1}{2}$. Other quotients of integers and their graphs are shown to the left.

These numbers, each of which can be represented as a quotient of integers, are examples of **rational numbers.** It's not possible to list the set of rational numbers using the notation that we have been using. For this reason, we will use a different notation.

Rational Numbers

$$\left\{\frac{a}{b}\,\middle|\, a \text{ and } b \text{ are integers and } b \neq 0\right\}$$

We read this set as "the set of all numbers $\frac{a}{b}$ such that a and b are integers and ***b* is not equal to 0.**" Notice that every integer is also a rational number since each integer can be expressed as a quotient of integers. For example, the integer 5 is also a rational number since $5 = \frac{5}{1}$.

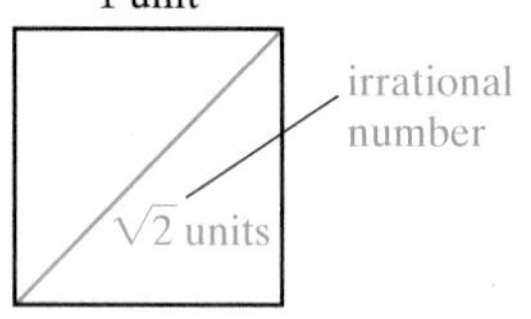

The number line also contains points that cannot be expressed as quotients of integers. These numbers are called **irrational numbers** because they cannot be represented by rational numbers. For example, $\sqrt{2}$ and π are irrational numbers.

Irrational Numbers

The set of **irrational numbers** is

{Nonrational numbers that correspond to points on the number line}.

That is, an irrational number is a number that cannot be expressed as a quotient of integers.

Both rational numbers and irrational numbers can be written as decimal numbers. The decimal equivalent of a rational number will either terminate or repeat in a pattern. For example, upon dividing we find that

Rational Numbers:

$\frac{3}{4} = 0.75$ (decimal number terminates or ends)

$\frac{2}{3} = 0.66666\ldots$ (decimal number repeats in a pattern)

The decimal representation of an irrational number will neither terminate nor repeat. For example, the decimal representations of irrational numbers $\sqrt{2}$ and π are

Irrational Numbers:

$\sqrt{2} = 1.414213562\ldots$ (decimal number does not terminate or repeat in a pattern)

$\pi = 3.141592653\ldots$ (decimal number does not terminate or repeat in a pattern)

(For further review of decimals, see the Appendix.)

Combining the rational numbers with the irrational numbers gives the set of **real numbers.** One and only one point on a number line corresponds to each real number.

Real Numbers

The set of **real numbers** is

{All numbers that correspond to points on the number line}

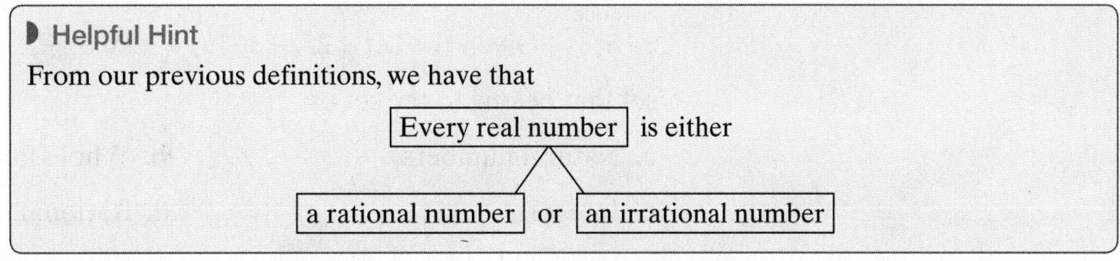

Helpful Hint
From our previous definitions, we have that
Every real number is either
a rational number or an irrational number

On the following number line, we see that real numbers can be positive, negative, or 0. Numbers to the left of 0 are called **negative numbers;** numbers to the right of 0 are called **positive numbers.** Positive and negative numbers are also called **signed numbers.**

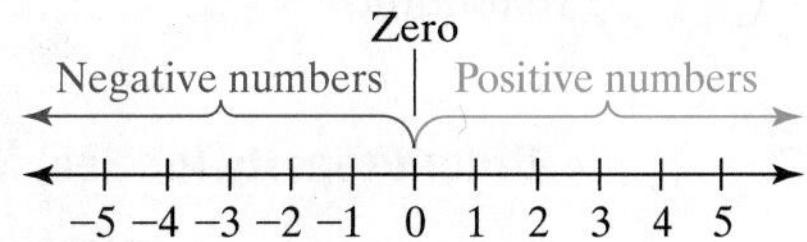

Several different sets of numbers have been discussed in this section. The following diagram shows the relationships among these sets of real numbers.

Common Sets of Numbers

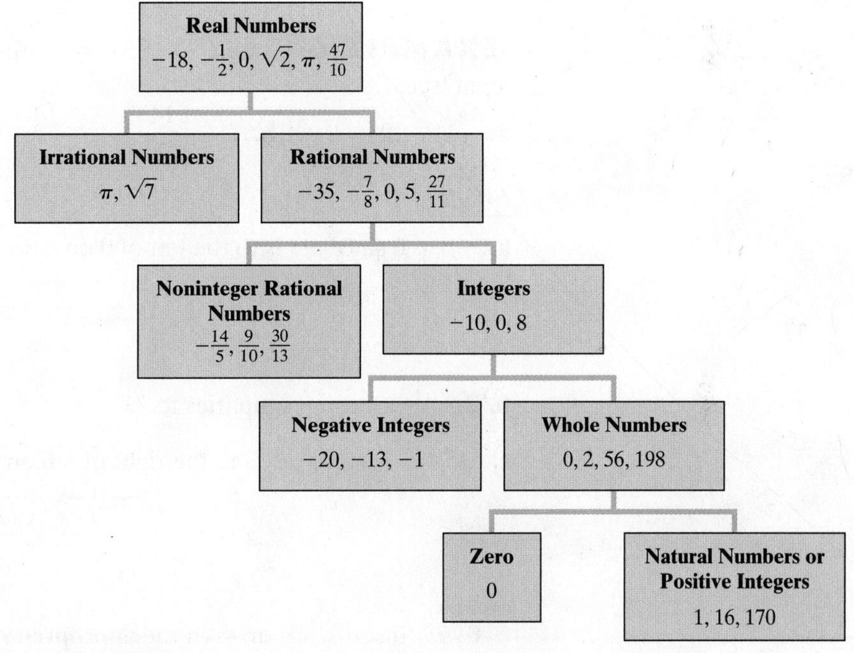

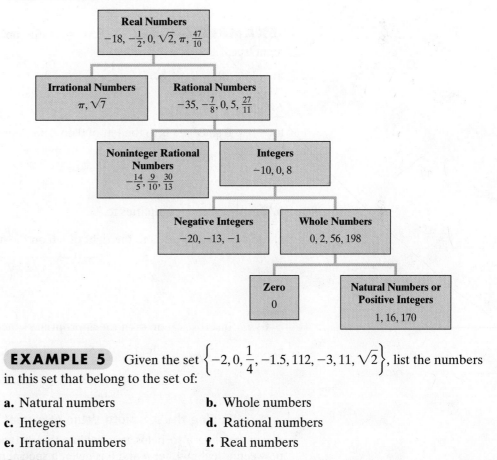

EXAMPLE 5 Given the set $\left\{-2, 0, \frac{1}{4}, -1.5, 112, -3, 11, \sqrt{2}\right\}$, list the numbers in this set that belong to the set of:

a. Natural numbers

b. Whole numbers

c. Integers

d. Rational numbers

e. Irrational numbers

f. Real numbers

Solution

a. The natural numbers are 11 and 112.

b. The whole numbers are 0, 11, and 112.

c. The integers are −3, −2, 0, 11, and 112.

d. Recall that integers are rational numbers also. The rational numbers are −3, −2, −1.5, 0, $\frac{1}{4}$, 11, and 112.

e. The irrational number is $\sqrt{2}$.

f. The real numbers are all numbers in the given set. □

PRACTICE
5 Given the set $\left\{25, \frac{7}{3}, -15, \frac{-3}{4}, \sqrt{5}, -3.7, 8.8, -99\right\}$, list the numbers in this set that belong to the set of:

a. Natural numbers **b.** Whole numbers
c. Integers **d.** Rational numbers
e. Irrational numbers **f.** Real numbers

We now extend the meaning and use of inequality symbols such as < and > to all real numbers.

Order Property for Real Numbers

For any two real numbers a and b, a is less than b if a is to the left of b on a number line.

a b

$a < b$ or also $b > a$

EXAMPLE 6 Insert <, >, or = in the appropriate space to make each statement true.

a. $-1 \quad 0$ **b.** $7 \quad \frac{14}{2}$ **c.** $-5 \quad -6$

Solution

a. $-1 < 0$ since -1 is to the left of 0 on a number line.

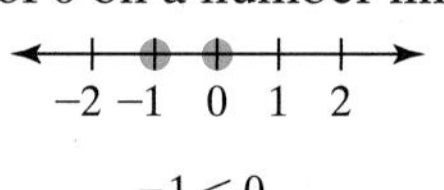

$-1 < 0$

b. $7 = \frac{14}{2}$ since $\frac{14}{2}$ simplifies to 7.

c. $-5 > -6$ since -5 is to the right of -6 on the number line.

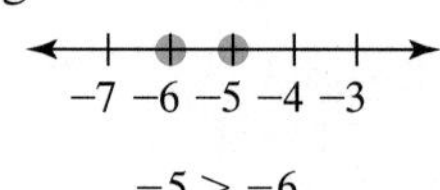

$-5 > -6$

PRACTICE
6 Insert<, >, or = in the appropriate space to make each statement true.

a. $0 \quad 3$ **b.** $15 \quad -5$ **c.** $3 \quad \frac{12}{4}$

OBJECTIVE

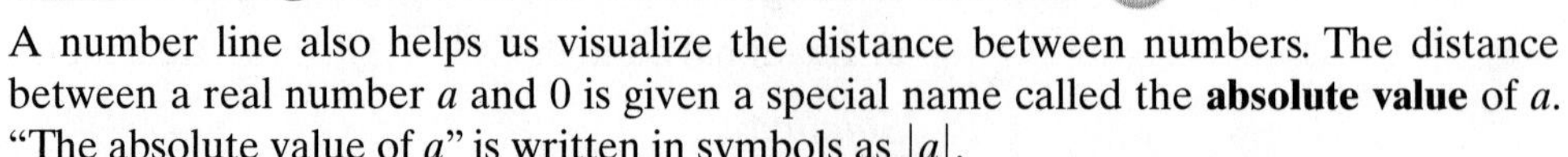

4 Finding the Absolute Value of a Real Number

A number line also helps us visualize the distance between numbers. The distance between a real number a and 0 is given a special name called the **absolute value** of a. "The absolute value of a" is written in symbols as $|a|$.

Absolute Value

The absolute value of a real number a, denoted by $|a|$, is the distance between a and 0 on a number line.

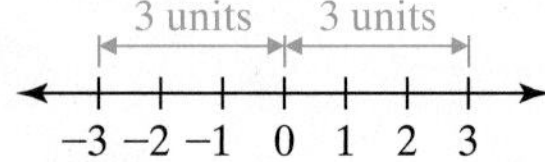

For example, $|3| = 3$ and $|-3| = 3$ since both 3 and -3 are a distance of 3 units from 0 on a number line.

Helpful Hint

Since $|a|$ is a distance, $|a|$ is always either positive or 0, never negative. That is, **for any real number** a, $|a| \geq 0$.

EXAMPLE 7 Find the absolute value of each number.

a. $|4|$ **b.** $|-5|$ **c.** $|0|$ **d.** $\left|-\frac{1}{2}\right|$ **e.** $|5.6|$

Solution

a. $|4| = 4$ since 4 is 4 units from 0 on a number line.

b. $|-5| = 5$ since -5 is 5 units from 0 on a number line.

c. $|0| = 0$ since 0 is 0 units from 0 on a number line.

d. $\left|-\frac{1}{2}\right| = \frac{1}{2}$ since $-\frac{1}{2}$ is $\frac{1}{2}$ unit from 0 on a number line.

e. $|5.6| = 5.6$ since 5.6 is 5.6 units from 0 on a number line. □

PRACTICE 7 Find the absolute value of each number.

a. $|-8|$ **b.** $|9|$ **c.** $|-2.5|$ **d.** $\left|\frac{5}{11}\right|$ **e.** $|\sqrt{3}|$

EXAMPLE 8 Insert <, >, or = in the appropriate space to make each statement true.

a. $|0| \quad 2$ **b.** $|-5| \quad 5$ **c.** $|-3| \quad |-2|$ **d.** $|5| \quad |6|$ **e.** $|-7| \quad |6|$

Solution

a. $|0| < 2$ since $|0| = 0$ and $0 < 2$.

b. $|-5| = 5$ since $5 = 5$.

c. $|-3| > |-2|$ since $3 > 2$.

d. $|5| < |6|$ since $5 < 6$.

e. $|-7| > |6|$ since $7 > 6$. □

PRACTICE 8 Insert <, >, or = in the appropriate space to make each statement true.

a. $|8| \quad |-8|$ **b.** $|-3| \quad 0$ **c.** $|-7| \quad |-11|$ **d.** $|3| \quad |2|$ **e.** $|0| \quad |-4|$

Vocabulary, Readiness & Video Check

Use the choices below to fill in each blank.

real	natural	whole	irrational
$\lvert b \rvert$	inequality	integers	rational

1. The _______ numbers are $\{0, 1, 2, 3, 4, \dots\}$.
2. The _______ numbers are $\{1, 2, 3, 4, 5, \dots\}$.
3. The symbols $\neq, \leq,$ and $>$ are called _______ symbols.
4. The _______ are $\{\dots, -3, -2, -1, 0, 1, 2, 3, \dots\}$.
5. The _______ numbers are {all numbers that correspond to points on the number line}.
6. The _______ numbers are $\left\{\frac{a}{b} \middle| a \text{ and } b \text{ are integers, } b \neq 0\right\}$.
7. The _______ numbers are {nonrational numbers that correspond to points on the number line}.
8. The distance between a number b and 0 on a number line is _______.

Martin-Gay Interactive Videos

See Video 1.2

Watch the section lecture video and answer the following questions.

OBJECTIVE 1

9. In Example 2, why is the symbol < inserted between the two numbers?

OBJECTIVE 2

10. Write the sentence given in Example 4 and translate it to a mathematical statement, using symbols.

OBJECTIVE 3

11. Which sets of numbers does the number in Example 6 belong to? Why is this number not an irrational number?

OBJECTIVE 4

12. Complete this statement based on the lecture given before Example 8. The ______ of a real number a, denoted by $|a|$, is the distance between a and 0 on a number line.

1.2 Exercise Set

MyMathLab®

Insert <, >, or = in the appropriate space to make the statement true. See Example 1.

1. 7   3

2. 9   15

3. 6.26   6.26

4. 2.13   1.13

5. 0   7

6. 20   0

7. −2   2

8. −4   −6

9. The freezing point of water is 32° Fahrenheit. The boiling point of water is 212° Fahrenheit. Write an inequality statement using < or > comparing the numbers 32 and 212.

10. The freezing point of water is 0° Celsius. The boiling point of water is 100° Celsius. Write an inequality statement using < or > comparing the numbers 0 and 100.

△ **11.** An angle measuring 30° is shown and an angle measuring 45° is shown. Use the inequality symbol ≤ or ≥ to write a statement comparing the numbers 30 and 45.

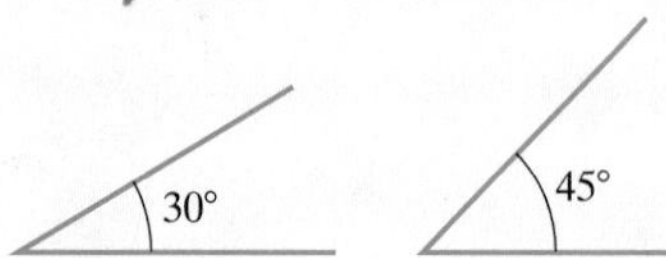

△ **12.** The sum of the measures of the angles of a triangle is 180°. The sum of the measures of the angles of a parallelogram is 360°. Use the inequality symbol ≤ or ≥ to write a statement comparing the numbers 360 and 180.

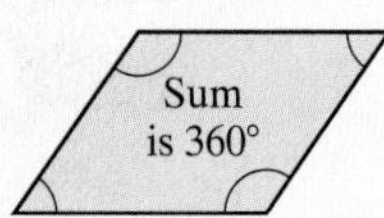

Are the following statements true or false? See Examples 2 and 6.

13. $11 \leq 11$

14. $4 \geq 7$

15. $10 > 11$

16. $17 > 16$

17. $3 + 8 \geq 3(8)$

18. $8 \cdot 8 \leq 8 \cdot 7$

19. $9 > 0$

20. $4 < 7$

21. $-6 > -2$

22. $0 < -15$

TRANSLATING

Write each sentence as a mathematical statement. See Example 3.

23. Eight is less than twelve.

24. Fifteen is greater than five.

25. Five is greater than or equal to four.

26. Negative ten is less than or equal to thirty-seven.

27. Fifteen is not equal to negative two.

28. Negative seven is not equal to seven.

Use integers to represent the values in each statement. See Example 4.

29. The highest elevation in California is Mt. Whitney, with an altitude of 14,494 feet. The lowest elevation in California is Death Valley, with an altitude of 282 feet below sea level. (*Source:* U.S. Geological Survey)

30. Driskill Mountain, in Louisiana, has an altitude of 535 feet. New Orleans, Louisiana, lies 8 feet below sea level. (*Source:* U.S. Geological Survey)

31. The number of graduate students at the University of Texas at Austin is 28,000 fewer than the number of undergraduate students. (*Source:* University of Texas at Austin)

32. The number of students admitted to the class of 2011 at UCLA is 38,792 fewer students than the number that had applied. (*Source:* UCLA)

33. Aaron Miller deposited \$350 in his savings account. He later withdrew \$126.

34. Aris Peña was deep-sea diving. During her dive, she ascended 30 feet and later descended 50 feet.

Tell which set or sets each number belongs to: natural numbers, whole numbers, integers, rational numbers, irrational numbers, and real numbers. See Example 5.

35. 0

36. $\frac{1}{4}$

37. −2

38. $-\frac{1}{2}$

39. 6
40. 5
41. $\frac{2}{3}$
42. $\sqrt{3}$
43. $-\sqrt{5}$
44. $-1\frac{5}{9}$

Tell whether each statement is true or false.

45. Every rational number is also an integer.
46. Every negative number is also a rational number.
47. Every natural number is positive.
48. Every rational number is also a real number.
49. 0 is a real number.
50. Every real number is also a rational number.
51. Every whole number is an integer.
52. $\frac{1}{2}$ is an integer.
53. A number can be both rational and irrational.
54. Every whole number is positive.

Insert $<$, $>$, or $=$ in the appropriate space to make a true statement. See Examples 6 through 8.

55. $-10 \quad -100$
56. $-200 \quad -20$
57. $32 \quad 5.2$
58. $7.1 \quad -7$
59. $\frac{18}{3} \quad \frac{24}{3}$
60. $\frac{8}{2} \quad \frac{12}{3}$
61. $-51 \quad -50$
62. $|-20| \quad -200$
63. $|-5| \quad -4$
64. $0 \quad |0|$
65. $|-1| \quad |1|$
66. $\left|\frac{2}{5}\right| \quad \left|-\frac{2}{5}\right|$
67. $|-2| \quad |-3|$
68. $-500 \quad |-50|$
69. $|0| \quad |-8|$
70. $|-12| \quad \frac{24}{2}$

CONCEPT EXTENSIONS

The graph below is called a bar graph. This particular graph shows the annual numbers of recreational visitors to U.S. National Parks. Each bar represents a different year, and the height of the bar represents the number of visitors (in millions) in that year.

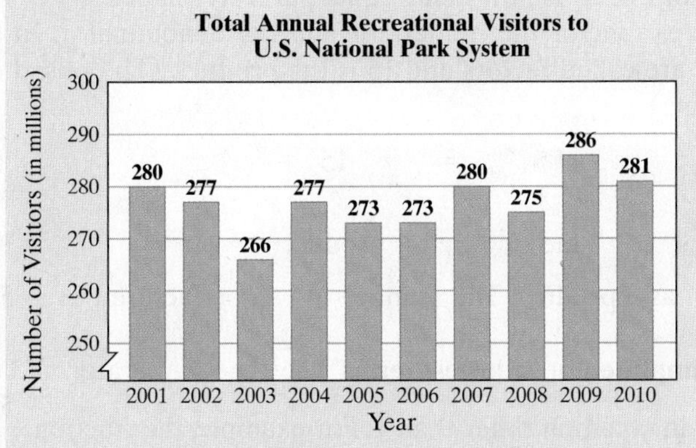

(*Note:* The ⚡ symbol means that some numbers are missing. Along the vertical data line, notice the numbers between 0 and 250 are missing or not shown.)

71. In which year(s) was the number of visitors the greatest?
72. What was the greatest number of visitors shown?
73. In what year(s) was the number of visitors greater than 280 million?
74. In what year(s) was the number of visitors less than 270 million?
75. Write an inequality statement comparing the number of annual visitors in 2001 and 2010.
76. Do you notice any trends shown by this bar graph?

The bar graph shows cranberry production from the top five cranberry-producing states.

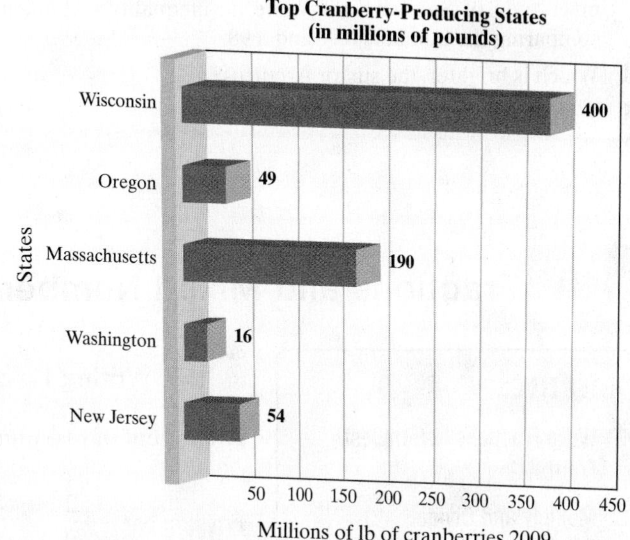

Data from National Agricultural Statistics Service

77. Write an inequality comparing the 2009 cranberry production in Oregon with the 2009 cranberry production in Washington.
78. Write an inequality comparing the 2009 cranberry production in Massachusetts with the 2009 cranberry production in Wisconsin.
79. Determine the difference between the 2009 cranberry production in Washington and the 2009 cranberry production in New Jersey.
80. According to the bar graph, which two states had almost equal 2009 cranberry crops?

The apparent magnitude of a star is the measure of its brightness as seen by someone on Earth. The smaller the apparent magnitude, the brighter the star. Use the apparent magnitudes in the table on page 16 to answer Exercises 81 through 86.

81. The apparent magnitude of the sun is -26.7. The apparent magnitude of the star Arcturus is -0.04. Write an inequality statement comparing the numbers -0.04 and -26.7.

Star	Apparent Magnitude	Star	Apparent Magnitude
Arcturus	−0.04	Spica	0.98
Sirius	−1.46	Rigel	0.12
Vega	0.03	Regulus	1.35
Antares	0.96	Canopus	−0.72
Sun	−26.7	Hadar	0.61

(Data from *Norton's Star Atlas and Reference Handbook*, 20th Edition, edited by Ian Ridpath. © 2004 Pearson Education, Inc.)

82. The apparent magnitude of Antares is 0.96. The apparent magnitude of Spica is 0.98. Write an inequality statement comparing the numbers 0.96 and 0.98.

83. Which is brighter, the sun or Arcturus?

84. Which is dimmer, Antares or Spica?

85. Which star listed is the brightest?

86. Which star listed is the dimmest?

Rewrite the following inequalities so that the inequality symbol points in the opposite direction and the resulting statement has the same meaning as the given one.

87. $25 \geq 20$

88. $-13 \leq 13$

89. $0 < 6$

90. $5 > 3$

91. $-10 > -12$

92. $-4 < -2$

93. In your own words, explain how to find the absolute value of a number.

94. Give an example of a real-life situation that can be described with integers but not with whole numbers.

1.3 Fractions and Mixed Numbers 

OBJECTIVES

1 Write Fractions in Simplest Form.

2 Multiply and Divide Fractions.

3 Add and Subtract Fractions.

4 Perform Operations on Mixed Numbers.

OBJECTIVE

1 Writing Fractions in Simplest Form 

A quotient of two numbers such as $\frac{2}{9}$ is called a **fraction.** The parts of a fraction are:

$$\text{Fraction bar} \rightarrow \frac{2}{9} \begin{array}{l} \leftarrow \text{Numerator} \\ \leftarrow \text{Denominator} \end{array}$$

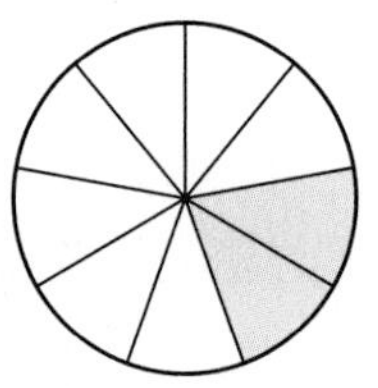

$\frac{2}{9}$ of the circle is shaded.

A fraction may be used to refer to part of a whole. For example, $\frac{2}{9}$ of the circle above is shaded. The denominator 9 tells us how many equal parts the whole circle is divided into, and the numerator 2 tells us how many equal parts are shaded.

To simplify fractions, we can factor the numerator and the denominator. In the statement $3 \cdot 5 = 15$, 3 and 5 are called **factors** and 15 is the **product.** (The raised dot symbol indicates multiplication.)

$$\underset{\text{factor}}{\underset{\uparrow}{3}} \cdot \underset{\text{factor}}{\underset{\uparrow}{5}} = \underset{\text{product}}{\underset{\uparrow}{15}}$$

To **factor** 15 means to write it as a product. The number 15 can be factored as $3 \cdot 5$ or as $1 \cdot 15$.

A fraction is said to be **simplified** or in **lowest terms** when the numerator and the denominator have no factors in common other than 1. For example, the fraction $\frac{5}{11}$ is in lowest terms since 5 and 11 have no common factors other than 1.

To help us simplify fractions, we write the numerator and the denominator as a product of **prime numbers.**

Prime Number and Composite Number

A **prime number** is a natural number, other than 1, whose only factors are 1 and itself. The first few prime numbers are

2, 3, 5, 7, 11, 13, 17, 19, 23, 29, and so on.

A natural number, other than 1, that is not a prime number is called a **composite number.**

> **Helpful Hint**
> The natural number 1 is neither prime nor composite.

Every composite number can be written as a product of prime numbers. We call this product of prime numbers the **prime factorization** of the composite number.

EXAMPLE 1 Write each of the following numbers as a product of primes.

a. 40 **b.** 63

Solution

a. First, write 40 as the product of any two whole numbers other than 1.

$$40 = 4 \cdot 10$$

Next, factor each of these numbers. Continue this process until all of the factors are prime numbers.

$$\begin{aligned} 40 &= 4 \cdot 10 \\ &\quad\downarrow\searrow \quad \downarrow\searrow \\ &= 2 \cdot 2 \cdot 2 \cdot 5 \end{aligned}$$

All the factors are now prime numbers. Then 40 written as a product of primes is

$$40 = 2 \cdot 2 \cdot 2 \cdot 5$$

b. $$\begin{aligned} 63 &= 9 \cdot 7 \\ &\quad\downarrow\searrow \quad \downarrow \\ &= 3 \cdot 3 \cdot 7 \end{aligned}$$ □

PRACTICE
1 Write each of the following numbers as a product of primes.

a. 36 **b.** 200

To use prime factors to write a fraction in lowest terms (or simplified form), apply the fundamental principle of fractions.

Fundamental Principle of Fractions

If $\frac{a}{b}$ is a fraction and c is a nonzero real number, then

$$\frac{a \cdot c}{b \cdot c} = \frac{a}{b}$$

To understand why this is true, we use the fact that since c is not zero, $\frac{c}{c} = 1$.

$$\frac{a \cdot c}{b \cdot c} = \frac{a}{b} \cdot \frac{c}{c} = \frac{a}{b} \cdot 1 = \frac{a}{b}$$

We will call this process dividing out the common factor of c.

EXAMPLE 2 Simplify each fraction (write it in lowest terms).

a. $\frac{42}{49}$ **b.** $\frac{11}{27}$ **c.** $\frac{88}{20}$

Solution

a. Write the numerator and the denominator as products of primes; then apply the fundamental principle to the common factor 7.

$$\frac{42}{49} = \frac{2 \cdot 3 \cdot 7}{7 \cdot 7} = \frac{2 \cdot 3}{7} \cdot \frac{7}{7} = \frac{2 \cdot 3}{7} = \frac{6}{7}$$

b. $\frac{11}{27} = \frac{11}{3 \cdot 3 \cdot 3}$

There are no common factors other than 1, so $\frac{11}{27}$ is already in simplest form.

c. $\frac{88}{20} = \frac{2 \cdot 2 \cdot 2 \cdot 11}{2 \cdot 2 \cdot 5} = \frac{2}{2} \cdot \frac{2}{2} \cdot \frac{2 \cdot 11}{5} = \frac{22}{5}$ □

PRACTICE
2 Write each fraction in lowest terms.

a. $\frac{63}{72}$ **b.** $\frac{64}{12}$ **c.** $\frac{7}{25}$

✓CONCEPT CHECK

Explain the error in the following steps.

a. $\frac{15}{55} = \frac{15}{55} = \frac{1}{5}$

b. $\frac{6}{7} = \frac{5+1}{5+2} = \frac{1}{2}$

OBJECTIVE

2 Multiplying and Dividing Fractions 

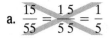

To multiply two fractions, multiply numerator times numerator to obtain the numerator of the product; multiply denominator times denominator to obtain the denominator of the product.

Multiplying Fractions

$$\frac{a}{b} \cdot \frac{c}{d} = \frac{a \cdot c}{b \cdot d} \quad \text{if } b \neq 0 \text{ and } d \neq 0$$

EXAMPLE 3 Multiply $\frac{2}{15}$ and $\frac{5}{13}$. Simplify the product if possible.

Solution

$$\frac{2}{15} \cdot \frac{5}{13} = \frac{2 \cdot 5}{15 \cdot 13}$$ Multiply numerators. Multiply denominators.

Next, simplify the product by dividing the numerator and the denominator by any common factors.

$$= \frac{2 \cdot \overset{1}{\cancel{5}}}{3 \cdot \underset{1}{\cancel{5}} \cdot 13}$$

$$= \frac{2}{39}$$ □

PRACTICE
3 Multiply $\frac{3}{8}$ and $\frac{7}{9}$. Simplify the product if possible.

Answers to Concept Check:
answers may vary

Before dividing fractions, we first define **reciprocals.** Two fractions are reciprocals of each other if their product is 1.

For example:

The reciprocal of $\frac{2}{3}$ is $\frac{3}{2}$ because $\frac{2}{3}\cdot\frac{3}{2} = \frac{6}{6} = 1$.

The reciprocal of 5 is $\frac{1}{5}$ because $5\cdot\frac{1}{5} = \frac{5}{1}\cdot\frac{1}{5} = \frac{5}{5} = 1$.

To divide fractions, multiply the first fraction by the reciprocal of the second fraction.

Dividing Fractions

$$\frac{a}{b} \div \frac{c}{d} = \frac{a}{b}\cdot\frac{d}{c}, \qquad \text{if } b \neq 0, d \neq 0, \text{and } c \neq 0$$

EXAMPLE 4 Divide. Simplify all quotients if possible.

a. $\frac{4}{5} \div \frac{5}{16}$ **b.** $\frac{7}{10} \div 14$ **c.** $\frac{3}{8} \div \frac{3}{10}$

Solution

a. $\frac{4}{5} \div \frac{5}{16} = \frac{4}{5}\cdot\frac{16}{5} = \frac{4\cdot16}{5\cdot5} = \frac{64}{25}$ — The numerator and denominator have no common factors.

b. $\frac{7}{10} \div 14 = \frac{7}{10} \div \frac{14}{1} = \frac{7}{10}\cdot\frac{1}{14} = \frac{\overset{1}{\cancel{7}}\cdot 1}{2\cdot5\cdot2\cdot\underset{1}{\cancel{7}}} = \frac{1}{20}$

c. $\frac{3}{8} \div \frac{3}{10} = \frac{3}{8}\cdot\frac{10}{3} = \frac{\overset{1}{\cancel{3}}\cdot\overset{1}{\cancel{2}}\cdot5}{\underset{1}{\cancel{2}}\cdot2\cdot2\cdot\underset{1}{\cancel{3}}} = \frac{5}{4}$ □

PRACTICE 4 Divide. Simplify all quotients if possible.

a. $\frac{3}{4} \div \frac{4}{9}$ **b.** $\frac{5}{12} \div 15$ **c.** $\frac{7}{6} \div \frac{7}{15}$

OBJECTIVE 3

Adding and Subtracting Fractions

To add or subtract fractions with the same denominator, combine numerators and place the sum or difference over the common denominator.

Adding and Subtracting Fractions with the Same Denominator

$$\frac{a}{b} + \frac{c}{b} = \frac{a+c}{b}, \qquad \text{if } b \neq 0$$

$$\frac{a}{b} - \frac{c}{b} = \frac{a-c}{b}, \qquad \text{if } b \neq 0$$

EXAMPLE 5 Add or subtract as indicated. Simplify each result if possible.

a. $\frac{2}{7} + \frac{4}{7}$ **b.** $\frac{3}{10} + \frac{2}{10}$ **c.** $\frac{9}{7} - \frac{2}{7}$ **d.** $\frac{5}{3} - \frac{1}{3}$

Solution

a. $\frac{2}{7} + \frac{4}{7} = \frac{2+4}{7} = \frac{6}{7}$

b. $\frac{3}{10} + \frac{2}{10} = \frac{3+2}{10} = \frac{5}{10} = \frac{\overset{1}{\cancel{5}}}{2\cdot\underset{1}{\cancel{5}}} = \frac{1}{2}$

c. $\frac{9}{7} - \frac{2}{7} = \frac{9-2}{7} = \frac{7}{7} = 1$

d. $\frac{5}{3} - \frac{1}{3} = \frac{5-1}{3} = \frac{4}{3}$ □

PRACTICE

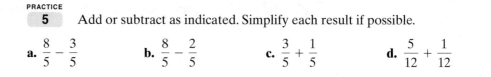

5 Add or subtract as indicated. Simplify each result if possible.

a. $\frac{8}{5} - \frac{3}{5}$ **b.** $\frac{8}{5} - \frac{2}{5}$ **c.** $\frac{3}{5} + \frac{1}{5}$ **d.** $\frac{5}{12} + \frac{1}{12}$

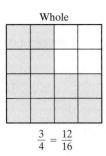

Whole

$\frac{3}{4} = \frac{12}{16}$

To add or subtract fractions without the same denominator, first write the fractions as equivalent fractions with a common denominator. **Equivalent fractions** are fractions that represent the same quantity. For example,

$$\frac{3}{4} \text{ and } \frac{12}{16} \text{ are equivalent fractions}$$

since they represent the same portion of a whole, as the diagram shows. Count the larger squares, and the shaded portion is $\frac{3}{4}$. Count the smaller squares, and the shaded portion is $\frac{12}{16}$. Thus, $\frac{3}{4} = \frac{12}{16}$.

We can write equivalent fractions by multiplying a given fraction by 1, as shown in the next example. Multiplying a fraction by 1 does not change the value of the fraction.

EXAMPLE 6 Write $\frac{2}{5}$ as an equivalent fraction with a denominator of 20.

Solution Since $5 \cdot 4 = 20$, multiply the fraction by $\frac{4}{4}$. Multiplying by $\frac{4}{4} = 1$ does not change the value of the fraction.

Multiply by $\frac{4}{4}$ or 1.

$$\frac{2}{5} = \frac{2}{5} \cdot \frac{4}{4} = \frac{2 \cdot 4}{5 \cdot 4} = \frac{8}{20}$$

Thus, $\frac{2}{5} = \frac{8}{20}$. □

PRACTICE

6 Write $\frac{2}{3}$ as an equivalent fraction with a denominator of 21.

To add or subtract with different denominators, we first write the fractions as **equivalent fractions** with the same denominator. We use the smallest or **least common denominator,** or **LCD.** (The LCD is the same as the least common multiple of the denominators.)

EXAMPLE 7 Add or subtract as indicated. Write each answer in simplest form.

a. $\frac{2}{5} + \frac{1}{4}$ **b.** $\frac{19}{6} - \frac{23}{12}$ **c.** $\frac{1}{2} + \frac{17}{22} - \frac{2}{11}$

Solution

a. Fractions must have a common denominator before they can be added or subtracted. Since 20 is the smallest number that both 5 and 4 divide into evenly, 20 is the **least common denominator** (LCD). Write both fractions as equivalent fractions with denominators of 20. Since

$$\frac{2}{5} \cdot \frac{4}{4} = \frac{2 \cdot 4}{5 \cdot 4} = \frac{8}{20} \quad \text{and} \quad \frac{1}{4} \cdot \frac{5}{5} = \frac{1 \cdot 5}{4 \cdot 5} = \frac{5}{20}$$

then

$$\frac{2}{5} + \frac{1}{4} = \frac{8}{20} + \frac{5}{20} = \frac{13}{20}$$

b. The LCD is 12. We write both fractions as equivalent fractions with denominators of 12.

$$\frac{19}{6} - \frac{23}{12} = \frac{38}{12} - \frac{23}{12}$$

$$= \frac{15}{12} = \frac{\overset{1}{\cancel{3}} \cdot 5}{2 \cdot 2 \cdot \underset{1}{\cancel{3}}} = \frac{5}{4}$$

c. The LCD for denominators 2, 22, and 11 is 22. First, write each fraction as an equivalent fraction with a denominator of 22. Then add or subtract from left to right.

$$\frac{1}{2} = \frac{1}{2} \cdot \frac{11}{11} = \frac{11}{22}, \quad \frac{17}{22} = \frac{17}{22}, \quad \text{and} \quad \frac{2}{11} = \frac{2}{11} \cdot \frac{2}{2} = \frac{4}{22}$$

Then

$$\frac{1}{2} + \frac{17}{22} - \frac{2}{11} = \frac{11}{22} + \frac{17}{22} - \frac{4}{22} = \frac{24}{22} = \frac{12}{11}$$

□

PRACTICE

7 Add or subtract as indicated. Write answers in simplest form.

a. $\frac{5}{11} + \frac{1}{7}$ **b.** $\frac{5}{21} - \frac{1}{6}$ **c.** $\frac{1}{3} + \frac{29}{30} - \frac{4}{5}$

OBJECTIVE

4 Performing Operations on Mixed Numbers 

To multiply or divide mixed numbers, first write each mixed number as an improper fraction. To recall how this is done, let's write $3\frac{1}{5}$ as an improper fraction.

$$3\frac{1}{5} = 3 + \frac{1}{5} = \frac{15}{5} + \frac{1}{5} = \frac{16}{5}$$

Because of the steps above, notice that we can use a shortcut process for writing a mixed number as an improper fraction.

$$3\frac{1}{5} = \frac{5 \cdot 3 + 1}{5} = \frac{16}{5}$$

EXAMPLE 8 Divide: $2\frac{1}{8} \div 1\frac{2}{3}$

Solution First write each mixed number as an improper fraction.

$$2\frac{1}{8} = \frac{8 \cdot 2 + 1}{8} = \frac{17}{8}; \qquad 1\frac{2}{3} = \frac{3 \cdot 1 + 2}{3} = \frac{5}{3}$$

Now divide as usual.

$$2\frac{1}{8} \div 1\frac{2}{3} = \frac{17}{8} \div \frac{5}{3} = \frac{17}{8} \cdot \frac{3}{5} = \frac{51}{40}$$

The fraction $\frac{51}{40}$ is improper. To write it as an equivalent mixed number, remember that the fraction bar means division and divide.

$$\begin{array}{r} 1\frac{11}{40} \\ 40\overline{)51} \\ -40 \\ \hline 11 \end{array}$$

Thus, the quotient is $\frac{51}{40}$ or $1\frac{11}{40}$. □

PRACTICE

8 Multiply: $5\frac{1}{6} \cdot 4\frac{2}{5}$

As a general rule, if the original exercise contains mixed numbers, write the result as a mixed number if possible.

When adding or subtracting mixed numbers, you might want to use the following method.

EXAMPLE 9 Subtract: $50\frac{1}{6} - 38\frac{1}{3}$

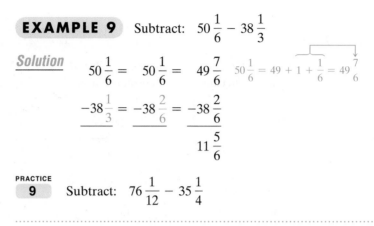

Solution

$$50\frac{1}{6} = 50\frac{1}{6} = 49\frac{7}{6} \qquad 50\frac{1}{6} = 49 + 1 + \frac{1}{6} = 49\frac{7}{6}$$

$$-38\frac{1}{3} = -38\frac{2}{6} = -38\frac{2}{6}$$

$$11\frac{5}{6}$$

□

PRACTICE 9 Subtract: $76\frac{1}{12} - 35\frac{1}{4}$

Vocabulary, Readiness & Video Check

Use the choices below to fill in each blank. Some choices may be used more than once.

simplified	reciprocals	equivalent	denominator
product	factors	fraction	numerator

1. A quotient of two numbers, such as $\frac{5}{8}$, is called a(n) ______________.
2. In the fraction $\frac{3}{11}$, the number 3 is called the ______________ and the number 11 is called the ______________.
3. To factor a number means to write it as a(n) ______________.
4. A fraction is said to be ______________ when the numerator and the denominator have no common factors other than 1.
5. In $7 \cdot 3 = 21$, the numbers 7 and 3 are called ______________ and the number 21 is called the ______________.
6. The fractions $\frac{2}{9}$ and $\frac{9}{2}$ are called ______________.
7. Fractions that represent the same quantity are called ______________ fractions.

Martin-Gay Interactive Videos

See Video 1.3

Watch the section lecture video and answer the following questions.

OBJECTIVE 1

8. What is the common factor in the numerator and denominator of Example 1? What principle is used to simplify this fraction?

OBJECTIVE 2

9. During the solving of Example 3, what two things change in the first step?

OBJECTIVE 3

10. What is the first step needed in order to subtract the fractions in Example 6 and why?

OBJECTIVE 4

11. For Example 7, why is the sum not left as $4\frac{7}{6}$?

1.3 Exercise Set

MyMathLab®

Represent the shaded part of each geometric figure by a fraction.

1.

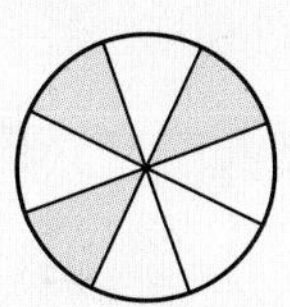

2.

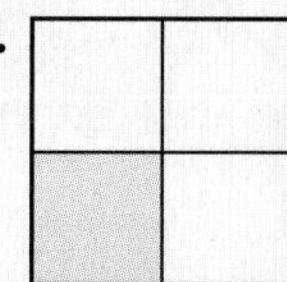

3.

4.

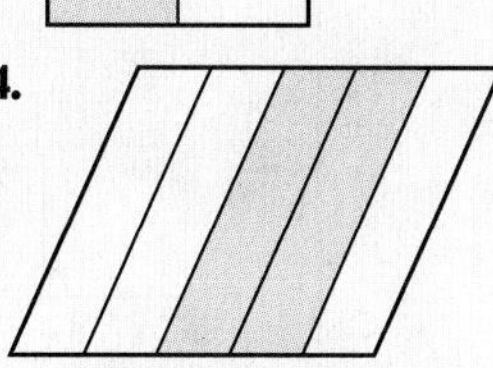

Write each number as a product of primes. See Example 1.

5. 33 **6.** 60 **7.** 98

8. 27 **9.** 20 **10.** 56

11. 75 **12.** 32 **13.** 45

14. 24

Write the fraction in lowest terms. See Example 2.

15. $\frac{2}{4}$ **16.** $\frac{3}{6}$ **17.** $\frac{10}{15}$

18. $\frac{15}{20}$ **19.** $\frac{3}{7}$ **20.** $\frac{5}{9}$

21. $\frac{18}{30}$ **22.** $\frac{42}{45}$ **23.** $\frac{120}{244}$

24. $\frac{360}{700}$

Multiply or divide as indicated. Simplify the answer if possible. See Examples 3 and 4.

25. $\frac{1}{2} \cdot \frac{3}{4}$ **26.** $\frac{7}{11} \cdot \frac{3}{5}$ **27.** $\frac{2}{3} \cdot \frac{3}{4}$

28. $\frac{7}{8} \cdot \frac{3}{21}$ **29.** $\frac{1}{2} \div \frac{7}{12}$ **30.** $\frac{7}{12} \div \frac{1}{2}$

31. $\frac{3}{4} \div \frac{1}{20}$ **32.** $\frac{3}{5} \div \frac{9}{10}$ **33.** $\frac{7}{10} \cdot \frac{5}{21}$

34. $\frac{3}{35} \cdot \frac{10}{63}$ **35.** $\frac{25}{9} \cdot \frac{1}{3}$ **36.** $\frac{1}{4} \cdot \frac{19}{6}$

The area of a plane figure is a measure of the amount of surface of the figure. Find the area of each figure below. (The area of a rectangle is the product of its length and width. The area of a triangle is $\frac{1}{2}$ the product of its base and height.)

37.

△ **38.**

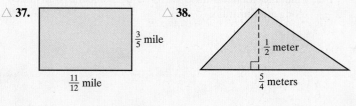

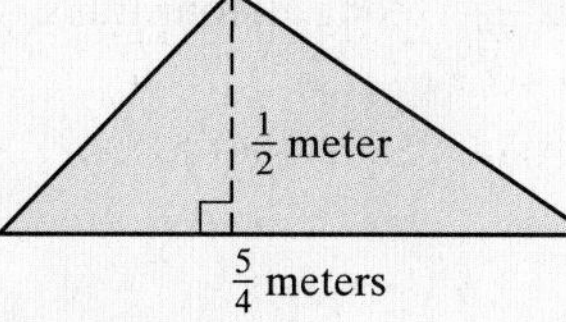

△ **39.**

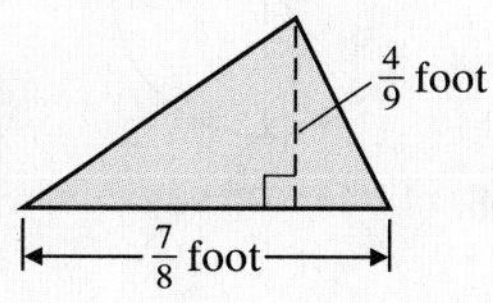

△ **40.**

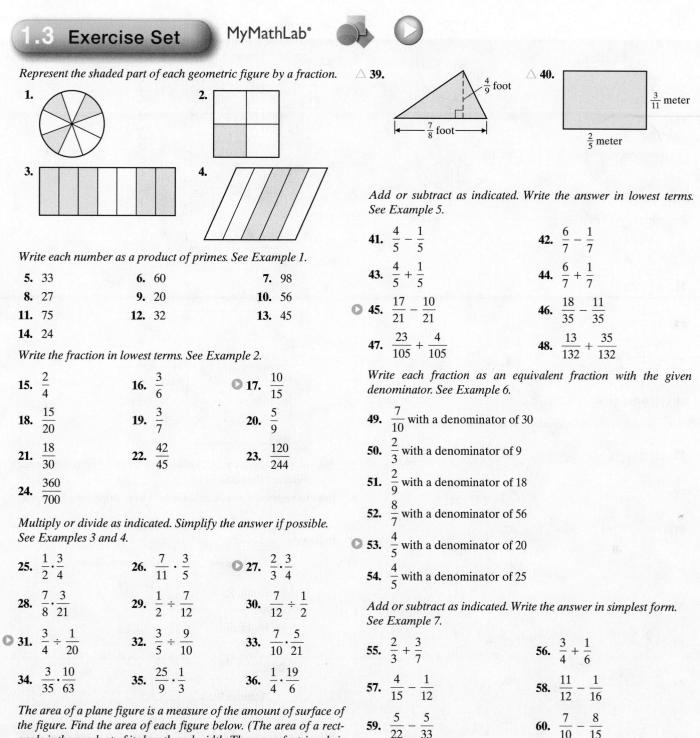

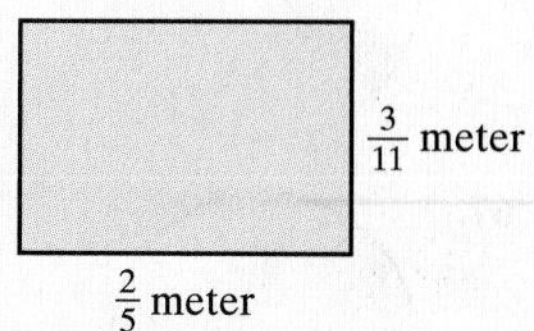

Add or subtract as indicated. Write the answer in lowest terms. See Example 5.

41. $\frac{4}{5} - \frac{1}{5}$ **42.** $\frac{6}{7} - \frac{1}{7}$

43. $\frac{4}{5} + \frac{1}{5}$ **44.** $\frac{6}{7} + \frac{1}{7}$

45. $\frac{17}{21} - \frac{10}{21}$ **46.** $\frac{18}{35} - \frac{11}{35}$

47. $\frac{23}{105} + \frac{4}{105}$ **48.** $\frac{13}{132} + \frac{35}{132}$

Write each fraction as an equivalent fraction with the given denominator. See Example 6.

49. $\frac{7}{10}$ with a denominator of 30

50. $\frac{2}{3}$ with a denominator of 9

51. $\frac{2}{9}$ with a denominator of 18

52. $\frac{8}{7}$ with a denominator of 56

53. $\frac{4}{5}$ with a denominator of 20

54. $\frac{4}{5}$ with a denominator of 25

Add or subtract as indicated. Write the answer in simplest form. See Example 7.

55. $\frac{2}{3} + \frac{3}{7}$ **56.** $\frac{3}{4} + \frac{1}{6}$

57. $\frac{4}{15} - \frac{1}{12}$ **58.** $\frac{11}{12} - \frac{1}{16}$

59. $\frac{5}{22} - \frac{5}{33}$ **60.** $\frac{7}{10} - \frac{8}{15}$

61. $\frac{12}{5} - 1$ **62.** $2 - \frac{3}{8}$

Each circle in Exercises 63–68 represents a whole, or 1. Use subtraction to determine the unknown part of the circle.

63.

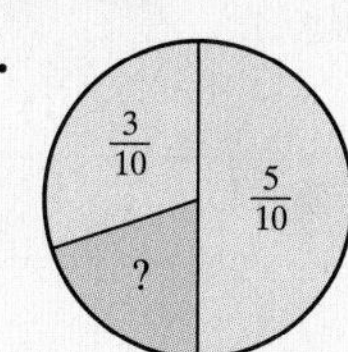

64.

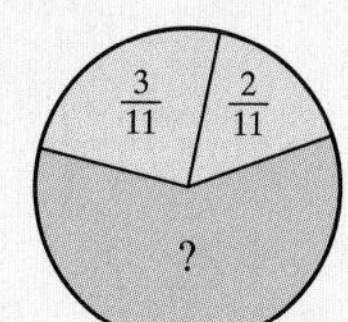

65.

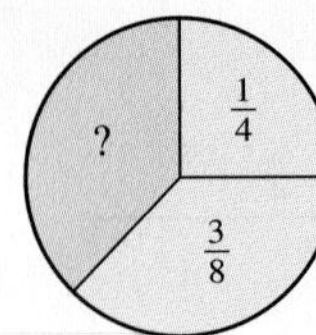

66.

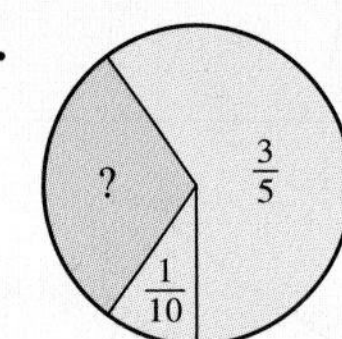

67.

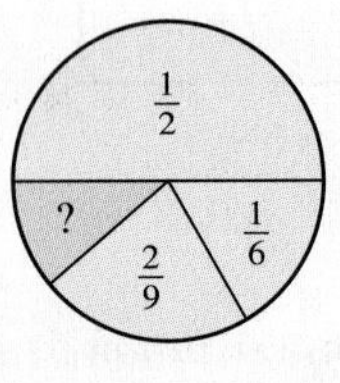

68.

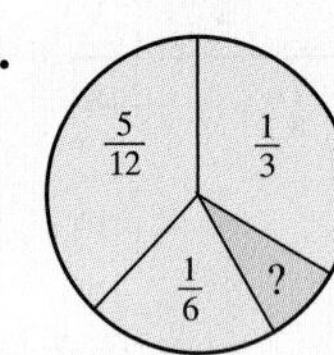

Perform the indicated operations. See Examples 8 and 9.

69. $5\frac{1}{9} \cdot 3\frac{2}{3}$

70. $2\frac{3}{4} \cdot 1\frac{7}{8}$

71. $8\frac{3}{5} \div 2\frac{9}{10}$

72. $1\frac{7}{8} \div 3\frac{8}{9}$

73. $17\frac{2}{5} + 30\frac{2}{3}$

74. $26\frac{11}{20} + 40\frac{7}{10}$

75. $8\frac{11}{12} - 1\frac{5}{6}$

76. $4\frac{7}{8} - 2\frac{3}{16}$

MIXED PRACTICE

Perform the following operations. Write answers in simplest form.

77. $\frac{10}{21} + \frac{5}{21}$

78. $\frac{11}{35} + \frac{3}{35}$

79. $\frac{10}{3} - \frac{5}{21}$

80. $\frac{11}{7} - \frac{3}{35}$

81. $\frac{2}{3} \cdot \frac{3}{5}$

82. $\frac{3}{4} \cdot \frac{7}{12}$

83. $\frac{2}{3} \div \frac{3}{5}$

84. $\frac{3}{4} \div \frac{7}{12}$

85. $5 + \frac{2}{3}$

86. $7 + \frac{1}{10}$

87. $7\frac{2}{5} \div \frac{1}{5}$

88. $9\frac{5}{6} \div \frac{1}{6}$

89. $\frac{1}{2} - \frac{14}{33}$

90. $\frac{7}{15} - \frac{7}{25}$

91. $\frac{23}{105} - \frac{2}{105}$

92. $\frac{57}{132} - \frac{13}{132}$

93. $1\frac{1}{2} + 3\frac{2}{3}$

94. $2\frac{3}{5} + 4\frac{7}{10}$

95. $\frac{2}{3} - \frac{5}{9} + \frac{5}{6}$

96. $\frac{8}{11} - \frac{1}{4} + \frac{1}{2}$

The perimeter of a plane figure is the total distance around the figure. Find the perimeter of each figure in Exercises 97 and 98.

△ **97.**

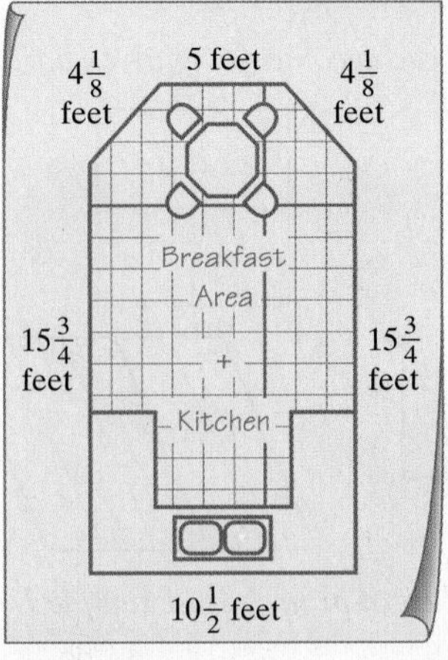

△ **98.**

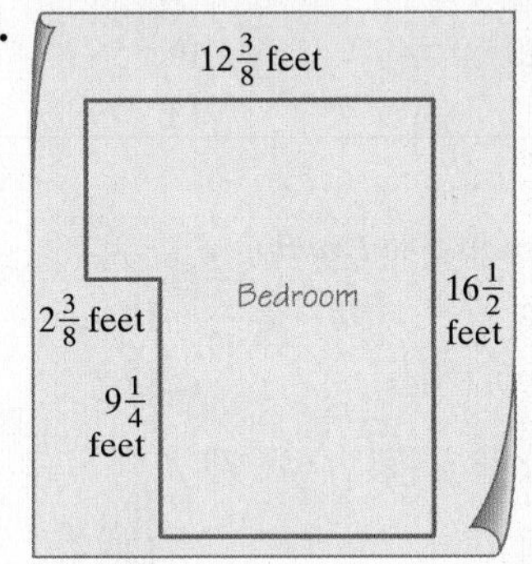

99. In your own words, explain how to add two fractions with different denominators.

100. In your own words, explain how to multiply two fractions.

The following trail chart is given to visitors at the Lakeview Forest Preserve.

Trail Name	*Distance (miles)*
Robin Path	$3\frac{1}{2}$
Red Falls	$5\frac{1}{2}$
Green Way	$2\frac{1}{8}$
Autumn Walk	$1\frac{3}{4}$

101. How much longer is Red Falls Trail than Green Way Trail?

102. Find the total distance traveled by someone who hiked along all four trails.

CONCEPT EXTENSIONS

The graph shown is called a circle graph or a pie chart. Use the graph to answer Exercises 103 through 106.

Fraction of U.S. Screens by Theater Type

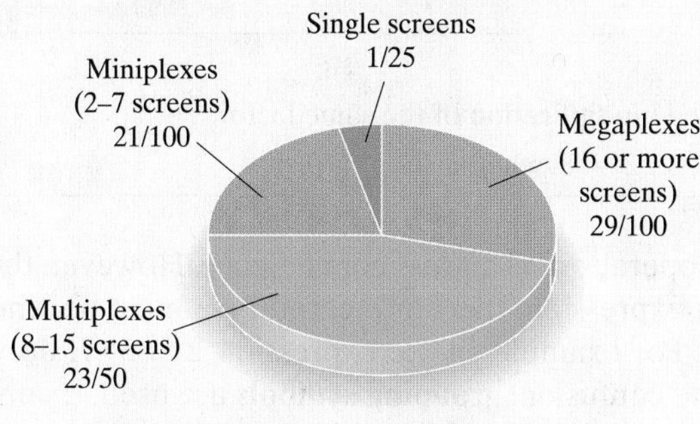

Data from Motion Picture Association of America

103. What fraction of U.S. movie screens are in miniplexes?

104. What fraction of U.S. movie screens are in single-screen theaters or multiplexes?

105. What theater type has the greatest fraction of screens?

106. What fraction of U.S. movie screens are in megaplexes or miniplexes?

For Exercises 107 through 110, determine whether the work is correct or incorrect. If incorrect, find the error and correct. See the Concept Check in this section.

107. $\frac{12}{24} = \frac{2+4+6}{2+4+6+12} = \frac{1}{12}$

108. $\frac{30}{60} = \frac{2\cdot3\cdot5}{2\cdot2\cdot3\cdot5} = \frac{1}{2}$

109. $\frac{2}{7} + \frac{9}{7} = \frac{11}{14}$

110. $\frac{16}{28} = \frac{2\cdot5+6\cdot1}{2\cdot5+6\cdot3} = \frac{1}{3}$

1.4 Exponents, Order of Operations, Variable Expressions, and Equations

OBJECTIVES

1 Define and Use Exponents and the Order of Operations.

2 Evaluate Algebraic Expressions, Given Replacement Values for Variables.

3 Determine Whether a Number Is a Solution of a Given Equation.

4 Translate Phrases into Expressions and Sentences into Statements.

OBJECTIVE

1 Using Exponents and the Order of Operations

Frequently in algebra, products occur that contain repeated multiplication of the same factor. For example, the volume of a cube whose sides each measure 2 centimeters is $(2\cdot2\cdot2)$ cubic centimeters. We may use **exponential notation** to write such products in a more compact form. For example,

$$2\cdot2\cdot2 \quad \textit{may be written as} \quad 2^3.$$

The 2 in 2^3 is called the **base;** it is the repeated factor. The 3 in 2^3 is called the **exponent** and is the number of times the base is used as a factor. The expression 2^3 is called an **exponential expression.**

exponent

base $2^3 = 2\cdot2\cdot2 = 8$

2 is a factor 3 times

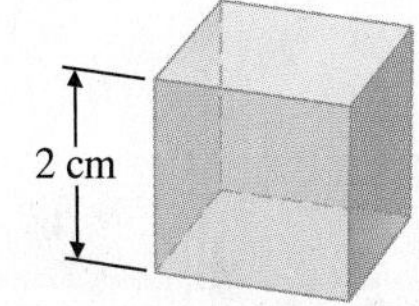

Volume is $(2\cdot2\cdot2)$ cubic centimeters.

EXAMPLE 1 Evaluate the following:

a. 3^2 [read as "3 squared" or as "3 to the second power"]

b. 5^3 [read as "5 cubed" or as "5 to the third power"]

c. 2^4 [read as "2 to the fourth power"]

d. 7^1 **e.** $\left(\frac{3}{7}\right)^2$

Solution

a. $3^2 = 3\cdot3 = 9$

b. $5^3 = 5\cdot5\cdot5 = 125$

c. $2^4 = 2\cdot2\cdot2\cdot2 = 16$

d. $7^1 = 7$

e. $\left(\frac{3}{7}\right)^2 = \left(\frac{3}{7}\right)\left(\frac{3}{7}\right) = \frac{9}{49}$ □

PRACTICE 1 Evaluate:

a. 1^3 **b.** 5^2 **c.** $\left(\frac{1}{10}\right)^2$ **d.** 9^1 **e.** $\left(\frac{2}{5}\right)^3$

> **Helpful Hint**
> $2^3 \neq 2 \cdot 3$ since 2^3 indicates repeated **multiplication** of the same factor.
> $$2^3 = 2 \cdot 2 \cdot 2 = 8, \text{ whereas } 2 \cdot 3 = 6.$$

Using symbols for mathematical operations is a great convenience. However, the more operation symbols present in an expression, the more careful we must be when performing the indicated operation. For example, in the expression $2 + 3 \cdot 7$, do we add first or multiply first? To eliminate confusion, **grouping symbols** are used. Examples of grouping symbols are parentheses ( ), brackets [ ], braces { }, and the fraction bar. If we wish $2 + 3 \cdot 7$ to be simplified by adding first, we enclose $2 + 3$ in parentheses.

$$(2 + 3) \cdot 7 = 5 \cdot 7 = 35$$

If we wish to multiply first, $3 \cdot 7$ may be enclosed in parentheses.

$$2 + (3 \cdot 7) = 2 + 21 = 23$$

To eliminate confusion when no grouping symbols are present, use the following agreed-upon order of operations.

> **Order of Operations**
>
> Simplify expressions using the order below. If grouping symbols such as parentheses are present, simplify expressions within those first, starting with the innermost set. If fraction bars are present, simplify the numerator and the denominator separately.
>
> **1.** Evaluate exponential expressions.
>
> **2.** Perform multiplications or divisions in order from left to right.
>
> **3.** Perform additions or subtractions in order from left to right.

Now simplify $2 + 3 \cdot 7$. There are no grouping symbols and no exponents, so we multiply and then add.

$$2 + 3 \cdot 7 = 2 + 21 \quad \text{Multiply.}$$
$$= 23 \quad \text{Add.}$$

EXAMPLE 2 Simplify each expression.

a. $6 \div 3 + 5^2$ **b.** $20 \div 5 \cdot 4$ **c.** $\frac{2(12 + 3)}{|-15|}$ **d.** $3 \cdot 4^2$ **e.** $\frac{3}{2} \cdot \frac{1}{2} - \frac{1}{2}$

Solution

a. Evaluate 5^2 first.

$$6 \div 3 + 5^2 = 6 \div 3 + 25$$

Next divide, then add.

$$= 2 + 25 \quad \text{Divide.}$$
$$= 27 \quad \text{Add.}$$

b. $20 \div 5 \cdot 4 = 4 \cdot 4$
$= 16$

> **Helpful Hint**
> Remember to multiply or divide in order from left to right.

c. First, simplify the numerator and the denominator separately.

$$\frac{2(12+3)}{|-15|} = \frac{2(15)}{15} \quad \text{Simplify numerator and denominator separately.}$$
$$= \frac{30}{15}$$
$$= 2 \quad \text{Simplify.}$$

d. In this example, only the 4 is squared. The factor of 3 is not part of the base because no grouping symbol includes it as part of the base.

$$3 \cdot 4^2 = 3 \cdot 16 \quad \text{Evaluate the exponential expression.}$$
$$= 48 \quad \text{Multiply.}$$

e. The order of operations applies to operations with fractions in exactly the same way as it applies to operations with whole numbers.

$$\frac{3}{2} \cdot \frac{1}{2} - \frac{1}{2} = \frac{3}{4} - \frac{1}{2} \quad \text{Multiply.}$$
$$= \frac{3}{4} - \frac{2}{4} \quad \text{The least common denominator is 4.}$$
$$= \frac{1}{4} \quad \text{Subtract.}$$

□

PRACTICE 2 Simplify each expression

a. $6 + 3 \cdot 9$ **b.** $4^3 \div 8 + 3$ **c.** $\left(\frac{2}{3}\right)^2 \cdot |-8|$

d. $\frac{9(14-6)}{|-2|}$ **e.** $\frac{7}{4} \cdot \frac{1}{4} - \frac{1}{4}$

Helpful Hint

Be careful when evaluating an exponential expression. In $3 \cdot 4^2$, the exponent 2 applies only to the base 4. In $(3 \cdot 4)^2$, we multiply first because of parentheses, so the exponent 2 applies to the product $3 \cdot 4$.

$$3 \cdot 4^2 = 3 \cdot 16 = 48 \qquad (3 \cdot 4)^2 = (12)^2 = 144$$

Expressions that include many grouping symbols can be confusing. When simplifying these expressions, keep in mind that grouping symbols separate the expression into distinct parts. Each is then simplified separately.

EXAMPLE 3 Simplify $\frac{3 + |4-3| + 2^2}{6-3}$.

Solution The fraction bar serves as a grouping symbol and separates the numerator and denominator. Simplify each separately. Also, the absolute value bars here serve as a grouping symbol. We begin in the numerator by simplifying within the absolute value bars.

$$\frac{3 + |4-3| + 2^2}{6-3} = \frac{3 + |1| + 2^2}{6-3} \quad \text{Simplify the expression inside the absolute value bars.}$$
$$= \frac{3 + 1 + 2^2}{3} \quad \text{Find the absolute value and simplify the denominator.}$$
$$= \frac{3 + 1 + 4}{3} \quad \text{Evaluate the exponential expression.}$$
$$= \frac{8}{3} \quad \text{Simplify the numerator.}$$

□

PRACTICE 3 Simplify $\frac{6^2 - 5}{3 + |6-5| \cdot 8}$.

EXAMPLE 4 Simplify $3[4 + 2(10 - 1)]$.

Solution Notice that both parentheses and brackets are used as grouping symbols. Start with the innermost set of grouping symbols.

> **Helpful Hint**
> Be sure to follow order of operations and resist the temptation to incorrectly add 4 and 2 first.

$$\begin{aligned} 3[4 + 2(10 - 1)] &= 3[4 + 2(9)] && \text{Simplify the expression in parentheses.} \\ &= 3[4 + 18] && \text{Multiply.} \\ &= 3[22] && \text{Add.} \\ &= 66 && \text{Multiply.} \end{aligned}$$

□

PRACTICE 4 Simplify $4[25 - 3(5 + 3)]$.

EXAMPLE 5 Simplify $\dfrac{8 + 2 \cdot 3}{2^2 - 1}$.

Solution

$$\frac{8 + 2 \cdot 3}{2^2 - 1} = \frac{8 + 6}{4 - 1} = \frac{14}{3}$$

□

PRACTICE 5 Simplify $\dfrac{36 \div 9 + 5}{5^2 - 3}$.

OBJECTIVE

2 Evaluating Algebraic Expressions 

In algebra, we use symbols, usually letters such as x, y, or z, to represent unknown numbers. A symbol that is used to represent a number is called a **variable.** An **algebraic expression** is a collection of numbers, variables, operation symbols, and grouping symbols. For example,

$$2x, \quad -3, \quad 2x + 10, \quad 5(p^2 + 1), \quad \text{and} \quad \frac{3y^2 - 6y + 1}{5}$$

are algebraic expressions.

Expression	*Meaning*
$2x$	$2 \cdot x$
$5(p^2 + 1)$	$5 \cdot (p^2 + 1)$
$3y^2$	$3 \cdot y^2$
xy	$x \cdot y$

If we give a specific value to a variable, we can **evaluate an algebraic expression.** To evaluate an algebraic expression means to find its numerical value once we know the values of the variables.

Algebraic expressions are often used in problem solving. For example, the expression

$$16t^2$$

gives the distance in feet (neglecting air resistance) that an object will fall in t seconds.

EXAMPLE 6 Evaluate each expression if $x = 3$ and $y = 2$.

a. $2x - y$ **b.** $\frac{3x}{2y}$ **c.** $\frac{x}{y} + \frac{y}{2}$ **d.** $x^2 - y^2$

Solution

a. Replace x with 3 and y with 2.

$$\begin{aligned} 2x - y &= 2(3) - 2 && \text{Let } x = 3 \text{ and } y = 2. \\ &= 6 - 2 && \text{Multiply.} \\ &= 4 && \text{Subtract.} \end{aligned}$$

b. $\frac{3x}{2y} = \frac{3 \cdot 3}{2 \cdot 2} = \frac{9}{4}$ Let $x = 3$ and $y = 2$.

c. Replace x with 3 and y with 2. Then simplify.

$$\frac{x}{y} + \frac{y}{2} = \frac{3}{2} + \frac{2}{2} = \frac{5}{2}$$

d. Replace x with 3 and y with 2.

$$x^2 - y^2 = 3^2 - 2^2 = 9 - 4 = 5$$

□

PRACTICE 6 Evaluate each expression if $x = 2$ and $y = 5$.

a. $2x + y$ **b.** $\frac{4x}{3y}$ **c.** $\frac{3}{x} + \frac{x}{y}$ **d.** $x^3 + y^2$

OBJECTIVE 3

Determining Whether a Number Is a Solution of an Equation

Many times, a problem-solving situation is modeled by an equation. An **equation** is a mathematical statement that two expressions have equal value. The equal symbol "=" is used to equate the two expressions. For example,

$$3 + 2 = 5,\quad 7x = 35,\quad \frac{2(x - 1)}{3} = 0, \text{ and } I = PRT \text{ are all equations.}$$

> **Helpful Hint**
> An equation contains the equal symbol "=". An algebraic expression does not.

✓CONCEPT CHECK
Which of the following are equations? Which are expressions?

a. $5x = 8$ **b.** $5x - 8$ **c.** $12y + 3x$ **d.** $12y = 3x$

When an equation contains a variable, deciding which values of the variable make an equation a true statement is called **solving** an equation for the variable. A **solution** of an equation is a value for the variable that makes the equation true. For example, 3 is a solution of the equation $x + 4 = 7$ because if x is replaced with 3, the statement is true.

$$x + 4 = 7$$
$$\downarrow$$
$$3 + 4 = 7 \quad \text{Replace } x \text{ with 3.}$$
$$7 = 7 \quad \text{True}$$

Similarly, 1 is not a solution of the equation $x + 4 = 7$ because $1 + 4 = 7$ is **not** a true statement.

Answers to Concept Check:
equations: a, d; expressions: b, c.

EXAMPLE 7 Decide whether 2 is a solution of $3x + 10 = 8x$.

Solution Replace x with 2 and see if a true statement results.

$$3x + 10 = 8x \quad \text{Original equation}$$
$$3(2) + 10 \stackrel{?}{=} 8(2) \quad \text{Replace } x \text{ with 2.}$$
$$6 + 10 \stackrel{?}{=} 16 \quad \text{Simplify each side.}$$
$$16 = 16 \quad \text{True}$$

Since we arrived at a true statement after replacing x with 2 and simplifying both sides of the equation, 2 is a solution of the equation. □

PRACTICE 7 Decide whether 4 is a solution of $9x - 6 = 7x$.

OBJECTIVE 4

Translating Phrases to Expressions and Sentences to Statements

Now that we know how to represent an unknown number by a variable, let's practice translating phrases into algebraic expressions and sentences into statements. Oftentimes, solving problems requires the ability to translate word phrases and sentences into symbols. Below is a list of some key words and phrases to help us translate.

Helpful Hint
Order matters when subtracting and dividing, so be especially careful with these translations.

Addition (+)	*Subtraction* (−)	*Multiplication* (·)	*Division* (÷)	*Equality* (=)
Sum	Difference of	Product	Quotient	Equals
Plus	Minus	Times	Divide	Gives
Added to	Subtracted from	Multiply	Into	Is/was/should be
More than	Less than	Twice	Ratio	Yields
Increased by	Decreased by	Of	Divided by	Amounts to
Total	Less			Represents/ Is the same as

EXAMPLE 8 Write an algebraic expression that represents each phrase. Let the variable x represent the unknown number.

a. The sum of a number and 3
b. The product of 3 and a number
c. Twice a number
d. 10 decreased by a number
e. 5 times a number, increased by 7

Solution

a. $x + 3$ since "sum" means to add
b. $3 \cdot x$ and $3x$ are both ways to denote the product of 3 and x
c. $2 \cdot x$ or $2x$
d. $10 - x$ because "decreased by" means to subtract
e. $\underbrace{5x}_{\text{5 times a number}} + 7$ □

PRACTICE 8 Write an algebraic expression that represents each phase. Let the variable x represent the unknown number.

a. Six times a number
b. A number decreased by 8
c. The product of a number and 9
d. Two times a number, plus 3
e. The sum of 7 and a number

Helpful Hint

Make sure you understand the difference when translating phrases containing "decreased by," "subtracted from," and "less than."

Phrase	***Translation***	
A number decreased by 10	$x - 10$	Notice the order.
A number subtracted from 10	$10 - x$	
10 less than a number	$x - 10$	
A number less 10	$x - 10$	

Now let's practice translating sentences into equations.

EXAMPLE 9 Write each sentence as an equation or inequality. Let x represent the unknown number.

a. The quotient of 15 and a number is 4.

b. Three subtracted from 12 is a number.

c. Four times a number, added to 17, is not equal to 21.

d. Triple a number is less than 48.

Solution

a. In words: the quotient of 15 and a number | is | 4

Translate: $\frac{15}{x} = 4$

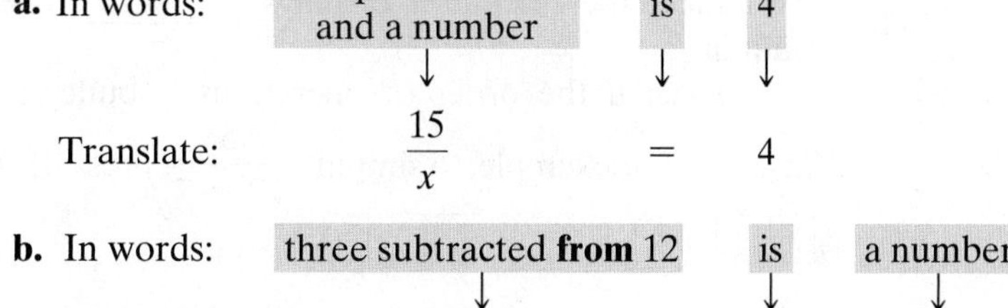

b. In words: three subtracted **from** 12 | is | a number

Translate: $12 - 3 = x$

Care must be taken when the operation is subtraction. The expression $3 - 12$ would be incorrect. Notice that $3 - 12 \neq 12 - 3$.

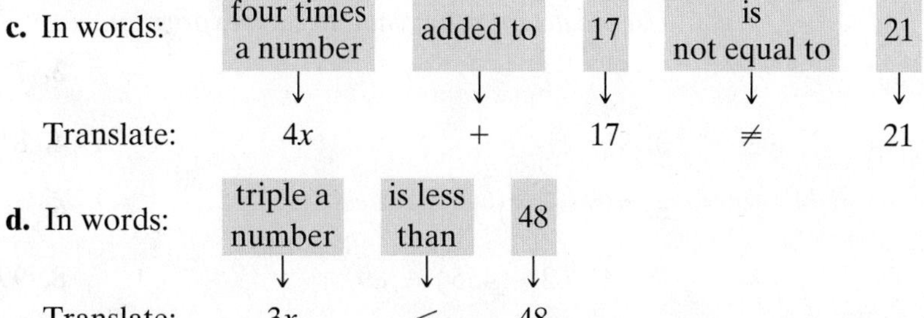

c. In words: four times a number | added to | 17 | is not equal to | 21

Translate: $4x + 17 \neq 21$

d. In words: triple a number | is less than | 48

Translate: $3x < 48$ □

PRACTICE

9 Write each sentence as an equation or inequality. Let x represent the unknown number.

a. A number increased by 7 is equal to 13.

b. Two less than a number is 11.

c. Double a number, added to 9, is not equal to 25.

d. Five times 11 is greater than or equal to an unknown number.

Calculator Explorations

Exponents

To evaluate exponential expressions on a scientific calculator, find the key marked $\boxed{y^x}$ or $\boxed{\wedge}$. To evaluate, for example, 3^5, press the following keys: $\boxed{3}$ $\boxed{y^x}$ $\boxed{5}$ $\boxed{=}$ or $\boxed{3}$ $\boxed{\wedge}$ $\boxed{5}$ $\boxed{=}$.

↕ or

$\boxed{\text{ENTER}}$

The display should read $\boxed{243}$ or $\boxed{\begin{matrix}3\wedge 5\\ 243\end{matrix}}$

Order of Operations

Some calculators follow the order of operations, and others do not. To see whether your calculator has the order of operations built in, use your calculator to find $2 + 3 \cdot 4$. To do this, press the following sequence of keys:

$\boxed{2}$ $\boxed{+}$ $\boxed{3}$ $\boxed{\times}$ $\boxed{4}$ $\boxed{=}$.

↕ or

$\boxed{\text{ENTER}}$

The correct answer is 14 because the order of operations is to multiply before we add. If the calculator displays $\boxed{14}$, then it has the order of operations built in.

Even if the order of operations is built in, parentheses must sometimes be inserted. For example, to simplify $\frac{5}{12 - 7}$, press the keys

$\boxed{5}$ $\boxed{\div}$ $\boxed{(}$ $\boxed{1}$ $\boxed{2}$ $\boxed{-}$ $\boxed{7}$ $\boxed{)}$ $\boxed{=}$.

↕ or

$\boxed{\text{ENTER}}$

The display should read $\boxed{1}$ or $\boxed{\begin{matrix}5/(12 - 7)\\ 1\end{matrix}}$

Use a calculator to evaluate each expression.

1. 5^4

2. 7^4

3. 9^5

4. 8^6

5. $2(20 - 5)$

6. $3(14 - 7) + 21$

7. $24(862 - 455) + 89$

8. $99 + (401 + 962)$

9. $\frac{4623 + 129}{36 - 34}$

10. $\frac{956 - 452}{89 - 86}$

Vocabulary, Readiness & Video Check

Use the choices below to fill in each blank.

equation	variable	base	grouping
expression	solution	solving	exponent

1. In the expression 5^2, the 5 is called the __________ and the 2 is called the __________.

2. The symbols ( ), [ ], and { } are examples of __________ symbols.

3. A symbol that is used to represent a number is called a(n) __________.

4. A collection of numbers, variables, operation symbols, and grouping symbols is called a(n) __________.

5. A mathematical statement that two expressions are equal is called a(n) __________.

6. A value for the variable that makes an equation a true statement is called a(n) __________.

7. Deciding what values of a variable make an equation a true statement is called __________ the equation.

Martin-Gay Interactive Videos

See Video 1.4

Watch the section lecture video and answer the following questions.

OBJECTIVE 1 — **8.** In Example 3 and the lecture before, what is the main point made about the order of operations?

OBJECTIVE 2 — **9.** What happens with the replacement value for z in Example 6 and why?

OBJECTIVE 3 — **10.** Is the value 0 a solution of the equation given in Example 9? How is this determined?

OBJECTIVE 4 — **11.** Earlier in this video the point was made that equations have =, while expressions do not. In the lecture before Example 10, translating from English to math is discussed and another difference between expressions and equations is explained. What is it?

1.4 Exercise Set

MyMathLab®

Evaluate. See Example 1.

1. 3^5

2. 2^5

3. 3^3

4. 4^4

5. 1^5

6. 1^8

7. 5^1

8. 8^1

9. 7^2

10. 9^2

11. $\left(\frac{2}{3}\right)^4$

12. $\left(\frac{6}{11}\right)^2$

13. $\left(\frac{1}{5}\right)^3$

14. $\left(\frac{1}{2}\right)^5$

15. $(1.2)^2$

16. $(1.5)^2$

17. $(0.04)^3$

18. $(0.03)^3$

MIXED PRACTICE

Simplify each expression. See Examples 2 through 5.

19. $5 + 6 \cdot 2$

20. $8 + 5 \cdot 3$

21. $4 \cdot 8 - 6 \cdot 2$

22. $12 \cdot 5 - 3 \cdot 6$

23. $2(8 - 3)$

24. $5(6 - 2)$

25. $2 + (5 - 2) + 4^2$

26. $6 - 2 \cdot 2 + 2^5$

27. $5 \cdot 3^2$

28. $2 \cdot 5^2$

29. $\frac{1}{4} \cdot \frac{2}{3} - \frac{1}{6}$

30. $\frac{3}{4} \cdot \frac{1}{2} + \frac{2}{3}$

31. $2[5 + 2(8 - 3)]$

32. $3[4 + 3(6 - 4)]$

33. $\frac{19 - 3 \cdot 5}{6 - 4}$

34. $\frac{4 \cdot 3 + 2}{4 + 3 \cdot 2}$

35. $\frac{|6 - 2| + 3}{8 + 2 \cdot 5}$

36. $\frac{15 - |3 - 1|}{12 - 3 \cdot 2}$

37. $\frac{3 + 3(5 + 3)}{3^2 + 1}$

38. $\frac{3 + 6(8 - 5)}{4^2 + 2}$

39. $\frac{6 + |8 - 2| + 3^2}{18 - 3}$

40. $\frac{16 + |13 - 5| + 4^2}{17 - 5}$

41. $2 + 3[10(4 \cdot 5 - 16) - 30]$

42. $3 + 4[8(5 \cdot 5 - 20) - 41]$

43. $\left(\frac{2}{3}\right)^3 + \frac{1}{9} + \frac{1}{3} \cdot \frac{4}{3}$

44. $\left(\frac{3}{8}\right)^2 + \frac{1}{4} + \frac{1}{8} \cdot \frac{3}{2}$

For Exercises 45 and 46, match each expression in the first column with its value in the second column.

45.

a. $(6 + 2) \cdot (5 + 3)$	19
b. $(6 + 2) \cdot 5 + 3$	22
c. $6 + 2 \cdot 5 + 3$	64
d. $6 + 2 \cdot (5 + 3)$	43

46.

a. $(1 + 4) \cdot 6 - 3$	15
b. $1 + 4 \cdot (6 - 3)$	13
c. $1 + 4 \cdot 6 - 3$	27
d. $(1 + 4) \cdot (6 - 3)$	22

Evaluate each expression when $x = 1$, $y = 3$, and $z = 5$. See Example 6.

47. $3y$

48. $4x$

49. $\frac{z}{5x}$

50. $\frac{y}{2z}$

51. $3x - 2$

52. $6y - 8$

53. $|2x + 3y|$

54. $|5z - 2y|$

55. $xy + z$

56. $yz - x$

57. $5y^2$

58. $2z^2$

Evaluate each expression if $x = 12$, $y = 8$, and $z = 4$. See Example 6.

59. $\frac{x}{z} + 3y$

60. $\frac{y}{z} + 8x$

61. $x^2 - 3y + x$

62. $y^2 - 3x + y$

63. $\frac{x^2 + z}{y^2 + 2z}$

64. $\frac{y^2 + x}{x^2 + 3y}$

Neglecting air resistance, the expression $16t^2$ gives the distance in feet an object will fall in t seconds.

65. Complete the chart below. To evaluate $16t^2$, remember to first find t^2, then multiply by 16.

Time t (in seconds)	*Distance $16t^2$ (in feet)*
1	
2	
3	
4	

66. Does an object fall the same distance *during* each second? Why or why not? (See Exercise 65.)

Decide whether the given number is a solution of the given equation. See Example 7.

67. Is 5 a solution of $3x + 30 = 9x$?

68. Is 6 a solution of $2x + 7 = 3x$?

69. Is 0 a solution of $2x + 6 = 5x - 1$?

70. Is 2 a solution of $4x + 2 = x + 8$?

71. Is 8 a solution of $2x - 5 = 5$?

72. Is 6 a solution of $3x - 10 = 8$?

73. Is 2 a solution of $x + 6 = x + 6$?

74. Is 10 a solution of $x + 6 = x + 6$?

75. Is 0 a solution of $x = 5x + 15$?

76. Is 1 a solution of $4 = 1 - x$?

TRANSLATING

Write each phrase as an algebraic expression. Let x represent the unknown number. See Example 8.

77. Fifteen more than a number

78. A number increased by 9

79. Five subtracted from a number

80. Five decreased by a number

81. The ratio of a number and 4

82. The quotient of a number and 9

83. Three times a number, increased by 22

84. Twice a number, decreased by 72

TRANSLATING

Write each sentence as an equation or inequality. Use x to represent any unknown number. See Example 9.

85. One increased by two equals the quotient of nine and three.

86. Four subtracted from eight is equal to two squared.

87. Three is not equal to four divided by two.

88. The difference of sixteen and four is greater than ten.

89. The sum of 5 and a number is 20.

90. Seven subtracted from a number is 0.

91. The product of 7.6 and a number is 17.

92. 9.1 times a number equals 4.

93. Thirteen minus three times a number is 13.

94. Eight added to twice a number is 42.

CONCEPT EXTENSIONS

Fill in each blank with one of the following:
add subtract multiply divide

95. To simplify the expression $1 + 3 \cdot 6$, first ________.

96. To simplify the expression $(1 + 3) \cdot 6$, first ________.

97. To simplify the expression $(20 - 4) \cdot 2$, first ________.

98. To simplify the expression $20 - 4 \div 2$, first ________.

99. Are parentheses necessary in the expression $2 + (3 \cdot 5)$? Explain your answer.

100. Are parentheses necessary in the expression $(2 + 3) \cdot 5$? Explain your answer.

△ *Recall that perimeter measures the distance around a plane figure and area measures the amount of surface of a plane figure. The expression $2l + 2w$ gives the perimeter of the rectangle below (measured in units), and the expression lw gives its area (measured in square units). Complete the chart below for the given lengths and widths. Be sure to include units.*

	Length: l	*Width: w*	*Perimeter of Rectangle: $2l + 2w$*	*Area of Rectangle: lw*
101.	4 in.	3 in.		
102.	6 in.	1 in.		
103.	5.3 in.	1.7 in.		
104.	4.6 in.	2.4 in.		

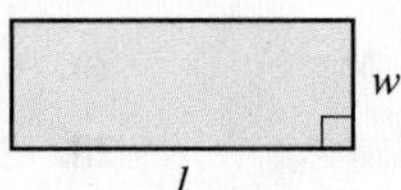

105. Study the perimeters and areas found in the chart on the previous page. Do you notice any trends?

106. In your own words, explain the difference between an expression and an equation.

107. Insert one set of parentheses so that the following expression simplifies to 32.

$$20 - 4 \cdot 4 \div 2$$

108. Insert one set of parentheses so that the following expression simplifies to 28.

$$2 \cdot 5 + 3^2$$

Determine whether each is an expression or an equation. See the Concept Check in this section.

109. a. $5x + 6$
b. $2a = 7$
c. $3a + 2 = 9$
d. $4x + 3y - 8z$
e. $5^2 - 2(6 - 2)$

110. a. $3x^2 - 26$
b. $3x^2 - 26 = 1$
c. $2x - 5 = 7x - 5$
d. $9y + x - 8$
e. $3^2 - 4(5 - 3)$

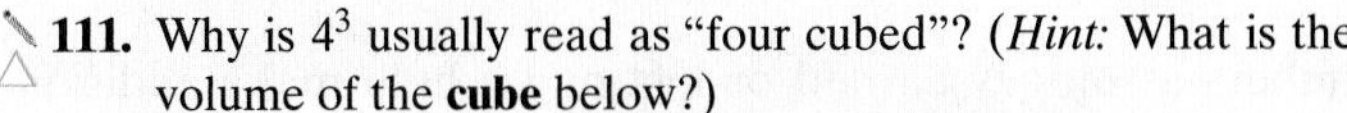

111. Why is 4^3 usually read as "four cubed"? (*Hint:* What is the volume of the **cube** below?)

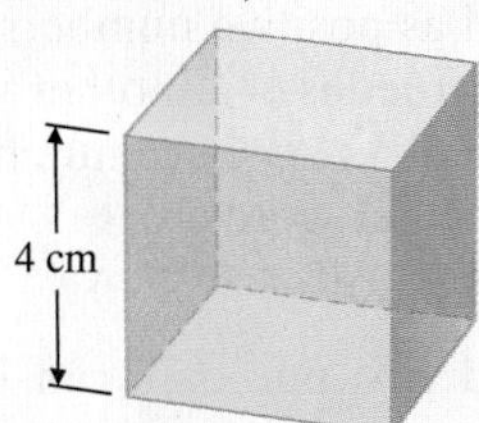

112. Why is 8^2 usually read as "eight squared"? (*Hint:* What is the area of the **square** below?)

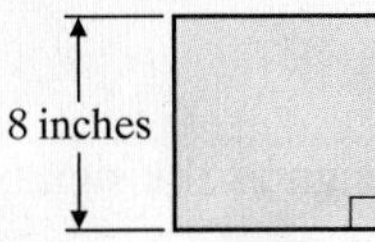

113. Write any expression, using 3 or more numbers, that simplifies to −11.

114. Write any expression, using 4 or more numbers, that simplifies to 7.

115. The area of a figure is the total enclosed surface of the figure. Area is measured in square units. The expression lw represents the area of a rectangle when l is its length and w is its width. Find the area of the following rectangular-shaped lot.

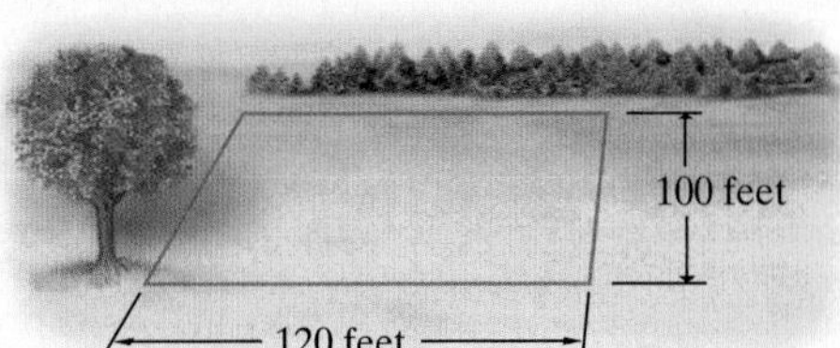

116. A trapezoid is a four-sided figure with exactly one pair of parallel sides. The expression $\frac{1}{2}h(B + b)$ represents its area, when B and b are the lengths of the two parallel sides and h is the height between these sides. Find the area if $B = 15$ inches, $b = 7$ inches, and $h = 5$ inches.

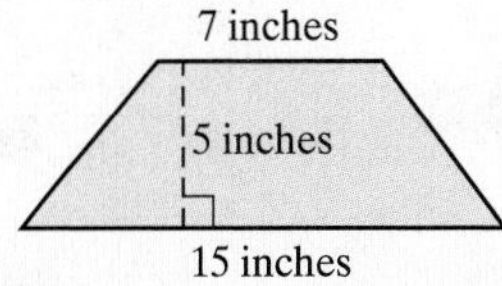

117. The expression $\frac{d}{t}$ represents the average speed in miles per hour if a distance of d miles is traveled in t hours. Find the speed to the nearest whole number if the distance between Dallas, Texas, and Kaw City, Oklahoma, is 432 miles, and it takes Peter Callac 8.5 hours to drive the distance.

118. The expression $\frac{I}{PT}$ represents the rate of interest being charged if a loan of P dollars for T years required I dollars in interest to be paid. Find the interest rate if a \$650 loan for 3 years to buy a used IBM personal computer requires \$126.75 in interest to be paid.

1.5 Adding Real Numbers

OBJECTIVES

1 Add Real Numbers.
2 Solve Applications That Involve Addition of Real Numbers.
3 Find the Opposite of a Number.

OBJECTIVE

1 Adding Real Numbers 

Real numbers can be added, subtracted, multiplied, divided, and raised to powers, just as whole numbers can. We use a number line to help picture the addition of real numbers. We begin by adding numbers with the same sign.

EXAMPLE 1 Add: $3 + 2$

Solution Recall that 3 and 2 are called **addends.** We start at 0 on a number line and draw an arrow representing the addend 3. This arrow is three units long and points to the right since 3 is positive. From the tip of this arrow, we draw another arrow, representing the addend 2. The number below the tip of this arrow is the sum, 5.

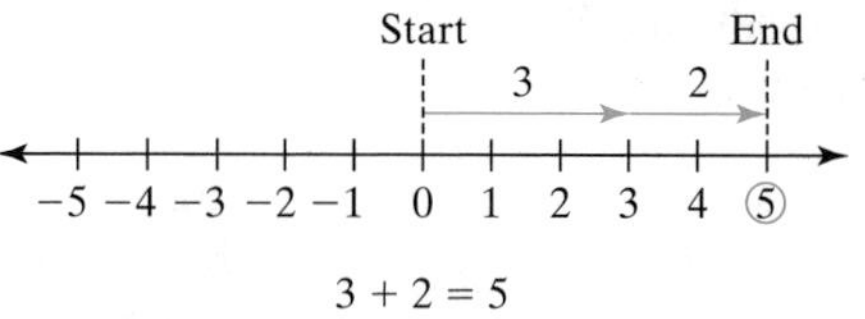

$3 + 2 = 5$

PRACTICE
1 Add using a number line: $2 + 4$.

EXAMPLE 2 Add: $-1 + (-2)$

Solution Here, -1 and -2 are addends. We start at 0 on a number line and draw an arrow representing -1. This arrow is one unit long and points to the left since -1 is negative. From the tip of this arrow, we draw another arrow, representing -2. The number below the tip of this arrow is the sum, -3.

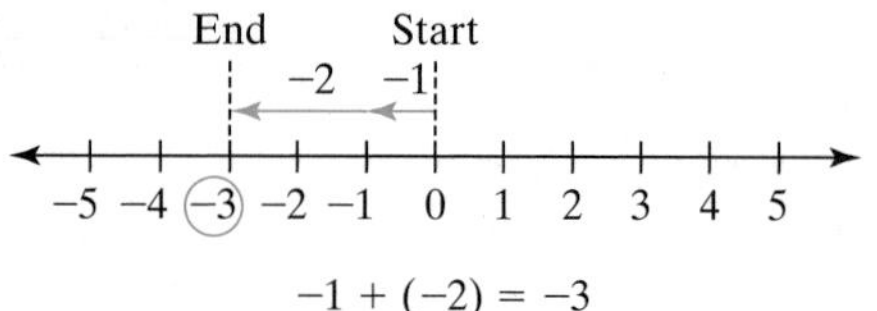

$-1 + (-2) = -3$

PRACTICE
2 Add using a number line: $-2 + (-3)$.

Thinking of signed numbers as money earned or lost might help make addition more meaningful. Earnings can be thought of as positive numbers. If \$1 is earned and later another \$3 is earned, the total amount earned is \$4. In other words, $1 + 3 = 4$.

On the other hand, losses can be thought of as negative numbers. If \$1 is lost and later another \$3 is lost, a total of \$4 is lost. In other words,

$$(-1) + (-3) = -4.$$

Using a number line each time we add two numbers can be time consuming. Instead, we can notice patterns in the previous examples and write rules for adding signed numbers. When adding two numbers with the same sign, notice that the sign of the sum is the same as the sign of the addends.

Adding Two Numbers with the Same Sign

Add their absolute values. Use their common sign as the sign of the sum.

EXAMPLE 3 Add.

a. $-3 + (-7)$ **b.** $-1 + (-20)$ **c.** $-2 + (-10)$

Solution Notice that each time, we are adding numbers with the same sign.

a. $-3 + (-7) = -10$ ← Add their absolute values: $3 + 7 = 10$.
↑ Use their common sign.

b. $-1 + (-20) = -21$ ← Add their absolute values: $1 + 20 = 21$.
↑ Common sign.

c. $-2 + (-10) = -12$ ← Add their absolute values.
↑ Common sign.

PRACTICE
3 Add. **a.** $-5 + (-8)$ **b.** $-31 + (-1)$

Adding numbers whose signs are not the same can also be pictured on a number line.

EXAMPLE 4 Add: $-4 + 6$

Solution

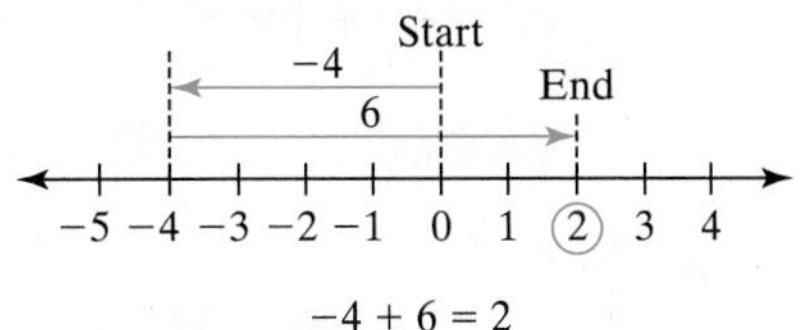

$-4 + 6 = 2$ □

PRACTICE 4 Add using a number line: $-3 + 8$.

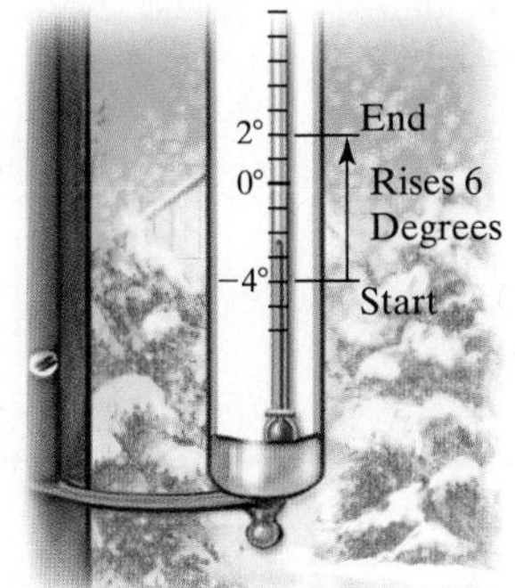

Using temperature as an example, if the thermometer registers 4 degrees below 0 degrees and then rises 6 degrees, the new temperature is 2 degrees above 0 degrees. Thus, it is reasonable that $-4 + 6 = 2$.

Once again, we can observe a pattern: when adding two numbers with different signs, the sign of the sum is the same as the sign of the addend whose absolute value is larger.

Adding Two Numbers with Different Signs

Subtract the smaller absolute value from the larger absolute value. Use the sign of the number whose absolute value is larger as the sign of the sum.

EXAMPLE 5 Add.

a. $3 + (-7)$ **b.** $-2 + 10$ **c.** $0.2 + (-0.5)$

Solution Notice that each time, we are adding numbers with different signs.

a. $3 + (-7) = -4$ ← Subtract their absolute values: $7 - 3 = 4$.
↑ The negative number, -7, has the larger absolute value so the sum is negative.

b. $-2 + 10 = 8$ ← Subtract their absolute values: $10 - 2 = 8$.
↑ The positive number, 10, has the larger absolute value so the sum is positive.

c. $0.2 + (-0.5) = -0.3$ ← Subtract their absolute values: $0.5 - 0.2 = 0.3$.
↑ The negative number, -0.5, has the larger absolute value so the sum is negative. □

PRACTICE 5 Add.

a. $15 + (-18)$ **b.** $-19 + 20$ **c.** $-0.6 + 0.4$

In general, we have the following:

Adding Real Numbers

To add two real numbers

1. with the *same sign,* add their absolute values. Use their common sign as the sign of the answer.
2. with *different signs,* subtract their absolute values. Give the answer the same sign as the number with the larger absolute value.

EXAMPLE 6 Add.

a. $-8 + (-11)$ b. $-5 + 35$ c. $0.6 + (-1.1)$

d. $-\frac{7}{10} + \left(-\frac{1}{10}\right)$ e. $11.4 + (-4.7)$ f. $-\frac{3}{8} + \frac{2}{5}$

Solution

a. $-8 + (-11) = -19$ Same sign. Add absolute values and use the common sign.

b. $-5 + 35 = 30$ Different signs. Subtract absolute values and use the sign of the number with the larger absolute value.

c. $0.6 + (-1.1) = -0.5$ Different signs.

d. $-\frac{7}{10} + \left(-\frac{1}{10}\right) = -\frac{8}{10} = -\frac{4}{5}$ Same sign.

e. $11.4 + (-4.7) = 6.7$

f. $-\frac{3}{8} + \frac{2}{5} = -\frac{15}{40} + \frac{16}{40} = \frac{1}{40}$ □

Helpful Hint
Don't forget that a common denominator is needed when adding or subtracting fractions. The common denominator here is 40.

PRACTICE
6 Add.

a. $-\frac{3}{5} + \left(-\frac{2}{5}\right)$ b. $3 + (-9)$

c. $2.2 + (-1.7)$ d. $-\frac{2}{7} + \frac{3}{10}$

EXAMPLE 7 Add.

a. $3 + (-7) + (-8)$ b. $[7 + (-10)] + [-2 + |-4|]$

Solution

a. Perform the additions from left to right.

$3 + (-7) + (-8) = -4 + (-8)$ Adding numbers with different signs.
$= -12$ Adding numbers with like signs.

b. Simplify inside brackets first.

$[7 + (-10)] + [-2 + |-4|] = [-3] + [-2 + 4]$
$= [-3] + [2]$
$= -1$ Add. □

Helpful Hint
Don't forget that brackets are grouping symbols. We simplify within them first.

PRACTICE
7 Add.

a. $8 + (-5) + (-9)$ b. $[-8 + 5] + [-5 + |-2|]$

✓CONCEPT CHECK
What is wrong with the following calculation?
$5 + (-22) = 17$

Answer to Concept Check:
$5 + (-22) = -17$

OBJECTIVE
2 Solving Applications by Adding Real Numbers 

Positive and negative numbers are often used in everyday life. Stock market returns show gains and losses as positive and negative numbers. Temperatures in cold climates often dip into the negative range, commonly referred to as "below zero" temperatures. Bank statements report deposits and withdrawals as positive and negative numbers.

EXAMPLE 8 **Calculating Temperature**

In Philadelphia, Pennsylvania, the record extreme high temperature is 104°F. Decrease this temperature by 111 degrees, and the result is the record extreme low temperature. Find this temperature. (*Source:* National Climatic Data Center)

Solution:

In words:	extreme low temperature	=	extreme high temperature	+	decrease of 111°
	↓		↓		↓
Translate:	extreme low temperature	=	104	+	(−111)
		= −7			

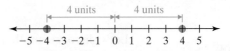

The record extreme low temperature in Philadelphia, Pennsylvania, is −7°F. □

PRACTICE

8 If the temperature was −7° Fahrenheit at 6 a.m., and it rose 4 degrees by 7 a.m. and then rose another 7 degrees in the hour from 7 a.m. to 8 a.m., what was the temperature at 8 a.m.?

OBJECTIVE

3 Finding the Opposite of a Number

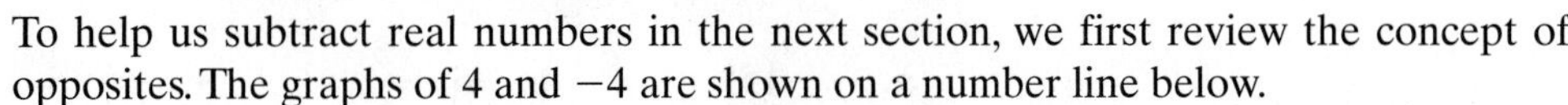

To help us subtract real numbers in the next section, we first review the concept of opposites. The graphs of 4 and −4 are shown on a number line below.

4 units 4 units

−5 −4 −3 −2 −1 0 1 2 3 4 5

Notice that 4 and −4 lie on opposite sides of 0, and each is 4 units away from 0.

This relationship between −4 and +4 is an important one. Such numbers are known as **opposites** or **additive inverses** of each other.

Opposites or Additive Inverses

Two numbers that are the same distance from 0 but lie on opposite sides of 0 are called **opposites** or **additive inverses** of each other.

EXAMPLE 9 Find the opposite or additive inverse of each number.

a. 5 **b.** −6 **c.** $\frac{1}{2}$ **d.** −4.5

Solution

a. The opposite of 5 is −5. Notice that 5 and −5 are on opposite sides of 0 when plotted on a number line and are equal distances away.

b. The opposite of −6 is 6.

c. The opposite of $\frac{1}{2}$ is $-\frac{1}{2}$.

d. The opposite of −4.5 is 4.5. □

PRACTICE

9 Find the opposite or additive inverse of each number.

a. $-\frac{5}{9}$ **b.** 8 **c.** 6.2 **d.** −3

We use the symbol "−" to represent the phrase "the opposite of" or "the additive inverse of." In general, if a is a number, we write the opposite or additive inverse of a as $-a$. We know that the opposite of -3 is 3. Notice that this translates as

the opposite of	−3	is	3
↓	↓	↓	↓
$-$	(-3)	$=$	3

This is true in general.

> If a is a number, then $-(-a) = a$.

EXAMPLE 10 Simplify each expression.

a. $-(-10)$ **b.** $-\left(-\frac{1}{2}\right)$ **c.** $-(-2x)$ **d.** $-|-6|$

Solution

a. $-(-10) = 10$ **b.** $-\left(-\frac{1}{2}\right) = \frac{1}{2}$ **c.** $-(-2x) = 2x$

d. Since $|-6| = 6$, then $-|-6| = -6$. □

PRACTICE 10 Simplify each expression.

a. $-|-15|$ **b.** $-\left(-\frac{3}{5}\right)$ **c.** $-(-5y)$ **d.** $-(-8)$

Let's discover another characteristic about opposites. Notice that the sum of a number and its opposite is 0.

$$10 + (-10) = 0$$
$$-3 + 3 = 0$$
$$\frac{1}{2} + \left(-\frac{1}{2}\right) = 0$$

In general, we can write the following:

> The sum of a number a and its opposite $-a$ is 0.
>
> $$a + (-a) = 0$$

This is why opposites are also called additive inverses. Notice that this also means that the opposite of 0 is then 0 since $0 + 0 = 0$.

Vocabulary, Readiness & Video Check

Use the choices below to fill in each blank.

a positive number	n	opposites
a negative number	0	$-n$

1. Two numbers that are the same distance from 0 but lie on opposite sides of 0 are called ______.

2. If n is a number, then $n + (-n) =$ ______.

3. If n is a number, then $-(-n) =$ ______.

4. The sum of two negative numbers is always ______.

Martin-Gay Interactive Videos

See Video 1.5

Watch the section lecture video and answer the following questions.

OBJECTIVE 1
5. Complete this statement based on the lecture given before Example 1. To add two numbers with the same sign, add their ________ and use their common sign as the sign of the sum.

OBJECTIVE 1
6. What is the sign of the sum in Example 6 and why?

OBJECTIVE 2
7. What is the real life application of negative numbers used in Example 9? The answer to Example 9 is –6. What does this number mean in the context of the problem?

OBJECTIVE 3
8. Example 12 illustrates the idea that if *a* is a real number, the opposite of –*a* is *a*. Example 13 looks similar to Example 12, but it's actually quite different. Explain the difference.

1.5 Exercise Set

MyMathLab®

MIXED PRACTICE

Add. See Examples 1 through 7.

1. $6 + 3$
2. $9 + (-12)$
3. $-6 + (-8)$
4. $-6 + (-14)$
5. $8 + (-7)$
6. $6 + (-4)$
7. $-14 + 2$
8. $-10 + 5$
9. $-2 + (-3)$
10. $-7 + (-4)$
11. $-9 + (-3)$
12. $7 + (-5)$
13. $-7 + 3$
14. $-5 + 9$
15. $10 + (-3)$
16. $8 + (-6)$
17. $5 + (-7)$
18. $3 + (-6)$
19. $-16 + 16$
20. $23 + (-23)$
21. $27 + (-46)$
22. $53 + (-37)$
23. $-18 + 49$
24. $-26 + 14$
25. $-33 + (-14)$
26. $-18 + (-26)$
27. $6.3 + (-8.4)$
28. $9.2 + (-11.4)$
29. $|-8| + (-16)$
30. $|-6| + (-61)$
31. $117 + (-79)$
32. $144 + (-88)$
33. $-9.6 + (-3.5)$
34. $-6.7 + (-7.6)$
35. $-\frac{3}{8} + \frac{5}{8}$
36. $-\frac{5}{12} + \frac{7}{12}$
37. $-\frac{7}{16} + \frac{1}{4}$
38. $-\frac{5}{9} + \frac{1}{3}$
39. $-\frac{7}{10} + \left(-\frac{3}{5}\right)$
40. $-\frac{5}{6} + \left(-\frac{2}{3}\right)$
41. $-15 + 9 + (-2)$
42. $-9 + 15 + (-5)$
43. $-21 + (-16) + (-22)$
44. $-18 + (-6) + (-40)$
45. $-23 + 16 + (-2)$
46. $-14 + (-3) + 11$
47. $|5 + (-10)|$
48. $|7 + (-17)|$
49. $6 + (-4) + 9$
50. $8 + (-2) + 7$
51. $[-17 + (-4)] + [-12 + 15]$
52. $[-2 + (-7)] + [-11 + 22]$
53. $|9 + (-12)| + |-16|$
54. $|43 + (-73)| + |-20|$
55. $-1.3 + [0.5 + (-0.3) + 0.4]$
56. $-3.7 + [0.1 + (-0.6) + 8.1]$

Solve. See Example 8.

57. The low temperature in Anoka, Minnesota, was −15° last night. During the day, it rose only 9°. Find the high temperature for the day.
58. On January 2, 1943, the temperature was −4° at 7:30 a.m. in Spearfish, South Dakota. Incredibly, it got 49° warmer in the next 2 minutes. To what temperature did it rise by 7:32?
59. The lowest point in Africa is −512 feet at Lake Assal in Djibouti. If you are standing at a point 658 feet above Lake Assal, what is your elevation? (*Source:* Microsoft Encarta)

60. The lowest elevation in Australia is −52 feet at Lake Eyre. If you are standing at a point 439 feet above Lake Eyre, what is your elevation? (*Source:* National Geographic Society)

A negative net income results when a company spends more money than it brings in.

61. Johnson Outdoors Inc. had the following quarterly net incomes during its 2009 fiscal year. (Data from Yahoo Finance)

Quarter of Fiscal 2009	*Net Income (in millions)*
First	2.5
Second	9
Third	−14.2
Fourth	−4.2

What was the total net income for fiscal year 2009?

62. LeapFrog Enterprises Inc. had the following quarterly net incomes during its 2009 fiscal year. (Data from Yahoo Finance)

Quarter of Fiscal 2009	*Net Income (in millions)*
First	−27.1
Second	−12.2
Third	7.2
Fourth	29.4

What was the total net income for fiscal year 2009?

In golf, scores that are under par for the entire round are shown as negative scores; positive scores are shown for scores that are over par, and 0 is par.

63. Sandra Gal was the winner of the 2011 LPGA Kia Classic in California. Her scores were −6, −5, −3, and −2. What was her overall score? (*Source:* Ladies Professional Golf Association)

64. During the 2011 PGA Waste Management Phoenix Open, Mark Wilson won with scores of −6, −7, −3, and −2. What was his overall score? (*Source:* Professional Golf Association)

Find each additive inverse or opposite. See Example 9.

65. 6 **66.** 4 **67.** −2

68. −8 **69.** 0 **70.** $-\frac{1}{4}$

71. $|-6|$ **72.** $|-11|$

Simplify each of the following. See Example 10.

73. $-|-2|$ **74.** $-(-3)$ **75.** $-|0|$

76. $\left|-\frac{2}{3}\right|$ **77.** $-\left|-\frac{2}{3}\right|$ **78.** $-(-7)$

Decide whether the given number is a solution of the given equation.

79. Is −4 a solution of $x + 9 = 5$?

80. Is 10 a solution of $7 = -x + 3$?

81. Is −1 a solution of $y + (-3) = -7$?

82. Is −6 a solution of $1 = y + 7$?

CONCEPT EXTENSIONS

The following bar graph shows each month's average daily low temperature in degrees Fahrenheit for Barrow, Alaska. Use this graph to answer Exercises 83 through 88.

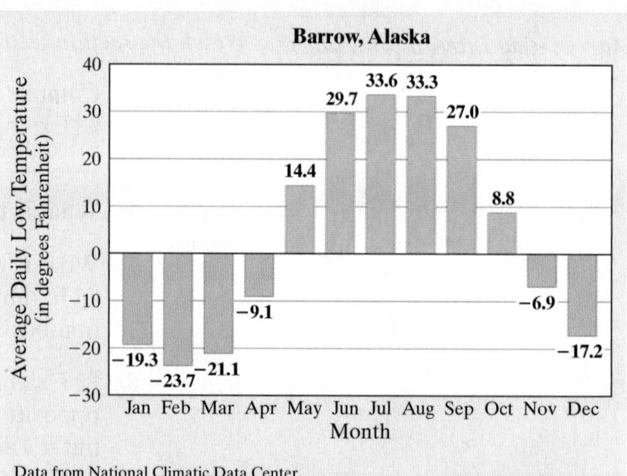

83. For what month is the graphed temperature the highest?

84. For what month is the graphed temperature the lowest?

85. For what month is the graphed temperature positive *and* closest to 0°?

86. For what month is the graphed temperature negative *and* closest to 0°?

87. Find the average of the temperatures shown for the months of April, May, and October. (To find the average of three temperatures, find their sum and divide by 3.)

88. Find the average of the temperatures shown for the months of January, September, and October.

Each calculation below is incorrect. Find the error and correct it. See the Concept Check in this section.

89. $7 + (-10) \stackrel{?}{=} 17$ **90.** $-4 + 14 \stackrel{?}{=} -18$

91. $-10 + (-12) \stackrel{?}{=} -120$ **92.** $-15 + (-17) \stackrel{?}{=} 32$

If a is a positive number and b is a negative number, fill in the blanks with the words positive or negative.

93. $-a$ is ______. **94.** $-b$ is ______.

95. $a + a$ is ______. **96.** $b + b$ is ______.

For Exercises 97 through 100, determine whether each statement is true or false.

97. The sum of two negative numbers is always a negative number.

98. The sum of two positive numbers is always a positive number.

99. The sum of a positive number and a negative number is always a negative number.

100. The sum of zero and a negative number is always a negative number.

101. In your own words, explain how to find the opposite of a number.

102. In your own words, explain why 0 is the only number that is its own opposite.

103. Explain why adding a negative number to another negative number always gives a negative sum.

104. When a positive and a negative number are added, sometimes the sum is positive, sometimes it is zero, and sometimes it is negative. Explain why and when this happens.

1.6 Subtracting Real Numbers

OBJECTIVES

1. Subtract Real Numbers.
2. Add and Subtract Real Numbers.
3. Evaluate Algebraic Expressions Using Real Numbers.
4. Solve Applications That Involve Subtraction of Real Numbers.
5. Find Complementary and Supplementary Angles.

OBJECTIVE

1 Subtracting Real Numbers 

Now that addition of signed numbers has been discussed, we can explore subtraction. We know that $9 - 7 = 2$. Notice that $9 + (-7) = 2$ also. This means that

$$9 - 7 = 9 + (-7)$$

Notice that the difference of 9 and 7 is the same as the sum of 9 and the opposite of 7. In general, we have the following.

Subtracting Two Real Numbers

If a and b are real numbers, then $a - b = a + (-b)$.

In other words, to find the difference of two numbers, add the first number to the opposite of the second number.

EXAMPLE 1 Subtract.

a. $-13 - 4$ **b.** $5 - (-6)$ **c.** $3 - 6$ **d.** $-1 - (-7)$

Solution

add

a. $-13 - 4 = -13 + (-4)$ Add −13 to the opposite of +4, which is −4.

opposite

$= -17$

add

b. $5 - (-6) = 5 + (6)$ Add 5 to the opposite of −6, which is 6.

opposite

$= 11$

c. $3 - 6 = 3 + (-6)$ Add 3 to the opposite of 6, which is −6.

$= -3$

d. $-1 - (-7) = -1 + (7) = 6$ □

PRACTICE

1 Subtract.

a. $-7 - 6$ **b.** $-8 - (-1)$ **c.** $9 - (-3)$ **d.** $5 - 7$

Helpful Hint

Study the patterns indicated.

No change — Change to addition. — Change to opposite.

$$5 - 11 = 5 + (-11) = -6$$
$$-3 - 4 = -3 + (-4) = -7$$
$$7 - (-1) = 7 + (1) = 8$$

EXAMPLE 2 Subtract.

a. $5.3 - (-4.6)$ **b.** $-\frac{3}{10} - \frac{5}{10}$ **c.** $-\frac{2}{3} - \left(-\frac{4}{5}\right)$

Solution

a. $5.3 - (-4.6) = 5.3 + (4.6) = 9.9$

b. $-\frac{3}{10} - \frac{5}{10} = -\frac{3}{10} + \left(-\frac{5}{10}\right) = -\frac{8}{10} = -\frac{4}{5}$

c. $-\frac{2}{3} - \left(-\frac{4}{5}\right) = -\frac{2}{3} + \left(\frac{4}{5}\right) = -\frac{10}{15} + \frac{12}{15} = \frac{2}{15}$ The common denominator is 15. □

PRACTICE 2 Subtract.

a. $8.4 - (-2.5)$ **b.** $-\frac{5}{8} - \left(-\frac{1}{8}\right)$ **c.** $-\frac{3}{4} - \frac{1}{5}$

EXAMPLE 3 Subtract 8 from -4.

Solution Be careful when interpreting this: The order of numbers in subtraction is important. 8 is to be subtracted **from** -4.

$$-4 - 8 = -4 + (-8) = -12$$ □

PRACTICE 3 Subtract 5 from -2.

OBJECTIVE

2 Adding and Subtracting Real Numbers 

If an expression contains additions and subtractions, just write the subtractions as equivalent additions. Then simplify from left to right.

EXAMPLE 4 Simplify each expression.

a. $-14 - 8 + 10 - (-6)$ **b.** $1.6 - (-10.3) + (-5.6)$

Solution

a. $-14 - 8 + 10 - (-6) = -14 + (-8) + 10 + 6$
$= -6$

b. $1.6 - (-10.3) + (-5.6) = 1.6 + 10.3 + (-5.6)$
$= 6.3$ □

PRACTICE 4 Simplify each expression.

a. $-15 - 2 - (-4) + 7$ **b.** $3.5 + (-4.1) - (-6.7)$

When an expression contains parentheses and brackets, remember the order of operations. Start with the innermost set of parentheses or brackets and work your way outward.

EXAMPLE 5 Simplify each expression.

a. $-3 + [(-2 - 5) - 2]$ **b.** $2^3 - |10| + [-6 - (-5)]$

Solution

a. Start with the innermost sets of parentheses. Rewrite $-2 - 5$ as a sum.

$$
\begin{aligned}
-3 + [(-2 - 5) - 2] &= -3 + [(-2 + (-5)) - 2] \\
&= -3 + [(-7) - 2] && \text{Add: } -2 + (-5). \\
&= -3 + [-7 + (-2)] && \text{Write } -7 - 2 \text{ as a sum.} \\
&= -3 + [-9] && \text{Add.} \\
&= -12 && \text{Add.}
\end{aligned}
$$

b. Start simplifying the expression inside the brackets by writing $-6 - (-5)$ as a sum.

$$
\begin{aligned}
2^3 - |10| + [-6 - (-5)] &= 2^3 - |10| + [-6 + 5] \\
&= 2^3 - |10| + [-1] && \text{Add.} \\
&= 8 - 10 + (-1) && \text{Evaluate } 2^3 \text{ and } |10|. \\
&= 8 + (-10) + (-1) && \text{Write } 8 - 10 \text{ as a sum.} \\
&= -2 + (-1) && \text{Add.} \\
&= -3 && \text{Add.}
\end{aligned}
$$

□

PRACTICE 5 Simplify each expression.

a. $-4 + [(-8 - 3) - 5]$ **b.** $|-13| - 3^2 + [2 - (-7)]$

OBJECTIVE 3

Evaluating Algebraic Expressions 

Knowing how to evaluate expressions for given replacement values is helpful when checking solutions of equations and when solving problems whose unknowns satisfy given expressions. The next example illustrates this.

EXAMPLE 6 Find the value of each expression when $x = 2$ and $y = -5$.

a. $\dfrac{x - y}{12 + x}$ **b.** $x^2 - 3y$

Solution

a. Replace x with 2 and y with -5. Be sure to put parentheses around -5 to separate signs. Then simplify the resulting expression.

$$
\begin{aligned}
\frac{x - y}{12 + x} &= \frac{2 - (-5)}{12 + 2} \\
&= \frac{2 + 5}{14} \\
&= \frac{7}{14} = \frac{1}{2}
\end{aligned}
$$

b. Replace x with 2 and y with -5 and simplify.

$$
\begin{aligned}
x^2 - 3y &= 2^2 - 3(-5) \\
&= 4 - 3(-5) \\
&= 4 - (-15) \\
&= 4 + 15 \\
&= 19
\end{aligned}
$$

□

PRACTICE
6 Find the value of each expression when $x = -3$ and $y = 4$.

a. $\frac{7 - x}{2y + x}$ **b.** $y^2 + x$

Helpful Hint

For additional help when replacing variables with replacement values, first place parentheses around any variables.

For Example 6b on the previous page, we have

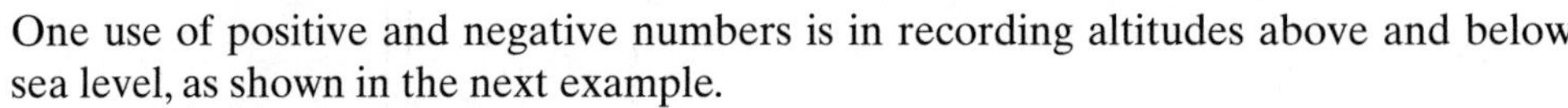

$$x^2 - 3y = \underbrace{(x)^2 - 3(y)}_{\text{Place parentheses around variables}} = \underbrace{(2)^2 - 3(-5)}_{\text{Replace variables with values}} = 4 - 3(-5) = 4 - (-15) = 4 + 15 = 19$$

OBJECTIVE
4 Solving Applications by Subtracting Real Numbers

One use of positive and negative numbers is in recording altitudes above and below sea level, as shown in the next example.

EXAMPLE 7 **Finding a Change in Elevation**

The highest point in the United States is the top of Mount McKinley, at a height of 20,320 feet above sea level. The lowest point is Death Valley, California, which is 282 feet below sea level. How much higher is Mount McKinley than Death Valley? (*Source:* U.S. Geological Survey)

Solution: To find "how much higher," we subtract. Don't forget that since Death Valley is 282 feet *below* sea level, we represent its height by −282. Draw a diagram to help visualize the problem.

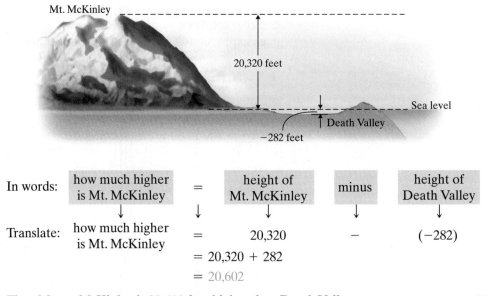

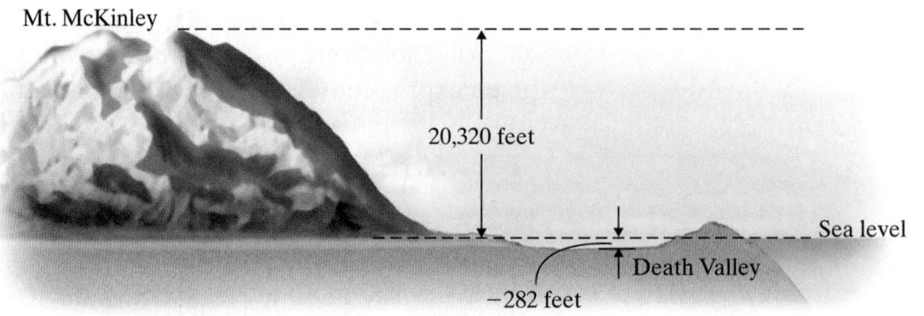

In words:	how much higher is Mt. McKinley	=	height of Mt. McKinley	minus	height of Death Valley
	↓	↓	↓	↓	↓
Translate:	how much higher is Mt. McKinley	=	20,320	−	(−282)
		=	20,320 + 282		
		=	20,602		

Thus, Mount McKinley is 20,602 feet higher than Death Valley. □

PRACTICE
7 On Tuesday morning, a bank account balance was \$282. On Thursday, the account balance had dropped to −\$75. Find the overall change in this account balance.

OBJECTIVE

5 Finding Complementary and Supplementary Angles

A knowledge of geometric concepts is needed by many professionals, such as doctors, carpenters, electronic technicians, gardeners, machinists, and pilots, just to name a few. With this in mind, we review the geometric concepts of **complementary** and **supplementary angles.**

Complementary and Supplementary Angles

Two angles are **complementary** if their sum is 90°.

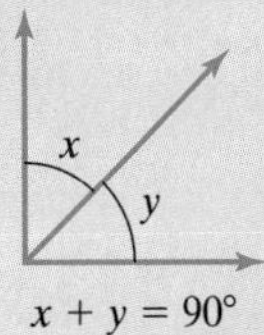

$x + y = 90°$

Two angles are **supplementary** if their sum is 180°.

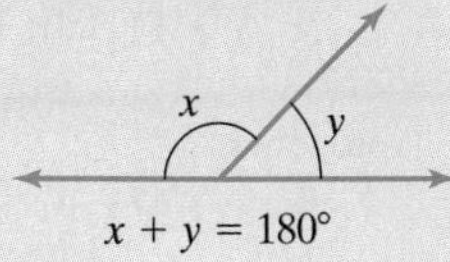

$x + y = 180°$

EXAMPLE 8 Find each unknown complementary or supplementary angle.

a.

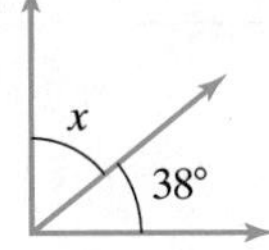

b.

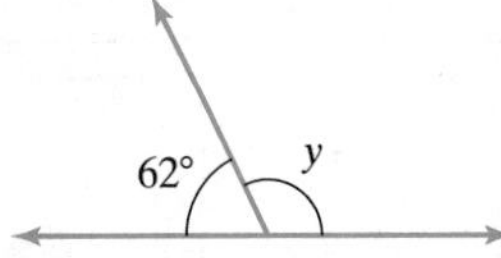

Solution

a. These angles are complementary, so their sum is 90°. This means that x is $90° - 38°$.

$$x = 90° - 38° = 52°$$

b. These angles are supplementary, so their sum is 180°. This means that y is $180° - 62°$.

$$y = 180° - 62° = 118°$$

PRACTICE

8 Find each unknown complementary or supplementary angle.

a.

x

62°

b.

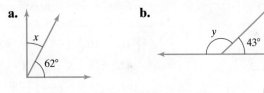

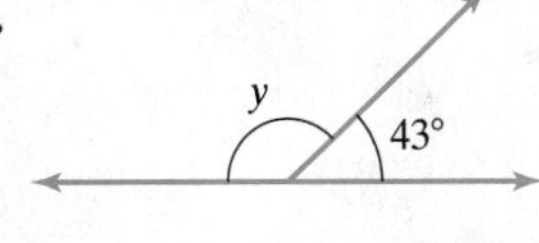

Vocabulary, Readiness & Video Check

Translate each phrase. Let x represent "a number." Use the choices below to fill in each blank.

$7 - x$ $\quad x - 7$

1. 7 minus a number ____________
2. 7 subtracted from a number ____________
3. A number decreased by 7 ____________
4. 7 less a number ____________
5. A number less than 7 ____________
6. A number subtracted from 7 ____________

Multiple choice: Select the correct lettered response following each exercise.

7. To evaluate $x - y$ for $x = -10$ and $y = -14$, we replace x with -10 and y with -14 and evaluate ___________.
a. $10 - 14$ **b.** $-10 - 14$ **c.** $-14 - 10$ **d.** $-10 - (-14)$

8. The expression $-5 - 10$ equals ___________.
a. $5 - 10$ **b.** $5 + 10$ **c.** $-5 + (-10)$ **d.** $10 - 5$

Martin-Gay Interactive Videos

See Video 1.6

Watch the section lecture video and answer the following questions.

OBJECTIVE 1
9. Complete this statement based on the lecture given before Example 1. To subtract two real numbers, change the operation to ___________ and take the ___________ of the second number.

OBJECTIVE 2
10. When simplifying Example 5, what is the result of the first step and why is the expression rewritten in this way?

OBJECTIVE 3
11. In Example 7, why are you told to be especially careful when working with the replacement value in the numerator?

OBJECTIVE 4
12. For Example 8, why is the overall vertical change represented as a negative number?

OBJECTIVE 5
13. The definition of supplementary angles is given just before Example 9. Explain how this definition is used to solve Example 9.

1.6 Exercise Set

MyMathLab®

MIXED PRACTICE

Subtract. See Examples 1 and 2.

1. $-6 - 4$
2. $-12 - 8$
3. $4 - 9$
4. $8 - 11$
5. $16 - (-3)$
6. $12 - (-5)$
7. $\frac{1}{2} - \frac{1}{3}$
8. $\frac{3}{4} - \frac{7}{8}$
9. $-16 - (-18)$
10. $-20 - (-48)$
11. $-6 - 5$
12. $-8 - 4$
13. $7 - (-4)$
14. $3 - (-6)$
15. $-6 - (-11)$
16. $-4 - (-16)$
17. $16 - (-21)$
18. $15 - (-33)$
19. $9.7 - 16.1$
20. $8.3 - 11.2$
21. $-44 - 27$
22. $-36 - 51$
23. $-21 - (-21)$
24. $-17 - (-17)$
25. $-2.6 - (-6.7)$
26. $-6.1 - (-5.3)$
27. $-\frac{3}{11} - \left(-\frac{5}{11}\right)$
28. $-\frac{4}{7} - \left(-\frac{1}{7}\right)$
29. $-\frac{1}{6} - \frac{3}{4}$
30. $-\frac{1}{10} - \frac{7}{8}$
31. $8.3 - (-0.62)$
32. $4.3 - (-0.87)$

TRANSLATING

Translate each phrase to an expression and simplify. See Example 3.

33. Subtract -5 from 8.
34. Subtract 3 from -2.
35. Subtract -1 from -6.
36. Subtract 17 from 1.
37. Subtract 8 from 7.
38. Subtract 9 from -4.
39. Decrease -8 by 15.
40. Decrease 11 by -14.

Simplify each expression. (Remember the order of operations.) See Examples 4 and 5.

41. $-10 - (-8) + (-4) - 20$
42. $-16 - (-3) + (-11) - 14$
43. $5 - 9 + (-4) - 8 - 8$
44. $7 - 12 + (-5) - 2 + (-2)$
45. $-6 - (2 - 11)$
46. $-9 - (3 - 8)$
47. $3^3 - 8 \cdot 9$
48. $2^3 - 6 \cdot 3$
49. $2 - 3(8 - 6)$
50. $4 - 6(7 - 3)$
51. $(3 - 6) + 4^2$
52. $(2 - 3) + 5^2$
53. $-2 + [(8 - 11) - (-2 - 9)]$
54. $-5 + [(4 - 15) - (-6) - 8]$
55. $|-3| + 2^2 + [-4 - (-6)]$
56. $|-2| + 6^2 + (-3 - 8)$

Evaluate each expression when $x = -5$, $y = 4$, and $t = 10$. See Example 6.

57. $x - y$

58. $y - x$

59. $|x| + 2t - 8y$

60. $|x + t - 7y|$

61. $\dfrac{9 - x}{y + 6}$

62. $\dfrac{15 - x}{y + 2}$

63. $y^2 - x$

64. $t^2 - x$

65. $\dfrac{|x - (-10)|}{2t}$

66. $\dfrac{|5y - x|}{6t}$

Solve. See Example 7.

67. Within 24 hours in 1916, the temperature in Browning, Montana, fell from 44°F to −56°F. How large a drop in temperature was this?

68. Much of New Orleans is below sea level. If George descends 12 feet from an elevation of 5 feet above sea level, what is his new elevation?

69. The coldest temperature ever recorded on Earth was −129°F in Antarctica. The warmest temperature ever recorded was 136°F in the Sahara Desert. How many degrees warmer is 136°F than −129°F? (*Source: Questions Kids Ask,* Grolier Limited, 1991, and *The World Almanac*)

70. The coldest temperature ever recorded in the United States was −80°F in Alaska. The warmest temperature ever recorded was 134°F in California. How many degrees warmer is 134°F than −80°F? (*Source: The World Almanac*)

71. Mauna Kea in Hawaii has an elevation of 13,796 feet above sea level. The Mid-America Trench in the Pacific Ocean has an elevation of 21,857 feet below sea level. Find the difference in elevation between those two points. (*Source:* National Geographic Society and Defense Mapping Agency)

72. A woman received a statement of her charge account at Old Navy. She spent $93 on purchases last month. She returned an $18 top because she didn't like the color. She also returned a $26 nightshirt because it was damaged. What does she actually owe on her account?

73. A commercial jetliner hits an air pocket and drops 250 feet. After climbing 120 feet, it drops another 178 feet. What is its overall vertical change?

74. In some card games, it is possible to have a negative score. Lavonne Schultz currently has a score of 15 points. She then loses 24 points. What is her new score?

75. The highest point in Africa is Mt. Kilimanjaro, Tanzania, at an elevation of 19,340 feet. The lowest point is Lake Assal, Djibouti, at 512 feet below sea level. How much higher is Mt. Kilimanjaro than Lake Assal? (*Source:* National Geographic Society)

76. The airport in Bishop, California, is at an elevation of 4101 feet above sea level. The nearby Furnace Creek Airport in Death Valley, California, is at an elevation of 226 feet below sea level. How much higher in elevation is the Bishop Airport than the Furnace Creek Airport? (*Source:* National Climatic Data Center)

Find each unknown complementary or supplementary angle. See Example 8.

77. **78.** **79.** **80.**

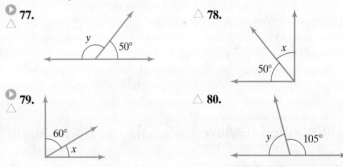

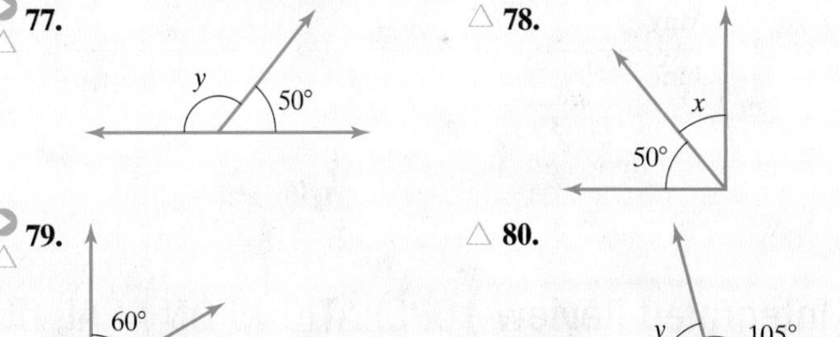

Decide whether the given number is a solution of the given equation.

81. Is −4 a solution of $x - 9 = 5$?

82. Is 3 a solution of $x - 10 = -7$?

83. Is −2 a solution of $-x + 6 = -x - 1$?

84. Is −10 a solution of $-x - 6 = -x - 1$?

85. Is 2 a solution of $-x - 13 = -15$?

86. Is 5 a solution of $4 = 1 - x$?

MIXED PRACTICE—TRANSLATING (*SECTIONS 1.5, 1.6*)

Translate each phrase to an algebraic expression. Use "x" to represent "a number."

87. The sum of −5 and a number.

88. The difference of −3 and a number.

89. Subtract a number from −20.

90. Add a number and −36.

CONCEPT EXTENSIONS

Recall the bar graph from Section 1.5. It shows each month's average daily low temperature in degrees Fahrenheit for Barrow, Alaska. Use this graph to answer Exercises 91 through 94.

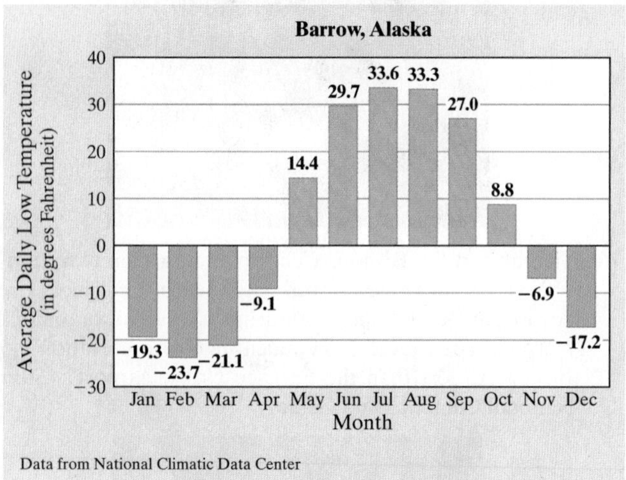

Data from National Climatic Data Center

91. Record the monthly increases and decreases in the low temperature from the previous month.

Month	*Monthly Increase or Decrease (from the previous month)*
February	
March	
April	
May	
June	

92. Record the monthly increases and decreases in the low temperature from the previous month.

Month	*Monthly Increase or Decrease (from the previous month)*
July	
August	
September	
October	
November	
December	

93. Which month had the greatest increase in temperature?

94. Which month had the greatest decrease in temperature?

95. Find two numbers whose difference is -5.

96. Find two numbers whose difference is -9.

Each calculation below is ***incorrect.*** *Find the error and correct it.*

97. $9 - (-7) \stackrel{?}{=} 2$

98. $-4 - 8 \stackrel{?}{=} 4$

99. $10 - 30 \stackrel{?}{=} 20$

100. $-3 - (-10) \stackrel{?}{=} -13$

If p is a positive number and n is a negative number, determine whether each statement is true or false. Explain your answer.

101. $p - n$ is always a positive number.

102. $n - p$ is always a negative number.

103. $|n| - |p|$ is always a positive number.

104. $|n - p|$ is always a positive number.

Without calculating, determine whether each answer is positive or negative. Then use a calculator to find the exact difference.

105. 56,875 − 87,262

106. 4.362 − 7.0086

Integrated Review OPERATIONS ON REAL NUMBERS

Sections 1.1–1.6

Answer the following with positive, negative, or 0.

1. The opposite of a positive number is a ______ number.
2. The sum of two negative numbers is a ______ number.
3. The absolute value of a negative number is a ______ number.
4. The absolute value of zero is ______.
5. The reciprocal of a positive number is a ______ number.
6. The sum of a number and its opposite is ______.
7. The absolute value of a positive number is a ______ number.
8. The opposite of a negative number is a ______ number.

Fill in the chart:

	Number	*Opposite*	*Absolute Value*
9.	$\frac{1}{7}$		
10.	$-\frac{12}{5}$		
11.		-3	
12.		$\frac{9}{11}$	

Perform each indicated operation and simplify.

13. $-19 + (-23)$

14. $7 - (-3)$

15. $-15 + 17$

16. $-8 - 10$

17. $18 + (-25)$

18. $-2 + (-37)$

19. $-14 - (-12)$

20. $5 - 14$

21. $4.5 - 7.9$

22. $-8.6 - 1.2$

23. $-\frac{3}{4} - \frac{1}{7}$

24. $\frac{2}{3} - \frac{7}{8}$

25. $-9 - (-7) + 4 - 6$

26. $11 - 20 + (-3) - 12$

27. $24 - 6(14 - 11)$

28. $30 - 5(10 - 8)$

29. $(7 - 17) + 4^2$

30. $9^2 + (10 - 30)$

31. $|-9| + 3^2 + (-4 - 20)$

32. $|-4 - 5| + 5^2 + (-50)$

33. $-7 + [(1 - 2) + (-2 - 9)]$

34. $-6 + [(-3 + 7) + (4 - 15)]$

35. Subtract 5 from 1.

36. Subtract -2 from -3.

37. Subtract $-\frac{2}{5}$ from $\frac{1}{4}$.

38. Subtract $\frac{1}{10}$ from $-\frac{5}{8}$.

39. $2(19 - 17)^3 - 3(-7 + 9)^2$

40. $3(10 - 9)^2 + 6(20 - 19)^3$

Evaluate each expression when $x = -2$, $y = -1$, and $z = 9$.

41. $x - y$

42. $x + y$

43. $y + z$

44. $z - y$

45. $\frac{|5z - x|}{y - x}$

46. $\frac{|-x - y + z|}{2z}$

1.7 Multiplying and Dividing Real Numbers

OBJECTIVES

1. Multiply Real Numbers.
2. Find the Reciprocal of a Real Number.
3. Divide Real Numbers.
4. Evaluate Expressions Using Real Numbers.
5. Solve Applications That Involve Multiplication or Division of Real Numbers.

OBJECTIVE

1 Multiplying Real Numbers

In this section, we discover patterns for multiplying and dividing real numbers. To discover sign rules for multiplication, recall that multiplication is repeated addition. Thus $3 \cdot 2$ means that 2 is an addend 3 times. That is,

$$2 + 2 + 2 = 3 \cdot 2$$

which equals 6. Similarly, $3 \cdot (-2)$ means -2 is an addend 3 times. That is,

$$(-2) + (-2) + (-2) = 3 \cdot (-2)$$

Since $(-2) + (-2) + (-2) = -6$, $3 \cdot (-2) = -6$. This suggests that the product of a positive number and a negative number is a negative number.

What about the product of two negative numbers? To find out, consider the following pattern.

Factor decreases by 1 each time

$$\left.\begin{aligned} -3 \cdot 2 &= -6 \\ -3 \cdot 1 &= -3 \\ -3 \cdot 0 &= 0 \end{aligned}\right\} \text{Product increases by 3 each time.}$$

This pattern continues as

Factor decreases by 1 each time

$$\left.\begin{aligned} -3 \cdot -1 &= 3 \\ -3 \cdot -2 &= 6 \end{aligned}\right\} \text{Product increases by 3 each time.}$$

This suggests that the product of two negative numbers is a positive number.

Multiplying Real Numbers

1. The product of two numbers with the *same* sign is a positive number.

2. The product of two numbers with *different* signs is a negative number.

EXAMPLE 1 Multiply.

a. $(-8)(4)$ **b.** $14(-1)$ **c.** $-9(-10)$

Solution

a. $-8(4) = -32$ **b.** $14(-1) = -14$ **c.** $-9(-10) = 90$ □

PRACTICE 1 Multiply.

a. $8(-5)$ **b.** $(-3)(-4)$ **c.** $(-6)(9)$

We know that every whole number multiplied by zero equals zero. This remains true for real numbers.

Zero as a Factor

If b is a real number, then $b \cdot 0 = 0$. Also, $0 \cdot b = 0$.

EXAMPLE 2 Perform the indicated operations.

a. $(7)(0)(-6)$ **b.** $(-2)(-3)(-4)$ **c.** $(-1)(5)(-9)$ **d.** $(-4)(-11) - (5)(-2)$

Solution

a. By the order of operations, we multiply from left to right. Notice that, because one of the factors is 0, the product is 0.

$$(7)(0)(-6) = 0(-6) = 0$$

b. Multiply two factors at a time, from left to right.

$$(-2)(-3)(-4) = (6)(-4) \quad \text{Multiply } (-2)(-3).$$
$$= -24$$

c. Multiply from left to right.

$$(-1)(5)(-9) = (-5)(-9) \quad \text{Multiply } (-1)(5).$$
$$= 45$$

d. Follow the rules for order of operation.

$$(-4)(-11) - (5)(-2) = 44 - (-10) \quad \text{Find each product.}$$
$$= 44 + 10 \quad \text{Add 44 to the opposite of } -10.$$
$$= 54 \quad \text{Add.}$$
□

PRACTICE 2 Perform the indicated operations.

a. $(-1)(-5)(-6)$ **b.** $(-3)(-2)(4)$

c. $(-4)(0)(5)$ **d.** $(-2)(-3) - (-4)(5)$

Helpful Hint

You may have noticed from the example that if we multiply:

- an *even* number of negative numbers, the product is *positive*.
- an *odd* number of negative numbers, the product is *negative*.

Answer to Concept Check
negative

✓CONCEPT CHECK

What is the sign of the product of five negative numbers? Explain.

Multiplying signed decimals or fractions is carried out exactly the same way as multiplying integers.

EXAMPLE 3 Multiply.

a. $(-1.2)(0.05)$ **b.** $\frac{2}{3} \cdot \left(-\frac{7}{10}\right)$ **c.** $\left(-\frac{4}{5}\right)(-20)$

Solution

a. The product of two numbers with different signs is negative.

$$(-1.2)(0.05) = -[(1.2)(0.05)]$$
$$= -0.06$$

b. $\frac{2}{3} \cdot \left(-\frac{7}{10}\right) = -\frac{2 \cdot 7}{3 \cdot 10} = -\frac{2 \cdot 7}{3 \cdot 2 \cdot 5} = -\frac{7}{15}$

c. $\left(-\frac{4}{5}\right)(-20) = \frac{4 \cdot 20}{5 \cdot 1} = \frac{4 \cdot 4 \cdot 5}{5 \cdot 1} = \frac{16}{1}$ or 16 □

PRACTICE 3 Multiply.

a. $(0.23)(-0.2)$ **b.** $\left(-\frac{3}{5}\right) \cdot \left(\frac{4}{9}\right)$ **c.** $\left(-\frac{7}{12}\right)(-24)$

Now that we know how to multiply positive and negative numbers, let's see how we find the values of $(-4)^2$ and -4^2, for example. Although these two expressions look similar, the difference between the two is the parentheses. In $(-4)^2$, the parentheses tell us that the base, or repeated factor, is -4. In -4^2, only 4 is the base. Thus,

$(-4)^2 = (-4)(-4) = 16$ The base is -4.

$-4^2 = -(4 \cdot 4) = -16$ The base is 4.

EXAMPLE 4 Evaluate.

a. $(-2)^3$ **b.** -2^3 **c.** $(-3)^2$ **d.** -3^2

Solution

a. $(-2)^3 = (-2)(-2)(-2) = -8$ The base is -2.

b. $-2^3 = -(2 \cdot 2 \cdot 2) = -8$ The base is 2.

c. $(-3)^2 = (-3)(-3) = 9$ The base is -3.

d. $-3^2 = -(3 \cdot 3) = -9$ The base is 3. □

PRACTICE 4 Evaluate.

a. $(-6)^2$ **b.** -6^2 **c.** $(-4)^3$ **d.** -4^3

Helpful Hint

Be careful when identifying the base of an exponential expression.

$(-3)^2$	-3^2
Base is -3	Base is 3
$(-3)^2 = (-3)(-3) = 9$	$-3^2 = -(3 \cdot 3) = -9$

OBJECTIVE

2 Finding Reciprocals 

Just as every difference of two numbers $a - b$ can be written as the sum $a + (-b)$, so too every quotient of two numbers can be written as a product. For example, the quotient $6 \div 3$ can be written as $6 \cdot \frac{1}{3}$. Recall that the pair of numbers 3 and $\frac{1}{3}$ has a special relationship. Their product is 1 and they are called reciprocals or **multiplicative inverses** of each other.

Reciprocals or Multiplicative Inverses

Two numbers whose product is 1 are called reciprocals or multiplicative inverses of each other.

Notice that **0 has no multiplicative inverse** since 0 multiplied by any number is never 1 but always 0.

EXAMPLE 5 Find the reciprocal of each number.

a. 22 **b.** $\frac{3}{16}$ **c.** -10 **d.** $-\frac{9}{13}$

Solution

a. The reciprocal of 22 is $\frac{1}{22}$ since $22 \cdot \frac{1}{22} = 1$.

b. The reciprocal of $\frac{3}{16}$ is $\frac{16}{3}$ since $\frac{3}{16} \cdot \frac{16}{3} = 1$.

c. The reciprocal of -10 is $-\frac{1}{10}$.

d. The reciprocal of $-\frac{9}{13}$ is $-\frac{13}{9}$. □

PRACTICE

5 Find the reciprocal of each number.

a. $\frac{8}{3}$ **b.** 15 **c.** $-\frac{2}{7}$ **d.** -5

OBJECTIVE

3 Dividing Real Numbers 

We may now write a quotient as an equivalent product.

Quotient of Two Real Numbers

If a and b are real numbers and b is not 0, then

$$a \div b = \frac{a}{b} = a \cdot \frac{1}{b}$$

In other words, the quotient of two real numbers is the product of the first number and the multiplicative inverse or reciprocal of the second number.

EXAMPLE 6 Use the definition of the quotient of two numbers to divide.

a. $-18 \div 3$ **b.** $\frac{-14}{-2}$ **c.** $\frac{20}{-4}$

Solution

a. $-18 \div 3 = -18 \cdot \frac{1}{3} = -6$

b. $\frac{-14}{-2} = -14 \cdot -\frac{1}{2} = 7$

c. $\frac{20}{-4} = 20 \cdot -\frac{1}{4} = -5$ □

PRACTICE
6 Use the definition of the quotient of two numbers to divide.

a. $\frac{16}{-2}$ **b.** $24 \div (-6)$ **c.** $\frac{-35}{-7}$

Since the quotient $a \div b$ can be written as the product $a \cdot \frac{1}{b}$, it follows that sign patterns for dividing two real numbers are the same as sign patterns for multiplying two real numbers.

Multiplying and Dividing Real Numbers

1. The product or quotient of two numbers with the *same* sign is a positive number.
2. The product or quotient of two numbers with *different* signs is a negative number.

EXAMPLE 7 Divide.

a. $\frac{-24}{-4}$ **b.** $\frac{-36}{3}$ **c.** $\frac{2}{3} \div \left(-\frac{5}{4}\right)$ **d.** $-\frac{3}{2} \div 9$

Solution

a. $\frac{-24}{-4} = 6$ **b.** $\frac{-36}{3} = -12$ **c.** $\frac{2}{3} \div \left(-\frac{5}{4}\right) = \frac{2}{3} \cdot \left(-\frac{4}{5}\right) = -\frac{8}{15}$

d. $-\frac{3}{2} \div 9 = -\frac{3}{2} \cdot \frac{1}{9} = -\frac{3 \cdot 1}{2 \cdot 9} = -\frac{3 \cdot 1}{2 \cdot 3 \cdot 3} = -\frac{1}{6}$

PRACTICE
7 Divide.

a. $\frac{-18}{-6}$ **b.** $\frac{-48}{3}$ **c.** $\frac{3}{5} \div \left(-\frac{1}{2}\right)$ **d.** $-\frac{4}{9} \div 8$

✓CONCEPT CHECK
What is wrong with the following calculation?

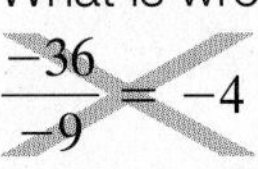

The definition of the quotient of two real numbers does not allow for division by 0 because 0 does not have a multiplicative inverse. There is no number we can multiply 0 by to get 1. How then do we interpret $\frac{3}{0}$? We say that division by 0 is not allowed or not defined and that $\frac{3}{0}$ does not represent a real number. The denominator of a fraction can never be 0.

Can the numerator of a fraction be 0? Can we divide 0 by a number? Yes. For example,

$$\frac{0}{3} = 0 \cdot \frac{1}{3} = 0$$

In general, the quotient of 0 and any nonzero number is 0.

Zero as a Divisor or Dividend

1. The quotient of any nonzero real number and 0 is undefined. In symbols, if $a \neq 0$, $\frac{a}{0}$ **is undefined.**
2. The quotient of 0 and any real number except 0 is 0. In symbols, if $a \neq 0$, $\frac{0}{a} = 0$.

Answer to Concept Check
$\frac{-36}{-9} = 4$

EXAMPLE 8 Perform the indicated operations.

a. $\frac{1}{0}$ **b.** $\frac{0}{-3}$ **c.** $\frac{0(-8)}{2}$

Solution

a. $\frac{1}{0}$ is undefined **b.** $\frac{0}{-3} = 0$ **c.** $\frac{0(-8)}{2} = \frac{0}{2} = 0$ □

PRACTICE 8 Perform the indicated operations.

a. $\frac{0}{-2}$ **b.** $\frac{-4}{0}$ **c.** $\frac{-5}{6(0)}$

Notice that $\frac{12}{-2} = -6$, $-\frac{12}{2} = -6$, and $\frac{-12}{2} = -6$. This means that

$$\frac{12}{-2} = -\frac{12}{2} = \frac{-12}{2}$$

In words, a single negative sign in a fraction can be written in the denominator, in the numerator, or in front of the fraction without changing the value of the fraction. Thus,

$$\frac{1}{-7} = \frac{-1}{7} = -\frac{1}{7}$$

In general, if a and b are real numbers, $b \neq 0$, then $\frac{a}{-b} = \frac{-a}{b} = -\frac{a}{b}$.

OBJECTIVE

4 Evaluating Expressions 

Examples combining basic arithmetic operations along with the principles of order of operations help us review these concepts.

EXAMPLE 9 Simplify each expression.

a. $\frac{(-12)(-3) + 3}{-7 - (-2)}$ **b.** $\frac{2(-3)^2 - 20}{-5 + 4}$

Solution

a. First, simplify the numerator and denominator separately, then divide.

$$\frac{(-12)(-3) + 3}{-7 - (-2)} = \frac{36 + 3}{-7 + 2}$$

$$= \frac{39}{-5} \text{ or } -\frac{39}{5}$$

b. Simplify the numerator and denominator separately, then divide.

$$\frac{2(-3)^2 - 20}{-5 + 4} = \frac{2 \cdot 9 - 20}{-5 + 4} = \frac{18 - 20}{-5 + 4} = \frac{-2}{-1} = 2$$ □

PRACTICE 9 Simplify each expression.

a. $\frac{(-8)(-11) - 4}{-9 - (-4)}$ **b.** $\frac{3(-2)^3 - 9}{-6 + 3}$

Using what we have learned about multiplying and dividing real numbers, we continue to practice evaluating algebraic expressions.

EXAMPLE 10 If $x = -2$ and $y = -4$, evaluate each expression.

a. $5x - y$ **b.** $x^4 - y^2$ **c.** $\frac{3x}{2y}$

Solution

a. Replace x with -2 and y with -4 and simplify.

$$5x - y = 5(-2) - (-4) = -10 - (-4) = -10 + 4 = -6$$

b. Replace x with -2 and y with -4.

$$x^4 - y^2 = (-2)^4 - (-4)^2 \quad \text{Substitute the given values for the variables.}$$
$$= 16 - (16) \quad \text{Evaluate exponential expressions.}$$
$$= 0 \quad \text{Subtract.}$$

c. Replace x with -2 and y with -4 and simplify.

$$\frac{3x}{2y} = \frac{3(-2)}{2(-4)} = \frac{-6}{-8} = \frac{3}{4}$$

□

PRACTICE 10 If $x = -5$ and $y = -2$, evaluate each expression.

a. $7y - x$ **b.** $x^2 - y^3$ **c.** $\frac{2x}{3y}$

OBJECTIVE 5 Solving Applications That Involve Multiplying or Dividing Numbers 

Many real-life problems involve multiplication and division of numbers.

EXAMPLE 11 Calculating a Total Golf Score

A professional golfer finished seven strokes under par (-7) for each of three days of a tournament. What was her total score for the tournament?

Solution Although the key word is "total," since this is repeated addition of the same number, we multiply.

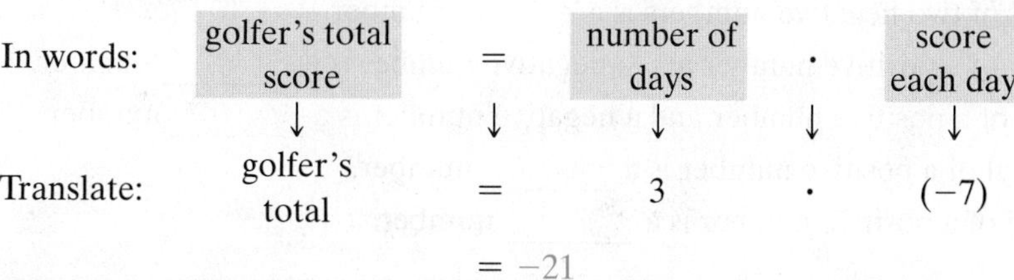

In words:
golfer's total score
=
number of days
·
score each day
Translate:
golfer's total
=
3
·
(−7)
= −21

Thus, the golfer's total score was -21, or 21 strokes under par. □

PRACTICE 11 A card player had a score of -13 for each of four games. Find the total score.

Calculator Explorations 

Entering Negative Numbers on a Scientific Calculator

To enter a negative number on a scientific calculator, find a key marked [+/−]. (On some calculators, this key is marked [CHS] for "change sign.") To enter -8, for example, press the keys [8] [+/−]. The display will read [−8].

(Continued on next page)

Entering Negative Numbers on a Graphing Calculator

To enter a negative number on a graphing calculator, find a key marked [(−)]. Do not confuse this key with the key [−], which is used for subtraction. To enter −8, for example, press the keys [(−)] [8]. The display will read [−8].

Operations with Real Numbers

To evaluate $-2(7-9)-20$ on a calculator, press the keys

[2] [+/−] [×] [(] [7] [−] [9] [)] [−] [2] [0] [=] or

[(−)] [2] [(] [7] [−] [9] [)] [−] [2] [0] [ENTER]

The display will read [−16] or [−2(7 − 9) − 20 / −16].

Use a calculator to simplify each expression.

1. $-38(26-27)$
2. $-59(-8)+1726$
3. $134+25(68-91)$
4. $45(32)-8(218)$
5. $\dfrac{-50(294)}{175-265}$
6. $\dfrac{-444-444.8}{-181-324}$
7. 9^5-4550
8. 5^8-6259
9. $(-125)^2$ (Be careful.)
10. -125^2 (Be careful.)

Vocabulary, Readiness & Video Check

Use the choices below to fill in each blank.

positive   0   negative   undefined

1. If n is a real number, then $n \cdot 0 =$ _______ and $0 \cdot n =$ _______.
2. If n is a real number, but not 0, then $\dfrac{0}{n} =$ _______ and we say $\dfrac{n}{0}$ is _______.
3. The product of two negative numbers is a _______ number.
4. The quotient of two negative numbers is a _______ number.
5. The quotient of a positive number and a negative number is a _______ number.
6. The product of a positive number and a negative number is a _______ number.
7. The reciprocal of a positive number is a _______ number.
8. The opposite of a positive number is a _______ number.

Martin-Gay Interactive Videos

See Video 1.7

Watch the section lecture video and answer the following questions.

OBJECTIVE 1
9. Explain the significance of the use of parentheses when comparing Examples 6 and 7.

OBJECTIVE 2
10. In Example 9, why is the reciprocal equal to $\frac{3}{2}$ and not $-\frac{3}{2}$?

OBJECTIVE 3
11. Before Example 11, the sign rules for division of real numbers are discussed. Are the sign rules for division the same as for multiplication? Why or why not?

OBJECTIVE 4
12. In Example 17, the importance of placing the replacement values in parentheses when evaluating is emphasized. Why?

OBJECTIVE 5
13. In Example 18, explain why each loss of 4 yards is represented by −4 and not 4.

1.7 Exercise Set

MyMathLab®

Multiply. See Examples 1 through 3.

1. $-6(4)$
2. $-8(5)$
3. $2(-1)$
4. $7(-4)$
5. $-5(-10)$
6. $-6(-11)$
7. $-3 \cdot 4$
8. $-2 \cdot 8$
9. $-7 \cdot 0$
10. $-6 \cdot 0$
11. $2(-9)$
12. $3(-5)$
13. $-\frac{1}{2}\left(-\frac{3}{5}\right)$
14. $-\frac{1}{8}\left(-\frac{1}{3}\right)$
15. $-\frac{3}{4}\left(-\frac{8}{9}\right)$
16. $-\frac{5}{6}\left(-\frac{3}{10}\right)$
17. $5(-1.4)$
18. $6(-2.5)$
19. $-0.2(-0.7)$
20. $-0.5(-0.3)$
21. $-10(80)$
22. $-20(60)$
23. $4(-7)$
24. $5(-9)$
25. $(-5)(-5)$
26. $(-7)(-7)$
27. $\frac{2}{3}\left(-\frac{4}{9}\right)$
28. $\frac{2}{7}\left(-\frac{2}{11}\right)$
29. $-11(11)$
30. $-12(12)$
31. $-\frac{20}{25}\left(\frac{5}{16}\right)$
32. $-\frac{25}{36}\left(\frac{6}{15}\right)$
33. $(-1)(2)(-3)(-5)$
34. $(-2)(-3)(-4)(-2)$

Perform the indicated operations. See Example 2.

35. $(-2)(5) - (-11)(3)$
36. $8(-3) - 4(-5)$
37. $(-6)(-1)(-2) - (-5)$
38. $20 - (-4)(3)(-2)$

Decide whether each statement is true or false.

39. The product of three negative integers is negative.
40. The product of three positive integers is positive.
41. The product of four negative integers is negative.
42. The product of four positive integers is positive.

Evaluate. See Example 4.

43. $(-2)^4$
44. -2^4
45. -1^5
46. $(-1)^5$
47. $(-5)^2$
48. -5^2
49. -7^2
50. $(-7)^2$

Find each reciprocal or multiplicative inverse. See Example 5.

51. 9
52. 100
53. $\frac{2}{3}$
54. $\frac{1}{7}$
55. -14
56. -8
57. $-\frac{3}{11}$
58. $-\frac{6}{13}$
59. 0.2
60. 1.5
61. $\frac{1}{-6.3}$
62. $\frac{1}{-8.9}$

Divide. See Examples 6 through 8.

63. $\frac{18}{-2}$
64. $\frac{20}{-10}$
65. $\frac{-16}{-4}$
66. $\frac{-18}{-6}$
67. $\frac{-48}{12}$
68. $\frac{-60}{5}$
69. $\frac{0}{-4}$
70. $\frac{0}{-9}$
71. $-\frac{15}{3}$
72. $-\frac{24}{8}$
73. $\frac{5}{0}$
74. $\frac{3}{0}$
75. $\frac{-12}{-4}$
76. $\frac{-45}{-9}$
77. $\frac{30}{-2}$
78. $\frac{14}{-2}$
79. $\frac{6}{7} \div \left(-\frac{1}{3}\right)$
80. $\frac{4}{5} \div \left(-\frac{1}{2}\right)$
81. $-\frac{5}{9} \div \left(-\frac{3}{4}\right)$
82. $-\frac{1}{10} \div \left(-\frac{8}{11}\right)$
83. $-\frac{4}{9} \div \frac{4}{9}$
84. $-\frac{5}{12} \div \frac{5}{12}$

MIXED PRACTICE

Simplify. See Examples 1 through 9.

85. $\frac{-9(-3)}{-6}$
86. $\frac{-6(-3)}{-4}$
87. $\frac{12}{9 - 12}$
88. $\frac{-15}{1 - 4}$
89. $\frac{-6^2 + 4}{-2}$
90. $\frac{3^2 + 4}{5}$
91. $\frac{8 + (-4)^2}{4 - 12}$
92. $\frac{6 + (-2)^2}{4 - 9}$
93. $\frac{22 + (3)(-2)}{-5 - 2}$
94. $\frac{-20 + (-4)(3)}{1 - 5}$
95. $\frac{-3 - 5^2}{2(-7)}$
96. $\frac{-2 - 4^2}{3(-6)}$
97. $\frac{6 - 2(-3)}{4 - 3(-2)}$
98. $\frac{8 - 3(-2)}{2 - 5(-4)}$
99. $\frac{-3 - 2(-9)}{-15 - 3(-4)}$
100. $\frac{-4 - 8(-2)}{-9 - 2(-3)}$
101. $\frac{|5 - 9| + |10 - 15|}{|2(-3)|}$
102. $\frac{|-3 + 6| + |-2 + 7|}{|-2 \cdot 2|}$

If $x = -5$ and $y = -3$, evaluate each expression. See Example 10.

103. $3x + 2y$
104. $4x + 5y$
105. $2x^2 - y^2$
106. $x^2 - 2y^2$
107. $x^3 + 3y$
108. $y^3 + 3x$
109. $\frac{2x - 5}{y - 2}$
110. $\frac{2y - 12}{x - 4}$
111. $\frac{-3 - y}{x - 4}$
112. $\frac{4 - 2x}{y + 3}$

TRANSLATING

Translate each phrase to an expression. Use x to represent "a number." See Example 11.

113. The product of -71 and a number
114. The quotient of -8 and a number
115. Subtract a number from -16.
116. The sum of a number and -12
117. -29 increased by a number
118. The difference of a number and -10
119. Divide a number by -33.
120. Multiply a number by -17.

Solve. See Example 11.

121. A football team lost four yards on each of three consecutive plays. Represent the total loss as a product of signed numbers and find the total loss.
122. Joe Norstrom lost $400 on each of seven consecutive days in the stock market. Represent his total loss as a product of signed numbers and find his total loss.
123. A deep-sea diver must move up or down in the water in short steps to keep from getting a physical condition called the "bends." Suppose a diver moves down from the surface in five steps of 20 feet each. Represent his total movement as a product of signed numbers and find the total depth.
124. A weather forecaster predicts that the temperature will drop five degrees each hour for the next six hours. Represent this drop as a product of signed numbers and find the total drop in temperature.

Decide whether the given number is a solution of the given equation.

125. Is 7 a solution of $-5x = -35$?
126. Is -4 a solution of $2x = x - 1$?
127. Is -20 a solution of $\frac{x}{10} = 2$?
128. Is -3 a solution of $\frac{45}{x} = -15$?
129. Is 5 a solution of $-3x - 5 = -20$?
130. Is -4 a solution of $2x + 4 = x + 8$?

CONCEPT EXTENSIONS

Study the bar graph below showing the average surface temperatures of planets. Use Exercises 131 and 132 to complete the planet temperatures on the graph. (Pluto is now classified as a dwarf planet.)

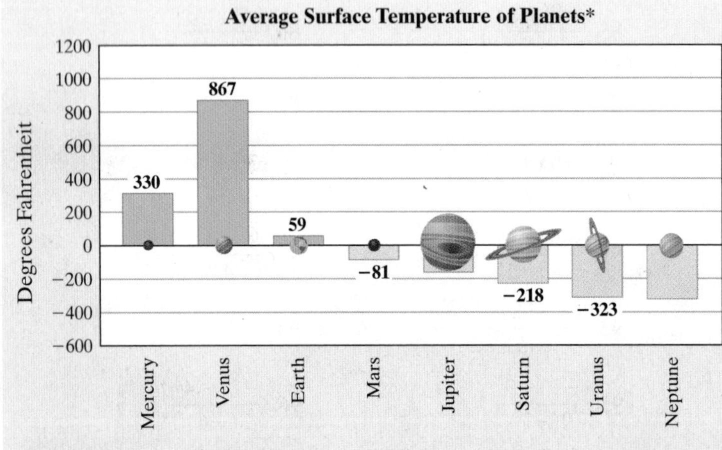

*For some planets, the temperature given is the temperature where the atmosphere pressure equals 1 Earth atmosphere; data from *The World Almanac*

131. The surface temperature of Jupiter is twice the temperature of Mars. Find this temperature.
132. The surface temperature of Neptune is equal to the temperature of Mercury divided by -1. Find this temperature.
133. Explain why the product of an even number of negative numbers is a positive number.
134. If a and b are any real numbers, is the statement $a \cdot b = b \cdot a$ always true? Why or why not?
135. Find any real numbers that are their own reciprocal.
136. Explain why 0 has no reciprocal.

If q is a negative number, r is a negative number, and t is a positive number, determine whether each expression simplifies to a positive or negative number. If it is not possible to determine, state so.

137. $\frac{q}{r \cdot t}$
138. $q^2 \cdot r \cdot t$
139. $q + t$
140. $t + r$
141. $t(q + r)$
142. $r(q - t)$

Write each of the following as an expression and evaluate.

143. The sum of -2 and the quotient of -15 and 3
144. The sum of 1 and the product of -8 and -5
145. Twice the sum of -5 and -3
146. 7 subtracted from the quotient of 0 and 5

1.8 Properties of Real Numbers 

OBJECTIVES

1 Use the Commutative and Associative Properties.

2 Use the Distributive Property.

3 Use the Identity and Inverse Properties.

OBJECTIVE

1 Using the Commutative and Associative Properties 

In this section, we give names to properties of real numbers with which we are already familiar. Throughout this section, the variables *a*, *b*, and *c* represent real numbers.

We know that order does not matter when adding numbers. For example, we know that $7 + 5$ is the same as $5 + 7$. This property is given a special name—the **commutative property of addition.** We also know that order does not matter when multiplying numbers. For example, we know that $-5(6) = 6(-5)$. This property means that multiplication is commutative also and is called the **commutative property of multiplication.**

Commutative Properties

Addition:	$a + b = b + a$
Multiplication:	$a \cdot b = b \cdot a$

These properties state that the *order* in which any two real numbers are added or multiplied does not change their sum or product. For example, if we let $a = 3$ and $b = 5$, then the commutative properties guarantee that

$$3 + 5 = 5 + 3 \quad \text{and} \quad 3 \cdot 5 = 5 \cdot 3$$

Helpful Hint

Is subtraction also commutative? Try an example. Does $3 - 2 = 2 - 3$? **No!** The left side of this statement equals 1; the right side equals -1. There is no commutative property of subtraction. Similarly, there is no commutative property for division. For example, $10 \div 2$ does not equal $2 \div 10$.

EXAMPLE 1 Use a commutative property to complete each statement.

a. $x + 5 =$ ________ **b.** $3 \cdot x =$ ________

Solution

a. $x + 5 = 5 + x$ By the commutative property of addition

b. $3 \cdot x = x \cdot 3$ By the commutative property of multiplication

PRACTICE

1 Use a commutative property to complete each statement.

a. $x \cdot 8 =$ ___ **b.** $x + 17 =$ _____

✓CONCEPT CHECK

Which of the following pairs of actions are commutative?

a. "raking the leaves" and "bagging the leaves"
b. "putting on your left glove" and "putting on your right glove"
c. "putting on your coat" and "putting on your shirt"
d. "reading a novel" and "reading a newspaper"

Let's now discuss grouping numbers. We know that when we add three numbers, the way in which they are grouped or associated does not change their sum. For example, we know that $2 + (3 + 4) = 2 + 7 = 9$. This result is the same if we group the numbers differently. In other words, $(2 + 3) + 4 = 5 + 4 = 9$ also. Thus, $2 + (3 + 4) = (2 + 3) + 4$. This property is called the **associative property of addition.**

We also know that changing the grouping of numbers when multiplying does not change their product. For example, $2 \cdot (3 \cdot 4) = (2 \cdot 3) \cdot 4$ (check it). This is the **associative property of multiplication.**

Answers to Concept Check:
b, d

Associative Properties

Addition: $(a + b) + c = a + (b + c)$

Multiplication: $(a \cdot b) \cdot c = a \cdot (b \cdot c)$

These properties state that the way in which three numbers are *grouped* does not change their sum or their product.

EXAMPLE 2 Use an associative property to complete each statement.

a. $5 + (4 + 6) =$ _______ **b.** $(-1 \cdot 2) \cdot 5 =$ _______

Solution

a. $5 + (4 + 6) = (5 + 4) + 6$ By the associative property of addition

b. $(-1 \cdot 2) \cdot 5 = -1 \cdot (2 \cdot 5)$ By the associative property of multiplication

PRACTICE 2 Use an associative property to complete each statement.

a. $(2 + 9) + 7 =$ _______ **b.** $-4 \cdot (2 \cdot 7) =$ _______

Helpful Hint

Remember the difference between the commutative properties and the associative properties. The commutative properties have to do with the *order* of numbers, and the associative properties have to do with the *grouping* of numbers.

Let's now illustrate how these properties can help us simplify expressions.

EXAMPLE 3 Simplify each expression.

a. $10 + (x + 12)$ **b.** $-3(7x)$

Solution

a. $10 + (x + 12) = 10 + (12 + x)$ By the commutative property of addition

$= (10 + 12) + x$ By the associative property of addition

$= 22 + x$ Add.

b. $-3(7x) = (-3 \cdot 7)x$ By the associative property of multiplication

$= -21x$ Multiply.

PRACTICE 3 Simplify each expression.

a. $(5 + x) + 9$ **b.** $5(-6x)$

OBJECTIVE

2 Using the Distributive Property

The **distributive property of multiplication over addition** is used repeatedly throughout algebra. It is useful because it allows us to write a product as a sum or a sum as a product.

We know that $7(2 + 4) = 7(6) = 42$. Compare that with $7(2) + 7(4) = 14 + 28 = 42$. Since both original expressions equal 42, they must equal each other, or

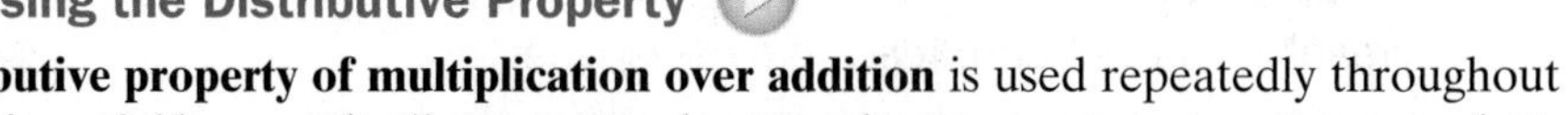

$$7(2 + 4) = 7(2) + 7(4)$$

This is an example of the distributive property. The product on the left side of the equal sign is equal to the sum on the right side. We can think of the 7 as being distributed to each number inside the parentheses.

Distributive Property of Multiplication Over Addition

$$a(b + c) = ab + ac$$

Since multiplication is commutative, this property can also be written as

$$(b + c)a = ba + ca$$

The distributive property can also be extended to more than two numbers inside the parentheses. For example,

$$\begin{aligned} 3(x + y + z) &= 3(x) + 3(y) + 3(z) \\ &= 3x + 3y + 3z \end{aligned}$$

Since we define subtraction in terms of addition, the distributive property is also true for subtraction. For example

$$\begin{aligned} 2(x - y) &= 2(x) - 2(y) \\ &= 2x - 2y \end{aligned}$$

EXAMPLE 4 Use the distributive property to write each expression without parentheses. Then simplify if possible.

a. $2(x + y)$ **b.** $-5(-3 + 2z)$ **c.** $5(x + 3y - z)$

d. $-1(2 - y)$ **e.** $-(3 + x - w)$ **f.** $\frac{1}{2}(6x + 14) + 10$

Solution

a. $2(x + y) = 2 \cdot x + 2 \cdot y$
$= 2x + 2y$

b. $-5(-3 + 2z) = -5(-3) + (-5)(2z)$
$= 15 - 10z$

c. $5(x + 3y - z) = 5(x) + 5(3y) - 5(z)$
$= 5x + 15y - 5z$

d. $-1(2 - y) = (-1)(2) - (-1)(y)$
$= -2 + y$

e. $-(3 + x - w) = -1(3 + x - w)$
$= (-1)(3) + (-1)(x) - (-1)(w)$
$= -3 - x + w$

Helpful Hint
Notice in part **(e)** that $-(3 + x - w)$ is first rewritten as $-1(3 + x - w)$.

f. $\frac{1}{2}(6x + 14) + 10 = \frac{1}{2}(6x) + \frac{1}{2}(14) + 10$ Apply the distributive property.
$= 3x + 7 + 10$ Multiply.
$= 3x + 17$ Add. □

PRACTICE

4 Use the distributive property to write each expression without parentheses. Then simplify if possible.

a. $5(x - y)$ **b.** $-6(4 + 2t)$

c. $2(3x - 4y - z)$ **d.** $(3 - y) \cdot (-1)$

e. $-(x - 7 + 2s)$ **f.** $\frac{1}{2}(2x + 4) + 9$

We can use the distributive property in reverse to write a sum as a product.

EXAMPLE 5 Use the distributive property to write each sum as a product.

a. $8 \cdot 2 + 8 \cdot x$ **b.** $7s + 7t$

Solution

a. $8 \cdot 2 + 8 \cdot x = 8(2 + x)$ **b.** $7s + 7t = 7(s + t)$

PRACTICE

5 Use the distributive property to write each sum as a product.

a. $5 \cdot w + 5 \cdot 3$ **b.** $9w + 9z$

OBJECTIVE

3 Using the Identity and Inverse Properties 

Next, we look at the **identity properties.**

The number 0 is called the identity for addition because when 0 is added to any real number, the result is the same real number. In other words, the *identity* of the real number is not changed.

The number 1 is called the identity for multiplication because when a real number is multiplied by 1, the result is the same real number. In other words, the *identity* of the real number is not changed.

Identities for Addition and Multiplication

0 is the identity element for addition.

$$a + 0 = a \quad \text{and} \quad 0 + a = a$$

1 is the identity element for multiplication.

$$a \cdot 1 = a \quad \text{and} \quad 1 \cdot a = a$$

Notice that 0 is the *only* number that can be added to any real number with the result that the sum is the same real number. Also, 1 is the *only* number that can be multiplied by any real number with the result that the product is the same real number.

Additive inverses or **opposites** were introduced in Section 1.5. Two numbers are called additive inverses or opposites if their sum is 0. The additive inverse or opposite of 6 is -6 because $6 + (-6) = 0$. The additive inverse or opposite of -5 is 5 because $-5 + 5 = 0$.

Reciprocals or **multiplicative inverses** were introduced in Section 1.3. Two nonzero numbers are called reciprocals or multiplicative inverses if their product is 1. The reciprocal or multiplicative inverse of $\frac{2}{3}$ is $\frac{3}{2}$ because $\frac{2}{3} \cdot \frac{3}{2} = 1$. Likewise, the reciprocal of -5 is $-\frac{1}{5}$ because $-5\left(-\frac{1}{5}\right) = 1$.

✓CONCEPT CHECK

Which of the following, $1, -\frac{10}{3}, \frac{3}{10}, 0, \frac{10}{3}, -\frac{3}{10}$, is the

a. opposite of $-\frac{3}{10}$?

b. reciprocal of $-\frac{3}{10}$?

Additive or Multiplicative Inverses

The numbers a and $-a$ are additive inverses or opposites of each other because their sum is 0; that is,

$$a + (-a) = 0$$

The numbers b and $\frac{1}{b}$ (for $b \neq 0$) are reciprocals or multiplicative inverses of each other because their product is 1; that is,

$$b \cdot \frac{1}{b} = 1$$

EXAMPLE 6 Name the property or properties illustrated by each true statement.

Solution

a. $3 \cdot y = y \cdot 3$ — Commutative property of multiplication (order changed)

b. $(x + 7) + 9 = x + (7 + 9)$ — Associative property of addition (grouping changed)

c. $(b + 0) + 3 = b + 3$ — Identity element for addition

d. $0.2 \cdot (z \cdot 5) = 0.2 \cdot (5 \cdot z)$ — Commutative property of multiplication (order changed)

e. $-2 \cdot \left(-\frac{1}{2}\right) = 1$ — Multiplicative inverse property

f. $-2 + 2 = 0$ — Additive inverse property

g. $-6 \cdot (y \cdot 2) = (-6 \cdot 2) \cdot y$ — Commutative and associative properties of multiplication (order and grouping changed) □

PRACTICE

6 Name the property or properties illustrated by each true statement.

a. $(7 \cdot 3x) \cdot 4 = (3x \cdot 7) \cdot 4$ — Commutative property of multiplication

b. $6 + (3 + y) = (6 + 3) + y$ — Associative property of addition

c. $8 + (t + 0) = 8 + t$ — Identity element for addition

d. $-\frac{3}{4} \cdot \left(-\frac{4}{3}\right) = 1$ — Multiplicative inverse property

e. $(2 + x) + 5 = 5 + (2 + x)$ — Commutative property of addition

f. $3 + (-3) = 0$ — Additive inverse property

g. $(-3b) \cdot 7 = (-3 \cdot 7) \cdot b$ — Commutative and associative properties of multiplication

Answers to Concept Check:

a. $\frac{3}{10}$ **b.** $-\frac{10}{3}$

Vocabulary, Readiness & Video Check

Use the choices below to fill in each blank.

distributive property	associative property of multiplication	commutative property of addition
opposites or additive inverses	associative property of addition	
reciprocals or multiplicative inverses	commutative property of multiplication	

1. $x + 5 = 5 + x$ is a true statement by the ______________________.
2. $x \cdot 5 = 5 \cdot x$ is a true statement by the ______________________.
3. $3(y + 6) = 3 \cdot y + 3 \cdot 6$ is true by the ______________________.
4. $2 \cdot (x \cdot y) = (2 \cdot x) \cdot y$ is a true statement by the ______________________.
5. $x + (7 + y) = (x + 7) + y$ is a true statement by the ______________________.
6. The numbers $-\frac{2}{3}$ and $-\frac{3}{2}$ are called ______________________.
7. The numbers $-\frac{2}{3}$ and $\frac{2}{3}$ are called ______________________.

Martin-Gay Interactive Videos

See Video 1.8

Watch the section lecture video and answer the following questions.

OBJECTIVE 1

8. The commutative properties are discussed in Examples 1 and 2 and the associative properties are discussed in Examples 3–7. What's the one word used again and again to describe the commutative property? The associative property?

OBJECTIVE 2

9. In Example 10, what point is made about the term 2?

OBJECTIVE 3

10. Complete these statements based on the lecture given before Example 12.
    - The identify element for addition is _____ because if we add _____ to any real number, the result is that real number.
    - The identify element for multiplication is _____ because any real number times _____ gives a result of that original real number.

1.8 Exercise Set

MyMathLab®

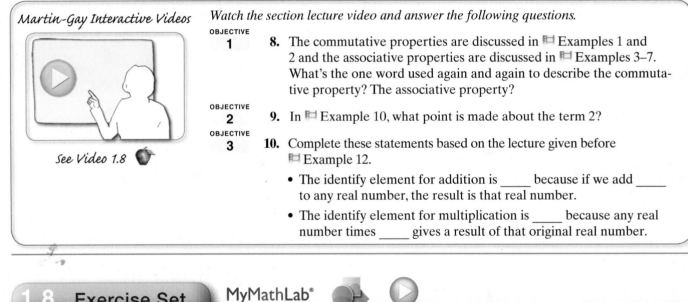

Use a commutative property to complete each statement. See Example 1.

1. $x + 16 =$ ______
2. $4 + y =$ ______
3. $-4 \cdot y =$ ______
4. $-2 \cdot x =$ ______
5. $xy =$ ______
6. $ab =$ ______
7. $2x + 13 =$ ______
8. $19 + 3y =$ ______

Use an associative property to complete each statement. See Example 2.

9. $(xy) \cdot z =$ ________
10. $3 \cdot (xy) =$ ________
11. $2 + (a + b) =$ ________
12. $(y + 4) + z =$ ________
13. $4 \cdot (ab) =$ ________
14. $(-3y) \cdot z =$ ________
15. $(a + b) + c =$ ________
16. $6 + (r + s) =$ ________

Use the commutative and associative properties to simplify each expression. See Example 3.

17. $8 + (9 + b)$
18. $(r + 3) + 11$
19. $4(6y)$
20. $2(42x)$
21. $\frac{1}{5}(5y)$
22. $\frac{1}{8}(8z)$
23. $(13 + a) + 13$
24. $7 + (x + 4)$
25. $-9(8x)$
26. $-3(12y)$
27. $\frac{3}{4}\left(\frac{4}{3}s\right)$
28. $\frac{2}{7}\left(\frac{7}{2}r\right)$
29. $\frac{2}{3} + \left(\frac{4}{3} + x\right)$
30. $\frac{7}{9} + \left(\frac{2}{9} + y\right)$

Use the distributive property to write each expression without parentheses. Then simplify the result. See Example 4.

31. $4(x + y)$
32. $7(a + b)$
33. $9(x - 6)$
34. $11(y - 4)$
35. $2(3x + 5)$
36. $5(7 + 8y)$
37. $7(4x - 3)$
38. $3(8x - 1)$
39. $3(6 + x)$
40. $2(x + 5)$
41. $-2(y - z)$
42. $-3(z - y)$
43. $-7(3y + 5)$
44. $-5(2r + 11)$
45. $5(x + 4m + 2)$
46. $8(3y + z - 6)$
47. $-4(1 - 2m + n)$
48. $-4(4 + 2p + 5q)$
49. $-(5x + 2)$
50. $-(9r + 5)$
51. $-(r - 3 - 7p)$
52. $-(q - 2 + 6r)$
53. $\frac{1}{2}(6x + 8)$
54. $\frac{1}{4}(4x - 2)$
55. $-\frac{1}{3}(3x - 9y)$
56. $-\frac{1}{5}(10a - 25b)$
57. $3(2r + 5) - 7$
58. $10(4s + 6) - 40$
59. $-9(4x + 8) + 2$
60. $-11(5x + 3) + 10$
61. $-4(4x + 5) - 5$
62. $-6(2x + 1) - 1$

Use the distributive property to write each sum as a product. See Example 5.

63. $4 \cdot 1 + 4 \cdot y$
64. $14 \cdot z + 14 \cdot 5$
65. $11x + 11y$
66. $9a + 9b$
67. $(-1) \cdot 5 + (-1) \cdot x$
68. $(-3)a + (-3)b$
69. $30a + 30b$
70. $25x + 25y$

Name the properties illustrated by each true statement. See Example 6.

71. $3 \cdot 5 = 5 \cdot 3$
72. $4(3 + 8) = 4 \cdot 3 + 4 \cdot 8$
73. $2 + (x + 5) = (2 + x) + 5$
74. $(x + 9) + 3 = (9 + x) + 3$
75. $9(3 + 7) = 9 \cdot 3 + 9 \cdot 7$
76. $1 \cdot 9 = 9$
77. $(4 \cdot y) \cdot 9 = 4 \cdot (y \cdot 9)$
78. $6 \cdot \frac{1}{6} = 1$
79. $0 + 6 = 6$
80. $(a + 9) + 6 = a + (9 + 6)$
81. $-4(y + 7) = -4 \cdot y + (-4) \cdot 7$
82. $(11 + r) + 8 = (r + 11) + 8$
83. $-4 \cdot (8 \cdot 3) = (8 \cdot -4) \cdot 3$
84. $r + 0 = r$

CONCEPT EXTENSIONS

Fill in the table with the opposite (additive inverse), and the reciprocal (multiplicative inverse). Assume that the value of each expression is not 0.

	Expression	*Opposite*	*Reciprocal*
85.	8		
86.	$-\frac{2}{3}$		
87.	x		
88.	$4y$		
89.			$\frac{1}{2x}$
90.		$7x$	

Decide whether each statement is true or false. See the second Concept Check in this section.

91. The opposite of $-\frac{a}{2}$ is $-\frac{2}{a}$.

92. The reciprocal of $-\frac{a}{2}$ is $\frac{a}{2}$.

Determine which pairs of actions are commutative. See the first Concept Check in this section.

93. "taking a test" and "studying for the test"
94. "putting on your shoes" and "putting on your socks"
95. "putting on your left shoe" and "putting on your right shoe"
96. "reading the sports section" and "reading the comics section"
97. "mowing the lawn" and "trimming the hedges"
98. "baking a cake" and "eating the cake"
99. "dialing a number" and "turning on the cell phone"
100. "feeding the dog" and "feeding the cat"

Name the property illustrated by each step.

101. a. $\triangle + (\square + \bigcirc) = (\square + \bigcirc) + \triangle$

b. $= (\bigcirc + \square) + \triangle$

c. $= \bigcirc + (\square + \triangle)$

102. a. $(x + y) + z = x + (y + z)$

b. $= (y + z) + x$

c. $= (z + y) + x$

103. Explain why 0 is called the identity element for addition.

104. Explain why 1 is called the identity element for multiplication.

105. Write an example that shows that division is not commutative.

106. Write an example that shows that subtraction is not commutative.

Chapter 1 Vocabulary Check

Fill in each blank with one of the words or phrases listed below.

set	inequality symbols	opposites	absolute value	numerator
denominator	grouping symbols	exponent	base	reciprocals
variable	equation	solution		

1. The symbols $\neq$, $<$, and $>$ are called ______________.
2. A mathematical statement that two expressions are equal is called a(n) ______________.
3. The ______________ of a number is the distance between that number and 0 on the number line.
4. A symbol used to represent a number is called a(n) ______________.
5. Two numbers that are the same distance from 0 but lie on opposite sides of 0 are called ______________.
6. The number in a fraction above the fraction bar is called the ______________.
7. A(n) ______________ of an equation is a value for the variable that makes the equation a true statement.
8. Two numbers whose product is 1 are called ______________.
9. In 2^3, the 2 is called the ______________ and the 3 is called the ______________.
10. The number in a fraction below the fraction bar is called the ______________.
11. Parentheses and brackets are examples of ______________.
12. A(n) ______________ is a collection of objects.

Chapter 1 Highlights

DEFINITIONS AND CONCEPTS	EXAMPLES
Section 1.2 Symbols and Sets of Numbers	
A **set** is a collection of objects, called **elements,** enclosed in braces.	$\{a, c, e\}$
Natural Numbers: $\{1, 2, 3, 4, \ldots\}$ **Whole Numbers:** $\{0, 1, 2, 3, 4, \ldots\}$	Given the set $\left\{-3.4, \sqrt{3}, 0, \frac{2}{3}, 5, -4\right\}$, list the numbers that belong to the set of
Integers: $\{\ldots, -3, -2, -1, 0, 1, 2, 3, \ldots\}$	Natural numbers: 5 Whole numbers: 0, 5
Rational Numbers: {real numbers that can be expressed as a quotient of integers}	Integers: $-4, 0, 5$ Rational numbers: $-4, -3.4, 0, \frac{2}{3}, 5$
Irrational Numbers: {real numbers that cannot be expressed as a quotient of integers}	Irrational Numbers: $\sqrt{3}$
Real Numbers: {all numbers that correspond to a point on the number line}	Real numbers: $-4, -3.4, 0, \frac{2}{3}, \sqrt{3}, 5$

DEFINITIONS AND CONCEPTS | **EXAMPLES**

Section 1.2 Symbols and Sets of Numbers (continued)

A line used to picture numbers is called a **number line.**

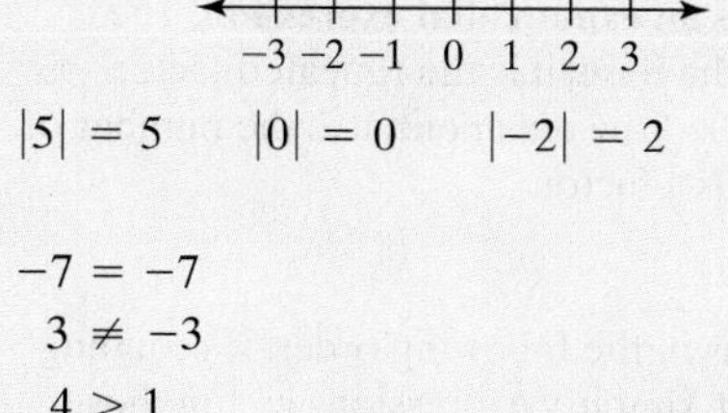

The **absolute value** of a real number a, denoted by $|a|$, is the distance between a and 0 on the number line.

$|5| = 5 \quad |0| = 0 \quad |-2| = 2$

Symbols:
- $=$ is equal to — $-7 = -7$
- $\neq$ is not equal to — $3 \neq -3$
- $>$ is greater than — $4 > 1$
- $<$ is less than — $1 < 4$
- $\leq$ is less than or equal to — $6 \leq 6$
- $\geq$ is greater than or equal to — $18 \geq -\frac{1}{3}$

Order Property for Real Numbers

For any two real numbers a and b, a is less than b if a is to the left of b on a number line.

−3 −2 −1 0 1 2 3

$-3 < 0 \quad 0 > -3 \quad 0 < 2.5 \quad 2.5 > 0$

Section 1.3 Fractions and Mixed Numbers

A quotient of two integers is called a **fraction.** The **numerator** of a fraction is the top number. The **denominator** of a fraction is the bottom number.

13 ← numerator
17 ← denominator

If $a \cdot b = c$, then a and b are **factors** and c is the **product.**

$7 \cdot 9 = 63$ (7: factor, 9: factor, 63: product)

A fraction is in **lowest terms** or **simplest form** when the numerator and the denominator have no factors in common other than 1.

$\frac{13}{17}$ is in simplest form.

To write a fraction in simplest form, factor the numerator and the denominator; then apply the fundamental principle.

Write in simplest form.

$$\frac{6}{14} = \frac{2 \cdot 3}{2 \cdot 7} = \frac{3}{7}$$

Two fractions are **reciprocals** if their product is 1. The reciprocal of $\frac{a}{b}$ is $\frac{b}{a}$.

The reciprocal of $\frac{6}{25}$ is $\frac{25}{6}$.

To multiply fractions, numerator times numerator is the numerator of the product and denominator times denominator is the denominator of the product.

Perform the indicated operations.

$$\frac{2}{5} \cdot \frac{3}{7} = \frac{6}{35}$$

To divide fractions, multiply the first fraction by the reciprocal of the second fraction.

$$\frac{5}{9} \div \frac{2}{7} = \frac{5}{9} \cdot \frac{7}{2} = \frac{35}{18}$$

To add fractions with the same denominator, add the numerators and place the sum over the common denominator.

$$\frac{5}{11} + \frac{3}{11} = \frac{8}{11}$$

To subtract fractions with the same denominator, subtract the numerators and place the difference over the common denominator.

$$\frac{13}{15} - \frac{3}{15} = \frac{10}{15} = \frac{2}{3}$$

Fractions that represent the same quantity are called **equivalent fractions.**

$$\frac{1}{5} = \frac{1 \cdot 4}{5 \cdot 4} = \frac{4}{20}$$

$\frac{1}{5}$ and $\frac{4}{20}$ are equivalent fractions.

DEFINITIONS AND CONCEPTS	EXAMPLES
Section 1.4 Exponents, Order of Operations, Variable Expressions, and Equations	
The expression a^n is an **exponential expression.** The number a is called the **base;** it is the repeated factor. The number n is called the **exponent;** it is the number of times that the base is a factor.	$4^3 = 4 \cdot 4 \cdot 4 = 64$ $7^2 = 7 \cdot 7 = 49$
Order of Operations Simplify expressions in the following order. If grouping symbols are present, simplify expressions within those first, starting with the innermost set. Also, simplify the numerator and the denominator of a fraction separately. **1.** Simplify exponential expressions. **2.** Multiply or divide in order from left to right. **3.** Add or subtract in order from left to right.	$\frac{8^2 + 5(7 - 3)}{3 \cdot 7} = \frac{8^2 + 5(4)}{21}$ $= \frac{64 + 5(4)}{21}$ $= \frac{64 + 20}{21}$ $= \frac{84}{21}$ $= 4$
A symbol used to represent a number is called a **variable.**	Examples of variables are: q, x, z
An **algebraic expression** is a collection of numbers, variables, operation symbols, and grouping symbols.	Examples of algebraic expressions are: $5x, 2(y - 6), \frac{q^2 - 3q + 1}{6}$
To evaluate an algebraic expression containing a variable, substitute a given number for the variable and simplify.	Evaluate $x^2 - y^2$ if $x = 5$ and $y = 3$. $x^2 - y^2 = (5)^2 - 3^2$ $= 25 - 9$ $= 16$
A mathematical statement that two expressions are equal is called an **equation.**	Examples of equations are: $3x - 9 = 20$ $A = \pi r^2$
A **solution** of an equation is a value for the variable that makes the equation a true statement.	Determine whether 4 is a solution of $5x + 7 = 27$. $5x + 7 = 27$ $5(4) + 7 \stackrel{?}{=} 27$ $20 + 7 \stackrel{?}{=} 27$ $27 = 27$ True 4 is a solution.
Section 1.5 Adding Real Numbers	
To Add Two Numbers with the Same Sign **1.** Add their absolute values. **2.** Use their common sign as the sign of the sum.	Add. $10 + 7 = 17$ $-3 + (-8) = -11$
To Add Two Numbers with Different Signs **1.** Subtract their absolute values. **2.** Use the sign of the number whose absolute value is larger as the sign of the sum.	$-25 + 5 = -20$ $14 + (-9) = 5$

DEFINITIONS AND CONCEPTS	EXAMPLES
Section 1.5 Adding Real Numbers (continued)	
Two numbers that are the same distance from 0 but lie on opposite sides of 0 are called **opposites** or **additive inverses.** The opposite of a number a is denoted by $-a$.	The opposite of -7 is 7. The opposite of 123 is -123.
The sum of a number a and its opposite, $-a$, is 0. $a + (-a) = 0$ If a is a number, then $-(-a) = a$.	$-4 + 4 = 0$ $12 + (-12) = 0$ $-(-8) = 8$ $-(-14) = 14$
Section 1.6 Subtracting Real Numbers	
To subtract two numbers a and b, add the first number a to the opposite of the second number b. $a - b = a + (-b)$	Subtract. $3 - (-44) = 3 + 44 = 47$ $-5 - 22 = -5 + (-22) = -27$ $-30 - (-30) = -30 + 30 = 0$
Section 1.7 Multiplying and Dividing Real Numbers	
Quotient of two real numbers $\frac{a}{b} = a \cdot \frac{1}{b}$	Multiply or divide. $\frac{42}{2} = 42 \cdot \frac{1}{2} = 21$
Multiplying and Dividing Real Numbers The product or quotient of two numbers with the same sign is a positive number. The product or quotient of two numbers with different signs is a negative number.	$7 \cdot 8 = 56 \quad -7 \cdot (-8) = 56$ $-2 \cdot 4 = -8 \quad 2 \cdot (-4) = -8$ $\frac{90}{10} = 9 \quad \frac{-90}{-10} = 9$ $\frac{42}{-6} = -7 \quad \frac{-42}{6} = -7$
Products and Quotients Involving Zero The product of 0 and any number is 0. $b \cdot 0 = 0$ and $0 \cdot b = 0$	$-4 \cdot 0 = 0 \quad 0 \cdot \left(-\frac{3}{4}\right) = 0$
The quotient of a nonzero number and 0 is undefined. $\frac{b}{0}$ is undefined.	$\frac{-85}{0}$ is undefined.
The quotient of 0 and any nonzero number is 0. $\frac{0}{b} = 0$	$\frac{0}{18} = 0 \quad \frac{0}{-47} = 0$
Section 1.8 Properties of Real Numbers	
Commutative Properties Addition: $a + b = b + a$ Multiplication: $a \cdot b = b \cdot a$	$3 + (-7) = -7 + 3$ $-8 \cdot 5 = 5 \cdot (-8)$
Associative Properties Addition: $(a + b) + c = a + (b + c)$ Multiplication: $(a \cdot b) \cdot c = a \cdot (b \cdot c)$	$(5 + 10) + 20 = 5 + (10 + 20)$ $(-3 \cdot 2) \cdot 11 = -3 \cdot (2 \cdot 11)$

(continued)

DEFINITIONS AND CONCEPTS	EXAMPLES
Section 1.8 Properties of Real Numbers (continued)	
Two numbers whose product is 1 are called **multiplicative inverses** or **reciprocals.** The reciprocal of a nonzero number a is $\frac{1}{a}$ because $a \cdot \frac{1}{a} = 1$.	The reciprocal of 3 is $\frac{1}{3}$. The reciprocal of $-\frac{2}{5}$ is $-\frac{5}{2}$.
Distributive Property $a(b + c) = a \cdot b + a \cdot c$	$5(6 + 10) = 5 \cdot 6 + 5 \cdot 10$ $-2(3 + x) = -2 \cdot 3 + (-2)(x)$
Identities $a + 0 = a$ $\quad 0 + a = a$ $a \cdot 1 = a$ $\quad 1 \cdot a = a$	$5 + 0 = 5$ $\quad 0 + (-2) = -2$ $-14 \cdot 1 = -14$ $\quad 1 \cdot 27 = 27$
Inverses Additive or opposite: $a + (-a) = 0$ Multiplicative or reciprocal: $b \cdot \frac{1}{b} = 1$	 $7 + (-7) = 0$ $3 \cdot \frac{1}{3} = 1$

Chapter 1 Review

(1.2) Insert $<, >,$ or $=$ in the appropriate space to make the following statements true.

1. 8 10

2. 7 2

3. −4 −5

4. $\frac{12}{2}$ −8

5. $|-7|$ $|-8|$

6. $|-9|$ −9

7. $-|-1|$ −1

8. $|-14|$ $-(-14)$

9. 1.2 1.02

10. $-\frac{3}{2}$ $-\frac{3}{4}$

TRANSLATING

Translate each statement into symbols.

11. Four is greater than or equal to negative three.

12. Six is not equal to five.

13. 0.03 is less than 0.3.

14. New York City has 155 museums and 400 art galleries. Write an inequality comparing the numbers 155 and 400. (*Source:* Absolute Trivia.com)

Given the following sets of numbers, list the numbers in each set that also belong to the set of:

a. Natural numbers

b. Whole numbers

c. Integers

d. Rational numbers

e. Irrational numbers

f. Real numbers

15. $\left\{-6, 0, 1, 1\frac{1}{2}, 3, \pi, 9.62\right\}$

16. $\left\{-3, -1.6, 2, 5, \frac{11}{2}, 15.1, \sqrt{5}, 2\pi\right\}$

The following chart shows the gains and losses in dollars of Density Oil and Gas stock for a particular week.

Day	*Gain or Loss in Dollars*
Monday	+1
Tuesday	−2
Wednesday	+5
Thursday	+1
Friday	−4

17. Which day showed the greatest loss?

18. Which day showed the greatest gain?

(1.3) Write the number as a product of prime factors.

19. 36

20. 120

Perform the indicated operations. Write results in lowest terms.

21. $\frac{8}{15} \cdot \frac{27}{30}$

22. $\frac{7}{8} \div \frac{21}{32}$

23. $\frac{7}{15} + \frac{5}{6}$

24. $\frac{3}{4} - \frac{3}{20}$

25. $2\frac{3}{4} + 6\frac{5}{8}$

26. $7\frac{1}{6} - 2\frac{2}{3}$

27. $5 \div \frac{1}{3}$

28. $2 \cdot 8\frac{3}{4}$

Each circle represents a whole, or 1. Determine the unknown part of the circle.

29.

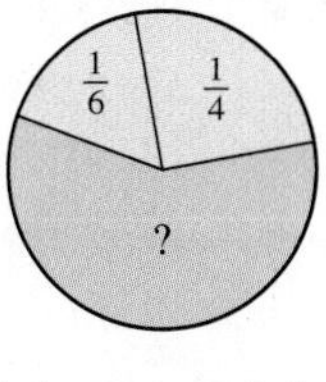

30.

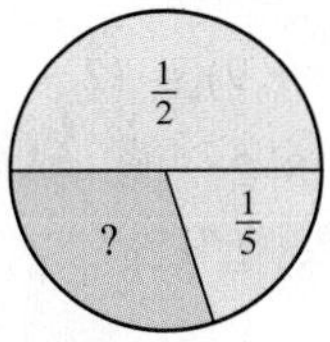

Find the area and the perimeter of each figure.

△ **31.**

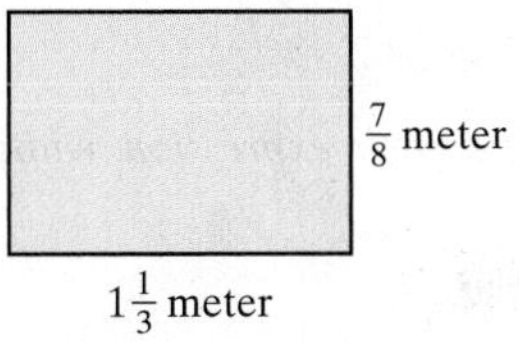

△ **32.**

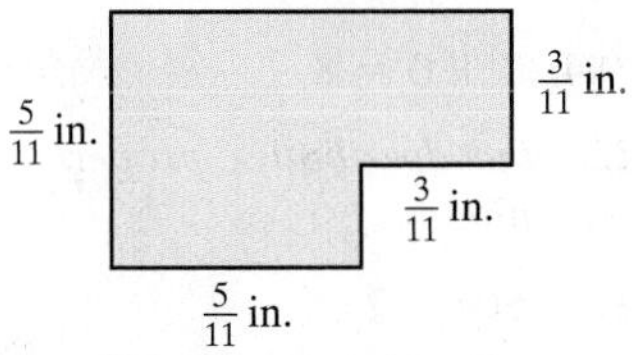

Octuplets were born in the U.S. in 2009. The following chart gives the octuplets' birthweights. The babies are listed in order of birth.

Baby	*Gender*	*Birthweight (pounds)*
Baby A	boy	$2\frac{1}{2}$
Baby B	girl	$2\frac{1}{8}$
Baby C	boy	$3\frac{1}{16}$
Baby D	girl	$2\frac{3}{16}$
Baby E	boy	$1\frac{3}{4}$
Baby F	boy	$2\frac{9}{16}$
Baby G	boy	$1\frac{13}{16}$
Baby H	boy	$2\frac{7}{16}$

33. What was the total weight of the boy octuplets?

34. What was the total weight of the girl octuplets?

35. Find the combined weight of all eight octuplets.

36. Which baby weighed the most?

37. Which baby weighed the least?

38. How much more did the heaviest baby weigh than the lightest baby?

(1.4) *Choose the correct answer for each statement.*

39. The expression $6 \cdot 3^2 + 2 \cdot 8$ simplifies to

a. −52 **b.** 448 **c.** 70 **d.** 64

40. The expression $68 - 5 \cdot 2^3$ simplifies to

a. −232 **b.** 28 **c.** 38 **d.** 504

Simplify each expression.

41. $\left(\frac{2}{7}\right)^2$

42. $\left(\frac{3}{4}\right)^3$

43. $3(1 + 2 \cdot 5) + 4$

44. $8 + 3(2 \cdot 6 - 1)$

45. $\dfrac{4 + |6 - 2| + 8^2}{4 + 6 \cdot 4}$

46. $5[3(2 + 5) - 5]$

TRANSLATING

Translate each word statement to symbols.

47. The difference of twenty and twelve is equal to the product of two and four.

48. The quotient of nine and two is greater than negative five.

Evaluate each expression if $x = 6$, $y = 2$, and $z = 8$.

49. $2x + 3y$

50. $x(y + 2z)$

51. $\dfrac{x}{y} + \dfrac{z}{2y}$

52. $x^2 - 3y^2$

△ **53.** The expression $180 - a - b$ represents the measure of the unknown angle of the given triangle. Replace a with 37 and b with 80 to find the measure of the unknown angle.

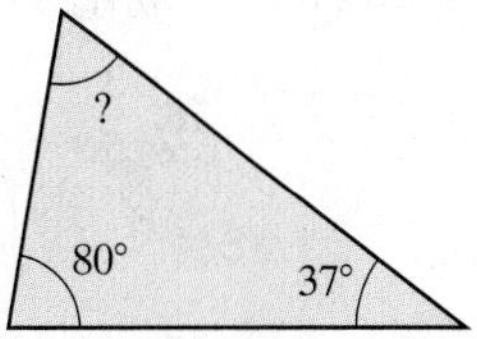

△ **54.** The expression $360 - a - b - c$ represents the measure of the unknown angle of the given quadrilateral. Replace a with 93, b with 80, and c with 82 to find the measure of the unknown angle.

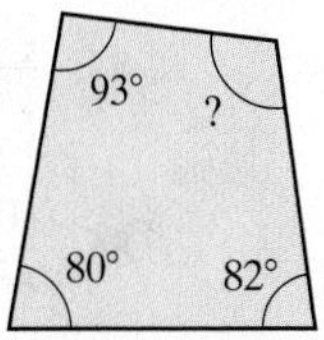

Decide whether the given number is a solution to the given equation.

55. Is $x = 3$ a solution of $7x - 3 = 18$?

56. Is $x = 1$ a solution of $3x^2 + 4 = x - 1$?

(1.5) *Find the additive inverse or the opposite.*

57. −9

58. $\frac{2}{3}$

59. $|-2|$

60. $-|-7|$

Find the following sums.

61. $-15 + 4$

62. $-6 + (-11)$

63. $\frac{1}{16} + \left(-\frac{1}{4}\right)$

64. $-8 + |-3|$

65. $-4.6 + (-9.3)$

66. $-2.8 + 6.7$

(1.6) *Perform the indicated operations.*

67. $6 - 20$

68. $-3.1 - 8.4$

69. $-6 - (-11)$

70. $4 - 15$

71. $-21 - 16 + 3(8 - 2)$

72. $\dfrac{11 - (-9) + 6(8 - 2)}{2 + 3 \cdot 4}$

Evaluate each expression for $x = 3$, $y = -6$, and $z = -9$. Then choose the correct evaluation.

73. $2x^2 - y + z$

a. 15 **b.** 3 **c.** 27 **d.** −3

74. $\dfrac{|y - 4x|}{2x}$

a. 3 **b.** 1 **c.** −1 **d.** −3

75. At the beginning of the week, the price of Density Oil and Gas stock from Exercises 17 and 18 is $50 per share. Find the price of a share of stock at the end of the week.

76. Find the price of a share of stock by the end of the day on Wednesday.

(1.7) *Find the multiplicative inverse or reciprocal.*

77. -6

78. $\dfrac{3}{5}$

Simplify each expression.

79. $6(-8)$

80. $(-2)(-14)$

81. $\dfrac{-18}{-6}$

82. $\dfrac{42}{-3}$

83. $\dfrac{4(-3) + (-8)}{2 + (-2)}$

84. $\dfrac{3(-2)^2 - 5}{-14}$

85. $\dfrac{-6}{0}$

86. $\dfrac{0}{-2}$

87. $-4^2 - (-3 + 5) \div (-1) \cdot 2$

88. $-5^2 - (2 - 20) \div (-3) \cdot 3$

If $x = -5$ and $y = -2$, evaluate each expression.

89. $x^2 - y^4$

90. $x^2 - y^3$

TRANSLATING

Translate each phrase to an expression. Use x to represent a number.

91. The product of −7 and a number

92. The quotient of a number and −13

93. Subtract a number from −20

94. The sum of −1 and a number

(1.8) *Name the property illustrated.*

95. $-6 + 5 = 5 + (-6)$

96. $6 \cdot 1 = 6$

97. $3(8 - 5) = 3 \cdot 8 - 3 \cdot (5)$

98. $4 + (-4) = 0$

99. $2 + (3 + 9) = (2 + 3) + 9$

100. $2 \cdot 8 = 8 \cdot 2$

101. $6(8 + 5) = 6 \cdot 8 + 6 \cdot 5$

102. $(3 \cdot 8) \cdot 4 = 3 \cdot (8 \cdot 4)$

103. $4 \cdot \dfrac{1}{4} = 1$

104. $8 + 0 = 8$

Use the distributive property to write each expression without parentheses.

105. $5(y - 2)$

106. $-3(z + y)$

107. $-(7 - x + 4z)$

108. $\dfrac{1}{2}(6z - 10)$

109. $-4(3x + 5) - 7$

110. $-8(2y + 9) - 1$

MIXED REVIEW

Insert $<$, $>$, or $=$ in the space between each pair of numbers.

111. $-|-11| \quad |11.4|$

112. $-1\frac{1}{2} \quad -2\frac{1}{2}$

Perform the indicated operations.

113. $-7.2 + (-8.1)$

114. $14 - 20$

115. $4(-20)$

116. $\dfrac{-20}{4}$

117. $-\dfrac{4}{5}\left(\dfrac{5}{16}\right)$

118. $-0.5(-0.3)$

119. $8 \div 2 \cdot 4$

120. $(-2)^4$

121. $\dfrac{-3 - 2(-9)}{-15 - 3(-4)}$

122. $5 + 2[(7 - 5)^2 + (1 - 3)]$

123. $-\dfrac{5}{8} \div \dfrac{3}{4}$

124. $\dfrac{-15 + (-4)^2 + |-9|}{10 - 2 \cdot 5}$

△ **125.** A trim carpenter needs a piece of quarter round molding $6\frac{1}{8}$ feet long for a bathroom. She finds a piece $7\frac{1}{2}$ feet long. How long a piece does she need to cut from the $7\frac{1}{2}$-foot-long molding in order to use it in the bathroom?

Chapter 1 Test

MyMathLab®

Translate the statement into symbols.

1. The absolute value of negative seven is greater than five.

2. The sum of nine and five is greater than or equal to four.

Simplify the expression.

3. $-13 + 8$

4. $-13 - (-2)$

5. $12 \div 4 \cdot 3 - 6 \cdot 2$

6. $(13)(-3)$

7. $(-6)(-2)$

8. $\frac{|-16|}{-8}$

9. $\frac{-8}{0}$

10. $\frac{|-6| + 2}{5 - 6}$

11. $\frac{1}{2} - \frac{5}{6}$

12. $5\frac{3}{4} - 1\frac{1}{8}$

13. $-0.6 + 1.875$

14. $3(-4)^2 - 80$

15. $6[5 + 2(3 - 8) - 3]$

16. $\frac{-12 + 3 \cdot 8}{4}$

17. $\frac{(-2)(0)(-3)}{-6}$

Insert $<$, $>$, or $=$ in the appropriate space to make each of the following statements true.

18. $-3 \quad -7$

19. $4 \quad -8$

20. $2 \quad |-3|$

21. $|-2| \quad -1 - (-3)$

22. In the state of Massachusetts, there are 2221 licensed child care centers and 10,993 licensed home-based child care providers. Write an inequality statement comparing the numbers 2221 and 10,993. (*Source:* Children's Foundation)

23. Given $\left\{-5, -1, 0, \frac{1}{4}, 1, 7, 11.6, \sqrt{7}, 3\pi\right\}$, list the numbers in this set that also belong to the set of:

a. Natural numbers

b. Whole numbers

c. Integers

d. Rational numbers

e. Irrational numbers

f. Real numbers

If $x = 6$, $y = -2$, and $z = -3$, evaluate each expression.

24. $x^2 + y^2$

25. $x + yz$

26. $2 + 3x - y$

27. $\frac{y + z - 1}{x}$

Identify the property illustrated by each expression.

28. $8 + (9 + 3) = (8 + 9) + 3$

29. $6 \cdot 8 = 8 \cdot 6$

30. $-6(2 + 4) = -6 \cdot 2 + (-6) \cdot 4$

31. $\frac{1}{6}(6) = 1$

32. Find the opposite of -9.

33. Find the reciprocal of $-\frac{1}{3}$.

The New Orleans Saints were 22 yards from the goal when the following series of gains and losses occurred.

	Gains and Losses in Yards
First Down	5
Second Down	−10
Third Down	−2
Fourth Down	29

34. During which down did the greatest loss of yardage occur?

35. Was a touchdown scored?

36. The temperature at the Winter Olympics was a frigid 14 degrees below zero in the morning, but by noon it had risen 31 degrees. What was the temperature at noon?

37. A health insurance provider had net incomes of \$356 million, \$460 million, and −\$166 million in 3 consecutive years. What was the health insurance provider's total net income for these three years?

38. A stockbroker decided to sell 280 shares of stock, which decreased in value by \$1.50 per share yesterday. How much money did she lose?

CHAPTER 2 Equations, Inequalities, and Problem Solving

Much of mathematics relates to deciding which statements are true and which are false. For example, the statement $x + 7 = 15$ is an equation stating that the sum $x + 7$ has the same value as 15. Is this statement true or false? It is false for some values of x and true for just one value of x, namely 8. Our purpose in this chapter is to learn ways of deciding which values make an equation or an inequality true.

Top 10 Countries by Number of Internet-Crime Complaints

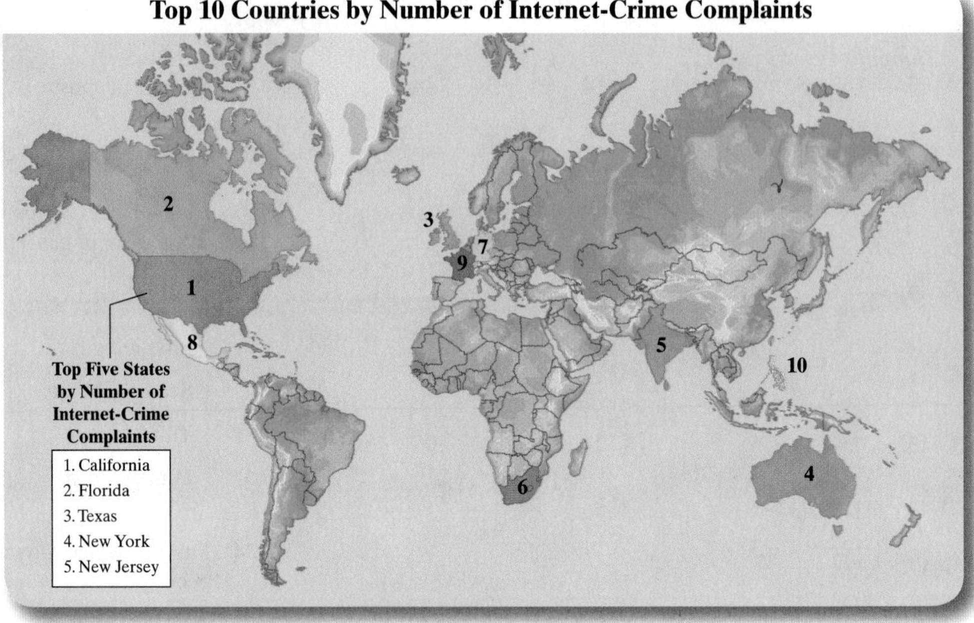

Internet Crime

The Internet Crime Complaint Center (IC3) is a joint operation between the FBI and the National White-Collar Crime Center. The IC3 receives and refers criminal complaints occurring on the Internet. Of course, nondelivery of merchandise or payment are the highest reported offenses.

In Section 2.7, Exercises 15 and 16, we analyze a bar graph on the yearly number of complaints received by the IC3.

Ages of Persons Filing Complaints

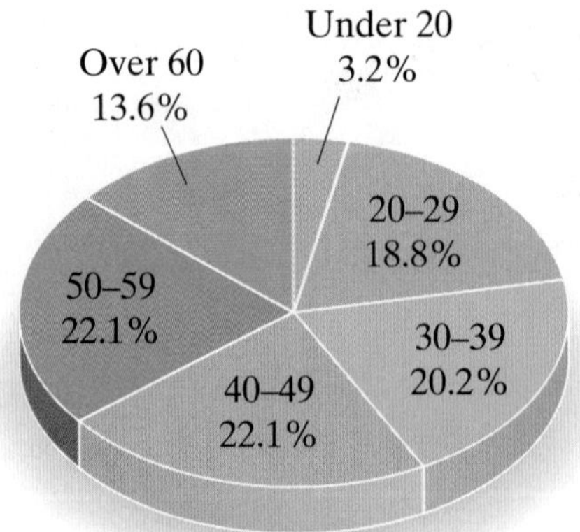

2.1 Simplifying Algebraic Expressions 

OBJECTIVES

1 Identify Terms, Like Terms, and Unlike Terms.

2 Combine Like Terms.

3 Use the Distributive Property to Remove Parentheses.

4 Write Word Phrases as Algebraic Expressions.

As we explore in this section, an expression such as $3x + 2x$ is not as simple as possible because—even without replacing x by a value—we can perform the indicated addition.

OBJECTIVE

1 Identifying Terms, Like Terms, and Unlike Terms 

Before we practice simplifying expressions, some new language of algebra is presented. A **term** is a number or the product of a number and variables raised to powers.

Terms

$$-y, \quad 2x^3, \quad -5, \quad 3xz^2, \quad \frac{2}{y}, \quad 0.8z$$

The **numerical coefficient** (sometimes also simply called the **coefficient**) of a term is the numerical factor. The numerical coefficient of $3x$ is 3. Recall that $3x$ means $3 \cdot x$.

Term	*Numerical Coefficient*
$3x$	3
$\frac{y^3}{5}$	$\frac{1}{5}$ since $\frac{y^3}{5}$ means $\frac{1}{5} \cdot y^3$
$0.7ab^3c^5$	0.7
z	1
$-y$	-1
-5	-5

Helpful Hint

The term $-y$ means $-1y$ and thus has a numerical coefficient of -1.

The term z means $1z$ and thus has a numerical coefficient of 1.

EXAMPLE 1 Identify the numerical coefficient in each term.

a. $-3y$ **b.** $22z^4$ **c.** y **d.** $-x$ **e.** $\frac{x}{7}$

Solution

a. The numerical coefficient of $-3y$ is -3.

b. The numerical coefficient of $22z^4$ is 22.

c. The numerical coefficient of y is 1, since y is $1y$.

d. The numerical coefficient of $-x$ is -1, since $-x$ is $-1x$.

e. The numerical coefficient of $\frac{x}{7}$ is $\frac{1}{7}$, since $\frac{x}{7}$ means $\frac{1}{7} \cdot x$.

PRACTICE

1 Identify the numerical coefficient in each term.

a. t **b.** $-7x$ **c.** $-\frac{w}{5}$ **d.** $43x^4$ **e.** $-b$

Terms with the same variables raised to exactly the same powers are called **like terms.** Terms that aren't like terms are called **unlike terms.**

Like Terms	*Unlike Terms*	*Reason*
$3x, 2x$	$5x, 5x^2$	Why? Same variable x but different powers x and x^2
$-6x^2y, 2x^2y, 4x^2y$	$7y, 3z, 8x^2$	Why? Different variables
$2ab^2c^3, ac^3b^2$	$6abc^3, 6ab^2$	Why? Different variables and different powers

Helpful Hint

In like terms, each variable and its exponent must match exactly, but these factors don't need to be in the same order.

$2x^2y$ and $3yx^2$ are like terms.

EXAMPLE 2 Determine whether the terms are like or unlike.

a. $2x, 3x^2$ **b.** $4x^2y, x^2y, -2x^2y$ **c.** $-2yz, -3zy$ **d.** $-x^4, x^4$

Solution

a. Unlike terms, since the exponents on x are not the same.

b. Like terms, since each variable and its exponent match.

c. Like terms, since $zy = yz$ by the commutative property.

d. Like terms.

PRACTICE 2 Determine whether the terms are like or unlike.

a. $-4xy, 5yx$

b. $5q, -3q^2$

c. $3ab^2, -2ab^2, 43ab^2$

d. $y^5, \frac{y^5}{2}$

OBJECTIVE 2

Combining Like Terms 

An algebraic expression containing the sum or difference of like terms can be simplified by applying the distributive property. For example, by the distributive property, we rewrite the sum of the like terms $3x + 2x$ as

$$3x + 2x = (3 + 2)x = 5x$$

Also,

$$-y^2 + 5y^2 = -1y^2 + 5y^2 = (-1 + 5)y^2 = 4y^2$$

Simplifying the sum or difference of like terms is called **combining like terms.**

EXAMPLE 3 Simplify each expression by combining like terms.

a. $7x - 3x$ **b.** $10y^2 + y^2$ **c.** $8x^2 + 2x - 3x$ **d.** $9n^2 - 5n^2 + n^2$

Solution

a. $7x - 3x = (7 - 3)x = 4x$

b. $10y^2 + y^2 = 10y^2 + 1y^2 = (10 + 1)y^2 = 11y^2$

c. $8x^2 + 2x - 3x = 8x^2 + (2 - 3)x = 8x^2 - x$

d. $9n^2 - 5n^2 + n^2 = (9 - 5 + 1)n^2 = 5n^2$

PRACTICE 3 Simplify each expression by combining like terms.

a. $-3y + 11y$

b. $4x^2 + x^2$

c. $5x - 3x^2 + 8x^2$

d. $20y^2 + 2y^2 - y^2$

The previous example suggests the following:

Combining Like Terms

To **combine like terms,** add the numerical coefficients and multiply the result by the common variable factors.

EXAMPLE 4 Simplify each expression by combining like terms.

a. $2x + 3x + 5 + 2$ **b.** $-5a - 3 + a + 2$ **c.** $4y - 3y^2$

d. $2.3x + 5x - 6$ **e.** $-\frac{1}{2}b + b$

Solution Use the distributive property to combine like terms.

a. $2x + 3x + 5 + 2 = (2 + 3)x + (5 + 2)$
$= 5x + 7$

b. $-5a - 3 + a + 2 = -5a + 1a + (-3 + 2)$
$= (-5 + 1)a + (-3 + 2)$
$= -4a - 1$

c. $4y - 3y^2$ These two terms cannot be combined because they are unlike terms.

d. $2.3x + 5x - 6 = (2.3 + 5)x - 6$
$= 7.3x - 6$

e. $-\frac{1}{2}b + b = -\frac{1}{2}b + 1b = \left(-\frac{1}{2} + 1\right)b = \frac{1}{2}b$ □

PRACTICE 4 Use the distributive property to combine like terms.

a. $3y + 8y - 7 + 2$ **b.** $6x - 3 - x - 3$ **c.** $\frac{3}{4}t - t$

d. $9y + 3.2y + 10 + 3$ **e.** $5z - 3z^4$

OBJECTIVE 3 Using the Distributive Property

Simplifying expressions makes frequent use of the distributive property to also remove parentheses.

It may be helpful to study the examples below.

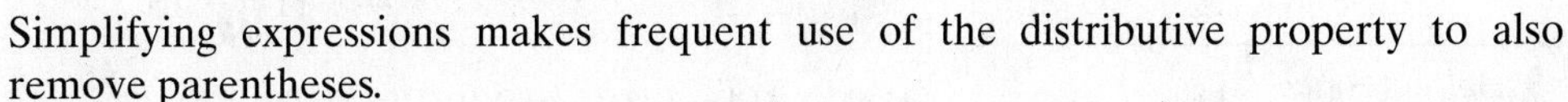

$+(3a + 2) = +1(3a + 2) = +1(3a) + (+1)(2) = 3a + 2$
(means)

$-(3a + 2) = -1(3a + 2) = -1(3a) + (-1)(2) = -3a - 2$
(means)

EXAMPLE 5 Find each product by using the distributive property to remove parentheses.

a. $5(3x + 2)$ **b.** $-2(y + 0.3z - 1)$ **c.** $-(9x + y - 2z + 6)$

Solution

a. $5(3x + 2) = 5 \cdot 3x + 5 \cdot 2$ Apply the distributive property.
$= 15x + 10$ Multiply.

b. $-2(y + 0.3z - 1) = -2(y) + (-2)(0.3z) + (-2)(-1)$ Apply the distributive property.
$= -2y - 0.6z + 2$ Multiply.

c. $-(9x + y - 2z + 6) = -1(9x + y - 2z + 6)$ Distribute −1 over each term.
$= -1(9x) - 1(y) - 1(-2z) - 1(6)$
$= -9x - y + 2z - 6$ □

PRACTICE
5 Find each product by using the distributive property to remove parentheses.

a. $3(2x - 7)$ **b.** $-5(x - 0.5z - 5)$
c. $-(2x - y + z - 2)$

Helpful Hint

If a "−" sign precedes parentheses, the sign of each term inside the parentheses is changed when the distributive property is applied to remove parentheses.

Examples:

$-(2x + 1) = -2x - 1$ $-(-5x + y - z) = 5x - y + z$
$-(x - 2y) = -x + 2y$ $-(-3x - 4y - 1) = 3x + 4y + 1$

When simplifying an expression containing parentheses, we often use the distributive property in both directions—first to remove parentheses and then again to combine any like terms.

EXAMPLE 6 Simplify each expression.

a. $3(2x - 5) + 1$ **b.** $-2(4x + 7) - (3x - 1)$ **c.** $9 + 3(4x - 10)$

Solution

a. $3(2x - 5) + 1 = 6x - 15 + 1$ Apply the distributive property.
$= 6x - 14$ Combine like terms.

b. $-2(4x + 7) - (3x - 1) = -8x - 14 - 3x + 1$ Apply the distributive property.
$= -11x - 13$ Combine like terms.

c. $9 + 3(4x - 10) = 9 + 12x - 30$ Apply the distributive property.
$= -21 + 12x$ Combine like terms.
or $12x - 21$ □

Helpful Hint

Don't forget to use the distributive property and multiply before adding or subtracting like terms.

PRACTICE
6 Simplify each expression.

a. $4(9x + 1) + 6$ **b.** $-7(2x - 1) - (6 - 3x)$ **c.** $8 - 5(6x + 5)$

EXAMPLE 7 Write the phrase below as an algebraic expression. Then simplify if possible.

"Subtract $4x - 2$ from $2x - 3$."

Solution "Subtract $4x - 2$ **from** $2x - 3$" translates to $(2x - 3) - (4x - 2)$. Next, simplify the algebraic expression.

$(2x - 3) - (4x - 2) = 2x - 3 - 4x + 2$ Apply the distributive property.
$= -2x - 1$ Combine like terms. □

PRACTICE 7 Write the phrase below as an algebraic expression. Then simplify if possible.

"Subtract $7x - 1$ from $2x + 3$."

OBJECTIVE 4 **Writing Word Phrases as Algebraic Expressions** 

Next, we practice writing word phrases as algebraic expressions.

EXAMPLE 8 Write the following phrases as algebraic expressions and simplify if possible. Let x represent the unknown number.

a. Twice a number, plus 6

b. The difference of a number and 4, divided by 7

c. Five added to triple the sum of a number and 1

d. The sum of twice a number, 3 times the number, and 5 times the number

Solution

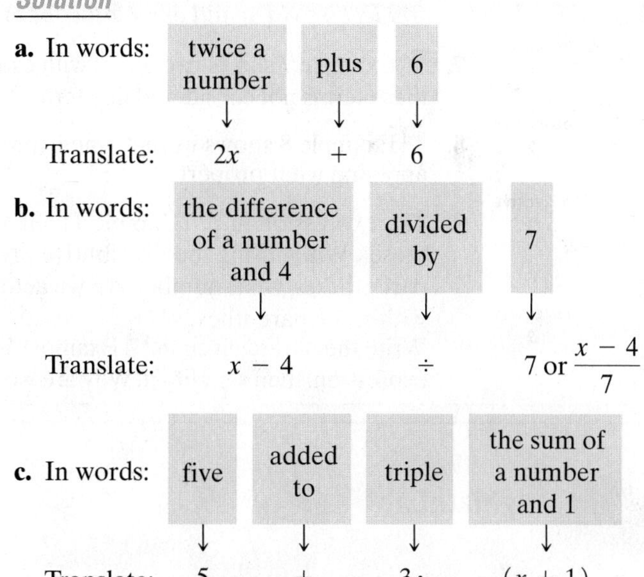

Next, we simplify this expression.

$$5 + 3(x + 1) = 5 + 3x + 3 \quad \text{Use the distributive property.}$$
$$= 8 + 3x \quad \text{Combine like terms.}$$

d. The phrase "the sum of" means that we add.

In words:	twice a number	added to	3 times the number	added to	5 times the number
	↓	↓	↓	↓	↓
Translate:	$2x$	$+$	$3x$	$+$	$5x$

Now let's simplify.

$$2x + 3x + 5x = 10x \quad \text{Combine like terms.}$$

PRACTICE 8 Write the following phrases as algebraic expressions and simplify if possible. Let x represent the unknown number.

a. Three added to double a number

b. Six subtracted from the sum of 5 and a number

c. Two times the sum of 3 and a number, increased by 4

d. The sum of a number, half the number, and 5 times the number

Vocabulary, Readiness & Video Check

Use the choices below to fill in each blank. Some choices may be used more than once.

like	numerical coefficient	term	distributive
unlike	combine like terms	expression	

1. $23y^2 + 10y - 6$ is called a(n) ________________ while $23y^2$, $10y$, and -6 are each called a(n) ________________.
2. To simplify $x + 4x$, we ________________.
3. The term y has an understood ________________ of 1.
4. The terms $7z$ and $7y$ are ________________ terms and the terms $7z$ and $-z$ are ________________ terms.
5. For the term $-\frac{1}{2}xy^2$, the number $-\frac{1}{2}$ is the ________________.
6. $5(3x - y)$ equals $15x - 5y$ by the ________________ property.

Martin-Gay Interactive Videos

See Video 2.1

Watch the section lecture video and answer the following questions.

OBJECTIVE 1
7. Example 7 shows two terms with exactly the same variables. Why are these terms not considered like terms?

OBJECTIVE 2
8. Example 8 shows us that when combining like terms, we are actually applying what property?

OBJECTIVE 3
9. The expression in Example 11 shows a minus sign before parentheses. When using the distributive property to multiply and remove parentheses, what number are we actually distributing to each term within the parentheses?

OBJECTIVE 4
10. Write the phrase given in Example 14, translate it to an algebraic expression, then simplify it. Why are we able to simplify it?

2.1 Exercise Set

MyMathLab®

Identify the numerical coefficient of each term. See Example 1.

1. $-7y$
2. $3x$
3. x
4. $-y$
5. $17x^2y$
6. $1.2xyz$

Indicate whether the terms in each list are like or unlike. See Example 2.

7. $5y, -y$
8. $-2x^2y, 6xy$
9. $2z, 3z^2$
10. $ab^2, -7ab^2$
11. $8wz, \frac{1}{7}zw$
12. $7.4p^3q^2, 6.2p^3q^2r$

Simplify each expression by combining any like terms. See Examples 3 and 4.

13. $7y + 8y$
14. $3x + 2x$
15. $8w - w + 6w$
16. $c - 7c + 2c$
17. $3b - 5 - 10b - 4$
18. $6g + 5 - 3g - 7$
19. $m - 4m + 2m - 6$
20. $a + 3a - 2 - 7a$
21. $5g - 3 - 5 - 5g$
22. $8p + 4 - 8p - 15$
23. $6.2x - 4 + x - 1.2$
24. $7.9y - 0.7 - y + 0.2$
25. $6x - 5x + x - 3 + 2x$
26. $8h + 13h - 6 + 7h - h$
27. $7x^2 + 8x^2 - 10x^2$
28. $8x^3 + x^3 - 11x^3$
29. $6x + 0.5 - 4.3x - 0.4x + 3$
30. $0.4y - 6.7 + y - 0.3 - 2.6y$

Simplify each expression. First use the distributive property to remove any parentheses. See Examples 5 and 6.

31. $5(y - 4)$
32. $7(r - 3)$
33. $-2(x + 2)$
34. $-4(y + 6)$
35. $7(d - 3) + 10$
36. $9(z + 7) - 15$
37. $-5(2x - 3y + 6)$
38. $-2(4x - 3z - 1)$

39. $-(3x - 2y + 1)$

40. $-(y + 5z - 7)$

41. $5(x + 2) - (3x - 4)$

42. $4(2x - 3) - 2(x + 1)$

Write each of the following as an algebraic expression. Simplify if possible. See Example 7.

43. Add $6x + 7$ to $4x - 10$.

44. Add $3y - 5$ to $y + 16$.

45. Subtract $7x + 1$ from $3x - 8$.

46. Subtract $4x - 7$ from $12 + x$.

47. Subtract $5m - 6$ from $m - 9$.

48. Subtract $m - 3$ from $2m - 6$.

MIXED PRACTICE

Simplify each expression. See Examples 3 through 7.

49. $2k - k - 6$

50. $7c - 8 - c$

51. $-9x + 4x + 18 - 10x$

52. $5y - 14 + 7y - 20y$

53. $-4(3y - 4) + 12y$

54. $-3(2x + 5) - 6x$

55. $3(2x - 5) - 5(x - 4)$

56. $2(6x - 1) - (x - 7)$

57. $-2(3x - 4) + 7x - 6$

58. $8y - 2 - 3(y + 4)$

59. $5k - (3k - 10)$

60. $-11c - (4 - 2c)$

61. Subtract $6x - 1$ from $3x + 4$

62. Subtract $4 + 3y$ from $8 - 5y$

63. $3.4m - 4 - 3.4m - 7$

64. $2.8w - 0.9 - 0.5 - 2.8w$

65. $\frac{1}{3}(7y - 1) + \frac{1}{6}(4y + 7)$

66. $\frac{1}{5}(9y + 2) + \frac{1}{10}(2y - 1)$

67. $2 + 4(6x - 6)$

68. $8 + 4(3x - 4)$

69. $0.5(m + 2) + 0.4m$

70. $0.2(k + 8) - 0.1k$

71. $10 - 3(2x + 3y)$

72. $14 - 11(5m + 3n)$

73. $6(3x - 6) - 2(x + 1) - 17x$

74. $7(2x + 5) - 4(x + 2) - 20x$

75. $\frac{1}{2}(12x - 4) - (x + 5)$

76. $\frac{1}{3}(9x - 6) - (x - 2)$

TRANSLATING

Write each phrase as an algebraic expression and simplify if possible. Let x represent the unknown number. See Examples 7 and 8.

77. Twice a number, decreased by four

78. The difference of a number and two, divided by five

79. Seven added to double a number

80. Eight more than triple a number

81. Three-fourths of a number, increased by twelve

82. Eleven, increased by two-thirds of a number

83. The sum of 5 times a number and -2, added to 7 times a number

84. The sum of 3 times a number and 10, subtracted from 9 times a number

85. Eight times the sum of a number and six

86. Six times the difference of a number and five

87. Double a number, minus the sum of the number and ten

88. Half a number, minus the product of the number and eight

89. The sum of 2, three times a number, -9, and four times a number

90. The sum of twice a number, -1, five times a number, and -12

REVIEW AND PREVIEW

Evaluate the following expressions for the given values. See Section 1.7.

91. If $x = -1$ and $y = 3$, find $y - x^2$.

92. If $g = 0$ and $h = -4$, find $gh - h^2$.

93. If $a = 2$ and $b = -5$, find $a - b^2$.

94. If $x = -3$, find $x^3 - x^2 + 4$.

95. If $y = -5$ and $z = 0$, find $yz - y^2$.

96. If $x = -2$, find $x^3 - x^2 - x$.

CONCEPT EXTENSIONS

△ 97. Recall that the perimeter of a figure is the total distance around the figure. Given the following rectangle, express the perimeter as an algebraic expression containing the variable x.

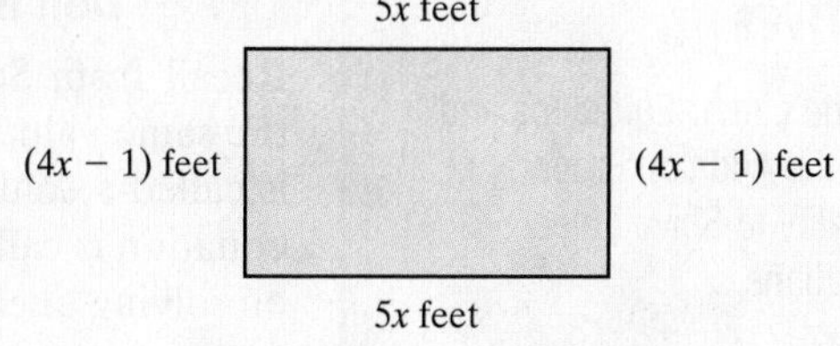

△ **98.** Given the following triangle, express its perimeter as an algebraic expression containing the variable x.

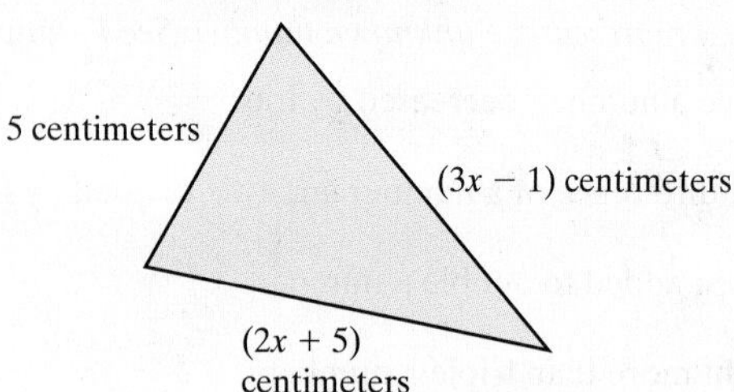

Given the following two rules, determine whether each scale in Exercises 99 through 102 is balanced.

1 cone balances 1 cube

1 cylinder balances 2 cubes

99.

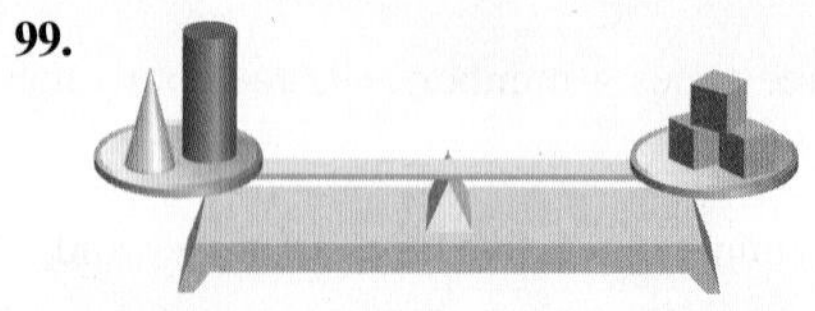

100.

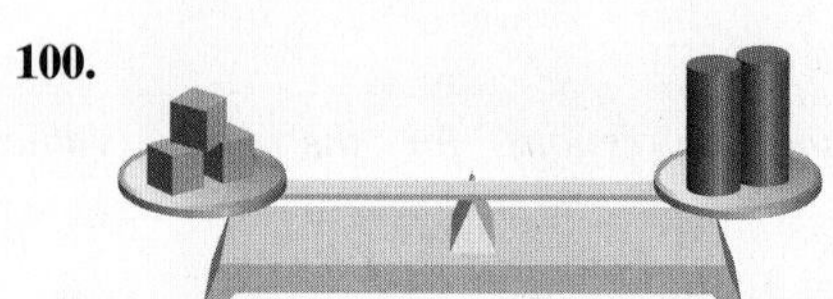

101.

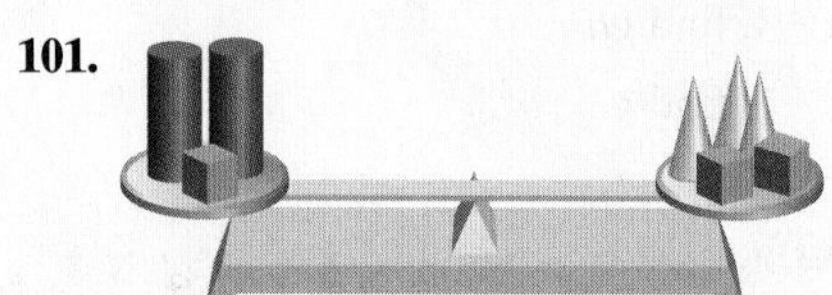

102.

Write each algebraic expression described.

103. Write an expression with 4 terms that simplifies to $3x - 4$.

104. Write an expression of the form ____ (____ + ____) whose product is $6x + 24$.

105. To convert from feet to inches, we multiply by 12. For example, the number of inches in 2 feet is $12 \cdot 2$ inches. If one board has a length of $(x + 2)$ *feet* and a second board has a length of $(3x - 1)$ *inches*, express their total length in inches as an algebraic expression.

106. The value of 7 nickels is $5 \cdot 7$ cents. Likewise, the value of x nickels is $5x$ cents. If the money box in a drink machine contains x *nickels*, $3x$ *dimes*, and $(30x - 1)$ *quarters*, express their total value in cents as an algebraic expression.

107. In your own words, explain how to combine like terms.

108. Do like terms always contain the same numerical coefficients? Explain your answer.

For Exercises 109 through 114, see the example below.

Example

Simplify $-3xy + 2x^2y - (2xy - 1)$.

Solution

$$-3xy + 2x^2y - (2xy - 1)$$
$$= -3xy + 2x^2y - 2xy + 1 = -5xy + 2x^2y + 1$$

Simplify each expression.

109. $5b^2c^3 + 8b^3c^2 - 7b^3c^2$

110. $4m^4p^2 + m^4p^2 - 5m^2p^4$

111. $3x - (2x^2 - 6x) + 7x^2$

112. $9y^2 - (6xy^2 - 5y^2) - 8xy^2$

113. $-(2x^2y + 3z) + 3z - 5x^2y$

114. $-(7c^3d - 8c) - 5c - 4c^3d$

2.2 The Addition Property of Equality

OBJECTIVES

1 Define Linear Equations and Use the Addition Property of Equality to Solve Linear Equations.

2 Write Word Phrases as Algebraic Expressions.

OBJECTIVE

1 Defining Linear Equations and Using the Addition Property

Recall from Section 1.4 that an equation is a statement that two expressions have the same value. Also, a value of the variable that makes an equation a true statement is called a solution or root of the equation. The process of finding the solution of an equation is called **solving** the equation for the variable. In this section, we concentrate on solving **linear equations** in one variable.

Linear Equation in One Variable

A linear equation in one variable can be written in the form

$$ax + b = c$$

where a, b, and c are real numbers and $a \neq 0$.

Evaluating a linear equation for a given value of the variable, as we did in Section 1.4, can tell us whether that value is a solution, but we can't rely on evaluating an equation as our method of solving it.

Instead, to solve a linear equation in x, we write a series of simpler equations, all *equivalent* to the original equation, so that the final equation has the form

$$x = \textbf{number} \qquad \textbf{or} \qquad \textbf{number} = x$$

Equivalent equations are equations that have the same solution. This means that the "number" above is the solution to the original equation.

The first property of equality that helps us write simpler equivalent equations is the **addition property of equality.**

Addition Property of Equality

If a, b, and c are real numbers, then

$$a = b \qquad \text{and} \qquad a + c = b + c$$

are equivalent equations.

This property guarantees that adding the same number to both sides of an equation does not change the solution of the equation. Since subtraction is defined in terms of addition, we may also **subtract the same number from both sides** without changing the solution.

A good way to picture a true equation is as a balanced scale. Since it is balanced, each side of the scale weighs the same.

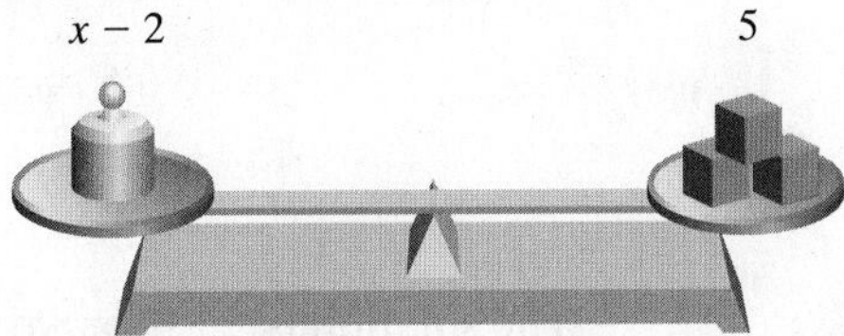

If the same weight is added to or subtracted from each side, the scale remains balanced.

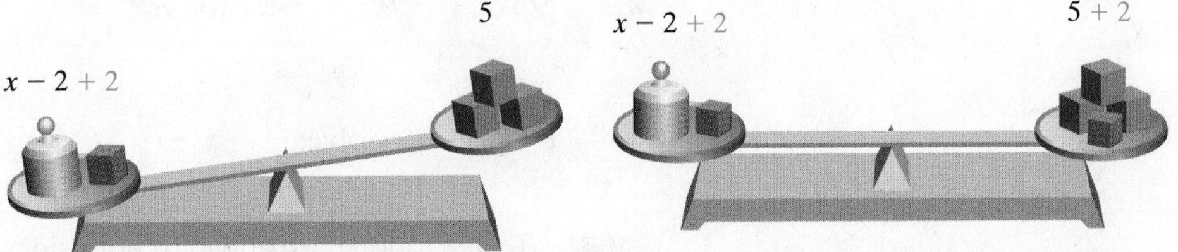

We use the addition property of equality to write equivalent equations until the variable is by itself on one side of the equation, and the equation looks like "$x =$ number" or "number $= x$."

EXAMPLE 1 Solve $x - 7 = 10$ for x.

Solution To solve for x, we want x alone on one side of the equation. To do this, we add 7 to both sides of the equation.

$$x - 7 = 10$$

$$x - 7 + 7 = 10 + 7 \quad \text{Add 7 to both sides.}$$

$$x = 17 \quad \text{Simplify.}$$

The solution of the equation $x = 17$ is obviously 17. Since we are writing equivalent equations, the solution of the equation $x - 7 = 10$ is also 17.

Check: To check, replace x with 17 in the original equation.

$$x - 7 = 10$$
$$17 - 7 \stackrel{?}{=} 10 \quad \text{Replace } x \text{ with 17 in the original equation.}$$
$$10 = 10 \quad \text{True}$$

Since the statement is true, 17 is the solution, or we can say that the solution set is $\{17\}$. □

PRACTICE
1 Solve: $x + 3 = -5$ for x.

✓CONCEPT CHECK

Use the addition property to fill in the blank so that the middle equation simplifies to the last equation.

$$x - 5 = 3$$
$$x - 5 + \_\_ = 3 + \_\_$$
$$x = 8$$

EXAMPLE 2 Solve $y + 0.6 = -1.0$ for y.

Solution To get y alone on one side of the equation, subtract 0.6 from both sides of the equation.

$$y + 0.6 = -1.0$$
$$y + 0.6 - 0.6 = -1.0 - 0.6 \quad \text{Subtract 0.6 from both sides.}$$
$$y = -1.6 \quad \text{Combine like terms.}$$

Check: To check the proposed solution, -1.6, replace y with -1.6 in the original equation.

$$y + 0.6 = -1.0$$
$$-1.6 + 0.6 \stackrel{?}{=} -1.0 \quad \text{Replace } y \text{ with } -1.6 \text{ in the original equation.}$$
$$-1.0 = -1.0 \quad \text{True}$$

The solution is -1.6, or we can say that the solution set is $\{-1.6\}$. □

PRACTICE
2 Solve: $y - 0.3 = -2.1$ for y.

EXAMPLE 3 Solve: $\frac{1}{2} = x - \frac{3}{4}$

Solution To get x alone, we add $\frac{3}{4}$ to both sides.

$$\frac{1}{2} = x - \frac{3}{4}$$
$$\frac{1}{2} + \frac{3}{4} = x - \frac{3}{4} + \frac{3}{4} \quad \text{Add } \frac{3}{4} \text{ to both sides.}$$
$$\frac{1}{2} \cdot \frac{2}{2} + \frac{3}{4} = x \quad \text{The LCD is 4.}$$
$$\frac{2}{4} + \frac{3}{4} = x \quad \text{Add the fractions.}$$
$$\frac{5}{4} = x$$

Answer to Concept Check: 5

Check:

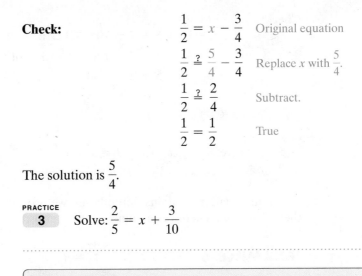

$$\frac{1}{2} = x - \frac{3}{4} \quad \text{Original equation}$$

$$\frac{1}{2} \stackrel{?}{=} \frac{5}{4} - \frac{3}{4} \quad \text{Replace } x \text{ with } \frac{5}{4}.$$

$$\frac{1}{2} \stackrel{?}{=} \frac{2}{4} \quad \text{Subtract.}$$

$$\frac{1}{2} = \frac{1}{2} \quad \text{True}$$

The solution is $\frac{5}{4}$. □

PRACTICE **3** Solve: $\frac{2}{5} = x + \frac{3}{10}$

> **Helpful Hint**
> We may solve an equation so that the variable is alone on *either* side of the equation. For example, $\frac{5}{4} = x$ is equivalent to $x = \frac{5}{4}$.

EXAMPLE 4 Solve $5t - 5 = 6t + 2$ for t.

Solution To solve for t, we first want all terms containing t on one side of the equation and all other terms on the other side of the equation. To do this, first subtract $5t$ from both sides of the equation.

$$5t - 5 = 6t + 2$$

$$5t - 5 - 5t = 6t + 2 - 5t \quad \text{Subtract } 5t \text{ from both sides.}$$

$$-5 = t + 2 \quad \text{Combine like terms.}$$

Next, subtract 2 from both sides, and the variable t will be isolated.

$$-5 = t + 2$$

$$-5 - 2 = t + 2 - 2 \quad \text{Subtract 2 from both sides.}$$

$$-7 = t$$

Check: Check the solution, -7, in the original equation. The solution is -7. □

PRACTICE **4** Solve: $4t + 7 = 5t - 3$

Many times, it is best to simplify one or both sides of an equation before applying the addition property of equality.

EXAMPLE 5 Solve: $2x + 3x - 5 + 7 = 10x + 3 - 6x - 4$

Solution First we simplify both sides of the equation.

$$2x + 3x - 5 + 7 = 10x + 3 - 6x - 4$$

$$5x + 2 = 4x - 1 \quad \text{Combine like terms on each side of the equation.}$$

Next, we want all terms with a variable on one side of the equation and all numbers on the other side.

$$5x + 2 - 4x = 4x - 1 - 4x \quad \text{Subtract } 4x \text{ from both sides.}$$

$$x + 2 = -1 \quad \text{Combine like terms.}$$

$$x + 2 - 2 = -1 - 2 \quad \text{Subtract 2 from both sides to get } x \text{ alone.}$$

$$x = -3 \quad \text{Combine like terms.}$$

Check:

$$2x + 3x - 5 + 7 = 10x + 3 - 6x - 4 \quad \text{Original equation}$$
$$2(-3) + 3(-3) - 5 + 7 \stackrel{?}{=} 10(-3) + 3 - 6(-3) - 4 \quad \text{Replace } x \text{ with } -3.$$
$$-6 - 9 - 5 + 7 \stackrel{?}{=} -30 + 3 + 18 - 4 \quad \text{Multiply.}$$
$$-13 = -13 \quad \text{True}$$

The solution is −3. □

PRACTICE 5 Solve: $8x - 5x - 3 + 9 = x + x + 3 - 7$

If an equation contains parentheses, we use the distributive property to remove them, as before. Then we combine any like terms.

EXAMPLE 6 Solve: $6(2a - 1) - (11a + 6) = 7$

Solution

$$6(2a - 1) - 1(11a + 6) = 7$$
$$6(2a) + 6(-1) - 1(11a) - 1(6) = 7 \quad \text{Apply the distributive property.}$$
$$12a - 6 - 11a - 6 = 7 \quad \text{Multiply.}$$
$$a - 12 = 7 \quad \text{Combine like terms.}$$
$$a - 12 + 12 = 7 + 12 \quad \text{Add 12 to both sides.}$$
$$a = 19 \quad \text{Simplify.}$$

Check: Check by replacing a with 19 in the original equation. □

PRACTICE 6 Solve: $4(2a - 3) - (7a + 4) = 2$

EXAMPLE 7 Solve: $3 - x = 7$

Solution First we subtract 3 from both sides.

$$3 - x = 7$$
$$3 - x - 3 = 7 - 3 \quad \text{Subtract 3 from both sides.}$$
$$-x = 4 \quad \text{Simplify.}$$

We have not yet solved for x since x is not alone. However, this equation does say that the opposite of x is 4. If the opposite of x is 4, then x is the opposite of 4, or $x = -4$. If $-x = 4$, then $x = -4$.

Check:

$$3 - x = 7 \quad \text{Original equation}$$
$$3 - (-4) \stackrel{?}{=} 7 \quad \text{Replace } x \text{ with } -4.$$
$$3 + 4 \stackrel{?}{=} 7 \quad \text{Add.}$$
$$7 = 7 \quad \text{True}$$

The solution is −4. □

PRACTICE 7 Solve: $12 - x = 20$

OBJECTIVE 2

Writing Word Phrases as Algebraic Expressions 

Next, we practice writing word phrases as algebraic expressions.

EXAMPLE 8

a. The sum of two numbers is 8. If one number is 3, find the other number.

b. The sum of two numbers is 8. If one number is x, write an expression representing the other number.

c. An 8-foot board is cut into two pieces. If one piece is x feet, express the length of the other piece in terms of x.

Solution

a. If the sum of two numbers is 8 and one number is 3, we find the other number by subtracting 3 from 8. The other number is $8 - 3$ or 5.

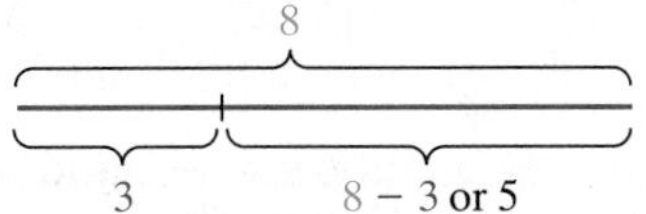

b. If the sum of two numbers is 8 and one number is x, we find the other number by subtracting x from 8. The other number is represented by $8 - x$.

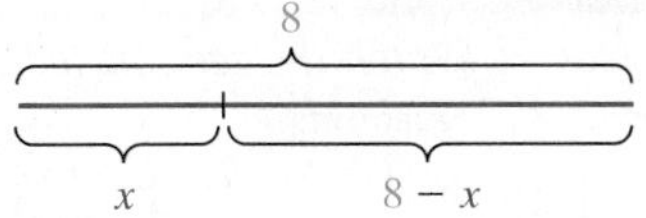

c. If an 8-foot board is cut into two pieces and one piece is x feet, we find the other length by subtracting x from 8. The other piece is $(8 - x)$ feet.

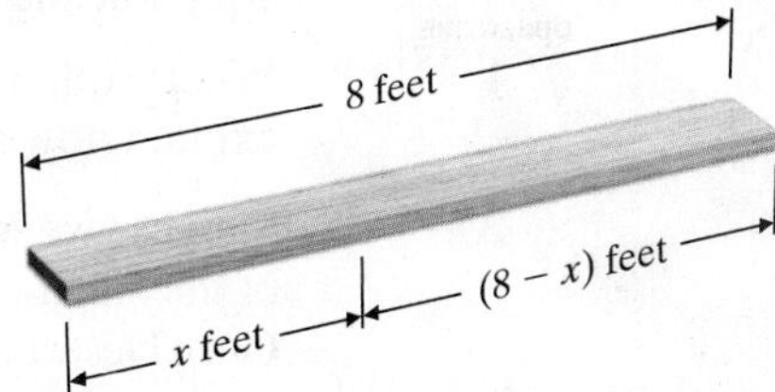

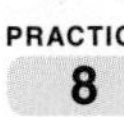

□

PRACTICE 8

a. The sum of two numbers is 9. If one number is 2, find the other number.

b. The sum of two numbers is 9. If one number is x, write an expression representing the other number.

c. A 9-foot rope is cut into two pieces. If one piece is x feet, express the length of the other piece in terms of x.

EXAMPLE 9 The Verrazano-Narrows Bridge in New York City is the longest suspension bridge in North America. The Golden Gate Bridge in San Francisco is 60 feet shorter than the Verrazano-Narrows Bridge. If the length of the Verrazano-Narrows Bridge is m feet, express the length of the Golden Gate Bridge as an algebraic expression in m. (*Source:* Survey of State Highway Engineers)

Solution Since the Golden Gate is 60 feet shorter than the Verrazano-Narrows Bridge, we have that its length is

In words:	length of Verrazano-Narrows Bridge	minus	60
	↓	↓	↓
Translate:	m	$-$	60

The Golden Gate Bridge is $(m - 60)$ feet long. □

PRACTICE 9 Currently, the fastest train in the world is the Japanese Maglev. The French TGV is 3.8 mph slower than the Maglev. If the speed of the Maglev is s miles per hour, express the speed of the French TGV as an algebraic expression in s.

Vocabulary, Readiness & Video Check

Use the choices below to fill in each blank. Not all choices will be used.

addition	solving	expression	true
equivalent	equation	solution	false

1. The difference between an equation and an expression is that a(n) ________ contains an equal sign, whereas a(n) ________ does not.
2. ________ equations are equations that have the same solution.
3. A value of the variable that makes the equation a true statement is called a(n) ________ of the equation.
4. The process of finding the solution of an equation is called ________ the equation for the variable.
5. By the ________ property of equality, $x = -2$ and $x + 10 = -2 + 10$ are equivalent equations.
6. True or false: The equations $x = \frac{1}{2}$ and $\frac{1}{2} = x$ are equivalent equations. ________

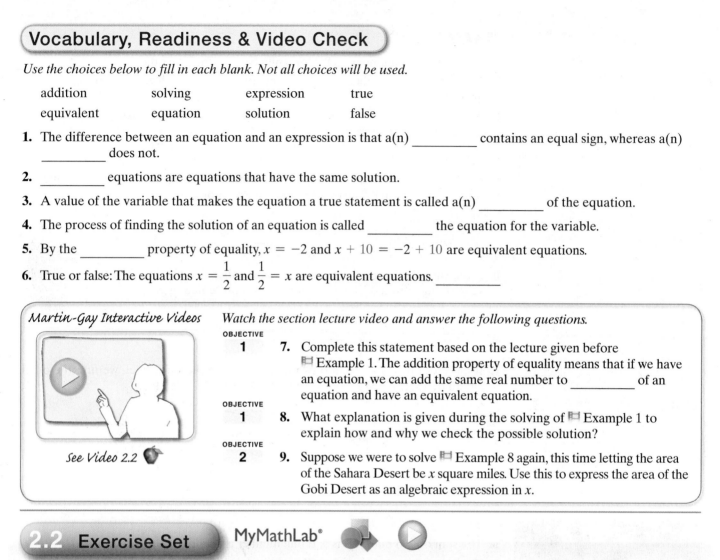

Martin-Gay Interactive Videos

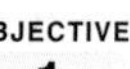

See Video 2.2

Watch the section lecture video and answer the following questions.

OBJECTIVE 1

7. Complete this statement based on the lecture given before Example 1. The addition property of equality means that if we have an equation, we can add the same real number to ________ of an equation and have an equivalent equation.

OBJECTIVE 1

8. What explanation is given during the solving of Example 1 to explain how and why we check the possible solution?

OBJECTIVE 2

9. Suppose we were to solve Example 8 again, this time letting the area of the Sahara Desert be x square miles. Use this to express the area of the Gobi Desert as an algebraic expression in x.

2.2 Exercise Set

MyMathLab®

Solve each equation. Check each solution. See Examples 1 through 3.

1. $x + 7 = 10$
2. $x + 14 = 25$
3. $x - 2 = -4$
4. $y - 9 = 1$
5. $-2 = t - 5$
6. $-17 = x + 3$
7. $r - 8.6 = -8.1$
8. $t - 9.2 = -6.8$
9. $\frac{3}{4} = \frac{1}{3} + f$
10. $\frac{3}{8} = c + \frac{1}{6}$
11. $5b - 0.7 = 6b$
12. $9x + 5.5 = 10x$
13. $7x - 3 = 6x$
14. $18x - 9 = 19x$
15. $y + \frac{11}{25} = -\frac{3}{25}$
16. $z + \frac{9}{19} = -\frac{2}{19}$

Solve each equation. Don't forget to first simplify each side of the equation if possible. Check each solution. See Examples 4 through 7.

17. $7x + 2x = 8x - 3$
18. $3n + 2n = 7 + 4n$
19. $\frac{5}{6}x + \frac{1}{6}x = -9$
20. $\frac{13}{11}y - \frac{2}{11}y = -3$
21. $2y + 10 = 5y - 4y$
22. $4x - 4 = 10x - 7x$
23. $-5(n - 2) = 8 - 4n$
24. $-4(z - 3) = 2 - 3z$
25. $\frac{3}{7}x + 2 = -\frac{4}{7}x - 5$
26. $\frac{1}{5}x - 1 = -\frac{4}{5}x - 13$
27. $5x - 6 = 6x - 5$
28. $2x + 7 = x - 10$
29. $8y + 2 - 6y = 3 + y - 10$
30. $4p - 11 - p = 2 + 2p - 20$
31. $-3(x - 4) = -4x$
32. $-2(x - 1) = -3x$
33. $\frac{3}{8}x - \frac{1}{6} = -\frac{5}{8}x - \frac{2}{3}$
34. $\frac{2}{5}x - \frac{1}{12} = -\frac{3}{5}x - \frac{3}{4}$
35. $2(x - 4) = x + 3$
36. $3(y + 7) = 2y - 5$
37. $3(n - 5) - (6 - 2n) = 4n$
38. $5(3 + z) - (8z + 9) = -4z$
39. $-2(x + 6) + 3(2x - 5) = 3(x - 4) + 10$
40. $-5(x + 1) + 4(2x - 3) = 2(x + 2) - 8$

MIXED PRACTICE

Solve. See Examples 1 through 7.

41. $-11 = 3 + x$
42. $-8 = 8 + z$
43. $x - \frac{2}{5} = -\frac{3}{20}$
44. $y - \frac{4}{7} = -\frac{3}{14}$
45. $3x - 6 = 2x + 5$
46. $7y + 2 = 6y + 2$

47. $13x - 9 + 2x - 5 = 12x - 1 + 2x$

48. $15x + 20 - 10x - 9 = 25x + 8 - 21x - 7$

49. $7(6 + w) = 6(2 + w)$

50. $6(5 + c) = 5(c - 4)$

51. $n + 4 = 3.6$

52. $m + 2 = 7.1$

53. $10 - (2x - 4) = 7 - 3x$

54. $15 - (6 - 7k) = 2 + 6k$

55. $\frac{1}{3} = x + \frac{2}{3}$

56. $\frac{1}{11} = y + \frac{10}{11}$

57. $-6.5 - 4x - 1.6 - 3x = -6x + 9.8$

58. $-1.4 - 7x - 3.6 - 2x = -8x + 4.4$

59. $-3\left(x - \frac{1}{4}\right) = -4x$

60. $-2\left(x - \frac{1}{7}\right) = -3x$

61. $7(m - 2) - 6(m + 1) = -20$

62. $-4(x - 1) - 5(2 - x) = -6$

63. $0.8t + 0.2(t - 0.4) = 1.75$

64. $0.6v + 0.4(0.3 + v) = 2.34$

TRANSLATING

See Examples 8 and 9.

65. Two numbers have a sum of 20. If one number is p, express the other number in terms of p.

66. Two numbers have a sum of 13. If one number is y, express the other number in terms of y.

67. A 10-foot board is cut into two pieces. If one piece is x feet long, express the other length in terms of x.

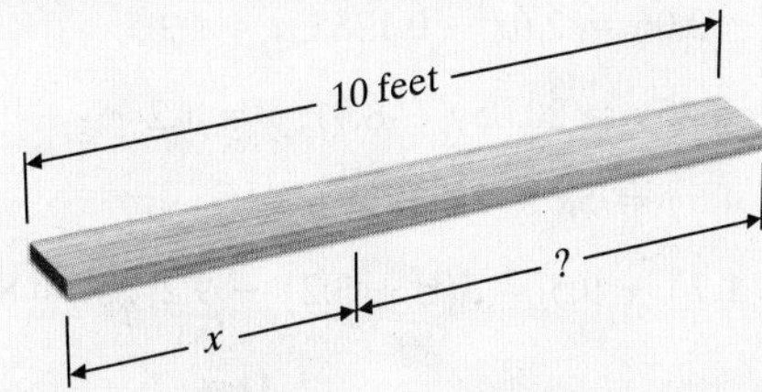

68. A 5-foot piece of string is cut into two pieces. If one piece is x feet long, express the other length in terms of x.

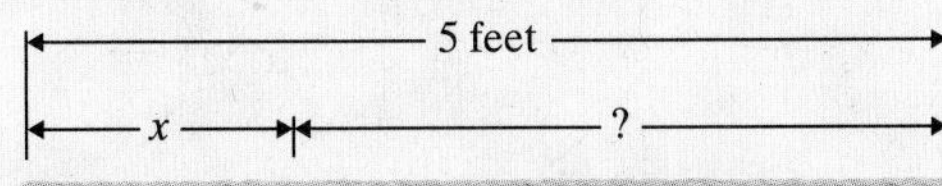

69. Two angles are *supplementary* if their sum is 180°. If one angle measures $x°$, express the measure of its supplement in terms of x.

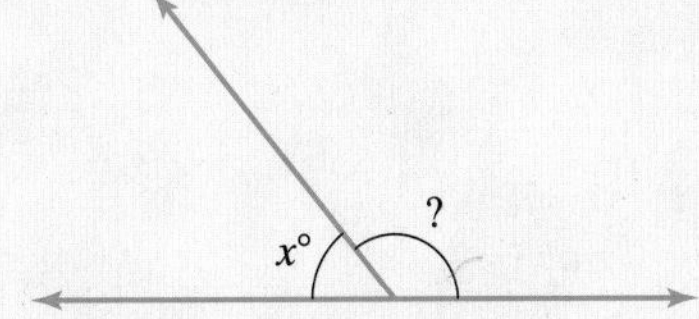

70. Two angles are *complementary* if their sum is 90°. If one angle measures $x°$, express the measure of its complement in terms of x.

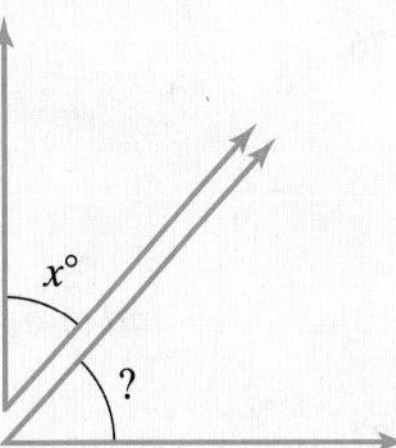

71. In a recent year, the number of graduate students at the University of Texas at Austin was approximately 28,000 fewer than the number of undergraduate students. If the number of undergraduate students was n, how many graduate students attend UT Austin? (*Source:* University of Texas at Austin)

72. The longest interstate highway in the U.S. is I-90, which connects Seattle, Washington, and Boston, Massachusetts. The second longest interstate highway, I-80 (connecting San Francisco, California, and Teaneck, New Jersey), is 178.5 miles shorter than I-90. If the length of I-80 is m miles, express the length of I-90 as an algebraic expression in m. (*Source:* U.S. Department of Transportation—Federal Highway Administration)

73. The area of the Sahara Desert in Africa is 7 times the area of the Gobi Desert in Asia. If the area of the Gobi Desert is x square miles, express the area of the Sahara Desert as an algebraic expression in x.

74. The largest meteorite in the world is the Hoba West located in Namibia. Its weight is 3 times the weight of the Armanty meteorite located in Outer Mongolia. If the weight of the Armanty meteorite is y kilograms, express the weight of the Hoba West meteorite as an algebraic expression in y.

REVIEW AND PREVIEW

Find the reciprocal or multiplicative inverse of each. See Section 1.7.

75. $\frac{5}{8}$ **76.** $\frac{7}{6}$ **77.** 2

78. 5 **79.** $-\frac{1}{9}$ **80.** $-\frac{3}{5}$

Perform each indicated operation and simplify. See Section 1.7.

81. $\frac{3x}{3}$ **82.** $\frac{-2y}{-2}$ **83.** $-5\left(-\frac{1}{5}y\right)$

84. $7\left(\frac{1}{7}r\right)$ **85.** $\frac{3}{5}\left(\frac{5}{3}x\right)$ **86.** $\frac{9}{2}\left(\frac{2}{9}x\right)$

CONCEPT EXTENSIONS

△ **87.** The sum of the angles of a triangle is 180°. If one angle of a triangle measures $x°$ and a second angle measures $(2x + 7)°$, express the measure of the third angle in terms of x. Simplify the expression.

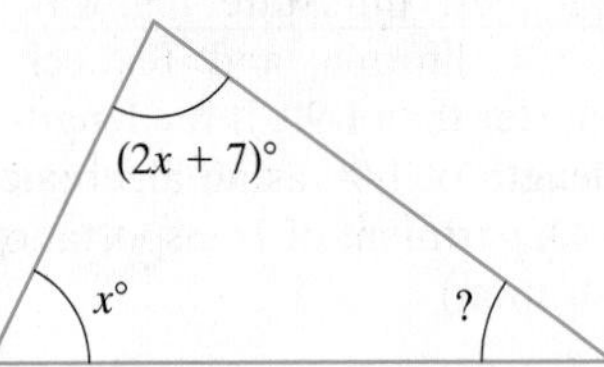

△ **88.** A quadrilateral is a four-sided figure like the one shown below whose angle sum is 360°. If one angle measures $x°$, a second angle measures $3x°$, and a third angle measures $5x°$, express the measure of the fourth angle in terms of x. Simplify the expression.

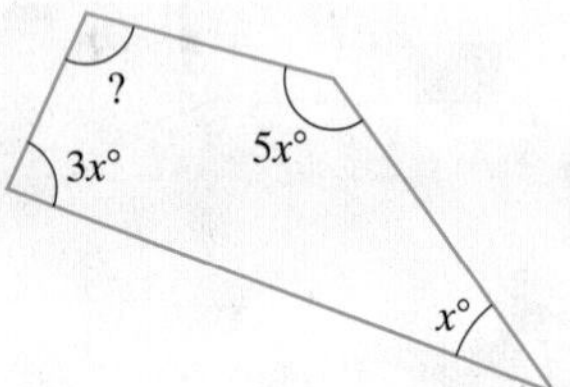

89. Write two terms whose sum is $-3x$.

90. Write four terms whose sum is $2y - 6$.

91. In your own words, explain what is meant by the solution of an equation.

92. In your own words, explain how to check a solution of an equation.

A nurses aide has two patients that are each to consume 1000 ml of fluid for the night. Use this information for Exercises 93 and 94.

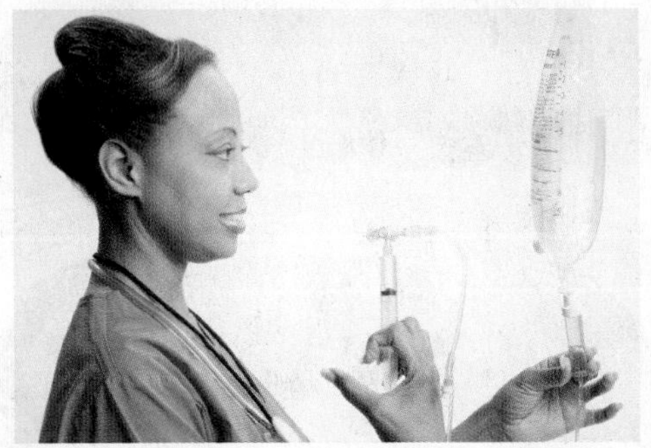

93. Patient intake thus far:

200 ml, 150 ml, and 400 ml.

To determine the remaining fluid needed for this patient, solve for x: $200 + 150 + 400 + x = 1000$

94. Patient intake thus far:

100 ml, 250 ml, and 500 ml.

To determine the remaining fluid needed for this patient, solve for x: $100 + 250 + 500 + x = 1000$

95. Let $x = 1$ and then $x = 2$ in the equation $x + 5 = x + 6$. Is either number a solution? How many solutions do you think this equation has? Explain your answer.

96. Let $x = 1$ and then $x = 2$ in the equation $x + 3 = x + 3$. Is either number a solution? How many solutions do you think this equation has? Explain your answer.

Use the addition property to fill in the blank so that the middle equation simplifies to the last equation. See the Concept Check in this section.

97.
$$x - 4 = -9$$
$$x - 4 + (\quad) = -9 + (\quad)$$
$$x = -5$$

98.
$$a + 9 = 15$$
$$a + 9 + (\quad) = 15 + (\quad)$$
$$a = 6$$

Fill in the blanks with numbers of your choice so that each equation has the given solution. Note: *Each blank may be replaced with a different number.*

99. ____ + x = ____; Solution: -3

100. x − ____ = ____; Solution: -10

Use a calculator to determine whether the given value is a solution of the given equation.

101. $1.23x - 0.06 = 2.6x - 0.1285; x = 0.05$

102. $8.13 + 5.85y = 20.05y - 8.91; y = 1.2$

103. $3(a + 4.6) = 5a + 2.5; a = 6.3$

104. $7(z - 1.7) + 9.5 = 5(z + 3.2) - 9.2; z = 4.8$

2.3 The Multiplication Property of Equality 

OBJECTIVES

1 Use the Multiplication Property of Equality to Solve Linear Equations.

2 Use Both the Addition and Multiplication Properties of Equality to Solve Linear Equations.

3 Write Word Phrases as Algebraic Expressions.

OBJECTIVE

1 Using the Multiplication Property 

As useful as the addition property of equality is, it cannot help us solve every type of linear equation in one variable. For example, adding or subtracting a value on both sides of the equation does not help solve

$$\frac{5}{2}x = 15.$$

Instead, we apply another important property of equality, the **multiplication property of equality.**

Multiplication Property of Equality

If a, b, and c are real numbers and $c \neq 0$, then

$$a = b \quad \text{and} \quad ac = bc$$

are equivalent equations.

This property guarantees that multiplying both sides of an equation by the same nonzero number does not change the solution of the equation. Since division is defined in terms of multiplication, we may also **divide both sides of the equation by the same nonzero number** without changing the solution.

Picturing again our balanced scale, if we multiply or divide the weight on each side by the same nonzero number, the scale (or equation) remains balanced.

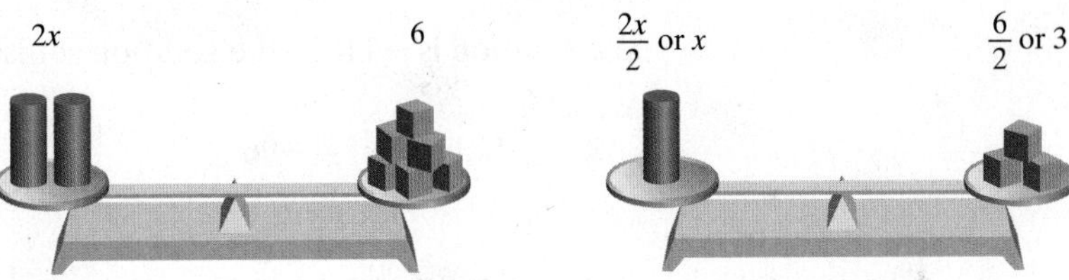

EXAMPLE 1 Solve: $\frac{5}{2}x = 15$.

Solution To get x alone, multiply both sides of the equation by the reciprocal of $\frac{5}{2}$, which is $\frac{2}{5}$.

$$\frac{5}{2}x = 15$$

$$\frac{2}{5} \cdot \frac{5}{2}x = \frac{2}{5} \cdot 15 \quad \text{Multiply both sides by } \frac{2}{5}.$$

$$\left(\frac{2}{5} \cdot \frac{5}{2}\right)x = \frac{2}{5} \cdot 15 \quad \text{Apply the associative property.}$$

$$1x = 6 \quad \text{Simplify.}$$

or

$$x = 6$$

Check: Replace x with 6 in the original equation.

$$\frac{5}{2}x = 15 \quad \text{Original equation}$$

$$\frac{5}{2}(6) \stackrel{?}{=} 15 \quad \text{Replace } x \text{ with 6.}$$

$$15 = 15 \quad \text{True}$$

The solution is 6, or we say that the solution set is $\{6\}$. □

PRACTICE 1 Solve: $\frac{4}{5}x = 16$

In the equation $\frac{5}{2}x = 15$, $\frac{5}{2}$ is the coefficient of x. When the coefficient of x is a *fraction,* we will get x alone by multiplying by the reciprocal. When the coefficient of x is an integer or a decimal, it is usually more convenient to divide both sides by the coefficient. (Dividing by a number is, of course, the same as multiplying by the reciprocal of the number.)

EXAMPLE 2 Solve: $-3x = 33$

Solution Recall that $-3x$ means $-3 \cdot x$. To get x alone, we divide both sides by the coefficient of x, that is, -3.

$$-3x = 33$$
$$\frac{-3x}{-3} = \frac{33}{-3} \quad \text{Divide both sides by } -3.$$
$$1x = -11 \quad \text{Simplify.}$$
$$x = -11$$

Check:

$$-3x = 33 \quad \text{Original equation}$$
$$-3(-11) \stackrel{?}{=} 33 \quad \text{Replace } x \text{ with } -11.$$
$$33 = 33 \quad \text{True}$$

The solution is -11, or the solution set is $\{-11\}$. □

PRACTICE 2 Solve: $8x = -96$

EXAMPLE 3 Solve: $\frac{y}{7} = 20$

Solution Recall that $\frac{y}{7} = \frac{1}{7}y$. To get y alone, we multiply both sides of the equation by 7, the reciprocal of $\frac{1}{7}$.

$$\frac{y}{7} = 20$$
$$\frac{1}{7}y = 20$$
$$7 \cdot \frac{1}{7}y = 7 \cdot 20 \quad \text{Multiply both sides by 7.}$$
$$1y = 140 \quad \text{Simplify.}$$
$$y = 140$$

Check:

$$\frac{y}{7} = 20 \quad \text{Original equation}$$
$$\frac{140}{7} \stackrel{?}{=} 20 \quad \text{Replace } y \text{ with 140.}$$
$$20 = 20 \quad \text{True}$$

The solution is 140. □

PRACTICE 3 Solve: $\frac{x}{5} = 13$

EXAMPLE 4 Solve: $3.1x = 4.96$

Solution

$$3.1x = 4.96$$

$$\frac{3.1x}{3.1} = \frac{4.96}{3.1} \quad \text{Divide both sides by 3.1.}$$

$$1x = 1.6 \quad \text{Simplify.}$$

$$x = 1.6$$

Check: Check by replacing x with 1.6 in the original equation. The solution is 1.6. □

PRACTICE 4 Solve: $2.7x = 4.05$

EXAMPLE 5 Solve: $-\frac{2}{3}x = -\frac{5}{2}$

Solution To get x alone, we multiply both sides of the equation by $-\frac{3}{2}$, the reciprocal of the coefficient of x.

$$-\frac{2}{3}x = -\frac{5}{2}$$

$$-\frac{3}{2} \cdot -\frac{2}{3}x = -\frac{3}{2} \cdot -\frac{5}{2} \quad \text{Multiply both sides by } -\frac{3}{2}, \text{ the reciprocal of } -\frac{2}{3}.$$

$$x = \frac{15}{4} \quad \text{Simplify.}$$

> **Helpful Hint**
> Don't forget to multiply *both sides* by $-\frac{3}{2}$.

Check: Check by replacing x with $\frac{15}{4}$ in the original equation. The solution is $\frac{15}{4}$. □

PRACTICE 5 Solve: $-\frac{5}{3}x = \frac{4}{7}$

OBJECTIVE

2 Using Both the Addition and Multiplication Properties

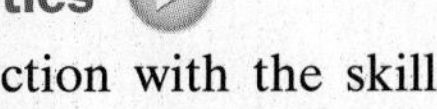

We are now ready to combine the skills learned in the last section with the skills learned from this section to solve equations by applying more than one property.

EXAMPLE 6 Solve: $-z - 4 = 6$

Solution First, get $-z$, the term containing the variable, alone on one side. To do so, add 4 to both sides of the equation.

$$-z - 4 + 4 = 6 + 4 \quad \text{Add 4 to both sides.}$$

$$-z = 10 \quad \text{Simplify.}$$

Next, recall that $-z$ means $-1 \cdot z$. To get z alone, either multiply or divide both sides of the equation by -1. In this example, we divide.

$$-z = 10$$

$$\frac{-z}{-1} = \frac{10}{-1} \quad \text{Divide both sides by the coefficient } -1.$$

$$z = -10 \quad \text{Simplify.}$$

Check: To check, replace z with -10 in the original equation. The solution is -10. □

PRACTICE 6 Solve: $-y + 3 = -8$

Don't forget to simplify one or both sides of an equation if possible.

EXAMPLE 7 Solve: $12a - 8a = 10 + 2a - 13 - 7$

Solution First, simplify both sides of the equation by combining like terms.

$$12a - 8a = 10 + 2a - 13 - 7$$
$$4a = 2a - 10 \qquad \text{Combine like terms.}$$

To get all terms containing a variable on one side, subtract $2a$ from both sides.

$$4a - 2a = 2a - 10 - 2a \qquad \text{Subtract } 2a \text{ from both sides.}$$
$$2a = -10 \qquad \text{Simplify.}$$
$$\frac{2a}{2} = \frac{-10}{2} \qquad \text{Divide both sides by 2.}$$
$$a = -5 \qquad \text{Simplify.}$$

Check: Check by replacing a with -5 in the original equation. The solution is -5. □

PRACTICE 7 Solve: $6b - 11b = 18 + 2b - 6 + 9$

EXAMPLE 8 Solve: $7x - 3 = 5x + 9$

Solution To get x alone, let's first use the addition property to get variable terms on one side of the equation and numbers on the other side. One way to get variable terms on one side is to subtract $5x$ from both sides.

$$7x - 3 = 5x + 9$$
$$7x - 3 - 5x = 5x + 9 - 5x \qquad \text{Subtract } 5x \text{ from both sides.}$$
$$2x - 3 = 9 \qquad \text{Simplify.}$$

Now, to get numbers on the other side, let's add 3 to both sides.

$$2x - 3 + 3 = 9 + 3 \qquad \text{Add 3 to both sides.}$$
$$2x = 12 \qquad \text{Simplify.}$$

Use the multiplication property to get x alone.

$$\frac{2x}{2} = \frac{12}{2} \qquad \text{Divide both sides by 2.}$$
$$x = 6 \qquad \text{Simplify.}$$

Check: To check, replace x with 6 in the original equation to see that a true statement results. The solution is 6. □

PRACTICE 8 Solve: $10x - 4 = 7x + 14$

If an equation has parentheses, don't forget to use the distributive property to remove them. Then combine any like terms.

EXAMPLE 9 Solve: $5(2x + 3) = -1 + 7$

Solution

$$5(2x + 3) = -1 + 7$$

$$5(2x) + 5(3) = -1 + 7 \quad \text{Apply the distributive property.}$$

$$10x + 15 = 6 \quad \text{Multiply and write } -1 + 7 \text{ as } 6.$$

$$10x + 15 - 15 = 6 - 15 \quad \text{Subtract 15 from both sides.}$$

$$10x = -9 \quad \text{Simplify.}$$

$$\frac{10x}{10} = -\frac{9}{10} \quad \text{Divide both sides by 10.}$$

$$x = -\frac{9}{10} \quad \text{Simplify.}$$

Check: To check, replace x with $-\frac{9}{10}$ in the original equation to see that a true statement results. The solution is $-\frac{9}{10}$. □

PRACTICE

9 Solve: $4(3x - 2) = -1 + 4$

OBJECTIVE

3 Writing Word Phrases as Algebraic Expressions 

Next, we continue to sharpen our problem-solving skills by writing word phrases as algebraic expressions.

EXAMPLE 10 If x is the first of three consecutive integers, express the sum of the three integers in terms of x. Simplify if possible.

Solution An example of three consecutive integers is

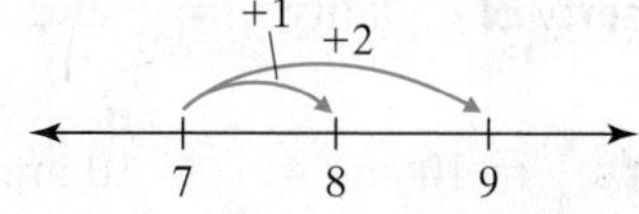

The second consecutive integer is always 1 more than the first, and the third consecutive integer is 2 more than the first. If x is the first of three consecutive integers, the three consecutive integers are

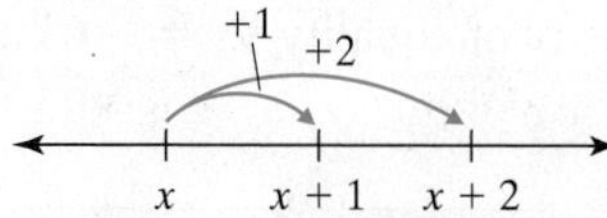

Their sum is

In words:	first integer	+	second integer	+	third integer
	↓		↓		↓
Translate:	x	+	$(x + 1)$	+	$(x + 2)$

which simplifies to $3x + 3$. □

PRACTICE

10 If x is the first of three consecutive *even* integers, express their sum in terms of x.

Below are examples of consecutive even and odd integers.

Consecutive Even Integers:

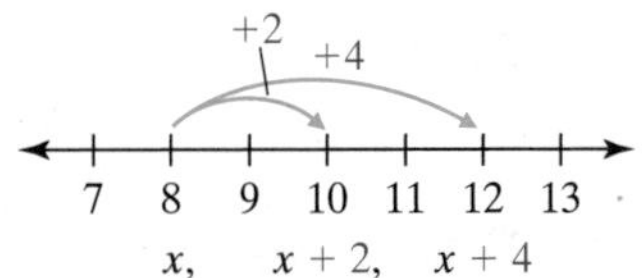

Consecutive Odd Integers:

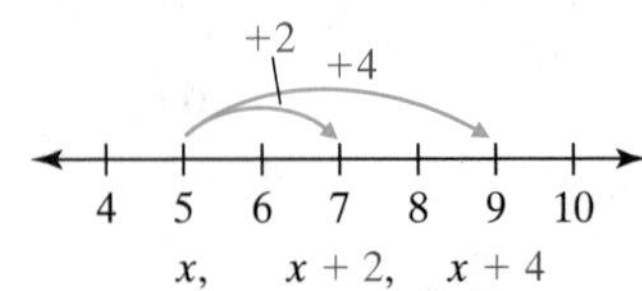

Helpful Hint

If x is an odd integer, then $x + 2$ is the next odd integer. This 2 simply means that odd integers are always 2 units from each other. (The same is true for even integers. They are always 2 units from each other.)

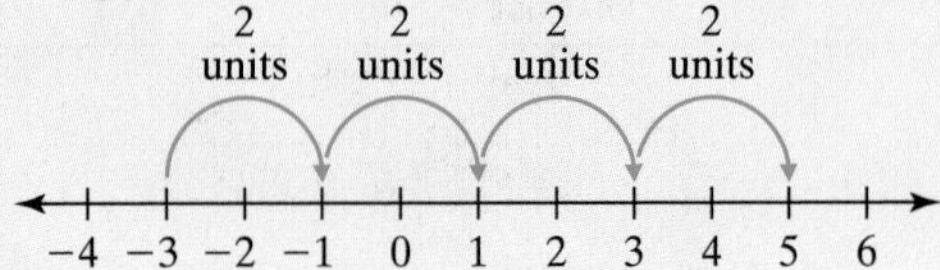

Vocabulary, Readiness & Video Check

Use the choices below to fill in each blank. Not all choices will be used.

true addition false multiplication

1. By the __________ property of equality, $y = \frac{1}{2}$ and $5 \cdot y = 5 \cdot \frac{1}{2}$ are equivalent equations.

2. True or false: The equations $\frac{z}{4} = 10$ and $4 \cdot \frac{z}{4} = 10$ are equivalent equations. __________

3. True or false: The equations $-7x = 30$ and $\frac{-7x}{-7} = \frac{30}{7}$ are equivalent equations. __________

4. By the __________ property of equality, $9x = -63$ and $\frac{9x}{9} = \frac{-63}{9}$ are equivalent equations.

Martin-Gay Interactive Videos *Watch the section lecture video and answer the following questions.*

See Video 2.3

OBJECTIVE 1

5. Complete this statement based on the lecture given before Example 1. We can multiply both sides of an equation by the ______ nonzero number and have an equivalent equation.

OBJECTIVE 2

6. Both the addition and multiplication properties of equality are used to solve Examples 4–6. In each of these exercises, what property is applied first? What property is applied last? What conclusion, if any, can you make?

OBJECTIVE 3

7. Let x be the first of four consecutive integers, as in Example 8. Now express the sum of the second integer and the fourth integer as an algebraic expression containing x.

2.3 Exercise Set

MyMathLab®

Solve each equation. Check each solution. See Examples 1 through 5.

1. $-5x = -20$
2. $-7x = -49$
3. $3x = 0$
4. $2x = 0$
5. $-x = -12$
6. $-y = 8$
7. $\frac{2}{3}x = -8$
8. $\frac{3}{4}n = -15$
9. $\frac{1}{2} = \frac{1}{6}d$
10. $\frac{1}{4} = \frac{1}{8}v$
11. $\frac{a}{2} = 1$
12. $\frac{d}{15} = 2$
13. $\frac{k}{-7} = 0$
14. $\frac{f}{-5} = 0$
15. $10.71 = 1.7x$
16. $19.55 = 8.5y$

Solve each equation. Check each solution. See Examples 6 and 7.

17. $2x - 4 = 16$
18. $3x - 1 = 26$
19. $-x + 2 = 22$
20. $-x + 4 = -24$
21. $6a + 3 = 3$
22. $8t + 5 = 5$
23. $\frac{x}{3} - 2 = -5$
24. $\frac{b}{4} - 1 = -7$
25. $6z - z = -2 + 2z - 1 - 6$
26. $4a + a = -1 + 3a - 1 - 2$
27. $1 = 0.4x - 0.6x - 5$
28. $19 = 0.4x - 0.9x - 6$
29. $\frac{2}{3}y - 11 = -9$
30. $\frac{3}{5}x - 14 = -8$
31. $\frac{3}{4}t - \frac{1}{2} = \frac{1}{3}$
32. $\frac{2}{7}z - \frac{1}{5} = \frac{1}{2}$

Solve each equation. See Examples 8 and 9.

33. $8x + 20 = 6x + 18$
34. $11x + 13 = 9x + 9$
35. $3(2x + 5) = -18 + 9$
36. $2(4x + 1) = -12 + 6$
37. $2x - 5 = 20x + 4$
38. $6x - 4 = -2x - 10$
39. $2 + 14 = -4(3x - 4)$
40. $8 + 4 = -6(5x - 2)$
41. $-6y - 3 = -5y - 7$
42. $-17z - 4 = -16z - 20$
43. $\frac{1}{2}(2x - 1) = -\frac{1}{7} - \frac{3}{7}$
44. $\frac{1}{3}(3x - 1) = -\frac{1}{10} - \frac{2}{10}$
45. $-10z - 0.5 = -20z + 1.6$
46. $-14y - 1.8 = -24y + 3.9$
47. $-4x + 20 = 4x - 20$
48. $-3x + 15 = 3x - 15$

MIXED PRACTICE

See Examples 1 through 9.

49. $42 = 7x$
50. $81 = 3x$
51. $4.4 = -0.8x$
52. $6.3 = -0.6x$
53. $6x + 10 = -20$
54. $10y + 15 = -5$
55. $5 - 0.3k = 5$
56. $2 - 0.4p = 2$
57. $13x - 5 = 11x - 11$
58. $20x - 20 = 16x - 40$
59. $9(3x + 1) = 4x - 5x$
60. $7(2x + 1) = 18x - 19x$
61. $-\frac{3}{7}p = -2$
62. $-\frac{4}{5}r = -5$
63. $-\frac{4}{3}x = 12$
64. $-\frac{10}{3}x = 30$
65. $-2x - \frac{1}{2} = \frac{7}{2}$
66. $-3n - \frac{1}{3} = \frac{8}{3}$
67. $10 = 2x - 1$
68. $12 = 3j - 4$
69. $10 - 3x - 6 - 9x = 7$
70. $12x + 30 + 8x - 6 = 10$
71. $z - 5z = 7z - 9 - z$
72. $t - 6t = -13 + t - 3t$
73. $-x - \frac{4}{5} = x + \frac{1}{2} + \frac{2}{5}$
74. $x + \frac{3}{7} = -x + \frac{1}{3} + \frac{4}{7}$
75. $-15 + 37 = -2(x + 5)$
76. $-19 + 74 = -5(x + 3)$

TRANSLATING

Write each algebraic expression described. Simplify if possible. See Example 10.

77. If x represents the first of two consecutive odd integers, express the sum of the two integers in terms of x.

78. If x is the first of four consecutive even integers, write their sum as an algebraic expression in x.

79. If x is the first of four consecutive integers, express the sum of the first integer and the third integer as an algebraic expression containing the variable x.

80. If x is the first of two consecutive integers, express the sum of 20 and the second consecutive integer as an algebraic expression containing the variable x.

81. Classrooms on one side of the science building are all numbered with consecutive even integers. If the first room on this side of the building is numbered x, write an expression in x for the sum of five classroom numbers in a row. Then simplify this expression.

82. Two sides of a quadrilateral have the same length, x, while the other two sides have the same length, both being the next consecutive odd integer. Write the sum of these lengths. Then simplify this expression.

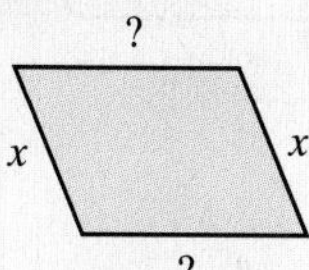

REVIEW AND PREVIEW

Simplify each expression. See Section 2.1.

83. $5x + 2(x - 6)$

84. $-7y + 2y - 3(y + 1)$

85. $6(2z + 4) + 20$

86. $8(z - 6) + 7z - 1$

87. $-(3a - 3) + 2a - 6$

88. $-(x - 1) + x$

Insert <, >, *or* = *in the appropriate space to make each statement true. See Sections 1.2 and 1.7.*

89. $(-3)^2 \quad -3^2$

90. $(-2)^4 \quad -2^4$

91. $(-2)^3 \quad -2^3$

92. $(-4)^3 \quad -4^3$

CONCEPT EXTENSIONS

For Exercises 93 and 94, fill in the blank with a number so that each equation has the given solution.

93. $6x =$ ____; solution: -8

94. ___ $x = 10$; solution: $\frac{1}{2}$

95. The equation $3x + 6 = 2x + 10 + x - 4$ is true for all real numbers. Substitute a few real numbers for x to see that this is so and then try solving the equation. Describe what happens.

96. The equation $6x + 2 - 2x = 4x + 1$ has no solution. Try solving this equation for x and describe what happens.

97. From the results of Exercises 95 and 96, when do you think an equation has all real numbers as its solutions?

98. From the results of Exercises 95 and 96, when do you think an equation has no solution?

99. A licensed nurse practitioner is instructed to give a patient 2100 milligrams of an antibiotic over a period of 36 hours. If the antibiotic is to be given every 4 hours starting immediately, how much antibiotic should be given in each dose? To answer this question, solve the equation $9x = 2100$.

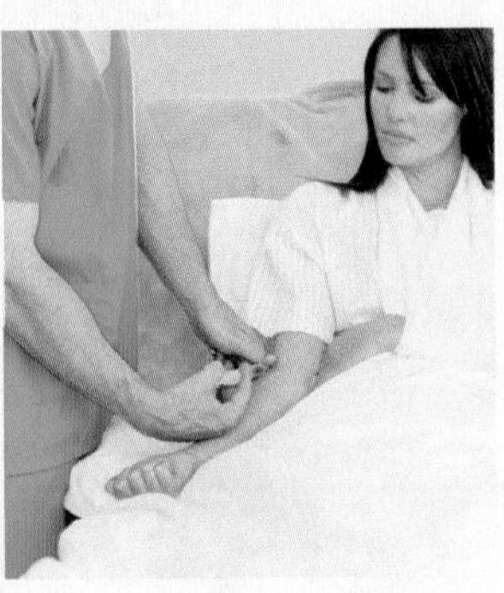

100. Suppose you are a pharmacist and a customer asks you the following question. His child is to receive 13.5 milliliters of a nausea medicine over a period of 54 hours. If the nausea medicine is to be administered every 6 hours starting immediately, how much medicine should be given in each dose? (*Hint:* See the last sentence for Exercise 99.)

Solve each equation.

101. $-3.6x = 10.62$

102. $4.95y = -31.185$

103. $7x - 5.06 = -4.92$

104. $0.06y + 2.63 = 2.5562$

2.4 Solving Linear Equations 

OBJECTIVES

1. Apply a General Strategy for Solving a Linear Equation.
2. Solve Equations Containing Fractions.
3. Solve Equations Containing Decimals.
4. Recognize Identities and Equations with No Solution.

OBJECTIVE

1 Apply a General Strategy for Solving a Linear Equation 

We now present a general strategy for solving linear equations. One new piece of strategy is a suggestion to "clear an equation of fractions" as a first step. Doing so makes the equation less tedious, since operating on integers is usually more convenient than operating on fractions.

Solving Linear Equations in One Variable

Step 1. Multiply on both sides by the LCD to clear the equation of fractions if they occur.

Step 2. Use the distributive property to remove parentheses if they occur.

Step 3. Simplify each side of the equation by combining like terms.

Step 4. Get all variable terms on one side and all numbers on the other side by using the addition property of equality.

Step 5. Get the variable alone by using the multiplication property of equality.

Step 6. Check the solution by substituting it into the original equation.

EXAMPLE 1 Solve: $4(2x - 3) + 7 = 3x + 5$

Solution There are no fractions, so we begin with Step 2.

$$4(2x - 3) + 7 = 3x + 5$$

Step 2. $8x - 12 + 7 = 3x + 5$ Apply the distributive property.

Step 3. $8x - 5 = 3x + 5$ Combine like terms.

Step 4. Get all variable terms on the same side of the equation by subtracting $3x$ from both sides, then adding 5 to both sides.

$8x - 5 - 3x = 3x + 5 - 3x$ Subtract $3x$ from both sides.

$5x - 5 = 5$ Simplify.

$5x - 5 + 5 = 5 + 5$ Add 5 to both sides.

$5x = 10$ Simplify.

Step 5. Use the multiplication property of equality to get x alone.

$\frac{5x}{5} = \frac{10}{5}$ Divide both sides by 5.

$x = 2$ Simplify.

Step 6. Check.

> **Helpful Hint**
> When checking solutions, remember to use the original written equation.

$4(2x - 3) + 7 = 3x + 5$ Original equation

$4[2(2) - 3] + 7 \stackrel{?}{=} 3(2) + 5$ Replace x with 2.

$4(4 - 3) + 7 \stackrel{?}{=} 6 + 5$

$4(1) + 7 \stackrel{?}{=} 11$

$4 + 7 \stackrel{?}{=} 11$

$11 = 11$ True

The solution is 2 or the solution set is $\{2\}$. □

PRACTICE

1 Solve: $2(4a - 9) + 3 = 5a - 6$

EXAMPLE 2 Solve: $8(2 - t) = -5t$

Solution First, we apply the distributive property.

$$8(2 - t) = -5t$$

Step 2. $16 - 8t = -5t$ Use the distributive property.

Step 4. $16 - 8t + 8t = -5t + 8t$ To get variable terms on one side, add $8t$ to both sides.

$16 = 3t$ Combine like terms.

Step 5. $\frac{16}{3} = \frac{3t}{3}$ Divide both sides by 3.

$\frac{16}{3} = t$ Simplify.

Step 6. Check.

$8(2 - t) = -5t$ Original equation

$8\left(2 - \frac{16}{3}\right) \stackrel{?}{=} -5\left(\frac{16}{3}\right)$ Replace t with $\frac{16}{3}$.

$8\left(\frac{6}{3} - \frac{16}{3}\right) \stackrel{?}{=} -\frac{80}{3}$ The LCD is 3.

$$8\left(-\frac{10}{3}\right) \stackrel{?}{=} -\frac{80}{3} \quad \text{Subtract fractions.}$$

$$-\frac{80}{3} = -\frac{80}{3} \quad \text{True}$$

The solution is $\frac{16}{3}$. □

PRACTICE 2 Solve: $7(x - 3) = -6x$

OBJECTIVE 2

Solving Equations Containing Fractions 

If an equation contains fractions, we can clear the equation of fractions by multiplying both sides by the LCD of all denominators. By doing this, we avoid working with time-consuming fractions.

EXAMPLE 3 Solve: $\frac{x}{2} - 1 = \frac{2}{3}x - 3$

Solution We begin by clearing fractions. To do this, we multiply both sides of the equation by the LCD of 2 and 3, which is 6.

$$\frac{x}{2} - 1 = \frac{2}{3}x - 3$$

Step 1. $6\left(\frac{x}{2} - 1\right) = 6\left(\frac{2}{3}x - 3\right)$ Multiply both sides by the LCD, 6.

Step 2. $6\left(\frac{x}{2}\right) - 6(1) = 6\left(\frac{2}{3}x\right) - 6(3)$ Apply the distributive property.

$3x - 6 = 4x - 18$ Simplify.

> **Helpful Hint**
> Don't forget to multiply *each* term by the LCD.

There are no longer grouping symbols and no like terms on either side of the equation, so we continue with Step 4.

$3x - 6 = 4x - 18$

Step 4. $3x - 6 - 3x = 4x - 18 - 3x$ To get variable terms on one side, subtract $3x$ from both sides.

$-6 = x - 18$ Simplify.

$-6 + 18 = x - 18 + 18$ Add 18 to both sides.

$12 = x$ Simplify.

Step 5. The variable is now alone, so there is no need to apply the multiplication property of equality.

Step 6. Check.

$\frac{x}{2} - 1 = \frac{2}{3}x - 3$ Original equation

$\frac{12}{2} - 1 \stackrel{?}{=} \frac{2}{3} \cdot 12 - 3$ Replace x with 12.

$6 - 1 \stackrel{?}{=} 8 - 3$ Simplify.

$5 = 5$ True

The solution is 12. □

PRACTICE 3 Solve: $\frac{3}{5}x - 2 = \frac{2}{3}x - 1$

EXAMPLE 4 Solve: $\frac{2(a+3)}{3} = 6a + 2$

Solution We clear the equation of fractions first.

$$\frac{2(a+3)}{3} = 6a + 2$$

Step 1. $3 \cdot \frac{2(a+3)}{3} = 3(6a+2)$ Clear the fraction by multiplying both sides by the LCD, 3.

$$2(a+3) = 3(6a+2)$$

Step 2. Next, we use the distributive property and remove parentheses.

$2a + 6 = 18a + 6$ Apply the distributive property.

Step 4. $2a + 6 - 6 = 18a + 6 - 6$ Subtract 6 from both sides.

$$2a = 18a$$

$2a - 18a = 18a - 18a$ Subtract $18a$ from both sides.

$$-16a = 0$$

Step 5. $\frac{-16a}{-16} = \frac{0}{-16}$ Divide both sides by -16.

$a = 0$ Write the fraction in simplest form.

Step 6. To check, replace a with 0 in the original equation. The solution is 0. □

PRACTICE 4 Solve: $\frac{4(y+3)}{3} = 5y - 7$

> **Helpful Hint**
> Remember: When solving an equation, it makes no difference on which side of the equation variable terms lie. Just make sure that constant terms lie on the other side.

OBJECTIVE

3 Solving Equations Containing Decimals 

When solving a problem about money, you may need to solve an equation containing decimals. If you choose, you may multiply to clear the equation of decimals.

EXAMPLE 5 Solve: $0.25x + 0.10(x - 3) = 0.05(22)$

Solution First we clear this equation of decimals by multiplying both sides of the equation by 100. Recall that multiplying a decimal number by 100 has the effect of moving the decimal point 2 places to the right.

> **Helpful Hint**
> By the distributive property, 0.10 is multiplied by x and -3. Thus to multiply each term here by 100, we only need to multiply 0.10 by 100.

$$0.25x + 0.10(x - 3) = 0.05(22)$$

Step 1. $0.25x + 0.10(x - 3) = 0.05(22)$ Multiply both sides by 100.

$$25x + 10(x - 3) = 5(22)$$

Step 2. $25x + 10x - 30 = 110$ Apply the distributive property.

Step 3. $35x - 30 = 110$ Combine like terms.

Step 4. $35x - 30 + 30 = 110 + 30$ Add 30 to both sides.

$35x = 140$ Combine like terms.

Step 5. $\frac{35x}{35} = \frac{140}{35}$ Divide both sides by 35.

$x = 4$

Step 6. To check, replace x with 4 in the original equation. The solution is 4. □

PRACTICE 5 Solve: $0.35x + 0.09(x + 4) = 0.03(12)$

OBJECTIVE 4

Recognizing Identities and Equations with No Solution 

So far, each equation that we have solved has had a single solution. However, not every equation in one variable has a single solution. Some equations have no solution, while others have an infinite number of solutions. For example,

$$x + 5 = x + 7$$

has no solution since no matter which **real number** we replace x with, the equation is false.

real number + 5 = same **real number** + 7 **FALSE**

On the other hand,

$$x + 6 = x + 6$$

has infinitely many solutions since x can be replaced by any real number and the equation is always true.

real number + 6 = same **real number** + 6 **TRUE**

The equation $x + 6 = x + 6$ is called an **identity.** The next two examples illustrate special equations like these.

EXAMPLE 6 Solve: $-2(x - 5) + 10 = -3(x + 2) + x$

Solution

$$-2(x - 5) + 10 = -3(x + 2) + x$$

$$-2x + 10 + 10 = -3x - 6 + x \quad \text{Apply the distributive property on both sides.}$$

$$-2x + 20 = -2x - 6 \quad \text{Combine like terms.}$$

$$-2x + 20 + 2x = -2x - 6 + 2x \quad \text{Add } 2x \text{ to both sides.}$$

$$20 = -6 \quad \text{Combine like terms.}$$

The final equation contains no variable terms, and there is no value for x that makes $20 = -6$ a true equation. We conclude that there is **no solution** to this equation. In set notation, we can indicate that there is no solution with the empty set, { }, or use the empty set or null set symbol, $\varnothing$. In this chapter, we will simply write *no solution*. □

PRACTICE 6 Solve: $4(x + 4) - x = 2(x + 11) + x$

EXAMPLE 7 Solve: $3(x - 4) = 3x - 12$

Solution

$$3(x - 4) = 3x - 12$$

$$3x - 12 = 3x - 12 \quad \text{Apply the distributive property.}$$

The left side of the equation is now identical to the right side. Every real number may be substituted for x and a true statement will result. We arrive at the same conclusion if we continue.

$$3x - 12 = 3x - 12$$
$$3x - 12 + 12 = 3x - 12 + 12 \quad \text{Add 12 to both sides.}$$
$$3x = 3x \quad \text{Combine like terms.}$$
$$3x - 3x = 3x - 3x \quad \text{Subtract } 3x \text{ from both sides.}$$
$$0 = 0$$

Again, one side of the equation is identical to the other side. Thus, $3(x - 4) = 3x - 12$ is an **identity** and **all real numbers** are solutions. In set notation, this is {all real numbers}. □

PRACTICE

7 Solve: $12x - 18 = 9(x - 2) + 3x$

✓CONCEPT CHECK

Suppose you have simplified several equations and obtain the following results. What can you conclude about the solutions to the original equation?

a. $7 = 7$ b. $x = 0$ c. $7 = -4$

Calculator Explorations

Checking Equations

We can use a calculator to check possible solutions of equations. To do this, replace the variable by the possible solution and evaluate both sides of the equation separately.

Equation: $3x - 4 = 2(x + 6)$ *Solution:* $x = 16$

$$3x - 4 = 2(x + 6) \quad \text{Original equation}$$
$$3(16) - 4 \stackrel{?}{=} 2(16 + 6) \quad \text{Replace } x \text{ with 16.}$$

Now evaluate each side with your calculator.

Evaluate left side:

[3] [×] [16] [−] [4] [=] or [ENTER] Display: [44] or [3*16 − 4 / 44]

Evaluate right side:

[2] [(] [16] [+] [6] [)] [=] or [ENTER] Display: [44] or [2(16 + 6) / 44]

Since the left side equals the right side, the solution checks.

Use a calculator to check the possible solutions to each equation.

1. $2x = 48 + 6x;\ x = -12$

2. $-3x - 7 = 3x - 1;\ x = -1$

3. $5x - 2.6 = 2(x + 0.8);\ x = 4.4$

4. $-1.6x - 3.9 = -6.9x - 25.6;\ x = 5$

5. $\frac{564x}{4} = 200x - 11(649);\ x = 121$

6. $20(x - 39) = 5x - 432;\ x = 23.2$

Answers to Concept Check:
a. Every real number is a solution.
b. The solution is 0.
c. There is no solution.

Vocabulary, Readiness & Video Check

Throughout algebra, it is important to be able to identify equations and expressions.

Remember,
- an equation contains an equal sign and
- an expression does not.

Also,
- we solve equations and
- we simplify or perform operations on expressions.

Identify each as an equation or an expression.

1. $x = -7$ ________ **2.** $x - 7$ ________ **3.** $4y - 6 + 9y + 1$ ________

4. $4y - 6 = 9y + 1$ ________ **5.** $\frac{1}{x} - \frac{x-1}{8}$ ________ **6.** $\frac{1}{x} - \frac{x-1}{8} = 6$ ________

7. $0.1x + 9 = 0.2x$ ________ **8.** $0.1x^2 + 9y - 0.2x^2$ ________

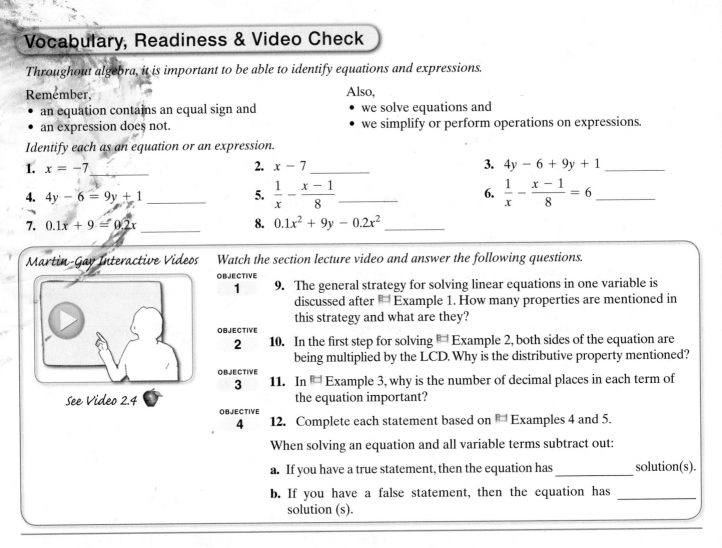

Martin-Gay Interactive Videos

See Video 2.4

Watch the section lecture video and answer the following questions.

OBJECTIVE 1
9. The general strategy for solving linear equations in one variable is discussed after Example 1. How many properties are mentioned in this strategy and what are they?

OBJECTIVE 2
10. In the first step for solving Example 2, both sides of the equation are being multiplied by the LCD. Why is the distributive property mentioned?

OBJECTIVE 3
11. In Example 3, why is the number of decimal places in each term of the equation important?

OBJECTIVE 4
12. Complete each statement based on Examples 4 and 5.

When solving an equation and all variable terms subtract out:

a. If you have a true statement, then the equation has ________ solution(s).

b. If you have a false statement, then the equation has ________ solution (s).

2.4 Exercise Set

MyMathLab®

Solve each equation. See Examples 1 and 2.

1. $-4y + 10 = -2(3y + 1)$

2. $-3x + 1 = -2(4x + 2)$

3. $15x - 8 = 10 + 9x$

4. $15x - 5 = 7 + 12x$

5. $-2(3x - 4) = 2x$

6. $-(5x - 10) = 5x$

7. $5(2x - 1) - 2(3x) = 1$

8. $3(2 - 5x) + 4(6x) = 12$

9. $-6(x - 3) - 26 = -8$

10. $-4(n - 4) - 23 = -7$

11. $8 - 2(a + 1) = 9 + a$

12. $5 - 6(2 + b) = b - 14$

13. $4x + 3 = -3 + 2x + 14$

14. $6y - 8 = -6 + 3y + 13$

15. $-2y - 10 = 5y + 18$

16. $-7n + 5 = 8n - 10$

Solve each equation. See Examples 3 through 5.

17. $\frac{2}{3}x + \frac{4}{3} = -\frac{2}{3}$

18. $\frac{4}{5}x - \frac{8}{5} = -\frac{16}{5}$

19. $\frac{3}{4}x - \frac{1}{2} = 1$

20. $\frac{2}{9}x - \frac{1}{3} = 1$

21. $0.50x + 0.15(70) = 35.5$

22. $0.40x + 0.06(30) = 9.8$

23. $\frac{2(x + 1)}{4} = 3x - 2$

24. $\frac{3(y + 3)}{5} = 2y + 6$

25. $x + \frac{7}{6} = 2x - \frac{7}{6}$

26. $\frac{5}{2}x - 1 = x + \frac{1}{4}$

27. $0.12(y - 6) + 0.06y = 0.08y - 0.70$

28. $0.60(z - 300) + 0.05z = 0.70z - 205$

Solve each equation. See Examples 6 and 7.

29. $4(3x + 2) = 12x + 8$
30. $14x + 7 = 7(2x + 1)$
31. $\frac{x}{4} + 1 = \frac{x}{4}$
32. $\frac{x}{3} - 2 = \frac{x}{3}$
33. $3x - 7 = 3(x + 1)$
34. $2(x - 5) = 2x + 10$
35. $-2(6x - 5) + 4 = -12x + 14$
36. $-5(4y - 3) + 2 = -20y + 17$

MIXED PRACTICE

Solve. See Examples 1 through 7.

37. $\frac{6(3 - z)}{5} = -z$
38. $\frac{4(5 - w)}{3} = -w$
39. $-3(2t - 5) + 2t = 5t - 4$
40. $-(4a - 7) - 5a = 10 + a$
41. $5y + 2(y - 6) = 4(y + 1) - 2$
42. $9x + 3(x - 4) = 10(x - 5) + 7$
43. $\frac{3(x - 5)}{2} = \frac{2(x + 5)}{3}$
44. $\frac{5(x - 1)}{4} = \frac{3(x + 1)}{2}$
45. $0.7x - 2.3 = 0.5$
46. $0.9x - 4.1 = 0.4$
47. $5x - 5 = 2(x + 1) + 3x - 7$
48. $3(2x - 1) + 5 = 6x + 2$
49. $4(2n + 1) = 3(6n + 3) + 1$
50. $4(4y + 2) = 2(1 + 6y) + 8$
51. $x + \frac{5}{4} = \frac{3}{4}x$
52. $\frac{7}{8}x + \frac{1}{4} = \frac{3}{4}x$
53. $\frac{x}{2} - 1 = \frac{x}{5} + 2$
54. $\frac{x}{5} - 7 = \frac{x}{3} - 5$
55. $2(x + 3) - 5 = 5x - 3(1 + x)$
56. $4(2 + x) + 1 = 7x - 3(x - 2)$
57. $0.06 - 0.01(x + 1) = -0.02(2 - x)$
58. $-0.01(5x + 4) = 0.04 - 0.01(x + 4)$
59. $\frac{9}{2} + \frac{5}{2}y = 2y - 4$
60. $3 - \frac{1}{2}x = 5x - 8$
61. $-2y - 10 = 5y + 18$
62. $7n + 5 = 10n - 10$
63. $0.6x - 0.1 = 0.5x + 0.2$
64. $0.2x - 0.1 = 0.6x - 2.1$
65. $0.02(6t - 3) = 0.12(t - 2) + 0.18$
66. $0.03(2m + 7) = 0.06(5 + m) - 0.09$

TRANSLATING

Write each phrase as an algebraic expression. Use x for the unknown number. See Section 2.1.

67. A number subtracted from -8
68. Three times a number
69. The sum of -3 and twice a number
70. The difference of 8 and twice a number
71. The product of 9 and the sum of a number and 20
72. The quotient of -12 and the difference of a number and 3

See Section 2.1.

73. A plot of land is in the shape of a triangle. If one side is x meters, a second side is $(2x - 3)$ meters, and a third side is $(3x - 5)$ meters, express the perimeter of the lot as a simplified expression in x.

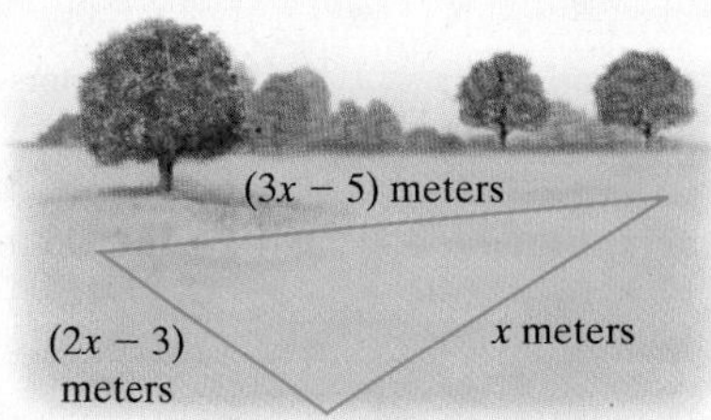

74. A portion of a board has length x feet. The other part has length $(7x - 9)$ feet. Express the total length of the board as a simplified expression in x.

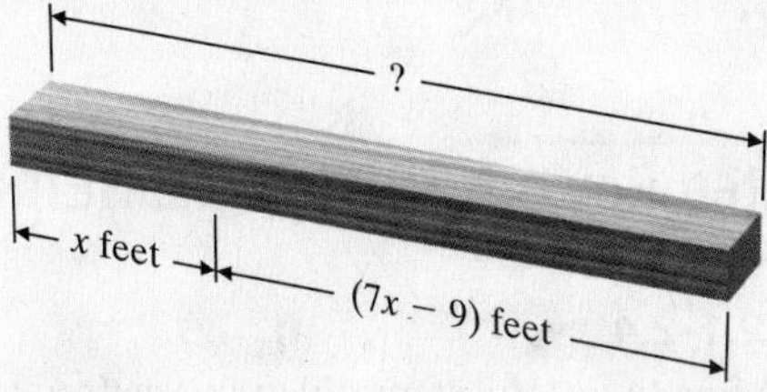

CONCEPT EXTENSIONS

See the Concept Check in this section.

75. **a.** Solve: $x + 3 = x + 3$
    b. If you simplify an equation and get $0 = 0$, what can you conclude about the solution(s) of the original equation?
    c. On your own, construct an equation for which every real number is a solution.
76. **a.** Solve: $x + 3 = x + 5$
    b. If you simplify an equation and get $3 = 5$, what can you conclude about the solution(s) of the original equation?
    c. On your own, construct an equation that has no solution.

For Exercises 77 through 82, match each equation in the first column with its solution in the second column. Items in the second column may be used more than once.

77. $5x + 1 = 5x + 1$

78. $3x + 1 = 3x + 2$

79. $2x - 6x - 10 = -4x + 3 - 10$

80. $x - 11x - 3 = -10x - 1 - 2$

81. $9x - 20 = 8x - 20$

82. $-x + 15 = x + 15$

A. all real numbers

B. no solution

C. 0

83. Explain the difference between simplifying an expression and solving an equation.

84. On your own, write an expression and then an equation. Label each.

For Exercises 85 and 86, **a.** *Write an equation for perimeter.* **b.** *Solve the equation in part (a).* **c.** *Find the length of each side.*

85. The perimeter of a geometric figure is the sum of the lengths of its sides. The perimeter of the following pentagon (five-sided figure) is 28 centimeters.

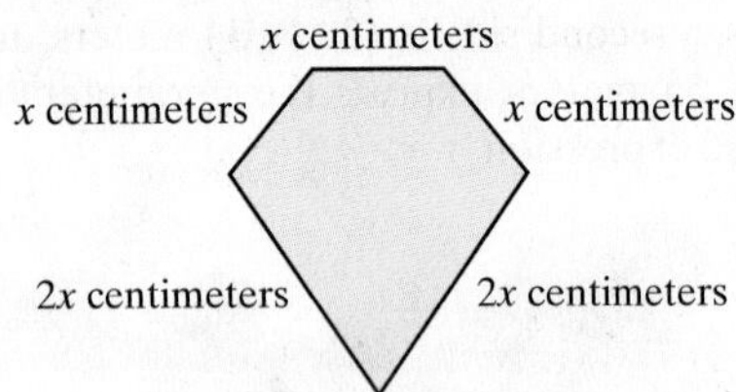

86. The perimeter of the following triangle is 35 meters.

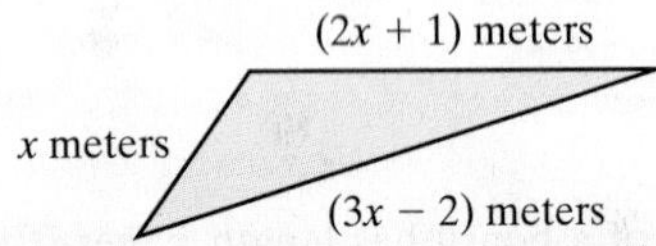

Fill in the blanks with numbers of your choice so that each equation has the given solution. Note: Each blank may be replaced by a different number.

87. $x + \_\_\_ = 2x - \_\_\_$; solution: 9

88. $-5x - \_\_\_ = \_\_\_$; solution: 2

Solve.

89. $1000(7x - 10) = 50(412 + 100x)$

90. $1000(x + 40) = 100(16 + 7x)$

91. $0.035x + 5.112 = 0.010x + 5.107$

92. $0.127x - 2.685 = 0.027x - 2.38$

For Exercises 93 through 96, see the example below.

Example

Solve: $t(t + 4) = t^2 - 2t + 6$.

Solution

$$t(t + 4) = t^2 - 2t + 6$$
$$t^2 + 4t = t^2 - 2t + 6$$
$$t^2 + 4t - t^2 = t^2 - 2t + 6 - t^2$$
$$4t = -2t + 6$$
$$4t + 2t = -2t + 6 + 2t$$
$$6t = 6$$
$$t = 1$$

Solve each equation.

93. $x(x - 3) = x^2 + 5x + 7$

94. $t^2 - 6t = t(8 + t)$

95. $2z(z + 6) = 2z^2 + 12z - 8$

96. $y^2 - 4y + 10 = y(y - 5)$

Integrated Review SOLVING LINEAR EQUATIONS

Sections 2.1–2.4

Solve. Feel free to use the steps given in Section 2.4.

1. $x - 10 = -4$

2. $y + 14 = -3$

3. $9y = 108$

4. $-3x = 78$

5. $-6x + 7 = 25$

6. $5y - 42 = -47$

7. $\frac{2}{3}x = 9$

8. $\frac{4}{5}z = 10$

9. $\frac{r}{-4} = -2$

10. $\frac{y}{-8} = 8$

11. $6 - 2x + 8 = 10$

12. $-5 - 6y + 6 = 19$

13. $2x - 7 = 2x - 27$

14. $3 + 8y = 8y - 2$

15. $-3a + 6 + 5a = 7a - 8a$

16. $4b - 8 - b = 10b - 3b$

17. $-\frac{2}{3}x = \frac{5}{9}$

18. $-\frac{3}{8}y = -\frac{1}{16}$

19. $10 = -6n + 16$

20. $-5 = -2m + 7$

21. $3(5c - 1) - 2 = 13c + 3$

22. $4(3t + 4) - 20 = 3 + 5t$

23. $\frac{2(z + 3)}{3} = 5 - z$

24. $\frac{3(w + 2)}{4} = 2w + 3$

25. $-2(2x - 5) = -3x + 7 - x + 3$

26. $-4(5x - 2) = -12x + 4 - 8x + 4$

27. $0.02(6t - 3) = 0.04(t - 2) + 0.02$

28. $0.03(m + 7) = 0.02(5 - m) + 0.03$

29. $-3y = \frac{4(y-1)}{5}$

30. $-4x = \frac{5(1-x)}{6}$

31. $\frac{5}{3}x - \frac{7}{3} = x$

32. $\frac{7}{5}n + \frac{3}{5} = -n$

33. $9(3x - 1) = -4 + 49$

34. $12(2x + 1) = -6 + 66$

35. $\frac{1}{10}(3x - 7) = \frac{3}{10}x + 5$

36. $\frac{1}{7}(2x - 5) = \frac{2}{7}x + 1$

37. $5 + 2(3x - 6) = -4(6x - 7)$

38. $3 + 5(2x - 4) = -7(5x + 2)$

2.5 An Introduction to Problem Solving 

OBJECTIVES

Apply the steps for problem solving as we

1. Solve Problems Involving Direct Translations.
2. Solve Problems Involving Relationships Among Unknown Quantities.
3. Solve Problems Involving Consecutive Integers.

OBJECTIVE

1 Solving Direct Translation Problems 

In previous sections, we practiced writing word phrases and sentences as algebraic expressions and equations to help prepare for problem solving. We now use these translations to help write equations that model a problem. The problem-solving steps given next may be helpful.

General Strategy for Problem Solving

1. UNDERSTAND the problem. During this step, become comfortable with the problem. Some ways of doing this are to:

Read and reread the problem.

Choose a variable to represent the unknown.

Construct a drawing whenever possible.

Propose a solution and check. Pay careful attention to how you check your proposed solution. This will help when writing an equation to model the problem.

2. TRANSLATE the problem into an equation.

3. SOLVE the equation.

4. INTERPRET the results: *Check* the proposed solution in the stated problem and state your conclusion.

Much of problem solving involves a direct translation from a sentence to an equation.

EXAMPLE 1 **Finding an Unknown Number**

Twice a number, added to seven, is the same as three subtracted from the number. Find the number.

Solution Translate the sentence into an equation and solve.

In words:	twice a number	added to	seven	is the same as	three subtracted from the number
	↓	↓	↓	↓	↓
Translate:	$2x$	$+$	7	$=$	$x - 3$

▶ Helpful Hint

Order matters when subtracting (and dividing), so be especially careful with these translations.

To solve, begin by subtracting x from both sides to get all variable terms on one side.

$$2x + 7 = x - 3$$

$$2x + 7 - x = x - 3 - x \quad \text{Subtract } x \text{ from both sides.}$$

$$x + 7 = -3 \quad \text{Combine like terms.}$$

$$x + 7 - 7 = -3 - 7 \quad \text{Subtract 7 from both sides.}$$

$$x = -10 \quad \text{Combine like terms.}$$

Check the solution in the problem as it was originally stated. To do so, replace "number" in the sentence with -10. Twice "-10" added to 7 is the same as 3 subtracted from "-10."

$$2(-10) + 7 = -10 - 3$$
$$-13 = -13$$

The unknown number is -10. □

PRACTICE

1 Three times a number, minus 6, is the same as two times a number, plus 3. Find the number.

Helpful Hint

When checking solutions, go back to the original stated problem, rather than to your equation, in case errors have been made in translating to an equation.

EXAMPLE 2 **Finding an Unknown Number**

Twice the sum of a number and 4 is the same as four times the number, decreased by 12. Find the number.

Solution

1. UNDERSTAND. Read and reread the problem. If we let

$x =$ the unknown number, then

"the sum of a number and 4" translates to "$x + 4$" and "four times the number" translates to "$4x$."

2. TRANSLATE.

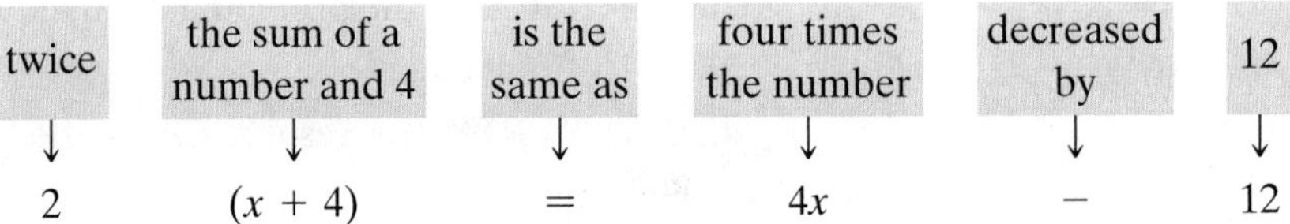

twice	the sum of a number and 4	is the same as	four times the number	decreased by	12
↓	↓	↓	↓	↓	↓
2	$(x + 4)$	$=$	$4x$	$-$	12

3. SOLVE.

$$2(x + 4) = 4x - 12$$
$$2x + 8 = 4x - 12 \quad \text{Apply the distributive property.}$$
$$2x + 8 - 4x = 4x - 12 - 4x \quad \text{Subtract } 4x \text{ from both sides.}$$
$$-2x + 8 = -12$$
$$-2x + 8 - 8 = -12 - 8 \quad \text{Subtract 8 from both sides.}$$
$$-2x = -20$$
$$\frac{-2x}{-2} = \frac{-20}{-2} \quad \text{Divide both sides by } -2.$$
$$x = 10$$

4. INTERPRET.

Check: Check this solution in the problem as it was originally stated. To do so, replace "number" with 10. Twice the sum of "10" and 4 is 28, which is the same as 4 times "10" decreased by 12.

State: The number is 10. □

PRACTICE

2 Three times a number, decreased by 4, is the same as double the difference of the number and 1.

OBJECTIVE

2 Solving Problems Involving Relationships Among Unknown Quantities

The next three examples have to do with relationships among unknown quantities.

EXAMPLE 3 **Finding the Length of a Board**

Balsa wood sticks are commonly used for building models (for example, bridge models). A 48-inch balsa wood stick is to be cut into two pieces so that the longer piece is 3 times the shorter. Find the length of each piece.

Solution

1. UNDERSTAND the problem. To do so, read and reread the problem. You may also want to propose a solution. For example, if 10 inches represents the length of the shorter piece, then $3(10) = 30$ inches is the length of the longer piece, since it is 3 times the length of the shorter piece. This guess gives a total stick length of 10 inches + 30 inches = 40 inches, too short. However, the purpose of proposing a solution is not to guess correctly but to help understand the problem better and how to model it.

   Since the length of the longer piece is given in terms of the length of the shorter piece, let's let

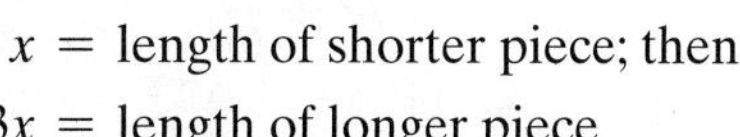

x = length of shorter piece; then
$3x$ = length of longer piece

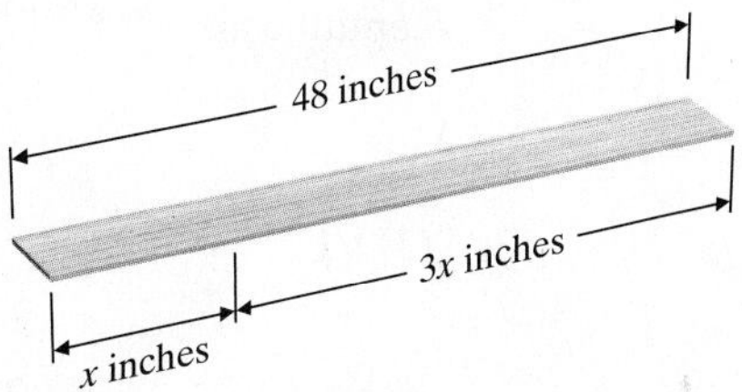

2. TRANSLATE the problem. First, we write the equation in words.

length of shorter piece	added to	length of longer piece	equals	total length of stick
↓	↓	↓	↓	↓
x	$+$	$3x$	$=$	48

3. SOLVE.

$$x + 3x = 48$$
$$4x = 48 \quad \text{Combine like terms.}$$
$$\frac{4x}{4} = \frac{48}{4} \quad \text{Divide both sides by 4.}$$
$$x = 12$$

4. INTERPRET.

Check: Check the solution in the stated problem. If the shorter piece of stick is 12 inches, the longer piece is $3 \cdot (12 \text{ inches}) = 36$ inches, and the sum of the two pieces is 12 inches + 36 inches = 48 inches.

State: The shorter piece of balsa wood is 12 inches, and the longer piece of balsa wood is 36 inches. □

> **Helpful Hint**
> Make sure that units are included in your answer, if appropriate.

PRACTICE

3 A 45-inch board is to be cut into two pieces so that the longer piece is 4 times the shorter. Find the length of each piece.

EXAMPLE 4 **Finding the Number of Republican and Democratic Representatives**

The 112th Congress began on January 3, 2011, and had a total of 435 Democrat and Republican representatives. There were 49 fewer Democratic representatives than Republican. Find the number of representatives from each party. (*Source:* Wikipedia.com)

Solution:

1. UNDERSTAND the problem. Read and reread the problem. Let's suppose that there were 200 Republican representatives. Since there were 49 fewer Democrats than Republicans, there must have been $200 - 49 = 151$ Democrats. The total number of Republicans and Democrats was then $200 + 151 = 351$. This is incorrect since the total should be 435, but we now have a better understanding of the problem.

   In general, if we let

$$x = \text{number of Republicans, then}$$
$$x - 49 = \text{number of Democrats}$$

2. TRANSLATE the problem. First, we write the equation in words.

number of Republicans	added to	number of Democrats	equals	435
↓	↓	↓	↓	↓
x	$+$	$(x - 49)$	$=$	435

3. SOLVE.

$$x + (x - 49) = 435$$
$$2x - 49 = 435 \quad \text{Combine like terms.}$$
$$2x - 49 + 49 = 435 + 49 \quad \text{Add 49 to both sides.}$$
$$2x = 484$$
$$\frac{2x}{2} = \frac{484}{2} \quad \text{Divide both sides by 2.}$$
$$x = 242$$

4. INTERPRET.

Check: If there were 242 Republican representatives, then there were $242 - 49 = 193$ Democratic representatives. The total number of representatives was then $242 + 193 = 435$. The results check.

State: There were 242 Republican and 193 Democratic representatives at the beginning of the 112th Congress. □

PRACTICE

4 In 2011, there were 9 fewer Democratic State Governors than Republican State Governors. Find the number of State Governors from each party. Rhode Island had an independent governor, so use a total of 49, representing the other 49 states. (*Source:* National Conference of State Legislatures).

△ **EXAMPLE 5** **Finding Angle Measures**

If the two walls of the Vietnam Veterans Memorial in Washington, D.C., were connected, an isosceles triangle would be formed. The measure of the third angle is 97.5° more than the measure of either of the other two equal angles. Find the measure of the third angle. (*Source:* National Park Service)

Solution

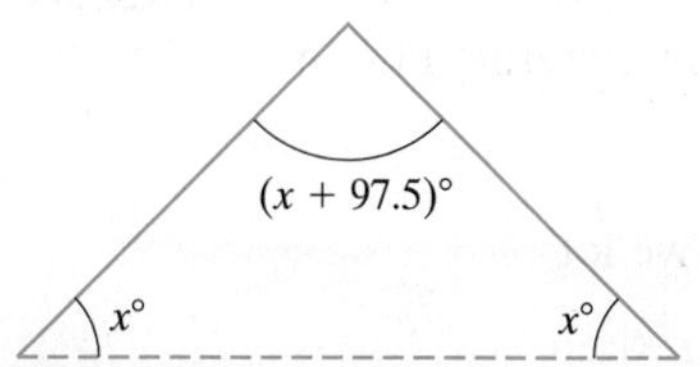

1. UNDERSTAND. Read and reread the problem. We then draw a diagram (recall that an isosceles triangle has two angles with the same measure) and let

$$x = \text{degree measure of one angle}$$
$$x = \text{degree measure of the second equal angle}$$
$$x + 97.5 = \text{degree measure of the third angle}$$

2. TRANSLATE. Recall that the sum of the measures of the angles of a triangle equals 180.

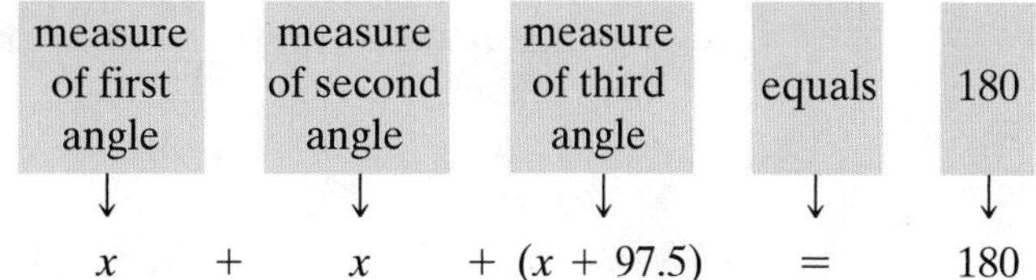

$$x \quad + \quad x \quad + (x + 97.5) \quad = \quad 180$$

3. SOLVE.

$$x + x + (x + 97.5) = 180$$
$$3x + 97.5 = 180 \qquad \text{Combine like terms.}$$
$$3x + 97.5 - 97.5 = 180 - 97.5 \qquad \text{Subtract 97.5 from both sides.}$$
$$3x = 82.5$$
$$\frac{3x}{3} = \frac{82.5}{3} \qquad \text{Divide both sides by 3.}$$
$$x = 27.5$$

4. INTERPRET.

Check: If $x = 27.5$, then the measure of the third angle is $x + 97.5 = 125$. The sum of the angles is then $27.5 + 27.5 + 125 = 180$, the correct sum.

State: The third angle measures 125°.* □

PRACTICE

5 The second angle of a triangle measures three times as large as the first. If the third angle measures 55° more than the first, find the measures of all three angles.

OBJECTIVE

3 Solving Consecutive Integer Problems 

The next example has to do with consecutive integers. Recall what we have learned thus far about these integers.

	Example	*General Representation*
Consecutive Integers	11, 12, 13 (+1, +1)	Let x be an integer. x, $x + 1$, $x + 2$ (+1, +1)
Consecutive Even Integers	38, 40, 42 (+2, +2)	Let x be an even integer. x, $x + 2$, $x + 4$ (+2, +2)
Consecutive Odd Integers	57, 59, 61 (+2, +2)	Let x be an odd integer. x, $x + 2$, $x + 4$ (+2, +2)

*The two walls actually meet at an angle of 125 degrees 12 minutes. The measurement of 97.5° given in the problem is an approximation.

EXAMPLE 6 Some states have a single area code for the entire state. Two such states have area codes that are consecutive odd integers. If the sum of these integers is 1208, find the two area codes. (*Source:* North American Numbering Plan Administration)

Solution:

1. UNDERSTAND. Read and reread the problem. If we let

$x =$ the first odd integer, then

$x + 2 =$ the next odd integer

> **Helpful Hint**
> Remember, the 2 here means that odd integers are 2 units apart, for example, the odd integers 13 and $13 + 2 = 15$.

2. TRANSLATE.

first odd integer	the sum of	next odd integer	is	1208
↓	↓	↓	↓	↓
x	$+$	$(x + 2)$	$=$	1208

3. SOLVE.

$$x + x + 2 = 1208$$
$$2x + 2 = 1208$$
$$2x + 2 - 2 = 1208 - 2$$
$$2x = 1206$$
$$\frac{2x}{2} = \frac{1206}{2}$$
$$x = 603$$

4. INTERPRET.

Check: If $x = 603$, then the next odd integer $x + 2 = 603 + 2 = 605$. Notice their sum, $603 + 605 = 1208$, as needed.

State: The area codes are 603 and 605.

Note: New Hampshire's area code is 603 and South Dakota's area code is 605. □

PRACTICE

6 The sum of three consecutive even integers is 144. Find the integers.

Vocabulary, Readiness & Video Check

Fill in the table.

1.	A number: → x	Double the number:	→ Double the number, decreased by 31:
2.	A number: → x	Three times the number:	→ Three times the number, increased by 17:
3.	A number: → x	The sum of the number and 5:	→ Twice the sum of the number and 5:
4.	A number: → x	The difference of the number and 11:	→ Seven times the difference of a number and 11:
5.	A number: → y	The difference of 20 and the number:	→ The difference of 20 and the number, divided by 3:
6.	A number: → y	The sum of -10 and the number:	→ The sum of -10 and the number, divided by 9:

Martin-Gay Interactive Videos *Watch the section lecture video and answer the following questions.*

OBJECTIVE 1

7. At the end of  Example 1, where are you told is the best place to check the solution of an application problem?

OBJECTIVE 2

8. The solution of the equation for  Example 3 is $x = 43$. Why is this not the solution to the application?

OBJECTIVE 3

9. What are two things that should be checked to make sure the solution of Example 4 is correct?

See Video 2.5

2.5 Exercise Set

MyMathLab®

TRANSLATING

Write each of the following as an equation. Then solve. See Examples 1 and 2.

1. The sum of six times a number, and 1, is equal to five times the number. Find the number.
2. The difference of three times a number, and 1, is the same as twice the number. Find the number.
3. Three times a number, minus 6, is equal to two times the number, plus 8. Find the number.
4. The sum of 4 times a number, and -2, is equal to the sum of 5 times the number, and -2. Find the number.
5. Twice the difference of a number and 8 is equal to three times the sum of the number and 3. Find the number.
6. Five times the sum of a number and -1 is the same as 6 times the difference of the number and 5. Find the number.
7. Twice the sum of -2 and a number is the same as the number decreased by $\frac{1}{2}$. Find the number.
8. If the difference of a number and four is doubled, the result is $\frac{1}{4}$ less than the number. Find the number.

Solve. For Exercises 9 and 10, the solutions have been started for you. See Examples 3 through 5.

9. A 25-inch piece of steel is cut into three pieces so that the second piece is twice as long as the first piece, and the third piece is one inch more than five times the length of the first piece. Find the lengths of the pieces.

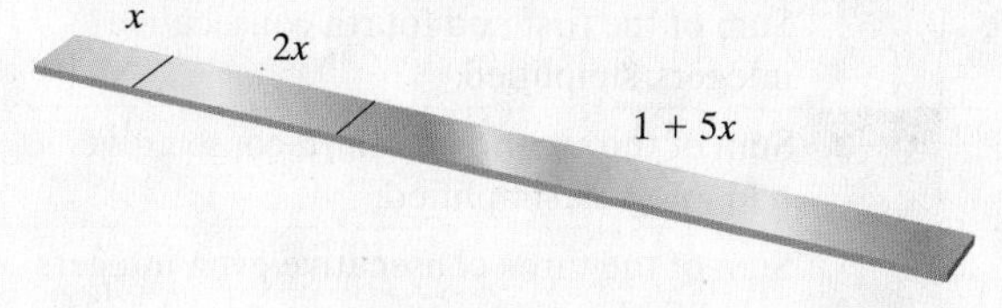

Start the solution:

1. UNDERSTAND the problem. Reread it as many times as needed.
2. TRANSLATE into an equation. (Fill in the blanks below.)

total length of steel	equals	length of first piece	plus	length of second piece	plus	length of third piece
↓	↓	↓	↓	↓	↓	↓
25	=	___	+	___	+	___

Finish with:

3. SOLVE and 4. INTERPRET

10. A 46-foot piece of rope is cut into three pieces so that the second piece is three times as long as the first piece, and the third piece is two feet more than seven times the length of the first piece. Find the lengths of the pieces.

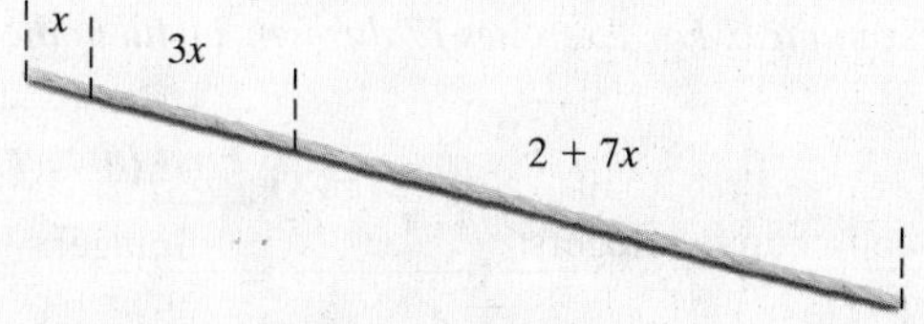

Start the solution:

1. UNDERSTAND the problem. Reread it as many times as needed.
2. TRANSLATE into an equation. (Fill in the blanks below.)

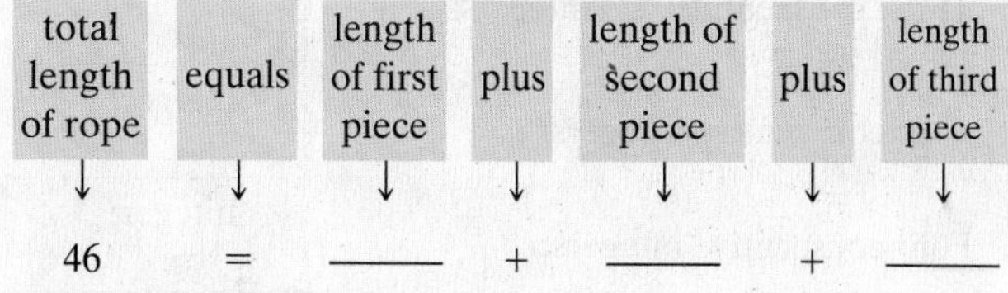

total length of rope	equals	length of first piece	plus	length of second piece	plus	length of third piece
↓	↓	↓	↓	↓	↓	↓
46	=	___	+	___	+	___

Finish with:

3. SOLVE and 4. INTERPRET

11. A 40-inch board is to be cut into three pieces so that the second piece is twice as long as the first piece, and the third piece is 5 times as long as the first piece. If x represents the length of the first piece, find the lengths of all three pieces.

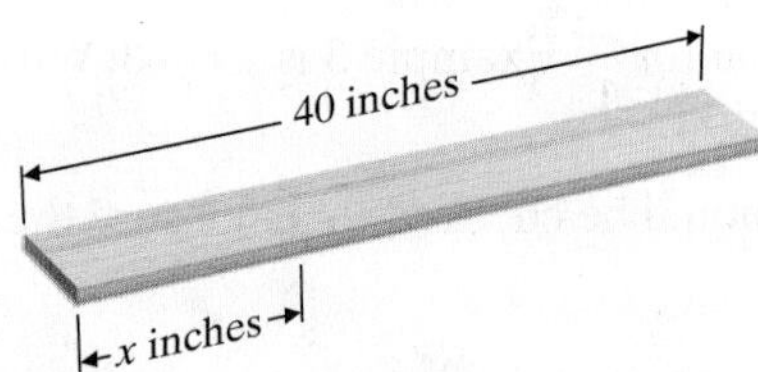

12. A 21-foot beam is to be cut into three pieces so that the second and the third piece are each 3 times the length of the first piece. If x represents the length of the shorter piece, find the lengths of all three pieces.

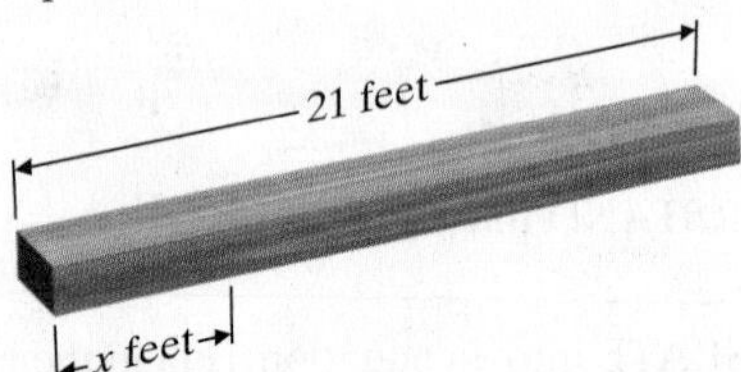

For Exercises 13 and 14, use each table to find the value of x. Then write a sentence to explain, in words, the meaning and the value of x. (Source: IHS Screen Digest)

13. The fastest growing type of movie screen 2010 is 3D.

Type of Screen	*Number of Screens*
3D	x
Non-3D	23,873 more than x
Total screens	39,547

14. The majority of movie screens in the United States in 2010 were located at sites with five or more screens.

Type of Site	*Number of Screens*
Smaller (1–4 screens)	x
Larger (5 or more screens)	22,857 more than x
Total screens	39,547

15. The flag of Equatorial Guinea contains an isosceles triangle. (Recall that an isosceles triangle contains two angles with the same measure.) If the measure of the third angle of the triangle is 30° more than twice the measure of either of the other two angles, find the measure of each angle of the triangle. (*Hint:* Recall that the sum of the measures of the angles of a triangle is 180°.)

16. The flag of Brazil contains a parallelogram. One angle of the parallelogram is 15° less than twice the measure of the angle next to it. Find the measure of each angle of the parallelogram. (*Hint:* Recall that opposite angles of a parallelogram have the same measure and that the sum of the measures of the angles is 360°.)

Solve. See Example 6. For Exercises 17 through 24, fill in the table. Most of the first row has been completed for you.

	First Integer →	*Next Integers*			→ *Indicated Sum*
17. Three consecutive integers:	Integer: x	$x + 1$	$x + 2$		Sum of the three consecutive integers, simplified:
18. Three consecutive integers:	Integer: x				Sum of the second and third consecutive integers, simplified:
19. Three consecutive even integers:	Even integer: x				Sum of the first and third even consecutive integers, simplified:
20. Three consecutive odd integers:	Odd integer: x				Sum of the three consecutive odd integers, simplified:
21. Four consecutive integers:	Integer: x				Sum of the four consecutive integers, simplified:
22. Four consecutive integers:	Integer: x				Sum of the first and fourth consecutive integers, simplified:
23. Three consecutive odd integers:	Odd integer: x				Sum of the second and third consecutive odd integers, simplified:
24. Three consecutive even integers:	Even integer: x				Sum of the three consecutive even integers, simplified:

The graph below shows the best-selling albums of all time. Use this graph for Exercises 59 through 64.

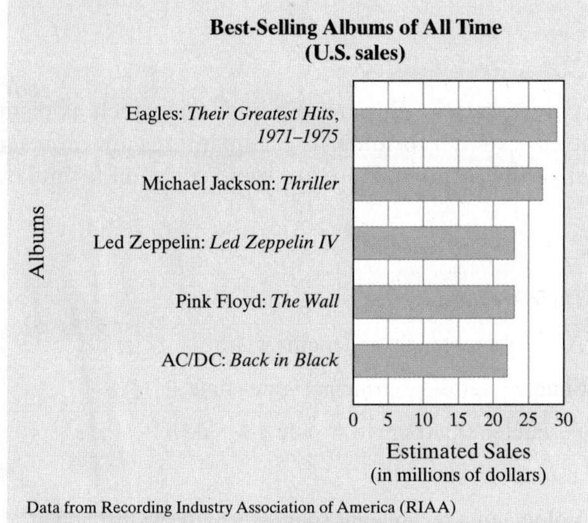

(*Note:* These numbers are before the death of Michael Jackson.)

59. Which album is the best-selling album of all time?

60. Which albums had total sales between \$20 million and \$25 million?

61. *Thriller* and *The Wall* had sales worth a total of \$50 million. *Thriller* brought in \$4 million more than *The Wall*. Find the amount of sales that each album brought in.

62. Eagles: *Their Greatest Hits, 1971–1975*, and AC/DC: *Back in Black* had sales worth \$51 million. Eagles: *Their Greatest Hits, 1971–1975*, sold \$7 million more than AC/DC: *Back in Black*. Find the amount of sales for each album.

Compare the lengths of the bars in the graph with your results for the exercises below. Are your answers reasonable?

63. Exercise 61 **64.** Exercise 62

REVIEW AND PREVIEW

Evaluate each expression for the given values. See Section 1.4.

65. $2W + 2L;\quad W = 7$ and $L = 10$

66. $\frac{1}{2}Bh;\quad B = 14$ and $h = 22$

67. $\pi r^2;\quad r = 15$

68. $r \cdot t;\quad r = 15$ and $t = 2$

CONCEPT EXTENSIONS

69. A golden rectangle is a rectangle whose length is approximately 1.6 times its width. The early Greeks thought that a rectangle with these dimensions was the most pleasing to the eye, and examples of the golden rectangle are found in many early works of art. For example, the Parthenon in Athens contains many examples of golden rectangles.

Mike Hallahan would like to plant a rectangular garden in the shape of a golden rectangle. If he has 78 feet of fencing available, find the dimensions of the garden.

70. Dr. Dorothy Smith gave the students in her geometry class at the University of New Orleans the following question. Is it possible to construct a triangle such that the second angle of the triangle has a measure that is twice the measure of the first angle and the measure of the third angle is 5 times the measure of the first? If so, find the measure of each angle. (*Hint:* Recall that the sum of the measures of the angles of a triangle is 180°.)

71. Only male crickets chirp. They chirp at different rates depending on their species and the temperature of their environment. Suppose a certain species is currently chirping at a rate of 90 chirps per minute. At this rate, how many chirps occur in one hour? In one 24-hour day? In one year?

72. The human eye blinks once every 5 seconds on average. How many times does the average eye blink in one hour? In one 16-hour day while awake? In one year while awake?

73. In your own words, explain why a solution of a word problem should be checked using the original wording of the problem and not the equation written from the wording.

74. Give an example of how you recently solved a problem using mathematics.

Recall from Exercise 69 that a golden rectangle is a rectangle whose length is approximately 1.6 times its width.

75. It is thought that for about 75% of adults, a rectangle in the shape of the golden rectangle is the most pleasing to the eye. Draw three rectangles, one in the shape of the golden rectangle, and poll your class. Do the results agree with the percentage given above?

76. Examples of golden rectangles can be found today in architecture and manufacturing packaging. Find an example of a golden rectangle in your home. A few suggestions: the front face of a book, the floor of a room, the front of a box of food.

For Exercises 77 and 78, measure the dimensions of each rectangle and decide which one best approximates the shape of a golden rectangle.

77.

78.

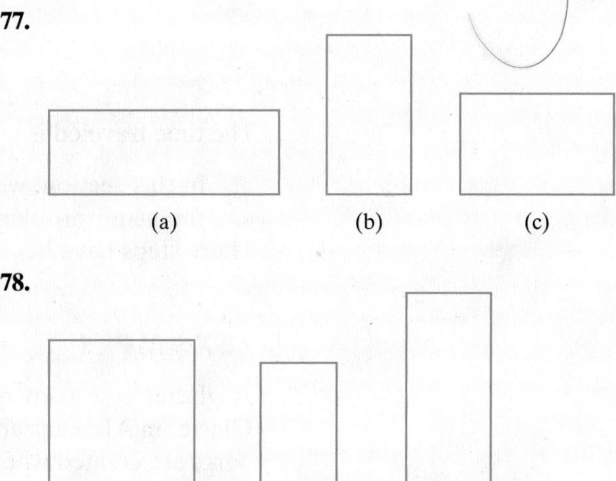

2.6 Formulas and Problem Solving 

OBJECTIVES

1 Use Formulas to Solve Problems.

2 Solve a Formula or Equation for One of Its Variables.

OBJECTIVE

1 Using Formulas to Solve Problems 

An equation that describes a known relationship among quantities, such as distance, time, volume, weight, and money, is called a **formula.** These quantities are represented by letters and are thus variables of the formula. Here are some common formulas and their meanings.

Formulas	*Their Meanings*
$A = lw$	Area of a rectangle = length · width
$I = PRT$	Simple interest = principal · rate · time
$P = a + b + c$	Perimeter of a triangle = side a + side b + side c
$d = rt$	distance = rate · time
$V = lwh$	Volume of a rectangular solid = length · width · height
$F = \left(\frac{9}{5}\right)C + 32$ or $F = 1.8C + 32$	degrees Fahrenheit = $\left(\frac{9}{5}\right)$ · degrees Celsius + 32

Formulas are valuable tools because they allow us to calculate measurements as long as we know certain other measurements. For example, if we know we traveled a distance of 100 miles at a rate of 40 miles per hour, we can replace the variables d and r in the formula $d = rt$ and find our time, t.

$$d = rt \quad \text{Formula.}$$

$$100 = 40t \quad \text{Replace } d \text{ with 100 and } r \text{ with 40.}$$

This is a linear equation in one variable, t. To solve for t, divide both sides of the equation by 40.

$$\frac{100}{40} = \frac{40t}{40} \quad \text{Divide both sides by 40.}$$

$$\frac{5}{2} = t \quad \text{Simplify.}$$

The time traveled is $\frac{5}{2}$ hours, or $2\frac{1}{2}$ hours, or 2.5 hours.

In this section, we solve problems that can be modeled by known formulas. We use the same problem-solving steps that were introduced in the previous section. These steps have been slightly revised to include formulas.

EXAMPLE 1 **Finding Time Given Rate and Distance**

A glacier is a giant mass of rocks and ice that flows downhill like a river. Portage Glacier in Alaska is about 6 miles, or 31,680 *feet,* long and moves 400 *feet* per year. Icebergs are created when the front end of the glacier flows into Portage Lake. How long does it take for ice at the head (beginning) of the glacier to reach the lake?

Solution

1. UNDERSTAND. Read and reread the problem. The appropriate formula needed to solve this problem is the distance formula, $d = rt$. To become familiar with this formula, let's find the distance that ice traveling at a rate of 400 feet per year travels in 100 years. To do so, we let time t be 100 years and rate r be the given 400 feet per year and substitute these values into the formula $d = rt$. We then have that distance $d = 400(100) = 40{,}000$ feet. Since we are interested in finding how long it takes ice to travel 31,680 feet, we now know that it is less than 100 years.

   Since we are using the formula $d = rt$, we let

   $t =$ the time in years for ice to reach the lake

   $r =$ rate or speed of ice

   $d =$ distance from beginning of glacier to lake

2. TRANSLATE. To translate to an equation, we use the formula $d = rt$ and let distance $d = 31{,}680$ feet and rate $r = 400$ feet per year.

$$d = r \cdot t$$

$$31{,}680 = 400 \cdot t \quad \text{Let } d = 31{,}680 \text{ and } r = 400.$$

3. SOLVE. Solve the equation for t. To solve for t, divide both sides by 400.

$$\frac{31{,}680}{400} = \frac{400 \cdot t}{400} \quad \text{Divide both sides by 400.}$$

$$79.2 = t \quad \text{Simplify.}$$

4. INTERPRET.

Check: To check, substitute 79.2 for t and 400 for r in the distance formula and check to see that the distance is 31,680 feet.

State: It takes 79.2 years for the ice at the head of Portage Glacier to reach the lake. □

> **Helpful Hint**
> Don't forget to include units if appropriate.

PRACTICE

1 The Stromboli Volcano, in Italy, began erupting in 2002 after a dormant period of over 17 years. In 2007, a volcanologist measured the lava flow to be moving at 5 meters/second. If the path the lava follows to the sea is 580 meters long, how long does it take the lava to reach the sea? *(Source:* Thorsten Boeckel and CNN*)*

EXAMPLE 2 Calculating the Length of a Garden

Charles Pecot can afford enough fencing to enclose a rectangular garden with a perimeter of 140 feet. If the width of his garden must be 30 feet, find the length.

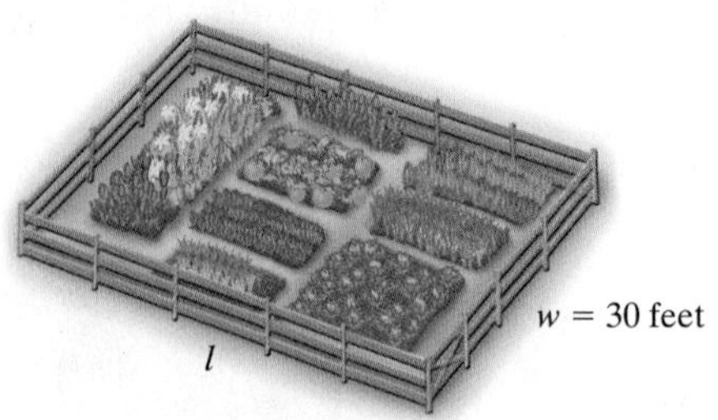

Solution

1. UNDERSTAND. Read and reread the problem. The formula needed to solve this problem is the formula for the perimeter of a rectangle, $P = 2l + 2w$. Before continuing, let's become familiar with this formula.

$l =$ the length of the rectangular garden
$w =$ the width of the rectangular garden
$P =$ perimeter of the garden

2. TRANSLATE. To translate to an equation, we use the formula $P = 2l + 2w$ and let perimeter $P = 140$ feet and width $w = 30$ feet.

$$P = 2l + 2w$$
$$\downarrow \qquad \searrow$$
$$140 = 2l + 2(30) \quad \text{Let } P = 140 \text{ and } w = 30.$$

3. SOLVE.

$$140 = 2l + 2(30)$$
$$140 = 2l + 60 \quad \text{Multiply } 2(30).$$
$$140 - 60 = 2l + 60 - 60 \quad \text{Subtract 60 from both sides.}$$
$$80 = 2l \quad \text{Combine like terms.}$$
$$40 = l \quad \text{Divide both sides by 2.}$$

4. INTERPRET.

Check: Substitute 40 for l and 30 for w in the perimeter formula and check to see that the perimeter is 140 feet.

State: The length of the rectangular garden is 40 feet. □

PRACTICE

2 Evelyn Gryk fenced in part of her backyard for a dog run. The dog run was 40 feet in length and used 98 feet of fencing. Find the width of the dog run.

EXAMPLE 3 Finding an Equivalent Temperature

The average minimum temperature for July in Shanghai, China, is 77° Fahrenheit. Find the equivalent temperature in degrees Celsius.

Solution

1. UNDERSTAND. Read and reread the problem. A formula that can be used to solve this problem is the formula for converting degrees Celsius to degrees Fahrenheit, $F = \frac{9}{5}C + 32$. Before continuing, become familiar with this formula. Using this formula, we let

$C =$ temperature in degrees Celsius, and
$F =$ temperature in degrees Fahrenheit.

2. TRANSLATE. To translate to an equation, we use the formula $F = \frac{9}{5}C + 32$ and let degrees Fahrenheit $F = 77$.

$$\text{Formula:} \quad F = \frac{9}{5}C + 32$$

$$\text{Substitute:} \quad 77 = \frac{9}{5}C + 32 \quad \text{Let } F = 77.$$

3. SOLVE.

$$77 = \frac{9}{5}C + 32$$

$$77 - 32 = \frac{9}{5}C + 32 - 32 \quad \text{Subtract 32 from both sides.}$$

$$45 = \frac{9}{5}C \quad \text{Combine like terms.}$$

$$\frac{5}{9} \cdot 45 = \frac{5}{9} \cdot \frac{9}{5}C \quad \text{Multiply both sides by } \frac{5}{9}.$$

$$25 = C \quad \text{Simplify.}$$

4. INTERPRET.

Check: To check, replace C with 25 and F with 77 in the formula and see that a true statement results.

State: Thus, 77° Fahrenheit is equivalent to 25° Celsius.

Note: There is a formula for directly converting degrees Fahrenheit to degrees Celsius. It is $C = \frac{5}{9}(F - 32)$, as we shall see in Example 8. □

PRACTICE

3 The average minimum temperature for July in Sydney, Australia, is 8° Celsius. Find the equivalent temperature in degrees Fahrenheit.

In the next example, we again use the formula for perimeter of a rectangle as in Example 2. In Example 2, we knew the width of the rectangle. In this example, both the length and width are unknown.

△ EXAMPLE 4 Finding Road Sign Dimensions

The length of a rectangular road sign is 2 feet less than three times its width. Find the dimensions if the perimeter is 28 feet.

Solution

1. UNDERSTAND. Read and reread the problem. Recall that the formula for the perimeter of a rectangle is $P = 2l + 2w$. Draw a rectangle and guess the solution. If the width of the rectangular sign is 5 feet, its length is 2 feet less than 3 times the width or $3(5 \text{ feet}) - 2 \text{ feet} = 13 \text{ feet}$. The perimeter P of the rectangle is then $2(13 \text{ feet}) + 2(5 \text{ feet}) = 36 \text{ feet}$, too much. We now know that the width is less than 5 feet.

Proposed rectangle: 5 feet, 13 feet

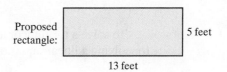

Let

$w =$ the width of the rectangular sign; then
$3w - 2 =$ the length of the sign.

w | $3w - 2$

Draw a rectangle and label it with the assigned variables.

2. TRANSLATE.

Formula: $P = 2l + 2w$ or
Substitute: $28 = 2(3w - 2) + 2w$.

3. SOLVE.

$$\begin{aligned} 28 &= 2(3w - 2) + 2w \\ 28 &= 6w - 4 + 2w && \text{Apply the distributive property.} \\ 28 &= 8w - 4 \\ 28 + 4 &= 8w - 4 + 4 && \text{Add 4 to both sides.} \\ 32 &= 8w \\ \frac{32}{8} &= \frac{8w}{8} && \text{Divide both sides by 8.} \\ 4 &= w \end{aligned}$$

4. INTERPRET.

Check: If the width of the sign is 4 feet, the length of the sign is 3(4 feet) − 2 feet = 10 feet. This gives a perimeter of $P = 2(4 \text{ feet}) + 2(10 \text{ feet}) = 28$ feet, the correct perimeter.

State: The width of the sign is 4 feet, and the length of the sign is 10 feet. □

PRACTICE

4 The new street signs along Route 114 have a length that is 3 inches more than 5 times the width. Find the dimensions of the signs if the perimeter of the signs is 66 inches.

OBJECTIVE

2 Solving a Formula for One of Its Variables

We say that the formula

$$F = \frac{9}{5}C + 32$$

is solved for F because F is alone on one side of the equation, and the other side of the equation contains no F's. Suppose that we need to convert many Fahrenheit temperatures to equivalent degrees Celsius. In this case, it is easier to perform this task by solving the formula $F = \frac{9}{5}C + 32$ for C. (See Example 8.) For this reason, it is important to be able to solve an equation for any one of its specified variables. For example, the formula $d = rt$ is solved for d in terms of r and t. We can also solve $d = rt$ for t in terms of d and r. To solve for t, divide both sides of the equation by r.

$$\begin{aligned} d &= rt \\ \frac{d}{r} &= \frac{rt}{r} && \text{Divide both sides by } r. \\ \frac{d}{r} &= t && \text{Simplify.} \end{aligned}$$

To solve a formula or an equation for a specified variable, we use the same steps as for solving a linear equation. These steps are listed next.

Solving Equations for a Specified Variable

Step 1. Multiply on both sides to clear the equation of fractions if they occur.

Step 2. Use the distributive property to remove parentheses if they occur.

Step 3. Simplify each side of the equation by combining like terms.

Step 4. Get all terms containing the specified variable on one side and all other terms on the other side by using the addition property of equality.

Step 5. Get the specified variable alone by using the multiplication property of equality.

EXAMPLE 5 Solve $V = lwh$ for l.

Solution This formula is used to find the volume of a box. To solve for l, divide both sides by wh.

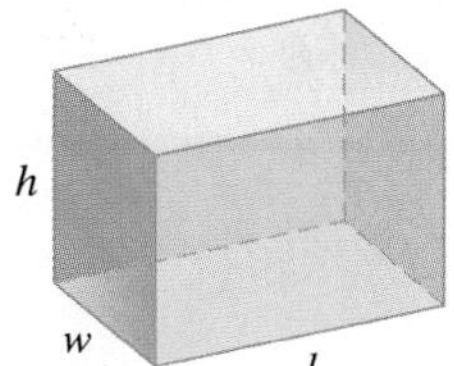

$$V = lwh$$

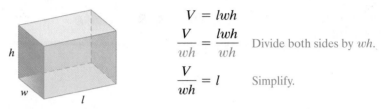

$$\frac{V}{wh} = \frac{lwh}{wh} \quad \text{Divide both sides by } wh.$$

$$\frac{V}{wh} = l \quad \text{Simplify.}$$

Since we have l alone on one side of the equation, we have solved for l in terms of V, w, and h. Remember that it does not matter on which side of the equation we isolate the variable. □

PRACTICE 5 Solve $I = Prt$ for r.

EXAMPLE 6 Solve $y = mx + b$ for x.

Solution The term containing the variable we are solving for, mx, is on the right side of the equation. Get mx alone by subtracting b from both sides.

$$y = mx + b$$

$$y - b = mx + b - b \quad \text{Subtract } b \text{ from both sides.}$$

$$y - b = mx \quad \text{Combine like terms.}$$

Next, solve for x by dividing both sides by m.

$$\frac{y-b}{m} = \frac{mx}{m}$$

$$\frac{y-b}{m} = x \quad \text{Simplify.}$$ □

PRACTICE 6 Solve $H = 5as + 10a$ for s.

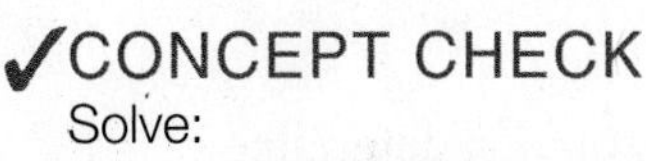

✓CONCEPT CHECK

Solve:

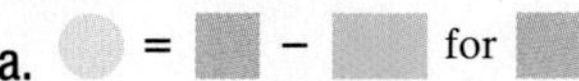

a. ● = ■ − ■ for ■

b. ● = ■ · ▲ − ■ for ■

EXAMPLE 7 Solve $P = 2l + 2w$ for w.

Solution This formula relates the perimeter of a rectangle to its length and width. Find the term containing the variable w. To get this term, $2w$, alone subtract $2l$ from both sides.

Answers to Concept Check:

a. ■ = ● + ■ b. ■ = (● + ■) / ▲

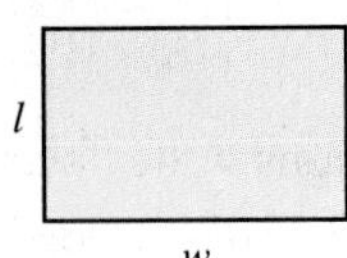

> **Helpful Hint**
> The 2's may *not* be divided out here. Although 2 is a factor of the denominator, 2 is *not* a factor of the numerator since it is not a factor of both terms in the numerator.

$$P = 2l + 2w$$

$$P - 2l = 2l + 2w - 2l \quad \text{Subtract } 2l \text{ from both sides.}$$

$$P - 2l = 2w \quad \text{Combine like terms.}$$

$$\frac{P - 2l}{2} = \frac{2w}{2} \quad \text{Divide both sides by 2.}$$

$$\frac{P - 2l}{2} = w \quad \text{Simplify.}$$

□

PRACTICE 7 Solve $N = F + d(n - 1)$ for d.

The next example has an equation containing a fraction. We will first clear the equation of fractions and then solve for the specified variable.

EXAMPLE 8 Solve $F = \frac{9}{5}C + 32$ for C.

Solution

$$F = \frac{9}{5}C + 32$$

$$5(F) = 5\left(\frac{9}{5}C + 32\right) \quad \text{Clear the fraction by multiplying both sides by the LCD.}$$

$$5F = 9C + 160 \quad \text{Distribute the 5.}$$

$$5F - 160 = 9C + 160 - 160 \quad \text{To get the term containing the variable } C \text{ alone, subtract 160 from both sides.}$$

$$5F - 160 = 9C \quad \text{Combine like terms.}$$

$$\frac{5F - 160}{9} = \frac{9C}{9} \quad \text{Divide both sides by 9.}$$

$$\frac{5F - 160}{9} = C \quad \text{Simplify.}$$

Note: An equivalent way to write this formula is $C = \frac{5}{9}(F - 32)$. □

PRACTICE 8 Solve $A = \frac{1}{2}a(b + B)$ for B.

Vocabulary, Readiness & Video Check

Martin-Gay Interactive Videos

See Video 2.6

Watch the section lecture video and answer the following questions.

OBJECTIVE 1
1. Complete this statement based on the lecture given before Example 1. A formula is an equation that describes known ________ among quantities.

OBJECTIVE 1
2. In Example 2, how are the units for the solution determined?

OBJECTIVE 2
3. During Example 4, why is the equation $5x = 30$?

2.6 Exercise Set

MyMathLab®

Substitute the given values into each given formula and solve for the unknown variable. If necessary, round to one decimal place. See Examples 1 through 4.

1. $A = bh$; $A = 45, b = 15$ (Area of a parallelogram)
2. $d = rt$; $d = 195, t = 3$ (Distance formula)
3. $S = 4lw + 2wh$; $S = 102, l = 7, w = 3$ (Surface area of a special rectangular box)
4. $V = lwh$; $l = 14, w = 8, h = 3$ (Volume of a rectangular box)
5. $A = \frac{1}{2}h(B + b)$; $A = 180, B = 11, b = 7$ (Area of a trapezoid)
6. $A = \frac{1}{2}h(B + b)$; $A = 60, B = 7, b = 3$ (Area of a trapezoid)
7. $P = a + b + c$; $P = 30, a = 8, b = 10$ (Perimeter of a triangle)
8. $V = \frac{1}{3}Ah$; $V = 45, h = 5$ (Volume of a pyramid)
9. $C = 2\pi r$; $C = 15.7$ (use the approximation 3.14 or a calculator approximation for π) (Circumference of a circle)
10. $A = \pi r^2$; $r = 4.5$ (use the approximation 3.14 or a calculator approximation for π) (Area of a circle)
11. $I = PRT$; $I = 3750, P = 25{,}000, R = 0.05$ (Simple interest formula)
12. $I = PRT$; $I = 1{,}056{,}000, R = 0.055, T = 6$ (Simple interest formula)
13. $V = \frac{1}{3}\pi r^2 h$; $V = 565.2, r = 6$ (use a calculator approximation for π) (Volume of a cone)
14. $V = \frac{4}{3}\pi r^3$; $r = 3$ (use a calculator approximation for π) (Volume of a sphere)

Solve each formula for the specified variable. See Examples 5 through 8.

15. $f = 5gh$ for h
16. $A = \pi ab$ for b
17. $V = lwh$ for w
18. $T = mnr$ for n
19. $3x + y = 7$ for y
20. $-x + y = 13$ for y
21. $A = P + PRT$ for R
22. $A = P + PRT$ for T
23. $V = \frac{1}{3}Ah$ for A
24. $D = \frac{1}{4}fk$ for k
25. $P = a + b + c$ for a
26. $PR = x + y + z + w$ for z
27. $S = 2\pi rh + 2\pi r^2$ for h
28. $S = 4lw + 2wh$ for h

Solve. For Exercises 29 and 30, the solutions have been started for you. See Examples 1 through 4.

29. The iconic NASDAQ sign in New York's Times Square has a width of 84 feet and an area of 10,080 square feet. Find the height (or length) of the sign. (*Source:* livedesignonline.com)

Start the solution:

1. UNDERSTAND the problem. Reread it as many times as needed.
2. TRANSLATE into an equation. (Fill in the blanks below.)

Area	=	length	times	width
↓	↓	↓	↓	↓
___	=	x	·	___

Finish with:

3. SOLVE and 4. INTERPRET

30. The world's largest sign for Coca-Cola is located in Arica, Chile. The rectangular sign has a length of 400 feet and an area of 52,400 square feet. Find the width of the sign. (*Source:* Fabulous Facts about Coca-Cola, Atlanta, GA)

Start the solution:

1. UNDERSTAND the problem. Reread it as many times as needed.
2. TRANSLATE into an equation. (Fill in the blanks below.)

Area	=	length	times	width
↓	↓	↓	↓	↓
___	=	___	·	x

Finish with:

3. SOLVE and 4. INTERPRET

31. For the purpose of purchasing new baseboard and carpet,
    a. Find the area and perimeter of the room below (neglecting doors).
    b. Identify whether baseboard has to do with area or perimeter and the same with carpet.

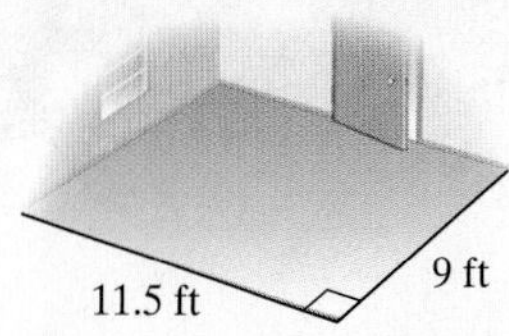

32. For the purpose of purchasing lumber for a new fence and seed to plant grass,
   a. Find the area and perimeter of the yard below.
   b. Identify whether a fence has to do with area or perimeter and the same with grass seed.

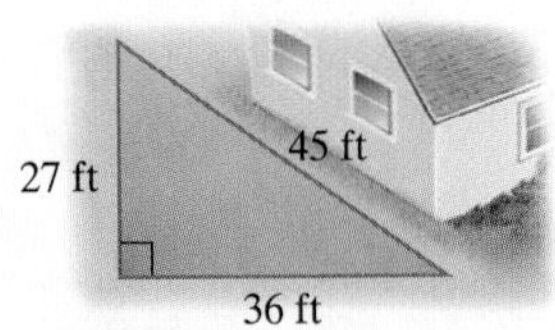

33. A frame shop charges according to both the amount of framing needed to surround the picture and the amount of glass needed to cover the picture.
   a. Find the area and perimeter of the trapezoid-shaped framed picture below.
   b. Identify whether the amount of framing has to do with perimeter or area and the same with the amount of glass.

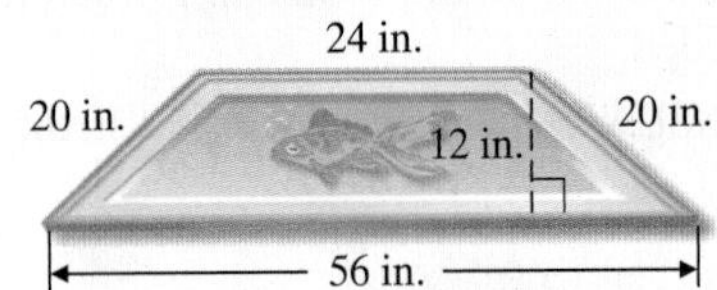

34. A decorator is painting and placing a border completely around the parallelogram-shaped wall.
   a. Find the area and perimeter of the wall below.
   b. Identify whether the border has to do with perimeter or area and the same with paint.

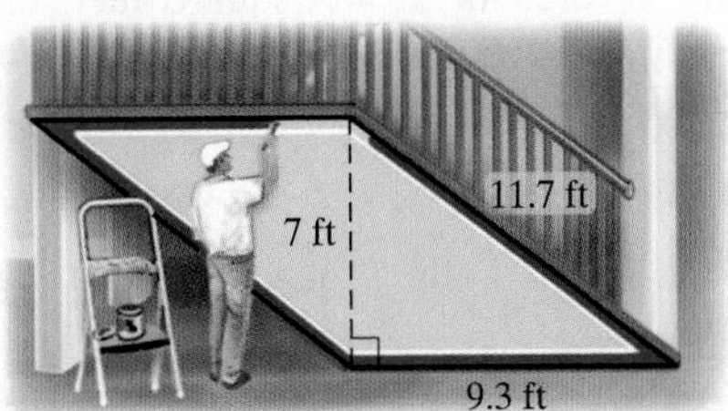

35. Convert Nome, Alaska's 14°F high temperature to Celsius.

36. Convert Paris, France's low temperature of −5°C to Fahrenheit.

△ 37. An architect designs a rectangular flower garden such that the width is exactly two-thirds of the length. If 260 feet of antique picket fencing are to be used to enclose the garden, find the dimensions of the garden.

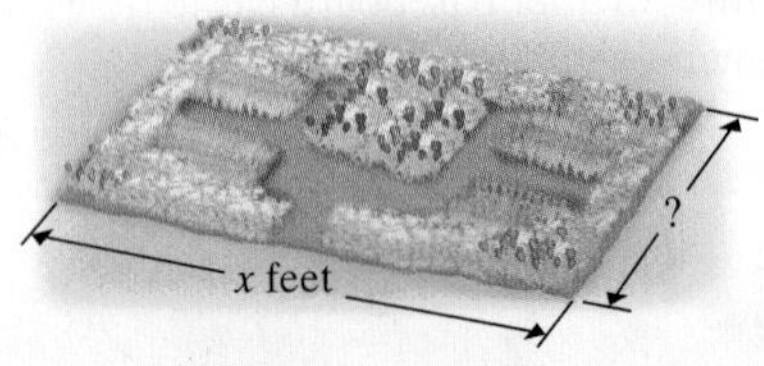

△ 38. If the length of a rectangular parking lot is 10 meters less than twice its width, and the perimeter is 400 meters, find the length of the parking lot.

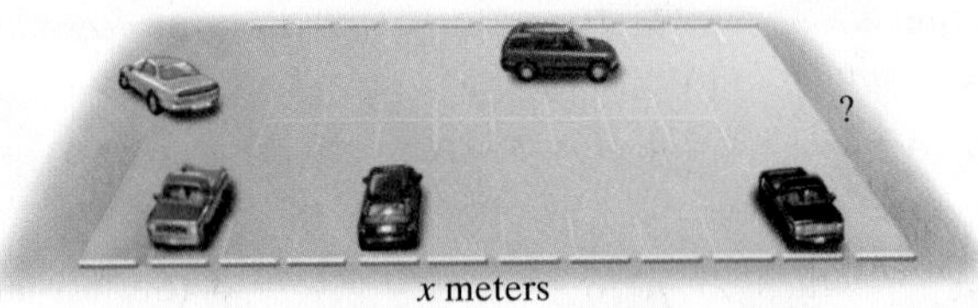

△ 39. A flower bed is in the shape of a triangle with one side twice the length of the shortest side, and the third side is 30 feet more than the length of the shortest side. Find the dimensions if the perimeter is 102 feet.

△ 40. The perimeter of a yield sign in the shape of an isosceles triangle is 22 feet. If the shortest side is 2 feet less than the other two sides, find the length of the shortest side. (*Hint:* An isosceles triangle has two sides the same length.)

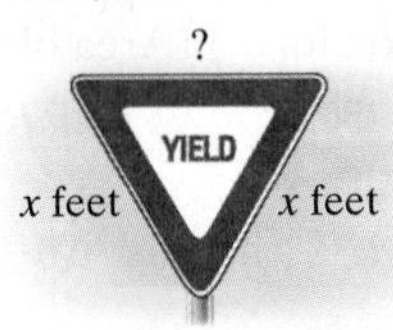

41. The Cat is a high-speed catamaran auto ferry that operates between Bar Harbor, Maine, and Yarmouth, Nova Scotia. The Cat can make the 138-mile trip in about $2\frac{1}{2}$ hours. Find the catamaran speed for this trip. (*Source:* Bay Ferries)

42. A family is planning their vacation to Disney World. They will drive from a small town outside New Orleans, Louisiana, to Orlando, Florida, a distance of 700 miles. They plan to average a rate of 55 mph. How long will this trip take?

△ **43.** Piranha fish require 1.5 cubic feet of water per fish to maintain a healthy environment. Find the maximum number of piranhas you could put in a tank measuring 8 feet by 3 feet by 6 feet.

△ **44.** Find the maximum number of goldfish you can put in a cylindrical tank whose diameter is 8 meters and whose height is 3 meters if each goldfish needs 2 cubic meters of water.

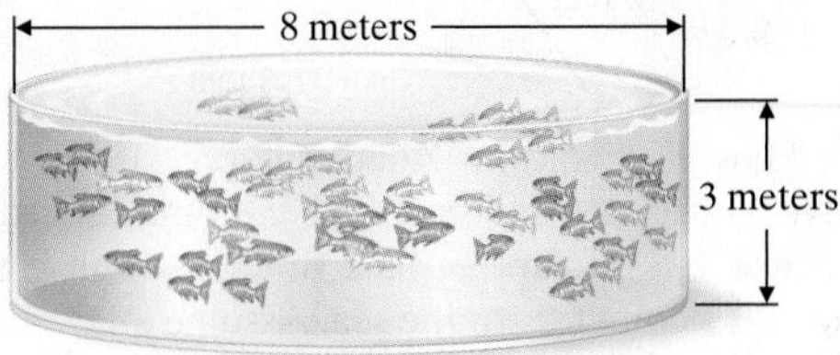

Dolbear's Law states the relationship between the rate at which Snowy Tree crickets chirp and the air temperature of their environment. The formula is

$$T = 50 + \frac{N - 40}{4}, \textit{ where } \quad \begin{array}{l} T = \text{temperature in degrees Fahrenheit and} \\ N = \text{number of chirps per minute} \end{array}$$

45. If $N = 86$, find the temperature in degrees Fahrenheit, T.

46. If $N = 94$, find the temperature in degrees Fahrenheit, T.

47. If $T = 55°\text{F}$, find the number of chirps per minute.

48. If $T = 65°\text{F}$, find the number of chirps per minute.

Use the results of Exercises 45–48 to complete each sentence with "increases" or "decreases."

49. As the number of cricket chirps per minute increases, the air temperature of their environment _______.

50. As the air temperature of their environment decreases, the number of cricket chirps per minute _______.

△ **51.** A lawn is in the shape of a trapezoid with a height of 60 feet and bases of 70 feet and 130 feet. How many whole bags of fertilizer must be purchased to cover the lawn if each bag covers 4000 square feet?

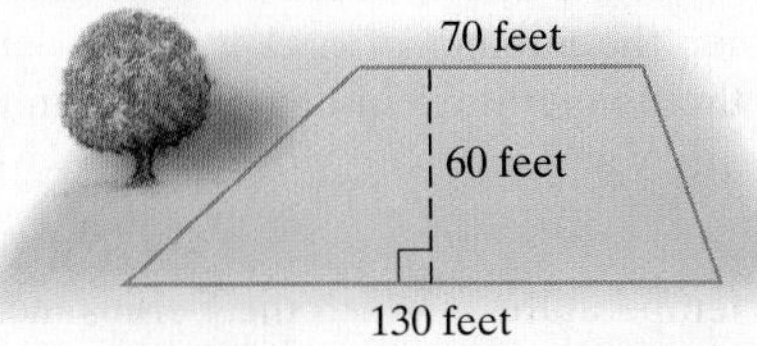

△ **52.** If the area of a right-triangularly shaped sail is 20 square feet and its base is 5 feet, find the height of the sail.

△ **53.** Maria's Pizza sells one 16-inch cheese pizza or two 10-inch cheese pizzas for $9.99. Determine which size gives more pizza.

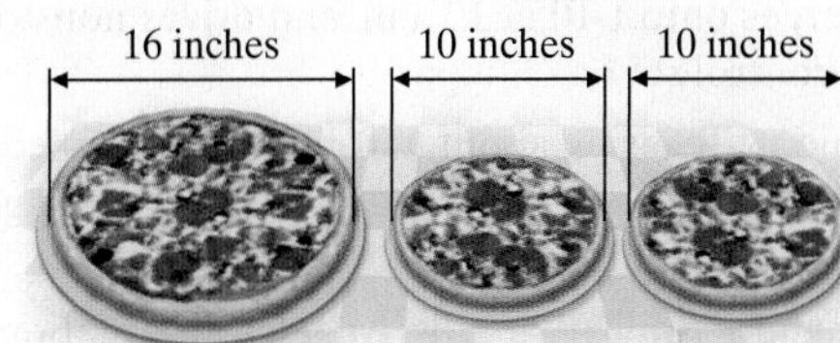

△ **54.** Find how much rope is needed to wrap around the earth at the equator if the radius of the earth is 4000 miles. (*Hint:* Use 3.14 for π and the formula for circumference.)

△ **55.** The perimeter of a geometric figure is the sum of the lengths of its sides. If the perimeter of the following pentagon (five-sided figure) is 48 meters, find the length of each side.

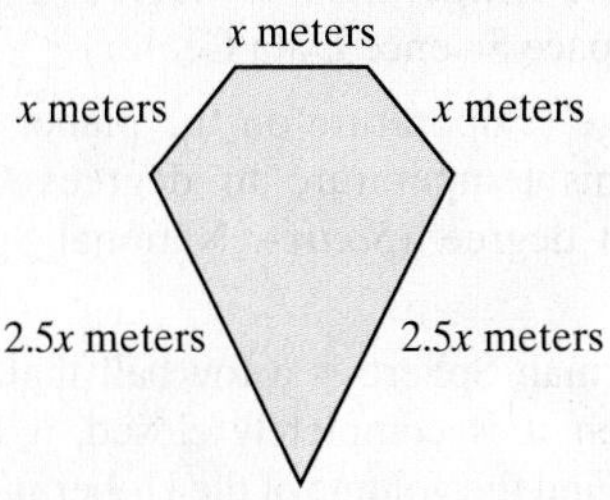

△ **56.** The perimeter of the following triangle is 82 feet. Find the length of each side.

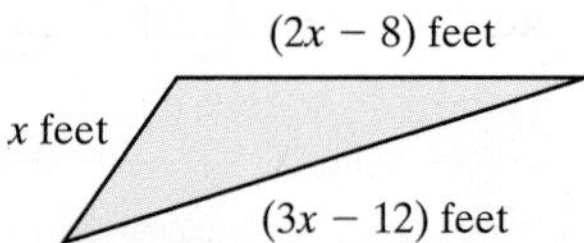

57. The Hawaiian volcano Kilauea is one of the world's most active volcanoes and has had continuous eruptive activity since 1983. Erupting lava flows through a tube system about 11 kilometers to the sea. Assume a lava flow speed of 0.5 kilometer per hour and calculate how long it takes to reach the sea.

58. The world's largest pink ribbon, the sign of the fight against breast cancer, was erected out of pink Post-it® notes on a billboard in New York City in October 2004. If the area of the rectangular billboard covered by the ribbon was approximately 3990 square feet, and the width of the billboard was approximately 57 feet, what was the height of this billboard?

△ **59.** The perimeter of an equilateral triangle is 7 inches more than the perimeter of a square, and the side of the triangle is 5 inches longer than the side of the square. Find the side of the triangle. (*Hint:* An equilateral triangle has three sides the same length.)

△ **60.** A square animal pen and a pen shaped like an equilateral triangle have equal perimeters. Find the length of the sides of each pen if the sides of the triangular pen are fifteen less than twice a side of the square pen.

61. Find how long it takes a person to drive 135 miles on I-10 if she merges onto I-10 at 10 a.m. and drives nonstop with her cruise control set on 60 mph.

62. Beaumont, Texas, is about 150 miles from Toledo Bend. If Leo Miller leaves Beaumont at 4 a.m. and averages 45 mph, when should he arrive at Toledo Bend?

△ **63.** The longest runway at Los Angeles International Airport has the shape of a rectangle and an area of 1,813,500 square feet. This runway is 150 feet wide. How long is the runway? (*Source:* Los Angeles World Airports)

64. Normal room temperature is about 78°F. Convert this temperature to Celsius.

65. The highest temperature ever recorded in Europe was 122°F in Seville, Spain, in August 1881. Convert this record high temperature to Celsius. (*Source:* National Climatic Data Center)

66. The lowest temperature ever recorded in Oceania was −10°C at the Haleakala Summit in Maui, Hawaii, in January 1961. Convert this record low temperature to Fahrenheit. (*Source:* National Climatic Data Center)

67. The average temperature on the planet Mercury is 167°C. Convert this temperature to degrees Fahrenheit. (*Source:* National Space Science Data Center)

68. The average temperature on the planet Jupiter is −227°F. Convert this temperature to degrees Celsius. Round to the nearest degree. (*Source:* National Space Science Data Center)

△ **69.** The Hoberman Sphere is a toy ball that expands and contracts. When it is completely closed, it has a diameter of 9.5 inches. Find the volume of the Hoberman Sphere when it is completely closed. Use 3.14 for π. Round to the nearest whole cubic inch. (*Hint:* Volume of a sphere $= \frac{4}{3}\pi r^3$. *Source:* Hoberman Designs, Inc.)

△ **70.** When the Hoberman Sphere (see Exercise 69) is completely expanded, its diameter is 30 inches. Find the volume of the Hoberman Sphere when it is completely expanded. Use 3.14 for π. (*Source:* Hoberman Designs, Inc.)

REVIEW AND PREVIEW

Write each percent as a decimal.

71. 32%

72. 8%

73. 200%

74. 0.5%

Write each decimal as a percent.

75. 0.17

76. 0.03

77. 7.2

78. 5

CONCEPT EXTENSIONS

△ **79.** The formula $V = lwh$ is used to find the volume of a box. If the length of a box is doubled, the width is doubled, and the height is doubled, how does this affect the volume? Explain your answer.

△ **80.** The formula $A = bh$ is used to find the area of a parallelogram. If the base of a parallelogram is doubled and its height is doubled, how does this affect the area? Explain your answer.

81. Use the Dolbear's Law formula for Exercises 45–48 and calculate when the number of cricket chirps per minute is the same as the temperature in degrees Fahrenheit. (*Hint:* Replace T with N and solve for N or replace N with T and solve for T.)

82. Find the temperature at which the Celsius measurement and the Fahrenheit measurement are the same number.

Solve.

83. $N = R + \frac{V}{G}$ for V (Urban forestry: tree plantings per year)

84. $B = \frac{F}{P - V}$ for V (Business: break-even point)

Solve. See the Concept Check in this section.

85. ■ − ● · ▮ = ▲ for ●

86. ⬟ · ■ + ▲ = ● for ■

87. The distance from the sun to the earth is approximately 93,000,000 miles. If light travels at a rate of 186,000 miles per second, how long does it take light from the sun to reach us?

88. Light travels at a rate of 186,000 miles per second. If our moon is 238,860 miles from the earth, how long does it take light from the moon to reach us? (Round to the nearest tenth of a second.)

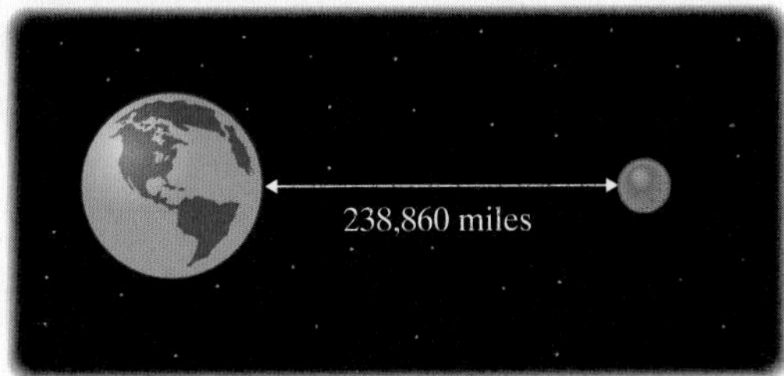

89. A glacier is a giant mass of rocks and ice that flows downhill like a river. Exit Glacier, near Seward, Alaska, moves at a rate of 20 inches a day. Find the distance in feet the glacier moves in a year. (Assume 365 days a year. Round to 2 decimal places.)

90. Flying fish do not *actually* fly, but glide. They have been known to travel a distance of 1300 feet at a rate of 20 miles per hour. How many seconds does it take to travel this distance? (*Hint:* First convert miles per hour to feet per second. Recall that 1 mile = 5280 feet. Round to the nearest tenth of a second.)

91. A Japanese "bullet" train set a new world record for train speed at 581 kilometers per hour during a manned test run on the Yamanashi Maglev Test Line in 2003. The Yamanashi Maglev Test Line is 42.8 kilometers long. How many *minutes* would a test run on the Yamanashi Line last at this record-setting speed? Round to the nearest hundredth of a minute. (*Source:* Japan Railways Central Co.)

92. The Boeing X-51 is an unmanned demonstration aircraft for hypersonic flight testing. In May 2010, it successfully completed a free flight at about 3800 mph. Neglecting altitude, if the circumference of Earth is approximately 25,000 miles, how long would it take for the X-51 to travel around Earth? Give your answer in hours and minutes rounded to the nearest whole minute.

93. In the United States, a notable hang glider flight was a 303-mile, $8\frac{1}{2}$-hour flight from New Mexico to Kansas. What was the average rate during this flight?

94. Stalactites join stalagmites to form columns. A column found at Natural Bridge Caverns near San Antonio, Texas, rises 15 feet and has a *diameter* of only 2 inches. Find the volume of this column in cubic inches. (*Hint:* Use the formula for volume of a cylinder and use a calculator approximation for π. Round to the nearest tenth of an inch.)

2.7 Percent and Mixture Problem Solving

OBJECTIVES

1 Solve Percent Equations.

2 Solve Discount and Mark-up Problems.

3 Solve Percent Increase and Percent Decrease Problems.

4 Solve Mixture Problems.

This section is devoted to solving problems in the categories listed. The same problem-solving steps used in previous sections are also followed in this section. They are listed below for review.

General Strategy for Problem Solving

1. UNDERSTAND the problem. During this step, become comfortable with the problem. Some ways of doing this are as follows:
   Read and reread the problem.
   Choose a variable to represent the unknown.
   Construct a drawing whenever possible.
   Propose a solution and check. Pay careful attention to how you check your proposed solution. This will help writing an equation to model the problem.
2. TRANSLATE the problem into an equation.
3. SOLVE the equation.
4. INTERPRET the results: *Check* the proposed solution in the stated problem and *state* your conclusion.

OBJECTIVE

1 Solving Percent Equations 

Many of today's statistics are given in terms of percent: a basketball player's free throw percent, current interest rates, stock market trends, and nutrition labeling, just to name a few. In this section, we first explore percent, percent equations, and applications involving percents. See Appendix E if a further review of percents is needed.

EXAMPLE 1 The number 63 is what percent of 72?

Solution

1. UNDERSTAND. Read and reread the problem. Next, let's suppose that the percent is 80%. To check, we find 80% of 72.

$$80\% \text{ of } 72 = 0.80(72) = 57.6$$

This is close but not 63. At this point, though, we have a better understanding of the problem, we know the correct answer is close to and greater than 80%, and we know how to check our proposed solution later.

Let $x =$ the unknown percent.

2. TRANSLATE. Recall that "is" means "equals" and "of" signifies multiplying. Let's translate the sentence directly.

the number 63	is	what percent	of	72
↓	↓	↓	↓	↓
63	=	x	·	72

3. SOLVE.

$$63 = 72x$$
$$0.875 = x \quad \text{Divide both sides by 72.}$$
$$87.5\% = x \quad \text{Write as a percent.}$$

4. INTERPRET.

Check: Verify that 87.5% of 72 is 63.

State: The number 63 is 87.5% of 72. □

PRACTICE

1 The number 35 is what percent of 56?

EXAMPLE 2 The number 120 is 15% of what number?

Solution

1. UNDERSTAND. Read and reread the problem.

Let $x =$ the unknown number.

2. TRANSLATE.

the number 120	is	15%	of	what number
↓	↓	↓	↓	↓
120	=	15%	·	x

3. SOLVE.

$$120 = 0.15x \quad \text{Write 15\% as 0.15.}$$
$$800 = x \quad \text{Divide both sides by 0.15.}$$

4. INTERPRET.

Check: Check the proposed solution by finding 15% of 800 and verifying that the result is 120.

State: Thus, 120 is 15% of 800. □

PRACTICE
2 The number 198 is 55% of what number?

Helpful Hint
The percents in a circle graph should have a sum of 100%.

The next example contains a circle graph. This particular circle graph shows percents of pets owned in the United States. Since the circle graph represents all pets owned in the United States, the percents should add to 100%.

EXAMPLE 3 The circle graph below shows the breakdown of total pets owned in the United States. Use this graph to answer the questions.

Pets Owned in the United States

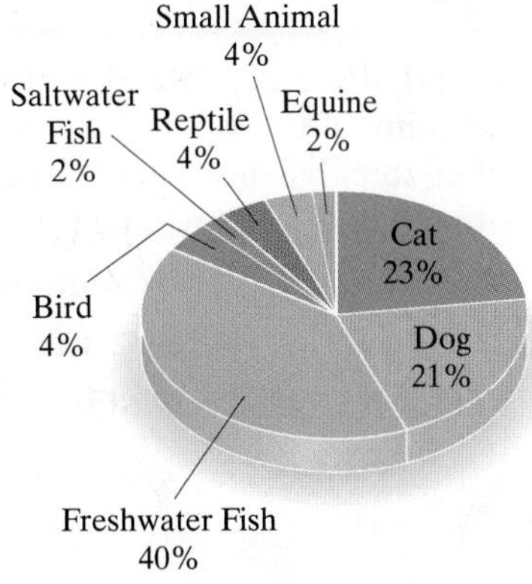

Data from American Pet Products Association's Industry Statistics and Trends results

a. What percent of pets owned in the United States are cats or dogs?

b. What percent of pets owned in the United States are not birds?

c. Currently, 377.41 million pets are owned in the United States. How many of these would be cats? (Round to the nearest tenth of a million.)

Solution

a. From the circle graph, we see that 23% of pets owned are cats and 21% are dogs; thus,

$$23\% + 21\% = 44\%.$$

44% of pets owned are cats or dogs.

b. The circle graph percents have a sum of 100%; thus, the percent of pets owned in the United States that are not birds is

$$100\% - 4\% = 96\%.$$

c. To find the number of cats owned, we find

$$\begin{aligned} 23\% \text{ of } 377.41 &= 0.23(377.41) \\ &= 86.8043 \\ &\approx 86.8 \quad \text{Rounded to the nearest tenth of a million.} \end{aligned}$$

Thus, about 86.8 million cats are owned in the United States. □

PRACTICE
3 Use the Example 3 circle graph to answer each question.

a. What percent of pets owned in the United States are freshwater fish or saltwater fish?

b. What percent of pets owned in the United States are not equines (horses, ponies, etc.)?

c. Currently, 377.41 million pets are owned in the United States. How many of these would be dogs? (Round to the nearest tenth of a million.)

OBJECTIVE

2 Solving Discount and Mark-up Problems 

The next example has to do with discounting the price of a cell phone.

EXAMPLE 4 Cell Phones Unlimited recently reduced the price of a $140 phone by 20%. What are the discount and the new price?

Solution

1. UNDERSTAND. Read and reread the problem. Make sure you understand the meaning of the word *discount*. Discount is the amount of money by which the cost of an item has been decreased. To find the discount, we simply find 20% of $140. In other words, we have the formulas,

$$\text{discount} = \text{percent} \cdot \text{original price} \quad \text{Then}$$
$$\text{new price} = \text{original price} - \text{discount}$$

2, 3. TRANSLATE and SOLVE.

$$\begin{aligned} \text{discount} &= \text{percent} \cdot \text{original price} \\ &\quad\;\downarrow \qquad\qquad \downarrow \\ &= 20\% \cdot \$140 \\ &= 0.20 \cdot \$140 \\ &= \$28 \end{aligned}$$

Thus, the discount in price is $28.

$$\begin{aligned} \text{new price} &= \text{original price} - \text{discount} \\ &\qquad\quad \downarrow \qquad\qquad\quad \downarrow \\ &= \$140 - \$28 \\ &= \$112 \end{aligned}$$

4. INTERPRET.

Check: Check your calculations in the formulas; see also whether our results are reasonable. They are.

State: The discount in price is $28, and the new price is $112. □

PRACTICE

4 A used treadmill, originally purchased for $480, was sold at a garage sale at a discount of 85% of the original price. What were the discount and the new price?

A concept similar to discount is mark-up. What is the difference between the two? A discount is subtracted from the original price while a mark-up is added to the original price. For mark-ups,

$$\text{mark-up} = \text{percent} \cdot \text{original price}$$
$$\text{new price} = \text{original price} + \text{mark-up}$$

> **Helpful Hint**
> Discounts are subtracted from the original price, while mark-ups are added.

Mark-up exercises can be found in Exercise Set 2.7 in the form of calculating tips.

OBJECTIVE

3 Solving Percent Increase and Percent Decrease Problems

Percent increase or percent decrease is a common way to describe how some measurement has increased or decreased. For example, crime increased by 8%, teachers received a 5.5% increase in salary, or a company decreased its number of employees by 10%. The next example is a review of percent increase.

EXAMPLE 5 **Calculating the Percent Increase of Attending College**

The tuition and fees cost of attending a public four-year college rose from \$4020 in 1996 to \$7610 in 2011. Find the percent increase. Round to the nearest tenth of a percent. (*Source:* The College Board)

Solution

1. UNDERSTAND. Read and reread the problem. Notice that the new tuition, \$7610, is almost double the old tuition of \$4020. Because of that, we know that the percent increase is close to 100%. To see this, let's guess that the percent increase is 100%. To check, we find 100% of \$4020 to find the *increase* in cost. Then we add this increase to \$4020 to find the *new cost.* In other words, 100%(\$4020) = 1.00(\$4020) = \$4020, the *increase* in cost. The *new cost* would be old cost + increase = \$4020 + \$4020 = \$8040, close to the actual new cost of \$7610. We now know that the increase is close to, but less than, 100% and we know how to check our proposed solution.

   Let $x =$ the percent increase.

2. TRANSLATE. First, find the **increase** and then the **percent increase.** The increase in cost is found by:

In words:	increase	=	new cost	−	old cost	or
Translate:	increase	=	\$7610	−	\$4020	
		= \$3590				

   Next, find the percent increase. The percent increase or percent decrease is always a percent of the original number or, in this case, the old cost.

In words:	increase	is	what percent	of	old cost
Translate:	\$3590	=	x	·	\$4020

3. SOLVE.

$3590 = 4020x$

$0.893 \approx x$ Divide both sides by 4020 and round to 3 decimal places.

$89.3\% \approx x$ Write as a percent.

4. INTERPRET.

Check: Check the proposed solution

State: The percent increase in cost is approximately 89.3%. □

PRACTICE

5 The tuition and fees cost of attending a public two-year college rose from \$1900 in 1996 to \$2710 in 2011. Find the percent increase. Round to the nearest tenth of a percent. (*Source:* The College Board)

Percent decrease is found using a method similar to that in Example 5. First find the decrease, then determine what percent of the original or first amount is that decrease.

Read the next example carefully. For Example 5, we were asked to find percent increase. In Example 6, we are given the percent increase and asked to find the number before the increase.

EXAMPLE 6 Most of the movie screens globally project analog film, but the number of cinemas using digital is increasing. Find the number of digital screens worldwide last year if, after a 122% increase, the number this year is 36,208. Round to the nearest whole number. (*Source:* Motion Picture Association of America)

Solution

1. UNDERSTAND. Read and reread the problem. Let's guess a solution and see how we would check our guess. If the number of digital screens worldwide last year was 20,000, we would see if 20,000 plus the increase is 36,208; that is,

$$20{,}000 + 122\%(20{,}000) = 20{,}000 + 1.22(20{,}000) = 2.22(20{,}000) = 44{,}400$$

Since 44,400 is too large, we know that our guess of 20,000 is too large. We also have a better understanding of the problem. Let

$$x = \text{number of digital screens last year}$$

2. TRANSLATE. To translate an equation, we remember that

In words:	number of digital screens last year	plus	increase	equals	number of digital screens this year
	↓	↓	↓	↓	↓
Translate:	x	$+$	$1.22x$	$=$	36,208

3. SOLVE.

$$2.22x = 36{,}208 \quad \text{Add like terms.}$$

$$x = \frac{36{,}208}{2.22}$$

$$x \approx 16{,}310$$

4. INTERPRET.

Check: Recall that x represents the number of digital screens worldwide last year. If this number is approximately 16,310, let's see if 16,310 plus the increase is close to 36,208. (We use the word *close* since 16,310 is rounded.)

$$16{,}310 + 122\%(16{,}310) = 16{,}310 + 1.22(16{,}310) = 2.22(16{,}310) = 36{,}208.2$$

which is close to 36,208.

State: There were approximately 16,310 digital screens worldwide last year. □

PRACTICE

6 The fastest-growing sector of digital theater screens is 3D. Find the number of digital 3D screens in the United States and Canada last year if after a 138% increase, the number this year is 8459. Round to the nearest whole. (*Source:* MPAA)

OBJECTIVE

4 Solving Mixture Problems 

Mixture problems involve two or more quantities being combined to form a new mixture. These applications range from Dow Chemical's need to form a chemical mixture of a required strength to Planter's Peanut Company's need to find the correct mixture of peanuts and cashews, given taste and price constraints.

EXAMPLE 7 **Calculating Percent for a Lab Experiment**

A chemist working on his doctoral degree at Massachusetts Institute of Technology needs 12 liters of a 50% acid solution for a lab experiment. The stockroom has only 40% and 70% solutions. How much of each solution should be mixed together to form 12 liters of a 50% solution?

Solution:

1. UNDERSTAND. First, read and reread the problem a few times. Next, guess a solution. Suppose that we need 7 liters of the 40% solution. Then we need $12 - 7 = 5$ liters of the 70% solution. To see if this is indeed the solution, find the amount of pure acid in 7 liters of the 40% solution, in 5 liters of the 70% solution, and in 12 liters of a 50% solution, the required amount and strength.

number of liters	×	acid strength	=	amount of pure acid
↓		↓		↓
7 liters	×	40%		7(0.40) or 2.8 liters
5 liters	×	70%		5(0.70) or 3.5 liters
12 liters	×	50%		12(0.50) or 6 liters

Since 2.8 liters + 3.5 liters = 6.3 liters and not 6, our guess is incorrect, but we have gained some valuable insight into how to model and check this problem.

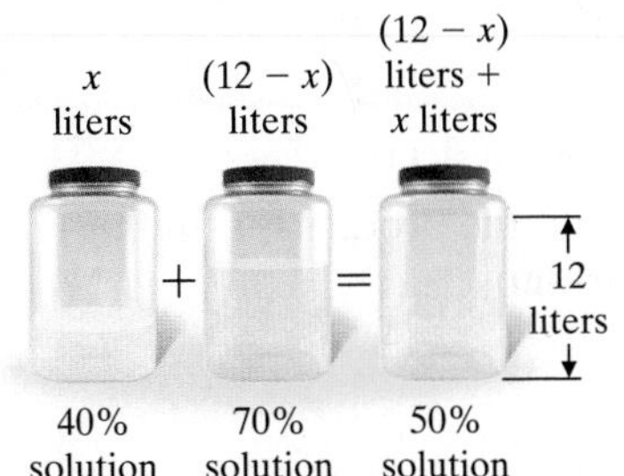

Let

$$x = \text{number of liters of 40\% solution; then}$$
$$12 - x = \text{number of liters of 70\% solution.}$$

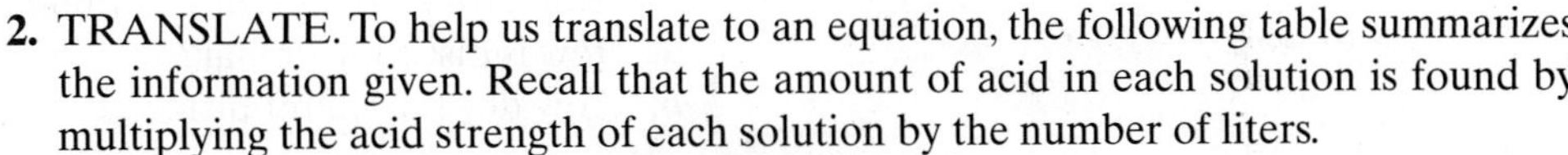

2. TRANSLATE. To help us translate to an equation, the following table summarizes the information given. Recall that the amount of acid in each solution is found by multiplying the acid strength of each solution by the number of liters.

	No. of Liters ·	***Acid Strength*** =	***Amount of Acid***
40% Solution	x	40%	$0.40x$
70% Solution	$12 - x$	70%	$0.70(12 - x)$
50% Solution Needed	12	50%	$0.50(12)$

The amount of acid in the final solution is the sum of the amounts of acid in the two beginning solutions.

In words:	acid in 40% solution	+	acid in 70% solution	=	acid in 50% mixture
	↓		↓		↓
Translate:	$0.40x$	+	$0.70(12 - x)$	=	$0.50(12)$

3. SOLVE.

$$0.40x + 0.70(12 - x) = 0.50(12)$$
$$0.4x + 8.4 - 0.7x = 6 \qquad \text{Apply the distributive property.}$$
$$-0.3x + 8.4 = 6 \qquad \text{Combine like terms.}$$
$$-0.3x = -2.4 \qquad \text{Subtract 8.4 from both sides.}$$
$$x = 8 \qquad \text{Divide both sides by } -0.3.$$

4. INTERPRET.

Check: To check, recall how we checked our guess.

State: If 8 liters of the 40% solution are mixed with $12 - 8$ or 4 liters of the 70% solution, the result is 12 liters of a 50% solution. □

PRACTICE

7 Hamida Barash was responsible for refilling the eye wash stations in the chemistry lab. She needed 6 liters of 3% strength eyewash to refill the dispensers. The supply room only had 2% and 5% eyewash in stock. How much of each solution should she mix to produce the needed 3% strength eyewash?

Vocabulary, Readiness & Video Check

Tell whether the percent labels in the circle graphs are correct.

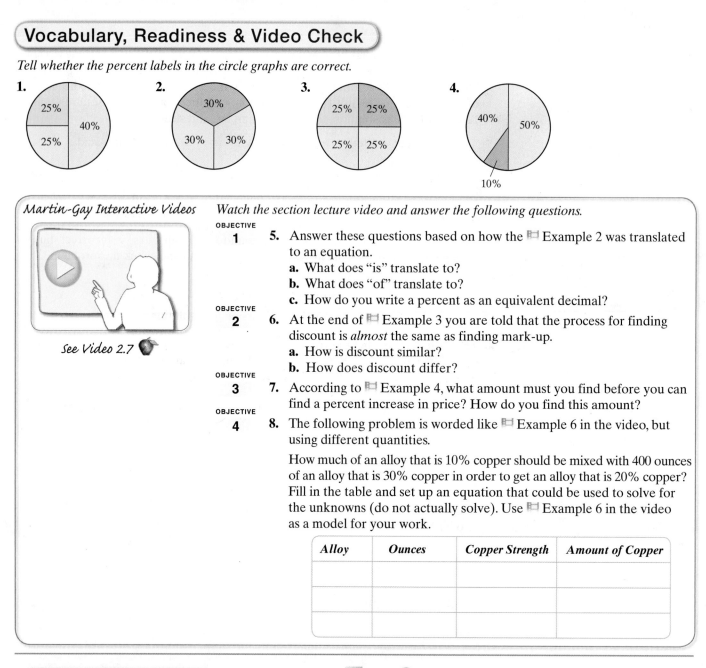

1.

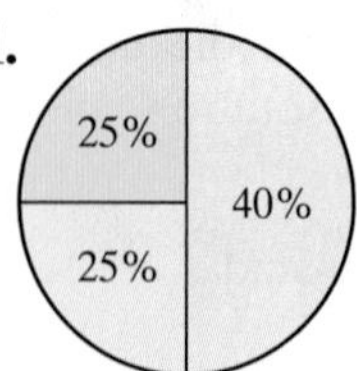

2.

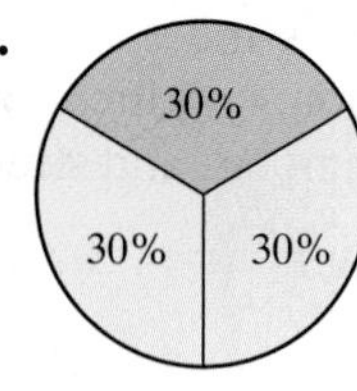

3.

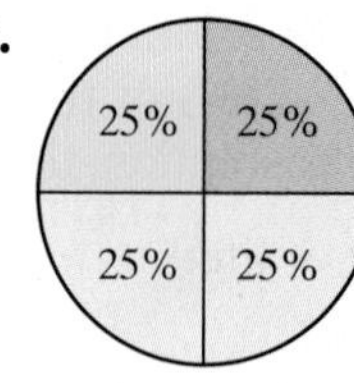

4.

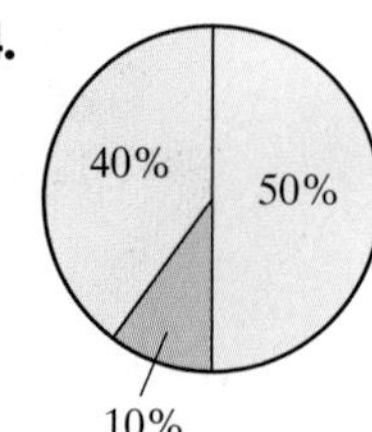

Martin-Gay Interactive Videos

See Video 2.7

Watch the section lecture video and answer the following questions.

OBJECTIVE 1

5. Answer these questions based on how the Example 2 was translated to an equation.
a. What does "is" translate to?
b. What does "of" translate to?
c. How do you write a percent as an equivalent decimal?

OBJECTIVE 2

6. At the end of Example 3 you are told that the process for finding discount is *almost* the same as finding mark-up.
a. How is discount similar?
b. How does discount differ?

OBJECTIVE 3

7. According to Example 4, what amount must you find before you can find a percent increase in price? How do you find this amount?

OBJECTIVE 4

8. The following problem is worded like Example 6 in the video, but using different quantities.

How much of an alloy that is 10% copper should be mixed with 400 ounces of an alloy that is 30% copper in order to get an alloy that is 20% copper? Fill in the table and set up an equation that could be used to solve for the unknowns (do not actually solve). Use Example 6 in the video as a model for your work.

Alloy	*Ounces*	*Copper Strength*	*Amount of Copper*

2.7 Exercise Set

MyMathLab®

Find each number described. For Exercises 1 and 2, the solutions have been started for you. See Examples 1 and 2.

1. What number is 16% of 70?

Start the solution:

1. UNDERSTAND the problem. Reread it as many times as needed.

2. TRANSLATE into an equation. (Fill in the blanks below.)

What number	is	16%	of	70
↓	↓	↓	↓	↓
x	___	0.16	___	70

Finish with:

3. SOLVE and **4.** INTERPRET

2. What number is 88% of 1000?

Start the solution:

1. UNDERSTAND the problem. Reread it as many times as needed.

2. TRANSLATE into an equation. (Fill in the blanks below.)

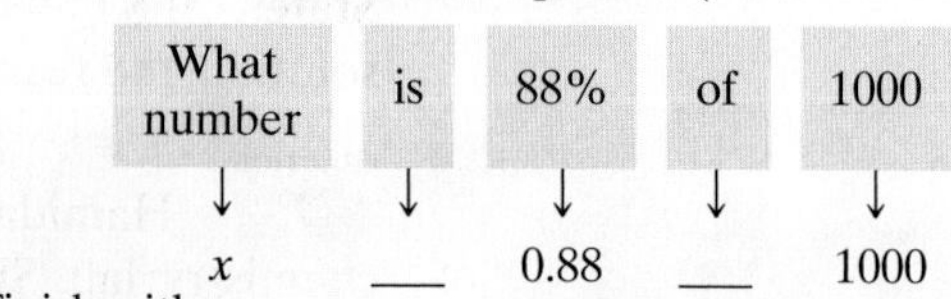

What number	is	88%	of	1000
↓	↓	↓	↓	↓
x	___	0.88	___	1000

Finish with:

3. SOLVE and **4.** INTERPRET

3. The number 28.6 is what percent of 52?

4. The number 87.2 is what percent of 436?

5. The number 45 is 25% of what number?
6. The number 126 is 35% of what number?

The circle graph below shows the number of minutes that adults spend on their home phone each day. Use this graph for Exercises 7 through 10. See Example 3.

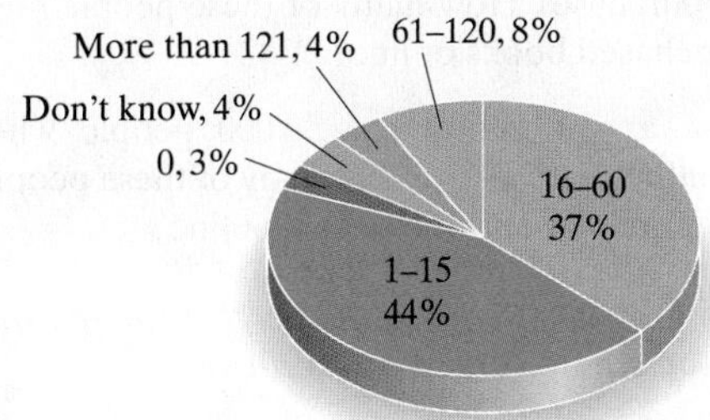

Data from Bruskin/Goldring Research for Sony Electronics

7. What percent of adults spend more than 121 minutes on the phone each day?
8. What percent of adults spend no time on the phone each day?
9. Florence is a town in Alabama whose adult population is approximately 27,000. How many of these adults might you expect to talk 16–60 minutes on the phone each day?
10. Columbus is a town in Indiana whose adult population is approximately 29,250. How many of these adults might you expect to talk 61–120 minutes on the phone each day?

Solve. If needed, round answers to the nearest cent. See Example 4.

11. A used automobile dealership recently reduced the price of a used compact car by 8%. If the price of the car before discount was $18,500, find the discount and the new price.
12. A music store is advertising a 25%-off sale on all new releases. Find the discount and the sale price of a newly released CD that regularly sells for $12.50.
13. A birthday celebration meal is $40.50 including tax. Find the total cost if a 15% tip is added to the given cost.
14. A retirement dinner for two is $65.40 including tax. Find the total cost if a 20% tip is added to the given cost.

Solve. See Example 5.

Use the graph below for Exercises 15 and 16.

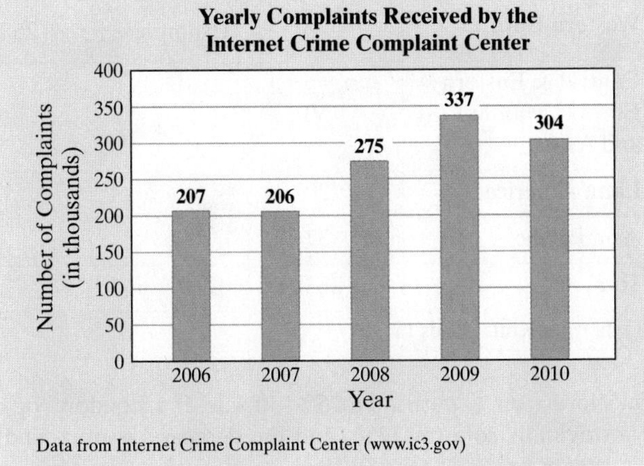

15. The number of Internet-crime complaints decreased from 2009 to 2010. Find the percent decrease. Round to the nearest tenth of a percent.
16. The number of Internet-crime complaints decreased from 2006 to 2007. Find the percent decrease. Round to the nearest tenth of a percent.
17. By decreasing each dimension by 1 unit, the area of a rectangle decreased from 40 square feet (on the left) to 28 square feet (on the right). Find the percent decrease in area.

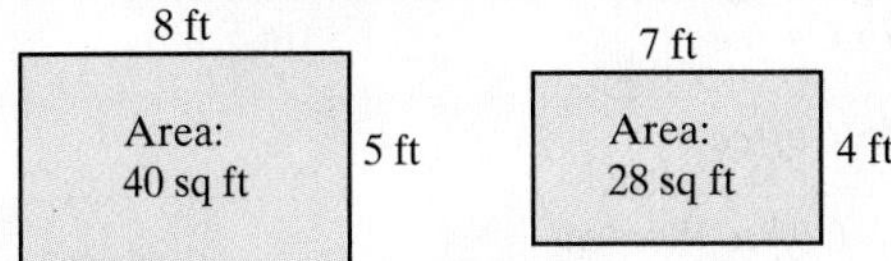

18. By decreasing the length of the side by one unit, the area of a square decreased from 100 square meters to 81 square meters. Find the percent decrease in area.

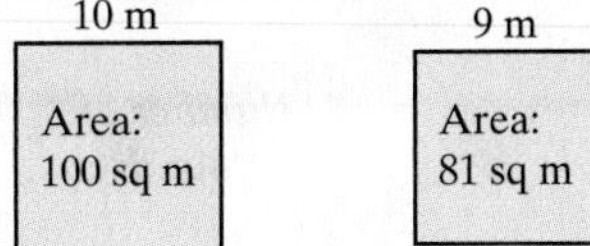

Solve. See Example 6.

19. Find the original price of a pair of shoes if the sale price is $78 after a 25% discount.
20. Find the original price of a popular pair of shoes if the increased price is $80 after a 25% increase.
21. Find last year's salary if, after a 4% pay raise, this year's salary is $44,200.
22. Find last year's salary if after a 3% pay raise, this year's salary is $55,620.

Solve. For each exercise, a table is given for you to complete and use to write an equation that models the situation. See Example 7.

23. How much pure acid should be mixed with 2 gallons of a 40% acid solution in order to get a 70% acid solution?

	Number of Gallons ·	*Acid Strength* =	*Amount of Acid*
Pure Acid		100%	
40% Acid Solution			
70% Acid Solution Needed			

24. How many cubic centimeters (cc) of a 25% antibiotic solution should be added to 10 cubic centimeters of a 60% antibiotic solution to get a 30% antibiotic solution?

	Number of Cubic cm ·	*Antibiotic Strength* =	*Amount of Antibiotic*
25% Antibiotic Solution			
60% Antibiotic Solution			
30% Antibiotic Solution Needed			

25. Community Coffee Company wants a new flavor of Cajun coffee. How many pounds of coffee worth \$7 a pound should be added to 14 pounds of coffee worth \$4 a pound to get a mixture worth \$5 a pound?

	Number of Pounds	· *Cost per pound*	= *Value*
\$7 per lb Coffee			
\$4 per lb Coffee			
\$5 per lb Coffee Wanted			

26. Planter's Peanut Company wants to mix 20 pounds of peanuts worth \$3 a pound with cashews worth \$5 a pound in order to make an experimental mix worth \$3.50 a pound. How many pounds of cashews should be added to the peanuts?

	Number of Pounds	· *Cost per pound*	= *Value*
\$3 per lb Peanuts			
\$5 per lb Cashews			
\$3.50 per lb Mixture Wanted			

MIXED PRACTICE

Solve. If needed, round money amounts to two decimal places and all other amounts to one decimal place. See Examples 1 through 7.

27. Find 23% of 20.

28. Find 140% of 86.

29. The number 40 is 80% of what number?

30. The number 56.25 is 45% of what number?

31. The number 144 is what percent of 480?

32. The number 42 is what percent of 35?

The graph below is the result of a survey of U.S. citizens who have made an online purchase in the past 3 months. Use this graph for Exercises 33 through 36.

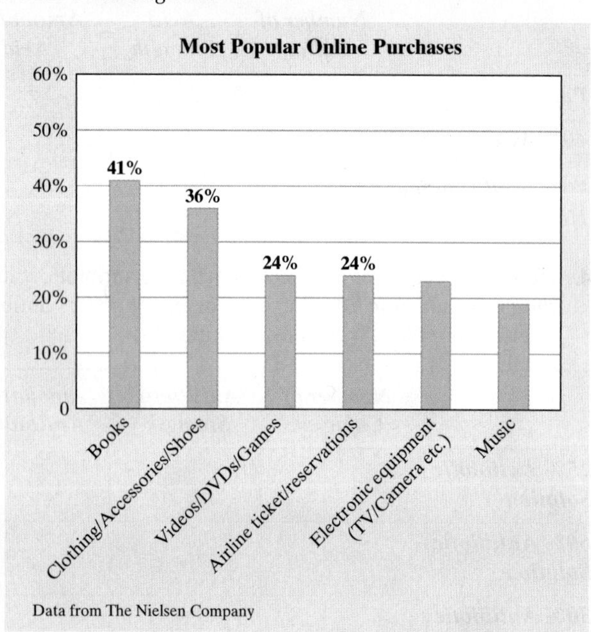

33. Estimate the percent of online purchases in the category of electronic equipment.

34. Estimate the percent of online purchases in the category of music.

35. Suppose a city has 220,500 people who regularly make online purchases. How many of these people might we predict purchased books online?

36. Suppose a community has 50,000 people who regularly make online purchases. How many of these people might we predict purchased airline tickets online?

For Exercises 37 and 38, fill in the percent column in each table. Each table contains a worked-out example.

37.

Top Cranberry-Producing States Forecast in 2010 (in millions of pounds)

	Millions of Pounds	*Percent of Total (rounded to nearest percent)*
Wisconsin	435	
Oregon	39	
Massachusetts	195	
Washington	14	
New Jersey	53	Example: $\frac{53}{736} \approx 7\%$
Total	736	

Data from National Agricultural Statistics Service

38.

Kraft Foods Number of Manufacturing Facilities

Region	*Volume (in pounds)*	*Percent of Total (round to nearest percent)*
U.S.	46	
Canada	11	
Western Europe	59	Example: $\frac{59}{223} \approx 26\%$
Central & Eastern Europe, Middle East and Africa	50	
Latin America	20	
Asia Pacific	37	
Total	223	

Data from Kraft Foods Inc.

39. Nordstrom advertised a 25%-off sale. If a London Fog coat originally sold for \$256, find the decrease in price and the sale price.

40. A gasoline station decreased the price of a \$0.95 cola by 15%. Find the decrease in price and the new price.

41. Although the consumption of washed leaf lettuce is increasing, the consumption of head lettuce is decreasing. The head lettuce consumption per capita in 2000 was 23.5 pounds, and in 2010 the consumption dropped to 17.1 pounds. Find the percent decrease. Round to the nearest tenth of a percent. (*Source:* USDA)

42. Iceberg lettuce is grown and shipped to stores for about 40 cents a head, and consumers purchase it for about 86 cents a head. Find the percent increase. (*Source: Statistical Abstract of the United States*)

43. The number of registered vehicles on the road in the United States is constantly increasing. In 2009, there were approximately 246 million registered vehicles. This represents a 3% increase over 2002. How many registered vehicles were there in the United States in 2002? Round to the nearest million. (*Source:* Federal Highway Administration)

44. A student at the University of New Orleans makes money by buying and selling used cars. Charles bought a used car and later sold it for a 20% profit. If he sold it for \$4680, how much did Charles pay for the car?

45. By doubling each dimension, the area of a parallelogram increased from 36 square centimeters to 144 square centimeters. Find the percent increase in area.

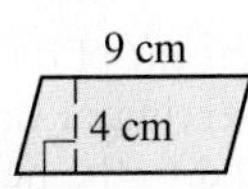

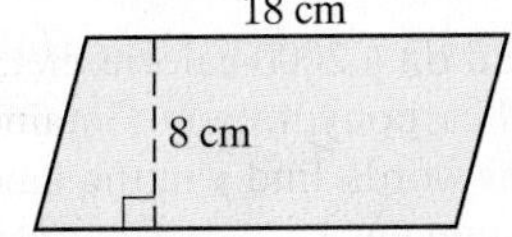

46. By doubling each dimension, the area of a triangle increased from 6 square miles to 24 square miles. Find the percent increase in area.

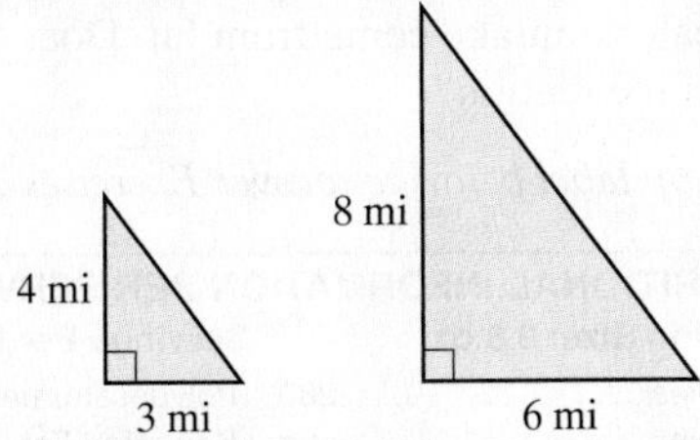

47. How much of an alloy that is 20% copper should be mixed with 200 ounces of an alloy that is 50% copper to get an alloy that is 30% copper?

48. How much water should be added to 30 gallons of a solution that is 70% antifreeze to get a mixture that is 60% antifreeze?

49. The number of farms in the United States is decreasing. In 1940, there were approximately 6.3 million farms, while in 2009 there were only 2.1 million farms. Find the percent decrease in the number of farms. Round to the nearest tenth of a percent.

50. During the 1982–1983 term, the Supreme Court made 151 decisions, while during the 2007–2008 term, they made only 72. Find the percent decrease in number of decisions. Round to the nearest tenth of a percent.

51. A company recently downsized its number of employees by 35%. If there are still 78 employees, how many employees were there prior to the layoffs?

52. The average number of children born to each U.S. woman has decreased by 44% since 1920. If this average is now 1.9, find the average in 1920. Round to the nearest tenth.

53. A recent survey showed that 42% of recent college graduates named flexible hours as their most desired employment benefit. In a graduating class of 860 college students, how many would you expect to rank flexible hours as their top priority in job benefits? (Round to the nearest whole.) (*Source:* JobTrak.com)

54. A recent survey showed that 64% of U.S. colleges have Internet access in their classrooms. There are approximately 9800 post-secondary institutions in the United States. How many of these would you expect to have Internet access in their classrooms? (*Source:* Market Data Retrieval, National Center for Education Statistics)

55. A new self-tanning lotion for everyday use is to be sold. First, an experimental lotion mixture is made by mixing 800 ounces of everyday moisturizing lotion worth \$0.30 an ounce with self-tanning lotion worth \$3 per ounce. If the experimental lotion is to cost \$1.20 per ounce, how many ounces of the self-tanning lotion should be in the mixture?

56. The owner of a local chocolate shop wants to develop a new trail mix. How many pounds of chocolate-covered peanuts worth \$5 a pound should be mixed with 10 pounds of granola bites worth \$2 a pound to get a mixture worth \$3 per pound?

57. Scoville units are used to measure the hotness of a pepper. Measuring 577 thousand Scoville units, the "Red Savina" habañero pepper was known as the hottest chili pepper. That has recently changed with the discovery of the Naga Jolokia pepper from India. It measures 48% hotter than the habañero. Find the measure of the Naga Jolokia pepper. Round to the nearest thousand units.

58. As of this writing, the women's record for throwing a disc (like a heavy Frisbee) was set by Valarie Jenkins of the United States in 2008. Her throw was 148.00 meters. The men's world record was set by Christian Sandstrom of Sweden in 2002. His throw was 68.9% farther than Valarie's. Find the distance of his throw. Round to the nearest meter. (*Source:* World Flying Disc Federation)

REVIEW AND PREVIEW

Place $<$, $>$, or $=$ in the appropriate space to make each a true statement. See Sections 1.2, 1.3, and 1.5.

59. $-5 \quad -7$

60. $\frac{12}{3} \quad 2^2$

61. $|-5| \quad -(-5)$

62. $-3^3 \quad (-3)^3$

63. $(-3)^2 \quad -3^2$

64. $|-2| \quad -|-2|$

CONCEPT EXTENSIONS

65. Is it possible to mix a 10% acid solution and a 40% acid solution to obtain a 60% acid solution? Why or why not?

66. Must the percents in a circle graph have a sum of 100%? Why or why not?

67. A trail mix is made by combining peanuts worth \$3 a pound, raisins worth \$2 a pound, and M&M's worth \$4 a pound. Would it make good business sense to sell the trail mix for \$1.98 a pound? Why or why not?

68. **a.** Can an item be marked up by more than 100%? Why or why not?

b. Can an item be discounted by more than 100%? Why or why not?

Standardized nutrition labels like the one below have been displayed on food items since 1994. The percent column on the right shows the percent of daily values (based on a 2000-calorie diet) shown at the bottom of the label. For example, a serving of this food contains 4 grams of total fat when the recommended daily fat based on a 2000-calorie diet is less than 65 grams of fat. This means that $\frac{4}{65}$ or approximately 6% (as shown) of your daily recommended fat is taken in by eating a serving of this food. Use this nutrition label to answer Exercises 69 through 71.

Nutrition Facts

Serving Size	18 Crackers (31g)
Servings Per Container	About 9
Amount Per Serving	
Calories 130	Calories from Fat 35
	% Daily Value*
Total Fat 4g	6%
Saturated Fat 0.5g	3%
Polyunsaturated Fat 0g	
Monounsaturated Fat 1.5g	
Cholesterol 0mg	0%
Sodium 230mg	*x*
Total Carbohydrate 23g	*y*
Dietary Fiber 2g	8%
Sugars 3g	
Protein 2g	
Vitamin A 0% • Vitamin C 0%	
Calcium 2% • Iron 6%	

* Percent Daily Values are based on a 2,000 calorie diet. Your daily values may be higher or lower depending on your calorie needs.

	Calories	2,000	2,500
Total Fat	Less than	65g	80g
Sat. Fat	Less than	20g	25g
Cholesterol	Less than	300mg	300mg
Sodium	Less than	2400mg	2400mg
Total Carbohydrate		300g	375g
Dietary Fiber		25g	30g

69. Based on a 2000-calorie diet, what percent of daily value of sodium is contained in a serving of this food? In other words, find x in the label. (Round to the nearest tenth of a percent.)

70. Based on a 2000-calorie diet, what percent of daily value of total carbohydrate is contained in a serving of this food? In other words, find y in the label. (Round to the nearest tenth of a percent.)

71. Notice on the nutrition label that one serving of this food contains 130 calories and 35 of these calories are from fat. Find the percent of calories from fat. (Round to the nearest tenth of a percent.) It is recommended that no more than 30% of calorie intake come from fat. Does this food satisfy this recommendation?

Use the nutrition label below to answer Exercises 72 through 74.

NUTRITIONAL INFORMATION PER SERVING

Serving Size: 9.8 oz		Servings Per Container: 1	
Calories	280	Polyunsaturated Fat	1g
Protein	12g	Saturated Fat	3g
Carbohydrate	45g	Cholesterol	20mg
Fat	6g	Sodium	520mg
Percent of Calories from Fat	?	Potassium	220mg

72. If fat contains approximately 9 calories per gram, find the percent of calories from fat in one serving of this food. (Round to the nearest tenth of a percent.)

73. If protein contains approximately 4 calories per gram, find the percent of calories from protein from one serving of this food. (Round to the nearest tenth of a percent.)

74. Find a food that contains more than 30% of its calories per serving from fat. Analyze the nutrition label and verify that the percents shown are correct.

2.8 Further Problem Solving 

OBJECTIVES

1 Solve Problems Involving Distance.

2 Solve Problems Involving Money.

3 Solve Problems Involving Interest

This section is devoted to solving problems in the categories listed. The same problem-solving steps used in previous sections are also followed in this section. They are listed below for review.

General Strategy for Problem Solving

1. UNDERSTAND the problem. During this step, become comfortable with the problem. Some ways of doing this are to:

   Read and reread the problem.

   Choose a variable to represent the unknown.

   Construct a drawing whenever possible.

   Propose a solution and check. Pay careful attention to how you check your proposed solution. This will help writing an equation to model the problem.
2. TRANSLATE the problem into an equation.
3. SOLVE the equation.
4. INTERPRET the results: *Check* the proposed solution in the stated problem and *state* your conclusion.

OBJECTIVE

1 Solving Distance Problems

Our first example involves distance. For a review of the distance formula, $d = r \cdot t$, see Section 2.6, Example 1, and the table before the example.

EXAMPLE 1 Finding Time Given Rate and Distance

Marie Antonio, a bicycling enthusiast, rode her 21-speed at an average speed of 18 miles per hour on level roads and then slowed down to an average of 10 mph on the hilly roads of the trip. If she covered a distance of 98 miles, how long did the entire trip take if traveling the level roads took the same time as traveling the hilly roads?

Solution

1. UNDERSTAND the problem. To do so, read and reread the problem. The formula $d = r \cdot t$ is needed. At this time, let's guess a solution. Suppose that she spent 2 hours traveling on the level roads. This means that she also spent 2 hours traveling on the hilly roads, since the times spent were the same. What is her total distance? Her distance on the level road is rate · time $= 18(2) = 36$ miles. Her distance on the hilly roads is rate · time $= 10(2) = 20$ miles. This gives a total distance of 36 miles + 20 miles = 56 miles, not the correct distance of 98 miles. Remember that the purpose of guessing a solution is not to guess correctly (although this may happen) but to help understand the problem better and how to model it with an equation. We are looking for the length of the entire trip, so we begin by letting

   $x =$ the time spent on level roads.

   Because the same amount of time is spent on hilly roads, then also

   $x =$ the time spent on hilly roads.

2. TRANSLATE. To help us translate to an equation, we now summarize the information from the problem on the following chart. Fill in the rates given and the variables used to represent the times and use the formula $d = r \cdot t$ to fill in the distance column.

	Rate ·	*Time* =	*Distance*
Level	18	x	$18x$
Hilly	10	x	$10x$

Since the entire trip covered 98 miles, we have that

In words:	total distance	=	level distance	+	hilly distance
	↓		↓		↓
Translate:	98	=	$18x$	+	$10x$

3. SOLVE.

$$98 = 28x \quad \text{Add like terms.}$$

$$\frac{98}{28} = \frac{28x}{28} \quad \text{Divide both sides by 28.}$$

$$3.5 = x$$

4. INTERPRET the results.

Check: Recall that x represents the time spent on the level portion of the trip and the time spent on the hilly portion. If Marie rides for 3.5 hours at 18 mph, her distance is $18(3.5) = 63$ miles. If Marie rides for 3.5 hours at 10 mph, her distance is $10(3.5) = 35$ miles. The total distance is 63 miles + 35 miles = 98 miles, the required distance.

State: The time of the entire trip is then 3.5 hours + 3.5 hours or 7 hours. □

PRACTICE

1 Sat Tranh took a short hike with his friends up Mt. Wachusett. They hiked uphill at a steady pace of 1.5 miles per hour and downhill at a rate of 4 miles per hour. If the time to climb the mountain took an hour more than the time to hike down, how long was the entire hike?

EXAMPLE 2 Finding Train Speeds

The Kansas City Southern Railway operates in 10 states and Mexico. Suppose two trains leave Neosho, Missouri, at the same time. One travels north and the other travels south at a speed that is 15 miles per hour faster. In 2 hours, the trains are 230 miles apart. Find the speed of each train.

Kansas City Southern Railway

Solution

1. UNDERSTAND the problem. Read and reread the problem. Guess a solution and check. Let's let

$$x = \text{speed of train traveling north}$$

Because the train traveling south is 15 mph faster, we have

$$x + 15 = \text{speed of train traveling south}$$

2. TRANSLATE. Just as for Example 1, let's summarize our information on a chart. Use the formula $d = r \cdot t$ to fill in the distance column.

	r ·	t =	d
North Train	x	2	$2x$
South Train	$x + 15$	2	$2(x + 15)$

Since the total distance between the trains is 230 miles, we have

In words:	north train distance	+	south train distance	=	total distance
	↓		↓		↓
Translate:	$2x$	+	$2(x + 15)$	=	230

3. SOLVE.

$$\begin{aligned} 2x + 2x + 30 &= 230 && \text{Use the distributive property.} \\ 4x + 30 &= 230 && \text{Combine like terms.} \\ 4x &= 200 && \text{Subtract 30 from both sides.} \\ \frac{4x}{4} &= \frac{200}{4} && \text{Divide both sides by 4.} \\ x &= 50 && \text{Simplify.} \end{aligned}$$

4. INTERPRET the results.

Check: Recall that x is the speed of the train traveling north, or 50 mph. In 2 hours, this train travels a distance of $2(50) = 100$ miles. The speed of the train traveling south is $x + 15$ or $50 + 15 = 65$ mph. In 2 hours, this train travels $2(65) = 130$ miles. The total distance of the trains is 100 miles + 130 miles = 230 miles, the required distance.

State: The northbound train's speed is 50 mph and the southbound train's speed is 65 mph. □

PRACTICE

2 The Kansas City Southern Railway has a station in Mexico City, Mexico. Suppose two trains leave Mexico City at the same time. One travels east and the other west at a speed that is 10 mph slower. In 1.5 hours, the trains are 171 miles apart. Find the speed of each train.

OBJECTIVE

2 Solving Money Problems

The next example has to do with finding an unknown number of a certain denomination of coin or bill. These problems are extremely useful in that they help you understand the difference between the number of coins or bills and the total value of the money.

For example, suppose there are seven \$5-bills. The *number* of \$5-bills is 7 and the *total value* of the money is $\$5(7) = \$35$.

Study the table below for more examples.

Denomination of Coin or Bill	*Number of Coins or Bills*	*Value of Coins or Bills*
20-dollar bills	17	$\$20(17) = \$340$
nickels	31	$\$0.05(31) = \$1.55$
quarters	x	$\$0.25(x) = \$0.25x$

EXAMPLE 3 Finding Numbers of Denominations

Part of the proceeds from a local talent show was \$2420 worth of \$10 and \$20 bills. If there were 37 more \$20 bills than \$10 bills, find the number of each denomination.

Solution

1. UNDERSTAND the problem. To do so, read and reread the problem. If you'd like, let's guess a solution. Suppose that there are 25 \$10 bills. Since there are 37 more \$20 bills, we have $25 + 37 = 62$ \$20 bills. The total amount of money is $\$10(25) + \$20(62) = \$1490$, below the given amount of \$2420. Remember that our purpose for guessing is to help us understand the problem better.

We are looking for the number of each denomination, so we let

$$x = \text{number of \$10 bills}$$

There are 37 more \$20 bills, so

$$x + 37 = \text{number of \$20 bills}$$

2. TRANSLATE. To help us translate to an equation, study the table below.

Denomination	***Number of Bills***	***Value of Bills (in dollars)***
\$10 bills	x	$10x$
\$20 bills	$x + 37$	$20(x + 37)$

Since the total value of these bills is \$2420, we have

In words:	value of \$10 bills	plus	value of \$20 bills	is	2420
	↓	↓	↓	↓	↓
Translate:	$10x$	$+$	$20(x + 37)$	$=$	2420

3. SOLVE:

$$10x + 20x + 740 = 2420 \quad \text{Use the distributive property.}$$
$$30x + 740 = 2420 \quad \text{Add like terms.}$$
$$30x = 1680 \quad \text{Subtract 740 from both sides.}$$
$$\frac{30x}{30} = \frac{1680}{30} \quad \text{Divide both sides by 30.}$$
$$x = 56$$

4. INTERPRET the results.

Check: Since x represents the number of \$10 bills, we have 56 \$10 bills and $56 + 37$, or 93 \$20 bills. The total amount of these bills is $\$10(56) + \$20(93) = \$2420$, the correct total.

State: There are 56 \$10 bills and 93 \$20 bills. □

PRACTICE

3 A stack of \$5 and \$20 bills was counted by the treasurer of an organization. The total value of the money was \$1710 and there were 47 more \$5 bills than \$20 bills. Find the number of each type of bill.

OBJECTIVE

3 Solving Interest Problems

The next example is an investment problem. For a review of the simple interest formula, $I = PRT$, see the table at the beginning of Section 2.6 and Exercises 11 and 12 in that exercise set.

EXAMPLE 4 Finding the Investment Amount

Rajiv Puri invested part of his \$20,000 inheritance in a mutual funds account that pays 7% simple interest yearly and the rest in a certificate of deposit that pays 9% simple interest yearly. At the end of one year, Rajiv's investments earned \$1550. Find the amount he invested at each rate.

Solution

1. UNDERSTAND. Read and reread the problem. Next, guess a solution. Suppose that Rajiv invested \$8000 in the 7% fund and the rest, \$12,000, in the fund paying 9%. To check, find his interest after one year. Recall the formula $I = PRT$, so the interest from the 7% fund $= \$8000(0.07)(1) = \$560$. The interest from the 9% fund $= \$12{,}000(0.09)(1) = \$1080$. The sum of the interests is $\$560 + \$1080 = \$1640$. Our guess is incorrect, since the sum of the interests is not \$1550, but we now have a better understanding of the problem.

Let

$x =$ amount of money in the account paying 7%.

The rest of the money is \$20,000 less x or

$20{,}000 - x =$ amount of money in the account paying 9%.

2. TRANSLATE. We apply the simple interest formula $I = PRT$ and organize our information in the following chart. Since there are two rates of interest and two amounts invested, we apply the formula twice.

	Principal ·	*Rate* ·	*Time* =	*Interest*
7% Fund	x	0.07	1	$x(0.07)(1)$ or $0.07x$
9% Fund	$20{,}000 - x$	0.09	1	$(20{,}000 - x)(0.09)(1)$ or $0.09(20{,}000 - x)$
Total	20,000			1550

The total interest earned, \$1550, is the sum of the interest earned at 7% and the interest earned at 9%.

In words:	interest at 7%	+	interest at 9%	=	total interest
	↓		↓		↓
Translate:	$0.07x$	+	$0.09(20{,}000 - x)$	=	1550

3. SOLVE.

$$0.07x + 0.09(20{,}000 - x) = 1550$$

$$0.07x + 1800 - 0.09x = 1550 \quad \text{Apply the distributive property.}$$

$$1800 - 0.02x = 1550 \quad \text{Combine like terms.}$$

$$-0.02x = -250 \quad \text{Subtract 1800 from both sides.}$$

$$x = 12{,}500 \quad \text{Divide both sides by } -0.02.$$

4. INTERPRET.

Check: If $x = 12{,}500$, then $20{,}000 - x = 20{,}000 - 12{,}500$ or 7500. These solutions are reasonable, since their sum is \$20,000 as required. The annual interest on \$12,500 at 7% is \$875; the annual interest on \$7500 at 9% is \$675, and \$875 + \$675 = \$1550.

State: The amount invested at 7% was \$12,500. The amount invested at 9% was \$7500. □

PRACTICE

4 Suzanne Scarpulla invested \$30,000, part of it in a high-risk venture that yielded 11.5% per year and the rest in a secure mutual fund paying interest of 6% per year. At the end of one year, Suzanne's investments earned \$2790. Find the amount she invested at each rate.

Vocabulary, Readiness & Video Check

Martin-Gay Interactive Videos

See Video 2.8

Watch the section lecture video and answer the following questions.

OBJECTIVE 1

1. The following problem is worded like Example 1 but using different quantities.

How long will it take a bus traveling at 55 miles per hour to overtake a car traveling at 50 mph if the car had a 3 hour head start? Fill in the table and set up an equation that could be used to solve for the unknown (do not actually solve). Use Example 1 in the video as a model for your work.

	r ·	*t* =	*d*
bus			
car			

OBJECTIVE 2

2. In the lecture before Example 3, what important point are you told to remember when working with applications that have to do with money?

OBJECTIVE 3

3. The following problem is worded like Example 4 in the video, but using different quantities.

How can $36,000 be invested, part at 6% annual simple interest and the remainder at 4% annual simple interest, so that the annual simple interest earned by the two accounts is equal? Fill in the table and set up an equation that could be used to solve for the unknowns (do not actually solve). Use Example 4 in the video as a model for your work.

P ·	*R* ·	*T* =	*I*

2.8 Exercise Set

MyMathLab®

Solve. See Examples 1 and 2.

1. A jet plane traveling at 500 mph overtakes a propeller plane traveling at 200 mph that had a 2-hour head start. How far from the starting point are the planes?

2. How long will it take a bus traveling at 60 miles per hour to overtake a car traveling at 40 mph if the car had a 1.5-hour head start?

3. A bus traveled on a level road for 3 hours at an average speed 20 miles per hour faster than it traveled on a winding road. The time spent on the winding road was 4 hours. Find the average speed on the level road if the entire trip was 305 miles.

4. The Jones family drove to Disneyland at 50 miles per hour and returned on the same route at 40 mph. Find the distance to Disneyland if the total driving time was 7.2 hours.

Complete the table. The first and sixth rows have been completed for you. See Example 3.

		Number of Coins or Bills	Value of Coins or Bills (in dollars)
	pennies	x	$0.01x$
5.	dimes	y	
6.	quarters	z	
7.	nickels	$(x + 7)$	
8.	half-dollars	$(20 - z)$	
	$5 bills	$9x$	$5(9x)$
9.	$20 bills	$4y$	
10.	$100 bills	$97z$	
11.	$50 bills	$(35 - x)$	
12.	$10 bills	$(15 - y)$	

13. Part of the proceeds from a garage sale was $280 worth of $5 and $10 bills. If there were 20 more $5 bills than $10 bills, find the number of each denomination.

	Number of Bills	Value of Bills
$5 bills		
$10 bills		
Total		

14. A bank teller is counting $20- and $50-dollar bills. If there are six times as many $20 bills as $50 bills and the total amount of money is $3910, find the number of each denomination.

	Number of Bills	Value of Bills
$20 bills		
$50 bills		
Total		

Solve. See Example 4.

15. Zoya Lon invested part of her $25,000 advance at 8% annual simple interest and the rest at 9% annual simple interest. If her total yearly interest from both accounts was $2135, find the amount invested at each rate.

16. Karen Waugtal invested some money at 9% annual simple interest and $250 more than that amount at 10% annual simple interest. If her total yearly interest was $101, how much was invested at each rate?

17. Sam Mathius invested part of his $10,000 bonus in a fund that paid an 11% profit and invested the rest in stock that suffered a 4% loss. Find the amount of each investment if his overall net profit was $650.

18. Bruce Blossum invested a sum of money at 10% annual simple interest and invested twice that amount at 12% annual simple interest. If his total yearly income from both investments was $2890, how much was invested at each rate?

19. The Concordia Theatre contains 500 seats, and the ticket prices for a recent play were $43 for adults and $28 for children. For one sold-out matinee, if the total proceeds were $16,805, how many of each type of ticket were sold?

20. A zoo in Oklahoma charged \$22 for adults and \$15 for children. During a summer day, 732 zoo tickets were sold, and the total receipts were \$12,912. How many child and how many adult tickets were sold?

MIXED PRACTICE

21. Two cars leave Richmond, Virginia, at the same time after visiting the nearby Richmond International Speedway. The cars travel in opposite directions, one traveling north at 56 mph and one traveling south at 47 mph. When will the two cars be 206 miles apart?

22. Two cars leave Las Vegas, Nevada, at the same time after visiting the Las Vegas Motor Speedway. The cars travel in opposite directions, one traveling northeast at 65 mph and one traveling southwest at 41 mph. When will the two cars be 530 miles apart?

23. How can \$54,000 be invested, part at 8% annual simple interest and the remainder at 10% annual simple interest, so that the interest earned by the two accounts will be equal?

24. Ms. Mills invested her \$20,000 bonus in two accounts. She took a 4% loss on one investment and made a 12% profit on another investment but ended up breaking even. How much was invested in each account?

25. Alan and Dave Schaferkötter leave from the same point driving in opposite directions, Alan driving at 55 miles per hour and Dave at 65 mph. Alan has a one-hour head start. How long will they be able to talk on their car phones if the phones have a 250-mile range?

26. Kathleen and Cade Williams leave simultaneously from the same point, hiking in opposite directions, Kathleen walking at 4 miles per hour and Cade at 5 mph. How long can they talk on their walkie-talkies if the walkie-talkies have a 20-mile radius?

27. Suppose two trains leave Corpus Christi, Texas, at the same time, traveling in opposite directions. One train travels 10 mph faster than the other. In 2.5 hours, the trains are 205 miles apart. Find the speed of each train.

28. Suppose two trains leave Edmonton, Canada, at the same time, traveling in opposite directions. One train travels 8 mph faster than the other. In 1.5 hours, the trains are 162 miles apart. Find the speed of each train.

29. A youth organization collected nickels and dimes for a charity drive. By the end of the 1-day drive, the youth had collected \$56.35. If there were three times as many dimes as nickels, how many of each type of coin was collected?

30. A collection of dimes and quarters is retrieved from a soft drink machine. There are five times as many dimes as quarters and the total value of the coins is \$27.75. Find the number of dimes and the number of quarters.

31. A truck and a van leave the same location at the same time and travel in opposite directions. The truck's speed is 52 mph and the van's speed is 63 mph. When will the truck and the van be 460 miles apart?

32. Two cars leave the same location at the same time and travel in opposite directions. One car's speed is 65 mph and the other car's speed is 45 mph. When will the two cars be 330 miles apart?

33. Two cars leave Pecos, Texas, at the same time and both travel east on Interstate 20. The first car's speed is 70 mph and the second car's speed is 58 mph. When will the cars be 30 miles apart?

34. Two cars leave Savannah, Georgia, at the same time and both travel north on Interstate 95. The first car's speed is 40 mph and the second car's speed is 50 mph. When will the cars be 20 miles apart?

35. If \$3000 is invested at 6% annual simple interest, how much should be invested at 9% annual simple interest so that the total yearly income from both investments is \$585?

36. Trudy Waterbury, a financial planner, invested a certain amount of money at 9% annual simple interest, twice that amount at 10% annual simple interest, and three times that amount at 11% annual simple interest. Find the amount invested at each rate if her total yearly income from the investments was \$2790.

37. Two hikers are 11 miles apart and walking toward each other. They meet in 2 hours. Find the rate of each hiker if one hiker walks 1.1 mph faster than the other.

38. Nedra and Latonya Dominguez are 12 miles apart hiking toward each other. How long will it take them to meet if Nedra walks at 3 mph and Latonya walks 1 mph faster?

39. Mark Martin can row upstream at 5 mph and downstream at 11 mph. If Mark starts rowing upstream until he gets tired and then rows downstream to his starting point, how far did Mark row if the entire trip took 4 hours?

40. On a 255-mile trip, Gary Alessandrini traveled at an average speed of 70 mph, got a speeding ticket, and then traveled at 60 mph for the remainder of the trip. If the entire trip took 4.5 hours and the speeding ticket stop took 30 minutes, how long did Gary speed before getting stopped?

REVIEW AND PREVIEW

Perform the indicated operations. See Sections 1.5 and 1.6.

41. $3 + (-7)$

42. $(-2) + (-8)$

43. $\frac{3}{4} - \frac{3}{16}$

44. $-11 + 2.9$

45. $-5 - (-1)$

46. $-12 - 3$

CONCEPT EXTENSIONS

47. A stack of \$20, \$50, and \$100 bills was retrieved as part of an FBI investigation. There were 46 more \$50 bills than \$100 bills. Also, the number of \$20 bills was 7 times the number of \$100 bills. If the total value of the money was \$9550, find the number of each type of bill.

48. A man places his pocket change in a jar every day. The jar is full and his children have counted the change. The total value is \$44.86. Let x represent the number of quarters and use the information below to find the number of each type of coin.

There are: 136 more dimes than quarters

8 times as many nickels as quarters

32 more than 16 times as many pennies as quarters

To "break even" in a manufacturing business, revenue R (income) ***must equal*** *the cost C of production, or* $R = C$.

49. The cost C to produce x number of skateboards is given by $C = 100 + 20x$. The skateboards are sold wholesale for \$24 each, so revenue R is given by $R = 24x$. Find how many skateboards the manufacturer needs to produce and sell to break even. (*Hint:* Set the expression for R equal to the expression for C, then solve for x.)

50. The revenue R from selling x number of computer boards is given by $R = 60x$, and the cost C of producing them is given by $C = 50x + 5000$. Find how many boards must be sold to break even. Find how much money is needed to produce the break-even number of boards.

51. The cost C of producing x number of paperback books is given by $C = 4.50x + 2400$. Income R from these books is given by $R = 7.50x$. Find how many books should be produced and sold to break even.

52. Find the break-even quantity for a company that makes x number of computer monitors at a cost C given by $C = 875 + 70x$ and receives revenue R given by $R = 105x$.

53. Exercises 49 through 52 involve finding the break-even point for manufacturing. Discuss what happens if a company makes and sells fewer products than the break-even point. Discuss what happens if more products than the break-even point are made and sold.

2.9 Solving Linear Inequalities

OBJECTIVES

1. Define Linear Inequality in One Variable, Graph Solution Sets on a Number Line, and Use Interval Notation.
2. Solve Linear Inequalities.
3. Solve Compound Inequalities.
4. Solve Inequality Applications.

OBJECTIVE

1 Graphing Solution Sets to Linear Inequalities and Using Interval Notation

In Chapter 1, we reviewed these inequality symbols and their meanings:

$<$ means "is less than" $\leq$ means "is less than or equal to"

$>$ means "is greater than" $\geq$ means "is greater than or equal to"

Equations	*Inequalities*
$x = 3$	$x \leq 3$
$5n - 6 = 14$	$5n - 6 < 14$
$12 = 7 - 3y$	$12 \geq 7 - 3y$
$\frac{x}{4} - 6 = 1$	$\frac{x}{4} - 6 > 1$

A linear inequality is similar to a linear equation except that the equality symbol is replaced with an inequality symbol.

Linear Inequality in One Variable

A **linear inequality in one variable** is an inequality that can be written in the form

$$ax + b < c$$

where a, b, and c are real numbers and a is not 0.

This definition and all other definitions, properties, and steps in this section also hold true for the inequality symbols $>$, $\geq$, and $\leq$.

A **solution of an inequality** is a value of the variable that makes the inequality a true statement. The solution set is the set of all solutions. For the inequality $x < 3$, replacing x with any number less than 3, that is, to the left of 3 on a number line, makes the resulting inequality true. This means that any number less than 3 is a solution of the inequality $x < 3$.

Since there are infinitely many such numbers, we cannot list all the solutions of the inequality. We *can* use set notation and write

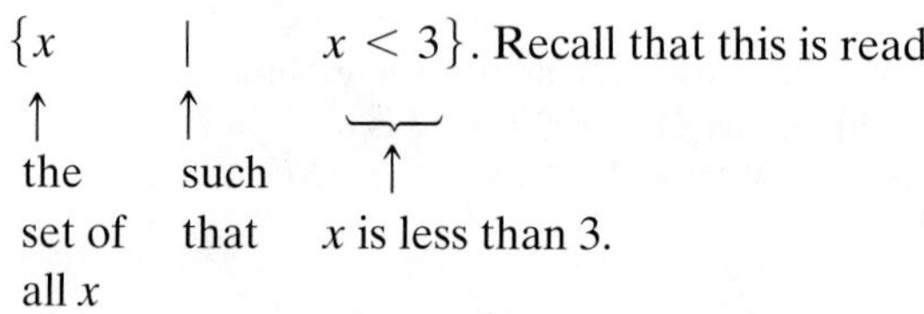

We can also picture the solutions on a number line. If we use open/closed-circle notation, the graph of $\{x|x < 3\}$ looks like the following.

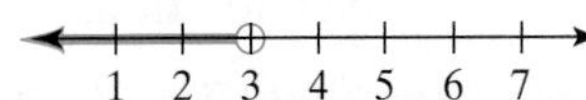

In this text, a convenient notation, called **interval notation,** will be used to write solution sets of inequalities. To help us understand this notation, a different graphing notation will be used. Instead of an open circle, we use a parenthesis; instead of a closed circle, we use a bracket. With this new notation, the graph of $\{x|x < 3\}$ now looks like

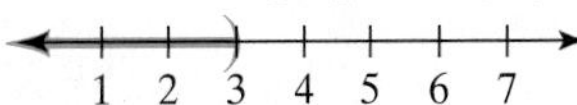

and can be represented in interval notation as $(-\infty, 3)$. The symbol $-\infty$, read as "negative infinity," does not indicate a number but does indicate that the shaded arrow to the left never ends. In other words, the interval $(-\infty, 3)$ includes *all* numbers less than 3.

Picturing the solutions of an inequality on a number line is called **graphing** the solutions or graphing the inequality, and the picture is called the **graph** of the inequality.

To graph $\{x|x \leq 3\}$ or simply $x \leq 3$, shade the numbers to the left of 3 and place a bracket at 3 on the number line. The bracket indicates that 3 **is** a solution: 3 **is** less than or equal to 3. In interval notation, we write $(-\infty, 3]$.

> **Helpful Hint**
>
> When writing an inequality in interval notation, it may be easier to graph the inequality first, then write it in interval notation. To help, think of the number line as approaching $-\infty$ to the left and $+\infty$ or ∞ to the right. Then simply write the interval notation by following your shading from left to right.
>
> $x > 5$ ∞ — 3 4 5 6 7 — $(5, \infty)$
>
> $-\infty$ $x \leq -7$ — −9 −8 −7 −6 −5 — $(-\infty, -7]$

EXAMPLE 1 Graph $x \geq -1$. Then write the solutions in interval notation.

Solution We place a bracket at -1 since the inequality symbol is $\geq$ and -1 is greater than or equal to -1. Then we shade to the right of -1.

−4 −3 −2 −1 0 1 2 3

In interval notation, this is $[-1, \infty)$. □

PRACTICE

1 Graph $x < 5$. Then write the solutions in interval notation.

OBJECTIVE

2 Solving Linear Inequalities 

When solutions of a linear inequality are not immediately obvious, they are found through a process similar to the one used to solve a linear equation. Our goal is to get the variable alone, and we use properties of inequality similar to properties of equality.

> **Addition Property of Inequality**
>
> If a, b, and c are real numbers, then
>
> $$a < b \quad \text{and} \quad a + c < b + c$$
>
> are equivalent inequalities.

This property also holds true for subtracting values, since subtraction is defined in terms of addition. In other words, adding or subtracting the same quantity from both sides of an inequality does not change the solution of the inequality.

EXAMPLE 2 Solve $x + 4 \le -6$ for x. Graph the solution set and write it in interval notation.

Solution To solve for x, subtract 4 from both sides of the inequality.

$x + 4 \le -6$	Original inequality
$x + 4 - 4 \le -6 - 4$	Subtract 4 from both sides.
$x \le -10$	Simplify.

−12 −11 −10 −9 −8 −7 −6

The solution set is $(-\infty, -10]$. □

PRACTICE

2 Solve $x + 11 \ge 6$. Graph the solution set and write it in interval notation.

Helpful Hint

Notice that any number less than or equal to -10 is a solution to $x \le -10$. For example, solutions include

$$-10,\ -200,\ -11\frac{1}{2},\ -7\pi,\ -\sqrt{130},\ -50.3$$

An important difference between linear equations and linear inequalities is shown when we multiply or divide both sides of an inequality by a nonzero real number. For example, start with the true statement $6 < 8$ and multiply both sides by 2. As we see below, the resulting inequality is also true.

$6 < 8$	True
$2(6) < 2(8)$	Multiply both sides by 2.
$12 < 16$	True

But if we start with the same true statement $6 < 8$ and multiply both sides by -2, the resulting inequality is not a true statement.

$6 < 8$	True
$-2(6) < -2(8)$	Multiply both sides by -2.
$-12 < -16$	False

Notice, however, that if we reverse the direction of the inequality symbol, the resulting inequality is true.

$-12 < -16$	False
$-12 > -16$	True

This demonstrates the multiplication property of inequality.

Multiplication Property of Inequality

1. If a, b, and c are real numbers, and c is **positive,** then

$$a < b \quad \text{and} \quad ac < bc$$

are equivalent inequalities.

2. If a, b, and c are real numbers, and c is **negative,** then

$$a < b \quad \text{and} \quad ac > bc$$

are equivalent inequalities.

Because division is defined in terms of multiplication, this property also holds true when dividing both sides of an inequality by a nonzero number. If we multiply or divide both sides of an inequality by a negative number, **the direction of the inequality symbol must be reversed for the inequalities to remain equivalent.**

> **Helpful Hint**
> Whenever both sides of an inequality are multiplied or divided by a negative number, the direction of the inequality symbol **must be** reversed to form an equivalent inequality.

EXAMPLE 3 Solve $-2x \le -4$. Graph the solution set and write it in interval notation.

Solution Remember to reverse the direction of the inequality symbol when dividing by a negative number.

> **Helpful Hint**
> Don't forget to reverse the direction of the inequality symbol.

$$-2x \le -4$$

$$\frac{-2x}{-2} \ge \frac{-4}{-2} \quad \text{Divide both sides by } -2 \text{ and reverse the direction of the inequality symbol.}$$

$$x \ge 2 \quad \text{Simplify.}$$

The solution set $[2, \infty)$ is graphed as shown.

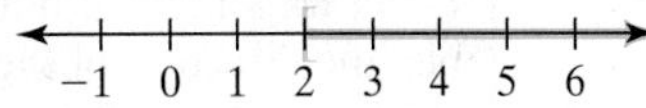

PRACTICE 3 Solve $-5x \ge -15$. Graph the solution set and write it in interval notation.

EXAMPLE 4 Solve $2x < -4$. Graph the solution set and write it in interval notation.

Solution

> **Helpful Hint**
> Do not reverse the inequality sign.

$$2x < -4$$

$$\frac{2x}{2} < \frac{-4}{2} \quad \text{Divide both sides by 2. Do not reverse the direction of the inequality sign.}$$

$$x < -2 \quad \text{Simplify.}$$

The solution set $(-\infty, -2)$ is graphed as shown.

−4 −3 −2 −1 0 1 2

PRACTICE 4 Solve $3x > -9$. Graph the solution set and write it in interval notation.

✓CONCEPT CHECK

Fill in the blank with $<$, $>$, $\le$, or $\ge$.

a. Since $-8 < -4$, then $3(-8)$ ____ $3(-4)$.

b. Since $5 \ge -2$, then $\frac{5}{-7}$ ____ $\frac{-2}{-7}$.

c. If $a < b$, then $2a$ ____ $2b$.

d. If $a \ge b$, then $\frac{a}{-3}$ ____ $\frac{b}{-3}$.

Answers to Concept Check:
a. $<$ **b.** $\le$ **c.** $<$ **d.** $\le$

The following steps may be helpful when solving inequalities. Notice that these steps are similar to the ones given in Section 2.4 for solving equations.

Solving Linear Inequalities in One Variable

Step 1. Clear the inequality of fractions by multiplying both sides of the inequality by the lowest common denominator (LCD) of all fractions in the inequality.

Step 2. Remove grouping symbols such as parentheses by using the distributive property.

Step 3. Simplify each side of the inequality by combining like terms.

Step 4. Write the inequality with variable terms on one side and numbers on the other side by using the addition property of inequality.

Step 5. Get the variable alone by using the multiplication property of inequality.

Helpful Hint

Don't forget that if both sides of an inequality are multiplied or divided by a negative number, the direction of the inequality symbol must be reversed.

EXAMPLE 5 Solve $-4x + 7 \geq -9$. Graph the solution set and write it in interval notation.

Solution

$$-4x + 7 \geq -9$$

$$-4x + 7 - 7 \geq -9 - 7 \quad \text{Subtract 7 from both sides.}$$

$$-4x \geq -16 \quad \text{Simplify.}$$

$$\frac{-4x}{-4} \leq \frac{-16}{-4} \quad \text{Divide both sides by } -4 \text{ and reverse the direction of the inequality symbol.}$$

$$x \leq 4 \quad \text{Simplify.}$$

The solution set $(-\infty, 4]$ is graphed as shown.

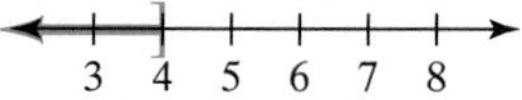

□

PRACTICE 5 Solve $45 - 7x \leq -4$. Graph the solution set and write it in interval notation.

EXAMPLE 6 Solve $2x + 7 \leq x - 11$. Graph the solution set and write it in interval notation.

Solution

$$2x + 7 \leq x - 11$$

$$2x + 7 - x \leq x - 11 - x \quad \text{Subtract } x \text{ from both sides.}$$

$$x + 7 \leq -11 \quad \text{Combine like terms.}$$

$$x + 7 - 7 \leq -11 - 7 \quad \text{Subtract 7 from both sides.}$$

$$x \leq -18 \quad \text{Combine like terms.}$$

The graph of the solution set $(-\infty, -18]$ is shown.

−20 −19 −18 −17 −16 −15 −14

□

PRACTICE 6 Solve $3x + 20 \leq 2x + 13$. Graph the solution set and write it in interval notation.

EXAMPLE 7 Solve $-5x + 7 < 2(x - 3)$. Graph the solution set and write it in interval notation.

Solution

$-5x + 7 < 2(x - 3)$	
$-5x + 7 < 2x - 6$	Apply the distributive property.
$-5x + 7 - 2x < 2x - 6 - 2x$	Subtract $2x$ from both sides.
$-7x + 7 < -6$	Combine like terms.
$-7x + 7 - 7 < -6 - 7$	Subtract 7 from both sides.
$-7x < -13$	Combine like terms.
$\frac{-7x}{-7} > \frac{-13}{-7}$	Divide both sides by -7 and reverse the direction of the inequality symbol.
$x > \frac{13}{7}$	Simplify.

The graph of the solution set $\left(\frac{13}{7}, \infty\right)$ is shown.

$\frac{13}{7}$

−2 −1 0 1 2 3 4

□

PRACTICE 7 Solve $6 - 5x > 3(x - 4)$. Graph the solution set and write it in interval notation.

EXAMPLE 8 Solve $2(x - 3) - 5 \le 3(x + 2) - 18$. Graph the solution set and write it in interval notation.

Solution

$2(x - 3) - 5 \le 3(x + 2) - 18$	
$2x - 6 - 5 \le 3x + 6 - 18$	Apply the distributive property.
$2x - 11 \le 3x - 12$	Combine like terms.
$-x - 11 \le -12$	Subtract $3x$ from both sides.
$-x \le -1$	Add 11 to both sides.
$\frac{-x}{-1} \ge \frac{-1}{-1}$	Divide both sides by -1 and reverse the direction of the inequality symbol.
$x \ge 1$	Simplify.

The graph of the solution set $[1, \infty)$ is shown.

−3 −2 −1 0 1 2 3

□

PRACTICE 8 Solve $3(x - 4) - 5 \le 5(x - 1) - 12$. Graph the solution set and write it in interval notation.

OBJECTIVE 3

Solving Compound Inequalities

Inequalities containing one inequality symbol are called **simple inequalities,** while inequalities containing two inequality symbols are called **compound inequalities.** A compound inequality is really two simple inequalities in one. The compound inequality

$$3 < x < 5 \quad \text{means} \quad 3 < x \textbf{ and } x < 5$$

This can be read "x is greater than 3 and less than 5."

A solution of a compound inequality is a value that is a solution of both of the simple inequalities that make up the compound inequality. For example,

$$4\frac{1}{2} \text{ is a solution of } 3 < x < 5 \text{ since } 3 < 4\frac{1}{2} \textbf{ and } 4\frac{1}{2} < 5.$$

To graph $3 < x < 5$, place parentheses at both 3 and 5 and shade between.

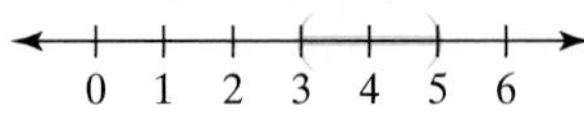

EXAMPLE 9 Graph $2 < x \le 4$. Write the solutions in interval notation.

Solution Graph all numbers greater than 2 and less than or equal to 4. Place a parenthesis at 2, a bracket at 4, and shade between.

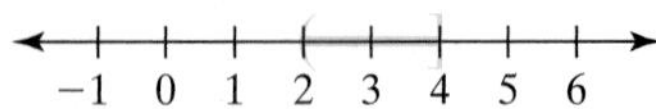

In interval notation, this is $(2, 4]$. □

PRACTICE

9 Graph $-3 \le x < 1$. Write the solutions in interval notation.

When we solve a simple inequality, we isolate the variable on one side of the inequality. When we solve a compound inequality, we isolate the variable in the middle part of the inequality. Also, when solving a compound inequality, we must perform the same operation to all **three** parts of the inequality: left, middle, and right.

EXAMPLE 10 Solve $-1 \le 2x - 3 < 5$. Graph the solution set and write it in interval notation.

Solution

$$-1 \le 2x - 3 < 5$$

$$-1 + 3 \le 2x - 3 + 3 < 5 + 3 \quad \text{Add 3 to all three parts.}$$

$$2 \le 2x < 8 \quad \text{Combine like terms.}$$

$$\frac{2}{2} \le \frac{2x}{2} < \frac{8}{2} \quad \text{Divide all three parts by 2.}$$

$$1 \le x < 4 \quad \text{Simplify.}$$

The graph of the solution set $[1, 4)$ is shown.

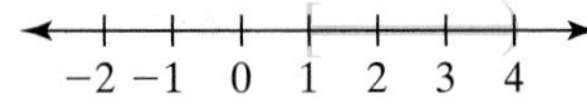

□

PRACTICE

10 Solve $-4 < 3x + 2 \le 8$. Graph the solution set and write it in interval notation.

EXAMPLE 11 Solve $3 \le \frac{3x}{2} + 4 \le 5$. Graph the solution set and write it in interval notation.

Solution

$$3 \le \frac{3x}{2} + 4 \le 5$$

$$2(3) \le 2\left(\frac{3x}{2} + 4\right) \le 2(5) \quad \text{Multiply all three parts by 2 to clear the fraction.}$$

$$6 \le 3x + 8 \le 10 \quad \text{Distribute.}$$

$$-2 \le 3x \le 2 \quad \text{Subtract 8 from all three parts.}$$

$$\frac{-2}{3} \le \frac{3x}{3} \le \frac{2}{3} \quad \text{Divide all three parts by 3.}$$

$$-\frac{2}{3} \le x \le \frac{2}{3} \quad \text{Simplify.}$$

The graph of the solution set $\left[-\frac{2}{3}, \frac{2}{3}\right]$ is shown.

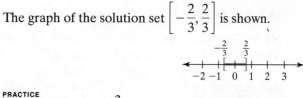

PRACTICE
11 Solve $1 < \frac{3}{4}x + 5 < 6$. Graph the solution set and write it in interval notation.

OBJECTIVE
4 Solving Inequality Applications

Problems containing words such as "at least," "at most," "between," "no more than," and "no less than" usually indicate that an inequality should be solved instead of an equation. In solving applications involving linear inequalities, use the same procedure you use to solve applications involving linear equations.

Some Inequality Translations

≥	≤	<	>
at least	at most	is less than	is greater than
no less than	no more than		

EXAMPLE 12 12 subtracted from 3 times a number is less than 21. Find all numbers that make this statement true.

Solution

1. UNDERSTAND. Read and reread the problem. This is a direct translation problem, and let's let
$x =$ the unknown number
2. TRANSLATE.

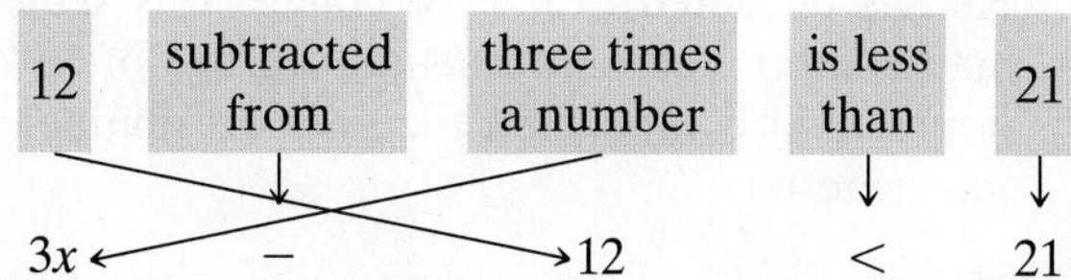

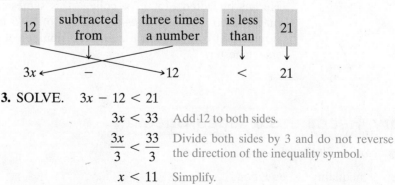

3. SOLVE. $3x - 12 < 21$

$3x < 33$ Add 12 to both sides.

$\frac{3x}{3} < \frac{33}{3}$ Divide both sides by 3 and do not reverse the direction of the inequality symbol.

$x < 11$ Simplify.

4. INTERPRET.

Check: Check the translation; then let's choose a number less than 11 to see if it checks. For example, let's check 10. 12 subtracted from 3 times 10 is 12 subtracted from 30, or 18. Since 18 is less than 21, the number 10 checks.

State: All numbers less than 11 make the original statement true.

PRACTICE
12 Twice a number, subtracted from 35, is greater than 15. Find all numbers that make this true.

EXAMPLE 13 **Staying within Budget**

Marie Chase and Jonathan Edwards are having their wedding reception at the Gallery Reception Hall. They may spend at most \$2000 for the reception. If the reception hall charges a \$100 cleanup fee plus \$36 per person, find the greatest number of people that they can invite and still stay within their budget.

Solution

1. UNDERSTAND. Read and reread the problem. Next, guess a solution. If 40 people attend the reception, the cost is $\$100 + \$36(40) = \$100 + \$1440 = \$1540$. Let $x =$ the number of people who attend the reception.

2. TRANSLATE.

In words:	cleanup fee	+	cost per person	must be less than or equal to	2000
	↓		↓	↓	↓
Translate:	100	+	$36x$	$\leq$	2000

3. SOLVE.

$$100 + 36x \leq 2000$$

$$36x \leq 1900 \quad \text{Subtract 100 from both sides.}$$

$$x \leq 52\frac{7}{9} \quad \text{Divide both sides by 36.}$$

4. INTERPRET.

Check: Since x represents the number of people, we round down to the nearest whole, or 52. Notice that if 52 people attend, the cost is

$\$100 + \$36(52) = \$1972$. If 53 people attend, the cost is

$\$100 + \$36(53) = \$2008$, which is more than the given 2000.

State: Marie Chase and Jonathan Edwards can invite at most 52 people to the reception. □

PRACTICE

13 Kasonga is eager to begin his education at his local community college. He has budgeted \$1500 for college this semester. His local college charges a \$300 matriculation fee and costs an average of \$375 for tuition, fees, and books for each three-credit course. Find the greatest number of classes Kasonga can afford to take this semester.

Vocabulary, Readiness & Video Check

Use the choices below to fill in each blank.

expression inequality equation

1. $6x - 7(x + 9)$ ________

2. $6x = 7(x + 9)$ ________

3. $6x < 7(x + 9)$ ________

4. $5y - 2 \geq -38$ ________

Decide which number listed is not a solution to each given inequality.

5. $x \geq -3;\quad -3, 0, -5, \pi$ ________

6. $x < 6;\quad -6, |-6|, 0, -3.2$ ________

Martin-Gay Interactive Videos

See Video 2.9

Watch the section lecture video and answer the following questions.

OBJECTIVE 1

7. Using Example 1 from the video as a reference, explain the connection between the graph of an inequality and interval notation.

OBJECTIVE 2

8. The steps for solving a linear inequality in one variable are discussed in the lecture before Example 6. Why are you told to be very careful when you use Step 5?

OBJECTIVE 3

9. For Example 8, explain how the solving would change if the compound inequality simplified to $0 < -3x < 14$ instead of $0 < 3x < 14$.

OBJECTIVE 4

10. What is the phrase in Example 9 that tells you to translate to an *inequality*? What does this phrase translate to?

2.9 Exercise Set

MyMathLab®

Graph each set of numbers given in interval notation. Then write an inequality statement in x describing the numbers graphed.

1. $[2, \infty)$

2. $(-3, \infty)$

3. $(-\infty, -5)$

4. $(-\infty, 4]$

Graph each inequality on a number line. Then write the solutions in interval notation. See Example 1.

5. $x \leq -1$

6. $y < 0$

7. $x < \frac{1}{2}$

8. $z < -\frac{2}{3}$

9. $y \geq 5$

10. $x > 3$

Solve each inequality. Graph the solution set and write it in interval notation. See Examples 2 through 4.

11. $2x < -6$

12. $3x > -9$

13. $x - 2 \geq -7$

14. $x + 4 \leq 1$

15. $-8x \leq 16$

16. $-5x < 20$

Solve each inequality. Graph the solution set and write it in interval notation. See Examples 5 and 6.

17. $3x - 5 > 2x - 8$

18. $3 - 7x \geq 10 - 8x$

19. $4x - 1 \leq 5x - 2x$

20. $7x + 3 < 9x - 3x$

Solve each inequality. Graph the solution set and write it in interval notation. See Examples 7 and 8.

21. $x - 7 < 3(x + 1)$

22. $3x + 9 \leq 5(x - 1)$

23. $-6x + 2 \geq 2(5 - x)$

24. $-7x + 4 > 3(4 - x)$

25. $4(3x - 1) \leq 5(2x - 4)$

26. $3(5x - 4) \leq 4(3x - 2)$

27. $3(x + 2) - 6 > -2(x - 3) + 14$

28. $7(x - 2) + x \leq -4(5 - x) - 12$

MIXED PRACTICE

Solve the following inequalities. Graph each solution set and write it in interval notation.

29. $-2x \leq -40$

30. $-7x > 21$

31. $-9 + x > 7$

32. $y - 4 \leq 1$

33. $3x - 7 < 6x + 2$

34. $2x - 1 \geq 4x - 5$

35. $5x - 7x \geq x + 2$

36. $4 - x < 8x + 2x$

37. $\frac{3}{4}x > 2$

38. $\frac{5}{6}x \geq -8$

39. $3(x - 5) < 2(2x - 1)$

40. $5(x + 4) < 4(2x + 3)$

41. $4(2x + 1) < 4$

42. $6(2 - x) \geq 12$

43. $-5x + 4 \geq -4(x - 1)$

44. $-6x + 2 < -3(x + 4)$

45. $-2(x - 4) - 3x < -(4x + 1) + 2x$

46. $-5(1 - x) + x \leq -(6 - 2x) + 6$

47. $\frac{1}{4}(x + 4) < \frac{1}{5}(2x + 3)$

48. $\frac{1}{3}(3x - 1) < \frac{1}{2}(x + 4)$

Graph each inequality. Then write the solutions in interval notation. See Example 9.

49. $-1 < x < 3$

50. $2 \leq y \leq 3$

51. $0 \leq y < 2$

52. $-1 \leq x \leq 4$

Solve each inequality. Graph the solution set and write it in interval notation. See Examples 10 and 11.

53. $-3 < 3x < 6$

54. $-5 < 2x < -2$

55. $2 \leq 3x - 10 \leq 5$

56. $4 \leq 5x - 6 \leq 19$

57. $-4 < 2(x - 3) \leq 4$

58. $0 < 4(x + 5) \leq 8$

59. $-2 < 3x - 5 < 7$

60. $1 < 4 + 2x \leq 7$

61. $-6 < 3(x - 2) \leq 8$

62. $-5 \leq 2(x + 4) < 8$

Solve the following. For Exercises 65 and 66, the solutions have been started for you. See Examples 12 and 13.

63. Six more than twice a number is greater than negative fourteen. Find all numbers that make this statement true.

64. One more than five times a number is less than or equal to ten. Find all such numbers.

65. The perimeter of a rectangle is to be no greater than 100 centimeters and the width must be 15 centimeters. Find the maximum length of the rectangle.

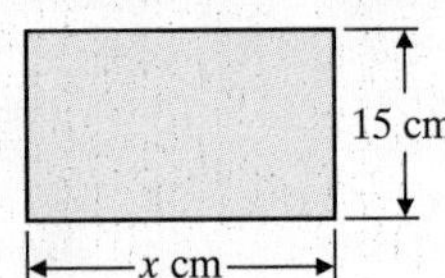

Start the solution:

1. UNDERSTAND the problem. Reread it as many times as needed.
2. TRANSLATE into an equation. (Fill in the blanks below.)

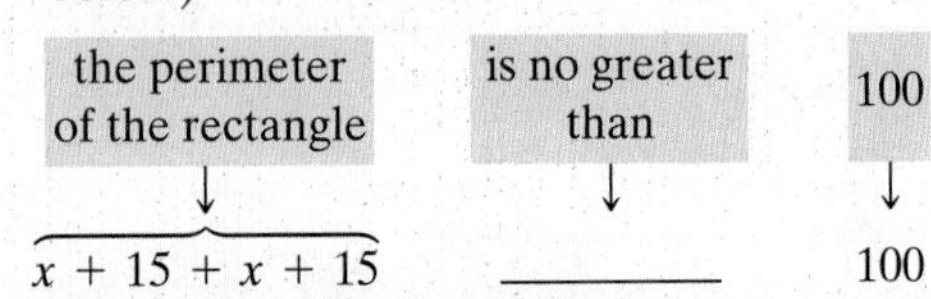

Finish with:

3. SOLVE and **4.** INTERPRET

66. One side of a triangle is three times as long as another side, and the third side is 12 inches long. If the perimeter can be no longer than 32 inches, find the maximum lengths of the other two sides.

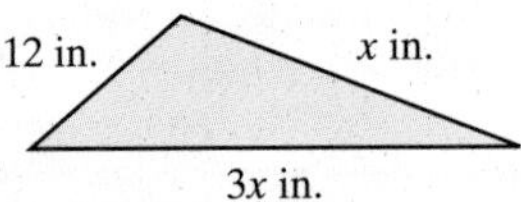

Start the solution:

1. UNDERSTAND the problem. Reread it as many times as needed.
2. TRANSLATE into an equation. (Fill in the blanks below.)

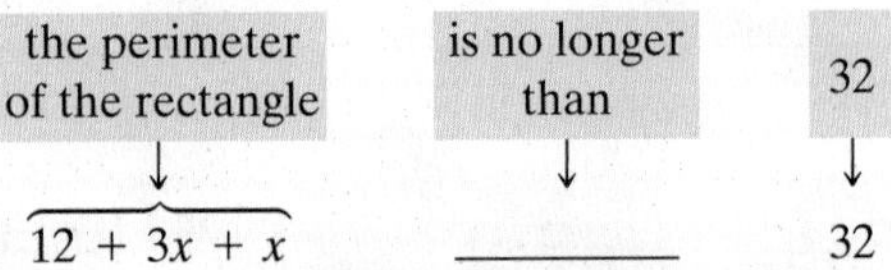

Finish with:

3. SOLVE and **4.** INTERPRET

67. Ben Holladay bowled 146 and 201 in his first two games. What must he bowl in his third game to have an average of at least 180? (*Hint:* The average of a list of numbers is their sum divided by the number of numbers in the list.)

68. On an NBA team, the two forwards measure 6′8″ and 6′6″ tall and the two guards measure 6′0″ and 5′9″ tall. How tall should the center be if they wish to have a starting team average height of at least 6′5″?

69. Dennis and Nancy Wood are celebrating their 30th wedding anniversary by having a reception at Tiffany Oaks reception hall. They have budgeted \$3000 for their reception. If the reception hall charges a \$50.00 cleanup fee plus \$34 per person, find the greatest number of people that they may invite and still stay within their budget.

70. A surprise retirement party is being planned for Pratap Puri. A total of \$860 has been collected for the event, which is to be held at a local reception hall. This reception hall charges a cleanup fee of \$40 and \$15 per person for drinks and light snacks. Find the greatest number of people that may be invited and still stay within the \$860 budget.

71. A 150-pound person uses 5.8 calories per minute when walking at a speed of 4 mph. How long must a person walk at this speed to use at least 200 calories? Round up to the nearest minute. (*Source:* Home & Garden Bulletin No. 72)

72. A 170-pound person uses 5.3 calories per minute when bicycling at a speed of 5.5 mph. How long must a person ride a bike at this speed to use at least 200 calories? Round up to the nearest minute. (*Source:* Same as Exercise 71)

73. Twice a number, increased by one, is between negative five and seven. Find all such numbers.

74. Half a number, decreased by four, is between two and three. Find all such numbers.

REVIEW AND PREVIEW

Evaluate the following. See Section 1.4.

75. $(2)^3$

76. $(3)^3$

77. $(1)^{12}$ **78.** 0^5

79. $\left(\frac{4}{7}\right)^2$ **80.** $\left(\frac{2}{3}\right)^3$

CONCEPT EXTENSIONS

Fill in the box with $<, >, \leq,$ or $\geq$. See the Concept Check in this section.

81. Since $3 < 5$, then $3(-4) \square 5(-4)$.

82. If $m \leq n$, then $2m \square 2n$.

83. If $m \leq n$, then $-2m \square -2n$.

84. If $-x < y$, then $x \square -y$.

85. When solving an inequality, when must you reverse the direction of the inequality symbol?

86. If both sides of the inequality $-3x < 30$ are divided by -3, do you reverse the direction of the inequality symbol? Why or why not?

Solve.

87. Eric Daly has scores of 75, 83, and 85 on his history tests. Use an inequality to find the scores he can make on his final exam to receive a B in the class. The final exam counts as **two** tests, and a B is received if the final course average is greater than or equal to 80.

88. Maria Lipco has scores of 85, 95, and 92 on her algebra tests. Use an inequality to find the scores she can make on her final exam to receive an A in the course. The final exam counts as **three** tests, and an A is received if the final course average is greater than or equal to 90. Round to one decimal place.

89. Explain how solving a linear inequality is similar to solving a linear equation.

90. Explain how solving a linear inequality is different from solving a linear equation.

91. Explain how solving a linear inequality is different from solving a compound inequality.

92. Explain how solving a linear inequality is similar to solving a compound inequality.

93. The formula $C = 3.14d$ can be used to approximate the circumference of a circle given its diameter. Waldo Manufacturing manufactures and sells a certain washer with an outside circumference of 3 centimeters. The company has decided that a washer whose actual circumference is in the interval $2.9 \leq C \leq 3.1$ centimeters is acceptable. Use a compound inequality and find the corresponding interval for diameters of these washers. (Round to 3 decimal places.)

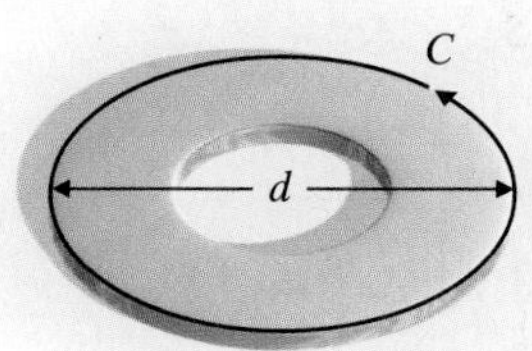

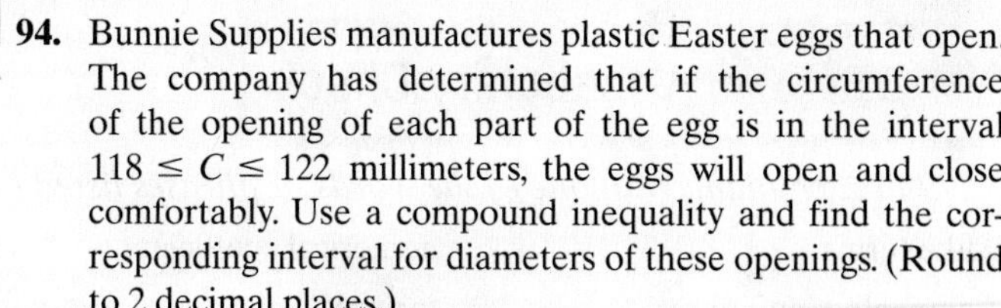

94. Bunnie Supplies manufactures plastic Easter eggs that open. The company has determined that if the circumference of the opening of each part of the egg is in the interval $118 \leq C \leq 122$ millimeters, the eggs will open and close comfortably. Use a compound inequality and find the corresponding interval for diameters of these openings. (Round to 2 decimal places.)

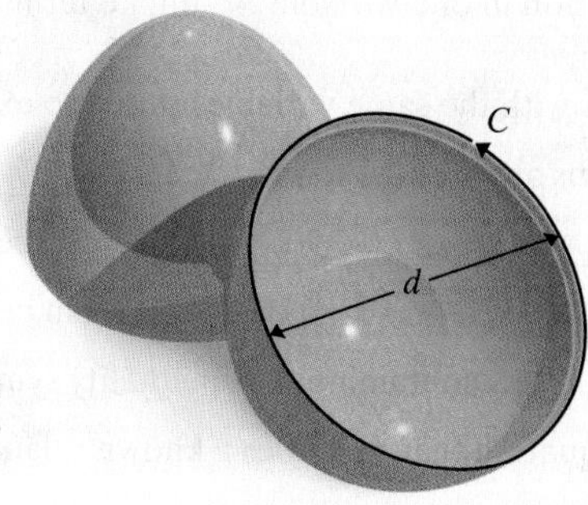

For Exercises 95 through 98, see the example below.

Solve $x(x - 6) > x^2 - 5x + 6$. Graph the solution set and write it in interval notation.

Solution

$$x(x-6) > x^2 - 5x + 6$$
$$x^2 - 6x > x^2 - 5x + 6$$
$$x^2 - 6x - x^2 > x^2 - 5x + 6 - x^2$$
$$-6x > -5x + 6$$
$$-x > 6$$
$$\frac{-x}{-1} < \frac{6}{-1}$$
$$x < -6$$

The solution set $(-\infty, -6)$ is graphed as shown.

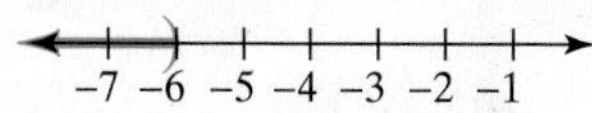

Solve each inequality. Graph the solution set and write it in interval notation.

95. $x(x + 4) > x^2 - 2x + 6$

96. $x(x - 3) \geq x^2 - 5x - 8$

97. $x^2 + 6x - 10 < x(x - 10)$

98. $x^2 - 4x + 8 < x(x + 8)$

Chapter 2 Vocabulary Check

Fill in each blank with one of the words or phrases listed below.

like terms	numerical coefficient	linear inequality in one variable	no solution
equivalent equations	formula	compound inequalities	reversed
linear equation in one variable	unlike terms	all real numbers	the same

1. Terms with the same variables raised to exactly the same powers are called ________________.
2. If terms are not like terms, they are ________________.
3. A(n) ________________ can be written in the form $ax + b = c$.
4. A(n) ________________ can be written in the form $ax + b < c$ (or $>, \leq, \geq$).
5. Inequalities containing two inequality symbols are called ________________.
6. An equation that describes a known relationship among quantities is called a(n) ________________.
7. The ________________ of a term is its numerical factor.
8. Equations that have the same solution are called ________________.
9. The solution(s) to the equation $x + 5 = x + 5$ is/are ________________.
10. The solution(s) to the equation $x + 5 = x + 4$ is/are ________________.
11. If both sides of an inequality are multiplied or divided by the same positive number, the direction of the inequality symbol is ________________.
12. If both sides of an inequality are multiplied by the same negative number, the direction of the inequality symbol is ________________.

Chapter 2 Highlights

DEFINITIONS AND CONCEPTS	EXAMPLES
Section 2.1 Simplifying Algebraic Expressions	
The **numerical coefficient** of a **term** is its numerical factor.	***Term*** — ***Numerical Coefficient*** $-7y$ — -7 x — 1 $\frac{1}{5}a^2b$ — $\frac{1}{5}$
Terms with the same variables raised to exactly the same powers are **like terms.**	***Like Terms*** — ***Unlike Terms*** $12x, -x$ — $3y, 3y^2$ $-2xy, 5yx$ — $7a^2b, -2ab^2$
To combine like terms, add the numerical coefficients and multiply the result by the common variable factor.	$9y + 3y = 12y$ $-4z^2 + 5z^2 - 6z^2 = -5z^2$
To remove parentheses, apply the distributive property.	$-4(x + 7) + 10(3x - 1)$ $= -4x - 28 + 30x - 10$ $= 26x - 38$

DEFINITIONS AND CONCEPTS	EXAMPLES
Section 2.2 The Addition Property of Equality	
A **linear equation in one variable** can be written in the form $ax + b = c$ where a, b, and c are real numbers and $a \neq 0$.	***Linear Equations*** $-3x + 7 = 2$ $3(x - 1) = -8(x + 5) + 4$
Equivalent equations are equations that have the same solution.	$x - 7 = 10$ and $x = 17$ are equivalent equations.
Addition Property of Equality Adding the same number to or subtracting the same number from both sides of an equation does not change its solution.	$y + 9 = 3$ $y + 9 - 9 = 3 - 9$ $y = -6$
Section 2.3 The Multiplication Property of Equality	
Multiplication Property of Equality Multiplying both sides or dividing both sides of an equation by the same nonzero number does not change its solution.	$\frac{2}{3}a = 18$ $\frac{3}{2}\left(\frac{2}{3}a\right) = \frac{3}{2}(18)$ $a = 27$
Section 2.4 Solving Linear Equations	
To Solve Linear Equations	Solve: $\frac{5(-2x + 9)}{6} + 3 = \frac{1}{2}$
1. Clear the equation of fractions.	**1.** $6 \cdot \frac{5(-2x + 9)}{6} + 6 \cdot 3 = 6 \cdot \frac{1}{2}$ $5(-2x + 9) + 18 = 3$
2. Remove any grouping symbols such as parentheses.	**2.** $-10x + 45 + 18 = 3$ Distributive property
3. Simplify each side by combining like terms.	**3.** $-10x + 63 = 3$ Combine like terms.
4. Write variable terms on one side and numbers on the other side using the addition property of equality.	**4.** $-10x + 63 - 63 = 3 - 63$ Subtract 63. $-10x = -60$
5. Get the variable alone using the multiplication property of equality.	**5.** $\frac{-10x}{-10} = \frac{-60}{-10}$ Divide by -10. $x = 6$
6. Check by substituting in the original equation.	**6.** $\frac{5(-2x + 9)}{6} + 3 = \frac{1}{2}$ $\frac{5(-2 \cdot 6 + 9)}{6} + 3 \stackrel{?}{=} \frac{1}{2}$ $\frac{5(-3)}{6} + 3 \stackrel{?}{=} \frac{1}{2}$ $-\frac{5}{2} + \frac{6}{2} \stackrel{?}{=} \frac{1}{2}$ $\frac{1}{2} = \frac{1}{2}$ True

DEFINITIONS AND CONCEPTS	EXAMPLES
Section 2.5 An Introduction to Problem Solving	
Problem-Solving Steps	The height of the Hudson volcano in Chile is twice the height of the Kiska volcano in the Aleutian Islands. If the sum of their heights is 12,870 feet, find the height of each.
1. UNDERSTAND the problem.	**1.** Read and reread the problem. Guess a solution and check your guess. Let x be the height of the Kiska volcano. Then $2x$ is the height of the Hudson volcano. x Kiska $2x$ Hudson
2. TRANSLATE the problem.	**2.** In word: height of Kiska → x; added to → $+$; height of Hudson → $2x$; is → $=$; 12,870 → 12,870 Translate: $x + 2x = 12{,}870$
3. SOLVE.	**3.** $x + 2x = 12{,}870$ $3x = 12{,}870$ $x = 4290$
4. INTERPRET the results.	**4.** *Check:* If x is 4290, then $2x$ is 2(4290) or 8580. Their sum is $4290 + 8580$ or 12,870, the required amount. *State:* The Kiska volcano is 4290 feet high and the Hudson volcano is 8580 feet high.
Section 2.6 Formulas and Problem Solving	
An equation that describes a known relationship among quantities is called a **formula.**	***Formulas*** $A = lw$ (area of a rectangle) $I = PRT$ (simple interest)
If all values for the variables in a formula are known except for one, this unknown value may be found by substituting in the known values and solving.	If $d = 182$ miles and $r = 52$ miles per hour in the formula $d = r \cdot t$, find t. $d = r \cdot t$ $182 = 52 \cdot t$ Let $d = 182$ and $r = 52$. $3.5 = t$ The time is 3.5 hours.
To solve a formula for a specified variable, use the same steps as for solving a linear equation. Treat the specified variable as the only variable of the equation.	Solve: $P = 2l + 2w$ for l. $P = 2l + 2w$ $P - 2w = 2l + 2w - 2w$ Subtract $2w$. $P - 2w = 2l$ $\frac{P - 2w}{2} = \frac{2l}{2}$ Divide by 2. $\frac{P - 2w}{2} = l$ Simplify.

DEFINITIONS AND CONCEPTS	EXAMPLES

Section 2.7 Percent and Mixture Problem Solving

Use the same problem-solving steps to solve a problem containing percents.

32% of what number is 36.8?

1. UNDERSTAND.

1. Read and reread. Propose a solution and check. Let $x =$ the unknown number.

2. TRANSLATE.

2.

32%	of	what number	is	36.8
↓	↓	↓	↓	↓
32%	·	x	=	36.8

3. SOLVE.

3. *Solve:*

$$32\% \cdot x = 36.8$$
$$0.32x = 36.8$$
$$\frac{0.32x}{0.32} = \frac{36.8}{0.32} \quad \text{Divide by 0.32.}$$
$$x = 115 \quad \text{Simplify.}$$

4. INTERPRET.

4. *Check, then state:* 32% of 115 is 36.8.

How many liters of a 20% acid solution must be mixed with a 50% acid solution to obtain 12 liters of a 30% solution?

1. UNDERSTAND.

1. Read and reread. Guess a solution and check.

Let $x =$ number of liters of 20% solution.

Then $12 - x =$ number of liters of 50% solution.

2. TRANSLATE.

2.

	No. of Liters ·	*Acid Strength* =	*Amount of Acid*
20% Solution	x	20%	$0.20x$
50% Solution	$12 - x$	50%	$0.50(12 - x)$
30% Solution Needed	12	30%	$0.30(12)$

In words:

	acid in 20% solution	+	acid in 50% solution	=	acid in 30% solution
	↓		↓		↓
Translate:	$0.20x$	+	$0.50(12 - x)$	=	$0.30(12)$

3. SOLVE.

3. Solve: $0.20x + 0.50(12 - x) = 0.30(12)$

$$0.20x + 6 - 0.50x = 3.6 \quad \text{Apply the distributive property.}$$
$$-0.30x + 6 = 3.6$$
$$-0.30x = -2.4 \quad \text{Subtract 6.}$$
$$x = 8 \quad \text{Divide by } -0.30.$$

4. INTERPRET.

4. *Check, then state:*

If 8 liters of a 20% acid solution are mixed with 12 − 8 or 4 liters of a 50% acid solution, the result is 12 liters of a 30% solution.

DEFINITIONS AND CONCEPTS	EXAMPLES

Section 2.8 Further Problem Solving

Problem-Solving Steps

A collection of dimes and quarters has a total value of \$19.55. If there are three times as many quarters as dimes, find the number of quarters.

1. UNDERSTAND.

1. Read and reread. Propose a solution and check.

Let x = number of dimes and

$3x$ = number of quarters.

2. TRANSLATE.

2. In words: value of dimes + value of quarters = 19.55

Translate: $0.10x + 0.25(3x) = 19.55$

3. SOLVE.

3. Solve: $0.10x + 0.75x = 19.55$ Multiply.

$0.85x = 19.55$ Add like terms.

$x = 23$ Divide by 0.85.

4. INTERPRET.

4. *Check, then state:*

The number of dimes is 23 and the number of quarters is 3(23) or 69. The total value of this money is

$0.10(23) + 0.25(69) = 19.55$, so our result checks.

The number of quarters is 69.

Section 2.9 Solving Linear Inequalities

A **linear inequality in one variable** is an inequality that can be written in one of the forms:

$$ax + b < c \qquad ax + b \le c$$
$$ax + b > c \qquad ax + b \ge c$$

where a, b, and c are real numbers and a is not 0.

Linear Inequalities

$$2x + 3 < 6 \qquad 5(x - 6) \ge 10$$
$$\frac{x-2}{5} > \frac{5x+7}{2} \qquad \frac{-(x+8)}{9} \le \frac{-2x}{11}$$

Addition Property of Inequality

Adding the same number to or subtracting the same number from both sides of an inequality does not change the solutions.

$$y + 4 \le -1$$
$$y + 4 - 4 \le -1 - 4 \quad \text{Subtract 4.}$$
$$y \le -5$$

$(-\infty, -5)$

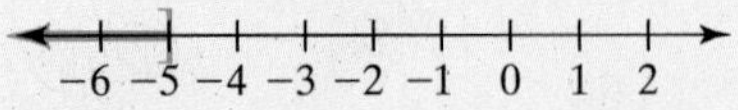

Multiplication Property of Inequality

Multiplying or dividing both sides of an inequality by the same positive number does not change its solutions.

$$\frac{1}{3}x > -2$$
$$3\left(\frac{1}{3}x\right) > 3 \cdot -2 \quad \text{Multiply by 3.}$$
$$x > -6 \quad (-6, \infty)$$

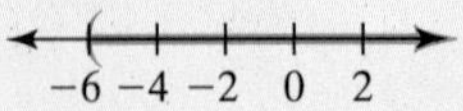

Multiplying or dividing both sides of an inequality by the same **negative number and reversing the direction of the inequality symbol** does not change its solutions.

$$-2x \le 4$$
$$\frac{-2x}{-2} \ge \frac{4}{-2} \quad \text{Divide by −2, reverse inequality symbol.}$$
$$x \ge -2 \quad [-2, \infty)$$

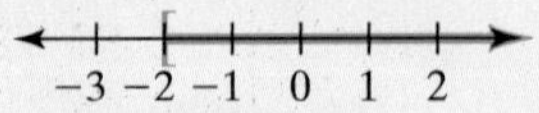

DEFINITIONS AND CONCEPTS	EXAMPLES
Section 2.9 Solving Linear Inequalities (continued)	
To Solve Linear Inequalities	Solve: $3(x+2) \le -2 + 8$
1. Clear the equation of fractions.	**1.** No fractions to clear. $3(x+2) \le -2+8$
2. Remove grouping symbols.	**2.** $3x + 6 \le -2 + 8$ Distributive property
3. Simplify each side by combining like terms.	**3.** $3x + 6 \le 6$ Combine like terms.
4. Write variable terms on one side and numbers on the other side, using the addition property of inequality.	**4.** $3x + 6 - 6 \le 6 - 6$ Subtract 6. $3x \le 0$
5. Get the variable alone, using the multiplication property of inequality.	**5.** $\frac{3x}{3} \le \frac{0}{3}$ Divide by 3. $x \le 0$ $(-\infty, 0)$ −2 −1 0 1 2
Inequalities containing two inequality symbols are called **compound inequalities.**	***Compound Inequalities*** $-2 < x < 6$ $5 \le 3(x-6) < \frac{20}{3}$
To solve a compound inequality, isolate the variable in the middle part of the inequality. Perform the same operation to all three parts of the inequality: left, middle, right.	Solve: $-2 < 3x + 1 < 7$ $-2 - 1 < 3x + 1 - 1 < 7 - 1$ Subtract 1. $-3 < 3x < 6$ $\frac{-3}{3} < \frac{3x}{3} < \frac{6}{3}$ Divide by 3. $-1 < x < 2$ $(-1, 2)$ −2 −1 0 1 2 3

Chapter 2 Review

(2.1) Simplify the following expressions.

1. $5x - x + 2x$
2. $0.2z - 4.6x - 7.4z$
3. $\frac{1}{2}x + 3 + \frac{7}{2}x - 5$
4. $\frac{4}{5}y + 1 + \frac{6}{5}y + 2$
5. $2(n-4) + n - 10$
6. $3(w+2) - (12 - w)$
7. Subtract $7x - 2$ from $x + 5$.
8. Subtract $1.4y - 3$ from $y - 0.7$.

Write each of the following as algebraic expressions.

9. Three times a number decreased by 7
10. Twice the sum of a number and 2.8 added to 3 times the number

(2.2) Solve each equation.

11. $8x + 4 = 9x$
12. $5y - 3 = 6y$
13. $\frac{2}{7}x + \frac{5}{7}x = 6$
14. $3x - 5 = 4x + 1$
15. $2x - 6 = x - 6$
16. $4(x+3) = 3(1+x)$
17. $6(3+n) = 5(n-1)$
18. $5(2+x) - 3(3x+2) = -5(x-6) + 2$

Use the addition property to fill in the blank so that the middle equation simplifies to the last equation.

19. $x - 5 = 3$
$x - 5 + \_\_ = 3 + \_\_$
$x = 8$

20. $x + 9 = -2$
$x + 9 - \_\_ = -2 - \_\_$
$x = -11$

Choose the correct algebraic expression.

21. The sum of two numbers is 10. If one number is x, express the other number in terms of x.
    a. $x - 10$ **b.** $10 - x$
    c. $10 + x$ **d.** $10x$

22. Mandy is 5 inches taller than Melissa. If x inches represents the height of Mandy, express Melissa's height in terms of x.
    a. $x - 5$ **b.** $5 - x$
    c. $5 + x$ **d.** $5x$

△ 23. If one angle measures $x°$, express the measure of its complement in terms of x.
    a. $(180 - x)°$ **b.** $(90 - x)°$
    c. $(x - 180)°$ **d.** $(x - 90)°$

△ **24.** If one angle measures $(x + 5)°$, express the measure of its supplement in terms of x.

a. $(185 + x)°$
b. $(95 + x)°$
c. $(175 - x)°$
d. $(x - 170)°$

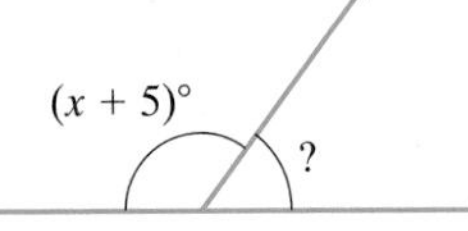

***(2.3)** Solve each equation.*

25. $\frac{3}{4}x = -9$

26. $\frac{x}{6} = \frac{2}{3}$

27. $-5x = 0$

28. $-y = 7$

29. $0.2x = 0.15$

30. $\frac{-x}{3} = 1$

31. $-3x + 1 = 19$

32. $5x + 25 = 20$

33. $7(x - 1) + 9 = 5x$

34. $7x - 6 = 5x - 3$

35. $-5x + \frac{3}{7} = \frac{10}{7}$

36. $5x + x = 9 + 4x - 1 + 6$

37. Write the sum of three consecutive integers as an expression in x. Let x be the first integer.

38. Write the sum of the first and fourth of four consecutive even integers. Let x be the first even integer.

***(2.4)** Solve each equation.*

39. $\frac{5}{3}x + 4 = \frac{2}{3}x$

40. $\frac{7}{8}x + 1 = \frac{5}{8}x$

41. $-(5x + 1) = -7x + 3$

42. $-4(2x + 1) = -5x + 5$

43. $-6(2x - 5) = -3(9 + 4x)$

44. $3(8y - 1) = 6(5 + 4y)$

45. $\frac{3(2 - z)}{5} = z$

46. $\frac{4(n + 2)}{5} = -n$

47. $0.5(2n - 3) - 0.1 = 0.4(6 + 2n)$

48. $-9 - 5a = 3(6a - 1)$

49. $\frac{5(c + 1)}{6} = 2c - 3$

50. $\frac{2(8 - a)}{3} = 4 - 4a$

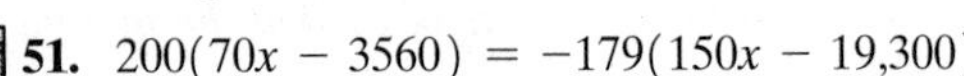

51. $200(70x - 3560) = -179(150x - 19{,}300)$

52. $1.72y - 0.04y = 0.42$

***(2.5)** Solve each of the following.*

53. The height of the Washington Monument is 50.5 inches more than 10 times the length of a side of its square base. If the sum of these two dimensions is 7327 inches, find the height of the Washington Monument. (*Source:* National Park Service)

54. A 12-foot board is to be divided into two pieces so that one piece is twice as long as the other. If x represents the length of the shorter piece, find the length of each piece.

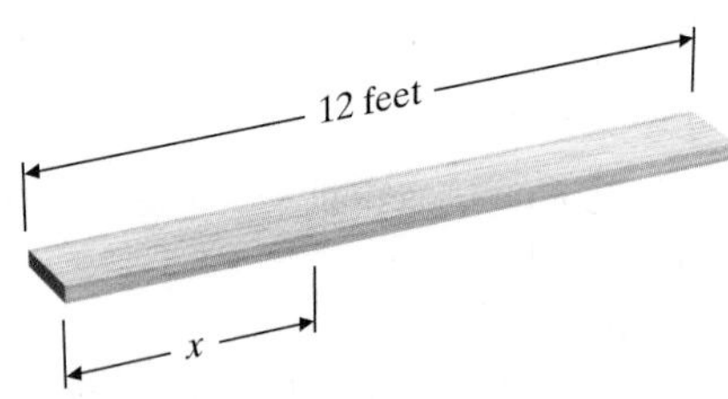

55. In a recent year, Kellogg Company acquired Keebler Foods Company. After the merger, the total number of Kellogg and Keebler manufacturing plants was 53. The number of Kellogg plants was one less than twice the number of Keebler plants. How many of each type of plant were there? (*Source: Kellogg Company 2000 Annual Report*)

56. Find three consecutive integers whose sum is -114.

57. The quotient of a number and 3 is the same as the difference of the number and two. Find the number.

58. Double the sum of a number and 6 is the opposite of the number. Find the number.

***(2.6)** Substitute the given values into the given formulas and solve for the unknown variable.*

59. $P = 2l + 2w$; $P = 46, l = 14$

60. $V = lwh$; $V = 192, l = 8, w = 6$

Solve each equation as indicated.

61. $y = mx + b$ for m

62. $r = vst - 5$ for s

63. $2y - 5x = 7$ for x

64. $3x - 6y = -2$ for y

△ **65.** $C = \pi D$ for π

△ **66.** $C = 2\pi r$ for π

△ **67.** A swimming pool holds 900 cubic meters of water. If its length is 20 meters and its height is 3 meters, find its width.

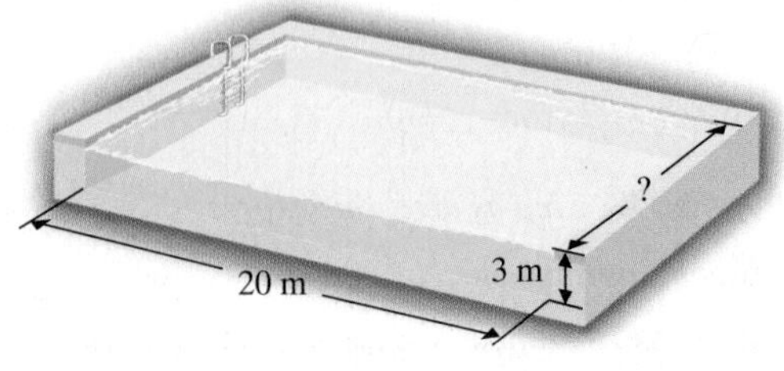

68. The perimeter of a rectangular billboard is 60 feet and has a length 6 feet longer than its width. Find the dimensions of the billboard.

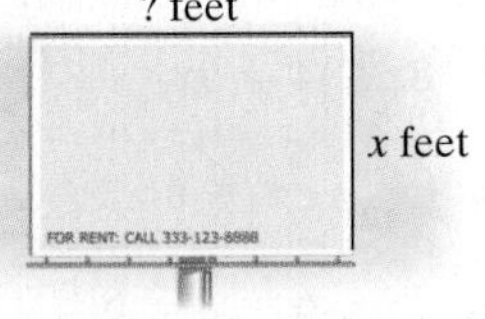

69. A charity 10K race is given annually to benefit a local hospice organization. How long will it take to run/walk a 10K race (10 kilometers or 10,000 meters) if your average pace is 125 **meters** per minute? Give your time in hours and minutes.

70. On April 28, 2001, the highest temperature recorded in the United States was 104°F, which occurred in Death Valley, California. Convert this temperature to degrees Celsius. (*Source:* National Weather Service)

(2.7) Find each of the following.

71. The number 9 is what percent of 45?

72. The number 59.5 is what percent of 85?

73. The number 137.5 is 125% of what number?

74. The number 768 is 60% of what number?

75. The price of a small diamond ring was recently increased by 11%. If the ring originally cost $1900, find the mark-up and the new price of the ring.

76. A recent survey found that 79% of Americans use the Internet. If a city has a population of 76,000 how many people in that city would you expect to use the Internet? (*Source:* PEW)

77. Thirty gallons of a 20% acid solution is needed for an experiment. Only 40% and 10% acid solutions are available. How much of each should be mixed to form the needed solution?

78. The ACT Assessment is a college entrance exam taken by about 60% of college-bound students. The national average score was 20.7 in 1993 and rose to 21.0 in 2008. Find the percent increase. (Round to the nearest hundredth of a percent.)

The graph below shows the percent(s) of cell phone users who have engaged in various behaviors while driving and talking on their cell phones. Use this graph to answer Exercises 79 through 82.

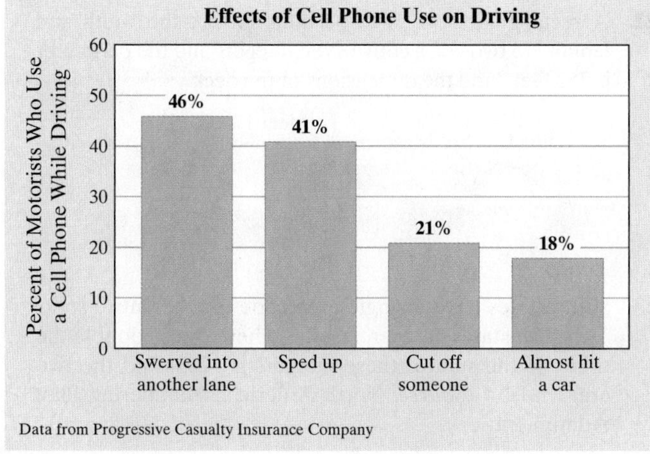

79. What percent of motorists who use a cell phone while driving have almost hit another car?

80. What is the most common effect of cell phone use on driving?

Suppose that a cell-phone service has an estimated 4600 customers who use their cell phones while driving. Use this information for Exercises 81 and 82.

81. How many of these customers would you expect to have cut someone off while driving and talking on their cell phones?

82. How many of these customers would you expect to have sped up while driving and talking on their cell phones?

83. If a price decreases from $250 to $170, what is the percent increase?

84. Find the original price of a DVD if the sale price is $19.20 after a 20% discount.

85. In 2005, Lance Armstrong incredibly won his seventh Tour de France, the first man in history to win more than five Tour de France Championships. Suppose he rides a bicycle up a category 2 climb at 10 km/hr and rides down the same distance at a speed of 50 km/hr. Find the distance traveled if the total time on the mountain was 3 hours.

(2.8) Solve.

86. A $50,000 retirement pension is to be invested into two accounts: a money market fund that pays 8.5% and a certificate of deposit that pays 10.5%. How much should be invested at each rate to provide a yearly interest income of $4550?

87. A pay phone is holding its maximum number of 500 coins consisting of nickels, dimes, and quarters. The number of quarters is twice the number of dimes. If the value of all the coins is $88.00, how many nickels were in the pay phone?

88. How long will it take an Amtrak passenger train to catch up to a freight train if their speeds are 60 and 45 mph and the freight train had an hour and a half head start?

(2.9) Solve and graph the solutions of each of the following inequalities.

89. $x > 0$

90. $x \leq -2$

91. $0.5 \leq y < 1.5$

92. $-1 < x < 1$

93. $-3x > 12$

94. $-2x \geq -20$

95. $x + 4 \geq 6x - 16$

96. $5x - 7 > 8x + 5$

97. $-3 < 4x - 1 < 2$

98. $2 \leq 3x - 4 < 6$

99. $4(2x - 5) \leq 5x - 1$

100. $-2(x - 5) > 2(3x - 2)$

101. Tina earns $175 per week plus a 5% commission on all her sales. Find the minimum amount of sales to ensure that she earns at least $300 per week.

102. Ellen Catarella shot rounds of 76, 82, and 79 golfing. What must she shoot on her next round so that her average will be below 80?

MIXED REVIEW

Solve each equation.

103. $6x + 2x - 1 = 5x + 11$

104. $2(3y - 4) = 6 + 7y$

105. $4(3 - a) - (6a + 9) = -12a$

106. $\frac{x}{3} - 2 = 5$

107. $2(y + 5) = 2y + 10$

108. $7x - 3x + 2 = 2(2x - 1)$

Solve.

109. The sum of six and twice a number is equal to seven less than the number. Find the number.

110. A 23-inch piece of string is to be cut into two pieces so that the length of the longer piece is three more than four times the shorter piece. If x represents the length of the shorter piece, find the lengths of both pieces.

Solve for the specified variable.

111. $V = \frac{1}{3}Ah$ for h

112. What number is 26% of 85?

113. The number 72 is 45% of what number?

114. A company recently increased its number of employees from 235 to 282. Find the percent increase.

Solve each inequality. Graph the solution set.

115. $4x - 7 > 3x + 2$

116. $-5x < 20$

117. $-3(1 + 2x) + x \geq -(3 - x)$

Chapter 2 Test

MyMathLab®

YouTube

Simplify each of the following expressions.

1. $2y - 6 - y - 4$

2. $2.7x + 6.1 + 3.2x - 4.9$

3. $4(x - 2) - 3(2x - 6)$

4. $7 + 2(5y - 3)$

Solve each of the following equations.

5. $-\frac{4}{5}x = 4$

6. $4(n - 5) = -(4 - 2n)$

7. $5y - 7 + y = -(y + 3y)$

8. $4z + 1 - z = 1 + z$

9. $\frac{2(x + 6)}{3} = x - 5$

10. $\frac{1}{2} - x + \frac{3}{2} = x - 4$

11. $-0.3(x - 4) + x = 0.5(3 - x)$

12. $-4(a + 1) - 3a = -7(2a - 3)$

13. $-2(x - 3) = x + 5 - 3x$

14. Find the value of x if $y = -14$, $m = -2$, and $b = -2$ in the formula $y = mx + b$.

Solve each of the following equations for the indicated variable.

15. $V = \pi r^2 h$ for h

16. $3x - 4y = 10$ for y

Solve each of the following inequalities. Graph each solution set and write it in interval notation.

17. $3x - 5 \geq 7x + 3$

18. $x + 6 > 4x - 6$

19. $-2 < 3x + 1 < 8$

20. $\frac{2(5x + 1)}{3} > 2$

Solve each of the following applications.

21. A number increased by two-thirds of the number is 35. Find the number.

22. A rectangular deck is to be built so that the width and length are two consecutive even integers and the perimeter is 252 feet. Find the dimensions of the deck.

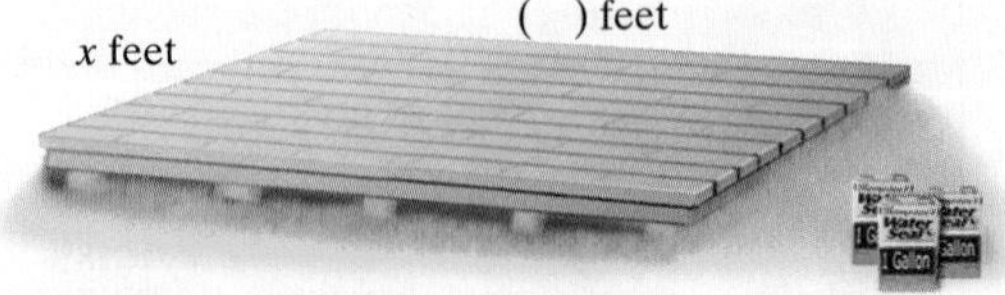

23. Some states have a single area code for the entire state. Two such states have area codes where one is double the other. If the sum of these integers is 1203, find the two area codes. *(Source:* North American Numbering Plan Administration*)*

24. Sedric Angell invested an amount of money in Amoxil stock that earned an annual 10% return, and then he invested twice that amount in IBM stock that earned an annual 12% return. If his total return from both investments was \$2890, find how much he invested in each stock.

The following graph shows the breakdown of tornadoes occurring in the United States by strength. The corresponding Fujita Tornado Scale categories are shown in parentheses. Use this graph to answer Exercise 26.

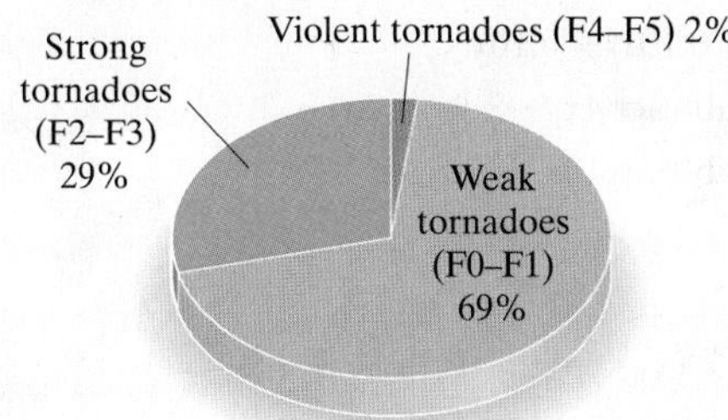

Data from National Climatic Data Center

25. Two trains leave Los Angeles simultaneously traveling on the same track in opposite directions at speeds of 50 and 64 mph. How long will it take before they are 285 miles apart?

26. According to the National Climatic Data Center, in an average year, about 800 tornadoes are reported in the United States. How many of these would you expect to be classified as "weak" tornadoes?

27. The number 72 is what percent of 180?

28. The number of employees of a company decreased from 225 to 189. Find this percent decrease.

Chapter 2 Cumulative Review

1. Given the set $\left\{-2, 0, \frac{1}{4}, -1.5, 112, -3, 11, \sqrt{2}\right\}$, list the numbers in this set that belong to the set of:

a. Natural numbers
b. Whole numbers
c. Integers
d. Rational numbers
e. Irrational numbers
f. Real numbers

2. Given the set $\left\{7, 2, -\frac{1}{5}, 0, \sqrt{3}, -185, 8\right\}$, list the numbers in this set that belong to the set of:

a. Natural numbers
b. Whole numbers
c. Integers
d. Rational numbers
e. Irrational numbers
f. Real numbers

3. Find the absolute value of each number.

a. $|4|$ **b.** $|-5|$
c. $|0|$ **d.** $\left|-\frac{1}{2}\right|$
e. $|5.6|$

4. Find the absolute value of each number.

a. $|5|$ **b.** $|-8|$ **c.** $\left|-\frac{2}{3}\right|$

5. Write each of the following numbers as a product of primes.

a. 40 **b.** 63

6. Write each number as a product of primes.

a. 44 **b.** 90

7. Write $\frac{2}{5}$ as an equivalent fraction with a denominator of 20.

8. Write $\frac{2}{3}$ as an equivalent fraction with a denominator of 24.

9. Simplify $3[4 + 2(10 - 1)]$.

10. Simplify $5[16 - 4(2 + 1)]$.

11. Decide whether 2 is a solution of $3x + 10 = 8x$.

12. Decide whether 3 is a solution of $5x - 2 = 4x$.

Add.

13. $-1 + (-2)$

14. $(-2) + (-8)$

15. $-4 + 6$

16. $-3 + 10$

17. Simplify each expression.

a. $-(-10)$ **b.** $-\left(-\frac{1}{2}\right)$
c. $-(-2x)$ **d.** $-|-6|$

18. Simplify each expression.

a. $-(-5)$ **b.** $-\left(-\frac{2}{3}\right)$
c. $-(-a)$ **d.** $-|-3|$

19. Subtract.

a. $5.3 - (-4.6)$ **b.** $-\frac{3}{10} - \frac{5}{10}$
c. $-\frac{2}{3} - \left(-\frac{4}{5}\right)$

20. Subtract

a. $-2.7 - 8.4$ **b.** $-\frac{4}{5} - \left(-\frac{3}{5}\right)$

c. $\frac{1}{4} - \left(-\frac{1}{2}\right)$

21. Find each unknown complementary or supplementary angle.

a.

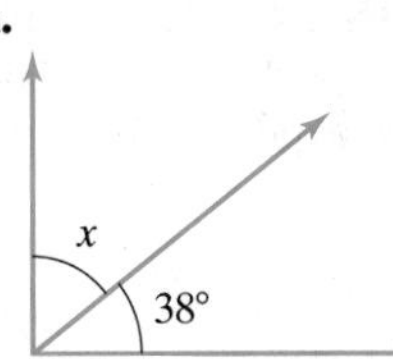

b.

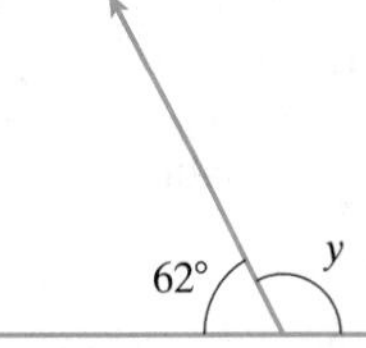

22. Find each unknown complementary or supplementary angle.

a.

b.

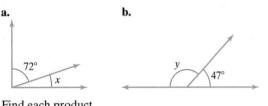

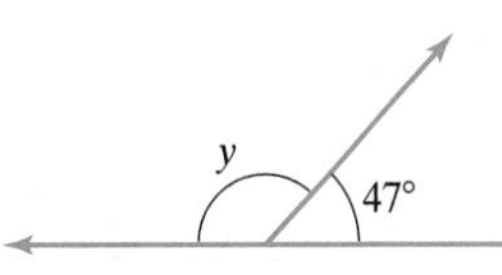

23. Find each product.

a. $(-1.2)(0.05)$ **b.** $\frac{2}{3} \cdot \left(-\frac{7}{10}\right)$

c. $\left(-\frac{4}{5}\right)(-20)$

24. Find each product.

a. $(4.5)(-0.08)$ **b.** $-\frac{3}{4} \cdot \left(-\frac{8}{17}\right)$

25. Divide.

a. $\frac{-24}{-4}$ **b.** $\frac{-36}{3}$

c. $\frac{2}{3} \div \left(-\frac{5}{4}\right)$ **d.** $-\frac{3}{2} \div 9$

26. Divide.

a. $\frac{-32}{8}$ **b.** $\frac{-108}{-12}$

c. $\frac{-5}{7} \div \left(\frac{-9}{2}\right)$

27. Use a commutative property to complete each statement.

a. $x + 5 =$ _____

b. $3 \cdot x =$ _____

28. Use a commutative property to complete each statement.

a. $y + 1 =$ _____

b. $y \cdot 4 =$ _____

29. Use the distributive property to write each sum as a product.

a. $8 \cdot 2 + 8 \cdot x$

b. $7s + 7t$

30. Use the distributive property to write each sum as a product.

a. $4 \cdot y + 4 \cdot \frac{1}{3}$

b. $0.10x + 0.10y$

31. Subtract $4x - 2$ from $2x - 3$.

32. Subtract $10x + 3$ from $-5x + 1$.

Solve.

33. $\frac{1}{2} = x - \frac{3}{4}$

34. $\frac{5}{6} + x = \frac{2}{3}$

35. $6(2a - 1) - (11a + 6) = 7$

36. $-3x + 1 - (-4x - 6) = 10$

37. $\frac{y}{7} = 20$

38. $\frac{x}{4} = 18$

39. $4(2x - 3) + 7 = 3x + 5$

40. $6x + 5 = 4(x + 4) - 1$

41. Twice the sum of a number and 4 is the same as four times the number, decreased by 12. Find the number.

42. A number increased by 4 is the same as 3 times the number decreased by 8. Find the number.

43. Solve $V = lwh$ for l.

44. Solve $C = 2\pi r$ for r.

45. Solve $x + 4 \leq -6$ for x. Graph the solution set and write it in interval notation.

46. Solve $x - 3 > 2$ for x. Graph the solution set and write it in interval notation.

CHAPTER 3

Graphing

In the previous chapter, we learned to solve and graph the solutions of linear equations and inequalities in one variable. Now we define and present techniques for solving and graphing linear equations and inequalities in two variables.

Tourism 2020 Vision is the World Tourism Organization's long-term forecast of world tourism through 2020. The broken-line graph below shows the forecasts. In Section 3.1, Exercises 1 through 6, we read a bar graph showing the top tourist destinations by country.

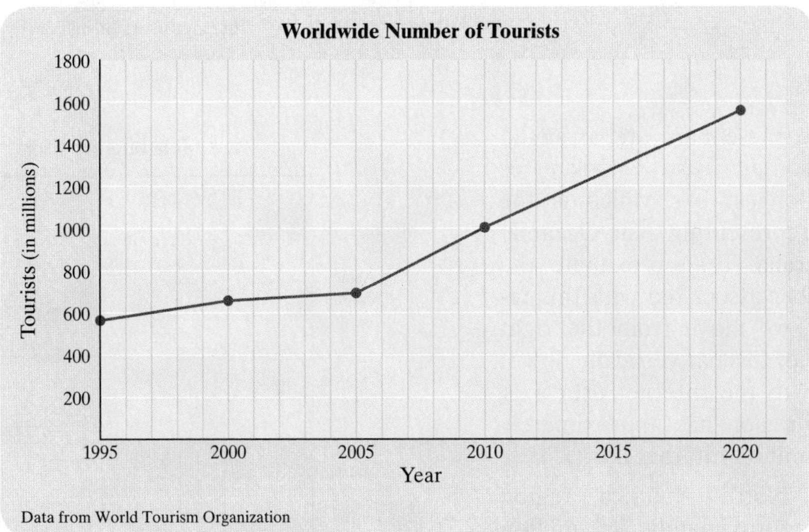

3.1 Reading Graphs and the Rectangular Coordinate System 

OBJECTIVES

1. Read Bar and Line Graphs.
2. Define the Rectangular Coordinate System and Plot Ordered Pairs of Numbers.
3. Graph Paired Data to Create a Scatter Diagram.
4. Determine Whether an Ordered Pair Is a Solution of an Equation in Two Variables.
5. Find the Missing Coordinate of an Ordered Pair Solution, Given One Coordinate of the Pair.

In today's world, where the exchange of information must be fast and entertaining, graphs are becoming increasingly popular. They provide a quick way of making comparisons, drawing conclusions, and approximating quantities.

OBJECTIVE

1 Reading Bar and Line Graphs 

A **bar graph** consists of a series of bars arranged vertically or horizontally. The bar graph in Example 1 shows a comparison of worldwide Internet users by region. The names of the regions are listed vertically and a bar is shown for each region. Corresponding to the length of the bar for each region is a number along a horizontal axis. These horizontal numbers are numbers of Internet users in millions.

EXAMPLE 1 The following bar graph shows the estimated number of Internet users worldwide by region as of a recent year.

a. Find the region that has the most Internet users and approximate the number of users.

b. How many more users are in the North America region than the Latin America/Caribbean region?

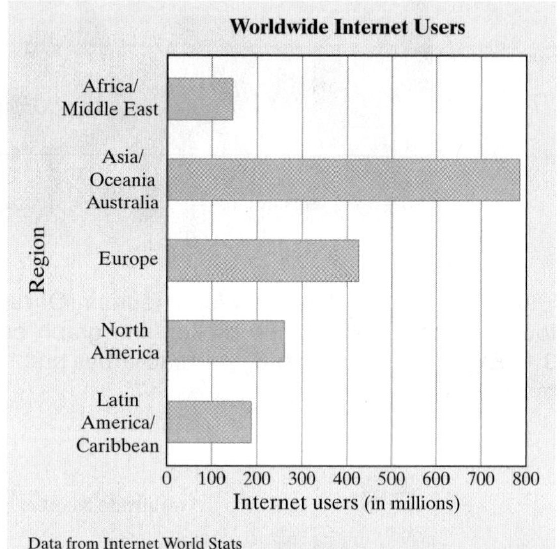

Solution

a. Since these bars are arranged horizontally, we look for the longest bar, which is the bar representing Asia/Oceania/Australia. To approximate the number associated with this region, we move from the right edge of this bar vertically downward to the Internet user axis. This region has approximately 785 million Internet users.

b. The North America region has approximately 260 million Internet users. The Latin America/Caribbean region has approximately 187 million Internet users. To find how many more users are in the North America region, we subtract $260 - 187 = 73$ million more Internet users. □

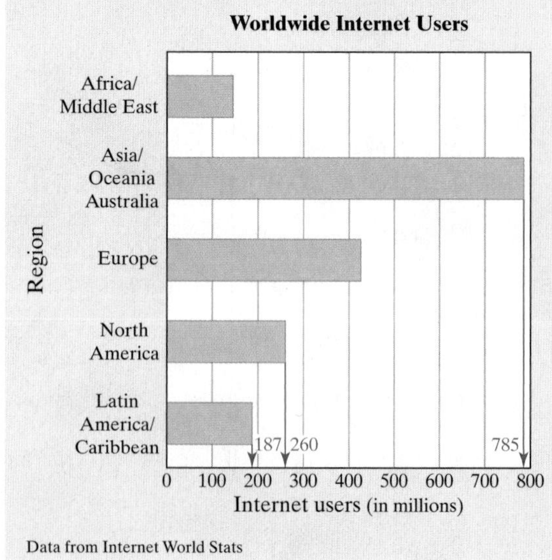

PRACTICE

1 Use the graph from Example 1 to answer the following.

a. Find the region with the fewest Internet users and approximate the number of users.

b. How many more users are in the Asia/Oceania/Australia region than in the Africa/Middle East region?

A **line graph** consists of a series of points connected by a line. The next graph is an example of a line graph. It is also sometimes called a **broken-line graph.**

EXAMPLE 2 The line graph shows the relationship between time spent smoking a cigarette and pulse rate. Time is recorded along the horizontal axis in minutes, with 0 minutes being the moment a smoker lights a cigarette. Pulse is recorded along the vertical axis in heartbeats per minute.

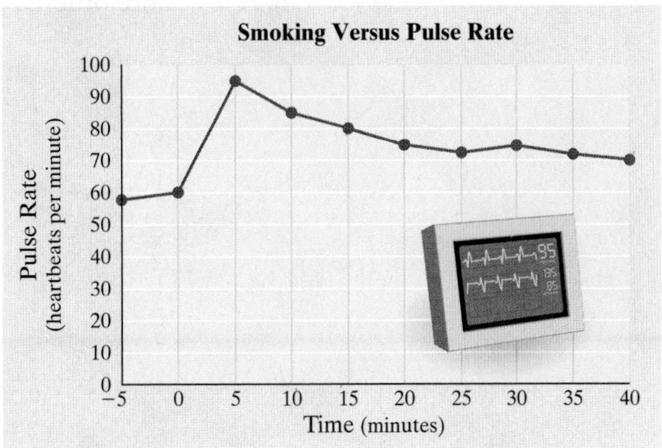

a. What is the pulse rate 15 minutes after a cigarette is lit?

b. When is the pulse rate the lowest?

c. When does the pulse rate show the greatest change?

Solution

a. We locate the number 15 along the time axis and move vertically upward until the line is reached. From this point on the line, we move horizontally to the left until the pulse rate axis is reached. Reading the number of beats per minute, we find that the pulse rate is 80 beats per minute 15 minutes after a cigarette is lit.

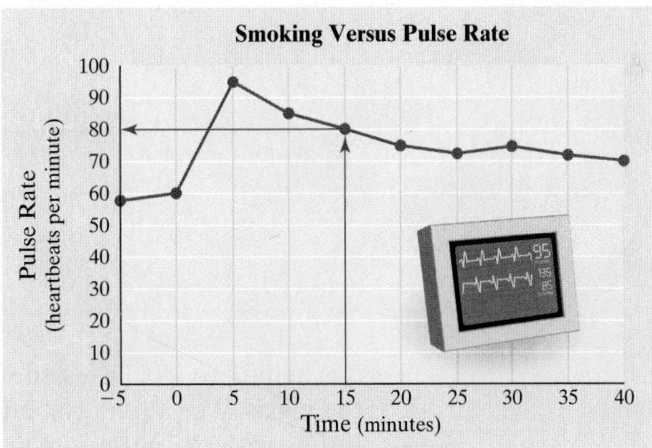

b. We find the lowest point of the line graph, which represents the lowest pulse rate. From this point, we move vertically downward to the time axis. We find that the pulse rate is the lowest at −5 minutes, which means 5 minutes *before* lighting a cigarette.

c. The pulse rate shows the greatest change during the 5 minutes between 0 and 5. Notice that the line graph is *steepest* between 0 and 5 minutes. □

PRACTICE

2 Use the graph from Example 2 to answer the following.

a. What is the pulse rate 40 minutes after lighting a cigarette?

b. What is the pulse rate when the cigarette is being lit?

c. When is the pulse rate the highest?

OBJECTIVE

2 Defining the Rectangular Coordinate System and Plotting Ordered Pairs of Numbers

Notice in the previous graph that two numbers are associated with each point of the graph. For example, we discussed earlier that 15 minutes after lighting a cigarette, the pulse rate is 80 beats per minute. If we agree to write the time first and the pulse rate second, we can say there is a point on the graph corresponding to the **ordered pair** of numbers (15, 80). A few more ordered pairs are listed alongside their corresponding points.

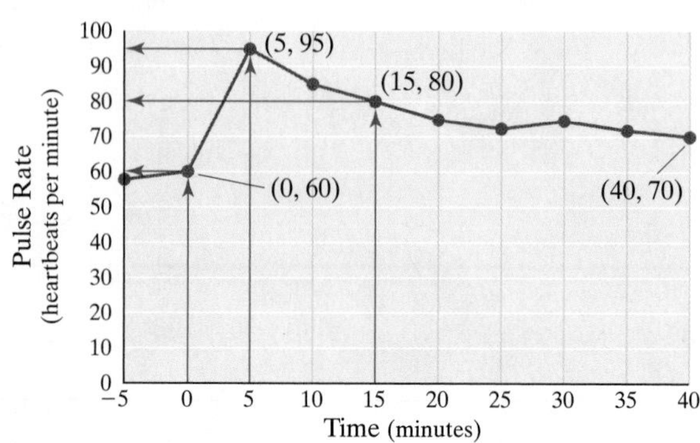

In general, we use this same ordered pair idea to describe the location of a point in a plane (such as a piece of paper). We start with a horizontal and a vertical axis. Each axis is a number line and, for the sake of consistency, we construct our axes to intersect at the 0 coordinate of both. This point of intersection is called the **origin.** Notice that these two number lines or axes divide the plane into four regions called **quadrants.** The quadrants are usually numbered with Roman numerals as shown. The axes are not considered to be in any quadrant.

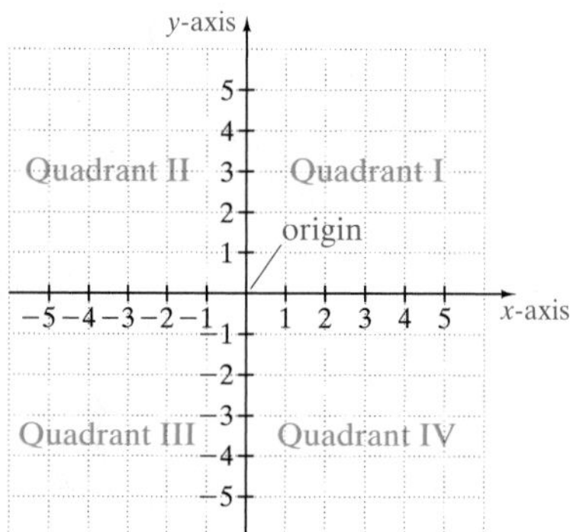

It is helpful to label axes, so we label the horizontal axis the ***x*-axis** and the vertical axis the ***y*-axis.** We call the system described above the **rectangular coordinate system.**

Just as with the pulse rate graph, we can then describe the locations of points by ordered pairs of numbers. We list the horizontal ***x*-axis** measurement first and the vertical ***y*-axis** measurement second.

To plot or graph the point corresponding to the ordered pair

$$(a, b)$$

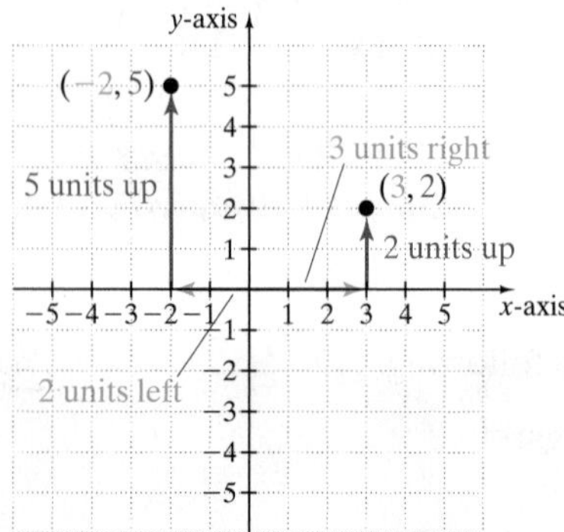

we start at the origin. We then move a units left or right (right if a is positive, left if a is negative). From there, we move b units up or down (up if b is positive, down if b is negative). For example, to plot the point corresponding to the ordered pair (3, 2), we start at the origin, move 3 units right and from there move 2 units up. (See the figure to the left.) The x-value, 3, is called the ***x*-coordinate** and the y-value, 2, is called the ***y*-coordinate.** From now on, we will call the point with coordinates (3, 2) simply the point (3, 2). The point $(-2, 5)$ is graphed to the left also.

Does the order in which the coordinates are listed matter? Yes! Notice below that the point corresponding to the ordered pair (2, 3) is in a different location than the point corresponding to (3, 2). These two ordered pairs of numbers describe two different points of the plane.

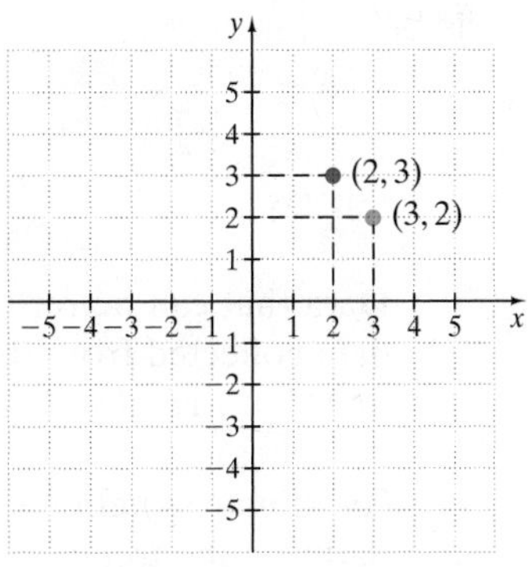

✓CONCEPT CHECK

Is the graph of the point $(-5, 1)$ in the same location as the graph of the point $(1, -5)$? Explain.

Helpful Hint

Don't forget that **each ordered pair corresponds to exactly one point in the plane and that each point in the plane corresponds to exactly one ordered pair.**

EXAMPLE 3 On a single coordinate system, plot each ordered pair. State in which quadrant, if any, each point lies.

a. $(5, 3)$ **b.** $(-5, 3)$ **c.** $(-2, -4)$ **d.** $(1, -2)$ **e.** $(0, 0)$

f. $(0, 2)$ **g.** $(-5, 0)$ **h.** $\left(0, -5\frac{1}{2}\right)$ **i.** $\left(4\frac{2}{3}, -3\right)$

Solution

a. Point $(5, 3)$ lies in quadrant I.

b. Point $(-5, 3)$ lies in quadrant II.

c. Point $(-2, -4)$ lies in quadrant III.

d. Point $(1, -2)$ lies in quadrant IV.

e.–h. Points $(0, 0)$, $(0, 2)$, $(-5, 0)$, and $\left(0, -5\frac{1}{2}\right)$ lie on axes, so they are not in any quadrant.

i. Point $\left(4\frac{2}{3}, -3\right)$ lies in quadrant IV.

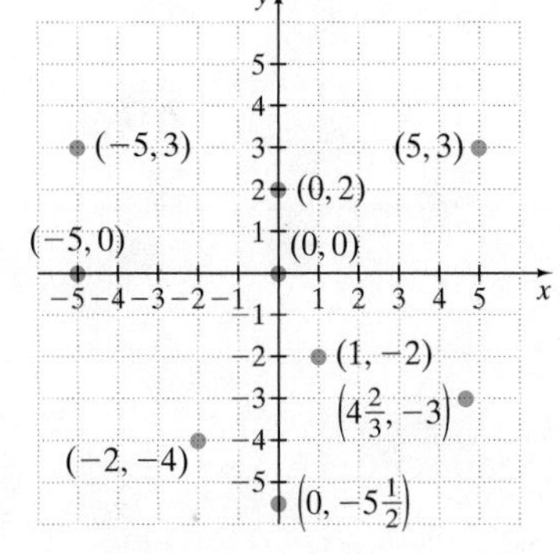

From Example 3, notice that the y-coordinate of any point on the x-axis is 0. For example, the point $(-5, 0)$ lies on the x-axis. Also, the x-coordinate of any point on the y-axis is 0. For example, the point $(0, 2)$ lies on the y-axis. □

PRACTICE

3 On a single coordinate system, plot each ordered pair. State in which quadrant, if any, each point lies.

a. $(4, -3)$ **b.** $(-3, 5)$ **c.** $(0, 4)$ **d.** $(-6, 1)$

e. $(-2, 0)$ **f.** $(5, 5)$ **g.** $\left(3\frac{1}{2}, 1\frac{1}{2}\right)$ **h.** $(-4, -5)$

Answer to Concept Check:
The graph of point $(-5, 1)$ lies in quadrant II and the graph of point $(1, -5)$ lies in quadrant IV. They are *not* in the same location.

✓CONCEPT CHECK

For each description of a point in the rectangular coordinate system, write an ordered pair that represents it.

a. Point A is located three units to the left of the y-axis and five units above the x-axis.

b. Point B is located six units below the origin.

OBJECTIVE

3 Graphing Paired Data 

Data that can be represented as an ordered pair is called **paired data.** Many types of data collected from the real world are paired data. For instance, the annual measurement of a child's height can be written as an ordered pair of the form (year, height in inches) and is paired data. The graph of paired data as points in the rectangular coordinate system is called a **scatter diagram.** Scatter diagrams can be used to look for patterns and trends in paired data.

EXAMPLE 4 The table gives the annual net sales (in billions of dollars) for Target stores for the years shown. (Data from www.retailsales.com)

Year	*Target Net Sales (in billions of dollars)*
2006	58
2007	61
2008	63
2009	63
2010	66

a. Write this paired data as a set of ordered pairs of the form (year, sales in billions of dollars).

b. Create a scatter diagram of the paired data.

c. What trend in the paired data does the scatter diagram show?

Solution

a. The ordered pairs are (2006, 58), (2007, 61), (2008, 63), (2009, 63), (2010, 66).

b. We begin by plotting the ordered pairs. Because the x-coordinate in each ordered pair is a year, we label the x-axis "Year" and mark the horizontal axis with the years given. Then we label the y-axis or vertical axis "Net Sales (in billions of dollars)." In this case we can mark the vertical axis in multiples of 2. Since no net sale is less than 58, we use the notation ⚡ to skip to 58, then proceed by multiples of 2.

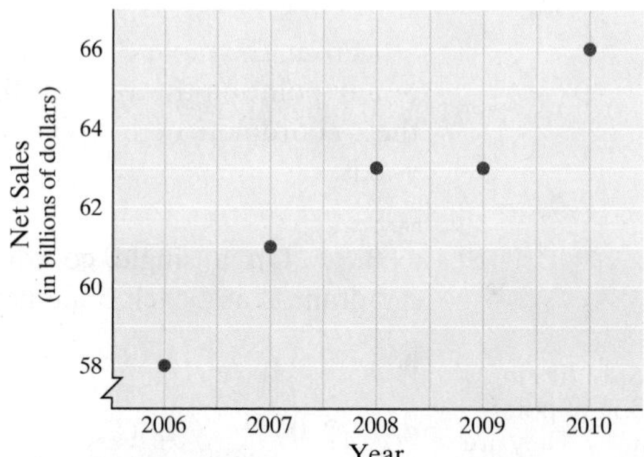

Answers to Concept Check:
a. $(-3, 5)$ **b.** $(0, -6)$

c. The scatter diagram shows that Target net sales were constant or steadily increased over the years 2006–2010. □

PRACTICE

4 The table gives the approximate annual number of wildfires (in thousands) that have occurred in the United States for the years shown. (Data from National Interagency Fire Center)

Year	*Wildfires (in thousands)*
2004	65
2005	67
2006	96
2007	86
2008	79
2009	79
2010	72

a. Write this paired data as a set of ordered pairs of the form (year, number of wildfires in thousands).

b. Create a scatter diagram of the paired data.

OBJECTIVE

4 Determining Whether an Ordered Pair Is a Solution 

Let's see how we can use ordered pairs to record solutions of equations containing two variables. An equation in one variable such as $x + 1 = 5$ has one solution, which is 4: The number 4 is the value of the variable x that makes the equation true.

An equation in two variables, such as $2x + y = 8$, has solutions consisting of two values, one for x and one for y. For example, $x = 3$ and $y = 2$ is a solution of $2x + y = 8$ because, if x is replaced with 3 and y with 2, we get a true statement.

$$\begin{aligned} 2x + y &= 8 \\ 2(3) + 2 &= 8 \\ 8 &= 8 \quad \text{True} \end{aligned}$$

The solution $x = 3$ and $y = 2$ can be written as $(3, 2)$, an **ordered pair** of numbers. The first number, 3, is the x-value and the second number, 2, is the y-value.

In general, an ordered pair is a **solution** of an equation in two variables if replacing the variables by the values of the ordered pair results in a true statement.

EXAMPLE 5 Determine whether each ordered pair is a solution of the equation $x - 2y = 6$.

a. $(6, 0)$ **b.** $(0, 3)$ **c.** $\left(1, -\frac{5}{2}\right)$

Solution

a. Let $x = 6$ and $y = 0$ in the equation $x - 2y = 6$.

$$\begin{aligned} x - 2y &= 6 \\ 6 - 2(0) &= 6 \quad \text{Replace } x \text{ with 6 and } y \text{ with 0.} \\ 6 - 0 &= 6 \quad \text{Simplify.} \\ 6 &= 6 \quad \text{True} \end{aligned}$$

$(6, 0)$ is a solution, since $6 = 6$ is a true statement.

b. Let $x = 0$ and $y = 3$.

$$\begin{aligned} x - 2y &= 6 \\ 0 - 2(3) &= 6 \quad \text{Replace } x \text{ with 0 and } y \text{ with 3.} \\ 0 - 6 &= 6 \\ -6 &= 6 \quad \text{False} \end{aligned}$$

(0, 3) is *not* a solution, since $-6 = 6$ is a false statement.

c. Let $x = 1$ and $y = -\frac{5}{2}$ in the equation.

$$\begin{aligned} x - 2y &= 6 \\ 1 - 2\left(-\frac{5}{2}\right) &= 6 \quad \text{Replace } x \text{ with 1 and } y \text{ with } -\frac{5}{2}. \\ 1 + 5 &= 6 \\ 6 &= 6 \quad \text{True} \end{aligned}$$

$\left(1, -\frac{5}{2}\right)$ is a solution, since $6 = 6$ is a true statement. □

PRACTICE

5 Determine whether each ordered pair is a solution of the equation $x + 3y = 6$.

a. (3, 1) **b.** (6, 0) **c.** $\left(-2, \frac{2}{3}\right)$

OBJECTIVE

5 Completing Ordered Pair Solutions 

If one value of an ordered pair solution of an equation is known, the other value can be determined. To find the unknown value, replace one variable in the equation by its known value. Doing so results in an equation with just one variable that can be solved for the variable using the methods of Chapter 2.

EXAMPLE 6 Complete the following ordered pair solutions for the equation $3x + y = 12$.

a. (0,) **b.** (, 6) **c.** (−1,)

Solution

a. In the ordered pair (0,), the x-value is 0. Let $x = 0$ in the equation and solve for y.

$$\begin{aligned} 3x + y &= 12 \\ 3(0) + y &= 12 \quad \text{Replace } x \text{ with 0.} \\ 0 + y &= 12 \\ y &= 12 \end{aligned}$$

The completed ordered pair is (0, 12).

b. In the ordered pair (, 6), the y-value is 6. Let $y = 6$ in the equation and solve for x.

$$\begin{aligned} 3x + y &= 12 \\ 3x + 6 &= 12 \quad \text{Replace } y \text{ with 6.} \\ 3x &= 6 \quad \text{Subtract 6 from both sides.} \\ x &= 2 \quad \text{Divide both sides by 3.} \end{aligned}$$

The ordered pair is (2, 6).

c. In the ordered pair $(-1,\ \ )$, the x-value is -1. Let $x = -1$ in the equation and solve for y.

$$3x + y = 12$$
$$3(-1) + y = 12 \quad \text{Replace } x \text{ with } -1.$$
$$-3 + y = 12$$
$$y = 15 \quad \text{Add 3 to both sides.}$$

The ordered pair is $(-1, 15)$. □

PRACTICE

6 Complete the following ordered pair solutions for the equation $2x - y = 8$.

a. $(0,\ \ )$ **b.** $(\ \ , 4)$ **c.** $(-3,\ \ )$

Solutions of equations in two variables can also be recorded in a **table of values,** as shown in the next example.

EXAMPLE 7 Complete the table for the equation $y = 3x$.

	x	y
a.	−1	
b.		0
c.		−9

Solution

a. Replace x with -1 in the equation and solve for y.

$$y = 3x$$
$$y = 3(-1) \quad \text{Let } x = -1.$$
$$y = -3$$

The ordered pair is $(-1, -3)$.

b. Replace y with 0 in the equation and solve for x.

$$y = 3x$$
$$0 = 3x \quad \text{Let } y = 0.$$
$$0 = x \quad \text{Divide both sides by 3.}$$

The ordered pair is $(0, 0)$.

c. Replace y with -9 in the equation and solve for x.

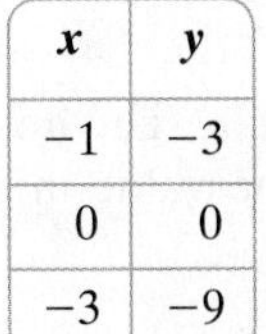

x	y
−1	−3
0	0
−3	−9

$$y = 3x$$
$$-9 = 3x \quad \text{Let } y = -9.$$
$$-3 = x \quad \text{Divide both sides by 3.}$$

The ordered pair is $(-3, -9)$. The completed table is shown to the left. □

PRACTICE

7 Complete the table for the equation $y = -4x$.

	x	y
a.	−2	
b.		−12
c.	0	

EXAMPLE 8 Complete the table for the equation

$$y = \frac{1}{2}x - 5.$$

	x	y
a.	−2	
b.	0	
c.		0

Solution

a. Let $x = -2$.

$$y = \frac{1}{2}x - 5$$
$$y = \frac{1}{2}(-2) - 5$$
$$y = -1 - 5$$
$$y = -6$$

b. Let $x = 0$.

$$y = \frac{1}{2}x - 5$$
$$y = \frac{1}{2}(0) - 5$$
$$y = 0 - 5$$
$$y = -5$$

c. Let $y = 0$.

$$y = \frac{1}{2}x - 5$$
$$0 = \frac{1}{2}x - 5 \quad \text{Now, solve for } x.$$
$$5 = \frac{1}{2}x \quad \text{Add 5.}$$
$$10 = x \quad \text{Multiply by 2.}$$

Ordered Pairs: $(-2, -6)$ $(0, -5)$ $(10, 0)$

The completed table is

x	y
−2	−6
0	−5
10	0

□

PRACTICE

8 Complete the table for the equation $y = \frac{1}{5}x - 2$.

	x	y
a.	−10	
b.	0	
c.		0

EXAMPLE 9 **Finding the Value of a Computer**

A computer was recently purchased for a small business for $2000. The business manager predicts that the computer will be used for 5 years and the value in dollars y of the computer in x years is $y = -300x + 2000$. Complete the table.

x	0	1	2	3	4	5
y						

Solution To find the value of y when x is 0, replace x with 0 in the equation. We use this same procedure to find y when x is 1 and when x is 2.

When x = 0,

$$y = -300x + 2000$$
$$y = -300 \cdot 0 + 2000$$
$$y = 0 + 2000$$
$$y = 2000$$

When x = 1,

$$y = -300x + 2000$$
$$y = -300 \cdot 1 + 2000$$
$$y = -300 + 2000$$
$$y = 1700$$

When x = 2,

$$y = -300x + 2000$$
$$y = -300 \cdot 2 + 2000$$
$$y = -600 + 2000$$
$$y = 1400$$

We have the ordered pairs (0, 2000), (1, 1700), and (2, 1400). This means that in 0 years, the value of the computer is $2000, in 1 year the value of the computer is $1700, and in

2 years the value is \$1400. To complete the table of values, we continue the procedure for $x = 3$, $x = 4$, and $x = 5$.

When $x = 3$,	***When*** $x = 4$,	***When*** $x = 5$,
$y = -300x + 2000$	$y = -300x + 2000$	$y = -300x + 2000$
$y = -300 \cdot 3 + 2000$	$y = -300 \cdot 4 + 2000$	$y = -300 \cdot 5 + 2000$
$y = -900 + 2000$	$y = -1200 + 2000$	$y = -1500 + 2000$
$y = 1100$	$y = 800$	$y = 500$

The completed table is

x	0	1	2	3	4	5
y	2000	1700	1400	1100	800	500

□

PRACTICE

9 A college student purchased a used car for \$12,000. The student predicted that she would need to use the car for four years and the value in dollars y of the car in x years is $y = -1800x + 12{,}000$. Complete this table.

x	0	1	2	3	4
y					

The ordered pair solutions recorded in the completed table for the example above are graphed below. Notice that the graph gives a visual picture of the decrease in value of the computer.

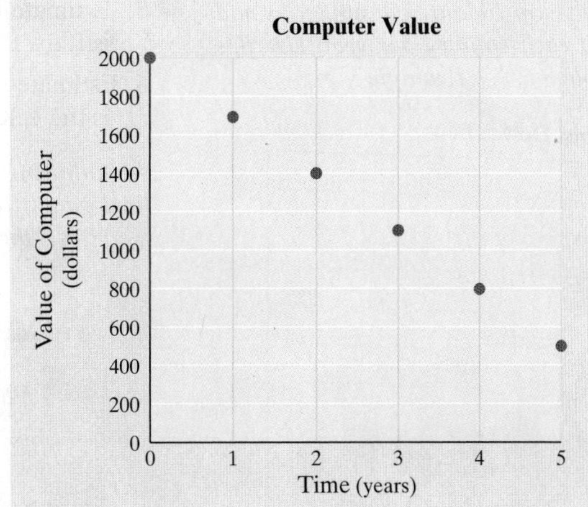

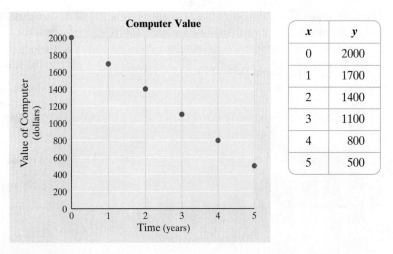

x	y
0	2000
1	1700
2	1400
3	1100
4	800
5	500

Vocabulary, Readiness & Video Check

Use the choices below to fill in each blank. The exercises below all have to do with the rectangular coordinate system.

origin	x-coordinate	x-axis	one	four
quadrants	y-coordinate	y-axis	solution	

1. The horizontal axis is called the __________ and the vertical axis is called the __________.
2. The intersection of the horizontal axis and the vertical axis is a point called the __________.
3. The axes divide the plane into regions called __________. There are __________ of these regions.
4. In the ordered pair of numbers $(-2, 5)$, the number -2 is called the __________ and the number 5 is called the __________.
5. Each ordered pair of numbers corresponds to __________ point in the plane.
6. An ordered pair is a(n) __________ of an equation in two variables if replacing the variables by the coordinates of the ordered pair results in a true statement.

Martin-Gay Interactive Videos *Watch the section lecture video and answer the following questions.*

See Video 3.1

OBJECTIVE 1

7. Examples 1–3 ask you to answer questions about a bar graph. What information is provided on the horizontal axis of this bar graph? On the vertical axis?

OBJECTIVE 2

8. Several points are plotted in Examples 4–11. Where do you always start when plotting a point? How does the 1st coordinate tell you to move? How does the 2nd coordinate tell you to move?

OBJECTIVE 3

9. In the lecture before Example 12, what connection is made between data and graphing?

OBJECTIVE 4

10. An ordered pair is a solution of an equation if, when the variables are replaced with their values, a true statement results. In Example 13, three ordered pairs are tested. What are the last two points to be tested? What lesson can be learned by the results of testing these two points and why?

OBJECTIVE 5

11. In Example 14, when one variable of a linear equation in two variables is replaced by a replacement value, what type of equation results?

3.1 Exercise Set

MyMathLab®

The following bar graph shows the top 10 tourist destinations and the number of tourists that visit each country per year. Use this graph to answer Exercises 1 through 6. See Example 1.

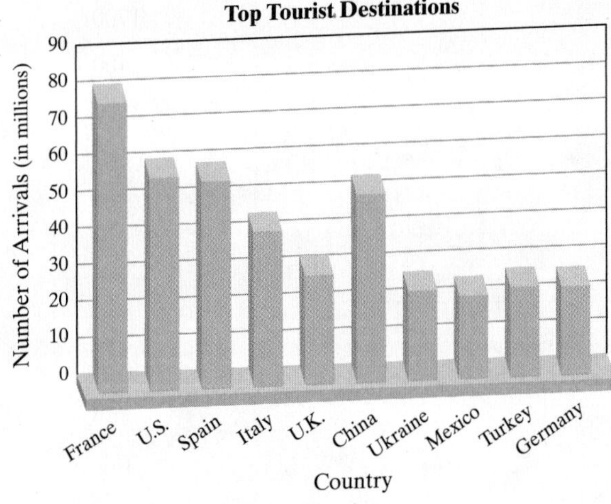

Data from MapXL (www.mapsofworld.com)

1. Which country shown is the most popular tourist destination?
2. Which country shown is the least popular tourist destination?
3. Which countries shown have more than 40 million tourists per year?
4. Which countries shown have fewer than 30 million tourists per year?
5. Estimate the number of tourists per year whose destination is Italy.
6. Estimate the number of tourists per year whose destination is the U.K.

The following line graph shows the paid attendance at each Super Bowl game from 2005 through 2011. Use this graph to answer Exercises 7 through 10. See Example 2.

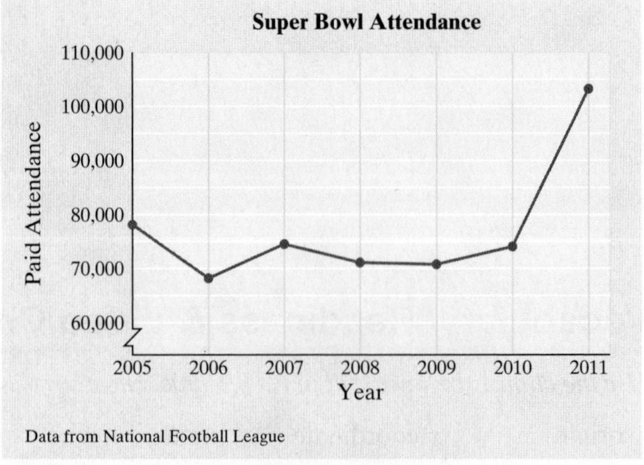

Data from National Football League

7. Estimate the Super Bowl attendance in 2009.
8. Estimate the Super Bowl attendance in 2010.
9. Find the year on the graph with the greatest Super Bowl attendance and approximate that attendance.
10. Find the year on the graph with the least Super Bowl attendance and approximate that attendance.

The line graph below shows the number of students per teacher in U.S. public elementary and secondary schools. Use this graph for Exercises 11 through 16. See Example 2.

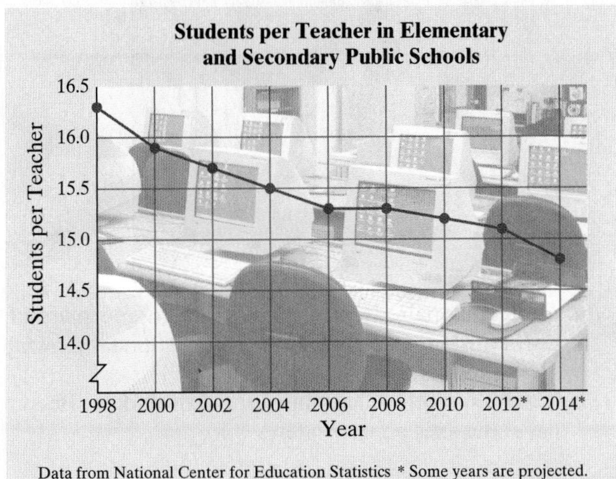

11. Approximate the number of students per teacher in 2000.
12. Approximate the number of students per teacher in 2012.
13. Between what years shown did the greatest decrease in number of students per teacher occur?
14. What was the first year shown that the number of students per teacher fell below 16?
15. What was the first year shown that the number of students per teacher fell below 15?
16. Discuss any trends shown by this line graph.

Plot each ordered pair. State in which quadrant or on which axis each point lies. See Example 3.

17. **a.** $(1, 5)$ **b.** $(-5, -2)$
 c. $(-3, 0)$ **d.** $(0, -1)$
 e. $(2, -4)$ **f.** $\left(-1, 4\frac{1}{2}\right)$
 g. $(3.7, 2.2)$ **h.** $\left(\frac{1}{2}, -3\right)$

18. **a.** $(2, 4)$ **b.** $(0, 2)$
 c. $(-2, 1)$ **d.** $(-3, -3)$
 e. $\left(3\frac{3}{4}, 0\right)$ **f.** $(5, -4)$
 g. $(-3.4, 4.8)$ **h.** $\left(\frac{1}{3}, -5\right)$

Find the x- and y-coordinates of each labeled point. See Example 3.

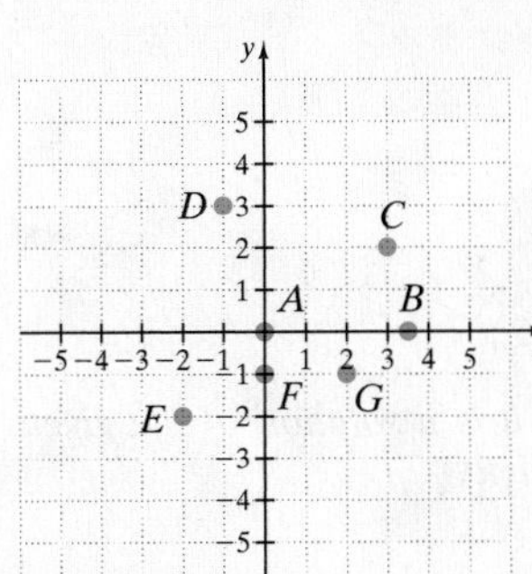

19. A
20. B
21. C
22. D
23. E
24. F
25. G

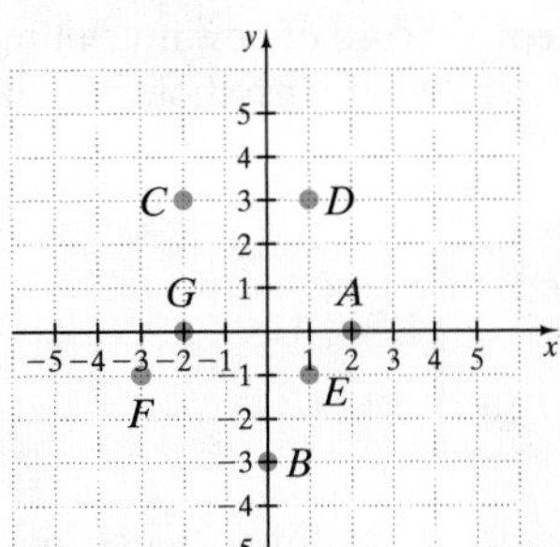

26. A
27. B
28. C
29. D
30. E
31. F
32. G

Solve. See Example 4.

33. The table shows the worldwide box office (in billions of dollars) for the movie industry during the years shown. (Data from Motion Picture Association of America)

Year	Box Office (in billions of dollars)
2006	25.5
2007	26.3
2008	27.7
2009	29.4
2010	31.8

a. Write this paired data as a set of ordered pairs of the form (year, box office).
b. In your own words, write the meaning of the ordered pair (2010, 31.8).
c. Create a scatter diagram of the paired data. Be sure to label the axes appropriately.
d. What trend in the paired data does the scatter diagram show?

34. The table shows the amount of money (in billions of dollars) Americans spent on their pets for the years shown. (Data from American Pet Products Manufacturers Association)

Year	Pet-Related Expenditures (in billions of dollars)
2008	43.2
2009	45.5
2010	48.4
2011	50.8

a. Write this paired data as a set of ordered pairs of the form (year, pet-related expenditures).
b. In your own words, write the meaning of the ordered pair (2011, 50.8).
c. Create a scatter diagram of the paired data. Be sure to label the axes appropriately.
d. What trend in the paired data does the scatter diagram show?

35. Minh, a psychology student, kept a record of how much time she spent studying for each of her 20-point psychology quizzes and her score on each quiz.

Hours Spent Studying	0.50	0.75	1.00	1.25	1.50	1.50	1.75	2.00
Quiz Score	10	12	15	16	18	19	19	20

a. Write the data as ordered pairs of the form (hours spent studying, quiz score).

b. In your own words, write the meaning of the ordered pair (1.25, 16).

c. Create a scatter diagram of the paired data. Be sure to label the axes appropriately.

d. What might Minh conclude from the scatter diagram?

36. A local lumberyard uses quantity pricing. The table shows the price per board for different amounts of lumber purchased.

Price per Board (in dollars)	*Number of Boards Purchased*
8.00	1
7.50	10
6.50	25
5.00	50
2.00	100

a. Write the data as ordered pairs of the form (price per board, number of boards purchased).

b. In your own words, write the meaning of the ordered pair (2.00, 100).

c. Create a scatter diagram of the paired data. Be sure to label the axes appropriately.

d. What trend in the paired data does the scatter diagram show?

37. The table shows the distance from the equator (in miles) and the average annual snowfall (in inches) for each of eight selected U.S. cities. (Data from National Climatic Data Center, Wake Forest University Albatross Project)

City	*Distance from Equator (in miles)*	*Average Annual Snowfall (in inches)*
1. Atlanta, GA	2313	2
2. Austin, TX	2085	1
3. Baltimore, MD	2711	21
4. Chicago, IL	2869	39
5. Detroit, MI	2920	42
6. Juneau, AK	4038	99
7. Miami, FL	1783	0
8. Winston-Salem, NC	2493	9

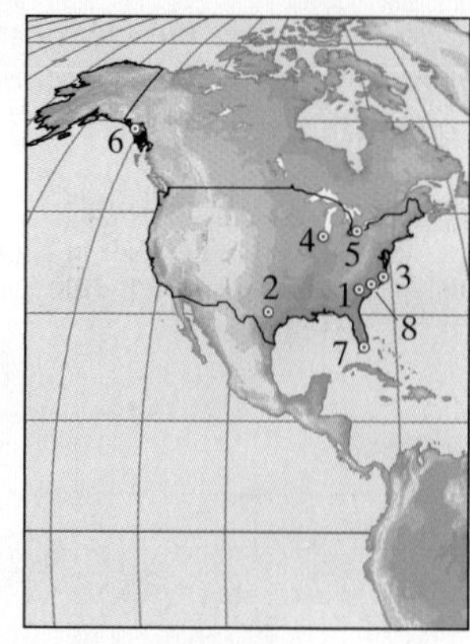

a. Write this paired data as a set of ordered pairs of the form (distance from equator, average annual snowfall).

b. Create a scatter diagram of the paired data. Be sure to label the axes appropriately.

c. What trend in the paired data does the scatter diagram show?

38. The table shows the average farm size (in acres) in the United States during the years shown. (Data from National Agricultural Statistics Service)

Year	*Average Farm Size (in acres)*
2001	438
2002	440
2003	441
2004	443
2005	445
2006	446

a. Write this paired data as a set of ordered pairs of the form (year, average farm size).

b. Create a scatter diagram of the paired data. Be sure to label the axes appropriately.

Determine whether each ordered pair is a solution of the given linear equation. See Example 5.

39. $2x + y = 7$; $(3, 1), (7, 0), (0, 7)$

40. $3x + y = 8$; $(2, 3), (0, 8), (8, 0)$

41. $x = -\frac{1}{3}y$; $(0, 0), (3, -9)$

42. $y = -\frac{1}{2}x$; $(0, 0), (4, 2)$

43. $x = 5$; $(4, 5), (5, 4), (5, 0)$

44. $y = -2$; $(-2, 2), (2, -2), (0, -2)$

Complete each ordered pair so that it is a solution of the given linear equation. See Examples 6 through 8.

45. $x - 4y = 4$; $(\quad, -2), (4, \quad)$

46. $x - 5y = -1$; $(\quad, -2), (4, \quad)$

47. $y = \frac{1}{4}x - 3$; $(-8,\ \ )$, $(\ \ , 1)$

48. $y = \frac{1}{5}x - 2$; $(-10,\ \ )$, $(\ \ , 1)$

Complete the table of ordered pairs for each linear equation. See Examples 6 through 8.

49. $y = -7x$

x	y
0	
−1	
	2

50. $y = -9x$

x	y
	0
−3	
	2

51. $y = -x + 2$

x	y
0	
	0
−3	

52. $x = -y + 4$

x	y
	0
0	
	−3

53. $y = \frac{1}{2}x$

x	y
0	
−6	
	1

54. $y = \frac{1}{3}x$

x	y
0	
−6	
	1

55. $x + 3y = 6$

x	y
0	
	0
	1

56. $2x + y = 4$

x	y
	4
2	
	2

57. $y = 2x - 12$

x	y
0	
	−2
3	

58. $y = 5x + 10$

x	y
	0
	5
0	

59. $2x + 7y = 5$

x	y
0	
	0
	1

60. $x - 6y = 3$

x	y
0	
1	
	−1

MIXED PRACTICE

Complete the table of ordered pairs for each equation. Then plot the ordered pair solutions. See Examples 1 through 7.

61. $x = -5y$

x	y
	0
	1
10	

62. $y = -3x$

x	y
0	
−2	
	9

63. $y = \frac{1}{3}x + 2$

x	y
0	
−3	
	0

64. $y = \frac{1}{2}x + 3$

x	y
0	
−4	
	0

Solve. See Example 9.

65. The cost in dollars y of producing x computer desks is given by $y = 80x + 5000$.

a. Complete the table.

x	100	200	300
y			

b. Find the number of computer desks that can be produced for \$8600. (*Hint:* Find x when $y = 8600$.)

66. The hourly wage y of an employee at a certain production company is given by $y = 0.25x + 9$ where x is the number of units produced by the employee in an hour.

a. Complete the table.

x	0	1	5	10
y				

b. Find the number of units that an employee must produce each hour to earn an hourly wage of \$12.25. (*Hint:* Find x when $y = 12.25$.)

67. The average annual cinema admission price y (in dollars) from 2001 through 2010 is given by $y = 0.24x + 5.28$. In this equation, x represents the number of years after 2000. (*Source:* Motion Picture Association of America)

a. Complete the table.

x	1	3	5
y			

b. Find the year in which the average cinema admission price was approximately \$7.50.
(*Hint:* Find x when $y = 7.50$ and round to the nearest whole number.)

c. Use the given equation to predict when the cinema admission price might be \$9.00. (Use the hint for part b.)

d. In your own words, write the meaning of the ordered pair (5, 6.48).

68. The average amount of money y spent per person on recorded music from 2002 to 2009 is given by $y = -4.22x + 65.70$. In this equation, x represents the number of years after 2000. (*Source:* edgeqld.org)

a. Complete the table.

x	1	5	9
y			

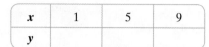

b. Find the year in which the yearly average amount of money per person spent on recorded music was approximately \$30. (*Hint:* Find x when $y = 30$ and round to the nearest whole number.)

The graph below shows the number of U.S. Walmart stores for each year. Use this graph to answer Exercises 69 through 72.

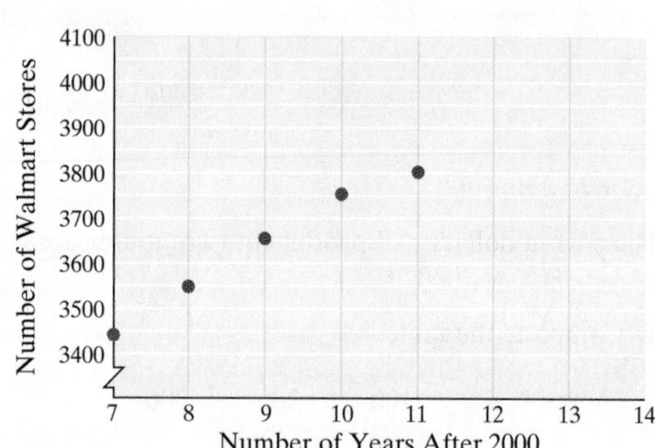

Data from Walmart

69. The ordered pair (10, 3755) is a point of the graph. Write a sentence describing the meaning of this ordered pair.

70. The ordered pair (11, 3804) is a point of the graph. Write a sentence describing the meaning of this ordered pair.

71. Estimate the increase in Walmart stores for years 8, 9, and 10.

72. Use a straightedge or ruler and this graph to predict the number of Walmart stores in the year 2014.

73. Describe what is similar about the coordinates of points whose graph lies on the x-axis.

74. Describe what is similar about the coordinates of points whose graph lies on the y-axis.

REVIEW AND PREVIEW

Solve each equation for y. See Section 2.6.

75. $x + y = 5$

76. $x - y = 3$

77. $2x + 4y = 5$

78. $5x + 2y = 7$

79. $10x = -5y$

80. $4y = -8x$

81. $x - 3y = 6$

82. $2x - 9y = -20$

CONCEPT EXTENSIONS

Answer each exercise with true or false.

83. Point $(-1, 5)$ lies in quadrant IV.

84. Point (3, 0) lies on the y-axis.

85. For the point $\left(-\frac{1}{2}, 1.5\right)$, the first value, $-\frac{1}{2}$, is the x-coordinate and the second value, 1.5, is the y-coordinate.

86. The ordered pair $\left(2, \frac{2}{3}\right)$ is a solution of $2x - 3y = 6$.

For Exercises 87 through 91, fill in each blank with "0," "positive," or "negative." For Exercises 92 and 93, fill in each blank with "x" or "y."

	Point	*Location*
87.	(_______, _______)	quadrant III
88.	(_______, _______)	quadrant I
89.	(_______, _______)	quadrant IV
90.	(_______, _______)	quadrant II
91.	(_______, _______)	origin
92.	(number, 0)	__-axis
93.	(0, number)	__-axis

94. Give an example of an ordered pair whose location is in (or on)

a. quadrant I **b.** quadrant II
c. quadrant III **d.** quadrant IV
e. x-axis **f.** y-axis

Solve. See the first Concept Check in this section.

95. Is the graph of (3, 0) in the same location as the graph of (0, 3)? Explain why or why not.

96. Give the coordinates of a point such that if the coordinates are reversed, their location is the same.

97. In general, what points can have coordinates reversed and still have the same location?

98. In your own words, describe how to plot or graph an ordered pair of numbers.

Write an ordered pair for each point described. See the second Concept Check in this section.

99. Point C is four units to the right of the y-axis and seven units below the x-axis.

100. Point D is three units to the left of the origin.

Solve.

101. Find the perimeter of the rectangle whose vertices are the points with coordinates $(-1, 5)$, $(3, 5)$, $(3, -4)$, and $(-1, -4)$.

102. Find the area of the rectangle whose vertices are the points with coordinates $(5, 2)$, $(5, -6)$, $(0, -6)$, and $(0, 2)$.

103. Three vertices of a rectangle are $(-2, -3)$, $(-7, -3)$, and $(-7, 6)$.

a. Find the coordinates of the fourth vertex of the rectangle.
b. Find the perimeter of the rectangle.
c. Find the area of the rectangle.

104. Three vertices of a square are $(-4, -1)$, $(-4, 8)$, and $(5, 8)$.

a. Find the coordinates of the fourth vertex of the square.
b. Find the perimeter of the square.
c. Find the area of the square.

3.2 Graphing Linear Equations 

OBJECTIVES

1 Identify Linear Equations.

2 Graph a Linear Equation by Finding and Plotting Ordered Pair Solutions.

OBJECTIVE

1 Identifying Linear Equations 

In the previous section, we found that equations in two variables may have more than one solution. For example, both (6, 0) and (2, −2) are solutions of the equation $x - 2y = 6$. In fact, this equation has an infinite number of solutions. Other solutions include $(0, -3)$, $(4, -1)$, and $(-2, -4)$. If we graph these solutions, notice that a pattern appears.

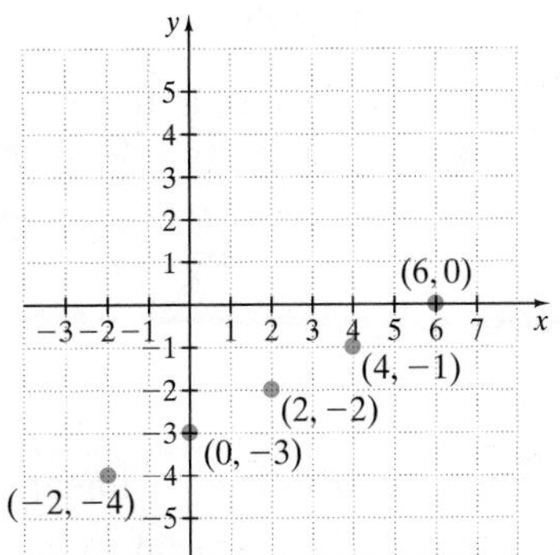

These solutions all appear to lie on the same line, which has been filled in below. It can be shown that every ordered pair solution of the equation corresponds to a point on this line, and every point on this line corresponds to an ordered pair solution. Thus, we say that this line is the **graph of the equation** $x - 2y = 6$.

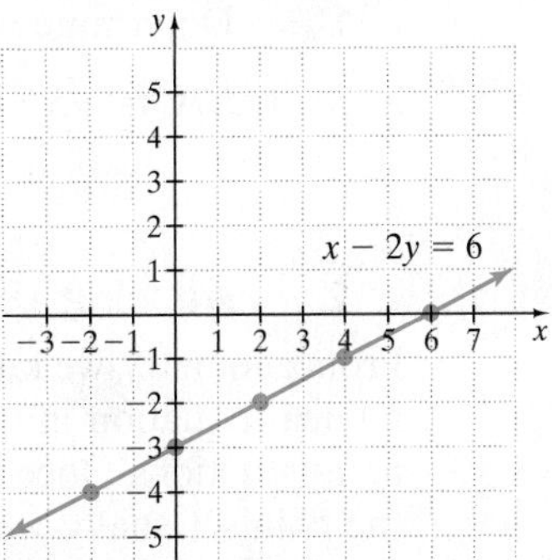

▶ Helpful Hint

Notice that we can only show a part of a line on a graph. The arrowheads on each end of the line remind us that the line actually extends indefinitely in both directions.

The equation $x - 2y = 6$ is called a **linear equation in two variables** and **the graph of every linear equation in two variables is a line.**

Linear Equation in Two Variables

A linear equation in two variables is an equation that can be written in the form

$$Ax + By = C$$

where A, B, and C are real numbers and A and B are not both 0. **The graph of a linear equation in two variables is a straight line.**

The form $Ax + By = C$ is called **standard form.**

▶ Helpful Hint

Notice in the form $Ax + By = C$, the understood exponent on both x and y is 1.

Examples of Linear Equations in Two Variables

$$2x + y = 8 \qquad -2x = 7y \qquad y = \frac{1}{3}x + 2 \qquad y = 7$$
(Standard Form)

Before we graph linear equations in two variables, let's practice identifying these equations.

EXAMPLE 1 Determine whether each equation is a linear equation in two variables.

a. $x - 1.5y = -1.6$ **b.** $y = -2x$ **c.** $x + y^2 = 9$ **d.** $x = 5$

Solution

a. This is a linear equation in two variables because it is written in the form $Ax + By = C$ with $A = 1$, $B = -1.5$, and $C = -1.6$.

b. This is a linear equation in two variables because it can be written in the form $Ax + By = C$.

$$y = -2x$$
$$2x + y = 0 \quad \text{Add } 2x \text{ to both sides.}$$

c. This is *not* a linear equation in two variables because y is squared.

d. This is a linear equation in two variables because it can be written in the form $Ax + By = C$.

$$x = 5$$
$$x + 0y = 5 \quad \text{Add } 0 \cdot y.$$

PRACTICE

1 Determine whether each equation is a linear equation in two variables.

a. $3x + 2.7y = -5.3$ **b.** $x^2 + y = 8$ **c.** $y = 12$ **d.** $5x = -3y$

OBJECTIVE

2 Graphing Linear Equations by Plotting Ordered Pair Solutions

From geometry, we know that a straight line is determined by just two points. Graphing a linear equation in two variables, then, requires that we find just two of its infinitely many solutions. Once we do so, we plot the solution points and draw the line connecting the points. Usually, we find a third solution as well, as a check.

EXAMPLE 2 Graph the linear equation $2x + y = 5$.

Solution Find three ordered pair solutions of $2x + y = 5$. To do this, choose a value for one variable, x or y, and solve for the other variable. For example, let $x = 1$. Then $2x + y = 5$ becomes

$$2x + y = 5$$
$$2(1) + y = 5 \quad \text{Replace } x \text{ with 1.}$$
$$2 + y = 5 \quad \text{Multiply.}$$
$$y = 3 \quad \text{Subtract 2 from both sides.}$$

Since $y = 3$ when $x = 1$, the ordered pair $(1, 3)$ is a solution of $2x + y = 5$. Next, let $x = 0$.

$$2x + y = 5$$
$$2(0) + y = 5 \quad \text{Replace } x \text{ with 0.}$$
$$0 + y = 5$$
$$y = 5$$

The ordered pair $(0, 5)$ is a second solution.

The two solutions found so far allow us to draw the straight line that is the graph of all solutions of $2x + y = 5$. However, we find a third ordered pair as a check. Let $y = -1$.

$$\begin{aligned} 2x + y &= 5 \\ 2x + (-1) &= 5 \quad \text{Replace } y \text{ with } -1. \\ 2x - 1 &= 5 \\ 2x &= 6 \quad \text{Add 1 to both sides.} \\ x &= 3 \quad \text{Divide both sides by 2.} \end{aligned}$$

The third solution is $(3, -1)$. These three ordered pair solutions are listed in table form as shown. The graph of $2x + y = 5$ is the line through the three points.

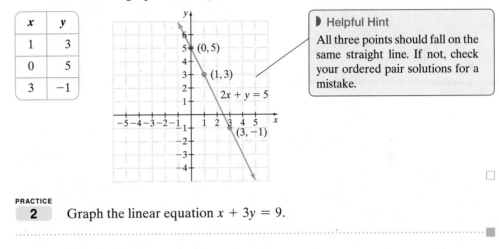

x	y
1	3
0	5
3	−1

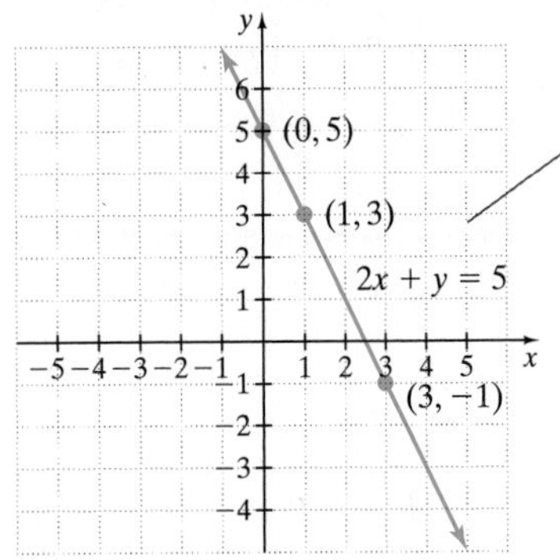

Helpful Hint

All three points should fall on the same straight line. If not, check your ordered pair solutions for a mistake.

PRACTICE

2 Graph the linear equation $x + 3y = 9$.

EXAMPLE 3 Graph the linear equation $-5x + 3y = 15$.

Solution Find three ordered pair solutions of $-5x + 3y = 15$.

Let $x = 0$.

$$\begin{aligned} -5x + 3y &= 15 \\ -5 \cdot 0 + 3y &= 15 \\ 0 + 3y &= 15 \\ 3y &= 15 \\ y &= 5 \end{aligned}$$

Let $y = 0$.

$$\begin{aligned} -5x + 3y &= 15 \\ -5x + 3 \cdot 0 &= 15 \\ -5x + 0 &= 15 \\ -5x &= 15 \\ x &= -3 \end{aligned}$$

Let $x = -2$.

$$\begin{aligned} -5x + 3y &= 15 \\ -5(-2) + 3y &= 15 \\ 10 + 3y &= 15 \\ 3y &= 5 \\ y &= \frac{5}{3} \end{aligned}$$

The ordered pairs are $(0, 5)$, $(-3, 0)$, and $\left(-2, \frac{5}{3}\right)$. The graph of $-5x + 3y = 15$ is the line through the three points.

x	y
0	5
−3	0
−2	$\frac{5}{3} = 1\frac{2}{3}$

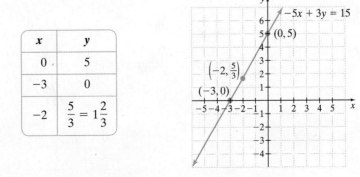

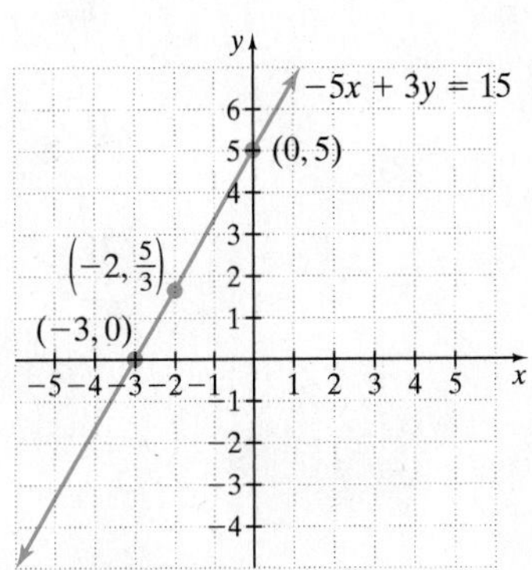

PRACTICE

3 Graph the linear equation $3x - 4y = 12$.

EXAMPLE 4 Graph the linear equation $y = 3x$.

Solution To graph this linear equation, we find three ordered pair solutions. Since this equation is solved for y, choose three x values.

If $x = 2, y = 3 \cdot 2 = 6$.
If $x = 0, y = 3 \cdot 0 = 0$.
If $x = -1, y = 3 \cdot -1 = -3$.

x	y
2	6
0	0
−1	−3

Next, graph the ordered pair solutions listed in the table above and draw a line through the plotted points as shown below. The line is the graph of $y = 3x$. Every point on the graph represents an ordered pair solution of the equation and every ordered pair solution is a point on this line.

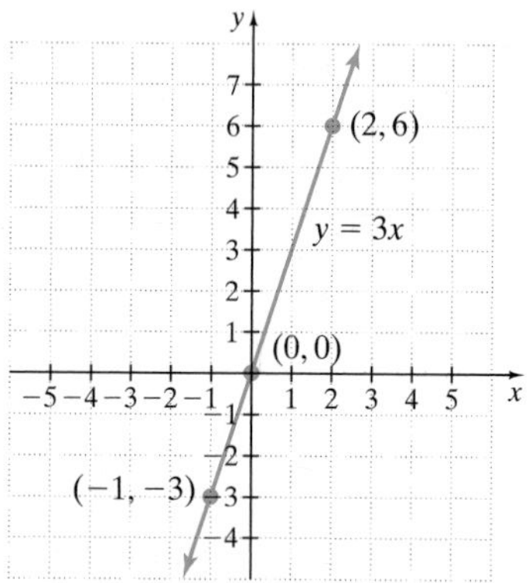

□

PRACTICE

4 Graph the linear equation $y = -2x$.

Helpful Hint

When graphing a linear equation in two variables, if it is

- solved for y, it may be easier to find ordered pair solutions by choosing x-values. If it is
- solved for x, it may be easier to find ordered pair solutions by choosing y-values.

EXAMPLE 5 Graph the linear equation $y = -\frac{1}{3}x + 2$.

Solution Find three ordered pair solutions, graph the solutions, and draw a line through the plotted solutions. To avoid fractions, choose x values that are multiples of 3 to substitute in the equation. When a multiple of 3 is multiplied by $-\frac{1}{3}$, the result is an integer. See the calculations shown above the table.

If $x = 6$, then $y = -\frac{1}{3} \cdot 6 + 2 = -2 + 2 = 0$

If $x = 0$, then $y = -\frac{1}{3} \cdot 0 + 2 = 0 + 2 = 2$

If $x = -3$, then $y = -\frac{1}{3} \cdot -3 + 2 = 1 + 2 = 3$

x	y
6	0
0	2
−3	3

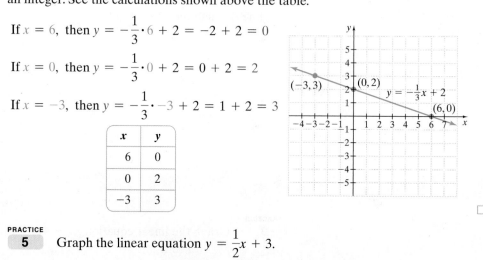

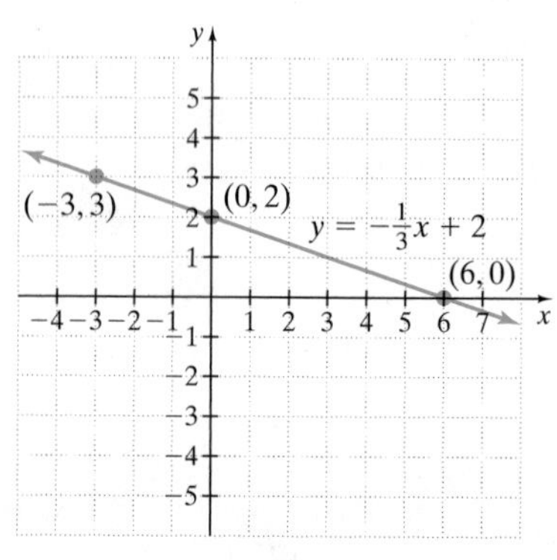

□

PRACTICE

5 Graph the linear equation $y = \frac{1}{2}x + 3$.

Let's compare the graphs in Examples 4 and 5. The graph of $y = 3x$ tilts upward (as we follow the line from left to right) and the graph of $y = -\frac{1}{3}x + 2$ tilts downward (as we follow the line from left to right). We will learn more about the tilt, or slope, of a line in Section 3.4.

EXAMPLE 6 Graph the linear equation $y = -2$.

Solution: The equation $y = -2$ can be written in standard form as $0x + y = -2$. No matter what value we replace x with, y is always -2.

x	y
0	−2
3	−2
−2	−2

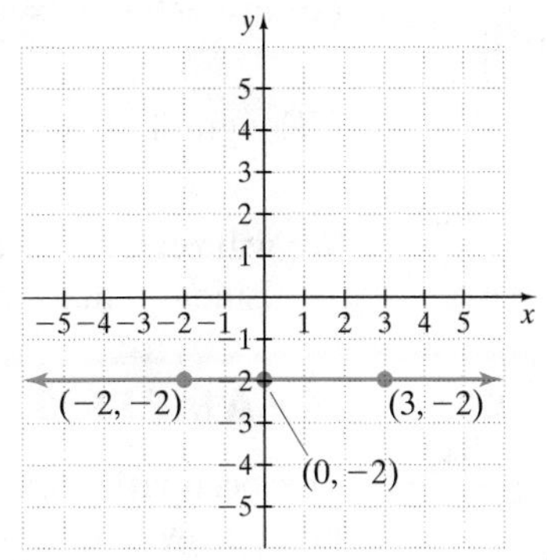

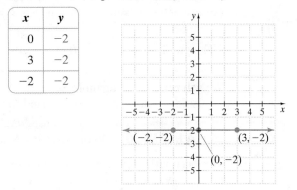

Notice that the graph of $y = -2$ is a horizontal line. □

PRACTICE 6 Graph the linear equation $x = -2$.

EXAMPLE 7 Graph the linear equation $y = 3x + 6$ and compare this graph with the graph of $y = 3x$ in Example 4.

Solution Find ordered pair solutions, graph the solutions, and draw a line through the plotted solutions. We choose x values and substitute in the equation $y = 3x + 6$.

If $x = -3$, then $y = 3(-3) + 6 = -3$.
If $x = 0$, then $y = 3(0) + 6 = 6$.
If $x = 1$, then $y = 3(1) + 6 = 9$.

x	y
−3	−3
0	6
1	9

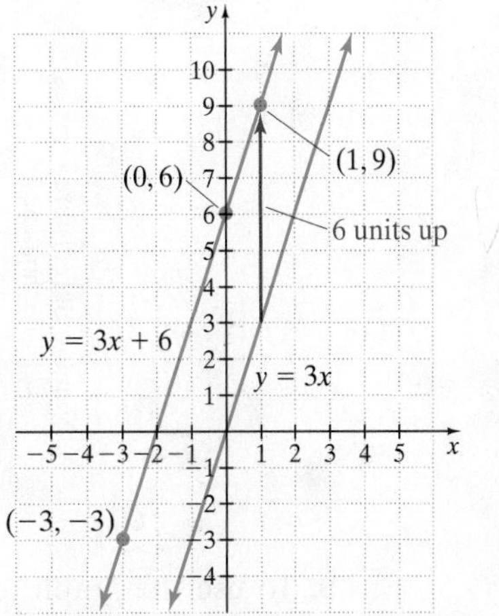

The most startling similarity is that both graphs appear to have the same upward tilt as we move from left to right. Also, the graph of $y = 3x$ crosses the y-axis at the origin, while the graph of $y = 3x + 6$ crosses the y-axis at 6. In fact, the graph of $y = 3x + 6$ is the same as the graph of $y = 3x$ moved vertically upward 6 units. □

PRACTICE

7 Graph the linear equation $y = -2x + 3$ and compare this graph with the graph of $y = -2x$ in Practice 4.

Notice that the graph of $y = 3x + 6$ crosses the y-axis at 6. This happens because when $x = 0$, $y = 3x + 6$ becomes $y = 3 \cdot 0 + 6 = 6$. The graph contains the point $(0, 6)$, which is on the y-axis.

In general, if a linear equation in two variables is solved for y, we say that it is written in the form $y = mx + b$. The graph of this equation contains the point $(0, b)$ because when $x = 0$, $y = mx + b$ is $y = m \cdot 0 + b = b$.

The graph of $y = mx + b$ crosses the y-axis at $(0, b)$.

We will review this again in Section 3.5.

Linear equations are often used to model real data as seen in the next example.

EXAMPLE 8 Estimating the Number of Registered Nurses

The occupation expected to have the most employment growth in the next few years is registered nurse. The number of people y (in thousands) employed as registered nurses in the United States can be estimated by the linear equation $y = 58.1x + 2619$, where x is the number of years after the year 2008. (Data from U.S. Bureau of Labor Statistics)

a. Graph the equation.

b. Use the graph to predict the number of registered nurses in the year 2018.

Solution:

a. To graph $y = 58.1x + 2619$, choose x-values and substitute in the equation.

If $x = 0$, then $y = 58.1(0) + 2619 = 2619$.
If $x = 2$, then $y = 58.1(2) + 2619 = 2735.2$.
If $x = 5$, then $y = 58.1(5) + 2619 = 2909.5$.

x	y
0	2619
2	2735.2
5	2909.5

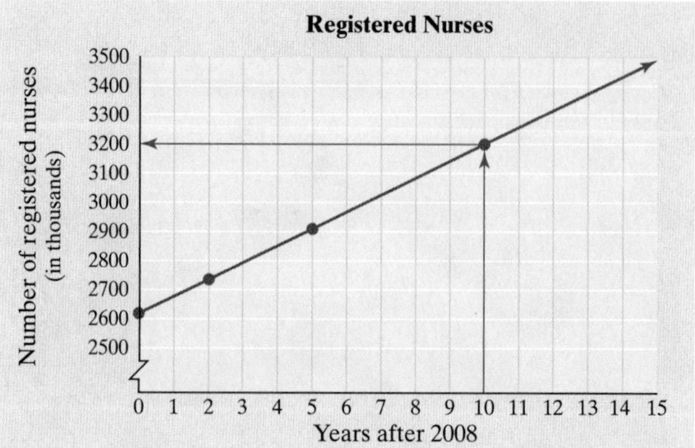

b. To use the graph to *predict* the number of registered nurses in the year 2018, we need to find the y-coordinate that corresponds to $x = 10$. (10 years after 2008 is the year 2018.) To do so, find 10 on the x-axis. Move vertically upward to the graphed line and then horizontally to the left. We approximate the number on the y-axis to be 3200. Thus, in the year 2018, we predict that there will be 3200 thousand registered nurses. (The actual value, using 10 for x, is 3200.) □

PRACTICE

8 One of the occupations expected to have a large growth in employment in the next few years is computer software application engineers. The number of people y (in thousands) employed as computer software application engineers in the United States can be estimated by the linear equation $y = 17.5x + 515$, where x is the number of years after 2008. (*Source:* Based on data from the Bureau of Labor Statistics)

a. Graph the equation.

b. Use the graph to predict the number of computer software application engineers in the year 2020.

Helpful Hint

Make sure you understand that models are mathematical approximations of the data for the known years. (For example, see the model in Example 8.) Any number of unknown factors can affect future years, so be cautious when using models to predict.

Graphing Calculator Explorations

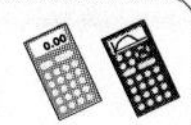

In this section, we begin an optional study of graphing calculators and graphing software packages for computers. These graphers use the same point plotting technique that was introduced in this section. The advantage of this graphing technology is, of course, that graphing calculators and computers can find and plot ordered pair solutions much faster than we can. Note, however, that the features described in these boxes may not be available on all graphing calculators.

The rectangular screen where a portion of the rectangular coordinate system is displayed is called a **window.** We call it a **standard window** for graphing when both the x- and y-axes show coordinates between -10 and 10. This information is often displayed in the window menu on a graphing calculator as

Xmin = −10
Xmax = 10
Xscl = 1 The scale on the x-axis is one unit per tick mark.
Ymin = −10
Ymax = 10
Yscl = 1 The scale on the y-axis is one unit per tick mark.

To use a graphing calculator to graph the equation $y = 2x + 3$, press the [Y=] key and enter the keystrokes [2] [x] [+] [3]. The top row should now read $Y_1 = 2x + 3$. Next press the [GRAPH] key, and the display should look like this:

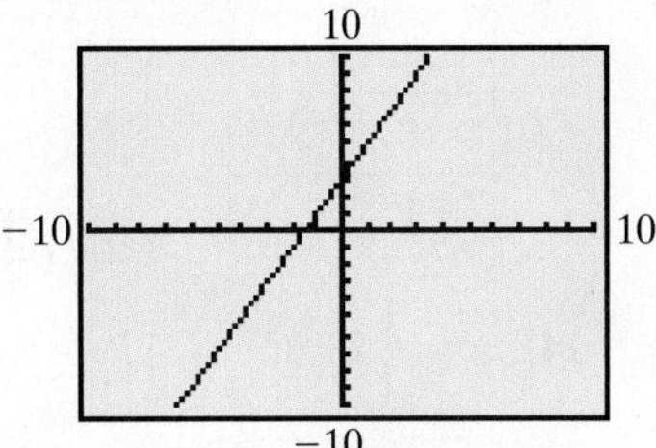

Use a standard window and graph the following linear equations. (Unless otherwise stated, use a standard window when graphing.)

1. $y = -3x + 7$

2. $y = -x + 5$

3. $y = 2.5x - 7.9$

4. $y = -1.3x + 5.2$

5. $y = -\frac{3}{10}x + \frac{32}{5}$

6. $y = \frac{2}{9}x - \frac{22}{3}$

Vocabulary, Readiness & Video Check

Martin-Gay Interactive Videos

See Video 3.2

Watch the section lecture video and answer the following questions.

OBJECTIVE 1

1. Exponents aren't mentioned in the definition of a linear equation in two variables. However, in determining whether Example 3 is a linear equation in two variables, the exponents or powers on the variables are discussed. Explain.

OBJECTIVE 2

2. In the lecture before Example 5, it's mentioned that you need only two points to determine a line. Why then are three ordered pair solutions found in Examples 5–7?

OBJECTIVE 2

3. What does a graphed line represent as discussed at the end of Examples 5 and 7?

3.2 Exercise Set

MyMathLab®

Determine whether each equation is a linear equation in two variables. See Example 1.

1. $-x = 3y + 10$
2. $y = x - 15$
3. $x = y$
4. $x = y^3$
5. $x^2 + 2y = 0$
6. $0.01x - 0.2y = 8.8$
7. $y = -1$
8. $x = 25$

For each equation, find three ordered pair solutions by completing the table. Then use the ordered pairs to graph the equation. See Examples 2 through 7.

9. $x - y = 6$

x	y
	0
4	
	−1

10. $x - y = 4$

x	y
0	
	2
−1	

11. $y = -4x$

x	y
1	
0	
−1	

12. $y = -5x$

x	y
1	
0	
−1	

13. $y = \frac{1}{3}x$

x	y
0	
6	
−3	

14. $y = \frac{1}{2}x$

x	y
0	
−4	
2	

15. $y = -4x + 3$

x	y
0	
1	
2	

16. $y = -5x + 2$

x	y
0	
1	
2	

MIXED PRACTICE

Graph each linear equation. See Examples 2 through 7.

17. $x + y = 1$
18. $x + y = 7$
19. $x - y = -2$
20. $-x + y = 6$
21. $x - 2y = 6$
22. $-x + 5y = 5$
23. $y = 6x + 3$
24. $y = -2x + 7$
25. $x = -4$
26. $y = 5$
27. $y = 3$
28. $x = -1$
29. $y = x$
30. $y = -x$
31. $x = -3y$
32. $x = -5y$
33. $x + 3y = 9$
34. $2x + y = 2$
35. $y = \frac{1}{2}x + 2$
36. $y = \frac{1}{4}x + 3$
37. $3x - 2y = 12$
38. $2x - 7y = 14$
39. $y = -3.5x + 4$
40. $y = -1.5x - 3$

Graph each pair of linear equations on the same set of axes. Discuss how the graphs are similar and how they are different. See Example 7.

41. $y = 5x; y = 5x + 4$
42. $y = 2x; y = 2x + 5$
43. $y = -2x; y = -2x - 3$
44. $y = x; y = x - 7$
45. $y = \frac{1}{2}x; y = \frac{1}{2}x + 2$
46. $y = -\frac{1}{4}x; y = -\frac{1}{4}x + 3$

The graph of $y = 5x$ *is given below as well as Figures a–d. For Exercises 47 through 50, match each equation with its graph. Hint: Recall that if an equation is written in the form* $y = mx + b$, *its graph crosses the y-axis at* $(0, b)$.

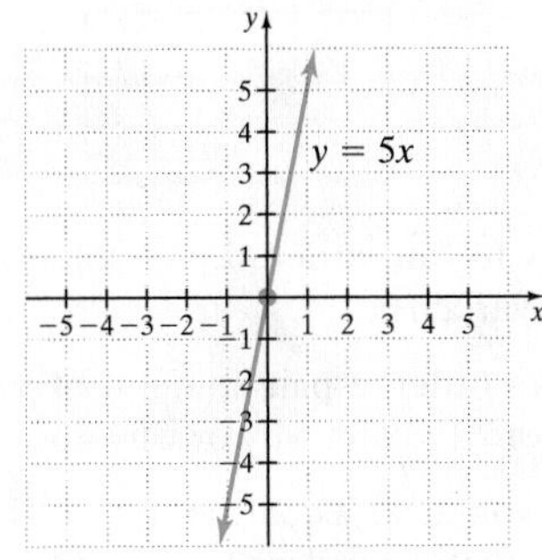

47. $y = 5x + 5$

48. $y = 5x - 4$

49. $y = 5x - 1$

50. $y = 5x + 2$

a.

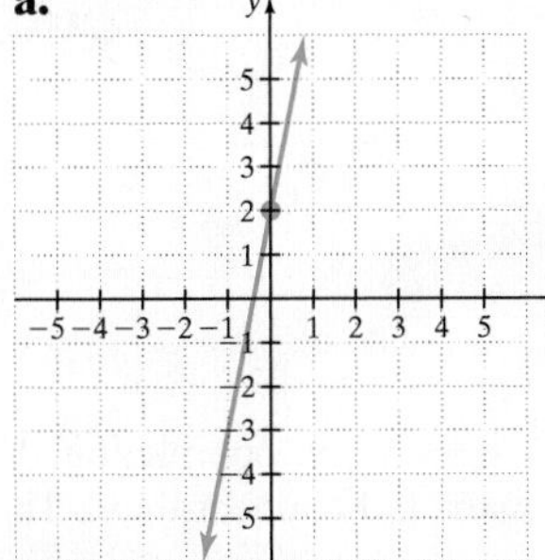

b.

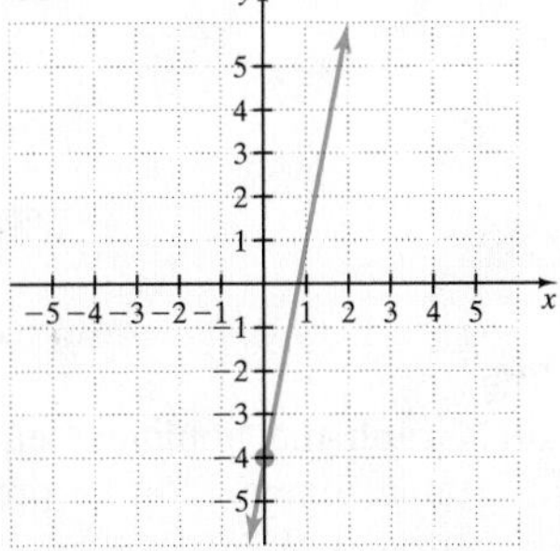

c.

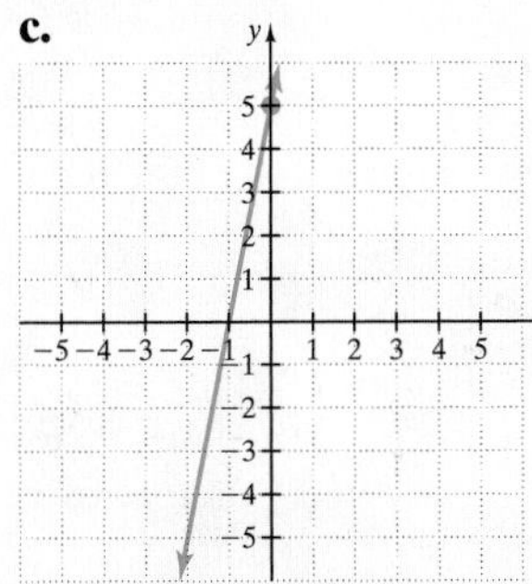

d.

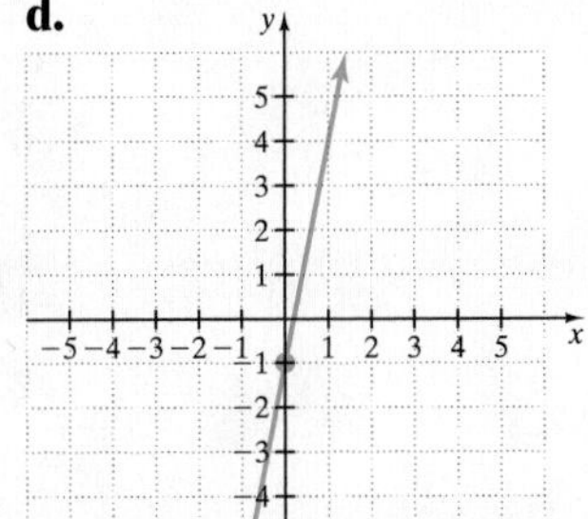

Solve. See Example 8.

51. Jogging is one of the few sports that has been consistently increasing over the past few years. The number of people jogging (in millions) from the years 2000 to 2009 is given by the equation $y = x + 23$, where x is the number of years after 2000. (*Source:* Based on data from the National Sporting Goods Association)

a. Use this equation or a graph of it to complete the ordered pair (8,).

b. Write a sentence explaining the meaning of the answer to part (a).

c. If this trend continues, how many joggers will there be in 2017?

52. The revenue y (in billions of dollars) for Home Depot stores during the years 2006 through 2010 is given by the equation $y = -3x + 79$, where x is the number of years after 2006. (*Source:* Based on data from Home Depot stores)

a. Use this equation or a graph of it to complete the ordered pair (4,).

b. Write a sentence explaining the meaning of the answer to part (a).

c. If this trend continues, predict the revenue for Home Depot stores for the year 2015.

53. One American rite of passage is a driver's license. The number of people y (in millions) who have a driver's license can be estimated by the linear equation $y = 2.2x + 190$, where x is the number of years after 2000. (Data from Federal Highway Administration)

a. Use this equation to complete the ordered pair (8,).

b. Write a sentence explaining the meaning of the ordered pair in part (a).

c. If this trend continues, predict the number of people with driver's licenses in 2016.

54. The percent of U.S. households y with at least one computer can be approximated by the linear equation $y = 4.3x + 51$, where x is the number of years since 2000. (Data from *Statistical Abstract of the United States*)

(Continued on next page)

a. Use the equation to complete the ordered pair (9,).

b. Write a sentence explaining the meaning of the ordered pair found in part (a).

c. If this trend continues, predict the percent of U.S. households that have at least one computer in 2012.

d. Explain any issues with your answer to part (c).

REVIEW AND PREVIEW

55. The coordinates of three vertices of a rectangle are $(-2, 5)$, $(4, 5)$, and $(-2, -1)$. Find the coordinates of the fourth vertex. See Section 3.1.

56. The coordinates of two vertices of a square are $(-3, -1)$ and $(2, -1)$. Find the coordinates of two pairs of points possible for the third and fourth vertices. See Section 3.1.

Complete each table. See Section 3.1.

57. $x - y = -3$

x	y
0	
	0

58. $y - x = 5$

x	y
0	
	0

59. $y = 2x$

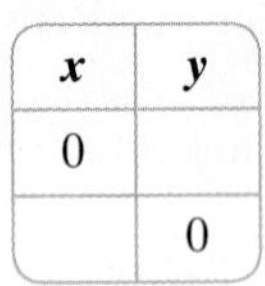

x	y
0	
	0

60. $x = -3y$

x	y
0	
	0

CONCEPT EXTENSIONS

Write each statement as an equation in two variables. Then graph the equation.

61. The y-value is 5 more than the x-value.

62. The y-value is twice the x-value.

63. Two times the x-value, added to three times the y-value is 6.

64. Five times the x-value, added to twice the y-value is -10.

65. The perimeter of the trapezoid below is 22 centimeters. Write a linear equation in two variables for the perimeter. Find y if x is 3 cm.

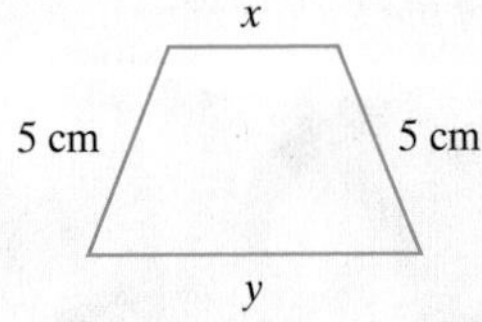

66. The perimeter of the rectangle below is 50 miles. Write a linear equation in two variables for this perimeter. Use this equation to find x when y is 20.

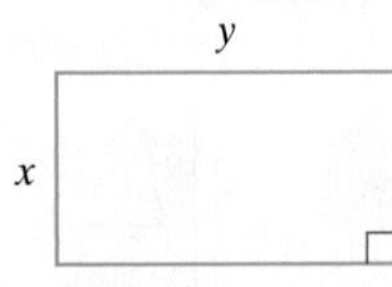

67. Explain how to find ordered pair solutions of linear equations in two variables.

68. If (a, b) is an ordered pair solution of $x + y = 5$, is (b, a) also a solution? Explain why or why not.

69. Graph the nonlinear equation $y = x^2$ by completing the table shown. Plot the ordered pairs and connect them with a smooth curve.

x	y
0	
1	
-1	
2	
-2	

70. Graph the nonlinear equation $y = |x|$ by completing the table shown. Plot the ordered pairs and connect them. This curve is "V" shaped.

$y = |x|$

x	y
0	
1	
-1	
2	
-2	

3.3 Intercepts

OBJECTIVES

1 Identify Intercepts of a Graph.

2 Graph a Linear Equation by Finding and Plotting Intercepts.

3 Identify and Graph Vertical and Horizontal Lines.

OBJECTIVE

1 Identifying Intercepts 

In this section, we graph linear equations in two variables by identifying intercepts. For example, the graph of $y = 4x - 8$ is shown on the right. Notice that this graph crosses the y-axis at the point $(0, -8)$. This point is called the ***y*-intercept.** Likewise, the graph crosses the x-axis at $(2, 0)$, and this point is called the ***x*-intercept.**

The intercepts are $(2, 0)$ and $(0, -8)$.

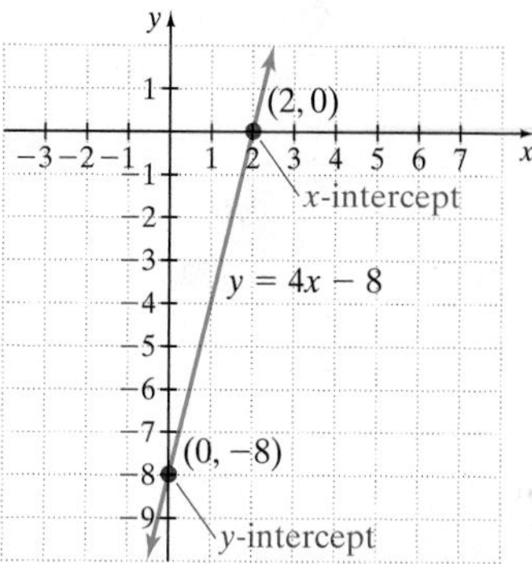

Helpful Hint

If a graph crosses the x-axis at $(-3, 0)$ and the y-axis at $(0, 7)$, then

$$\underbrace{(-3, 0)}_{x\text{-intercept}} \qquad \underbrace{(0, 7)}_{y\text{-intercept}}$$

Notice that for the y-intercept, the x-value is 0 and for the x-intercept, the y-value is 0.
Note: Sometimes in mathematics, you may see just the number 7 stated as the y-intercept, and -3 stated as the x-intercept.

EXAMPLES Identify the x- and y-intercepts.

1.

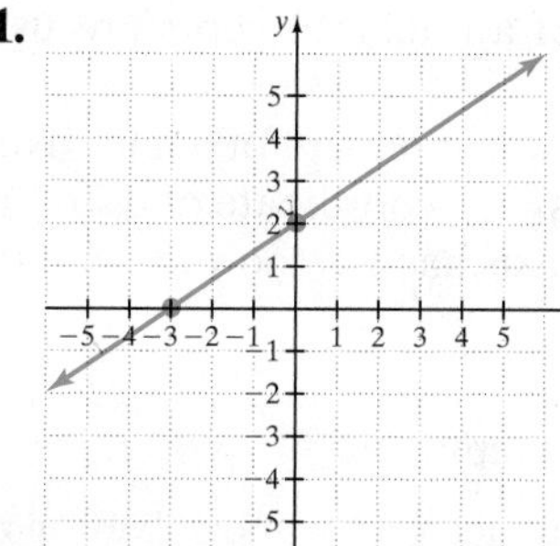

Solution

x-intercept: $(-3, 0)$
y-intercept: $(0, 2)$

2.

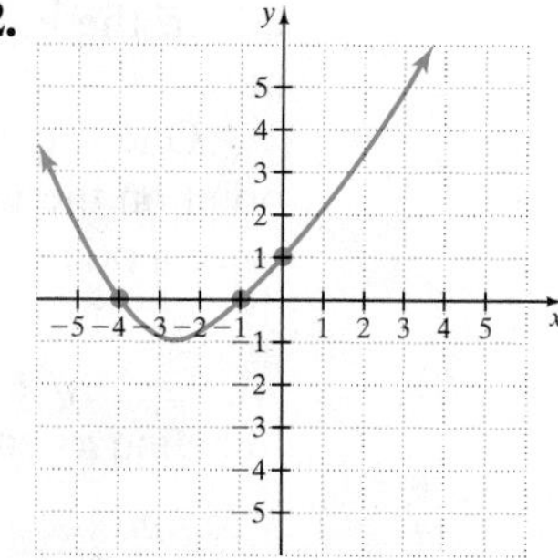

Solution

x-intercepts: $(-4, 0)$, $(-1, 0)$
y-intercept: $(0, 1)$

Helpful Hint

Notice that any time $(0, 0)$ is a point of a graph, then it is an x-intercept and a y-intercept. Why? It is the only point that lies on both axes.

3.

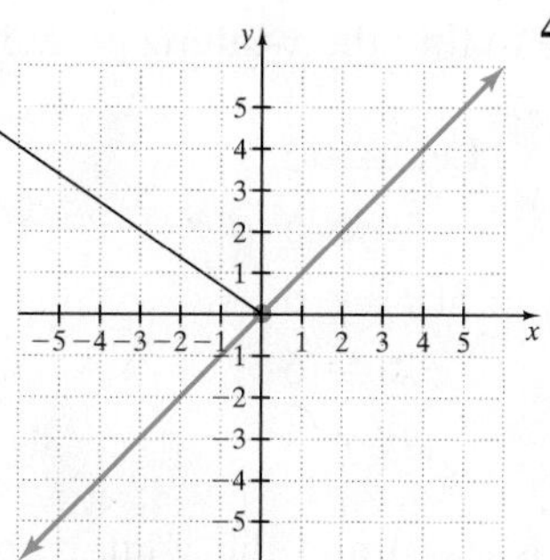

Solution

x-intercept: $(0, 0)$
y-intercept: $(0, 0)$

4.

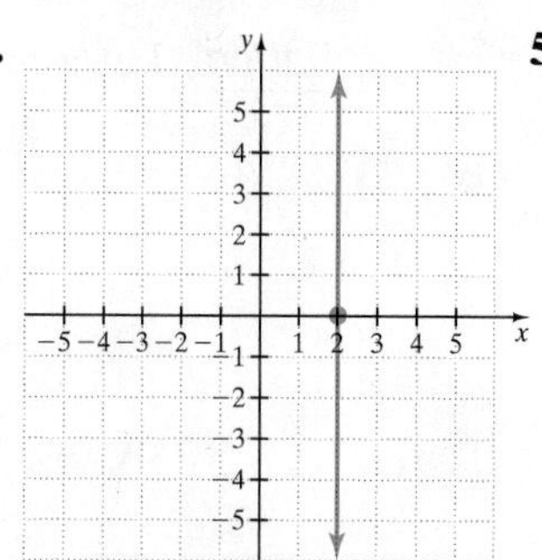

Solution

x-intercept: $(2, 0)$
y-intercept: none

5.

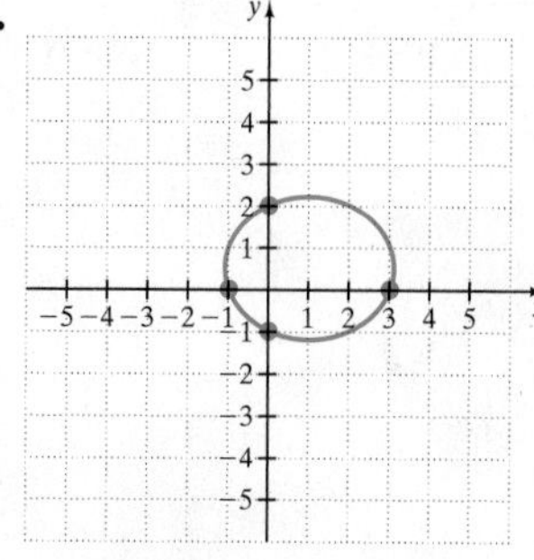

Solution

x-intercepts:
$(-1, 0)$, $(3, 0)$
y-intercepts:
$(0, 2)$, $(0, -1)$

□

PRACTICE 1–5 Identify the x- and y-intercepts.

1.

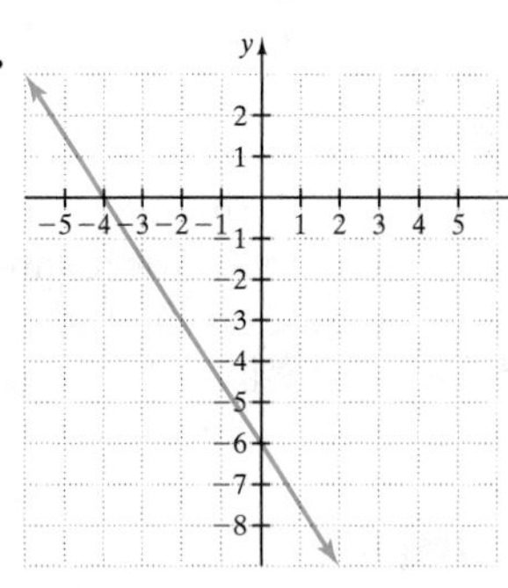

2.

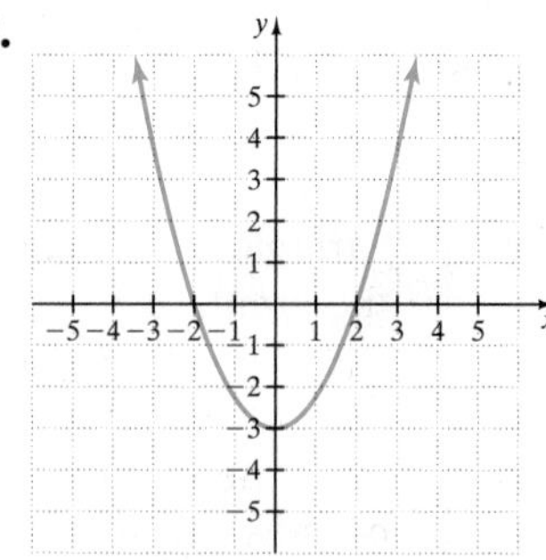

3.

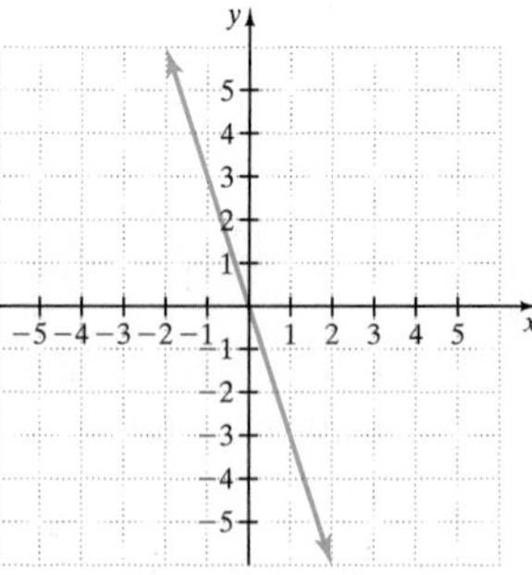

4.

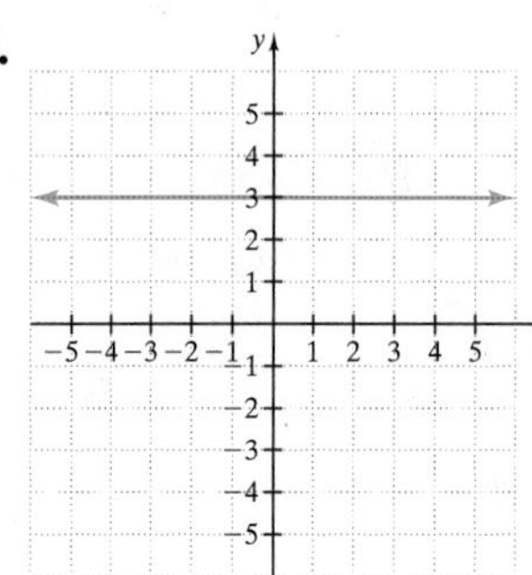

5.

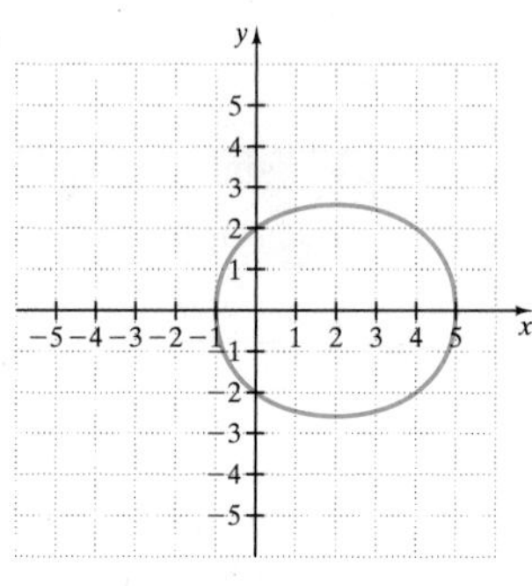

OBJECTIVE

2 Using Intercepts to Graph a Linear Equation 

Given the equation of a line, intercepts are usually easy to find since one coordinate is 0.

One way to find the y-intercept of a line, given its equation, is to let $x = 0$, since a point on the y-axis has an x-coordinate of 0. To find the x-intercept of a line, let $y = 0$, since a point on the x-axis has a y-coordinate of 0.

Finding x- and y-intercepts

To find the x-intercept, let $y = 0$ and solve for x.
To find the y-intercept, let $x = 0$ and solve for y.

EXAMPLE 6 Graph $x - 3y = 6$ by finding and plotting intercepts.

Solution Let $y = 0$ to find the x-intercept and let $x = 0$ to find the y-intercept.

Let $y = 0$	Let $x = 0$
$x - 3y = 6$	$x - 3y = 6$
$x - 3(0) = 6$	$0 - 3y = 6$
$x - 0 = 6$	$-3y = 6$
$x = 6$	$y = -2$

The x-intercept is $(6, 0)$ and the y-intercept is $(0, -2)$. We find a third ordered pair solution to check our work. If we let $y = -1$, then $x = 3$. Plot the points $(6, 0)$, $(0, -2)$, and $(3, -1)$. The graph of $x - 3y = 6$ is the line drawn through these points, as shown.

x	y
6	0
0	−2
3	−1

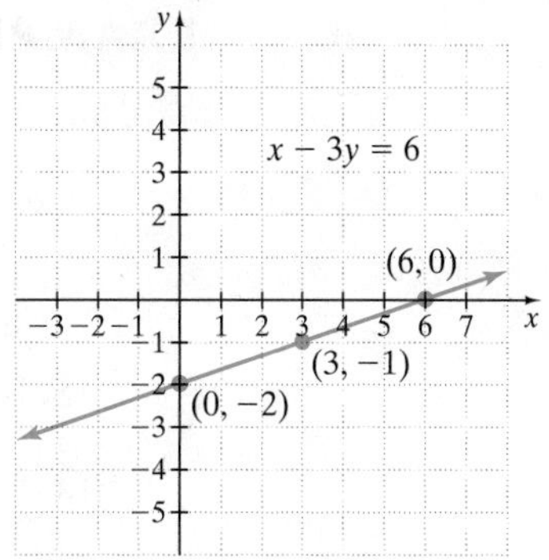

PRACTICE 6 Graph $x + 2y = -4$ by finding and plotting intercepts.

EXAMPLE 7 Graph $x = -2y$ by plotting intercepts.

Solution Let $y = 0$ to find the x-intercept and $x = 0$ to find the y-intercept.

Let $y = 0$	Let $x = 0$
$x = -2y$	$x = -2y$
$x = -2(0)$	$0 = -2y$
$x = 0$	$0 = y$

Both the x-intercept and y-intercept are $(0, 0)$. In other words, when $x = 0$, then $y = 0$, which gives the ordered pair $(0, 0)$. Also, when $y = 0$, then $x = 0$, which gives the same ordered pair $(0, 0)$. This happens when the graph passes through the origin. Since two points are needed to determine a line, we must find at least one more ordered pair that satisfies $x = -2y$. We will let $y = -1$ to find a second ordered pair solution and let $y = 1$ as a checkpoint.

Let $y = -1$	Let $y = 1$
$x = -2(-1)$	$x = -2(1)$
$x = 2$	$x = -2$

The ordered pairs are $(0, 0)$, $(2, -1)$, and $(-2, 1)$. Plot these points to graph $x = -2y$.

x	y
0	0
2	−1
−2	1

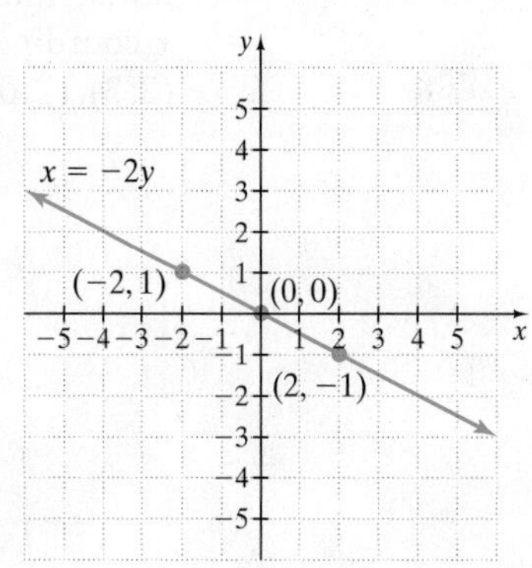

PRACTICE 7 Graph $x = 3y$ by plotting intercepts.

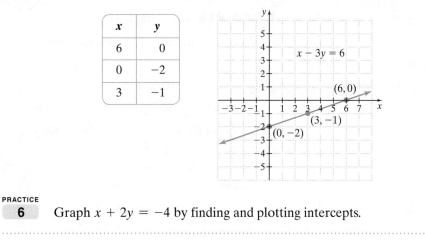

EXAMPLE 8 Graph $4x = 3y - 9$.

Solution Find the x- and y-intercepts, and then choose $x = 2$ to find a checkpoint.

Let $y = 0$	Let $x = 0$	Let $x = 2$
$4x = 3(0) - 9$	$4 \cdot 0 = 3y - 9$	$4(2) = 3y - 9$
$4x = -9$	$9 = 3y$	$8 = 3y - 9$
Solve for x.	Solve for y.	Solve for y.
$x = -\frac{9}{4}$ or $-2\frac{1}{4}$	$3 = y$	$17 = 3y$
		$\frac{17}{3} = y$ or $y = 5\frac{2}{3}$

The ordered pairs are $\left(-2\frac{1}{4}, 0\right)$, $(0, 3)$, and $\left(2, 5\frac{2}{3}\right)$. The equation $4x = 3y - 9$ is graphed as follows.

x	y
$-2\frac{1}{4}$	0
0	3
2	$5\frac{2}{3}$

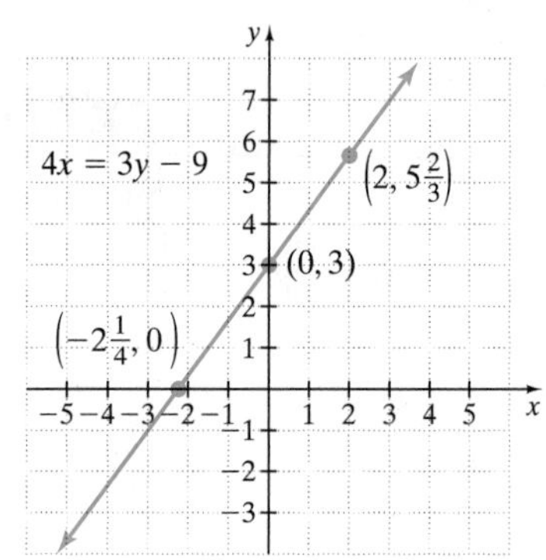

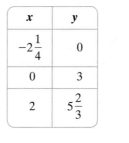

PRACTICE 8 Graph $3x = 2y + 4$.

OBJECTIVE 3 **Graphing Vertical and Horizontal Lines**

The equation $x = c$, where c is a real number constant, is a linear equation in two variables because it can be written in the form $x + 0y = c$. The graph of this equation is a vertical line as shown in the next example.

EXAMPLE 9 Graph $x = 2$.

Solution The equation $x = 2$ can be written as $x + 0y = 2$. For any y-value chosen, notice that x is 2. No other value for x satisfies $x + 0y = 2$. Any ordered pair whose x-coordinate is 2 is a solution of $x + 0y = 2$. We will use the ordered pair solutions $(2, 3)$, $(2, 0)$, and $(2, -3)$ to graph $x = 2$.

x	y
2	3
2	0
2	−3

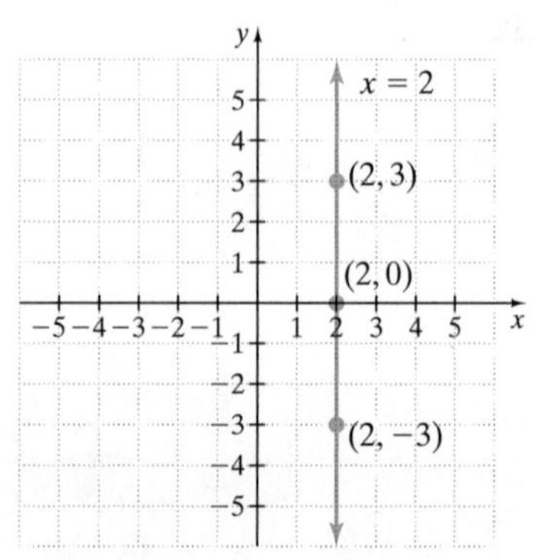

The graph is a vertical line with x-intercept $(2, 0)$. Note that this graph has no y-intercept because x is never 0. □

PRACTICE

9 Graph $x = -2$.

Vertical Lines

The graph of $x = c$, where c is a real number, is a vertical line with x-intercept $(c, 0)$.

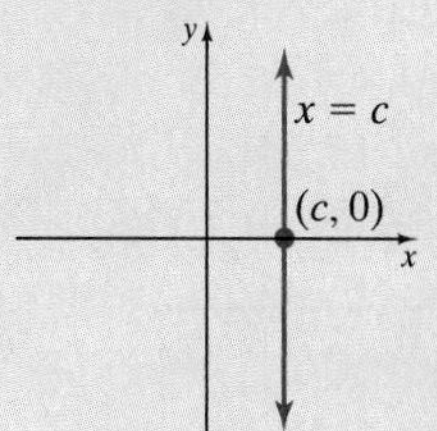

EXAMPLE 10 Graph $y = -3$.

Solution The equation $y = -3$ can be written as $0x + y = -3$. For any x-value chosen, y is -3. If we choose 4, 1, and -2 as x-values, the ordered pair solutions are $(4, -3)$, $(1, -3)$, and $(-2, -3)$. Use these ordered pairs to graph $y = -3$. The graph is a horizontal line with y-intercept $(0, -3)$ and no x-intercept.

x	y
4	-3
1	-3
-2	-3

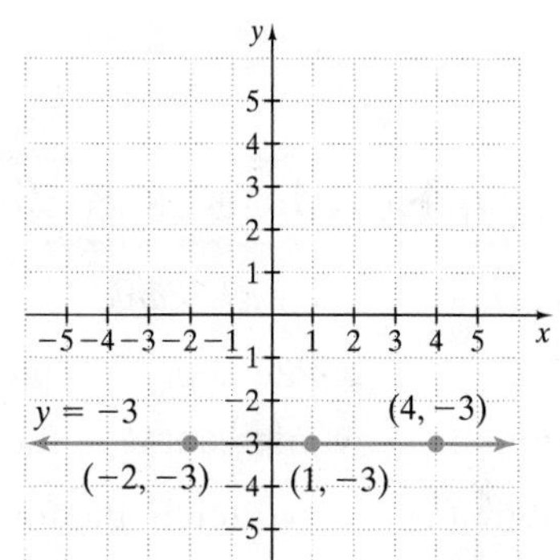

□

PRACTICE

10 Graph $y = 2$.

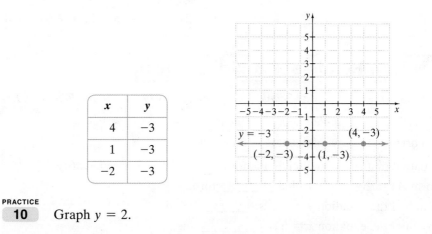

Horizontal Lines

The graph of $y = c$, where c is a real number, is a horizontal line with y-intercept $(0, c)$.

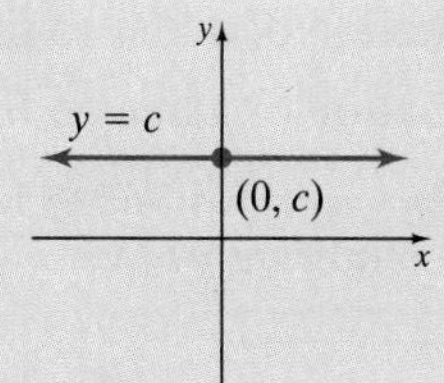

Graphing Calculator Explorations 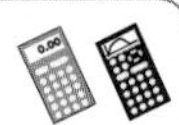

You may have noticed that to use the $\boxed{Y=}$ key on a grapher to graph an equation, the equation must be solved for y. For example, to graph $2x + 3y = 7$, we solve this equation for y.

$$2x + 3y = 7$$

$$3y = -2x + 7 \quad \text{Subtract } 2x \text{ from both sides.}$$

$$\frac{3y}{3} = -\frac{2x}{3} + \frac{7}{3} \quad \text{Divide both sides by 3.}$$

$$y = -\frac{2}{3}x + \frac{7}{3} \quad \text{Simplify.}$$

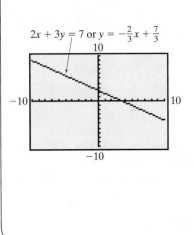

$2x + 3y = 7$ or $y = -\frac{2}{3}x + \frac{7}{3}$

10

−10 10

−10

To graph $2x + 3y = 7$ or $y = -\frac{2}{3}x + \frac{7}{3}$, press the $\boxed{Y=}$ key and enter

$$Y_1 = -\frac{2}{3}x + \frac{7}{3}$$

Graph each linear equation.

1. $x = 3.78y$

2. $-2.61y = x$

3. $3x + 7y = 21$

4. $-4x + 6y = 21$

5. $-2.2x + 6.8y = 15.5$

6. $5.9x - 0.8y = -10.4$

Vocabulary, Readiness & Video Check

Use the choices below to fill in each blank. Some choices may be used more than once. Exercises 1 and 2 come from Section 3.2.

x	vertical	x-intercept	linear
y	horizontal	y-intercept	standard

1. An equation that can be written in the form $Ax + By = C$ is called a(n) ________ equation in two variables.

2. The form $Ax + By = C$ is called ________ form.

3. The graph of the equation $y = -1$ is a ________ line.

4. The graph of the equation $x = 5$ is a ________ line.

5. A point where a graph crosses the y-axis is called a(n) ________.

6. A point where a graph crosses the x-axis is called a(n) ________.

7. Given an equation of a line, to find the x-intercept (if there is one), let ________ = 0 and solve for ________.

8. Given an equation of a line, to find the y-intercept (if there is one), let ________ = 0 and solve for ________.

Martin-Gay Interactive Videos

See Video 3.3

Watch the section lecture video and answer the following questions.

OBJECTIVE 1

9. At the end of Example 2, patterns are discussed. What reason is given for why x-intercepts have y-values of 0? For why y-intercepts have x-values of 0?

OBJECTIVE 2

10. In Example 3, the goal is to use the x- and y-intercepts to graph a line. Yet once the two intercepts are found, a third point is also found before the line is graphed. Why do think this practice of finding a third point is continued?

OBJECTIVE 3

11. From Examples 5 and 6, what can you say about the coefficient of x when the equation of a horizontal line is written as $Ax + By = C$? What about the coefficient of y when the equation of a vertical line is written as $Ax + By = C$?

3.3 Exercise Set

MyMathLab®

Identify the intercepts. See Examples 1 through 5.

1.

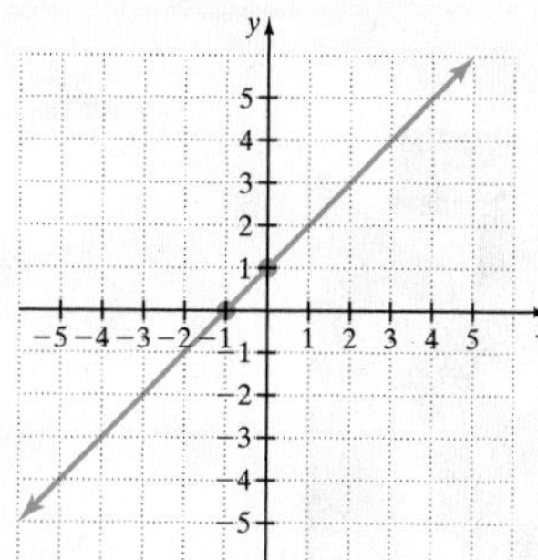

2.

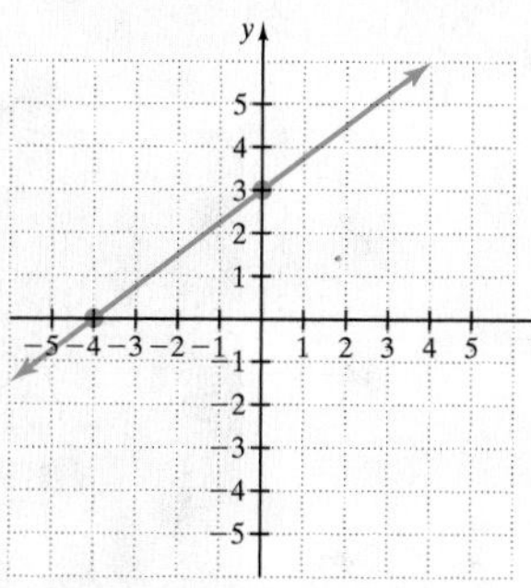

3.

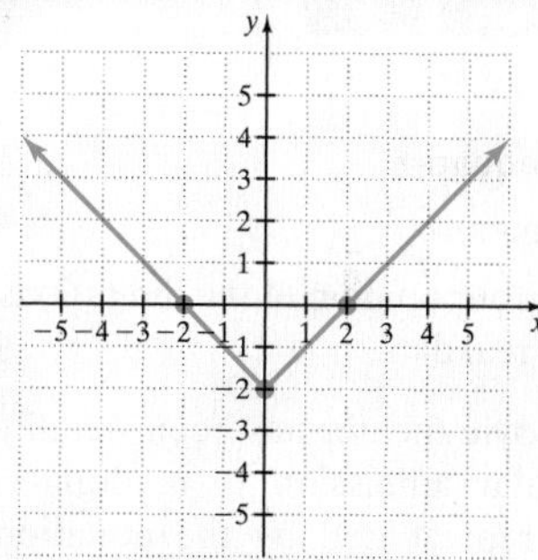

4.

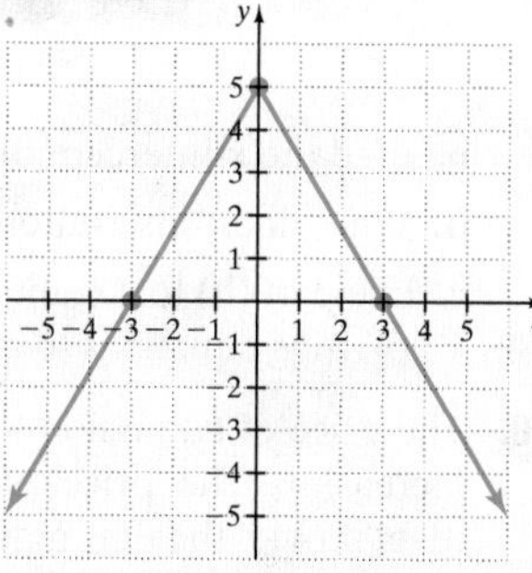

5.

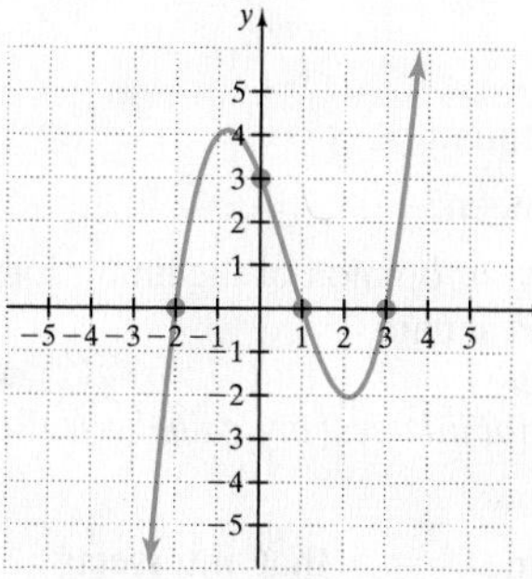

6.

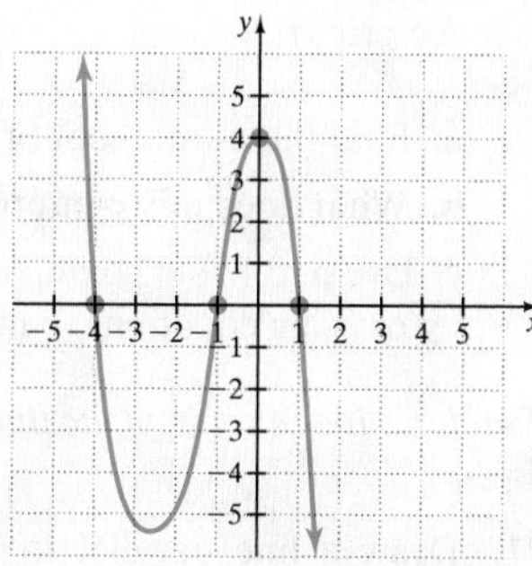

7.

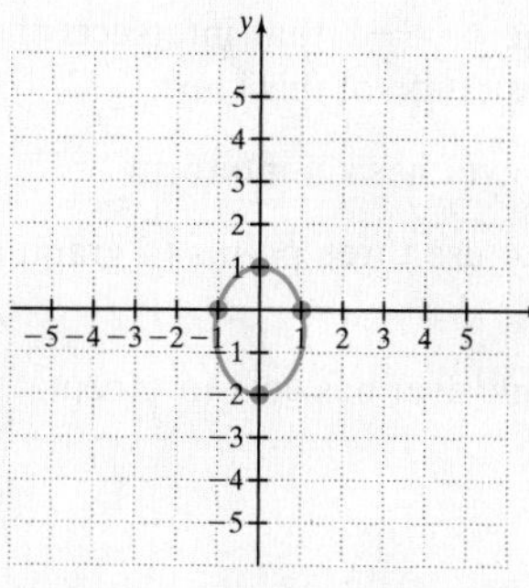

8.

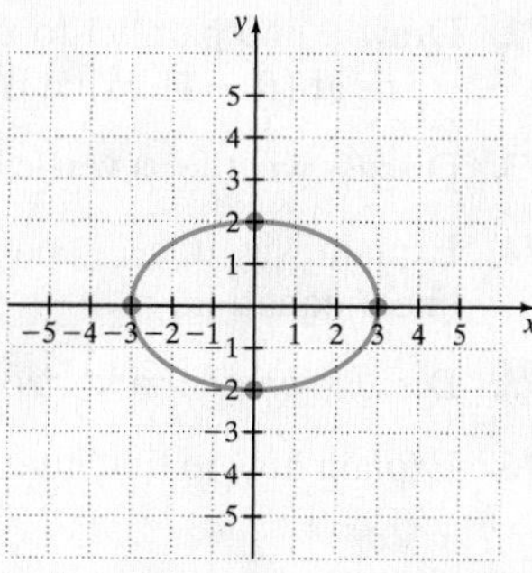

Solve. See Examples 1 through 5.

9. What is the greatest number of intercepts for a line?

10. What is the least number of intercepts for a line?

11. What is the least number of intercepts for a circle?

12. What is the greatest number of intercepts for a circle?

Graph each linear equation by finding and plotting its intercepts. See Examples 6 through 8.

13. $x - y = 3$ **14.** $x - y = -4$ **15.** $x = 5y$

16. $x = 2y$ **17.** $-x + 2y = 6$ **18.** $x - 2y = -8$

19. $2x - 4y = 8$ **20.** $2x + 3y = 6$ **21.** $y = 2x$

22. $y = -2x$ **23.** $y = 3x + 6$ **24.** $y = 2x + 10$

Graph each linear equation. See Examples 9 and 10.

25. $x = -1$ **26.** $y = 5$ **27.** $y = 0$

28. $x = 0$ **29.** $y + 7 = 0$ **30.** $x - 2 = 0$

31. $x + 3 = 0$ **32.** $y - 6 = 0$

MIXED PRACTICE

Graph each linear equation. See Examples 6 through 10.

33. $x = y$ **34.** $x = -y$

35. $x + 8y = 8$ **36.** $x + 3y = 9$

37. $5 = 6x - y$ **38.** $4 = x - 3y$

39. $-x + 10y = 11$ **40.** $-x + 9y = 10$

41. $x = -4\frac{1}{2}$ **42.** $x = -1\frac{3}{4}$

43. $y = 3\frac{1}{4}$ **44.** $y = 2\frac{1}{2}$

45. $y = -\frac{2}{3}x + 1$ **46.** $y = -\frac{3}{5}x + 3$

47. $4x - 6y + 2 = 0$ **48.** $9x - 6y + 3 = 0$

For Exercises 49 through 54, match each equation with its graph. See Graphs A.–F. below and on the next page.

49. $y = 3$ **50.** $y = 2x + 2$

51. $x = -1$ **52.** $x = 3$

53. $y = 2x + 3$ **54.** $y = -2x$

A.

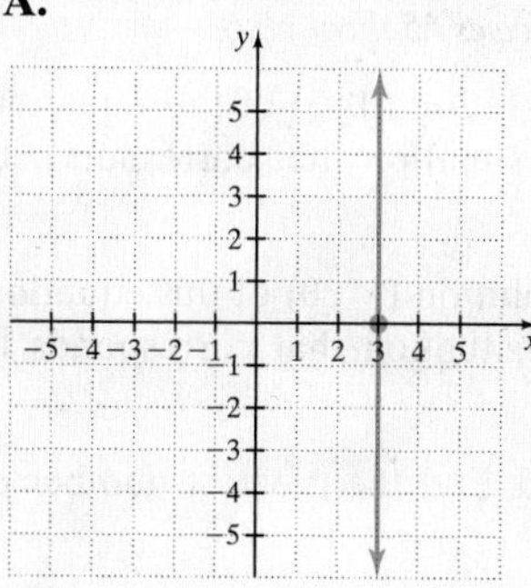

B.

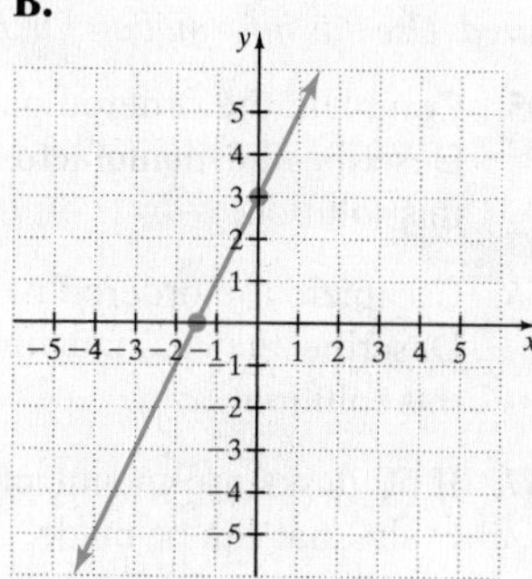

C.

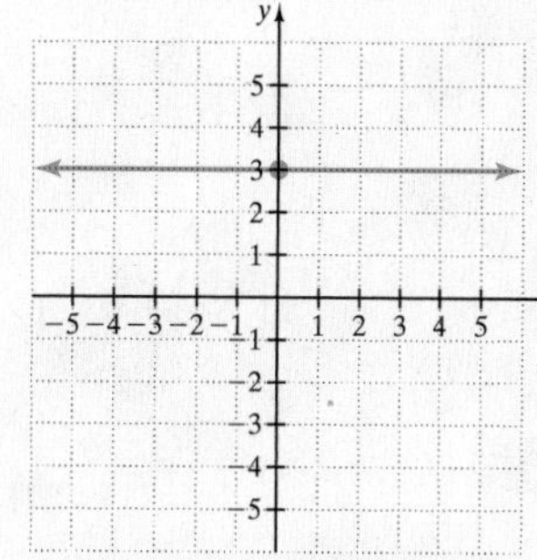

D.

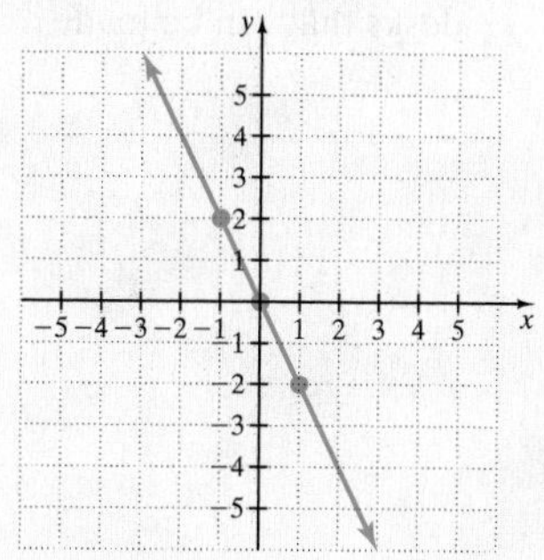

E.

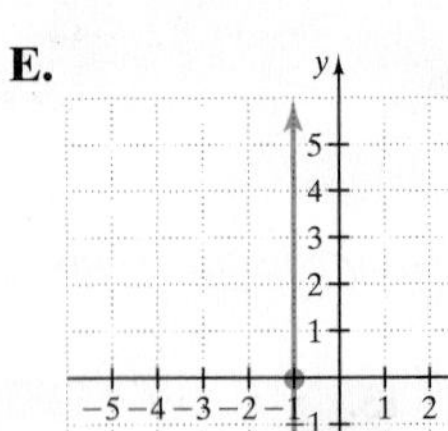

F.

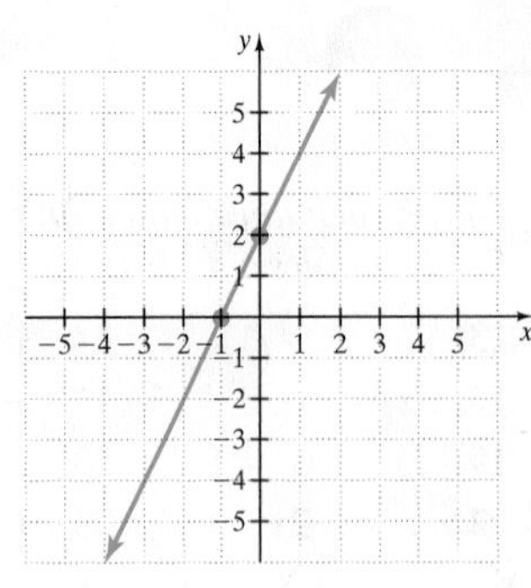

REVIEW AND PREVIEW

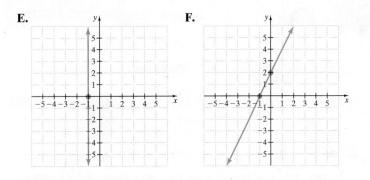

Simplify. See Sections 1.4 through 1.7.

55. $\frac{-6 - 3}{2 - 8}$

56. $\frac{4 - 5}{-1 - 0}$

57. $\frac{-8 - (-2)}{-3 - (-2)}$

58. $\frac{12 - 3}{10 - 9}$

59. $\frac{0 - 6}{5 - 0}$

60. $\frac{2 - 2}{3 - 5}$

CONCEPT EXTENSIONS

Answer the following true or false.

61. All lines have an x-intercept *and* a y-intercept.

62. The graph of $y = 4x$ contains the point $(0, 0)$.

63. The graph of $x + y = 5$ has an x-intercept of $(5, 0)$ and a y-intercept of $(0, 5)$.

64. The graph of $y = 5x$ contains the point $(5, 1)$.

The production supervisor at Alexandra's Office Products finds that it takes 3 hours to manufacture a particular office chair and 6 hours to manufacture an office desk. A total of 1200 hours is available to produce office chairs and desks of this style. The linear equation that models this situation is $3x + 6y = 1200$, where x represents the number of chairs produced and y the number of desks manufactured. Use this information for Exercises 65 through 68.

65. Complete the ordered pair solution $(0, \quad)$ of this equation. Describe the manufacturing situation that corresponds to this solution.

66. Complete the ordered pair solution $(\quad, 0)$ of this equation. Describe the manufacturing situation that corresponds to this solution.

67. If 50 desks are manufactured, find the greatest number of chairs that can be made.

68. If 50 chairs are manufactured, find the greatest number of desks that can be made.

69. Since 2002, admissions at movie theaters in the U.S. and Canada have mostly been in a decline. The number of people y (in billions) who go to movie theaters each year can be estimated by the equation $y = -0.025x + 1.55$, where x represents the number of years since 2002. (*Source:* Based on data from Motion Picture Association of America)

a. Find the x-intercept of this equation.

b. What does this x-intercept mean?

c. Use part (b) to comment on your opinion of the limitations of using equations to model real data.

70. The price of admission to a movie theater has been steadily increasing. The price of regular admission y (in dollars) to a movie theater may be represented by the equation $y = 0.24x + 5.28$, where x is the number of years after 2000. (*Source:* Based on data from Motion Picture Association of America)

a. Find the x-intercept of this equation.

b. What does this x-intercept mean?

c. Use part (b) to comment on your opinion of the limitations of using equations to model real data.

Two lines in the same plane that do not intersect are called **parallel lines.**

△ **71.** Draw a line parallel to the line $x = 5$ that intersects the x-axis at $(1, 0)$. What is the equation of this line?

△ **72.** Draw a line parallel to the line $y = -1$ that intersects the y-axis at $(0, -4)$. What is the equation of this line?

73. Discuss whether a vertical line ever has a y-intercept.

74. Explain why it is a good idea to use three points to graph a linear equation.

75. Discuss whether a horizontal line ever has an x-intercept.

76. Explain how to find intercepts.

3.4 Slope and Rate of Change 

OBJECTIVES

1. Find the Slope of a Line Given Two Points of the Line.
2. Find the Slope of a Line Given Its Equation.
3. Find the Slopes of Horizontal and Vertical Lines.
4. Compare the Slopes of Parallel and Perpendicular Lines.
5. Interpret Slope as a Rate of Change.

OBJECTIVE

1 Finding the Slope of a Line Given Two Points of the Line 

Thus far, much of this chapter has been devoted to graphing lines. You have probably noticed by now that a key feature of a line is its slant or steepness. In mathematics, the slant or steepness of a line is formally known as its **slope.** We measure the slope of a line by the ratio of vertical change to the corresponding horizontal change as we move along the line.

On the line below, for example, suppose that we begin at the point (1, 2) and move to the point (4, 6). The vertical change is the change in y-coordinates: $6 - 2$ or 4 units. The corresponding horizontal change is the change in x-coordinates: $4 - 1 = 3$ units. The ratio of these changes is

$$\text{slope} = \frac{\text{change in } y \text{ (vertical change)}}{\text{change in } x \text{ (horizontal change)}} = \frac{4}{3}$$

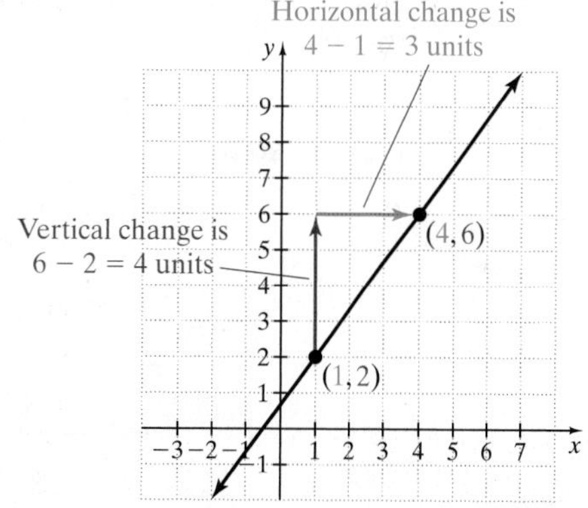

The slope of this line, then, is $\frac{4}{3}$. This means that for every 4 units of change in y-coordinates, there is a corresponding change of 3 units in x-coordinates.

Helpful Hint

It makes no difference what two points of a line are chosen to find its slope. The slope of a line is the same everywhere on the line.

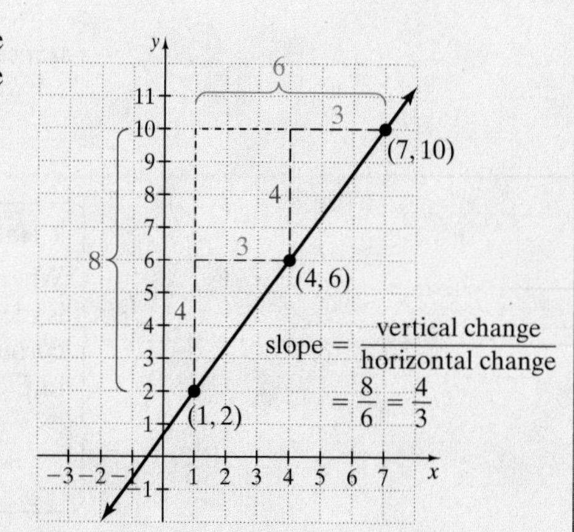

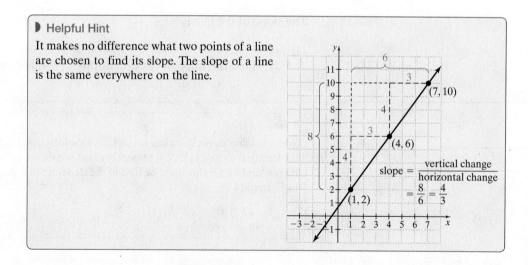

To find the slope of a line, then, choose two points of the line. Label the two x-coordinates of two points x_1 and x_2 (read "x sub one" and "x sub two"), and label the corresponding y-coordinates y_1 and y_2.

The vertical change or **rise** between these points is the difference in the y-coordinates: $y_2 - y_1$. The horizontal change or **run** between the points is the difference of the x-coordinates: $x_2 - x_1$. The slope of the line is the ratio of $y_2 - y_1$ to $x_2 - x_1$, and we traditionally use the letter m to denote slope: $m = \dfrac{y_2 - y_1}{x_2 - x_1}$.

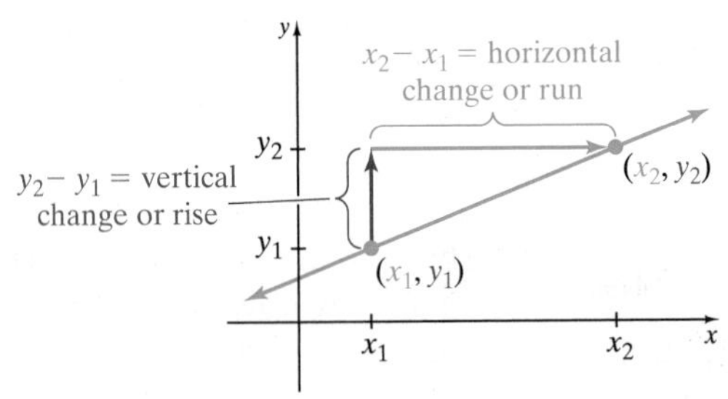

Slope of a Line

The slope m of the line containing the points (x_1, y_1) and (x_2, y_2) is given by

$$m = \frac{\text{rise}}{\text{run}} = \frac{\text{change in } y}{\text{change in } x} = \frac{y_2 - y_1}{x_2 - x_1}, \quad \text{as long as } x_2 \neq x_1$$

EXAMPLE 1 Find the slope of the line through $(-1, 5)$ and $(2, -3)$. Graph the line.

Solution If we let (x_1, y_1) be $(-1, 5)$, then $x_1 = -1$ and $y_1 = 5$. Also, let (x_2, y_2) be $(2, -3)$ so that $x_2 = 2$ and $y_2 = -3$. Then, by the definition of slope,

$$m = \frac{y_2 - y_1}{x_2 - x_1} = \frac{-3 - 5}{2 - (-1)} = \frac{-8}{3} = -\frac{8}{3}$$

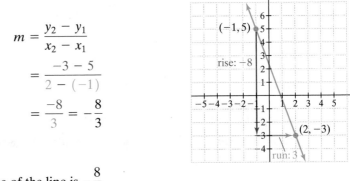

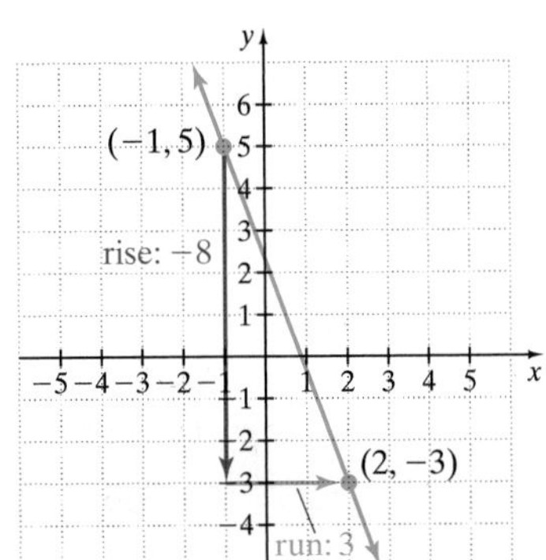

The slope of the line is $-\frac{8}{3}$. □

PRACTICE

1 Find the slope of the line through $(-4, 11)$ and $(2, 5)$.

Helpful Hint

When finding slope, it makes no difference which point is identified as (x_1, y_1) and which is identified as (x_2, y_2). Just remember that whatever y-value is first in the numerator, its corresponding x-value must be first in the denominator. Another way to calculate the slope in Example 1 is:

$$m = \frac{y_2 - y_1}{x_2 - x_1} = \frac{5 - (-3)}{-1 - 2} = \frac{8}{-3} \text{ or } -\frac{8}{3} \quad \leftarrow \text{Same slope as found in Example 1}$$

✓**CONCEPT CHECK**

The points $(-2, -5)$, $(0, -2)$, $(4, 4)$, and $(10, 13)$ all lie on the same line. Work with a partner and verify that the slope is the same no matter which points are used to find slope.

Answer to Concept Check:
$m = \frac{3}{2}$

EXAMPLE 2 Find the slope of the line through $(-1, -2)$ and $(2, 4)$. Graph the line.

Solution Let (x_1, y_1) be $(2, 4)$ and (x_2, y_2) be $(-1, -2)$.

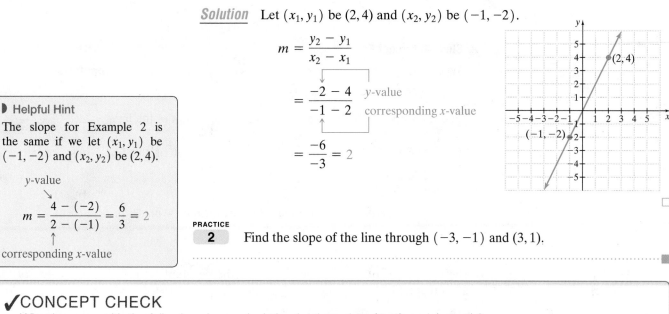

$$m = \frac{y_2 - y_1}{x_2 - x_1}$$

$$= \frac{-2 - 4}{-1 - 2} \quad \begin{array}{l} y\text{-value} \\ \text{corresponding } x\text{-value} \end{array}$$

$$= \frac{-6}{-3} = 2$$

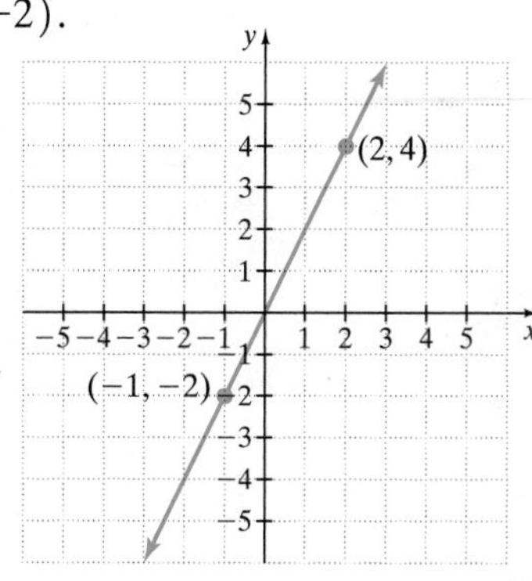

> **Helpful Hint**
>
> The slope for Example 2 is the same if we let (x_1, y_1) be $(-1, -2)$ and (x_2, y_2) be $(2, 4)$.
>
> y-value
>
> $$m = \frac{4 - (-2)}{2 - (-1)} = \frac{6}{3} = 2$$
>
> corresponding x-value

PRACTICE

2 Find the slope of the line through $(-3, -1)$ and $(3, 1)$.

✓CONCEPT CHECK

What is wrong with the following slope calculation for the points $(3, 5)$ and $(-2, 6)$?

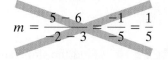

$$m = \frac{5 - 6}{-2 - 3} = \frac{-1}{-5} = \frac{1}{5}$$

Notice that the slope of the line in Example 1 is negative, whereas the slope of the line in Example 2 is positive. Let your eye follow the line with negative slope from left to right and notice that the line "goes down." Following the line with positive slope from left to right, notice that the line "goes up." This is true in general.

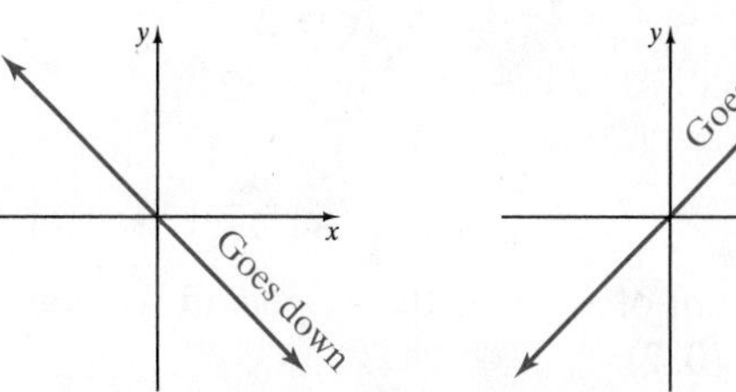

Negative slope Positive slope

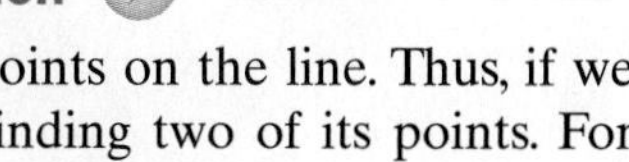

> **Helpful Hint**
>
> To decide whether a line "goes up" or "goes down," always follow the line from left to right.

OBJECTIVE

2 Finding the Slope of a Line Given Its Equation

As we have seen, the slope of a line is defined by two points on the line. Thus, if we know the equation of a line, we can find its slope by finding two of its points. For example, let's find the slope of the line

$$y = 3x + 2$$

To find two points, we can choose two values for x and substitute to find corresponding y-values. If $x = 0$, for example, $y = 3 \cdot 0 + 2$ or $y = 2$. If $x = 1$, $y = 3 \cdot 1 + 2$ or $y = 5$. This gives the ordered pairs $(0, 2)$ and $(1, 5)$. Using the definition for slope, we have

$$m = \frac{5 - 2}{1 - 0} = \frac{3}{1} = 3 \quad \text{The slope is 3.}$$

Notice that the slope, 3, is the same as the coefficient of x in the equation $y = 3x + 2$.

Also, recall from Section 3.2 that the graph of an equation of the form $y = mx + b$ has y-intercept $(0, b)$.

Answer to Concept Check:
The order in which the x- and y-values are used must be the same.

$$m = \frac{5 - 6}{3 - (-2)} = \frac{-1}{5} = -\frac{1}{5}$$

This means that the y-intercept of the graph of $y = 3x + 2$ is $(0, 2)$. This is true in general and the form $y = mx + b$ is appropriately called the **slope-intercept form.**

$$y = \underset{\text{slope}}{m}x + \underset{\substack{y\text{-intercept} \\ (0,\, b)}}{b}$$

Slope-Intercept Form

When a linear equation in two variables is written in slope-intercept form,

$$y = mx + b$$

m is the slope of the line and $(0, b)$ is the y-intercept of the line.

EXAMPLE 3 Find the slope and y-intercept of the line whose equation is $y = \frac{3}{4}x + 6$

Solution The equation is in slope-intercept form, $y = mx + b$.

$$y = \frac{3}{4}x + 6$$

The coefficient of x, $\frac{3}{4}$, is the slope and the constant term, 6, is the y-value of the y-intercept, $(0, 6)$. □

PRACTICE 3 Find the slope and y-intercept of the line whose equation is $y = \frac{2}{3}x - 2$.

EXAMPLE 4 Find the slope and the y-intercept of the line whose equation is $-y = 5x - 2$.

Solution: Remember, the equation must be solved for y (not $-y$) in order for it to be written in slope-intercept form.

To solve for y, let's divide both sides of the equation by -1.

$$-y = 5x - 2$$

$$\frac{-y}{-1} = \frac{5x}{-1} - \frac{2}{-1} \quad \text{Divide both sides by } -1.$$

$$y = -5x + 2 \quad \text{Simplify.}$$

The coefficient of x, -5, is the slope and the constant term, 2, is the y-value of the y-intercept, $(0, 2)$. □

PRACTICE 4 Find the slope and y-intercept of the line whose equation is $-y = -6x + 5$.

EXAMPLE 5 Find the slope and the y-intercept of the line whose equation is $3x - 4y = 4$.

Solution Write the equation in slope-intercept form by solving for y.

$$3x - 4y = 4$$

$$-4y = -3x + 4 \quad \text{Subtract } 3x \text{ from both sides.}$$

$$\frac{-4y}{-4} = \frac{-3x}{-4} + \frac{4}{-4} \quad \text{Divide both sides by } -4.$$

$$y = \frac{3}{4}x - 1 \quad \text{Simplify.}$$

The coefficient of x, $\frac{3}{4}$, is the slope, and the y-intercept is $(0, -1)$. □

PRACTICE 5 Find the slope and the y-intercept of the line whose equation is $5x + 2y = 8$.

OBJECTIVE 3

Finding Slopes of Horizontal and Vertical Lines 

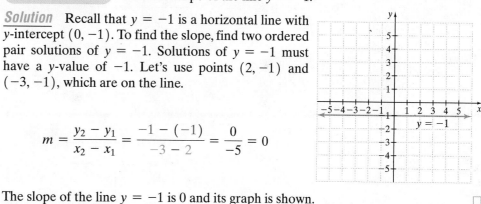

Recall that if a line tilts upward from left to right, its slope is positive. If a line tilts downward from left to right, its slope is negative. Let's now find the slopes of two special lines, horizontal and vertical lines.

EXAMPLE 6 Find the slope of the line $y = -1$.

Solution Recall that $y = -1$ is a horizontal line with y-intercept $(0, -1)$. To find the slope, find two ordered pair solutions of $y = -1$. Solutions of $y = -1$ must have a y-value of -1. Let's use points $(2, -1)$ and $(-3, -1)$, which are on the line.

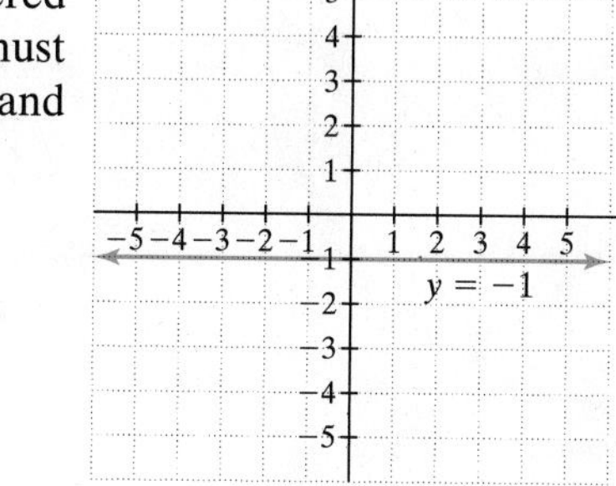

$$m = \frac{y_2 - y_1}{x_2 - x_1} = \frac{-1 - (-1)}{-3 - 2} = \frac{0}{-5} = 0$$

The slope of the line $y = -1$ is 0 and its graph is shown. □

PRACTICE 6 Find the slope of the line $y = 3$.

Any two points of a horizontal line will have the same y-values. This means that the y-values will always have a difference of 0 for all horizontal lines. Thus, **all horizontal lines have a slope 0.**

EXAMPLE 7 Find the slope of the line $x = 5$.

Solution Recall that the graph of $x = 5$ is a vertical line with x-intercept $(5, 0)$.

To find the slope, find two ordered pair solutions of $x = 5$. Solutions of $x = 5$ must have an x-value of 5. Let's use points $(5, 0)$ and $(5, 4)$, which are on the line.

$$m = \frac{y_2 - y_1}{x_2 - x_1} = \frac{4 - 0}{5 - 5} = \frac{4}{0}$$

Since $\frac{4}{0}$ is undefined, we say the slope of the vertical line $x = 5$ is undefined, and its graph is shown.

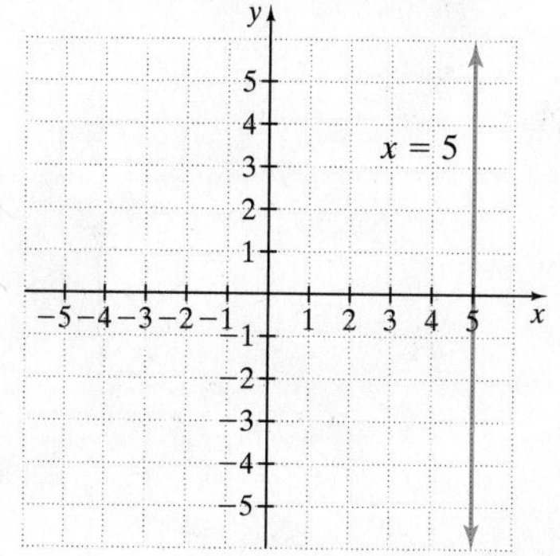

□

PRACTICE 7 Find the slope of the line $x = -4$.

Any two points of a vertical line will have the same x-values. This means that the x-values will always have a difference of 0 for all vertical lines. Thus **all vertical lines have undefined slope.**

> **Helpful Hint**
>
> Slope of 0 and undefined slope are not the same. Vertical lines have undefined slope or no slope, while horizontal lines have a slope of 0.

Here is a general review of slope.

Summary of Slope

Slope m of the line through (x_1, y_1) and (x_2, y_2) is given by the equation

$$m = \frac{y_2 - y_1}{x_2 - x_1}.$$

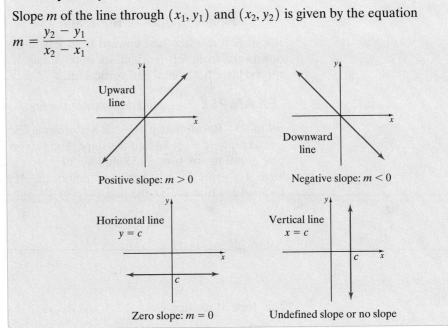

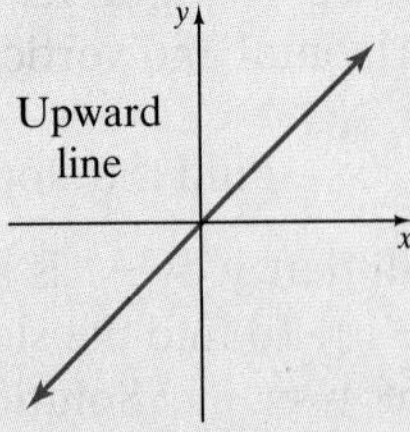

Positive slope: $m > 0$

Downward line

Negative slope: $m < 0$

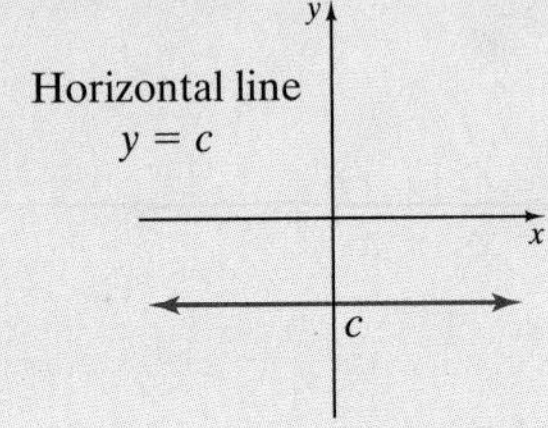

Zero slope: $m = 0$

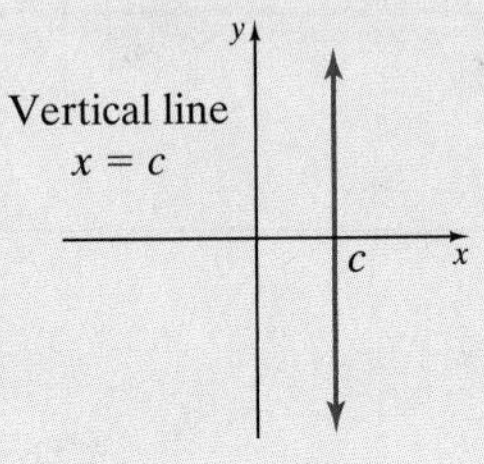

Undefined slope or no slope

OBJECTIVE 4 Slopes of Parallel and Perpendicular Lines

Two lines in the same plane are **parallel** if they do not intersect. Slopes of lines can help us determine whether lines are parallel. Parallel lines have the same steepness, so it follows that they have the same slope.

For example, the graphs of

$$y = -2x + 4$$

and

$$y = -2x - 3$$

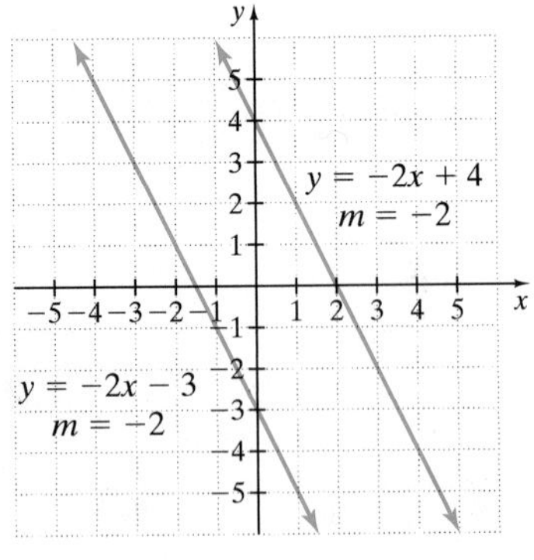

are shown. These lines have the same slope, −2. They also have different y-intercepts, so the lines are distinct and parallel. (If the y-intercepts were the same also, the lines would be the same.)

Parallel Lines

Nonvertical parallel lines have the same slope and different y-intercepts.

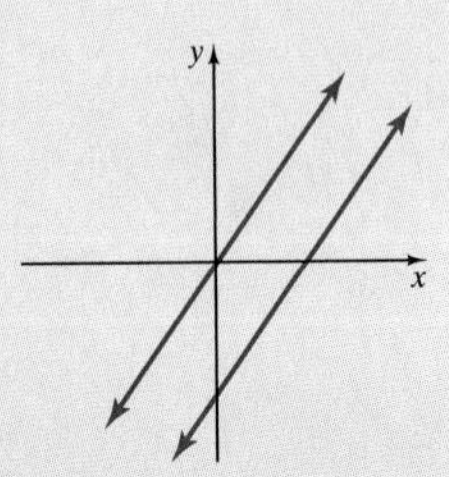

Two lines are **perpendicular** if they lie in the same plane and meet at a 90° (right) angle. How do the slopes of perpendicular lines compare? The product of the slopes of two perpendicular lines is −1.

For example, the graphs of

$$y = 4x + 1$$

and

$$y = -\frac{1}{4}x - 3$$

are shown. The slopes of the lines are 4 and $-\frac{1}{4}$. Their product is $4\left(-\frac{1}{4}\right) = -1$, so the lines are perpendicular.

Perpendicular Lines

If the product of the slopes of two lines is -1, then the lines are perpendicular.

(Two nonvertical lines are perpendicular if the slope of one is the negative reciprocal of the slope of the other.)

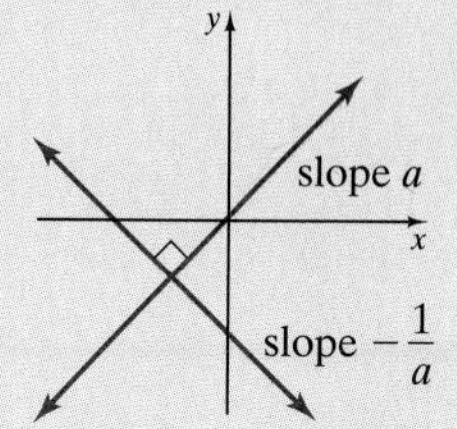

Helpful Hint

Here are examples of numbers that are negative (opposite) reciprocals.

Number	*Negative Reciprocal*	*Their Product Is* **−1.**
$\frac{2}{3}$	$-\frac{3}{2}$	$\frac{2}{3}\cdot -\frac{3}{2} = -\frac{6}{6} = -1$
-5 or $-\frac{5}{1}$	$\frac{1}{5}$	$-5\cdot\frac{1}{5} = -\frac{5}{5} = -1$

Helpful Hint

Here are a few important facts about vertical and horizontal lines.

- Two distinct vertical lines are parallel.
- Two distinct horizontal lines are parallel.
- A horizontal line and a vertical line are always perpendicular.

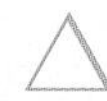

EXAMPLE 8 Determine whether each pair of lines is parallel, perpendicular, or neither.

a. $y = -\frac{1}{5}x + 1$
$2x + 10y = 3$

b. $x + y = 3$
$-x + y = 4$

c. $3x + y = 5$
$2x + 3y = 6$

Solution

a. The slope of the line $y = -\frac{1}{5}x + 1$ is $-\frac{1}{5}$. We find the slope of the second line by solving its equation for y.

$$2x + 10y = 3$$

$$10y = -2x + 3 \quad \text{Subtract } 2x \text{ from both sides.}$$

$$y = \frac{-2}{10}x + \frac{3}{10} \quad \text{Divide both sides by 10.}$$

$$y = -\frac{1}{5}x + \frac{3}{10} \quad \text{Simplify.}$$

The slope of this line is $-\frac{1}{5}$ also. Since the lines have the same slope and different y-intercepts, they are parallel, as shown in the figure on the next page.

b. To find each slope, we solve each equation for y.

$$x + y = 3 \qquad\qquad -x + y = 4$$
$$y = -x + 3 \qquad\qquad y = x + 4$$

The slope is -1. The slope is 1.

The slopes are not the same, so the lines are not parallel. Next we check the product of the slopes: $(-1)(1) = -1$. Since the product is -1, the lines are perpendicular, as shown in the figure.

a.

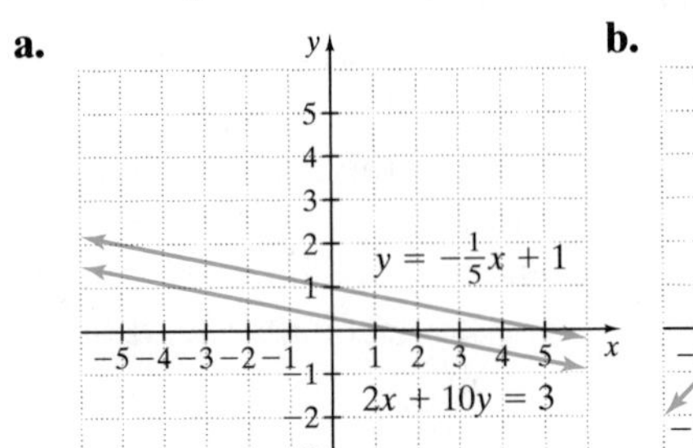

b.

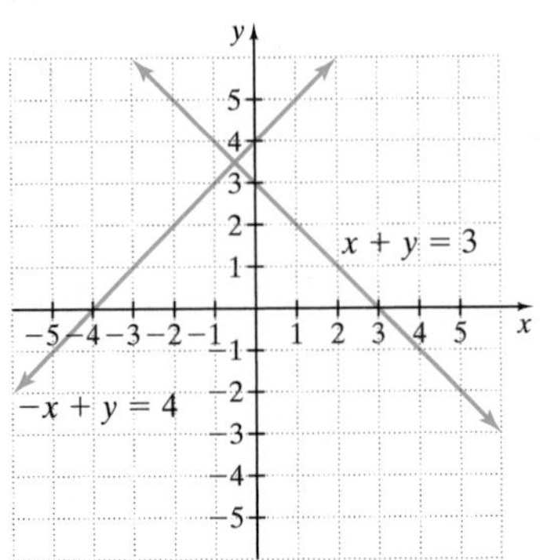

c. We solve each equation for y to find each slope. The slopes are -3 and $-\frac{2}{3}$. The slopes are not the same and their product is not -1. Thus, the lines are neither parallel nor perpendicular.

PRACTICE

8 Determine whether each pair of lines is parallel, perpendicular, or neither.

a. $y = -5x + 1$
$x - 5y = 10$

b. $x + y = 11$
$2x + y = 11$

c. $2x + 3y = 21$
$6y = -4x - 2$

✓CONCEPT CHECK

Consider the line $-6x + 2y = 1$.

a. Write the equations of two lines parallel to this line.

b. Write the equations of two lines perpendicular to this line.

OBJECTIVE

5 Slope as a Rate of Change 

Slope can also be interpreted as a rate of change. In other words, slope tells us how fast y is changing with respect to x. To see this, let's look at a few of the many real-world applications of slope. For example, the pitch of a roof, used by builders and architects, is its slope. The pitch of the roof on the left is $\frac{7}{10}\left(\frac{\text{rise}}{\text{run}}\right)$. This means that the roof rises vertically 7 feet for every horizontal 10 feet. The rate of change for the roof is 7 vertical feet (y) per 10 horizontal feet (x).

The grade of a road is its slope written as a percent. A 7% grade, as shown below, means that the road rises (or falls) 7 feet for every horizontal 100 feet. $\left(\text{Recall that } 7\% = \frac{7}{100}.\right)$ Here, the slope of $\frac{7}{100}$ gives us the rate of change. The road rises (in our diagram) 7 vertical feet (y) for every 100 horizontal feet (x).

Answers to Concept Check:

a. any two lines with $m = 3$ and y-intercept not $\left(0, \frac{1}{2}\right)$

b. any two lines with $m = -\frac{1}{3}$

EXAMPLE 9 **Finding the Grade of a Road**

At one part of the road to the summit of Pikes Peak, the road rises at a rate of 15 vertical feet for a horizontal distance of 250 feet. Find the grade of the road.

Solution Recall that the grade of a road is its slope written as a percent.

$$\text{grade} = \frac{\text{rise}}{\text{run}} = \frac{15}{250} = 0.06 = 6\%$$

15 feet

250 feet

The grade is 6%. □

PRACTICE

9 One part of the Mt. Washington (New Hampshire) cog railway rises about 1794 feet over a horizontal distance of 7176 feet. Find the grade of this part of the railway.

EXAMPLE 10 **Finding the Slope of a Line**

The following graph shows annual food and drink sales y (in billions of dollars) for year x. Find the slope of the line and attach the proper units for the rate of change. Then write a sentence explaining the meaning of the slope for this application.

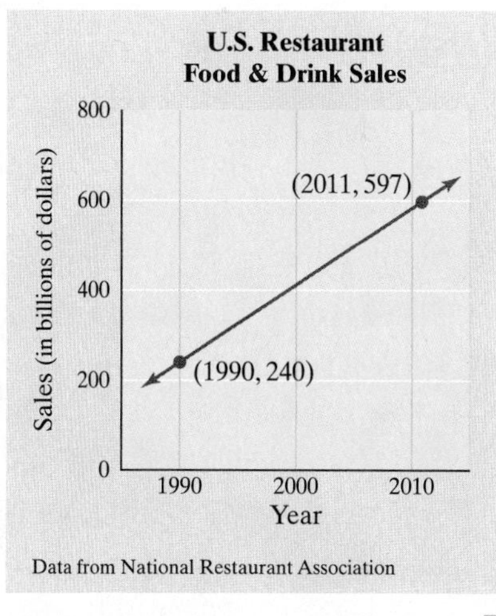

Solution Use (1990, 240) and (2011, 597) to calculate slope.

$$m = \frac{597 - 240}{2011 - 1990} = \frac{357}{21} = \frac{17 \text{ billion dollars}}{1 \text{ year}}$$

This means that the rate of change of restarurant food and drink sales increases by 17 billion dollars every 1 year, or $17 billion per year. □

PRACTICE

10 The following graph shows the cost y (in dollars) of having laundry done at the Wash-n-Fold, where x is the number of pounds of laundry. Find the slope of the line, and attach the proper units for the rate of change.

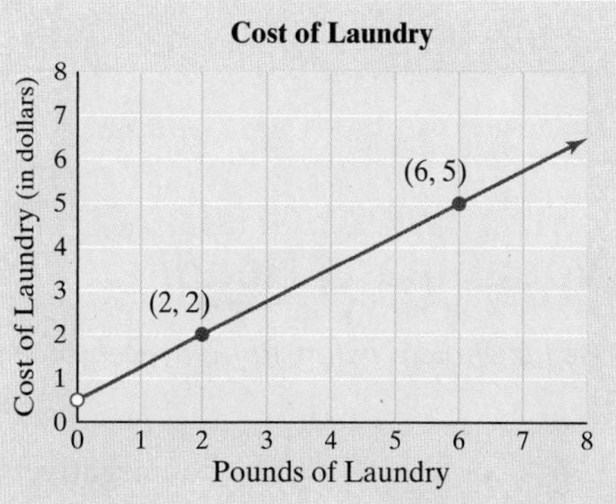

Graphing Calculator Explorations 

It is possible to use a grapher to sketch the graph of more than one equation on the same set of axes. This feature can be used to confirm our findings from Section 3.2 when we learned that the graph of an equation written in the form $y = mx + b$ has a y-intercept of b. For example, graph the equations $y = \frac{2}{5}x$, $y = \frac{2}{5}x + 7$, and

$y = \frac{2}{5}x - 4$ on the same set of axes. To do so, press the $\boxed{Y=}$ key and enter the equations on the first three lines.

$$Y_1 = \left(\frac{2}{5}\right)x$$

$$Y_2 = \left(\frac{2}{5}\right)x + 7$$

$$Y_3 = \left(\frac{2}{5}\right)x - 4$$

The screen should look like:

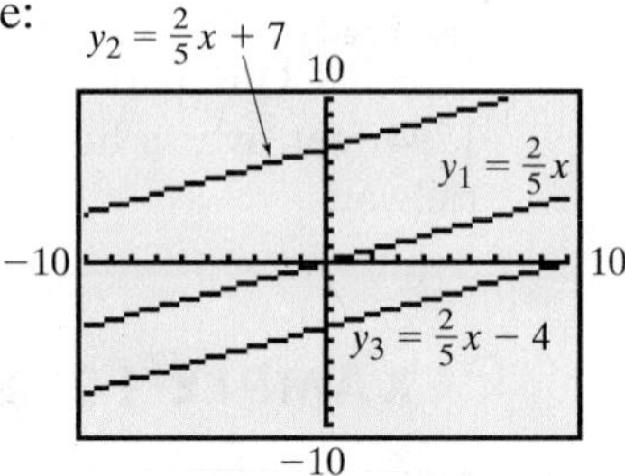

Notice that all three graphs appear to have the same positive slope. The graph of $y = \frac{2}{5}x + 7$ is the graph of $y = \frac{2}{5}x$ moved 7 units upward with a y-intercept of 7. Also, the graph of $y = \frac{2}{5}x - 4$ is the graph of $y = \frac{2}{5}x$ moved 4 units downward with a y-intercept of -4.

Graph the equations on the same set of axes. Describe the similarities and differences in their graphs.

1. $y = 3.8x, y = 3.8x - 3, y = 3.8x + 9$

2. $y = -4.9x, y = -4.9x + 1, y = -4.9x + 8$

3. $y = \frac{1}{4}x; y = \frac{1}{4}x + 5, y = \frac{1}{4}x - 8$

4. $y = -\frac{3}{4}x, y = -\frac{3}{4}x - 5, y = -\frac{3}{4}x + 6$

Vocabulary, Readiness & Video Check

Use the choices below to fill in each blank. Not all choices will be used.

m	x	0	positive	undefined
b	y	slope	negative	

1. The measure of the steepness or tilt of a line is called __________.

2. If an equation is written in the form $y = mx + b$, the value of the letter __________ is the value of the slope of the graph.

3. The slope of a horizontal line is __________.

4. The slope of a vertical line is __________.

5. If the graph of a line moves upward from left to right, the line has __________ slope.

6. If the graph of a line moves downward from left to right, the line has __________ slope.

7. Given two points of a line, slope $= \frac{\text{change in \_\_}}{\text{change in \_\_}}$.

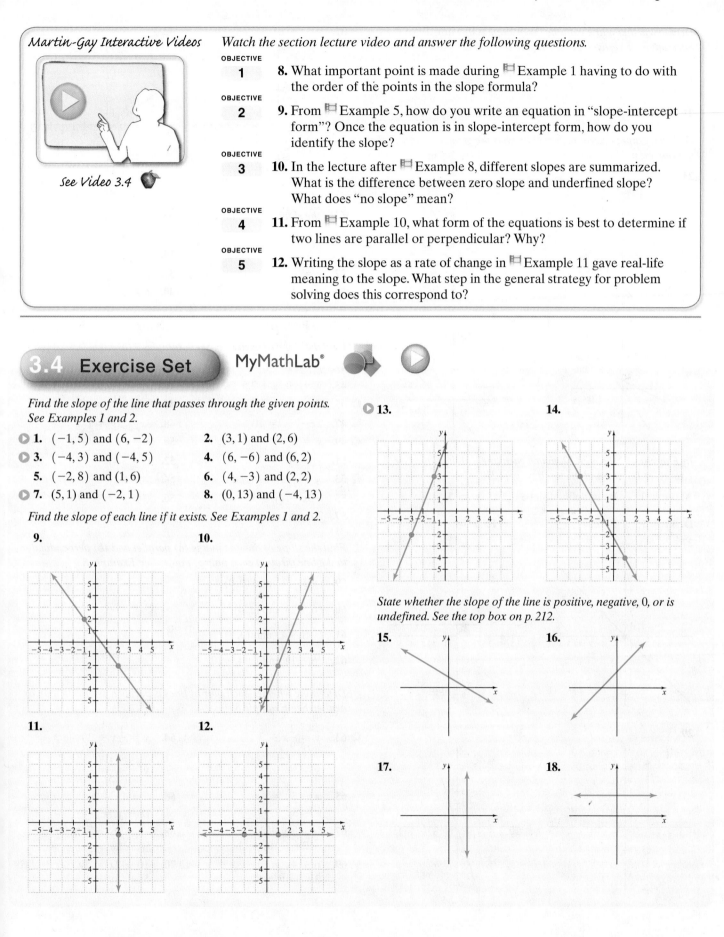

Martin-Gay Interactive Videos

See Video 3.4

Watch the section lecture video and answer the following questions.

OBJECTIVE 1

8. What important point is made during Example 1 having to do with the order of the points in the slope formula?

OBJECTIVE 2

9. From Example 5, how do you write an equation in "slope-intercept form"? Once the equation is in slope-intercept form, how do you identify the slope?

OBJECTIVE 3

10. In the lecture after Example 8, different slopes are summarized. What is the difference between zero slope and underfined slope? What does "no slope" mean?

OBJECTIVE 4

11. From Example 10, what form of the equations is best to determine if two lines are parallel or perpendicular? Why?

OBJECTIVE 5

12. Writing the slope as a rate of change in Example 11 gave real-life meaning to the slope. What step in the general strategy for problem solving does this correspond to?

3.4 Exercise Set

MyMathLab®

Find the slope of the line that passes through the given points. See Examples 1 and 2.

1. $(-1, 5)$ and $(6, -2)$

2. $(3, 1)$ and $(2, 6)$

3. $(-4, 3)$ and $(-4, 5)$

4. $(6, -6)$ and $(6, 2)$

5. $(-2, 8)$ and $(1, 6)$

6. $(4, -3)$ and $(2, 2)$

7. $(5, 1)$ and $(-2, 1)$

8. $(0, 13)$ and $(-4, 13)$

Find the slope of each line if it exists. See Examples 1 and 2.

9.

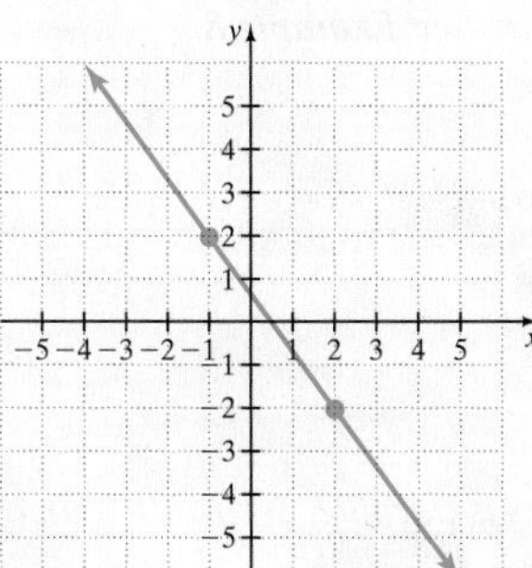

10.

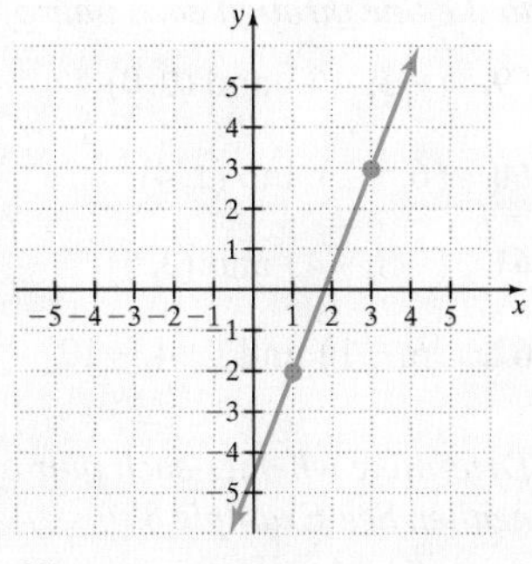

11.

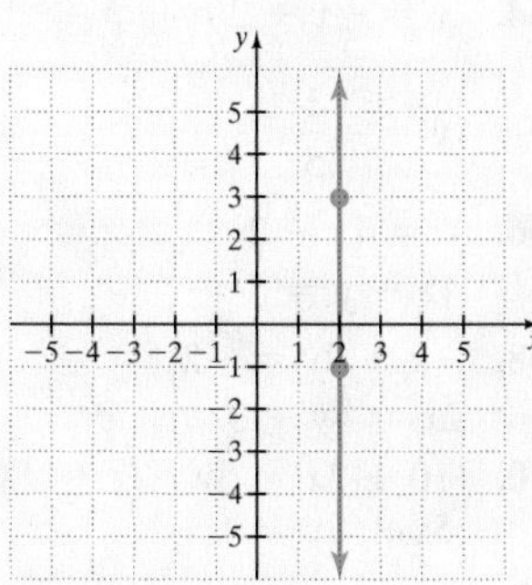

12.

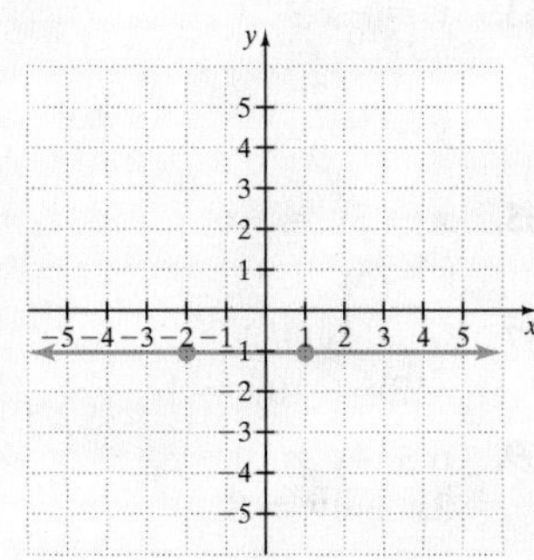

13.

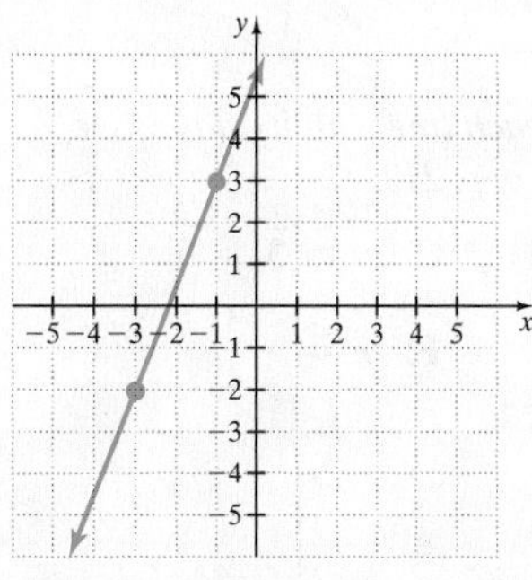

14.

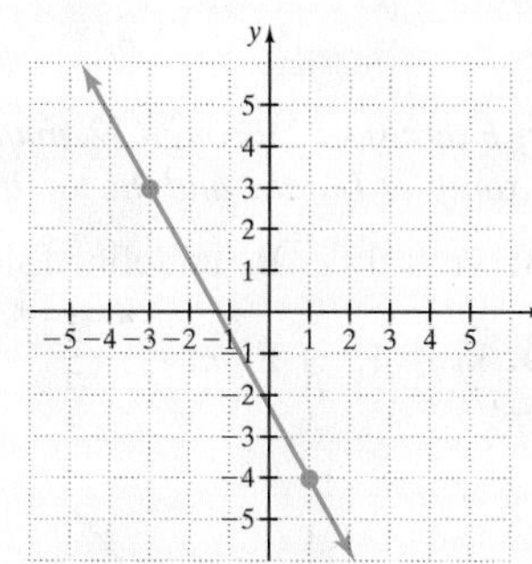

State whether the slope of the line is positive, negative, 0, or is undefined. See the top box on p. 212.

15.

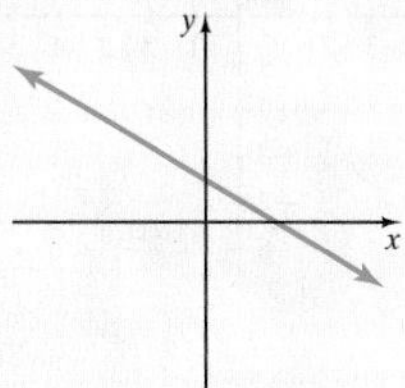

16.

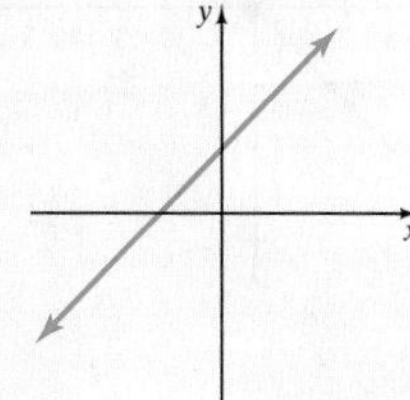

17.

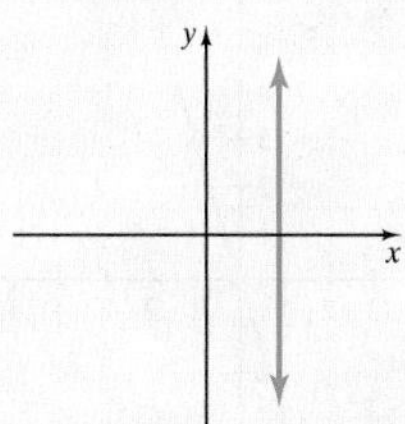

18.

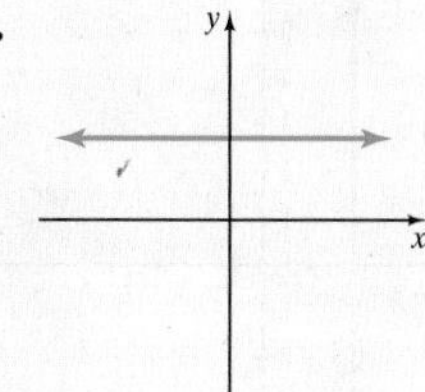

Decide whether a line with the given slope is upward, downward, horizontal, or vertical. See the top box on p. 212.

19. $m = \frac{7}{6}$ ________ **20.** $m = -3$ ________

21. $m = 0$ ________ **22.** m is undefined. ________

For each graph, determine which line has the greater slope. See the top box on p. 212.

23.

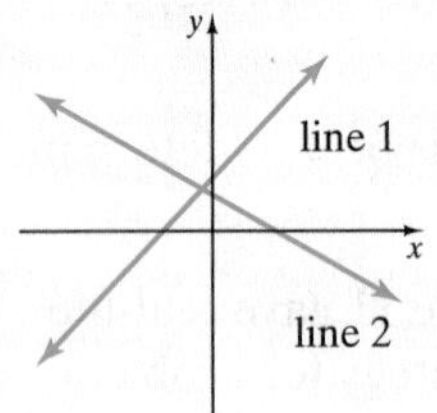

24.

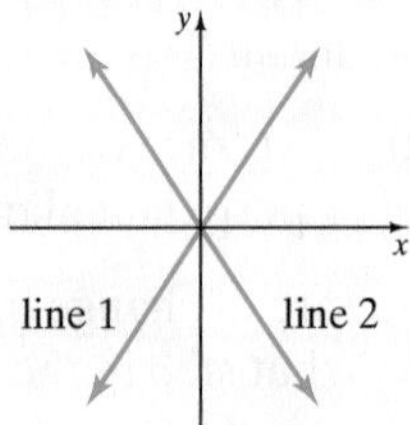

25.

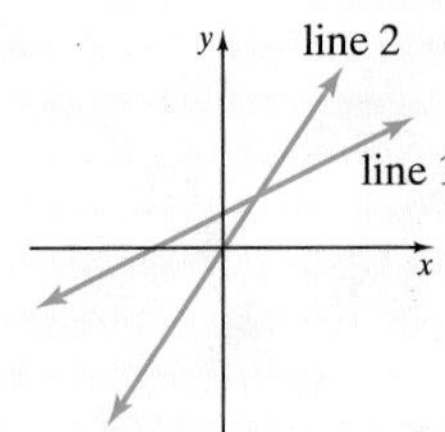

26.

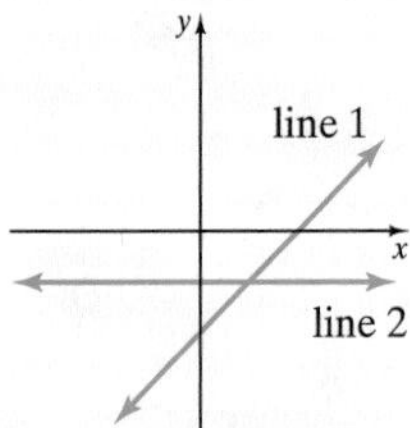

In Exercises 27 through 32, match each line with its slope. See Examples 1 and 2 and the top box on p. 212.

A. $m = 0$ **B.** undefined slope **C.** $m = 3$

D. $m = 1$ **E.** $m = -\frac{1}{2}$ **F.** $m = -\frac{3}{4}$

27.

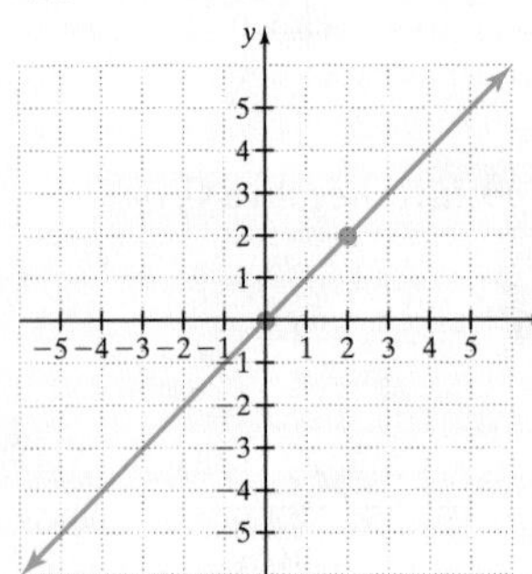

28.

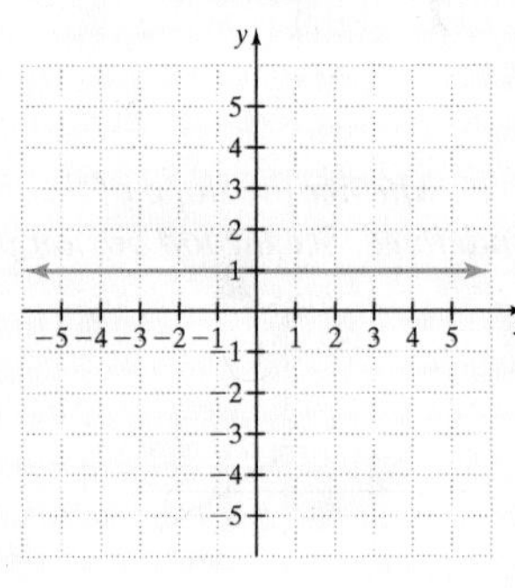

29.

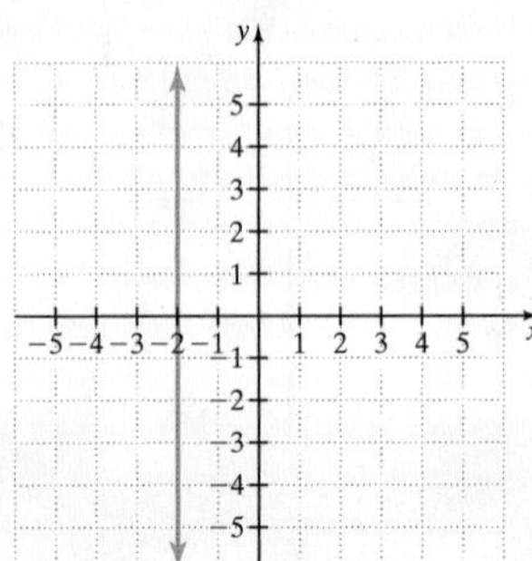

30.

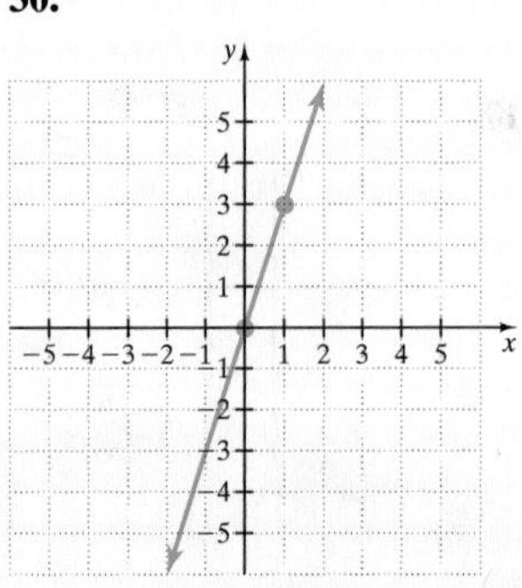

31.

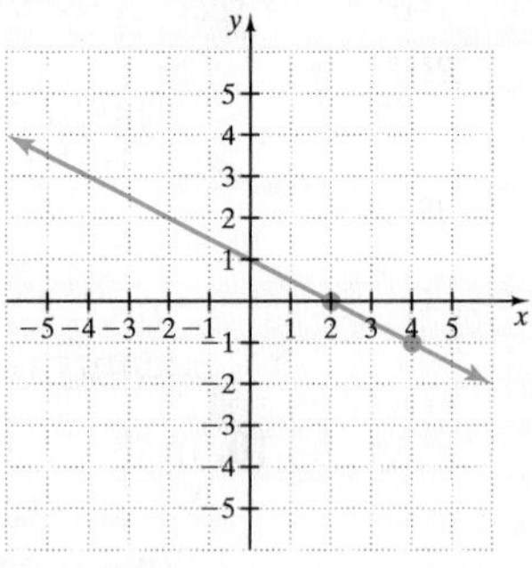

32.

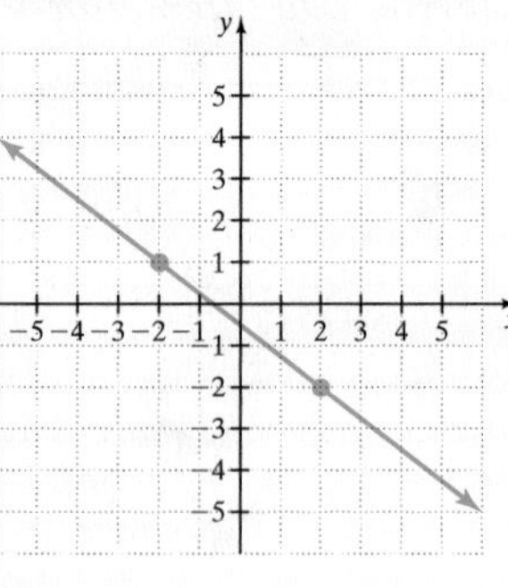

Find the slope of each line. See Examples 6 and 7.

33. $x = 6$ **34.** $y = 4$

35. $y = -4$ **36.** $x = 2$

37. $x = -3$ **38.** $y = -11$

39. $y = 0$ **40.** $x = 0$

MIXED PRACTICE

Find the slope of each line. See Examples 3 through 7.

41. $y = 5x - 2$ **42.** $y = -2x + 6$

43. $y = -0.3x + 2.5$ **44.** $y = -7.6x - 0.1$

45. $2x + y = 7$ **46.** $-5x + y = 10$

47. $2x - 3y = 10$ **48.** $3x - 5y = 1$

49. $x = 1$ **50.** $y = -2$

51. $x = 2y$ **52.** $x = -4y$

53. $y = -3$ **54.** $x = 5$

55. $-3x - 4y = 6$ **56.** $-4x - 7y = 9$

57. $20x - 5y = 1.2$ **58.** $24x - 3y = 5.7$

△ *Find the slope of the line that is **(a)** parallel and **(b)** perpendicular to the line through each pair of points. See Example 8.*

59. $(-3, -3)$ and $(0, 0)$

60. $(6, -2)$ and $(1, 4)$

61. $(-8, -4)$ and $(3, 5)$

62. $(6, -1)$ and $(-4, -10)$

△ *Determine whether each pair of lines is parallel, perpendicular, or neither. See Example 8.*

63. $y = \frac{2}{9}x + 3$
$y = -\frac{2}{9}x$

64. $y = \frac{1}{5}x + 20$
$y = -\frac{1}{5}x$

65. $x - 3y = -6$
$y = 3x - 9$

66. $y = 4x - 2$
$4x + y = 5$

67. $6x = 5y + 1$
$-12x + 10y = 1$

68. $-x + 2y = -2$
$2x = 4y + 3$

69. $6 + 4x = 3y$
$3x + 4y = 8$

70. $10 + 3x = 5y$
$5x + 3y = 1$

The pitch of a roof is its slope. Find the pitch of each roof shown. See Example 9.

71.

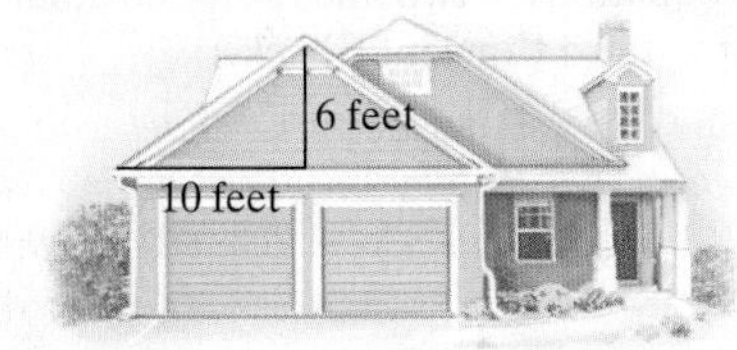

72.

The grade of a road is its slope written as a percent. Find the grade of each road shown. See Example 9.

73.

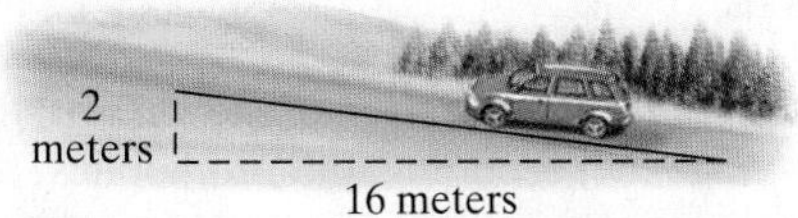

74.

75. One of Japan's superconducting "bullet" trains is researched and tested at the Yamanashi Maglev Test Line near Otsuki City. The steepest section of the track has a rise of 2580 meters for a horizontal distance of 6450 meters. What is the grade of this section of track? (*Source:* Japan Railways Central Co.)

76. Professional plumbers suggest that a sewer pipe should rise 0.25 inch for every horizontal foot. Find the recommended slope for a sewer pipe. Round to the nearest hundredth.

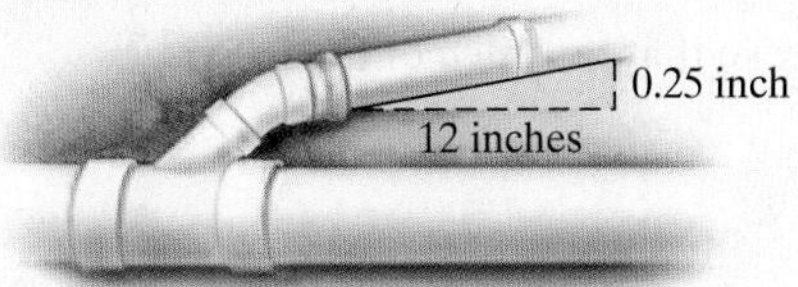

77. There has been controversy over the past few years about the world's steepest street. The *Guinness Book of Records* actually listed Baldwin Street, in Dunedin, New Zealand, as the world's steepest street, but Canton Avenue in the Pittsburgh neighborhood of Beechview may be steeper. Calculate each grade to the nearest percent.

		Grade (%)
Canton Avenue	for every 30 meters of horizontal distance, the vertical change is 11 meters	
Baldwin Street	for every 2.86 meters of horizontal distance, the vertical change is 1 meter	

78. According to federal regulations, a wheelchair ramp should rise no more than 1 foot for a horizontal distance of 12 feet. Write the slope as a grade. Round to the nearest tenth of a percent.

Find the slope of each line and write the slope as a rate of change. Don't forget to attach the proper units. See Example 10.

79. This graph approximates the number of U.S. households that have televisions y (in millions) for year x.

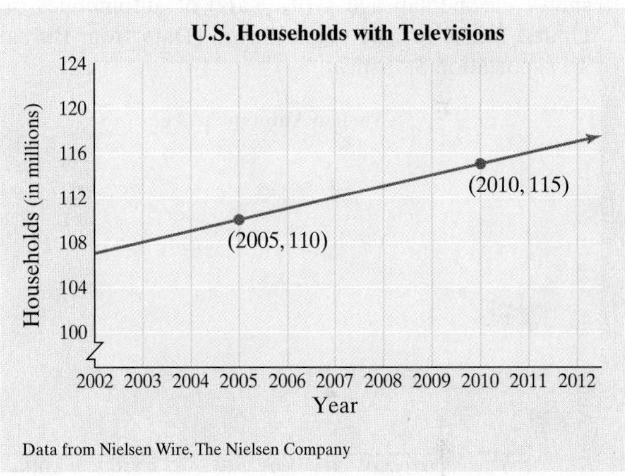

80. The graph approximates the amount of money y (in billions of dollars) spent worldwide on tourism for year x. (Data from World Tourism Organization)

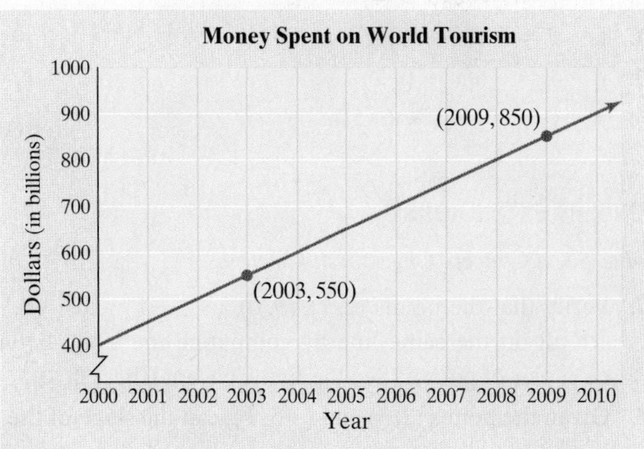

For Exercises 81 and 82, write the slope as a decimal.

81. The graph below shows the total cost y (in dollars) of owning and operating a compact car where x is the number of miles driven.

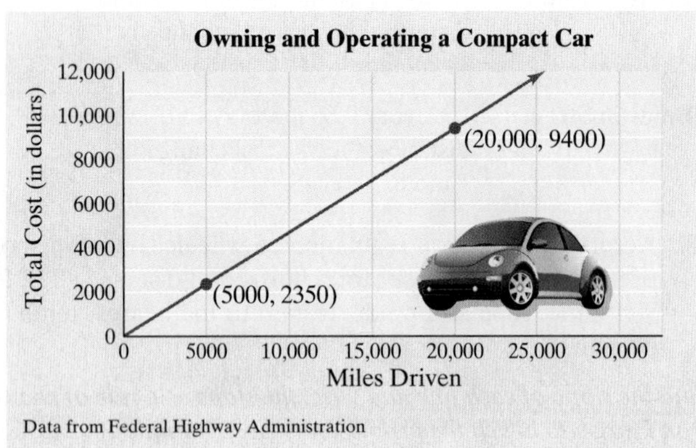

82. Americans are keeping their cars longer. The graph below shows the median age y (in years) of automobiles in the United States for the years shown. (Data from Bureau of Transportation Statistics)

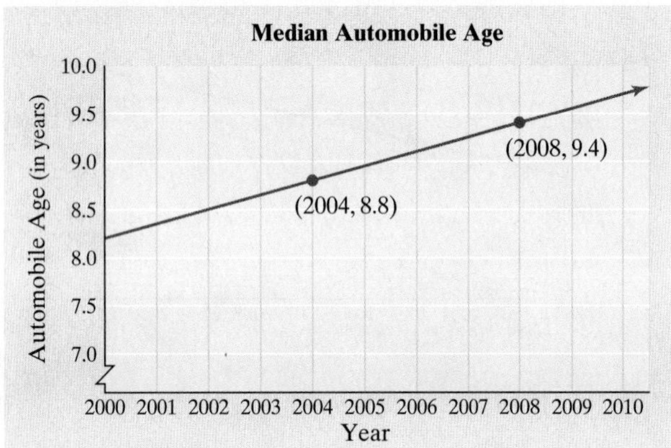

REVIEW AND PREVIEW

Solve each equation for y. See Section 2.5.

83. $y - (-6) = 2(x - 4)$

84. $y - 7 = -9(x - 6)$

85. $y - 1 = -6(x - (-2))$

86. $y - (-3) = 4(x - (-5))$

CONCEPT EXTENSIONS

Solve. See a Concept Check in this section.

87. Verify that the points $(2, 1)$, $(0, 0)$, $(-2, -1)$ and $(-4, -2)$ are all on the same line by computing the slope between each pair of points. (See the first Concept Check.)

88. Given the points $(2, 3)$ and $(-5, 1)$, can the slope of the line through these points be calculated by $\frac{1 - 3}{2 - (-5)}$? Why or why not? (See the second Concept Check.)

89. Write the equations of three lines parallel to $10x - 5y = -7$. (See the third Concept Check.)

90. Write the equations of two lines perpendicular to $10x - 5y = -7$. (See the third Concept Check.)

The following line graph shows the average fuel economy (in miles per gallon) of passenger automobiles produced during each of the model years shown. Use this graph to answer Exercises 91 through 96.

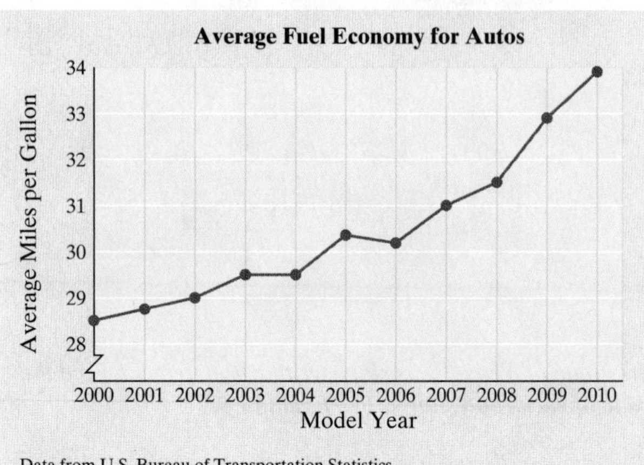

91. Between what two years shown was there a decrease in average fuel economy for automobiles?

92. What was the average fuel economy (in miles per gallon) for automobiles produced during 2008?

93. During which of the model years shown was average fuel economy the lowest?
What was the average fuel economy for that year?

94. During which of the model years shown was average fuel economy the highest?
What was the average fuel economy for that year?

95. Of the following line segments, which has the greatest slope: from 2002 to 2003, from 2006 to 2007, or from 2008 to 2009?

96. Which line segment has a slope of 0?

Solve.

97. Find x so that the pitch of the roof is $\frac{1}{3}$.

98. Find x so that the pitch of the roof is $\frac{2}{5}$.

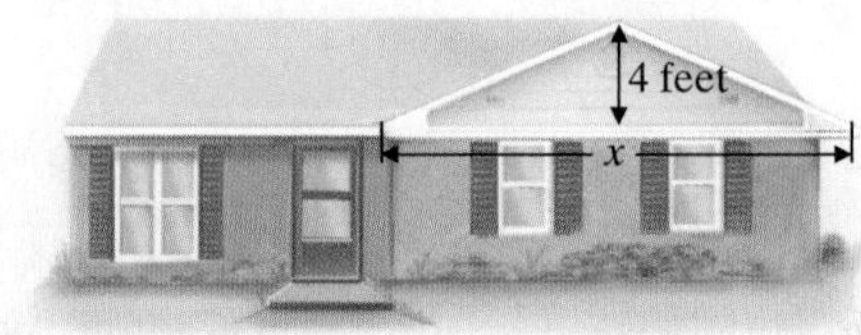

99. Approximately 2207 heart transplants were performed in the United States in 2007. In 2010, the number of heart transplants in the United States rose to 2333. (*Source:* Organ Procurement and Transplantation Network)

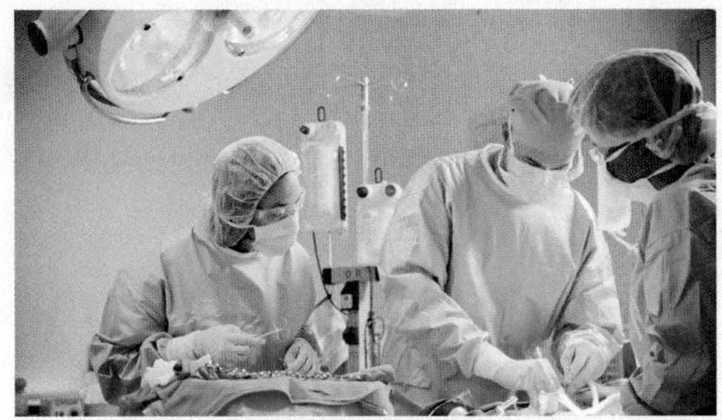

a. Write two ordered pairs of the form (year, number of heart transplants).

b. Find the slope of the line between the two points.

c. Write a sentence explaining the meaning of the slope as a rate of change.

100. The average price of an acre of U.S. farmland was $1210 in 2002. In 2008, the price of an acre rose to $2350. (*Source:* National Agricultural Statistics Services)

a. Write two ordered pairs of the form (year, price of an acre).

b. Find the slope of the line through the two points.

c. Write a sentence explaining the meaning of the slope as a rate of change.

101. Show that a triangle with vertices at the points $(1, 1)$, $(-4, 4)$, and $(-3, 0)$ is a right triangle.

102. Show that the quadrilateral with vertices $(1, 3)$, $(2, 1)$, $(-4, 0)$, and $(-3, -2)$ is a parallelogram.

Find the slope of the line through the given points.

103. $(2.1, 6.7)$ and $(-8.3, 9.3)$

104. $(-3.8, 1.2)$ and $(-2.2, 4.5)$

105. $(2.3, 0.2)$ and $(7.9, 5.1)$

106. $(14.3, -10.1)$ and $(9.8, -2.9)$

107. The graph of $y = -\frac{1}{3}x + 2$ has a slope of $-\frac{1}{3}$. The graph of $y = -2x + 2$ has a slope of -2. The graph of $y = -4x + 2$ has a slope of -4. Graph all three equations on a single coordinate system. As the absolute value of the slope becomes larger, how does the steepness of the line change?

108. The graph of $y = \frac{1}{2}x$ has a slope of $\frac{1}{2}$. The graph of $y = 3x$ has a slope of 3. The graph of $y = 5x$ has a slope of 5. Graph all three equations on a single coordinate system. As slope becomes larger, how does the steepness of the line change?

Integrated Review SUMMARY ON SLOPE & GRAPHING LINEAR EQUATIONS

Sections 3.1–3.4

Find the slope of each line.

1.

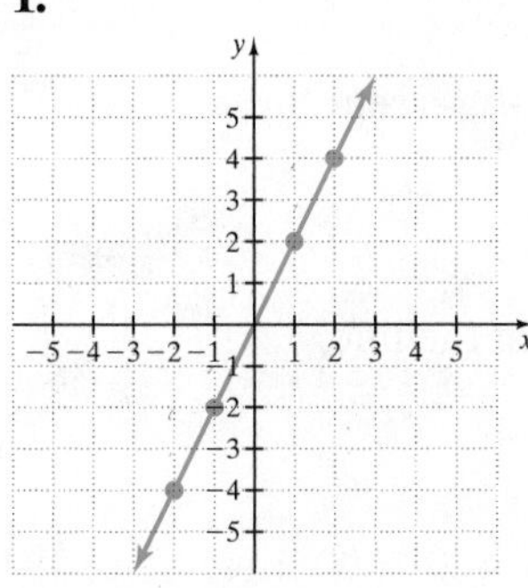

2.

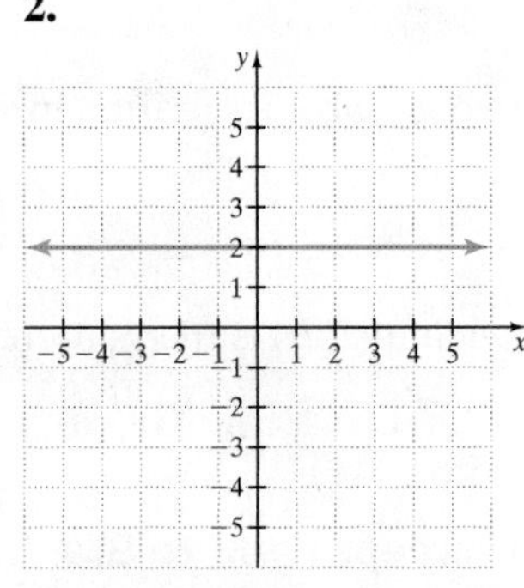

3.

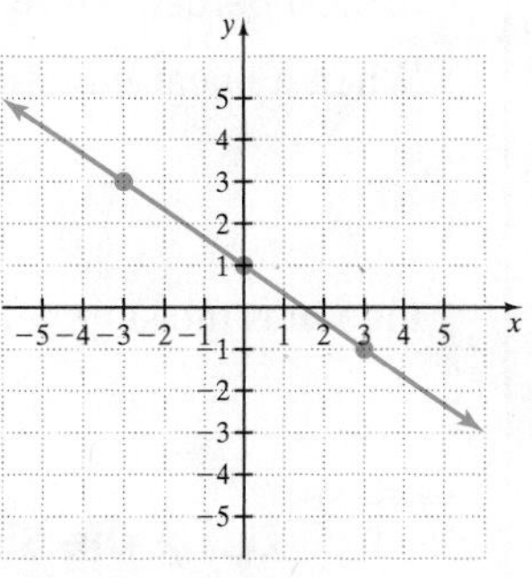

4.

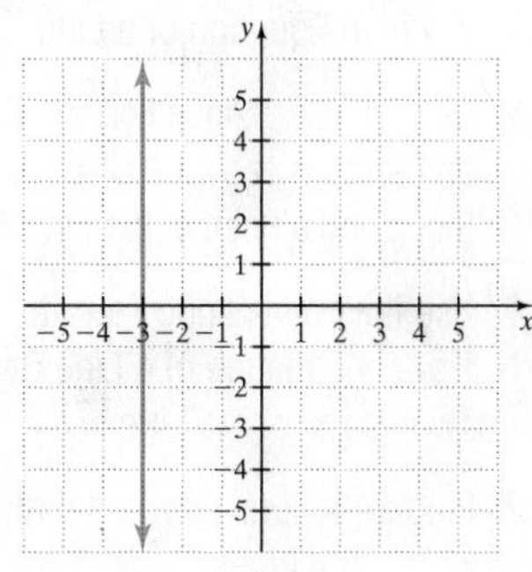

Graph each linear equation.

5. $y = -2x$

6. $x + y = 3$

7. $x = -1$

8. $y = 4$

9. $x - 2y = 6$

10. $y = 3x + 2$

11. $5x + 3y = 15$

12. $2x - 4y = 8$

Determine whether the lines through the points are parallel, perpendicular, or neither.

13. $y = -\frac{1}{5}x + \frac{1}{3}$
$3x = -15y$

14. $x - y = \frac{1}{2}$
$3x - y = \frac{1}{2}$

15. In the years 2000 through 2009 the number of bridges on public roads (in thousands) in the United States can be modeled by the linear equation $y = 1.7x + 587$ where x is years after 2000 and y is number of bridges (in thousands). (*Source:* U.S. Dept. of Transportation)

a. Find the y-intercept of this line.
b. Write a sentence explaining the meaning of this intercept.
c. Find the slope of this line.
d. Write a sentence explaining the meaning of the slope as a rate of change.

16. Online advertising is a means of promoting products and services using the Internet. The revenue (in billions of dollars) for online advertising for the years 2003 through 2010 is given by $y = 3.3x - 3.1$, where x is the number of years after 2000.

a. Use this equation to complete the ordered pair (9,   ).
b. Write a sentence explaining the meaning of the answer to part (a).

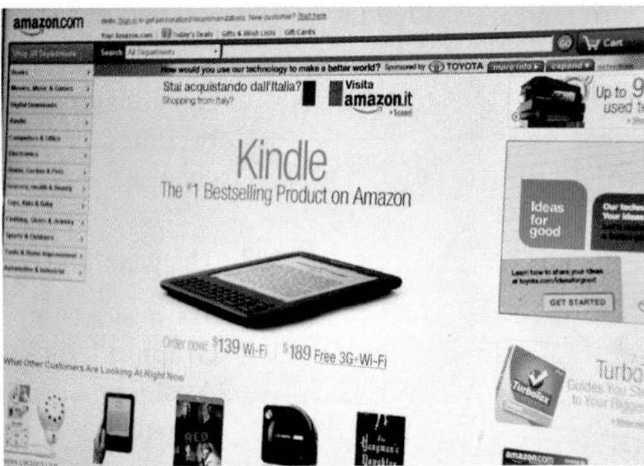

3.5 Equations of Lines

OBJECTIVES

1. Use the Slope-Intercept Form to Graph a Linear Equation.
2. Use the Slope-Intercept Form to Write an Equation of a Line.
3. Use the Point-Slope Form to Find an Equation of a Line Given Its Slope and a Point of the Line.
4. Use the Point-Slope Form to Find an Equation of a Line Given Two Points of the Line.
5. Find Equations of Vertical and Horizontal Lines.
6. Use the Point-Slope Form to Solve Problems.

Recall that the form $y = mx + b$ is appropriately called the *slope-intercept form* of a linear equation.

m ↑ slope; b ↑ y-intercept is $(0, b)$

Slope-Intercept Form

When a linear equation in two variables is written in **slope-intercept form,**

$$y = mx + b$$

m ↑ slope; b ↑ $(0, b)$, y-intercept

then m is the slope of the line and $(0, b)$ is the y-intercept of the line.

OBJECTIVE

1 Using the Slope-Intercept Form to Graph an Equation 

We can use the slope-intercept form of the equation of a line to graph a linear equation.

EXAMPLE 1 Use the slope-intercept form to graph the equation

$$y = \frac{3}{5}x - 2$$

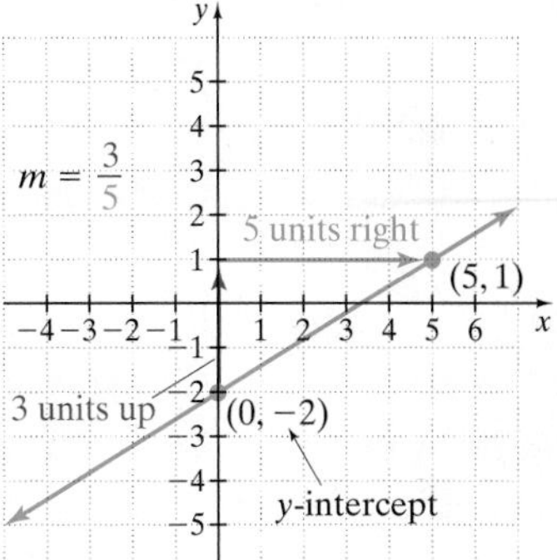

Solution Since the equation $y = \frac{3}{5}x - 2$ is written in slope-intercept form $y = mx + b$, the slope of its graph is $\frac{3}{5}$ and the y-intercept is $(0, -2)$. To graph this equation, we begin by plotting the point $(0, -2)$. From this point, we can find another point of the graph by using the slope $\frac{3}{5}$ and recalling that slope is $\frac{\text{rise}}{\text{run}}$. We start at the y-intercept and move 3 units up since the numerator of the slope is 3; then we move 5 units to the right since the denominator of the slope is 5. We stop at the point $(5, 1)$. The line through $(0, -2)$ and $(5, 1)$ is the graph of $y = \frac{3}{5}x - 2$. □

PRACTICE 1 Graph $y = \frac{2}{3}x - 5$.

EXAMPLE 2 Use the slope-intercept form to graph the equation $4x + y = 1$.

Solution First we write the given equation in slope-intercept form.

$$4x + y = 1$$
$$y = -4x + 1$$

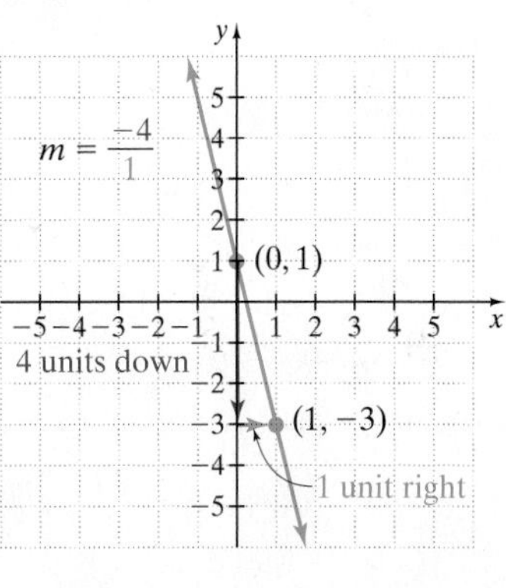

The graph of this equation will have slope -4 and y-intercept $(0, 1)$. To graph this line, we first plot the point $(0, 1)$. To find another point of the graph, we use the slope -4, which can be written as $\frac{-4}{1}\left(\frac{4}{-1} \text{ could also be used}\right)$. We start at the point $(0, 1)$ and move 4 units down (since the numerator of the slope is -4) and then 1 unit to the right (since the denominator of the slope is 1).

We arrive at the point $(1, -3)$. The line through $(0, 1)$ and $(1, -3)$ is the graph of $4x + y = 1$. □

> **Helpful Hint**
> In Example 2, if we interpret the slope of -4 as $\frac{4}{-1}$, we arrive at $(-1, 5)$ for a second point. Notice that this point is also on the line.

PRACTICE 2 Use the slope-intercept form to graph the equation $3x - y = 2$.

OBJECTIVE

2 Using the Slope-Intercept Form to Write an Equation

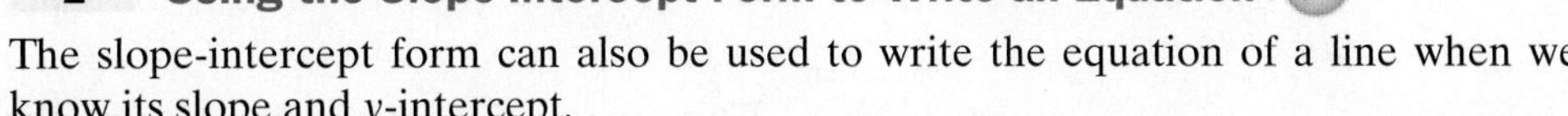

The slope-intercept form can also be used to write the equation of a line when we know its slope and y-intercept.

EXAMPLE 3 Find an equation of the line with y-intercept $(0, -3)$ and slope of $\frac{1}{4}$.

Solution We are given the slope and the y-intercept. We let $m = \frac{1}{4}$ and $b = -3$ and write the equation in slope-intercept form, $y = mx + b$.

$$y = mx + b$$
$$y = \frac{1}{4}x + (-3) \quad \text{Let } m = \frac{1}{4} \text{ and } b = -3.$$
$$y = \frac{1}{4}x - 3 \quad \text{Simplify.}$$

□

PRACTICE 3 Find an equation of the line with y-intercept (0, 7) and slope of $\frac{1}{2}$.

OBJECTIVE 3

Writing an Equation Given Slope and a Point 

Thus far, we have seen that we can write an equation of a line if we know its slope and y-intercept. We can also write an equation of a line if we know its slope and any point on the line. To see how we do this, let m represent slope and (x_1, y_1) represent the point on the line. Then if (x, y) is any other point of the line, we have that

$$\frac{y - y_1}{x - x_1} = m$$

$$y - y_1 = m(x - x_1) \quad \text{Multiply both sides by } (x - x_1).$$

(↑ slope)

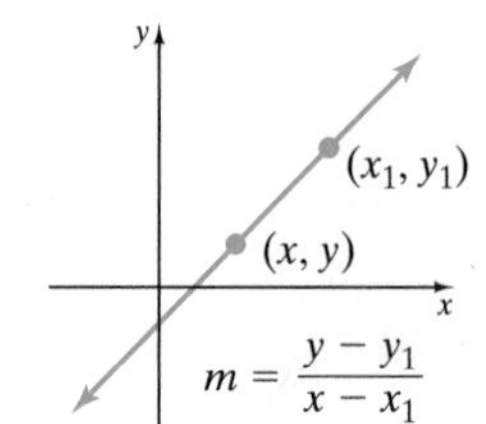

This is the *point-slope form* of the equation of a line.

Point-Slope Form of the Equation of a Line

The **point-slope form** of the equation of a line is

$$y - y_1 = m(x - x_1)$$

(↑ slope; ↑ (x_1, y_1) point on the line)

where m is the slope of the line and (x_1, y_1) is a point on the line.

EXAMPLE 4 Find an equation of the line with slope -2 that passes through $(-1, 5)$. Write the equation in slope-intercept form, $y = mx + b$, and in standard form, $Ax + By = C$.

Solution Since the slope and a point on the line are given, we use point-slope form $y - y_1 = m(x - x_1)$ to write the equation. Let $m = -2$ and $(-1, 5) = (x_1, y_1)$.

$$y - y_1 = m(x - x_1)$$

$$y - 5 = -2[x - (-1)] \quad \text{Let } m = -2 \text{ and } (x_1, y_1) = (-1, 5).$$

$$y - 5 = -2(x + 1) \quad \text{Simplify.}$$

$$y - 5 = -2x - 2 \quad \text{Use the distributive property.}$$

To write the equation in slope-intercept form, $y = mx + b$, we simply solve the equation for y. To do this, we add 5 to both sides.

$$y - 5 = -2x - 2$$

$$y = -2x + 3 \quad \text{Slope-intercept form.}$$

$$2x + y = 3 \quad \text{Add } 2x \text{ to both sides and we have standard form.}$$

□

PRACTICE 4 Find an equation of the line passing through (2, 3) with slope 4. Write the equation in standard form: $Ax + By = C$.

OBJECTIVE 4

Writing an Equation Given Two Points 

We can also find an equation of a line when we are given any two points of the line.

EXAMPLE 5 Find an equation of the line through $(2, 5)$ and $(-3, 4)$. Write the equation in standard form.

Solution First, use the two given points to find the slope of the line.

$$m = \frac{4 - 5}{-3 - 2} = \frac{-1}{-5} = \frac{1}{5}$$

Next we use the slope $\frac{1}{5}$ and either one of the given points to write the equation in point-slope form. We use $(2, 5)$. Let $x_1 = 2$, $y_1 = 5$, and $m = \frac{1}{5}$.

$$y - y_1 = m(x - x_1) \quad \text{Use point-slope form.}$$

$$y - 5 = \frac{1}{5}(x - 2) \quad \text{Let } x_1 = 2, y_1 = 5, \text{ and } m = \frac{1}{5}.$$

$$5(y - 5) = 5 \cdot \frac{1}{5}(x - 2) \quad \text{Multiply both sides by 5 to clear fractions.}$$

$$5y - 25 = x - 2 \quad \text{Use the distributive property and simplify.}$$

$$-x + 5y - 25 = -2 \quad \text{Subtract } x \text{ from both sides.}$$

$$-x + 5y = 23 \quad \text{Add 25 to both sides.}$$

□

PRACTICE

5 Find an equation of the line through $(-1, 6)$ and $(3, 1)$. Write the equation in standard form.

Helpful Hint

Multiply both sides of the equation $-x + 5y = 23$ by -1, and it becomes $x - 5y = -23$. Both $-x + 5y = 23$ and $x - 5y = -23$ are in standard form, and they are equations of the same line.

OBJECTIVE

5 Finding Equations of Vertical and Horizontal Lines

Recall from Section 3.3 that:

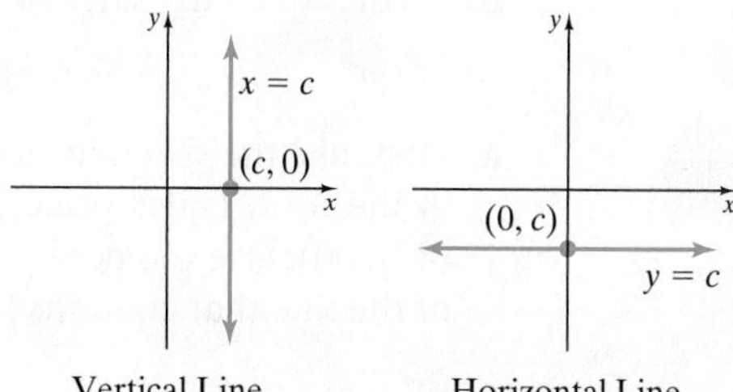

Vertical Line Horizontal Line

EXAMPLE 6 Find an equation of the vertical line through $(-1, 5)$.

Solution The equation of a vertical line can be written in the form $x = c$, so an equation for a vertical line passing through $(-1, 5)$ is $x = -1$.

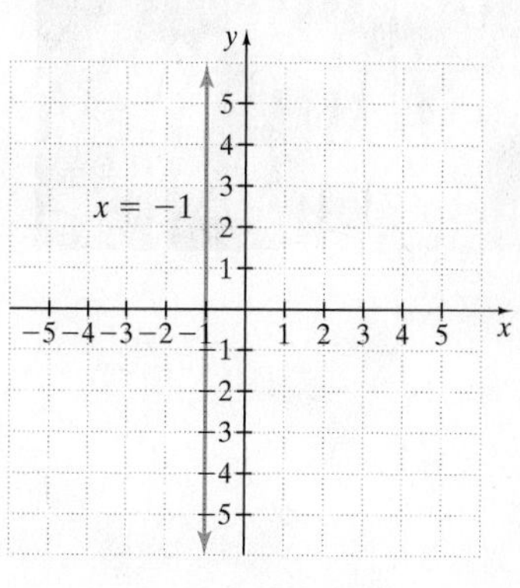

□

PRACTICE

6 Find an equation of the vertical line through $(3, -2)$.

EXAMPLE 7 Find an equation of the line parallel to the line $y = 5$ and passing through $(-2, -3)$.

Solution Since the graph of $y = 5$ is a horizontal line, any line parallel to it is also horizontal. The equation of a horizontal line can be written in the form $y = c$. An equation for the horizontal line passing through

$$(-2, -3) \text{ is } y = -3.$$

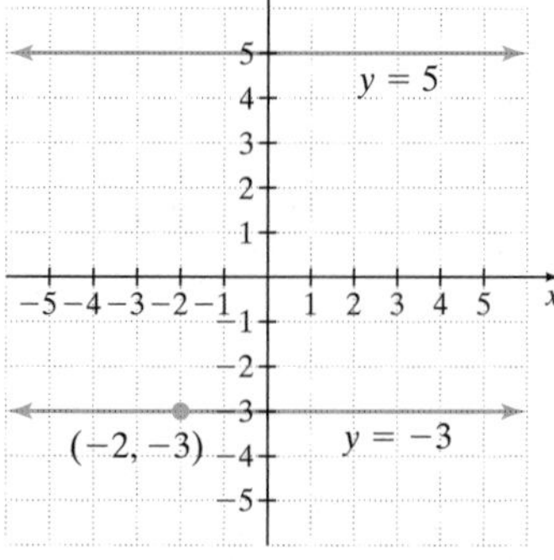

□

PRACTICE

7 Find an equation of the line parallel to the line $y = -2$ and passing through $(4, 3)$.

OBJECTIVE

6 Using the Point-Slope Form to Solve Problems

Problems occurring in many fields can be modeled by linear equations in two variables. The next example is from the field of marketing and shows how consumer demand of a product depends on the price of the product.

EXAMPLE 8 **Predicting the Sales of T-Shirts**

A web-based T-shirt company has learned that by pricing a clearance-sale T-shirt at $6, sales will reach 2000 T-shirts per day. Raising the price to $8 will cause the sales to fall to 1500 T-shirts per day.

a. Assume that the relationship between sales price and number of T-shirts sold is linear and write an equation describing this relationship. Write the equation in slope-intercept form.

b. Predict the daily sales of T-shirts if the price is $7.50.

Solution

a. First, use the given information and write two ordered pairs. Ordered pairs will be in the form (sales price, number sold) so that our ordered pairs are $(6, 2000)$ and $(8, 1500)$. Use the point-slope form to write an equation. To do so, we find the slope of the line that contains these points.

$$m = \frac{2000 - 1500}{6 - 8} = \frac{500}{-2} = -250$$

Next, use the slope and either one of the points to write the equation in point-slope form. We use $(6, 2000)$.

$$y - y_1 = m(x - x_1) \quad \text{Use point-slope form.}$$
$$y - 2000 = -250(x - 6) \quad \text{Let } x_1 = 6, y_1 = 2000, \text{ and } m = -250.$$
$$y - 2000 = -250x + 1500 \quad \text{Use the distributive property.}$$
$$y = -250x + 3500 \quad \text{Write in slope-intercept form.}$$

b. To predict the sales if the price is $7.50, we find y when $x = 7.50$.

$$y = -250x + 3500$$
$$y = -250(7.50) + 3500 \quad \text{Let } x = 7.50.$$
$$y = -1875 + 3500$$
$$y = 1625$$

If the price is $7.50, sales will reach 1625 T-shirts per day. □

PRACTICE

8 The new *Camelot* condos were selling at a rate of 30 per month when they were priced at $150,000 each. Lowering the price to $120,000 caused the sales to rise to 50 condos per month.

a. Assume that the relationship between number of condos sold and price is linear, and write an equation describing this relationship. Write the equation in slope-intercept form.

b. How should the condos be priced if the developer wishes to sell 60 condos per month?

The preceding example may also be solved by using ordered pairs of the form (number sold, sales price).

Forms of Linear Equations

$Ax + By = C$	**Standard form** of a linear equation. A and B are not both 0.
$y = mx + b$	**Slope-intercept form** of a linear equation. The slope is m and the y-intercept is $(0, b)$.
$y - y_1 = m(x - x_1)$	**Point-slope form** of a linear equation. The slope is m and (x_1, y_1) is a point on the line.
$y = c$	**Horizontal line** The slope is 0 and the y-intercept is $(0, c)$.
$x = c$	**Vertical line** The slope is undefined and the x-intercept is $(c, 0)$.

Parallel and Perpendicular Lines

Nonvertical parallel lines have the same slope.

The product of the slopes of two nonvertical perpendicular lines is -1.

Graphing Calculator Explorations

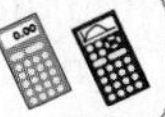

A grapher is a very useful tool for discovering patterns. To discover the change in the graph of a linear equation caused by a change in slope, try the following. Use a standard window and graph a linear equation in the form $y = mx + b$. Recall that the graph of such an equation will have slope m and y-intercept b.

First, graph $y = x + 3$. To do so, press the [Y =] key and enter $Y_1 = x + 3$. Notice that this graph has slope 1 and that the y-intercept is 3. Next, on the same set of axes, graph $y = 2x + 3$ and $y = 3x + 3$ by pressing [Y =] and entering $Y_2 = 2x + 3$ and $Y_3 = 3x + 3$.

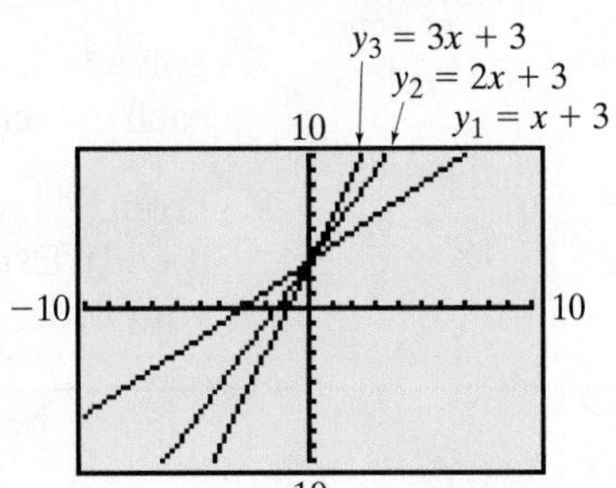

Notice the difference in the graph of each equation as the slope changes from 1 to 2 to 3. How would the graph of $y = 5x + 3$ appear? To see the change in the graph caused by a change in negative slope, try graphing $y = -x + 3$, $y = -2x + 3$, and $y = -3x + 3$ on the same set of axes.

Use a grapher to graph the following equations. For each exercise, graph the first equation and use its graph to predict the appearance of the other equations. Then graph the other equations on the same set of axes and check your prediction.

1. $y = x; y = 6x, y = -6x$

2. $y = -x; y = -5x, y = -10x$

3. $y = \frac{1}{2}x + 2; y = \frac{3}{4}x + 2, y = x + 2$

4. $y = x + 1; y = \frac{5}{4}x + 1, y = \frac{5}{2}x + 1$

5. $y = -7x + 5; y = 7x + 5$

6. $y = 3x - 1; y = -3x - 1$

Vocabulary, Readiness & Video Check

Use the choices below to fill in each blank. Some choices may be used more than once and some not at all.

b	(y_1, x_1)	point-slope	vertical	standard
m	(x_1, y_1)	slope-intercept	horizontal	

1. The form $y = mx + b$ is called ____________ form. When a linear equation in two variables is written in this form, ____________ is the slope of its graph and (0, ____________) is its y-intercept.

2. The form $y - y_1 = m(x - x_1)$ is called ____________ form. When a linear equation in two variables is written in this form, ____________ is the slope of its graph and ____________ is a point on the graph.

Martin-Gay Interactive Videos

See Video 3.5

Watch the section lecture video and answer the following questions.

OBJECTIVE 1

3. We can use the slope-intercept form to graph a line. Complete these statements based on Example 1. Start by graphing the ______________.

Find another point by applying the slope to this point—rewrite the slope as a ______________ if necessary.

OBJECTIVE 2

4. In Example 3, what is the y-intercept?

OBJECTIVE 3

5. In Example 4 we use the point-slope form to find the equation of a line given the slope and a point. How do we then write this equation in standard form?

OBJECTIVE 4

6. The lecture before Example 5 discusses how to find the equation of a line given two points. Is there any circumstance when you might want to use the slope-intercept form to find the equation of a line given two points? If so, when?

OBJECTIVE 5

7. Solve Examples 6 and 7 again, this time using the point $(-2, -3)$ in each exercise.

OBJECTIVE 6

8. From Example 8, we are told to use ordered pairs of the form (time, speed). Explain why it is important to keep track of how you define your ordered pairs and/or define your variables?

3.5 Exercise Set

MyMathLab®

Use the slope-intercept form to graph each equation. See Examples 1 and 2.

1. $y = 2x + 1$
2. $y = -4x - 1$
3. $y = \frac{2}{3}x + 5$
4. $y = \frac{1}{4}x - 3$
5. $y = -5x$
6. $y = -6x$
7. $4x + y = 6$
8. $-3x + y = 2$
9. $4x - 7y = -14$
10. $3x - 4y = 4$
11. $x = \frac{5}{4}y$
12. $x = \frac{3}{2}y$

Write an equation of the line with each given slope, m, and y-intercept, (0, b). See Example 3.

13. $m = 5, b = 3$
14. $m = -3, b = -3$
15. $m = -4, b = -\frac{1}{6}$
16. $m = 2, b = \frac{3}{4}$
17. $m = \frac{2}{3}, b = 0$
18. $m = -\frac{4}{5}, b = 0$
19. $m = 0, b = -8$
20. $m = 0, b = -2$
21. $m = -\frac{1}{5}, b = \frac{1}{9}$
22. $m = \frac{1}{2}, b = -\frac{1}{3}$

Find an equation of each line with the given slope that passes through the given point. Write the equation in the form $Ax + By = C$. See Example 4.

23. $m = 6;\ (2, 2)$
24. $m = 4;\ (1, 3)$
25. $m = -8;\ (-1, -5)$
26. $m = -2;\ (-11, -12)$
27. $m = \frac{3}{2};\ (5, -6)$
28. $m = \frac{2}{3};\ (-8, 9)$
29. $m = -\frac{1}{2};\ (-3, 0)$
30. $m = -\frac{1}{5};\ (4, 0)$

Find an equation of the line passing through each pair of points. Write the equation in the form $Ax + By = C$. See Example 5.

31. $(3, 2)$ and $(5, 6)$
32. $(6, 2)$ and $(8, 8)$
33. $(-1, 3)$ and $(-2, -5)$
34. $(-4, 0)$ and $(6, -1)$
35. $(2, 3)$ and $(-1, -1)$
36. $(7, 10)$ and $(-1, -1)$
37. $(0, 0)$ and $\left(-\frac{1}{8}, \frac{1}{13}\right)$
38. $(0, 0)$ and $\left(-\frac{1}{2}, \frac{1}{3}\right)$

Find an equation of each line. See Example 6.

39. Vertical line through $(0, 2)$
40. Horizontal line through $(1, 4)$
41. Horizontal line through $(-1, 3)$
42. Vertical line through $(-1, 3)$
43. Vertical line through $\left(-\frac{7}{3}, -\frac{2}{5}\right)$
44. Horizontal line through $\left(\frac{2}{7}, 0\right)$

Find an equation of each line. See Example 7.

45. △ Parallel to $y = 5$, through $(1, 2)$
46. △ Perpendicular to $y = 5$, through $(1, 2)$
47. △ Perpendicular to $x = -3$, through $(-2, 5)$
48. △ Parallel to $y = -4$, through $(0, -3)$
49. △ Parallel to $x = 0$, through $(6, -8)$
50. △ Perpendicular to $x = 7$, through $(-5, 0)$

MIXED PRACTICE

See Examples 1 through 7. Find an equation of each line described. Write each equation in slope-intercept form (solved for y), when possible.

51. With slope $-\frac{1}{2}$, through $\left(0, \frac{5}{3}\right)$
52. With slope $\frac{5}{7}$, through $(0, -3)$
53. Through $(10, 7)$ and $(7, 10)$
54. Through $(5, -6)$ and $(-6, 5)$
55. With undefined slope, through $\left(-\frac{3}{4}, 1\right)$
56. With slope 0, through $(6.7, 12.1)$
57. Slope 1, through $(-7, 9)$
58. Slope 5, through $(6, -8)$
59. Slope -5, y-intercept $(0, 7)$
60. Slope -2; y-intercept $(0, -4)$
61. Through $(6, 7)$, parallel to the x-axis
62. Through $(1, -5)$, parallel to the y-axis
63. Through $(2, 3)$ and $(0, 0)$
64. Through $(4, 7)$ and $(0, 0)$
65. Through $(-2, -3)$, perpendicular to the y-axis
66. Through $(0, 12)$, perpendicular to the x-axis
67. Slope $-\frac{4}{7}$, through $(-1, -2)$
68. Slope $-\frac{3}{5}$, through $(4, 4)$

Solve. Assume each exercise describes a linear relationship. Write the equations in slope-intercept form. See Example 8.

69. A rock is dropped from the top of a 400-foot cliff. After 1 second, the rock is traveling 32 feet per second. After 3 seconds, the rock is traveling 96 feet per second.

a. Assume that the relationship between time and speed is linear and write an equation describing this relationship. Use ordered pairs of the form (time, speed).

b. Use this equation to determine the speed of the rock 4 seconds after it was dropped.

70. A Hawaiian fruit company is studying the sales of a pineapple sauce to see if this product is to be continued. At the end of its first year, profits on this product amounted to \$30,000. At the end of the fourth year, profits were \$66,000.

a. Assume that the relationship between years on the market and profit is linear and write an equation describing this relationship. Use ordered pairs of the form (years on the market, profit).

b. Use this equation to predict the profit at the end of 7 years.

71. Sales of hybrid (gas-electric) vehicles have been decreasing for the past few years. In 2007, 356 thousand new gasoline-electric hybrid vehicles were sold in the United States. In 2009, only were 290 thousand were sold. (*Source:* U.S. Energy Information Administration)

a. Write an equation describing the relationship between time and number of registered gasoline-electric hybrid vehicles sold. Use ordered pairs of the form (years past 2007, number of hybrids sold).

b. Use this equation to predict the number of gasoline-electric hybrids sold in the year 2015.

72. In 2011, there were 960 thousand eating establishments in the United States. In 2006, there were 935 thousand eating establishments. (*Source:* National Restaurant Association)

a. Write an equation describing the relationship between time and number of eating establishments. Use ordered pairs of the form (years past 2006, number of eating establishments in thousands).

b. Use this equation to predict the number of eating establishments in 2018.

73. In 2006, the U.S. population per square mile of land area was 85. In 2011, the person per square mile population was 88.

a. Write an equation describing the relationship between year and persons per square mile. Use ordered pairs of the form (years past 2006, persons per square mile).

b. Use this equation to predict the person per square mile population in 2020.

74. In 2005, there were a total of 150 thousand apparel and accessory stores. In 2009, there were a total of 157 thousand apparel and accessory stores. (*Source:* U.S. Bureau of the Census. *County Business Patterns*)

a. Write an equation describing this relationship. Use ordered pairs of the form (years past 2005, numbers of stores in thousands).

b. Use this equation to predict the number of apparel and accessory stores in 2015.

75. It has been said that newspapers are disappearing, replaced by various electronic media. In 2001, newspaper circulation (number of copies it distributed in a day) was about 60 million. In 2009, this circulation dropped to 46 million. (*Source:* Newspaper Asso. of America)

a. Write an equation describing the relationship between time and circulation. Use ordered pairs of the form (years past 2001, number of newspapers circulated).

b. Use this equation to predict the newspaper circulation in 2017.

76. A certain chain of book stores is slowly closing down stores. Suppose that in 2006 there were 3991 stores and in 2010 there were 3200 stores.

a. Write an equation describing the relationship between time and number of store locations. Use ordered pairs of the form (years past 2006, number of stores).

b. Use this equation to predict the number of stores in 2018.

77. The Pool Fun Company has learned that, by pricing a newly released Fun Noodle at \$3, sales will reach 10,000 Fun Noodles per day during the summer. Raising the price to \$5 will cause sales to fall to 8000 Fun Noodles per day.

a. Assume that the relationship between price and number of Fun Noodles sold is linear and write an equation describing this relationship. Use ordered pairs of the form (price, number sold).

b. Predict the daily sales of Fun Noodles if the price is \$3.50.

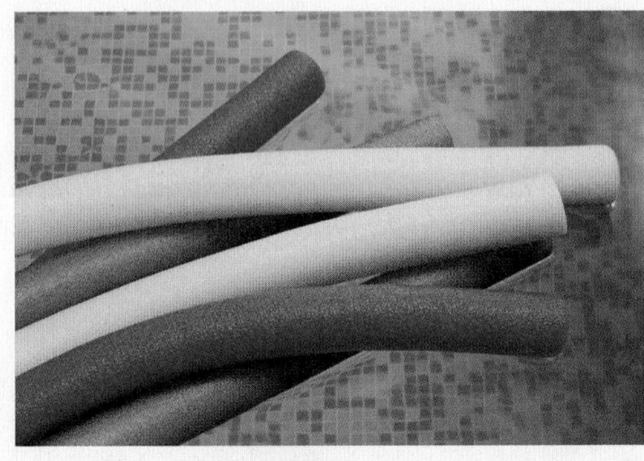

78. The value of a building bought in 2000 may be depreciated (or decreased) as time passes for income tax purposes. Seven years after the building was bought, this value was \$225,000 and 12 years after it was bought, this value was \$195,000.

a. If the relationship between number of years past 2000 and the depreciated value of the building is linear, write an equation describing this relationship. Use ordered pairs of the form (years past 2000, value of building).

b. Use this equation to estimate the depreciated value of the building in 2018.

REVIEW AND PREVIEW

Find the value of $x^2 - 3x + 1$ for each given value of x. See Section 1.7.

79. 2 **80.** 5 **81.** -1 **82.** -3

For each graph, determine whether any x-values correspond to two or more y-values. See Section 3.1.

83.

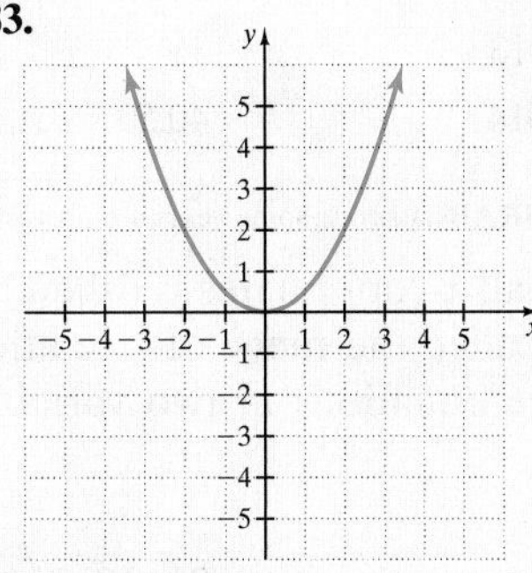

84.

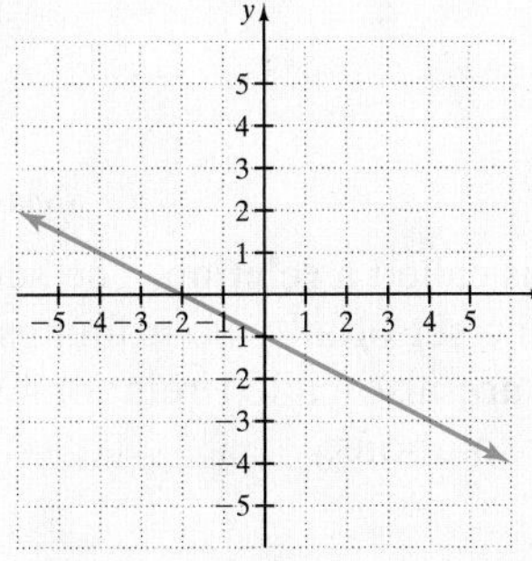

85.

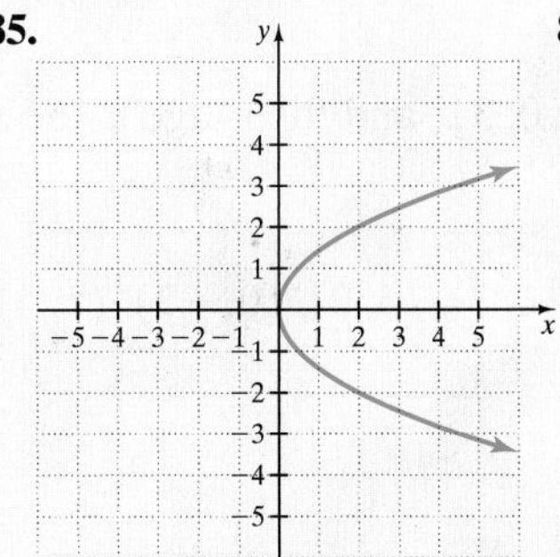

86.

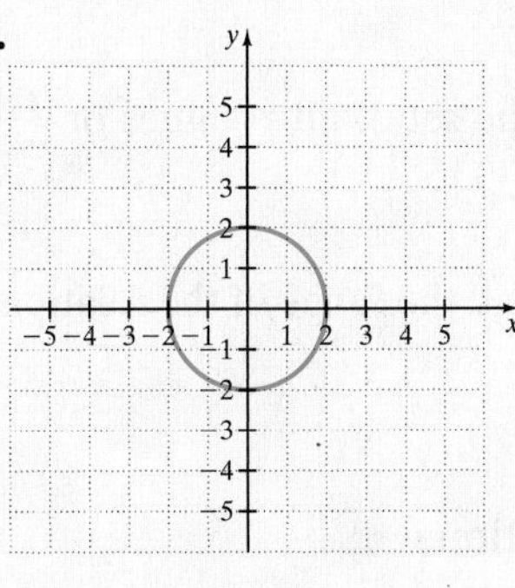

CONCEPT EXTENSIONS

For Exercises 87 through 90, identify the form that the linear equation in two variables is written in. For Exercises 91 and 92, identify the appearance of the graph of the equation.

87. $y - 7 = 4(x + 3)$; __________ form

88. $5x - 9y = 11$; __________ form

89. $y = \frac{3}{4}x - \frac{1}{3}$; __________ form

90. $y + 2 = \frac{-1}{3}(x - 2)$; __________ form

91. $y = \frac{1}{2}$; __________ line **92.** $x = -17$; __________ line

93. Given the equation of a nonvertical line, explain how to find the slope without finding two points on the line.

94. Given two points on a nonvertical line, explain how to use the point-slope form to find the equation of the line.

95. Write an equation in standard form of the line that contains the point $(-1, 2)$ and is

a. parallel to the line $y = 3x - 1$.

b. perpendicular to the line $y = 3x - 1$.

96. Write an equation in standard form of the line that contains the point $(4, 0)$ and is

a. parallel to the line $y = -2x + 3$.

b. perpendicular to the line $y = -2x + 3$.

97. Write an equation in standard form of the line that contains the point $(3, -5)$ and is

a. parallel to the line $3x + 2y = 7$.

b. perpendicular to the line $3x + 2y = 7$.

98. Write an equation in standard form of the line that contains the point $(-2, 4)$ and is

a. parallel to the line $x + 3y = 6$.

b. perpendicular to the line $x + 3y = 6$.

3.6 Functions

OBJECTIVES

1 Identify Relations, Domains, and Ranges.

2 Identify Functions.

3 Use the Vertical Line Test.

4 Use Function Notation.

OBJECTIVE

1 Identifying Relations, Domains, and Ranges 

In previous sections, we have discussed the relationships between two quantities. For example, the relationship between the length of the side of a square x and its area y is described by the equation $y = x^2$. Ordered pairs can be used to write down solutions of this equation. For example, (2, 4) is a solution of $y = x^2$, and this notation tells us that the x-value 2 is related to the y-value 4 for this equation. In other words, when the length of the side of a square is 2 units, its area is 4 square units.

Examples of Relationships Between Two Quantities

Area of Square: $y = x^2$		*Equation of Line:* $y = x + 2$		*U.S. Online Advertising Revenue*	
Some Ordered Pairs		*Some Ordered Pairs*		*Ordered Pairs*	
x	y	x	y	*Year*	*Billions of Dollars*
2	4	–3	–1	2010	25.8
5	25	0	2	2011	28.5
7	49	2	4	2012	32.6
12	144	9	11	2013	36
				2014	40.5

(*Source:* GRABstats.com; some years are projected)

A set of ordered pairs is called a **relation.** The set of all x-coordinates is called the **domain** of a relation, and the set of all y-coordinates is called the **range** of a relation. Equations such as $y = x^2$ are also called relations since equations in two variables define a set of ordered pair solutions.

EXAMPLE 1 Find the domain and the range of the relation $\{(0, 2), (3, 3), (-1, 0), (3, -2)\}$.

Solution The domain is the set of all x-values or $\{-1, 0, 3\}$, and the range is the set of all y-values, or $\{-2, 0, 2, 3\}$. □

PRACTICE
1 Find the domain and the range of the relation $\{(1, 3), (5, 0), (0, -2), (5, 4)\}$.

OBJECTIVE
2 Identifying Functions 

Some relations are also functions.

Function
A function is a set of ordered pairs that assigns to each x-value exactly one y-value.

In other words, a function cannot have two ordered pairs with the same x-coordinate but different y-coordinates.

EXAMPLE 2 Determine whether each relation is also a function.

a. $\{(-1, 1), (2, 3), (7, 3), (8, 6)\}$ **b.** $\{(0, -2), (1, 5), (0, 3), (7, 7)\}$

Solution

a. Although the ordered pairs (2, 3) and (7, 3) have the same y-value, each x-value is assigned to only one y-value, so this set of ordered pairs is a function.

b. The x-value 0 is assigned to two y-values, -2 and 3, so this set of ordered pairs is not a function. □

PRACTICE 2 Determine whether each relation is also a function.

a. $\{(4, 1), (3, -2), (8, 5), (-5, 3)\}$ **b.** $\{(1, 2), (-4, 3), (0, 8), (1, 4)\}$

Relations and functions can be described by a graph of their ordered pairs.

EXAMPLE 3 Determine whether each graph is the graph of a function.

a.

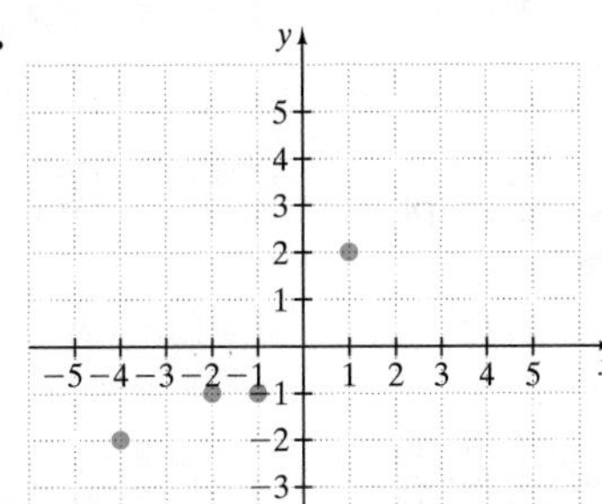

b.

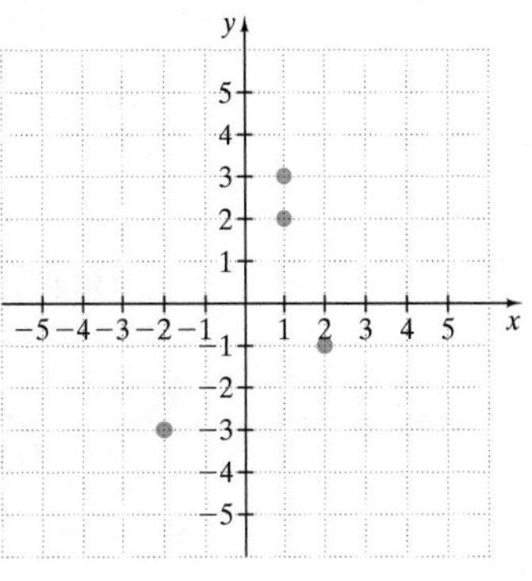

Solution

a. This is the graph of the relation $\{(-4, -2), (-2, -1), (-1, -1), (1, 2)\}$. Each x-coordinate has exactly one y-coordinate, so this is the graph of a function.

b. This is the graph of the relation $\{(-2, -3), (1, 2), (1, 3), (2, -1)\}$. The x-coordinate 1 is paired with two y-coordinates, 2 and 3, so this is not the graph of a function. □

PRACTICE 3 Determine whether each graph is the graph of a function.

a.

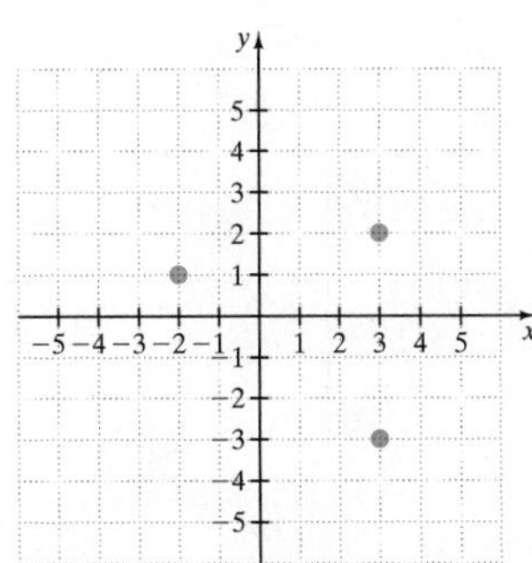

b.

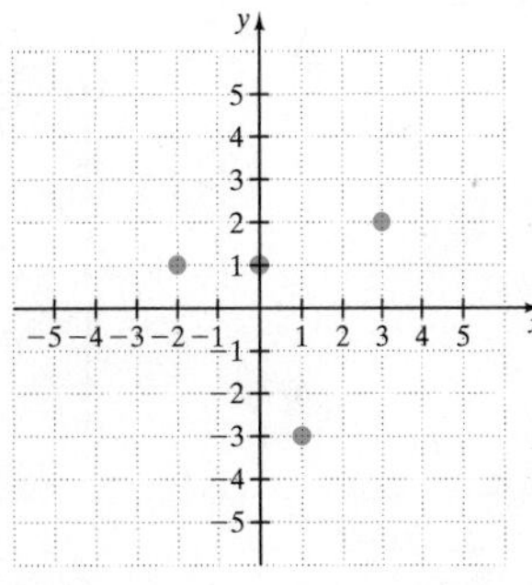

x-coordinate 1 paired with two y-coordinates, 2 and 3.

(1, 3)

(1, 2)

Not the graph of a function.

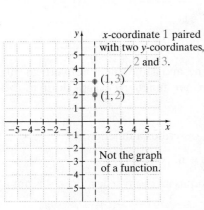

OBJECTIVE 3

Using the Vertical Line Test

The graph in Example 3(b) was not the graph of a function because the x-coordinate 1 was paired with two y-coordinates, 2 and 3. Notice that when an x-coordinate is paired with more than one y-coordinate, a vertical line can be drawn that will intersect the graph at more than one point. We can use this fact to determine whether a relation is also a function. We call this the **vertical line test.**

Vertical Line Test

If a vertical line can be drawn so that it intersects a graph more than once, the graph is not the graph of a function.

This vertical line test works for all types of graphs on the rectangular coordinate system.

EXAMPLE 4 Use the vertical line test to determine whether each graph is the graph of a function.

a.

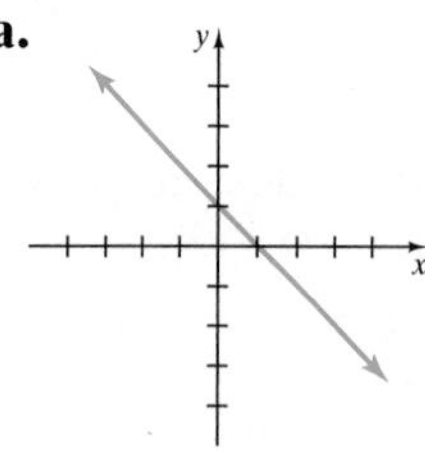

b.

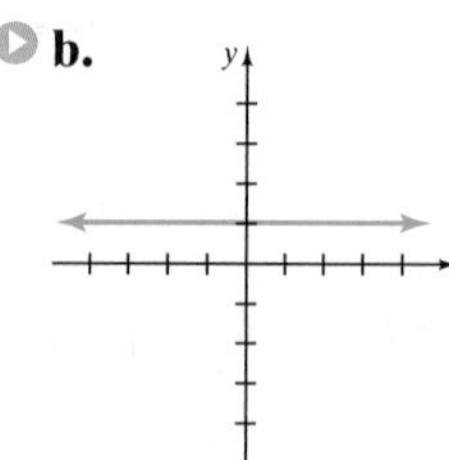

c.

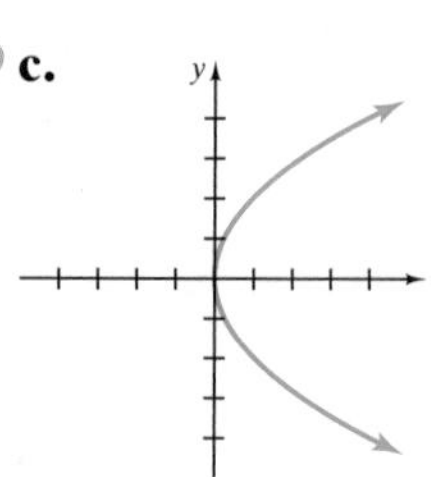

d.

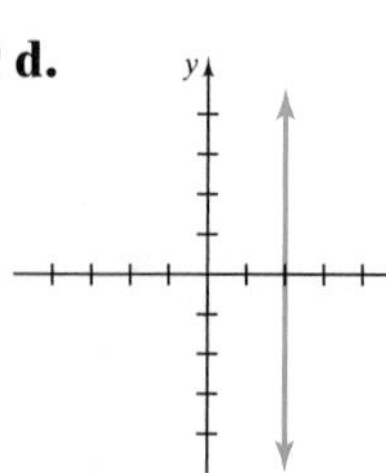

Solution

a. This graph is the graph of a function since no vertical line will intersect this graph more than once.

b. This graph is also the graph of a function; no vertical line will intersect it more than once.

c. This graph is not the graph of a function. Vertical lines can be drawn that intersect the graph in two points. An example of one is shown.

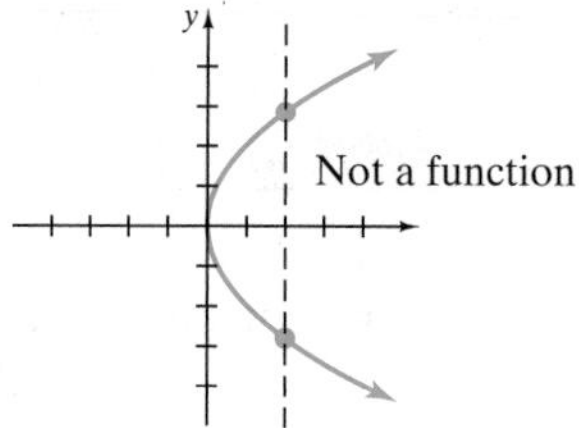

d. This graph is not the graph of a function. A vertical line can be drawn that intersects this line at every point. □

PRACTICE

4 Use the vertical line test to determine whether each graph is the graph of a function.

a.

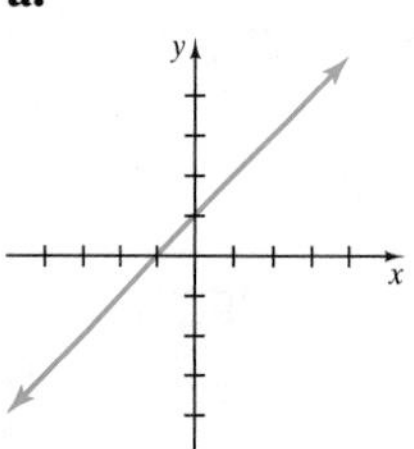

b.

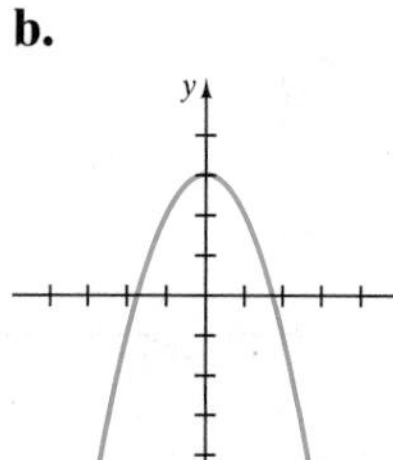

c.

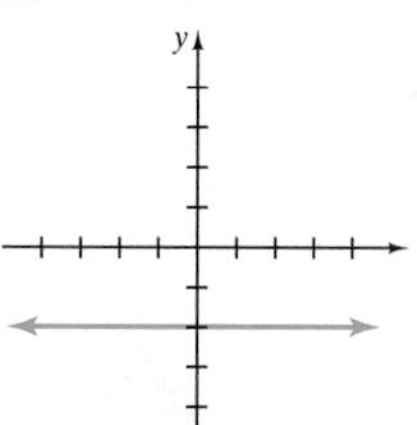

d.

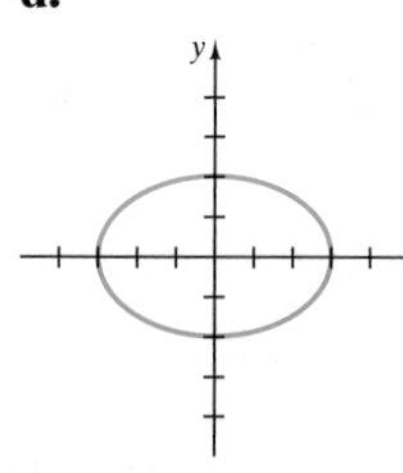

Recall that the graph of a linear equation is a line, and a line that is not vertical will pass the vertical line test. **Thus, all linear equations are functions except those of the form $x = c$, which are vertical lines.**

EXAMPLE 5 Decide whether the equation describes a function.

a. $y = x$ **b.** $y = 2x + 1$ **c.** $y = 5$ **d.** $x = -1$

Solution **a, b,** and **c** are functions because their graphs are nonvertical lines. **d** is not a function because its graph is a vertical line. □

PRACTICE

5 Decide whether the equation describes a function.

a. $y = 2x$ **b.** $y = -3x - 1$ **c.** $y = 8$ **d.** $x = 2$

Examples of functions can often be found in magazines, newspapers, books, and other printed material in the form of tables or graphs such as that in Example 6.

EXAMPLE 6 The graph shows the sunrise time for Indianapolis, Indiana, for the year. Use this graph to answer the questions.

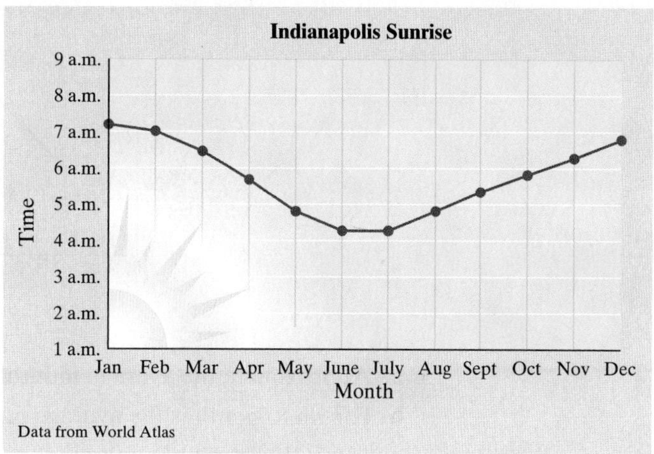

a. Approximate the time of sunrise on February 1.

b. Approximately when does the sun rise at 5 a.m.?

c. Is this the graph of a function?

Solution

a. To approximate the time of sunrise on February 1, we find the mark on the horizontal axis that corresponds to February 1. From this mark, we move vertically upward until the graph is reached. From that point on the graph, we move horizontally to the left until the vertical axis is reached. The vertical axis there reads 7 a.m.

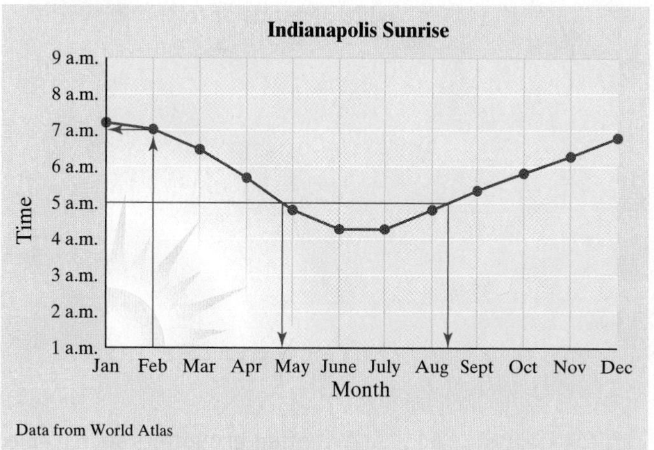

b. To approximate when the sun rises at 5 a.m., we find 5 a.m. on the time axis and move horizontally to the right. Notice that we will reach the graph twice, corresponding to two dates for which the sun rises at 5 a.m. We follow both points on the graph vertically downward until the horizontal axis is reached. The sun rises at 5 a.m. at approximately the end of the month of April and the middle of the month of August.

c. The graph is the graph of a function since it passes the vertical line test. In other words, for every day of the year in Indianapolis, there is exactly one sunrise time. □

PRACTICE

6 The graph shows the average monthly temperature for Chicago, Illinois, for the year. Use this graph to answer the questions.

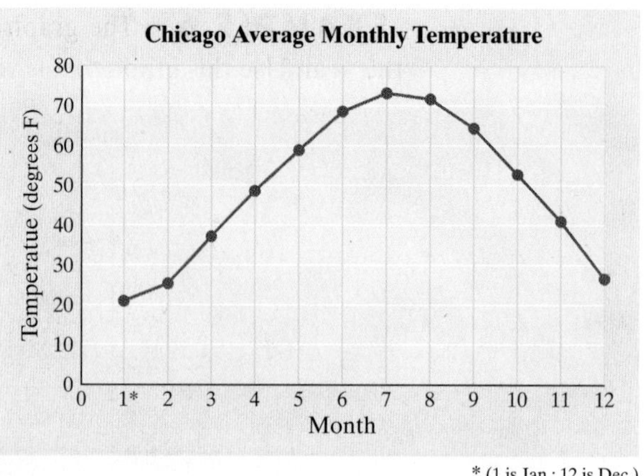

* (1 is Jan.; 12 is Dec.)

a. Approximate the average monthly temperature for June.

b. For what month is the average monthly temperature 26°?

c. Is this the graph of a function?

OBJECTIVE

4 Using Function Notation 

The graph of the linear equation $y = 2x + 1$ passes the vertical line test, so we say that $y = 2x + 1$ is a function. In other words, $y = 2x + 1$ gives us a rule for writing ordered pairs where every x-coordinate is paired with one and only one y-coordinate.

The variable y is a function of the variable x. For each value of x, there is only one value of y. Thus, we say the variable x is the **independent variable** because any value in the domain can be assigned to x. The variable y is the **dependent variable** because its value depends on x.

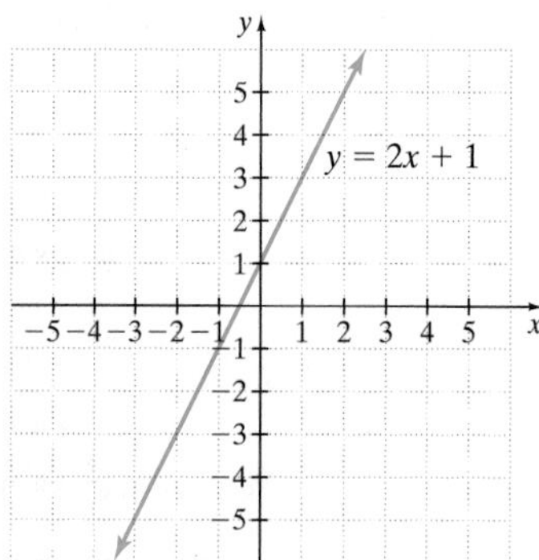

We often use letters such as f, g, and h to name functions. For example, the symbol $f(x)$ means *function of x* and is read "f of x." This notation is called **function notation.** The equation $y = 2x + 1$ can be written as $f(x) = 2x + 1$ using function notation, and these equations mean the same thing. In other words, $y = f(x)$.

The notation $f(1)$ means to replace x with 1 and find the resulting y or function value. Since

$$f(x) = 2x + 1$$

then

$$f(1) = 2(1) + 1$$
$$= 3$$

This means that, when $x = 1$, y or $f(x) = 3$, and we have the ordered pair $(1, 3)$. Now let's find $f(2)$, $f(0)$, and $f(-1)$.

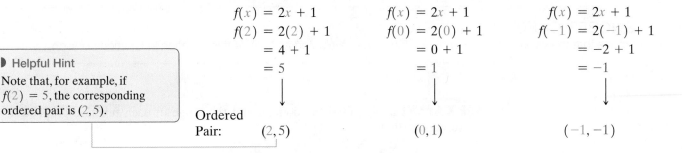

> **Helpful Hint**
> Note that, for example, if $f(2) = 5$, the corresponding ordered pair is $(2, 5)$.

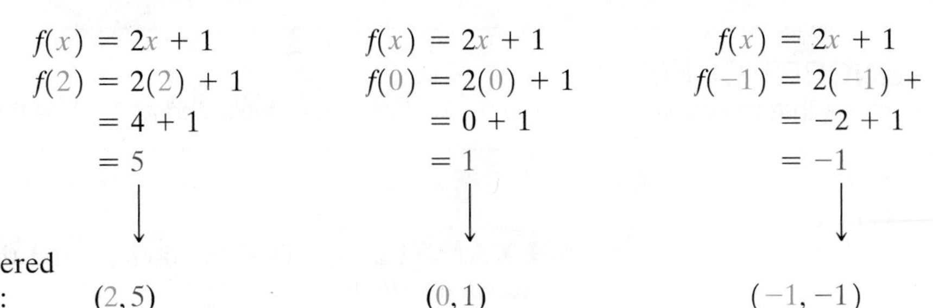

$$f(x) = 2x + 1 \quad f(2) = 2(2) + 1 = 4 + 1 = 5$$

$$f(x) = 2x + 1 \quad f(0) = 2(0) + 1 = 0 + 1 = 1$$

$$f(x) = 2x + 1 \quad f(-1) = 2(-1) + 1 = -2 + 1 = -1$$

Ordered Pair: $(2, 5)$ $\quad$ $(0, 1)$ $\quad$ $(-1, -1)$

> **Helpful Hint**
> Note that $f(x)$ is a special symbol in mathematics used to denote a function. The symbol $f(x)$ is read "f of x." It does **not** mean $f \cdot x$ (f times x).

EXAMPLE 7 Given $g(x) = x^2 - 3$, find the following. Then write the corresponding ordered pairs generated.

a. $g(2)$ **b.** $g(-2)$ **c.** $g(0)$

Solution

a. $g(x) = x^2 - 3$

$g(2) = 2^2 - 3 = 4 - 3 = 1$

b. $g(x) = x^2 - 3$

$g(-2) = (-2)^2 - 3 = 4 - 3 = 1$

c. $g(x) = x^2 - 3$

$g(0) = 0^2 - 3 = 0 - 3 = -3$

Ordered Pairs:	$g(2) = 1$ gives $(2, 1)$	$g(-2) = 1$ gives $(-2, 1)$	$g(0) = -3$ gives $(0, -3)$

□

PRACTICE 7 Given $h(x) = x^2 + 5$, find the following. Then write the corresponding ordered pairs generated.

a. $h(2)$ **b.** $h(-5)$ **c.** $h(0)$

We now practice finding the domain and the range of a function. The domains of our functions will be the set of all possible real numbers that x can be replaced by. The range is the set of corresponding y-values.

EXAMPLE 8 Find the domain of each function.

a. $g(x) = \frac{1}{x}$ **b.** $f(x) = 2x + 1$

Solution

a. Recall that we cannot divide by 0, so the domain of $g(x)$ is the set of all real numbers except 0. In interval notation, we can write $(-\infty, 0) \cup (0, \infty)$.

b. In this function, x can be any real number. The domain of $f(x)$ is the set of all real numbers, or $(-\infty, \infty)$ in interval notation. □

PRACTICE 8 Find the domain of each function.

a. $h(x) = 6x + 3$ **b.** $f(x) = \frac{1}{x^2}$

✓CONCEPT CHECK
Suppose that the value of f is -7 when the function is evaluated at 2. Write this situation in function notation.

EXAMPLE 9 Find the domain and the range of each function graphed. Use interval notation.

a.

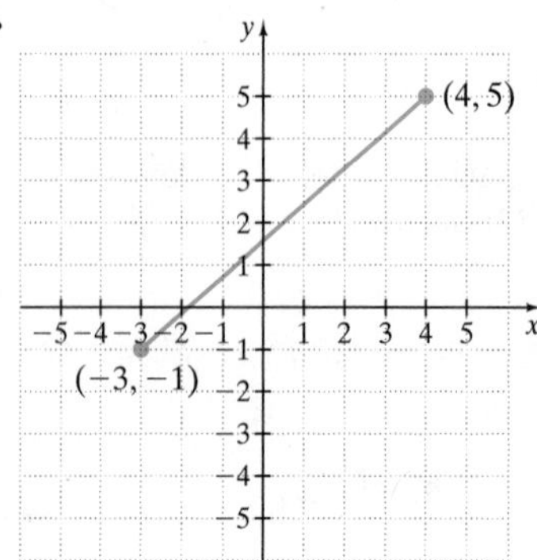

b.

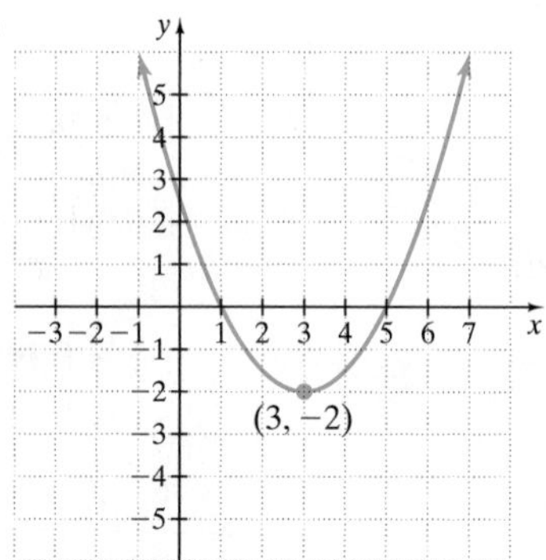

Solution

a.

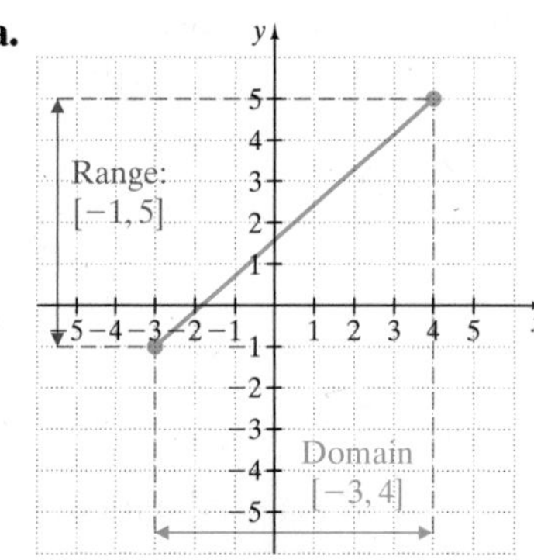

b.

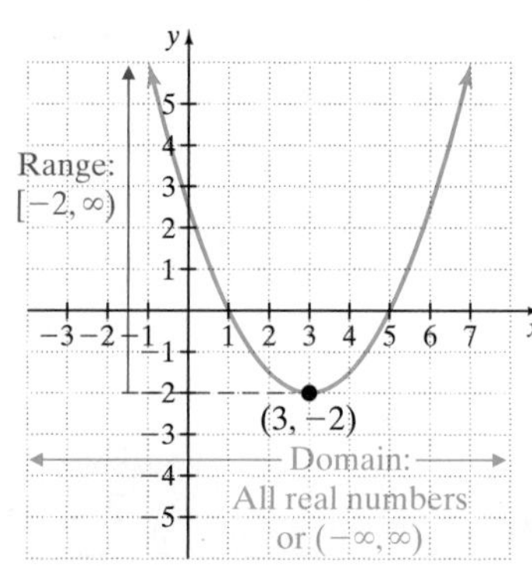

□

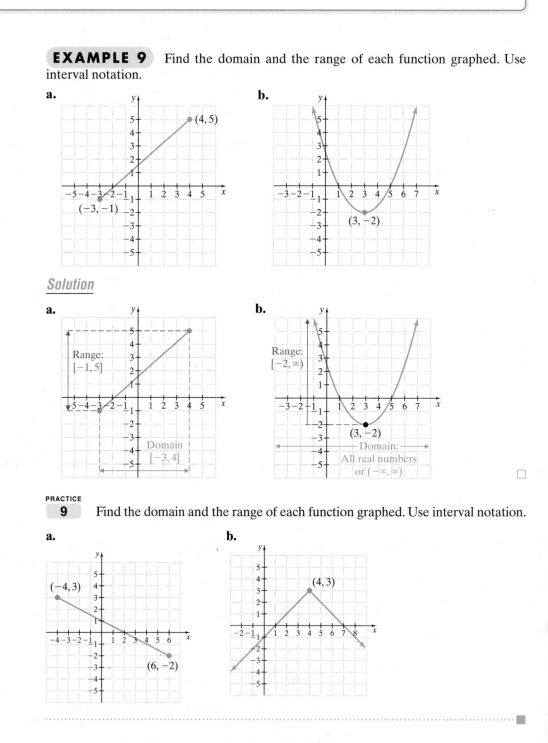

PRACTICE 9 Find the domain and the range of each function graphed. Use interval notation.

a.

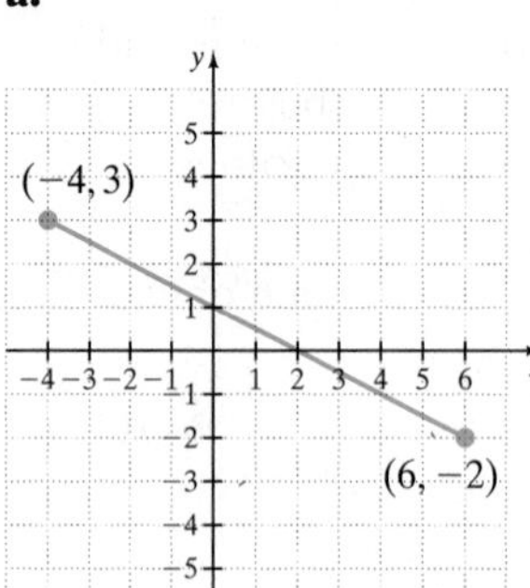

b.

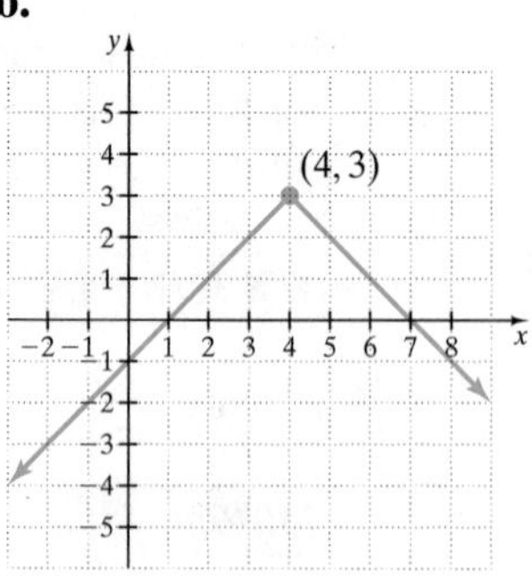

Answer to Concept Check:
$f(2) = -7$

Vocabulary, Readiness & Video Check

Use the choices below to fill in each blank. Some choices may not be used.

$x = c$	horizontal	domain	relation	$(7, 3)$
$y = c$	vertical	range	function	$(3, 7)$

1. A set of ordered pairs is called a(n) ______.
2. A set of ordered pairs that assigns to each x-value exactly one y-value is called a(n) ______.
3. The set of all y-coordinates of a relation is called the ______.
4. The set of all x-coordinates of a relation is called the ______.
5. All linear equations are functions except those whose graphs are ______ lines.
6. All linear equations are functions except those whose equations are of the form ______.

Martin-Gay Interactive Videos

See Video 3.6

Watch the section lecture video and answer the following questions.

OBJECTIVE 1

7. In the lecture before Example 1, relations are discussed. Why can an equation in two variables define a relation?

OBJECTIVE 2

8. Based on Examples 2 and 3, can a set of ordered pairs with no repeated x-values, but with repeated y-values be a function? For example: $\{(0, 4), (-3, 4), (2, 4)\}$.

OBJECTIVE 3

9. After reviewing Example 8, explain why the vertical line test works.

OBJECTIVE 4

10. Using Example 10, write the three function values found and their corresponding ordered pairs. For example, $f(0) = 2$ corresponds to $(0, 2)$.

3.6 Exercise Set

MyMathLab®

Find the domain and the range of each relation. See Example 1.

1. $\{(2, 4), (0, 0), (-7, 10), (10, -7)\}$
2. $\{(3, -6), (1, 4), (-2, -2)\}$
3. $\{(0, -2), (1, -2), (5, -2)\}$
4. $\{(5, 0), (5, -3), (5, 4), (5, 3)\}$

Determine whether each relation is also a function. See Example 2.

5. $\{(1, 1), (2, 2), (-3, -3), (0, 0)\}$
6. $\{(11, 6), (-1, -2), (0, 0), (3, -2)\}$
7. $\{(-1, 0), (-1, 6), (-1, 8)\}$
8. $\{(1, 2), (3, 2), (1, 4)\}$

MIXED PRACTICE

Determine whether each graph is the graph of a function. See Examples 3 and 4.

9.

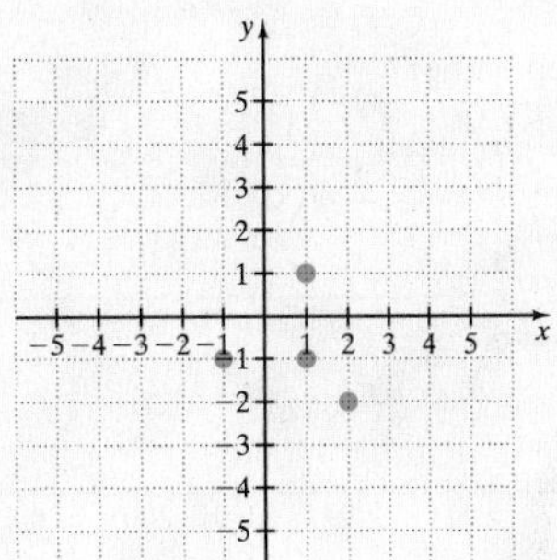

10.

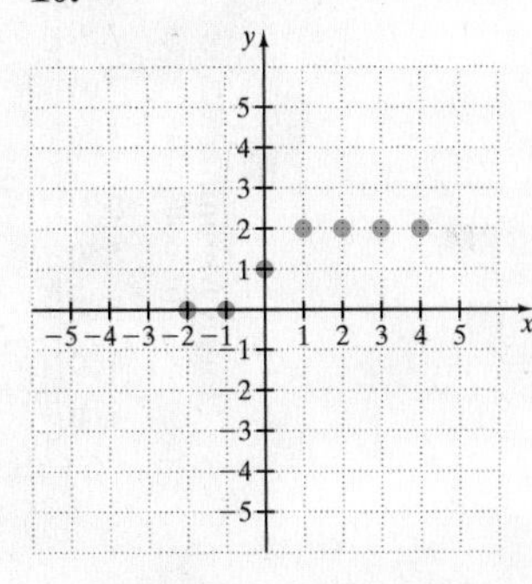

11.

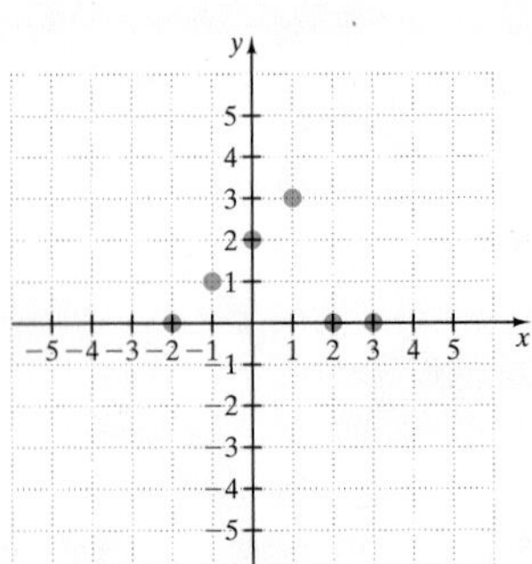

12.

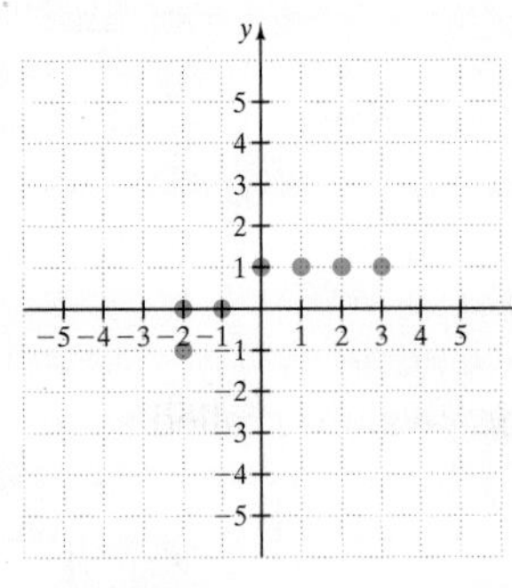
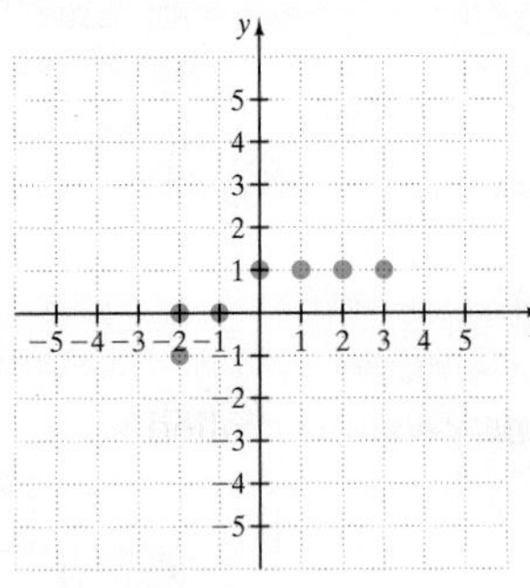

13.

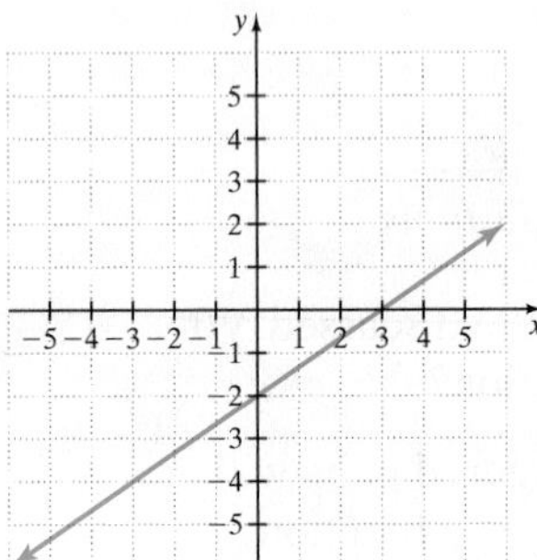

14.

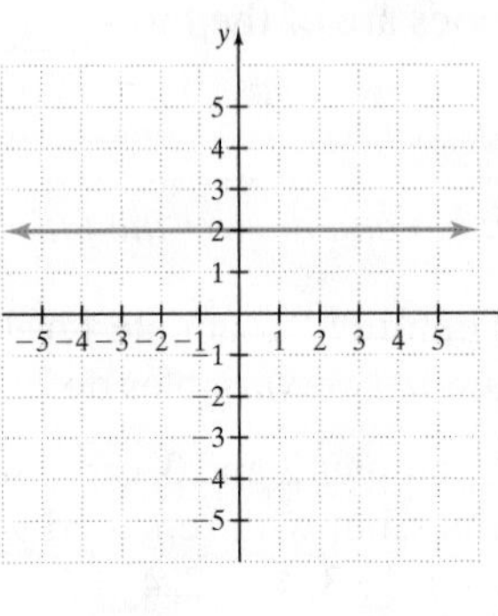

15.

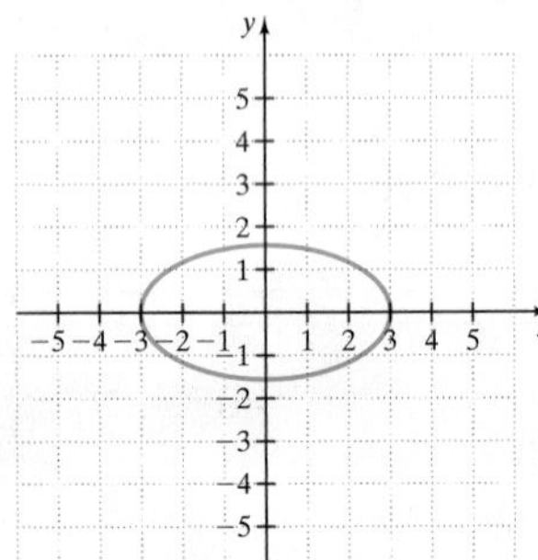

16.

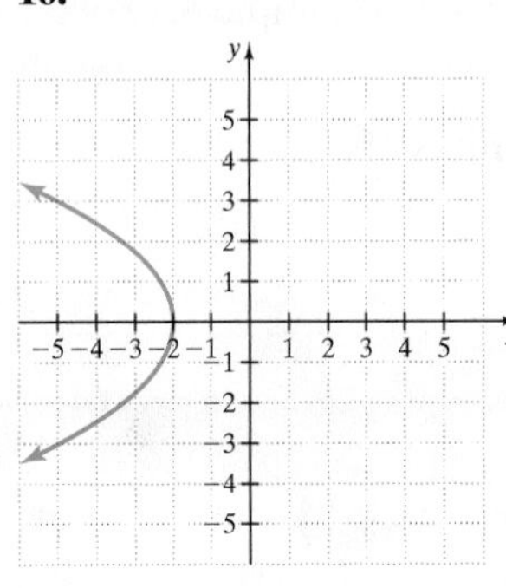

Decide whether the equation describes a function. See Example 5.

17. $y = x + 1$

18. $y = x - 1$

19. $y - x = 7$

20. $2x - 3y = 9$

21. $y = 6$

22. $x = 3$

23. $x = -2$

24. $y = -9$

25. $x = y^2$

26. $y = x^2 - 3$

The graph shows the sunset times for Seward, Alaska. Use this graph to answer Exercises 27 through 32. See Example 6.

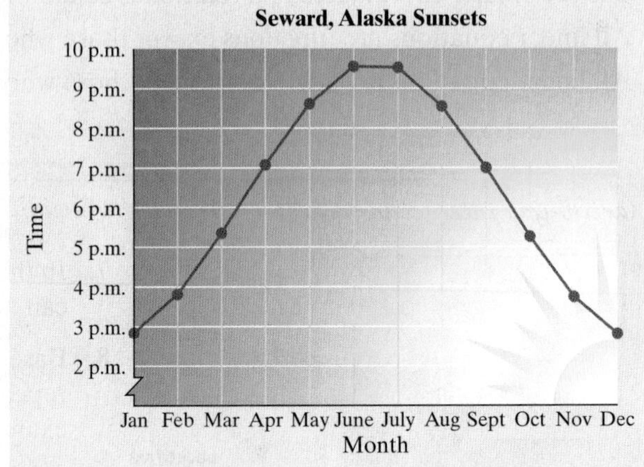

27. Approximate the time of sunset on June 1.

28. Approximate the time of sunset on November 1.

29. Approximate when the sunset is at 3 p.m.

30. Approximate when the sunset is at 9 p.m.

31. Is this graph the graph of a function? Why or why not?

32. Do you think a graph of sunset times for any location will always be a function? Why or why not?

This graph shows the U.S. hourly minimum wage for each year shown. Use this graph to answer Exercises 33 through 38. See Example 6.

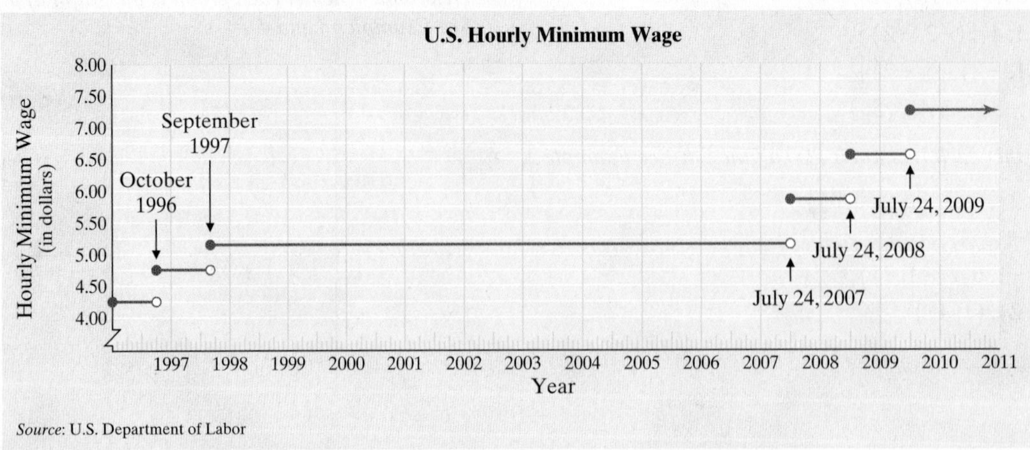

33. Approximate the minimum wage before October 1996.

34. Approximate the minimum wage in 2006.

35. Approximate the year when the minimum wage increased to over \$7.00 per hour.

36. According to the graph, what hourly wage was in effect for the greatest number of years?

37. Is this graph the graph of a function? Why or why not?

38. Do you think that a similar graph of your hourly wage on January 1 of every year (whether you are working or not) will be the graph of a function? Why or why not?

This graph shows the cost of mailing a large envelope through the U.S. Postal Service by weight. (Note: These costs will increase.) Use this graph to answer Exercises 39 through 44. See Example 6.

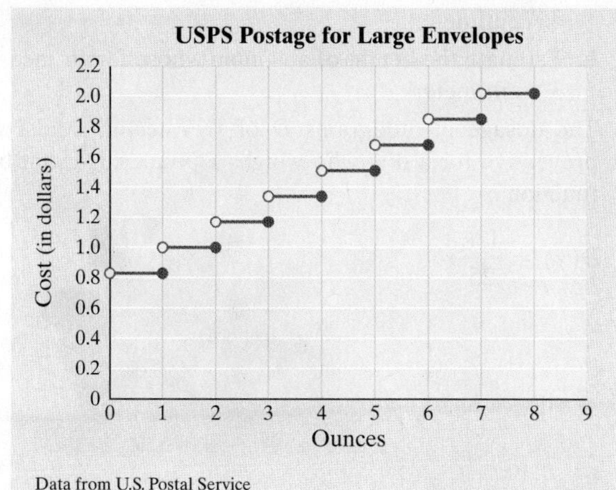

39. Approximate the postage to mail a large envelope weighing more than 4 ounces but not more than 5 ounces.

40. Approximate the postage to mail a large envelope weighing more than 7 ounces but not more than 8 ounces.

41. Give the weight of a large envelope that costs $1 to mail.

42. If you have $1.25, what is the weight of the largest envelope you can mail for that amount of money?

43. Is this graph a function? Why or why not?

44. Do you think that a similar graph of postage to mail a first-class letter will be the graph of a function? Why or why not?

Find $f(-2)$, $f(0)$, and $f(3)$ for each function. See Example 7.

45. $f(x) = 2x - 5$
46. $f(x) = 3 - 7x$
47. $f(x) = x^2 + 2$
48. $f(x) = x^2 - 4$
49. $f(x) = 3x$
50. $f(x) = -3x$
51. $f(x) = |x|$
52. $f(x) = |2 - x|$

Find $h(-1)$, $h(0)$, and $h(4)$ for each function. See Example 7.

53. $h(x) = -5x$
54. $h(x) = -3x$
55. $h(x) = 2x^2 + 3$
56. $h(x) = 3x^2$

For each given function value, write a corresponding ordered pair.

57. $f(3) = 6$
58. $f(7) = -2$
59. $g(0) = -\frac{1}{2}$
60. $g(0) = -\frac{7}{8}$
61. $h(-2) = 9$
62. $h(-10) = 1$

Find the domain of each function. See Example 8.

63. $f(x) = 3x - 7$
64. $g(x) = 5 - 2x$
65. $h(x) = \frac{1}{x + 5}$
66. $f(x) = \frac{1}{x - 6}$
67. $g(x) = |x + 1|$
68. $h(x) = |2x|$

Find the domain and the range of each relation graphed. See Example 9.

69.

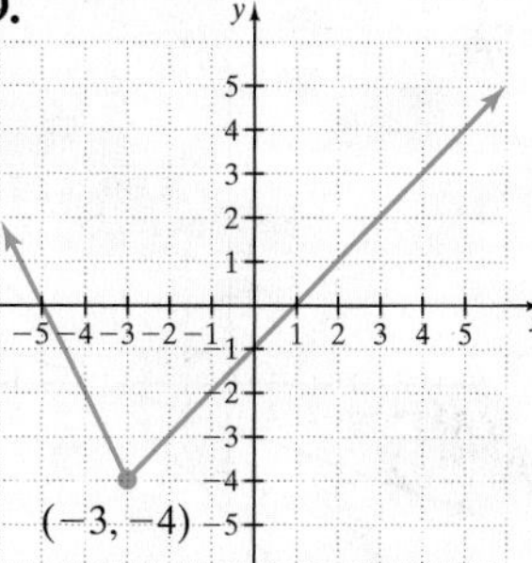

70.

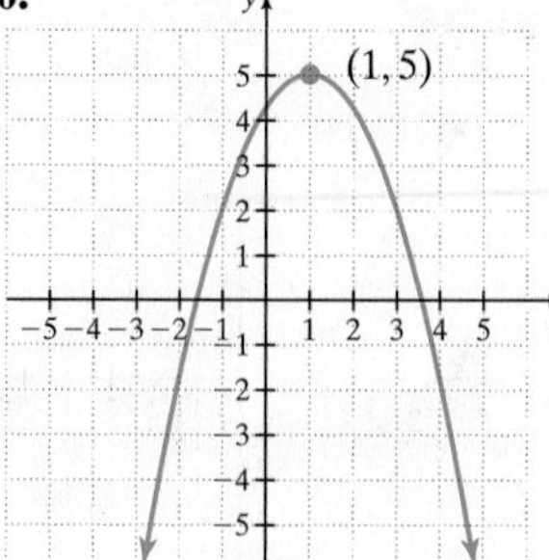

71.

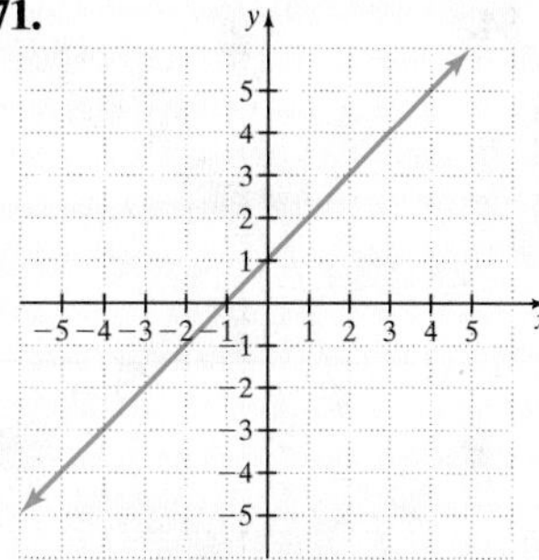

72.

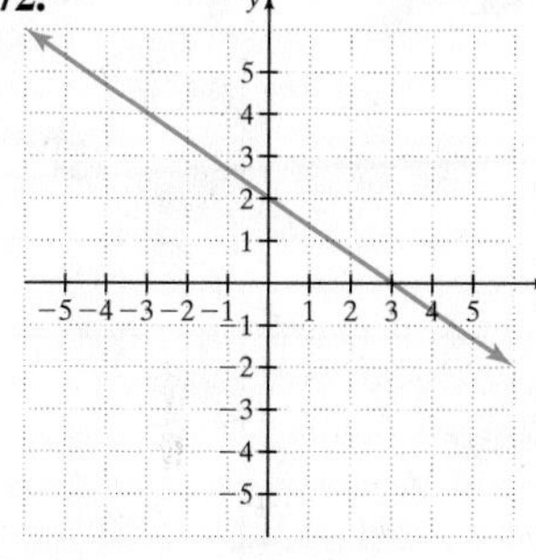

73.

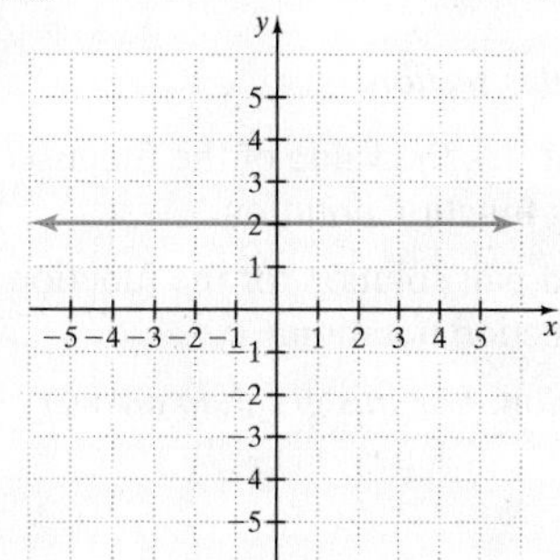

74.

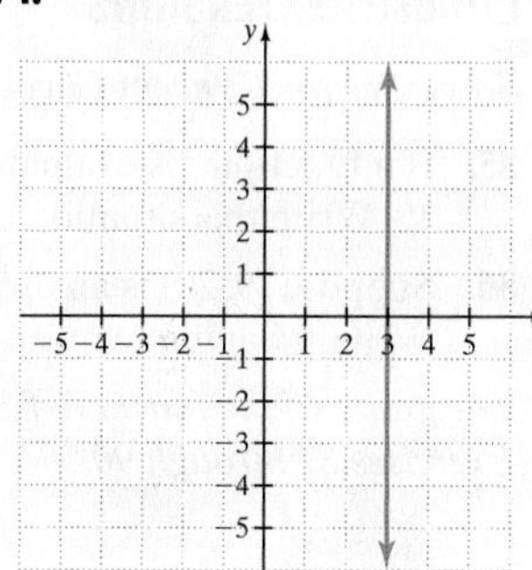

Use the graph of f below to answer Exercises 75 through 80.

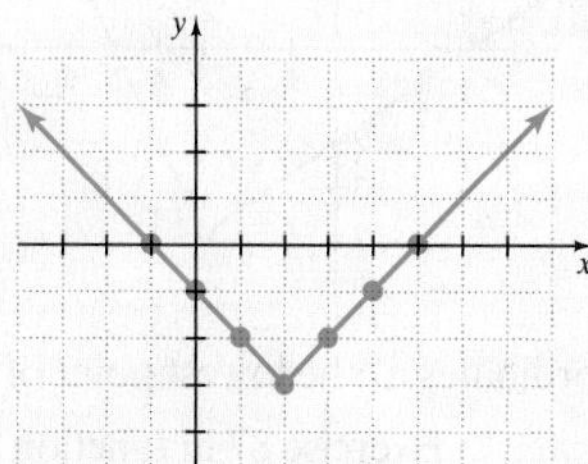

75. Complete the ordered pair solution for *f*. (0,)

76. Complete the ordered pair solution for *f*. (3,)

77. $f(0) =$ _____?

78. $f(3) =$ _____?

79. If $f(x) = 0$, find the value(s) of x.

80. If $f(x) = -1$, find the value(s) of x.

REVIEW AND PREVIEW

Find the coordinates of the point of intersection. See Section 3.1.

81.

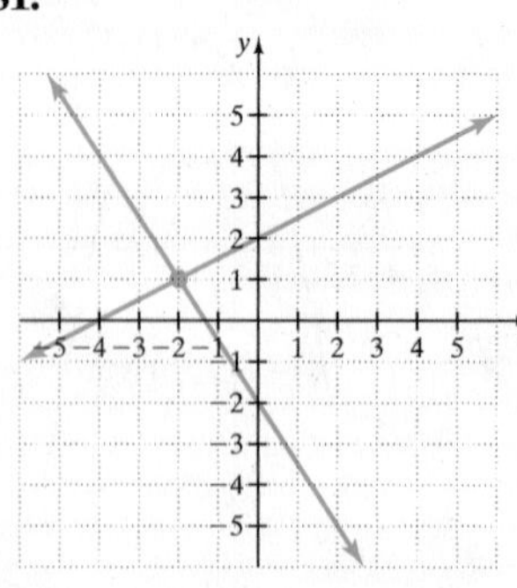

82.

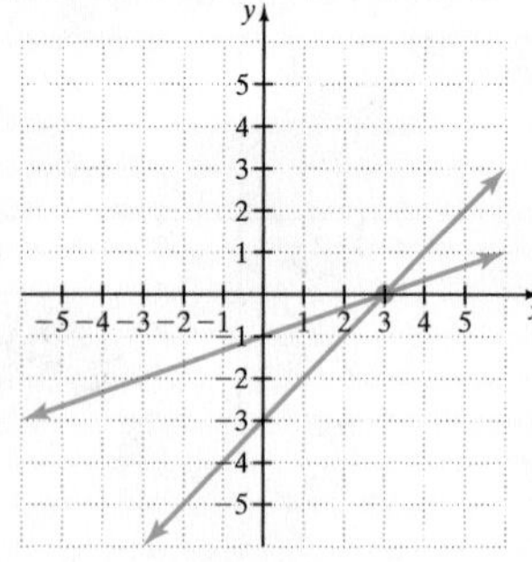

83.

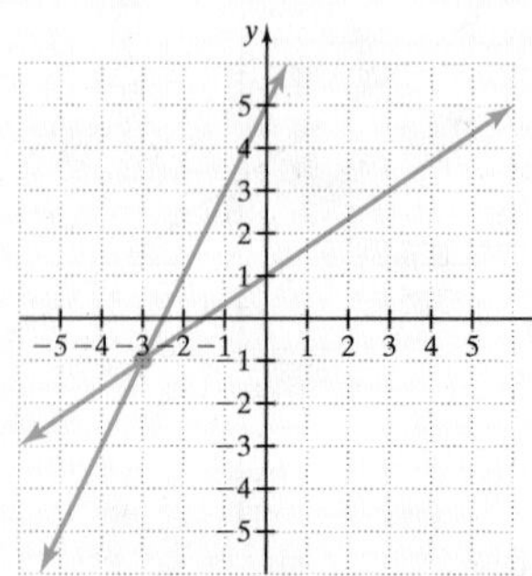

84.

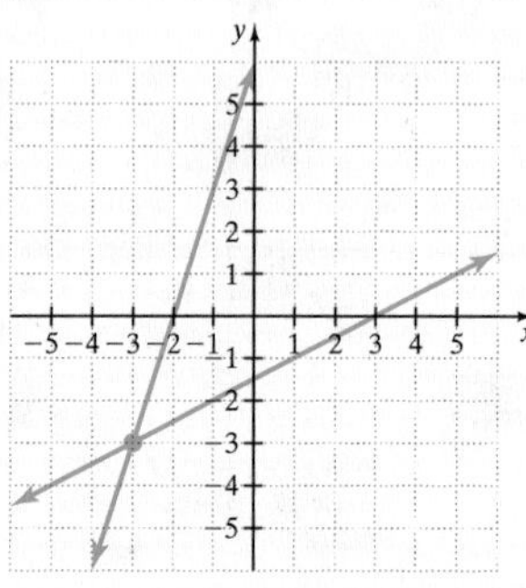

CONCEPT EXTENSIONS

Solve. See the Concept Check in this section.

85. If a function f is evaluated at -5, the value of the function is 12. Write this situation using function notation.

86. Suppose $(9, 20)$ is an ordered-pair solution for the function g. Write this situation using function notation.

The graph of the function, f, is below. Use this graph to answer Exercises 87 through 90.

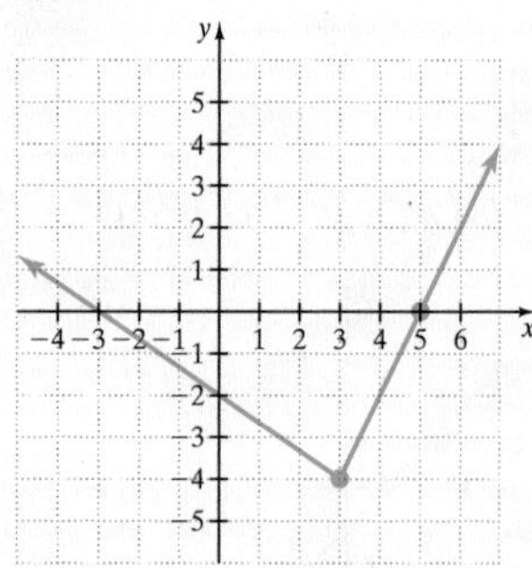

87. Write the coordinates of the lowest point of the graph.

88. Write the answer to Exercise 87 in function notation.

89. An x-intercept of this graph is $(5, 0)$. Write this using function notation.

90. Write the other x-intercept of this graph (see Exercise 89) using function notation.

91. Forensic scientists use the function

$$H(x) = 2.59x + 47.24$$

to estimate the height of a woman in centimeters given the length x of her femur bone.

a. Estimate the height of a woman whose femur measures 46 centimeters.

b. Estimate the height of a woman whose femur measures 39 centimeters.

92. The dosage in milligrams D of Ivermectin, a heartworm preventive for a dog who weighs x pounds, is given by the function

$$D(x) = \frac{136}{25}x$$

a. Find the proper dosage for a dog that weighs 35 pounds.

b. Find the proper dosage for a dog that weighs 70 pounds.

93. In your own words, define **(a)** function; **(b)** domain; **(c)** range.

94. Explain the vertical line test and how it is used.

95. Since $y = x + 7$ is a function, rewrite the equation using function notation.

96. Since $y = 3$ is a function, rewrite this equation using function notation.

See the example below for Exercises 97 through 100.

Example

If $f(x) = x^2 + 2x + 1$, find $f(\pi)$.

Solution:

$$f(x) = x^2 + 2x + 1$$
$$f(\pi) = \pi^2 + 2\pi + 1$$

Given the following functions, find the indicated values.

97. $f(x) = 2x + 7$

a. $f(2)$ **b.** $f(a)$

98. $g(x) = -3x + 12$

a. $g(s)$ **b.** $g(r)$

99. $h(x) = x^2 + 7$

a. $h(3)$ **b.** $h(a)$

100. $f(x) = x^2 - 12$

a. $f(12)$ **b.** $f(a)$

Chapter 3 Vocabulary Check

Fill in each blank with one of the words listed below.

relation	function	domain	range	standard	slope-intercept
y-axis	x-axis	solution	linear	slope	point-slope
x-intercept	y-intercept	y	x		

1. An ordered pair is a(n) ______________ of an equation in two variables if replacing the variables by the coordinates of the ordered pair results in a true statement.
2. The vertical number line in the rectangular coordinate system is called the ______________.
3. A(n) ______________ equation can be written in the form $Ax + By = C$.
4. A(n) ______________ is a point of the graph where the graph crosses the x-axis.
5. The form $Ax + By = C$ is called ______________ form.
6. A(n) ______________ is a point of the graph where the graph crosses the y-axis.
7. The equation $y = 7x - 5$ is written in ______________ form.
8. The equation $y + 1 = 7(x - 2)$ is written in ______________ form.
9. To find an x-intercept of a graph, let ______________ = 0.
10. The horizontal number line in the rectangular coordinate system is called the ______________.
11. To find a y-intercept of a graph, let ______________ = 0.
12. The ______________ of a line measures the steepness or tilt of a line.
13. A set of ordered pairs that assigns to each x-value exactly one y-value is called a(n) ______________.
14. The set of all x-coordinates of a relation is called the ______________ of the relation.
15. The set of all y-coordinates of a relation is called the ______________ of the relation.
16. A set of ordered pairs is called a(n) ______________.

Chapter 3 Highlights

DEFINITIONS AND CONCEPTS | **EXAMPLES**

Section 3.1 Reading Graphs and the Rectangular Coordinate System

The **rectangular coordinate system** consists of a plane and a vertical and a horizontal number line intersecting at their 0 coordinates. The vertical number line is called the **y-axis** and the horizontal number line is called the **x-axis.** The point of intersection of the axes is called the **origin.**

To **plot** or **graph** an ordered pair means to find its corresponding point on a rectangular coordinate system.

To plot or graph an ordered pair such as $(3, -2)$, start at the origin. Move 3 units to the right and, from there, 2 units down.

To plot or graph $(-3, 4)$ start at the origin. Move 3 units to the left and, from there, 4 units up.

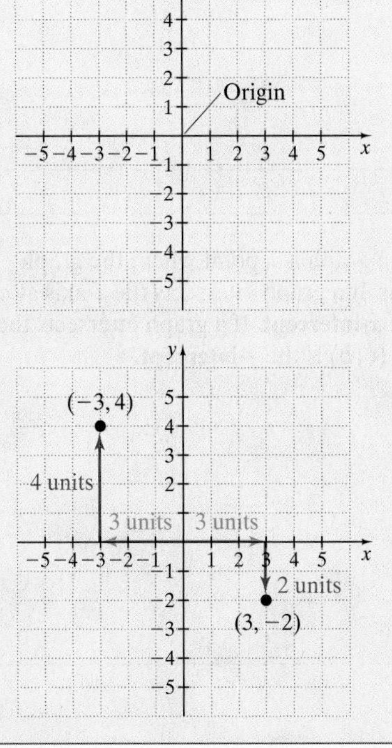

(continued)

DEFINITIONS AND CONCEPTS	EXAMPLES

Section 3.1 Reading Graphs and the Rectangular Coordinate System (continued)

An ordered pair is a **solution** of an equation in two variables if replacing the variables by the coordinates of the ordered pair results in a true statement.

Determine whether $(-1, 5)$ is a solution of $2x + 3y = 13$.

$$2x + 3y = 13$$
$$2(-1) + 3 \cdot 5 = 13 \quad \text{Let } x = -1, y = 5$$
$$-2 + 15 = 13$$
$$13 = 13 \quad \text{True}$$

If one coordinate of an ordered pair solution is known, the other value can be determined by substitution.

Complete the ordered pair solution (0,) for the equation $x - 6y = 12$.

$$x - 6y = 12$$
$$0 - 6y = 12 \quad \text{Let } x = 0.$$
$$\frac{-6y}{-6} = \frac{12}{-6} \quad \text{Divide by } -6.$$
$$y = -2$$

The ordered pair solution is $(0, -2)$.

Section 3.2 Graphing Linear Equations

A **linear equation in two variables** is an equation that can be written in the form $Ax + By = C$ where A and B are not both 0. The form $Ax + By = C$ is called **standard form.**

Linear Equations

$$3x + 2y = -6 \qquad x = -5$$
$$y = 3 \qquad y = -x + 10$$

$x + y = 10$ is in standard form.

To graph a linear equation in two variables, find three ordered pair solutions. Plot the solution points and draw the line connecting the points.

Graph $x - 2y = 5$.

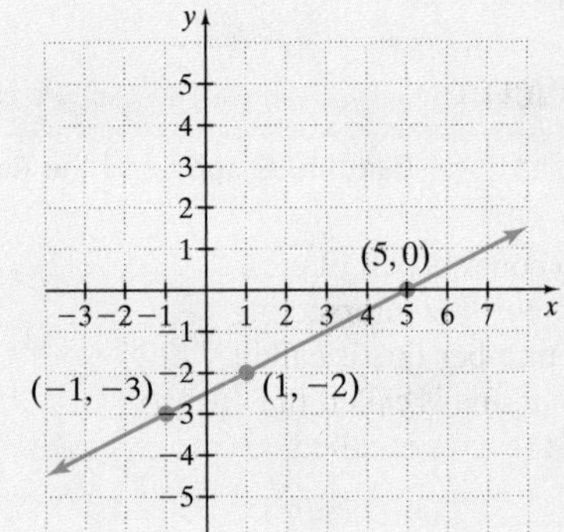

x	y
5	0
1	−2
−1	−3

Section 3.3 Intercepts

An **intercept** of a graph is a point where the graph intersects an axis. If a graph intersects the x-axis at a, then $(a, 0)$ is the ***x*-intercept.** If a graph intersects the y-axis at b, then $(0, b)$ is the ***y*-intercept.**

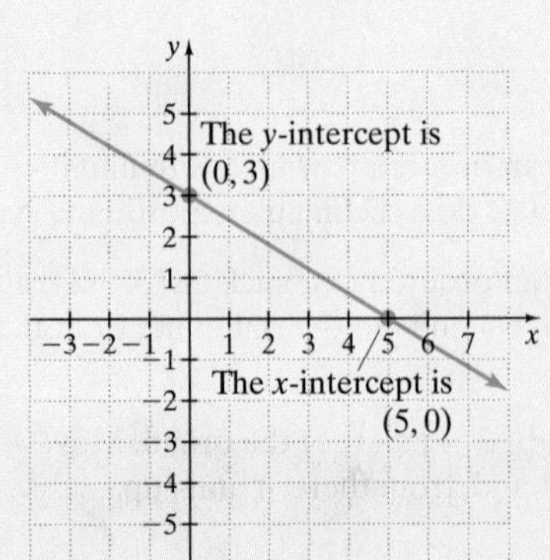

DEFINITIONS AND CONCEPTS	EXAMPLES

Section 3.3 Intercepts (continued)

To find the x-intercept, let $y = 0$ and solve for x.

To find the y-intercept, let $x = 0$ and solve for y.

Graph $2x - 5y = -10$ by finding intercepts.

If $y = 0$, then	If $x = 0$, then
$2x - 5 \cdot 0 = -10$	$2 \cdot 0 - 5y = -10$
$2x = -10$	$-5y = -10$
$\frac{2x}{2} = \frac{-10}{2}$	$\frac{-5y}{-5} = \frac{-10}{-5}$
$x = -5$	$y = 2$

The x-intercept is $(-5, 0)$. The y-intercept is $(0, 2)$.

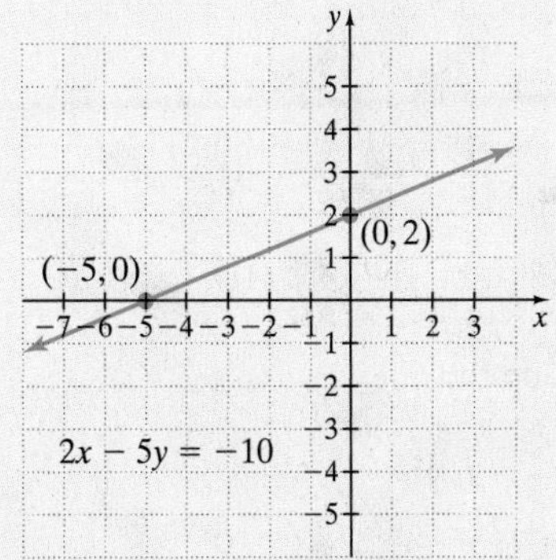

The graph of $x = c$ is a vertical line with x-intercept $(c, 0)$.

The graph of $y = c$ is a horizontal line with y-intercept $(0, c)$.

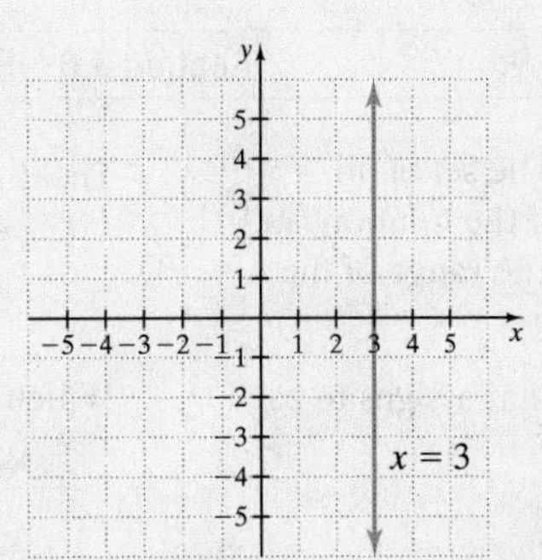

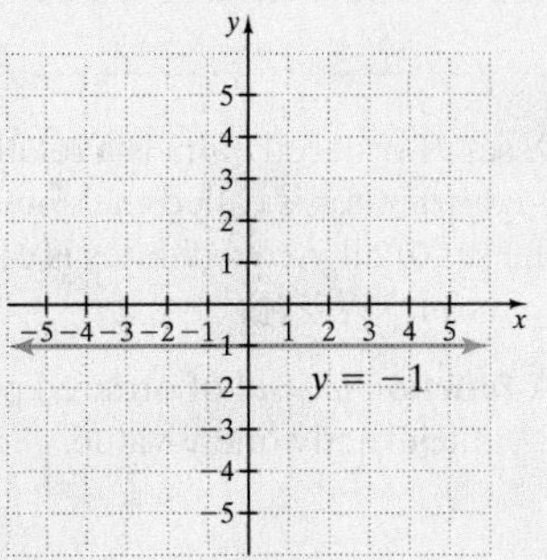

Section 3.4 Slope and Rate of Change

The **slope m** of the line through points (x_1, y_1) and (x_2, y_2) is given by

$$m = \frac{y_2 - y_1}{x_2 - x_1} \quad \text{as long as } x_2 \neq x_1$$

The slope of the line through points $(-1, 6)$ and $(-5, 8)$ is

$$m = \frac{y_2 - y_1}{x_2 - x_1} = \frac{8 - 6}{-5 - (-1)} = \frac{2}{-4} = -\frac{1}{2}$$

A horizontal line has slope 0.
The slope of a vertical line is undefined.

The slope of the line $y = -5$ is 0.
The line $x = 3$ has undefined slope.

Nonvertical parallel lines have the same slope.

Two nonvertical lines are perpendicular if the slope of one is the negative reciprocal of the slope of the other.

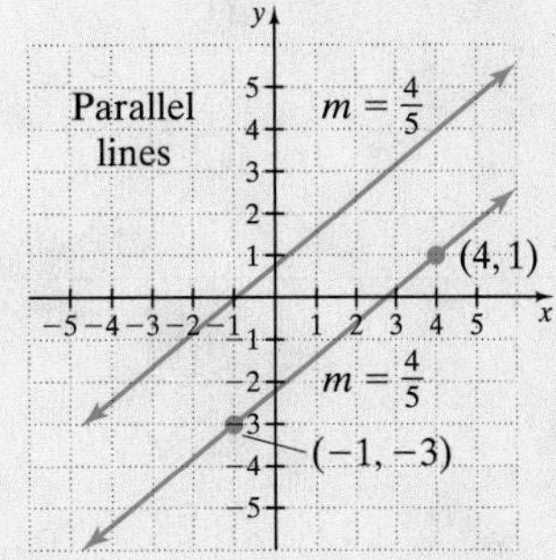

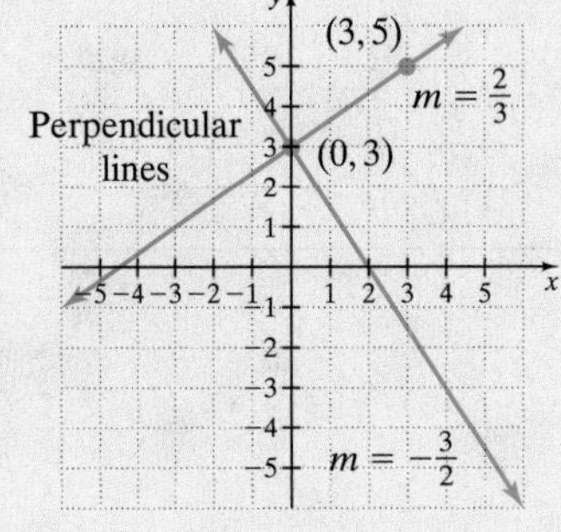

DEFINITIONS AND CONCEPTS	EXAMPLES
Section 3.5 Equations of Lines	
Slope-Intercept Form $$y = mx + b$$ m is the slope of the line. $(0, b)$ is the y-intercept.	Find the slope and the y-intercept of the line whose equation is $2x + 3y = 6$. Solve for y: $2x + 3y = 6$; $3y = -2x + 6$ Subtract $2x$.; $y = -\frac{2}{3}x + 2$ Divide by 3. The slope of the line is $-\frac{2}{3}$ and the y-intercept is $(0, 2)$. Find an equation of the line with slope 3 and y-intercept $(0, -1)$. The equation is $y = 3x - 1$.
Point-Slope Form $$y - y_1 = m(x - x_1)$$ m is the slope. (x_1, y_1) is a point on the line.	Find an equation of the line with slope $\frac{3}{4}$ that contains the point $(-1, 5)$. $y - 5 = \frac{3}{4}[x - (-1)]$; $4(y - 5) = 3(x + 1)$ Multiply by 4.; $4y - 20 = 3x + 3$ Distribute.; $-3x + 4y = 23$ Subtract $3x$ and add 20.
Section 3.6 Functions	
A set of ordered pairs is a **relation.** The set of all x-coordinates is called the **domain** of the relation, and the set of all y-coordinates is called the **range** of the relation.	The domain of the relation $\{(0, 5), (2, 5), (4, 5), (5, -2)\}$ is $\{0, 2, 4, 5\}$. The range is $\{-2, 5\}$.
A **function** is a set of ordered pairs that assigns to each x-value exactly one y-value. ***Vertical Line Test*** If a vertical line can be drawn so that it intersects a graph more than once, the graph is not the graph of a function.	Which are graphs of functions? 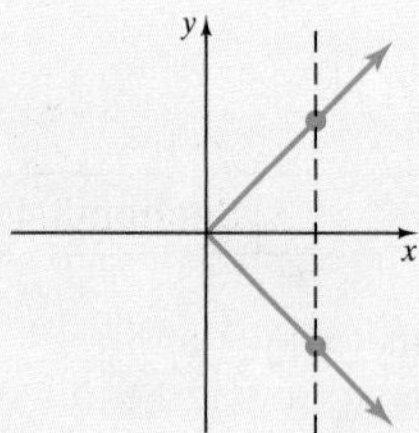 This graph is not the graph of a function. 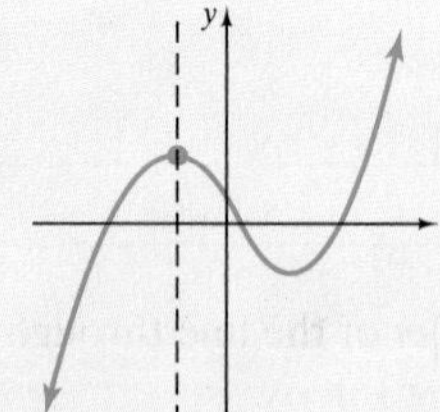 This graph is the graph of a function.
The symbol $f(x)$ means **function of *x*.** This notation is called **function** notation.	If $f(x) = 2x^2 + 6x - 1$, find $f(3)$. $f(3) = 2(3)^2 + 6 \cdot 3 - 1$; $= 2 \cdot 9 + 18 - 1$; $= 18 + 18 - 1$; $= 35$

Chapter 3 Review

(3.1) *Plot the following ordered pairs on a Cartesian coordinate system.*

1. $(-7, 0)$ **2.** $\left(0, 4\frac{4}{5}\right)$

3. $(-2, -5)$ **4.** $(1, -3)$

5. $(0.7, 0.7)$ **6.** $(-6, 4)$

7. A local office supply store uses quantity pricing. The table shows the price per box of #10 security envelopes for different numbers of envelopes in a box purchased.

Price per Box of Envelopes (in dollars)	*Number of Envelopes in Box*
5.00	50
8.50	100
20.00	250
27.00	500

a. Write each paired data as an ordered pair of the form (price per box of envelopes, number of envelopes in box).

b. Create a scatter diagram of the paired data. Be sure to label the axes appropriately.

8. The table shows the annual overnight stays in national parks. (Data from National Park Service)

Year	*Overnight Stays in National Parks (in millions)*
2005	13.8
2006	13.6
2007	14.1
2008	13.9
2009	14.6
2010	14.6

a. Write each paired data as an ordered pair of the form (year, number of overnight stays).

b. Create a scatter diagram of the paired data. Be sure to label the axes properly.

Determine whether each ordered pair is a solution of the given equation.

9. $7x - 8y = 56$; $(0, 56), (8, 0)$

10. $-2x + 5y = 10$; $(-5, 0), (1, 1)$

11. $x = 13$; $(13, 5), (13, 13)$

12. $y = 2$; $(7, 2), (2, 7)$

Complete the ordered pairs so that each is a solution of the given equation.

13. $-2 + y = 6x$; $(7, \quad)$ **14.** $y = 3x + 5$; $(\quad, -8)$

Complete the table of values for each given equation; then plot the ordered pairs. Use a single coordinate system for each exercise.

15. $9 = -3x + 4y$

x	y
	0
	3
9	

16. $x = 2y$

x	y
	0
	5
	−5

The cost in dollars of producing x compact disc holders is given by $y = 5x + 2000$. Use this equation for Exercises 17 and 18.

17. Complete the following table.

x	y
1	
100	
1000	

18. Find the number of compact disc holders that can be produced for $6430.

(3.2) *Graph each linear equation.*

19. $x - y = 1$ **20.** $x + y = 6$

21. $x - 3y = 12$ **22.** $5x - y = -8$

23. $x = 3y$ **24.** $y = -2x$

25. $2x - 3y = 6$ **26.** $4x - 3y = 12$

(3.3) *Identify the intercepts.*

27.

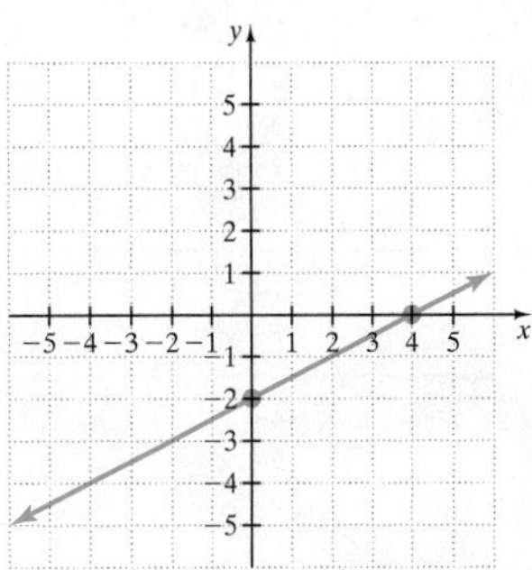

28.

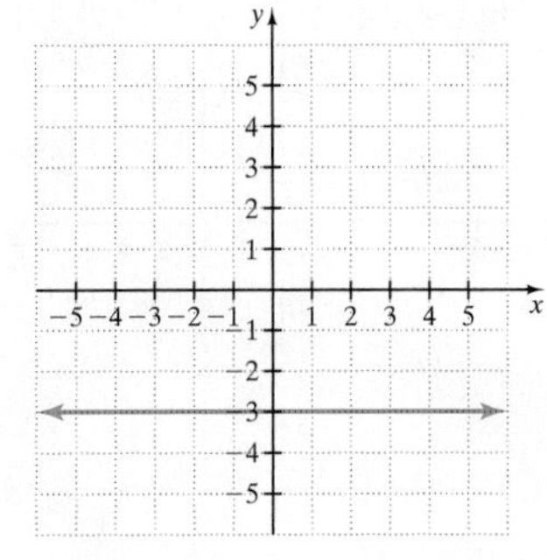

29.

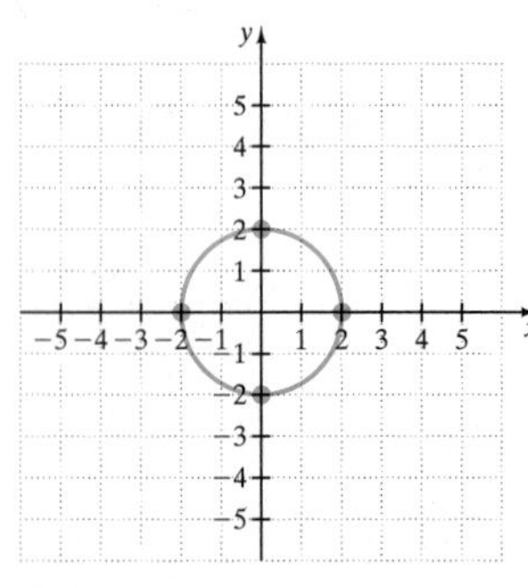

30.

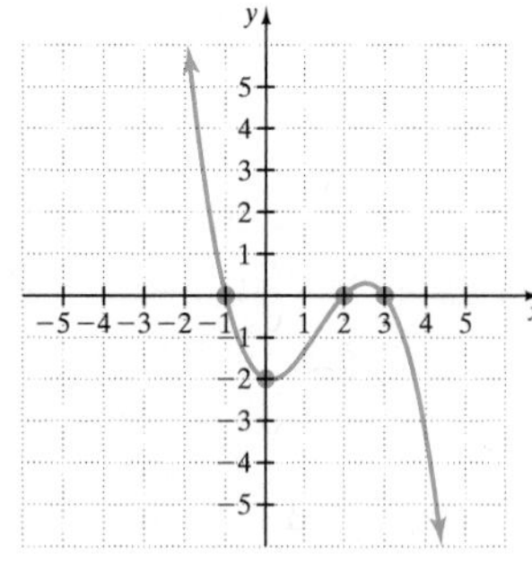

Graph each linear equation by finding its intercepts.

31. $x - 3y = 12$

32. $-4x + y = 8$

33. $y = -3$

34. $x = 5$

35. $y = -3x$

36. $x = 5y$

37. $x - 2 = 0$

38. $y + 6 = 0$

***(3.4)** Find the slope of each line.*

39.

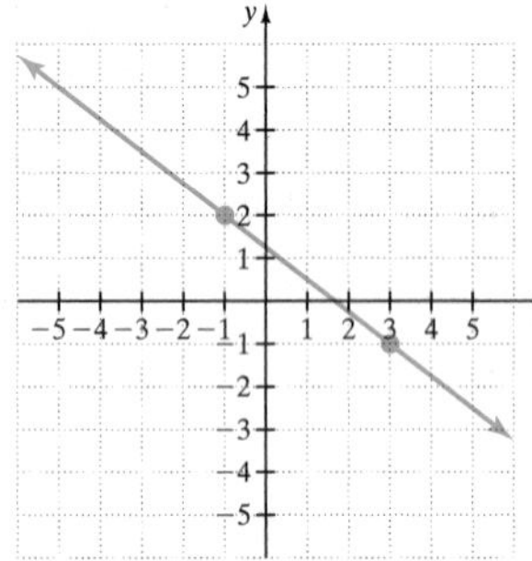

40.

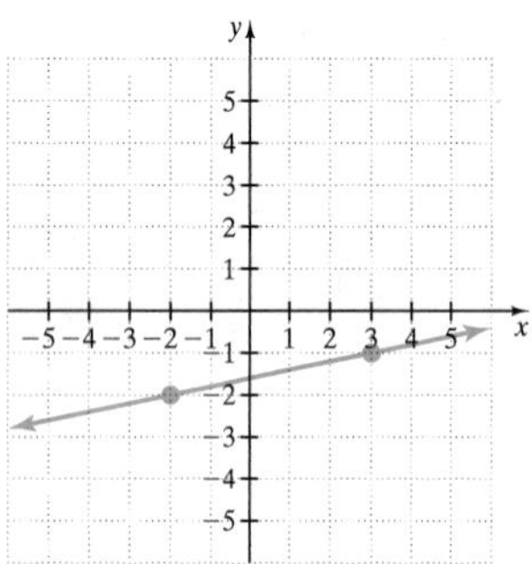

In Exercises 41 through 44, match each line with its slope.

a.

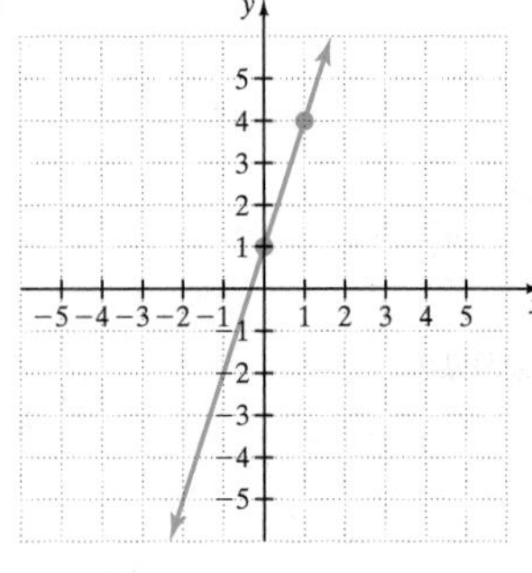

b.

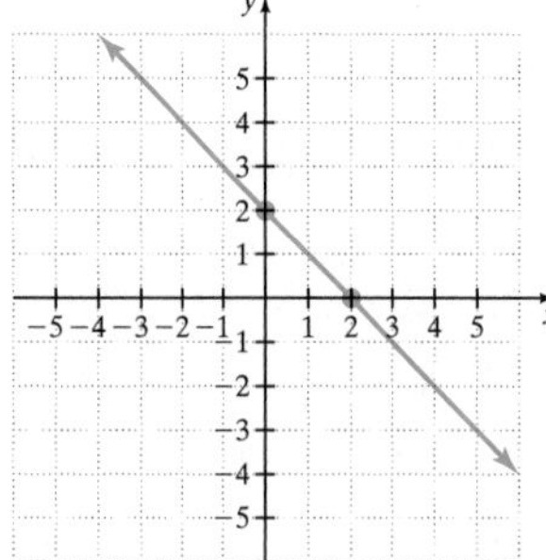

c.

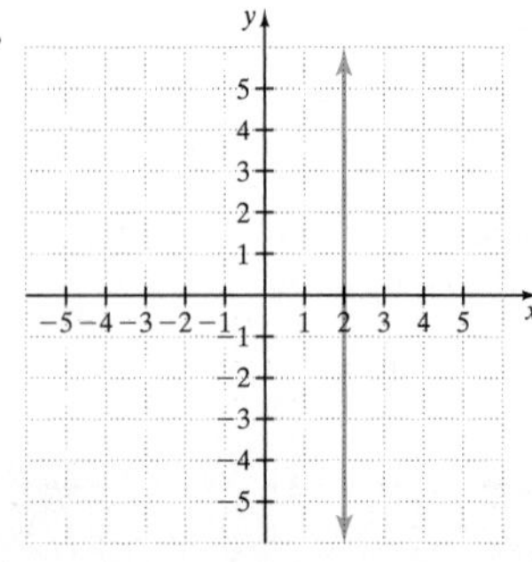

d.

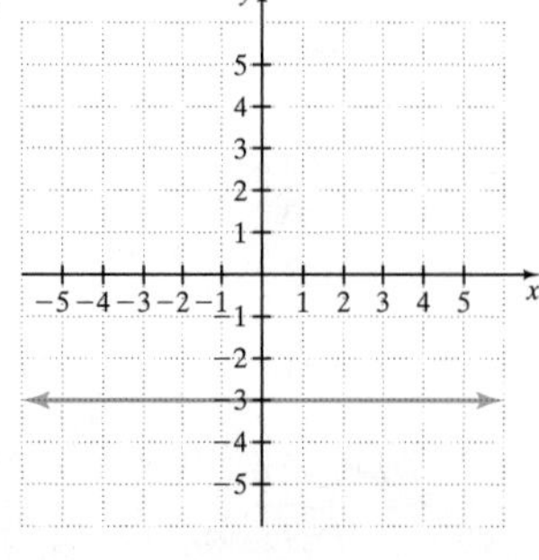

41. $m = 0$

42. $m = -1$

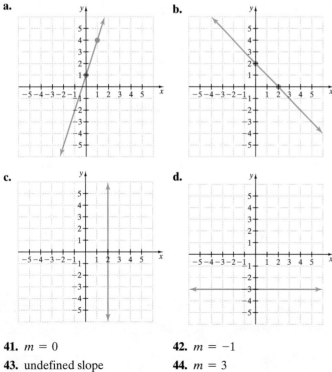

43. undefined slope

44. $m = 3$

Find the slope of the line that goes through the given points.

45. $(2, 5)$ and $(6, 8)$

46. $(4, 7)$ and $(1, 2)$

47. $(1, 3)$ and $(-2, -9)$

48. $(-4, 1)$ and $(3, -6)$

Find the slope of each line.

49. $y = 3x + 7$

50. $x - 2y = 4$

51. $y = -2$

52. $x = 0$

Determine whether each pair of lines is parallel, perpendicular, or neither.

53. $x - y = -6$
$x + y = 3$

54. $3x + y = 7$
$-3x - y = 10$

55. $y = 4x + \frac{1}{2}$
$4x + 2y = 1$

56. $x = 4$
$y = -2$

Find the slope of each line and write the slope as a rate of change. Don't forget to attach the proper units.

57. The graph below approximates the number of U.S. college students (in thousands) earning an associate's degree for each year x.

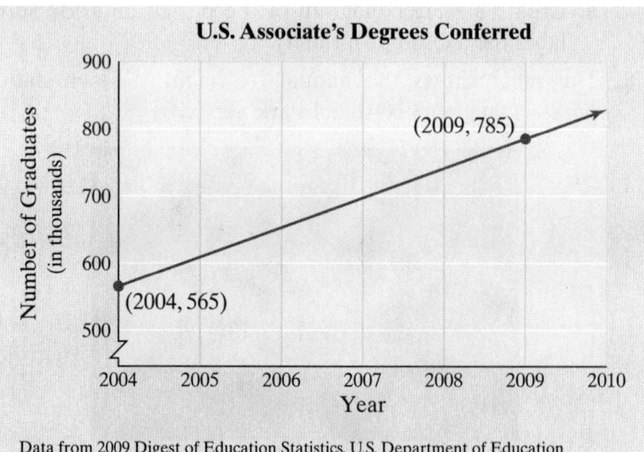

58. The graph below approximates the number of kidney transplants y in the United States for year x.

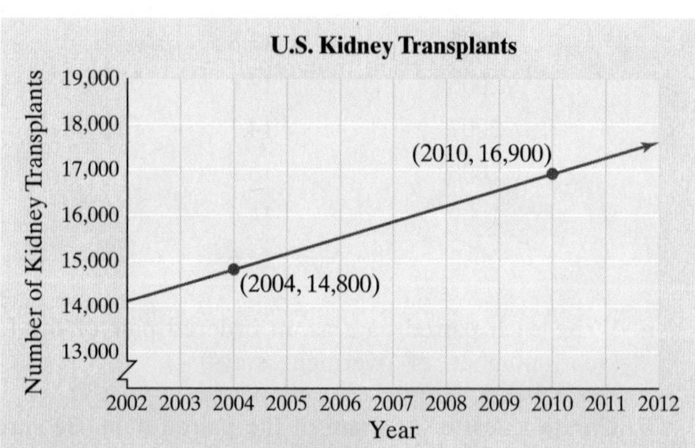

(3.5) Determine the slope and the y-intercept of the graph of each equation.

59. $3x + y = 7$ **60.** $x - 6y = -1$

61. $y = 2$ **62.** $x = -5$

Use the slope-intercept form to graph each equation.

63. $y = 3x - 1$ **64.** $y = -3x$

65. $5x - 3y = 15$ **66.** $-x + 2y = 8$

Write an equation of each line in slope-intercept form.

67. slope -5; y-intercept $\left(0, \frac{1}{2}\right)$

68. slope $\frac{2}{3}$; y-intercept $(0, 6)$

Match each equation with its graph.

69. $y = -4x$ **70.** $y = 2x$

71. $y = 2x - 1$ **72.** $y = -2x + 1$

a. 

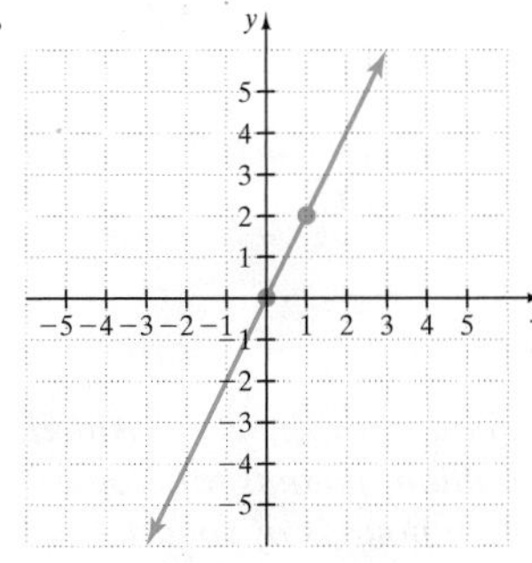

b.

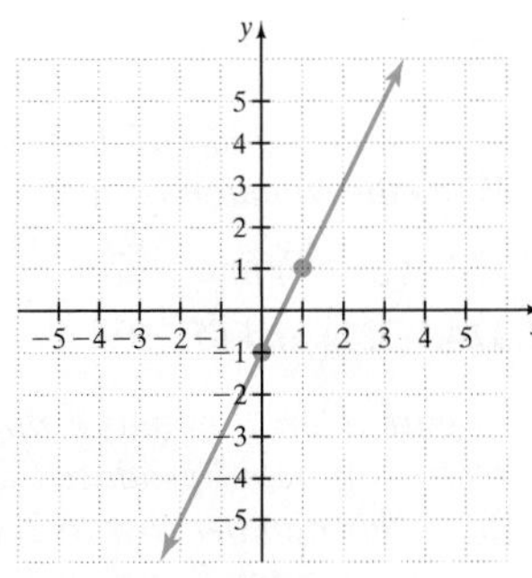

c.

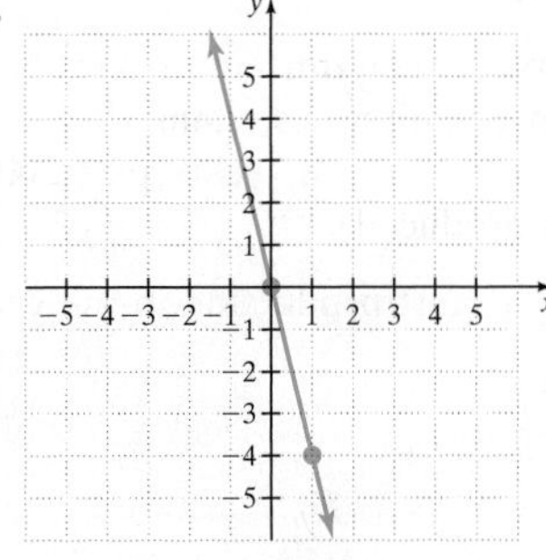

d.

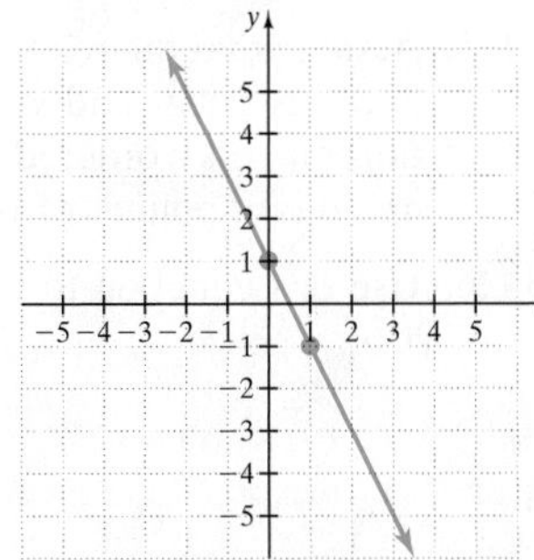

College is getting more expensive every year. The average cost for tuition and fees at a public two-year college y from 1995 through 2011 can be approximated by the linear equation $y = 56x, + 1859$, where x is the number of years after 1995. Use this information for Exercises 73 and 74. (Source: The College Board: Trends in College Pricing 2010)

73. Find the y-intercept of this equation.

74. What does the y-intercept mean?

Write an equation of each line. Write each equation in standard form, $Ax + By = C$, or $x = c$ or $y = c$ form.

75. With slope -3, through $(0, -5)$

76. With slope $\frac{1}{2}$, through $\left(0, -\frac{7}{2}\right)$

77. With slope 0, through $(-2, -3)$

78. With 0 slope, through the origin

79. With slope -6, through $(2, -1)$

80. With slope 12, through $\left(\frac{1}{2}, 5\right)$

81. Through $(0, 6)$ and $(6, 0)$

82. Through $(0, -4)$ and $(-8, 0)$

83. Vertical line, through $(5, 7)$

84. Horizontal line, through $(-6, 8)$

85. Through $(6, 0)$, perpendicular to $y = 8$

86. Through $(10, 12)$, perpendicular to $x = -2$

(3.6) Determine whether each of the following are functions.

87. $\{(7, 1), (7, 5), (2, 6)\}$

88. $\{(0, -1), (5, -1), (2, 2)\}$

89. $7x - 6y = 1$ **90.** $y = 7$

91. $x = 2$ **92.** $y = x^3$

93. **94.**

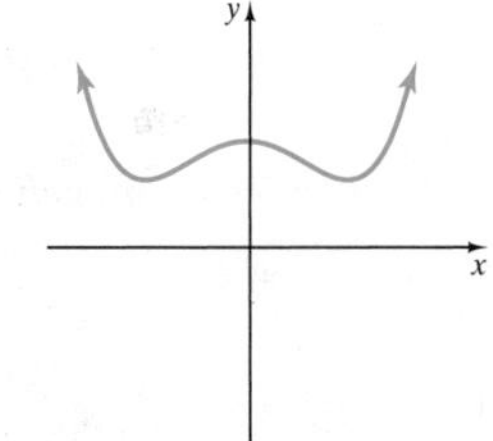

Given the following functions, find the indicated function values.

95. Given $f(x) = -2x + 6$, find

a. $f(0)$ **b.** $f(-2)$ **c.** $f\left(\frac{1}{2}\right)$

96. Given $h(x) = -5 - 3x$, find

a. $h(2)$ **b.** $h(-3)$ **c.** $h(0)$

97. Given $g(x) = x^2 + 12x$, find

a. $g(3)$ **b.** $g(-5)$ **c.** $g(0)$

98. Given $h(x) = 6 - |x|$, find

a. $h(-1)$ **b.** $h(1)$ **c.** $h(-4)$

Find the domain of each function.

99. $f(x) = 2x + 7$ **100.** $g(x) = \frac{7}{x - 2}$

Find the domain and the range of each function graphed.

101.

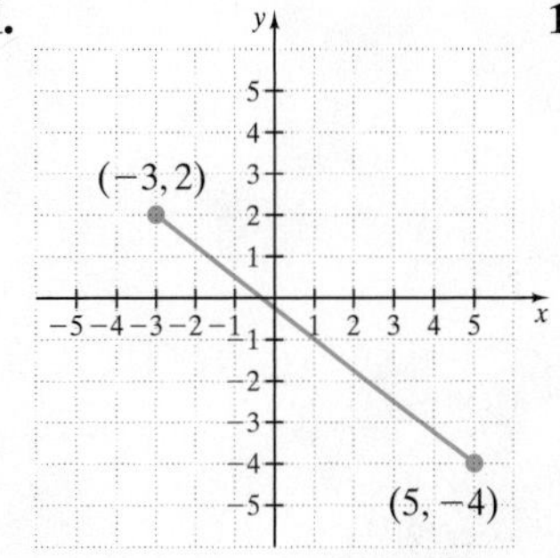

102.

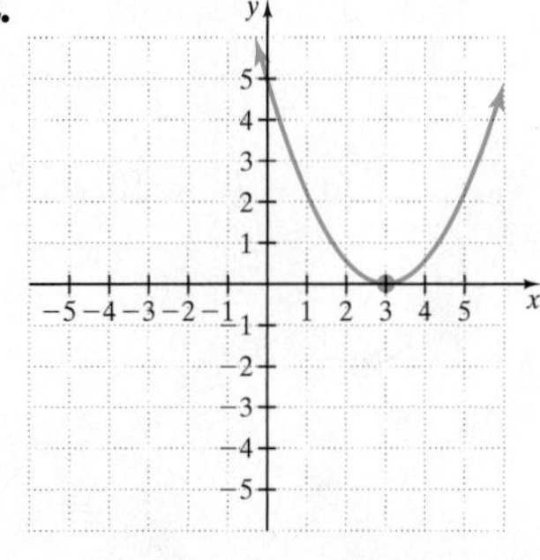

103.

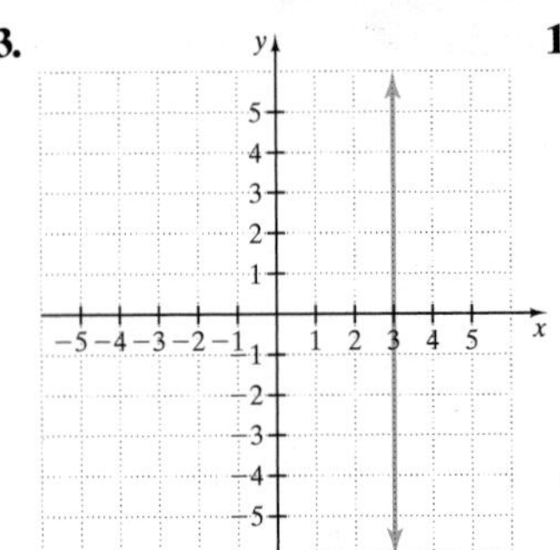

104.

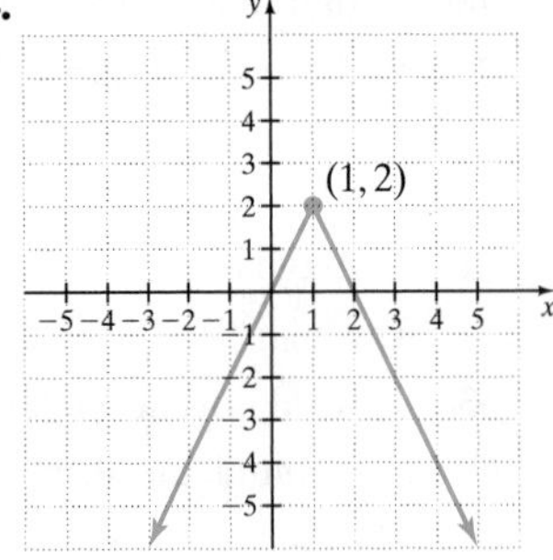

MIXED REVIEW

Complete the table of values for each given equation.

105. $2x - 5y = 9$

x	y
	1
2	
	−3

106. $x = -3y$

x	y
0	
	1
6	

Find the intercepts for each equation.

107. $2x - 3y = 6$

108. $-5x + y = 10$

Graph each linear equation.

109. $x - 5y = 10$

110. $x + y = 4$

111. $y = -4x$

112. $2x + 3y = -6$

113. $x = 3$

114. $y = -2$

Find the slope of the line that passes through each pair of points.

115. $(3, -5)$ and $(-4, 2)$

116. $(1, 3)$ and $(-6, -8)$

Find the slope of each line.

117.

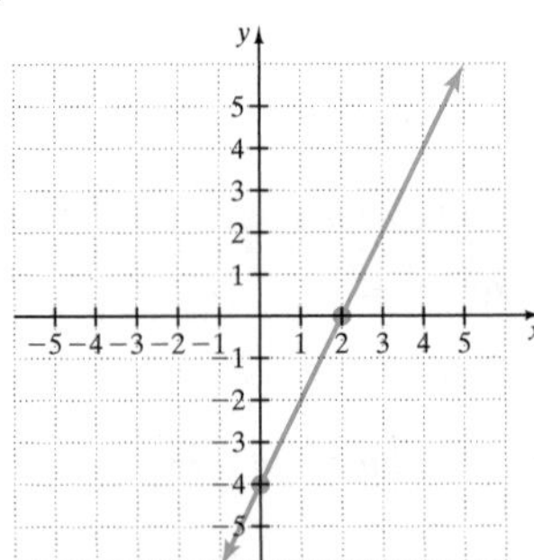

118.

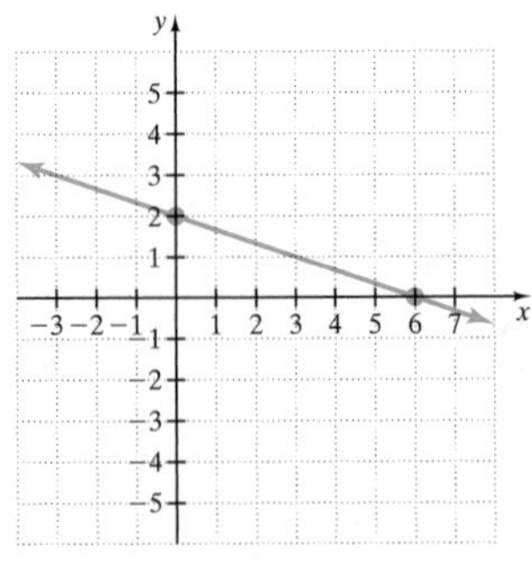

Determine the slope and y-intercept of the graph of each equation.

119. $-2x + 3y = -15$

120. $6x + y - 2 = 0$

Write an equation of the line with the given slope that passes through the given point. Write the equation in the form $Ax + By = C$.

121. $m = -5; (3, -7)$

122. $m = 3; (0, 6)$

Write an equation of the line passing through each pair of points. Write the equation in the form $Ax + By = C$.

123. $(-3, 9)$ and $(-2, 5)$

124. $(3, 1)$ and $(5, -9)$

Yogurt is an ever more popular food item. In 2002, American Dairy affiliates produced 2134 million pounds of yogurt. In 2009, this number rose to 3800 million pounds of yogurt. Use this information for Exercises 125 and 126.

125. Assume that the relationship between time and yogurt produced is linear and write an equation describing this relationship. Use ordered pairs of the form (years past 2002, millions of pounds of yogurt produced).

126. Use this equation to predict yogurt production in the year 2014.

Chapter 3 Test

MyMathLab®

YouTube

Graph the following.

1. $y = \frac{1}{2}x$

2. $2x + y = 8$

3. $5x - 7y = 10$

4. $y = -1$

5. $x - 3 = 0$

For Exercises 6 through 10, find the slopes of the lines.

6.

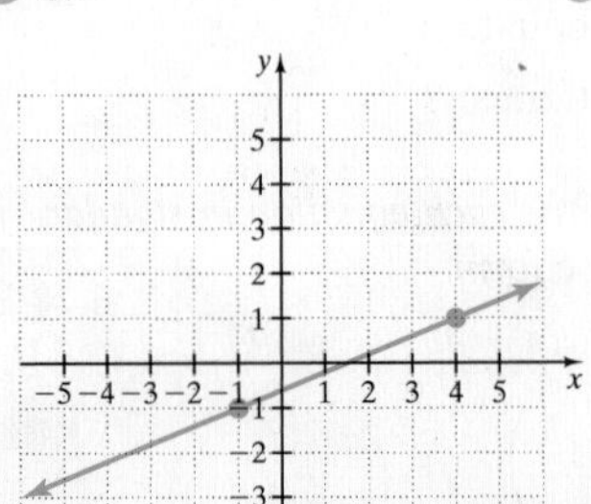

7.

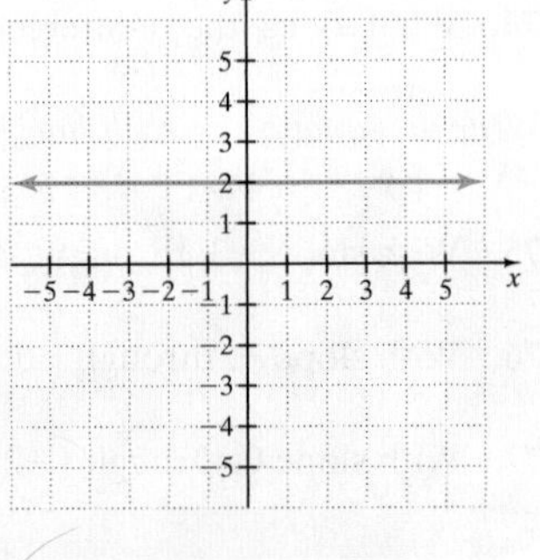

8. Through $(6, -5)$ and $(-1, 2)$

9. $-3x + y = 5$

10. $x = 6$

11. Determine the slope and the y-intercept of the graph of $7x - 3y = 2$.

12. Determine whether the graphs of $y = 2x - 6$ and $-4x = 2y$ are parallel lines, perpendicular lines, or neither.

Find equations of the following lines. Write the equation in standard form.

13. With slope of $-\frac{1}{4}$, through $(2, 2)$

14. Through the origin and $(6, -7)$

15. Through $(2, -5)$ and $(1, 3)$

16. Through $(-5, -1)$ and parallel to $x = 7$

17. With slope $\frac{1}{8}$ and y-intercept $(0, 12)$

Determine whether each graph is the graph of a function.

18.  **19.**

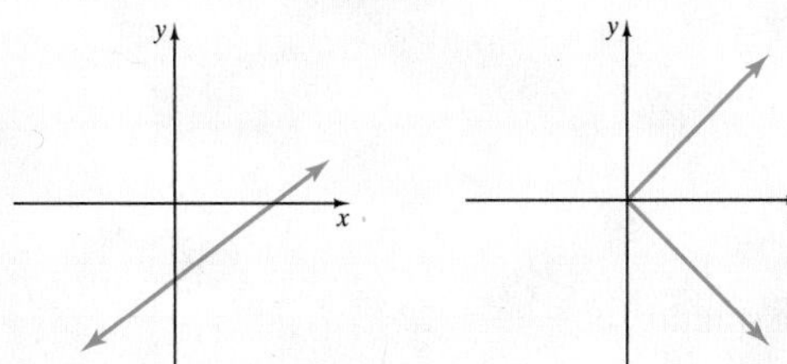

Given the following functions, find the indicated function values.

20. $h(x) = x^3 - x$

a. $h(-1)$ **b.** $h(0)$ **c.** $h(4)$

21. Find the domain of $y = \frac{1}{x + 1}$.

For Exercises 22 and 23,

***a.** Identify the x- and y-intercepts.*

***b.** Find the domain and the range of each function graphed.*

22. **23.**

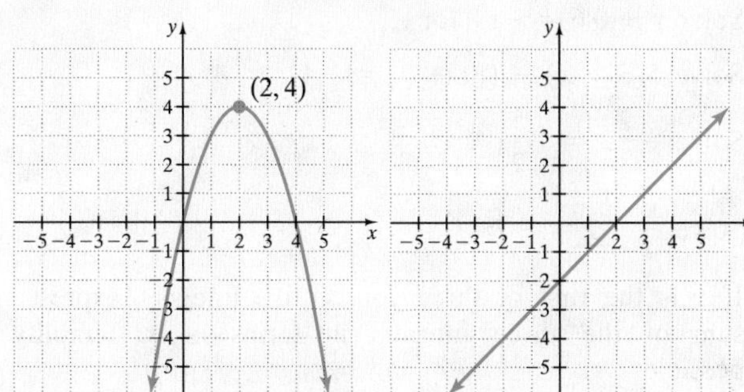

24. If $f(7) = 20$, write the corresponding ordered pair.

Use the bar graph below to answer Exercises 25 and 26.

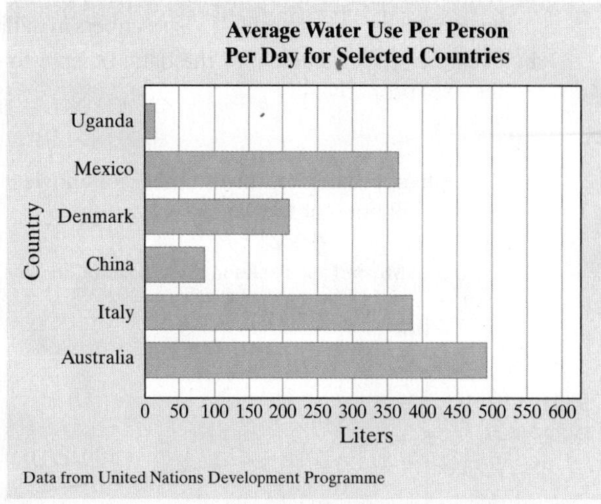

25. Estimate the average water use per person per day in Denmark.

26. Estimate the average water use per person per day in Australia.

Use this graph to answer Exercises 27 through 29.

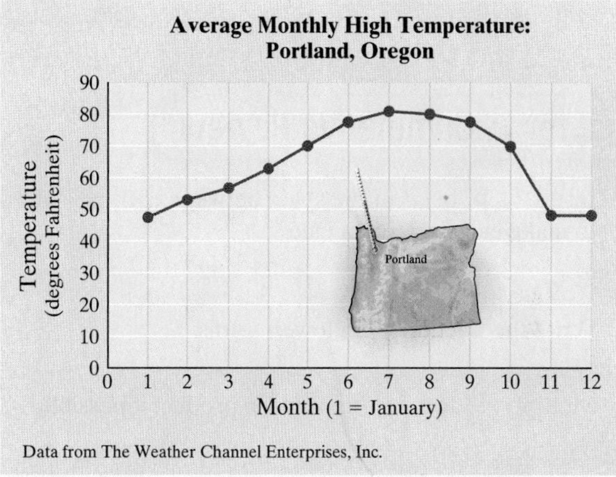

27. During what month is the average high temperature the greatest?

28. Approximate the average high temperature for the month of April.

29. During what month(s) is the average high temperature below 60°F?

30. The table gives the number of basic cable TV subscribers (in millions) for the years shown. (Data from National Cable and Telecommunications Association)

Year	*Basic Cable TV Subscribers (in millions)*
2003	66.0
2004	65.4
2005	65.4
2006	65.6
2007	64.9
2008	63.7
2009	62.1

a. Write this data as a set of ordered pairs of the form (year, number of basic cable TV subscribers in millions).

b. Create a scatter diagram of the data. Be sure to label the axes properly.

31. This graph approximates the number of movie ticket sales y (in millions) for the year x.

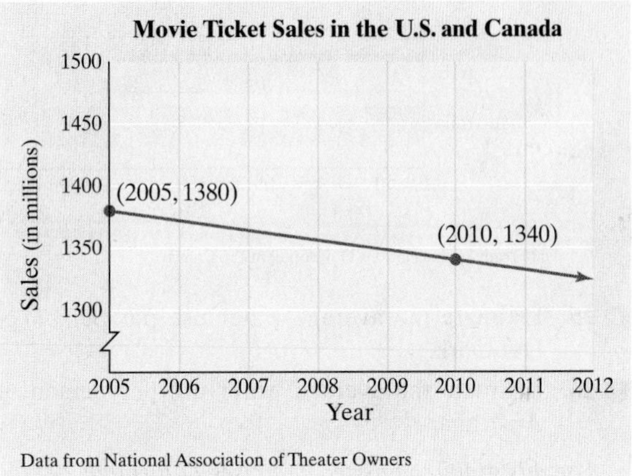

a. Find the slope of the line and write the slope as a rate of change. Don't forget to attach the proper units.

b. Write two ordered pairs of the form (years past 2005, number of tickets sold in millions).

c. Use the two ordered pairs from part (b) to write a linear equation. Write the equation in slope-intercept form.

d. Use the equation from part (c) to predict the number of movie tickets sold in 2015.

Chapter 3 Cumulative Review

1. Insert <, >, or = in the space between each pair of numbers to make each statement true.

a. 2 3 **b.** 7 4 **c.** 72 27

2. Write the fraction $\frac{56}{64}$ in lowest terms.

3. Multiply $\frac{2}{15}$ and $\frac{5}{13}$. Simplify the product if possible.

4. Add: $\frac{10}{3} + \frac{5}{21}$

5. Simplify: $\frac{3 + |4 - 3| + 2^2}{6 - 3}$

6. Simplify: $16 - 3 \cdot 3 + 2^4$

7. Add.

a. $-8 + (-11)$ **b.** $-5 + 35$

c. $0.6 + (-1.1)$ **d.** $-\frac{7}{10} + \left(-\frac{1}{10}\right)$

e. $11.4 + (-4.7)$ **f.** $-\frac{3}{8} + \frac{2}{5}$

8. Simplify: $|9 + (-20)| + |-10|$

9. Simplify each expression.

a. $-14 - 8 + 10 - (-6)$

b. $1.6 - (-10.3) + (-5.6)$

10. Simplify: $-9 - (3 - 8)$

11. If $x = -2$ and $y = -4$, evaluate each expression.

a. $5x - y$ **b.** $x^4 - y^2$

c. $\frac{3x}{2y}$

12. Is -20 a solution of $\frac{x}{-10} = 2$?

13. Simplify each expression.

a. $10 + (x + 12)$

b. $-3(7x)$

14. Simplify: $(12 + x) - (4x - 7)$

15. Identify the numerical coefficient in each term.

a. $-3y$ **b.** $22z^4$

c. y **d.** $-x$

e. $\frac{x}{7}$

16. Multiply: $-5(x - 7)$

17. Solve $y + 0.6 = -1.0$ for y.

18. Solve: $5(3 + z) - (8z + 9) = -4$

19. Solve: $-\frac{2}{3}x = -\frac{5}{2}$

20. Solve: $\frac{x}{4} - 1 = -7$

21. If x is the first of three consecutive integers, express the sum of the three integers in terms of x. Simplify if possible.

22. Solve: $\frac{x}{3} - 2 = \frac{x}{3}$

23. Solve: $\frac{2(a + 3)}{3} = 6a + 2$

24. Solve: $x + 2y = 6$ for y.

25. The 112th Congress began on January 3, 2011, and had a total of 435 Democratic and Republican representatives. There were 49 fewer Democratic representatives than Republican. Find the number of representatives from each party. (*Source*: Wikipedia.com).

26. Solve $5(x + 4) \geq 4(2x + 3)$. Write the solution set in interval notation.

27. Charles Pecot can afford enough fencing to enclose a rectangular garden with a perimeter of 140 feet. If the width of his garden must be 30 feet, find the length.

28. Solve $-3 < 4x - 1 \leq 2$. Write the solution set in interval notation.

29. Solve $y = mx + b$ for x.

30. Complete the table for $y = -5x$.

x	y
0	
-1	
	-10

31. A chemist working on his doctoral degree at Massachusetts Institute of Technology needs 12 liters of a 50% acid solution for a lab experiment. The stockroom has only 40% and 70% solutions. How much of each solution should be mixed together to form 12 liters of a 50% solution?

32. Graph: $y = -3x + 5$

33. Graph $x \geq -1$. Then write the solutions in interval notation.

34. Find the x- and y-intercepts of $2x + 4y = -8$.

35. Solve $-1 \leq 2x - 3 < 5$. Graph the solution set and write it in interval notation.

36. Graph $x = 2$ on a rectangular coordinate system.

37. Determine whether each ordered pair is a solution of the equation $x - 2y = 6$.

a. $(6, 0)$ **b.** $(0, 3)$

c. $\left(1, -\frac{5}{2}\right)$

38. Find the slope of the line through $(0, 5)$ and $(-5, 4)$.

39. Determine whether each equation is a linear equation in two variables.

a. $x - 1.5y = -1.6$ **b.** $y = -2x$

c. $x + y^2 = 9$ **d.** $x = 5$

40. Find the slope of the line $x = -10$.

41. Find the slope of the line $y = -1$.

42. Find the slope and y-intercept of the line whose equation is $2x - 5y = 10$.

43. Find an equation of the line with y-intercept $(0, -3)$ and slope of $\frac{1}{4}$.

44. Write an equation of the line through $(2, 3)$ and $(0, 0)$. Write the equation in standard form.

CHAPTER 4

Solving Systems of Linear Equations and Inequalities

In Chapter 3, we graphed equations containing two variables. Equations like these are often needed to represent relationships between two different values. There are also many real-life opportunities to compare and contrast two such equations, called a system of equations. This chapter presents linear systems and ways we solve these systems and apply them to real-life situations.

According to the Bureau of Labor Statistics, workers age 55 years and over are projected to have the greatest increase in the labor force during the years 2008 to 2018. These workers are anticipated to increase from 18.1% to 23.9% of the labor force during this period. Workers age 25 to 34 years are projected to increase also, but not at the same rate. In Section 4.3, Exercise 73, we will predict when these two age groups have the same percent of the labor force.

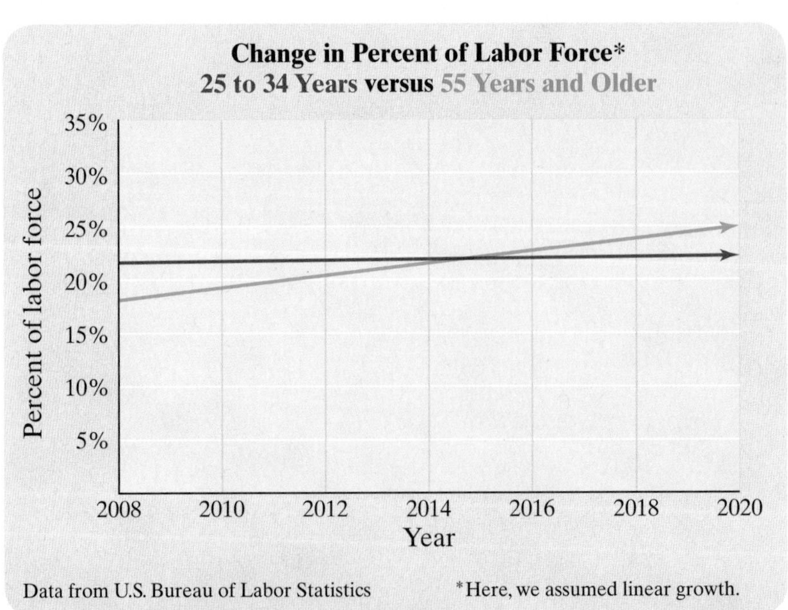

4.1 Solving Systems of Linear Equations by Graphing 

OBJECTIVES

1. Determine if an Ordered Pair Is a Solution of a System of Equations in Two Variables.
2. Solve a System of Linear Equations by Graphing.
3. Without Graphing, Determine the Number of Solutions of a System.

A **system of linear equations** consists of two or more linear equations. In this section, we focus on solving systems of linear equations containing two equations in two variables. Examples of such linear systems are

$$\begin{cases} 3x - 3y = 0 \\ x = 2y \end{cases} \quad \begin{cases} x - y = 0 \\ 2x + y = 10 \end{cases} \quad \begin{cases} y = 7x - 1 \\ y = 4 \end{cases}$$

OBJECTIVE

1 Deciding Whether an Ordered Pair is a Solution 

A **solution** of a system of two equations in two variables is an ordered pair of numbers that is a solution of both equations in the system.

EXAMPLE 1 Determine whether (12, 6) is a solution of the system

$$\begin{cases} 2x - 3y = 6 \\ x = 2y \end{cases}$$

Solution To determine whether (12, 6) is a solution of the system, we replace x with 12 and y with 6 in both equations.

$2x - 3y = 6$	First equation	$x = 2y$	Second equation
$2(12) - 3(6) \stackrel{?}{=} 6$	Let $x = 12$ and $y = 6$.	$12 \stackrel{?}{=} 2(6)$	Let $x = 12$ and $y = 6$.
$24 - 18 \stackrel{?}{=} 6$	Simplify.	$12 = 12$	True
$6 = 6$	True		

Since (12, 6) is a solution of both equations, it is a solution of the system. □

PRACTICE

1 Determine whether (4, 12) is a solution of the system.

$$\begin{cases} 4x - y = 2 \\ y = 3x \end{cases}$$

EXAMPLE 2 Determine whether $(-1, 2)$ is a solution of the system

$$\begin{cases} x + 2y = 3 \\ 4x - y = 6 \end{cases}$$

Solution We replace x with -1 and y with 2 in both equations.

$x + 2y = 3$	First equation	$4x - y = 6$	Second equation
$-1 + 2(2) \stackrel{?}{=} 3$	Let $x = -1$ and $y = 2$.	$4(-1) - 2 \stackrel{?}{=} 6$	Let $x = -1$ and $y = 2$.
$-1 + 4 \stackrel{?}{=} 3$	Simplify.	$-4 - 2 \stackrel{?}{=} 6$	Simplify.
$3 = 3$	True	$-6 = 6$	False

$(-1, 2)$ is not a solution of the second equation, $4x - y = 6$, so it is not a solution of the system. □

PRACTICE

2 Determine whether $(-4, 1)$ is a solution of the system.

$$\begin{cases} x - 3y = -7 \\ 2x + 9y = 1 \end{cases}$$

OBJECTIVE

2 Solving Systems of Equations by Graphing 

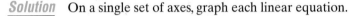

Since a solution of a system of two equations in two variables is a solution common to both equations, it is also a point common to the graphs of both equations. Let's practice finding solutions of both equations in a system—that is, solutions of a system—by graphing and identifying points of intersection.

EXAMPLE 3 Solve the system of equations by graphing.

$$\begin{cases} -x + 3y = 10 \\ x + y = 2 \end{cases}$$

Solution On a single set of axes, graph each linear equation.

$-x + 3y = 10$

x	y
0	$\frac{10}{3}$
−4	2
2	4

$x + y = 2$

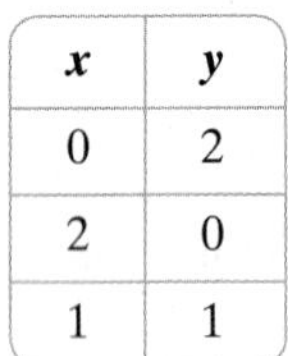

x	y
0	2
2	0
1	1

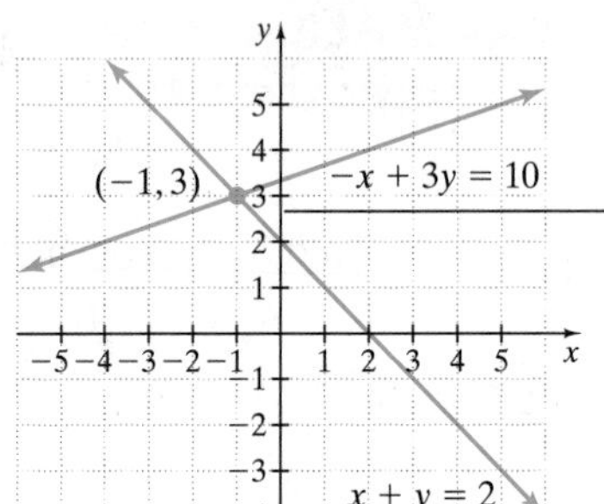

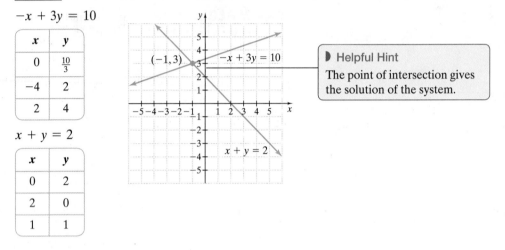

Helpful Hint
The point of intersection gives the solution of the system.

The two lines appear to intersect at the point $(-1, 3)$. To check, we replace x with -1 and y with 3 in both equations.

$-x + 3y = 10$	First equation	$x + y = 2$	Second equation
$-(-1) + 3(3) \stackrel{?}{=} 10$	Let $x = -1$ and $y = 3$.	$-1 + 3 \stackrel{?}{=} 2$	Let $x = -1$ and $y = 3$.
$1 + 9 \stackrel{?}{=} 10$	Simplify.	$2 = 2$	True
$10 = 10$	True		

$(-1, 3)$ checks, so it is the solution of the system. □

PRACTICE

3 Solve the system of equations by graphing:

$$\begin{cases} x - y = 3 \\ x + 2y = 18 \end{cases}$$

Helpful Hint
Neatly drawn graphs can help when you are estimating the solution of a system of linear equations by graphing.

In the example above, notice that the two lines intersected in a point. This means that the system has 1 solution.

EXAMPLE 4 Solve the system of equations by graphing.

$$\begin{cases} 2x + 3y = -2 \\ x = 2 \end{cases}$$

Solution We graph each linear equation on a single set of axes.

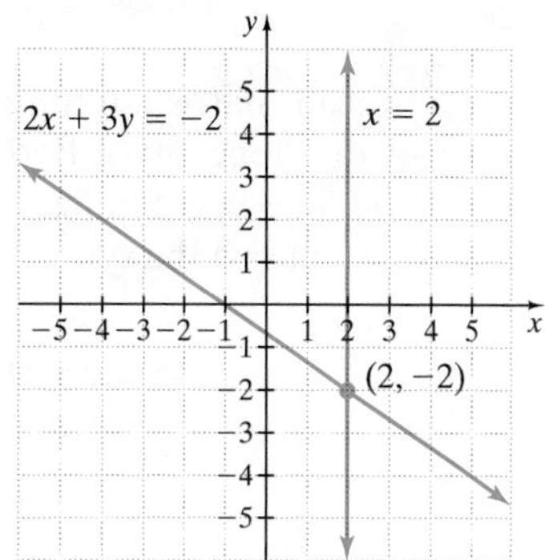

The two lines appear to intersect at the point $(2, -2)$. To determine whether $(2, -2)$ is the solution, we replace x with 2 and y with -2 in both equations.

$2x + 3y = -2$	First equation	$x = 2$	Second equation
$2(2) + 3(-2) \stackrel{?}{=} -2$	Let $x = 2$ and $y = -2$.	$2 \stackrel{?}{=} 2$	Let $x = 2$.
$4 + (-6) \stackrel{?}{=} -2$	Simplify.	$2 = 2$	True
$-2 = -2$	True		

Since a true statement results in both equations, $(2, -2)$ is the solution of the system. □

PRACTICE 4 Solve the system of equations by graphing.

$$\begin{cases} -4x + 3y = -3 \\ y = -5 \end{cases}$$

A system of equations that has at least one solution as in Examples 3 and 4 is said to be a **consistent system.** A system that has no solution is said to be an **inconsistent system.**

EXAMPLE 5 Solve the following system of equations by graphing.

$$\begin{cases} 2x + y = 7 \\ 2y = -4x \end{cases}$$

Solution Graph the two lines in the system.

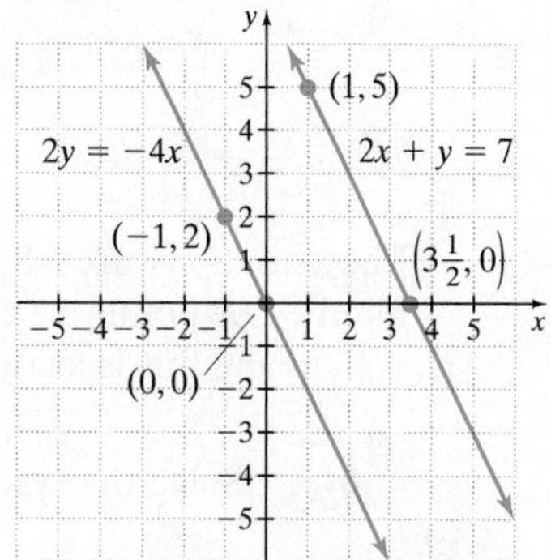

The lines **appear** to be parallel. To confirm this, write both equations in slope-intercept form by solving each equation for y.

$2x + y = 7$ First equation

$\mathbf{y = -2x + 7}$ Subtract $2x$ from both sides.

$2y = -4x$ Second equation

$\frac{2y}{2} = \frac{-4x}{2}$ Divide both sides by 2.

$\mathbf{y = -2x}$

Recall that when an equation is written in slope-intercept form, the coefficient of x is the slope. Since both equations have the same slope, -2, but different y-intercepts, the lines are parallel and have no points in common. Thus, there is no solution of the system and the system is inconsistent. □

PRACTICE
5 Solve the system of equations by graphing.

$$\begin{cases} 3y = 9x \\ 6x - 2y = 12 \end{cases}$$

In Examples 3, 4, and 5, the graphs of the two linear equations of each system are different. When this happens, we call these equations **independent equations.** If the graphs of the two equations in a system are identical, we call the equations **dependent equations.**

EXAMPLE 6 Solve the system of equations by graphing.

$$\begin{cases} x - y = 3 \\ -x + y = -3 \end{cases}$$

Solution Graph each line.

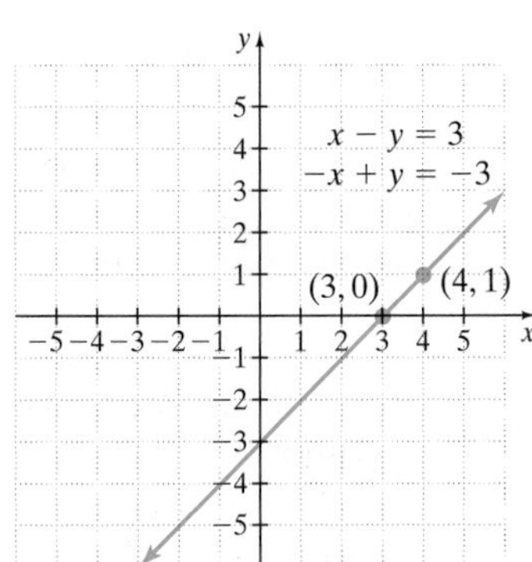

These graphs **appear** to be identical. To confirm this, write each equation in slope-intercept form.

$x - y = 3$ First equation

$-y = -x + 3$ Subtract x from both sides.

$\frac{-y}{-1} = \frac{-x}{-1} + \frac{3}{-1}$ Divide both sides by -1.

$\mathbf{y = x - 3}$

$-x + y = -3$ Second equation

$\mathbf{y = x - 3}$ Add x to both sides.

The equations are identical and so must be their graphs. The lines have an infinite number of points in common. Thus, there is an infinite number of solutions of the system and this is a consistent system. The equations are dependent equations. □

PRACTICE
6 Solve the system of equations by graphing.

$$\begin{cases} x - y = 4 \\ -2x + 2y = -8 \end{cases}$$

As we have seen, three different situations can occur when graphing the two lines associated with the two equations in a linear system:

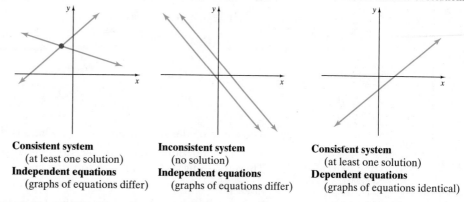

One point of intersection: one solution

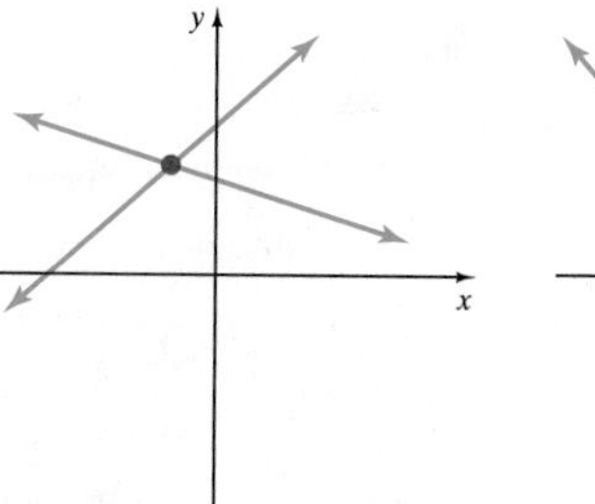

Consistent system
(at least one solution)
Independent equations
(graphs of equations differ)

Parallel lines: no solution

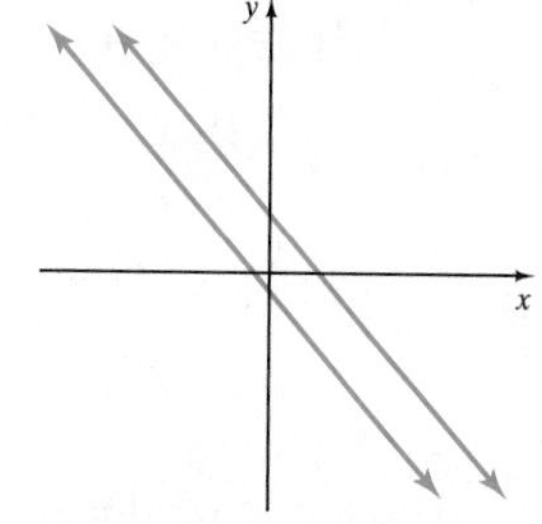

Inconsistent system
(no solution)
Independent equations
(graphs of equations differ)

Same line: infinite number of solutions

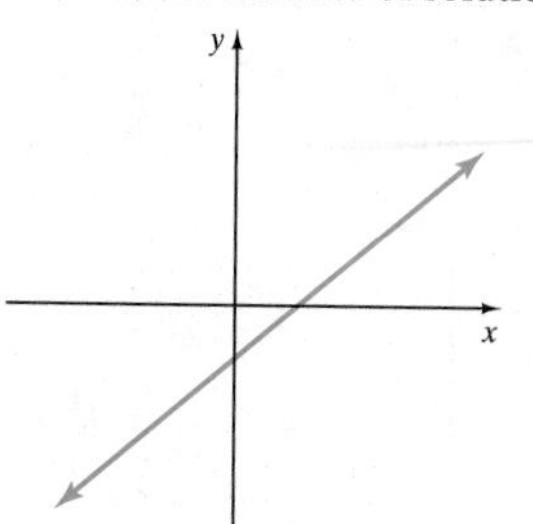

Consistent system
(at least one solution)
Dependent equations
(graphs of equations identical)

OBJECTIVE

3 Finding the Number of Solutions of a System without Graphing

You may have suspected by now that graphing alone is not an accurate way to solve a system of linear equations. For example, a solution of $\left(\frac{1}{2}, \frac{2}{9}\right)$ is unlikely to be read correctly from a graph. The next two sections present two accurate methods of solving these systems. In the meantime, we can decide how many solutions a system has by writing each equation in the slope-intercept form.

EXAMPLE 7 Without graphing, determine the number of solutions of the system.

$$\begin{cases} \frac{1}{2}x - y = 2 \\ x = 2y + 5 \end{cases}$$

Solution First write each equation in slope-intercept form.

$\frac{1}{2}x - y = 2$ First equation

$\frac{1}{2}x = y + 2$ Add y to both sides.

$\frac{1}{2}x - 2 = y$ Subtract 2 from both sides.

$x = 2y + 5$ Second equation

$x - 5 = 2y$ Subtract 5 from both sides.

$\frac{x}{2} - \frac{5}{2} = \frac{2y}{2}$ Divide both sides by 2.

$\frac{1}{2}x - \frac{5}{2} = y$ Simplify.

The slope of each line is $\frac{1}{2}$, but they have different y-intercepts. This tells us that the lines representing these equations are parallel. Since the lines are parallel, the system has no solution and is inconsistent. □

PRACTICE

7 Without graphing, determine the number of solutions of the system.

$$\begin{cases} 5x + 4y = 6 \\ x - y = 3 \end{cases}$$

EXAMPLE 8 Without graphing, determine the number of solutions of the system.

$$\begin{cases} 3x - y = 4 \\ x + 2y = 8 \end{cases}$$

Solution Once again, the slope-intercept form helps determine how many solutions this system has.

$3x - y = 4$	First equation	$x + 2y = 8$	Second equation
$3x = y + 4$	Add y to both sides.	$x = -2y + 8$	Subtract $2y$ from both sides.
$3x - 4 = y$	Subtract 4 from both sides.	$x - 8 = -2y$	Subtract 8 from both sides.
		$\frac{x}{-2} - \frac{8}{-2} = \frac{-2y}{-2}$	Divide both sides by -2.
		$-\frac{1}{2}x + 4 = y$	Simplify.

The slope of the second line is $-\frac{1}{2}$, whereas the slope of the first line is 3. Since the slopes are not equal, the two lines are neither parallel nor identical and must intersect. Therefore, this system has one solution and is consistent. □

PRACTICE 8 Without graphing, determine the number of solutions of the system.

$$\begin{cases} -\frac{2}{3}x + y = 6 \\ 3y = 2x + 5 \end{cases}$$

Graphing Calculator Explorations

A graphing calculator may be used to approximate solutions of systems of equations. For example, to approximate the solution of the system

$$\begin{cases} y = -3.14x - 1.35 \\ y = 4.88x + 5.25, \end{cases}$$

first graph each equation on the same set of axes. Then use the intersect feature of your calculator to approximate the point of intersection.

The approximate point of intersection is $(-0.82, 1.23)$.

Solve each system of equations. Approximate the solutions to two decimal places.

1. $\begin{cases} y = -2.68x + 1.21 \\ y = 5.22x - 1.68 \end{cases}$
2. $\begin{cases} y = 4.25x + 3.89 \\ y = -1.88x + 3.21 \end{cases}$
3. $\begin{cases} 4.3x - 2.9y = 5.6 \\ 8.1x + 7.6y = -14.1 \end{cases}$
4. $\begin{cases} -3.6x - 8.6y = 10 \\ -4.5x + 9.6y = -7.7 \end{cases}$

Vocabulary, Readiness & Video Check

Fill in each blank with one of the words or phrases listed below.

system of linear equations	solution	consistent
dependent	inconsistent	independent

1. In a system of linear equations in two variables, if the graphs of the equations are the same, the equations are ______________ equations.
2. Two or more linear equations are called a(n) ______________.
3. A system of equations that has at least one solution is called a(n) ______________ system.
4. A(n) ______________ of a system of two equations in two variables is an ordered pair of numbers that is a solution of both equations in the system.
5. A system of equations that has no solution is called a(n) ______________ system.
6. In a system of linear equations in two variables, if the graphs of the equations are different, the equations are ______________ equations.

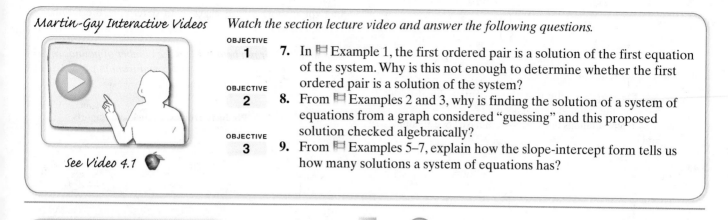

Martin-Gay Interactive Videos

See Video 4.1

Watch the section lecture video and answer the following questions.

OBJECTIVE 1

7. In Example 1, the first ordered pair is a solution of the first equation of the system. Why is this not enough to determine whether the first ordered pair is a solution of the system?

OBJECTIVE 2

8. From Examples 2 and 3, why is finding the solution of a system of equations from a graph considered "guessing" and this proposed solution checked algebraically?

OBJECTIVE 3

9. From Examples 5–7, explain how the slope-intercept form tells us how many solutions a system of equations has?

4.1 Exercise Set

MyMathLab®

Each rectangular coordinate system shows the graph of the equations in a system of equations. Use each graph to determine the number of solutions for each associated system. If the system has only one solution, give its coordinates.

1.

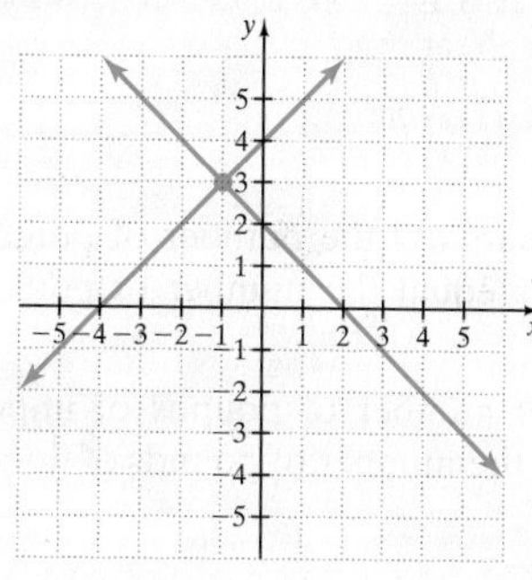

2.

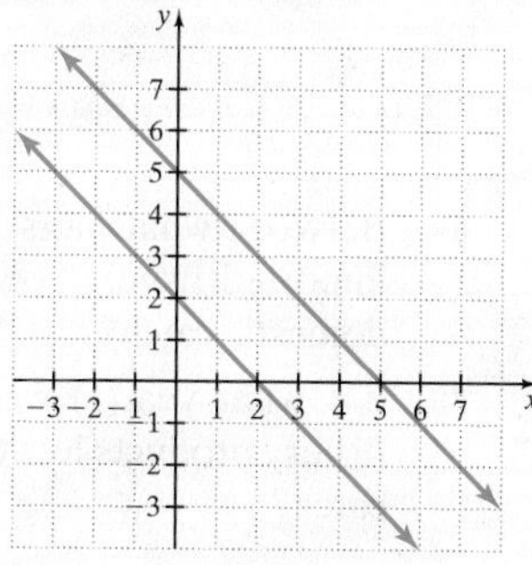

3.

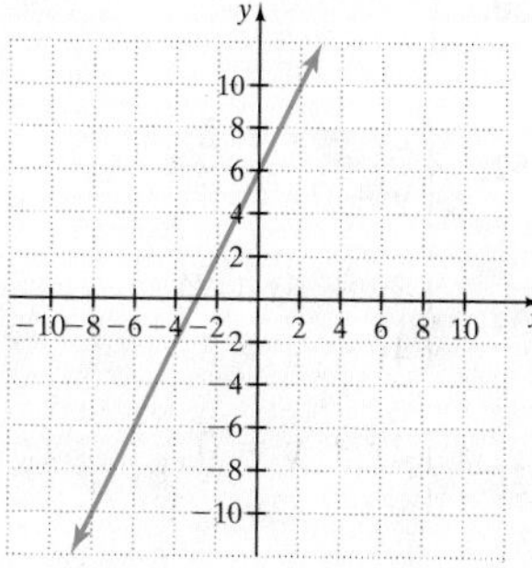

4.

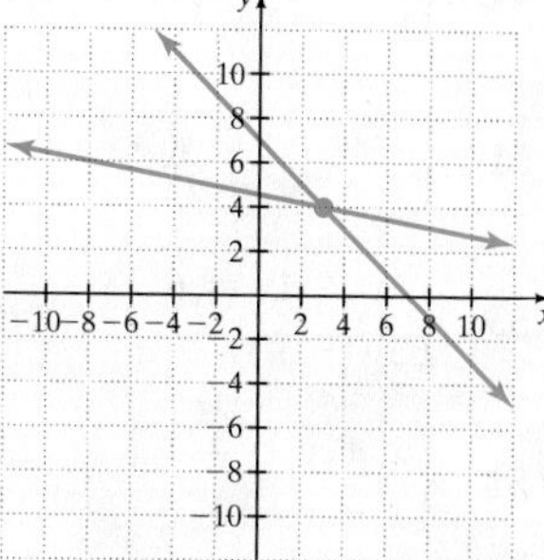

Determine whether each ordered pair is a solution of the system of linear equations. See Examples 1 and 2.

5. $\begin{cases} x + y = 8 \\ 3x + 2y = 21 \end{cases}$
   a. $(2, 4)$
   b. $(5, 3)$

6. $\begin{cases} 2x + y = 5 \\ x + 3y = 5 \end{cases}$
   a. $(5, 0)$
   b. $(2, 1)$

7. $\begin{cases} 3x - y = 5 \\ x + 2y = 11 \end{cases}$
   a. $(3, 4)$
   b. $(0, -5)$

8. $\begin{cases} 2x - 3y = 8 \\ x - 2y = 6 \end{cases}$
   a. $(-2, -4)$
   b. $(7, 2)$

9. $\begin{cases} 2y = 4x + 6 \\ 2x - y = -3 \end{cases}$
   a. $(-3, -3)$
   b. $(0, 3)$

10. $\begin{cases} x + 5y = -4 \\ -2x = 10y + 8 \end{cases}$
   a. $(-4, 0)$
   b. $(6, -2)$

11. $\begin{cases} -2 = x - 7y \\ 6x - y = 13 \end{cases}$
   a. $(-2, 0)$
   b. $\left(\frac{1}{2}, \frac{5}{14}\right)$

12. $\begin{cases} 4x = 1 - y \\ x - 3y = -8 \end{cases}$
   a. $(0, 1)$
   b. $\left(\frac{1}{6}, \frac{1}{3}\right)$

MIXED PRACTICE

Solve each system of linear equations by graphing. See Examples 3 through 6.

13. $\begin{cases} x + y = 4 \\ x - y = 2 \end{cases}$

14. $\begin{cases} x + y = 3 \\ x - y = 5 \end{cases}$

15. $\begin{cases} x + y = 6 \\ -x + y = -6 \end{cases}$

16. $\begin{cases} x + y = 1 \\ -x + y = -3 \end{cases}$

17. $\begin{cases} y = 2x \\ 3x - y = -2 \end{cases}$

18. $\begin{cases} y = -3x \\ 2x - y = -5 \end{cases}$

19. $\begin{cases} y = x + 1 \\ y = 2x - 1 \end{cases}$

20. $\begin{cases} y = 3x - 4 \\ y = x + 2 \end{cases}$

21. $\begin{cases} 2x + y = 0 \\ 3x + y = 1 \end{cases}$

22. $\begin{cases} 2x + y = 1 \\ 3x + y = 0 \end{cases}$

23. $\begin{cases} y = -x - 1 \\ y = 2x + 5 \end{cases}$

24. $\begin{cases} y = x - 1 \\ y = -3x - 5 \end{cases}$

25. $\begin{cases} x + y = 5 \\ x + y = 6 \end{cases}$

26. $\begin{cases} x - y = 4 \\ x - y = 1 \end{cases}$

27. $\begin{cases} 2x - y = 6 \\ y = 2 \end{cases}$

28. $\begin{cases} x + y = 5 \\ x = 4 \end{cases}$

29. $\begin{cases} x - 2y = 2 \\ 3x + 2y = -2 \end{cases}$

30. $\begin{cases} x + 3y = 7 \\ 2x - 3y = -4 \end{cases}$

31. $\begin{cases} 2x + y = 4 \\ 6x = -3y + 6 \end{cases}$

32. $\begin{cases} y + 2x = 3 \\ 4x = 2 - 2y \end{cases}$

33. $\begin{cases} y - 3x = -2 \\ 6x - 2y = 4 \end{cases}$

34. $\begin{cases} x - 2y = -6 \\ -2x + 4y = 12 \end{cases}$

35. $\begin{cases} x = 3 \\ y = -1 \end{cases}$

36. $\begin{cases} x = -5 \\ y = 3 \end{cases}$

37. $\begin{cases} y = x - 2 \\ y = 2x + 3 \end{cases}$

38. $\begin{cases} y = x + 5 \\ y = -2x - 4 \end{cases}$

39. $\begin{cases} 2x - 3y = -2 \\ -3x + 5y = 5 \end{cases}$

40. $\begin{cases} 4x - y = 7 \\ 2x - 3y = -9 \end{cases}$

41. $\begin{cases} 6x - y = 4 \\ \frac{1}{2}y = -2 + 3x \end{cases}$

42. $\begin{cases} 3x - y = 6 \\ \frac{1}{3}y = -2 + x \end{cases}$

Without graphing, decide.

a. Are the graphs of the equations identical lines, parallel lines, or lines intersecting at a single point?

b. How many solutions does the system have? See Examples 7 and 8.

43. $\begin{cases} 4x + y = 24 \\ x + 2y = 2 \end{cases}$

44. $\begin{cases} 3x + y = 1 \\ 3x + 2y = 6 \end{cases}$

45. $\begin{cases} 2x + y = 0 \\ 2y = 6 - 4x \end{cases}$

46. $\begin{cases} 3x + y = 0 \\ 2y = -6x \end{cases}$

47. $\begin{cases} 6x - y = 4 \\ \frac{1}{2}y = -2 + 3x \end{cases}$

48. $\begin{cases} 3x - y = 2 \\ \frac{1}{3}y = -2 + 3x \end{cases}$

49. $\begin{cases} x = 5 \\ y = -2 \end{cases}$

50. $\begin{cases} y = 3 \\ x = -4 \end{cases}$

51. $\begin{cases} 3y - 2x = 3 \\ x + 2y = 9 \end{cases}$

52. $\begin{cases} 2y = x + 2 \\ y + 2x = 3 \end{cases}$

53. $\begin{cases} 6y + 4x = 6 \\ 3y - 3 = -2x \end{cases}$

54. $\begin{cases} 8y + 6x = 4 \\ 4y - 2 = 3x \end{cases}$

55. $\begin{cases} x + y = 4 \\ x + y = 3 \end{cases}$

56. $\begin{cases} 2x + y = 0 \\ y = -2x + 1 \end{cases}$

REVIEW AND PREVIEW

Solve each equation. See Section 2.4.

57. $5(x - 3) + 3x = 1$

58. $-2x + 3(x + 6) = 17$

59. $4\left(\frac{y + 1}{2}\right) + 3y = 0$

60. $-y + 12\left(\frac{y - 1}{4}\right) = 3$

61. $8a - 2(3a - 1) = 6$

62. $3z - (4z - 2) = 9$

CONCEPT EXTENSIONS

63. Draw a graph of two linear equations whose associated system has the solution $(-1, 4)$.

64. Draw a graph of two linear equations whose associated system has the solution $(3, -2)$.

65. Draw a graph of two linear equations whose associated system has no solution.

66. Draw a graph of two linear equations whose associated system has an infinite number of solutions.

67. The ordered pair $(-2, 3)$ is a solution of all three independent equations:

$$x + y = 1$$
$$2x - y = -7$$
$$x + 3y = 7$$

Describe the graph of all three equations on the same axes.

68. Explain how to use a graph to determine the number of solutions of a system.

The double-line graph below shows the number of pounds of fishery products from U.S. domestic catch and from imports. Use this graph to answer Exercises 69 and 70.

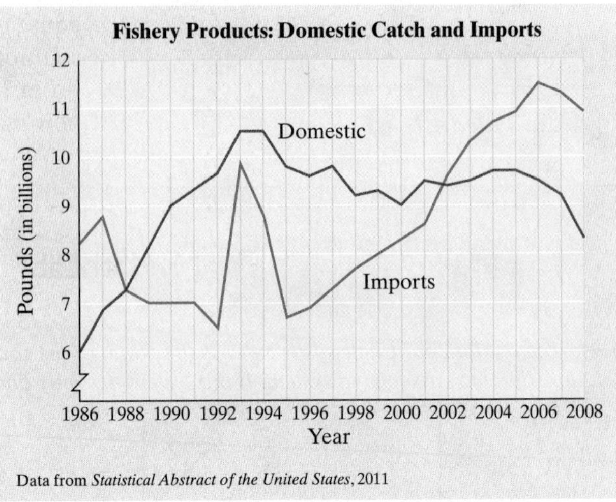

Data from *Statistical Abstract of the United States*, 2011

69. Between what pairs of years did the number of pounds of imported fishery products equal the number of pounds of domestic catch?

70. For what year(s) was the number of pounds of imported fishery products less than the number of pounds of domestic catch?

The double-line graph below shows the average attendance per game for the years shown for the Minnesota Twins and the Texas Rangers baseball teams. Use this for Exercises 71 and 72.

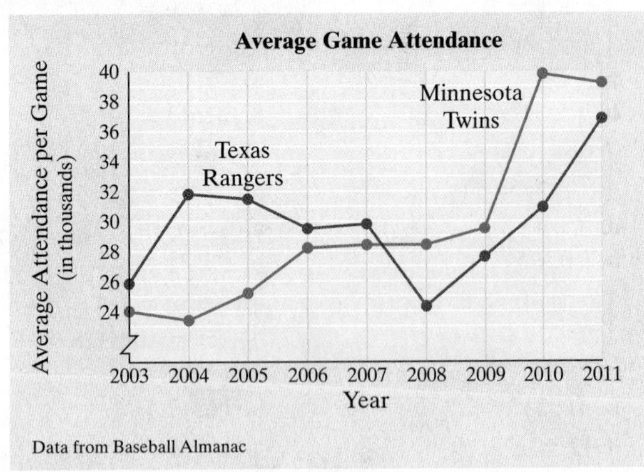

Data from Baseball Almanac

71. In what year(s) was the average attendance per game for the Texas Rangers greater than the average attendance per game for the Minnesota Twins?

72. In what year was the average attendance per game for the Texas Rangers closest to the average attendance per game for the Minnesota Twins, 2006 or 2009?

73. Construct a system of two linear equations that has $(1, 3)$ as a solution.

74. Construct a system of two linear equations that has $(0, 7)$ as a solution.

75. Below are tables of values for two linear equations.

a. Find a solution of the corresponding system.

b. Graph several ordered pairs from each table and sketch the two lines.

x	y
1	3
2	5
3	7
4	9
5	11

x	y
1	6
2	7
3	8
4	9
5	10

c. Does your graph confirm the solution from part (a)?

76. Tables of values for two linear equations are shown.

a. Find a solution of the corresponding system.

b. Graph several ordered pairs from each table and sketch the two lines.

x	y
−3	5
−1	1
0	−1
1	−3
2	−5

x	y
−3	7
−1	1
0	−2
1	−5
2	−8

c. Does your graph confirm the solution from part (a)?

77. Explain how writing each equation in a linear system in slope-intercept form helps determine the number of solutions of a system.

78. Is it possible for a system of two linear equations in two variables to be inconsistent but with dependent equations? Why or why not?

4.2 Solving Systems of Linear Equations by Substitution 

OBJECTIVE

1 Use the Substitution Method to Solve a System of Linear Equations.

OBJECTIVE

1 Using the Substitution Method 

As we stated in the preceding section, graphing alone is not an accurate way to solve a system of linear equations. In this section, we discuss a second, more accurate method for solving systems of equations. This method is called the **substitution method** and is introduced in the next example.

EXAMPLE 1 Solve the system:

$$\begin{cases} 2x + y = 10 & \text{First equation} \\ x = y + 2 & \text{Second equation} \end{cases}$$

Solution The second equation in this system is $x = y + 2$. This tells us that x and $y + 2$ have the same value. This means that we may substitute $y + 2$ for x in the first equation.

$$2x + y = 10 \quad \text{First equation}$$

$$2(y + 2) + y = 10 \quad \text{Substitute } y + 2 \text{ for } x \text{ since } x = y + 2.$$

Notice that this equation now has one variable, y. Let's now solve this equation for y.

Helpful Hint
Don't forget the distributive property.

$$2(y + 2) + y = 10$$

$$2y + 4 + y = 10 \quad \text{Use the distributive property.}$$

$$3y + 4 = 10 \quad \text{Combine like terms.}$$

$$3y = 6 \quad \text{Subtract 4 from both sides.}$$

$$y = 2 \quad \text{Divide both sides by 3.}$$

Now we know that the y-value of the ordered pair solution of the system is 2. To find the corresponding x-value, we replace y with 2 in the equation $x = y + 2$ and solve for x.

$$x = y + 2$$

$$x = 2 + 2 \quad \text{Let } y = 2.$$

$$x = 4$$

The solution of the system is the ordered pair (4, 2). Since an ordered pair solution must satisfy both linear equations in the system, we could have chosen the equation $2x + y = 10$ to find the corresponding x-value. The resulting x-value is the same.

Check: We check to see that (4, 2) satisfies both equations of the original system.

First Equation		***Second Equation***	
$2x + y = 10$		$x = y + 2$	
$2(4) + 2 \stackrel{?}{=} 10$		$4 \stackrel{?}{=} 2 + 2$	Let $x = 4$ and $y = 2$.
$10 = 10$	True	$4 = 4$	True

The solution of the system is (4, 2).

A graph of the two equations shows the two lines intersecting at the point (4, 2).

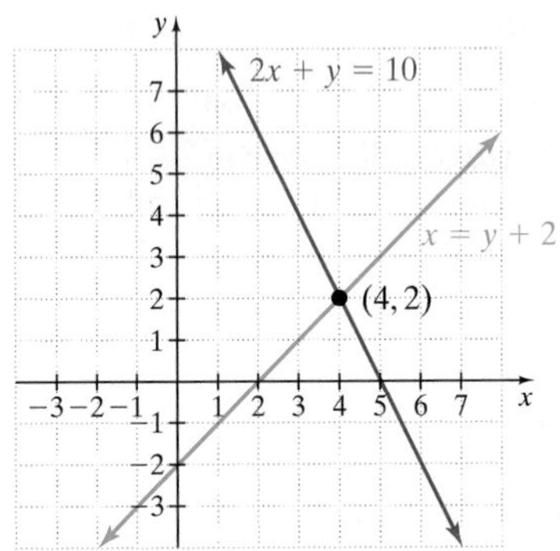

□

PRACTICE 1 Solve the system:

$$\begin{cases} 2x - y = 9 \\ x = y + 1 \end{cases}$$

EXAMPLE 2 Solve the system:

$$\begin{cases} 5x - y = -2 \\ y = 3x \end{cases}$$

Solution The second equation is solved for y in terms of x. We substitute $3x$ for y in the first equation.

$$5x - y = -2 \quad \text{First equation}$$

$$5x - (3x) = -2 \quad \text{Substitute } 3x \text{ for } y.$$

Now we solve for x.

$$5x - 3x = -2$$

$$2x = -2 \quad \text{Combine like terms.}$$

$$x = -1 \quad \text{Divide both sides by 2.}$$

The x-value of the ordered pair solution is -1. To find the corresponding y-value, we replace x with -1 in the second equation $y = 3x$.

$$y = 3x \quad \text{Second equation}$$

$$y = 3(-1) \quad \text{Let } x = -1.$$

$$y = -3$$

Check to see that the solution of the system is $(-1, -3)$. □

PRACTICE 2 Solve the system:

$$\begin{cases} 7x - y = -15 \\ y = 2x \end{cases}$$

To solve a system of equations by substitution, we first need an equation solved for one of its variables, as in Examples 1 and 2. If neither equation in a system is solved for x or y, this will be our first step.

EXAMPLE 3 Solve the system:

$$\begin{cases} x + 2y = 7 \\ 2x + 2y = 13 \end{cases}$$

Solution We choose one of the equations and solve for x or y. We will solve the first equation for x by subtracting $2y$ from both sides.

$$x + 2y = 7 \quad \text{First equation}$$
$$x = 7 - 2y \quad \text{Subtract } 2y \text{ from both sides.}$$

Since $x = 7 - 2y$, we now substitute $7 - 2y$ for x in the second equation and solve for y.

$$2x + 2y = 13 \quad \text{Second equation}$$
$$2(7 - 2y) + 2y = 13 \quad \text{Let } x = 7 - 2y.$$
$$14 - 4y + 2y = 13 \quad \text{Use the distributive property.}$$
$$14 - 2y = 13 \quad \text{Simplify.}$$
$$-2y = -1 \quad \text{Subtract 14 from both sides.}$$
$$y = \frac{1}{2} \quad \text{Divide both sides by } -2.$$

Helpful Hint
Don't forget to insert parentheses when substituting $7 - 2y$ for x.

To find x, we let $y = \frac{1}{2}$ in the equation $x = 7 - 2y$.

$$x = 7 - 2y$$
$$x = 7 - 2\left(\frac{1}{2}\right) \quad \text{Let } y = \frac{1}{2}.$$
$$x = 7 - 1$$
$$x = 6$$

Helpful Hint
To find x, any equation in two variables equivalent to the original equations of the system may be used. We used this equation since it is solved for x.

The solution is $\left(6, \frac{1}{2}\right)$. Check the solution in both equations of the original system. □

PRACTICE 3 Solve the system: $\begin{cases} x + 3y = 6 \\ 2x + 3y = 10 \end{cases}$

The following steps may be used to solve a system of equations by the substitution method.

Solving a System of Two Linear Equations by the Substitution Method

Step 1. Solve one of the equations for one of its variables.

Step 2. Substitute the expression for the variable found in Step 1 into the other equation.

Step 3. Solve the equation from Step 2 to find the value of one variable.

Step 4. Substitute the value found in Step 3 in any equation containing both variables to find the value of the other variable.

Step 5. Check the proposed solution in the original system.

✓CONCEPT CHECK

As you solve the system $\begin{cases} 2x + y = -5 \\ x - y = 5 \end{cases}$ you find that $y = -5$. Is this the solution of the system?

EXAMPLE 4 Solve the system:

$$\begin{cases} 7x - 3y = -14 \\ -3x + y = 6 \end{cases}$$

Solution To avoid introducing fractions, we will solve the second equation for y.

$$-3x + y = 6 \quad \text{Second equation}$$
$$y = 3x + 6$$

Next, substitute $3x + 6$ for y in the first equation.

$$7x - 3y = -14 \quad \text{First equation}$$
$$7x - 3(3x + 6) = -14 \quad \text{Let } y = 3x + 6.$$
$$7x - 9x - 18 = -14 \quad \text{Use the distributive property.}$$
$$-2x - 18 = -14 \quad \text{Simplify.}$$
$$-2x = 4 \quad \text{Add 18 to both sides.}$$
$$x = -2 \quad \text{Divide both sides by } -2.$$

To find the corresponding y-value, substitute -2 for x in the equation $y = 3x + 6$. Then $y = 3(-2) + 6$ or $y = 0$. The solution of the system is $(-2, 0)$. Check this solution in both equations of the system. □

PRACTICE 4 Solve the system:

$$\begin{cases} 5x + 3y = -9 \\ -2x + y = 8 \end{cases}$$

Helpful Hint

When solving a system of equations by the substitution method, begin by solving an equation for one of its variables. If possible, solve for a variable that has a coefficient of 1 or -1. This way, we avoid working with time-consuming fractions.

✓CONCEPT CHECK

To avoid fractions, which of the equations below would you use to solve for x?

a. $3x - 4y = 15$ **b.** $14 - 3y = 8x$ **c.** $7y + x = 12$

Answer to Concept Check:
No, the solution will be an ordered pair.

Answer to Concept Check:
c

EXAMPLE 5 Solve the system:

$$\begin{cases} \frac{1}{2}x - y = 3 \\ x = 6 + 2y \end{cases}$$

Solution The second equation is already solved for x in terms of y. Thus we substitute $6 + 2y$ for x in the first equation and solve for y.

$$\frac{1}{2}x - y = 3 \quad \text{First equation}$$

$$\frac{1}{2}(6 + 2y) - y = 3 \quad \text{Let } x = 6 + 2y.$$

$$3 + y - y = 3 \quad \text{Use the distributive property.}$$

$$3 = 3 \quad \text{Simplify.}$$

Arriving at a true statement such as $3 = 3$ indicates that the two linear equations in the original system are equivalent. This means that their graphs are identical and there is an infinite number of solutions of the system. Any solution of one equation is also a solution of the other.

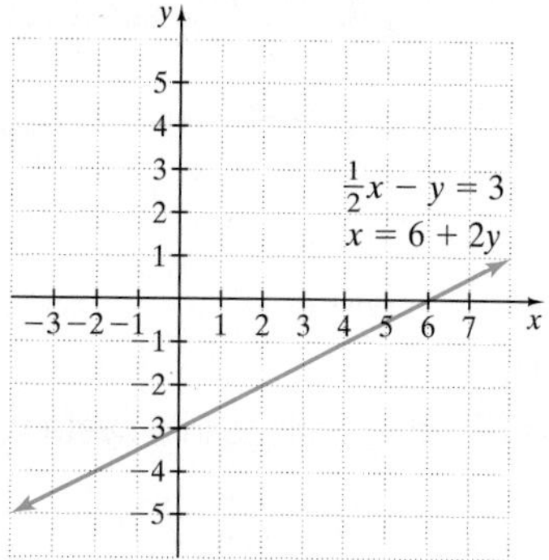

□

PRACTICE 5 Solve the system:

$$\begin{cases} \frac{1}{4}x - y = 2 \\ x = 4y + 8 \end{cases}$$

EXAMPLE 6 Use substitution to solve the system.

$$\begin{cases} 6x + 12y = 5 \\ -4x - 8y = 0 \end{cases}$$

Solution Choose the second equation and solve for y.

$$-4x - 8y = 0 \quad \text{Second equation}$$

$$-8y = 4x \quad \text{Add } 4x \text{ to both sides.}$$

$$\frac{-8y}{-8} = \frac{4x}{-8} \quad \text{Divide both sides by } -8.$$

$$y = -\frac{1}{2}x \quad \text{Simplify.}$$

Now replace y with $-\frac{1}{2}x$ in the first equation.

$$6x + 12y = 5 \quad \text{First equation}$$

$$6x + 12\left(-\frac{1}{2}x\right) = 5 \quad \text{Let } y = -\frac{1}{2}x.$$

$$6x + (-6x) = 5 \quad \text{Simplify.}$$

$$0 = 5 \quad \text{Combine like terms.}$$

The false statement $0 = 5$ indicates that this system has no solution and is inconsistent. The graph of the linear equations in the system is a pair of parallel lines.

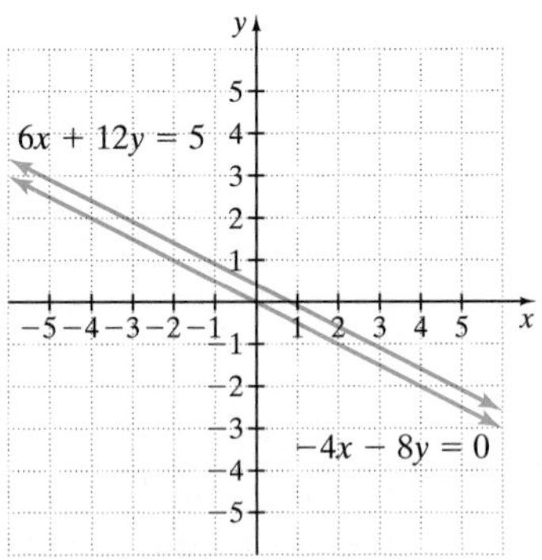

PRACTICE 6 Use substitution to solve the system.

$$\begin{cases} 4x - 3y = 12 \\ -8x + 6y = -30 \end{cases}$$

Answers to Concept Check:
a. parallel lines
b. intersect at one point
c. identical graphs

✓CONCEPT CHECK

Describe how the graphs of the equations in a system appear if the system has

a. no solution
b. one solution
c. an infinite number of solutions

Vocabulary, Readiness & Video Check

Give the solution of each system. If the system has no solution or an infinite number of solutions, say so. If the system has one solution, find it.

1. $\begin{cases} y = 4x \\ -3x + y = 1 \end{cases}$
When solving, you obtain $x = 1$

2. $\begin{cases} 4x - y = 17 \\ -8x + 2y = 0 \end{cases}$
When solving, you obtain $0 = 34$

3. $\begin{cases} 4x - y = 17 \\ -8x + 2y = -34 \end{cases}$
When solving, you obtain $0 = 0$

4. $\begin{cases} 5x + 2y = 25 \\ x = y + 5 \end{cases}$
When solving, you obtain $y = 0$

5. $\begin{cases} x + y = 0 \\ 7x - 7y = 0 \end{cases}$
When solving, you obtain $x = 0$

6. $\begin{cases} y = -2x + 5 \\ 4x + 2y = 10 \end{cases}$
When solving, you obtain $0 = 0$

Martin-Gay Interactive Videos

See Video 4.2

Watch the section lecture video and answer the following question.

OBJECTIVE 1

7. The systems in Examples 2–4 all need one of their equations solved for a variable as a first step. What important part of the substitution method is emphasized in each example?

4.2 Exercise Set

MyMathLab®

Solve each system of equations by the substitution method. See Examples 1 and 2.

1. $\begin{cases} x + y = 3 \\ x = 2y \end{cases}$

2. $\begin{cases} x + y = 20 \\ x = 3y \end{cases}$

3. $\begin{cases} x + y = 6 \\ y = -3x \end{cases}$

4. $\begin{cases} x + y = 6 \\ y = -4x \end{cases}$

5. $\begin{cases} y = 3x + 1 \\ 4y - 8x = 12 \end{cases}$

6. $\begin{cases} y = 2x + 3 \\ 5y - 7x = 18 \end{cases}$

7. $\begin{cases} y = 2x + 9 \\ y = 7x + 10 \end{cases}$

8. $\begin{cases} y = 5x - 3 \\ y = 8x + 4 \end{cases}$

MIXED PRACTICE

Solve each system of equations by the substitution method. See Examples 1 through 6.

9. $\begin{cases} 3x - 4y = 10 \\ y = x - 3 \end{cases}$

10. $\begin{cases} 4x - 3y = 10 \\ y = x - 5 \end{cases}$

11. $\begin{cases} x + 2y = 6 \\ 2x + 3y = 8 \end{cases}$

12. $\begin{cases} x + 3y = -5 \\ 2x + 2y = 6 \end{cases}$

13. $\begin{cases} 3x + 2y = 16 \\ x = 3y - 2 \end{cases}$

14. $\begin{cases} 2x + 3y = 18 \\ x = 2y - 5 \end{cases}$

15. $\begin{cases} 2x - 5y = 1 \\ 3x + y = -7 \end{cases}$

16. $\begin{cases} 3y - x = 6 \\ 4x + 12y = 0 \end{cases}$

17. $\begin{cases} 4x + 2y = 5 \\ -2x = y + 4 \end{cases}$

18. $\begin{cases} 2y = x + 2 \\ 6x - 12y = 0 \end{cases}$

19. $\begin{cases} 4x + y = 11 \\ 2x + 5y = 1 \end{cases}$

20. $\begin{cases} 3x + y = -14 \\ 4x + 3y = -22 \end{cases}$

21. $\begin{cases} x + 2y + 5 = -4 + 5y - x \\ 2x + x = y + 4 \end{cases}$
(*Hint:* First simplify each equation.)

22. $\begin{cases} 5x + 4y - 2 = -6 + 7y - 3x \\ 3x + 4x = y + 3 \end{cases}$
(*Hint:* See Exercise 21.)

23. $\begin{cases} 6x - 3y = 5 \\ x + 2y = 0 \end{cases}$

24. $\begin{cases} 10x - 5y = -21 \\ x + 3y = 0 \end{cases}$

25. $\begin{cases} 3x - y = 1 \\ 2x - 3y = 10 \end{cases}$

26. $\begin{cases} 2x - y = -7 \\ 4x - 3y = -11 \end{cases}$

27. $\begin{cases} -x + 2y = 10 \\ -2x + 3y = 18 \end{cases}$

28. $\begin{cases} -x + 3y = 18 \\ -3x + 2y = 19 \end{cases}$

29. $\begin{cases} 5x + 10y = 20 \\ 2x + 6y = 10 \end{cases}$

30. $\begin{cases} 6x + 3y = 12 \\ 9x + 6y = 15 \end{cases}$

31. $\begin{cases} 3x + 6y = 9 \\ 4x + 8y = 16 \end{cases}$

32. $\begin{cases} 2x + 4y = 6 \\ 5x + 10y = 16 \end{cases}$

33. $\begin{cases} \frac{1}{3}x - y = 2 \\ x - 3y = 6 \end{cases}$

34. $\begin{cases} \frac{1}{4}x - 2y = 1 \\ x - 8y = 4 \end{cases}$

35. $\begin{cases} x = \frac{3}{4}y - 1 \\ 8x - 5y = -6 \end{cases}$

36. $\begin{cases} x = \frac{5}{6}y - 2 \\ 12x - 5y = -9 \end{cases}$

Solve each system by the substitution method. First, simplify each equation by combining like terms.

37. $\begin{cases} -5y + 6y = 3x + 2(x - 5) - 3x + 5 \\ 4(x + y) - x + y = -12 \end{cases}$

38. $\begin{cases} 5x + 2y - 4x - 2y = 2(2y + 6) - 7 \\ 3(2x - y) - 4x = 1 + 9 \end{cases}$

REVIEW AND PREVIEW

Write equivalent equations by multiplying both sides of the given equation by the given nonzero number. See Section 2.3.

39. $3x + 2y = 6$ by -2

40. $-x + y = 10$ by 5

41. $-4x + y = 3$ by 3

42. $5a - 7b = -4$ by -4

Add the expressions by combining any like terms. See Section 2.1.

43. $\begin{array}{r} 3n + 6m \\ \underline{2n - 6m} \end{array}$

44. $\begin{array}{r} -2x + 5y \\ \underline{2x + 11y} \end{array}$

45. $\begin{array}{r} -5a - 7b \\ \underline{5a - 8b} \end{array}$

46. $\begin{array}{r} 9q + p \\ \underline{-9q - p} \end{array}$

CONCEPT EXTENSIONS

47. Explain how to identify a system with no solution when using the substitution method.

48. Occasionally, when using the substitution method, we obtain the equation $0 = 0$. Explain how this result indicates that the graphs of the equations in the system are identical.

Solve. See a Concept Check in this section.

49. As you solve the system $\begin{cases} 3x - y = -6 \\ -3x + 2y = 7 \end{cases}$, you find that $y = 1$. Is this the solution to the system?

50. As you solve the system $\begin{cases} x = 5y \\ y = 2x \end{cases}$, you find that $x = 0$ and $y = 0$. What is the solution to this system?

51. To avoid fractions, which of the equations below would you use if solving for y? Explain why.

a. $\frac{1}{2}x - 4y = \frac{3}{4}$ **b.** $8x - 5y = 13$

c. $7x - y = 19$

52. Give the number of solutions for a system if the graphs of the equations in the system are

a. lines intersecting in one point

b. parallel lines

c. same line

53. The number of men and women receiving bachelor's degrees each year has been steadily increasing. For the years 1970 through the projection of 2014, the number of men receiving degrees (in thousands) is given by the equation $y = 3.9x + 443$, and for women, the equation is $y = 14.2x + 314$ where x is the number of years after 1970. (*Source:* National Center for Education Statistics)

(Continued on next page.)

a. Use the substitution method to solve this system of equations. (Round your final results to the nearest whole numbers.)

b. Explain the meaning of your answer to part (a).

c. Sketch a graph of the system of equations. Write a sentence describing the trends for men and women receiving bachelor's degrees.

54. The number of Adult Contemporary Music radio stations in the United States from 2000 to 2009 is given by the equation $y = -11x + 734$, where x is the number of years after 2000. The number of Spanish radio stations is given by $y = 31x + 542$ for the same time period. (*Source:* M Street Corporation)

a. Use the substitution method to solve this system of equations. (Round your numbers to the nearest tenth.)

b. Explain the meaning of your answer to part (a).

c. Sketch a graph of the system of equations. Write a sentence describing the trends in the popularity of these two types of music formats.

Solve each system by substitution. When necessary, round answers to the nearest hundredth.

55. $\begin{cases} y = 5.1x + 14.56 \\ y = -2x - 3.9 \end{cases}$

56. $\begin{cases} y = 3.1x - 16.35 \\ y = -9.7x + 28.45 \end{cases}$

57. $\begin{cases} 3x + 2y = 14.05 \\ 5x + y = 18.5 \end{cases}$

58. $\begin{cases} x + y = -15.2 \\ -2x + 5y = -19.3 \end{cases}$

4.3 Solving Systems of Linear Equations by Addition

OBJECTIVE

1 Use the Addition Method to Solve a System of Linear Equations.

OBJECTIVE

1 Using the Addition Method 

We have seen that substitution is an accurate way to solve a linear system. Another method for solving a system of equations accurately is the **addition** or **elimination method.** The addition method is based on the addition property of equality: adding equal quantities to both sides of an equation does not change the solution of the equation. In symbols,

$$\text{if } A = B \text{ and } C = D, \text{ then } A + C = B + D.$$

EXAMPLE 1 Solve the system:

$$\begin{cases} x + y = 7 \\ x - y = 5 \end{cases}$$

Solution Since the left side of each equation is equal to the right side, we add equal quantities by adding the left sides of the equations together and the right sides of the equations together. This adding eliminates the variable y and gives us an equation in one variable, x. We can then solve for x.

> **Helpful Hint**
> Notice in Example 1 that our goal when solving a system of equations by the addition method is to eliminate a variable when adding the equations.

$$\begin{aligned} x + y &= 7 && \text{First equation} \\ x - y &= 5 && \text{Second equation} \\ 2x \quad &= 12 && \text{Add the equations.} \\ x &= 6 && \text{Divide both sides by 2.} \end{aligned}$$

The x-value of the solution is 6. To find the corresponding y-value, let $x = 6$ in either equation of the system. We will use the first equation.

$$\begin{aligned} x + y &= 7 && \text{First equation} \\ 6 + y &= 7 && \text{Let } x = 6. \\ y &= 7 - 6 && \text{Solve for } y. \\ y &= 1 && \text{Simplify.} \end{aligned}$$

Check: The solution is (6, 1). Check this in both equations.

First Equation		*Second Equation*	
$x + y = 7$		$x - y = 5$	
$6 + 1 \stackrel{?}{=} 7$		$6 - 1 \stackrel{?}{=} 5$	Let $x = 6$ and $y = 1$.
$7 = 7$	True	$5 = 5$	True

Thus, the solution of the system is (6, 1) and the graphs of the two equations intersect at the point (6, 1) as shown next.

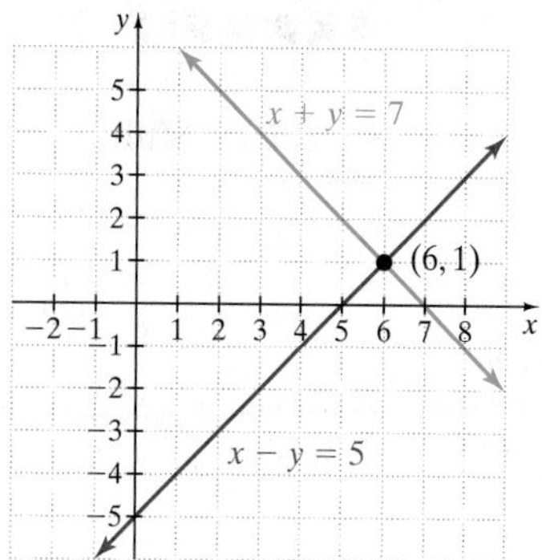

□

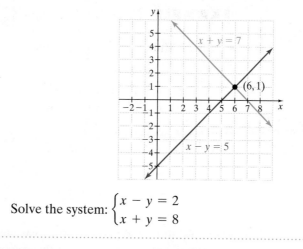

PRACTICE 1 Solve the system: $\begin{cases} x - y = 2 \\ x + y = 8 \end{cases}$

EXAMPLE 2 Solve the system: $\begin{cases} -2x + y = 2 \\ -x + 3y = -4 \end{cases}$

Solution If we simply add the two equations, the result is still an equation in two variables. However, remember from Example 1 that our goal is to eliminate one of the variables. Notice what happens if we multiply *both sides* of the first equation by −3, which we are allowed to do by the multiplication property of equality. The system

$$\begin{cases} -3(-2x + y) = -3(2) \\ -x + 3y = -4 \end{cases} \quad \text{simplifies to} \quad \begin{cases} 6x - 3y = -6 \\ -x + 3y = -4 \end{cases}$$

Now add the resulting equations and the y-variable is eliminated.

$$\begin{aligned} 6x - 3y &= -6 \\ -x + 3y &= -4 \\ \hline 5x \quad\quad &= -10 \quad \text{Add.} \\ x &= -2 \quad \text{Divide both sides by 5.} \end{aligned}$$

To find the corresponding y-value, let $x = -2$ in any of the preceding equations containing both variables. We use the first equation of the original system.

$$\begin{aligned} -2x + y &= 2 \quad \text{First equation} \\ -2(-2) + y &= 2 \quad \text{Let } x = -2. \\ 4 + y &= 2 \\ y &= -2 \quad \text{Subtract 4 from both sides.} \end{aligned}$$

The solution is (−2, −2). Check this ordered pair in both equations of the original system. □

PRACTICE 2 Solve the system: $\begin{cases} x - 2y = 11 \\ 3x - y = 13 \end{cases}$

Helpful Hint

When finding the second value of an ordered pair solution, any equation equivalent to one of the original equations in the system may be used.

In Example 2, the decision to multiply the first equation by -3 was no accident. **To eliminate a variable** when adding two equations, **the coefficient of the variable in one equation must be the opposite of its coefficient in the other equation.**

> **Helpful Hint**
> Be sure to multiply *both sides* of an equation by a chosen number when solving by the addition method. A common mistake is to multiply only the side containing the variables.

EXAMPLE 3 Solve the system: $\begin{cases} 2x - y = 7 \\ 8x - 4y = 1 \end{cases}$

Solution Multiply both sides of the first equation by -4 and the resulting coefficient of x is -8, the opposite of 8, the coefficient of x in the second equation. The system becomes

> **Helpful Hint**
> Don't forget to multiply **both** sides by -4.

$$\begin{cases} -4(2x - y) = -4(7) \\ 8x - 4y = 1 \end{cases} \quad \text{simplifies to} \quad \begin{cases} -8x + 4y = -28 \\ 8x - 4y = 1 \end{cases}$$

Now add the resulting equations.

$$\begin{aligned} -8x + 4y &= -28 \\ 8x - 4y &= 1 \quad \text{Add the equations.} \\ \hline 0 &= -27 \quad \text{False} \end{aligned}$$

When we add the equations, both variables are eliminated and we have $0 = -27$, a false statement. This means that the system has no solution. The graphs of these equations are parallel lines. □

PRACTICE 3 Solve the system: $\begin{cases} x - 3y = 5 \\ 2x - 6y = -3 \end{cases}$

EXAMPLE 4 Solve the system: $\begin{cases} 3x - 2y = 2 \\ -9x + 6y = -6 \end{cases}$

Solution First we multiply both sides of the first equation by 3, then we add the resulting equations.

$$\begin{cases} 3(3x - 2y) = 3(2) \\ -9x + 6y = -6 \end{cases} \quad \text{simplifies to} \quad \begin{cases} 9x - 6y = 6 \\ -9x + 6y = -6 \end{cases} \quad \text{Add the equations.}$$

$$0 = 0 \quad \text{True}$$

Both variables are eliminated and we have $0 = 0$, a true statement. Whenever you eliminate a variable and get the equation $0 = 0$, the system has an infinite number of solutions. The graphs of these equations are identical. □

PRACTICE 4 Solve the system: $\begin{cases} 4x - 3y = 5 \\ -8x + 6y = -10 \end{cases}$

Answer to Concept Check: b

> **✓CONCEPT CHECK**
> Suppose you are solving the system
>
> $$\begin{cases} 3x + 8y = -5 \\ 2x - 4y = 3 \end{cases}$$
>
> You decide to use the addition method and begin by multiplying both sides of the first equation by -2. In which of the following was the multiplication performed correctly? Explain.
>
> **a.** $-6x - 16y = -5$ **b.** $-6x - 16y = 10$

EXAMPLE 5 Solve the system: $\begin{cases} 3x + 4y = 13 \\ 5x - 9y = 6 \end{cases}$

Solution We can eliminate the variable y by multiplying the first equation by 9 and the second equation by 4.

$$\begin{cases} 9(3x + 4y) = 9(13) \\ 4(5x - 9y) = 4(6) \end{cases} \text{ simplifies to } \begin{cases} 27x + 36y = 117 \\ 20x - 36y = 24 \end{cases}$$

$$47x = 141 \quad \text{Add the equations.}$$

$$x = 3 \quad \text{Divide both sides by 47.}$$

To find the corresponding y-value, we let $x = 3$ in any equation in this example containing two variables. Doing so in any of these equations will give $y = 1$. The solution to this system is (3, 1). Check to see that (3, 1) satisfies each equation in the original system. □

PRACTICE 5 Solve the system: $\begin{cases} 4x + 3y = 14 \\ 3x - 2y = 2 \end{cases}$

If we had decided to eliminate x instead of y in Example 5, the first equation could have been multiplied by 5 and the second by -3. Try solving the original system this way to check that the solution is (3, 1).

The following steps summarize how to solve a system of linear equations by the addition method.

Solving a System of Two Linear Equations by the Addition Method

Step 1. Rewrite each equation in standard form $Ax + By = C$.

Step 2. If necessary, multiply one or both equations by a nonzero number so that the coefficients of a chosen variable in the system are opposites.

Step 3. Add the equations.

Step 4. Find the value of one variable by solving the resulting equation from Step 3.

Step 4. Find the value of the second variable by substituting the value found in Step 4 into either of the original equations.

Step 6. Check the proposed solution in the original system.

✓CONCEPT CHECK

Suppose you are solving the system

$$\begin{cases} -4x + 7y = 6 \\ x + 2y = 5 \end{cases}$$

by the addition method.

a. What step(s) should you take if you wish to eliminate x when adding the equations?

b. What step(s) should you take if you wish to eliminate y when adding the equations?

EXAMPLE 6 Solve the system: $\begin{cases} -x - \dfrac{y}{2} = \dfrac{5}{2} \\ \dfrac{x}{6} - \dfrac{y}{2} = 0 \end{cases}$

Solution: We begin by clearing each equation of fractions. To do so, we multiply both sides of the first equation by the LCD, 2, and both sides of the second equation by the LCD, 6. Then the system

Answers to Concept Check:

a. multiply the second equation by 4

b. possible answer: multiply the first equation by -2 and the second equation by 7

$$\begin{cases} 2\left(-x - \dfrac{y}{2}\right) = 2\left(\dfrac{5}{2}\right) \\ 6\left(\dfrac{x}{6} - \dfrac{y}{2}\right) = 6(0) \end{cases} \quad \text{simplifies to} \quad \begin{cases} -2x - y = 5 \\ x - 3y = 0 \end{cases}$$

We can now eliminate the variable x by multiplying the second equation by 2.

$$\begin{cases} -2x - y = 5 \\ 2(x - 3y) = 2(0) \end{cases} \quad \text{simplifies to} \quad \begin{cases} -2x - y = 5 \\ 2x - 6y = 0 \end{cases}$$

$$-7y = 5 \qquad \text{Add the equations.}$$

$$y = -\frac{5}{7} \qquad \text{Solve for } y.$$

To find x, we could replace y with $-\frac{5}{7}$ in one of the equations with two variables. Instead, let's go back to the simplified system and multiply by appropriate factors to eliminate the variable y and solve for x. To do this, we multiply the first equation by -3. Then the system

$$\begin{cases} -3(-2x - y) = -3(5) \\ x - 3y = 0 \end{cases} \quad \text{simplifies to} \quad \begin{cases} 6x + 3y = -15 \\ x - 3y = 0 \end{cases}$$

$$7x = -15 \qquad \text{Add the equations.}$$

$$x = -\frac{15}{7} \qquad \text{Solve for } x.$$

Check the ordered pair $\left(-\frac{15}{7}, -\frac{5}{7}\right)$ in both equations of the original system. The solution is $\left(-\frac{15}{7}, -\frac{5}{7}\right)$. □

PRACTICE 6 Solve the system: $\begin{cases} -2x + \dfrac{3y}{2} = 5 \\ -\dfrac{x}{2} - \dfrac{y}{4} = \dfrac{1}{2} \end{cases}$

Vocabulary, Readiness & Video Check

Given the system $\begin{cases} 3x - 2y = -9 \\ x + 5y = 14 \end{cases}$ *read each row (Step 1, Step 2, and Result). Then answer whether the result is true or false.*

	Step 1	*Step 2*	*Result*	*True or False?*
1.	Multiply 2nd equation through by -3.	Add the resulting equation to the 1st equation.	The y's are eliminated.	
2.	Multiply 2nd equation through by -3.	Add the resulting equation to the 1st equation.	The x's are eliminated.	
3.	Multiply 1st equation by 5 and 2nd equation by 2.	Add the two new equations.	The y's are eliminated.	
4.	Multiply 1st equation by 5 and 2nd equation by -2.	Add the two new equations.	The y's are eliminated.	

Martin-Gay Interactive Videos

See Video 4.3

Watch the section lecture video and answer the following question.

OBJECTIVE 1

5. For the addition/elimination methods, sometimes you need to multiply an equation through by a nonzero number so that the coefficients of a variable are opposites, as is shown in Example 2. What property allows us to do this? What important reminder is made at this step?

4.3 Exercise Set

MyMathLab®

Solve each system of equations by the addition method. See Example 1.

1. $\begin{cases} 3x + y = 5 \\ 6x - y = 4 \end{cases}$

2. $\begin{cases} 4x + y = 13 \\ 2x - y = 5 \end{cases}$

3. $\begin{cases} x - 2y = 8 \\ -x + 5y = -17 \end{cases}$

4. $\begin{cases} x - 2y = -11 \\ -x + 5y = 23 \end{cases}$

Solve each system of equations by the addition method. If a system contains fractions or decimals, you may want to clear each equation of fractions or decimals first. See Examples 2 through 6.

5. $\begin{cases} 3x + y = -11 \\ 6x - 2y = -2 \end{cases}$

6. $\begin{cases} 4x + y = -13 \\ 6x - 3y = -15 \end{cases}$

7. $\begin{cases} 3x + 2y = 11 \\ 5x - 2y = 29 \end{cases}$

8. $\begin{cases} 4x + 2y = 2 \\ 3x - 2y = 12 \end{cases}$

9. $\begin{cases} x + 5y = 18 \\ 3x + 2y = -11 \end{cases}$

10. $\begin{cases} x + 4y = 14 \\ 5x + 3y = 2 \end{cases}$

11. $\begin{cases} x + y = 6 \\ x - y = 6 \end{cases}$

12. $\begin{cases} x - y = 1 \\ -x + 2y = 0 \end{cases}$

13. $\begin{cases} 2x + 3y = 0 \\ 4x + 6y = 3 \end{cases}$

14. $\begin{cases} 3x + y = 4 \\ 9x + 3y = 6 \end{cases}$

15. $\begin{cases} -x + 5y = -1 \\ 3x - 15y = 3 \end{cases}$

16. $\begin{cases} 2x + y = 6 \\ 4x + 2y = 12 \end{cases}$

17. $\begin{cases} 3x - 2y = 7 \\ 5x + 4y = 8 \end{cases}$

18. $\begin{cases} 6x - 5y = 25 \\ 4x + 15y = 13 \end{cases}$

19. $\begin{cases} 8x = -11y - 16 \\ 2x + 3y = -4 \end{cases}$

20. $\begin{cases} 10x + 3y = -12 \\ 5x = -4y - 16 \end{cases}$

21. $\begin{cases} 4x - 3y = 7 \\ 7x + 5y = 2 \end{cases}$

22. $\begin{cases} -2x + 3y = 10 \\ 3x + 4y = 2 \end{cases}$

23. $\begin{cases} 4x - 6y = 8 \\ 6x - 9y = 12 \end{cases}$

24. $\begin{cases} 9x - 3y = 12 \\ 12x - 4y = 18 \end{cases}$

25. $\begin{cases} 2x - 5y = 4 \\ 3x - 2y = 4 \end{cases}$

26. $\begin{cases} 6x - 5y = 7 \\ 4x - 6y = 7 \end{cases}$

27. $\begin{cases} \dfrac{x}{3} + \dfrac{y}{6} = 1 \\ \dfrac{x}{2} - \dfrac{y}{4} = 0 \end{cases}$

28. $\begin{cases} \dfrac{x}{2} + \dfrac{y}{8} = 3 \\ x - \dfrac{y}{4} = 0 \end{cases}$

29. $\begin{cases} \dfrac{10}{3}x + 4y = -4 \\ 5x + 6y = -6 \end{cases}$

30. $\begin{cases} \dfrac{3}{2}x + 4y = 1 \\ 9x + 24y = 5 \end{cases}$

31. $\begin{cases} x - \dfrac{y}{3} = -1 \\ -\dfrac{x}{2} + \dfrac{y}{8} = \dfrac{1}{4} \end{cases}$

32. $\begin{cases} 2x - \dfrac{3y}{4} = -3 \\ x + \dfrac{y}{9} = \dfrac{13}{3} \end{cases}$

33. $\begin{cases} -4(x + 2) = 3y \\ 2x - 2y = 3 \end{cases}$

34. $\begin{cases} -9(x + 3) = 8y \\ 3x - 3y = 8 \end{cases}$

35. $\begin{cases} \dfrac{x}{3} - y = 2 \\ -\dfrac{x}{2} + \dfrac{3y}{2} = -3 \end{cases}$

36. $\begin{cases} \dfrac{x}{2} + \dfrac{y}{4} = 1 \\ -\dfrac{x}{4} - \dfrac{y}{8} = 1 \end{cases}$

37. $\begin{cases} \dfrac{3}{5}x - y = -\dfrac{4}{5} \\ 3x + \dfrac{y}{2} = -\dfrac{9}{5} \end{cases}$

38. $\begin{cases} 3x + \dfrac{7}{2}y = \dfrac{3}{4} \\ -\dfrac{x}{2} + \dfrac{5}{3}y = -\dfrac{5}{4} \end{cases}$

39. $\begin{cases} 3.5x + 2.5y = 17 \\ -1.5x - 7.5y = -33 \end{cases}$

40. $\begin{cases} -2.5x - 6.5y = 47 \\ 0.5x - 4.5y = 37 \end{cases}$

41. $\begin{cases} 0.02x + 0.04y = 0.09 \\ -0.1x + 0.3y = 0.8 \end{cases}$

42. $\begin{cases} 0.04x - 0.05y = 0.105 \\ 0.2x - 0.6y = 1.05 \end{cases}$

MIXED PRACTICE

Solve each system by either the addition method or the substitution method.

43. $\begin{cases} 2x - 3y = -11 \\ y = 4x - 3 \end{cases}$

44. $\begin{cases} 4x - 5y = 6 \\ y = 3x - 10 \end{cases}$

45. $\begin{cases} x + 2y = 1 \\ 3x + 4y = -1 \end{cases}$

46. $\begin{cases} x + 3y = 5 \\ 5x + 6y = -2 \end{cases}$

47. $\begin{cases} 2y = x + 6 \\ 3x - 2y = -6 \end{cases}$

48. $\begin{cases} 3y = x + 14 \\ 2x - 3y = -16 \end{cases}$

49. $\begin{cases} y = 2x - 3 \\ y = 5x - 18 \end{cases}$

50. $\begin{cases} y = 6x - 5 \\ y = 4x - 11 \end{cases}$

51. $\begin{cases} x + \frac{1}{6}y = \frac{1}{2} \\ 3x + 2y = 3 \end{cases}$

52. $\begin{cases} x + \frac{1}{3}y = \frac{5}{12} \\ 8x + 3y = 4 \end{cases}$

53. $\begin{cases} \frac{x+2}{2} = \frac{y+11}{3} \\ \frac{x}{2} = \frac{2y+16}{6} \end{cases}$

54. $\begin{cases} \frac{x+5}{2} = \frac{y+14}{4} \\ \frac{x}{3} = \frac{2y+2}{6} \end{cases}$

55. $\begin{cases} 2x + 3y = 14 \\ 3x - 4y = -69.1 \end{cases}$

56. $\begin{cases} 5x - 2y = -19.8 \\ -3x + 5y = -3.7 \end{cases}$

REVIEW AND PREVIEW

Translating *Rewrite the following sentences using mathematical symbols. Do not solve the equations. See Sections 2.4 and 2.5.*

57. Twice a number, added to 6, is 3 less than the number.

58. The sum of three consecutive integers is 66.

59. Three times a number, subtracted from 20, is 2.

60. Twice the sum of 8 and a number is the difference of the number and 20.

61. The product of 4 and the sum of a number and 6 is twice the number.

62. If the quotient of twice a number and 7 is subtracted from the reciprocal of the number, the result is 2.

CONCEPT EXTENSIONS

Solve. See a Concept Check in this section.

63. To solve this system by the addition method and eliminate the variable y,

$$\begin{cases} 4x + 2y = -7 \\ 3x - y = -12 \end{cases}$$

by what value would you multiply the second equation? What do you get when you complete the multiplication?

64. Given the system of linear equations $\begin{cases} 3x - y = -8 \\ 5x + 3y = 2 \end{cases}$, use the addition method and

a. Solve the system by eliminating x.

b. Solve the system by eliminating y.

65. Suppose you are solving the system

$$\begin{cases} 3x + 8y = -5 \\ 2x - 4y = 3. \end{cases}$$

You decide to use the addition method by multiplying both sides of the second equation by 2. In which of the following was the multiplication performed correctly? Explain.

a. $4x - 8y = 3$

b. $4x - 8y = 6$

66. Suppose you are solving the system

$$\begin{cases} -2x - y = 0 \\ -2x + 3y = 6. \end{cases}$$

You decide to use the addition method by multiplying both sides of the first equation by 3, then adding the resulting equation to the second equation. Which of the following is the correct sum? Explain.

a. $-8x = 6$

b. $-8x = 9$

67. When solving a system of equations by the addition method, how do we know when the system has no solution?

68. Explain why the addition method might be preferred over the substitution method for solving the system $\begin{cases} 2x - 3y = 5 \\ 5x + 2y = 6. \end{cases}$

69. Use the system of linear equations below to answer the questions.

$$\begin{cases} x + y = 5 \\ 3x + 3y = b \end{cases}$$

a. Find the value of b so that the system has an infinite number of solutions.

b. Find a value of b so that there are no solutions to the system.

70. Use the system of linear equations below to answer the questions.

$$\begin{cases} x + y = 4 \\ 2x + by = 8 \end{cases}$$

a. Find the value of b so that the system has an infinite number of solutions.

b. Find a value of b so that the system has a single solution.

Solve each system by the addition method.

71. $\begin{cases} 1.2x + 3.4y = 27.6 \\ 7.2x - 1.7y = -46.56 \end{cases}$

72. $\begin{cases} 5.1x - 2.4y = 3.15 \\ -15.3x + 1.2y = 27.75 \end{cases}$

73. Two age groups predicted to increase in percent of the workforce are those age 25–34 and those age 55 and older. The percent of workers age 25–34 predicted for 2008 through 2018 can be approximated by $0.05x - y = -21.6$. The percent of workers age 55 and older for the same years can be approximated by $0.58x - y = -18.1$. For both equations, x is the number of years since 2008 and y is the percent of the workforce. (*Source:* Bureau of Labor Statistics)

a. Use the addition method to solve the system:

$$\begin{cases} 0.05x - y = -21.6 \\ 0.58x - y = -18.1 \end{cases}$$

(Round answer to the nearest whole.)

b. Use your result from part (a) and estimate the year in which both percents are the same.

c. Use your results from part (a) and part (b) to estimate the percent of workers age 25–34 and those age 55 and older.

74. Two occupations predicted to greatly increase in the number of jobs are pharmacy technicians and network system analysts. The number of pharmacy technician jobs predicted for 2006 through 2016 can be approximated by $9.1x - y = -295$. The number of network system analyst jobs predicted for 2006 through 2016 can be approximated by $14x - y = -262$. For both equations, x is the number of years since 2006, and y is the number of jobs in the thousands. (*Source:* Bureau of Labor Statistics)

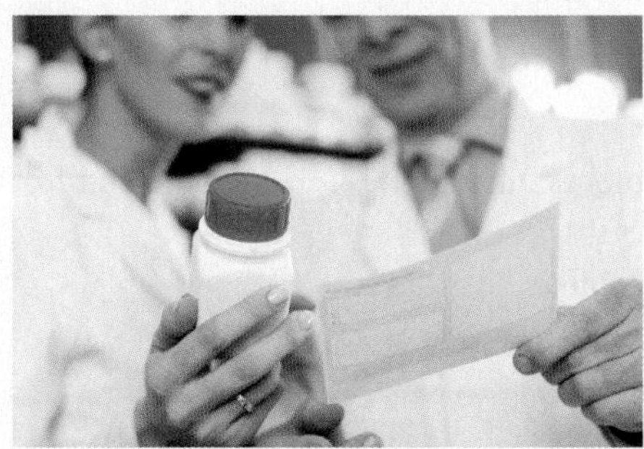

a. Use the addition method to solve this system of equations.

$$\begin{cases} 9.1x - y = -295 \\ 14x - y = -262 \end{cases}$$

(Eliminate y first and solve for x. Round this result to the nearest whole.)

b. Interpret your solution from part (a).

c. Using the year in your answer to part (b), estimate the number of pharmacy technician jobs and network system analyst jobs in that year.

Integrated Review SOLVING SYSTEMS OF EQUATIONS

Sections 4.1–4.3

Solve each system by either the addition method or the substitution method.

1. $\begin{cases} 2x - 3y = -11 \\ y = 4x - 3 \end{cases}$

2. $\begin{cases} 4x - 5y = 6 \\ y = 3x - 10 \end{cases}$

3. $\begin{cases} x + y = 3 \\ x - y = 7 \end{cases}$

4. $\begin{cases} x - y = 20 \\ x + y = -8 \end{cases}$

5. $\begin{cases} x + 2y = 1 \\ 3x + 4y = -1 \end{cases}$

6. $\begin{cases} x + 3y = 5 \\ 5x + 6y = -2 \end{cases}$

7. $\begin{cases} y = x + 3 \\ 3x - 2y = -6 \end{cases}$

8. $\begin{cases} y = -2x \\ 2x - 3y = -16 \end{cases}$

9. $\begin{cases} y = 2x - 3 \\ y = 5x - 18 \end{cases}$

10. $\begin{cases} y = 6x - 5 \\ y = 4x - 11 \end{cases}$

11. $\begin{cases} x + \frac{1}{6}y = \frac{1}{2} \\ 3x + 2y = 3 \end{cases}$

12. $\begin{cases} x + \frac{1}{3}y = \frac{5}{12} \\ 8x + 3y = 4 \end{cases}$

13. $\begin{cases} x - 5y = 1 \\ -2x + 10y = 3 \end{cases}$

14. $\begin{cases} -x + 2y = 3 \\ 3x - 6y = -9 \end{cases}$

15. $\begin{cases} 0.2x - 0.3y = -0.95 \\ 0.4x + 0.1y = 0.55 \end{cases}$

16. $\begin{cases} 0.08x - 0.04y = -0.11 \\ 0.02x - 0.06y = -0.09 \end{cases}$

17. $\begin{cases} x = 3y - 7 \\ 2x - 6y = -14 \end{cases}$

18. $\begin{cases} y = \frac{x}{2} - 3 \\ 2x - 4y = 0 \end{cases}$

19. $\begin{cases} 2x + 5y = -1 \\ 3x - 4y = 33 \end{cases}$

20. $\begin{cases} 7x - 3y = 2 \\ 6x + 5y = -21 \end{cases}$

21. Which method, substitution or addition, would you prefer to use to solve the system below? Explain your reasoning.

$$\begin{cases} 3x + 2y = -2 \\ y = -2x \end{cases}$$

22. Which method, substitution or addition, would you prefer to use to solve the system below? Explain your reasoning.

$$\begin{cases} 3x - 2y = -3 \\ 6x + 2y = 12 \end{cases}$$

4.4 Systems of Linear Equations and Problem Solving 

OBJECTIVE

1 Use a System of Equations to Solve Problems.

OBJECTIVE

1 Using a System of Equations for Problem Solving 

Many of the word problems solved earlier using one-variable equations can also be solved using two equations in **two** variables. We use the same problem-solving steps that have been used throughout this text. The only difference is that two variables are assigned to represent the two unknown quantities and that the stated problem is translated into **two** equations.

Problem-Solving Steps

Step 1. UNDERSTAND the problem. During this step, become comfortable with the problem. Some ways of doing this are to

Read and reread the problem.

Choose two variables to represent the two unknowns.

Construct a drawing if possible.

Propose a solution and check. Pay careful attention to how you check your proposed solution. This will help when writing equations to model the problem.

Step 2. TRANSLATE the problem into two equations.

Step 3. SOLVE the system of equations.

Step 4. INTERPRET the results: **Check** the proposed solution in the stated problem and **state** your conclusion.

EXAMPLE 1 **Finding Unknown Numbers**

Find two numbers whose sum is 37 and whose difference is 21.

Solution

1. UNDERSTAND. Read and reread the problem. Suppose that one number is 20. If their sum is 37, the other number is 17 because $20 + 17 = 37$. Is their difference 21? No; $20 - 17 = 3$. Our proposed solution is incorrect, but we now have a better understanding of the problem.

   Since we are looking for two numbers, we let

   $x =$ first number

   $y =$ second number

2. TRANSLATE. Since we have assigned two variables to this problem, we translate our problem into two equations.

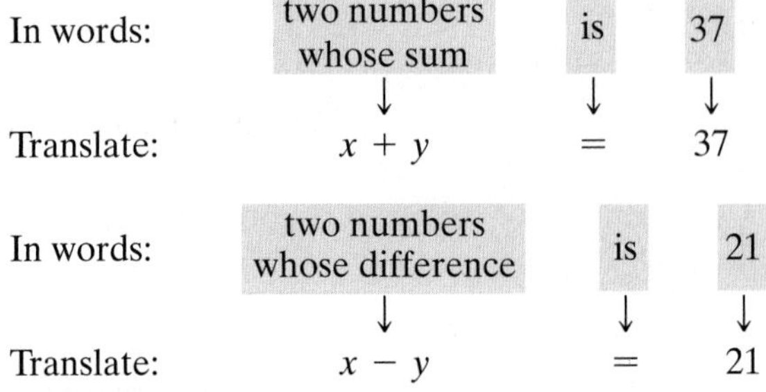

3. SOLVE. Now we solve the system

$$\begin{cases} x + y = 37 \\ x - y = 21 \end{cases}$$

Notice that the coefficients of the variable y are opposites. Let's then solve by the addition method and begin by adding the equations.

$$\begin{aligned} x + y &= 37 \\ x - y &= 21 \\ \hline 2x \quad &= 58 \qquad \text{Add the equations.} \\ x &= \frac{58}{2} = 29 \qquad \text{Divide both sides by 2.} \end{aligned}$$

Now we let $x = 29$ in the first equation to find y.

$$\begin{aligned} x + y &= 37 \qquad \text{First equation} \\ 29 + y &= 37 \\ y &= 8 \qquad \text{Subtract 29 from both sides.} \end{aligned}$$

4. INTERPRET. The solution of the system is (29, 8).

Check: Notice that the sum of 29 and 8 is $29 + 8 = 37$, the required sum. Their difference is $29 - 8 = 21$, the required difference.

State: The numbers are 29 and 8. □

PRACTICE

1 Find two numbers whose sum is 30 and whose difference is 6.

EXAMPLE 2 Solving a Problem about Prices

The Cirque du Soleil show Varekai is performing locally. Matinee admission for 4 adults and 2 children is \$374, while admission for 2 adults and 3 children is \$285.

a. What is the price of an adult's ticket?

b. What is the price of a child's ticket?

c. Suppose that a special rate of \$1000 is offered for groups of 20 persons. Should a group of 4 adults and 16 children use the group rate? Why or why not?

Solution

1. UNDERSTAND. Read and reread the problem and guess a solution. Let's suppose that the price of an adult's ticket is \$50 and the price of a child's ticket is \$40. To check our proposed solution, let's see if admission for 4 adults and 2 children is \$374. Admission for 4 adults is 4(\$50) or \$200 and admission for 2 children is 2(\$40) or \$80. This gives a total admission of \$200 + \$80 = \$280, not the required \$374. Again, though, we have accomplished the purpose of this process: We have a better understanding of the problem. To continue, we let

A = the price of an adult's ticket and

C = the price of a child's ticket

2. TRANSLATE. We translate the problem into two equations using both variables.

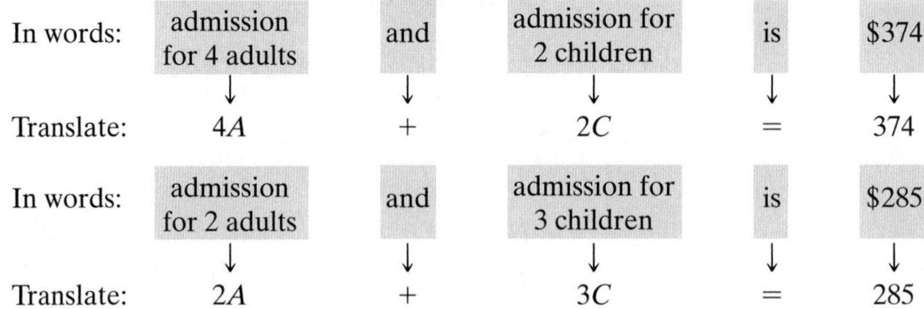

In words:	admission for 4 adults	and	admission for 2 children	is	\$374
	↓	↓	↓	↓	↓
Translate:	$4A$	$+$	$2C$	$=$	374

In words:	admission for 2 adults	and	admission for 3 children	is	\$285
	↓	↓	↓	↓	↓
Translate:	$2A$	$+$	$3C$	$=$	285

3. SOLVE. We solve the system.

$$\begin{cases} 4A + 2C = 374 \\ 2A + 3C = 285 \end{cases}$$

Since both equations are written in standard form, we solve by the addition method. First we multiply the second equation by -2 so that when we add the equations, we eliminate the variable A. Then the system

$$\begin{cases} 4A + 2C = 374 \\ -2(2A + 3C) = -2(285) \end{cases} \quad \text{simplifies to} \quad \begin{cases} 4A + 2C = 374 \\ -4A - 6C = -570 \end{cases}$$

Add the equations.

$$-4C = -196$$

$C = 49$ or \$49, the children's ticket price.

To find A, we replace C with 49 in the first equation.

$$4A + 2C = 374 \quad \text{First equation}$$
$$4A + 2(49) = 374 \quad \text{Let } C = 49$$
$$4A + 98 = 374$$
$$4A = 276$$

$A = 69$ or \$69, the adult's ticket price

4. INTERPRET.

Check: Notice that 4 adults and 2 children will pay $4(\$69) + 2(\$49) = \$276 + \$98 = \$374$, the required amount. Also, the price for 2 adults and 3 children is $2(\$69) + 3(\$49) = \$138 + \$147 = \$285$, the required amount.

State: Answer the three original questions.

a. Since $A = 69$, the price of an adult's ticket is \$69.

b. Since $C = 49$, the price of a child's ticket is \$49.

c. The regular admission price for 4 adults and 16 children is

$$4(\$69) + 16(\$49) = \$276 + \$784$$
$$= \$1060$$

This is \$60 more than the special group rate of \$1000, so they should request the group rate. □

PRACTICE

2 It is considered a premium game when the Red Sox or the Yankees come to Texas to play the Rangers. Admission for one of these games for three adults and three children under 14 is \$75, while admission for two adults and four children is \$62. (*Source:* MLB.com, Texas Rangers)

a. What is the price of an adult admission at Ameriquest Park?

b. What is the price of a child's admission?

c. Suppose that a special rate of \$200 is offered for groups of 20 persons. Should a group of 5 adults and 15 children use the group rate? Why or why not?

EXAMPLE 3 Finding Rates

As part of an exercise program, two students, Albert and Louis, started walking each morning. They live 15 miles away from each other and decided to meet one day by walking toward one another. After 2 hours, they meet. If Louis walks one mile per hour faster than Albert, find both walking speeds.

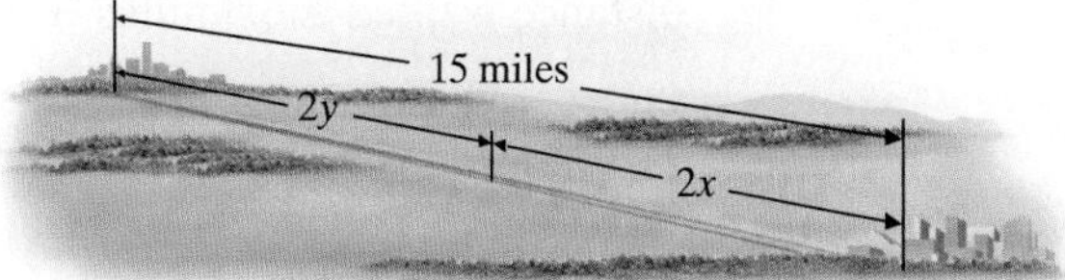

Solution

1. UNDERSTAND. Read and reread the problem. Let's propose a solution and use the formula $d = r \cdot t$ to check. Suppose that Louis's rate is 4 miles per hour. Since Louis's rate is 1 mile per hour faster, Albert's rate is 3 miles per hour. To check, see if they can walk a total of 15 miles in 2 hours. Louis's distance is rate · time $= 4(2) = 8$ miles and Albert's distance is rate · time $= 3(2) = 6$ miles. Their total distance is 8 miles + 6 miles = 14 miles, not the required 15 miles. Now that we have a better understanding of the problem, let's model it with a system of equations.

First, we let

x = Albert's rate in miles per hour
y = Louis's rate in miles per hour

Now we use the facts stated in the problem and the formula $d = rt$ to fill in the following chart.

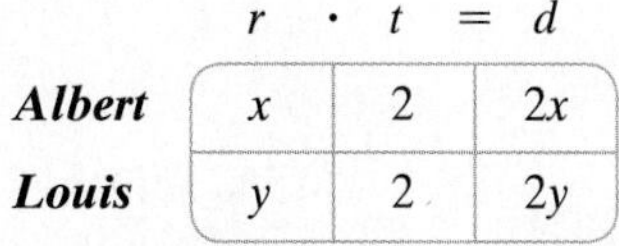

	r	$\cdot$ t	$= d$
Albert	x	2	$2x$
Louis	y	2	$2y$

2. TRANSLATE. We translate the problem into two equations using both variables.

In words:	Albert's distance	+	Louis's distance	=	15
	↓		↓		↓
Translate:	$2x$	+	$2y$	=	15

In words:	Louis's rate	is	1 mile per hour faster than Albert's
	↓	↓	↓
Translate:	y	=	$x + 1$

3. SOLVE. The system of equations we are solving is

$$\begin{cases} 2x + 2y = 15 \\ y = x + 1 \end{cases}$$

Let's use substitution to solve the system since the second equation is solved for y.

$$2x + 2y = 15 \quad \text{First equation}$$

$$2x + 2(x + 1) = 15 \quad \text{Replace } y \text{ with } x + 1.$$

$$2x + 2x + 2 = 15$$

$$4x = 13$$

$$x = 3.25 \quad \text{Divide both sides by 4 and write the result as a decimal.}$$

$$y = x + 1 = 3.25 + 1 = 4.25$$

4. INTERPRET. Albert's proposed rate is 3.25 miles per hour and Louis's proposed rate is 4.25 miles per hour.

Check: Use the formula $d = rt$ and find that in 2 hours, Albert's distance is (3.25)(2) miles or 6.5 miles. In 2 hours, Louis's distance is (4.25)(2) miles or 8.5 miles. The total distance walked is 6.5 miles + 8.5 miles or 15 miles, the given distance.

State: Albert walks at a rate of 3.25 miles per hour and Louis walks at a rate of 4.25 miles per hour. □

PRACTICE

3 Two hikers on a straight trail are 22 miles apart and walking toward each other. After 4 hours, they meet. If one hiker is a nature lover and walks 2 miles per hour slower than the other hiker, find both walking speeds.

EXAMPLE 4 Finding Amounts of Solutions

A chemistry teaching assistant needs 10 liters of a 20% saline solution (salt water) for his 2 p.m. laboratory class. Unfortunately, the only mixtures on hand are a 5% saline solution and a 25% saline solution. How much of each solution should he mix to produce the 20% solution?

Solution

1. UNDERSTAND. Read and reread the problem. Suppose that we need 4 liters of the 5% solution. Then we need 10 − 4 = 6 liters of the 25% solution. To see if this gives us 10 liters of a 20% saline solution, let's find the amount of pure salt in each solution.

	concentration rate	×	amount of solution	=	amount of pure salt
	↓		↓		↓
5% solution:	0.05	×	4 liters	=	0.2 liters
25% solution:	0.25	×	6 liters	=	1.5 liters
20% solution:	0.20	×	10 liters	=	2 liters

Since 0.2 liters + 1.5 liters = 1.7 liters, not 2 liters, our proposed solution is incorrect. But we have gained some insight into how to model and check this problem.

We let

x = number of liters of 5% solution

y = number of liters of 25% solution

Now we use a table to organize the given data.

	Concentration Rate	*Liters of Solution*	*Liters of Pure Salt*
First Solution	5%	x	$0.05x$
Second Solution	25%	y	$0.25y$
Mixture Needed	20%	10	$(0.20)(10)$

2. TRANSLATE. We translate into two equations using both variables.

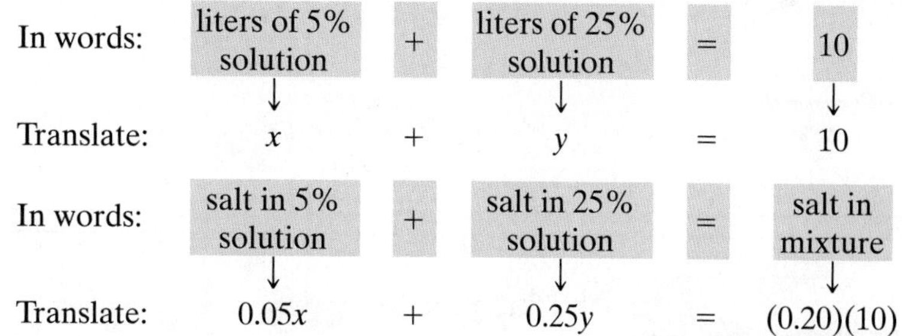

In words: liters of 5% solution + liters of 25% solution = 10
Translate: x + y = 10
In words: salt in 5% solution + salt in 25% solution = salt in mixture
Translate: 0.05x + 0.25y = (0.20)(10)

3. SOLVE. Here we solve the system

$$\begin{cases} x + y = 10 \\ 0.05x + 0.25y = 2 \end{cases}$$

To solve by the addition method, we first multiply the first equation by -25 and the second equation by 100. Then the system

$$\begin{cases} -25(x + y) = -25(10) \\ 100(0.05x + 0.25y) = 100(2) \end{cases} \text{ simplifies to } \begin{cases} -25x - 25y = -250 \\ 5x + 25y = 200 \end{cases}$$

$$-20x = -50 \quad \text{Add.}$$

$$x = 2.5 \quad \text{Divide by } -20.$$

To find y, we let $x = 2.5$ in the first equation of the original system.

$$x + y = 10$$
$$2.5 + y = 10 \quad \text{Let } x = 2.5.$$
$$y = 7.5$$

4. INTERPRET. Thus, we propose that he needs to mix 2.5 liters of 5% saline solution with 7.5 liters of 25% saline solution.

Check: Notice that $2.5 + 7.5 = 10$, the required number of liters. Also, the sum of the liters of salt in the two solutions equals the liters of salt in the required mixture:

$$0.05(2.5) + 0.25(7.5) = 0.20(10)$$
$$0.125 + 1.875 = 2$$

State: The teaching assistant needs 2.5 liters of the 5% saline solution and 7.5 liters of the 25% solution. □

PRACTICE

4 Jemima Juarez owns the Sola Café in southern California. She is known for her interesting coffee blends. To create a new blend, she has decided to use Hawaiian Kona and Jamaica Blue Mountain coffee. To test her new blend, she intends to create 20 pounds of the mix. If the Hawaiian Kona costs \$20 per pound and the Jamaica Blue Mountain costs \$28 per pound, how much of each coffee type should she use to create a new blend that costs \$22 per pound?

Answer to Concept Check:
b

✓CONCEPT CHECK

Suppose you mix an amount of a 30% acid solution with an amount of a 50% acid solution. Which of the following acid strengths would be possible for the resulting acid mixture?

a. 22% **b.** 44% **c.** 63%

Vocabulary, Readiness & Video Check

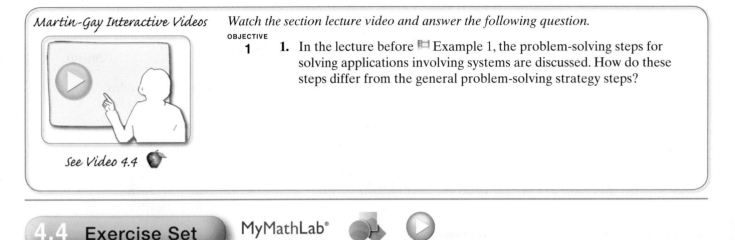

Martin-Gay Interactive Videos

See Video 4.4 

Watch the section lecture video and answer the following question.

OBJECTIVE 1

1. In the lecture before Example 1, the problem-solving steps for solving applications involving systems are discussed. How do these steps differ from the general problem-solving strategy steps?

4.4 Exercise Set

MyMathLab®

Without actually solving each problem, choose each correct solution by deciding which choice satisfies the given conditions.

△ **1.** The length of a rectangle is 3 feet longer than the width. The perimeter is 30 feet. Find the dimensions of the rectangle.

a. length = 8 feet; width = 5 feet
b. length = 8 feet; width = 7 feet
c. length = 9 feet; width = 6 feet

△ **2.** An isosceles triangle, a triangle with two sides of equal length, has a perimeter of 20 inches. Each of the equal sides is one inch longer than the third side. Find the lengths of the three sides.

a. 6 inches, 6 inches, and 7 inches
b. 7 inches, 7 inches, and 6 inches
c. 6 inches, 7 inches, and 8 inches

3. Two computer disks and three notebooks cost \$17. However, five computer disks and four notebooks cost \$32. Find the price of each.

a. notebook = \$4; computer disk = \$3
b. notebook = \$3; computer disk = \$4
c. notebook = \$5; computer disk = \$2

4. Two music CDs and four music cassette tapes cost a total of \$40. However, three music CDs and five cassette tapes cost \$55. Find the price of each.

a. CD = \$12; cassette = \$4
b. CD = \$15; cassette = \$2
c. CD = \$10; cassette = \$5

5. Kesha has a total of 100 coins, all of which are either dimes or quarters. The total value of the coins is \$13.00. Find the number of each type of coin.

a. 80 dimes; 20 quarters **b.** 20 dimes; 44 quarters
c. 60 dimes; 40 quarters

6. Samuel has 28 gallons of saline solution available in two large containers at his pharmacy. One container holds three times as much as the other container. Find the capacity of each container.

a. 15 gallons; 5 gallons **b.** 20 gallons; 8 gallons
c. 21 gallons; 7 gallons

TRANSLATING

Write a system of equations in x and y describing each situation. Do not solve the system. See Example 1.

7. A smaller number and a larger number add up to 15 and have a difference of 7. (Let x be the larger number.)

8. The total of two numbers is 16. The first number plus 2 more than 3 times the second equals 18. (Let x be the first number.)

9. Keiko has a total of \$6500, which she has invested in two accounts. The larger account is \$800 greater than the smaller account. (Let x be the amount of money in the larger account.)

10. Dominique has four times as much money in his savings account as in his checking account. The total amount is \$2300. (Let x be the amount of money in his checking account.)

MIXED PRACTICE

Solve. See Examples 1 through 4.

11. Two numbers total 83 and have a difference of 17. Find the two numbers.

12. The sum of two numbers is 76 and their difference is 52. Find the two numbers.

13. A first number plus twice a second number is 8. Twice the first number, plus the second totals 25. Find the numbers.

14. One number is 4 more than twice the second number. Their total is 25. Find the numbers.

15. Miguel Cabrera of the Detroit Tigers led Major League Baseball in runs batted in for the 2010 regular season. Alex Rodriguez of the New York Yankees, who came in second to Cabrera, had 1 fewer run batted in for the 2010 regular season. Together, these two players brought home 251 runs during the 2010 regular season. How many runs batted in did each player account for? (*Source:* Major League Baseball)

16. The highest scorer during the WNBA 2010 regular season was Cappie Pondexter of the New York Liberty. Over the season, Pondexter scored 13 more points than the second-highest scorer, Angel McCoughtry of the Atlanta Dream. Together, Pondexter and McCoughtry scored 1445 points during the 2010 regular season. How many points did each player score over the course of the season? (*Source:* Women's National Basketball Association)

17. Ann Marie Jones has been pricing Amtrak train fares for a group trip to New York. Three adults and four children must pay \$159. Two adults and three children must pay \$112. Find the price of an adult's ticket and find the price of a child's ticket.

18. Last month, Jerry Papa purchased two DVDs and five CDs at Wall-to-Wall Sound for \$65. This month, he bought four DVDs and three CDs for \$81. Find the price of each DVD and find the price of each CD.

19. Johnston and Betsy Waring have a jar containing 80 coins, all of which are either quarters or nickels. The total value of the coins is \$14.60. How many of each type of coin do they have?

20. Sarah and Keith Robinson purchased 40 stamps, a mixture of 44¢ and 20¢ stamps. Find the number of each type of stamp if they spent \$16.40.

21. Norman and Suzanne Scarpulla own 35 shares of McDonald's stock and 69 shares of The Ohio Art Company stock (makers of Etch A Sketch and other toys). On a particular day in 2011, their stock portfolio consisting of these two stocks was worth \$2814. The McDonald's stock was \$70 more per share than The Ohio Art Company stock. What was the price of each stock on that day? (*Source:* Yahoo finance)

22. Saralee Rose has investments in Google and Nintendo stock. During a particular day in 2011, Google stock was at \$540 per share, and Nintendo stock was at \$30 per share. Saralee's portfolio made up of these two stocks was worth \$20,610 at that time. If Saralee owns 16 more shares of Google stock than she owns of Nintendo stock, how many shares of each type of stock does she own?

23. Twice last month, Judy Carter rented a car from Enterprise in Fresno, California, and traveled around the Southwest on business. Enterprise rents this car for a daily fee plus an additional charge per mile driven. Judy recalls that her first trip lasted 4 days, she drove 450 miles, and the rental cost her \$240.50. On her second business trip, she drove the same model of car a distance of 200 miles in 3 days and paid \$146.00 for the rental. Find the daily fee and the mileage charge.

24. Joan Gundersen rented the same car model twice from Hertz, which rents this car model for a daily fee plus an additional charge per mile driven. Joan recalls that the car rented for 5 days and driven for 300 miles cost her \$178, while the same model car rented for 4 days and driven for 500 miles cost \$197. Find the daily fee and find the mileage charge.

25. Pratap Puri rowed 18 miles down the Delaware River in 2 hours, but the return trip took him $4\frac{1}{2}$ hours. Find the rate Pratap can row in still water and find the rate of the current.

Let x = rate Pratap can row in still water and
y = rate of the current

	d	$=$ r	$\cdot$ t
Downstream		$x + y$	
Upstream		$x - y$	

26. The Jonathan Schultz family took a canoe 10 miles down the Allegheny River in $1\frac{1}{4}$ hours. After lunch, it took them 4 hours to return. Find the rate of the current.

Let x = rate the family can row in still water and
y = rate of the current

	d	$=$ r	$\cdot$ t
Downstream		$x + y$	
Upstream		$x - y$	

27. Dave and Sandy Hartranft are frequent flyers with Delta Airlines. They often fly from Philadelphia to Chicago, a distance of 780 miles. On one particular trip, they fly into the wind, and the flight takes 2 hours. The return trip, with the wind behind them, only takes $1\frac{1}{2}$ hours. If the wind speed is the same on each trip, find the speed of the wind and find the speed of the plane in still air.

28. With a strong wind behind it, a United Airlines jet flies 2400 miles from Los Angeles to Orlando in $4\frac{3}{4}$ hours. The return trip takes 6 hours because the plane flies into the wind. If the wind speed is the same on each trip, find the speed of the plane in still air and find the wind speed to the nearest tenth of a mile per hour.

29. Kevin Briley began a 114-mile bicycle trip to build up stamina for a triathlete competition. Unfortunately, his bicycle chain broke, so he finished the trip walking. The whole trip took 6 hours. If Kevin walks at a rate of 4 miles per hour and rides at 24 miles per hour, find the amount of time he spent on the bicycle.

30. In Canada, eastbound and westbound trains travel along the same track, with sidings to pull onto to avoid accidents. Two trains are now 150 miles apart, with the westbound train traveling twice as fast as the eastbound train. A warning must be issued to pull one train onto a siding or else the trains will crash in $1\frac{1}{4}$ hours. Find the speed of the eastbound train and the speed of the westbound train.

31. Doreen Schmidt is a chemist with Gemco Pharmaceutical. She needs to prepare 12 liters of a 9% hydrochloric acid solution. Find the amount of a 4% solution and the amount of a 12% solution she should mix to get this solution.

Concentration Rate	*Liters of Solution*	*Liters of Pure Acid*
0.04	x	$0.04x$
0.12	y	?
0.09	12	?

32. Elise Everly is preparing 15 liters of a 25% saline solution. Elise has two other saline solutions, with strengths of 40% and 10%. Find the amount of 40% solution and the amount of 10% solution she should mix to get 15 liters of a 25% solution.

Concentration Rate	*Liters of Solution*	*Liters of Pure Salt*
0.40	x	$0.40x$
0.10	y	?
0.25	15	?

33. Wayne Osby blends coffee for a local coffee café. He needs to prepare 200 pounds of blended coffee beans selling for \$3.95 per pound. He intends to do this by blending together a high-quality bean costing \$4.95 per pound and a cheaper bean costing \$2.65 per pound. To the nearest pound, find how much high-quality coffee bean and how much cheaper coffee bean he should blend.

34. Macadamia nuts cost an astounding \$16.50 per pound, but research by an independent firm says that mixed nuts sell better if macadamias are included. The standard mix costs \$9.25 per pound. Find how many pounds of macadamias and how many pounds of the standard mix should be combined to produce 40 pounds that will cost \$10 per pound. Find the amounts to the nearest tenth of a pound.

35. Recall that two angles are complementary if the sum of their measures is 90°. Find the measures of two complementary angles if one angle is twice the other.

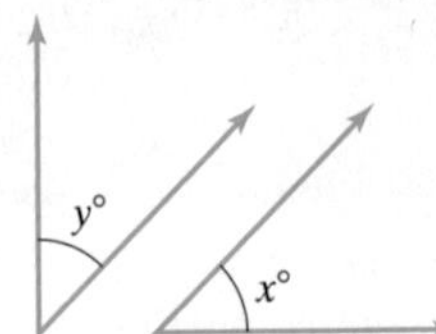

36. Recall that two angles are supplementary if the sum of their measures is 180°. Find the measures of two supplementary angles if one angle is 20° more than four times the other. (See art on top of next column.)

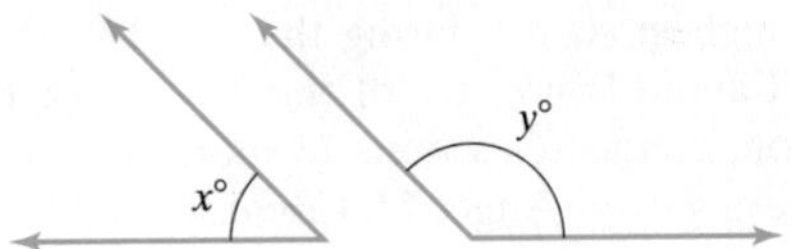

37. Find the measures of two complementary angles if one angle is 10° more than three times the other.

38. Find the measures of two supplementary angles if one angle is 18° more than twice the other.

39. Kathi and Robert Hawn had a pottery stand at the annual Skippack Craft Fair. They sold some of their pottery at the original price of \$9.50 each but later decreased the price of each by \$2. If they sold all 90 pieces and took in \$721, find how many they sold at the original price and how many they sold at the reduced price.

40. A charity fund-raiser consisted of a spaghetti supper where a total of 387 people were fed. They charged \$6.80 for adults and half price for children. If they took in \$2444.60, find how many adults and how many children attended the supper.

41. The Santa Fe National Historic Trail is approximately 1200 miles between Old Franklin, Missouri, and Santa Fe, New Mexico. Suppose that a group of hikers start from each town and walk the trail toward each other. They meet after a total hiking time of 240 hours. If one group travels $\frac{1}{2}$ mile per hour slower than the other group, find the rate of each group. (*Source:* National Park Service)

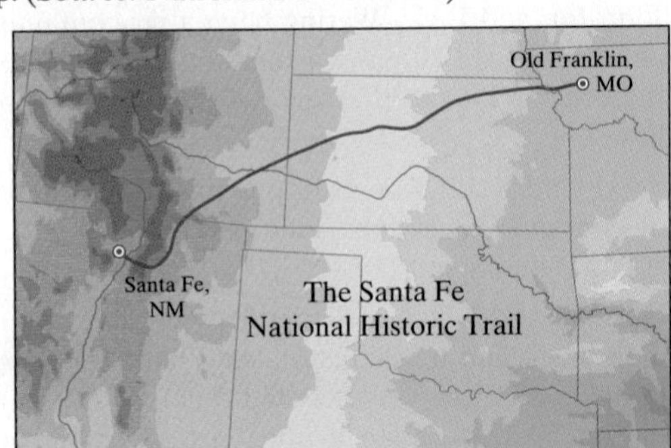

42. California 1 South is a historic highway that stretches 123 miles along the coast from Monterey to Morro Bay. Suppose that two antique cars start driving this highway, one from each town. They meet after 3 hours. Find the rate of each car if one car travels 1 mile per hour faster than the other car. (*Source:* National Geographic)

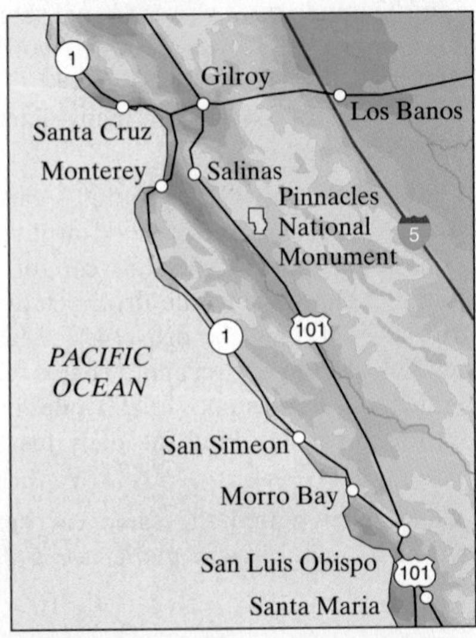

43. A 30% solution of fertilizer is to be mixed with a 60% solution of fertilizer to get 150 gallons of a 50% solution. How many gallons of the 30% solution and 60% solution should be mixed?

44. A 10% acid solution is to be mixed with a 50% acid solution to get 120 ounces of a 20% acid solution. How many ounces of the 10% solution and 50% solution should be mixed?

45. Traffic signs are regulated by the *Manual on Uniform Traffic Control Devices* (MUTCD). According to this manual, if the sign below is placed on a freeway, its perimeter must be 144 inches. Also, its length is 12 inches longer than its width. Find the dimensions of this sign.

46. According to the MUTCD (see Exercise 45), this sign must have a perimeter of 60 inches. Also, its length must be 6 inches longer than its width. Find the dimensions of this sign.

REVIEW AND PREVIEW

Solve each linear inequality. Write your solution in interval notation. See Section 2.9.

47. $-3x < -9$

48. $2x - 7 \leq 5x + 11$

49. $4(2x - 1) \geq 0$

50. $\frac{2}{3}x < \frac{1}{3}$

CONCEPT EXTENSIONS

Solve. See the Concept Check in this section.

51. Suppose you mix an amount of candy costing \$0.49 a pound with candy costing \$0.65 a pound. Which of the following costs per pound could result?

a. \$0.58 **b.** \$0.72 **c.** \$0.29

52. Suppose you mix a 50% acid solution with pure acid (100%). Which of the following acid strengths are possible for the resulting acid mixture?

a. 25% **b.** 150% **c.** 62% **d.** 90%

△ **53.** Dale and Sharon Mahnke have decided to fence off a garden plot behind their house, using their house as the "fence" along one side of the garden. The length (which runs parallel to the house) is 3 feet less than twice the width. Find the dimensions if 33 feet of fencing is used along the three sides requiring it.

△ **54.** Judy McElroy plans to erect 152 feet of fencing around her rectangular horse pasture. A river bank serves as one side length of the rectangle. If each width is 4 feet longer than half the length, find the dimensions.

55. The percent of viewers who watch nightly network news can be approximated by the equation $y = 0.82x + 17.2$, where x is the years of age over 18 of the viewer. The percent of viewers who watch cable TV news is approximated by the equation $y = 0.33x + 30.5$ where x is also the years of age over 18 of the viewer. (*Source:* The Pew Research Center for the People & the Press)

a. Solve the system of equations: $\begin{cases} y = 0.82x + 17.2 \\ y = 0.33x + 30.5 \end{cases}$

Round x and y to the nearest tenth.

b. Explain what the point of intersection means in terms of the context of the exercise.

c. Look at the slopes of both equations of the system. What type of news attracts older viewers more? What type of news attracts younger viewers more?

56. In the triangle below, the measure of angle x is 6 times the measure of angle y. Find the measure of x and y by writing a system of two equations in two unknowns and solving the system.

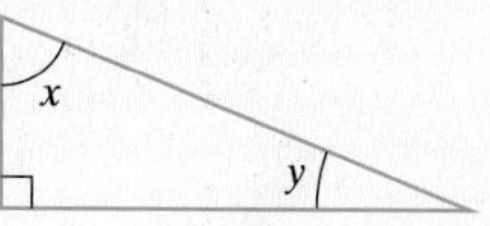

4.5 Graphing Linear Inequalities

OBJECTIVE

1 Determine Whether an Ordered Pair Is a Solution of a Linear Inequality in Two Variables.

2 Graph a Linear Inequality in Two Variables.

In the next section, we continue our work with systems by solving systems of linear inequalities. Before that section, we first need to learn to graph a single linear inequality in two variables.

Recall that a linear equation in two variables is an equation that can be written in the form $Ax + By = C$ where A, B, and C are real numbers and A and B are not both 0. The definition of a linear inequality is the same except that the equal sign is replaced with an inequality symbol.

A **linear inequality in two variables** is an inequality that can be written in one of the forms:

$$Ax + By < C \qquad Ax + By \leq C$$
$$Ax + By > C \qquad Ax + By \geq C$$

where A, B, and C are real numbers and A and B are not both 0.

OBJECTIVE

1 Determining Solutions of Linear Inequalities in Two Variables

Just as for linear equations in x and y, an ordered pair is a **solution** of an inequality in x and y if replacing the variables by coordinates of the ordered pair results in a true statement.

EXAMPLE 1 Determine whether each ordered pair is a solution of the inequality $2x - y < 6$.

a. $(5, -1)$ **b.** $(2, 7)$

Solution

a. We replace x with 5 and y with -1 and see if a true statement results.

$$2x - y < 6$$
$$2(5) - (-1) < 6 \quad \text{Replace } x \text{ with 5 and } y \text{ with } -1.$$
$$10 + 1 < 6$$
$$11 < 6 \quad \text{False}$$

The ordered pair $(5, -1)$ is not a solution since $11 < 6$ is a false statement.

b. We replace x with 2 and y with 7 and see if a true statement results.

$$2x - y < 6$$
$$2(2) - (7) < 6 \quad \text{Replace } x \text{ with 2 and } y \text{ with 7.}$$
$$4 - 7 < 6$$
$$-3 < 6 \quad \text{True}$$

The ordered pair $(2, 7)$ is a solution since $-3 < 6$ is a true statement. □

PRACTICE

1 Determine whether each ordered pair is a solution of $x - 4y > 8$.

a. $(-3, 2)$ **b.** $(9, 0)$

OBJECTIVE

2 Graphing Linear Inequalities in Two Variables

The linear equation $x - y = 1$ is graphed next. Recall that all points on the line correspond to ordered pairs that satisfy the equation $x - y = 1$.

Notice the line defined by $x - y = 1$ divides the rectangular coordinate system plane into 2 sides. All points on one side of the line satisfy the inequality $x - y < 1$ and all points on the other side satisfy the inequality $x - y > 1$. The graph on page 289 shows a few examples of this.

Check	$x - y < 1$
(1, 3)	$1 - 3 < 1$ True
(−2, 1)	$-2 - 1 < 1$ True
(−4, −4)	$-4 - (-4) < 1$ True

Check	$x - y > 1$
(4, 1)	$4 - 1 > 1$ True
(2, −2)	$2 - (-2) > 1$ True
(0, −4)	$0 - (-4) > 1$ True

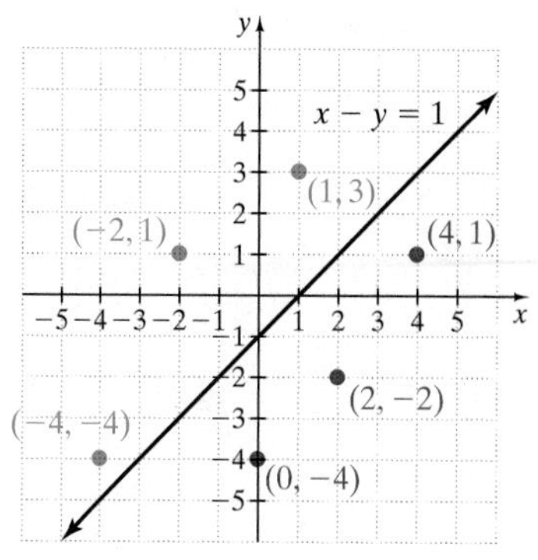

The graph of $x - y < 1$ is the region shaded blue, and the graph of $x - y > 1$ is the region shaded red below.

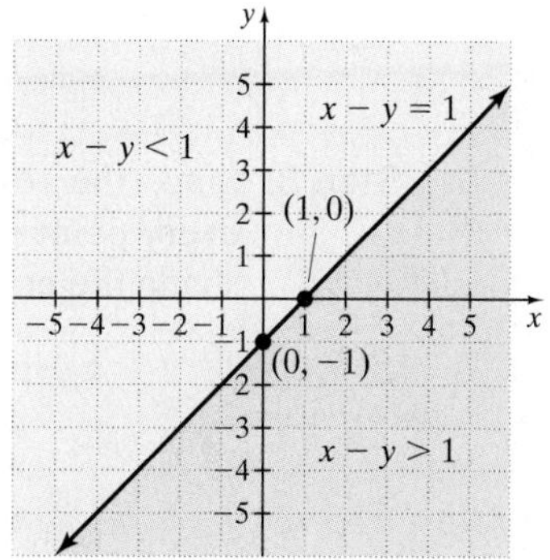

The region to the left of the line and the region to the right of the line are called **half-planes.** Every line divides the plane (similar to a sheet of paper extending indefinitely in all directions) into two half-planes; the line is called the **boundary.**

Recall that the inequality $x - y \le 1$ means

$$x - y = 1 \quad \text{or} \quad x - y < 1$$

Thus, the graph of $x - y \le 1$ is the half-plane $x - y < 1$ along with the boundary line $x - y = 1$.

Graphing a Linear Inequality in Two Variables

Step 1. Graph the boundary line found by replacing the inequality sign with an equal sign. If the inequality sign is $>$ or $<$, graph a dashed boundary line (indicating that the points on the line are not solutions of the inequality). If the inequality sign is $\ge$ or $\le$, graph a solid boundary line (indicating that the points on the line are solutions of the inequality).

Step 2. Choose a point, *not* on the boundary line, as a test point. Substitute the coordinates of this test point into the *original* inequality.

Step 3. If a true statement is obtained in Step 2, shade the half-plane that contains the test point. If a false statement is obtained, shade the half-plane that does not contain the test point.

EXAMPLE 2 Graph: $x + y < 7$

Solution

Step 1. First we graph the boundary line by graphing the equation $x + y = 7$. We graph this boundary as a dashed line because the inequality sign is $<$, and thus the points on the line are not solutions of the inequality $x + y < 7$.

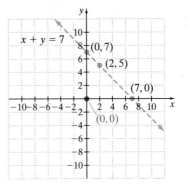

Step 2. Next, we choose a test point, being careful not to choose a point on the boundary line. We choose (0, 0). Substitute the coordinates of (0, 0) into $x + y < 7$.

$$x + y < 7 \quad \text{Original inequality}$$
$$0 + 0 \stackrel{?}{<} 7 \quad \text{Replace } x \text{ with 0 and } y \text{ with 0.}$$
$$0 < 7 \quad \text{True}$$

Step 3. Since the result is a true statement, (0, 0) is a solution of $x + y < 7$, and every point in the same half-plane as (0, 0) is also a solution. To indicate this, shade the entire half-plane containing (0, 0), as shown.

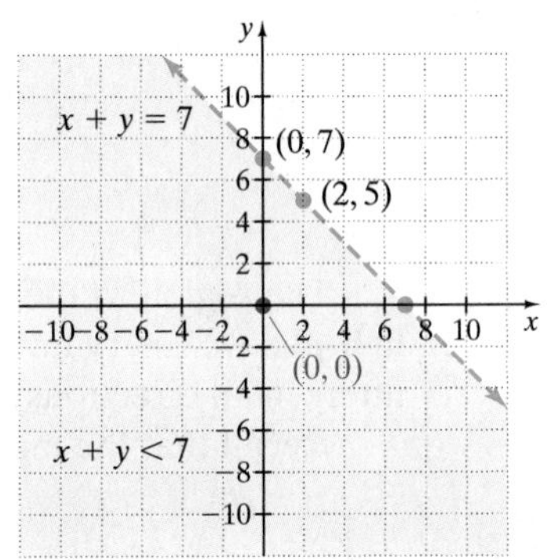

□

PRACTICE 2 Graph: $x + y > 5$

✓CONCEPT CHECK

Determine whether (0, 0) is included in the graph of

a. $y \geq 2x + 3$ **b.** $x < 7$ **c.** $2x - 3y < 6$

EXAMPLE 3 Graph: $2x - y \geq 3$

Solution

Step 1. We graph the boundary line by graphing $2x - y = 3$. We draw this line as a solid line because the inequality sign is $\geq$, and thus the points on the line are solutions of $2x - y \geq 3$.

Step 2. Once again, (0, 0) is a convenient test point since it is not on the boundary line. We substitute 0 for x and 0 for y into the original inequality.

$$2x - y \geq 3$$
$$2(0) - 0 \geq 3 \quad \text{Let } x = 0 \text{ and } y = 0.$$
$$0 \geq 3 \quad \text{False}$$

Answers to Concept Check:
a. no **b.** yes **c.** yes

Step 3. Since the statement is false, no point in the half-plane containing (0, 0) is a solution. Therefore, we shade the half-plane that does not contain (0, 0). Every point in the shaded half-plane and every point on the boundary line is a solution of $2x - y \geq 3$.

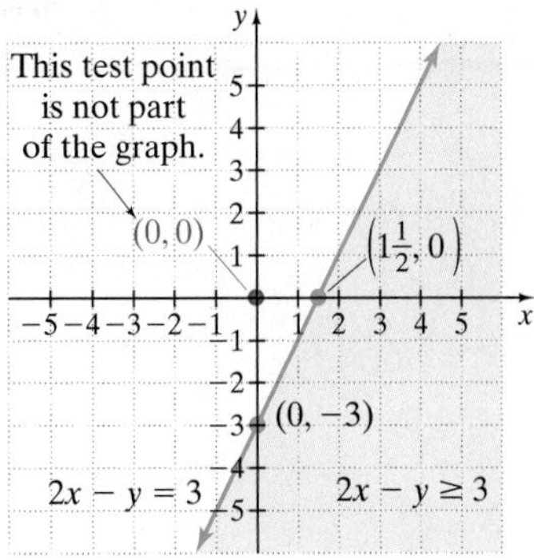

PRACTICE **3** Graph: $3x - y \geq 4$

> **Helpful Hint**
> When graphing an inequality, make sure the test point is substituted into the **original inequality.** For Example 3, we substituted the test point (0, 0) into the **original inequality** $2x - y \geq 3$, *not* $2x - y = 3$.

EXAMPLE 4 Graph: $x > 2y$

Solution

Step 1. We find the boundary line by graphing $x = 2y$. The boundary line is a dashed line since the inequality symbol is $>$.

Step 2. We cannot use (0, 0) as a test point because it is a point on the boundary line. We choose instead (0, 2).

$x > 2y$

$0 > 2(2)$ Let $x = 0$ and $y = 2$.

$0 > 4$ False

Step 3. Since the statement is false, we shade the half-plane that does not contain the test point (0, 2), as shown.

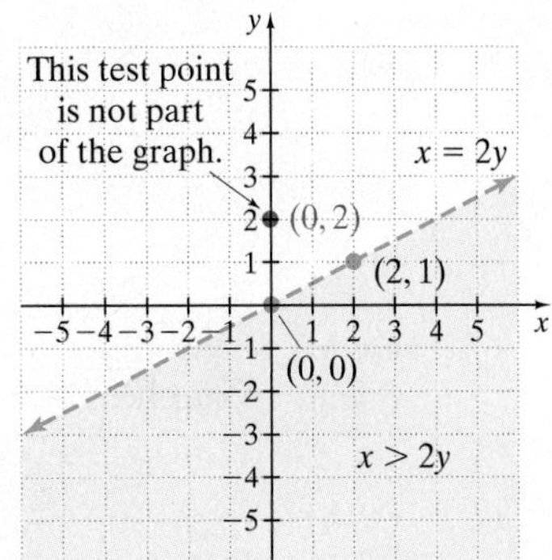

PRACTICE **4** Graph: $x > 3y$

EXAMPLE 5 Graph: $5x + 4y \le 20$

Solution We graph the solid boundary line $5x + 4y = 20$ and choose $(0, 0)$ as the test point.

$$5x + 4y \le 20$$
$$5(0) + 4(0) \overset{?}{\le} 20 \quad \text{Let } x = 0 \text{ and } y = 0.$$
$$0 \le 20 \quad \text{True}$$

We shade the half-plane that contains $(0, 0)$, as shown.

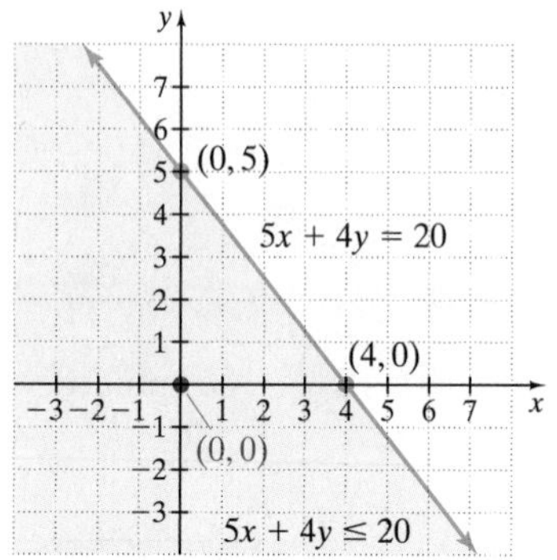

PRACTICE 5 Graph: $3x + 4y \ge 12$

EXAMPLE 6 Graph: $y > 3$

Solution We graph the dashed boundary line $y = 3$ and choose $(0, 0)$ as the test point. (Recall that the graph of $y = 3$ is a horizontal line with y-intercept 3.)

$$y > 3$$
$$0 \overset{?}{>} 3 \quad \text{Let } y = 0.$$
$$0 > 3 \quad \text{False}$$

We shade the half-plane that does not contain $(0, 0)$, as shown.

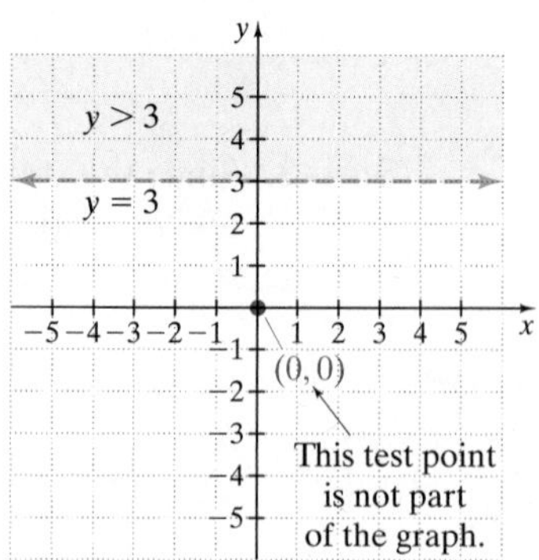

PRACTICE 6 Graph: $x > 3$

EXAMPLE 7 Graph: $y \le \frac{2}{3}x - 4$

Solution Graph the solid boundary line $y = \frac{2}{3}x - 4$. This equation is in slope-intercept form with slope $\frac{2}{3}$ and y-intercept -4.

We use this information to graph the line. Then we choose (0, 0) as our test point.

$$y \le \frac{2}{3}x - 4$$

$$0 \overset{?}{\le} \frac{2}{3} \cdot 0 - 4$$

$$0 \le -4 \quad \text{False}$$

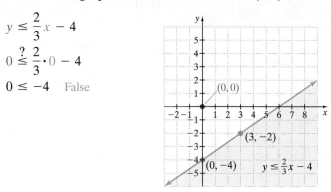

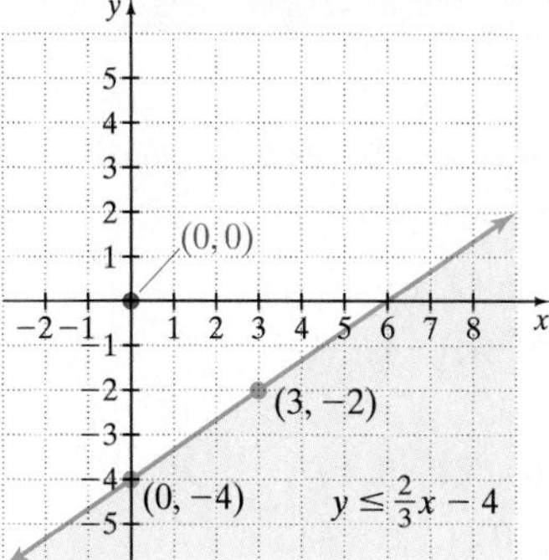

We shade the half-plane that does not contain (0, 0), as shown. □

PRACTICE 7 Graph $y < \frac{3}{4}x - 1$.

Vocabulary, Readiness & Video Check

Use the choices below to fill in each blank. Some choices may be used more than once and some not at all.

true	$x < 3$	$y < 3$	half-planes	yes
false	$x \le 3$	$y \le 3$	linear inequality in two variables	no

1. The statement $5x - 6y < 7$ is an example of a(n) ______________________.
2. A boundary line divides a plane into two regions called ______________________.
3. True or false: The graph of $5x - 6y < 7$ includes its corresponding boundary line. ______________________
4. True or false: When graphing a linear inequality, to determine which side of the boundary line to shade, choose a point *not* on the boundary line. ______________________
5. True or false: The boundary line for the inequality $5x - 6y < 7$ is the graph of $5x - 6y = 7$.

   ____
6. The graph of ______________________ is

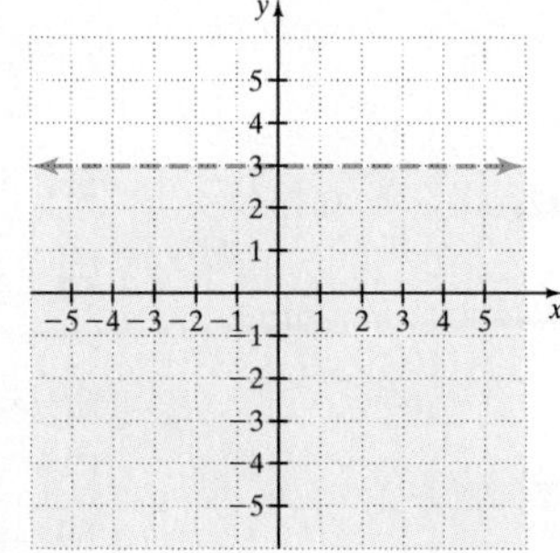

Martin-Gay Interactive Videos

See Video 4.5

Watch the section lecture video and answer the following questions.

OBJECTIVE 1

7. From Example 1, how do you determine whether an ordered pair in x and y is a solution of an inequality in x and y?

OBJECTIVE 2

8. From Example 3, how do you find the equation of the boundary line? How do you determine if the points on the boundary line are solutions of the inequality?

4.5 Exercise Set

MyMathLab®

Determine which ordered pairs given are solutions of the linear inequality in two variables. See Example 1.

1. $x - y > 3; (2, -1), (5, 1)$
2. $y - x < -2; (2, 1), (5, -1)$
3. $3x - 5y \le -4; (-1, -1), (4, 0)$
4. $2x + y \ge 10; (-1, -4), (5, 0)$
5. $x < -y; (0, 2), (-5, 1)$
6. $y > 3x; (0, 0), (-1, -4)$

MIXED PRACTICE

Graph each inequality. See Examples 2 through 7.

7. $x + y \le 1$
8. $x + y \ge -2$
9. $2x + y > -4$
10. $x + 3y \le 3$
11. $x + 6y \le -6$
12. $7x + y > -14$
13. $2x + 5y > -10$
14. $5x + 2y \le 10$
15. $x + 2y \le 3$
16. $2x + 3y > -5$
17. $2x + 7y > 5$
18. $3x + 5y \le -2$
19. $x - 2y \ge 3$
20. $4x + y \le 2$
21. $5x + y < 3$
22. $x + 2y > -7$
23. $4x + y < 8$
24. $9x + 2y \ge -9$
25. $y \ge 2x$
26. $x < 5y$
27. $x \ge 0$
28. $y \le 0$
29. $y \le -3$
30. $x > -\frac{2}{3}$
31. $2x - 7y > 0$
32. $5x + 2y \le 0$
33. $3x - 7y \ge 0$
34. $-2x - 9y > 0$
35. $x > y$
36. $x \le -y$
37. $x - y \le 6$
38. $x - y > 10$
39. $-\frac{1}{4}y + \frac{1}{3}x > 1$
40. $\frac{1}{2}x - \frac{1}{3}y \le -1$
41. $-x < 0.4y$
42. $0.3x \ge 0.1y$

In Exercises 43 through 48, match each inequality with its graph.

a. $x > 2$ b. $y < 2$ c. $y < 2x$

d. $y \le -3x$ e. $2x + 3y < 6$ f. $3x + 2y > 6$

43.

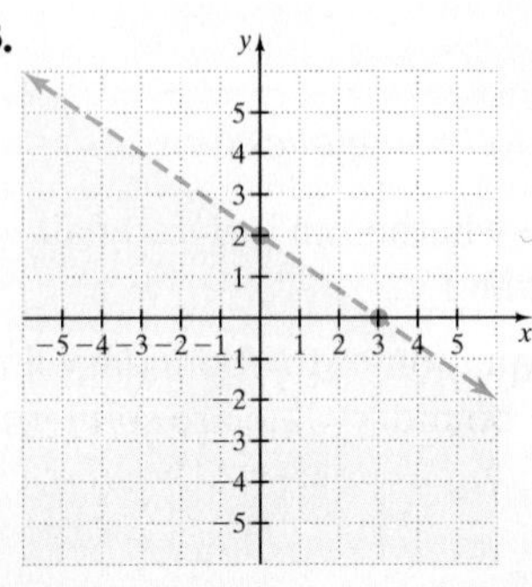

44.

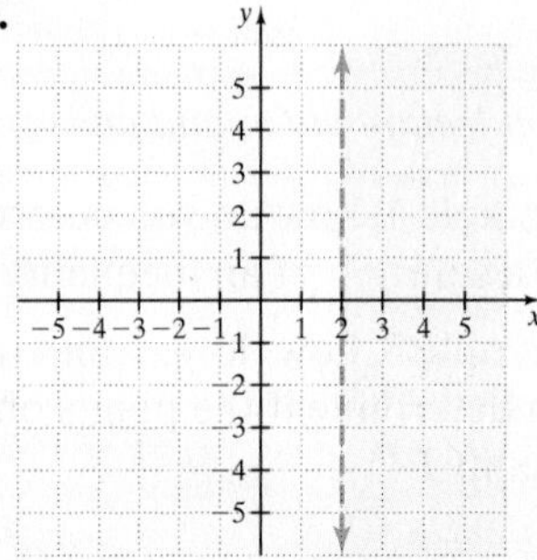

45.

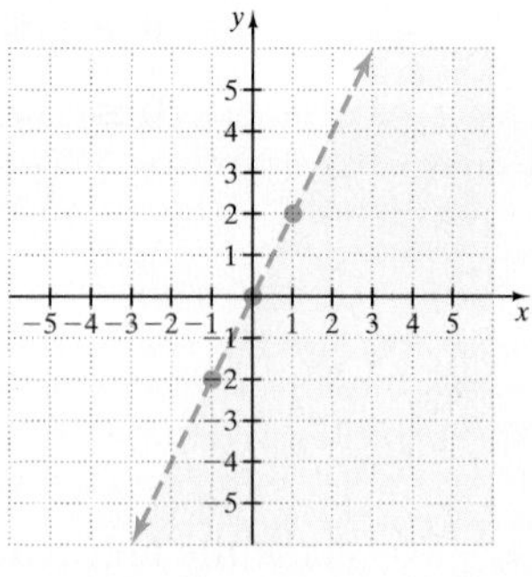

46.

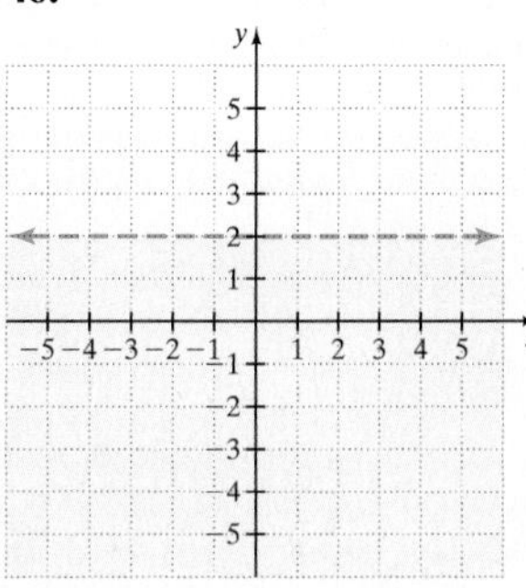

47.

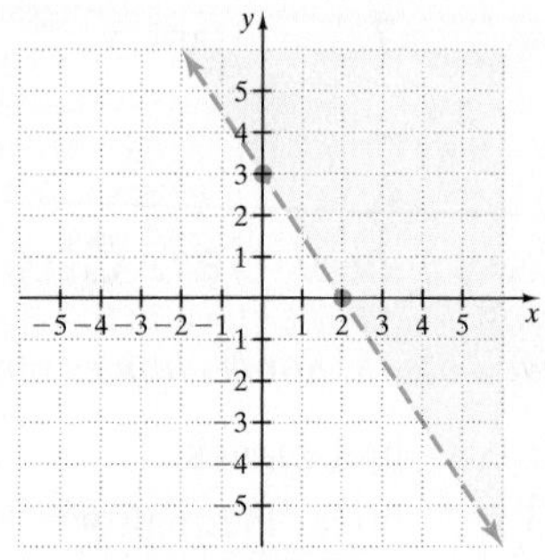

48.

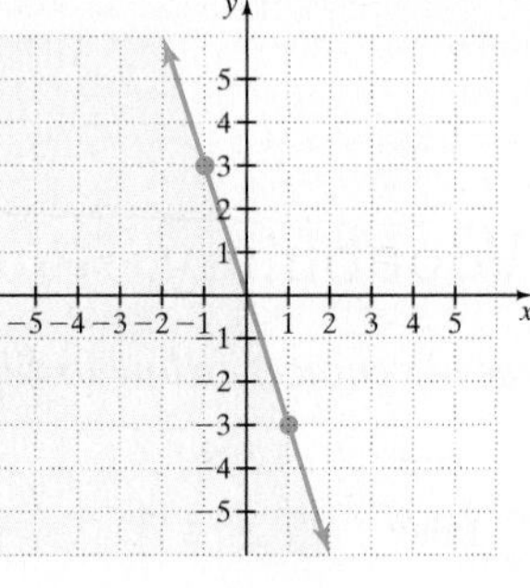

REVIEW AND PREVIEW

Evaluate. See Section 1.4.

49. 2^3
50. 3^4
51. $(-2)^5$
52. -2^5
53. $3 \cdot 4^2$
54. $4 \cdot 3^3$

Evaluate each expression for the given replacement value. See Section 1.4.

55. x^2 if x is -5
56. x^3 if x is -5
57. $2x^3$ if x is -1
58. $3x^2$ if x is -1

State whether the graph of each inequality includes its corresponding boundary line.

59. $y \ge x + 4$
60. $x - y > -7$
61. $y \ge x$
62. $x > 0$

CONCEPT EXTENSIONS

Determine whether (1, 1) is included in each graph. See the Concept Check in this section.

63. $3x + 4y < 8$
64. $y > 5x$
65. $y \ge -\frac{1}{2}x$
66. $x > 3$

67. Write an inequality whose solutions are all pairs of numbers x and y whose sum is at least 13. Graph the inequality.

68. Write an inequality whose solutions are all the pairs of numbers x and y whose sum is at most -4. Graph the inequality.

69. Explain why a point on the boundary line should not be chosen as the test point.

70. Describe the graph of a linear inequality.

71. The price for a taxi cab in a small city is \$2.50 per mile, x, while traveling, and \$0.25 every minute, y, while waiting. If you have \$20 to spend on a cab ride, the inequality

$$2.5x + 0.25y \le 20$$

represents your situation. Graph this inequality in the first quadrant only.

72. A word processor charges \$22 per hour, x, for typing a first draft, and \$15 per hour, y, for making changes and editing to create a second draft. If you need a document created (first and second drafts) and have \$100, the inequality

$$22x + 15y \le 100$$

represents your situation. Graph the inequality in the first quadrant only.

73. In Exercises 71 and 72, why were you instructed to graph each inequality in the first quadrant only?

74. It's the end of the budgeting period for Dennis Fernandes, and he has \$500 left in his budget for car rental expenses. He plans to spend this budget on a sales trip throughout southern Texas. He will rent a car that costs \$30 per day and \$0.15 per mile, and he can spend no more than \$500.

a. Write an inequality describing this situation. Let $x =$ number of days and let $y =$ number of miles.

b. Graph this inequality.

c. Why should we be interested in only quadrant I of this graph?

75. Scott Sambracci and Sara Thygeson are planning their wedding. They have calculated that they want the cost of their wedding ceremony x plus the cost of their reception y to be no more than \$5000.

a. Write an inequality describing this relationship.

b. Graph this inequality.

c. Why should we be interested in only quadrant I of this graph?

4.6 Systems of Linear Inequalities 

OBJECTIVE

1 Solve a System of Linear Inequalities.

OBJECTIVE

1 Solving Systems of Linear Inequalities

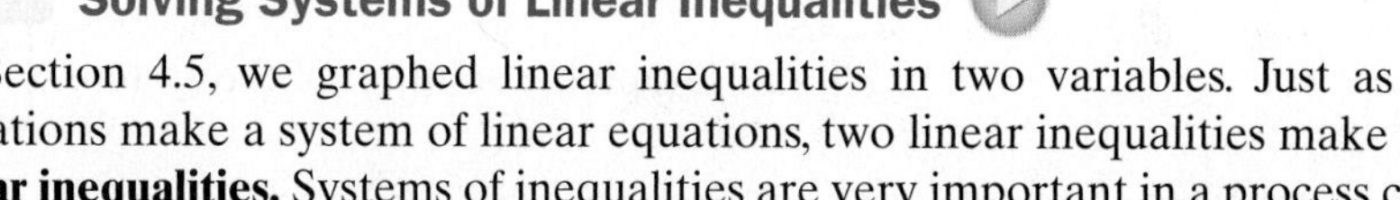

In Section 4.5, we graphed linear inequalities in two variables. Just as two linear equations make a system of linear equations, two linear inequalities make a **system of linear inequalities.** Systems of inequalities are very important in a process called linear programming. Many businesses use linear programming to find the most profitable way to use limited resources such as employees, machines, or buildings.

A **solution of a system of linear inequalities** is an ordered pair that satisfies each inequality in the system. The set of all such ordered pairs is the solution set of the system. Graphing this set gives us a picture of the solution set. We can graph a system of inequalities by graphing each inequality in the system and identifying the region of overlap.

EXAMPLE 1 Graph the solution of the system: $\begin{cases} 3x \ge y \\ x + 2y \le 8 \end{cases}$

Solution We begin by graphing each inequality on the same set of axes. The graph of the solution of the system is the region contained in the graphs of both inequalities. It is their intersection.

First, graph $3x \ge y$. The boundary line is the graph of $3x = y$. Sketch a solid boundary line since the inequality $3x \ge y$ means $3x > y$ or $3x = y$. The test point $(1, 0)$ satisfies the inequality, so shade the half-plane that includes $(1, 0)$. (This is shown on the next page.)

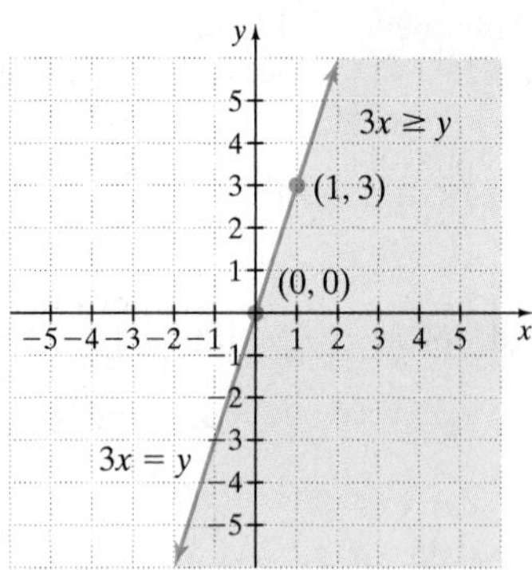

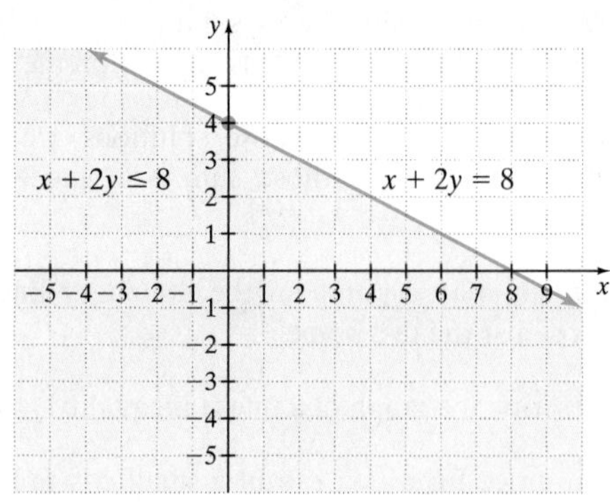

Next, graph $x + 2y \leq 8$ on the same set of axes. (For clarity, the graph of $x + 2y \leq 8$ is shown on a separate set of axes.) Sketch a solid boundary line $x + 2y = 8$. The test point $(0, 0)$ satisfies the inequality $x + 2y \leq 8$, so shade the half-plane that includes $(0, 0)$.

An ordered pair solution of the system must satisfy both inequalities. These solutions are points that lie in both shaded regions. The solution of the system is the purple shaded region as seen below. This solution includes parts of both boundary lines.

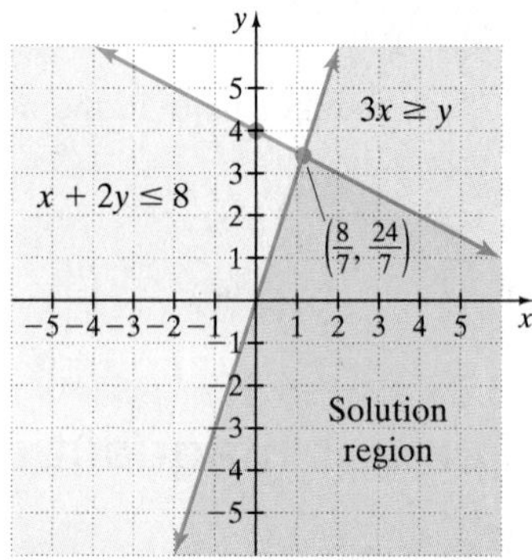

□

PRACTICE 1 Graph the solution of the system: $\begin{cases} 4x \leq y \\ x + 3y \geq 9 \end{cases}$

In linear programming, it is sometimes necessary to find the coordinates of the **corner point:** the point at which the two boundary lines intersect. To find the point of intersection, solve the related linear system

$$\begin{cases} 3x = y \\ x + 2y = 8 \end{cases}$$

by the substitution method or the addition method. The lines intersect at $\left(\frac{8}{7}, \frac{24}{7}\right)$, the corner point of the graph.

Graphing the Solution of a System of Linear Inequalities

Step 1. Graph each inequality in the system on the same set of axes.

Step 2. The solutions of the system are the points common to the graphs of all the inequalities in the system.

EXAMPLE 2 Graph the solution of the system: $\begin{cases} x - y < 2 \\ x + 2y > -1 \end{cases}$

Solution Graph both inequalities on the same set of axes. Both boundary lines are dashed lines since the inequality symbols are < and >. The solution of the system is the region shown by the purple shading. In this example, the boundary lines are not a part of the solution.

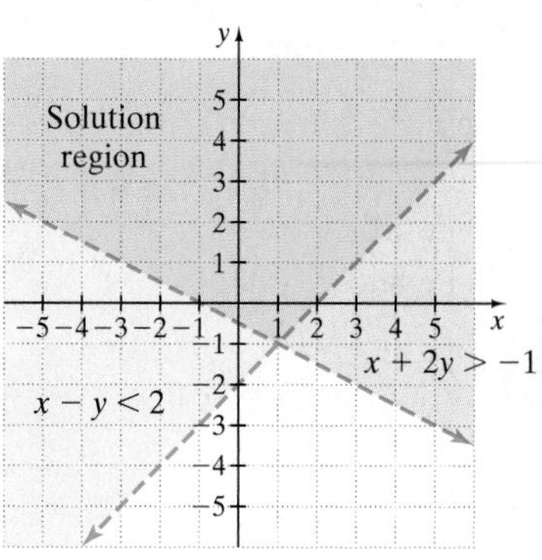

□

PRACTICE 2 Graph the solution of the system: $\begin{cases} x - y > 4 \\ x + 3y < -4 \end{cases}$

EXAMPLE 3 Graph the solution of the system: $\begin{cases} -3x + 4y < 12 \\ x \geq 2 \end{cases}$

Solution Graph both inequalities on the same set of axes.

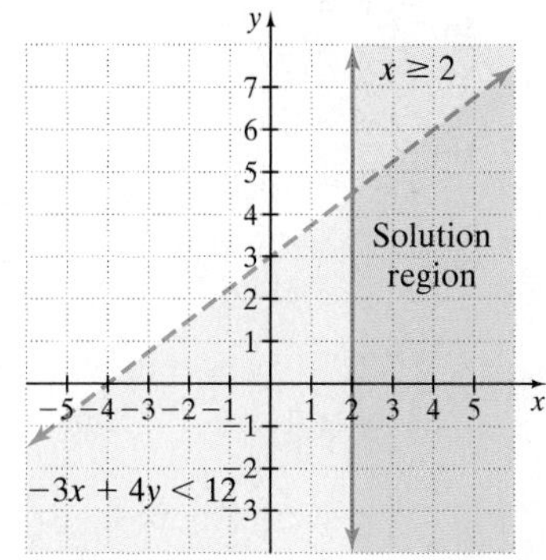

The solution of the system is the purple shaded region, including a portion of the line $x = 2$. □

PRACTICE 3 Graph the solution of the system: $\begin{cases} y \leq 6 \\ -2x + 5y > 10 \end{cases}$

Vocabulary, Readiness & Video Check

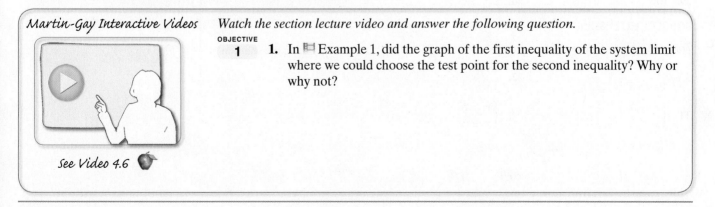

Martin-Gay Interactive Videos

See Video 4.6

Watch the section lecture video and answer the following question.

OBJECTIVE 1

1. In Example 1, did the graph of the first inequality of the system limit where we could choose the test point for the second inequality? Why or why not?

4.6 Exercise Set

MyMathLab®

MIXED PRACTICE

Graph the solution of each system of linear inequalities. See Examples 1 through 3.

1. $\begin{cases} y \geq x + 1 \\ y \geq 3 - x \end{cases}$

2. $\begin{cases} y \geq x - 3 \\ y \geq -1 - x \end{cases}$

3. $\begin{cases} y < 3x - 4 \\ y \leq x + 2 \end{cases}$

4. $\begin{cases} y \leq 2x + 1 \\ y > x + 2 \end{cases}$

5. $\begin{cases} y \leq -2x - 2 \\ y \geq x + 4 \end{cases}$

6. $\begin{cases} y \leq 2x + 4 \\ y \geq -x - 5 \end{cases}$

7. $\begin{cases} y \geq -x + 2 \\ y \leq 2x + 5 \end{cases}$

8. $\begin{cases} y \geq x - 5 \\ y \leq -3x + 3 \end{cases}$

9. $\begin{cases} x \geq 3y \\ x + 3y \leq 6 \end{cases}$

10. $\begin{cases} -2x < y \\ x + 2y < 3 \end{cases}$

11. $\begin{cases} y + 2x \geq 0 \\ 5x - 3y \leq 12 \end{cases}$

12. $\begin{cases} y + 2x \leq 0 \\ 5x + 3y \geq -2 \end{cases}$

13. $\begin{cases} 3x - 4y \geq -6 \\ 2x + y \leq 7 \end{cases}$

14. $\begin{cases} 4x - y \geq -2 \\ 2x + 3y \leq -8 \end{cases}$

15. $\begin{cases} x \leq 2 \\ y \geq -3 \end{cases}$

16. $\begin{cases} x \geq -3 \\ y \geq -2 \end{cases}$

17. $\begin{cases} y \geq 1 \\ x < -3 \end{cases}$

18. $\begin{cases} y > 2 \\ x \geq -1 \end{cases}$

19. $\begin{cases} 2x + 3y < -8 \\ x \geq -4 \end{cases}$

20. $\begin{cases} 3x + 2y \leq 6 \\ x < 2 \end{cases}$

21. $\begin{cases} 2x - 5y \leq 9 \\ y \leq -3 \end{cases}$

22. $\begin{cases} 2x + 5y \leq -10 \\ y \geq 1 \end{cases}$

23. $\begin{cases} y \geq \frac{1}{2}x + 2 \\ y \leq \frac{1}{2}x - 3 \end{cases}$

24. $\begin{cases} y \geq \frac{-3}{2}x + 3 \\ y < \frac{-3}{2}x + 6 \end{cases}$

REVIEW AND PREVIEW

Find the square of each expression. For example, the square of 7 is 7^2 or 49. The square of 5x is $(5x)^2$ or $25x^2$. See Section 1.4.

25. 4

26. 3

27. $6x$

28. $11y$

29. $10y^3$

30. $8x^5$

CONCEPT EXTENSIONS

For each system of inequalities, choose the corresponding graph.

31. $\begin{cases} y < 5 \\ x > 3 \end{cases}$

32. $\begin{cases} y > 5 \\ x < 3 \end{cases}$

33. $\begin{cases} y \leq 5 \\ x < 3 \end{cases}$

34. $\begin{cases} y > 5 \\ x \geq 3 \end{cases}$

A.

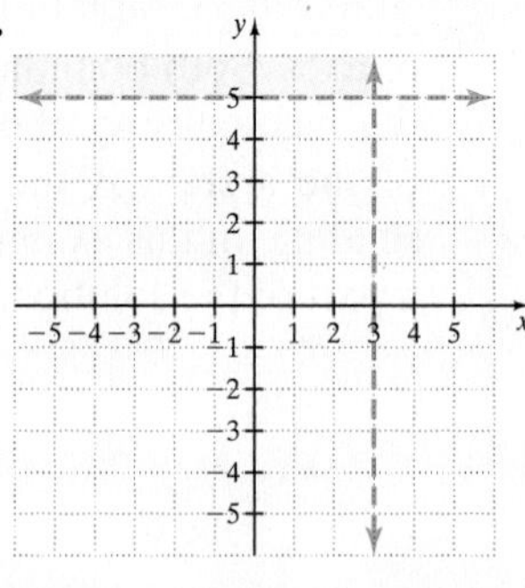

B.

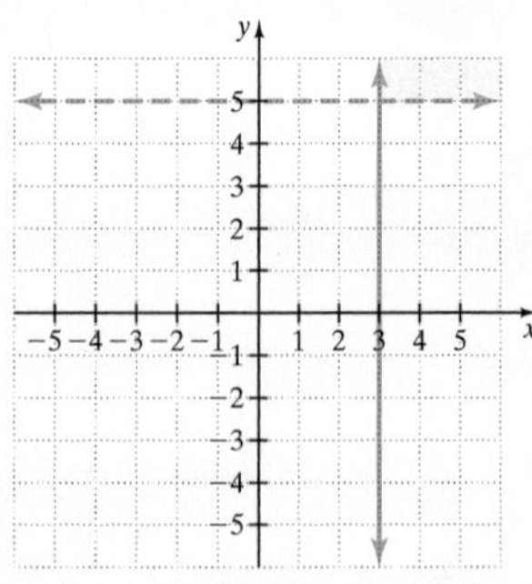

C.

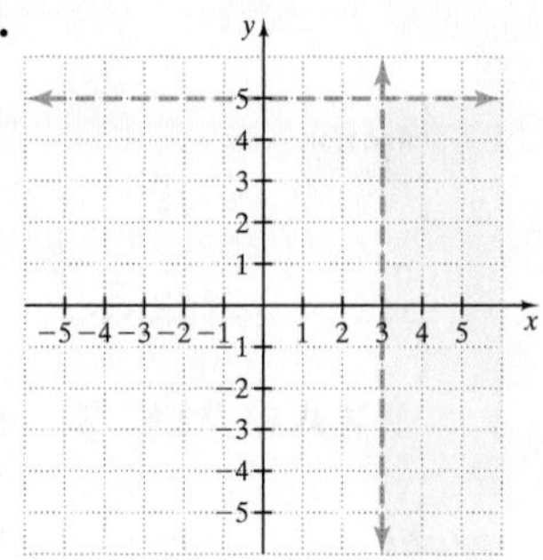

D.

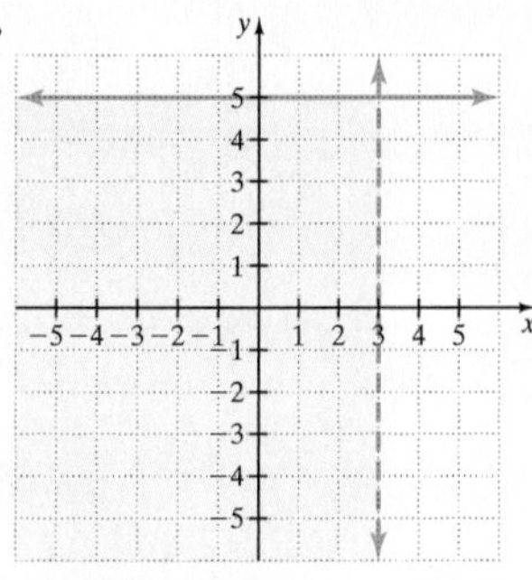

35. Explain how to decide which region to shade to show the solution region of the following system.

$$\begin{cases} x \geq 3 \\ y \geq -2 \end{cases}$$

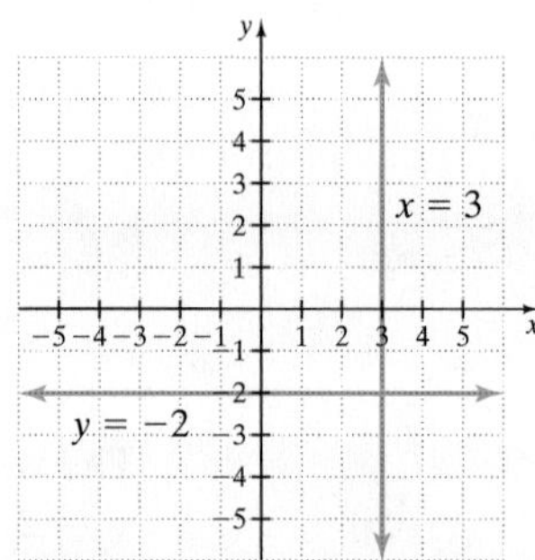

36. Describe the location of the solution region of the system

$$\begin{cases} x > 0 \\ y > 0. \end{cases}$$

37. Graph the solution of $\begin{cases} 2x - y \leq 6 \\ x \geq 3 \\ y > 2 \end{cases}$

38. Graph the solution of $\begin{cases} x + y < 5 \\ y < 2x \\ x \geq 0 \\ y \geq 0 \end{cases}$

Chapter 4 Vocabulary Check

Fill in each blank with one of the words or phrases listed below.

system of linear equations	solution	consistent	independent
dependent	inconsistent	substitution	addition
system of linear inequalities			

1. In a system of linear equations in two variables, if the graphs of the equations are the same, the equations are ____________________ equations.
2. Two or more linear equations are called a(n) ____________________.
3. A system of equations that has at least one solution is called a(n) ____________________ system.
4. A(n) ____________________ of a system of two equations in two variables is an ordered pair of numbers that is a solution of both equations in the system.
5. Two algebraic methods for solving systems of equations are ____________________ and ____________________.
6. A system of equations that has no solution is called a(n) ____________________ system.
7. In a system of linear equations in two variables, if the graphs of the equations are different, the equations are ____________________ equations.
8. Two or more linear inequalities are called a(n) ____________________.

Chapter 4 Highlights

DEFINITIONS AND CONCEPTS | **EXAMPLES**

Section 4.1 Solving Systems of Linear Equations by Graphing

A **solution** of a system of two equations in two variables is an ordered pair of numbers that is a solution of both equations in the system.

Determine whether $(-1, 3)$ is a solution of the system:

$$\begin{cases} 2x - y = -5 \\ x = 3y - 10 \end{cases}$$

Replace x with -1 and y with 3 in both equations.

$$\begin{array}{ll} 2x - y = -5 & x = 3y - 10 \\ 2(-1) - 3 \stackrel{?}{=} -5 & -1 \stackrel{?}{=} 3 \cdot 3 - 10 \\ -5 = -5 \quad \text{True} & -1 = -1 \quad \text{True} \end{array}$$

$(-1, 3)$ is a solution of the system.

Graphically, a solution of a system is a point common to the graphs of both equations.

Solve by graphing. $\begin{cases} 3x - 2y = -3 \\ x + y = 4 \end{cases}$

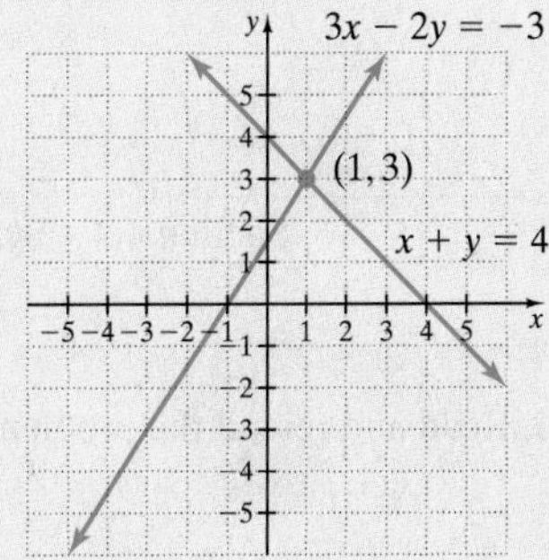

A system of equations with at least one solution is a **consistent system.** A system that has no solution is an **inconsistent system.**

If the graphs of two linear equations are identical, the equations are **dependent.** If their graphs are different, the equations are **independent.**

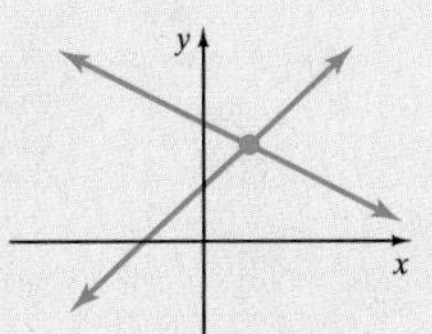
Consistent and independent

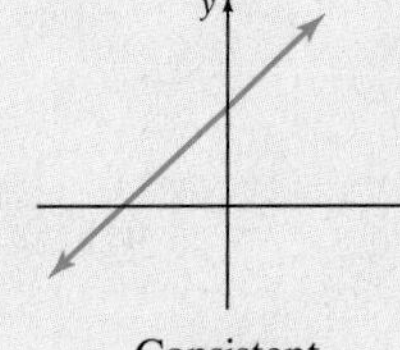
Consistent and dependent

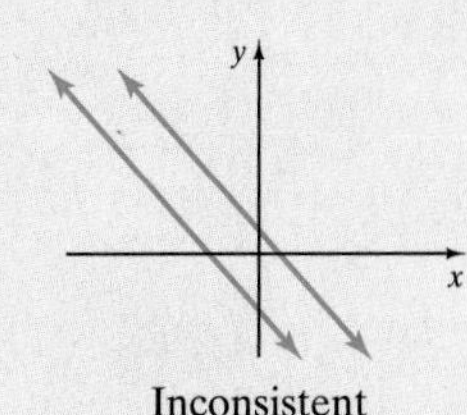
Inconsistent and independent

DEFINITIONS AND CONCEPTS	EXAMPLES
Section 4.2 Solving Systems of Linear Equations by Substitution	
To solve a system of linear equations by the substitution method: **Step 1.** Solve one equation for a variable. **Step 2.** Substitute the expression for the variable into the other equation. **Step 3.** Solve the equation from Step 2 to find the value of one variable. **Step 4.** Substitute the value from Step 3 in either original equation to find the value of the other variable. **Step 5.** Check the solution in both equations.	Solve by substitution. $\begin{cases} 3x + 2y = 1 \\ x = y - 3 \end{cases}$ Substitute $y - 3$ for x in the first equation. $3x + 2y = 1$ $3(y - 3) + 2y = 1$ $3y - 9 + 2y = 1$ $5y = 10$ $y = 2$ Divide by 5. To find x, substitute 2 for y in $x = y - 3$ so that $x = 2 - 3$ or -1. The solution $(-1, 2)$ checks.
Section 4.3 Solving Systems of Linear Equations by Addition	
To solve a system of linear equations by the addition method: **Step 1.** Rewrite each equation in standard form $Ax + By = C$. **Step 2.** Multiply one or both equations by a nonzero number so that the coefficients of a variable are opposites. **Step 3.** Add the equations. **Step 4.** Find the value of one variable by solving the resulting equation. **Step 5.** Substitute the value from Step 4 into either original equation to find the value of the other variable. **Step 6.** Check the solution in both equations.	Solve by addition. $\begin{cases} x - 2y = 8 \\ 3x + y = -4 \end{cases}$ Multiply both sides of the first equation by -3. $\begin{cases} -3x + 6y = -24 \\ 3x + y = -4 \end{cases}$ $7y = -28$ Add. $y = -4$ Divide by 7. To find x, let $y = -4$ in an original equation. $x - 2(-4) = 8$ First equation $x + 8 = 8$ $x = 0$ The solution $(0, -4)$ checks.
If solving a system of linear equations by substitution or addition yields a true statement such as $-2 = -2$, then the graphs of the equations in the system are identical and there is an infinite number of solutions of the system.	Solve: $\begin{cases} 2x - 6y = -2 \\ x = 3y - 1 \end{cases}$ Substitute $3y - 1$ for x in the first equation. $2(3y - 1) - 6y = -2$ $6y - 2 - 6y = -2$ $-2 = -2$ True The system has an infinite number of solutions.
Section 4.4 Systems of Linear Equations and Problem Solving	
Problem-solving steps **1.** UNDERSTAND. Read and reread the problem.	Two angles are supplementary if their sum is 180°. The larger of two supplementary angles is three times the smaller, decreased by twelve. Find the measure of each angle. Let $x =$ measure of smaller angle $y =$ measure of larger angle 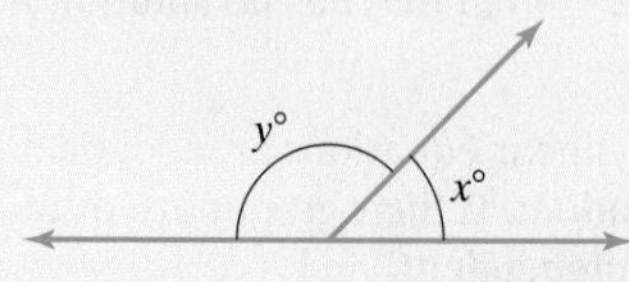

DEFINITIONS AND CONCEPTS	EXAMPLES
Section 4.4 Systems of Linear Equations and Problem Solving (continued)	

2. TRANSLATE.

In words:

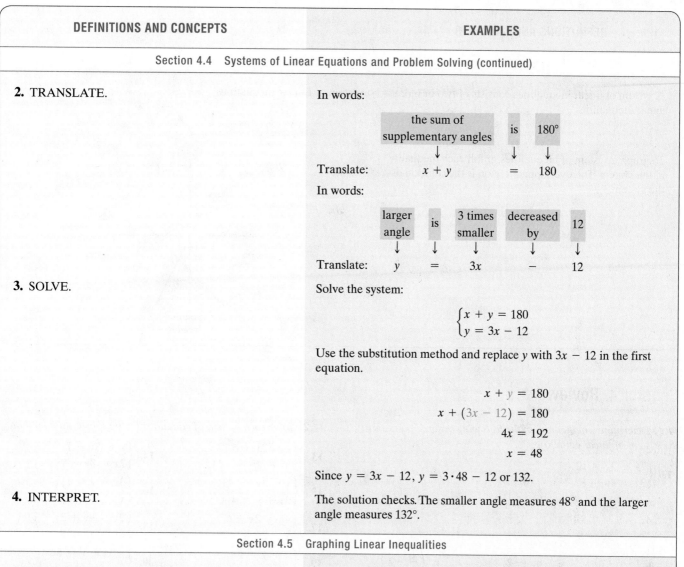

the sum of supplementary angles	is	180°
↓	↓	↓
$x + y$	$=$	180

Translate: $x + y = 180$

In words:

larger angle	is	3 times smaller	decreased by	12
↓	↓	↓	↓	↓
y	$=$	$3x$	$-$	12

Translate: $y = 3x - 12$

3. SOLVE.

Solve the system:

$$\begin{cases} x + y = 180 \\ y = 3x - 12 \end{cases}$$

Use the substitution method and replace y with $3x - 12$ in the first equation.

$$\begin{aligned} x + y &= 180 \\ x + (3x - 12) &= 180 \\ 4x &= 192 \\ x &= 48 \end{aligned}$$

Since $y = 3x - 12$, $y = 3 \cdot 48 - 12$ or 132.

4. INTERPRET.

The solution checks. The smaller angle measures 48° and the larger angle measures 132°.

Section 4.5 Graphing Linear Inequalities

A **linear inequality in two variables** is an inequality that can be written in one of the forms:

$$Ax + By < C \qquad Ax + By \le C$$
$$Ax + By > C \qquad Ax + By \ge C$$

To graph a linear inequality:

1. Graph the boundary line by graphing the related equation. Draw the line solid if the inequality symbol is $\le$ or $\ge$. Draw the line dashed if the inequality symbol is $<$ or $>$.
2. Choose a test point not on the line. Substitute its coordinates into the original inequality.
3. If the resulting inequality is true, shade the half-plane that contains the test point. If the inequality is not true, shade the half-plane that does not contain the test point.

Linear Inequalities

$$2x - 5y < 6 \qquad x \ge -5$$
$$y > -8x \qquad y \le 2$$

Graph $2x - y \le 4$.

1. Graph $2x - y = 4$. Draw a solid line because the inequality symbol is $\le$.
2. Check the test point $(0, 0)$ in the inequality $2x - y \le 4$.

$$2 \cdot 0 - 0 \le 4 \quad \text{Let } x = 0 \text{ and } y = 0.$$
$$0 \le 4 \quad \text{True}$$

3. The inequality is true, so we shade the half-plane containing $(0, 0)$.

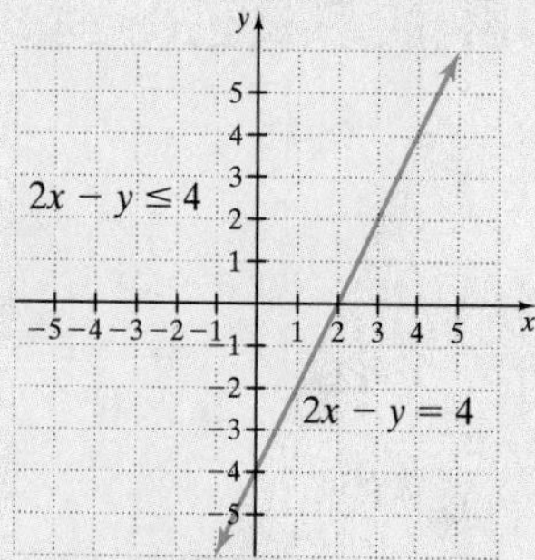

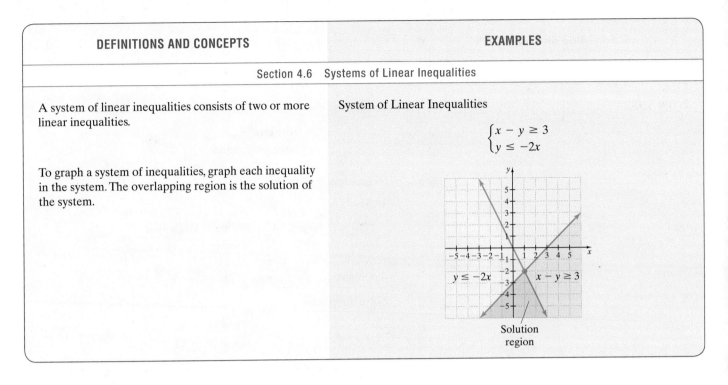

DEFINITIONS AND CONCEPTS	EXAMPLES
Section 4.6 Systems of Linear Inequalities	
A system of linear inequalities consists of two or more linear inequalities.	System of Linear Inequalities $\begin{cases} x - y \geq 3 \\ y \leq -2x \end{cases}$
To graph a system of inequalities, graph each inequality in the system. The overlapping region is the solution of the system.	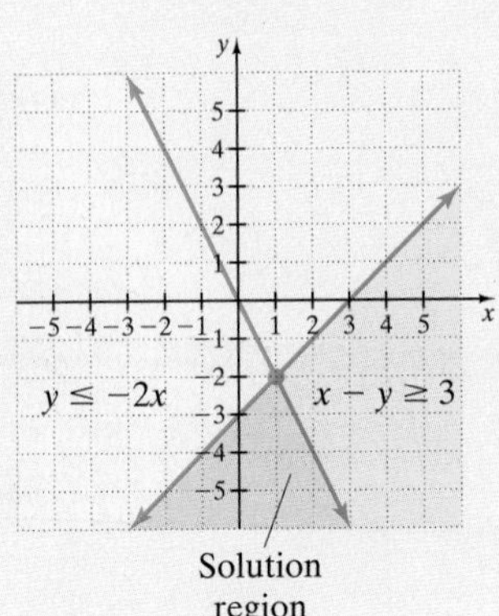

Chapter 4 Review

(4.1) Determine whether any of the following ordered pairs satisfy the system of linear equations.

1. $\begin{cases} 2x - 3y = 12 \\ 3x + 4y = 1 \end{cases}$

   a. $(12, 4)$ **b.** $(3, -2)$ **c.** $(-3, 6)$

2. $\begin{cases} 4x + y = 0 \\ -8x - 5y = 9 \end{cases}$

   a. $\left(\frac{3}{4}, -3\right)$ **b.** $(-2, 8)$ **c.** $\left(\frac{1}{2}, -2\right)$

3. $\begin{cases} 5x - 6y = 18 \\ 2y - x = -4 \end{cases}$

   a. $(-6, -8)$ **b.** $\left(3, \frac{5}{2}\right)$ **c.** $\left(3, -\frac{1}{2}\right)$

4. $\begin{cases} 2x + 3y = 1 \\ 3y - x = 4 \end{cases}$

   a. $(2, 2)$ **b.** $(-1, 1)$ **c.** $(2, -1)$

Solve each system of equations by graphing.

5. $\begin{cases} x + y = 5 \\ x - y = 1 \end{cases}$

6. $\begin{cases} x + y = 3 \\ x - y = -1 \end{cases}$

7. $\begin{cases} x = 5 \\ y = -1 \end{cases}$

8. $\begin{cases} x = -3 \\ y = 2 \end{cases}$

9. $\begin{cases} 2x + y = 5 \\ x = -3y \end{cases}$

10. $\begin{cases} 3x + y = -2 \\ y = -5x \end{cases}$

11. $\begin{cases} y = 3x \\ -6x + 2y = 6 \end{cases}$

12. $\begin{cases} x - 2y = 2 \\ -2x + 4y = -4 \end{cases}$

(4.2) Solve each system of equations by the substitution method.

13. $\begin{cases} y = 2x + 6 \\ 3x - 2y = -11 \end{cases}$

14. $\begin{cases} y = 3x - 7 \\ 2x - 3y = 7 \end{cases}$

15. $\begin{cases} x + 3y = -3 \\ 2x + y = 4 \end{cases}$

16. $\begin{cases} 3x + y = 11 \\ x + 2y = 12 \end{cases}$

17. $\begin{cases} 4y = 2x + 6 \\ x - 2y = -3 \end{cases}$

18. $\begin{cases} 9x = 6y + 3 \\ 6x - 4y = 2 \end{cases}$

19. $\begin{cases} x + y = 6 \\ y = -x - 4 \end{cases}$

20. $\begin{cases} -3x + y = 6 \\ y = 3x + 2 \end{cases}$

(4.3) Solve each system of equations by the addition method.

21. $\begin{cases} 2x + 3y = -6 \\ x - 3y = -12 \end{cases}$

22. $\begin{cases} 4x + y = 15 \\ -4x + 3y = -19 \end{cases}$

23. $\begin{cases} 2x - 3y = -15 \\ x + 4y = 31 \end{cases}$

24. $\begin{cases} x - 5y = -22 \\ 4x + 3y = 4 \end{cases}$

25. $\begin{cases} 2x - 6y = -1 \\ -x + 3y = \frac{1}{2} \end{cases}$

26. $\begin{cases} 0.6x - 0.3y = -1.5 \\ 0.04x - 0.02y = -0.1 \end{cases}$

27. $\begin{cases} \frac{3}{4}x + \frac{2}{3}y = 2 \\ x + \frac{y}{3} = 6 \end{cases}$

28. $\begin{cases} 10x + 2y = 0 \\ 3x + 5y = 33 \end{cases}$

(4.4) Solve each problem by writing and solving a system of linear equations.

29. The sum of two numbers is 16. Three times the larger number decreased by the smaller number is 72. Find the two numbers.

30. The Forrest Theater can seat a total of 360 people. It takes in \$15,150 when every seat is sold. If orchestra section tickets cost \$45 and balcony tickets cost \$35, find the number of seats in the orchestra section and the number of seats in the balcony.

31. A riverboat can head 340 miles upriver in 19 hours, but the return trip takes only 14 hours. Find the current of the river and find the speed of the riverboat in still water to the nearest tenth of a mile.

	d =	r ·	t
Upriver	340	$x - y$	19
Downriver	340	$x + y$	14

32. Find the amount of a 6% acid solution and the amount of a 14% acid solution Pat Mayfield should combine to prepare 50 cc (cubic centimeters) of a 12% solution.

33. A deli charges \$3.80 for a breakfast of three eggs and four strips of bacon. The charge is \$2.75 for two eggs and three strips of bacon. Find the cost of each egg and the cost of each strip of bacon.

34. An exercise enthusiast alternates between jogging and walking. He traveled 15 miles during the past 3 hours. He jogs at a rate of 7.5 miles per hour and walks at a rate of 4 miles per hour. Find how much time, to the nearest hundredth of an hour, he actually spent jogging and how much time he spent walking.

(4.5) Graph each inequality.

35. $5x + 4y < 20$

36. $x + 3y > 4$

37. $y \geq -7$

38. $y \leq -4$

39. $-x \leq y$

40. $x \geq -y$

(4.6) Graph the solutions of the following systems of linear inequalities.

41. $\begin{cases} y \geq 2x - 3 \\ y \leq -2x + 1 \end{cases}$

42. $\begin{cases} y \leq -3x - 3 \\ y \leq 2x + 7 \end{cases}$

43. $\begin{cases} -3x + 2y > -1 \\ y < -2 \end{cases}$

44. $\begin{cases} -2x + 3y > -7 \\ x \geq -2 \end{cases}$

MIXED REVIEW

Solve each system of equations by graphing.

45. $\begin{cases} x - 2y = 1 \\ 2x + 3y = -12 \end{cases}$

46. $\begin{cases} 3x - y = -4 \\ 6x - 2y = -8 \end{cases}$

Solve each system of equations.

47. $\begin{cases} x + 4y = 11 \\ 5x - 9y = -3 \end{cases}$

48. $\begin{cases} x + 9y = 16 \\ 3x - 8y = 13 \end{cases}$

49. $\begin{cases} y = -2x \\ 4x + 7y = -15 \end{cases}$

50. $\begin{cases} 3y = 2x + 15 \\ -2x + 3y = 21 \end{cases}$

51. $\begin{cases} 3x - y = 4 \\ 4y = 12x - 16 \end{cases}$

52. $\begin{cases} x + y = 19 \\ x - y = -3 \end{cases}$

53. $\begin{cases} x - 3y = -11 \\ 4x + 5y = -10 \end{cases}$

54. $\begin{cases} -x - 15y = 44 \\ 2x + 3y = 20 \end{cases}$

55. $\begin{cases} 2x + y = 3 \\ 6x + 3y = 9 \end{cases}$

56. $\begin{cases} -3x + y = 5 \\ -3x + y = -2 \end{cases}$

Solve each problem by writing and solving a system of linear equations.

57. The sum of two numbers is 12. Three times the smaller number increased by the larger number is 20. Find the numbers.

58. The difference of two numbers is −18. Twice the smaller decreased by the larger is −23. Find the two numbers.

59. Emma Hodges has a jar containing 65 coins, all of which are either nickels or dimes. The total value of the coins is \$5.30. How many of each type does she have?

60. Sarah and Owen Hebert purchased 26 stamps, a mixture of 13¢ and 22¢ stamps. Find the number of each type of stamp if they spent \$4.19.

Graph each inequality.

61. $x + 6y < 6$

62. $x + y > -2$

Chapter 4 **Test**

MyMathLab®  YouTube

Answer each question true or false.

1. A system of two linear equations in two variables can have exactly two solutions.

2. Although (1, 4) is not a solution of $x + 2y = 6$, it can still be a solution of the system $\begin{cases} x + 2y = 6 \\ x + y = 5 \end{cases}$

3. If the two equations in a system of linear equations are added and the result is $3 = 0$, the system has no solution.

4. If the two equations in a system of linear equations are added and the result is $3x = 0$, the system has no solution.

Is the ordered pair a solution of the given linear system?

5. $\begin{cases} 2x - 3y = 5 \\ 6x + y = 1 \end{cases}$; $(1, -1)$

6. $\begin{cases} 4x - 3y = 24 \\ 4x + 5y = -8 \end{cases}$; $(3, -4)$

Solve each system by graphing.

7. $\begin{cases} x - y = 2 \\ 3x - y = -2 \end{cases}$

8. $\begin{cases} y = -3x \\ 3x + y = 6 \end{cases}$

Solve each system by the substitution method.

9. $\begin{cases} 3x - 2y = -14 \\ y = x + 5 \end{cases}$

10. $\begin{cases} \frac{1}{2}x + 2y = -\frac{15}{4} \\ 4x = -y \end{cases}$

Solve each system by the addition method.

11. $\begin{cases} x + y = 28 \\ x - y = 12 \end{cases}$

12. $\begin{cases} 4x - 6y = 7 \\ -2x + 3y = 0 \end{cases}$

Solve each system using the substitution method or the addition method.

13. $\begin{cases} 3x + y = 7 \\ 4x + 3y = 1 \end{cases}$

14. $\begin{cases} 3(2x + y) = 4x + 20 \\ x - 2y = 3 \end{cases}$

15. $\begin{cases} \frac{x - 3}{2} = \frac{2 - y}{4} \\ \frac{7 - 2x}{3} = \frac{y}{2} \end{cases}$

16. $\begin{cases} 8x - 4y = 12 \\ y = 2x - 3 \end{cases}$

17. $\begin{cases} 0.01x - 0.06y = -0.23 \\ 0.2x + 0.4y = 0.2 \end{cases}$

18. $\begin{cases} x - \frac{2}{3}y = 3 \\ -2x + 3y = 10 \end{cases}$

Solve each problem by writing and using a system of linear equations.

19. Two numbers have a sum of 124 and a difference of 32. Find the numbers.

20. Find the amount of a 12% saline solution a lab assistant should add to 80 cc (cubic centimeters) of a 22% saline solution to have a 16% solution.

21. Texas and Missouri are the states with the most number of farms. Texas has 140 thousand more farms than Missouri and the total number of farms for these two states is 356 thousand. Find the number of farms for each state.

22. Two hikers start at opposite ends of the St. Tammany Trail and walk toward each other. The trail is 36 miles long and they meet in 4 hours. If one hiker is twice as fast as the other, find both hiking speeds.

Graph each inequality.

23. $x - y \geq -2$

24. $y > -4x$

25. $2x - 3y > -6$

Graph the solutions of the following systems of linear inequalities.

26. $\begin{cases} y + 2x \leq 4 \\ y \geq 2 \end{cases}$

27. $\begin{cases} 2y - x \geq 1 \\ x + y \geq -4 \end{cases}$

Chapter 4 Cumulative Review

1. Insert $<, >$, or $=$ in the appropriate space to make each statement true.

   a. $-1 \quad 0$ b. $7 \quad \frac{14}{2}$ c. $-5 \quad -6$

2. Evaluate.

   a. 5^2 b. 2^5

3. Name the property or properties illustrated by each true statement.

   a. $3 \cdot y = y \cdot 3$
   b. $(x + 7) + 9 = x + (7 + 9)$
   c. $(b + 0) + 3 = b + 3$
   d. $0.2 \cdot (z \cdot 5) = 0.2 \cdot (5 \cdot z)$
   e. $-2 \cdot \left(-\frac{1}{2}\right) = 1$
   f. $-2 + 2 = 0$
   g. $-6 \cdot (y \cdot 2) = (-6 \cdot 2) \cdot y$

4. Evaluate $y^2 - 3x$ for $x = 8$ and $y = 5$.

5. Subtract $4x - 2$ from $2x - 3$

6. Simplify: $7 - 12 + (-5) - 2 + (-2)$

7. Solve $5t - 5 = 6t + 2$ for t.

8. Evaluate $2y^2 - x^2$ for $x = -7$ and $y = -3$.

9. Solve: $\frac{5}{2}x = 15$

10. Simplify: $0.4y - 6.7 + y - 0.3 - 2.6y$

11. Solve: $\frac{x}{2} - 1 = \frac{2}{3}x - 3$

12. Solve: $7(x - 2) - 6(x + 1) = 20$

13. Twice the sum of a number and 4 is the same as four times the number, decreased 12. Find the number.

14. Solve: $5(y - 5) = 5y + 10$

15. Solve $y = mx + b$ for x.

16. Five times the sum of a number and -1 is the same as 6 times the number. Find the number.

17. Solve $-2x \leq -4$. Write the solution set in interval notation.

18. Solve $P = a + b + c$ for b.

19. Graph $x = -2y$ by plotting intercepts.

20. Solve $3x + 7 \geq x - 9$. Write the solution set in interval notation.

21. Find the slope of the line through $(-1, 5)$ and $(2, -3)$.

22. Complete the table of values for $x - 3y = 3$

x	y
	-1
3	
	2

23. Find the slope of the line whose equation is $y = \frac{3}{4}x + 6$.

24. Find the slope of a line parallel to the line passing through $(-1, 3)$ and $(2, -8)$.

25. Find the slope and the y-intercept of the line whose equation is $3x - 4y = 4$.

26. Find the slope and y-intercept of the line whose equation is $y = 7x$.

27. Find an equation of the line with slope -2 that passes through $(-1, 5)$ Write the equation in standard form: $Ax + By = C$.

28. Determine whether the lines are parallel, perpendicular, or neither.

$$y = 4x - 5$$
$$-4x + y = 7$$

29. Find an equation of the vertical line through $(-1, 5)$.

30. Write an equation of the line with slope -5, through $(-2, 3)$.

31. Find the domain and the range of the relation $\{(0, 2), (3, 3), (-1, 0), (3, -2)\}$.

32. If $f(x) = 5x^2 - 6$, find $f(0)$ and $f(-2)$.

33. Determine whether each relation is also a function.

a. $\{(-1, 1), (2, 3), (7, 3), (8, 6)\}$

b. $\{(0, -2), (1, 5), (0, 3), (7, 7)\}$

34. Determine whether each graph is the graph of a function.

a.

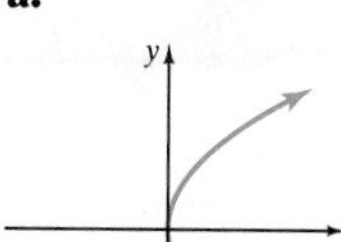

b.

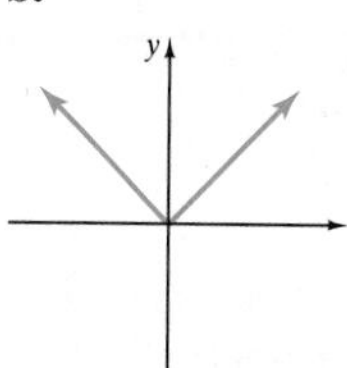

c.

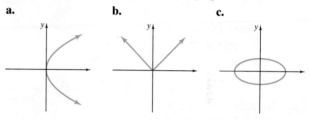

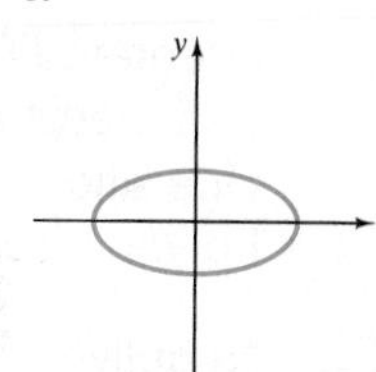

35. Without graphing, determine the number of solutions of the system.

$$\begin{cases} 3x - y = 4 \\ x + 2y = 8 \end{cases}$$

36. Determine whether each ordered pair satisfies the given system.

$$\begin{cases} 2x - y = 6 \\ 3x + 2y = -5 \end{cases}$$

a. $(1, -4)$ b. $(0, 6)$ c. $(3, 0)$

Solve each system.

37. $\begin{cases} x + 2y = 7 \\ 2x + 2y = 13 \end{cases}$

38. $\begin{cases} 3x - 4y = 10 \\ y = 2x \end{cases}$

39. $\begin{cases} x + y = 7 \\ x - y = 5 \end{cases}$

40. $\begin{cases} x = 5y - 3 \\ x = 8y + 4 \end{cases}$

41. Find two numbers whose sum is 37 and whose difference is 21.

42. Graph: $x > 1$

43. Graph: $2x - y \geq 3$

44. Graph the solution of the system: $\begin{cases} 2x + 3y < 6 \\ y < 2 \end{cases}$

CHAPTER 5 Exponents and Polynomials

Recall from Chapter 1 that an exponent is a shorthand notation for repeated factors. This chapter explores additional concepts about exponents and exponential expressions. An especially useful type of exponential expression is a polynomial. Polynomials model many real-world phenomena. In this chapter, we focus on polynomials and operations on polynomials.

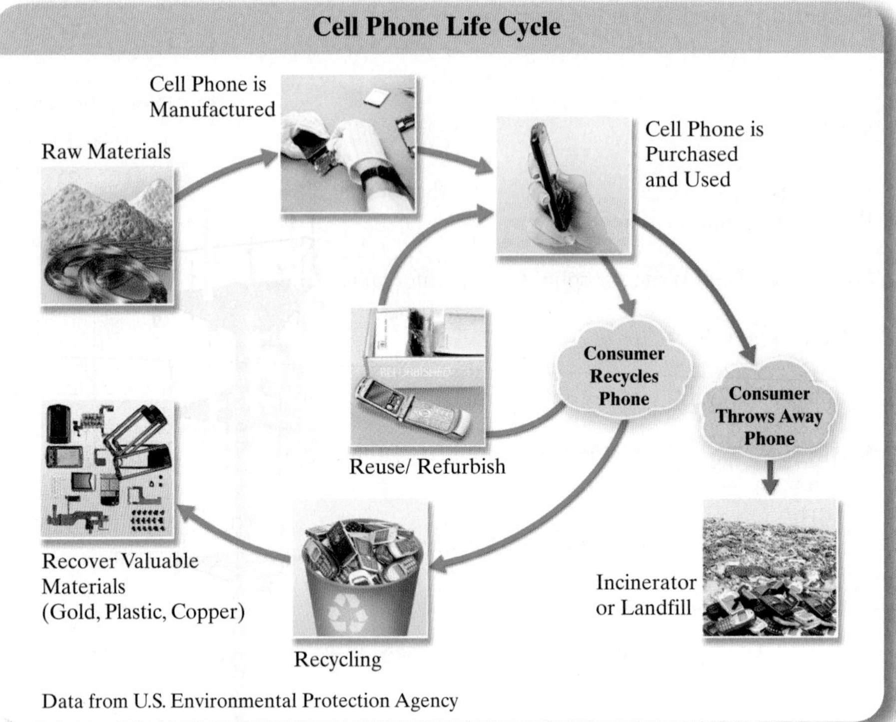

According to a recent survey, the average American adult owns 2.4 cell phones, and over 56% of those surveyed still have their old cell phones. In fact, it is estimated that 130 million cell phones will be retired this year. With new cell phone models having new features and increased technologies, this number of retired cell phones will probably only increase. The good news is that the number of cell phones being recycled is increasing. Below is a graph showing the growth of cell phones recycled by the largest recycler. In Exercises 103 and 104 of Section 5.2, you will explore some information about recycling cell phones.

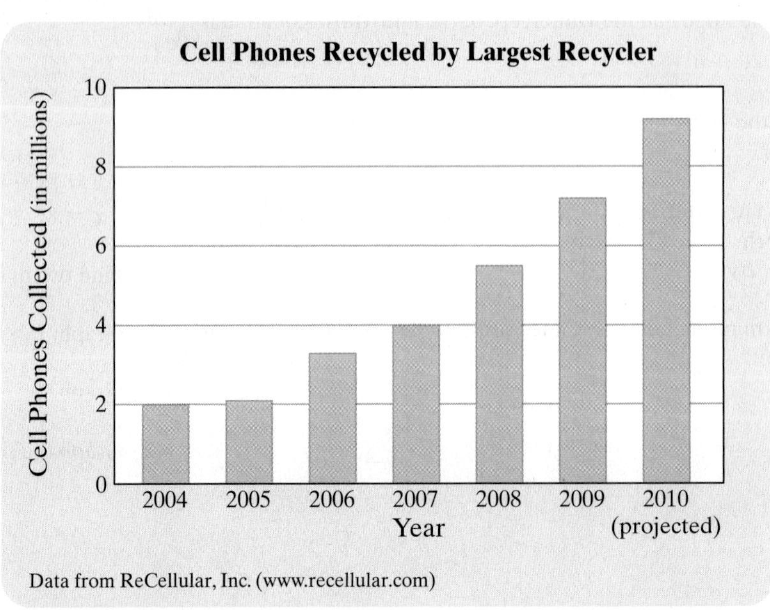

5.1 Exponents

OBJECTIVES

1. Evaluate Exponential Expressions.
2. Use the Product Rule for Exponents.
3. Use the Power Rule for Exponents.
4. Use the Power Rules for Products and Quotients.
5. Use the Quotient Rule for Exponents, and Define a Number Raised to the 0 Power.
6. Decide Which Rule(s) to Use to Simplify an Expression.

OBJECTIVE

1 Evaluating Exponential Expressions 

As we reviewed in Section 1.4, an exponent is a shorthand notation for repeated factors. For example, $2 \cdot 2 \cdot 2 \cdot 2 \cdot 2$ can be written as 2^5. The expression 2^5 is called an **exponential expression.** It is also called the fifth **power** of 2, or we say that 2 is **raised** to the fifth power.

$$5^6 = \underbrace{5 \cdot 5 \cdot 5 \cdot 5 \cdot 5 \cdot 5}_{6 \text{ factors; each factor is } 5} \quad \text{and} \quad (-3)^4 = \underbrace{(-3) \cdot (-3) \cdot (-3) \cdot (-3)}_{4 \text{ factors; each factor is } -3}$$

The **base** of an exponential expression is the repeated factor. The **exponent** is the number of times that the base is used as a factor.

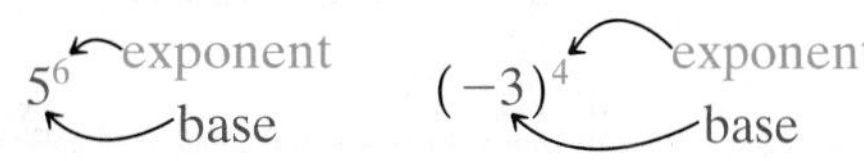

EXAMPLE 1 Evaluate each expression.

a. 2^3 **b.** 3^1 **c.** $(-4)^2$ **d.** -4^2 **e.** $\left(\frac{1}{2}\right)^4$ **f.** $(0.5)^3$ **g.** $4 \cdot 3^2$

Solution

a. $2^3 = 2 \cdot 2 \cdot 2 = 8$

b. To raise 3 to the first power means to use 3 as a factor only once. Therefore, $3^1 = 3$. Also, when no exponent is shown, the exponent is assumed to be 1.

c. $(-4)^2 = (-4)(-4) = 16$ **d.** $-4^2 = -(4 \cdot 4) = -16$

e. $\left(\frac{1}{2}\right)^4 = \frac{1}{2} \cdot \frac{1}{2} \cdot \frac{1}{2} \cdot \frac{1}{2} = \frac{1}{16}$ **f.** $(0.5)^3 = (0.5)(0.5)(0.5) = 0.125$

g. $4 \cdot 3^2 = 4 \cdot 9 = 36$ □

PRACTICE

1 Evaluate each expression.

a. 3^3 **b.** 4^1 **c.** $(-8)^2$ **d.** -8^2

e. $\left(\frac{3}{4}\right)^3$ **f.** $(0.3)^4$ **g.** $3 \cdot 5^2$

Notice how similar -4^2 is to $(-4)^2$ in the example above. The difference between the two is the parentheses. In $(-4)^2$, the parentheses tell us that the base, or repeated factor, is -4. In -4^2, only 4 is the base.

> **Helpful Hint**
>
> Be careful when identifying the base of an exponential expression. Pay close attention to the use of parentheses.
>
> | $(-3)^2$ | -3^2 | $2 \cdot 3^2$ |
> |---|---|---|
> | The base is -3. | The base is 3. | The base is 3. |
> | $(-3)^2 = (-3)(-3) = 9$ | $-3^2 = -(3 \cdot 3) = -9$ | $2 \cdot 3^2 = 2 \cdot 3 \cdot 3 = 18$ |

An exponent has the same meaning whether the base is a number or a variable. If x is a real number and n is a positive integer, then x^n is the product of n factors, each of which is x.

$$x^n = \underbrace{x \cdot x \cdot x \cdot x \cdot x \cdot \ldots \cdot x}_{n \text{ factors of } x}$$

EXAMPLE 2 Evaluate each expression for the given value of x.

a. $2x^3$; x is 5 **b.** $\frac{9}{x^2}$; x is -3

Solution **a.** If x is 5, $2x^3 = 2\cdot(5)^3$
$= 2\cdot(5\cdot5\cdot5)$
$= 2\cdot125$
$= 250$

b. If x is -3, $\frac{9}{x^2} = \frac{9}{(-3)^2}$
$= \frac{9}{(-3)(-3)}$
$= \frac{9}{9}$
$= 1$ □

PRACTICE
2 Evaluate each expression for the given value of x.

a. $3x^4$; x is 3 **b.** $\frac{6}{x^2}$; x is -4

OBJECTIVE
2 Using the Product Rule 

Exponential expressions can be multiplied, divided, added, subtracted, and themselves raised to powers. By our definition of an exponent,

$$5^4\cdot5^3 = \underbrace{(5\cdot5\cdot5\cdot5)}_{4 \text{ factors of } 5}\cdot\underbrace{(5\cdot5\cdot5)}_{3 \text{ factors of } 5}$$
$$= \underbrace{5\cdot5\cdot5\cdot5\cdot5\cdot5\cdot5}_{7 \text{ factors of } 5}$$
$$= 5^7$$

Also,

$$x^2\cdot x^3 = (x\cdot x)\cdot(x\cdot x\cdot x)$$
$$= x\cdot x\cdot x\cdot x\cdot x$$
$$= x^5$$

In both cases, notice that the result is exactly the same if the exponents are added.

$$5^4\cdot5^3 = 5^{4+3} = 5^7 \quad \text{and} \quad x^2\cdot x^3 = x^{2+3} = x^5$$

This suggests the following rule.

Product Rule for Exponents

If m and n are positive integers and a is a real number, then

$$a^m\cdot a^n = a^{m+n}$$ ← Add exponents. (↑ Keep common base.)

For example, $3^5\cdot3^7 = 3^{5+7} = 3^{12}$ ← Add exponents. (↑ Keep common base.)

Helpful Hint

Don't forget that

$3^5\cdot3^7 \neq 9^{12}$ ← Add exponents. (↑ **Common base *not* kept.**)

$$3^5\cdot3^7 = \underbrace{3\cdot3\cdot3\cdot3\cdot3}_{5 \text{ factors of } 3}\cdot\underbrace{3\cdot3\cdot3\cdot3\cdot3\cdot3\cdot3}_{7 \text{ factors of } 3}$$
$= 3^{12}$ 12 factors of 3, *not* 9

In other words, to multiply two exponential expressions with the **same base,** we keep the base and add the exponents. We call this **simplifying** the exponential expression.

EXAMPLE 3 Use the product rule to simplify.

a. $4^2 \cdot 4^5$ **b.** $x^4 \cdot x^6$ **c.** $y^3 \cdot y$ **d.** $y^3 \cdot y^2 \cdot y^7$ **e.** $(-5)^7 \cdot (-5)^8$ **f.** $a^2 \cdot b^2$

Solution

a. $4^2 \cdot 4^5 = 4^{2+5} = 4^7$ ← Add exponents.
↑ **Keep** common base.

b. $x^4 \cdot x^6 = x^{4+6} = x^{10}$

c. $y^3 \cdot y = y^3 \cdot y^1$
$= y^{3+1}$
$= y^4$

Helpful Hint
Don't forget that if no exponent is written, it is assumed to be 1.

d. $y^3 \cdot y^2 \cdot y^7 = y^{3+2+7} = y^{12}$

e. $(-5)^7 \cdot (-5)^8 = (-5)^{7+8} = (-5)^{15}$

f. $a^2 \cdot b^2$ Cannot be simplified because a and b are different bases. □

PRACTICE
3 Use the product rule to simplify.

a. $3^4 \cdot 3^6$ **b.** $y^3 \cdot y^2$
c. $z \cdot z^4$ **d.** $x^3 \cdot x^2 \cdot x^6$
e. $(-2)^5 \cdot (-2)^3$ **f.** $b^3 \cdot t^5$

✓**CONCEPT CHECK**
Where possible, use the product rule to simplify the expression.
a. $z^2 \cdot z^{14}$ b. $x^2 \cdot y^{14}$ c. $9^8 \cdot 9^3$ d. $9^8 \cdot 2^7$

EXAMPLE 4 Use the product rule to simplify $(2x^2)(-3x^5)$.

Solution Recall that $2x^2$ means $2 \cdot x^2$ and $-3x^5$ means $-3 \cdot x^5$.

$(2x^2)(-3x^5) = 2 \cdot x^2 \cdot -3 \cdot x^5$ Remove parentheses.
$= 2 \cdot -3 \cdot x^2 \cdot x^5$ Group factors with common bases.
$= -6x^7$ Simplify. □

PRACTICE
4 Use the product rule to simplify $(-5y^3)(-3y^4)$.

EXAMPLE 5 Simplify.

a. $(x^2y)(x^3y^2)$ **b.** $(-a^7b^4)(3ab^9)$

Solution

a. $(x^2y)(x^3y^2) = (x^2 \cdot x^3) \cdot (y^1 \cdot y^2)$ Group like bases and write y as y^1.
$= x^5 \cdot y^3$ or x^5y^3 Multiply.

b. $(-a^7b^4)(3ab^9) = (-1 \cdot 3) \cdot (a^7 \cdot a^1) \cdot (b^4 \cdot b^9)$
$= -3a^8b^{13}$ □

PRACTICE
5 Simplify.

a. $(y^7z^3)(y^5z)$ **b.** $(-m^4n^4)(7mn^{10})$

Answers to Concept Check:
a. z^{16} **b.** cannot be simplified
c. 9^{11} **d.** cannot be simplified

Helpful Hint

These examples will remind you of the difference between adding and multiplying terms.

Addition

$$5x^3 + 3x^3 = (5 + 3)x^3 = 8x^3 \quad \text{By the distributive property.}$$
$$7x + 4x^2 = 7x + 4x^2 \quad \text{Cannot be combined.}$$

Multiplication

$$(5x^3)(3x^3) = 5 \cdot 3 \cdot x^3 \cdot x^3 = 15x^{3+3} = 15x^6 \quad \text{By the product rule.}$$
$$(7x)(4x^2) = 7 \cdot 4 \cdot x \cdot x^2 = 28x^{1+2} = 28x^3 \quad \text{By the product rule.}$$

OBJECTIVE

3 Using the Power Rule 

Exponential expressions can themselves be raised to powers. Let's try to discover a rule that simplifies an expression like $(x^2)^3$. By definition,

$$(x^2)^3 = \underbrace{(x^2)(x^2)(x^2)}_{3 \text{ factors of } x^2}$$

which can be simplified by the product rule for exponents.

$$(x^2)^3 = (x^2)(x^2)(x^2) = x^{2+2+2} = x^6$$

Notice that the result is exactly the same if we multiply the exponents.

$$(x^2)^3 = x^{2 \cdot 3} = x^6$$

The following property states this result.

Power Rule for Exponents

If m and n are positive integers and a is a real number, then

$$(a^m)^n = a^{mn} \leftarrow \text{Multiply exponents.}$$
$$\uparrow \text{Keep common base.}$$

For example, $(7^2)^5 = 7^{2 \cdot 5} = 7^{10}$ ← Multiply exponents. ↑ Keep common base.

To raise a power to a power, keep the base and multiply the exponents.

EXAMPLE 6 Use the power rule to simplify.

a. $(y^8)^2$ **b.** $(8^4)^5$ **c.** $[(-5)^3]^7$

Solution

a. $(y^8)^2 = y^{8 \cdot 2} = y^{16}$ **b.** $(8^4)^5 = 8^{4 \cdot 5} = 8^{20}$ **c.** $[(-5)^3]^7 = (-5)^{21}$ □

PRACTICE

6 Use the power rule to simplify.

a. $(z^3)^7$ **b.** $(4^9)^2$ **c.** $[(-2)^3]^5$

Helpful Hint

Take a moment to make sure that you understand when to apply the product rule and when to apply the power rule.

Product Rule → Add Exponents	*Power Rule → Multiply Exponents*
$x^5 \cdot x^7 = x^{5+7} = x^{12}$	$(x^5)^7 = x^{5 \cdot 7} = x^{35}$
$y^6 \cdot y^2 = y^{6+2} = y^8$	$(y^6)^2 = y^{6 \cdot 2} = y^{12}$

OBJECTIVE

4 Using the Power Rules for Products and Quotients 

When the base of an exponential expression is a product, the definition of x^n still applies. To simplify $(xy)^3$, for example,

$$\begin{aligned}(xy)^3 &= (xy)(xy)(xy) && (xy)^3 \text{ means 3 factors of } (xy).\\ &= x\cdot x\cdot x\cdot y\cdot y\cdot y && \text{Group factors with common bases.}\\ &= x^3y^3 && \text{Simplify.}\end{aligned}$$

Notice that to simplify the expression $(xy)^3$, we raise each factor within the parentheses to a power of 3.

$$(xy)^3 = x^3y^3$$

In general, we have the following rule.

Power of a Product Rule

If n is a positive integer and a and b are real numbers, then

$$(ab)^n = a^nb^n$$

For example, $(3x)^5 = 3^5x^5$.

In other words, to raise a product to a power, we raise each factor to the power.

EXAMPLE 7 Simplify each expression.

a. $(st)^4$ **b.** $(2a)^3$ **c.** $\left(\frac{1}{3}mn^3\right)^2$ **d.** $(-5x^2y^3z)^2$

Solution

a. $(st)^4 = s^4\cdot t^4 = s^4t^4$ Use the power of a product rule.

b. $(2a)^3 = 2^3\cdot a^3 = 8a^3$ Use the power of a product rule.

c. $\left(\frac{1}{3}mn^3\right)^2 = \left(\frac{1}{3}\right)^2\cdot(m)^2\cdot(n^3)^2 = \frac{1}{9}m^2n^6$ Use the power of a product rule.

d. $(-5x^2y^3z)^2 = (-5)^2\cdot(x^2)^2\cdot(y^3)^2\cdot(z^1)^2$ Use the power of a product rule.

$= 25x^4y^6z^2$ Use the power rule for exponents. □

PRACTICE

7 Simplify each expression.

a. $(pr)^5$ **b.** $(6b)^2$ **c.** $\left(\frac{1}{4}x^2y\right)^3$ **d.** $(-3a^3b^4c)^4$

Let's see what happens when we raise a quotient to a power. To simplify $\left(\frac{x}{y}\right)^3$, for example,

$$\begin{aligned}\left(\frac{x}{y}\right)^3 &= \left(\frac{x}{y}\right)\left(\frac{x}{y}\right)\left(\frac{x}{y}\right) && \left(\frac{x}{y}\right)^3 \text{ means 3 factors of } \left(\frac{x}{y}\right)\\ &= \frac{x\cdot x\cdot x}{y\cdot y\cdot y} && \text{Multiply fractions.}\\ &= \frac{x^3}{y^3} && \text{Simplify.}\end{aligned}$$

Notice that to simplify the expression $\left(\frac{x}{y}\right)^3$, we raise both the numerator and the denominator to a power of 3.

$$\left(\frac{x}{y}\right)^3 = \frac{x^3}{y^3}$$

In general, we have the following.

Power of a Quotient Rule

If n is a positive integer and a and c are real numbers, then

$$\left(\frac{a}{c}\right)^n = \frac{a^n}{c^n}, \quad c \neq 0$$

For example, $\left(\frac{y}{7}\right)^4 = \frac{y^4}{7^4}$.

In other words, to raise a quotient to a power, we raise both the numerator and the denominator to the power.

EXAMPLE 8 Simplify each expression.

a. $\left(\frac{m}{n}\right)^7$ **b.** $\left(\frac{x^3}{3y^5}\right)^4$

Solution

a. $\left(\frac{m}{n}\right)^7 = \frac{m^7}{n^7}, n \neq 0$ Use the power of a quotient rule.

b. $\left(\frac{x^3}{3y^5}\right)^4 = \frac{(x^3)^4}{3^4 \cdot (y^5)^4}, y \neq 0$ Use the power of a product or quotient rule.

$= \frac{x^{12}}{81y^{20}}$ Use the power rule for exponents. □

PRACTICE

8 Simplify each expression.

a. $\left(\frac{x}{y^2}\right)^5$ **b.** $\left(\frac{2a^4}{b^3}\right)^5$

OBJECTIVE

5 Using the Quotient Rule and Defining the Zero Exponent 

Another pattern for simplifying exponential expressions involves quotients.

To simplify an expression like $\frac{x^5}{x^3}$, in which the numerator and the denominator have a common base, we can apply the fundamental principle of fractions and divide the numerator and the denominator by the common base factors. Assume for the remainder of this section that denominators are not 0.

$$\frac{x^5}{x^3} = \frac{x \cdot x \cdot x \cdot x \cdot x}{x \cdot x \cdot x}$$
$$= \frac{x \cdot x \cdot x \cdot x \cdot x}{x \cdot x \cdot x}$$
$$= x \cdot x$$
$$= x^2$$

Notice that the result is exactly the same if we subtract exponents of the common bases.

$$\frac{x^5}{x^3} = x^{5-3} = x^2$$

The quotient rule for exponents states this result in a general way.

Quotient Rule for Exponents

If m and n are positive integers and a is a real number, then

$$\frac{a^m}{a^n} = a^{m-n}$$

as long as a is not 0.

For example, $\frac{x^6}{x^2} = x^{6-2} = x^4$.

In other words, to divide one exponential expression by another with a common base, keep the base and subtract exponents.

EXAMPLE 9 Simplify each quotient.

a. $\frac{x^5}{x^2}$ **b.** $\frac{4^7}{4^3}$ **c.** $\frac{(-3)^5}{(-3)^2}$ **d.** $\frac{s^2}{t^3}$ **e.** $\frac{2x^5y^2}{xy}$

Solution

a. $\frac{x^5}{x^2} = x^{5-2} = x^3$ Use the quotient rule.

b. $\frac{4^7}{4^3} = 4^{7-3} = 4^4 = 256$ Use the quotient rule.

c. $\frac{(-3)^5}{(-3)^2} = (-3)^3 = -27$ Use the quotient rule.

d. $\frac{s^2}{t^3}$ Cannot be simplified because s and t are different bases.

e. Begin by grouping common bases.

$$\frac{2x^5y^2}{xy} = 2 \cdot \frac{x^5}{x^1} \cdot \frac{y^2}{y^1}$$

$$= 2 \cdot (x^{5-1}) \cdot (y^{2-1}) \quad \text{Use the quotient rule.}$$

$$= 2x^4y^1 \quad \text{or} \quad 2x^4y$$

□

PRACTICE

9 Simplify each quotient.

a. $\frac{z^8}{z^4}$ **b.** $\frac{(-5)^5}{(-5)^3}$ **c.** $\frac{8^8}{8^6}$ **d.** $\frac{q^5}{t^2}$ **e.** $\frac{6x^3y^7}{xy^5}$

✓CONCEPT CHECK

Suppose you are simplifying each expression. Tell whether you would *add* the exponents, *subtract* the exponents, *multiply* the exponents, *divide* the exponents, or *none of these*.

a. $(x^{63})^{21}$ b. $\frac{y^{15}}{y^3}$ c. $z^{16} + z^8$ d. $w^{45} \cdot w^9$

Let's now give meaning to an expression such as x^0. To do so, we will simplify $\frac{x^3}{x^3}$ in two ways and compare the results.

$\frac{x^3}{x^3} = x^{3-3} = x^0$ Apply the quotient rule.

$\frac{x^3}{x^3} = \frac{x \cdot x \cdot x}{x \cdot x \cdot x} = 1$ Apply the fundamental principle for fractions.

Answers to Concept Check:
a. multiply **b.** subtract
c. none of these **d.** add

Since $\frac{x^3}{x^3} = x^0$ and $\frac{x^3}{x^3} = 1$, we define that $x^0 = 1$ as long as x is not 0.

Zero Exponent

$a^0 = 1$, as long as a is not 0.

In other words, any base raised to the 0 power is 1 as long as the base is not 0.

EXAMPLE 10 Simplify each expression.

a. 3^0 **b.** $(5x^3y^2)^0$ **c.** $(-5)^0$ **d.** -5^0 **e.** $\left(\frac{3}{100}\right)^0$ **f.** $4x^0$

Solution

a. $3^0 = 1$

b. Assume that neither x nor y is zero.

$$(5x^3y^2)^0 = 1$$

c. $(-5)^0 = 1$

d. $-5^0 = -1 \cdot 5^0 = -1 \cdot 1 = -1$

e. $\left(\frac{3}{100}\right)^0 = 1$

f. $4x^0 = 4 \cdot x^0 = 4 \cdot 1 = 4$ □

PRACTICE
10 Simplify the following expressions.

a. -3^0 **b.** $(-3)^0$ **c.** 8^0 **d.** $(0.2)^0$
e. $(7a^2y^4)^0$ **f.** $7y^0$

OBJECTIVE
6 Deciding Which Rule to Use

Let's practice deciding which rule(s) to use to simplify. We will continue this discussion with more examples in Section 5.5.

EXAMPLE 11 Simplify each expression.

a. $x^7 \cdot x^4$ **b.** $\left(\frac{t}{2}\right)^4$ **c.** $(9y^5)^2$

Solution

a. Here, we have a product, so we use the product rule to simplify.
$x^7 \cdot x^4 = x^{7+4} = x^{11}$

b. This is a quotient raised to a power, so we use the power of a quotient rule.
$\left(\frac{t}{2}\right)^4 = \frac{t^4}{2^4} = \frac{t^4}{16}$

c. This is a product raised to a power, so we use the power of a product rule.
$(9y^5)^2 = 9^2(y^5)^2 = 81y^{10}$ □

PRACTICE
11

a. $\left(\frac{z}{12}\right)^2$ **b.** $(4x^6)^3$ **c.** $y^{10} \cdot y^3$

EXAMPLE 12 Simplify each expression.

a. $4^2 - 4^0$ **b.** $(x^0)^3 + (2^0)^5$ **c.** $\left(\frac{3y^7}{6x^5}\right)^2$ **d.** $\frac{(2a^3b^4)^3}{-8a^9b^2}$

Solution

a. $4^2 - 4^0 = 16 - 1 = 15$ Remember that $4^0 = 1$.

b. $(x^0)^3 + (2^0)^5 = 1^3 + 1^5 = 1 + 1 = 2$

c. $\left(\frac{3y^7}{6x^5}\right)^2 = \frac{3^2(y^7)^2}{6^2(x^5)^2} = \frac{9 \cdot y^{14}}{36 \cdot x^{10}} = \frac{y^{14}}{4x^{10}}$

d. $\frac{(2a^3b^4)^3}{-8a^9b^2} = \frac{2^3(a^3)^3(b^4)^3}{-8a^9b^2} = \frac{8a^9b^{12}}{-8a^9b^2} = -1 \cdot (a^{9-9}) \cdot (b^{12-2})$

$= -1 \cdot a^0 \cdot b^{10} = -1 \cdot 1 \cdot b^{10} = -b^{10}$ □

PRACTICE 12 Simplify each expression.

a. $8^2 - 8^0$ **b.** $(z^0)^6 + (4^0)^5$ **c.** $\left(\frac{5x^3}{15y^4}\right)^2$ **d.** $\frac{(2z^8x^5)^4}{-16z^2x^{20}}$

Vocabulary, Readiness & Video Check

Use the choices below to fill in each blank. Some choices may be used more than once.

0	base	add
1	exponent	multiply

1. Repeated multiplication of the same factor can be written using a(n) ________.
2. In 5^2, the 2 is called the ________ and the 5 is called the ________.
3. To simplify $x^2 \cdot x^7$, keep the base and ________ the exponents.
4. To simplify $(x^3)^6$, keep the base and ________ the exponents.
5. The understood exponent on the term y is ________.
6. If $x^{\square} = 1$, the exponent is ________.

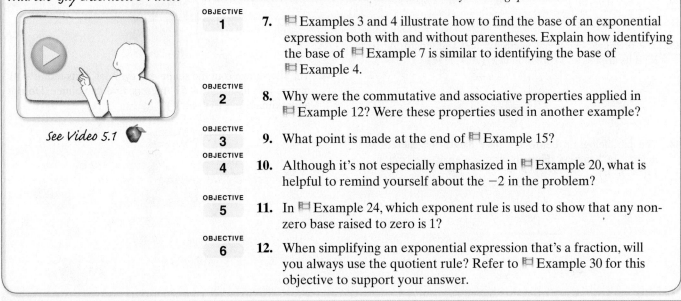

Martin-Gay Interactive Videos

See Video 5.1 

Watch the section lecture video and answer the following questions.

OBJECTIVE 1

7. Examples 3 and 4 illustrate how to find the base of an exponential expression both with and without parentheses. Explain how identifying the base of Example 7 is similar to identifying the base of Example 4.

OBJECTIVE 2

8. Why were the commutative and associative properties applied in Example 12? Were these properties used in another example?

OBJECTIVE 3

9. What point is made at the end of Example 15?

OBJECTIVE 4

10. Although it's not especially emphasized in Example 20, what is helpful to remind yourself about the −2 in the problem?

OBJECTIVE 5

11. In Example 24, which exponent rule is used to show that any nonzero base raised to zero is 1?

OBJECTIVE 6

12. When simplifying an exponential expression that's a fraction, will you always use the quotient rule? Refer to Example 30 for this objective to support your answer.

5.1 Exercise Set

MyMathLab®

For each of the following expressions, state the exponent shown and its corresponding base.

1. 3^2
2. $(-3)^6$
3. -4^2
4. $5 \cdot 3^4$
5. $5x^2$
6. $(5x)^2$

Evaluate each expression. See Example 1.

7. 7^2
8. -3^2
9. $(-5)^1$
10. $(-3)^2$
11. -2^4
12. -4^3
13. $(-2)^4$
14. $(-4)^3$
15. $(0.1)^5$
16. $(0.2)^5$
17. $\left(\frac{1}{3}\right)^4$
18. $\left(-\frac{1}{9}\right)^2$
19. $7 \cdot 2^5$
20. $9 \cdot 1^7$
21. $-2 \cdot 5^3$
22. $-4 \cdot 3^3$

Evaluate each expression given the replacement values for x. See Example 2.

23. $x^2; x = -2$
24. $x^3; x = -2$
25. $5x^3; x = 3$
26. $4x^2; x = -1$
27. $2xy^2; x = 3$ and $y = 5$
28. $-4x^2y^3; x = 2$ and $y = -1$
29. $\frac{2z^4}{5}; z = -2$
30. $\frac{10}{3y^3}; y = 5$

Use the product rule to simplify each expression. Write the results using exponents. See Examples 3 through 5.

31. $x^2 \cdot x^5$
32. $y^2 \cdot y$
33. $(-3)^3 \cdot (-3)^9$
34. $(-5)^7 \cdot (-5)^6$
35. $(5y^4)(3y)$
36. $(-2z^3)(-2z^2)$
37. $(x^9y)(x^{10}y^5)$
38. $(a^2b)(a^{13}b^{17})$
39. $(-8mn^6)(9m^2n^2)$
40. $(-7a^3b^3)(7a^{19}b)$
41. $(4z^{10})(-6z^7)(z^3)$
42. $(12x^5)(-x^6)(x^4)$

△ 43. The rectangle below has width $4x^2$ feet and length $5x^3$ feet. Find its area as an expression in x. $(A = l \cdot w)$

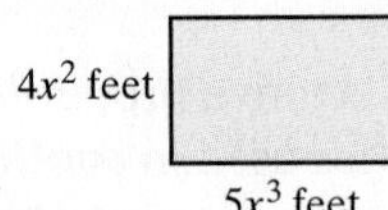

△ 44. The parallelogram below has base length $9y^7$ meters and height $2y^{10}$ meters. Find its area as an expression in y. $(A = b \cdot h)$

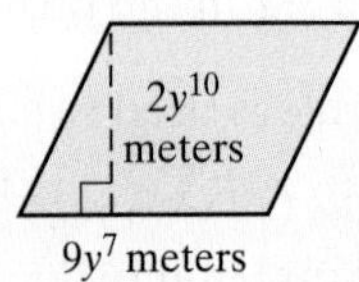

MIXED PRACTICE

Use the power rule and the power of a product or quotient rule to simplify each expression. See Examples 6 through 8.

45. $(x^9)^4$
46. $(y^7)^5$
47. $(pq)^8$
48. $(ab)^6$
49. $(2a^5)^3$
50. $(4x^6)^2$
51. $(x^2y^3)^5$
52. $(a^4b)^7$
53. $(-7a^2b^5c)^2$
54. $(-3x^7yz^2)^3$
55. $\left(\frac{r}{s}\right)^9$
56. $\left(\frac{q}{t}\right)^{11}$
57. $\left(\frac{mp}{n}\right)^5$
58. $\left(\frac{xy}{7}\right)^2$
59. $\left(\frac{-2xz}{y^5}\right)^2$
60. $\left(\frac{xy^4}{-3z^3}\right)^3$

△ 61. The square shown has sides of length $8z^5$ decimeters. Find its area. $(A = s^2)$

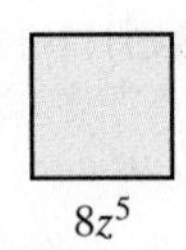

△ 62. Given the circle below with radius $5y$ centimeters, find its area. Do not approximate π. $(A = \pi r^2)$

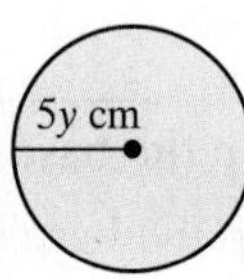

△ 63. The vault below is in the shape of a cube. If each side is $3y^4$ feet, find its volume. $(V = s^3)$

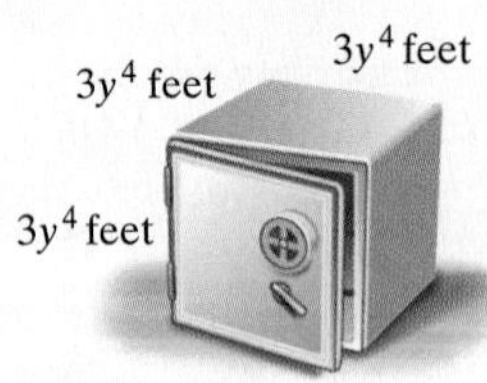

△ 64. The silo shown is in the shape of a cylinder. If its radius is $4x$ meters and its height is $5x^3$ meters, find its volume. Do not approximate π. $(V = \pi r^2 h)$

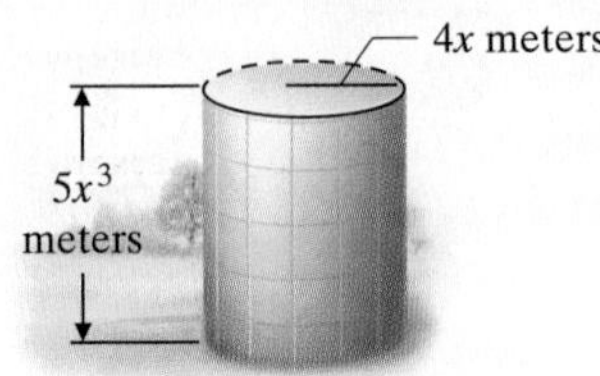

Use the quotient rule and simplify each expression. See Example 9.

65. $\frac{x^3}{x}$

66. $\frac{y^{10}}{y^9}$

67. $\frac{(-4)^6}{(-4)^3}$

68. $\frac{(-6)^{13}}{(-6)^{11}}$

69. $\frac{p^7q^{20}}{pq^{15}}$

70. $\frac{x^8y^6}{xy^5}$

71. $\frac{7x^2y^6}{14x^2y^3}$

72. $\frac{9a^4b^7}{27ab^2}$

Simplify each expression. See Example 10.

73. 7^0

74. 23^0

75. $(2x)^0$

76. $(4y)^0$

77. $-7x^0$

78. $-2x^0$

79. $5^0 + y^0$

80. $-3^0 + 4^0$

MIXED PRACTICE

Simplify each expression. See Examples 1 through 12.

81. -9^2

82. $(-9)^2$

83. $\left(\frac{1}{4}\right)^3$

84. $\left(\frac{2}{3}\right)^3$

85. b^4b^2

86. y^4y

87. $a^2a^3a^4$

88. $x^2x^{15}x^9$

89. $(2x^3)(-8x^4)$

90. $(3y^4)(-5y)$

91. $(a^7b^{12})(a^4b^8)$

92. $(y^2z^2)(y^{15}z^{13})$

93. $(-2mn^6)(-13m^8n)$

94. $(-3s^5t)(-7st^{10})$

95. $(z^4)^{10}$

96. $(t^5)^{11}$

97. $(4ab)^3$

98. $(2ab)^4$

99. $(-6xyz^3)^2$

100. $(-3xy^2a^3)^3$

101. $\frac{3x^5}{x^4}$

102. $\frac{5x^9}{x^3}$

103. $(9xy)^2$

104. $(2ab)^5$

105. $2^3 + 2^0$

106. $7^2 - 7^0$

107. $\left(\frac{3y^5}{6x^4}\right)^3$

108. $\left(\frac{2ab}{6yz}\right)^4$

109. $\frac{2x^3y^2z}{xyz}$

110. $\frac{x^{12}y^{13}}{x^5y^7}$

111. $(5^0)^3 + (y^0)^7$

112. $(9^0)^4 + (z^0)^5$

113. $\left(\frac{5x^9}{10y^{11}}\right)^2$

114. $\left(\frac{3a^4}{9b^5}\right)^2$

115. $\frac{(2a^5b^3)^4}{-16a^{20}b^7}$

116. $\frac{(2x^6y^2)^5}{-32x^{20}y^{10}}$

REVIEW AND PREVIEW

Simplify each expression by combining any like terms. Use the distributive property to remove any parentheses. See Section 2.1.

117. $y - 10 + y$

118. $-6z + 20 - 3z$

119. $7x + 2 - 8x - 6$

120. $10y - 14 - y - 14$

121. $2(x - 5) + 3(5 - x)$

122. $-3(w + 7) + 5(w + 1)$

CONCEPT EXTENSIONS

Solve. See the Concept Checks in this section. For Exercises 123 through 126, match the expression with the operation needed to simplify each. A letter may be used more than once and a letter may not be used at all.

123. $(x^{14})^{23}$

124. $x^{14} \cdot x^{23}$

125. $x^{14} + x^{23}$

126. $\frac{x^{35}}{x^{17}}$

a. Add the exponents
b. Subtract the exponents
c. Multiply the exponents
d. Divide the exponents
e. None of these

Fill in the boxes so that each statement is true. (More than one answer is possible for each exercise.)

127. $x^{\square} \cdot x^{\square} = x^{12}$

128. $(x^{\square})^{\square} = x^{20}$

129. $\frac{y^{\square}}{y^{\square}} = y^7$

130. $(y^{\square})^{\square} \cdot (y^{\square})^{\square} = y^{30}$

△ 131. The formula $V = x^3$ can be used to find the volume V of a cube with side length x. Find the volume of a cube with side length 7 meters. (Volume is measured in cubic units.)

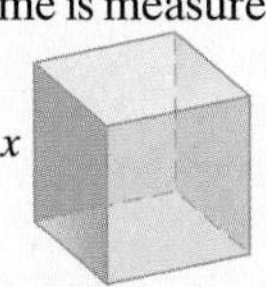

△ 132. The formula $S = 6x^2$ can be used to find the surface area S of a cube with side length x. Find the surface area of a cube with side length 5 meters. (Surface area is measured in square units.)

△ 133. To find the amount of water that a swimming pool in the shape of a cube can hold, do we use the formula for volume of the cube or surface area of the cube? (See Exercises 131 and 132.)

△ 134. To find the amount of material needed to cover an ottoman in the shape of a cube, do we use the formula for volume of the cube or surface area of the cube? (See Exercises 131 and 132.)

135. Explain why $(-5)^4 = 625$, while $-5^4 = -625$.

136. Explain why $5 \cdot 4^2 = 80$, while $(5 \cdot 4)^2 = 400$.

137. In your own words, explain why $5^0 = 1$.

138. In your own words, explain when $(-3)^n$ is positive and when it is negative.

Simplify each expression. Assume that variables represent positive integers.

139. $x^{5a}x^{4a}$

140. $b^{9a}b^{4a}$

141. $(a^b)^5$

142. $(2a^{4b})^4$

143. $\frac{x^{9a}}{x^{4a}}$

144. $\frac{y^{15b}}{y^{6b}}$

Solve. Round money amounts to 2 decimal places.

145. Suppose you borrow money for 6 months. If the interest is compounded monthly, the formula $A = P\left(1 + \frac{r}{12}\right)^6$ gives the total amount A to be repaid at the end of 6 months. For a loan of $P = \$1000$ and interest rate of 9% ($r = 0.09$), how much money is needed to pay off the loan?

146. Suppose you borrow money for 3 years. If the interest is compounded quarterly, the formula $A = P\left(1 + \frac{r}{4}\right)^{12}$ gives the total amount A to be repaid at the end of 3 years. For a loan of \$10,000 and interest rate of 8% ($r = 0.08$), how much money is needed to pay off the loan in 3 years?

5.2 Adding and Subtracting Polynomials 

OBJECTIVES

1 Define Polynomial, Monomial, Binomial, Trinomial, and Degree.

2 Find the Value of a Polynomial Given Replacement Values for the Variables.

3 Simplify a Polynomial by Combining Like Terms.

4 Add and Subtract Polynomials.

OBJECTIVE

1 Defining Polynomial, Monomial, Binomial, Trinomial, and Degree 

In this section, we introduce a special algebraic expression called a polynomial. Let's first review some definitions presented in Section 2.1.

Recall that a **term** is a number or the product of a number and variables raised to powers. The terms of the expression $4x^2 + 3x$ are $4x^2$ and $3x$.

The terms of the expression $9x^4 - 7x - 1$ are $9x^4$, $-7x$, and -1.

Expression	*Terms*
$4x^2 + 3x$	$4x^2, 3x$
$9x^4 - 7x - 1$	$9x^4, -7x, -1$
$7y^3$	$7y^3$
5	5

The **numerical coefficient** of a term, or simply the **coefficient**, is the numerical factor of each term. If no numerical factor appears in the term, then the coefficient is understood to be 1. If the term is a number only, it is called a **constant term** or simply a **constant.**

Term	*Coefficient*
x^5	1
$3x^2$	3
$-4x$	-4
$-x^2y$	-1
3 (constant)	3

Now we are ready to define a polynomial.

Polynomial

A **polynomial in *x*** is a finite sum of terms of the form ax^n, where a is a real number and n is a whole number.

For example,

$$x^5 - 3x^3 + 2x^2 - 5x + 1$$

is a polynomial. Notice that this polynomial is written in **descending powers** of x because the powers of x decrease from left to right. (Recall that the term 1 can be thought of as $1x^0$.)

On the other hand,

$$x^{-5} + 2x - 3$$

is **not** a polynomial because it contains an exponent, -5, that is not a whole number. (We study negative exponents in Section 5.5 of this chapter.)

Some polynomials are given special names.

Types of Polynomials

A **monomial** is a polynomial with exactly one term.

A **binomial** is a polynomial with exactly two terms.

A **trinomial** is a polynomial with exactly three terms.

The following are examples of monomials, binomials, and trinomials. Each of these examples is also a polynomial.

POLYNOMIALS

Monomials	*Binomials*	*Trinomials*	*More than Three Terms*
ax^2	$x + y$	$x^2 + 4xy + y^2$	$5x^3 - 6x^2 + 3x - 6$
$-3z$	$3p + 2$	$x^5 + 7x^2 - x$	$-y^5 + y^4 - 3y^3 - y^2 + y$
4	$4x^2 - 7$	$-q^4 + q^3 - 2q$	$x^6 + x^4 - x^3 + 1$

Each term of a polynomial has a **degree.**

Degree of a Term

The degree of a term is the sum of the exponents on the variables contained in the term.

EXAMPLE 1 Find the degree of each term.

a. $-3x^2$ **b.** $5x^3yz$ **c.** 2

Solution

a. The exponent on x is 2, so the degree of the term is 2.

b. $5x^3yz$ can be written as $5x^3y^1z^1$. The degree of the term is the sum of the exponents on its variables, so the degree is $3 + 1 + 1$ or 5.

c. The constant, 2, can be written as $2x^0$ (since $x^0 = 1$). The degree of 2 or $2x^0$ is 0. □

PRACTICE

1 Find the degree of each term.

a. $5y^3$ **b.** $-3a^2b^5c$ **c.** 8

From the preceding, we can say that **the degree of a constant is 0.**

Each polynomial also has a degree.

Degree of a Polynomial

The degree of a polynomial is the greatest degree of any term of the polynomial.

EXAMPLE 2 Find the degree of each polynomial and tell whether the polynomial is a monomial, binomial, trinomial, or none of these.

a. $-2t^2 + 3t + 6$ **b.** $15x - 10$ **c.** $7x + 3x^3 + 2x^2 - 1$

Solution

a. The degree of the trinomial $-2t^2 + 3t + 6$ is 2, the greatest degree of any of its terms.

b. The degree of the binomial $15x - 10$ or $15x^1 - 10$ is 1.

c. The degree of the polynomial $7x + 3x^3 + 2x^2 - 1$ is 3. □

PRACTICE

2 Find the degree of each polynomial and tell whether the polynomial is a monomial, binomial, trinomial, or none of these.

a. $5b^2 - 3b + 7$ **b.** $7t + 3$

c. $5x^2 + 3x - 6x^3 + 4$

EXAMPLE 3 Complete the table for the polynomial

$$7x^2y - 6xy + x^2 - 3y + 7$$

Use the table to give the degree of the polynomial.

Solution

Term	*Numerical Coefficient*	*Degree of Term*
$7x^2y$	7	3
$-6xy$	−6	2
x^2	1	2
$-3y$	−3	1
7	7	0

The degree of the polynomial is 3. ←

PRACTICE 3 Complete the table for the polynomial $-3x^3y^2 + 4xy^2 - y^2 + 3x - 2$.

Term	*Numerical Coefficient*	*Degree of Term*
$-3x^3y^2$		
$4xy^2$		
$-y^2$		
$3x$		
−2		

OBJECTIVE 2

Evaluating Polynomials

Polynomials have different values depending on replacement values for the variables.

EXAMPLE 4 Find the value of each polynomial when $x = -2$.

a. $-5x + 6$ **b.** $3x^2 - 2x + 1$

Solution

a. $-5x + 6 = -5(-2) + 6$ Replace x with −2.

$= 10 + 6$

$= 16$

b. $3x^2 - 2x + 1 = 3(-2)^2 - 2(-2) + 1$ Replace x with −2.

$= 3(4) + 4 + 1$

$= 12 + 4 + 1$

$= 17$

PRACTICE 4 Find the value of each polynomial when $x = -3$.

a. $-10x + 1$ **b.** $2x^2 - 5x + 3$

Many physical phenomena can be modeled by polynomials.

EXAMPLE 5 **Finding the Height of a Dropped Object**

The Swiss Re Building, in London, is a unique building. Londoners often refer to it as the "pickle building." The building is 592.1 feet tall. An object is dropped from the highest point of this building. Neglecting air resistance, the height in feet of the object above ground at time t seconds is given by the polynomial $-16t^2 + 592.1$. Find the height of the object when $t = 1$ second and when $t = 6$ seconds.

Solution To find each height, we evaluate the polynomial when $t = 1$ and when $t = 6$.

$$\begin{aligned} -16t^2 + 592.1 &= -16(1)^2 + 592.1 \quad \text{Replace } t \text{ with 1.} \\ &= -16(1) + 592.1 \\ &= -16 + 592.1 \\ &= 576.1 \end{aligned}$$

The height of the object at 1 second is 576.1 feet.

$$\begin{aligned} -16t^2 + 592.1 &= -16(6)^2 + 592.1 \quad \text{Replace } t \text{ with 6.} \\ &= -16(36) + 592.1 \\ &= -576 + 592.1 = 16.1 \end{aligned}$$

The height of the object at 6 seconds is 16.1 feet. □

PRACTICE

5 The cliff divers of Acapulco dive 130 feet into La Quebrada several times a day for the entertainment of the tourists. If a tourist is standing near the diving platform and drops his camera off the cliff, the height of the camera above the water at time t seconds is given by the polynomial $-16t^2 + 130$. Find the height of the camera when $t = 1$ second and when $t = 2$ seconds.

OBJECTIVE

3 Simplifying Polynomials by Combining Like Terms

Polynomials with like terms can be simplified by combining the like terms. Recall that like terms are terms that contain exactly the same variables raised to exactly the same powers.

Like Terms	*Unlike Terms*
$5x^2, -7x^2$	$3x, 3y$
$y, 2y$	$-2x^2, -5x$
$\frac{1}{2}a^2b, -a^2b$	$6st^2, 4s^2t$

Only like terms can be combined. We combine like terms by applying the distributive property.

EXAMPLE 6 Simplify each polynomial by combining any like terms.

a. $-3x + 7x$

b. $x + 3x^2$

c. $9x^3 + x^3$

d. $11x^2 + 5 + 2x^2 - 7$

e. $\frac{2}{5}x^4 + \frac{2}{3}x^3 - x^2 + \frac{1}{10}x^4 - \frac{1}{6}x^3$

Solution

a. $-3x + 7x = (-3 + 7)x = 4x$

b. $x + 3x^2$ These terms cannot be combined because x and $3x^2$ are not like terms.

c. $9x^3 + x^3 = 9x^3 + 1x^3 = 10x^3$

d. $11x^2 + 5 + 2x^2 - 7 = 11x^2 + 2x^2 + 5 - 7$

$= 13x^2 - 2$ Combine like terms.

e. $\frac{2}{5}x^4 + \frac{2}{3}x^3 - x^2 + \frac{1}{10}x^4 - \frac{1}{6}x^3$

$= \left(\frac{2}{5} + \frac{1}{10}\right)x^4 + \left(\frac{2}{3} - \frac{1}{6}\right)x^3 - x^2$

$= \left(\frac{4}{10} + \frac{1}{10}\right)x^4 + \left(\frac{4}{6} - \frac{1}{6}\right)x^3 - x^2$

$= \frac{5}{10}x^4 + \frac{3}{6}x^3 - x^2$

$= \frac{1}{2}x^4 + \frac{1}{2}x^3 - x^2$ □

PRACTICE

6 Simplify each polynomial by combining any like terms.

a. $-4y + 2y$ **b.** $z + 5z^3$ **c.** $15x^3 - x^3$

d. $7a^2 - 5 - 3a^2 - 7$ **e.** $\frac{3}{8}x^3 - x^2 + \frac{5}{6}x^4 + \frac{1}{12}x^3 - \frac{1}{2}x^4$

✓CONCEPT CHECK

When combining like terms in the expression $5x - 8x^2 - 8x$, which of the following is the proper result?

a. $-11x^2$ b. $-8x^2 - 3x$ c. $-11x$ d. $-11x^4$

EXAMPLE 7 Combine like terms to simplify.

$$-9x^2 + 3xy - 5y^2 + 7yx$$

Solution

$-9x^2 + 3xy - 5y^2 + 7yx = -9x^2 + (3 + 7)xy - 5y^2$

$= -9x^2 + 10xy - 5y^2$ □

Helpful Hint

This term can be written as $7yx$ or $7xy$.

PRACTICE

7 Combine like terms to simplify: $9xy - 3x^2 - 4yx + 5y^2$.

EXAMPLE 8 Write a polynomial that describes the total area of the squares and rectangles shown below. Then simplify the polynomial.

Solution

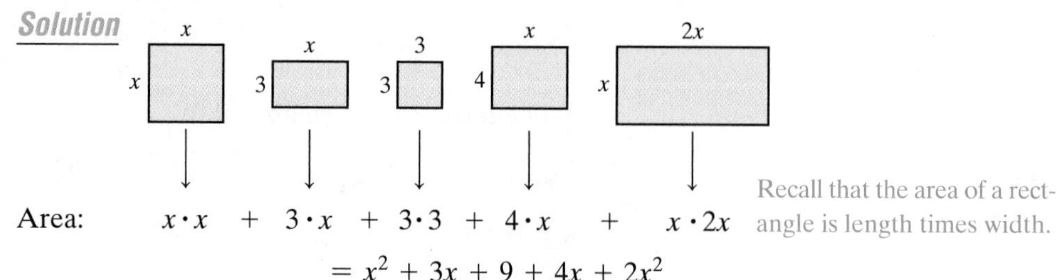

Recall that the area of a rectangle is length times width.

$= x^2 + 3x + 9 + 4x + 2x^2$

$= 3x^2 + 7x + 9$ Combine like terms. □

PRACTICE

8 Write a polynomial that describes the total area of the squares and rectangles shown below. Then simplify the polynomial.

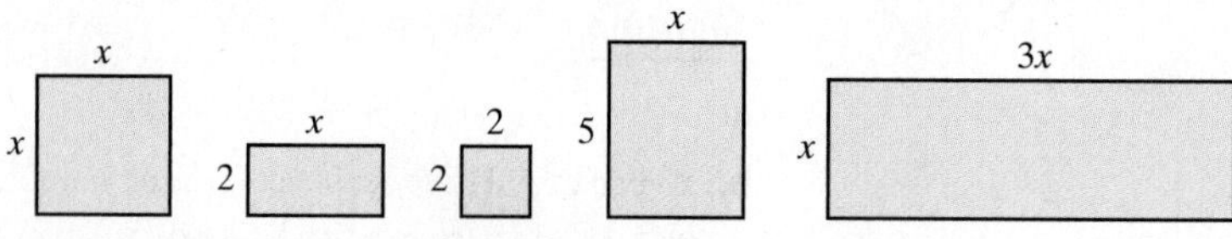

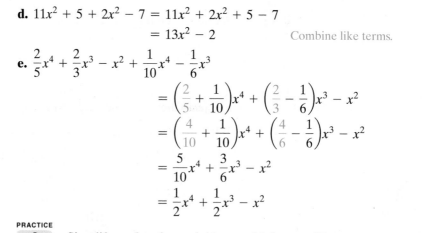

Answer to Concept Check:
b

OBJECTIVE

4 Adding and Subtracting Polynomials 

We now practice adding and subtracting polynomials.

Adding Polynomials

To add polynomials, combine all like terms.

EXAMPLE 9 Add $(-2x^2 + 5x - 1)$ and $(-2x^2 + x + 3)$.

Solution

$$\begin{aligned}(-2x^2 + 5x - 1) + (-2x^2 + x + 3) &= -2x^2 + 5x - 1 - 2x^2 + x + 3\\ &= (-2x^2 - 2x^2) + (5x + 1x) + (-1 + 3)\\ &= -4x^2 + 6x + 2\end{aligned}$$

□

PRACTICE

9 Add. $(-3x^2 - 4x + 9)$ and $(2x^2 - 2x)$.

EXAMPLE 10 Add: $(4x^3 - 6x^2 + 2x + 7) + (5x^2 - 2x)$.

Solution

$$\begin{aligned}(4x^3 - 6x^2 + 2x + 7) + (5x^2 - 2x) &= 4x^3 - 6x^2 + 2x + 7 + 5x^2 - 2x\\ &= 4x^3 + (-6x^2 + 5x^2) + (2x - 2x) + 7\\ &= 4x^3 - x^2 + 7\end{aligned}$$

□

PRACTICE

10 Add. $(-3x^3 + 7x^2 + 3x - 4) + (3x^2 - 9x)$.

Polynomials can be added vertically if we line up like terms underneath one another.

EXAMPLE 11 Add $(7y^3 - 2y^2 + 7)$ and $(6y^2 + 1)$ using the vertical format.

Solution Vertically line up like terms and add.

$$\begin{array}{r}7y^3 - 2y^2 + 7\\ 6y^2 + 1\\ \hline 7y^3 + 4y^2 + 8\end{array}$$

□

PRACTICE

11 Add $(5z^3 + 3z^2 + 4z)$ and $(5z^2 + 4z)$ using the vertical format.

To subtract one polynomial from another, recall the definition of subtraction. To subtract a number, we add its opposite: $a - b = a + (-b)$. To subtract a polynomial, we also add its opposite. Just as $-b$ is the opposite of b, $-(x^2 + 5)$ is the opposite of $(x^2 + 5)$.

EXAMPLE 12 Subtract: $(5x - 3) - (2x - 11)$.

Solution From the definition of subtraction, we have

$$\begin{aligned}(5x - 3) - (2x - 11) &= (5x - 3) + [-(2x - 11)] && \text{Add the opposite.}\\ &= (5x - 3) + (-2x + 11) && \text{Apply the distributive property.}\\ &= (5x - 2x) + (-3 + 11)\\ &= 3x + 8\end{aligned}$$

□

PRACTICE

12 Subtract: $(8x - 7) - (3x - 6)$.

Subtracting Polynomials

To subtract two polynomials, change the signs of the terms of the polynomial being subtracted and then add.

EXAMPLE 13 Subtract: $(2x^3 + 8x^2 - 6x) - (2x^3 - x^2 + 1)$.

Solution First, change the sign of each term of the second polynomial and then add.

> **Helpful Hint**
> Notice the sign of each term is changed.

$$(2x^3 + 8x^2 - 6x) - (2x^3 - x^2 + 1) = (2x^3 + 8x^2 - 6x) + (-2x^3 + x^2 - 1)$$
$$= 2x^3 - 2x^3 + 8x^2 + x^2 - 6x - 1$$
$$= 9x^2 - 6x - 1 \quad \text{Combine like terms.}$$ □

PRACTICE **13** Subtract: $(3x^3 - 5x^2 + 4x) - (x^3 - x^2 + 6)$.

EXAMPLE 14 Subtract $(5y^2 + 2y - 6)$ from $(-3y^2 - 2y + 11)$ using the vertical format.

Solution Arrange the polynomials in vertical format, lining up like terms.

$$\begin{array}{r} -3y^2 - 2y + 11 \\ \underline{-(5y^2 + 2y - 6)} \end{array} \qquad \begin{array}{r} -3y^2 - 2y + 11 \\ \underline{-5y^2 - 2y + 6} \\ -8y^2 - 4y + 17 \end{array}$$ □

PRACTICE **14** Subtract $(6z^2 + 3z - 7)$ from $(-2z^2 - 8z + 5)$ using the vertical format.

EXAMPLE 15 Subtract $(5z - 7)$ from the sum of $(8z + 11)$ and $(9z - 2)$.

Solution Notice that $(5z - 7)$ is to be subtracted **from** a sum. The translation is

$$[(8z + 11) + (9z - 2)] - (5z - 7)$$
$$= 8z + 11 + 9z - 2 - 5z + 7 \quad \text{Remove grouping symbols.}$$
$$= 8z + 9z - 5z + 11 - 2 + 7 \quad \text{Group like terms.}$$
$$= 12z + 16 \quad \text{Combine like terms.}$$ □

PRACTICE **15** Subtract $(3x + 5)$ from the sum of $(8x - 11)$ and $(2x + 5)$.

EXAMPLE 16 Add or subtract as indicated.

a. $(3x^2 - 6xy + 5y^2) + (-2x^2 + 8xy - y^2)$

b. $(9a^2b^2 + 6ab - 3ab^2) - (5b^2a + 2ab - 3 - 9b^2)$

Solution

a. $(3x^2 - 6xy + 5y^2) + (-2x^2 + 8xy - y^2)$

$= 3x^2 - 6xy + 5y^2 - 2x^2 + 8xy - y^2$

$= x^2 + 2xy + 4y^2$ Combine like terms.

b. $(9a^2b^2 + 6ab - 3ab^2) - (5b^2a + 2ab - 3 - 9b^2)$

$= 9a^2b^2 + 6ab - 3ab^2 - 5b^2a - 2ab + 3 + 9b^2$ Change the sign of each term of the polynomial being subtracted.

$= 9a^2b^2 + 4ab - 8ab^2 + 3 + 9b^2$ Combine like terms. □

Answers to Concept Check:
a. $3y$ **b.** $2y^2$ **c.** $-3y$ **d.** $2y^2$
e. cannot be simplified

PRACTICE 16 Add or subtract as indicated.

a. $(3a^2 - 4ab + 7b^2) + (-8a^2 + 3ab - b^2)$
b. $(5x^2y^2 - 6xy - 4xy^2) - (2x^2y^2 + 4xy - 5 + 6y^2)$

✓CONCEPT CHECK
If possible, simplify each expression by performing the indicated operation.
a. $2y + y$ b. $2y \cdot y$ c. $-2y - y$ d. $(-2y)(-y)$ e. $2x + y$

Vocabulary, Readiness & Video Check

Use the choices below to fill in each blank. Not all choices will be used.

least	monomial	trinomial	coefficient
greatest	binomial	constant	

1. A ____________ is a polynomial with exactly 2 terms.
2. A ____________ is a polynomial with exactly one term.
3. A ____________ is a polynomial with exactly three terms.
4. The numerical factor of a term is called the ____________.
5. A number term is also called a ____________.
6. The degree of a polynomial is the ____________ degree of any term of the polynomial.

Martin-Gay Interactive Videos

See Video 5.2

Watch the section lecture video and answer the following questions.

OBJECTIVE 1 **7.** For Example 2, why is the degree of each **term** found when the example asks for the degree of the **polynomial** only?

OBJECTIVE 2 **8.** From Example 3, what does the value of a polynomial depend on?

OBJECTIVE 3 **9.** When combining any like terms in a polynomial, as in Examples 4–6, what are we doing to the polynomial?

OBJECTIVE 4 **10.** From Example 7, when we simply remove parentheses and combine the like terms of two polynomials, what operation do we perform? Is this true of Examples 9–11?

5.2 Exercise Set

MyMathLab®

Find the degree of each of the following polynomials and determine whether it is a monomial, binomial, trinomial, or none of these. See Examples 1 through 3.

1. $x + 2$
2. $-6y + 4$
3. $9m^3 - 5m^2 + 4m - 8$
4. $a + 5a^2 + 3a^3 - 4a^4$
5. $12x^4y - x^2y^2 - 12x^2y^4$
6. $7r^2s^2 + 2rs - 3rs^5$
7. $3 - 5x^8$
8. $5y^7 + 2$

In the second column, write the degree of the polynomial in the first column. See Examples 1 through 3.

	Polynomial	Degree
9.	$3xy^2 - 4$	
10.	$8x^2y^2$	
11.	$5a^2 - 2a + 1$	
12.	$4z^6 + 3z^2$	

Find the value of each polynomial when **(a)** $x = 0$ *and* **(b)** $x = -1$. *See Examples 4 and 5.*

13. $5x - 6$

14. $2x - 10$

15. $x^2 - 5x - 2$

16. $x^2 + 3x - 4$

17. $-x^3 + 4x^2 - 15x + 1$

18. $-2x^3 + 3x^2 - 6x + 1$

The CN Tower in Toronto, Ontario, is 1821 feet tall and is the world's tallest self-supporting structure. An object is dropped from the Skypod of the Tower, which is at 1150 feet. Neglecting air resistance, the height of the object at time t seconds is given by the polynomial $-16t^2 + 1150$. *Find the height of the object at the given times.*

	Time, t (in seconds)	**Height** $-16t^2 + 1150$
19.	1	
20.	7	
21.	3	
22.	6	

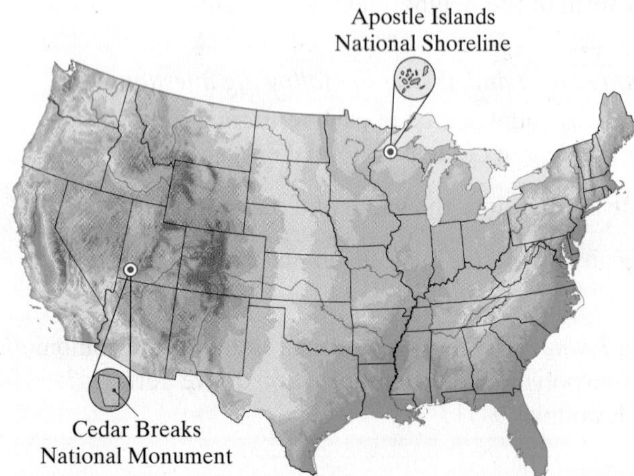

23. The polynomial $-7.5x^2 + 93x - 100$ models the yearly number of visitors (in thousands) x years after 2000 at Apostle Islands National Park. Use this polynomial to estimate the number of visitors to the park in 2008 ($x = 8$).

24. The polynomial $8x^2 - 90.6x + 752$ models the yearly number of visitors (in thousands) x years after 2000 at Cedar Breaks National Park. Use this polynomial to estimate the number of visitors to the park in 2007 ($x = 7$).

Simplify each of the following by combining like terms. See Examples 6 and 7.

25. $9x - 20x$

26. $14y - 30y$

27. $14x^2 + 9x^2$

28. $18x^3 - 4x^3$

29. $15x^2 - 3x^2 - y$

30. $12k^3 - 9k^3 + 11$

31. $8s - 5s + 4s$

32. $5y + 7y - 6y$

33. $0.1y^2 - 1.2y^2 + 6.7 - 1.9$

34. $7.6y + 3.2y^2 - 8y - 2.5y^2$

35. $\frac{2}{3}x^4 + 12x^3 + \frac{1}{6}x^4 - 19x^3 - 19$

36. $\frac{2}{5}x^4 - 23x^2 + \frac{1}{15}x^4 + 5x^2 - 5$

37. $\frac{2}{5}x^2 - \frac{1}{3}x^3 + x^2 - \frac{1}{4}x^3 + 6$

38. $\frac{1}{6}x^4 - \frac{1}{7}x^2 + 5 - \frac{1}{2}x^4 - \frac{3}{7}x^2 + 1$

39. $6a^2 - 4ab + 7b^2 - a^2 - 5ab + 9b^2$

40. $x^2y + xy - y + 10x^2y - 2y + xy$

Perform the indicated operations. See Examples 9 through 13.

41. $(2x + 5) - (3x - 9)$

42. $(4 + 5a) - (-a - 5)$

43. $(2x^2 + 5) - (3x^2 - 9)$

44. $(5x^2 + 4) - (-2y^2 + 4)$

45. $(-7x + 5) + (-3x^2 + 7x + 5)$

46. $(3x - 8) + (4x^2 - 3x + 3)$

47. $3x - (5x - 9)$

48. $4 - (-12y - 4)$

49. $(2x^2 + 3x - 9) - (-4x + 7)$

50. $(-7x^2 + 4x + 7) - (-8x + 2)$

Perform the indicated operations. See Examples 11, 14, and 15.

51. $\begin{array}{r} 3t^2 + 4 \\ \underline{+5t^2 - 8} \end{array}$

52. $\begin{array}{r} 7x^3 + 3 \\ \underline{+2x^3 + 1} \end{array}$

53. $\begin{array}{r} 4z^2 - 8z + 3 \\ \underline{-(6z^2 + 8z - 3)} \end{array}$

54. $\begin{array}{r} 7a^2 - 9a + 6 \\ \underline{-(11a^2 - 4a + 2)} \end{array}$

55. $\begin{array}{r} 5x^3 - 4x^2 + 6x - 2 \\ \underline{-(3x^3 - 2x^2 - x - 4)} \end{array}$

56. $\begin{array}{r} 5u^5 - 4u^2 + 3u - 7 \\ \underline{-(3u^5 + 6u^2 - 8u + 2)} \end{array}$

57. Subtract $(19x^2 + 5)$ from $(81x^2 + 10)$.

58. Subtract $(2x + xy)$ from $(3x - 9xy)$.

59. Subtract $(2x + 2)$ from the sum of $(8x + 1)$ and $(6x + 3)$.

60. Subtract $(-12x - 3)$ from the sum of $(-5x - 7)$ and $(12x + 3)$.

MIXED PRACTICE

Perform the indicated operations.

61. $(2y + 20) + (5y - 30)$

62. $(14y + 12) + (-3y - 5)$

63. $(x^2 + 2x + 1) - (3x^2 - 6x + 2)$

64. $(5y^2 - 3y - 1) - (2y^2 + y + 1)$

65. $(3x^2 + 5x - 8) + (5x^2 + 9x + 12) - (8x^2 - 14)$

66. $(2x^2 + 7x - 9) + (x^2 - x + 10) - (3x^2 - 30)$

67. $(-a^2 + 1) - (a^2 - 3) + (5a^2 - 6a + 7)$

68. $(-m^2 + 3) - (m^2 - 13) + (6m^2 - m + 1)$

TRANSLATING

Perform each indicated operation.

69. Subtract $4x$ from $(7x - 3)$.
70. Subtract y from $(y^2 - 4y + 1)$.
71. Add $(4x^2 - 6x + 1)$ and $(3x^2 + 2x + 1)$.
72. Add $(-3x^2 - 5x + 2)$ and $(x^2 - 6x + 9)$.
73. Subtract $(5x + 7)$ from $(7x^2 + 3x + 9)$.
74. Subtract $(5y^2 + 8y + 2)$ from $(7y^2 + 9y - 8)$.
75. Subtract $(4y^2 - 6y - 3)$ from the sum of $(8y^2 + 7)$ and $(6y + 9)$.
76. Subtract $(4x^2 - 2x + 2)$ from the sum of $(x^2 + 7x + 1)$ and $(7x + 5)$.

Find the area of each figure. Write a polynomial that describes the total area of the rectangles and squares shown in Exercises 77–78. Then simplify the polynomial. See Example 8.

△ 77. △ 78.

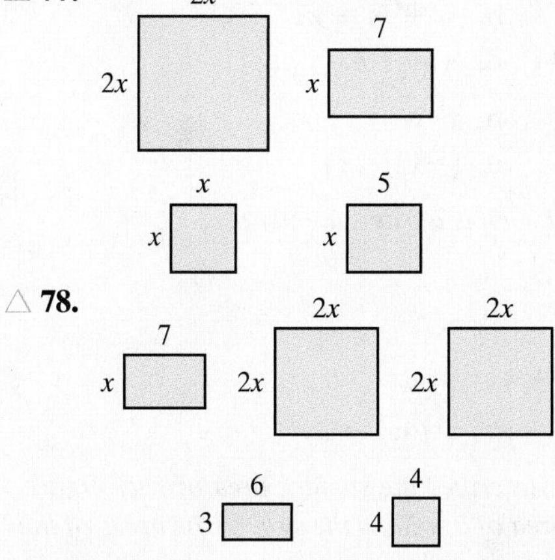

Add or subtract as indicated. See Example 16.

79. $(9a + 6b - 5) + (-11a - 7b + 6)$
80. $(3x - 2 + 6y) + (7x - 2 - y)$
81. $(4x^2 + y^2 + 3) - (x^2 + y^2 - 2)$
82. $(7a^2 - 3b^2 + 10) - (-2a^2 + b^2 - 12)$
83. $(x^2 + 2xy - y^2) + (5x^2 - 4xy + 20y^2)$
84. $(a^2 - ab + 4b^2) + (6a^2 + 8ab - b^2)$
85. $(11r^2s + 16rs - 3 - 2r^2s^2) - (3sr^2 + 5 - 9r^2s^2)$
86. $(3x^2y - 6xy + x^2y^2 - 5) - (11x^2y^2 - 1 + 5yx^2)$

Simplify each polynomial by combining like terms.

87. $7.75x + 9.16x^2 - 1.27 - 14.58x^2 - 18.34$
88. $1.85x^2 - 3.76x + 9.25x^2 + 10.76 - 4.21x$

Perform each indicated operation.

89. $[(7.9y^4 - 6.8y^3 + 3.3y) + (6.1y^3 - 5)] - (4.2y^4 + 1.1y - 1)$
90. $[(1.2x^2 - 3x + 9.1) - (7.8x^2 - 3.1 + 8)] + (1.2x - 6)$

REVIEW AND PREVIEW

Multiply. See Section 5.1.

91. $3x(2x)$
92. $-7x(x)$
93. $(12x^3)(-x^5)$
94. $6r^3(7r^{10})$
95. $10x^2(20xy^2)$
96. $-z^2y(11zy)$

CONCEPT EXTENSIONS

Recall that the perimeter of a figure is the sum of the lengths of its sides. For Exercises 97 through 100, find the perimeter of each polynomial.

△ 97. △ 98.

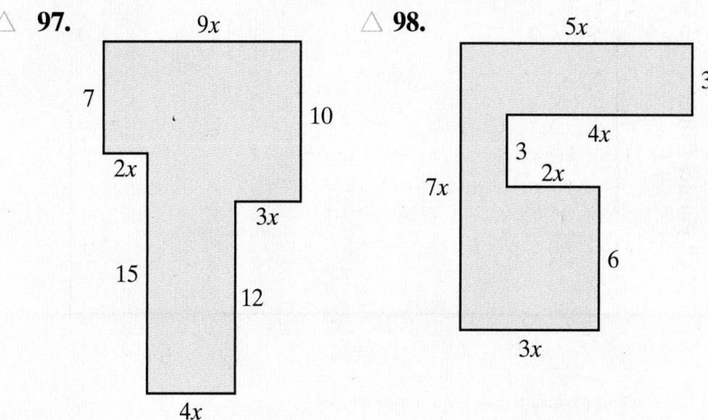

△ 99.

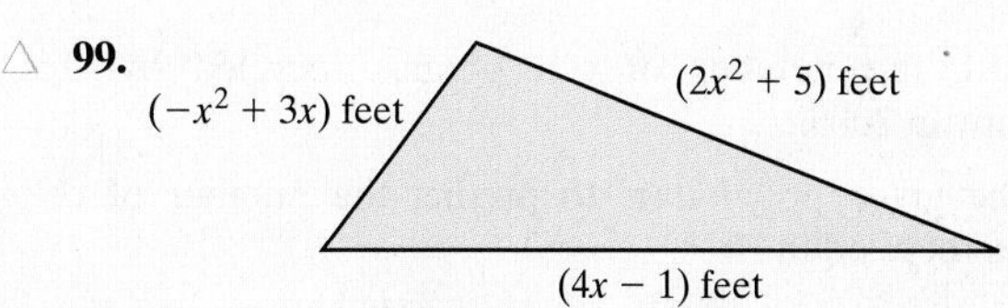

△ 100.

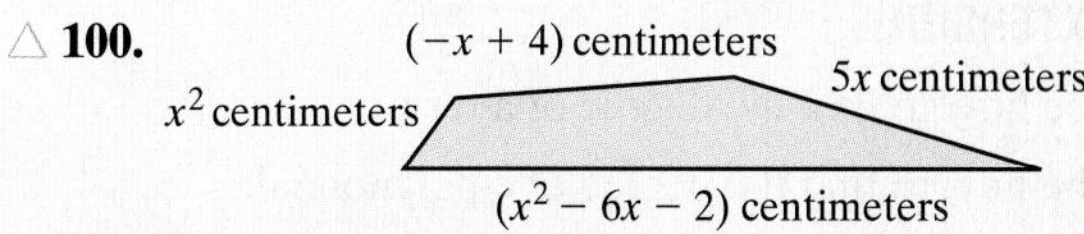

△ 101. A wooden beam is $(4y^2 + 4y + 1)$ meters long. If a piece $(y^2 - 10)$ meters is cut, express the length of the remaining piece of beam as a polynomial in y.

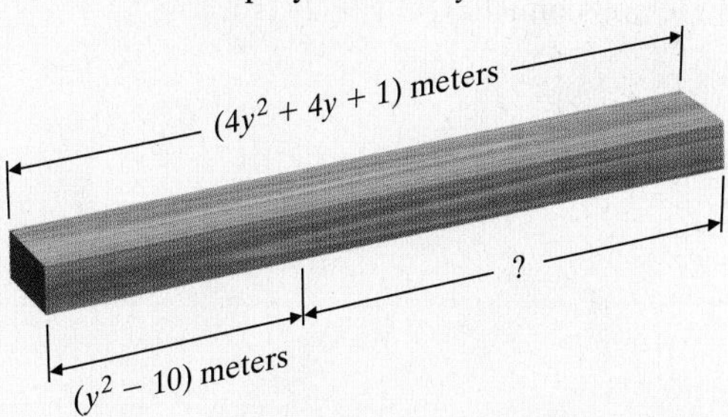

△ 102. A piece of quarter-round molding is $(13x - 7)$ inches long. If a piece $(2x + 2)$ inches is removed, express the length of the remaining piece of molding as a polynomial in x.

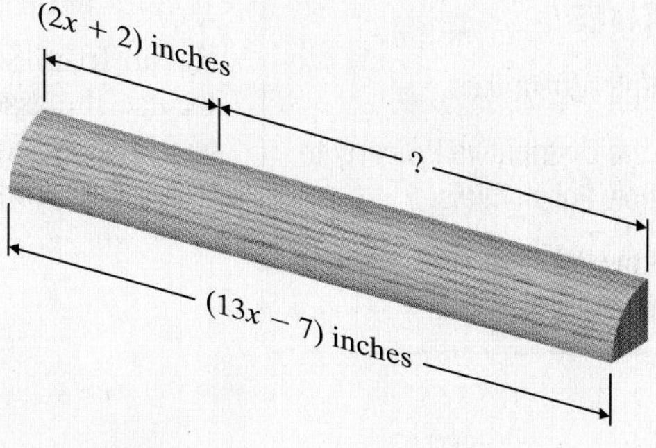

The number of cell phones recycled by the largest recycler x years after 2004 is given by the polynomial $0.16x^2 + 0.29x + 1.88$ *for the years 2004 through 2010. Use this polynomial for Exercises 103 and 104.*

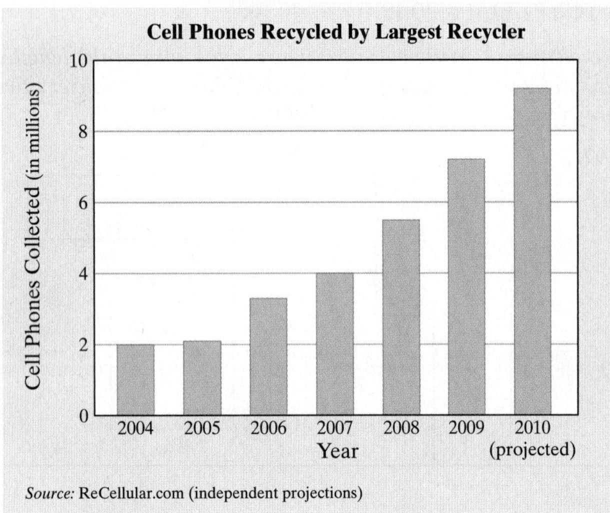

Source: ReCellular.com (independent projections)

103. Estimate the number of cell phones recycled by this company in 2009.

104. Use the given polynomial to predict the number of cell phones recycled in 2014.

CONCEPT EXTENSIONS

105. Describe how to find the degree of a term.

106. Describe how to find the degree of a polynomial.

107. Explain why xyz is a monomial while $x + y + z$ is a trinomial.

108. Explain why the degree of the term $5y^3$ is 3 and the degree of the polynomial $2y + y + 2y$ is 1.

Match each expression on the left with its simplification on the right. Not all letters on the right must be used, and a letter may be used more than once.

109. $10y - 6y^2 - y$ **a.** $3y$

110. $5x + 5x$ **b.** $9y - 6y^2$

111. $(5x - 3) + (5x - 3)$ **c.** $10x$

112. $(15x - 3) - (5x - 3)$ **d.** $25x^2$

e. $10x - 6$

f. none of these

Simplify each expression by performing the indicated operation. Explain how you arrived at each answer. See the last Concept Check in this section.

113. **a.** $z + 3z$ **b.** $z \cdot 3z$ **c.** $-z - 3z$ **d.** $(-z)(-3z)$

114. **a.** $2y + y$ **b.** $2y \cdot y$ **c.** $-2y - y$ **d.** $(-2y)(-y)$

115. **a.** $m \cdot m \cdot m$ **b.** $m + m + m$ **c.** $(-m)(-m)(-m)$ **d.** $-m - m - m$

116. **a.** $x + x$ **b.** $x \cdot x$ **c.** $-x - x$ **d.** $(-x)(-x)$

Fill in the squares so that each is a true statement.

117. $3x^{\square} + 4x^2 = 7x^{\square}$

118. $9y^7 + 3y^{\square} = 12y^7$

119. $2x^{\square} + 3x^{\square} - 5x^{\square} + 4x^{\square} = 6x^4 - 2x^3$

120. $3y^{\square} + 7y^{\square} - 2y^{\square} - y^{\square} = 10y^5 - 3y^2$

Write a polynomial that describes the surface area of each figure. (Recall that the surface area of a solid is the sum of the areas of the faces or sides of the solid.)

△ **121.**

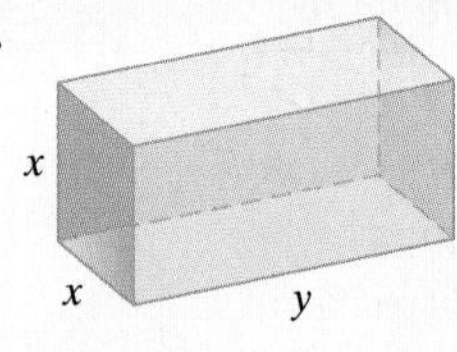

△ **122.**

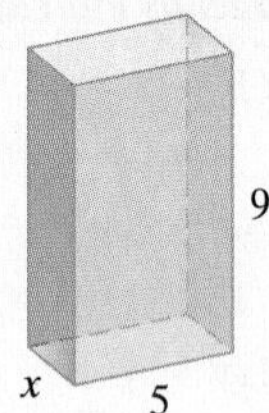

5.3 Multiplying Polynomials 

OBJECTIVES

1 Multiply Monomials. 

2 Use the Distributive Property to Multiply Polynomials. 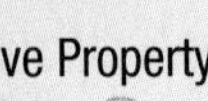

3 Multiply Polynomials Vertically.

OBJECTIVE

1 Multiplying Monomials 

Recall from Section 5.1 that to multiply two monomials such as $(-5x^3)$ and $(-2x^4)$, we use the associative and commutative properties and regroup. Remember, also, that to multiply exponential expressions with a common base, we use the product rule for exponents and add exponents.

$$(-5x^3)(-2x^4) = (-5)(-2)(x^3)(x^4) = 10x^7$$

EXAMPLES Multiply.

1. $6x \cdot 4x = (6 \cdot 4)(x \cdot x)$ Use the commutative and associative properties.
$= 24x^2$ Multiply.

2. $-7x^2 \cdot 0.2x^5 = (-7 \cdot 0.2)(x^2 \cdot x^5)$
$= -1.4x^7$

3. $\left(-\frac{1}{3}x^5\right)\left(-\frac{2}{9}x\right) = \left(-\frac{1}{3} \cdot -\frac{2}{9}\right) \cdot (x^5 \cdot x)$
$= \frac{2}{27}x^6$ □

PRACTICES 1–3 Multiply.

1. $5y \cdot 2y$ **2.** $(5z^3) \cdot (-0.4z^5)$ **3.** $\left(-\frac{1}{9}b^6\right)\left(-\frac{7}{8}b^3\right)$

✓CONCEPT CHECK
Simplify.
a. $3x \cdot 2x$ b. $3x + 2x$

OBJECTIVE
2 Using the Distributive Property to Multiply Polynomials 

To multiply polynomials that are not monomials, use the distributive property.

EXAMPLE 4 Use the distributive property to find each product.

a. $5x(2x^3 + 6)$ **b.** $-3x^2(5x^2 + 6x - 1)$

Solution

a. $5x(2x^3 + 6) = 5x(2x^3) + 5x(6)$ Use the distributive property.
$= 10x^4 + 30x$ Multiply.

b. $-3x^2(5x^2 + 6x - 1)$
$= (-3x^2)(5x^2) + (-3x^2)(6x) + (-3x^2)(-1)$ Use the distributive property.
$= -15x^4 - 18x^3 + 3x^2$ Multiply. □

PRACTICE 4 Use the distributive property to find each product.

a. $3x(9x^5 + 11)$ **b.** $-6x^3(2x^2 - 9x + 2)$

We also use the distributive property to multiply two binomials. To multiply $(x + 3)$ by $(x + 1)$, distribute the factor $(x + 3)$ first.

$(x + 3)(x + 1) = x(x + 1) + 3(x + 1)$ Distribute $(x + 3)$.
$= x(x) + x(1) + 3(x) + 3(1)$ Apply the distributive property a second time.
$= x^2 + x + 3x + 3$ Multiply.
$= x^2 + 4x + 3$ Combine like terms.

Answers to Concept Check:
a. $6x^2$ **b.** $5x$

This idea can be expanded so that we can multiply any two polynomials.

To Multiply Two Polynomials

Multiply each term of the first polynomial by each term of the second polynomial and then combine like terms.

EXAMPLE 5 Multiply $(3x + 2)(2x - 5)$.

Solution Multiply each term of the first binomial by each term of the second.

$$(3x + 2)(2x - 5) = 3x(2x) + 3x(-5) + 2(2x) + 2(-5)$$
$$= 6x^2 - 15x + 4x - 10 \quad \text{Multiply.}$$
$$= 6x^2 - 11x - 10 \quad \text{Combine like terms.}$$

PRACTICE

5 Multiply$(5x - 2)(2x + 3)$

EXAMPLE 6 Multiply $(2x - y)^2$.

Solution Recall that $a^2 = a \cdot a$, so $(2x - y)^2 = (2x - y)(2x - y)$. Multiply each term of the first polynomial by each term of the second.

$$(2x - y)(2x - y) = 2x(2x) + 2x(-y) + (-y)(2x) + (-y)(-y)$$
$$= 4x^2 - 2xy - 2xy + y^2 \quad \text{Multiply.}$$
$$= 4x^2 - 4xy + y^2 \quad \text{Combine like terms.}$$

PRACTICE

6 Multiply$(5x - 3y)^2$

✓CONCEPT CHECK

Square where indicated. Simplify if possible.

a. $(4a)^2 + (3b)^2$ b. $(4a + 3b)^2$

EXAMPLE 7 Multiply $(t + 2)$ by $(3t^2 - 4t + 2)$.

Solution Multiply each term of the first polynomial by each term of the second.

$$(t + 2)(3t^2 - 4t + 2) = t(3t^2) + t(-4t) + t(2) + 2(3t^2) + 2(-4t) + 2(2)$$
$$= 3t^3 - 4t^2 + 2t + 6t^2 - 8t + 4$$
$$= 3t^3 + 2t^2 - 6t + 4 \quad \text{Combine like terms.}$$

PRACTICE

7 Multiply $(y + 4)$ by $(2y^2 - 3y + 5)$

EXAMPLE 8 Multiply $(3a + b)^3$.

Solution Write $(3a + b)^3$ as $(3a + b)(3a + b)(3a + b)$.

$$(3a + b)(3a + b)(3a + b) = (9a^2 + 3ab + 3ab + b^2)(3a + b)$$
$$= (9a^2 + 6ab + b^2)(3a + b)$$
$$= 9a^2(3a + b) + 6ab(3a + b) + b^2(3a + b)$$
$$= 27a^3 + 9a^2b + 18a^2b + 6ab^2 + 3ab^2 + b^3$$
$$= 27a^3 + 27a^2b + 9ab^2 + b^3$$

PRACTICE

8 Multiply $(s + 2t)^3$.

Answers to Concept Check:
a. $16a^2 + 9b^2$
b. $16a^2 + 24ab + 9b^2$

OBJECTIVE

3 Multiplying Polynomials Vertically 

Another convenient method for multiplying polynomials is to use a vertical format similar to the format used to multiply real numbers. We demonstrate this method by multiplying $(3y^2 - 4y + 1)$ by $(y + 2)$.

EXAMPLE 9 Multiply $(3y^2 - 4y + 1)(y + 2)$. Use a vertical format.

Solution

$$\begin{array}{rl} 3y^2 - 4y + 1 & \\ \times \qquad y + 2 & \text{1st, Multiply } 3y^2 - 4y + 1 \text{ by } 2. \\ \hline 6y^2 - 8y + 2 & \text{2nd, Multiply } 3y^2 - 4y + 1 \text{ by } y. \\ 3y^3 - 4y^2 + y \qquad & \text{Line up like terms.} \\ \hline 3y^3 + 2y^2 - 7y + 2 & \text{3rd, Combine like terms.} \end{array}$$

Helpful Hint
Make sure like terms are lined up.

Thus, $(3y^2 - 4y + 1)(y + 2) = 3y^3 + 2y^2 - 7y + 2$. □

PRACTICE

9 Multiply $(5x^2 - 3x + 5)(x - 4)$.

When multiplying vertically, be careful if a power is missing; you may want to leave space in the partial products and take care that like terms are lined up.

EXAMPLE 10 Multiply $(2x^3 - 3x + 4)(x^2 + 1)$. Use a vertical format.

Solution

$$\begin{array}{rl} 2x^3 - 3x + 4 & \\ \times \qquad x^2 + 1 & \\ \hline 2x^3 \qquad - 3x + 4 & \text{Leave space for missing powers of } x. \\ 2x^5 - 3x^3 + 4x^2 \qquad\qquad & \leftarrow \text{Line up like terms.} \\ \hline 2x^5 - x^3 + 4x^2 - 3x + 4 & \text{Combine like terms.} \end{array}$$

□

PRACTICE

10 Multiply $(x^3 - 2x^2 + 1)(x^2 + 2)$ using a vertical format.

EXAMPLE 11 Find the product of $(2x^2 - 3x + 4)$ and $(x^2 + 5x - 2)$ using a vertical format.

Solution First, we arrange the polynomials in a vertical format. Then we multiply each term of the second polynomial by each term of the first polynomial.

$$\begin{array}{rl} 2x^2 - 3x + 4 & \\ \times \qquad x^2 + 5x - 2 & \\ \hline -4x^2 + 6x - 8 & \text{Multiply } 2x^2 - 3x + 4 \text{ by } -2. \\ 10x^3 - 15x^2 + 20x \qquad & \text{Multiply } 2x^2 - 3x + 4 \text{ by } 5x. \\ 2x^4 - 3x^3 + 4x^2 \qquad\qquad & \text{Multiply } 2x^2 - 3x + 4 \text{ by } x^2. \\ \hline 2x^4 + 7x^3 - 15x^2 + 26x - 8 & \text{Combine like terms.} \end{array}$$

□

PRACTICE

11 Find the product of $(5x^2 + 2x - 2)$ and $(x^2 - x + 3)$ using a vertical format.

Vocabulary, Readiness & Video Check

Fill in each blank with the correct choice.

1. The expression $5x(3x + 2)$ equals $5x \cdot 3x + 5x \cdot 2$ by the ______________ property.
a. commutative **b.** associative **c.** distributive

2. The expression $(x + 4)(7x - 1)$ equals $x(7x - 1) + 4(7x - 1)$ by the ______________ property.
a. commutative **b.** associative **c.** distributive

3. The expression $(5y - 1)^2$ equals ______________.
a. $2(5y - 1)$ **b.** $(5y - 1)(5y + 1)$ **c.** $(5y - 1)(5y - 1)$

4. The expression $9x \cdot 3x$ equals ______________.
a. $27x$ **b.** $27x^2$ **c.** $12x$ **d.** $12x^2$

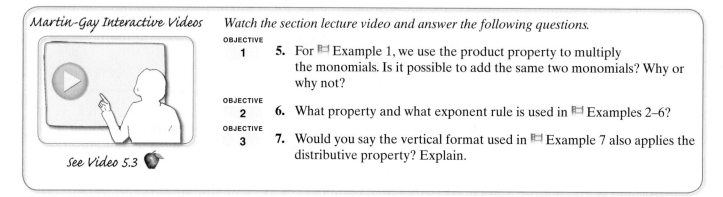

Martin-Gay Interactive Videos

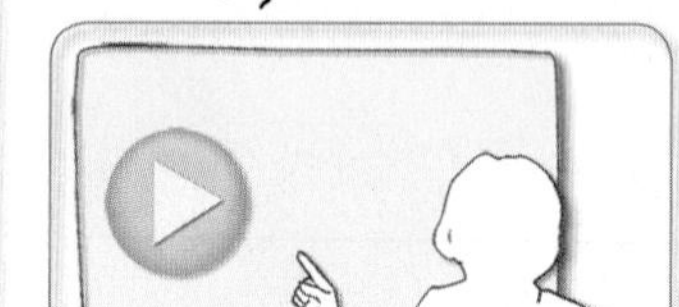

See Video 5.3

Watch the section lecture video and answer the following questions.

OBJECTIVE 1 **5.** For Example 1, we use the product property to multiply the monomials. Is it possible to add the same two monomials? Why or why not?

OBJECTIVE 2 **6.** What property and what exponent rule is used in Examples 2–6?

OBJECTIVE 3 **7.** Would you say the vertical format used in Example 7 also applies the distributive property? Explain.

5.3 Exercise Set

MyMathLab®

Multiply. See Examples 1 through 3.

1. $-4n^3 \cdot 7n^7$
2. $9t^6(-3t^5)$
3. $(-3.1x^3)(4x^9)$
4. $(-5.2x^4)(3x^4)$
5. $\left(-\frac{1}{3}y^2\right)\left(\frac{2}{5}y\right)$
6. $\left(-\frac{3}{4}y^7\right)\left(\frac{1}{7}y^4\right)$
7. $(2x)(-3x^2)(4x^5)$
8. $(x)(5x^4)(-6x^7)$

Multiply. See Example 4.

9. $3x(2x + 5)$
10. $2x(6x + 3)$
11. $-2a(a + 4)$
12. $-3a(2a + 7)$
13. $3x(2x^2 - 3x + 4)$
14. $4x(5x^2 - 6x - 10)$
15. $-2a^2(3a^2 - 2a + 3)$
16. $-4b^2(3b^3 - 12b^2 - 6)$
17. $-y(4x^3 - 7x^2y + xy^2 + 3y^3)$
18. $-x(6y^3 - 5xy^2 + x^2y - 5x^3)$
19. $\frac{1}{2}x^2(8x^2 - 6x + 1)$
20. $\frac{1}{3}y^2(9y^2 - 6y + 1)$

Multiply. See Examples 5 and 6.

21. $(x + 4)(x + 3)$
22. $(x + 2)(x + 9)$
23. $(a + 7)(a - 2)$
24. $(y - 10)(y + 11)$
25. $\left(x + \frac{2}{3}\right)\left(x - \frac{1}{3}\right)$
26. $\left(x + \frac{3}{5}\right)\left(x - \frac{2}{5}\right)$
27. $(3x^2 + 1)(4x^2 + 7)$
28. $(5x^2 + 2)(6x^2 + 2)$
29. $(2y - 4)^2$
30. $(6x - 7)^2$
31. $(4x - 3)(3x - 5)$
32. $(8x - 3)(2x - 4)$
33. $(3x^2 + 1)^2$
34. $(x^2 + 4)^2$

35. Perform the indicated operations.
a. $4y^2(-y^2)$
b. $4y^2 - y^2$
c. Explain the difference between the two expressions.

36. Perform the indicated operations.
a. $9x^2(-10x^2)$
b. $9x^2 - 10x^2$
c. Explain the difference between the two expressions.

Multiply. See Example 7.

37. $(x - 2)(x^2 - 3x + 7)$
38. $(x + 3)(x^2 + 5x - 8)$
39. $(x + 5)(x^3 - 3x + 4)$
40. $(a + 2)(a^3 - 3a^2 + 7)$
41. $(2a - 3)(5a^2 - 6a + 4)$
42. $(3 + b)(2 - 5b - 3b^2)$

Multiply. See Example 8.

43. $(x + 2)^3$
44. $(y - 1)^3$
45. $(2y - 3)^3$
46. $(3x + 4)^3$

Multiply vertically. See Examples 9 through 11.

47. $(2x - 11)(6x + 1)$
48. $(4x - 7)(5x + 1)$
49. $(5x + 1)(2x^2 + 4x - 1)$
50. $(4x - 5)(8x^2 + 2x - 4)$
51. $(x^2 + 5x - 7)(2x^2 - 7x - 9)$
52. $(3x^2 - x + 2)(x^2 + 2x + 1)$

MIXED PRACTICE

Multiply. See Examples 1 through 11.

53. $-1.2y(-7y^6)$
54. $-4.2x(-2x^5)$
55. $-3x(x^2 + 2x - 8)$
56. $-5x(x^2 - 3x + 10)$
57. $(x + 19)(2x + 1)$
58. $(3y + 4)(y + 11)$
59. $\left(x + \frac{1}{7}\right)\left(x - \frac{3}{7}\right)$
60. $\left(m + \frac{2}{9}\right)\left(m - \frac{1}{9}\right)$
61. $(3y + 5)^2$
62. $(7y + 2)^2$
63. $(a + 4)(a^2 - 6a + 6)$
64. $(t + 3)(t^2 - 5t + 5)$
65. $(2x - 5)^3$
66. $(3y - 1)^3$
67. $(4x + 5)(8x^2 + 2x - 4)$
68. $(5x + 4)(x^2 - x + 4)$
69. $(3x^2 + 2x - 4)(2x^2 - 4x + 3)$
70. $(a^2 + 3a - 2)(2a^2 - 5a - 1)$

Express as the product of polynomials. Then multiply.

△ 71. Find the area of the rectangle.

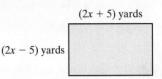

(2x + 5) yards

(2x − 5) yards

△ 72. Find the area of the square field.

△ 73. Find the area of the triangle.
(Triangle Area $= \frac{1}{2} \cdot$ base $\cdot$ height)

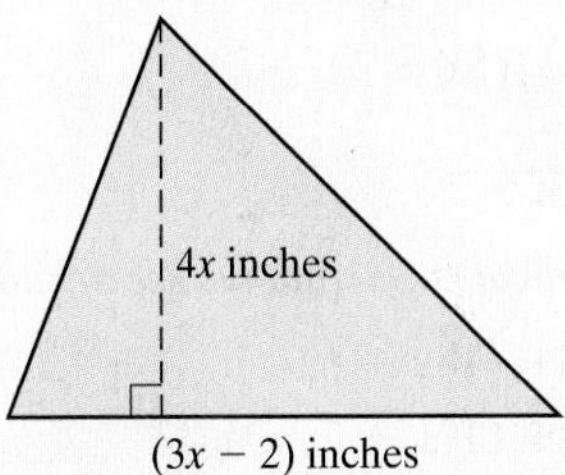

△ 74. Find the volume of the cube.
(Volume = length · width · height)

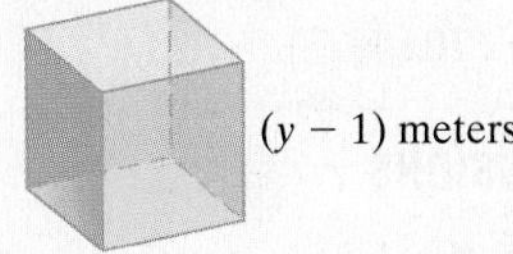

REVIEW AND PREVIEW

In this section, we review operations on monomials. Study the box below, then proceed. See Sections 1.8, 5.1, and 5.2. (Continued on next page.)

Operations on Monomials	
Multiply	Review the product rule for exponents.
Divide	Review the quotient rule for exponents.
Add or Subtract	Remember, we may only combine like terms.

Perform the operations on the monomials if possible. The first two rows have been completed for you.

	Monomials	Add	Subtract	Multiply	Divide
	$6x, 3x$	$6x + 3x = 9x$	$6x - 3x = 3x$	$6x \cdot 3x = 18x^2$	$\frac{6x}{3x} = 2$
	$-12x^2, 2x$	$-12x^2 + 2x$, can't be simplified	$-12x^2 - 2x$, can't be simplified	$-12x^2 \cdot 2x = -24x^3$	$\frac{-12x^2}{2x} = -6x$
75.	$5a, 15a$				
76.	$4y^7, 4y^3$				
77.	$-3y^5, 9y^4$				
78.	$-14x^2, 2x^2$				

CONCEPT EXTENSIONS

79. Perform each indicated operation. Explain the difference between the two expressions.
a. $(3x + 5) + (3x + 7)$
b. $(3x + 5)(3x + 7)$

80. Perform each indicated operation. Explain the difference between the two expressions.
a. $(8x - 3) - (5x - 2)$
b. $(8x - 3)(5x - 2)$

MIXED PRACTICE

Perform the indicated operations. See Sections 5.1 and 5.2.

81. $(3x - 1) + (10x - 6)$
82. $(2x - 1) + (10x - 7)$
83. $(3x - 1)(10x - 6)$
84. $(2x - 1)(10x - 7)$
85. $(3x - 1) - (10x - 6)$
86. $(2x - 1) - (10x - 7)$

CONCEPT EXTENSIONS

87. The area of the larger rectangle on the right is $x(x + 3)$. Find another expression for this area by finding the sum of the areas of the two smaller rectangles.

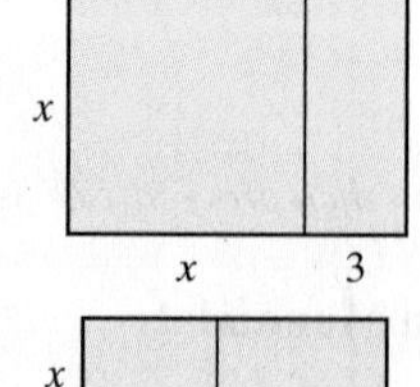

88. Write an expression for the area of the larger rectangle on the right in two different ways.

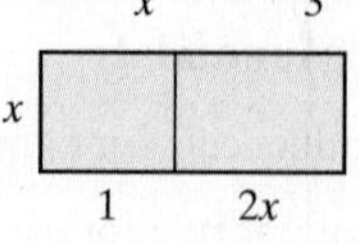

89. The area of the figure on the right is $(x + 2)(x + 3)$. Find another expression for this area by finding the sum of the areas of the four smaller rectangles.

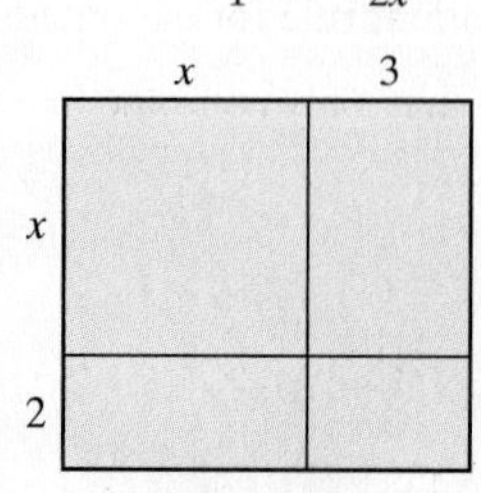

90. Write an expression for the area of the figure in two different ways.

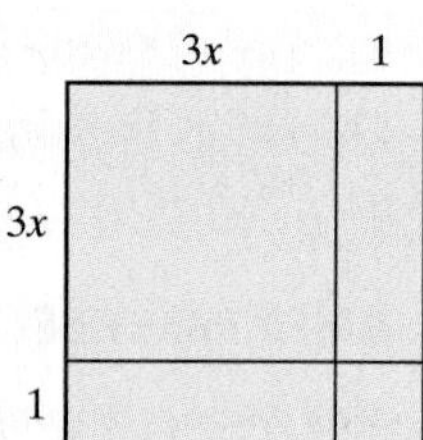

Simplify. See the Concept Checks in this section.

91. $5a + 6a$
92. $5a \cdot 6a$

Square where indicated. Simplify if possible.

93. $(5x)^2 + (2y)^2$
94. $(5x + 2y)^2$

95. Multiply each of the following polynomials.
a. $(a + b)(a - b)$
b. $(2x + 3y)(2x - 3y)$
c. $(4x + 7)(4x - 7)$
d. Can you make a general statement about all products of the form $(x + y)(x - y)$?

96. Evaluate each of the following.
a. $(2 + 3)^2; 2^2 + 3^2$
b. $(8 + 10)^2; 8^2 + 10^2$
c. Does $(a + b)^2 = a^2 + b^2$ no matter what the values of a and b are? Why or why not?

97. Write a polynomial that describes the area of the shaded region. (Find the area of the larger square minus the area of the smaller square.)

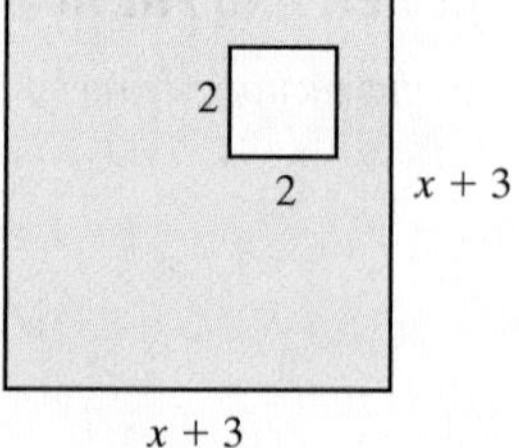

98. Write a polynomial that describes the area of the shaded region. (See Exercise 97.)

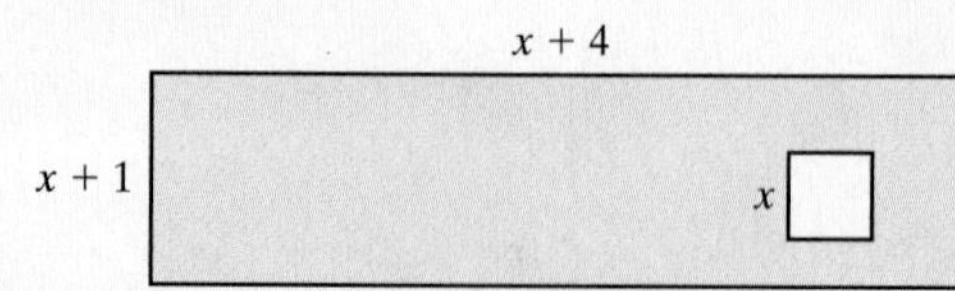

5.4 Special Products

OBJECTIVES

1. Multiply Two Binomials Using the FOIL Method.
2. Square a Binomial.
3. Multiply the Sum and Difference of Two Terms.
4. Use Special Products to Multiply Binomials.

OBJECTIVE

1 Using the FOIL Method

In this section, we multiply binomials using special products. First, a special order for multiplying binomials called the FOIL order or method is introduced. This method is demonstrated by multiplying $(3x + 1)$ by $(2x + 5)$ as shown below.

The FOIL Method

F stands for the product of the **First** terms. $(3x + 1)(2x + 5)$

$(3x)(2x) = 6x^2$ **F**

O stands for the product of the **Outer** terms. $(3x + 1)(2x + 5)$

$(3x)(5) = 15x$ **O**

I stands for the product of the **Inner** terms. $(3x + 1)(2x + 5)$

$(1)(2x) = 2x$ **I**

L stands for the product of the **Last** terms. $(3x + 1)(3x + 5)$

$(1)(5) = 5$ **L**

$$\begin{aligned}(3x + 1)(2x + 5) &= \overset{F}{6x^2} + \overset{O}{15x} + \overset{I}{2x} + \overset{L}{5}\\ &= 6x^2 + 17x + 5 \quad \text{Combine like terms.}\end{aligned}$$

✓CONCEPT CHECK

Multiply $(3x + 1)(2x + 5)$ using methods from the last section. Show that the product is still $6x^2 + 17x + 5$.

EXAMPLE 1 Multiply $(x - 3)(x + 4)$ by the FOIL method.

Solution

$$\begin{aligned}(x - 3)(x + 4) &= \overset{F}{(x)(x)} + \overset{O}{(x)(4)} + \overset{I}{(-3)(x)} + \overset{L}{(-3)(4)}\\ &= x^2 + 4x - 3x - 12\\ &= x^2 + x - 12 \quad \text{Combine like terms.}\end{aligned}$$

□

> **Helpful Hint**
> Remember that the FOIL order for multiplying can be used only for the product of 2 binomials.

PRACTICE

1 Multiply $(x + 2)(x - 5)$ by the FOIL method.

EXAMPLE 2 Multiply $(5x - 7)(x - 2)$ by the FOIL method.

Solution

$$\begin{aligned}(5x - 7)(x - 2) &= \overset{F}{5x(x)} + \overset{O}{5x(-2)} + \overset{I}{(-7)(x)} + \overset{L}{(-7)(-2)}\\ &= 5x^2 - 10x - 7x + 14\\ &= 5x^2 - 17x + 14 \quad \text{Combine like terms.}\end{aligned}$$

□

PRACTICE

2 Multiply $(4x - 9)(x - 1)$ by the FOIL method.

Answer to Concept Check:
Multiply and simplify:
$3x(2x + 5) + 1(2x + 5)$

EXAMPLE 3 Multiply $2(y + 6)(2y - 1)$.

Solution

$$2(y+6)(2y-1) = \overset{F}{}\overset{O}{}\overset{I}{}\overset{L}{}2(2y^2 - 1y + 12y - 6)$$

$$= 2(2y^2 + 11y - 6) \quad \text{Simplify inside parentheses.}$$

$$= 4y^2 + 22y - 12 \quad \text{Now use the distributive property.}$$

□

PRACTICE 3 Multiply $3(x + 5)(3x - 1)$.

OBJECTIVE 2

Squaring Binomials 

Now, try squaring a binomial using the FOIL method.

EXAMPLE 4 Multiply $(3y + 1)^2$.

Solution $(3y + 1)^2 = (3y + 1)(3y + 1)$

F O I L

$$= (3y)(3y) + (3y)(1) + 1(3y) + 1(1)$$

$$= 9y^2 + 3y + 3y + 1$$

$$= 9y^2 + 6y + 1$$

□

PRACTICE 4 Multiply $(4x - 1)^2$.

Notice the pattern that appears in Example 4.

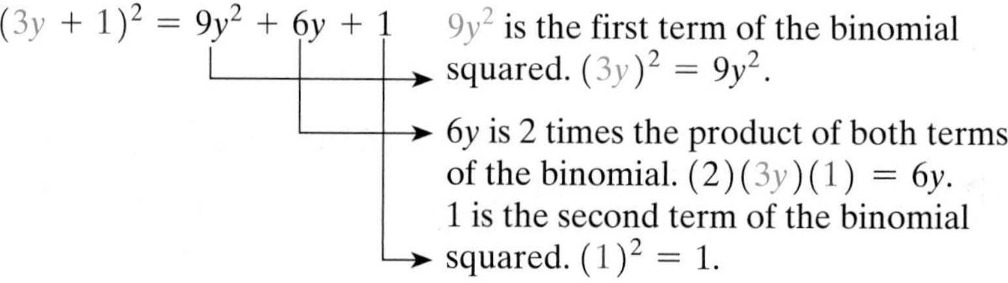

$(3y + 1)^2 = 9y^2 + 6y + 1$

$9y^2$ is the first term of the binomial squared. $(3y)^2 = 9y^2$.

$6y$ is 2 times the product of both terms of the binomial. $(2)(3y)(1) = 6y$.

1 is the second term of the binomial squared. $(1)^2 = 1$.

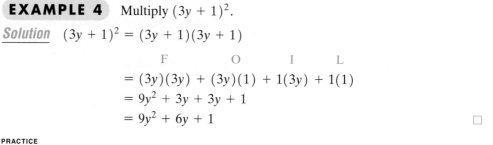

This pattern leads to the following, which can be used when squaring a binomial. We call these **special products.**

Squaring a Binomial

A binomial squared is equal to the square of the first term plus or minus twice the product of both terms plus the square of the second term.

$$(a + b)^2 = a^2 + 2ab + b^2$$

$$(a - b)^2 = a^2 - 2ab + b^2$$

This product can be visualized geometrically.

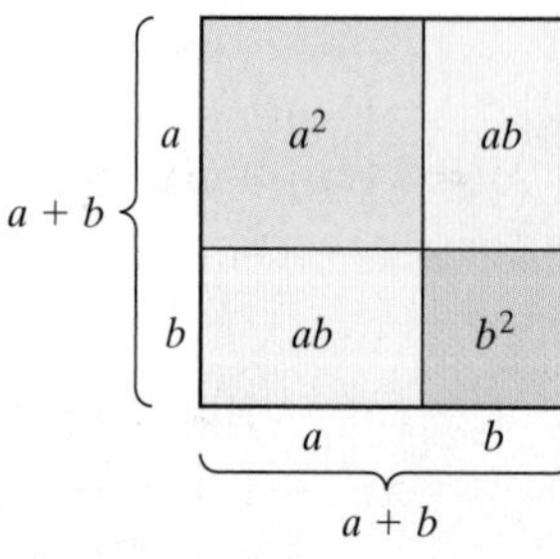

The area of the large square is side · side.

$$\text{Area} = (a + b)(a + b) = (a + b)^2$$

The area of the large square is also the sum of the areas of the smaller rectangles.

$$\text{Area} = a^2 + ab + ab + b^2 = a^2 + 2ab + b^2$$

Thus, $(a + b)^2 = a^2 + 2ab + b^2$.

EXAMPLE 5 Use a special product to square each binomial.

a. $(t + 2)^2$ **b.** $(p - q)^2$ **c.** $(2x + 5)^2$ **d.** $(x^2 - 7y)^2$

Solution

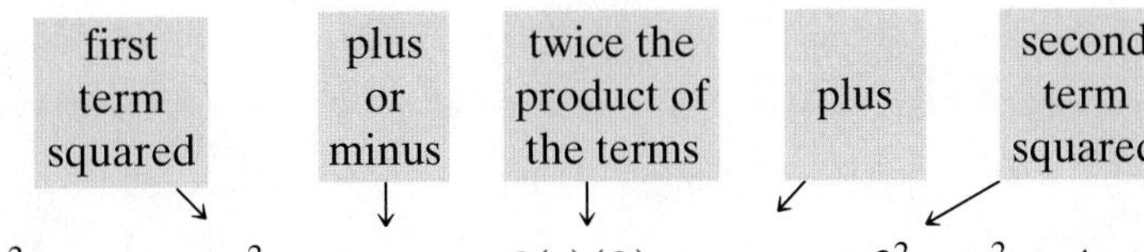

a. $(t + 2)^2 = t^2 + 2(t)(2) + 2^2 = t^2 + 4t + 4$

b. $(p - q)^2 = p^2 - 2(p)(q) + q^2 = p^2 - 2pq + q^2$

c. $(2x + 5)^2 = (2x)^2 + 2(2x)(5) + 5^2 = 4x^2 + 20x + 25$

d. $(x^2 - 7y)^2 = (x^2)^2 - 2(x^2)(7y) + (7y^2) = x^4 - 14x^2y + 49y^2$ □

PRACTICE

5 Use a special product to square each binomial.

a. $(b + 3)^2$ **b.** $(x - y)^2$

c. $(3y + 2)^2$ **d.** $(a^2 - 5b)^2$

Helpful Hint

Notice that

$$(a + b)^2 \neq a^2 + b^2 \quad \text{The middle term } 2ab \text{ is missing.}$$
$$(a + b)^2 = (a + b)(a + b) = a^2 + 2ab + b^2$$

Likewise,

$$(a - b)^2 \neq a^2 - b^2$$
$$(a - b)^2 = (a - b)(a - b) = a^2 - 2ab + b^2$$

OBJECTIVE

3 Multiplying the Sum and Difference of Two Terms 

Another special product is the product of the sum and difference of the same two terms, such as $(x + y)(x - y)$. Finding this product by the FOIL method, we see a pattern emerge.

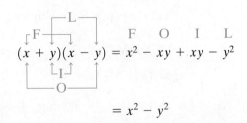

$$\overset{\text{F O I L}}{(x + y)(x - y) = x^2 - xy + xy - y^2}$$
$$= x^2 - y^2$$

Notice that the middle two terms subtract out. This is because the **O**uter product is the opposite of the **I**nner product. Only the **difference of squares** remains.

Multiplying the Sum and Difference of Two Terms

The product of the sum and difference of two terms is the square of the first term minus the square of the second term.

$$(a + b)(a - b) = a^2 - b^2$$

EXAMPLE 6 Use a special product to multiply.

a. $4(x + 4)(x - 4)$ **b.** $(6t + 7)(6t - 7)$ **c.** $\left(x - \frac{1}{4}\right)\left(x + \frac{1}{4}\right)$
d. $(2p - q)(2p + q)$ **e.** $(3x^2 - 5y)(3x^2 + 5y)$

Solution

first term squared ↓ minus ↓ second term squared ↙

a. $4(x + 4)(x - 4) = 4(x^2 - 4^2) = 4(x^2 - 16) = 4x^2 - 64$

b. $(6t + 7)(6t - 7) = (6t)^2 - 7^2 = 36t^2 - 49$

c. $\left(x - \frac{1}{4}\right)\left(x + \frac{1}{4}\right) = x^2 - \left(\frac{1}{4}\right)^2 = x^2 - \frac{1}{16}$

d. $(2p - q)(2p + q) = (2p)^2 - q^2 = 4p^2 - q^2$

e. $(3x^2 - 5y)(3x^2 + 5y) = (3x^2)^2 - (5y)^2 = 9x^4 - 25y^2$ □

PRACTICE 6 Use a special product to multiply.

a. $3(x + 5)(x - 5)$ **b.** $(4b - 3)(4b + 3)$
c. $\left(x + \frac{2}{3}\right)\left(x - \frac{2}{3}\right)$ **d.** $(5s + t)(5s - t)$
e. $(2y - 3z^2)(2y + 3z^2)$

✓CONCEPT CHECK

Match expression number 1 and number 2 to the equivalent expression or expressions in the list below.

1. $(a + b)^2$ **2.** $(a + b)(a - b)$

a. $(a + b)(a + b)$ **b.** $a^2 - b^2$ **c.** $a^2 + b^2$ **d.** $a^2 - 2ab + b^2$ **e.** $a^2 + 2ab + b^2$

OBJECTIVE 4 Using Special Products

Let's now practice multiplying polynomials in general. If possible, use a special product.

EXAMPLE 7 Use a special product to multiply, if possible.

a. $(x - 5)(3x + 4)$ **b.** $(7x + 4)^2$ **c.** $(y - 0.6)(y + 0.6)$
d. $\left(y^4 + \frac{2}{5}\right)\left(3y^2 - \frac{1}{5}\right)$ **e.** $(a - 3)(a^2 + 2a - 1)$

Solution

a. $(x - 5)(3x + 4) = 3x^2 + 4x - 15x - 20$ FOIL.
$= 3x^2 - 11x - 20$

b. $(7x + 4)^2 = (7x)^2 + 2(7x)(4) + 4^2$ Squaring a binomial.
$= 49x^2 + 56x + 16$

c. $(y - 0.6)(y + 0.6) = y^2 - (0.6)^2 = y^2 - 0.36$ Multiplying the sum and difference of 2 terms.

d. $\left(y^4 + \frac{2}{5}\right)\left(3y^2 - \frac{1}{5}\right) = 3y^6 - \frac{1}{5}y^4 + \frac{6}{5}y^2 - \frac{2}{25}$ FOIL.

e. I've inserted this product as a reminder that since it is not a binomial times a binomial, the FOIL order may not be used.

Answers to Concept Check:
1. a and e **2.** b

$$
\begin{aligned}
(a - 3)(a^2 + 2a - 1) &= a(a^2 + 2a - 1) - 3(a^2 + 2a - 1) \\
&= a^3 + 2a^2 - a - 3a^2 - 6a + 3 \\
&= a^3 - a^2 - 7a + 3
\end{aligned}
$$

Multiplying each term of the binomial by each term of the trinomial. □

PRACTICE 7 Use a special product to multiply, if possible.

a. $(4x + 3)(x - 6)$

b. $(7b - 2)^2$

c. $(x + 0.4)(x - 0.4)$

d. $\left(x^2 - \frac{3}{7}\right)\left(3x^4 + \frac{2}{7}\right)$

e. $(x + 1)(x^2 + 5x - 2)$

> **Helpful Hint**
> - When multiplying two binomials, you may always use the FOIL order or method.
> - When multiplying any two polynomials, you may always use the distributive property to find the product.

Vocabulary, Readiness & Video Check

Answer each exercise true or false.

1. $(x + 4)^2 = x^2 + 16$

2. For $(x + 6)(2x - 1)$ the product of the first terms is $2x^2$.

3. $(x + 4)(x - 4) = x^2 + 16$

4. The product $(x - 1)(x^3 + 3x - 1)$ is a polynomial of degree 5.

Martin-Gay Interactive Videos

See Video 5.4

Watch the section lecture video and answer the following questions.

OBJECTIVE 1 — **5.** From Examples 1–3, for what type of multiplication problem is the FOIL order of multiplication used?

OBJECTIVE 2 — **6.** Name at least one other method you can use to multiply Example 4.

OBJECTIVE 3 — **7.** From Example 5, why does multiplying the sum and difference of the same two terms always give you a binomial answer?

OBJECTIVE 4 — **8.** Why was the FOIL method not used for Example 10?

5.4 Exercise Set

MyMathLab®

Multiply using the FOIL method. See Examples 1 through 3.

1. $(x + 3)(x + 4)$

2. $(x + 5)(x - 1)$

3. $(x - 5)(x + 10)$

4. $(y - 12)(y + 4)$

5. $(5x - 6)(x + 2)$

6. $(3y - 5)(2y - 7)$

7. $5(y - 6)(4y - 1)$

8. $2(x - 11)(2x - 9)$

9. $(2x + 5)(3x - 1)$

10. $(6x + 2)(x - 2)$

11. $\left(x - \frac{1}{3}\right)\left(x + \frac{2}{3}\right)$

12. $\left(x - \frac{2}{5}\right)\left(x + \frac{1}{5}\right)$

Multiply. See Examples 4 and 5.

13. $(x + 2)^2$

14. $(x + 7)^2$

15. $(2x - 1)^2$

16. $(7x - 3)^2$

17. $(3a - 5)^2$

18. $(5a + 2)^2$

19. $(5x + 9)^2$

20. $(6s - 2)^2$

Multiply. See Example 6.

21. $(a - 7)(a + 7)$
22. $(b + 3)(b - 3)$
23. $(3x - 1)(3x + 1)$
24. $(4x - 5)(4x + 5)$
25. $\left(3x - \frac{1}{2}\right)\left(3x + \frac{1}{2}\right)$
26. $\left(10x + \frac{2}{7}\right)\left(10x - \frac{2}{7}\right)$
27. $(9x + y)(9x - y)$
28. $(2x - y)(2x + y)$
29. $(2x + 0.1)(2x - 0.1)$
30. $(5x - 1.3)(5x + 1.3)$

MIXED PRACTICE

Multiply. See Example 7.

31. $(a + 5)(a + 4)$
32. $(a - 5)(a - 7)$
33. $(a + 7)^2$
34. $(b - 2)^2$
35. $(4a + 1)(3a - 1)$
36. $(6a + 7)(6a + 5)$
37. $(x + 2)(x - 2)$
38. $(x - 10)(x + 10)$
39. $(3a + 1)^2$
40. $(4a - 2)^2$
41. $(x^2 + y)(4x - y^4)$
42. $(x^3 - 2)(5x + y)$
43. $(x + 3)(x^2 - 6x + 1)$
44. $(x - 2)(x^2 - 4x + 2)$
45. $(2a - 3)^2$
46. $(5b - 4x)^2$
47. $(5x - 6z)(5x + 6z)$
48. $(11x - 7y)(11x + 7y)$
49. $(x^5 - 3)(x^5 - 5)$
50. $(a^4 + 5)(a^4 + 6)$
51. $(x + 0.8)(x - 0.8)$
52. $(y - 0.9)(y + 0.9)$
53. $(a^3 + 11)(a^4 - 3)$
54. $(x^5 + 5)(x^2 - 8)$
55. $3(x - 2)^2$
56. $2(3b + 7)^2$
57. $(3b + 7)(2b - 5)$
58. $(3y - 13)(y - 3)$
59. $(7p - 8)(7p + 8)$
60. $(3s - 4)(3s + 4)$
61. $\left(\frac{1}{3}a^2 - 7\right)\left(\frac{1}{3}a^2 + 7\right)$
62. $\left(\frac{2}{3}a - b^2\right)\left(\frac{2}{3}a - b^2\right)$
63. $5x^2(3x^2 - x + 2)$
64. $4x^3(2x^2 + 5x - 1)$
65. $(2r - 3s)(2r + 3s)$
66. $(6r - 2x)(6r + 2x)$
67. $(3x - 7y)^2$
68. $(4s - 2y)^2$
69. $(4x + 5)(4x - 5)$
70. $(3x + 5)(3x - 5)$
71. $(8x + 4)^2$
72. $(3x + 2)^2$
73. $\left(a - \frac{1}{2}y\right)\left(a + \frac{1}{2}y\right)$
74. $\left(\frac{a}{2} + 4y\right)\left(\frac{a}{2} - 4y\right)$
75. $\left(\frac{1}{5}x - y\right)\left(\frac{1}{5}x + y\right)$
76. $\left(\frac{y}{6} - 8\right)\left(\frac{y}{6} + 8\right)$
77. $(a + 1)(3a^2 - a + 1)$
78. $(b + 3)(2b^2 + b - 3)$

Express each as a product of polynomials in x. Then multiply and simplify.

△ 79. Find the area of the square rug shown if its side is $(2x + 1)$ feet.

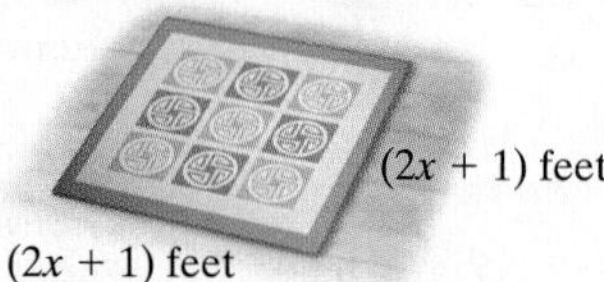

△ 80. Find the area of the rectangular canvas if its length is $(3x - 2)$ inches and its width is $(x - 4)$ inches.

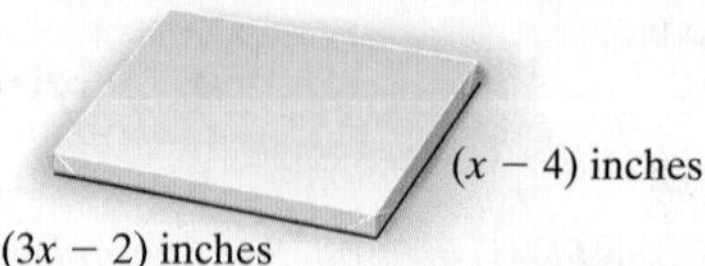

REVIEW AND PREVIEW

Simplify each expression. See Section 5.1.

81. $\frac{50b^{10}}{70b^5}$
82. $\frac{x^3y^6}{xy^2}$
83. $\frac{8a^{17}b^{15}}{-4a^7b^{10}}$
84. $\frac{-6a^8y}{3a^4y}$
85. $\frac{2x^4y^{12}}{3x^4y^4}$
86. $\frac{-48ab^6}{32ab^3}$

Find the slope of each line. See Section 3.4.

87.

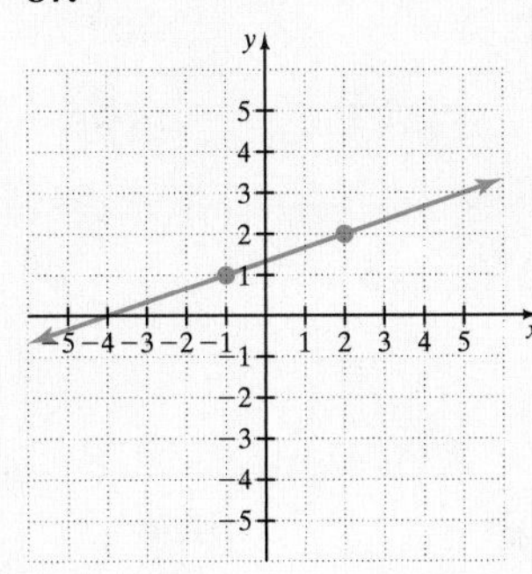

88.

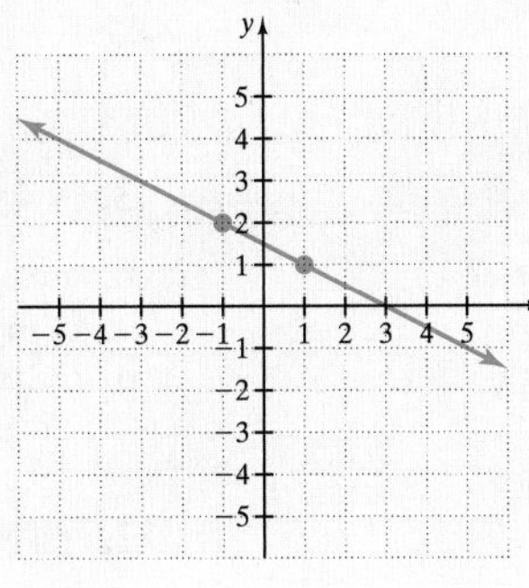

89.

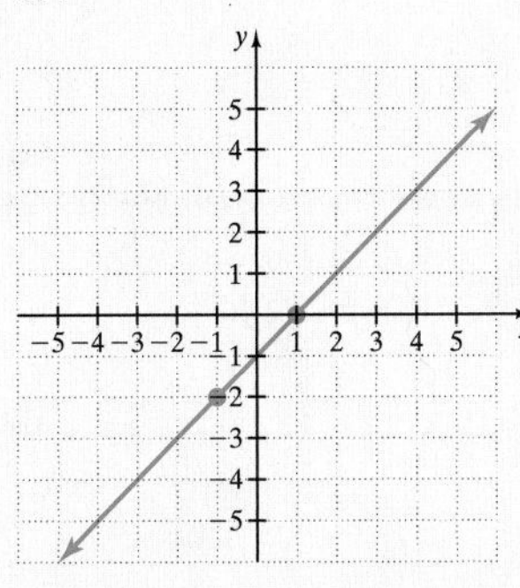

90.

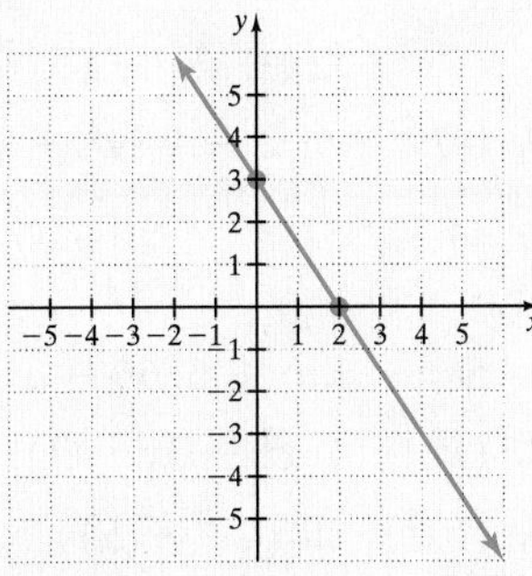

CONCEPT EXTENSIONS

Match each expression on the left to the equivalent expression on the right. See the second Concept Check in this section.

91. $(a - b)^2$ **a.** $a^2 - b^2$

92. $(a - b)(a + b)$ **b.** $a^2 + b^2$

93. $(a + b)^2$ **c.** $a^2 - 2ab + b^2$

94. $(a + b)^2(a - b)^2$ **d.** $a^2 + 2ab + b^2$

e. none of these

Fill in the squares so that a true statement forms.

95. $(x^{\square} + 7)(x^{\square} + 3) = x^4 + 10x^2 + 21$

96. $(5x^{\square} - 2)^2 = 25x^6 - 20x^3 + 4$

Find the area of the shaded figure. To do so, subtract the area of the smaller square(s) from the area of the larger geometric figure.

△ **97.**

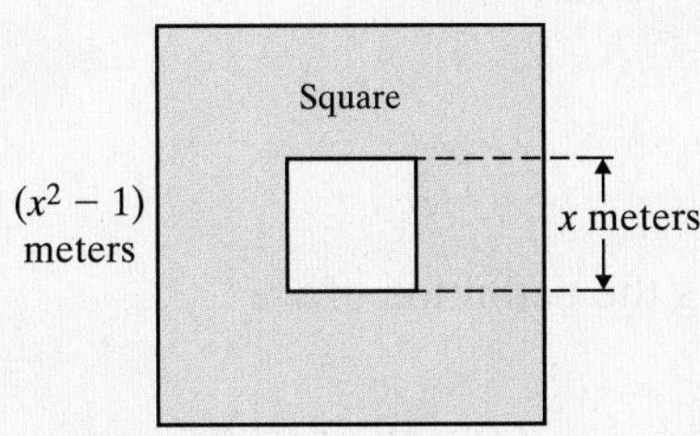

△ **98.**

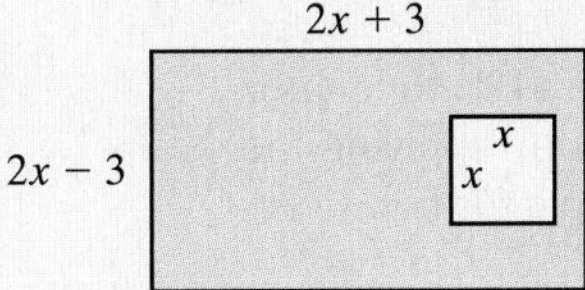

△ **99.**

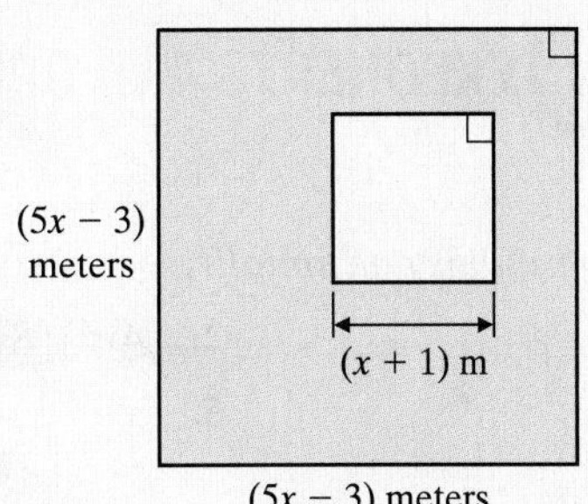

△ **100.**

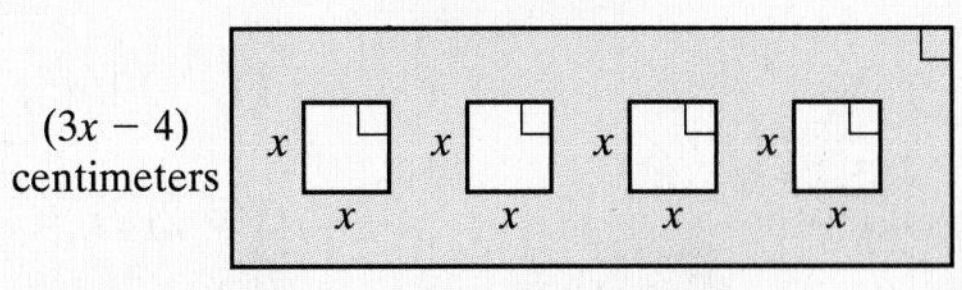

For Exercises 101 and 102, find the area of the shaded figure.

△ **101.**

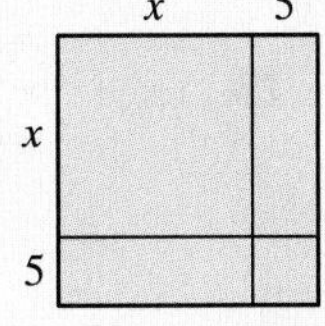

△ **102.**

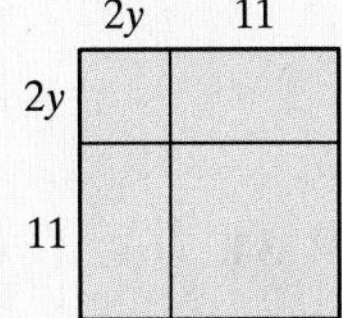

103. In your own words, describe the different methods that can be used to find the product: $(2x - 5)(3x + 1)$.

104. In your own words, describe the different methods that can be used to find the product: $(5x + 1)^2$.

105. Suppose that a classmate asked you why $(2x + 1)^2$ is **not** $(4x^2 + 1)$. Write down your response to this classmate.

106. Suppose that a classmate asked you why $(2x + 1)^2$ **is** $(4x^2 + 4x + 1)$. Write down your response to this classmate.

107. Using your own words, explain how to square a binomial such as $(a + b)^2$.

108. Explain how to find the product of two binomials using the FOIL method.

Find each product. For example,

$$[(a + b) - 2][(a + b) + 2] = (a + b)^2 - 2^2$$
$$= a^2 + 2ab + b^2 - 4$$

109. $[(x + y) - 3][(x + y) + 3]$

110. $[(a + c) - 5][(a + c) + 5]$

111. $[(a - 3) + b][(a - 3) - b]$

112. $[(x - 2) + y][(x - 2) - y]$

Integrated Review EXPONENTS AND OPERATIONS ON POLYNOMIALS

Sections 5.1–5.4

Perform the indicated operations and simplify.

1. $(5x^2)(7x^3)$
2. $(4y^2)(8y^7)$
3. -4^2
4. $(-4)^2$
5. $(x - 5)(2x + 1)$
6. $(3x - 2)(x + 5)$
7. $(x - 5) + (2x + 1)$
8. $(3x - 2) + (x + 5)$
9. $\dfrac{7x^9y^{12}}{x^3y^{10}}$
10. $\dfrac{20a^2b^8}{14a^2b^2}$
11. $(12m^7n^6)^2$
12. $(4y^9z^{10})^3$
13. $3(4y - 3)(4y + 3)$
14. $2(7x - 1)(7x + 1)$
15. $(x^7y^5)^9$
16. $(3^1x^9)^3$
17. $(7x^2 - 2x + 3) - (5x^2 + 9)$
18. $(10x^2 + 7x - 9) - (4x^2 - 6x + 2)$
19. $0.7y^2 - 1.2 + 1.8y^2 - 6y + 1$
20. $7.8x^2 - 6.8x + 3.3 + 0.6x^2 - 9$
21. $(x + 4y)^2$
22. $(y - 9z)^2$
23. $(x + 4y) + (x + 4y)$
24. $(y - 9z) + (y - 9z)$
25. $7x^2 - 6xy + 4(y^2 - xy)$
26. $5a^2 - 3ab + 6(b^2 - a^2)$
27. $(x - 3)(x^2 + 5x - 1)$
28. $(x + 1)(x^2 - 3x - 2)$
29. $(2x^3 - 7)(3x^2 + 10)$
30. $(5x^3 - 1)(4x^4 + 5)$
31. $(2x - 7)(x^2 - 6x + 1)$
32. $(5x - 1)(x^2 + 2x - 3)$

Perform exercises and simplify if possible.

33. $5x^3 + 5y^3$
34. $(5x^3)(5y^3)$
35. $(5x^3)^3$
36. $\dfrac{5x^3}{5y^3}$
37. $x + x$
38. $x \cdot x$

5.5 Negative Exponents and Scientific Notation

OBJECTIVES

1. Simplify Expressions Containing Negative Exponents.
2. Use All the Rules and Definitions for Exponents to Simplify Exponential Expressions.
3. Write Numbers in Scientific Notation.
4. Convert Numbers from Scientific Notation to Standard Form.
5. Perform Operations on Numbers Written in Scientific Notation.

OBJECTIVE

1 Simplifying Expressions Containing Negative Exponents

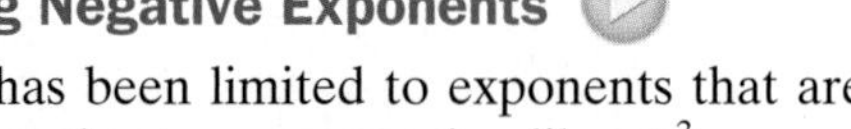

Our work with exponential expressions so far has been limited to exponents that are positive integers or 0. Here we expand to give meaning to an expression like x^{-3}.

Suppose that we wish to simplify the expression $\frac{x^2}{x^5}$. If we use the quotient rule for exponents, we subtract exponents:

$$\frac{x^2}{x^5} = x^{2-5} = x^{-3}, \quad x \neq 0$$

But what does x^{-3} mean? Let's simplify $\frac{x^2}{x^5}$ using the definition of x^n.

$$\frac{x^2}{x^5} = \frac{x \cdot x}{x \cdot x \cdot x \cdot x \cdot x}$$

$$= \frac{x \cdot x}{x \cdot x \cdot x \cdot x \cdot x}$$ Divide numerator and denominator by common factors by applying the fundamental principle for fractions.

$$= \frac{1}{x^3}$$

If the quotient rule is to hold true for negative exponents, then x^{-3} must equal $\frac{1}{x^3}$. From this example, we state the definition for negative exponents.

Negative Exponents

If a is a real number other than 0 and n is an integer, then

$$a^{-n} = \frac{1}{a^n}$$

For example, $x^{-3} = \frac{1}{x^3}$.

In other words, another way to write a^{-n} is to take its reciprocal and change the sign of its exponent.

EXAMPLE 1 Simplify by writing each expression with positive exponents only.

a. 3^{-2} **b.** $2x^{-3}$ **c.** $2^{-1} + 4^{-1}$ **d.** $(-2)^{-4}$ **e.** y^{-4}

Solution

a. $3^{-2} = \frac{1}{3^2} = \frac{1}{9}$ Use the definition of negative exponents.

b. $2x^{-3} = 2 \cdot \frac{1}{x^3} = \frac{2}{x^3}$ Use the definition of negative exponents.

> **Helpful Hint**
> Don't forget that since there are no parentheses, only x is the base for the exponent -3.

c. $2^{-1} + 4^{-1} = \frac{1}{2} + \frac{1}{4} = \frac{2}{4} + \frac{1}{4} = \frac{3}{4}$

d. $(-2)^{-4} = \frac{1}{(-2)^4} = \frac{1}{(-2)(-2)(-2)(-2)} = \frac{1}{16}$

e. $y^{-4} = \frac{1}{y^4}$ □

PRACTICE

1 Simplify by writing each expression with positive exponents only.

a. 5^{-3} **b.** $3y^{-4}$ **c.** $3^{-1} + 2^{-1}$ **d.** $(-5)^{-2}$ **e.** x^{-5}

> **Helpful Hint**
> A negative exponent *does not affect* the sign of its base.
> Remember: Another way to write a^{-n} is to take its reciprocal and change the sign of its exponent: $a^{-n} = \frac{1}{a^n}$. For example,
>
> $$x^{-2} = \frac{1}{x^2}, \qquad 2^{-3} = \frac{1}{2^3} \text{ or } \frac{1}{8}$$
>
> $$\frac{1}{y^{-4}} = \frac{1}{\frac{1}{y^4}} = y^4, \qquad \frac{1}{5^{-2}} = 5^2 \text{ or } 25$$

From the preceding Helpful Hint, we know that $x^{-2} = \frac{1}{x^2}$ and $\frac{1}{y^{-4}} = y^4$. We can use this to include another statement in our definition of negative exponents.

Negative Exponents

If a is a real number other than 0 and n is an integer, then

$$a^{-n} = \frac{1}{a^n} \quad \text{and} \quad \frac{1}{a^{-n}} = a^n$$

EXAMPLE 2 Simplify each expression. Write results using positive exponents only.

a. $\frac{1}{x^{-3}}$ **b.** $\frac{1}{3^{-4}}$ **c.** $\frac{p^{-4}}{q^{-9}}$ **d.** $\frac{5^{-3}}{2^{-5}}$

Solution

a. $\frac{1}{x^{-3}} = \frac{x^3}{1} = x^3$ **b.** $\frac{1}{3^{-4}} = \frac{3^4}{1} = 81$ **c.** $\frac{p^{-4}}{q^{-9}} = \frac{q^9}{p^4}$ **d.** $\frac{5^{-3}}{2^{-5}} = \frac{2^5}{5^3} = \frac{32}{125}$ □

PRACTICE 2 Simplify each expression. Write results using positive exponents only.

a. $\frac{1}{s^{-5}}$ **b.** $\frac{1}{2^{-3}}$ **c.** $\frac{x^{-7}}{y^{-5}}$ **d.** $\frac{4^{-3}}{3^{-2}}$

EXAMPLE 3 Simplify each expression. Write answers with positive exponents.

a. $\frac{y}{y^{-2}}$ **b.** $\frac{3}{x^{-4}}$ **c.** $\frac{x^{-5}}{x^7}$ **d.** $\left(\frac{2}{3}\right)^{-3}$

Solution

a. $\frac{y}{y^{-2}} = \frac{y^1}{y^{-2}} = y^{1-(-2)} = y^3$ Remember that $\frac{a^m}{a^n} = a^{m-n}$.

b. $\frac{3}{x^{-4}} = 3 \cdot \frac{1}{x^{-4}} = 3 \cdot x^4$ or $3x^4$

c. $\frac{x^{-5}}{x^7} = x^{-5-7} = x^{-12} = \frac{1}{x^{12}}$

d. $\left(\frac{2}{3}\right)^{-3} = \frac{2^{-3}}{3^{-3}} = \frac{3^3}{2^3} = \frac{27}{8}$ □

PRACTICE 3 Simplify each expression. Write answers with positive exponents.

a. $\frac{x^{-3}}{x^2}$ **b.** $\frac{5}{y^{-7}}$ **c.** $\frac{z}{z^{-4}}$ **d.** $\left(\frac{5}{9}\right)^{-2}$

OBJECTIVE

2 Simplifying Exponential Expressions 

All the previously stated rules for exponents apply for negative exponents also. Here is a summary of the rules and definitions for exponents.

Summary of Exponent Rules

If m and n are integers and a, b, and c are real numbers, then:

Product rule for exponents: $a^m \cdot a^n = a^{m+n}$

Power rule for exponents: $(a^m)^n = a^{m \cdot n}$

Power of a product: $(ab)^n = a^n b^n$

Power of a quotient: $\left(\frac{a}{c}\right)^n = \frac{a^n}{c^n}$, $c \neq 0$

Quotient rule for exponents: $\frac{a^m}{a^n} = a^{m-n}$, $a \neq 0$

Zero exponent: $a^0 = 1$, $a \neq 0$

Negative exponent: $a^{-n} = \frac{1}{a^n}$, $a \neq 0$

EXAMPLE 4 Simplify the following expressions. Write each result using positive exponents only.

a. $(y^{-3}z^6)^{-6}$ **b.** $\dfrac{(2x^3)^4x}{x^7}$ **c.** $\left(\dfrac{3a^2}{b}\right)^{-3}$ **d.** $\dfrac{4^{-1}x^{-3}y}{4^{-3}x^2y^{-6}}$ **e.** $\left(\dfrac{-2x^3y}{xy^{-1}}\right)^3$

Solution

a. $(y^{-3}z^6)^{-6} = y^{18} \cdot z^{-36} = \dfrac{y^{18}}{z^{36}}$

b. $\dfrac{(2x^3)^4x}{x^7} = \dfrac{2^4 \cdot x^{12} \cdot x}{x^7} = \dfrac{16 \cdot x^{12+1}}{x^7} = \dfrac{16x^{13}}{x^7} = 16x^{13-7} = 16x^6$ Use the power rule.

c. $\left(\dfrac{3a^2}{b}\right)^{-3} = \dfrac{3^{-3}(a^2)^{-3}}{b^{-3}}$ Raise each factor in the numerator and the denominator to the −3 power.

$= \dfrac{3^{-3}a^{-6}}{b^{-3}}$ Use the power rule.

$= \dfrac{b^3}{3^3a^6}$ Use the negative exponent rule.

$= \dfrac{b^3}{27a^6}$ Write 3^3 as 27.

d. $\dfrac{4^{-1}x^{-3}y}{4^{-3}x^2y^{-6}} = 4^{-1-(-3)}x^{-3-2}y^{1-(-6)} = 4^2x^{-5}y^7 = \dfrac{4^2y^7}{x^5} = \dfrac{16y^7}{x^5}$

e. $\left(\dfrac{-2x^3y}{xy^{-1}}\right)^3 = \dfrac{(-2)^3x^9y^3}{x^3y^{-3}} = \dfrac{-8x^9y^3}{x^3y^{-3}} = -8x^{9-3}y^{3-(-3)} = -8x^6y^6$ □

PRACTICE

4 Simplify the following expressions. Write each result using positive exponents only.

a. $(a^4b^{-3})^{-5}$ **b.** $\dfrac{x^2(x^5)^3}{x^7}$ **c.** $\left(\dfrac{5p^8}{q}\right)^{-2}$

d. $\dfrac{6^{-2}x^{-4}y^{-7}}{6^{-3}x^3y^{-9}}$ **e.** $\left(\dfrac{-3x^4y}{x^2y^{-2}}\right)^3$

OBJECTIVE

3 Writing Numbers in Scientific Notation 

Both very large and very small numbers frequently occur in many fields of science. For example, the distance between the sun and the dwarf planet Pluto is approximately 5,906,000,000 kilometers, and the mass of a proton is approximately 0.00000000000000000000000165 gram. It can be tedious to write these numbers in this standard decimal notation, so **scientific notation** is used as a convenient shorthand for expressing very large and very small numbers.

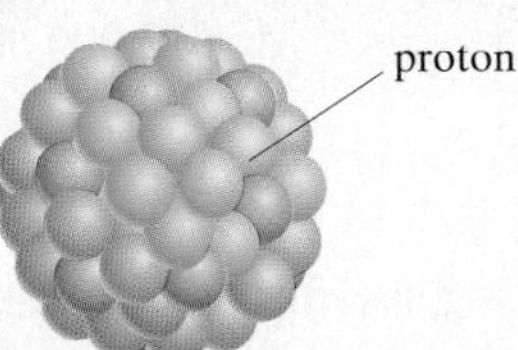

Mass of proton is approximately 0.000 000 000 000 000 000 000 001 65 gram

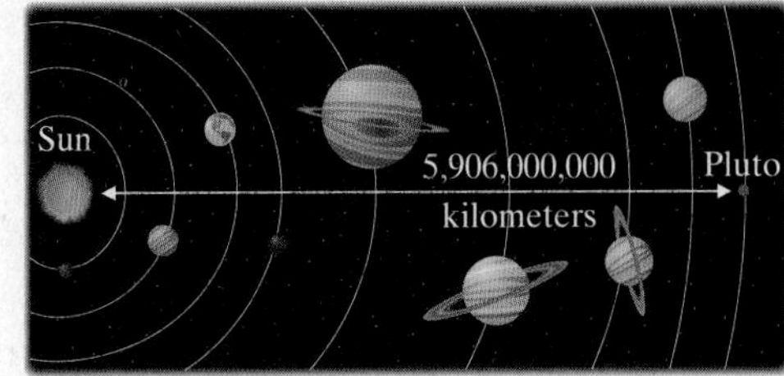

Scientific Notation

A positive number is written in scientific notation if it is written as the product of a number a, where $1 \le a < 10$, and an integer power r of 10:

$$a \times 10^r$$

The numbers below are written in scientific notation. The $\times$ sign for multiplication is used as part of the notation.

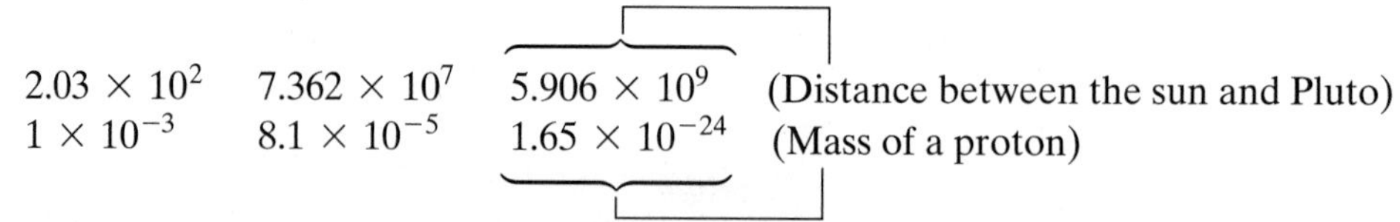

2.03×10^2 $\quad 7.362 \times 10^7$ $\quad 5.906 \times 10^9$ (Distance between the sun and Pluto)
1×10^{-3} $\quad 8.1 \times 10^{-5}$ $\quad 1.65 \times 10^{-24}$ (Mass of a proton)

The following steps are useful when writing numbers in scientific notation.

To Write a Number in Scientific Notation

Step 1. Move the decimal point in the original number to the left or right so that the new number has a value between 1 and 10.

Step 2. Count the number of decimal places the decimal point is moved in Step 1. If the original number is 10 or greater, the count is positive. If the original number is less than 1, the count is negative.

Step 3. Multiply the new number in Step 1 by 10 raised to an exponent equal to the count found in Step 2.

EXAMPLE 5 Write each number in scientific notation.

a. 367,000,000 **b.** 0.000003 **c.** 20,520,000,000 **d.** 0.00085

Solution

a. Step 1. Move the decimal point until the number is between 1 and 10.

367,000,000.
8 places

Step 2. The decimal point is moved 8 places, and the original number is 10 or greater, so the count is positive 8.

Step 3. $367{,}000{,}000 = 3.67 \times 10^8$.

b. Step 1. Move the decimal point until the number is between 1 and 10.

0.000003
6 places

Step 2. The decimal point is moved 6 places, and the original number is less than 1, so the count is -6.

Step 3. $0.000003 = 3.0 \times 10^{-6}$

c. $20{,}520{,}000{,}000 = 2.052 \times 10^{10}$

d. $0.00085 = 8.5 \times 10^{-4}$

PRACTICE 5 Write each number in scientific notation.

a. 0.000007 **b.** 20,700,000
c. 0.0043 **d.** 812,000,000

OBJECTIVE 4 Converting Numbers to Standard Form 

A number written in scientific notation can be rewritten in standard form. For example, to write 8.63×10^3 in standard form, recall that $10^3 = 1000$.

$$8.63 \times 10^3 = 8.63(1000) = 8630$$

Notice that the exponent on the 10 is positive 3, and we moved the decimal point 3 places to the right.

To write 7.29×10^{-3} in standard form, recall that $10^{-3} = \frac{1}{10^3} = \frac{1}{1000}$.

$$7.29 \times 10^{-3} = 7.29\left(\frac{1}{1000}\right) = \frac{7.29}{1000} = 0.00729$$

The exponent on the 10 is negative 3, and we moved the decimal to the left 3 places.

In general, **to write a scientific notation number in standard form,** move the decimal point the same number of places as the exponent on 10. If the exponent is positive, move the decimal point to the right; if the exponent is negative, move the decimal point to the left.

EXAMPLE 6 Write each number in standard notation, without exponents.

a. 1.02×10^5 **b.** 7.358×10^{-3} **c.** 8.4×10^7 **d.** 3.007×10^{-5}

Solution

a. Move the decimal point 5 places to the right.

$$1.02 \times 10^5 = 102{,}000.$$

b. Move the decimal point 3 places to the left.

$$7.358 \times 10^{-3} = 0.007358$$

c. $8.4 \times 10^7 = 84{,}000{,}000.$ 7 places to the right

d. $3.007 \times 10^{-5} = 0.00003007$ 5 places to the left

PRACTICE

6 Write each number in standard notation, without exponents.

a. 3.67×10^{-4} **b.** 8.954×10^6

c. 2.009×10^{-5} **d.** 4.054×10^3

✓CONCEPT CHECK

Which number in each pair is larger?

a. 7.8×10^3 or 2.1×10^5 b. 9.2×10^{-2} or 2.7×10^4 c. 5.6×10^{-4} or 6.3×10^{-5}

OBJECTIVE

5 Performing Operations with Scientific Notation

Performing operations on numbers written in scientific notation uses the rules and definitions for exponents.

EXAMPLE 7 Perform each indicated operation. Write each result in standard decimal notation.

a. $(8 \times 10^{-6})(7 \times 10^3)$ **b.** $\frac{12 \times 10^2}{6 \times 10^{-3}}$

Solution

a.
$$\begin{aligned}(8 \times 10^{-6})(7 \times 10^3) &= (8 \cdot 7) \times (10^{-6} \cdot 10^3)\\ &= 56 \times 10^{-3}\\ &= 0.056\end{aligned}$$

b. $\frac{12 \times 10^2}{6 \times 10^{-3}} = \frac{12}{6} \times 10^{2-(-3)} = 2 \times 10^5 = 200{,}000$

PRACTICE

7 Perform each indicated operation. Write each result in standard decimal notation.

a. $(5 \times 10^{-4})(8 \times 10^6)$ **b.** $\frac{64 \times 10^3}{32 \times 10^{-7}}$

Answers to Concept Check:
a. 2.1×10^5 **b.** 2.7×10^4
c. 5.6×10^{-4}

Calculator Explorations 

Scientific Notation

To enter a number written in scientific notation on a scientific calculator, locate the scientific notation key, which may be marked EE or EXP. To enter 3.1×10^7, press 3.1 EE 7. The display should read 3.1 07.

Enter each number written in scientific notation on your calculator.

1. 5.31×10^3
2. -4.8×10^{14}
3. 6.6×10^{-9}
4. -9.9811×10^{-2}

Multiply each of the following on your calculator. Notice the form of the result.

5. $3{,}000{,}000 \times 5{,}000{,}000$
6. $230{,}000 \times 1000$

Multiply each of the following on your calculator. Write the product in scientific notation.

7. $(3.26 \times 10^6)(2.5 \times 10^{13})$
8. $(8.76 \times 10^{-4})(1.237 \times 10^9)$

Vocabulary, Readiness & Video Check

Fill in each blank with the correct choice.

1. The expression x^{-3} equals ______.
   a. $-x^3$ b. $\frac{1}{x^3}$ c. $\frac{-1}{x^3}$ d. $\frac{1}{x^{-3}}$
2. The expression 5^{-4} equals ______.
   a. -20 b. -625 c. $\frac{1}{20}$ d. $\frac{1}{625}$
3. The number 3.021×10^{-3} is written in ______________.
   a. standard form b. expanded form
   c. scientific notation
4. The number 0.0261 is written in ______________.
   a. standard form b. expanded form
   c. scientific notation

Martin-Gay Interactive Videos

See Video 5.5

Watch the section lecture video and answer the following questions.

OBJECTIVE 1 5. What important reminder is made at the end of Example 1?

OBJECTIVE 2 6. Name all the rules and definitions used to simplify Example 8.

OBJECTIVE 3 7. From Examples 9 and 10, explain how the movement of the decimal point in step 1 suggests the sign of the exponent on the number 10.

OBJECTIVE 4 8. From Example 11, what part of a number written in scientific notation is key in telling you how to write the number in standard form?

OBJECTIVE 5 9. For Example 13, what exponent rules were needed to evaluate?

5.5 Exercise Set

MyMathLab®  

Simplify each expression. Write each result using positive exponents only. See Examples 1 through 3.

1. 4^{-3}
2. 6^{-2}
3. $(-3)^{-4}$
4. $(-3)^{-5}$
5. $7x^{-3}$
6. $(7x)^{-3}$
7. $\left(\frac{1}{2}\right)^{-5}$
8. $\left(\frac{1}{8}\right)^{-2}$
9. $\left(-\frac{1}{4}\right)^{-3}$
10. $\left(-\frac{1}{8}\right)^{-2}$
11. $3^{-1} + 5^{-1}$
12. $4^{-1} + 4^{-2}$

13. $\frac{1}{p^{-3}}$

14. $\frac{1}{q^{-5}}$

15. $\frac{p^{-5}}{q^{-4}}$

16. $\frac{r^{-5}}{s^{-2}}$

17. $\frac{x^{-2}}{x}$

18. $\frac{y}{y^{-3}}$

19. $\frac{z^{-4}}{z^{-7}}$

20. $\frac{x^{-4}}{x^{-1}}$

21. $3^{-2} + 3^{-1}$

22. $4^{-2} - 4^{-3}$

23. $\frac{-1}{p^{-4}}$

24. $\frac{-1}{y^{-6}}$

25. $-2^0 - 3^0$

26. $5^0 + (-5)^0$

MIXED PRACTICE

Simplify each expression. Write each result using positive exponents only. See Examples 1 through 4.

27. $\frac{x^2x^5}{x^3}$

28. $\frac{y^4y^5}{y^6}$

29. $\frac{p^2p}{p^{-1}}$

30. $\frac{y^3y}{y^{-2}}$

31. $\frac{(m^5)^4m}{m^{10}}$

32. $\frac{(x^2)^8x}{x^9}$

33. $\frac{r}{r^{-3}r^{-2}}$

34. $\frac{p}{p^{-3}q^{-5}}$

35. $(x^5y^3)^{-3}$

36. $(z^5x^5)^{-3}$

37. $\frac{(x^2)^3}{x^{10}}$

38. $\frac{(y^4)^2}{y^{12}}$

39. $\frac{(a^5)^2}{(a^3)^4}$

40. $\frac{(x^2)^5}{(x^4)^3}$

41. $\frac{8k^4}{2k}$

42. $\frac{27r^4}{3r^6}$

43. $\frac{-6m^4}{-2m^3}$

44. $\frac{15a^4}{-15a^5}$

45. $\frac{-24a^6b}{6ab^2}$

46. $\frac{-5x^4y^5}{15x^4y^2}$

47. $(-2x^3y^{-4})(3x^{-1}y)$

48. $(-5a^4b^{-7})(-a^{-4}b^3)$

49. $(a^{-5}b^2)^{-6}$

50. $(4^{-1}x^5)^{-2}$

51. $\left(\frac{x^{-2}y^4}{x^3y^7}\right)^2$

52. $\left(\frac{a^5b}{a^7b^{-2}}\right)^{-3}$

53. $\frac{4^2z^{-3}}{4^3z^{-5}}$

54. $\frac{3^{-1}x^4}{3^3x^{-7}}$

55. $\frac{2^{-3}x^{-4}}{2^2x}$

56. $\frac{5^{-1}z^7}{5^{-2}z^9}$

57. $\frac{7ab^{-4}}{7^{-1}a^{-3}b^2}$

58. $\frac{6^{-5}x^{-1}y^2}{6^{-2}x^{-4}y^4}$

59. $\left(\frac{a^{-5}b}{ab^3}\right)^{-4}$

60. $\left(\frac{r^{-2}s^{-3}}{r^{-4}s^{-3}}\right)^{-3}$

61. $\frac{(xy^3)^5}{(xy)^{-4}}$

62. $\frac{(rs)^{-3}}{(r^2s^3)^2}$

63. $\frac{(-2xy^{-3})^{-3}}{(xy^{-1})^{-1}}$

64. $\frac{(-3x^2y^2)^{-2}}{(xyz)^{-2}}$

65. $\frac{6x^2y^3}{-7xy^5}$

66. $\frac{-8xa^2b}{-5xa^5b}$

67. $\frac{(a^4b^{-7})^{-5}}{(5a^2b^{-1})^{-2}}$

68. $\frac{(a^6b^{-2})^4}{(4a^{-3}b^{-3})^3}$

Write each number in scientific notation. See Example 5.

69. 78,000

70. 9,300,000,000

71. 0.00000167

72. 0.00000017

73. 0.00635

74. 0.00194

75. 1,160,000

76. 700,000

77. More than 2,000,000,000 pencils are manufactured in the United States annually. Write this number in scientific notation. (*Source:* AbsoluteTrivia.com)

78. The temperature at the interior of the Earth is 20,000,000 degrees Celsius. Write 20,000,000 in scientific notation.

79. As of this writing, the world's largest optical telescope is the Gran Telescopio Canaris, located in La Palma, Canary Islands, Spain. The elevation of this telescope is 2400 meters above sea level. Write 2400 in scientific notation.

80. In January 2009, the twin Mars rovers, Spirit and Opportunity, celebrated their fifth anniversary of landing on Mars. These rovers, which were expected to last about 90 days, have defied all expectations and have been transmitting signals back to Earth from as far away as 250,000,000 miles. Write 250,000,000 in scientific notation. (*Source:* NASA)

Write each number in standard notation. See Example 6.

81. 8.673×10^{-10}

82. 9.056×10^{-4}

83. 3.3×10^{-2}

84. 4.8×10^{-6}

85. 2.032×10^{4}

86. 9.07×10^{10}

87. Each second, the Sun converts 7.0×10^{8} tons of hydrogen into helium and energy in the form of gamma rays. Write this number in standard notation. (*Source:* Students for the Exploration and Development of Space)

88. In chemistry, Avogadro's number is the number of atoms in one mole of an element. Avogadro's number is $6.02214199 \times 10^{23}$. Write this number in standard notation. (*Source:* National Institute of Standards and Technology)

89. The distance light travels in 1 year is 9.460×10^{12} kilometers. Write this number in standard notation.

90. The population of the world is 6.95×10^{9}. Write this number in standard notation. (*Source:* U.S. Bureau of the Census)

MIXED PRACTICE

See Examples 5 and 6. Below are some interesting facts about selected countries' national debts during a certain time period. If a number is written in standard form, write it in scientific notation. If a number is written in scientific notation, write it in standard form. (Data from CIA World Factbook)

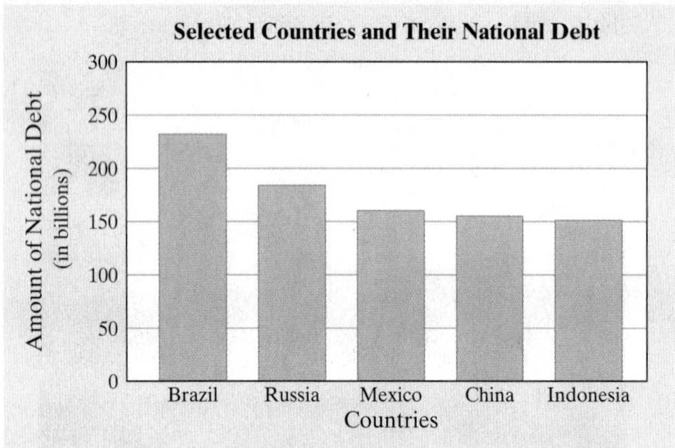

91. The national debt of Russia during a certain time period was \$184,000,000,000.

92. The amount by which Russia's debt is greater than Mexico's debt is \$24,000,000,000.

93. At a certain time period, China's national debt was $\$1.55 \times 10^{11}$.

94. At a certain time period, the national debt of the United States was $\$1.1 \times 10^{13}$.

95. At a certain time period, the estimated per person share of the United States' national debt was $\$3.5 \times 10^{4}$.

96. The bar graph shows the national debt of five countries. Estimate the height of the tallest bar and the shortest bar in standard notation. Then write each number in scientific notation.

Evaluate each expression using exponential rules. Write each result in standard notation. See Example 7.

97. $(1.2 \times 10^{-3})(3 \times 10^{-2})$

98. $(2.5 \times 10^{6})(2 \times 10^{-6})$

99. $(4 \times 10^{-10})(7 \times 10^{-9})$

100. $(5 \times 10^{6})(4 \times 10^{-8})$

101. $\dfrac{8 \times 10^{-1}}{16 \times 10^{5}}$

102. $\dfrac{25 \times 10^{-4}}{5 \times 10^{-9}}$

103. $\dfrac{1.4 \times 10^{-2}}{7 \times 10^{-8}}$

104. $\dfrac{0.4 \times 10^{5}}{0.2 \times 10^{11}}$

REVIEW AND PREVIEW

Simplify the following. See Section 5.1.

105. $\dfrac{5x^7}{3x^4}$

106. $\dfrac{27y^{14}}{3y^7}$

107. $\dfrac{15z^4y^3}{21zy}$

108. $\dfrac{18a^7b^{17}}{30a^7b}$

Use the distributive property and multiply. See Sections 5.3 and 5.5.

109. $\dfrac{1}{y}(5y^2 - 6y + 5)$

110. $\dfrac{2}{x}(3x^5 + x^4 - 2)$

CONCEPT EXTENSIONS

△ **111.** Find the volume of the cube.

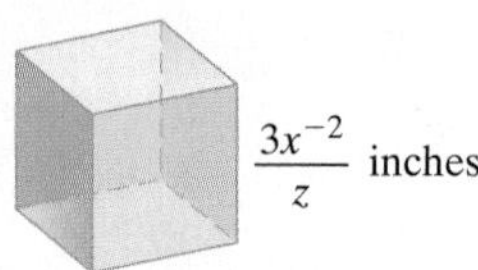

△ **112.** Find the area of the triangle.

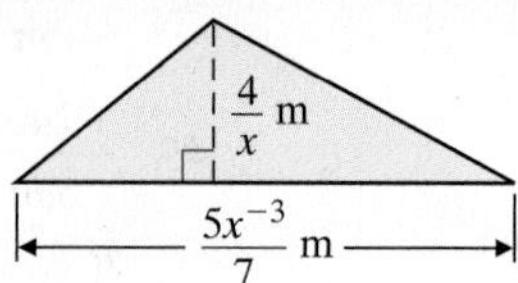

Simplify.

113. $(2a^3)^3a^4 + a^5a^8$

114. $(2a^3)^3a^{-3} + a^{11}a^{-5}$

Fill in the boxes so that each statement is true. (More than one answer may be possible for these exercises.)

115. $x^{\square} = \dfrac{1}{x^5}$

116. $7^{\square} = \dfrac{1}{49}$

117. $z^{\square} \cdot z^{\square} = z^{-10}$

118. $(x^{\square})^{\square} = x^{-15}$

119. Which is larger? See the Concept Check in this section.

a. 9.7×10^{-2} or 1.3×10^{1}

b. 8.6×10^{5} or 4.4×10^{7}

c. 6.1×10^{-2} or 5.6×10^{-4}

120. Determine whether each statement is true or false.

a. $5^{-1} < 5^{-2}$

b. $\left(\frac{1}{5}\right)^{-1} < \left(\frac{1}{5}\right)^{-2}$

c. $a^{-1} < a^{-2}$ for all nonzero numbers.

121. It was stated earlier that for an integer *n*,

$$x^{-n} = \frac{1}{x^n}, \quad x \neq 0$$

Explain why *x* may not equal 0.

122. The quotient rule states that

$$\frac{a^m}{a^n} = a^{m-n}, a \neq 0.$$

Explain why *a* may not equal 0.

Simplify each expression. Assume that variables represent positive integers.

123. $a^{-4m} \cdot a^{5m}$

124. $(x^{-3s})^3$

125. $(3y^{2z})^3$

126. $a^{4m+1} \cdot a^4$

Simplify each expression. Write each result in standard notation.

127. $(2.63 \times 10^{12})(-1.5 \times 10^{-10})$

128. $(6.785 \times 10^{-4})(4.68 \times 10^{10})$

Light travels at a rate of 1.86×10^5 *miles per second. Use this information and the distance formula* $d = r \cdot t$ *to answer Exercises 129 and 130.*

129. If the distance from the moon to Earth is 238,857 miles, find how long it takes the reflected light of the moon to reach Earth. (Round to the nearest tenth of a second.)

130. If the distance from the sun to Earth is 93,000,000 miles, find how long it takes the light of the sun to reach Earth.

5.6 Dividing Polynomials

OBJECTIVES

1 Divide a Polynomial by a Monomial.

2 Use Long Division to Divide a Polynomial by Another Polynomial.

OBJECTIVE

1 Dividing by a Monomial

Now that we know how to add, subtract, and multiply polynomials, we practice dividing polynomials.

To divide a polynomial by a monomial, recall addition of fractions. Fractions that have a common denominator are added by adding the numerators:

$$\frac{a}{c} + \frac{b}{c} = \frac{a+b}{c}$$

If we read this equation from right to left and let *a*, *b*, and *c* be monomials, $c \neq 0$, we have the following:

Dividing a Polynomial by a Monomial

Divide each term of the polynomial by the monomial.

$$\frac{a+b}{c} = \frac{a}{c} + \frac{b}{c}, \quad c \neq 0$$

Throughout this section, we assume that denominators are not 0.

EXAMPLE 1 Divide $6m^2 + 2m$ by $2m$.

Solution We begin by writing the quotient in fraction form. Then we divide each term of the polynomial $6m^2 + 2m$ by the monomial $2m$.

$$\frac{6m^2 + 2m}{2m} = \frac{6m^2}{2m} + \frac{2m}{2m}$$

$$= 3m + 1 \quad \text{Simplify.}$$

Check: We know that if $\frac{6m^2 + 2m}{2m} = 3m + 1$, then $2m \cdot (3m + 1)$ must equal $6m^2 + 2m$. Thus, to check, we multiply.

$$2m(3m + 1) = 2m(3m) + 2m(1) = 6m^2 + 2m$$

The quotient $3m + 1$ checks. □

PRACTICE

1 Divide $8t^3 + 4t^2$ by $4t^2$

EXAMPLE 2 Divide $\dfrac{9x^5 - 12x^2 + 3x}{3x^2}$.

Solution

$$\frac{9x^5 - 12x^2 + 3x}{3x^2} = \frac{9x^5}{3x^2} - \frac{12x^2}{3x^2} + \frac{3x}{3x^2} \quad \text{Divide each term by } 3x^2.$$

$$= 3x^3 - 4 + \frac{1}{x} \quad \text{Simplify.}$$

Notice that the quotient is not a polynomial because of the term $\frac{1}{x}$. This expression is called a rational expression—we will study rational expressions further in Chapter 7. Although the quotient of two polynomials is not always a polynomial, we may still check by multiplying.

Check: $3x^2\left(3x^3 - 4 + \frac{1}{x}\right) = 3x^2(3x^3) - 3x^2(4) + 3x^2\left(\frac{1}{x}\right)$

$= 9x^5 - 12x^2 + 3x$ □

PRACTICE 2 Divide $\dfrac{16x^6 + 20x^3 - 12x}{4x^2}$.

EXAMPLE 3 Divide $\dfrac{8x^2y^2 - 16xy + 2x}{4xy}$.

Solution

$$\frac{8x^2y^2 - 16xy + 2x}{4xy} = \frac{8x^2y^2}{4xy} - \frac{16xy}{4xy} + \frac{2x}{4xy} \quad \text{Divide each term by } 4xy.$$

$$= 2xy - 4 + \frac{1}{2y} \quad \text{Simplify.}$$

Check: $4xy\left(2xy - 4 + \frac{1}{2y}\right) = 4xy(2xy) - 4xy(4) + 4xy\left(\frac{1}{2y}\right)$

$= 8x^2y^2 - 16xy + 2x$ □

PRACTICE 3 Divide $\dfrac{15x^4y^4 - 10xy + y}{5xy}$.

✓CONCEPT CHECK

In which of the following is $\dfrac{x + 5}{5}$ simplified correctly?

a. $\dfrac{x}{5} + 1$ b. x c. $x + 1$

OBJECTIVE 2

Using Long Division to Divide by a Polynomial

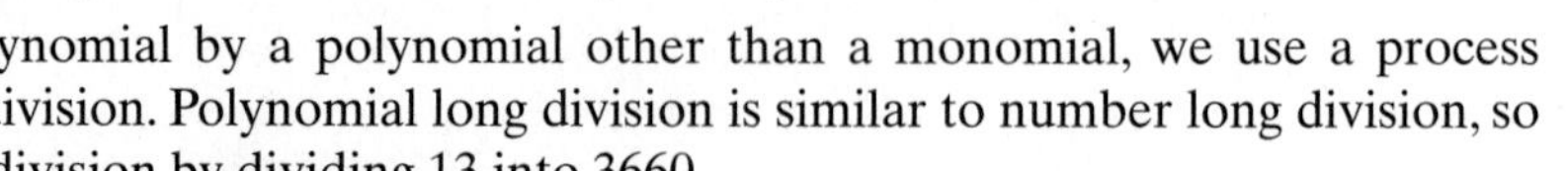

To divide a polynomial by a polynomial other than a monomial, we use a process known as long division. Polynomial long division is similar to number long division, so we review long division by dividing 13 into 3660.

Helpful Hint
Recall that 3660 is called the dividend.

```
   281
13)3660
   26↓      2·13 = 26
   106      Subtract and bring down the next digit in the dividend.
   104↓     8·13 = 104
     20     Subtract and bring down the next digit in the dividend.
     13     1·13 = 13
      7     Subtract. There are no more digits to bring down, so the remainder is 7.
```

Answer to Concept Check: a

The quotient is 281 R 7, which can be written as $281\frac{7}{13}$. ← remainder / ← divisor

Recall that division can be checked by multiplication. To check a division problem such as this one, we see that

$$13 \cdot 281 + 7 = 3660$$

Now we demonstrate long division of polynomials.

EXAMPLE 4 Divide $x^2 + 7x + 12$ by $x + 3$ using long division.

Solution

To subtract, change the signs of these terms and add.

$$\begin{array}{r} x \\ x + 3\overline{)x^2 + 7x + 12} \\ \underline{-x^2 - 3x } \downarrow \\ 4x + 12 \end{array}$$

How many times does x divide x^2? $\frac{x^2}{x} = x$.
Multiply: $x(x + 3)$.
Subtract and bring down the next term.

Now we repeat this process.

$$\begin{array}{r} x + 4 \\ x + 3\overline{)x^2 + 7x + 12} \\ \underline{-x^2 - 3x } \\ 4x + 12 \\ \underline{-4x - 12} \\ 0 \end{array}$$

How many times does x divide $4x$? $\frac{4x}{x} = 4$.

To subtract, change the signs of these terms and add.

Multiply: $4(x + 3)$.
Subtract. The remainder is 0.

The quotient is $x + 4$.

Check: We check by multiplying.

divisor	·	quotient	+	remainder	=	dividend
↓		↓		↓		↓
$(x + 3)$	·	$(x + 4)$	+	0	=	$x^2 + 7x + 12$

The quotient checks. □

PRACTICE
4 Divide $x^2 + 5x + 6$ by $x + 2$ using long division.

EXAMPLE 5 Divide $6x^2 + 10x - 5$ by $3x - 1$ using long division.

Solution

$$\begin{array}{r} 2x + 4 \\ 3x - 1\overline{)6x^2 + 10x - 5} \\ \underline{-6x^2 + 2x } \downarrow \\ 12x - 5 \\ \underline{-12x + 4} \\ -1 \end{array}$$

$\frac{6x^2}{3x} = 2x$, so $2x$ is a term of the quotient.
Multiply $2x(3x - 1)$.
Subtract and bring down the next term.
$\frac{12x}{3x} = 4$, multiply $4(3x - 1)$
Subtract. The remainder is -1.

Thus $(6x^2 + 10x - 5)$ divided by $(3x - 1)$ is $(2x + 4)$ with a remainder of -1. This can be written as

$$\frac{6x^2 + 10x - 5}{3x - 1} = 2x + 4 + \frac{-1}{3x - 1} \quad \begin{array}{l} \leftarrow \text{remainder} \\ \leftarrow \text{divisor} \end{array}$$

Check: To check, we multiply $(3x - 1)(2x + 4)$. Then we add the remainder, -1, to this product.

$$\begin{aligned} (3x - 1)(2x + 4) + (-1) &= (6x^2 + 12x - 2x - 4) - 1 \\ &= 6x^2 + 10x - 5 \end{aligned}$$

The quotient checks. □

PRACTICE
5 Divide $4x^2 + 8x - 7$ by $2x + 1$ using long division.

In Example 5, the degree of the divisor, $3x - 1$, is 1 and the degree of the remainder, -1, is 0. The division process is continued until the degree of the remainder polynomial is less than the degree of the divisor polynomial.

Writing the dividend and divisor in a form with descending order of powers and with no missing terms is helpful when dividing polynomials.

EXAMPLE 6 Divide $\frac{4x^2 + 7 + 8x^3}{2x + 3}$.

Solution Before we begin the division process, we rewrite

$$4x^2 + 7 + 8x^3 \quad \text{as} \quad 8x^3 + 4x^2 + 0x + 7$$

Notice that we have written the polynomial in descending order and have represented the missing x term by $0x$.

$$\begin{array}{r} 4x^2 - 4x + 6 \\ 2x + 3\overline{)8x^3 + 4x^2 + 0x + 7} \\ \underline{8x^3 \mp 12x^2} \qquad\qquad \\ -8x^2 + 0x \qquad \\ \underline{\mp 8x^2 \mp 12x} \qquad \\ 12x + 7 \\ \underline{12x \mp 18} \\ -11 \end{array} \quad \text{Remainder}$$

Thus, $\frac{4x^2 + 7 + 8x^3}{2x + 3} = 4x^2 - 4x + 6 + \frac{-11}{2x + 3}$. □

PRACTICE 6 Divide $\frac{11x - 3 + 9x^3}{3x + 2}$.

EXAMPLE 7 Divide $\frac{2x^4 - x^3 + 3x^2 + x - 1}{x^2 + 1}$.

Solution Before dividing, rewrite the divisor polynomial

$$x^2 + 1 \quad \text{as} \quad x^2 + 0x + 1$$

The $0x$ term represents the missing x^1 term in the divisor.

$$\begin{array}{r} 2x^2 - x + 1 \\ x^2 + 0x + 1\overline{)2x^4 - x^3 + 3x^2 + x - 1} \\ \underline{2x^4 \mp 0x^3 \mp 2x^2} \qquad\qquad \\ -x^3 + x^2 + x \qquad \\ \underline{\mp x^3 \mp 0x^2 \mp x} \qquad \\ x^2 + 2x - 1 \\ \underline{x^2 \mp 0x \mp 1} \\ 2x - 2 \end{array} \quad \text{Remainder}$$

Thus, $\frac{2x^4 - x^3 + 3x^2 + x - 1}{x^2 + 1} = 2x^2 - x + 1 + \frac{2x - 2}{x^2 + 1}$. □

PRACTICE 7 Divide $\frac{3x^4 - 2x^3 - 3x^2 + x + 4}{x^2 + 2}$.

EXAMPLE 8 Divide $x^3 - 8$ by $x - 2$.

Solution: Notice that the polynomial $x^3 - 8$ is missing an x^2-term and an x-term. We'll represent these terms by inserting $0x^2$ and $0x$.

$$\begin{array}{r} x^2 + 2x + 4 \\ x - 2\overline{)x^3 + 0x^2 + 0x - 8} \\ \underline{-x^3 + 2x^2} \quad\quad\quad\quad \\ 2x^2 + 0x \quad\quad \\ \underline{-2x^2 + 4x} \quad\quad \\ 4x - 8 \\ \underline{-4x + 8} \\ 0 \end{array}$$

Thus, $\frac{x^3 - 8}{x - 2} = x^2 + 2x + 4$. □

Check: To check, see that $(x^2 + 2x + 4)(x - 2) = x^3 - 8$.

PRACTICE 8 Divide $x^3 + 27$ by $x + 3$.

Vocabulary, Readiness & Video Check

Use the choices below to fill in each blank. Choices may be used more than once.

dividend divisor quotient

1. In $6\overline{)18}^{\,3}$, the 18 is the ________, the 3 is the ________ and the 6 is the ________.

2. In $x + 1\overline{)x^2 + 3x + 2}^{\,x+2}$, the $x + 1$ is the ________, the $x^2 + 3x + 2$ is the ________ and the $x + 2$ is the ________.

Simplify each expression mentally.

3. $\frac{a^6}{a^4}$ **4.** $\frac{p^8}{p^3}$ **5.** $\frac{y^2}{y}$ **6.** $\frac{a^3}{a}$

Martin-Gay Interactive Videos

See Video 5.6

Watch the section lecture video and answer the following questions.

OBJECTIVE 1 **7.** The lecture before Example 1 begins with adding two fractions with the same denominator. From there, the lecture continues to a method for dividing a polynomial by a monomial. What role does the monomial play in the fraction example?

OBJECTIVE 2 **8.** In Example 5, you're told that although you don't have to fill in missing powers in the divisor and the dividend, it really is a good idea to do so. Why?

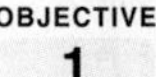

5.6 Exercise Set

MyMathLab®

Perform each division. See Examples 1 through 3.

1. $\frac{12x^4 + 3x^2}{x}$

2. $\frac{15x^2 - 9x^5}{x}$

3. $\frac{20x^3 - 30x^2 + 5x + 5}{5}$

4. $\frac{8x^3 - 4x^2 + 6x + 2}{2}$

5. $\dfrac{15p^3 + 18p^2}{3p}$

6. $\dfrac{14m^2 - 27m^3}{7m}$

7. $\dfrac{-9x^4 + 18x^5}{6x^5}$

8. $\dfrac{6x^5 + 3x^4}{3x^4}$

9. $\dfrac{-9x^5 + 3x^4 - 12}{3x^3}$

10. $\dfrac{6a^2 - 4a + 12}{-2a^2}$

11. $\dfrac{4x^4 - 6x^3 + 7}{-4x^4}$

12. $\dfrac{-12a^3 + 36a - 15}{3a}$

Find each quotient using long division. See Examples 4 and 5.

13. $\dfrac{x^2 + 4x + 3}{x + 3}$

14. $\dfrac{x^2 + 7x + 10}{x + 5}$

15. $\dfrac{2x^2 + 13x + 15}{x + 5}$

16. $\dfrac{3x^2 + 8x + 4}{x + 2}$

17. $\dfrac{2x^2 - 7x + 3}{x - 4}$

18. $\dfrac{3x^2 - x - 4}{x - 1}$

19. $\dfrac{9a^3 - 3a^2 - 3a + 4}{3a + 2}$

20. $\dfrac{4x^3 + 12x^2 + x - 14}{2x + 3}$

21. $\dfrac{8x^2 + 10x + 1}{2x + 1}$

22. $\dfrac{3x^2 + 17x + 7}{3x + 2}$

23. $\dfrac{2x^3 + 2x^2 - 17x + 8}{x - 2}$

24. $\dfrac{4x^3 + 11x^2 - 8x - 10}{x + 3}$

Find each quotient using long division. Don't forget to write the polynomials in descending order and fill in any missing terms. See Examples 6 through 8.

25. $\dfrac{x^2 - 36}{x - 6}$

26. $\dfrac{a^2 - 49}{a - 7}$

27. $\dfrac{x^3 - 27}{x - 3}$

28. $\dfrac{x^3 + 64}{x + 4}$

29. $\dfrac{1 - 3x^2}{x + 2}$

30. $\dfrac{7 - 5x^2}{x + 3}$

31. $\dfrac{-4b + 4b^2 - 5}{2b - 1}$

32. $\dfrac{-3y + 2y^2 - 15}{2y + 5}$

MIXED PRACTICE

Divide. If the divisor contains 2 or more terms, use long division. See Examples 1 through 8.

33. $\dfrac{a^2b^2 - ab^3}{ab}$

34. $\dfrac{m^3n^2 - mn^4}{mn}$

35. $\dfrac{8x^2 + 6x - 27}{2x - 3}$

36. $\dfrac{18w^2 + 18w - 8}{3w + 4}$

37. $\dfrac{2x^2y + 8x^2y^2 - xy^2}{2xy}$

38. $\dfrac{11x^3y^3 - 33xy + x^2y^2}{11xy}$

39. $\dfrac{2b^3 + 9b^2 + 6b - 4}{b + 4}$

40. $\dfrac{2x^3 + 3x^2 - 3x + 4}{x + 2}$

41. $\dfrac{5x^2 + 28x - 10}{x + 6}$

42. $\dfrac{2x^2 + x - 15}{x + 3}$

43. $\dfrac{10x^3 - 24x^2 - 10x}{10x}$

44. $\dfrac{2x^3 + 12x^2 + 16}{4x^2}$

45. $\dfrac{6x^2 + 17x - 4}{x + 3}$

46. $\dfrac{2x^2 - 9x + 15}{x - 6}$

47. $\dfrac{30x^2 - 17x + 2}{5x - 2}$

48. $\dfrac{4x^2 - 13x - 12}{4x + 3}$

49. $\dfrac{3x^4 - 9x^3 + 12}{-3x}$

50. $\dfrac{8y^6 - 3y^2 - 4y}{4y}$

51. $\dfrac{x^3 + 6x^2 + 18x + 27}{x + 3}$

52. $\dfrac{x^3 - 8x^2 + 32x - 64}{x - 4}$

53. $\dfrac{y^3 + 3y^2 + 4}{y - 2}$

54. $\frac{3x^3 + 11x + 12}{x + 4}$

55. $\frac{5 - 6x^2}{x - 2}$

56. $\frac{3 - 7x^2}{x - 3}$

Divide.

57. $\frac{x^5 + x^2}{x^2 + x}$

58. $\frac{x^6 - x^4}{x^3 + 1}$

REVIEW AND PREVIEW

Multiply each expression. See Section 5.3.

59. $2a(a^2 + 1)$
60. $-4a(3a^2 - 4)$
61. $2x(x^2 + 7x - 5)$
62. $4y(y^2 - 8y - 4)$
63. $-3xy(xy^2 + 7x^2y + 8)$
64. $-9xy(4xyz + 7xy^2z + 2)$
65. $9ab(ab^2c + 4bc - 8)$
66. $-7sr(6s^2r + 9sr^2 + 9rs + 8)$

CONCEPT EXTENSIONS

67. The perimeter of a square is $(12x^3 + 4x - 16)$ feet. Find the length of its side.

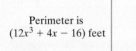

Perimeter is $(12x^3 + 4x - 16)$ feet

△ 68. The volume of the swimming pool shown is $(36x^5 - 12x^3 + 6x^2)$ cubic feet. If its height is $2x$ feet and its width is $3x$ feet, find its length.

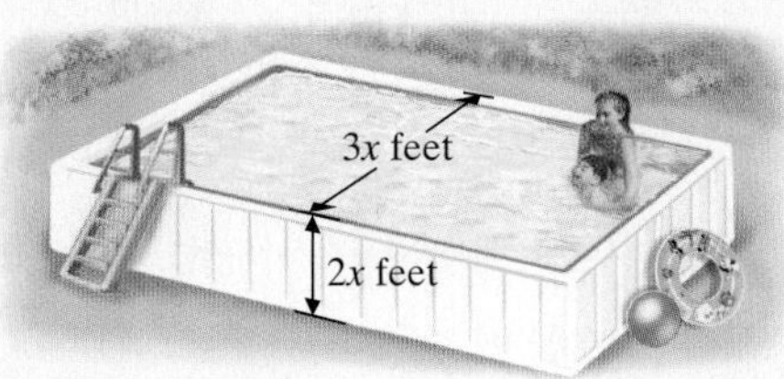

69. In which of the following is $\frac{a + 7}{7}$ simplified correctly? See the Concept Check in this section.

a. $a + 1$ **b.** a **c.** $\frac{a}{7} + 1$

70. In which of the following is $\frac{5x + 15}{5}$ simplified correctly? See the Concept Check in this section.

a. $x + 15$ **b.** $x + 3$ **c.** $x + 1$

71. Explain how to check a polynomial long division result when the remainder is 0.

72. Explain how to check a polynomial long division result when the remainder is not 0.

△ 73. The area of the following parallelogram is $(10x^2 + 31x + 15)$ square meters. If its base is $(5x + 3)$ meters, find its height.

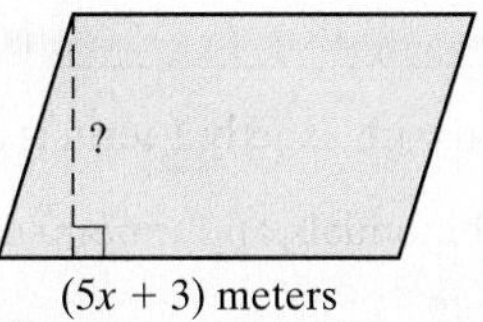

△ 74. The area of the top of the Ping-Pong table is $(49x^2 + 70x - 200)$ square inches. If its length is $(7x + 20)$ inches, find its width.

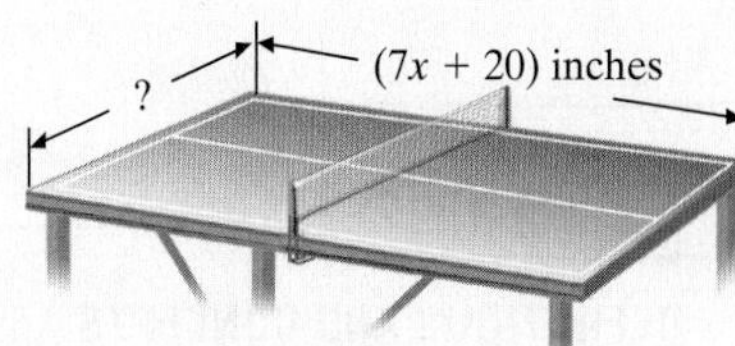

75. $(18x^{10a} - 12x^{8a} + 14x^{5a} - 2x^{3a}) \div 2x^{3a}$
76. $(25y^{11b} + 5y^{6b} - 20y^{3b} + 100y^{b}) \div 5y^{b}$

Chapter 5 Vocabulary Check

Fill in each blank with one of the words or phrases listed below.

term	coefficient	monomial	binomial	trinomial
polynomials	degree of a term	distributive	FOIL	degree of a polynomial

1. A ______________ is a number or the product of numbers and variables raised to powers.
2. The ______________ method may be used when multiplying two binomials.
3. A polynomial with exactly 3 terms is called a ______________.
4. The ______________ is the greatest degree of any term of the polynomial.
5. A polynomial with exactly 2 terms is called a ______________.
6. The ______________ of a term is its numerical factor.
7. The ______________ is the sum of the exponents on the variables in the term.
8. A polynomial with exactly 1 term is called a ______________.
9. Monomials, binomials, and trinomials are all examples of ______________.
10. The ______________ property is used to multiply $2x(x - 4)$.

Chapter 5 Highlights

DEFINITIONS AND CONCEPTS	EXAMPLES
Section 5.1 Exponents	
a^n means the product of n factors, each of which is a.	$3^2 = 3 \cdot 3 = 9$ $(-5)^3 = (-5)(-5)(-5) = -125$ $\left(\frac{1}{2}\right)^4 = \frac{1}{2}\cdot\frac{1}{2}\cdot\frac{1}{2}\cdot\frac{1}{2} = \frac{1}{16}$
If m and n are integers and no denominators are 0,	
Product Rule: $a^m \cdot a^n = a^{m+n}$	$x^2 \cdot x^7 = x^{2+7} = x^9$
Power Rule: $(a^m)^n = a^{mn}$	$(5^3)^8 = 5^{3\cdot 8} = 5^{24}$
Power of a Product Rule: $(ab)^n = a^n b^n$	$(7y)^4 = 7^4 y^4$
Power of a Quotient Rule: $\left(\frac{a}{b}\right)^n = \frac{a^n}{b^n}$	$\left(\frac{x}{8}\right)^3 = \frac{x^3}{8^3}$
Quotient Rule: $\frac{a^m}{a^n} = a^{m-n}$	$\frac{x^9}{x^4} = x^{9-4} = x^5$
Zero Exponent: $a^0 = 1, a \neq 0.$	$5^0 = 1, x^0 = 1, x \neq 0$

DEFINITIONS AND CONCEPTS	EXAMPLES
Section 5.2 Adding and Subtracting Polynomials	
A **term** is a number or the product of numbers and variables raised to powers.	***Terms*** $-5x, 7a^2b, \frac{1}{4}y^4, 0.2$
The **numerical coefficient** or **coefficient** of a term is its numerical factor.	***Term*** — ***Coefficient*** $7x^2$ — 7 y — 1 $-a^2b$ — -1
A **polynomial** is a term or a finite sum of terms in which variables may appear in the numerator raised to whole number powers only. A **monomial** is a polynomial with exactly 1 term. A **binomial** is a polynomial with exactly 2 terms. A **trinomial** is a polynomial with exactly 3 terms.	***Polynomials*** $3x^2 - 2x + 1$ (Trinomial) $-0.2a^2b - 5b^2$ (Binomial) $\frac{5}{6}y^3$ (Monomial)
The **degree of a term** is the sum of the exponents on the variables in the term.	***Term*** — ***Degree*** $-5x^3$ — 3 3 (or $3x^0$) — 0 $2a^2b^2c$ — 5
The **degree of a polynomial** is the greatest degree of any term of the polynomial.	***Polynomial*** — ***Degree*** $5x^2 - 3x + 2$ — 2 $7y + 8y^2z^3 - 12$ — $2 + 3 = 5$
To add polynomials, add or combine like terms.	Add: $(7x^2 - 3x + 2) + (-5x - 6) = 7x^2 - 3x + 2 - 5x - 6$ $= 7x^2 - 8x - 4$
To subtract two polynomials, change the signs of the terms of the second polynomial, then add.	Subtract: $(17y^2 - 2y + 1) - (-3y^3 + 5y - 6)$ $= (17y^2 - 2y + 1) + (3y^3 - 5y + 6)$ $= 17y^2 - 2y + 1 + 3y^3 - 5y + 6$ $= 3y^3 + 17y^2 - 7y + 7$
Section 5.3 Multiplying Polynomials	
To multiply two polynomials, multiply each term of one polynomial by each term of the other polynomial and then combine like terms.	Multiply: $(2x + 1)(5x^2 - 6x + 2)$ $= 2x(5x^2 - 6x + 2) + 1(5x^2 - 6x + 2)$ $= 10x^3 - 12x^2 + 4x + 5x^2 - 6x + 2$ $= 10x^3 - 7x^2 - 2x + 2$

DEFINITIONS AND CONCEPTS	EXAMPLES
Section 5.4 Special Products	
The **FOIL method** may be used when multiplying two binomials.	Multiply: $(5x - 3)(2x + 3)$ First, Last, Outer, Inner $(5x - 3)(2x + 3) = \overset{F}{(5x)(2x)} + \overset{O}{(5x)(3)} + \overset{I}{(-3)(2x)} + \overset{L}{(-3)(3)}$ $= 10x^2 + 15x - 6x - 9$ $= 10x^2 + 9x - 9$
Squaring a Binomial $(a + b)^2 = a^2 + 2ab + b^2$ $(a - b)^2 = a^2 - 2ab + b^2$	Square each binomial. $(x + 5)^2 = x^2 + 2(x)(5) + 5^2$ $= x^2 + 10x + 25$ $(3x - 2y)^2 = (3x)^2 - 2(3x)(2y) + (2y)^2$ $= 9x^2 - 12xy + 4y^2$
Multiplying the Sum and Difference of Two Terms $(a + b)(a - b) = a^2 - b^2$	Multiply: $(6y + 5)(6y - 5) = (6y)^2 - 5^2$ $= 36y^2 - 25$
Section 5.5 Negative Exponents and Scientific Notation	
If $a \neq 0$ and n is an integer, $a^{-n} = \frac{1}{a^n}$ Rules for exponents are true for positive and negative integers.	$3^{-2} = \frac{1}{3^2} = \frac{1}{9}; 5x^{-2} = \frac{5}{x^2}$ Simplify: $\left(\frac{x^{-2}y}{x^5}\right)^{-2} = \frac{x^4y^{-2}}{x^{-10}}$ $= x^{4-(-10)}y^{-2}$ $= \frac{x^{14}}{y^2}$
A positive number is written in scientific notation if it is written as the product of a number a, $1 \leq a < 10$, and an integer power r of 10. $a \times 10^r$	Write each number in scientific notation. $12{,}000 = 1.2 \times 10^4$ $0.00000568 = 5.68 \times 10^{-6}$
Section 5.6 Dividing Polynomials	
To divide a polynomial by a monomial: $\frac{a + b}{c} = \frac{a}{c} + \frac{b}{c}$	Divide: $\frac{15x^5 - 10x^3 + 5x^2 - 2x}{5x^2} = \frac{15x^5}{5x^2} - \frac{10x^3}{5x^2} + \frac{5x^2}{5x^2} - \frac{2x}{5x^2}$ $= 3x^3 - 2x + 1 - \frac{2}{5x}$
To divide a polynomial by a polynomial other than a monomial, use long division.	$5x - 1 + \frac{-4}{2x + 3}$ $2x + 3\overline{)10x^2 + 13x - 7}$ $\underline{10x^2 \not+ 15x}$ (signs changed: $-$, $-$) $-2x - 7$ $\underline{\not-2x \not- 3}$ (signs changed: $+$, $+$) -4

Chapter 5 Review

(5.1) State the base and the exponent for each expression.

1. 7^9
2. $(-5)^4$
3. -5^4
4. x^6

Evaluate each expression.

5. 8^3
6. $(-6)^2$
7. -6^2
8. $-4^3 - 4^0$
9. $(3b)^0$
10. $\frac{8b}{8b}$

Simplify each expression.

11. $y^2 \cdot y^7$
12. $x^9 \cdot x^5$
13. $(2x^5)(-3x^6)$
14. $(-5y^3)(4y^4)$
15. $(x^4)^2$
16. $(y^3)^5$
17. $(3y^6)^4$
18. $(2x^3)^3$
19. $\frac{x^9}{x^4}$
20. $\frac{z^{12}}{z^5}$
21. $\frac{a^5b^4}{ab}$
22. $\frac{x^4y^6}{xy}$
23. $\frac{12xy^6}{3x^4y^{10}}$
24. $\frac{2x^7y^8}{8xy^2}$
25. $5a^7(2a^4)^3$
26. $(2x)^2(9x)$
27. $(-5a)^0 + 7^0 + 8^0$
28. $8x^0 + 9^0$

Simplify the given expression and choose the correct result.

29. $\left(\frac{3x^4}{4y}\right)^3$

a. $\frac{27x^{64}}{64y^3}$ b. $\frac{27x^{12}}{64y^3}$ c. $\frac{9x^{12}}{12y^3}$ d. $\frac{3x^{12}}{4y^3}$

30. $\left(\frac{5a^6}{b^3}\right)^2$

a. $\frac{10a^{12}}{b^6}$ b. $\frac{25a^{36}}{b^9}$ c. $\frac{25a^{12}}{b^6}$ d. $25a^{12}b^6$

(5.2) Find the degree of each term.

31. $-5x^4y^3$
32. $10x^3y^2z$
33. $35a^5bc^2$
34. $95xyz$

Find the degree of each polynomial.

35. $y^5 + 7x - 8x^4$
36. $9y^2 + 30y + 25$
37. $-14x^2y - 28x^2y^3 - 42x^2y^2$
38. $6x^2y^2z^2 + 5x^2y^3 - 12xyz$
39. The Glass Bridge Skywalk is suspended 4000 feet over the Colorado River at the very edge of the Grand Canyon. Neglecting air resistance, the height of an object dropped from the Skywalk at time t seconds is given by the polynomial $-16t^2 + 4000$. Find the height of the object at the given times. (See top of next column.)

t	0 seconds	1 second	3 seconds	5 seconds
$-16t^2 + 4000$				

△ 40. The surface area of a box with a square base and a height of 5 units is given by the polynomial $2x^2 + 20x$. Fill in the table below by evaluating $2x^2 + 20x$ for the given values of x.

x	1	3	5.1	10
$2x^2 + 20x$				

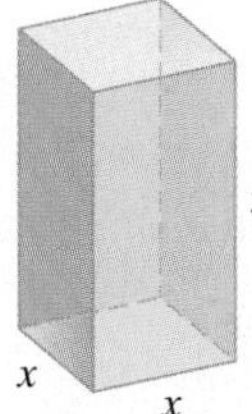

Combine like terms in each expression.

41. $7a^2 - 4a^2 - a^2$
42. $9y + y - 14y$
43. $6a^2 + 4a + 9a^2$
44. $21x^2 + 3x + x^2 + 6$
45. $4a^2b - 3b^2 - 8q^2 - 10a^2b + 7q^2$
46. $2s^{14} + 3s^{13} + 12s^{12} - s^{10}$

Add or subtract as indicated.

47. $(3x^2 + 2x + 6) + (5x^2 + x)$
48. $(2x^5 + 3x^4 + 4x^3 + 5x^2) + (4x^2 + 7x + 6)$
49. $(-5y^2 + 3) - (2y^2 + 4)$
50. $(3x^2 - 7xy + 7y^2) - (4x^2 - xy + 9y^2)$
51. $(-9x^2 + 6x + 2) + (4x^2 - x - 1)$
52. $(8x^6 - 5xy - 10y^2) - (7x^6 - 9xy - 12y^2)$

TRANSLATING

Perform the indicated operations.

53. Subtract $(3x - y)$ from $(7x - 14y)$.

54. Subtract $(4x^2 + 8x - 7)$ from the sum of $(x^2 + 7x + 9)$ and $(x^2 + 4)$.

(5.3) *Multiply each expression.*

55. $4(2a + 7)$

56. $9(6a - 3)$

57. $-7x(x^2 + 5)$

58. $-8y(4y^2 - 6)$

59. $(3a^3 - 4a + 1)(-2a)$

60. $(6b^3 - 4b + 2)(7b)$

61. $(2x + 2)(x - 7)$

62. $(2x - 5)(3x + 2)$

63. $(x - 9)^2$

64. $(x - 12)^2$

65. $(4a - 1)(a + 7)$

66. $(6a - 1)(7a + 3)$

67. $(5x + 2)^2$

68. $(3x + 5)^2$

69. $(x + 7)(x^3 + 4x - 5)$

70. $(x + 2)(x^5 + x + 1)$

71. $(x^2 + 2x + 4)(x^2 + 2x - 4)$

72. $(x^3 + 4x + 4)(x^3 + 4x - 4)$

73. $(x + 7)^3$

74. $(2x - 5)^3$

(5.4) *Use special products to multiply each of the following.*

75. $(x + 7)^2$

76. $(x - 5)^2$

77. $(3x - 7)^2$

78. $(4x + 2)^2$

79. $(5x - 9)^2$

80. $(5x + 1)(5x - 1)$

81. $(7x + 4)(7x - 4)$

82. $(a + 2b)(a - 2b)$

83. $(2x - 6)(2x + 6)$

84. $(4a^2 - 2b)(4a^2 + 2b)$

Express each as a product of polynomials in x. Then multiply and simplify.

△ **85.** Find the area of the square if its side is $(3x - 1)$ meters.

(3x − 1) meters

△ **86.** Find the area of the rectangle.

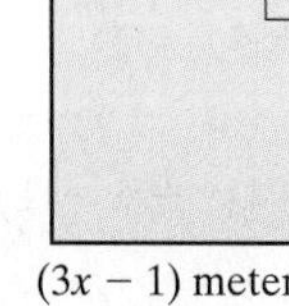

(5.5) *Simplify each expression.*

87. 7^{-2} **88.** -7^{-2} **89.** $2x^{-4}$ **90.** $(2x)^{-4}$

91. $\left(\frac{1}{5}\right)^{-3}$ **92.** $\left(\frac{-2}{3}\right)^{-2}$

93. $2^0 + 2^{-4}$ **94.** $6^{-1} - 7^{-1}$

Simplify each expression. Write each answer using positive exponents only.

95. $\frac{x^5}{x^{-3}}$ **96.** $\frac{z^4}{z^{-4}}$

97. $\frac{r^{-3}}{r^{-4}}$ **98.** $\frac{y^{-2}}{y^{-5}}$

99. $\left(\frac{bc^{-2}}{bc^{-3}}\right)^4$ **100.** $\left(\frac{x^{-3}y^{-4}}{x^{-2}y^{-5}}\right)^{-3}$

101. $\frac{x^{-4}y^{-6}}{x^2y^7}$ **102.** $\frac{a^5b^{-5}}{a^{-5}b^5}$

103. $a^{6m}a^{5m}$ **104.** $\frac{(x^{5+h})^3}{x^5}$

105. $(3xy^{2z})^3$ **106.** $a^{m+2}a^{m+3}$

Write each number in scientific notation.

107. 0.00027 **108.** 0.8868

109. 80,800,000 **110.** 868,000

111. Google.com is an Internet search engine that handles 91,000,000 searches every day. Write 91,000,000 in scientific notation. (*Source:* Google, Inc.)

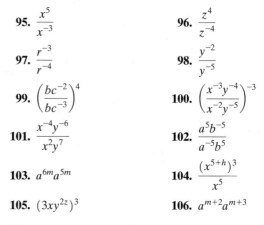

112. The approximate diameter of the Milky Way galaxy is 150,000 light years. Write this number in scientific notation. (*Source:* NASA IMAGE/POETRY Education and Public Outreach Program)

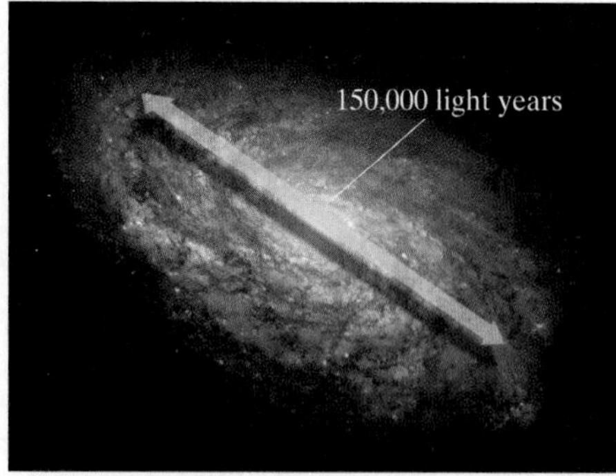

Write each number in standard form.

113. 8.67×10^5 **114.** 3.86×10^{-3}

115. 8.6×10^{-4} **116.** 8.936×10^5

117. The volume of the planet Jupiter is 1.43128×10^{15} cubic kilometers. Write this number in standard notation. (*Source:* National Space Science Data Center)

118. An angstrom is a unit of measure, equal to 1×10^{-10} meter, used for measuring wavelengths or the diameters of atoms. Write this number in standard notation. (*Source:* National Institute of Standards and Technology)

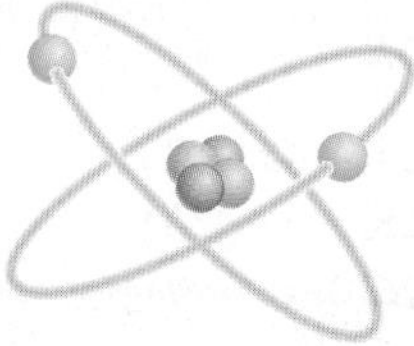

Simplify. Express each result in standard form.

119. $(8 \times 10^4)(2 \times 10^{-7})$

120. $\dfrac{8 \times 10^4}{2 \times 10^{-7}}$

(5.6) *Divide.*

121. $\dfrac{x^2 + 21x + 49}{7x^2}$

122. $\dfrac{5a^3b - 15ab^2 + 20ab}{-5ab}$

123. $(a^2 - a + 4) \div (a - 2)$

124. $(4x^2 + 20x + 7) \div (x + 5)$

125. $\dfrac{a^3 + a^2 + 2a + 6}{a - 2}$

126. $\dfrac{9b^3 - 18b^2 + 8b - 1}{3b - 2}$

127. $\dfrac{4x^4 - 4x^3 + x^2 + 4x - 3}{2x - 1}$

128. $\dfrac{-10x^2 - x^3 - 21x + 18}{x - 6}$

△ 129. The area of the rectangle below is $(15x^3 - 3x^2 + 60)$ square feet. If its length is $3x^2$ feet, find its width.

Area is $(15x^3 - 3x^2 + 60)$ sq feet

△ 130. The perimeter of the equilateral triangle below is $(21a^3b^6 + 3a - 3)$ units. Find the length of a side.

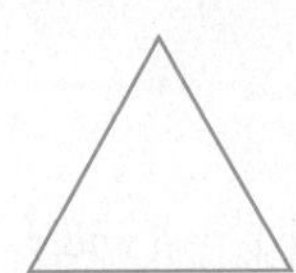

Perimeter is $(21a^3b^6 + 3a - 3)$ units

MIXED REVIEW

Evaluate.

131. $\left(-\dfrac{1}{2}\right)^3$

Simplify each expression. Write each answer using positive exponents only.

132. $(4xy^2)(x^3y^5)$

133. $\dfrac{18x^9}{27x^3}$

134. $\left(\dfrac{3a^4}{b^2}\right)^3$

135. $(2x^{-4}y^3)^{-4}$

136. $\dfrac{a^{-3}b^6}{9^{-1}a^{-5}b^{-2}}$

Perform the indicated operations and simplify.

137. $(6x + 2) + (5x - 7)$

138. $(-y^2 - 4) + (3y^2 - 6)$

139. $(8y^2 - 3y + 1) - (3y^2 + 2)$

140. $(5x^2 + 2x - 6) - (-x - 4)$

141. $4x(7x^2 + 3)$

142. $(2x + 5)(3x - 2)$

143. $(x - 3)(x^2 + 4x - 6)$

144. $(7x - 2)(4x - 9)$

Use special products to multiply.

145. $(5x + 4)^2$

146. $(6x + 3)(6x - 3)$

Divide.

147. $\dfrac{8a^4 - 2a^3 + 4a - 5}{2a^3}$

148. $\dfrac{x^2 + 2x + 10}{x + 5}$

149. $\dfrac{4x^3 + 8x^2 - 11x + 4}{2x - 3}$

Chapter 5 Test

MyMathLab® YouTube

Evaluate each expression.

1. 2^5
2. $(-3)^4$
3. -3^4
4. 4^{-3}

Simplify each exponential expression. Write the result using only positive exponents.

5. $(3x^2)(-5x^9)$
6. $\dfrac{y^7}{y^2}$
7. $\dfrac{r^{-8}}{r^{-3}}$
8. $\left(\dfrac{x^2y^3}{x^3y^{-4}}\right)^2$
9. $\dfrac{6^2x^{-4}y^{-1}}{6^3x^{-3}y^7}$

Express each number in scientific notation.

10. 563,000
11. 0.0000863

Write each number in standard form.

12. 1.5×10^{-3}

13. 6.23×10^{4}

14. Simplify. Write the answer in standard form.

$$(1.2 \times 10^{5})(3 \times 10^{-7})$$

15. a. Complete the table for the polynomial $4xy^2 + 7xyz + x^3y - 2$.

Term	Numerical Coefficient	Degree of Term
$4xy^2$		
$7xyz$		
x^3y		
-2		

b. What is the degree of the polynomial?

16. Simplify by combining like terms.

$$5x^2 + 4xy - 7x^2 + 11 + 8xy$$

Perform each indicated operation.

17. $(8x^3 + 7x^2 + 4x - 7) + (8x^3 - 7x - 6)$

18. $5x^3 + x^2 + 5x - 2 - (8x^3 - 4x^2 + x - 7)$

19. Subtract $(4x + 2)$ from the sum of $(8x^2 + 7x + 5)$ and $(x^3 - 8)$.

Multiply.

20. $(3x + 7)(x^2 + 5x + 2)$

21. $3x^2(2x^2 - 3x + 7)$

22. $(x + 7)(3x - 5)$

23. $\left(3x - \frac{1}{5}\right)\left(3x + \frac{1}{5}\right)$

24. $(4x - 2)^2$

25. $(8x + 3)^2$

26. $(x^2 - 9b)(x^2 + 9b)$

Solve.

27. The height of the Bank of China in Hong Kong is 1001 feet. Neglecting air resistance, the height of an object dropped from this building at time t seconds is given by the polynomial $-16t^2 + 1001$. Find the height of the object at the given times below.

t	0 seconds	1 second	3 seconds	5 seconds
$-16t^2 + 1001$				

28. Find the area of the top of the table. Express the area as a product, then multiply and simplify.

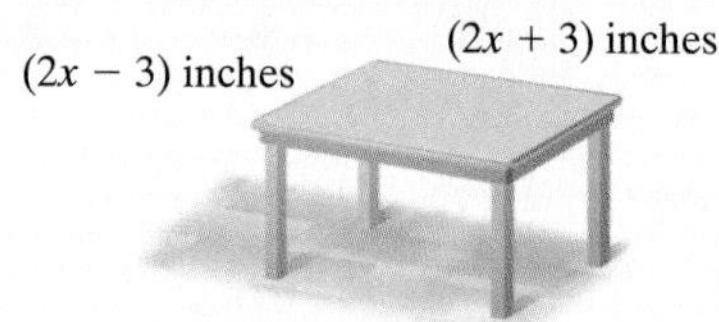

Divide.

29. $\dfrac{4x^2 + 24xy - 7x}{8xy}$

30. $(x^2 + 7x + 10) \div (x + 5)$

31. $\dfrac{27x^3 - 8}{3x + 2}$

Chapter 5 Cumulative Review

1. Tell whether each statement is true or false.

a. $8 \geq 8$ **b.** $8 \leq 8$

c. $23 \leq 0$ **d.** $23 \geq 0$

2. Find the absolute value of each number.

a. $|-7.2|$ **b.** $|0|$ **c.** $\left|-\frac{1}{2}\right|$

3. Divide. Simplify all quotients if possible.

a. $\frac{4}{5} \div \frac{5}{16}$ **b.** $\frac{7}{10} \div 14$ **c.** $\frac{3}{8} \div \frac{3}{10}$

4. Multiply. Write products in lowest terms.

a. $\frac{3}{4} \cdot \frac{7}{21}$ **b.** $\frac{1}{2} \cdot 4\frac{5}{6}$

5. Evaluate the following:

a. 3^2 **b.** 5^3 **c.** 2^4

d. 7^1 **e.** $\left(\frac{3}{7}\right)^2$

6. Evaluate $\frac{2x - 7y}{x^2}$ for $x = 5$ and $y = 1$.

7. Add.

a. $-3 + (-7)$ **b.** $-1 + (-20)$

c. $-2 + (-10)$

8. Simplify: $8 + 3(2 \cdot 6 - 1)$

9. Subtract 8 from -4.

10. Is $x = 1$ a solution of $5x^2 + 2 = x - 8$?

11. Find the reciprocal of each number.

a. 22 **b.** $\frac{3}{16}$

c. -10 **d.** $-\frac{9}{13}$

12. Subtract:

a. $7 - 40$ **b.** $-5 - (-10)$

13. Use an associative property to complete each statement.

a. $5 + (4 + 6) =$ _____

b. $(-1 \cdot 2) \cdot 5 =$ _____

14. Simplify: $\dfrac{4(-3) + (-8)}{5 + (-5)}$

15. Simplify each expression.

a. $10 + (x + 12)$

b. $-3(7x)$

16. Use the distributive property to write $-2(x + 3y - z)$ without parentheses.

17. Find each product by using the distributive property to remove parentheses.

a. $5(3x + 2)$

b. $-2(y + 0.3z - 1)$

c. $-(9x + y - 2z + 6)$

18. Simplify: $2(6x - 1) - (x - 7)$

19. Solve $x - 7 = 10$ for x.

20. Write the phrase as an algebraic expression: double a number, subtracted from the sum of a number and seven

21. Solve: $\frac{5}{2}x = 15$

22. Solve: $2x + \frac{1}{8} = x - \frac{3}{8}$

23. Solve: $4(2x - 3) + 7 = 3x + 5$

24. Solve: $10 = 5j - 2$

25. Twice a number, added to seven, is the same as three subtracted from the number. Find the number.

26. Solve: $\dfrac{7x + 5}{3} = x + 3$

27. The length of a rectangular road sign is 2 feet less than three times its width. Find the dimensions if the perimeter is 28 feet.

28. Graph $x < 5$ and write in interval notation.

29. Solve $F = \frac{9}{5}C + 32$ for C.

30. Find the slope of each line.

a. $x = -1$

b. $y = 7$

31. Graph $2 < x \leq 4$.

32. Recall that the grade of a road is its slope written as a percent. Find the grade of the road shown.

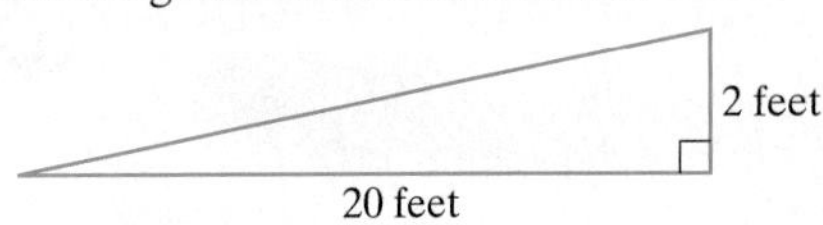

33. Complete the following ordered pair solutions for the equation $3x + y = 12$.

a. $(0, \ )$

b. $(\ , 6)$

c. $(-1, \ )$

34. Solve the system: $\begin{cases} 3x + 2y = -8 \\ 2x - 6y = -9 \end{cases}$

35. Graph the linear equation $2x + y = 5$.

36. Solve the system: $\begin{cases} x = -3y + 3 \\ 2x + 9y = 5 \end{cases}$

37. Graph $x = 2$.

38. Evaluate.

a. $(-5)^2$

b. -5^2

c. $2 \cdot 5^2$

39. Find the slope of the line $x = 5$.

40. Simplify: $\dfrac{(z^2)^3 \cdot z^7}{z^9}$

41. Graph $x + y < 7$.

42. Subtract $(5y^2 - 6) - (y^2 + 2)$.

43. Use the product rule to simplify $(2x^2)(-3x^5)$.

44. Find the value of $-x^2$ when

a. $x = 2$

b. $x = -2$

45. Add $(-2x^2 + 5x - 1)$ and $(-2x^2 + x + 3)$.

46. Multiply $(10x^2 - 3)(10x^2 + 3)$.

47. Multiply $(2x - y)^2$.

48. Multiply $(10x^2 + 3)^2$.

49. Divide $6m^2 + 2m$ by $2m$.

50. Evaluate.

a. 5^{-1}

b. 7^{-2}

CHAPTER 6 Factoring Polynomials

In Chapter 5, you learned how to multiply polynomials. This chapter deals with an operation that is the reverse process of multiplying, called *factoring*. Factoring is an important algebraic skill because this process allows us to write a sum as a product.

At the end of this chapter, we use factoring to help us solve equations other than linear equations, and in Chapter 7, we use factoring to simplify and perform arithmetic operations on rational expressions.

As shown on the circle graph below, games and music are the leading moneymakers in the digital market. Record companies' revenue from digital music channels grew 6% in a recent year to total \$4.6 billion in the United States alone. From 2004 to 2010, the digital music market increased 1000%.

In Section 6.1, Exercise 109, we explore digital music downloads further.

Global Digital Revenues

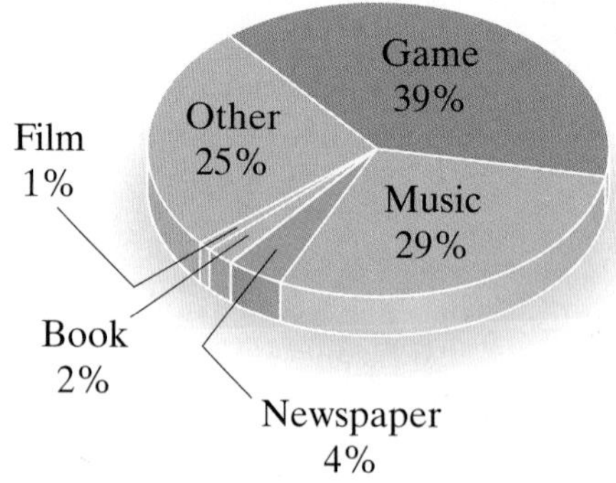

Data from NPD Group; Pollstar; Recording Industry Association of America (RIAA); U.S. Department of Labor; and Tera Consultants

6.1 The Greatest Common Factor and Factoring by Grouping 

OBJECTIVES

1. Find the Greatest Common Factor of a List of Integers.
2. Find the Greatest Common Factor of a List of Terms.
3. Factor Out the Greatest Common Factor from a Polynomial.
4. Factor a Polynomial by Grouping.

In the product $2 \cdot 3 = 6$, the numbers 2 and 3 are called **factors** of 6 and $2 \cdot 3$ is a **factored form** of 6. This is true of polynomials also. Since $(x + 2)(x + 3) = x^2 + 5x + 6$, $(x + 2)$ and $(x + 3)$ are factors of $x^2 + 5x + 6$, and $(x + 2)(x + 3)$ is a factored form of the polynomial.

a factored form of 6

$$\underbrace{2 \cdot 3}_{\text{factor}\ \ \text{factor}} = \underbrace{6}_{\text{product}}$$

a factored form of x^5

$$\underbrace{x^2 \cdot x^3}_{\text{factor}\ \ \text{factor}} = \underbrace{x^5}_{\text{product}}$$

a factored form of $x^2 + 5x + 6$

$$\underbrace{(x + 2)(x + 3)}_{\text{factor}\ \ \text{factor}} = \underbrace{x^2 + 5x + 6}_{\text{product}}$$

The process of writing a polynomial as a product is called **factoring** the polynomial.

Do you see that factoring is the reverse process of multiplying?

$$x^2 + 5x + 6 = (x + 2)(x + 3)$$

factoring (left to right); multiplying (right to left)

✓CONCEPT CHECK

Multiply: $2(x - 4)$

What do you think the result of factoring $2x - 8$ would be? Why?

OBJECTIVE

1 Finding the Greatest Common Factor of a List of Integers 

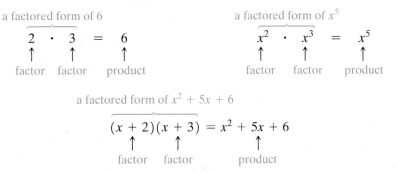

The first step in factoring a polynomial is to see whether the terms of the polynomial have a common factor. If there is one, we can write the polynomial as a product by **factoring out** the common factor. We will usually factor out the **greatest common factor (GCF).**

The GCF of a list of integers is the largest integer that is a factor of all the integers in the list. For example, the GCF of 12 and 20 is 4 because 4 is the largest integer that is a factor of both 12 and 20. With large integers, the GCF may not be found easily by inspection. When this happens, use the following steps.

Finding the GCF of a List of Integers

Step 1. Write each number as a product of prime numbers.

Step 2. Identify the common prime factors.

Step 3. The product of all common prime factors found in Step 2 is the greatest common factor. If there are no common prime factors, the greatest common factor is 1.

Recall from Section 1.3 that a prime number is a whole number other than 1 whose only factors are 1 and itself.

Answers to Concept Check:
$2x - 8$; The result would be $2(x - 4)$ because factoring is the reverse process of multiplying.

EXAMPLE 1 Find the GCF of each list of numbers.

a. 28 and 40 **b.** 55 and 21 **c.** 15, 18, and 66

Solution

a. Write each number as a product of primes.

$$28 = 2 \cdot 2 \cdot 7 = 2^2 \cdot 7$$
$$40 = 2 \cdot 2 \cdot 2 \cdot 5 = 2^3 \cdot 5$$

There are two common factors, each of which is 2, so the GCF is

$$\text{GCF} = 2 \cdot 2 = 4$$

b. $55 = 5 \cdot 11$

$21 = 3 \cdot 7$

There are no common prime factors; thus, the GCF is 1.

c. $15 = 3 \cdot 5$

$18 = 2 \cdot 3 \cdot 3 = 2 \cdot 3^2$

$66 = 2 \cdot 3 \cdot 11$

The only prime factor common to all three numbers is 3, so the GCF is

$$\text{GCF} = 3$$

□

PRACTICE

1 Find the GCF of each list of numbers.

a. 36 and 42 **b.** 35 and 44 **c.** 12, 16, and 40

OBJECTIVE

2 Finding the Greatest Common Factor of a List of Terms

The greatest common factor of a list of variables raised to powers is found in a similar way. For example, the GCF of x^2, x^3, and x^5 is x^2 because each term contains a factor of x^2 and no higher power of x is a factor of each term.

$$x^2 = x \cdot x$$
$$x^3 = x \cdot x \cdot x$$
$$x^5 = x \cdot x \cdot x \cdot x \cdot x$$

There are two common factors, each of which is x, so the GCF $= x \cdot x$ or x^2.

From this example, we see that **the GCF of a list of common variables raised to powers is the variable raised to the smallest exponent in the list.**

EXAMPLE 2 Find the GCF of each list of terms.

a. x^3, x^7, and x^5 **b.** y, y^4, and y^7

Solution

a. The GCF is x^3, since 3 is the smallest exponent to which x is raised.

b. The GCF is y^1 or y, since 1 is the smallest exponent on y. □

PRACTICE

2 Find the GCF of each list of terms.

a. y^7, y^4, and y^6 **b.** x, x^4, and x^2

In general, the **greatest common factor (GCF) of a list of terms** is the product of the GCF of the numerical coefficients and the GCF of the variable factors.

$$20x^2y^2 = 2 \cdot 2 \cdot 5 \cdot x \cdot x \cdot y \cdot y$$
$$6xy^3 = 2 \cdot 3 \cdot x \cdot y \cdot y \cdot y$$
$$\text{GCF} = 2 \cdot x \cdot y \cdot y = 2xy^2$$

Helpful Hint

Remember that the GCF of a list of terms contains the smallest exponent on each common variable.

The GCF of x^5y^6, x^2y^7, and x^3y^4 is x^2y^4. ← Smallest exponent on x; Smallest exponent on y

EXAMPLE 3 Find the GCF of each list of terms.

a. $6x^2$, $10x^3$, and $-8x$ **b.** $-18y^2$, $-63y^3$, and $27y^4$ **c.** a^3b^2, a^5b, and a^6b^2

Solution

a.
$$6x^2 = 2 \cdot 3 \cdot x^2$$
$$10x^3 = 2 \cdot 5 \cdot x^3$$
$$-8x = -1 \cdot 2 \cdot 2 \cdot 2 \cdot x^1$$
→ The GCF of x^2, x^3, and x^1 is x^1 or x.
$$\text{GCF} = 2 \cdot x^1 \quad \text{or} \quad 2x$$

b.
$$-18y^2 = -1 \cdot 2 \cdot 3 \cdot 3 \cdot y^2$$
$$-63y^3 = -1 \cdot 3 \cdot 3 \cdot 7 \cdot y^3$$
$$27y^4 = 3 \cdot 3 \cdot 3 \cdot y^4$$
→ The GCF of y^2, y^3, and y^4 is y^2.
$$\text{GCF} = 3 \cdot 3 \cdot y^2 \quad \text{or} \quad 9y^2$$

c. The GCF of a^3, a^5, and a^6 is a^3.

The GCF of b^2, b, and b^2 is b. Thus,

the GCF of a^3b^2, a^5b, and a^6b^2 is a^3b. □

PRACTICE

3 Find the GCF of each list of terms.

a. $5y^4$, $15y^2$, and $-20y^3$ **b.** $4x^2$, x^3, and $3x^8$ **c.** a^4b^2, a^3b^5, and a^2b^3

OBJECTIVE

3 Factoring Out the Greatest Common Factor

The first step in factoring a polynomial is to find the GCF of its terms. Once we do so, we can write the polynomial as a product by **factoring out** the GCF.

The polynomial $8x + 14$, for example, contains two terms: $8x$ and 14. The GCF of these terms is 2. We factor out 2 from each term by writing each term as a product of 2 and the term's remaining factors.

$$8x + 14 = 2 \cdot 4x + 2 \cdot 7$$

Using the distributive property, we can write

$$8x + 14 = 2 \cdot 4x + 2 \cdot 7$$
$$= 2(4x + 7)$$

Thus, a factored form of $8x + 14$ is $2(4x + 7)$. We can check by multiplying:

$$2(4x + 7) = 2 \cdot 4x + 2 \cdot 7 = 8x + 14.$$

Helpful Hint

A factored form of $8x + 14$ is *not*

$$2 \cdot 4x + 2 \cdot 7$$

Although the *terms* have been factored (written as a product), the *polynomial* $8x + 14$ has not been factored (written as a product).

A factored form of $8x + 14$ is the *product* $2(4x + 7)$.

✓**CONCEPT CHECK**

Which of the following is/are factored form(s) of $7t + 21$?

a. 7 b. $7 \cdot t + 7 \cdot 3$ c. $7(t + 3)$ d. $7(t + 21)$

EXAMPLE 4 Factor each polynomial by factoring out the GCF.

a. $6t + 18$ **b.** $y^5 - y^7$

Solution

a. The GCF of terms $6t$ and 18 is 6.

$$6t + 18 = 6 \cdot t + 6 \cdot 3$$
$$= 6(t + 3) \quad \text{Apply the distributive property.}$$

Our work can be checked by multiplying 6 and $(t + 3)$.

$$6(t + 3) = 6 \cdot t + 6 \cdot 3 = 6t + 18, \text{ the original polynomial.}$$

b. The GCF of y^5 and y^7 is y^5. Thus,

$$y^5 - y^7 = y^5(1) - y^5(y^2)$$
$$= y^5(1 - y^2)$$

Helpful Hint
Don't forget the 1.

PRACTICE 4 Factor each polynomial by factoring out the GCF.

a. $4t + 12$ **b.** $y^8 + y^4$

EXAMPLE 5 Factor: $-9a^5 + 18a^2 - 3a$

Solution

a. $-9a^5 + 18a^2 - 3a = (3a)(-3a^4) + (3a)(6a) + (3a)(-1)$
$= 3a(-3a^4 + 6a - 1)$

Helpful Hint
Don't forget the -1.

PRACTICE 5 Factor $-8b^6 + 16b^4 - 8b^2$.

In Example 5, we could have chosen to factor out a $-3a$ instead of $3a$. If we factor out a $-3a$, we have

$$-9a^5 + 18a^2 - 3a = (-3a)(3a^4) + (-3a)(-6a) + (-3a)(1)$$
$$= -3a(3a^4 - 6a + 1)$$

Helpful Hint
Notice the changes in signs when factoring out $-3a$.

EXAMPLES Factor.

6. $6a^4 - 12a = 6a(a^3 - 2)$

7. $\frac{3}{7}x^4 + \frac{1}{7}x^3 - \frac{5}{7}x^2 = \frac{1}{7}x^2(3x^2 + x - 5)$

8. $15p^2q^4 + 20p^3q^5 + 5p^3q^3 = 5p^2q^3(3q + 4pq^2 + p)$

PRACTICES 6–8 Factor.

6. $5x^4 - 20x$ **7.** $\frac{5}{9}z^5 + \frac{1}{9}z^4 - \frac{2}{9}z^3$ **8.** $8a^2b^4 - 20a^3b^3 + 12ab^3$

Answer to Concept Check: c

EXAMPLE 9 Factor: $5(x + 3) + y(x + 3)$

Solution The binomial $(x + 3)$ is the greatest common factor. Use the distributive property to factor out $(x + 3)$.

$$5(x + 3) + y(x + 3) = (x + 3)(5 + y)$$ □

PRACTICE 9 Factor $8(y - 2) + x(y - 2)$.

EXAMPLE 10 Factor: $3m^2n(a + b) - (a + b)$

Solution The greatest common factor is $(a + b)$.

$$3m^2n(a + b) - 1(a + b) = (a + b)(3m^2n - 1)$$ □

PRACTICE 10 Factor $7xy^3(p + q) - (p + q)$

OBJECTIVE 4

Factoring by Grouping

Once the GCF is factored out, we can often continue to factor the polynomial, using a variety of techniques. We discuss here a technique for factoring polynomials called **factoring by grouping.**

EXAMPLE 11 Factor $xy + 2x + 3y + 6$ by grouping. Check by multiplying.

Solution The GCF of the first two terms is x, and the GCF of the last two terms is 3.

$$xy + 2x + 3y + 6 = (xy + 2x) + (3y + 6) \quad \text{Group terms.}$$
$$= x(y + 2) + 3(y + 2) \quad \text{Factor out GCF from each grouping.}$$

Helpful Hint
Notice that this form, $x(y + 2) + 3(y + 2)$, is *not* a factored form of the original polynomial. It is a sum, not a product.

Next we factor out the common binomial factor, $(y + 2)$.

$$x(y + 2) + 3(y + 2) = (y + 2)(x + 3)$$

Now the result is a factored form because it is a product. We were able to write the polynomial as a product because of the common binomial factor, $(y + 2)$, that appeared. If this does not happen, try rearranging the terms of the original polynomial.

Check: Multiply $(y + 2)$ by $(x + 3)$.

$$(y + 2)(x + 3) = xy + 2x + 3y + 6,$$

the original polynomial.
Thus, the factored form of $xy + 2x + 3y + 6$ is the product $(y + 2)(x + 3)$. □

PRACTICE 11 Factor $xy + 3y + 4x + 12$ by grouping. Check by multiplying.

You may want to try these steps when factoring by grouping.

To Factor a Four-Term Polynomial by Grouping

Step 1. Group the terms in two groups of two terms so that each group has a common factor.

Step 2. Factor out the GCF from each group.

Step 3. If there is now a common binomial factor in the groups, factor it out.

Step 4. If not, rearrange the terms and try these steps again.

EXAMPLES Factor by grouping.

12. $15x^3 - 10x^2 + 6x - 4$

$= (15x^3 - 10x^2) + (6x - 4)$ Group the terms.

$= 5x^2(3x - 2) + 2(3x - 2)$ Factor each group.

$= (3x - 2)(5x^2 + 2)$ Factor out the common factor, $(3x - 2)$.

13. $3x^2 + 4xy - 3x - 4y$

$= (3x^2 + 4xy) + (-3x - 4y)$

$= x(3x + 4y) - 1(3x + 4y)$ Factor each group. A -1 is factored from the second pair of terms so that there is a common factor, $(3x + 4y)$.

$= (3x + 4y)(x - 1)$ Factor out the common factor, $(3x + 4y)$.

14. $2a^2 + 5ab + 2a + 5b$

$= (2a^2 + 5ab) + (2a + 5b)$ Factor each group. An understood 1 is written before $(2a + 5b)$ to help remember that $(2a + 5b)$ is $1(2a + 5b)$.

$= a(2a + 5b) + 1(2a + 5b)$

$= (2a + 5b)(a + 1)$ Factor out the common factor, $(2a + 5b)$. □

Helpful Hint
Notice the factor of 1 is written when $(2a + 5b)$ is factored out.

PRACTICES
12–14

12. Factor $40x^3 - 24x^2 + 15x - 9$ by grouping.
13. Factor $2xy + 3y^2 - 2x - 3y$ by grouping.
14. Factor $7a^3 + 5a^2 + 7a + 5$ by grouping.

EXAMPLES Factor by grouping.

15. $3xy + 2 - 3x - 2y$

Notice that the first two terms have no common factor other than 1. However, if we rearrange these terms, a grouping emerges that does lead to a common factor.

$3xy + 2 - 3x - 2y$

$= (3xy - 3x) + (-2y + 2)$

$= 3x(y - 1) - 2(y - 1)$ Factor -2 from the second group so that there is a common factor $(y - 1)$.

$= (y - 1)(3x - 2)$ Factor out the common factor, $(y - 1)$.

16. $5x - 10 + x^3 - x^2 = 5(x - 2) + x^2(x - 1)$

There is no common binomial factor that can now be factored out. No matter how we rearrange the terms, no grouping will lead to a common factor. Thus, this polynomial is not factorable by grouping. □

PRACTICES
15–16

15. Factor $4xy + 15 - 12x - 5y$ by grouping.
16. Factor $9y - 18 + y^3 - 4y^2$ by grouping.

Helpful Hint
One more reminder: When **factoring** a polynomial, make sure the polynomial is written as a **product.** For example, it is true that

$$3x^2 + 4xy - 3x - 4y = \underbrace{x(3x + 4y) - 1(3x + 4y)}_{\text{but is not a factored form}},$$

since it is a **sum (difference),** not a **product.** A factored form of $3x^2 + 4xy - 3x - 4y$ is the product $(3x + 4y)(x - 1)$.

Factoring out a greatest common factor first makes factoring by any method easier, as we see in the next example.

EXAMPLE 17 Factor: $4ax - 4ab - 2bx + 2b^2$

Solution First, factor out the common factor 2 from all four terms.

$4ax - 4ab - 2bx + 2b^2$

$= 2(2ax - 2ab - bx + b^2)$ Factor out 2 from all four terms.

$= 2[2a(x - b) - b(x - b)]$ Factor each pair of terms. A "$-b$" is factored from the second pair so that there is a common factor, $x - b$.

$= 2(x - b)(2a - b)$ Factor out the common binomial. □

PRACTICE 17 Factor $3xy - 3ay - 6ax + 6a^2$.

> **Helpful Hint**
> Throughout this chapter, we will be factoring polynomials. Even when the instructions do not so state, it is always a good idea to check your answers by multiplying.

Vocabulary, Readiness & Video Check

Use the choices below to fill in each blank. Some choices may be used more than once and some may not be used at all.

greatest common factor | factors | factoring | true | false | least | greatest

1. Since $5 \cdot 4 = 20$, the numbers 5 and 4 are called ________________ of 20.
2. The ________________ of a list of integers is the largest integer that is a factor of all the integers in the list.
3. The greatest common factor of a list of common variables raised to powers is the variable raised to the ________________ exponent in the list.
4. The process of writing a polynomial as a product is called ________________.
5. True or false: A factored form of $7x + 21 + xy + 3y$ is $7(x + 3) + y(x + 3)$. ________________
6. True or false: A factored form of $3x^3 + 6x + x^2 + 2$ is $3x(x^2 + 2)$. ________________

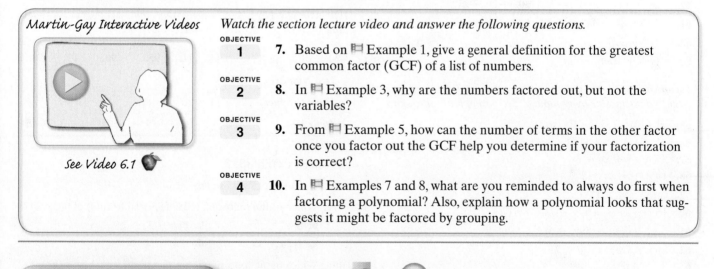

Martin-Gay Interactive Videos

See Video 6.1

Watch the section lecture video and answer the following questions.

OBJECTIVE 1

7. Based on Example 1, give a general definition for the greatest common factor (GCF) of a list of numbers.

OBJECTIVE 2

8. In Example 3, why are the numbers factored out, but not the variables?

OBJECTIVE 3

9. From Example 5, how can the number of terms in the other factor once you factor out the GCF help you determine if your factorization is correct?

OBJECTIVE 4

10. In Examples 7 and 8, what are you reminded to always do first when factoring a polynomial? Also, explain how a polynomial looks that suggests it might be factored by grouping.

6.1 Exercise Set

MyMathLab®

Find the GCF for each list. See Examples 1 through 3.

1. 32, 36
2. 36, 90
3. 18, 42, 84
4. 30, 75, 135
5. 24, 14, 21
6. 15, 25, 27
7. y^2, y^4, y^7
8. x^3, x^2, x^5
9. z^7, z^9, z^{11}
10. y^8, y^{10}, y^{12}
11. $x^{10}y^2, xy^2, x^3y^3$
12. p^7q, p^8q^2, p^9q^3
13. $14x, 21$
14. $20y, 15$
15. $12y^4, 20y^3$
16. $32x^5, 18x^2$
17. $-10x^2, 15x^3$
18. $-21x^3, 14x$
19. $12x^3, -6x^4, 3x^5$
20. $15y^2, 5y^7, -20y^3$

21. $-18x^2y, 9x^3y^3, 36x^3y$
22. $7x^3y^3, -21x^2y^2, 14xy^4$
23. $20a^6b^2c^8, 50a^7b$
24. $40x^7y^2z, 64x^9y$

Factor out the GCF from each polynomial. See Examples 4 through 10.

25. $3a + 6$
26. $18a + 12$
27. $30x - 15$
28. $42x - 7$
29. $x^3 + 5x^2$
30. $y^5 + 6y^4$
31. $6y^4 + 2y^3$
32. $5x^2 + 10x^6$
33. $4x - 8y + 4$
34. $7x + 21y - 7$
35. $6x^3 - 9x^2 + 12x$
36. $12x^3 + 16x^2 - 8x$
37. $a^7b^6 - a^3b^2 + a^2b^5 - a^2b^2$
38. $x^9y^6 + x^3y^5 - x^4y^3 + x^3y^3$
39. $8x^5 + 16x^4 - 20x^3 + 12$
40. $9y^6 - 27y^4 + 18y^2 + 6$
41. $\frac{1}{3}x^4 + \frac{2}{3}x^3 - \frac{4}{3}x^5 + \frac{1}{3}x$
42. $\frac{2}{5}y^7 - \frac{4}{5}y^5 + \frac{3}{5}y^2 - \frac{2}{5}y$
43. $y(x^2 + 2) + 3(x^2 + 2)$
44. $x(y^2 + 1) - 3(y^2 + 1)$
45. $z(y + 4) - 3(y + 4)$
46. $8(x + 2) - y(x + 2)$
47. $r(z^2 - 6) + (z^2 - 6)$
48. $q(b^3 - 5) + (b^3 - 5)$

Factor a negative number or a GCF with a negative coefficient from each polynomial. See Example 5.

49. $-2x - 14$
50. $-7y - 21$
51. $-2x^5 + x^7$
52. $-5y^3 + y^6$
53. $-6a^4 + 9a^3 - 3a^2$
54. $-5m^6 + 10m^5 - 5m^3$

Factor each four-term polynomial by grouping. If this is not possible, write "not factorable by grouping." See Examples 11 through 17.

55. $x^3 + 2x^2 + 5x + 10$
56. $x^3 + 4x^2 + 3x + 12$
57. $5x + 15 + xy + 3y$
58. $xy + y + 2x + 2$
59. $6x^3 - 4x^2 + 15x - 10$
60. $16x^3 - 28x^2 + 12x - 21$
61. $5m^3 + 6mn + 5m^2 + 6n$
62. $8w^2 + 7wv + 8w + 7v$
63. $2y - 8 + xy - 4x$
64. $6x - 42 + xy - 7y$
65. $2x^3 - x^2 + 8x - 4$
66. $2x^3 - x^2 - 10x + 5$
67. $3x - 3 + x^3 - 4x^2$
68. $7x - 21 + x^3 - 2x^2$
69. $4x^2 - 8xy - 3x + 6y$
70. $5xy - 15x - 6y + 18$
71. $5q^2 - 4pq - 5q + 4p$
72. $6m^2 - 5mn - 6m + 5n$
73. $2x^4 + 5x^3 + 2x^2 + 5x$
74. $4y^4 + y^2 + 20y^3 + 5y$
75. $12x^2y - 42x^2 - 4y + 14$
76. $90 + 15y^2 - 18x - 3xy^2$

MIXED PRACTICE

Factor. See Examples 4 through 17.

77. $32xy - 18x^2$
78. $10xy - 15x^2$
79. $y(x + 2) - 3(x + 2)$
80. $z(y - 4) + 3(y - 4)$
81. $14x^3y + 7x^2y - 7xy$
82. $5x^3y - 15x^2y + 10xy$
83. $28x^3 - 7x^2 + 12x - 3$
84. $15x^3 + 5x^2 - 6x - 2$
85. $-40x^8y^6 - 16x^9y^5$
86. $-21x^3y - 49x^2y^2$
87. $6a^2 + 9ab^2 + 6ab + 9b^3$
88. $16x^2 + 4xy^2 + 8xy + 2y^3$

REVIEW AND PREVIEW

Multiply. See Section 5.3.

89. $(x + 2)(x + 5)$
90. $(y + 3)(y + 6)$
91. $(b + 1)(b - 4)$
92. $(x - 5)(x + 10)$

Fill in the chart by finding two numbers that have the given product and sum. The first column is filled in for you.

		93.	94.	95.	96.	97.	98.
Two Numbers	4, 7						
Their Product	28	12	20	8	16	−10	−24
Their Sum	11	8	9	−9	−10	3	−5

CONCEPT EXTENSIONS

See the Concept Checks in this section.

99. Which of the following is/are factored form(s) of $8a - 24$?
 a. $8 \cdot a - 24$ **b.** $8(a - 3)$
 c. $4(2a - 12)$ **d.** $8 \cdot a - 2 \cdot 12$

100. Which of the following is/are factored form(s) of $-2x + 14$?
 a. $-2(x + 7)$ **b.** $-2 \cdot x + 14$
 c. $-2(x - 14)$ **d.** $-2(x - 7)$

Determine whether the following expressions are factored.

101. $(a + 6)(a + 2)$
102. $(x + 5)(x + y)$
103. $5(2y + z) - b(2y + z)$
104. $3x(a + 2b) + 2(a + 2b)$

105. Construct a binomial whose greatest common factor is $5a^3$. (*Hint:* Multiply $5a^3$ by a binomial whose terms contain no common factor other than 1. $5a^3(\square + \square)$.)

106. Construct a trinomial whose greatest common factor is $2x^2$. See the hint for Exercise 105.

107. Explain how you can tell whether a polynomial is written in factored form.

108. Construct a four-term polynomial that can be factored by grouping.

109. The number (in millions) of single digital downloads annually in the United States each year during 2004–2010 can be modeled by the polynomial $-20x^2 + 300x + 120$, where x is the number of years since 2004. (*Source:* Recording Industry Association of America)

a. Find the number of single digital downloads in 2010. To do so, let $x = 6$ and evaluate $-20x^2 + 300x + 120$.

b. Use this expression to predict the number of single digital downloads in 2014.

c. Factor the polynomial $-20x^2 + 300x + 120$ by factoring -20 from each term.

110. The number (in thousands) of students who graduated from U.S. public high schools each year during 2000–2009 can be modeled by $0.8x^2 + 44x + 2552$, where x is the number of years since 2000. (*Source:* Digest of Education Statistics)

a. Find the number of students who graduated from U.S. high schools in 2005. To do so, let $x = 5$ and evaluate $0.8x^2 + 44x + 2552$.

b. Use this expression to predict the number of students who will graduate from U.S. high schools in 2015.

c. Factor the polynomial $0.8x^2 + 44x + 2552$ by factoring out 0.8 from each term.

Write an expression for the area of each shaded region. Then write the expression as a factored polynomial.

△ **111.**

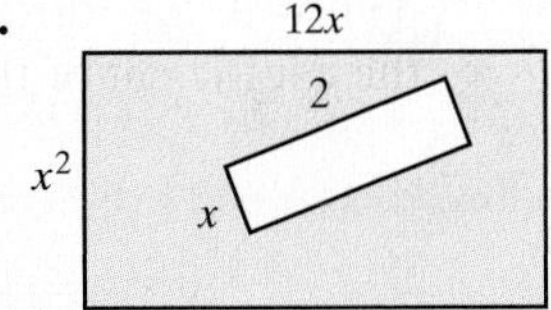

△ **112.**

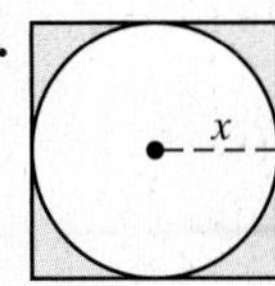

Write an expression for the length of each rectangle. (**Hint:** *Factor the area binomial and recall that Area = width · length.*)

△ **113.**

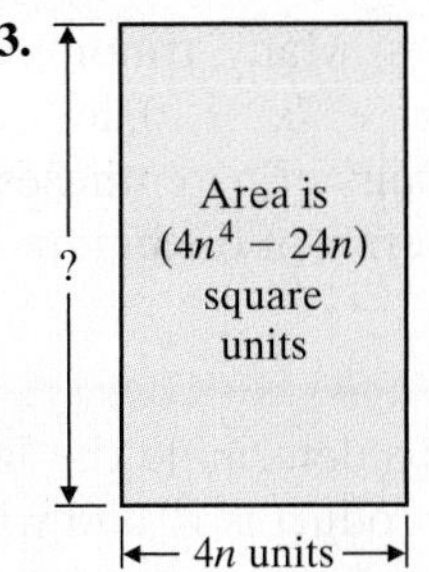

△ **114.**

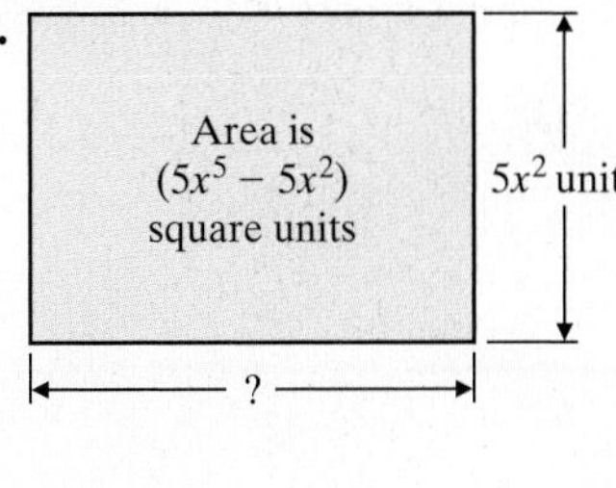

Factor each polynomial by grouping.

115. $x^{2n} + 2x^n + 3x^n + 6$
(**Hint:** Don't forget that $x^{2n} = x^n \cdot x^n$.)

116. $x^{2n} + 6x^n + 10x^n + 60$

117. $3x^{2n} + 21x^n - 5x^n - 35$

118. $12x^{2n} - 10x^n - 30x^n + 25$

6.2 Factoring Trinomials of the Form $x^2 + bx + c$

OBJECTIVES

1 Factor Trinomials of the Form $x^2 + bx + c$.

2 Factor Out the Greatest Common Factor and Then Factor a Trinomial of the Form $x^2 + bx + c$.

OBJECTIVE

1 Factoring Trinomials of the Form $x^2 + bx + c$

In this section, we factor trinomials of the form $x^2 + bx + c$, such as

$$x^2 + 4x + 3, \quad x^2 - 8x + 15, \quad x^2 + 4x - 12, \quad r^2 - r - 42$$

Notice that for these trinomials, the coefficient of the squared variable is 1.

Recall that factoring means to write as a product and that factoring and multiplying are reverse processes. Using the FOIL method of multiplying binomials, we have that

$$\begin{aligned}(x + 3)(x + 1) &= \overset{F}{x^2} + \overset{O}{1x} + \overset{I}{3x} + \overset{L}{3} \\ &= x^2 + 4x + 3\end{aligned}$$

Thus, a factored form of $x^2 + 4x + 3$ is $(x + 3)(x + 1)$.

Notice that the product of the first terms of the binomials is $x \cdot x = x^2$, the first term of the trinomial. Also, the product of the last two terms of the binomials is $3 \cdot 1 = 3$, the third term of the trinomial. The sum of these same terms is $3 + 1 = 4$, the coefficient of the middle term, x, of the trinomial.

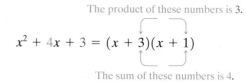

The product of these numbers is 3.

$$x^2 + 4x + 3 = (x + 3)(x + 1)$$

The sum of these numbers is 4.

Many trinomials, such as the one above, factor into two binomials. To factor $x^2 + 7x + 10$, let's assume that it factors into two binomials and begin by writing two pairs of parentheses. The first term of the trinomial is x^2, so we use x and x as the first terms of the binomial factors.

$$x^2 + 7x + 10 = (x + \square)(x + \square)$$

To determine the last term of each binomial factor, we look for two integers whose product is 10 and whose sum is 7. Since our numbers must have a positive product and a positive sum, we list pairs of positive integer factors of 10 only.

Positive Factors of 10	*Sum of Factors*
1, 10	$1 + 10 = 11$
2, 5	$2 + 5 = 7$

The correct pair of numbers is 2 and 5 because their product is 10 and their sum is 7. Now we can fill in the last terms of the binomial factors.

$$x^2 + 7x + 10 = (x + 2)(x + 5)$$

Check: To see if we have factored correctly, multiply.

$$\begin{aligned}(x + 2)(x + 5) &= x^2 + 5x + 2x + 10 \\ &= x^2 + 7x + 10 \quad \text{Combine like terms.}\end{aligned}$$

Helpful Hint

Since multiplication is commutative, the factored form of $x^2 + 7x + 10$ can be written as either $(x + 2)(x + 5)$ or $(x + 5)(x + 2)$.

Factoring a Trinomial of the Form $x^2 + bx + c$

The factored form of $x^2 + bx + c$ is

The product of these numbers is c.

$$x^2 + bx + c = (x + \square)(x + \square)$$

The sum of these numbers is b.

EXAMPLE 1 Factor: $x^2 + 7x + 12$

Solution We begin by writing the first terms of the binomial factors.

$$(x + \square)(x + \square)$$

Next we look for two numbers whose product is 12 and whose sum is 7. Since our numbers must have a positive product and a positive sum, we look at pairs of positive factors of 12 only.

Positive Factors of 12	*Sum of Factors*
1, 12	13
2, 6	8
3, 4	7

Correct sum, so the numbers are 3 and 4.

Thus, $x^2 + 7x + 12 = (x + 3)(x + 4)$

Check: $(x + 3)(x + 4) = x^2 + 4x + 3x + 12 = x^2 + 7x + 12.$ □

PRACTICE 1 Factor $x^2 + 5x + 6$.

EXAMPLE 2 Factor: $x^2 - 12x + 35$

Solution Again, we begin by writing the first terms of the binomials.

$$(x + \square)(x + \square)$$

Now we look for two numbers whose product is 35 and whose sum is -12. Since our numbers must have a positive product and a negative sum, we look at pairs of negative factors of 35 only.

Negative Factors of 35	*Sum of Factors*
−1, −35	−36
−5, −7	−12

Correct sum, so the numbers are −5 and −7.

Thus, $x^2 - 12x + 35 = (x - 5)(x - 7)$

Check: To check, multiply $(x - 5)(x - 7)$. □

PRACTICE 2 Factor $x^2 - 17x + 70$.

EXAMPLE 3 Factor: $x^2 + 4x - 12$

Solution $x^2 + 4x - 12 = (x + \square)(x + \square)$

We look for two numbers whose product is -12 and whose sum is 4. Since our numbers must have a negative product, we look at pairs of factors with opposite signs.

Factors of −12	*Sum of Factors*
−1, 12	11
1, −12	−11
−2, 6	4
2, −6	−4
−3, 4	1
3, −4	−1

Correct sum, so the numbers are −2 and 6.

Thus, $x^2 + 4x - 12 = (x - 2)(x + 6)$ □

PRACTICE 3 Factor $x^2 + 5x - 14$.

EXAMPLE 4 Factor: $r^2 - r - 42$

Solution Because the variable in this trinomial is r, the first term of each binomial factor is r.

$$r^2 - r - 42 = (r + \square)(r + \square)$$

Now we look for two numbers whose product is -42 and whose sum is -1, the numerical coefficient of r. The numbers are 6 and -7. Therefore,

$$r^2 - r - 42 = (r + 6)(r - 7)$$ □

PRACTICE 4 Factor $p^2 - 2p - 63$.

EXAMPLE 5 Factor: $a^2 + 2a + 10$

Solution Look for two numbers whose product is 10 and whose sum is 2. Neither 1 and 10 nor 2 and 5 give the required sum, 2. We conclude that $a^2 + 2a + 10$ is not factorable with integers. A polynomial such as $a^2 + 2a + 10$ is called a **prime polynomial.** □

PRACTICE 5 Factor $b^2 + 5b + 1$.

EXAMPLE 6 Factor: $x^2 + 7xy + 6y^2$

Solution

$$x^2 + 7xy + 6y^2 = (x + \square)(x + \square)$$

Recall that the middle term $7xy$ is the same as $7yx$. Thus, we can see that $7y$ is the "coefficient" of x. We then look for two terms whose product is $6y^2$ and whose sum is $7y$. The terms are $6y$ and $1y$ or $6y$ and y because $6y \cdot y = 6y^2$ and $6y + y = 7y$. Therefore,

$$x^2 + 7xy + 6y^2 = (x + 6y)(x + y)$$ □

PRACTICE 6 Factor $x^2 + 7xy + 12y^2$.

EXAMPLE 7 Factor: $x^4 + 5x^2 + 6$

Solution As usual, we begin by writing the first terms of the binomials. Since the greatest power of x in this polynomial is x^4, we write

$$(x^2 + \square)(x^2 + \square) \quad \text{since } x^2 \cdot x^2 = x^4$$

Now we look for two factors of 6 whose sum is 5. The numbers are 2 and 3. Thus,

$$x^4 + 5x^2 + 6 = (x^2 + 2)(x^2 + 3)$$ □

PRACTICE 7 Factor $x^4 + 13x^2 + 12$.

If the terms of a polynomial are not written in descending powers of the variable, you may want to do so before factoring.

EXAMPLE 8 Factor: $40 - 13t + t^2$

Solution First, we rearrange terms so that the trinomial is written in descending powers of t.

$$40 - 13t + t^2 = t^2 - 13t + 40$$

Next, try to factor.

$$t^2 - 13t + 40 = (t + \square)(t + \square)$$

Now we look for two factors of 40 whose sum is −13. The numbers are −8 and −5. Thus,

$$t^2 - 13t + 40 = (t - 8)(t - 5)$$ □

PRACTICE 8 Factor $48 - 14x + x^2$.

The following sign patterns may be useful when factoring trinomials.

> **Helpful Hint**
>
> A positive constant in a trinomial tells us to look for two numbers with the same sign. The sign of the coefficient of the middle term tells us whether the signs are both positive or both negative.
>
> both positive ↓, same sign ↙
>
> $$x^2 + 10x + 16 = (x + 2)(x + 8)$$
>
> both negative ↓, same sign ↙
>
> $$x^2 - 10x + 16 = (x - 2)(x - 8)$$
>
> A negative constant in a trinomial tells us to look for two numbers with opposite signs.
>
> opposite signs ↓
>
> $$x^2 + 6x - 16 = (x + 8)(x - 2)$$
>
> opposite signs ↓
>
> $$x^2 - 6x - 16 = (x - 8)(x + 2)$$

OBJECTIVE 2

Factoring Out the Greatest Common Factor 

Remember that the first step in factoring any polynomial is to factor out the greatest common factor (if there is one other than 1 or −1).

EXAMPLE 9 Factor: $3m^2 - 24m - 60$

Solution First we factor out the greatest common factor, 3, from each term.

$$3m^2 - 24m - 60 = 3(m^2 - 8m - 20)$$

Now we factor $m^2 - 8m - 20$ by looking for two factors of −20 whose sum is −8. The factors are −10 and 2. Therefore, the complete factored form is

$$3m^2 - 24m - 60 = 3(m + 2)(m - 10)$$ □

> **Helpful Hint**
>
> Remember to write the common factor 3 as part of the factored form.

PRACTICE 9 Factor $4x^2 - 24x + 36$.

EXAMPLE 10 Factor: $2x^4 - 26x^3 + 84x^2$

Solution

$$2x^4 - 26x^3 + 84x^2 = 2x^2(x^2 - 13x + 42) \quad \text{Factor out common factor, } 2x^2.$$

$$= 2x^2(x - 6)(x - 7) \quad \text{Factor } x^2 - 13x + 42.$$ □

PRACTICE 10 Factor $3y^4 - 18y^3 - 21y^2$.

Vocabulary, Readiness & Video Check

Fill in each blank with "true or false."

1. To factor $x^2 + 7x + 6$, we look for two numbers whose product is 6 and whose sum is 7. ______
2. We can write the factorization $(y + 2)(y + 4)$ also as $(y + 4)(y + 2)$. ______
3. The factorization $(4x - 12)(x - 5)$ is completely factored. ______
4. The factorization $(x + 2y)(x + y)$ may also be written as $(x + 2y)^2$. ______

Complete each factored form.

5. $x^2 + 9x + 20 = (x + 4)(x \quad)$
6. $x^2 + 12x + 35 = (x + 5)(x \quad)$
7. $x^2 - 7x + 12 = (x - 4)(x \quad)$
8. $x^2 - 13x + 22 = (x - 2)(x \quad)$
9. $x^2 + 4x + 4 = (x + 2)(x \quad)$
10. $x^2 + 10x + 24 = (x + 6)(x \quad)$

Martin-Gay Interactive Videos

See Video 6.2

Watch the section lecture video and answer the following questions.

OBJECTIVE 1

11. In  Example 2, why are only negative factors of 15 considered?

OBJECTIVE 2

12. In Example 5, we know we need a positive and a negative factor of -10. How do we determine which factor is negative?

6.2 Exercise Set

MyMathLab®

Factor each trinomial completely. If a polynomial can't be factored, write "prime." See Examples 1 through 8.

1. $x^2 + 7x + 6$
2. $x^2 + 6x + 8$
3. $y^2 - 10y + 9$
4. $y^2 - 12y + 11$
5. $x^2 - 6x + 9$
6. $x^2 - 10x + 25$
7. $x^2 - 3x - 18$
8. $x^2 - x - 30$
9. $x^2 + 3x - 70$
10. $x^2 + 4x - 32$
11. $x^2 + 5x + 2$
12. $x^2 - 7x + 5$
13. $x^2 + 8xy + 15y^2$
14. $x^2 + 6xy + 8y^2$
15. $a^4 - 2a^2 - 15$
16. $y^4 - 3y^2 - 70$
17. $13 + 14m + m^2$
18. $17 + 18n + n^2$
19. $10t - 24 + t^2$
20. $6q - 27 + q^2$
21. $a^2 - 10ab + 16b^2$
22. $a^2 - 9ab + 18b^2$

MIXED PRACTICE

Factor each trinomial completely. Some of these trinomials contain a greatest common factor (other than 1). Don't forget to factor out the GCF first. See Examples 1 through 10.

23. $2z^2 + 20z + 32$
24. $3x^2 + 30x + 63$
25. $2x^3 - 18x^2 + 40x$
26. $3x^3 - 12x^2 - 36x$
27. $x^2 - 3xy - 4y^2$
28. $x^2 - 4xy - 77y^2$
29. $x^2 + 15x + 36$
30. $x^2 + 19x + 60$
31. $x^2 - x - 2$
32. $x^2 - 5x - 14$
33. $r^2 - 16r + 48$
34. $r^2 - 10r + 21$
35. $x^2 + xy - 2y^2$

36. $x^2 - xy - 6y^2$
37. $3x^2 + 9x - 30$
38. $4x^2 - 4x - 48$
39. $3x^2 - 60x + 108$
40. $2x^2 - 24x + 70$
41. $x^2 - 18x - 144$
42. $x^2 + x - 42$
43. $r^2 - 3r + 6$
44. $x^2 + 4x - 10$
45. $x^2 - 8x + 15$
46. $x^2 - 9x + 14$
47. $6x^3 + 54x^2 + 120x$
48. $3x^3 + 3x^2 - 126x$
49. $4x^2y + 4xy - 12y$
50. $3x^2y - 9xy + 45y$
51. $x^2 - 4x - 21$
52. $x^2 - 4x - 32$
53. $x^2 + 7xy + 10y^2$
54. $x^2 - 2xy - 15y^2$
55. $64 + 24t + 2t^2$
56. $50 + 20t + 2t^2$
57. $x^3 - 2x^2 - 24x$
58. $x^3 - 3x^2 - 28x$
59. $2t^5 - 14t^4 + 24t^3$
60. $3x^6 + 30x^5 + 72x^4$
61. $5x^3y - 25x^2y^2 - 120xy^3$
62. $7a^3b - 35a^2b^2 + 42ab^3$
63. $162 - 45m + 3m^2$
64. $48 - 20n + 2n^2$
65. $-x^2 + 12x - 11$ (Factor out -1 first.)
66. $-x^2 + 8x - 7$ (Factor out -1 first.)
67. $\frac{1}{2}y^2 - \frac{9}{2}y - 11$ (Factor out $\frac{1}{2}$ first.)
68. $\frac{1}{3}y^2 - \frac{5}{3}y - 8$ (Factor out $\frac{1}{3}$ first.)
69. $x^3y^2 + x^2y - 20x$
70. $a^2b^3 + ab^2 - 30b$

REVIEW AND PREVIEW

Multiply. See Section 5.4.

71. $(2x + 1)(x + 5)$
72. $(3x + 2)(x + 4)$
73. $(5y - 4)(3y - 1)$
74. $(4z - 7)(7z - 1)$
75. $(a + 3b)(9a - 4b)$
76. $(y - 5x)(6y + 5x)$

CONCEPT EXTENSIONS

77. Write a polynomial that factors as $(x - 3)(x + 8)$.
78. To factor $x^2 + 13x + 42$, think of two numbers whose ______ is 42 and whose ______ is 13.

Complete each sentence in your own words.

79. If $x^2 + bx + c$ is factorable and c is negative, then the signs of the last-term factors of the binomials are opposite because…
80. If $x^2 + bx + c$ is factorable and c is positive, then the signs of the last-term factors of the binomials are the same because…

Remember that perimeter means distance around. Write the perimeter of each rectangle as a simplified polynomial. Then factor the polynomial.

81.

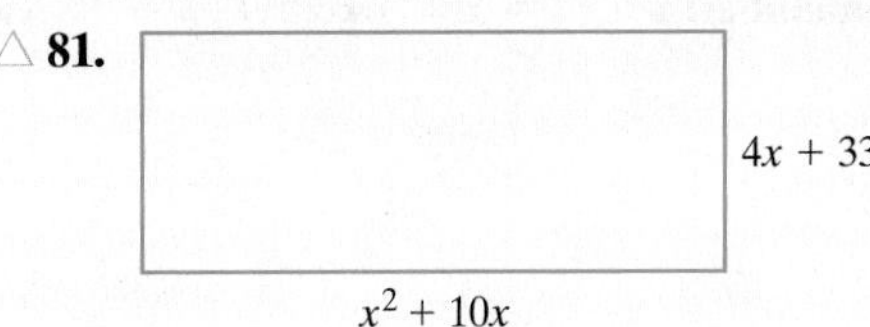

82.

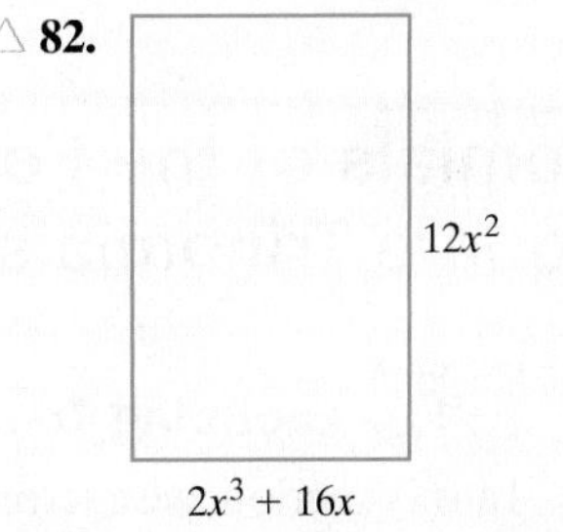

83. An object is thrown upward from the top of an 80-foot building with an initial velocity of 64 feet per second. Neglecting air resistance, the height of the object after t seconds is given by $-16t^2 + 64t + 80$. Factor this polynomial.

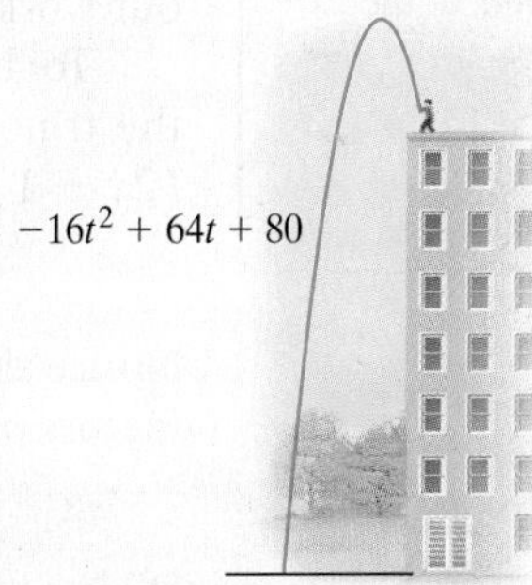

84. An object is thrown upward from the top of a 112-foot building with an initial velocity of 96 feet per second. Neglecting air resistance, the height of the object after t seconds is given by $-16t^2 + 96t + 112$. Factor this polynomial.

Factor each trinomial completely.

85. $x^2 + \frac{1}{2}x + \frac{1}{16}$

86. $x^2 + x + \frac{1}{4}$

87. $z^2(x + 1) - 3z(x + 1) - 70(x + 1)$

88. $y^2(x + 1) - 2y(x + 1) - 15(x + 1)$

Factor each trinomial. (**Hint:** *Notice that* $x^{2n} + 4x^n + 3$ *factors as* $(x^n + 1)(x^n + 3)$. **Remember:** $x^n \cdot x^n = x^{n+n}$ *or* x^{2n}.)

89. $x^{2n} + 8x^n - 20$

90. $x^{2n} + 5x^n + 6$

Find a positive value of c so that each trinomial is factorable.

91. $x^2 + 6x + c$

92. $t^2 + 8t + c$

93. $y^2 - 4y + c$

94. $n^2 - 16n + c$

Find a positive value of b so that each trinomial is factorable.

95. $x^2 + bx + 15$

96. $y^2 + by + 20$

97. $m^2 + bm - 27$

98. $x^2 + bx - 14$

6.3 Factoring Trinomials of the Form $ax^2 + bx + c$ and Perfect Square Trinomials

OBJECTIVES

1. Factor Trinomials of the Form $ax^2 + bx + c$, Where $a \neq 1$.
2. Factor Out a GCF Before Factoring a Trinomial of the Form $ax^2 + bx + c$.
3. Factor Perfect Square Trinomials.

OBJECTIVE

1 Factoring Trinomials of the Form $ax^2 + bx + c$

In this section, we factor trinomials of the form $ax^2 + bx + c$, such as

$$3x^2 + 11x + 6, \qquad 8x^2 - 22x + 5, \qquad \text{and} \qquad 2x^2 + 13x - 7$$

Notice that the coefficient of the squared variable in these trinomials is a number other than 1. We will factor these trinomials using a trial-and-check method based on our work in the last section.

To begin, let's review the relationship between the numerical coefficients of the trinomial and the numerical coefficients of its factored form. For example, since $(2x + 1)(x + 6) = 2x^2 + 13x + 6$,

$$\text{a factored form of } 2x^2 + 13x + 6 \text{ is } (2x + 1)(x + 6)$$

Notice that $2x$ and x are factors of $2x^2$, the first term of the trinomial. Also, 6 and 1 are factors of 6, the last term of the trinomial, as shown:

$$2x^2 + 13x + 6 = (2x + 1)(x + 6)$$

$2x \cdot x$

$1 \cdot 6$

Also notice that $13x$, the middle term, is the sum of the following products:

$$2x^2 + 13x + 6 = (2x + 1)(x + 6)$$

$1x$

$+12x$

$13x$ Middle term

Let's use this pattern to factor $5x^2 + 7x + 2$. First, we find factors of $5x^2$. Since all numerical coefficients in this trinomial are positive, we will use factors with positive numerical coefficients only. Thus, the factors of $5x^2$ are $5x$ and x. Let's try these factors as first terms of the binomials. Thus far, we have

$$5x^2 + 7x + 2 = (5x + \square)(x + \square)$$

Next, we need to find positive factors of 2. Positive factors of 2 are 1 and 2. Now we try possible combinations of these factors as second terms of the binomials until we obtain a middle term of $7x$.

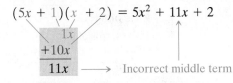

$$(5x + 1)(x + 2) = 5x^2 + 11x + 2$$

$1x$
$+10x$
$11x$ → Incorrect middle term

Let's try switching factors 2 and 1.

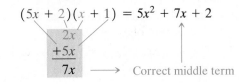

$$(5x + 2)(x + 1) = 5x^2 + 7x + 2$$

$2x$
$+5x$
$7x$ → Correct middle term

Thus the factored form of $5x^2 + 7x + 2$ is $(5x + 2)(x + 1)$. To check, we multiply $(5x + 2)$ and $(x + 1)$. The product is $5x^2 + 7x + 2$.

EXAMPLE 1 Factor: $3x^2 + 11x + 6$

Solution Since all numerical coefficients are positive, we use factors with positive numerical coefficients. We first find factors of $3x^2$.

$$\text{Factors of } 3x^2\text{:} \quad 3x^2 = 3x \cdot x$$

If factorable, the trinomial will be of the form

$$3x^2 + 11x + 6 = (3x + \square)(x + \square)$$

Next we factor 6.

$$\text{Factors of 6:} \quad 6 = 1 \cdot 6, \quad 6 = 2 \cdot 3$$

Now we try combinations of factors of 6 until a middle term of $11x$ is obtained. Let's try 1 and 6 first.

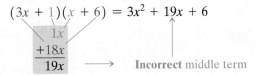

$$(3x + 1)(x + 6) = 3x^2 + 19x + 6$$

$1x$
$+18x$
$19x$ → **Incorrect** middle term

Now let's next try 6 and 1.

$$(3x + 6)(x + 1)$$

Before multiplying, notice that the terms of the factor $3x + 6$ have a common factor of 3. The terms of the original trinomial $3x^2 + 11x + 6$ have no common factor other than 1, so the terms of its factors will also contain no common factor other than 1. This means that $(3x + 6)(x + 1)$ is not a factored form.

Next let's try 2 and 3 as last terms.

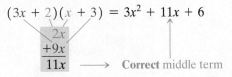

$$(3x + 2)(x + 3) = 3x^2 + 11x + 6$$

$2x$
$+9x$
$11x$ → **Correct** middle term

Thus a factored form of $3x^2 + 11x + 6$ is $(3x + 2)(x + 3)$. □

PRACTICE
1 Factor: $2x^2 + 11x + 15$.

Helpful Hint

If the terms of a trinomial have no common factor (other than 1), then the terms of neither of its binomial factors will contain a common factor (other than 1).

✓CONCEPT CHECK

Do the terms of $3x^2 + 29x + 18$ have a common factor? Without multiplying, decide which of the following factored forms could not be a factored form of $3x^2 + 29x + 18$.

a. $(3x + 18)(x + 1)$ b. $(3x + 2)(x + 9)$ c. $(3x + 6)(x + 3)$ d. $(3x + 9)(x + 2)$

EXAMPLE 2 Factor: $8x^2 - 22x + 5$

Solution Factors of $8x^2$: $8x^2 = 8x \cdot x$, $8x^2 = 4x \cdot 2x$

We'll try $8x$ and x.

$$8x^2 - 22x + 5 = (8x + \square)(x + \square)$$

Since the middle term, $-22x$, has a negative numerical coefficient, we factor 5 into negative factors.

Factors of 5: $5 = -1 \cdot -5$

Let's try -1 and -5.

$$(8x - 1)(x - 5) = 8x^2 - 41x + 5$$

$-1x$
$+(-40x)$
$-41x$ ⟶ Incorrect middle term

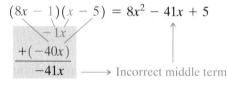

Now let's try -5 and -1.

$$(8x - 5)(x - 1) = 8x^2 - 13x + 5$$

$-5x$
$+(-8x)$
$-13x$ ⟶ Incorrect middle term

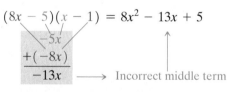

Don't give up yet! We can still try other factors of $8x^2$. Let's try $4x$ and $2x$ with -1 and -5.

$$(4x - 1)(2x - 5) = 8x^2 - 22x + 5$$

$-2x$
$+(-20x)$
$-22x$ ⟶ Correct middle term

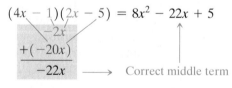

A factored form of $8x^2 - 22x + 5$ is $(4x - 1)(2x - 5)$. □

PRACTICE
2 Factor: $15x^2 - 22x + 8$.

EXAMPLE 3 Factor: $2x^2 + 13x - 7$

Solution Factors of $2x^2$: $2x^2 = 2x \cdot x$

Factors of -7: $-7 = -1 \cdot 7$, $-7 = 1 \cdot -7$

Answers to Concept Check:
no; a, c, d

We try possible combinations of these factors:

$$(2x + 1)(x - 7) = 2x^2 - 13x - 7 \quad \text{Incorrect middle term}$$
$$(2x - 1)(x + 7) = 2x^2 + 13x - 7 \quad \text{Correct middle term}$$

A factored form of $2x^2 + 13x - 7$ is $(2x - 1)(x + 7)$. □

PRACTICE 3 Factor: $4x^2 + 11x - 3$.

EXAMPLE 4 Factor: $10x^2 - 13xy - 3y^2$

Solution Factors of $10x^2$: $10x^2 = 10x \cdot x$, $10x^2 = 2x \cdot 5x$

Factors of $-3y^2$: $-3y^2 = -3y \cdot y$, $-3y^2 = 3y \cdot -y$

We try some combinations of these factors:

Correct ↓ Correct ↓

$$(10x - 3y)(x + y) = 10x^2 + 7xy - 3y^2$$
$$(x + 3y)(10x - y) = 10x^2 + 29xy - 3y^2$$
$$(5x + 3y)(2x - y) = 10x^2 + xy - 3y^2$$
$$(2x - 3y)(5x + y) = 10x^2 - 13xy - 3y^2 \quad \text{Correct middle term}$$

A factored form of $10x^2 - 13xy - 3y^2$ is $(2x - 3y)(5x + y)$. □

PRACTICE 4 Factor: $21x^2 + 11xy - 2y^2$.

EXAMPLE 5 Factor: $3x^4 - 5x^2 - 8$

Solution Factors of $3x^4$: $3x^4 = 3x^2 \cdot x^2$

Factors of -8: $-8 = -2 \cdot 4, 2 \cdot -4, -1 \cdot 8, 1 \cdot -8$

Try combinations of these factors:

Correct ↓ Correct ↓

$$(3x^2 - 2)(x^2 + 4) = 3x^4 + 10x^2 - 8$$
$$(3x^2 + 4)(x^2 - 2) = 3x^4 - 2x^2 - 8$$
$$(3x^2 + 8)(x^2 - 1) = 3x^4 + 5x^2 - 8 \quad \text{Incorrect sign on middle term, so switch signs in binomial factors.}$$
$$(3x^2 - 8)(x^2 + 1) = 3x^4 - 5x^2 - 8 \quad \text{Correct middle term.}$$

A factored form of $3x^4 - 5x^2 - 8$ is $(3x^2 - 8)(x^2 + 1)$. □

PRACTICE 5 Factor: $2x^4 - 5x^2 - 7$.

Helpful Hint

Study the last two lines of Example 5. If a factoring attempt gives you a middle term whose numerical coefficient is the opposite of the desired numerical coefficient, try switching the signs of the last terms in the binomials.

Switched signs

$$(3x^2 + 8)(x^2 - 1) = 3x^4 + 5x^2 - 8 \quad \text{Middle term: } +5x^2$$
$$(3x^2 - 8)(x^2 + 1) = 3x^4 - 5x^2 - 8 \quad \text{Middle term: } -5x^2$$

OBJECTIVE

2 Factoring Out the Greatest Common Factor 

Don't forget that the first step in factoring any polynomial is to look for a common factor to factor out.

EXAMPLE 6 Factor: $24x^4 + 40x^3 + 6x^2$

Solution Notice that all three terms have a common factor of $2x^2$. Thus we factor out $2x^2$ first.

$$24x^4 + 40x^3 + 6x^2 = 2x^2(12x^2 + 20x + 3)$$

Next we factor $12x^2 + 20x + 3$.

Factors of $12x^2$: $12x^2 = 4x \cdot 3x$, $12x^2 = 12x \cdot x$, $12x^2 = 6x \cdot 2x$

Since all terms in the trinomial have positive numerical coefficients, we factor 3 using positive factors only.

Factors of 3: $3 = 1 \cdot 3$

We try some combinations of the factors.

$$2x^2(4x + 3)(3x + 1) = 2x^2(12x^2 + 13x + 3)$$
$$2x^2(12x + 1)(x + 3) = 2x^2(12x^2 + 37x + 3)$$
$$2x^2(2x + 3)(6x + 1) = 2x^2(12x^2 + 20x + 3) \quad \text{Correct middle term}$$

A factored form of $24x^4 + 40x^3 + 6x^2$ is $2x^2(2x + 3)(6x + 1)$. □

> **Helpful Hint**
> Don't forget to include the common factor in the factored form.

PRACTICE

6 Factor: $3x^3 + 17x^2 + 10x$

When the term containing the squared variable has a negative coefficient, you may want to first factor out a common factor of -1.

EXAMPLE 7 Factor: $-6x^2 - 13x + 5$

Solution We begin by factoring out a common factor of -1.

$$-6x^2 - 13x + 5 = -1(6x^2 + 13x - 5) \quad \text{Factor out } -1.$$
$$= -1(3x - 1)(2x + 5) \quad \text{Factor } 6x^2 + 13x - 5.$$ □

PRACTICE

7 Factor: $-8x^2 + 2x + 3$

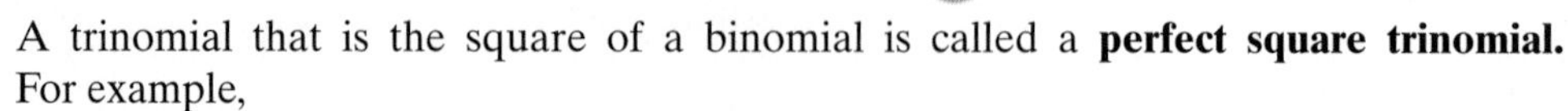

OBJECTIVE

3 Factoring Perfect Square Trinomials

A trinomial that is the square of a binomial is called a **perfect square trinomial.** For example,

$$(x + 3)^2 = (x + 3)(x + 3)$$
$$= x^2 + 6x + 9$$

Thus $x^2 + 6x + 9$ is a perfect square trinomial.

In Chapter 5, we discovered special product formulas for squaring binomials.

$$(a + b)^2 = a^2 + 2ab + b^2 \quad \text{and} \quad (a - b)^2 = a^2 - 2ab + b^2$$

Because multiplication and factoring are reverse processes, we can now use these special products to help us factor perfect square trinomials. If we reverse these equations, we have the following.

Factoring Perfect Square Trinomials

$$a^2 + 2ab + b^2 = (a + b)^2$$
$$a^2 - 2ab + b^2 = (a - b)^2$$

Helpful Hint

Notice that for both given forms of a perfect square trinomial, the last term is positive. This is because the last term is a square.

To use these equations to help us factor, we must first be able to recognize a perfect square trinomial. A trinomial is a perfect square when

1. two terms, a^2 and b^2, are squares and
2. the remaining term is $2 \cdot a \cdot b$ or $-2 \cdot a \cdot b$. That is, this term is twice the product of a and b, or its opposite.

When a trinomial fits this description, its factored form is $(a + b)^2$ or $(a - b)^2$.

EXAMPLE 8 Factor: $x^2 + 12x + 36$

Solution First, is this a perfect square trinomial?

$$x^2 + 12x + 36$$

1. $x^2 = (x)^2$ and $36 = 6^2$.
2. Is the middle term $2 \cdot x \cdot 6$? Yes, $2 \cdot x \cdot 6 = 12x$, the middle term.

Thus, $x^2 + 12x + 36$ factors as $(x + 6)^2$. □

PRACTICE 8 Factor $x^2 + 14x + 49$.

EXAMPLE 9 Factor: $25x^2 + 25xy + 4y^2$

Solution Is this a perfect square trinomial?

$$25x^2 + 25xy + 4y^2$$

1. $25x^2 = (5x)^2$ and $4y^2 = (2y)^2$.
2. Is the middle term $2 \cdot 5x \cdot 2y$? **No,** $2 \cdot 5x \cdot 2y = 20xy$, **not** the middle term $25xy$.

Therefore, $25x^2 + 25xy + 4y^2$ is not a perfect square trinomial. It is factorable, though. Using earlier techniques, we find that $25x^2 + 25xy + 4y^2$ factors as $(5x + 4y)(5x + y)$. □

Helpful Hint

A perfect square trinomial can also be factored by other methods.

PRACTICE 9 Factor $4x^2 + 20xy + 9y^2$.

EXAMPLE 10 Factor: $4m^4 - 4m^2 + 1$

Solution Is this a perfect square trinomial?

$$4m^4 - 4m^2 + 1$$

1. $4m^4 = (2m^2)^2$ and $1 = 1^2$.
2. Is the middle term $2 \cdot 2m^2 \cdot 1$ or $-2 \cdot 2m^2 \cdot 1$? Yes, $-2 \cdot 2m^2 \cdot 1 = -4m^2$, the middle term.

Thus, $4m^4 - 4m^2 + 1$ factors as $(2m^2 - 1)^2$. □

PRACTICE 10 Factor $36n^4 - 12n^2 + 1$.

EXAMPLE 11 Factor: $162x^3 - 144x^2 + 32x$

Solution Don't forget to look first for a common factor. There is a greatest common factor of $2x$ in this trinomial.

$$\begin{aligned}162x^3 - 144x^2 + 32x &= 2x(81x^2 - 72x + 16)\\ &= 2x[(9x)^2 - 2\cdot 9x\cdot 4 + 4^2]\\ &= 2x(9x - 4)^2\end{aligned}$$

PRACTICE 11 Factor $12x^3 - 84x^2 + 147x$.

Vocabulary, Readiness & Video Check

Use the choices below to fill in each blank. Some choices will be used more than once and some not used at all.

$5y^2$	$(x + 5y)^2$	perfect square trinomial
$(5y)^2$	$(x - 5y)^2$	perfect square binomial

1. A ________________ is a trinomial that is the square of a binomial.
2. The term $25y^2$ written as a square is ________________.
3. The expression $x^2 + 10xy + 25y^2$ is called a ________________.
4. The factorization $(x + 5y)(x + 5y)$ may also be written as ________________.

Complete each factorization.

5. $2x^2 + 5x + 3$ factors as $(2x + 3)(\ ?\ )$.
   a. $(x + 3)$ **b.** $(2x + 1)$ **c.** $(3x + 4)$ **d.** $(x + 1)$
6. $7x^2 + 9x + 2$ factors as $(7x + 2)(\ ?\ )$.
   a. $(3x + 1)$ **b.** $(x + 1)$ **c.** $(x + 2)$ **d.** $(7x + 1)$

Martin-Gay Interactive Videos

See Video 6.3

Watch the section lecture video and answer the following questions.

OBJECTIVE 1
7. From Example 1, explain in general terms how you would go about factoring a trinomial with a first-term coefficient $\neq 1$.

OBJECTIVE 2
8. From Examples 3 and 5, how can factoring the GCF from a trinomial help you save time when trying to factor the remaining trinomial?

OBJECTIVE 3
9. Describe in words the special patterns that the trinomials in Examples 7 and 8 have that identify them as perfect square trinomials.

6.3 Exercise Set

MyMathLab®

Complete each factored form. See Examples 1 through 5, and 8 through 10.

1. $5x^2 + 22x + 8 = (5x + 2)(\quad)$
2. $2y^2 + 27y + 25 = (2y + 25)(\quad)$
3. $50x^2 + 15x - 2 = (5x + 2)(\quad)$
4. $6y^2 + 11y - 10 = (2y + 5)(\quad)$
5. $25x^2 - 20x + 4 = (5x - 2)(\quad)$
6. $4y^2 - 20y + 25 = (2y - 5)(\quad)$

Factor completely. See Examples 1 through 5.

7. $2x^2 + 13x + 15$
8. $3x^2 + 8x + 4$
9. $8y^2 - 17y + 9$
10. $21x^2 - 31x + 10$
11. $2x^2 - 9x - 5$
12. $36r^2 - 5r - 24$
13. $20r^2 + 27r - 8$

14. $3x^2 + 20x - 63$

15. $10x^2 + 31x + 3$

16. $12x^2 + 17x + 5$

17. $2m^2 + 17m + 10$

18. $3n^2 + 20n + 5$

19. $6x^2 - 13xy + 5y^2$

20. $8x^2 - 14xy + 3y^2$

21. $15m^2 - 16m - 15$

22. $25n^2 - 5n - 6$

Factor completely. See Examples 1 through 7.

23. $12x^3 + 11x^2 + 2x$

24. $8a^3 + 14a^2 + 3a$

25. $21b^2 - 48b - 45$

26. $12x^2 - 14x - 10$

27. $7z + 12z^2 - 12$

28. $16t + 15t^2 - 15$

29. $6x^2y^2 - 2xy^2 - 60y^2$

30. $8x^2y + 34xy - 84y$

31. $4x^2 - 8x - 21$

32. $6x^2 - 11x - 10$

33. $-x^2 + 2x + 24$

34. $-x^2 + 4x + 21$

35. $4x^3 - 9x^2 - 9x$

36. $6x^3 - 31x^2 + 5x$

37. $24x^2 - 58x + 9$

38. $36x^2 + 55x - 14$

Factor each perfect square trinomial completely. See Examples 8 through 11.

39. $x^2 + 22x + 121$

40. $x^2 + 18x + 81$

41. $x^2 - 16x + 64$

42. $x^2 - 12x + 36$

43. $16a^2 - 24a + 9$

44. $25x^2 - 20x + 4$

45. $x^4 + 4x^2 + 4$

46. $m^4 + 10m^2 + 25$

47. $2n^2 - 28n + 98$

48. $3y^2 - 6y + 3$

49. $16y^2 + 40y + 25$

50. $9y^2 + 48y + 64$

MIXED PRACTICE

Factor each trinomial completely. See Examples 1 through 11 and Section 6.2.

51. $2x^2 - 7x - 99$

52. $2x^2 + 7x - 72$

53. $24x^2 + 41x + 12$

54. $24x^2 - 49x + 15$

55. $3a^2 + 10ab + 3b^2$

56. $2a^2 + 11ab + 5b^2$

57. $-9x + 20 + x^2$

58. $-7x + 12 + x^2$

59. $p^2 + 12pq + 36q^2$

60. $m^2 + 20mn + 100n^2$

61. $x^2y^2 - 10xy + 25$

62. $x^2y^2 - 14xy + 49$

63. $40a^2b + 9ab - 9b$

64. $24y^2x + 7yx - 5x$

65. $30x^3 + 38x^2 + 12x$

66. $6x^3 - 28x^2 + 16x$

67. $6y^3 - 8y^2 - 30y$

68. $12x^3 - 34x^2 + 24x$

69. $10x^4 + 25x^3y - 15x^2y^2$

70. $42x^4 - 99x^3y - 15x^2y^2$

71. $-14x^2 + 39x - 10$

72. $-15x^2 + 26x - 8$

73. $16p^4 - 40p^3 + 25p^2$

74. $9q^4 - 42q^3 + 49q^2$

75. $x + 3x^2 - 2$

76. $y + 8y^2 - 9$

77. $8x^2 + 6xy - 27y^2$

78. $54a^2 + 39ab - 8b^2$

79. $1 + 6x^2 + x^4$

80. $1 + 16x^2 + x^4$

81. $9x^2 - 24xy + 16y^2$

82. $25x^2 - 60xy + 36y^2$

83. $18x^2 - 9x - 14$

84. $42a^2 - 43a + 6$

85. $-27t + 7t^2 - 4$

86. $-3t + 4t^2 - 7$

87. $49p^2 - 7p - 2$

88. $3r^2 + 10r - 8$

89. $m^3 + 18m^2 + 81m$

90. $y^3 + 12y^2 + 36y$

91. $5x^2y^2 + 20xy + 1$

92. $3a^2b^2 + 12ab + 1$

93. $6a^5 + 37a^3b^2 + 6ab^4$

94. $5m^5 + 26m^3h^2 + 5mh^4$

REVIEW AND PREVIEW

Multiply the following. See Section 5.4.

95. $(x - 2)(x + 2)$

96. $(y^2 + 3)(y^2 - 3)$

97. $(a + 3)(a^2 - 3a + 9)$

98. $(z - 2)(z^2 + 2z + 4)$

The following graph shows the percent of text message users in each age group. See Section 3.1.

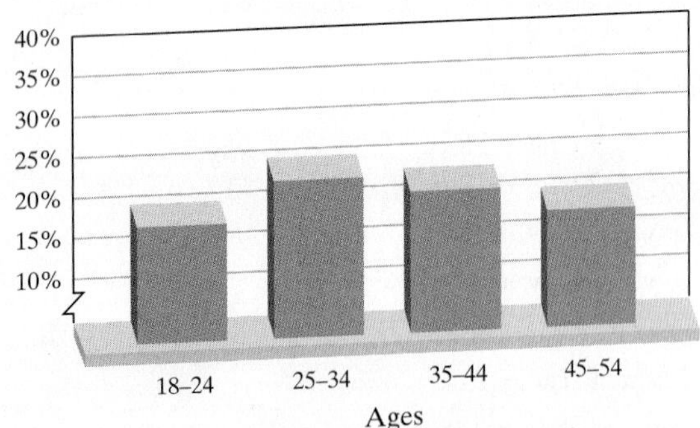

Data from CellSigns, Inc.

99. What range of ages shown has the highest percent of text message users?

100. What range of ages shown has the lowest percent of text message users?

101. Describe any trend you see.

102. Why don't the percents shown in the graph add to 100%?

CONCEPT EXTENSIONS

See the Concept Check in this section.

103. Do the terms of $4x^2 + 19x + 12$ have a common factor (other than 1)?

104. Without multiplying, decide which of the following factored forms is not a factored form of $4x^2 + 19x + 12$.

a. $(2x + 4)(2x + 3)$ **b.** $(4x + 4)(x + 3)$

c. $(4x + 3)(x + 4)$ **d.** $(2x + 2)(2x + 6)$

105. Describe a perfect square trinomial.

106. Write the perfect square trinomial that factors as $(x + 3y)^2$.

Write the perimeter of each figure as a simplified polynomial. Then factor the polynomial.

107.

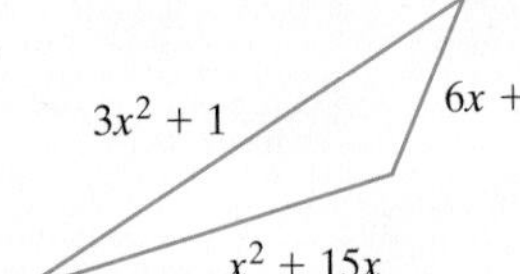

108.

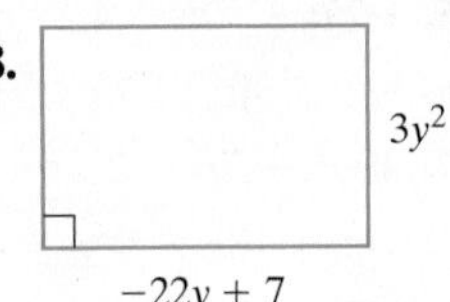

Factor each trinomial completely.

109. $4x^2 + 2x + \frac{1}{4}$

110. $27x^2 + 2x - \frac{1}{9}$

111. $4x^2(y - 1)^2 + 10x(y - 1)^2 + 25(y - 1)^2$

112. $3x^2(a + 3)^3 - 10x(a + 3)^3 + 25(a + 3)^3$

113. Fill in the blank so that $x^2 +$ ______ $x + 16$ is a perfect square trinomial.

114. Fill in the blank so that $9x^2 +$ ______ $x + 25$ is a perfect square trinomial.

The area of the largest square in the figure is $(a + b)^2$. Use this figure to answer Exercises 115 and 116.

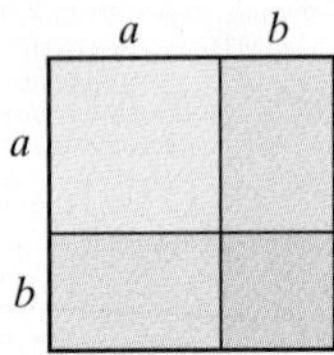

△ **115.** Write the area of the largest square as the sum of the areas of the smaller squares and rectangles.

△ **116.** What factoring formula from this section is visually represented by this square?

Find a positive value of b so that each trinomial is factorable.

117. $3x^2 + bx - 5$ **118.** $2y^2 + by + 3$

Find a positive value of c so that each trinomial is factorable.

119. $5x^2 + 7x + c$ **120.** $11y^2 - 40y + c$

Factor completely. Don't forget to first factor out the greatest common factor.

121. $-12x^3y^2 + 3x^2y^2 + 15xy^2$

122. $-12r^3x^2 + 38r^2x^2 + 14rx^2$

123. $4x^2(y - 1)^2 + 20x(y - 1)^2 + 25(y - 1)^2$

124. $3x^2(a + 3)^3 - 28x(a + 3)^3 + 25(a + 3)^3$

Factor.

125. $3x^{2n} + 17x^n + 10$

126. $2x^{2n} + 5x^n - 12$

127. In your own words, describe the steps you will use to factor a trinomial.

6.4 Factoring Trinomials of the Form $ax^2 + bx + c$ by Grouping

OBJECTIVE

1 Use the Grouping Method to Factor Trinomials of the Form $ax^2 + bx + c$.

OBJECTIVE

1 Using the Grouping Method 

There is an alternative method that can be used to factor trinomials of the form $ax^2 + bx + c, a \neq 1$. This method is called the **grouping method** because it uses factoring by grouping as we learned in Section 6.1.

To see how this method works, recall from Section 6.2 that to factor a trinomial such as $x^2 + 11x + 30$, we find two numbers such that

Product is 30

$$x^2 + 11x + 30$$

Sum is 11.

To factor a trinomial such as $2x^2 + 11x + 12$ by grouping, we use an extension of the method in Section 6.2. Here we look for two numbers such that

Product is $2 \cdot 12 = 24$

$$2x^2 + 11x + 12$$

Sum is 11.

This time, we use the two numbers to write

$$2x^2 + 11x + 12 \text{ as}$$
$$= 2x^2 + \square x + \square x + 12$$

Then we factor by grouping. Since we want a positive product, 24, and a positive sum, 11, we consider pairs of positive factors of 24 only.

Factors of 24	*Sum of Factors*	
1, 24	25	
2, 12	14	
3, 8	11	Correct sum

The factors are 3 and 8. Now we use these factors to write the middle term $11x$ as $3x + 8x$ (or $8x + 3x$). We replace $11x$ with $3x + 8x$ in the original trinomial and then we can factor by grouping.

$$\begin{aligned} 2x^2 + 11x + 12 &= 2x^2 + 3x + 8x + 12 \\ &= (2x^2 + 3x) + (8x + 12) && \text{Group the terms.} \\ &= x(2x + 3) + 4(2x + 3) && \text{Factor each group.} \\ &= (2x + 3)(x + 4) && \text{Factor out } (2x + 3). \end{aligned}$$

In general, we have the following procedure.

To Factor Trinomials by Grouping

Step 1. Factor out a greatest common factor if there is one other than 1.

Step 2. For the resulting trinomial $ax^2 + bx + c$, find two numbers whose product is $a \cdot c$ and whose sum is b.

Step 3. Write the middle term, bx, using the factors found in Step 2.

Step 4. Factor by grouping.

EXAMPLE 1 Factor $3x^2 + 31x + 10$ by grouping.

Solution

Step 1. The terms of this trinomial contain no greatest common factor other than 1 (or −1).

Step 2. In $3x^2 + 31x + 10$, $a = 3$, $b = 31$, and $c = 10$.

Let's find two numbers whose product is $a \cdot c$ or $3(10) = 30$ and whose sum is b or 31. The numbers are 1 and 30.

Factors of 30	Sum of Factors
5, 6	11
3, 10	13
2, 15	17
1, 30	31

Correct sum

Step 3. Write $31x$ as $1x + 30x$ so that $3x^2 + 31x + 10 = 3x^2 + 1x + 30x + 10$.

Step 4. Factor by grouping.

$$3x^2 + 1x + 30x + 10 = x(3x + 1) + 10(3x + 1)$$
$$= (3x + 1)(x + 10)$$

□

PRACTICE 1 Factor $5x^2 + 61x + 12$ by grouping.

EXAMPLE 2 Factor $8x^2 - 14x + 5$ by grouping.

Solution

Step 1. The terms of this trinomial contain no greatest common factor other than 1.

Step 2. This trinomial is of the form $ax^2 + bx + c$ with $a = 8$, $b = -14$, and $c = 5$. Find two numbers whose product is $a \cdot c$ or $8 \cdot 5 = 40$, and whose sum is b or -14.

The numbers are -4 and -10.

Factors of 40	Sum of Factors
−40, −1	−41
−20, −2	−22
−10, −4	−14

Correct sum

Step 3. Write $-14x$ as $-4x - 10x$ so that

$$8x^2 - 14x + 5 = 8x^2 - 4x - 10x + 5$$

Step 4. Factor by grouping.

$$8x^2 - 4x - 10x + 5 = 4x(2x - 1) - 5(2x - 1)$$
$$= (2x - 1)(4x - 5)$$

□

PRACTICE 2 Factor $12x^2 - 19x + 5$ by grouping.

EXAMPLE 3 Factor $6x^2 - 2x - 20$ by grouping.

Solution

Step 1. First factor out the greatest common factor, 2.

$$6x^2 - 2x - 20 = 2(3x^2 - x - 10)$$

Step 2. Next, notice that $a = 3$, $b = -1$, and $c = -10$ in the resulting trinomial. Find two numbers whose product is $a \cdot c$ or $3(-10) = -30$ and whose sum is b, -1. The numbers are -6 and 5.

Step 3. $3x^2 - x - 10 = 3x^2 - 6x + 5x - 10$

Step 4. $3x^2 - 6x + 5x - 10 = 3x(x - 2) + 5(x - 2)$
$$= (x - 2)(3x + 5)$$

The factored form of $6x^2 - 2x - 20 = 2(x - 2)(3x + 5)$.

↑ Don't forget to include the common factor of 2. □

PRACTICE 3 Factor $30x^2 - 14x - 4$ by grouping.

EXAMPLE 4 Factor $18y^4 + 21y^3 - 60y^2$ by grouping.

Solution

Step 1. First factor out the greatest common factor, $3y^2$.

$$18y^4 + 21y^3 - 60y^2 = 3y^2(6y^2 + 7y - 20)$$

Step 2. Notice that $a = 6, b = 7$, and $c = -20$ in the resulting trinomial. Find two numbers whose product is $a \cdot c$ or $6(-20) = -120$ and whose sum is 7. It may help to factor -120 as a product of primes and -1.

$$-120 = 2 \cdot 2 \cdot 2 \cdot 3 \cdot 5 \cdot (-1)$$

Then choose pairings of factors until you have a pairing whose sum is 7.

$$\underbrace{2 \cdot 2 \cdot 2}_{} \cdot 3 \cdot 5 \cdot (-1)$$

-8 (from $2 \cdot 2 \cdot 2 \cdot (-1)$), 15 (from $3 \cdot 5$)

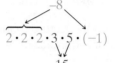

The numbers are -8 and 15.

Step 3. $6y^2 + 7y - 20 = 6y^2 - 8y + 15y - 20$

Step 4. $6y^2 - 8y + 15y - 20 = 2y(3y - 4) + 5(3y - 4)$
$$= (3y - 4)(2y + 5)$$

The factored form of $18y^4 + 21y^3 - 60y^2$ is $3y^2(3y - 4)(2y + 5)$.

↑ Don't forget to include the common factor of $3y^2$. □

PRACTICE 4 Factor $40m^4 + 5m^3 - 35m^2$ by grouping.

EXAMPLE 5 Factor $4x^2 + 20x + 25$ by grouping.

Solution

Step 1. The terms of this trinomial contain no greatest common factor other than 1 (or -1).

Step 2. In $4x^2 + 20x + 25$, $a = 4, b = 20$, and $c = 25$. Find two numbers whose product is $a \cdot c$ or $4 \cdot 25 = 100$ and whose sum is 20. The numbers are 10 and 10.

Step 3. Write $20x$ as $10x + 10x$ so that

$$4x^2 + 20x + 25 = 4x^2 + 10x + 10x + 25$$

Step 4. Factor by grouping.

$$4x^2 + 10x + 10x + 25 = 2x(2x + 5) + 5(2x + 5)$$
$$= (2x + 5)(2x + 5)$$

The factored form of $4x^2 + 20x + 25$ is $(2x + 5)(2x + 5)$ or $(2x + 5)^2$. □

PRACTICE 5 Factor $16x^2 + 24x + 9$ by grouping.

A trinomial that is the square of a binomial, such as the trinomial in Example 5, is called a **perfect square trinomial.** From Chapter 5, there are special product formulas we can use to help us recognize and factor these trinomials. To study these formulas further, see Section 6.3, Objective 3. **Remember:** A perfect square trinomial, such as the one in Example 5, may be factored by special product formulas or by other methods of factoring trinomials, such as by grouping.

Vocabulary, Readiness & Video Check

For each trinomial $ax^2 + bx + c$, choose two numbers whose product is $a \cdot c$ and whose sum is b.

1. $x^2 + 6x + 8$

a. 4, 2 **b.** 7, 1 **c.** 6, 2 **d.** 6, 8

2. $x^2 + 11x + 24$

a. 6, 4 **b.** 24, 1 **c.** 8, 3 **d.** 2, 12

3. $2x^2 + 13x + 6$

a. 2, 6 **b.** 12, 1 **c.** 13, 1 **d.** 3, 4

4. $4x^2 + 8x + 3$

a. 4, 3 **b.** 4, 4 **c.** 12, 1 **d.** 2, 6

Martin-Gay Interactive Videos

See Video 6.4

Watch the section lecture video and answer the following question.

OBJECTIVE 1

5. In the lecture following Example 1, why does writing a term as the sum or difference of two terms suggest we'd then try to factor by grouping?

6.4 Exercise Set

MyMathLab®

Factor each polynomial by grouping. Notice that Step 3 has already been done in these exercises. See Examples 1 through 5.

1. $x^2 + 3x + 2x + 6$

2. $x^2 + 5x + 3x + 15$

3. $y^2 + 8y - 2y - 16$

4. $z^2 + 10z - 7z - 70$

5. $8x^2 - 5x - 24x + 15$

6. $4x^2 - 9x - 32x + 72$

7. $5x^4 - 3x^2 + 25x^2 - 15$

8. $2y^4 - 10y^2 + 7y^2 - 35$

MIXED PRACTICE

Factor each trinomial by grouping. Exercises 9–12 are broken into parts to help you get started. See Examples 1 through 5.

9. $6x^2 + 11x + 3$

a. Find two numbers whose product is $6 \cdot 3 = 18$ and whose sum is 11.

b. Write $11x$ using the factors from part (a).

c. Factor by grouping.

10. $8x^2 + 14x + 3$

a. Find two numbers whose product is $8 \cdot 3 = 24$ and whose sum is 14.

b. Write $14x$ using the factors from part (a).

c. Factor by grouping.

11. $15x^2 - 23x + 4$

a. Find two numbers whose product is $15 \cdot 4 = 60$ and whose sum is -23.

b. Write $-23x$ using the factors from part (a).

c. Factor by grouping.

12. $6x^2 - 13x + 5$

a. Find two numbers whose product is $6 \cdot 5 = 30$ and whose sum is -13.

b. Write $-13x$ using the factors from part (a).

c. Factor by grouping.

13. $21y^2 + 17y + 2$

14. $15x^2 + 11x + 2$

15. $7x^2 - 4x - 11$
16. $8x^2 - x - 9$
17. $10x^2 - 9x + 2$
18. $30x^2 - 23x + 3$
19. $2x^2 - 7x + 5$
20. $2x^2 - 7x + 3$
21. $12x + 4x^2 + 9$
22. $20x + 25x^2 + 4$
23. $4x^2 - 8x - 21$
24. $6x^2 - 11x - 10$
25. $10x^2 - 23x + 12$
26. $21x^2 - 13x + 2$
27. $2x^3 + 13x^2 + 15x$
28. $3x^3 + 8x^2 + 4x$
29. $16y^2 - 34y + 18$
30. $4y^2 - 2y - 12$
31. $-13x + 6 + 6x^2$
32. $-25x + 12 + 12x^2$
33. $54a^2 - 9a - 30$
34. $30a^2 + 38a - 20$
35. $20a^3 + 37a^2 + 8a$
36. $10a^3 + 17a^2 + 3a$
37. $12x^3 - 27x^2 - 27x$
38. $30x^3 - 155x^2 + 25x$
39. $3x^2y + 4xy^2 + y^3$
40. $6r^2t + 7rt^2 + t^3$
41. $20z^2 + 7z + 1$
42. $36z^2 + 6z + 1$
43. $5x^2 + 50xy + 125y^2$
44. $3x^2 + 42xy + 147y^2$
45. $24a^2 - 6ab - 30b^2$
46. $30a^2 + 5ab - 25b^2$
47. $15p^4 + 31p^3q + 2p^2q^2$
48. $20s^4 + 61s^3t + 3s^2t^2$
49. $162a^4 - 72a^2 + 8$
50. $32n^4 - 112n^2 + 98$
51. $35 + 12x + x^2$
52. $33 + 14x + x^2$
53. $6 - 11x + 5x^2$
54. $5 - 12x + 7x^2$

REVIEW AND PREVIEW

Multiply. See Sections 5.3 and 5.4.

55. $(x - 2)(x + 2)$
56. $(y - 5)(y + 5)$
57. $(y + 4)(y + 4)$
58. $(x + 7)(x + 7)$
59. $(9z + 5)(9z - 5)$
60. $(8y + 9)(8y - 9)$
61. $(x - 3)(x^2 + 3x + 9)$
62. $(2z - 1)(4z^2 + 2z + 1)$

CONCEPT EXTENSIONS

Write the perimeter of each figure as a simplified polynomial. Then factor the polynomial.

63.

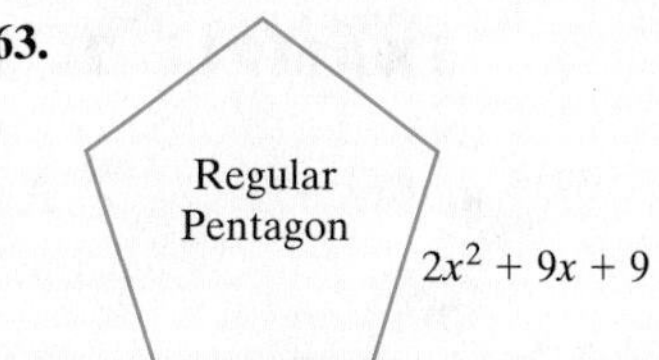

64.

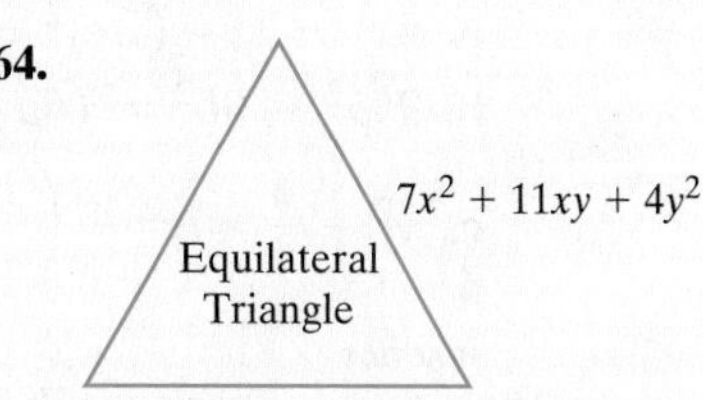

Factor each polynomial by grouping.

65. $x^{2n} + 2x^n + 3x^n + 6$
(***Hint:*** Don't forget that $x^{2n} = x^n \cdot x^n$.)
66. $x^{2n} + 6x^n + 10x^n + 60$
67. $3x^{2n} + 16x^n - 35$
68. $12x^{2n} - 40x^n + 25$
69. In your own words, explain how to factor a trinomial by grouping.

6.5 Factoring Binomials

OBJECTIVES

1 Factor the Difference of Two Squares.

2 Factor the Sum or Difference of Two Cubes.

OBJECTIVE

1 Factoring the Difference of Two Squares

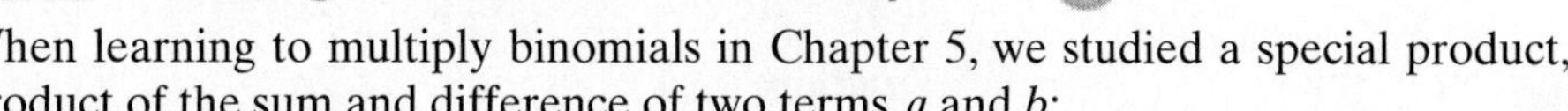

When learning to multiply binomials in Chapter 5, we studied a special product, the product of the sum and difference of two terms, a and b:

$$(a + b)(a - b) = a^2 - b^2$$

For example, the product of $x + 3$ and $x - 3$ is

$$(x + 3)(x - 3) = x^2 - 9$$

The binomial $x^2 - 9$ is called a **difference of squares**. In this section, we reverse the pattern for the product of a sum and difference to factor the binomial difference of squares.

Factoring the Difference of Two Squares

$$a^2 - b^2 = (a + b)(a - b)$$

Helpful Hint

Since multiplication is commutative, remember that the order of factors does not matter. In other words,

$$a^2 - b^2 = (a + b)(a - b) \text{ or } (a - b)(a + b)$$

EXAMPLE 1 Factor: $x^2 - 25$

Solution $x^2 - 25$ is the difference of two squares since $x^2 - 25 = x^2 - 5^2$. Therefore,

$$x^2 - 25 = x^2 - 5^2 = (x + 5)(x - 5)$$

Multiply to check. □

PRACTICE 1 Factor $x^2 - 81$.

EXAMPLE 2 Factor each difference of squares.

a. $4x^2 - 1$ **b.** $25a^2 - 9b^2$ **c.** $y^2 - \frac{4}{9}$

Solution

a. $4x^2 - 1 = (2x)^2 - 1^2 = (2x + 1)(2x - 1)$

b. $25a^2 - 9b^2 = (5a)^2 - (3b)^2 = (5a + 3b)(5a - 3b)$

c. $y^2 - \frac{4}{9} = y^2 - \left(\frac{2}{3}\right)^2 = \left(y + \frac{2}{3}\right)\left(y - \frac{2}{3}\right)$ □

PRACTICE 2 Factor each difference of squares.

a. $9x^2 - 1$ **b.** $36a^2 - 49b^2$ **c.** $p^2 - \frac{25}{36}$

EXAMPLE 3 Factor: $x^4 - y^6$

Solution This is a difference of squares since $x^4 = (x^2)^2$ and $y^6 = (y^3)^2$. Thus,

$$x^4 - y^6 = (x^2)^2 - (y^3)^2 = (x^2 + y^3)(x^2 - y^3)$$ □

PRACTICE 3 Factor $p^4 - q^{10}$.

EXAMPLE 4 Factor each binomial.

a. $y^4 - 16$ **b.** $x^2 + 4$

Solution

a. $y^4 - 16 = (y^2)^2 - 4^2$

$= (y^2 + 4)(y^2 - 4)$ Factor the difference of two squares.

This binomial ($y^2 - 4$) can be factored further since it is the difference of two squares.

$= (y^2 + 4)(y + 2)(y - 2)$ Factor the difference of two squares.

b. $x^2 + 4$

Note that the binomial $x^2 + 4$ is the *sum* of two squares since we can write $x^2 + 4$ as $x^2 + 2^2$. We might try to factor using $(x + 2)(x + 2)$ or $(x - 2)(x - 2)$. But when we multiply to check, we find that neither factoring is correct.

$$(x + 2)(x + 2) = x^2 + 4x + 4$$
$$(x - 2)(x - 2) = x^2 - 4x + 4$$

In both cases, the product is a trinomial, not the required binomial. In fact, $x^2 + 4$ is a prime polynomial. □

PRACTICE 4 Factor each binomial.

a. $z^4 - 81$ **b.** $m^2 + 49$

Helpful Hint

When factoring, don't forget:

- See whether the terms have a greatest common factor (GCF) (other than 1) that can be factored out.
- Other than a GCF, the **sum** of two squares cannot be factored using real numbers.
- Factor completely. Always check to see whether any factors can be factored further.

EXAMPLES Factor each binomial.

5. $4x^3 - 49x = x(4x^2 - 49)$ Factor out the common factor, x.
$= x[(2x)^2 - 7^2]$
$= x(2x + 7)(2x - 7)$ Factor the difference of two squares.

6. $162x^4 - 2 = 2(81x^4 - 1)$ Factor out the common factor, 2.
$= 2(9x^2 + 1)(9x^2 - 1)$ Factor the difference of two squares.
$= 2(9x^2 + 1)(3x + 1)(3x - 1)$ Factor the difference of two squares. □

PRACTICES 5–6 Factor each binomial.

5. $36y^3 - 25y$ **6.** $80y^4 - 5$

EXAMPLE 7 Factor: $-49x^2 + 16$

Solution Factor as is, or, if you like, rearrange terms.

Factor as is: $-49x^2 + 16 = -1(49x^2 - 16)$ Factor out -1.
$= -1(7x + 4)(7x - 4)$ Factor the difference of two squares.

Rewrite binomial: $-49x^2 + 16 = 16 - 49x^2 = 4^2 - (7x)^2$
$= (4 + 7x)(4 - 7x)$

Helpful Hint

When rearranging terms, keep in mind that the sign of a term is in front of the term.

Both factorizations are correct and are equal. To see this, factor -1 from $(4 - 7x)$ in the second factorization. □

PRACTICE 7 Factor $-9x^2 + 100$.

OBJECTIVE

2 Factoring the Sum or Difference of Two Cubes 

Although the sum of two squares usually does not factor, the sum or difference of two cubes can be factored and reveals factoring patterns. The pattern for the sum of cubes is illustrated by multiplying the binomial $x + y$ and the trinomial $x^2 - xy + y^2$.

$$\begin{array}{r} x^2 - xy + y^2 \\ x + y \\ \hline x^2y - xy^2 + y^3 \\ x^3 - x^2y + xy^2 \quad\quad \\ \hline x^3 \quad\quad\quad\quad + y^3 \end{array}$$

Thus, $(x + y)(x^2 - xy + y^2) = x^3 + y^3$ Sum of cubes

The pattern for the difference of two cubes is illustrated by multiplying the binomial $x - y$ by the trinomial $x^2 + xy + y^2$. The result is

$$(x - y)(x^2 + xy + y^2) = x^3 - y^3 \quad \text{Difference of cubes}$$

Factoring the Sum or Difference of Two Cubes

$$a^3 + b^3 = (a + b)(a^2 - ab + b^2)$$
$$a^3 - b^3 = (a - b)(a^2 + ab + b^2)$$

Recall that "factor" means "to write as a product." Above are patterns for writing sums and differences as products.

EXAMPLE 8 Factor: $x^3 + 8$

Solution First, write the binomial in the form $a^3 + b^3$.

$$x^3 + 8 = x^3 + 2^3 \quad \text{Write in the form } a^3 + b^3.$$

If we replace a with x and b with 2 in the formula above, we have

$$\begin{aligned} x^3 + 2^3 &= (x + 2)[x^2 - (x)(2) + 2^2] \\ &= (x + 2)(x^2 - 2x + 4) \end{aligned}$$

□

PRACTICE

8 Factor $x^3 + 64$.

Helpful Hint

When factoring sums or differences of cubes, notice the sign patterns.

same sign

$$x^3 + y^3 = (x + y)(x^2 - xy + y^2)$$

opposite signs always positive

same sign

$$x^3 - y^3 = (x - y)(x^2 + xy + y^2)$$

opposite signs always positive

EXAMPLE 9 Factor: $y^3 - 27$

Solution

$$\begin{aligned} y^3 - 27 &= y^3 - 3^3 \quad \text{Write in the form } a^3 - b^3. \\ &= (y - 3)[y^2 + (y)(3) + 3^2] \\ &= (y - 3)(y^2 + 3y + 9) \end{aligned}$$

□

PRACTICE

9 Factor $x^3 - 125$.

EXAMPLE 10 Factor: $64x^3 + 1$

Solution

$$\begin{aligned} 64x^3 + 1 &= (4x)^3 + 1^3 \\ &= (4x + 1)[(4x)^2 - (4x)(1) + 1^2] \\ &= (4x + 1)(16x^2 - 4x + 1) \end{aligned}$$

□

PRACTICE 10 Factor $27y^3 + 1$.

EXAMPLE 11 Factor: $54a^3 - 16b^3$

Solution Remember to factor out common factors first before using other factoring methods.

$$\begin{aligned} 54a^3 - 16b^3 &= 2(27a^3 - 8b^3) && \text{Factor out the GCF 2.} \\ &= 2[(3a)^3 - (2b)^3] && \text{Difference of two cubes} \\ &= 2(3a - 2b)[(3a)^2 + (3a)(2b) + (2b)^2] \\ &= 2(3a - 2b)(9a^2 + 6ab + 4b^2) \end{aligned}$$

□

PRACTICE 11 Factor $32x^3 - 500y^3$.

Calculator Explorations

Graphing

A graphing calculator is a convenient tool for evaluating an expression at a given replacement value. For example, let's evaluate $x^2 - 6x$ when $x = 2$. To do so, store the value 2 in the variable x and then enter and evaluate the algebraic expression.

```
2→X
                2
X²-6X
               -8
```

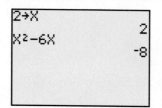

The value of $x^2 - 6x$ when $x = 2$ is -8. You may want to use this method for evaluating expressions as you explore the following.

We can use a graphing calculator to explore factoring patterns numerically. Use your calculator to evaluate $x^2 - 2x + 1$, $x^2 - 2x - 1$, and $(x - 1)^2$ for each value of x given in the table. What do you observe?

	$x^2 - 2x + 1$	$x^2 - 2x - 1$	$(x - 1)^2$
$x = 5$			
$x = -3$			
$x = 2.7$			
$x = -12.1$			
$x = 0$			

Notice in each case that $x^2 - 2x - 1 \neq (x - 1)^2$. Because for each x in the table the value of $x^2 - 2x + 1$ and the value of $(x - 1)^2$ are the same, we might guess that $x^2 - 2x + 1 = (x - 1)^2$. We can verify our guess algebraically with multiplication:

$$(x - 1)(x - 1) = x^2 - x - x + 1 = x^2 - 2x + 1$$

Vocabulary, Readiness & Video Check

Use the choices below to fill in each blank. Some choices may be used more than once and some choices may not be used at all.

true	difference of two squares	sum of two cubes
false	difference of two cubes	

1. The expression $x^3 - 27$ is called a ________________.

2. The expression $x^2 - 49$ is called a ________________.

3. The expression $z^3 + 1$ is called a ________________.

4. True or false: The binomial $y^2 + 9$ factors as $(y + 3)^2$. ________________

Write each term as a square.

5. $49x^2$ **6.** $25y^4$

Write each term as a cube.

7. $8y^3$ **8.** x^6

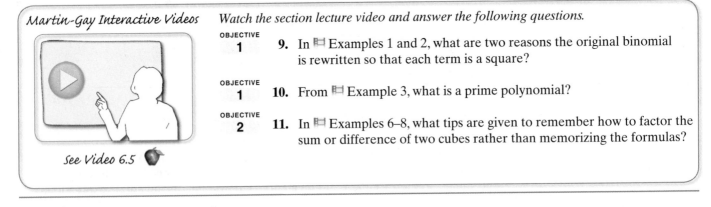

Martin-Gay Interactive Videos *Watch the section lecture video and answer the following questions.*

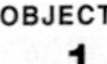

OBJECTIVE 1 **9.** In Examples 1 and 2, what are two reasons the original binomial is rewritten so that each term is a square?

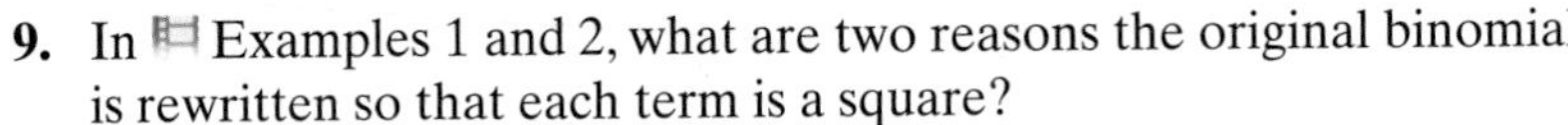

OBJECTIVE 1 **10.** From Example 3, what is a prime polynomial?

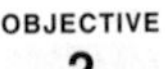

OBJECTIVE 2 **11.** In Examples 6–8, what tips are given to remember how to factor the sum or difference of two cubes rather than memorizing the formulas?

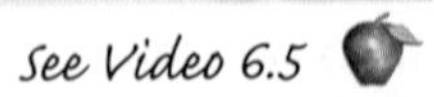

See Video 6.5

6.5 Exercise Set

MyMathLab®

Factor each binomial completely. See Examples 1 through 7.

1. $x^2 - 4$

2. $x^2 - 36$

3. $81p^2 - 1$

4. $49m^2 - 1$

5. $25y^2 - 9$

6. $49a^2 - 16$

7. $121m^2 - 100n^2$

8. $169a^2 - 49b^2$

9. $x^2y^2 - 1$

10. $a^2b^2 - 16$

11. $x^2 - \frac{1}{4}$

12. $y^2 - \frac{1}{16}$

13. $-4r^2 + 1$

14. $-9t^2 + 1$

15. $16r^2 + 1$

16. $49y^2 + 1$

17. $-36 + x^2$

18. $-1 + y^2$

19. $m^4 - 1$

20. $n^4 - 16$

21. $m^4 - n^{18}$

22. $n^4 - r^6$

Factor the sum or difference of two cubes. See Examples 8 through 11.

23. $x^3 + 125$

24. $p^3 + 1$

25. $8a^3 - 1$

26. $27y^3 - 1$

27. $m^3 + 27n^3$

28. $y^3 + 64z^3$

29. $5k^3 + 40$

30. $6r^3 + 162$

31. $x^3y^3 - 64$

32. $a^3b^3 - 8$

33. $250r^3 - 128t^3$

34. $24x^3 - 81y^3$

MIXED PRACTICE

Factor each binomial completely. See Examples 1 through 11.

35. $r^2 - 64$

36. $q^2 - 121$

37. $x^2 - 169y^2$

38. $x^2 - 225y^2$

39. $27 - t^3$

40. $125 - r^3$

41. $18r^2 - 8$

42. $32t^2 - 50$

43. $9xy^2 - 4x$

44. $36x^2y - 25y$

45. $8m^3 + 64$

46. $2x^3 + 54$

47. $xy^3 - 9xyz^2$

48. $x^3y - 4xy^3$

49. $36x^2 - 64y^2$

50. $225a^2 - 81b^2$

51. $144 - 81x^2$

52. $12x^2 - 27$

53. $x^3y^3 - z^6$

54. $a^3b^3 - c^9$

55. $49 - \frac{9}{25}m^2$

56. $100 - \frac{4}{81}n^2$

57. $t^3 + 343$

58. $s^3 + 216$

59. $n^3 + 49n$

60. $y^3 + 64y$

61. $x^6 - 81x^2$

62. $n^9 - n^5$

63. $64p^3q - 81pq^3$

64. $100x^3y - 49xy^3$

65. $27x^2y^3 + xy^2$

66. $8x^3y^3 + x^3y$

67. $125a^4 - 64ab^3$

68. $64m^4 - 27mn^3$

69. $16x^4 - 64x^2$

70. $25y^4 - 100y^2$

REVIEW AND PREVIEW

Solve each equation. See Section 2.3.

71. $x - 6 = 0$

72. $y + 5 = 0$

73. $2m + 4 = 0$

74. $3x - 9 = 0$

75. $5z - 1 = 0$

76. $4a + 2 = 0$

CONCEPT EXTENSIONS

Factor each expression completely.

77. $(x + 2)^2 - y^2$

78. $(y - 6)^2 - z^2$

79. $a^2(b - 4) - 16(b - 4)$

80. $m^2(n + 8) - 9(n + 8)$

81. $(x^2 + 6x + 9) - 4y^2$ (***Hint:*** Factor the trinomial in parentheses first.)

82. $(x^2 + 2x + 1) - 36y^2$

83. $x^{2n} - 100$

84. $x^{2n} - 81$

85. What binomial multiplied by $(x - 6)$ gives the difference of two squares?

86. What binomial multiplied by $(5 + y)$ gives the difference of two squares?

87. In your own words, explain how to tell whether a binomial is a difference of squares. Then explain how to factor a difference of squares.

88. In your own words, explain how to tell whether a binomial is a sum of cubes. Then explain how to factor a sum of cubes.

89. The Toroweap Overlook, on the North Rim of the Grand Canyon, lies 3000 vertical feet above the Colorado River. The view is spectacular, and the sheer drop is dramatic. A film crew creating a documentary about the Grand Canyon has suspended a camera platform 296 feet below the Overlook. A camera filter comes loose and falls to the river below. The height of the filter above the river after t seconds is given by the expression $2704 - 16t^2$.

a. Find the height of the filter above the river after 3 seconds.

b. Find the height of the filter above the river after 7 seconds.

c. To the nearest whole second, estimate when the filter lands in the river.

d. Factor $2704 - 16t^2$.

90. An object is dropped from the top of Pittsburgh's USX Tower, which is 841 feet tall. (*Source: World Almanac* research) The height of the object after t seconds is given by the expression $841 - 16t^2$.

a. Find the height of the object after 2 seconds.

b. Find the height of the object after 5 seconds.

c. To the nearest whole second, estimate when the object hits the ground.

d. Factor $841 - 16t^2$.

91. At this writing, the world's second tallest building is the Taipei 101 in Taipei, Taiwan, at a height of 1671 feet. (*Source:* Council on Tall Buildings and Urban Habitat) Suppose a worker is suspended 71 feet below the top of the pinnacle atop the building, at a height of 1600 feet above the ground. If the worker accidentally drops a bolt, the height of the bolt after t seconds is given by the expression $1600 - 16t^2$.

a. Find the height of the bolt after 3 seconds.

b. Find the height of the bolt after 7 seconds.

c. To the nearest whole second, estimate when the bolt hits the ground.

d. Factor $1600 - 16t^2$.

92. A performer with the Moscow Circus is planning a stunt involving a free fall from the top of the Moscow State University building, which is 784 feet tall. (*Source:* Council on Tall Buildings and Urban Habitat) Neglecting air resistance, the performer's height above gigantic cushions positioned at ground level after t seconds is given by the expression $784 - 16t^2$.

a. Find the performer's height after 2 seconds.

b. Find the performer's height after 5 seconds.

c. To the nearest whole second, estimate when the performer reaches the cushions positioned at ground level.

d. Factor $784 - 16t^2$.

Integrated Review CHOOSING A FACTORING STRATEGY

Sections 6.1–6.5

The following steps may be helpful when factoring polynomials.

Factoring a Polynomial

Step 1. Are there any common factors? If so, factor out the GCF.

Step 2. How many terms are in the polynomial?

a. If there are **two** terms, decide if one of the following can be applied.

i. Difference of two squares: $a^2 - b^2 = (a + b)(a - b)$.

ii. Difference of two cubes: $a^3 - b^3 = (a - b)(a^2 + ab + b^2)$.

iii. Sum of two cubes: $a^3 + b^3 = (a + b)(a^2 - ab + b^2)$.

b. If there are **three** terms, try one of the following.

i. Perfect square trinomial: $a^2 + 2ab + b^2 = (a + b)^2$
$a^2 - 2ab + b^2 = (a - b)^2$.

ii. If not a perfect square trinomial, factor using the methods presented in Sections 6.2 through 6.4.

c. If there are **four** or more terms, try factoring by grouping.

Step 3. See if any factors in the factored polynomial can be factored further.

Step 4. Check by multiplying.

Study the next five examples to help you use the steps above.

EXAMPLE 1 Factor $10t^2 - 17t + 3$.

Solution

Step 1. The terms of this polynomial have no common factor (other than 1).

Step 2. There are three terms, so this polynomial is a trinomial. This trinomial is not a perfect square trinomial, so factor using methods from earlier sections.

$$\text{Factors of } 10t^2\text{: } 10t^2 = 2t \cdot 5t, \quad 10t^2 = t \cdot 10t$$

Since the middle term, $-17t$, has a negative numerical coefficient, find negative factors of 3.

$$\text{Factors of 3: } 3 = -1 \cdot -3$$

Try different combinations of these factors. The correct combination is

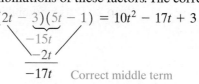

$$(2t - 3)(5t - 1) = 10t^2 - 17t + 3$$

$-15t$

$-2t$

$-17t$ Correct middle term

Step 3. No factor can be factored further, so we have factored completely.

Step 4. To check, multiply $2t - 3$ and $5t - 1$.

$$(2t - 3)(5t - 1) = 10t^2 - 2t - 15t + 3 = 10t^2 - 17t + 3$$

The factored form of $10t^2 - 17t + 3$ is $(2t - 3)(5t - 1)$. □

PRACTICE 1 Factor $6x^2 - 11x + 3$.

EXAMPLE 2 Factor $2x^3 + 3x^2 - 2x - 3$.

Solution

Step 1. There are no factors common to all terms.

Step 2. Try factoring by grouping since this polynomial has four terms.

$2x^3 + 3x^2 - 2x - 3 = x^2(2x + 3) - 1(2x + 3)$ Factor out the greatest common factor for each pair of terms.

$= (2x + 3)(x^2 - 1)$ Factor out $2x + 3$.

Step 3. The binomial $x^2 - 1$ can be factored further. It is the difference of two squares.

$= (2x + 3)(x + 1)(x - 1)$ Factor $x^2 - 1$ as a difference of squares.

Step 4. Check by finding the product of the three binomials. The polynomial factored completely is $(2x + 3)(x + 1)(x - 1)$. □

PRACTICE 2 Factor $3x^3 + x^2 - 12x - 4$.

EXAMPLE 3 Factor $12m^2 - 3n^2$.

Solution

Step 1. The terms of this binomial contain a greatest common factor of 3.

$12m^2 - 3n^2 = 3(4m^2 - n^2)$ Factor out the greatest common factor.

Step 2. The binomial $4m^2 - n^2$ is a difference of squares.

$= 3(2m + n)(2m - n)$ Factor the difference of squares.

Step 3. No factor can be factored further.

Step 4. We check by multiplying.

$$3(2m + n)(2m - n) = 3(4m^2 - n^2) = 12m^2 - 3n^2$$

The factored form of $12m^2 - 3n^2$ is $3(2m + n)(2m - n)$. □

PRACTICE 3 Factor $27x^2 - 3y^2$.

EXAMPLE 4 Factor $x^3 + 27y^3$.

Solution

Step 1. The terms of this binomial contain no common factor (other than 1).

Step 2. This binomial is the sum of two cubes.

$$\begin{aligned} x^3 + 27y^3 &= (x)^3 + (3y)^3 \\ &= (x + 3y)[x^2 - x(3y) + (3y)^2] \\ &= (x + 3y)(x^2 - 3xy + 9y^2) \end{aligned}$$

Step 3. No factor can be factored further.

Step 4. We check by multiplying.

$$\begin{aligned} (x + 3y)(x^2 - 3xy + 9y^2) &= x(x^2 - 3xy + 9y^2) + 3y(x^2 - 3xy + 9y^2) \\ &= x^3 - 3x^2y + 9xy^2 + 3x^2y - 9xy^2 + 27y^3 \\ &= x^3 + 27y^3 \end{aligned}$$

Thus, $x^3 + 27y^3$ factored completely is $(x + 3y)(x^2 - 3xy + 9y^2)$. □

PRACTICE 4 Factor $8a^3 + b^3$.

EXAMPLE 5 Factor $30a^2b^3 + 55a^2b^2 - 35a^2b$.

Solution

Step 1. $30a^2b^3 + 55a^2b^2 - 35a^2b = 5a^2b(6b^2 + 11b - 7)$ Factor out the GCF.

Step 2. $= 5a^2b(2b - 1)(3b + 7)$ Factor the resulting trinomial.

Step 3. No factor can be factored further.

Step 4. Check by multiplying.

The trinomial factored completely is $5a^2b(2b - 1)(3b + 7)$. □

PRACTICE 5 Factor $60x^3y^2 - 66x^2y^2 - 36xy^2$.

Factor the following completely.

1. $x^2 + 2xy + y^2$
2. $x^2 - 2xy + y^2$
3. $a^2 + 11a - 12$
4. $a^2 - 11a + 10$
5. $a^2 - a - 6$
6. $a^2 - 2a + 1$
7. $x^2 + 2x + 1$
8. $x^2 + x - 2$
9. $x^2 + 4x + 3$
10. $x^2 + x - 6$
11. $x^2 + 7x + 12$
12. $x^2 + x - 12$
13. $x^2 + 3x - 4$
14. $x^2 - 7x + 10$
15. $x^2 + 2x - 15$
16. $x^2 + 11x + 30$
17. $x^2 - x - 30$
18. $x^2 + 11x + 24$
19. $2x^2 - 98$
20. $3x^2 - 75$
21. $x^2 + 3x + xy + 3y$
22. $3y - 21 + xy - 7x$
23. $x^2 + 6x - 16$
24. $x^2 - 3x - 28$
25. $4x^3 + 20x^2 - 56x$
26. $6x^3 - 6x^2 - 120x$
27. $12x^2 + 34x + 24$
28. $8a^2 + 6ab - 5b^2$
29. $4a^2 - b^2$
30. $28 - 13x - 6x^2$
31. $20 - 3x - 2x^2$
32. $x^2 - 2x + 4$
33. $a^2 + a - 3$
34. $6y^2 + y - 15$
35. $4x^2 - x - 5$
36. $x^2y - y^3$
37. $4t^2 + 36$
38. $x^2 + x + xy + y$
39. $ax + 2x + a + 2$
40. $18x^3 - 63x^2 + 9x$
41. $12a^3 - 24a^2 + 4a$
42. $x^2 + 14x - 32$
43. $x^2 - 14x - 48$
44. $16a^2 - 56ab + 49b^2$
45. $25p^2 - 70pq + 49q^2$
46. $7x^2 + 24xy + 9y^2$
47. $125 - 8y^3$
48. $64x^3 + 27$

49. $-x^2 - x + 30$
50. $-x^2 + 6x - 8$
51. $14 + 5x - x^2$
52. $3 - 2x - x^2$
53. $3x^4y + 6x^3y - 72x^2y$
54. $2x^3y + 8x^2y^2 - 10xy^3$
55. $5x^3y^2 - 40x^2y^3 + 35xy^4$
56. $4x^4y - 8x^3y - 60x^2y$
57. $12x^3y + 243xy$
58. $6x^3y^2 + 8xy^2$
59. $4 - x^2$
60. $9 - y^2$
61. $3rs - s + 12r - 4$
62. $x^3 - 2x^2 + 3x - 6$
63. $4x^2 - 8xy - 3x + 6y$
64. $4x^2 - 2xy - 7yz + 14xz$
65. $6x^2 + 18xy + 12y^2$
66. $12x^2 + 46xy - 8y^2$
67. $xy^2 - 4x + 3y^2 - 12$
68. $x^2y^2 - 9x^2 + 3y^2 - 27$
69. $5(x + y) + x(x + y)$
70. $7(x - y) + y(x - y)$
71. $14t^2 - 9t + 1$
72. $3t^2 - 5t + 1$
73. $3x^2 + 2x - 5$
74. $7x^2 + 19x - 6$
75. $x^2 + 9xy - 36y^2$
76. $3x^2 + 10xy - 8y^2$
77. $1 - 8ab - 20a^2b^2$
78. $1 - 7ab - 60a^2b^2$
79. $9 - 10x^2 + x^4$
80. $36 - 13x^2 + x^4$
81. $x^4 - 14x^2 - 32$
82. $x^4 - 22x^2 - 75$
83. $x^2 - 23x + 120$
84. $y^2 + 22y + 96$
85. $6x^3 - 28x^2 + 16x$
86. $6y^3 - 8y^2 - 30y$
87. $27x^3 - 125y^3$
88. $216y^3 - z^3$
89. $x^3y^3 + 8z^3$
90. $27a^3b^3 + 8$
91. $2xy - 72x^3y$
92. $2x^3 - 18x$
93. $x^3 + 6x^2 - 4x - 24$
94. $x^3 - 2x^2 - 36x + 72$
95. $6a^3 + 10a^2$
96. $4n^2 - 6n$
97. $a^2(a + 2) + 2(a + 2)$
98. $a - b + x(a - b)$
99. $x^3 - 28 + 7x^2 - 4x$
100. $a^3 - 45 - 9a + 5a^2$

CONCEPT EXTENSIONS

Factor.

101. $(x - y)^2 - z^2$
102. $(x + 2y)^2 - 9$
103. $81 - (5x + 1)^2$
104. $b^2 - (4a + c)^2$
105. Explain why it makes good sense to factor out the GCF first, before using other methods of factoring.
106. The sum of two squares usually does not factor. Is the sum of two squares $9x^2 + 81y^2$ factorable?
107. Which of the following are equivalent to $(x + 10)(x - 7)$?

a. $(x - 7)(x + 10)$ **b.** $-1(x + 10)(x - 7)$
c. $-1(x + 10)(7 - x)$ **d.** $-1(-x - 10)(7 - x)$

108. Which of the following are equivalent to $(x - 2)(x - 5)$?

a. $-1(x + 2)(x + 5)$ **b.** $(x - 5)(x - 2)$
c. $(5 - x)(2 - x)$ **d.** $-1(x + 2)(x - 5)$

6.6 Solving Quadratic Equations by Factoring

OBJECTIVES

1. Solve Quadratic Equations by Factoring.
2. Solve Equations with Degree Greater than 2 by Factoring.
3. Find the x-Intercepts of the Graph of a Quadratic Equation in Two Variables.

In this section, we introduce a new type of equation—the **quadratic equation.**

Quadratic Equation

A quadratic equation is one that can be written in the form

$$ax^2 + bx + c = 0$$

where a, b, and c are real numbers and $a \neq 0$.

Some examples of quadratic equations are shown below.

$$x^2 - 9x - 22 = 0 \qquad 4x^2 - 28 = -49 \qquad x(2x - 7) = 4$$

The form $ax^2 + bx + c = 0$ is called the **standard form** of a quadratic equation. The quadratic equation $x^2 - 9x - 22 = 0$ is the only equation above that is in standard form.

Quadratic equations model many real-life situations. For example, let's suppose we want to know how long before a person diving from a 144-foot cliff reaches the ocean. The answer to this question is found by solving the quadratic equation $-16t^2 + 144 = 0$. (See Example 1 in Section 6.7.)

OBJECTIVE

1 Solving Quadratic Equations by Factoring 

Some quadratic equations can be solved by making use of factoring and the **zero factor property.**

> **Zero Factor Theorem**
>
> If a and b are real numbers and if $ab = 0$, then $a = 0$ or $b = 0$.

This theorem states that if the product of two numbers is 0 then at least one of the numbers must be 0.

EXAMPLE 1 Solve: $(x - 3)(x + 1) = 0$

Solution If this equation is to be a true statement, then either the factor $x - 3$ must be 0 or the factor $x + 1$ must be 0. In other words, either

$$x - 3 = 0 \quad \text{or} \quad x + 1 = 0$$

If we solve these two linear equations, we have

$$x = 3 \quad \text{or} \quad x = -1$$

Thus, 3 and -1 are both solutions of the equation $(x - 3)(x + 1) = 0$. To check, we replace x with 3 in the original equation. Then we replace x with -1 in the original equation.

Check: Let $x = 3$.

$(x - 3)(x + 1) = 0$

$(3 - 3)(3 + 1) \stackrel{?}{=} 0$ Replace x with 3.

$0(4) = 0$ True

Let $x = -1$.

$(x - 3)(x + 1) = 0$

$(-1 - 3)(-1 + 1) \stackrel{?}{=} 0$ Replace x with -1.

$(-4)(0) = 0$ True

The solutions are 3 and -1, or we say that the solution set is $\{-1, 3\}$. □

PRACTICE

1 Solve: $(x + 4)(x - 5) = 0$.

> **Helpful Hint**
>
> The zero factor property says that *if a product is 0, then a factor is 0.*
>
> If $a \cdot b = 0$, then $a = 0$ or $b = 0$.
> If $x(x + 5) = 0$, then $x = 0$ or $x + 5 = 0$.
> If $(x + 7)(2x - 3) = 0$, then $x + 7 = 0$ or $2x - 3 = 0$.
>
> Use this property only when the product is 0.
> For example, if $a \cdot b = 8$, we do not know the value of a or b. The values may be $a = 2, b = 4$ or $a = 8, b = 1$, or any other two numbers whose product is 8.

EXAMPLE 2 Solve: $(x - 5)(2x + 7) = 0$

Solution: The product is 0. By the zero factor property, this is true only when a factor is 0. To solve, we set each factor equal to 0 and solve the resulting linear equations.

$$(x - 5)(2x + 7) = 0$$

$$x - 5 = 0 \quad \text{or} \quad 2x + 7 = 0$$

$$x = 5 \qquad\qquad 2x = -7$$

$$x = -\frac{7}{2}$$

Check: Let $x = 5$.

$$(x - 5)(2x + 7) = 0$$
$$(5 - 5)(2 \cdot 5 + 7) \stackrel{?}{=} 0 \quad \text{Replace } x \text{ with 5.}$$
$$0 \cdot 17 \stackrel{?}{=} 0$$
$$0 = 0 \quad \text{True}$$

Let $x = -\frac{7}{2}$.

$$(x - 5)(2x + 7) = 0$$
$$\left(-\frac{7}{2} - 5\right)\left(2\left(-\frac{7}{2}\right) + 7\right) \stackrel{?}{=} 0 \quad \text{Replace } x \text{ with } -\frac{7}{2}.$$
$$\left(-\frac{17}{2}\right)(-7 + 7) \stackrel{?}{=} 0$$
$$\left(-\frac{17}{2}\right) \cdot 0 \stackrel{?}{=} 0$$
$$0 = 0 \quad \text{True}$$

The solutions are 5 and $-\frac{7}{2}$. □

PRACTICE

2 Solve: $(x - 12)(4x + 3) = 0$

EXAMPLE 3 Solve: $x(5x - 2) = 0$

Solution

$$x(5x - 2) = 0$$
$$x = 0 \quad \text{or} \quad 5x - 2 = 0 \quad \text{Use the zero factor property.}$$
$$5x = 2$$
$$x = \frac{2}{5}$$

Check: Let $x = 0$.

$$x(5x - 2) = 0$$
$$0(5 \cdot 0 - 2) \stackrel{?}{=} 0 \quad \text{Replace } x \text{ with 0.}$$
$$0(-2) \stackrel{?}{=} 0$$
$$0 = 0 \quad \text{True}$$

Let $x = \frac{2}{5}$.

$$x(5x - 2) = 0$$
$$\frac{2}{5}\left(5 \cdot \frac{2}{5} - 2\right) \stackrel{?}{=} 0 \quad \text{Replace } x \text{ with } \frac{2}{5}.$$
$$\frac{2}{5}(2 - 2) \stackrel{?}{=} 0$$
$$\frac{2}{5}(0) \stackrel{?}{=} 0$$
$$0 = 0 \quad \text{True}$$

The solutions are 0 and $\frac{2}{5}$. □

PRACTICE

3 Solve: $x(7x - 6) = 0$.

EXAMPLE 4 Solve: $x^2 - 9x - 22 = 0$

Solution One side of the equation is 0. However, to use the zero factor property, one side of the equation must be 0 *and* the other side must be written as a product (must be factored). Thus, we must first factor this polynomial.

$$x^2 - 9x - 22 = 0$$
$$(x - 11)(x + 2) = 0 \quad \text{Factor.}$$

Now we can apply the zero factor property.

$$x - 11 = 0 \quad \text{or} \quad x + 2 = 0$$
$$x = 11 \qquad\qquad x = -2$$

Check: Let $x = 11$.

$$x^2 - 9x - 22 = 0$$
$$11^2 - 9 \cdot 11 - 22 \stackrel{?}{=} 0$$
$$121 - 99 - 22 \stackrel{?}{=} 0$$
$$22 - 22 \stackrel{?}{=} 0$$
$$0 = 0 \quad \text{True}$$

Let $x = -2$.

$$x^2 - 9x - 22 = 0$$
$$(-2)^2 - 9(-2) - 22 \stackrel{?}{=} 0$$
$$4 + 18 - 22 \stackrel{?}{=} 0$$
$$22 - 22 \stackrel{?}{=} 0$$
$$0 = 0 \quad \text{True}$$

The solutions are 11 and -2. □

PRACTICE
4 Solve: $x^2 - 8x - 48 = 0$.

EXAMPLE 5 Solve: $4x^2 - 28x = -49$

Solution First we rewrite the equation in standard form so that one side is 0. Then we factor the polynomial.

$$4x^2 - 28x = -49$$
$$4x^2 - 28x + 49 = 0 \quad \text{Write in standard form by adding 49 to both sides.}$$
$$(2x - 7)(2x - 7) = 0 \quad \text{Factor.}$$

Next we use the zero factor property and set each factor equal to 0. Since the factors are the same, the related equations will give the same solution.

$$2x - 7 = 0 \quad \text{or} \quad 2x - 7 = 0 \quad \text{Set each factor equal to 0.}$$
$$2x = 7 \qquad\qquad 2x = 7 \quad \text{Solve.}$$
$$x = \frac{7}{2} \qquad\qquad x = \frac{7}{2}$$

Check: Although $\frac{7}{2}$ occurs twice, there is a single solution. Check this solution in the original equation. The solution is $\frac{7}{2}$. □

PRACTICE
5 Solve: $9x^2 - 24x = -16$.

The following steps may be used to solve a quadratic equation by factoring.

To Solve Quadratic Equations by Factoring

Step 1. Write the equation in standard form so that one side of the equation is 0.
Step 2. Factor the quadratic expression completely.
Step 3. Set each factor containing a variable equal to 0.
Step 4. Solve the resulting equations.
Step 5. Check each solution in the original equation.

Since it is not always possible to factor a quadratic polynomial, not all quadratic equations can be solved by factoring. Other methods of solving quadratic equations are presented in Chapter 9.

EXAMPLE 6 Solve: $x(2x - 7) = 4$

Solution First we write the equation in standard form; then we factor.

$$x(2x - 7) = 4$$
$$2x^2 - 7x = 4 \quad \text{Multiply.}$$
$$2x^2 - 7x - 4 = 0 \quad \text{Write in standard form.}$$
$$(2x + 1)(x - 4) = 0 \quad \text{Factor.}$$
$$2x + 1 = 0 \quad \text{or} \quad x - 4 = 0 \quad \text{Set each factor equal to zero.}$$
$$2x = -1 \qquad x = 4 \quad \text{Solve.}$$
$$x = -\frac{1}{2}$$

Check the solutions in the original equation. The solutions are $-\frac{1}{2}$ and 4. □

Helpful Hint

To solve the equation $x(2x - 7) = 4$, do **not** set each factor equal to 4. Remember that to apply the zero factor property, one side of the equation must be 0 and the other side of the equation must be in factored form.

PRACTICE 6 Solve: $x(3x + 7) = 6$.

✓CONCEPT CHECK

Explain the error and solve the equation correctly.

$$(x - 3)(x + 1) = 5$$
$$x - 3 = 5 \quad \text{or} \quad x + 1 = 5$$
$$x = 8 \quad \text{or} \quad x = 4$$

EXAMPLE 7 Solve: $-2x^2 - 4x + 30 = 0$.

Solution The equation is in standard form, so we begin by factoring out a common factor of -2.

$$-2x^2 - 4x + 30 = 0$$
$$-2(x^2 + 2x - 15) = 0 \quad \text{Factor out } -2.$$
$$-2(x + 5)(x - 3) = 0 \quad \text{Factor the quadratic.}$$

Next, set each factor **containing a variable** equal to 0.

$$x + 5 = 0 \quad \text{or} \quad x - 3 = 0 \quad \text{Set each factor containing a variable equal to 0.}$$
$$x = -5 \quad \text{or} \quad x = 3 \quad \text{Solve.}$$

Note: The factor -2 is a constant term containing no variables and can never equal 0. The solutions are -5 and 3. □

PRACTICE 7 Solve: $-3x^2 - 6x + 72 = 0$.

Answer to Concept Check:
To use the zero factor property, one side of the equation must be 0, not 5. Correctly, $(x - 3)(x + 1) = 5$, $x^2 - 2x - 3 = 5$, $x^2 - 2x - 8 = 0$, $(x - 4)(x + 2) = 0$, $x - 4 = 0$ or $x + 2 = 0$, $x = 4$ or $x = -2$.

OBJECTIVE

2 Solving Equations with Degree Greater than Two by Factoring

Some equations involving polynomials of degree higher than 2 may also be solved by factoring and then applying the zero factor theorem.

EXAMPLE 8 Solve: $3x^3 - 12x = 0$.

Solution Factor the left side of the equation. Begin by factoring out the common factor of $3x$.

$$3x^3 - 12x = 0$$

$$3x(x^2 - 4) = 0 \quad \text{Factor out the GCF } 3x.$$

$$3x(x + 2)(x - 2) = 0 \quad \text{Factor } x^2 - 4\text{, a difference of squares.}$$

$$3x = 0 \quad \text{or} \quad x + 2 = 0 \quad \text{or} \quad x - 2 = 0 \quad \text{Set each factor equal to 0.}$$

$$x = 0 \quad \text{or} \quad x = -2 \quad \text{or} \quad x = 2 \quad \text{Solve.}$$

Thus, the equation $3x^3 - 12x = 0$ has three solutions: 0, -2, and 2. To check, replace x with each solution in the original equation.

Let $x = 0$.

$$3(0)^3 - 12(0) \stackrel{?}{=} 0$$
$$0 = 0$$

Let $x = -2$.

$$3(-2)^3 - 12(-2) \stackrel{?}{=} 0$$
$$3(-8) + 24 \stackrel{?}{=} 0$$
$$0 = 0$$

Let $x = 2$.

$$3(2)^3 - 12(2) \stackrel{?}{=} 0$$
$$3(8) - 24 \stackrel{?}{=} 0$$
$$0 = 0$$

Substituting $0, -2$, or 2 into the original equation results each time in a true equation. The solutions are $0, -2$, and 2. □

PRACTICE
8 Solve: $7x^3 - 63x = 0$.

EXAMPLE 9 Solve: $(5x - 1)(2x^2 + 15x + 18) = 0$.

Solution

$$(5x - 1)(2x^2 + 15x + 18) = 0$$

$$(5x - 1)(2x + 3)(x + 6) = 0 \quad \text{Factor the trinomial.}$$

$$5x - 1 = 0 \quad \text{or} \quad 2x + 3 = 0 \quad \text{or} \quad x + 6 = 0 \quad \text{Set each factor equal to 0.}$$

$$5x = 1 \quad \text{or} \quad 2x = -3 \quad \text{or} \quad x = -6 \quad \text{Solve.}$$

$$x = \frac{1}{5} \quad \text{or} \quad x = -\frac{3}{2}$$

The solutions are $\frac{1}{5}, -\frac{3}{2}$, and -6. Check by replacing x with each solution in the original equation. The solutions are $-6, -\frac{3}{2}$, and $\frac{1}{5}$. □

PRACTICE
9 Solve: $(3x - 2)(2x^2 - 13x + 15) = 0$.

EXAMPLE 10 Solve: $2x^3 - 4x^2 - 30x = 0$.

Solution Begin by factoring out the GCF $2x$.

$$2x^3 - 4x^2 - 30x = 0$$

$$2x(x^2 - 2x - 15) = 0 \quad \text{Factor out the GCF } 2x.$$

$$2x(x - 5)(x + 3) = 0 \quad \text{Factor the quadratic.}$$

$$2x = 0 \quad \text{or} \quad x - 5 = 0 \quad \text{or} \quad x + 3 = 0 \quad \text{Set each factor containing a variable equal to 0.}$$

$$x = 0 \quad \text{or} \quad x = 5 \quad \text{or} \quad x = -3 \quad \text{Solve.}$$

Check by replacing x with each solution in the cubic equation. The solutions are -3, 0, and 5. □

PRACTICE
10 Solve: $5x^3 + 5x^2 - 30x = 0$.

OBJECTIVE

3 Finding x-Intercepts of the Graph of a Quadratic Equation 

In Chapter 3, we graphed linear equations in two variables, such as $y = 5x - 6$. Recall that to find the x-intercept of the graph of a linear equation, let $y = 0$ and solve for x. This is also how to find the x-intercepts of the graph of a **quadratic equation in two variables,** such as $y = x^2 - 5x + 4$.

EXAMPLE 11 Find the x-intercepts of the graph of $y = x^2 - 5x + 4$.

Solution Let $y = 0$ and solve for x.

$$y = x^2 - 5x + 4$$

$$0 = x^2 - 5x + 4 \quad \text{Let } y = 0.$$

$$0 = (x - 1)(x - 4) \quad \text{Factor.}$$

$$x - 1 = 0 \quad \text{or} \quad x - 4 = 0 \quad \text{Set each factor equal to 0.}$$

$$x = 1 \quad \text{or} \quad x = 4 \quad \text{Solve.}$$

The x-intercepts of the graph of $y = x^2 - 5x + 4$ are $(1, 0)$ and $(4, 0)$.
The graph of $y = x^2 - 5x + 4$ is shown in the margin. □

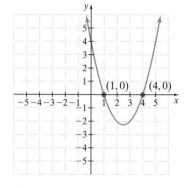

PRACTICE

11 Find the x-intercepts of the graph of $y = x^2 - 6x + 8$.

In general, a quadratic equation in two variables is one that can be written in the form $y = ax^2 + bx + c$ where $a \neq 0$. The graph of such an equation is called a **parabola** and will open up or down depending on the sign of a.

Notice that the x-intercepts of the graph of $y = ax^2 + bx + c$ are the real number solutions of $0 = ax^2 + bx + c$. Also, the real number solutions of $0 = ax^2 + bx + c$ are the x-intercepts of the graph of $y = ax^2 + bx + c$. We study more about graphs of quadratic equations in two variables in Chapter 9.

Graph of $y = ax^2 + bx + c$
x-intercepts are solutions of $0 = ax^2 + bx + c$

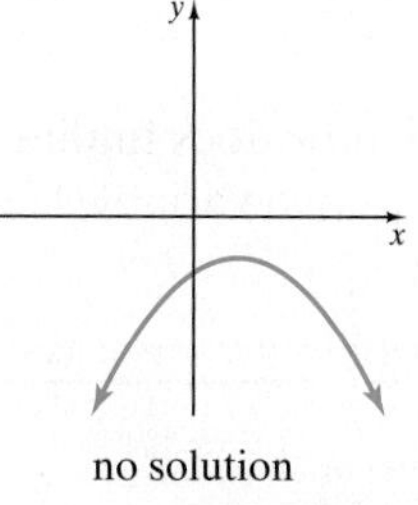

no solution

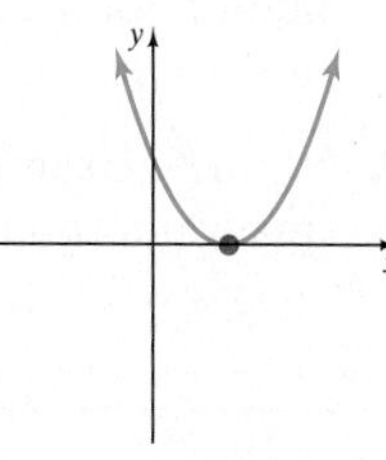

1 solution

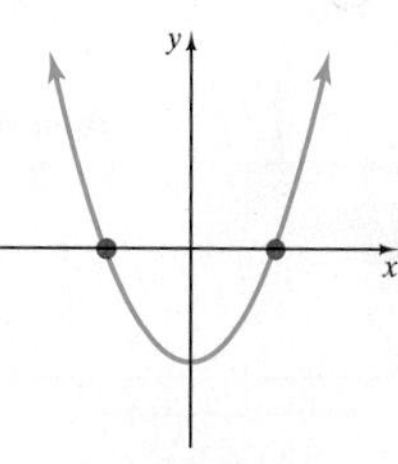

2 solutions

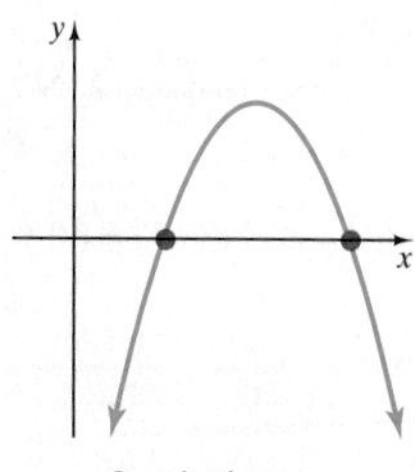

2 solutions

Graphing Calculator Explorations 

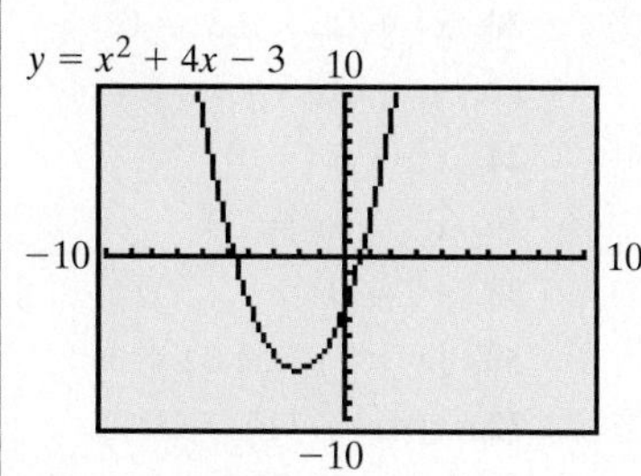

A grapher may be used to find solutions of a quadratic equation whether the related quadratic polynomial is factorable or not. For example, let's use a grapher to approximate the solutions of $0 = x^2 + 4x - 3$. To do so, graph $y_1 = x^2 + 4x - 3$. Recall that the x-intercepts of this graph are the solutions of $0 = x^2 + 4x - 3$.

Notice that the graph appears to have an x-intercept between -5 and -4 and one between 0 and 1. Many graphers contain a TRACE feature. This feature activates a graph cursor that can be used to *trace* along a graph while the corresponding x- and y-coordinates are shown on the screen. Use the TRACE feature to confirm that x-intercepts lie between -5 and -4 and between 0 and 1. To approximate the x-intercepts to the nearest tenth, use a ROOT or a ZOOM feature on your grapher or redefine the viewing window. (A ROOT feature calculates the x-intercept. A ZOOM feature magnifies the viewing window around a specific

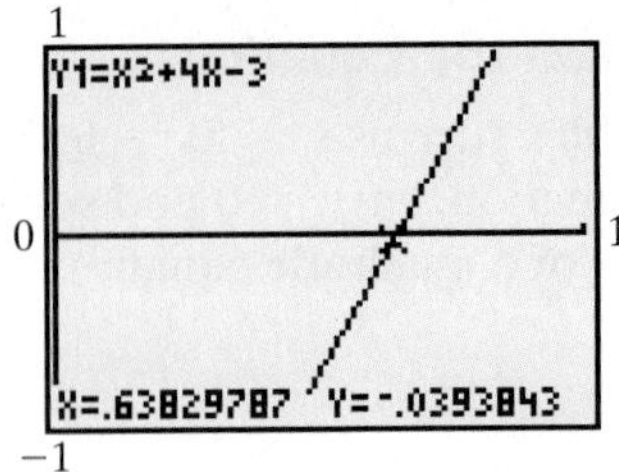

location such as the graph cursor.) If we redefine the window to $[0, 1]$ on the x-axis and $[-1, 1]$ on the y-axis, the graph to the left is generated.

By using the TRACE feature, we can conclude that one x-intercept is approximately 0.6 to the nearest tenth. By repeating these steps for the other x-intercept, we find that it is approximately -4.6.

Use a grapher to approximate the real number solutions to the nearest tenth. If an equation has no real number solution, state so.

1. $3x^2 - 4x - 6 = 0$

2. $x^2 - x - 9 = 0$

3. $2x^2 + x + 2 = 0$

4. $-4x^2 - 5x - 4 = 0$

5. $-x^2 + x + 5 = 0$

6. $10x^2 + 6x - 3 = 0$

Vocabulary, Readiness & Video Check

Use the choices below to fill in each blank. Not all choices will be used.

$-3, 5$	$a = 0$ or $b = 0$	0	linear
$3, -5$	quadratic	1	

1. An equation that can be written in the form $ax^2 + bx + c = 0$, with $a \neq 0$, is called a ________ equation.

2. If the product of two numbers is 0, then at least one of the numbers must be ________.

3. The solutions to $(x - 3)(x + 5) = 0$ are ________.

4. If $a \cdot b = 0$, then ____________.

Martin-Gay Interactive Videos

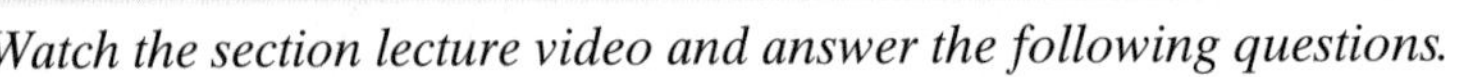

Watch the section lecture video and answer the following questions.

See Video 6.6

OBJECTIVE 1

5. As shown in Examples 1–3, what two things have to be true in order to use the zero factor theorem?

OBJECTIVE 2

6. Example 4 implies that the zero factor theorem can be used with any number of factors on one side of the equation so long as the other side of the equation is zero. Why do you think this is true?

OBJECTIVE 3

7. From Example 5, how does finding the x-intercepts of the graph of a quadratic equation in two variables lead to solving a quadratic equation?

6.6 Exercise Set

MyMathLab®

Solve each equation. See Examples 1 through 3.

1. $(x - 6)(x - 7) = 0$

2. $(x - 10)(x - 5) = 0$

3. $(x - 2)(x + 1) = 0$

4. $(x + 4)(x - 10) = 0$

5. $(x + 9)(x + 17) = 0$

6. $(x + 11)(x + 1) = 0$

7. $x(x + 6) = 0$

8. $x(x - 7) = 0$

9. $3x(x - 8) = 0$

10. $2x(x + 12) = 0$

11. $(2x + 3)(4x - 5) = 0$

12. $(3x - 2)(5x + 1) = 0$

13. $(2x - 7)(7x + 2) = 0$

14. $(9x + 1)(4x - 3) = 0$

15. $\left(x - \frac{1}{2}\right)\left(x + \frac{1}{3}\right) = 0$

16. $\left(x + \frac{2}{9}\right)\left(x - \frac{1}{4}\right) = 0$

17. $(x + 0.2)(x + 1.5) = 0$

18. $(x + 1.7)(x + 2.3) = 0$

Solve. See Examples 4 through 7.

19. $x^2 - 13x + 36 = 0$

20. $x^2 + 2x - 63 = 0$

21. $x^2 + 2x - 8 = 0$

22. $x^2 - 5x + 6 = 0$

23. $x^2 - 7x = 0$

24. $x^2 - 3x = 0$

25. $x^2 - 4x = 32$

26. $x^2 - 5x = 24$

27. $x^2 = 16$

28. $x^2 = 9$

29. $(x + 4)(x - 9) = 4x$

30. $(x + 3)(x + 8) = x$

31. $x(3x - 1) = 14$

32. $x(4x - 11) = 3$

33. $-3x^2 + 75 = 0$

34. $-2y^2 + 72 = 0$

35. $24x^2 + 44x = 8$

36. $6x^2 + 57x = 30$

Solve each equation. See Examples 8 through 10.

37. $x^3 - 12x^2 + 32x = 0$

38. $x^3 - 14x^2 + 49x = 0$

39. $(4x - 3)(16x^2 - 24x + 9) = 0$

40. $(2x + 5)(4x^2 + 20x + 25) = 0$

41. $4x^3 - x = 0$

42. $4y^3 - 36y = 0$

43. $32x^3 - 4x^2 - 6x = 0$

44. $15x^3 + 24x^2 - 63x = 0$

MIXED PRACTICE

Solve each equation. See Examples 1 through 10. (A few exercises are linear equations.)

45. $(x + 3)(x - 2) = 0$

46. $(x - 6)(x + 7) = 0$

47. $x^2 + 20x = 0$

48. $x^2 + 15x = 0$

49. $4(x - 7) = 6$

50. $5(3 - 4x) = 9$

51. $4y^2 - 1 = 0$

52. $4y^2 - 81 = 0$

53. $(2x + 3)(2x^2 - 5x - 3) = 0$

54. $(2x - 9)(x^2 + 5x - 36) = 0$

55. $x^2 - 15 = -2x$

56. $x^2 - 26 = -11x$

57. $30x^2 - 11x - 30 = 0$

58. $12x^2 + 7x - 12 = 0$

59. $5x^2 - 6x - 8 = 0$

60. $9x^2 + 7x = 2$

61. $6y^2 - 22y - 40 = 0$

62. $3x^2 - 6x - 9 = 0$

63. $(y - 2)(y + 3) = 6$

64. $(y - 5)(y - 2) = 28$

65. $3x^3 + 19x^2 - 72x = 0$

66. $36x^3 + x^2 - 21x = 0$

67. $x^2 + 14x + 49 = 0$

68. $x^2 + 22x + 121 = 0$

69. $12y = 8y^2$

70. $9y = 6y^2$

71. $7x^3 - 7x = 0$

72. $3x^3 - 27x = 0$

73. $3x^2 + 8x - 11 = 13 - 6x$

74. $2x^2 + 12x - 1 = 4 + 3x$

75. $3x^2 - 20x = -4x^2 - 7x - 6$

76. $4x^2 - 20x = -5x^2 - 6x - 5$

Find the x-intercepts of the graph of each equation. See Example 11.

77. $y = (3x + 4)(x - 1)$

78. $y = (5x - 3)(x - 4)$

79. $y = x^2 - 3x - 10$

80. $y = x^2 + 7x + 6$

81. $y = 2x^2 + 11x - 6$

82. $y = 4x^2 + 11x + 6$

For Exercises 83 through 88, match each equation with its graph. See Example 11.

83. $y = (x + 2)(x - 1)$

84. $y = (x - 5)(x + 2)$

85. $y = x(x + 3)$

86. $y = x(x - 4)$

87. $y = 2x^2 - 8$

88. $y = 2x^2 - 2$

a.

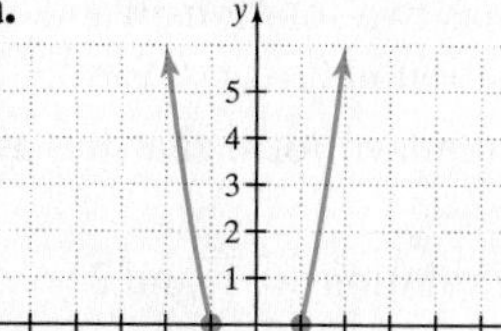

b.

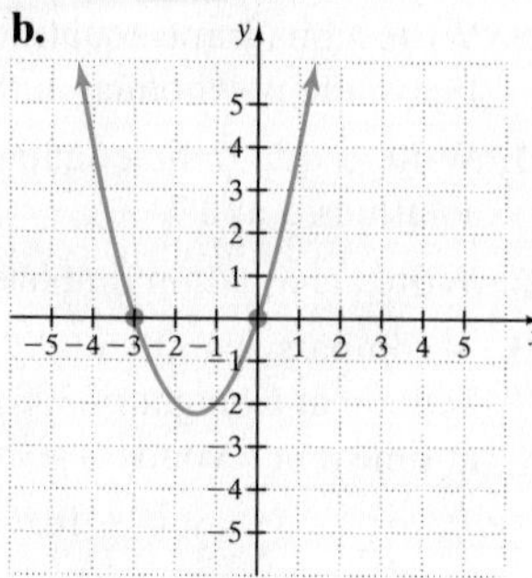

c.

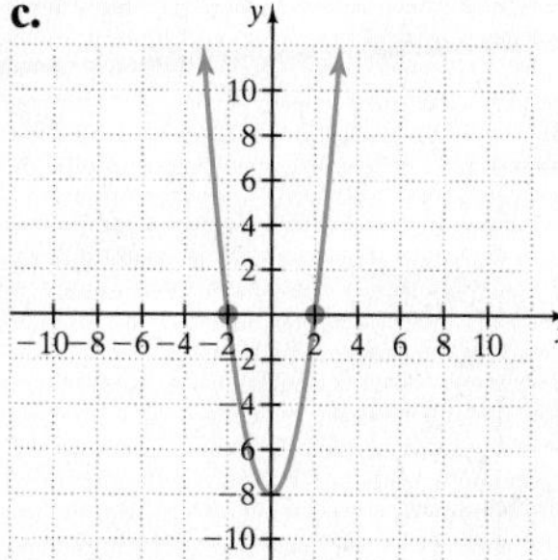

d.

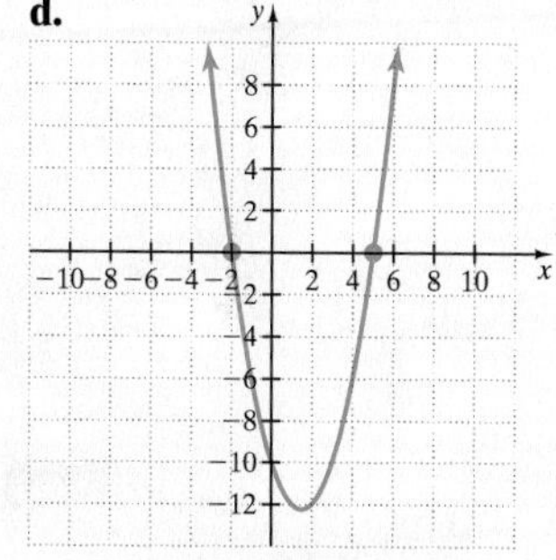

e.

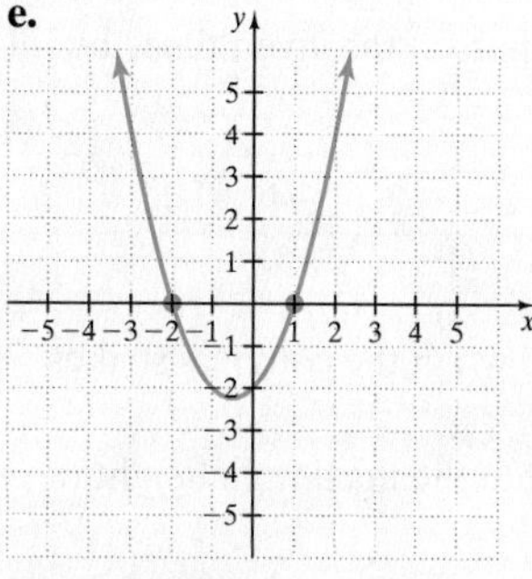

f.

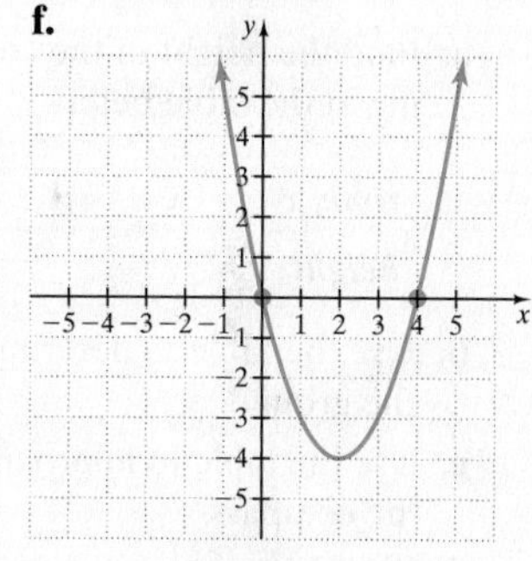

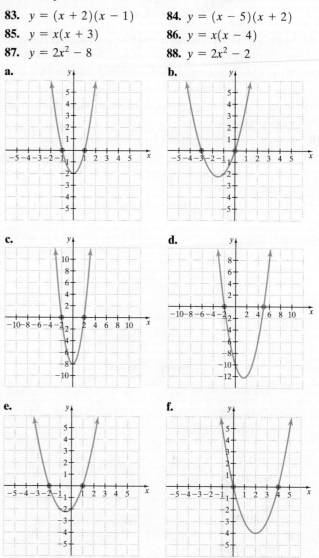

REVIEW AND PREVIEW

Perform the following operations. Write all results in lowest terms. See Section 1.3.

89. $\frac{3}{5} + \frac{4}{9}$

90. $\frac{2}{3} + \frac{3}{7}$

91. $\frac{7}{10} - \frac{5}{12}$

92. $\frac{5}{9} - \frac{5}{12}$

93. $\frac{7}{8} \div \frac{7}{15}$

94. $\frac{5}{12} - \frac{3}{10}$

95. $\frac{4}{5} \cdot \frac{7}{8}$

96. $\frac{3}{7} \cdot \frac{12}{17}$

CONCEPT EXTENSIONS

For Exercises 97 and 98, see the Concept Check in this section.

97. Explain the error and solve correctly:

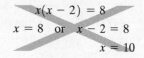

$$x(x - 2) = 8$$
$$x = 8 \quad \text{or} \quad x - 2 = 8$$
$$x = 10$$

98. Explain the error and solve correctly:

$$(x - 4)(x + 2) = 0$$
$$x = -4 \quad \text{or} \quad x = 2$$

99. Write a quadratic equation that has two solutions, 6 and −1. Leave the polynomial in the equation in factored form.

100. Write a quadratic equation that has two solutions, 0 and −2. Leave the polynomial in the equation in factored form.

101. Write a quadratic equation in standard form that has two solutions, 5 and 7.

102. Write an equation that has three solutions, 0, 1, and 2.

103. A compass is accidentally thrown upward and out of an air balloon at a height of 300 feet. The height, y, of the compass at time x in seconds is given by the equation

$$y = -16x^2 + 20x + 300$$

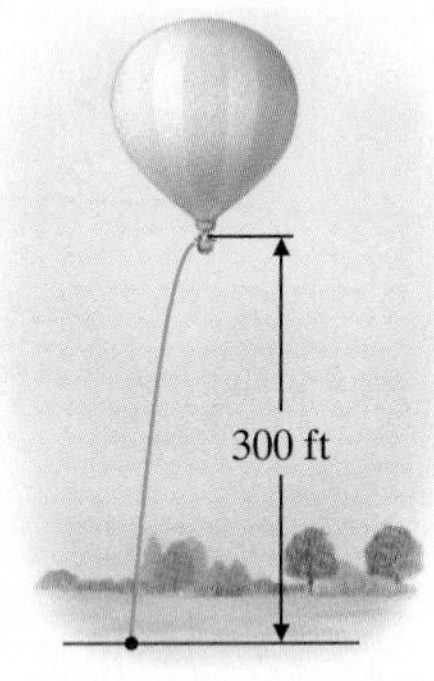

a. Find the height of the compass at the given times by filling in the table below.

time, x	0	1	2	3	4	5	6
height, y							

b. Use the table to determine when the compass strikes the ground.

c. Use the table to approximate the maximum height of the compass.

d. Plot the points (x, y) on a rectangular coordinate system and connect them with a smooth curve. Explain your results.

104. A rocket is fired upward from the ground with an initial velocity of 100 feet per second. The height, y, of the rocket at any time x is given by the equation

$$y = -16x^2 + 100x$$

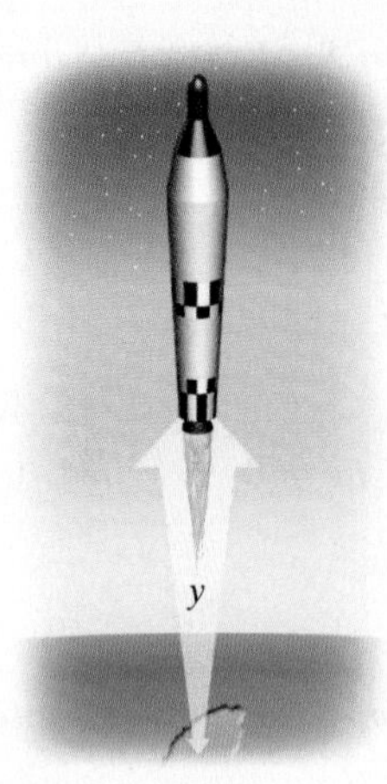

a. Find the height of the rocket at the given times by filling in the table below.

time, x	0	1	2	3	4	5	6	7
height, y								

b. Use the table to approximate when the rocket strikes the ground to the nearest second.

c. Use the table to approximate the maximum height of the rocket.

d. Plot the points (x, y) on a rectangular coordinate system and connect them with a smooth curve. Explain your results.

Solve each equation. First, multiply the binomial.

To solve $(x - 6)(2x - 3) = (x + 2)(x + 9)$, see below.

$$(x - 6)(2x - 3) = (x + 2)(x + 9)$$
$$2x^2 - 15x + 18 = x^2 + 11x + 18$$
$$x^2 - 26x = 0$$
$$x(x - 26) = 0$$
$$x = 0 \quad \text{or} \quad x - 26 = 0$$
$$x = 26$$

105. $(x - 3)(3x + 4) = (x + 2)(x - 6)$

106. $(2x - 3)(x + 6) = (x - 9)(x + 2)$

107. $(2x - 3)(x + 8) = (x - 6)(x + 4)$

108. $(x + 6)(x - 6) = (2x - 9)(x + 4)$

6.7 Quadratic Equations and Problem Solving

OBJECTIVE

1 Solve Problems That Can Be Modeled by Quadratic Equations.

OBJECTIVE

1 Solving Problems Modeled by Quadratic Equations

Some problems may be modeled by quadratic equations. To solve these problems, we use the same problem-solving steps that were introduced in Section 2.5. When solving these problems, keep in mind that a solution of an equation that models a problem may not be a solution to the problem. For example, a person's age or the length of a rectangle is always a positive number. Discard solutions that do not make sense as solutions of the problem.

EXAMPLE 1 **Finding Free-Fall Time**

Since the 1940s, one of the top tourist attractions in Acapulco, Mexico, is watching the cliff divers off the La Quebrada. The divers' platform is about 144 feet above the sea. These divers must time their descent just right, since they land in the crashing Pacific, in an inlet that is at most $9\frac{1}{2}$ feet deep. Neglecting air resistance, the height h in feet of a cliff diver above the ocean after t seconds is given by the quadratic equation $h = -16t^2 + 144$.

Find out how long it takes the diver to reach the ocean.

Solution

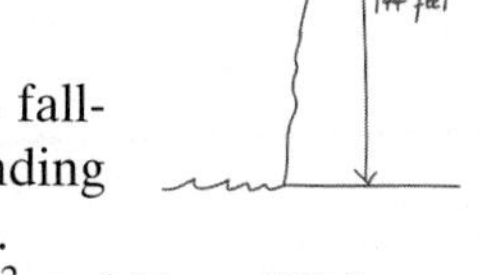

1. UNDERSTAND. Read and reread the problem. Then draw a picture of the problem.

   The equation $h = -16t^2 + 144$ models the height of the falling diver at time t. Familiarize yourself with this equation by finding the height of the diver at time $t = 1$ second and $t = 2$ seconds.

   When $t = 1$ second, the height of the diver is $h = -16(1)^2 + 144 = 128$ feet.
   When $t = 2$ seconds, the height of the diver is $h = -16(2)^2 + 144 = 80$ feet.
2. TRANSLATE. To find out how long it takes the diver to reach the ocean, we want to know the value of t for which $h = 0$.
3. SOLVE.

$$
\begin{aligned}
0 &= -16t^2 + 144 && \\
0 &= -16(t^2 - 9) && \text{Factor out } -16. \\
0 &= -16(t - 3)(t + 3) && \text{Factor completely.} \\
t - 3 = 0 \quad &\text{or} \quad t + 3 = 0 && \text{Set each factor containing a variable equal to 0.} \\
t = 3 \quad &\text{or} \quad t = -3 && \text{Solve.}
\end{aligned}
$$

4. INTERPRET. Since the time t cannot be negative, the proposed solution is 3 seconds.

Check: Verify that the height of the diver when t is 3 seconds is 0.

When $t = 3$ seconds, $h = -16(3)^2 + 144 = -144 + 144 = 0$.

State: It takes the diver 3 seconds to reach the ocean. □

PRACTICE

1 Cliff divers also frequent the falls at Waimea Falls Park in Oahu, Hawaii. One of the popular diving spots is 64 feet high. Neglecting air resistance, the height of a diver above the pool after t seconds is $h = -16t^2 + 64$. Find how long it takes a diver to reach the pool.

EXAMPLE 2 **Finding an Unknown Number**

The square of a number plus three times the number is 70. Find the number.

Solution

1. UNDERSTAND. Read and reread the problem. Suppose that the number is 5. The square of 5 is 5^2 or 25. Three times 5 is 15. Then $25 + 15 = 40$, not 70, so the number is not 5. Remember, the purpose of proposing a number, such as 5, is to understand the problem better. Now that we do, we will let $x =$ the number.

2. TRANSLATE.

the square of a number	plus	three times the number	is	70
↓	↓	↓	↓	↓
x^2	$+$	$3x$	$=$	70

3. SOLVE.

$$x^2 + 3x = 70$$
$$x^2 + 3x - 70 = 0 \quad \text{Subtract 70 from both sides.}$$
$$(x + 10)(x - 7) = 0 \quad \text{Factor.}$$
$$x + 10 = 0 \quad \text{or} \quad x - 7 = 0 \quad \text{Set each factor equal to 0.}$$
$$x = -10 \qquad x = 7 \quad \text{Solve.}$$

4. INTERPRET.

Check: The square of -10 is $(-10)^2$, or 100. Three times -10 is $3(-10)$ or -30. Then $100 + (-30) = 70$, the correct sum, so -10 checks.

The square of 7 is 7^2 or 49. Three times 7 is 3(7), or 21. Then $49 + 21 = 70$, the correct sum, so 7 checks.

State: There are two numbers. They are -10 and 7. □

PRACTICE

2 The square of a number minus eight times the number is equal to forty-eight. Find the number.

EXAMPLE 3 Finding the Dimensions of a Sail

The height of a triangular sail is 2 meters less than twice the length of the base. If the sail has an area of 30 square meters, find the length of its base and the height.

Solution

1. UNDERSTAND. Read and reread the problem. Since we are finding the length of the base and the height, we let

$x =$ the length of the base

and since the height is 2 meters less than twice the base,

$2x - 2 =$ the height

An illustration is shown to the right.

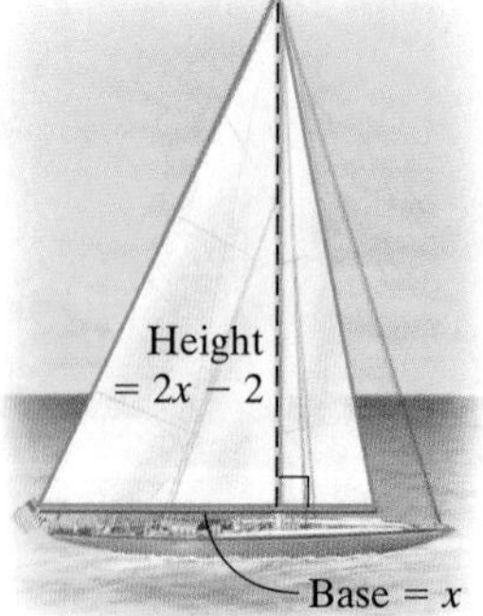

2. TRANSLATE. We are given that the area of the triangle is 30 square meters, so we use the formula for area of a triangle.

area of triangle	=	$\frac{1}{2}$	·	base	·	height
↓		↓		↓		↓
30	=	$\frac{1}{2}$	·	x	·	$(2x - 2)$

3. SOLVE. Now we solve the quadratic equation.

$$30 = \frac{1}{2}x(2x - 2)$$

$$30 = x^2 - x \quad \text{Multiply.}$$

$$x^2 - x - 30 = 0 \quad \text{Write in standard form.}$$

$$(x - 6)(x + 5) = 0 \quad \text{Factor.}$$

$$x - 6 = 0 \quad \text{or} \quad x + 5 = 0 \quad \text{Set each factor equal to 0.}$$

$$x = 6 \qquad\qquad x = -5$$

4. INTERPRET. Since x represents the length of the base, we discard the solution -5. The base of a triangle cannot be negative. The base is then 6 meters and the height is $2(6) - 2 = 10$ meters.

Check: To check this problem, we recall that $\frac{1}{2}$base · height = area, or

$$\frac{1}{2}(6)(10) = 30 \quad \text{The required area}$$

State: The base of the triangular sail is 6 meters and the height is 10 meters. □

PRACTICE

3 An engineering team from Georgia Tech earned second place in a flight competition, with their triangular shaped paper hang glider. The base of their prize-winning entry was 1 foot less than three times the height. If the area of the triangular glider wing was 210 square feet, find the dimensions of the wing. (*Source: The Technique* [Georgia Tech's newspaper], April 18, 2003)

Study the following diagrams for a review of consecutive integers.

Examples

If x is the first integer, then consecutive integers are $x, x + 1, x + 2, \ldots$

If x is the first even integer, then consecutive even integers are $x, x + 2, x + 4, \ldots$

If x is the first odd integer, then consecutive odd integers are $x, x + 2, x + 4, \ldots$

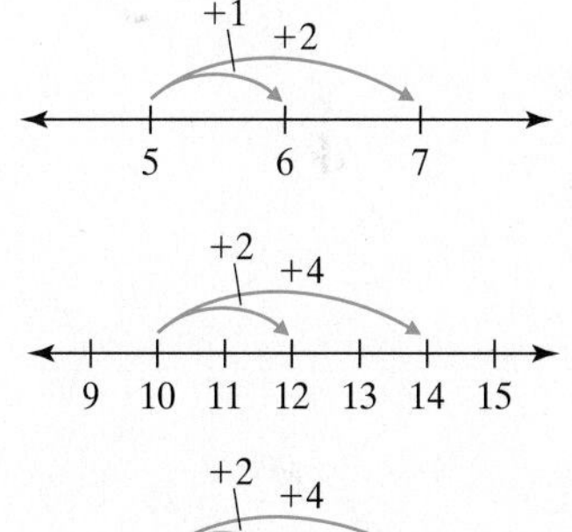

EXAMPLE 4 **Finding Consecutive Even Integers**

Find two consecutive even integers whose product is 34 more than their sum.

Solution

1. UNDERSTAND. Read and reread the problem. Let's just choose two consecutive even integers to help us better understand the problem. Let's choose 10 and 12. Their product is $10(12) = 120$ and their sum is $10 + 12 = 22$. The product is $120 - 22$, or 98 greater than the sum. Thus our guess is incorrect, but we have a better understanding of this example.

Let's let x and $x + 2$ be the consecutive even integers.

2. TRANSLATE.

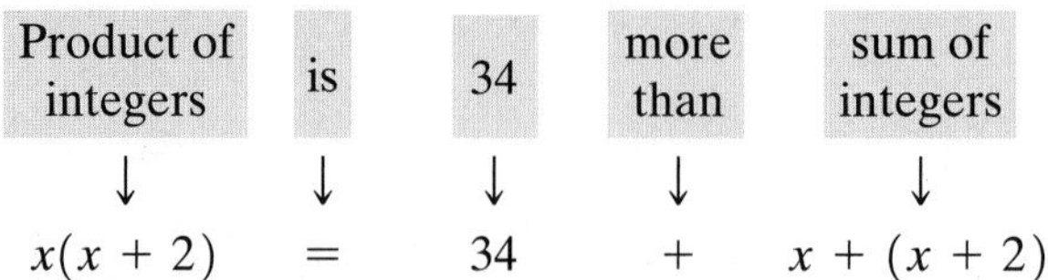

$$x(x + 2) \quad = \quad 34 \quad + \quad x + (x + 2)$$

3. SOLVE. Now we solve the equation.

$$x(x + 2) = 34 + x + (x + 2)$$

$$x^2 + 2x = 34 + x + x + 2 \qquad \text{Multiply.}$$

$$x^2 + 2x = 2x + 36 \qquad \text{Combine like terms.}$$

$$x^2 - 36 = 0 \qquad \text{Write in standard form.}$$

$$(x + 6)(x - 6) = 0 \qquad \text{Factor.}$$

$$x + 6 = 0 \quad \text{or} \quad x - 6 = 0 \qquad \text{Set each factor equal to 0.}$$

$$x = -6 \qquad x = 6 \qquad \text{Solve.}$$

4. INTERPRET. If $x = -6$, then $x + 2 = -6 + 2$, or -4.
If $x = 6$, then $x + 2 = 6 + 2$, or 8.

Check: $-6, -4$

$$-6(-4) \stackrel{?}{=} 34 + (-6) + (-4)$$

$$24 \stackrel{?}{=} 34 + (-10)$$

$$24 = 24 \qquad \text{True}$$

$6, 8$

$$6(8) \stackrel{?}{=} 34 + 6 + 8$$

$$48 \stackrel{?}{=} 34 + 14$$

$$48 = 48 \qquad \text{True}$$

State: The two consecutive even integers are -6 and -4 or 6 and 8. □

PRACTICE
4 Find two consecutive integers whose product is 41 more than their sum.

The next example uses the **Pythagorean theorem** and consecutive integers. Before we review this theorem, recall that a **right triangle** is a triangle that contains a 90° or right angle. The **hypotenuse** of a right triangle is the side opposite the right angle and is the longest side of the triangle. The **legs** of a right triangle are the other sides of the triangle.

Pythagorean Theorem

In a right triangle, the sum of the squares of the lengths of the two legs is equal to the square of the length of the hypotenuse.

$$(\text{leg})^2 + (\text{leg})^2 = (\text{hypotenuse})^2 \quad \text{or} \quad a^2 + b^2 = c^2$$

Helpful Hint
If you use this formula, don't forget that c represents the length of the hypotenuse.

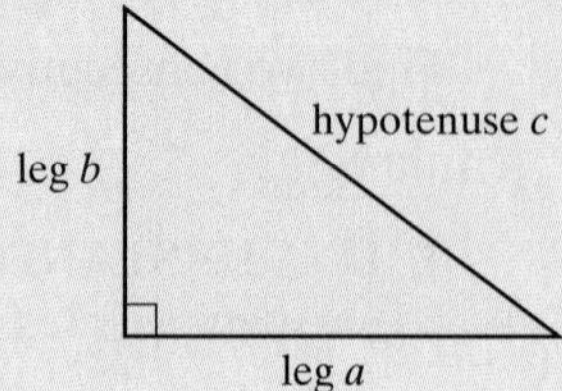

EXAMPLE 5 Finding the Dimensions of a Triangle

Find the lengths of the sides of a right triangle if the lengths can be expressed as three consecutive even integers.

Solution

1. UNDERSTAND. Read and reread the problem. Let's suppose that the length of one leg of the right triangle is 4 units. Then the other leg is the next even integer, or 6 units, and the hypotenuse of the triangle is the next even integer, or 8 units. Remember that the hypotenuse is the longest side. Let's see if a triangle with sides of these lengths forms a right triangle. To do this, we check to see whether the Pythagorean theorem holds true.

$$
\begin{aligned}
4^2 + 6^2 &\stackrel{?}{=} 8^2 \\
16 + 36 &\stackrel{?}{=} 64 \\
52 &= 64 \quad \text{False}
\end{aligned}
$$

4 units, 8 units, 6 units

Our proposed numbers do not check, but we now have a better understanding of the problem.

We let x, $x + 2$, and $x + 4$ be three consecutive even integers. Since these integers represent lengths of the sides of a right triangle, we have the following.

$$
\begin{aligned}
x &= \text{one leg} \\
x + 2 &= \text{other leg} \\
x + 4 &= \text{hypotenuse (longest side)}
\end{aligned}
$$

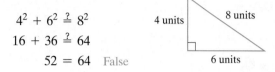

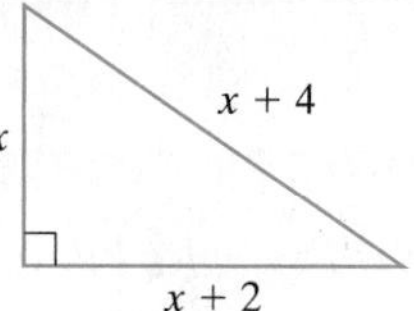

2. TRANSLATE. By the Pythagorean theorem, we have that

$$
\begin{aligned}
(\text{leg})^2 + (\text{leg})^2 &= (\text{hypotenuse})^2 \\
(x)^2 + (x + 2)^2 &= (x + 4)^2
\end{aligned}
$$

3. SOLVE. Now we solve the equation.

$$
\begin{aligned}
x^2 + (x + 2)^2 &= (x + 4)^2 \\
x^2 + x^2 + 4x + 4 &= x^2 + 8x + 16 && \text{Multiply.} \\
2x^2 + 4x + 4 &= x^2 + 8x + 16 && \text{Combine like terms.} \\
x^2 - 4x - 12 &= 0 && \text{Write in standard form.} \\
(x - 6)(x + 2) &= 0 && \text{Factor.}
\end{aligned}
$$

$$
\begin{aligned}
x - 6 = 0 \quad &\text{or} \quad x + 2 = 0 && \text{Set each factor equal to 0.} \\
x = 6 \quad & \quad\quad\; x = -2
\end{aligned}
$$

4. INTERPRET. We discard $x = -2$ since length cannot be negative. If $x = 6$, then $x + 2 = 8$ and $x + 4 = 10$.

Check: Verify that

$$
\begin{aligned}
(\text{leg})^2 + (\text{leg})^2 &= (\text{hypotenuse})^2 \\
6^2 + 8^2 &\stackrel{?}{=} 10^2 \\
36 + 64 &\stackrel{?}{=} 100 \\
100 &= 100 \quad \text{True}
\end{aligned}
$$

State: The sides of the right triangle have lengths 6 units, 8 units, and 10 units.

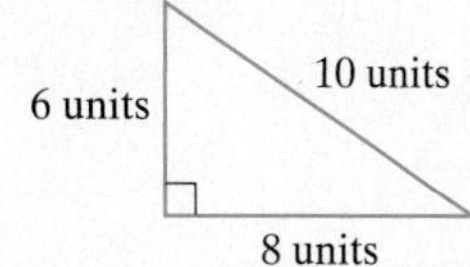

□

PRACTICE

5 Find the dimensions of a right triangle where the second leg is 1 unit less than double the first leg, and the hypotenuse is 1 unit more than double the length of the first leg.

Vocabulary, Readiness & Video Check

Martin-Gay Interactive Videos

See Video 6.7

Watch the section lecture video and answer the following question.

OBJECTIVE 1

1. In each of Examples 1–3, why aren't both solutions of the translated equation accepted as solutions to the application?

6.7 Exercise Set

MyMathLab®

MIXED PRACTICE

See Examples 1 through 5 for all exercises.

TRANSLATING

For Exercises 1 through 6, represent each given condition using a single variable, x.

△ 1. The length and width of a rectangle whose length is 4 centimeters more than its width

△ 2. The length and width of a rectangle whose length is twice its width

3. Two consecutive odd integers

4. Two consecutive even integers

△ 5. The base and height of a triangle whose height is one more than four times its base

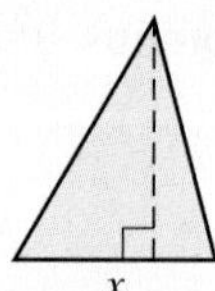

△ 6. The base and height of a trapezoid whose base is three less than five times its height

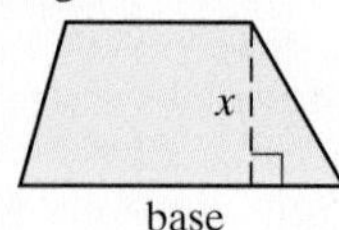

Use the information given to find the dimensions of each figure.

△ 7. The *area* of the square is 121 square units. Find the length of its sides.

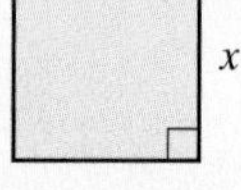

△ 8. The *area* of the rectangle is 84 square inches. Find its length and width.

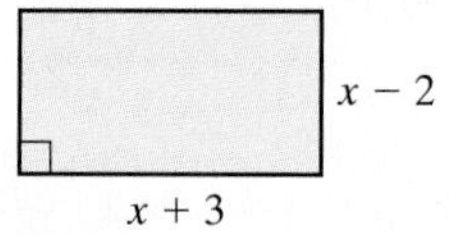

△ 9. The *perimeter* of the quadrilateral is 120 centimeters. Find the lengths of the sides.

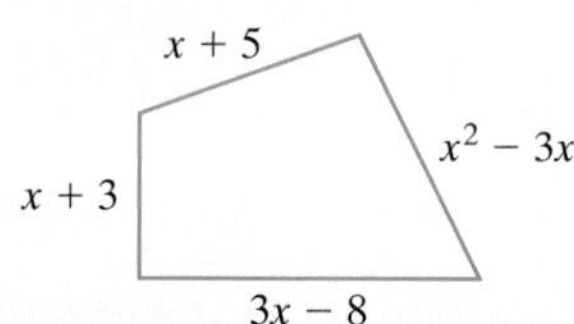

△ 10. The *perimeter* of the triangle is 85 feet. Find the lengths of its sides.

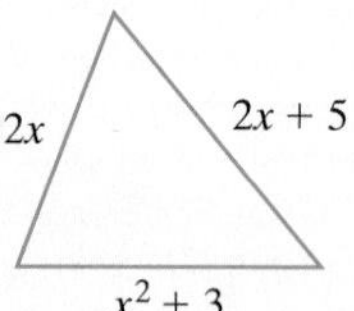

△ 11. The *area* of the parallelogram is 96 square miles. Find its base and height.

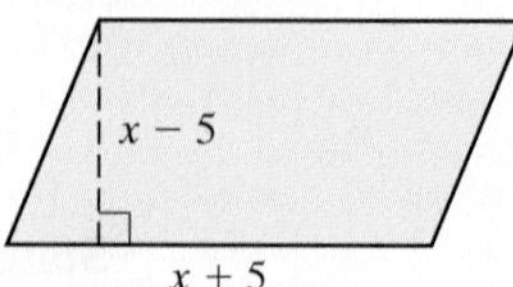

△ 12. The *area* of the circle is 25π square kilometers. Find its radius.

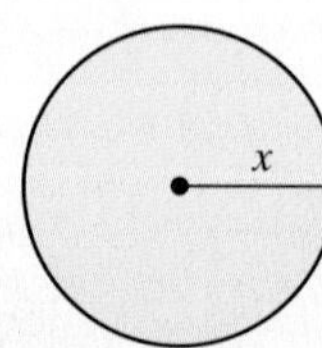

Solve.

13. An object is thrown upward from the top of an 80-foot building with an initial velocity of 64 feet per second. The height h of the object after t seconds is given by the quadratic equation $h = -16t^2 + 64t + 80$. When will the object hit the ground?

14. A hang glider pilot accidentally drops her compass from the top of a 400-foot cliff. The height h of the compass after t seconds is given by the quadratic equation $h = -16t^2 + 400$. When will the compass hit the ground?

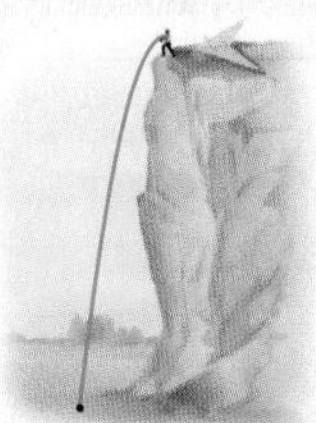

15. The width of a rectangle is 7 centimeters less than twice its length. Its area is 30 square centimeters. Find the dimensions of the rectangle.

16. The length of a rectangle is 9 inches more than its width. Its area is 112 square inches. Find the dimensions of the rectangle.

The equation $D = \frac{1}{2}n(n-3)$ gives the number of diagonals D for a polygon with n sides. For example, a polygon with 6 sides has $D = \frac{1}{2}\cdot 6(6-3)$ or $D = 9$ diagonals. (See if you can count all 9 diagonals. Some are shown in the figure.) Use this equation, $D = \frac{1}{2}n(n-3)$, for Exercises 17 through 20.

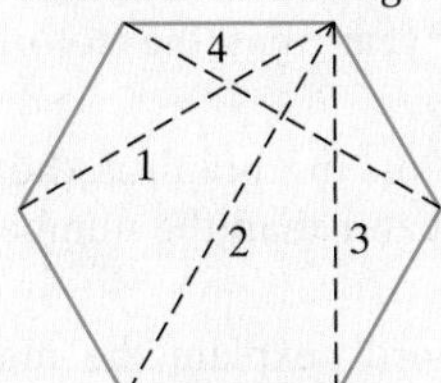

17. Find the number of diagonals for a polygon that has 12 sides.

18. Find the number of diagonals for a polygon that has 15 sides.

19. Find the number of sides n for a polygon that has 35 diagonals.

20. Find the number of sides n for a polygon that has 14 diagonals.

Solve.

21. The sum of a number and its square is 132. Find the number(s).

22. The sum of a number and its square is 182. Find the number(s).

23. The product of two consecutive room numbers is 210. Find the room numbers.

24. The product of two consecutive page numbers is 420. Find the page numbers.

25. A ladder is leaning against a building so that the distance from the ground to the top of the ladder is one foot less than the length of the ladder. Find the length of the ladder if the distance from the bottom of the ladder to the building is 5 feet.

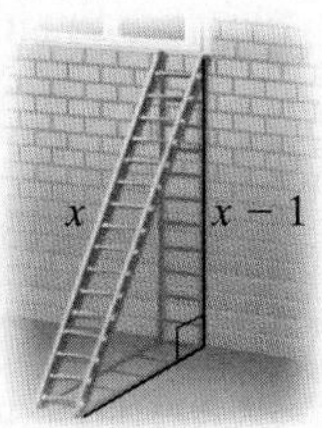

26. Use the given figure to find the length of the guy wire.

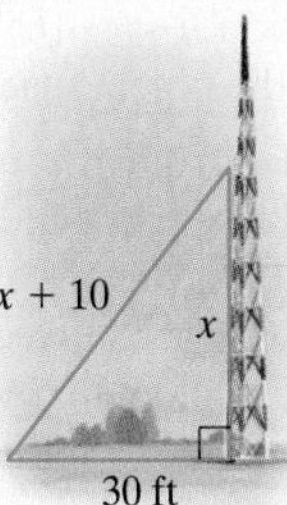

27. If the sides of a square are increased by 3 inches, the area becomes 64 square inches. Find the length of the sides of the original square.

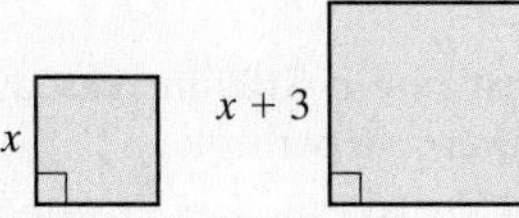

28. If the sides of a square are increased by 5 meters, the area becomes 100 square meters. Find the length of the sides of the original square.

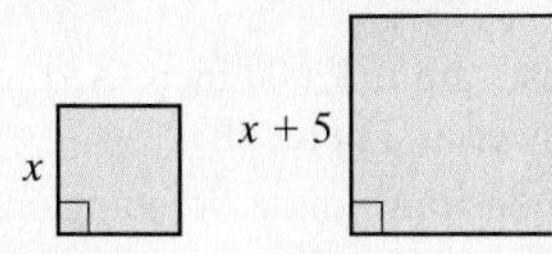

29. One leg of a right triangle is 4 millimeters longer than the smaller leg and the hypotenuse is 8 millimeters longer than the smaller leg. Find the lengths of the sides of the triangle.

30. One leg of a right triangle is 9 centimeters longer than the other leg and the hypotenuse is 45 centimeters. Find the lengths of the legs of the triangle.

31. The length of the base of a triangle is twice its height. If the area of the triangle is 100 square kilometers, find the height.

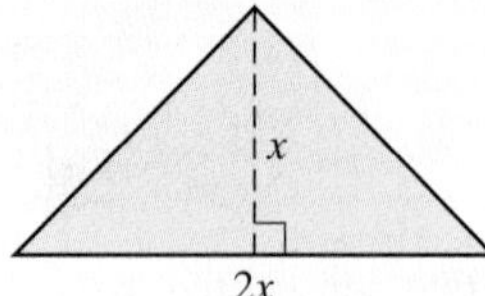

32. The height of a triangle is 2 millimeters less than the base. If the area is 60 square millimeters, find the base.

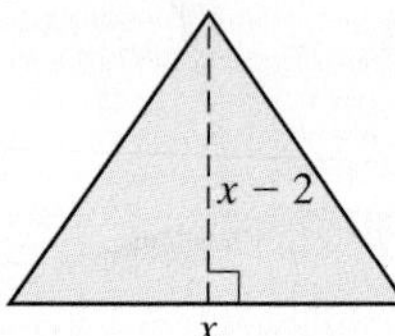

33. Find the length of the shorter leg of a right triangle if the longer leg is 12 feet more than the shorter leg and the hypotenuse is 12 feet less than twice the shorter leg.

34. Find the length of the shorter leg of a right triangle if the longer leg is 10 miles more than the shorter leg and the hypotenuse is 10 miles less than twice the shorter leg.

35. An object is dropped from 39 feet below the tip of the pinnacle atop one of the 1483-foot-tall Petronas Twin Towers in Kuala Lumpur, Malaysia. (*Source:* Council on Tall Buildings and Urban Habitat) The height h of the object after t seconds is given by the equation $h = -16t^2 + 1444$. Find how many seconds pass before the object reaches the ground.

36. An object is dropped from the top of 311 South Wacker Drive, a 961-foot-tall office building in Chicago. (*Source:* Council on Tall Buildings and Urban Habitat) The height h of the object after t seconds is given by the equation $h = -16t^2 + 961$. Find how many seconds pass before the object reaches the ground.

37. At the end of 2 years, P dollars invested at an interest rate r compounded annually increases to an amount, A dollars, given by

$$A = P(1 + r)^2$$

Find the interest rate if $100 increased to $144 in 2 years. Write your answer as a percent.

38. At the end of 2 years, P dollars invested at an interest rate r compounded annually increases to an amount, A dollars, given by

$$A = P(1 + r)^2$$

Find the interest rate if $2000 increased to $2420 in 2 years. Write your answer as a percent.

39. Find the dimensions of a rectangle whose width is 7 miles less than its length and whose area is 120 square miles.

40. Find the dimensions of a rectangle whose width is 2 inches less than half its length and whose area is 160 square inches.

41. If the cost, C, for manufacturing x units of a certain product is given by $C = x^2 - 15x + 50$, find the number of units manufactured at a cost of $9500.

42. If a switchboard handles n telephones, the number C of telephone connections it can make simultaneously is given by the equation $C = \frac{n(n-1)}{2}$. Find how many telephones are handled by a switchboard making 120 telephone connections simultaneously.

REVIEW AND PREVIEW

The following double-line graph shows a comparison of the number of annual visitors (in millions) to Glacier National Park and Gettysburg National Military Park for the years shown. Use this graph to answer Exercises 43 through 50. See Section 3.1.

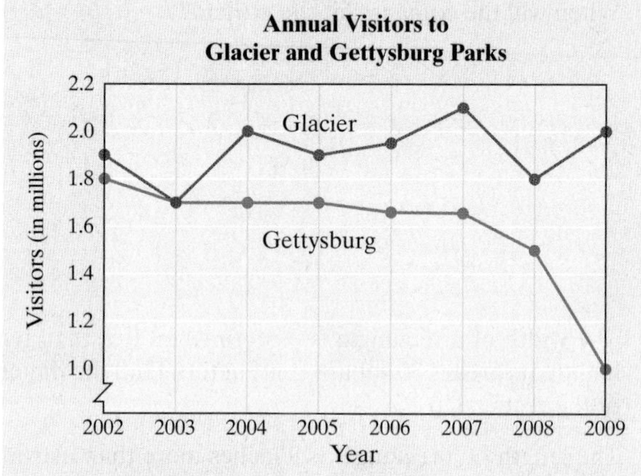

43. Approximate the number of visitors to Glacier National Park in 2009.

44. Approximate the number of visitors to Gettysburg National Military Park in 2009.

45. Approximate the number of visitors to Glacier National Park in 2005.

46. Approximate the number of visitors to Gettysburg National Military Park in 2005.

47. Determine the year that the colored lines in this graph intersect.

48. For what years on the graph is the number of visitors to Glacier Park greater than the number of visitors to Gettysburg Park?

49. In your own words, explain the meaning of the point of intersection in the graph.

50. Describe the trends shown in this graph and speculate as to why these trends have occurred.

Write each fraction in simplest form. See Section 1.3.

51. $\frac{20}{35}$

52. $\frac{24}{32}$

53. $\frac{27}{18}$

54. $\frac{15}{27}$

55. $\frac{14}{42}$

56. $\frac{45}{50}$

CONCEPT EXTENSIONS

△ **57.** Two boats travel at right angles to each other after leaving the same dock at the same time. One hour later, the boats are 17 miles apart. If one boat travels 7 miles per hour faster than the other boat, find the rate of each boat.

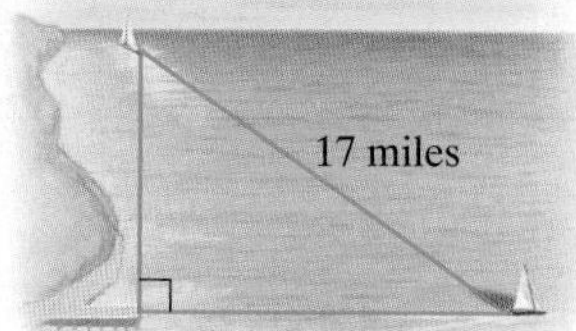

△ **58.** The side of a square equals the width of a rectangle. The length of the rectangle is 6 meters longer than its width. The sum of the areas of the square and the rectangle is 176 square meters. Find the side of the square.

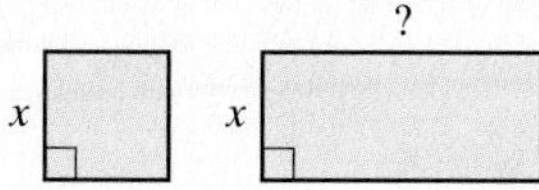

59. The sum of two numbers is 20, and the sum of their squares is 218. Find the numbers.

60. The sum of two numbers is 25, and the sum of their squares is 325. Find the numbers.

△ **61.** A rectangular pool is surrounded by a walk 4 meters wide. The pool is 6 meters longer than its width. If the total area of the pool and walk is 576 square meters more than the area of the pool, find the dimensions of the pool.

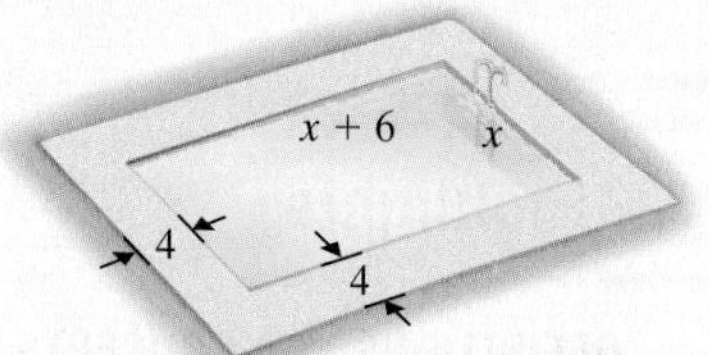

△ **62.** A rectangular garden is surrounded by a walk of uniform width. The area of the garden is 180 square yards. If the dimensions of the garden plus the walk are 16 yards by 24 yards, find the width of the walk.

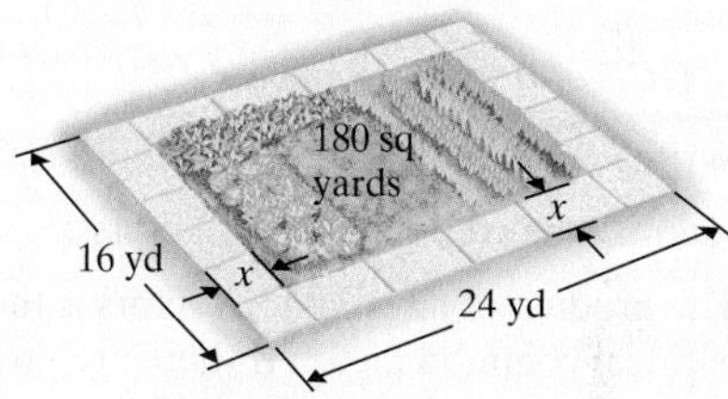

63. Write down two numbers whose sum is 10. Square each number and find the sum of the squares. Use this work to write a word problem like Exercise 59. Then give the word problem to a classmate to solve.

64. Write down two numbers whose sum is 12. Square each number and find the sum of the squares. Use this work to write a word problem like Exercise 60. Then give the word problem to a classmate to solve.

Chapter 6 Vocabulary Check

Fill in each blank with one of the words or phrases listed below. Not all choices will be used and some choices may be used more than once.

factoring	quadratic equation	perfect square trinomial	0
greatest common factor	hypotenuse	sum of two cubes	1
difference of two cubes	difference of two squares	triangle	leg

1. An equation that can be written in the form $ax^2 + bx + c = 0$ (with a not 0) is called a ________________.

2. ________________ is the process of writing an expression as a product.

3. The ________________ of a list of terms is the product of all common factors.

4. A trinomial that is the square of some binomial is called a ________________.

5. The expression $a^2 - b^2$ is called a ________________.

6. The expression $a^3 - b^3$ is called a ________________.

7. The expression $a^3 + b^3$ is called a ________________.

8. By the zero factor property, if the product of two numbers is 0, then at least one of the numbers must be ________________.

9. In a right triangle, the side opposite the right angle is called the ________________.

10. In a right triangle, each side adjacent to the right angle is called a ________________.

11. The Pythagorean theorem states that $(\text{leg})^2 + (\text{leg})^2 = (\_\_\_\_\_\_\_\_\_\_)^2$.

Chapter 6 Highlights

DEFINITIONS AND CONCEPTS	EXAMPLES
Section 6.1 The Greatest Common Factor and Factoring by Grouping	
Factoring is the process of writing an expression as a product.	Factor: $6 = 2 \cdot 3$ $x^2 + 5x + 6 = (x + 2)(x + 3)$
To Find the GCF of a List of Integers	Find the GCF of 12, 36, and 48.
Step 1. Write each number as a product of primes.	$12 = 2 \cdot 2 \cdot 3$
Step 2. Identify the common prime factors.	$36 = 2 \cdot 2 \cdot 3 \cdot 3$
Step 3. The product of all common factors is the greatest common factor. If there are no common prime factors, the GCF is 1.	$48 = 2 \cdot 2 \cdot 2 \cdot 2 \cdot 3$ $\text{GCF} = 2 \cdot 2 \cdot 3 = 12$
The GCF of a list of common variables raised to powers is the variable raised to the smallest exponent in the list.	The GCF of z^5, z^3, and z^{10} is z^3.
The GCF of a list of terms is the product of all common factors.	Find the GCF of $8x^2y$, $10x^3y^2$, and $26x^2y^3$. The GCF of 8, 10, and 26 is 2. The GCF of x^2, x^3, and x^2 is x^2. The GCF of y, y^2, and y^3 is y. The GCF of the terms is $2x^2y$.
To Factor by Grouping	Factor $10ax + 15a - 6xy - 9y$.
Step 1. Arrange the terms so that the first two terms have a common factor and the last two have a common factor.	**Step 1.** $10ax + 15a - 6xy - 9y$
Step 2. For each pair of terms, factor out the pair's GCF.	**Step 2.** $5a(2x + 3) - 3y(2x + 3)$
Step 3. If there is now a common binomial factor, factor it out.	**Step 3.** $(2x + 3)(5a - 3y)$
Step 4. If there is no common binomial factor, begin again, rearranging the terms differently. If no rearrangement leads to a common binomial factor, the polynomial cannot be factored.	
Section 6.2 Factoring Trinomials of the Form $x^2 + bx + c$	
The product of these numbers is c. $x^2 + bx + c = (x + \square)(x + \square)$ The sum of these numbers is b.	Factor: $x^2 + 7x + 12$ $3 + 4 = 7$ $\quad 3 \cdot 4 = 12$ $x^2 + 7x + 12 = (x + 3)(x + 4)$

DEFINITIONS AND CONCEPTS	EXAMPLES
Section 6.3 Factoring Trinomials of the Form $ax^2 + bx + c$ and Perfect Square Trinomials	
To factor $ax^2 + bx + c$, try various combinations of factors of ax^2 and c until a middle term of bx is obtained when checking.	Factor: $3x^2 + 14x - 5$ Factors of $3x^2$: $3x, x$ Factors of -5: $-1, 5$ and $1, -5$. 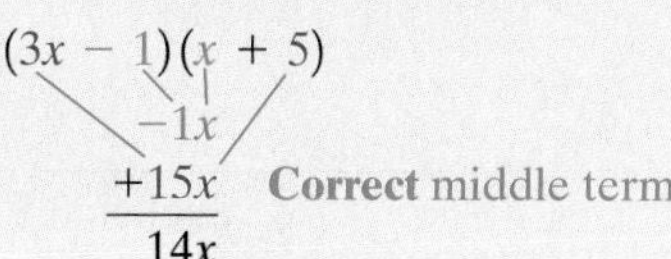 $(3x - 1)(x + 5)$ $-1x$ $+15x$ **Correct** middle term $14x$
A **perfect square trinomial** is a trinomial that is the square of some binomial.	Perfect square trinomial = square of binomial $x^2 + 4x + 4 = (x + 2)^2$ $25x^2 - 10x + 1 = (5x - 1)^2$
Factoring Perfect Square Trinomials: $a^2 + 2ab + b^2 = (a + b)^2$ $a^2 - 2ab + b^2 = (a - b)^2$	Factor: $x^2 + 6x + 9 = x^2 + 2 \cdot x \cdot 3 + 3^2 = (x + 3)^2$ $4x^2 - 12x + 9 = (2x)^2 - 2 \cdot 2x \cdot 3 + 3^2 = (2x - 3)^2$
Section 6.4 Factoring Trinomials of the Form $ax^2 + bx + c$ by Grouping	
To Factor $ax^2 + bx + c$ by Grouping	Factor: $3x^2 + 14x - 5$
Step 1. Find two numbers whose product is $a \cdot c$ and whose sum is b.	**Step 1.** Find two numbers whose product is $3 \cdot (-5)$ or -15 and whose sum is 14. They are 15 and -1.
Step 2. Rewrite bx, using the factors found in Step 1.	**Step 2.** $3x^2 + 14x - 5$ $= 3x^2 + 15x - 1x - 5$
Step 3. Factor by grouping.	**Step 3.** $= 3x(x + 5) - 1(x + 5)$ $= (x + 5)(3x - 1)$
Section 6.5 Factoring Binomials	
Difference of Squares $a^2 - b^2 = (a + b)(a - b)$	Factor: $x^2 - 9 = x^2 - 3^2 = (x + 3)(x - 3)$
Sum or Difference of Cubes $a^3 + b^3 = (a + b)(a^2 - ab + b^2)$ $a^3 - b^3 = (a - b)(a^2 + ab + b^2)$	$y^3 + 8 = y^3 + 2^3 = (y + 2)(y^2 - 2y + 4)$ $125z^3 - 1 = (5z)^3 - 1^3 = (5z - 1)(25z^2 + 5z + 1)$
Integrated Review—Choosing a Factoring Strategy	
To Factor a Polynomial,	Factor: $2x^4 - 6x^2 - 8$
Step 1. Factor out the GCF.	**Step 1.** $2x^4 - 6x^2 - 8 = 2(x^4 - 3x^2 - 4)$
Step 2. **a.** If two terms, **i.** $a^2 - b^2 = (a + b)(a - b)$ **ii.** $a^3 - b^3 = (a - b)(a^2 + ab + b^2)$ **iii.** $a^3 + b^3 = (a + b)(a^2 - ab + b^2)$	
b. If three terms, **i.** $a^2 + 2ab + b^2 = (a + b)^2$ **ii.** Methods in Sections 6.2 and 6.3	**Step 2.** **b. ii.** $= 2(x^2 + 1)(x^2 - 4)$
c. If four or more terms, try factoring by grouping.	*(continued)*

DEFINITIONS AND CONCEPTS	EXAMPLES
Integrated Review—Choosing a Factoring Strategy (continued)	
Step 3. See if any factors can be factored further.	**Step 3.** $= 2(x^2 + 1)(x + 2)(x - 2)$
Step 4. Check by multiplying.	**Step 4.** Check by multiplying. $2(x^2 + 1)(x + 2)(x - 2) = 2(x^2 + 1)(x^2 - 4)$ $= 2(x^4 - 3x^2 - 4)$ $= 2x^4 - 6x^2 - 8$
Section 6.6 Solving Quadratic Equations by Factoring	
A **quadratic equation** is an equation that can be written in the form $ax^2 + bx + c = 0$ with a not 0. The form $ax^2 + bx + c = 0$ is called the **standard form** of a quadratic equation.	***Quadratic Equation*** — ***Standard Form*** $x^2 = 16$ — $x^2 - 16 = 0$ $y = -2y^2 + 5$ — $2y^2 + y - 5 = 0$
Zero Factor Theorem If a and b are real numbers and if $ab = 0$, then $a = 0$ or $b = 0$.	If $(x + 3)(x - 1) = 0$, then $x + 3 = 0$ or $x - 1 = 0$
To solve quadratic equations by factoring,	Solve: $3x^2 = 13x - 4$
Step 1. Write the equation in standard form: $ax^2 + bx + c = 0$.	**Step 1.** $3x^2 - 13x + 4 = 0$
Step 2. Factor the quadratic.	**Step 2.** $(3x - 1)(x - 4) = 0$
Step 3. Set each factor containing a variable equal to 0.	**Step 3.** $3x - 1 = 0$ or $x - 4 = 0$
Step 4. Solve the equations.	**Step 4.** $3x = 1$ or $x = 4$ $x = \frac{1}{3}$
Step 5. Check in the original equation.	**Step 5.** Check both $\frac{1}{3}$ and 4 in the original equation.
Section 6.7 Quadratic Equations and Problem Solving	
Problem-Solving Steps	A garden is in the shape of a rectangle whose length is two feet more than its width. If the area of the garden is 35 square feet, find its dimensions.
1. UNDERSTAND the problem.	**1.** Read and reread the problem. Guess a solution and check your guess. Let x be the width of the rectangular garden. Then $x + 2$ is the length. 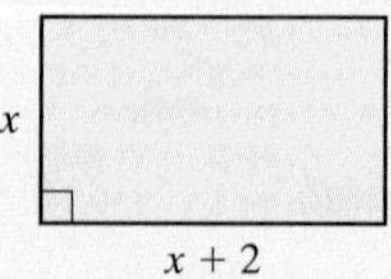
2. TRANSLATE.	**2.** In words: length · width = area ↓ ↓ ↓ Translate: $(x + 2) \cdot x = 35$

DEFINITIONS AND CONCEPTS	EXAMPLES
Section 6.7 Quadratic Equations and Problem Solving (continued)	
3. SOLVE.	**3.** $(x + 2)x = 35$ $x^2 + 2x - 35 = 0$ $(x - 5)(x + 7) = 0$ $x - 5 = 0$ or $x + 7 = 0$ $x = 5$ or $x = -7$
4. INTERPRET.	**4.** Discard the solution of -7 since x represents width. **Check:** If x is 5 feet then $x + 2 = 5 + 2 = 7$ feet. The area of a rectangle whose width is 5 feet and whose length is 7 feet is (5 feet)(7 feet) or 35 square feet. **State:** The garden is 5 feet by 7 feet.

Chapter 6 Review

***(6.1)** Complete the factoring.*

1. $6x^2 - 15x = 3x(\quad)$

2. $2x^3y + 6x^2y^2 + 8xy^3 = 2xy(\quad)$

Factor the GCF from each polynomial.

3. $20x^2 + 12x$

4. $6x^2y^2 - 3xy^3$

5. $3x(2x + 3) - 5(2x + 3)$

6. $5x(x + 1) - (x + 1)$

Factor each polynomial by grouping.

7. $3x^2 - 3x + 2x - 2$

8. $3a^2 + 9ab + 3b^2 + ab$

9. $10a^2 + 5ab + 7b^2 + 14ab$

10. $6x^2 + 10x - 3x - 5$

***(6.2)** Factor each trinomial.*

11. $x^2 + 6x + 8$

12. $x^2 - 11x + 24$

13. $x^2 + x + 2$

14. $x^2 - x + 2$

15. $x^2 + 4xy - 12y^2$

16. $x^2 + 8xy + 15y^2$

17. $72 - 18x - 2x^2$

18. $32 + 12x - 4x^2$

19. $10a^3 - 110a^2 + 100a$

20. $5y^3 - 50y^2 + 120y$

21. To factor $x^2 + 2x - 48$, think of two numbers whose product is ____ and whose sum is ____.

22. What is the first step in factoring $3x^2 + 15x + 30$?

***(6.3) or (6.4)** Factor each trinomial.*

23. $2x^2 + 13x + 6$

24. $4x^2 + 4x - 3$

25. $6x^2 + 5xy - 4y^2$

26. $18x^2 - 9xy - 20y^2$

27. $10y^3 + 25y^2 - 60y$

28. $60y^3 - 39y^2 + 6y$

29. $18x^2 - 60x + 50$

30. $4x^2 - 28xy + 49y^2$

***(6.5)** Factor each binomial.*

31. $4x^2 - 9$

32. $9t^2 - 25s^2$

33. $16x^2 + y^2$

34. $x^3 - 8y^3$

35. $8x^3 + 27$

36. $2x^3 + 8x$

37. $54 - 2x^3y^3$

38. $9x^2 - 4y^2$

39. $16x^4 - 1$

40. $x^4 + 16$

***(6.6)** Solve the following equations.*

41. $(x + 6)(x - 2) = 0$

42. $3x(x + 1)(7x - 2) = 0$

43. $4(5x + 1)(x + 3) = 0$

44. $x^2 + 8x + 7 = 0$

45. $x^2 - 2x - 24 = 0$

46. $x^2 + 10x = -25$

47. $x(x - 10) = -16$

48. $(3x - 1)(9x^2 - 6x + 1) = 0$

49. $56x^2 - 5x - 6 = 0$

50. $20x^2 - 7x - 6 = 0$

51. $5(3x + 2) = 4$

52. $6x^2 - 3x + 8 = 0$

53. $12 - 5t = -3$

54. $5x^3 + 20x^2 + 20x = 0$

55. $4t^3 - 5t^2 - 21t = 0$

56. Write a quadratic equation that has the two solutions 4 and 5.

(6.7) Use the given information to choose the correct dimensions.

57. The perimeter of a rectangle is 24 inches. The length is twice the width. Find the dimensions of the rectangle.

a. 5 inches by 7 inches
b. 5 inches by 10 inches
c. 4 inches by 8 inches
d. 2 inches by 10 inches

58. The area of a rectangle is 80 square meters. The length is one more than three times the width. Find the dimensions of the rectangle.

a. 8 meters by 10 meters
b. 4 meters by 13 meters
c. 4 meters by 20 meters
d. 5 meters by 16 meters

Use the given information to find the dimensions of each figure.

59. The *area* of the square is 81 square units. Find the length of a side.

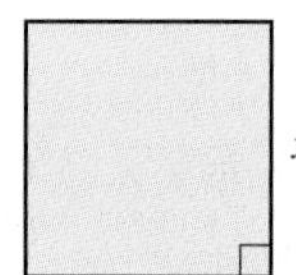

60. The *perimeter* of the quadrilateral is 47 units. Find the lengths of the sides.

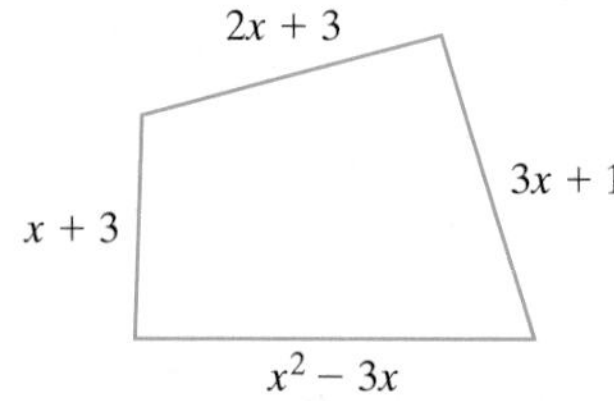

61. A flag for a local organization is in the shape of a rectangle whose length is 15 inches less than twice its width. If the area of the flag is 500 square inches, find its dimensions.

62. The base of a triangular sail is four times its height. If the area of the triangle is 162 square yards, find the base.

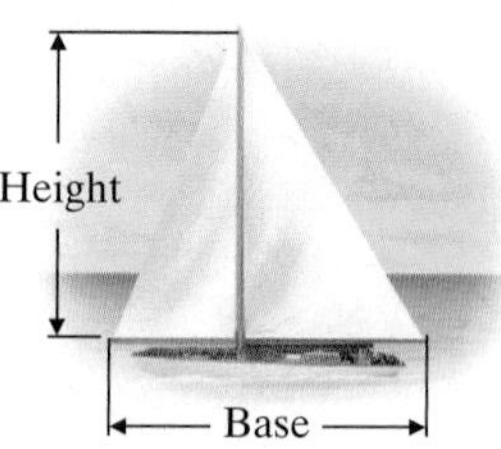

63. Find two consecutive positive integers whose product is 380.

64. Find two consecutive positive even integers whose product is 440.

65. A rocket is fired from the ground with an initial velocity of 440 feet per second. Its height h after t seconds is given by the equation

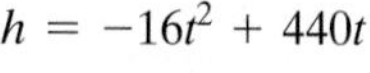
$$h = -16t^2 + 440t$$

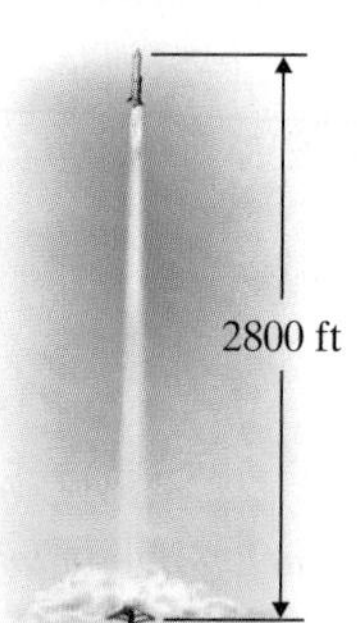

a. Find how many seconds pass before the rocket reaches a height of 2800 feet. Explain why two answers are obtained.

b. Find how many seconds pass before the rocket reaches the ground again.

66. An architect's squaring instrument is in the shape of a right triangle. Find the length of the longer leg of the right triangle if the hypotenuse is 8 centimeters longer than the longer leg and the shorter leg is 8 centimeters shorter than the longer leg.

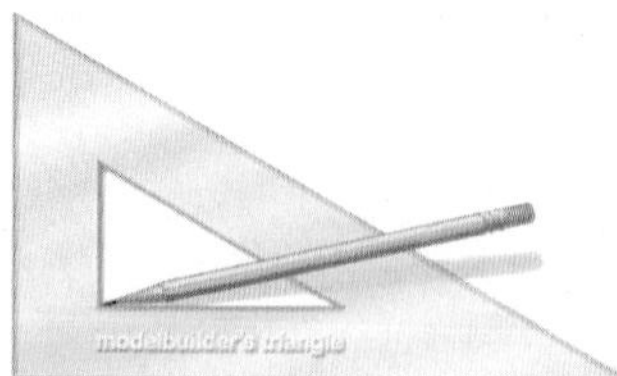

MIXED REVIEW

Factor completely.

67. $7x - 63$

68. $11x(4x - 3) - 6(4x - 3)$

69. $m^2 - \frac{4}{25}$

70. $3x^3 - 4x^2 + 6x - 8$

71. $xy + 2x - y - 2$

72. $2x^2 + 2x - 24$

73. $3x^3 - 30x^2 + 27x$

74. $4x^2 - 81$

75. $2x^2 - 18$

76. $16x^2 - 24x + 9$

77. $5x^2 + 20x + 20$

78. $2x^2 + 5x - 12$

79. $4x^2y - 6xy^2$

80. $125x^3 + 27$

81. $24x^2 - 3x - 18$

82. $(x + 7)^2 - y^2$

83. $x^2(x + 3) - 4(x + 3)$

84. $54a^3b - 2b$

Write the perimeter of each figure as a simplified polynomial. Then factor each polynomial.

△ **85.**

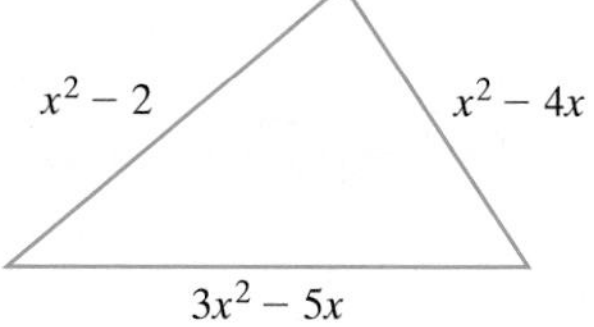

△ **86.**

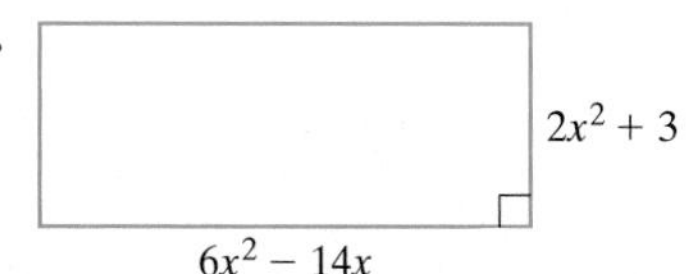

Solve.

87. $2x^2 - x - 28 = 0$

88. $x^2 - 2x = 15$

89. $2x(x + 7)(x + 4) = 0$

90. $x(x - 5) = -6$

91. $x^2 = 16x$

Solve.

△ **92.** The perimeter of the following triangle is 48 inches. Find the lengths of its sides.

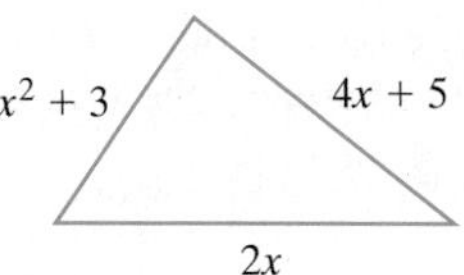

93. The width of a rectangle is 4 inches less than its length. Its area is 12 square inches. Find the dimensions of the rectangle.

94. A 6-foot-tall person drops an object from the top of the Westin Peachtree Plaza in Atlanta, Georgia. The Westin building is 723 feet tall. (*Source: World Almanac* research) The height h of the object after t seconds is given by the equation $h = -16t^2 + 729$. Find how many seconds pass before the object reaches the ground.

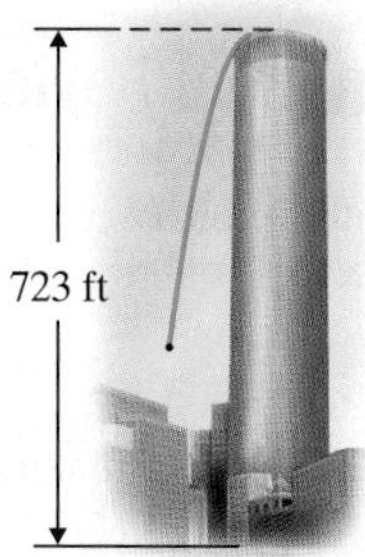

Write an expression for the area of the shaded region. Then write the expression as a factored polynomial.

△ **95.**

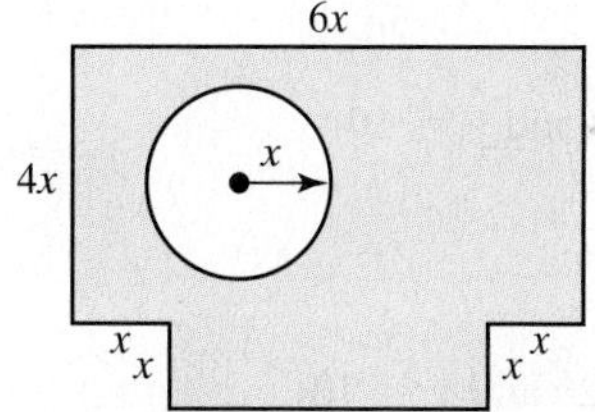

Chapter 6 Test

Factor each polynomial completely. If a polynomial cannot be factored, write "prime."

1. $x^2 + 11x + 28$

2. $49 - m^2$

3. $y^2 + 22y + 121$

4. $4(a + 3) - y(a + 3)$

5. $x^2 + 4$

6. $y^2 - 8y - 48$

7. $x^2 + x - 10$

8. $9x^3 + 39x^2 + 12x$

9. $3a^2 + 3ab - 7a - 7b$

10. $3x^2 - 5x + 2$

11. $x^2 + 14xy + 24y^2$

12. $180 - 5x^2$

13. $6t^2 - t - 5$

14. $xy^2 - 7y^2 - 4x + 28$

15. $x - x^5$

16. $-xy^3 - x^3y$

17. $64x^3 - 1$

18. $8y^3 - 64$

Solve each equation.

19. $(x - 3)(x + 9) = 0$

20. $x^2 + 5x = 14$

21. $x(x + 6) = 7$

22. $3x(2x - 3)(3x + 4) = 0$

23. $5t^3 - 45t = 0$

24. $t^2 - 2t - 15 = 0$

25. $6x^2 = 15x$

Solve each problem.

26. A deck for a home is in the shape of a triangle. The length of the base of the triangle is 9 feet longer than its altitude. If the area of the triangle is 68 square feet, find the length of the base.

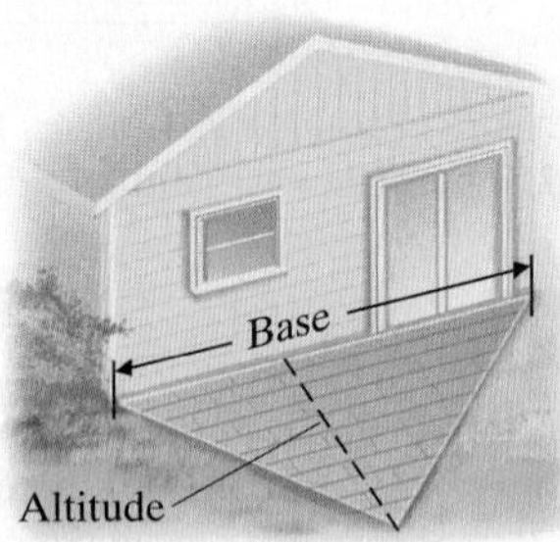

27. The sum of two numbers is 17 and the sum of their squares is 145. Find the numbers.

28. An object is dropped from the top of the Woolworth Building on Broadway in New York City. The height h of the object after t seconds is given by the equation

$$h = -16t^2 + 784$$

Find how many seconds pass before the object reaches the ground.

29. Find the lengths of the sides of a right triangle if the hypotenuse is 10 centimeters longer than the shorter leg and 5 centimeters longer than the longer leg.

Chapter 6 Cumulative Review

1. Translate each sentence into a mathematical statement.
   a. Nine is less than or equal to eleven.
   b. Eight is greater than one.
   c. Three is not equal to four.
2. Insert < or > in the space to make each statement true.
   a. $|-5| \quad |-3|$
   b. $|0| \quad |-2|$
3. Simplify each fraction (write it in lowest terms).
   a. $\frac{42}{49}$ **b.** $\frac{11}{27}$ **c.** $\frac{88}{20}$
4. Evaluate $\frac{x}{y} + 5x$ if $x = 20$ and $y = 10$.
5. Simplify: $\frac{8 + 2 \cdot 3}{2^2 - 1}$
6. Evaluate $\frac{x}{y} + 5x$ if $x = -20$ and $y = 10$.
7. Add.
   a. $3 + (-7) + (-8)$
   b. $[7 + (-10)] + [-2 + |-4|]$
8. Evaluate $\frac{x}{y} + 5x$ if $x = -20$ and $y = -10$.
9. Multiply.
   a. $(-8)(4)$
   b. $14(-1)$
   c. $(-9)(-10)$
10. Simplify: $5 - 2(3x - 7)$
11. Simplify each expression by combining like terms.
    a. $7x - 3x$ **b.** $10y^2 + y^2$
    c. $8x^2 + 2x - 3x$ **d.** $9n^2 - 5n^2 + n^2$
12. Solve: $0.8y + 0.2(y - 1) = 1.8$

Solve.

13. $3 - x = 7$
14. $\frac{x}{-7} = -4$
15. $-3x = 33$
16. $-\frac{2}{3}x = -22$
17. $8(2 - t) = -5t$
18. $-z = \frac{7z + 3}{5}$
19. Balsa wood sticks are commonly used to build models (for example, bridge models). A 48-inch balsa wood stick is to be cut into two pieces so that the longer piece is 3 times the shorter. Find the length of each piece.
20. Solve $3x + 9 \leq 5(x - 1)$. Write the solution set using interval notation.
21. Graph the linear equation $y = -\frac{1}{3}x + 2$.
22. Is the ordered pair $(-1, 2)$ a solution of $-7x - 8y = -9$?
23. Find the slope and the y-intercept of the line whose equation is $3x - 4y = 4$.
24. Find the slope of the line through $(5, -6)$ and $(5, 2)$.
25. Evaluate each expression for the given value of x.
    a. $2x^3$; x is 5 **b.** $\frac{9}{x^2}$; x is -3
26. Find the slope and y-intercept of the line whose equation is $7x - 3y = 2$.
27. Find the degree of each term.
    a. $-3x^2$ **b.** $5x^3yz$ **c.** 2
28. Find an equation of the vertical line through $(0, 7)$.
29. Subtract: $(2x^3 + 8x^2 - 6x) - (2x^3 - x^2 + 1)$

30. Find an equation of the line with slope 4 and y-intercept $\left(0, \frac{1}{2}\right)$. Write the equation in standard form.

31. Multiply $(3x + 2)(2x - 5)$.

32. Write an equation of the line through $(-4, 0)$ and $(6, -1)$. Write the equation in standard form.

33. Multiply $(3y + 1)^2$.

34. Solve the system: $\begin{cases} -x + 3y = 18 \\ -3x + 2y = 19 \end{cases}$

35. Simplify by writing each expression with positive exponents only.

a. 3^{-2} **b.** $2x^{-3}$

c. $2^{-1} + 4^{-1}$ **d.** $(-2)^{-4}$

e. y^{-4}

36. Simplify: $\dfrac{(5a^7)^2}{a^5}$

37. Write each number in scientific notation.

a. 367,000,000

b. 0.000003

c. 20,520,000,000

d. 0.00085

38. Multiply $(3x - 7y)^2$.

39. Divide $x^2 + 7x + 12$ by $x + 3$ using long division.

40. Simplify: $\dfrac{(xy)^{-3}}{(x^5y^6)^3}$

41. Find the GCF of each list of terms.

a. x^3, x^7, and x^5 **b.** y, y^4, and y^7

Factor.

42. $z^3 + 7z + z^2 + 7$

43. $x^2 + 7x + 12$

44. $2x^3 + 2x^2 - 84x$

45. $8x^2 - 22x + 5$

46. $-4x^2 - 23x + 6$

47. $25a^2 - 9b^2$

48. $9xy^2 - 16x$

49. Solve: $(x - 3)(x + 1) = 0$

50. Solve $x^2 - 13x = -36$.

CHAPTER 7 Rational Expressions

In this chapter, we expand our knowledge of algebraic expressions to include algebraic fractions, called *rational expressions.* We explore the operations of addition, subtraction, multiplication, and division using principles similar to the principles for numerical fractions.

Many sport statistics are calculated using formulas containing rational expressions. Below are a few examples:

Sport	***Formula***	***Explanation***
Baseball	$S = \dfrac{h + d + 2t + 3r}{b}$	A baseball player's slugging percentage S, where h = number of hits, d = number of doubles, t = number of triples, r = number of home runs, and b = number of at bats
NCAA Football	$R_{\text{NCAA}} = \dfrac{100C + 330T - 200I + 8.4Y}{A}$	A quarterback's rating in NCAA football R, where C = the number of completed passes, A = the number of attempted passes, T = the number of touchdown passes, Y = the number of yards in the completed passes, and I = the number of interceptions
NFL Football	$R_{\text{NFL}} = \dfrac{a + b + c + d}{6} \times 100$	A quarterback's rating in NFL football R, where $a, b, c,$ and d are each formulas containing rational expressions.

(*Source:* Wikipedia.org.)

In Section 7.1, Exercises 101 and 102, you will have the opportunity to calculate some sport statistics.

7.1 Simplifying Rational Expressions

OBJECTIVES

1 Find the Value of a Rational Expression Given a Replacement Number.

2 Identify Values for Which a Rational Expression Is Undefined.

3 Simplify or Write Rational Expressions in Lowest Terms.

4 Write Equivalent Rational Expressions of the Form $-\frac{a}{b}=\frac{-a}{b}=\frac{a}{-b}$.

OBJECTIVE

1 Evaluating Rational Expressions

As we reviewed in Chapter 1, a rational number is a number that can be written as a quotient of integers. A **rational expression** is also a quotient; it is a quotient of polynomials.

Rational Expression

A rational expression is an expression that can be written in the form

$$\frac{P}{Q},$$

where P and Q are polynomials and $Q \neq 0$.

Rational Expressions

$$\frac{2}{3} \qquad \frac{3y^3}{8} \qquad \frac{-4p}{p^3+2p+1} \qquad \frac{5x^2-3x+2}{3x+7}$$

Rational expressions have different values depending on what value replaces the variable. Next, we review the standard order of operations by finding values of rational expressions for given replacement values of the variable.

EXAMPLE 1 Find the value of $\frac{x+4}{2x-3}$ for the given replacement values.

a. $x = 5$ **b.** $x = -2$

Solution

a. Replace each x in the expression with 5 and then simplify.

$$\frac{x+4}{2x-3}=\frac{5+4}{2(5)-3}=\frac{9}{10-3}=\frac{9}{7}$$

b. Replace each x in the expression with -2 and then simplify.

$$\frac{x+4}{2x-3}=\frac{-2+4}{2(-2)-3}=\frac{2}{-7} \quad \text{or} \quad -\frac{2}{7}$$

□

PRACTICE

1 Find the value of $\frac{x+6}{3x-2}$ for the given replacement values.

a. $x = 3$ **b.** $x = -3$

In the example above, we wrote $\frac{2}{-7}$ as $-\frac{2}{7}$. For a negative fraction such as $\frac{2}{-7}$, recall from Section 1.7 that

$$\frac{2}{-7}=\frac{-2}{7}=-\frac{2}{7}$$

In general, for any fraction,

$$\frac{-a}{b}=\frac{a}{-b}=-\frac{a}{b}, \qquad b \neq 0$$

This is also true for rational expressions. For example,

$$\underbrace{\frac{-(x+2)}{x}}=\frac{x+2}{-x}=-\frac{x+2}{x}$$

Notice the parentheses.

> **Helpful Hint**
>
> **Do you recall why division by 0 is not defined?** Remember, for example, that $\frac{8}{4} = 2$ because $2 \cdot 4 = 8$. Thus, if $\frac{8}{0} =$ *a number*, then *the number* $\cdot 0 = 8$. There is no number that when multiplied by 0 equals 8; thus $\frac{8}{0}$ is undefined. This is true in general for fractions and rational expressions.

OBJECTIVE

2 Identifying When a Rational Expression Is Undefined 

In the definition of rational expression (first "box" in this section), notice that we wrote $Q \neq 0$ for the denominator Q. This is because the denominator of a rational expression must not equal 0 since division by 0 is not defined. (See the margin Helpful Hint.) This means we must be careful when replacing the variable in a rational expression by a number. For example, suppose we replace x with 5 in the rational expression $\frac{3 + x}{x - 5}$. The expression becomes

$$\frac{3 + x}{x - 5} = \frac{3 + 5}{5 - 5} = \frac{8}{0}$$

But division by 0 is undefined. Therefore, in this rational expression, we can allow x to be any real number *except* 5. **A rational expression is undefined for values that make the denominator 0.** Thus, to find values for which a rational expression is undefined, find values for which the denominator is 0.

EXAMPLE 2 Are there any values for x for which each rational expression is undefined?

a. $\frac{x}{x - 3}$ **b.** $\frac{x^2 + 2}{3x^2 - 5x + 2}$ **c.** $\frac{x^3 - 6x^2 - 10x}{3}$ **d.** $\frac{2}{x^2 + 1}$

Solution To find values for which a rational expression is undefined, find values that make the *denominator* 0.

a. The denominator of $\frac{x}{x - 3}$ is 0 when $x - 3 = 0$ or when $x = 3$. Thus, when $x = 3$, the expression $\frac{x}{x - 3}$ is undefined.

b. Set the denominator equal to zero.

$$3x^2 - 5x + 2 = 0$$

$$(3x - 2)(x - 1) = 0 \qquad \text{Factor.}$$

$$3x - 2 = 0 \quad \text{or} \quad x - 1 = 0 \qquad \text{Set each factor equal to zero.}$$

$$3x = 2 \quad \text{or} \quad x = 1 \qquad \text{Solve.}$$

$$x = \frac{2}{3}$$

Thus, when $x = \frac{2}{3}$ or $x = 1$, the denominator $3x^2 - 5x + 2$ is 0. So the rational expression $\frac{x^2 + 2}{3x^2 - 5x + 2}$ is undefined when $x = \frac{2}{3}$ or when $x = 1$.

c. The denominator of $\frac{x^3 - 6x^2 - 10x}{3}$ is never 0, so there are no values of x for which this expression is undefined.

d. No matter which real number x is replaced by, the denominator $x^2 + 1$ does not equal 0, so there are no real numbers for which this expression is undefined. □

PRACTICE

2 Are there any values of x for which each rational expression is undefined?

a. $\frac{x}{x + 6}$ **b.** $\frac{x^4 - 3x^2 + 7x}{7}$ **c.** $\frac{x^2 - 5}{x^2 + 6x + 8}$ **d.** $\frac{3}{x^4 + 5}$

Note: Unless otherwise stated, we will now assume that variables in rational expressions are replaced only by values for which the expressions are defined.

OBJECTIVE

3 Simplifying Rational Expressions

A fraction is said to be written in lowest terms or simplest form when the numerator and denominator have no common factors other than 1 (or −1). For example, the fraction $\frac{7}{10}$ is in lowest terms since the numerator and denominator have no common factors other than 1 (or −1).

The process of writing a rational expression in lowest terms or simplest form is called **simplifying** a rational expression.

Simplifying a rational expression is similar to simplifying a fraction. Recall that to simplify a fraction, we essentially "remove factors of 1." Our ability to do this comes from these facts:

- Any nonzero number over itself simplifies to 1

  $\left(\frac{5}{5} = 1, \frac{-7.26}{-7.26} = 1, \text{ or } \frac{c}{c} = 1 \text{ as long as } c \text{ is not } 0\right)$, and

- The product of any number and 1 is that number

  $\left(19 \cdot 1 = 19, -8.9 \cdot 1 = -8.9, \frac{a}{b} \cdot 1 = \frac{a}{b}\right)$.

In other words, we have the following:

$$\frac{a \cdot c}{b \cdot c} = \frac{a}{b} \cdot \frac{c}{c} = \frac{a}{b}$$

Since $\frac{a}{b} \cdot 1 = \frac{a}{b}$

Simplify: $\frac{15}{20}$

$\frac{15}{20} = \frac{3 \cdot 5}{2 \cdot 2 \cdot 5}$ Factor the numerator and the denominator.

$= \frac{3 \cdot 5}{2 \cdot 2 \cdot 5}$ Look for common factors.

$= \frac{3}{2 \cdot 2} \cdot \frac{5}{5}$ Common factors in the numerator and denominator form factors of 1.

$= \frac{3}{2 \cdot 2} \cdot 1$ Write $\frac{5}{5}$ as 1.

$= \frac{3}{2 \cdot 2} = \frac{3}{4}$ Multiply to remove a factor of 1.

Before we use the same technique to simplify a rational expression, remember that as long as the denominator is not 0, $\frac{a^3b}{a^3b} = 1$, $\frac{x+3}{x+3} = 1$, and $\frac{7x^2 + 5x - 100}{7x^2 + 5x - 100} = 1$.

Simplify: $\frac{x^2 - 9}{x^2 + x - 6}$

$\frac{x^2 - 9}{x^2 + x - 6} = \frac{(x-3)(x+3)}{(x-2)(x+3)}$ Factor the numerator and the denominator.

$= \frac{(x-3)(x+3)}{(x-2)(x+3)}$ Look for common factors.

$= \frac{x-3}{x-2} \cdot \frac{x+3}{x+3}$

$= \frac{x-3}{x-2} \cdot 1$ Write $\frac{x+3}{x+3}$ as 1.

$= \frac{x-3}{x-2}$ Multiply to remove a factor of 1.

Just as for numerical fractions, we can use a shortcut notation. Remember that as long as exact factors in both the numerator and denominator are divided out, we are "removing a factor of 1." We will use the following notation to show this:

$$\frac{x^2 - 9}{x^2 + x - 6} = \frac{(x - 3)(x + 3)}{(x - 2)(x + 3)} \quad \text{A factor of 1 is identified by the shading.}$$

$$= \frac{x - 3}{x - 2} \quad \text{Remove a factor of 1.}$$

Thus, the rational expression $\frac{x^2 - 9}{x^2 + x - 6}$ has the same value as the rational expression $\frac{x - 3}{x - 2}$ for all values of x except 2 and -3. (Remember that when x is 2, the denominator of both rational expressions is 0 and when x is -3, the original rational expression has a denominator of 0.)

As we simplify rational expressions, we will assume that the simplified rational expression is equal to the original rational expression for all real numbers except those for which either denominator is 0. The following steps may be used to simplify rational expressions.

To Simplify a Rational Expression

Step 1. Completely factor the numerator and denominator.

Step 2. Divide out factors common to the numerator and denominator. (This is the same as "removing a factor of 1.")

EXAMPLE 3 Simplify: $\frac{5x - 5}{x^3 - x^2}$

Solution To begin, we factor the numerator and denominator if possible. Then we look for common factors.

$$\frac{5x - 5}{x^3 - x^2} = \frac{5(x - 1)}{x^2(x - 1)} = \frac{5}{x^2}$$

□

PRACTICE 3 Simplify: $\frac{x^6 - x^5}{6x - 6}$

■

EXAMPLE 4 Simplify: $\frac{x^2 + 8x + 7}{x^2 - 4x - 5}$

Solution We factor the numerator and denominator and then look for common factors.

$$\frac{x^2 + 8x + 7}{x^2 - 4x - 5} = \frac{(x + 7)(x + 1)}{(x - 5)(x + 1)} = \frac{x + 7}{x - 5}$$

□

PRACTICE 4 Simplify: $\frac{x^2 + 5x + 4}{x^2 + 2x - 8}$

■

EXAMPLE 5 Simplify: $\frac{x^2 + 4x + 4}{x^2 + 2x}$

Solution We factor the numerator and denominator and then look for common factors.

$$\frac{x^2 + 4x + 4}{x^2 + 2x} = \frac{(x + 2)(x + 2)}{x(x + 2)} = \frac{x + 2}{x}$$

□

PRACTICE 5 Simplify: $\frac{x^3 + 9x^2}{x^2 + 18x + 81}$

■

Helpful Hint

When simplifying a rational expression, we look for **common *factors*, not common *terms*.**

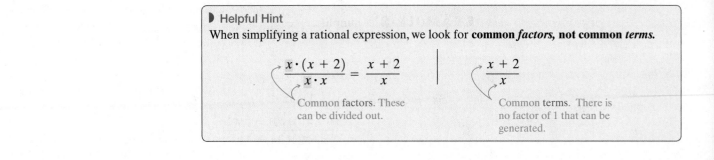

$$\frac{x \cdot (x+2)}{x \cdot x} = \frac{x+2}{x}$$

Common factors. These can be divided out.

$$\frac{x+2}{x}$$

Common terms. There is no factor of 1 that can be generated.

✓CONCEPT CHECK

Recall that we can remove only *factors* of 1. Which of the following are *not* true? Explain why.

a. $\frac{3-1}{3+5}$ simplifies to $-\frac{1}{5}$? **b.** $\frac{2x+10}{2}$ simplifies to $x+5$?

c. $\frac{37}{72}$ simplifies to $\frac{3}{2}$? **d.** $\frac{2x+3}{2}$ simplifies to $x+3$?

EXAMPLE 6 Simplify: $\frac{x+9}{x^2-81}$

Solution We factor and then divide out common factors.

$$\frac{x+9}{x^2-81} = \frac{x+9}{(x+9)(x-9)} = \frac{1}{x-9}$$

PRACTICE 6 Simplify: $\frac{x-7}{x^2-49}$

EXAMPLE 7 Simplify each rational expression.

a. $\frac{x+y}{y+x}$ **b.** $\frac{x-y}{y-x}$

Solution

a. The expression $\frac{x+y}{y+x}$ can be simplified by using the commutative property of addition to rewrite the denominator $y+x$ as $x+y$.

$$\frac{x+y}{y+x} = \frac{x+y}{x+y} = 1$$

b. The expression $\frac{x-y}{y-x}$ can be simplified by recognizing that $y-x$ and $x-y$ are opposites. In other words, $y-x = -1(x-y)$. We proceed as follows:

$$\frac{x-y}{y-x} = \frac{1 \cdot (x-y)}{-1 \cdot (x-y)} = \frac{1}{-1} = -1$$

PRACTICE 7 Simplify each rational expression.

a. $\frac{s-t}{t-s}$ **b.** $\frac{2c+d}{d+2c}$

Answers to Concept Check:
a, c, d

EXAMPLE 8 Simplify: $\dfrac{4 - x^2}{3x^2 - 5x - 2}$

Solution

$$\frac{4 - x^2}{3x^2 - 5x - 2} = \frac{(2 - x)(2 + x)}{(x - 2)(3x + 1)} \quad \text{Factor.}$$

$$= \frac{(-1)(x - 2)(2 + x)}{(x - 2)(3x + 1)} \quad \text{Write } 2 - x \text{ as } -1(x - 2).$$

$$= \frac{(-1)(2 + x)}{3x + 1} \text{ or } \frac{-2 - x}{3x + 1} \quad \text{Simplify.}$$

□

PRACTICE 8 Simplify: $\dfrac{2x^2 - 5x - 12}{16 - x^2}$

OBJECTIVE 4

Writing Equivalent Forms of Rational Expressions

From Example 7a, we have $y + x = x + y$. $y + x$ and $x + y$ are equivalent.
From Example 7b, we have $y - x = -1(x - y)$. $y - x$ and $x - y$ are opposites.

$$\text{Thus, } \frac{x + y}{y + x} = \frac{x + y}{x + y} = 1 \quad \text{and} \quad \frac{x - y}{y - x} = \frac{x - y}{-1(x - y)} = \frac{1}{-1} = -1.$$

When performing operations on rational expressions, equivalent forms of answers often result. For this reason, it is very important to be able to recognize equivalent answers.

EXAMPLE 9 List some equivalent forms of $-\dfrac{5x - 1}{x + 9}$.

Solution To do so, recall that $-\dfrac{a}{b} = \dfrac{-a}{b} = \dfrac{a}{-b}$. Thus

$$-\frac{5x - 1}{x + 9} = \frac{-(5x - 1)}{x + 9} = \frac{-5x + 1}{x + 9} \text{ or } \frac{1 - 5x}{x + 9}.$$

Also,

$$-\frac{5x - 1}{x + 9} = \frac{5x - 1}{-(x + 9)} = \frac{5x - 1}{-x - 9} \text{ or } \frac{5x - 1}{-9 - x}$$

Thus $-\dfrac{5x - 1}{x + 9} = \dfrac{-(5x - 1)}{x + 9} = \dfrac{-5x + 1}{x + 9} = \dfrac{5x - 1}{-(x + 9)} = \dfrac{5x - 1}{-x - 9}$ □

Helpful Hint

Remember, a negative sign in front of a fraction or rational expression may be moved to the numerator or the denominator, but *not* both.

PRACTICE 9 List some equivalent forms of $-\dfrac{x + 3}{6x - 11}$.

Keep in mind that many rational expressions may look different but in fact be equivalent.

Vocabulary, Readiness & Video Check

Use the choices below to fill in each blank. Not all choices will be used.

-1 0 simplifying $\frac{-a}{-b}$ $\frac{-a}{b}$ $\frac{a}{-b}$

1 2 rational expression

1. A ______________ is an expression that can be written in the form $\frac{P}{Q}$ where P and Q are polynomials and $Q \neq 0$.

2. The expression $\frac{x+3}{3+x}$ simplifies to ______________.

3. The expression $\frac{x-3}{3-x}$ simplifies to ______________.

4. A rational expression is undefined for values that make the denominator ______________.

5. The expression $\frac{7x}{x-2}$ is undefined for $x =$ ______________.

6. The process of writing a rational expression in lowest terms is called ______________.

7. For a rational expression, $-\frac{a}{b} = \underline{\quad} = \underline{\quad}$.

Martin-Gay Interactive Videos

See Video 7.1

Watch the section lecture video and answer the following questions.

OBJECTIVE 1

8. From the lecture before Example 1, what do the different values of a rational expression depend on? How are these different values found?

OBJECTIVE 2

9. Why can't the denominators of rational expressions be zero? How can you find the numbers for which a rational expression is undefined?

OBJECTIVE 3

10. In Example 7, why isn't a factor of x divided out of the expression at the end?

OBJECTIVE 4

11. From Example 9, if you move a negative sign from in front of a rational expression to either the numerator or denominator, when would you need to use parentheses and why?

7.1 Exercise Set

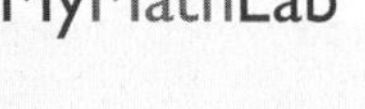

MyMathLab®

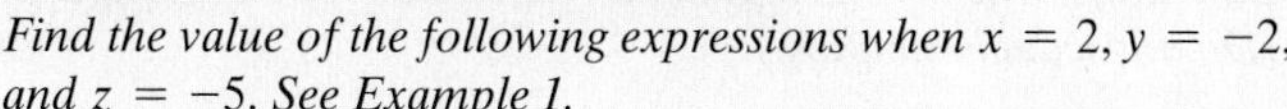

Find the value of the following expressions when $x = 2$, $y = -2$, and $z = -5$. See Example 1.

1. $\frac{x+5}{x+2}$

2. $\frac{x+8}{x+1}$

3. $\frac{4z-1}{z-2}$

4. $\frac{7y-1}{y-1}$

5. $\frac{y^3}{y^2-1}$

6. $\frac{z}{z^2-5}$

7. $\frac{x^2+8x+2}{x^2-x-6}$

8. $\frac{x+5}{x^2+4x-8}$

Find any numbers for which each rational expression is undefined. See Example 2.

9. $\frac{7}{2x}$

10. $\frac{3}{5x}$

11. $\frac{x+3}{x+2}$

12. $\frac{5x+1}{x-9}$

13. $\frac{x-4}{2x-5}$

14. $\frac{x+1}{5x-2}$

15. $\frac{x^2-5x-2}{4}$

16. $\frac{9y^5+y^3}{9}$

17. $\frac{3x^2+9}{x^2-5x-6}$

18. $\frac{11x^2+1}{x^2-5x-14}$

19. $\frac{9x^3+4}{x^2+36}$

20. $\frac{19x^3+2}{x^2+4}$

21. $\frac{x}{3x^2+13x+14}$

22. $\frac{x}{2x^2+15x+27}$

Study Example 9. Then list four equivalent forms for each rational expression.

23. $-\frac{x-10}{x+8}$

24. $-\frac{x+11}{x-4}$

25. $-\frac{5y-3}{y-12}$

26. $-\frac{8y-1}{y-15}$

Simplify each expression. See Examples 3 through 8.

27. $\frac{x+7}{7+x}$
28. $\frac{y+9}{9+y}$
29. $\frac{x-7}{7-x}$
30. $\frac{y-9}{9-y}$
31. $\frac{2}{8x+16}$
32. $\frac{3}{9x+6}$
33. $\frac{x-2}{x^2-4}$
34. $\frac{x+5}{x^2-25}$
35. $\frac{2x-10}{3x-30}$
36. $\frac{3x-9}{4x-16}$
37. $\frac{-5a-5b}{a+b}$
38. $\frac{-4x-4y}{x+y}$
39. $\frac{7x+35}{x^2+5x}$
40. $\frac{9x+99}{x^2+11x}$
41. $\frac{x+5}{x^2-4x-45}$
42. $\frac{x-3}{x^2-6x+9}$
43. $\frac{5x^2+11x+2}{x+2}$
44. $\frac{12x^2+4x-1}{2x+1}$
45. $\frac{x^3+7x^2}{x^2+5x-14}$
46. $\frac{x^4-10x^3}{x^2-17x+70}$
47. $\frac{14x^2-21x}{2x-3}$
48. $\frac{4x^2+24x}{x+6}$
49. $\frac{x^2+7x+10}{x^2-3x-10}$
50. $\frac{2x^2+7x-4}{x^2+3x-4}$
51. $\frac{3x^2+7x+2}{3x^2+13x+4}$
52. $\frac{4x^2-4x+1}{2x^2+9x-5}$
53. $\frac{2x^2-8}{4x-8}$
54. $\frac{5x^2-500}{35x+350}$
55. $\frac{4-x^2}{x-2}$
56. $\frac{49-y^2}{y-7}$
57. $\frac{x^2-1}{x^2-2x+1}$
58. $\frac{x^2-16}{x^2-8x+16}$
59. $\frac{m^2-6m+9}{m^2-m-6}$
60. $\frac{m^2-4m+4}{m^2+m-6}$
61. $\frac{11x^2-22x^3}{6x-12x^2}$
62. $\frac{24y^2-8y^3}{15y-5y^2}$

Simplify. These expressions contain 4-term polynomials and sums and differences of cubes.

63. $\frac{x^2+xy+2x+2y}{x+2}$
64. $\frac{ab+ac+b^2+bc}{b+c}$
65. $\frac{5x+15-xy-3y}{2x+6}$
66. $\frac{xy-6x+2y-12}{y^2-6y}$
67. $\frac{x^3+8}{x+2}$
68. $\frac{x^3+64}{x+4}$
69. $\frac{x^3-1}{1-x}$
70. $\frac{3-x}{x^3-27}$
71. $\frac{2xy+5x-2y-5}{3xy+4x-3y-4}$
72. $\frac{2xy+2x-3y-3}{2xy+4x-3y-6}$

MIXED PRACTICE

Simplify each expression. Then determine whether the given answer is correct. See Examples 3 through 9.

73. $\frac{9-x^2}{x-3}$; Answer: $-3-x$
74. $\frac{100-x^2}{x-10}$; Answer: $-10-x$
75. $\frac{7-34x-5x^2}{25x^2-1}$; Answer: $\frac{x+7}{-5x-1}$
76. $\frac{2-15x-8x^2}{64x^2-1}$; Answer: $\frac{x+2}{-8x-1}$

REVIEW AND PREVIEW

Perform each indicated operation. See Section 1.3.

77. $\frac{1}{3}\cdot\frac{9}{11}$
78. $\frac{5}{27}\cdot\frac{2}{5}$
79. $\frac{1}{3}\div\frac{1}{4}$
80. $\frac{7}{8}\div\frac{1}{2}$
81. $\frac{13}{20}\div\frac{2}{9}$
82. $\frac{8}{15}\div\frac{5}{8}$

CONCEPT EXTENSIONS

Which of the following are incorrect and why? See the Concept Check in this section.

83. $\frac{5a-15}{5}$ simplifies to $a-3$?
84. $\frac{7m-9}{7}$ simplifies to $m-9$?
85. $\frac{1+2}{1+3}$ simplifies to $\frac{2}{3}$?
86. $\frac{46}{54}$ simplifies to $\frac{6}{5}$?

Determine whether each rational expression can be simplified. (Do not actually simplify.)

87. $\frac{x}{x+7}$
88. $\frac{3+x}{x+3}$
89. $\frac{5-x}{x-5}$
90. $\frac{x+2}{x+8}$
91. Explain how to write a fraction in lowest terms.
92. Explain how to write a rational expression in lowest terms.
93. Explain why the denominator of a fraction or a rational expression must not equal 0.
94. Does $\frac{(x-3)(x+3)}{x-3}$ have the same value as $x+3$ for all real numbers? Explain why or why not.

95. The total revenue R from the sale of a popular music compact disc is approximately given by the equation

$$R = \frac{150x^2}{x^2 + 3}$$

where x is the number of years since the CD has been released and revenue R is in millions of dollars.

a. Find the total revenue generated by the end of the first year.

b. Find the total revenue generated by the end of the second year.

c. Find the total revenue generated in the second year only.

96. For a certain model fax machine, the manufacturing cost C per machine is given by the equation

$$C = \frac{250x + 10{,}000}{x}$$

where x is the number of fax machines manufactured and cost C is in dollars per machine.

a. Find the cost per fax machine when manufacturing 100 fax machines.

b. Find the cost per fax machine when manufacturing 1000 fax machines.

c. Does the cost per machine decrease or increase when more machines are manufactured? Explain why this is so.

Solve.

97. The dose of medicine prescribed for a child depends on the child's age A in years and the adult dose D for the medication. Young's Rule is a formula used by pediatricians that gives a child's dose C as

$$C = \frac{DA}{A + 12}$$

Suppose that an 8-year-old child needs medication, and the normal adult dose is 1000 mg. What size dose should the child receive?

98. Calculating body-mass index is a way to gauge whether a person should lose weight. Doctors recommend that body-mass index values fall between 18.5 and 25. The formula for body-mass index B is

$$B = \frac{703w}{h^2}$$

where w is weight in pounds and h is height in inches. Should a 148-pound person who is 5 feet 6 inches tall lose weight?

99. Anthropologists and forensic scientists use a measure called the cephalic index to help classify skulls. The cephalic index of a skull with width W and length L from front to back is given by the formula

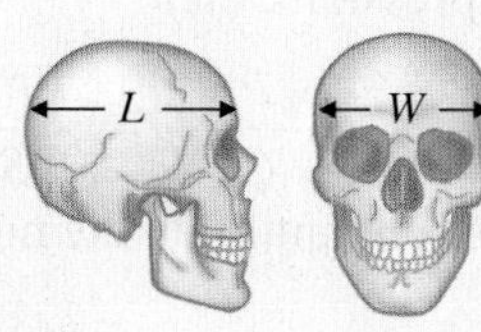

$$C = \frac{100W}{L}$$ (Cont. in next column)

A long skull has an index value less than 75, a medium skull has an index value between 75 and 85, and a broad skull has an index value over 85. Find the cephalic index of a skull that is 5 inches wide and 6.4 inches long. Classify the skull.

100. A company's gross profit margin P can be computed with the formula $P = \frac{R - C}{R}$, where R = the company's revenue and C = cost of goods sold. During a recent fiscal year, computer company Apple had revenues of \$32.5 billion and cost of goods sold \$21.3 billion. (*Source:* Apple, Inc.) What was Apple's gross profit margin in this year? Express the answer as a percent, rounded to the nearest tenth of a percent.

101. A baseball player's slugging percentage S can be calculated with the following formula:

$S = \frac{h + d + 2t + 3r}{b}$, where h = number of hits, d = number of doubles, t = number of triples, r = number of home runs, and b = number of at bats. In 2010, Albert Pujols of the St. Louis Cardinals led Major League Baseball in slugging percent. During the 2010 season, Pujols had 587 at bats, 183 hits, 39 doubles, 1 triple, and 42 home runs. (*Source:* Major League Baseball) Calculate Pujols' 2010 slugging percent. Round to the nearest tenth of a percent.

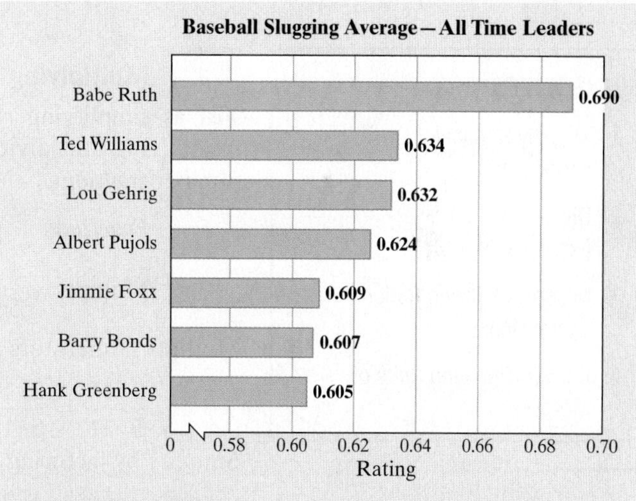

102. To calculate a quarterback's rating in NCAA football, you may use the formula $\frac{100C + 330T - 200I + 8.4Y}{A}$, where C = the number of completed passes, A = the number of attempted passes, T = the number of touchdown passes, Y = the number of yards in the completed passes, and I = the number of interceptions. Cam Newton of Auburn University was selected as the 2010 winner of the Heisman Memorial Trophy as the Most Outstanding Football Player. Newton, a junior quarterback, ended the season with 280 attempts, 185 completions, 2854 yards, 30 touchdowns, and only 7 interceptions. Calculate Newton's quarterback rating for the 2010 season. (*Source:* NCAA) Round the answer to the nearest tenth.

How does the graph of $y = \dfrac{x^2 - 9}{x - 3}$ *compare to the graph of* $y = x + 3$*? Recall that* $\dfrac{x^2 - 9}{x - 3} = \dfrac{(x + 3)(x - 3)}{x - 3} = x + 3$ *as long as x is not 3. This means that the graph of* $y = \dfrac{x^2 - 9}{x - 3}$ *is the same as the graph of* $y = x + 3$ *with* $x \neq 3$*. To graph* $y = \dfrac{x^2 - 9}{x - 3}$*, then, graph the linear equation* $y = x + 3$ *and place an open dot on the graph at 3. This open dot or interruption of the line at 3 means* $x \neq 3$*.*

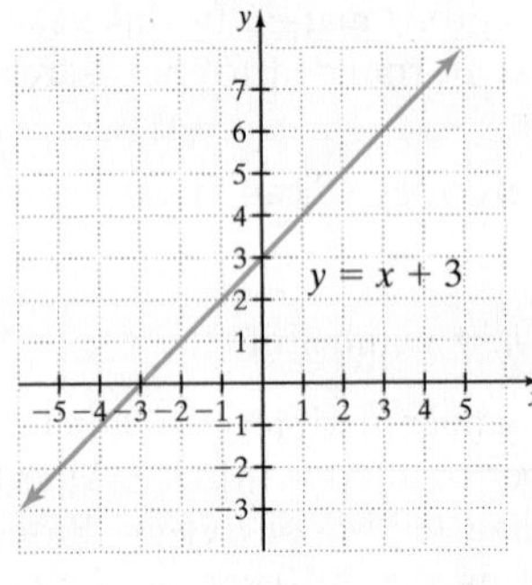

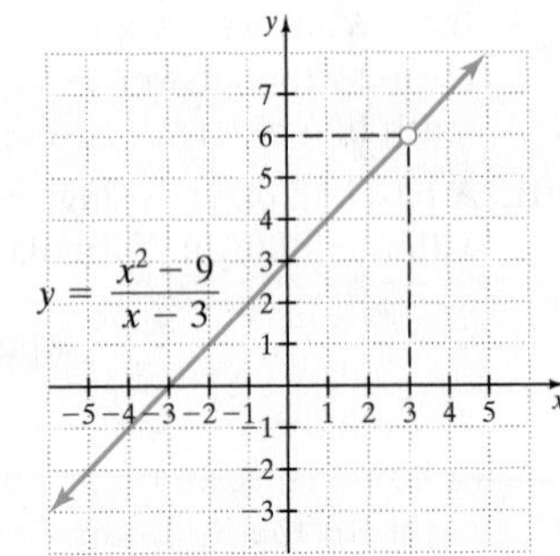

103. Graph $y = \dfrac{x^2 - 25}{x + 5}$.

104. Graph $y = \dfrac{x^2 - 16}{x - 4}$.

105. Graph $y = \dfrac{x^2 + x - 12}{x + 4}$.

106. Graph $y = \dfrac{x^2 - 6x + 8}{x - 2}$.

7.2 Multiplying and Dividing Rational Expressions 

OBJECTIVES

1. Multiply Rational Expressions.
2. Divide Rational Expressions.
3. Multiply or Divide Rational Expressions.
4. Convert Between Units of Measure.

OBJECTIVE

1 Multiplying Rational Expressions 

Just as simplifying rational expressions is similar to simplifying number fractions, multiplying and dividing rational expressions is similar to multiplying and dividing number fractions.

Fractions	*Rational Expressions*
Multiply: $\dfrac{3}{5} \cdot \dfrac{10}{11}$	Multiply: $\dfrac{x - 3}{x + 5} \cdot \dfrac{2x + 10}{x^2 - 9}$

Multiply numerators and multiply denominators.

$\dfrac{3}{5} \cdot \dfrac{10}{11} = \dfrac{3 \cdot 10}{5 \cdot 11}$	$\dfrac{x - 3}{x + 5} \cdot \dfrac{2x + 10}{x^2 - 9} = \dfrac{(x - 3) \cdot (2x + 10)}{(x + 5) \cdot (x^2 - 9)}$

Simplify by factoring numerators and denominators.

$= \dfrac{3 \cdot 2 \cdot 5}{5 \cdot 11}$	$= \dfrac{(x - 3) \cdot 2(x + 5)}{(x + 5)(x + 3)(x - 3)}$

Apply the fundamental principle.

$= \dfrac{3 \cdot 2}{11}$ or $\dfrac{6}{11}$	$= \dfrac{2}{x + 3}$

Multiplying Rational Expressions

If $\dfrac{P}{Q}$ and $\dfrac{R}{S}$ are rational expressions, then

$$\frac{P}{Q} \cdot \frac{R}{S} = \frac{PR}{QS}$$

To multiply rational expressions, multiply the numerators and multiply the denominators.

Note: Recall that for Sections 7.1 through 7.4, we assume variables in rational expressions have only those replacement values for which the expressions are defined.

EXAMPLE 1 Multiply.

a. $\frac{25x}{2} \cdot \frac{1}{y^3}$ **b.** $\frac{-7x^2}{5y} \cdot \frac{3y^5}{14x^2}$

Solution To multiply rational expressions, multiply the numerators and multiply the denominators of both expressions. Then simplify if possible.

a. $\frac{25x}{2} \cdot \frac{1}{y^3} = \frac{25x \cdot 1}{2 \cdot y^3} = \frac{25x}{2y^3}$

The expression $\frac{25x}{2y^3}$ is in simplest form.

b. $\frac{-7x^2}{5y} \cdot \frac{3y^5}{14x^2} = \frac{-7x^2 \cdot 3y^5}{5y \cdot 14x^2}$ Multiply.

The expression $\frac{-7x^2 \cdot 3y^5}{5y \cdot 14x^2}$ is not in simplest form, so we factor the numerator and the denominator and divide out common factors.

$$= \frac{-1 \cdot 7 \cdot 3 \cdot x^2 \cdot y \cdot y^4}{5 \cdot 2 \cdot 7 \cdot x^2 \cdot y}$$

$$= -\frac{3y^4}{10}$$

□

PRACTICE 1 Multiply.

a. $\frac{4a}{5} \cdot \frac{3}{b^2}$ **b.** $\frac{-3p^4}{q^2} \cdot \frac{2q^3}{9p^4}$

When multiplying rational expressions, it is usually best to factor each numerator and denominator. This will help us when we divide out common factors to write the product in lowest terms.

EXAMPLE 2 Multiply: $\frac{x^2 + x}{3x} \cdot \frac{6}{5x + 5}$

Solution $\frac{x^2 + x}{3x} \cdot \frac{6}{5x + 5} = \frac{x(x + 1)}{3x} \cdot \frac{2 \cdot 3}{5(x + 1)}$ Factor numerators and denominators.

$$= \frac{x(x + 1) \cdot 2 \cdot 3}{3x \cdot 5\,(x + 1)}$$ Multiply.

$$= \frac{2}{5}$$ Simplify by dividing out common factors.

□

PRACTICE 2 Multiply: $\frac{x^2 - x}{5x} \cdot \frac{15}{x^2 - 1}$

The following steps may be used to multiply rational expressions.

Multiplying Rational Expressions

Step 1. Completely factor numerators and denominators.

Step 2. Multiply numerators and multiply denominators.

Step 3. Simplify or write the product in lowest terms by dividing out common factors.

✓CONCEPT CHECK

Which of the following is a true statement?

a. $\frac{1}{3} \cdot \frac{1}{2} = \frac{1}{5}$ b. $\frac{2}{x} \cdot \frac{5}{x} = \frac{10}{x}$ c. $\frac{3}{x} \cdot \frac{1}{2} = \frac{3}{2x}$ d. $\frac{x}{7} \cdot \frac{x+5}{4} = \frac{2x+5}{28}$

EXAMPLE 3 Multiply: $\frac{3x+3}{5x-5x^2} \cdot \frac{2x^2+x-3}{4x^2-9}$

Solution

$$\frac{3x+3}{5x-5x^2} \cdot \frac{2x^2+x-3}{4x^2-9} = \frac{3(x+1)}{5x(1-x)} \cdot \frac{(2x+3)(x-1)}{(2x-3)(2x+3)}$$ Factor.

$$= \frac{3(x+1)(2x+3)(x-1)}{5x(1-x)(2x-3)(2x+3)}$$ Multiply.

$$= \frac{3(x+1)(x-1)}{5x(1-x)(2x-3)}$$ Divide out common factors.

Next, recall that $x - 1$ and $1 - x$ are opposites so that $x - 1 = -1(1 - x)$.

$$= \frac{3(x+1)(-1)(1-x)}{5x(1-x)(2x-3)}$$ Write $x - 1$ as $-1(1 - x)$.

$$= \frac{-3(x+1)}{5x(2x-3)} \quad \text{or} \quad -\frac{3(x+1)}{5x(2x-3)}$$ Divide out common factors. □

PRACTICE 3 Multiply: $\frac{6-3x}{6x+6x^2} \cdot \frac{3x^2-2x-5}{x^2-4}$.

OBJECTIVE

2 Dividing Rational Expressions

We can divide by a rational expression in the same way we divide by a fraction. To divide by a fraction, multiply by its reciprocal.

Helpful Hint

Don't forget how to find reciprocals. The reciprocal of $\frac{a}{b}$ is  $\frac{b}{a}$, $a \neq 0, b \neq 0$.

For example, to divide $\frac{3}{2}$ by $\frac{7}{8}$, multiply $\frac{3}{2}$ by $\frac{8}{7}$.

$$\frac{3}{2} \div \frac{7}{8} = \frac{3}{2} \cdot \frac{8}{7} = \frac{3 \cdot 4 \cdot 2}{2 \cdot 7} = \frac{12}{7}$$

Dividing Rational Expressions

If $\frac{P}{Q}$ and $\frac{R}{S}$ are rational expressions and $\frac{R}{S}$ is not 0, then

$$\frac{P}{Q} \div \frac{R}{S} = \frac{P}{Q} \cdot \frac{S}{R} = \frac{PS}{QR}$$

To divide two rational expressions, multiply the first rational expression by the reciprocal of the second rational expression.

Answer to Concept Check: c

EXAMPLE 4 Divide: $\frac{3x^3y^7}{40} \div \frac{4x^3}{y^2}$

Solution

$$\frac{3x^3y^7}{40} \div \frac{4x^3}{y^2} = \frac{3x^3y^7}{40} \cdot \frac{y^2}{4x^3} \quad \text{Multiply by the reciprocal of } \frac{4x^3}{y^2}.$$

$$= \frac{3x^3y^9}{160x^3}$$

$$= \frac{3y^9}{160} \quad \text{Simplify.}$$

□

PRACTICE 4 Divide: $\frac{5a^3b^2}{24} \div \frac{10a^5}{6}$

EXAMPLE 5 Divide: $\frac{(x+2)^2}{10} \div \frac{2x+4}{5}$

Solution:

$$\frac{(x+2)^2}{10} \div \frac{2x+4}{5} = \frac{(x+2)^2}{10} \cdot \frac{5}{2x+4} \quad \text{Multiply by the reciprocal of } \frac{2x+4}{5}.$$

$$= \frac{(x+2)(x+2)\cdot 5}{5\cdot 2\cdot 2\cdot (x+2)} \quad \text{Factor and multiply.}$$

$$= \frac{x+2}{4} \quad \text{Simplify.}$$

□

> **Helpful Hint**
> Remember, **to Divide by a Rational Expression**, multiply by its reciprocal.

PRACTICE 5 Divide $\frac{(x-5)^2}{3}$ by $\frac{4x-20}{9}$.

The following may be used to divide by a rational expression.

> **Dividing by a Rational Expression**
> Multiply by its reciprocal.

EXAMPLE 6 Divide: $\frac{6x+2}{x^2-1} \div \frac{3x^2+x}{x-1}$

Solution

$$\frac{6x+2}{x^2-1} \div \frac{3x^2+x}{x-1} = \frac{6x+2}{x^2-1} \cdot \frac{x-1}{3x^2+x} \quad \text{Multiply by the reciprocal.}$$

$$= \frac{2(3x+1)(x-1)}{(x+1)(x-1)\cdot x(3x+1)} \quad \text{Factor and multiply.}$$

$$= \frac{2}{x(x+1)} \quad \text{Simplify.}$$

□

PRACTICE 6 Divide $\frac{10x-2}{x^2-9} \div \frac{5x^2-x}{x+3}$.

EXAMPLE 7 Divide: $\frac{2x^2 - 11x + 5}{5x - 25} \div \frac{4x - 2}{10}$

Solution

$$\frac{2x^2 - 11x + 5}{5x - 25} \div \frac{4x - 2}{10} = \frac{2x^2 - 11x + 5}{5x - 25} \cdot \frac{10}{4x - 2}$$ Multiply by the reciprocal.

$$= \frac{(2x - 1)(x - 5) \cdot 2 \cdot 5}{5(x - 5) \cdot 2(2x - 1)}$$ Factor and multiply.

$$= \frac{1}{1} \text{ or } 1$$ Simplify. □

PRACTICE 7 Divide $\frac{3x^2 - 11x - 4}{2x - 8} \div \frac{9x + 3}{6}$.

OBJECTIVE 3 Multiplying or Dividing Rational Expressions 

Let's make sure that we understand the difference between multiplying and dividing rational expressions.

Rational Expressions	
Multiplication	Multiply the numerators and multiply the denominators.
Division	Multiply by the reciprocal of the divisor.

EXAMPLE 8 Multiply or divide as indicated.

a. $\frac{x - 4}{5} \cdot \frac{x}{x - 4}$ **b.** $\frac{x - 4}{5} \div \frac{x}{x - 4}$ **c.** $\frac{x^2 - 4}{2x + 6} \cdot \frac{x^2 + 4x + 3}{2 - x}$

Solution

a. $\frac{x - 4}{5} \cdot \frac{x}{x - 4} = \frac{(x - 4) \cdot x}{5 \cdot (x - 4)} = \frac{x}{5}$

b. $\frac{x - 4}{5} \div \frac{x}{x - 4} = \frac{x - 4}{5} \cdot \frac{x - 4}{x} = \frac{(x - 4)^2}{5x}$

c. $\frac{x^2 - 4}{2x + 6} \cdot \frac{x^2 + 4x + 3}{2 - x} = \frac{(x - 2)(x + 2) \cdot (x + 1)(x + 3)}{2(x + 3) \cdot (2 - x)}$ Factor and multiply.

$$= \frac{(x - 2)(x + 2) \cdot (x + 1)(x + 3)}{2(x + 3) \cdot (2 - x)}$$

$$= \frac{-1(x + 2)(x + 1)}{2}$$ Divide out common factors. Recall that $\frac{x - 2}{2 - x} = -1$

$$= -\frac{(x + 2)(x + 1)}{2}$$ □

PRACTICE 8 Multiply or divide as indicated.

a. $\frac{y + 9}{8x} \cdot \frac{y + 9}{2x}$ **b.** $\frac{y + 9}{8x} \div \frac{y + 9}{2}$ **c.** $\frac{35x - 7x^2}{x^2 - 25} \cdot \frac{x^2 + 3x - 10}{x^2 + 4x}$

OBJECTIVE

4 Converting Between Units of Measure 

How many square inches are in 1 square foot?

How many cubic feet are in a cubic yard?

If you have trouble answering these questions, this section will be helpful to you.

Now that we know how to multiply fractions and rational expressions, we can use this knowledge to help us convert between units of measure. To do so, we will use **unit fractions.** A unit fraction is a fraction that equals 1. For example, since 12 in. $=$ 1 ft, we have the unit fractions

$$\frac{12 \text{ in.}}{1 \text{ ft}} = 1 \quad \text{and} \quad \frac{1 \text{ ft}}{12 \text{ in.}} = 1$$

EXAMPLE 9 18 square feet $=$ ____ square yards

Solution: Let's multiply 18 square feet by a unit fraction that has square feet in the denominator and square yards in the numerator. From the diagram, you can see that

1 square yard $=$ 9 square feet

Thus,

$$18 \text{ sq ft} = \frac{18 \text{ sq ft}}{1} \cdot 1 = \frac{\overset{2}{\cancel{18}} \cancel{\text{sq ft}}}{1} \cdot \frac{1 \text{ sq yd}}{\underset{1}{\cancel{9}} \cancel{\text{sq ft}}}$$

$$= \frac{2 \cdot 1}{1 \cdot 1} \text{ sq yd} = 2 \text{ sq yd}$$

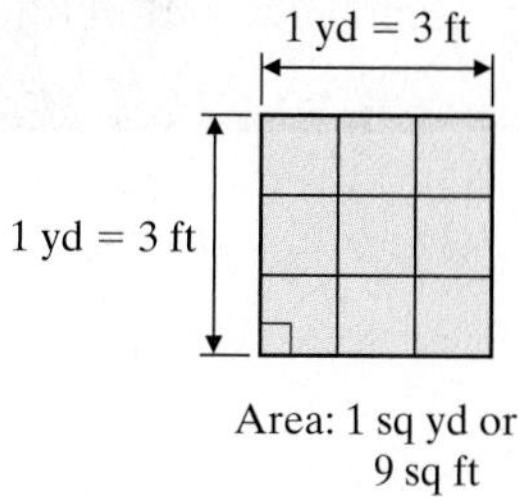

Area: 1 sq yd or 9 sq ft

Thus, 18 sq ft $=$ 2 sq yd.

Draw a diagram of 18 sq ft to help you see that this is reasonable. □

PRACTICE

9 288 square inches = ____ square feet

EXAMPLE 10 5.2 square yards $=$ ____ square feet

Solution:

$$5.2 \text{ sq yd} = \frac{5.2 \text{ sq yd}}{1} \cdot 1 = \frac{5.2 \cancel{\text{sq yd}}}{1} \cdot \frac{9 \text{ sq ft}}{1 \cancel{\text{sq yd}}} \begin{array}{l} \leftarrow \text{Units converting to} \\ \leftarrow \text{Units given} \end{array}$$

$$= \frac{5.2 \cdot 9}{1 \cdot 1} \text{ sq ft}$$

$$= 46.8 \text{ sq ft}$$

Thus, 5.2 sq yd $=$ 46.8 sq ft.

Draw a diagram to see that this is reasonable. □

PRACTICE

10 3.5 square feet $=$ ____ square inches

EXAMPLE 11 **Converting from Cubic Feet to Cubic Yards**

The largest building in the world by volume is The Boeing Company's Everett, Washington, factory complex, where Boeing's wide-body jetliners, the 747, 767, and 777, are built. The volume of this factory complex is 472,370,319 cubic feet. Find the volume of this Boeing facility in cubic yards. (*Source:* The Boeing Company) (Solution on next page)

Solution: There are 27 cubic feet in 1 cubic yard. (See the diagram.)

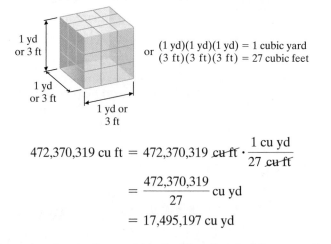

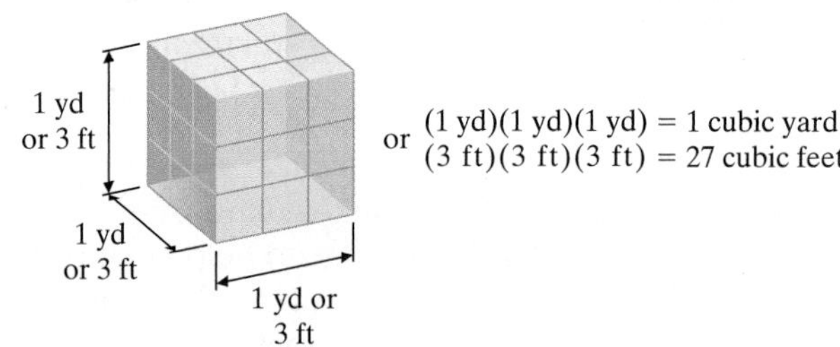

$$\begin{aligned} 472{,}370{,}319 \text{ cu ft} &= 472{,}370{,}319 \cancel{\text{cu ft}} \cdot \frac{1 \text{ cu yd}}{27 \cancel{\text{cu ft}}} \\ &= \frac{472{,}370{,}319}{27} \text{ cu yd} \\ &= 17{,}495{,}197 \text{ cu yd} \end{aligned}$$

PRACTICE

11 The largest casino in the world is the Venetian, in Macau, on the southern tip of China. The gaming area for this casino is approximately 61,000 *square yards*. Find the size of the gaming area in *square feet*. (*Source: USA Today*)

> **Helpful Hint**
>
> When converting among units of measurement, if possible write the unit fraction so that **the numerator contains the units you are converting to** and **the denominator contains the original units.**
>
> $$\begin{aligned} 48 \text{ in.} &= \frac{48 \cancel{\text{in.}}}{1} \cdot \underbrace{\frac{1 \text{ ft}}{12 \cancel{\text{in.}}}}_{\text{Unit fraction}} \begin{matrix} \leftarrow \text{Units converting to} \\ \leftarrow \text{Original units} \end{matrix} \\ &= \frac{48}{12} \text{ ft} = 4 \text{ ft} \end{aligned}$$

EXAMPLE 12 At the 2008 Summer Olympics, Jamaican athlete Usain Bolt won the gold medal in the men's 100-meter track event. He ran the distance at an average speed of 33.9 feet per second. Convert this speed to miles per hour. (*Source:* Beijing 2008 Olympics Committee)

Solution: Recall that 1 mile = 5280 feet and 1 hour = 3600 seconds $(60 \cdot 60)$.

$$\begin{aligned} 33.9 \text{ feet/second} &= \frac{33.9 \text{ feet}}{1 \text{ second}} \cdot \overbrace{\frac{3600 \text{ seconds}}{1 \text{ hour}} \cdot \frac{1 \text{ mile}}{5280 \text{ feet}}}^{\text{Unit fractions}} \\ &= \frac{33.9 \cdot 3600}{5280} \text{ miles/hour} \\ &\approx 23.1 \text{ miles/hour (rounded to the nearest tenth)} \end{aligned}$$

PRACTICE

12 The cheetah is the fastest land animal, being clocked at about 102.7 feet per second. Convert this to miles per hour. Round to the nearest tenth. (*Source: World Almanac and Book of Facts*)

Vocabulary, Readiness & Video Check

Use one of the choices below to fill in the blank.

opposites reciprocals

1. The expressions $\frac{x}{2y}$ and $\frac{2y}{x}$ are called ________.

Multiply or divide as indicated.

2. $\frac{a}{b} \cdot \frac{c}{d} =$ ________ **3.** $\frac{a}{b} \div \frac{c}{d} =$ ________ **4.** $\frac{x}{7} \cdot \frac{x}{6} =$ ________ **5.** $\frac{x}{7} \div \frac{x}{6} =$ ________

Martin-Gay Interactive Videos

See Video 7.2

Watch the section lecture video and answer the following questions.

OBJECTIVE 1

6. Would you say a person needs to be quite comfortable with factoring polynomials in order to be successful with multiplying rational expressions? Explain, referencing Example 2 in your answer.

OBJECTIVE 2

7. Based on the lecture before Example 3, complete the following statements. Dividing rational expressions is exactly like dividing ____. Therefore, to divide by a rational expression, multiply by its ____.

OBJECTIVE 3

8. In Examples 4 and 5, determining the operation is the first step in deciding how to simplify. Why do you think this is so?

OBJECTIVE 4

9. When converting between units of measurement, a unit fraction may be used. What units are used in the numerator and what units are used in the denominator of your unit fraction?

7.2 Exercise Set

MyMathLab®

Find each product and simplify if possible. See Examples 1 through 3.

1. $\frac{3x}{y^2} \cdot \frac{7y}{4x}$

2. $\frac{9x^2}{y} \cdot \frac{4y}{3x^3}$

3. $\frac{8x}{2} \cdot \frac{x^5}{4x^2}$

4. $\frac{6x^2}{10x^3} \cdot \frac{5x}{12}$

5. $-\frac{5a^2b}{30a^2b^2} \cdot b^3$

6. $-\frac{9x^3y^2}{18xy^5} \cdot y^3$

7. $\frac{x}{2x - 14} \cdot \frac{x^2 - 7x}{5}$

8. $\frac{4x - 24}{20x} \cdot \frac{5}{x - 6}$

9. $\frac{6x + 6}{5} \cdot \frac{10}{36x + 36}$

10. $\frac{x^2 + x}{8} \cdot \frac{16}{x + 1}$

11. $\frac{(m + n)^2}{m - n} \cdot \frac{m}{m^2 + mn}$

12. $\frac{(m - n)^2}{m + n} \cdot \frac{m}{m^2 - mn}$

13. $\frac{x^2 - 25}{x^2 - 3x - 10} \cdot \frac{x + 2}{x}$

14. $\frac{a^2 - 4a + 4}{a^2 - 4} \cdot \frac{a + 3}{a - 2}$

15. $\frac{x^2 + 6x + 8}{x^2 + x - 20} \cdot \frac{x^2 + 2x - 15}{x^2 + 8x + 16}$

16. $\frac{x^2 + 9x + 20}{x^2 - 15x + 44} \cdot \frac{x^2 - 11x + 28}{x^2 + 12x + 35}$

Find each quotient and simplify. See Examples 4 through 7.

17. $\frac{5x^7}{2x^5} \div \frac{15x}{4x^3}$

18. $\frac{9y^4}{6y} \div \frac{y^2}{3}$

19. $\frac{8x^2}{y^3} \div \frac{4x^2y^3}{6}$

20. $\frac{7a^2b}{3ab^2} \div \frac{21a^2b^2}{14ab}$

21. $\frac{(x - 6)(x + 4)}{4x} \div \frac{2x - 12}{8x^2}$

22. $\frac{(x + 3)^2}{5} \div \frac{5x + 15}{25}$

23. $\frac{3x^2}{x^2 - 1} \div \frac{x^5}{(x + 1)^2}$

24. $\frac{9x^5}{a^2 - b^2} \div \frac{27x^2}{3b - 3a}$

25. $\frac{m^2 - n^2}{m + n} \div \frac{m}{m^2 + nm}$

26. $\frac{(m - n)^2}{m + n} \div \frac{m^2 - mn}{m}$

27. $\frac{x + 2}{7 - x} \div \frac{x^2 - 5x + 6}{x^2 - 9x + 14}$

28. $\frac{x - 3}{2 - x} \div \frac{x^2 + 3x - 18}{x^2 + 2x - 8}$

29. $\frac{x^2 + 7x + 10}{x - 1} \div \frac{x^2 + 2x - 15}{x - 1}$

30. $\frac{x + 1}{(x + 1)(2x + 3)} \div \frac{20x + 100}{2x + 3}$

MIXED PRACTICE

Multiply or divide as indicated. See Examples 1 through 8.

31. $\frac{5x-10}{12} \div \frac{4x-8}{8}$
32. $\frac{6x+6}{5} \div \frac{9x+9}{10}$
33. $\frac{x^2+5x}{8} \cdot \frac{9}{3x+15}$
34. $\frac{3x^2+12x}{6} \cdot \frac{9}{2x+8}$
35. $\frac{7}{6p^2+q} \div \frac{14}{18p^2+3q}$
36. $\frac{3x+6}{20} \div \frac{4x+8}{8}$
37. $\frac{3x+4y}{x^2+4xy+4y^2} \cdot \frac{x+2y}{2}$
38. $\frac{x^2-y^2}{3x^2+3xy} \cdot \frac{3x^2+6x}{3x^2-2xy-y^2}$
39. $\frac{(x+2)^2}{x-2} \div \frac{x^2-4}{2x-4}$
40. $\frac{x+3}{x^2-9} \div \frac{5x+15}{(x-3)^2}$
41. $\frac{x^2-4}{24x} \div \frac{2-x}{6xy}$
42. $\frac{3y}{3-x} \div \frac{12xy}{x^2-9}$
43. $\frac{a^2+7a+12}{a^2+5a+6} \cdot \frac{a^2+8a+15}{a^2+5a+4}$
44. $\frac{b^2+2b-3}{b^2+b-2} \cdot \frac{b^2-4}{b^2+6b+8}$
45. $\frac{5x-20}{3x^2+x} \cdot \frac{3x^2+13x+4}{x^2-16}$
46. $\frac{9x+18}{4x^2-3x} \cdot \frac{4x^2-11x+6}{x^2-4}$
47. $\frac{8n^2-18}{2n^2-5n+3} \div \frac{6n^2+7n-3}{n^2-9n+8}$
48. $\frac{36n^2-64}{3n^2+10n+8} \div \frac{3n^2-13n+12}{n^2-5n-14}$
49. Find the quotient of $\frac{x^2-9}{2x}$ and $\frac{x+3}{8x^4}$.
50. Find the quotient of $\frac{4x^2+4x+1}{4x+2}$ and $\frac{4x+2}{16}$.

Multiply or divide as indicated. Some of these expressions contain 4-term polynomials and sums and differences of cubes. See Examples 1 through 8.

51. $\frac{a^2+ac+ba+bc}{a-b} \div \frac{a+c}{a+b}$
52. $\frac{x^2+2x-xy-2y}{x^2-y^2} \div \frac{2x+4}{x+y}$
53. $\frac{3x^2+8x+5}{x^2+8x+7} \cdot \frac{x+7}{x^2+4}$
54. $\frac{16x^2+2x}{16x^2+10x+1} \cdot \frac{1}{4x^2+2x}$
55. $\frac{x^3+8}{x^2-2x+4} \cdot \frac{4}{x^2-4}$
56. $\frac{9y}{3y-3} \cdot \frac{y^3-1}{y^3+y^2+y}$
57. $\frac{a^2-ab}{6a^2+6ab} \div \frac{a^3-b^3}{a^2-b^2}$
58. $\frac{x^3+27y^3}{6x} \div \frac{x^2-9y^2}{x^2-3xy}$

Convert as indicated. See Examples 9 through 12.

59. 10 square feet = ________ square inches.
60. 1008 square inches = ________ square feet.
61. 45 square feet = ________ square yards.
62. 2 square yards = ________ square inches.
63. 3 cubic yards = ________ cubic feet.
64. 2 cubic yards = ________ cubic inches.
65. 50 miles per hour = ________ feet per second (round to the nearest whole).
66. 10 feet per second = ________ miles per hour (round to the nearest tenth).
67. 6.3 square yards = ________ square feet.
68. 3.6 square yards = ________ square feet.
69. In January 2010, the Burj Khalifa Tower officially became the tallest building in the world. This tower has a curtain wall (the exterior skin of the building) that is approximately 133,500 square yards. Convert this to square feet. (*Source:* Burj Khalifa)
70. The Pentagon, headquarters for the Department of Defense, contains 3,705,793 square feet of office and storage space. Convert this to square yards. Round to the nearest square yard. (*Source:* U.S. Department of Defense)
71. On October 9, 2007, Russ Wicks set a new stock car world speed record of 359.2 feet/second on the Bonneville Salt Flats in Utah. Convert this speed to miles/hour. Round to the nearest tenth. (*Source:* RussWicks.com)
72. On October 4, 2004, the rocket plane *SpaceShipOne* shot to an altitude of more than 100 km for the second time inside a week to claim the $10 million Ansari X-Prize. At one point in its flight, *SpaceShipOne* was traveling past Mach 1, about 930 miles per hour. Find this speed in feet per second. (*Source:* Space.com)

REVIEW AND PREVIEW

Perform each indicated operation. See Section 1.3.

73. $\frac{1}{5} + \frac{4}{5}$
74. $\frac{3}{15} + \frac{6}{15}$
75. $\frac{9}{9} - \frac{19}{9}$
76. $\frac{4}{3} - \frac{8}{3}$
77. $\frac{6}{5} + \left(\frac{1}{5} - \frac{8}{5}\right)$
78. $-\frac{3}{2} + \left(\frac{1}{2} - \frac{3}{2}\right)$

Graph each linear equation. See Section 3.2.

79. $x - 2y = 6$
80. $5x - y = 10$

CONCEPT EXTENSIONS

Identify each statement as true or false. If false, correct the multiplication. See the Concept Check in this section.

81. $\frac{4}{a} \cdot \frac{1}{b} = \frac{4}{ab}$

82. $\frac{2}{3} \cdot \frac{2}{4} = \frac{2}{7}$

83. $\frac{x}{5} \cdot \frac{x+3}{4} = \frac{2x+3}{20}$

84. $\frac{7}{a} \cdot \frac{3}{a} = \frac{21}{a}$

△ **85.** Find the area of the rectangle.

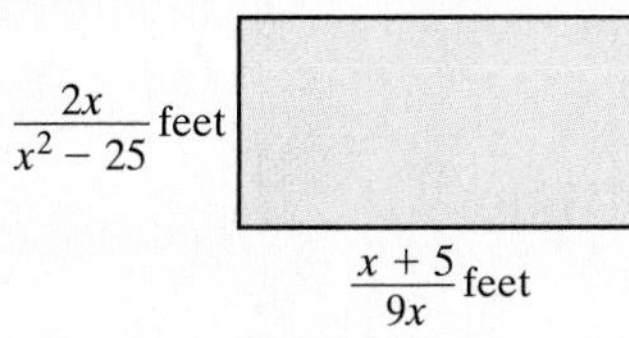

△ **86.** Find the area of the square.

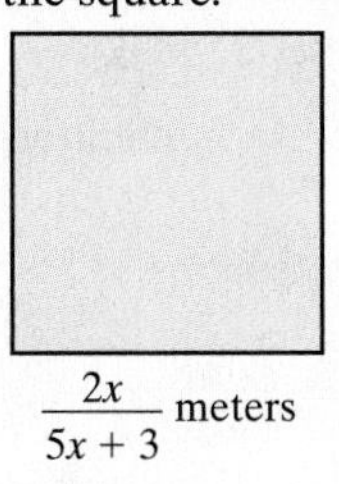

Multiply or divide as indicated.

87. $\left(\frac{x^2 - y^2}{x^2 + y^2} \div \frac{x^2 - y^2}{3x}\right) \cdot \frac{x^2 + y^2}{6}$

88. $\left(\frac{x^2 - 9}{x^2 - 1} \cdot \frac{x^2 + 2x + 1}{2x^2 + 9x + 9}\right) \div \frac{2x + 3}{1 - x}$

89. $\left(\frac{2a + b}{b^2} \cdot \frac{3a^2 - 2ab}{ab + 2b^2}\right) \div \frac{a^2 - 3ab + 2b^2}{5ab - 10b^2}$

90. $\left(\frac{x^2y^2 - xy}{4x - 4y} \div \frac{3y - 3x}{8x - 8y}\right) \cdot \frac{y - x}{8}$

91. In your own words, explain how you multiply rational expressions.

92. Explain how dividing rational expressions is similar to dividing rational numbers.

7.3 Adding and Subtracting Rational Expressions with Common Denominators and Least Common Denominator

OBJECTIVES

1. Add and Subtract Rational Expressions with the Same Denominator.
2. Find the Least Common Denominator of a List of Rational Expressions.
3. Write a Rational Expression as an Equivalent Expression Whose Denominator Is Given.

OBJECTIVE

1 Adding and Subtracting Rational Expressions with the Same Denominator

Like multiplication and division, addition and subtraction of rational expressions is similar to addition and subtraction of rational numbers. In this section, we add and subtract rational expressions with a common (or the same) denominator.

Add: $\frac{6}{5} + \frac{2}{5}$

Add: $\frac{9}{x+2} + \frac{3}{x+2}$

Add the numerators and place the sum over the common denominator.

$$\frac{6}{5} + \frac{2}{5} = \frac{6+2}{5}$$

$$= \frac{8}{5} \quad \text{Simplify.}$$

$$\frac{9}{x+2} + \frac{3}{x+2} = \frac{9+3}{x+2}$$

$$= \frac{12}{x+2} \quad \text{Simplify.}$$

Adding and Subtracting Rational Expressions with Common Denominators

If $\frac{P}{R}$ and $\frac{Q}{R}$ are rational expressions, then

$$\frac{P}{R} + \frac{Q}{R} = \frac{P+Q}{R} \quad \text{and} \quad \frac{P}{R} - \frac{Q}{R} = \frac{P-Q}{R}$$

To add or subtract rational expressions, add or subtract the numerators and place the sum or difference over the common denominator.

EXAMPLE 1 Add: $\frac{5m}{2n} + \frac{m}{2n}$

Solution $\frac{5m}{2n} + \frac{m}{2n} = \frac{5m + m}{2n}$ Add the numerators.

$= \frac{6m}{2n}$ Simplify the numerator by combining like terms.

$= \frac{3m}{n}$ Simplify by applying the fundamental principle. □

PRACTICE 1 Add: $\frac{7a}{4b} + \frac{a}{4b}$

EXAMPLE 2 Subtract: $\frac{2y}{2y - 7} - \frac{7}{2y - 7}$

Solution $\frac{2y}{2y - 7} - \frac{7}{2y - 7} = \frac{2y - 7}{2y - 7}$ Subtract the numerators.

$= \frac{1}{1}$ or 1 Simplify. □

PRACTICE 2 Subtract: $\frac{3x}{3x - 2} - \frac{2}{3x - 2}$

EXAMPLE 3 Subtract: $\frac{3x^2 + 2x}{x - 1} - \frac{10x - 5}{x - 1}$

Solution $\frac{3x^2 + 2x}{x - 1} - \frac{10x - 5}{x - 1} = \frac{(3x^2 + 2x) - (10x - 5)}{x - 1}$ Subtract the numerators. Notice the parentheses.

$= \frac{3x^2 + 2x - 10x + 5}{x - 1}$ Use the distributive property.

$= \frac{3x^2 - 8x + 5}{x - 1}$ Combine like terms.

$= \frac{(x - 1)(3x - 5)}{x - 1}$ Factor.

$= 3x - 5$ Simplify. □

Helpful Hint
Parentheses are inserted so that the entire numerator, $10x - 5$, is subtracted.

PRACTICE 3 Subtract: $\frac{4x^2 + 15x}{x + 3} - \frac{8x + 15}{x + 3}$

Helpful Hint
Notice how the numerator $10x - 5$ has been subtracted in Example 3.

This $-$ sign applies to the entire numerator of $10x - 5$.

So parentheses are inserted here to indicate this.

$$\frac{3x^2 + 2x}{x - 1} - \frac{10x - 5}{x - 1} = \frac{3x^2 + 2x - (10x - 5)}{x - 1}$$

OBJECTIVE

2 Finding the Least Common Denominator 

To add and subtract fractions with **unlike** denominators, first find a least common denominator (LCD) and then write all fractions as equivalent fractions with the LCD.

For example, suppose we add $\frac{8}{3}$ and $\frac{2}{5}$. The LCD of denominators 3 and 5 is 15, since 15 is the least common multiple (LCM) of 3 and 5. That is, 15 is the smallest number that both 3 and 5 divide into evenly.

Next, rewrite each fraction so that its denominator is 15.

$$\frac{8}{3} + \frac{2}{5} = \frac{8(5)}{3(5)} + \frac{2(3)}{5(3)} = \frac{40}{15} + \frac{6}{15} = \frac{40 + 6}{15} = \frac{46}{15}$$

↑ ↑

We are multiplying by 1.

To add or subtract rational expressions with unlike denominators, we also first find an LCD and then write all rational expressions as equivalent expressions with the LCD. The **least common denominator (LCD) of a list of rational expressions** is a polynomial of least degree whose factors include all the factors of the denominators in the list.

Finding the Least Common Denominator (LCD)

Step 1. Factor each denominator completely.

Step 2. The least common denominator (LCD) is the product of all unique factors found in Step 1, each raised to a power equal to the greatest number of times that the factor appears in any one factored denominator.

EXAMPLE 4 Find the LCD for each pair.

a. $\frac{1}{8}, \frac{3}{22}$ **b.** $\frac{7}{5x}, \frac{6}{15x^2}$

Solution

a. Start by finding the prime factorization of each denominator.

$$8 = 2 \cdot 2 \cdot 2 = 2^3 \quad \text{and}$$
$$22 = 2 \cdot 11$$

Next, write the product of all the unique factors, each raised to a power equal to the greatest number of times that the factor appears in any denominator.

The greatest number of times that the factor 2 appears is 3.

The greatest number of times that the factor 11 appears is 1.

$$\text{LCD} = 2^3 \cdot 11^1 = 8 \cdot 11 = 88$$

b. Factor each denominator.

$$5x = 5 \cdot x \quad \text{and}$$
$$15x^2 = 3 \cdot 5 \cdot x^2$$

The greatest number of times that the factor 5 appears is 1.

The greatest number of times that the factor 3 appears is 1.

The greatest number of times that the factor x appears is 2.

$$\text{LCD} = 3^1 \cdot 5^1 \cdot x^2 = 15x^2$$

□

PRACTICE

4 Find the LCD for each pair.

a. $\frac{3}{14}, \frac{5}{21}$ **b.** $\frac{4}{9y}, \frac{11}{15y^3}$

EXAMPLE 5 Find the LCD of

a. $\dfrac{7x}{x+2}$ and $\dfrac{5x^2}{x-2}$ **b.** $\dfrac{3}{x}$ and $\dfrac{6}{x+4}$

Solution

a. The denominators $x + 2$ and $x - 2$ are completely factored already. The factor $x + 2$ appears once and the factor $x - 2$ appears once.

$$\text{LCD} = (x+2)(x-2)$$

b. The denominators x and $x + 4$ cannot be factored further. The factor x appears once and the factor $x + 4$ appears once.

$$\text{LCD} = x(x+4)$$

□

PRACTICE 5 Find the LCD of

a. $\dfrac{16}{y-5}$ and $\dfrac{3y^3}{y-4}$ **b.** $\dfrac{8}{a}$ and $\dfrac{5}{a+2}$

EXAMPLE 6 Find the LCD of $\dfrac{6m^2}{3m+15}$ and $\dfrac{2}{(m+5)^2}$.

Solution We factor each denominator.

$$3m + 15 = 3(m+5)$$

$$(m+5)^2 = (m+5)^2 \quad \text{This denominator is already factored.}$$

The greatest number of times that the factor 3 appears is 1.

The greatest number of times that the factor $m + 5$ appears *in any one denominator* is 2.

$$\text{LCD} = 3(m+5)^2$$

□

PRACTICE 6 Find the LCD of $\dfrac{2x^3}{(2x-1)^2}$ and $\dfrac{5x}{6x-3}$.

✓CONCEPT CHECK

Choose the correct LCD of $\dfrac{x}{(x+1)^2}$ and $\dfrac{5}{x+1}$.

a. $x + 1$ b. $(x+1)^2$ c. $(x+1)^3$ d. $5x(x+1)^2$

EXAMPLE 7 Find the LCD of $\dfrac{t-10}{t^2-t-6}$ and $\dfrac{t+5}{t^2+3t+2}$.

Solution Start by factoring each denominator.

$$t^2 - t - 6 = (t-3)(t+2)$$

$$t^2 + 3t + 2 = (t+1)(t+2)$$

$$\text{LCD} = (t-3)(t+2)(t+1)$$

□

PRACTICE 7 Find the LCD of $\dfrac{x-5}{x^2+5x+4}$ and $\dfrac{x+8}{x^2-16}$.

Answer to Concept Check: b

EXAMPLE 8 Find the LCD of $\frac{2}{x-2}$ and $\frac{10}{2-x}$.

Solution The denominators $x - 2$ and $2 - x$ are opposites. That is, $2 - x = -1(x - 2)$. Use $x - 2$ or $2 - x$ as the LCD.

$$\text{LCD} = x - 2 \quad \text{or} \quad \text{LCD} = 2 - x$$ □

PRACTICE 8 Find the LCD of $\frac{5}{3-x}$ and $\frac{4}{x-3}$.

OBJECTIVE 3 Writing Equivalent Rational Expressions

Next we practice writing a rational expression as an equivalent rational expression with a given denominator. To do this, we multiply by a form of 1. Recall that multiplying an expression by 1 produces an equivalent expression. In other words,

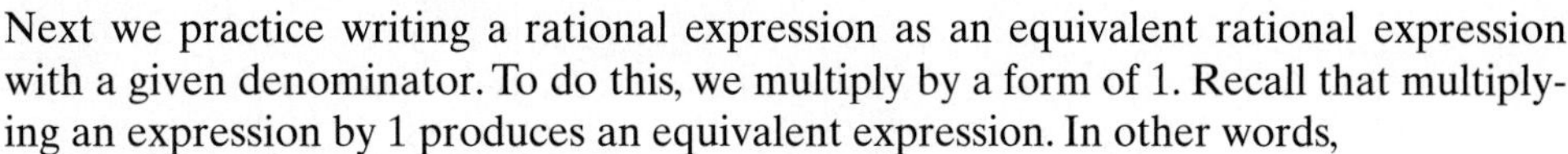

$$\frac{P}{Q} = \frac{P}{Q} \cdot 1 = \frac{P}{Q} \cdot \frac{R}{R} = \frac{PR}{QR}.$$

EXAMPLE 9 Write each rational expression as an equivalent rational expression with the given denominator.

a. $\frac{4b}{9a} = \frac{}{27a^2b}$ **b.** $\frac{7x}{2x+5} = \frac{}{6x+15}$

Solution

a. We can ask ourselves: "What do we multiply $9a$ by to get $27a^2b$?" The answer is $3ab$, since $9a(3ab) = 27a^2b$. So we multiply by 1 in the form of $\frac{3ab}{3ab}$.

$$\frac{4b}{9a} = \frac{4b}{9a} \cdot 1$$

$$= \frac{4b}{9a} \cdot \frac{3ab}{3ab}$$

$$= \frac{4b(3ab)}{9a(3ab)} = \frac{12ab^2}{27a^2b}$$

b. First, factor the denominator on the right.

$$\frac{7x}{2x+5} = \frac{}{3(2x+5)}$$

To obtain the denominator on the right from the denominator on the left, we multiply by 1 in the form of $\frac{3}{3}$.

$$\frac{7x}{2x+5} = \frac{7x}{2x+5} \cdot \frac{3}{3} = \frac{7x \cdot 3}{(2x+5) \cdot 3} = \frac{21x}{3(2x+5)} \text{ or } \frac{21x}{6x+15}$$ □

PRACTICE 9 Write each rational expression as an equivalent fraction with the given denominator.

a. $\frac{3x}{5y} = \frac{}{35xy^2}$ **b.** $\frac{9x}{4x+7} = \frac{}{8x+14}$

EXAMPLE 10 Write the rational expression as an equivalent rational expression with the given denominator.

$$\frac{5}{x^2-4} = \frac{}{(x-2)(x+2)(x-4)}$$

Solution First, factor the denominator $x^2 - 4$ as $(x-2)(x+2)$.

If we multiply the original denominator $(x-2)(x+2)$ by $x-4$, the result is the new denominator $(x-2)(x+2)(x-4)$. Thus, we multiply by 1 in the form of $\frac{x-4}{x-4}$.

$$\frac{5}{x^2-4} = \frac{5}{(x-2)(x+2)} = \frac{5}{(x-2)(x+2)} \cdot \frac{x-4}{x-4}$$

Factored denominator

$$= \frac{5(x-4)}{(x-2)(x+2)(x-4)}$$

$$= \frac{5x-20}{(x-2)(x+2)(x-4)}$$

PRACTICE 10 Write the rational expression as an equivalent rational expression with the given denominator.

$$\frac{3}{x^2-2x-15} = \frac{}{(x-2)(x+3)(x-5)}$$

Vocabulary, Readiness & Video Check

Use the choices below to fill in each blank. Not all choices will be used.

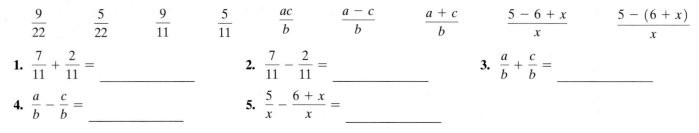

$\frac{9}{22}$ $\quad \frac{5}{22}$ $\quad \frac{9}{11}$ $\quad \frac{5}{11}$ $\quad \frac{ac}{b}$ $\quad \frac{a-c}{b}$ $\quad \frac{a+c}{b}$ $\quad \frac{5-6+x}{x}$ $\quad \frac{5-(6+x)}{x}$

1. $\frac{7}{11} + \frac{2}{11} =$ ________
2. $\frac{7}{11} - \frac{2}{11} =$ ________
3. $\frac{a}{b} + \frac{c}{b} =$ ________
4. $\frac{a}{b} - \frac{c}{b} =$ ________
5. $\frac{5}{x} - \frac{6+x}{x} =$ ________

Martin-Gay Interactive Videos

See Video 7.3

Watch the section lecture video and answer the following questions.

OBJECTIVE 1

6. In Example 3, why is it important to place parentheses around the second numerator when writing as one expression?

OBJECTIVE 2

7. In Examples 4 and 5, we factor the denominators completely. How does this help determine the LCD?

OBJECTIVE 3

8. Based on Example 6, complete the following statements. To write an equivalent rational expression, you multiply the ________ of a rational expression by the same expression as the denominator. This means you're multiplying the original rational expression by a factor of ________ and therefore not changing the ________ of the original expression.

7.3 Exercise Set

MyMathLab®

Add or subtract as indicated. Simplify the result if possible. See Examples 1 through 3.

1. $\frac{a+1}{13}+\frac{8}{13}$
2. $\frac{x+1}{7}+\frac{6}{7}$
3. $\frac{4m}{3n}+\frac{5m}{3n}$
4. $\frac{3p}{2q}+\frac{11p}{2q}$
5. $\frac{4m}{m-6}-\frac{24}{m-6}$
6. $\frac{8y}{y-2}-\frac{16}{y-2}$
7. $\frac{9}{3+y}+\frac{y+1}{3+y}$
8. $\frac{9}{y+9}+\frac{y-5}{y+9}$
9. $\frac{5x^2+4x}{x-1}-\frac{6x+3}{x-1}$
10. $\frac{x^2+9x}{x+7}-\frac{4x+14}{x+7}$
11. $\frac{4a}{a^2+2a-15}-\frac{12}{a^2+2a-15}$
12. $\frac{3y}{y^2+3y-10}-\frac{6}{y^2+3y-10}$
13. $\frac{2x+3}{x^2-x-30}-\frac{x-2}{x^2-x-30}$
14. $\frac{3x-1}{x^2+5x-6}-\frac{2x-7}{x^2+5x-6}$
15. $\frac{2x+1}{x-3}+\frac{3x+6}{x-3}$
16. $\frac{4p-3}{2p+7}+\frac{3p+8}{2p+7}$
17. $\frac{2x^2}{x-5}-\frac{25+x^2}{x-5}$
18. $\frac{6x^2}{2x-5}-\frac{25+2x^2}{2x-5}$
19. $\frac{5x+4}{x-1}-\frac{2x+7}{x-1}$
20. $\frac{7x+1}{x-4}-\frac{2x+21}{x-4}$

Find the LCD for each list of rational expressions. See Examples 4 through 8.

21. $\frac{19}{2x}, \frac{5}{4x^3}$
22. $\frac{17x}{4y^5}, \frac{2}{8y}$
23. $\frac{9}{8x}, \frac{3}{2x+4}$
24. $\frac{1}{6y}, \frac{3x}{4y+12}$
25. $\frac{2}{x+3}, \frac{5}{x-2}$
26. $\frac{-6}{x-1}, \frac{4}{x+5}$
27. $\frac{x}{x+6}, \frac{10}{3x+18}$
28. $\frac{12}{x+5}, \frac{x}{4x+20}$
29. $\frac{8x^2}{(x-6)^2}, \frac{13x}{5x-30}$
30. $\frac{9x^2}{7x-14}, \frac{6x}{(x-2)^2}$
31. $\frac{1}{3x+3}, \frac{8}{2x^2+4x+2}$
32. $\frac{19x+5}{4x-12}, \frac{3}{2x^2-12x+18}$
33. $\frac{5}{x-8}, \frac{3}{8-x}$
34. $\frac{2x+5}{3x-7}, \frac{5}{7-3x}$
35. $\frac{5x+1}{x^2+3x-4}, \frac{3x}{x^2+2x-3}$
36. $\frac{4}{x^2+4x+3}, \frac{4x-2}{x^2+10x+21}$
37. $\frac{2x}{3x^2+4x+1}, \frac{7}{2x^2-x-1}$
38. $\frac{3x}{4x^2+5x+1}, \frac{5}{3x^2-2x-1}$
39. $\frac{1}{x^2-16}, \frac{x+6}{2x^3-8x^2}$
40. $\frac{5}{x^2-25}, \frac{x+9}{3x^3-15x^2}$

Rewrite each rational expression as an equivalent rational expression with the given denominator. See Examples 9 and 10.

41. $\frac{3}{2x}=\frac{}{4x^2}$
42. $\frac{3}{9y^5}=\frac{}{72y^9}$
43. $\frac{6}{3a}=\frac{}{12ab^2}$
44. $\frac{5}{4y^2x}=\frac{}{32y^3x^2}$
45. $\frac{9}{2x+6}=\frac{}{2y(x+3)}$
46. $\frac{4x+1}{3x+6}=\frac{}{3y(x+2)}$
47. $\frac{9a+2}{5a+10}=\frac{}{5b(a+2)}$
48. $\frac{5+y}{2x^2+10}=\frac{}{4(x^2+5)}$
49. $\frac{x}{x^3+6x^2+8x}=\frac{}{x(x+4)(x+2)(x+1)}$
50. $\frac{5x}{x^3+2x^2-3x}=\frac{}{x(x-1)(x-5)(x+3)}$
51. $\frac{9y-1}{15x^2-30}=\frac{}{30x^2-60}$
52. $\frac{6m-5}{3x^2-9}=\frac{}{12x^2-36}$

MIXED PRACTICE

Perform the indicated operations.

53. $\frac{5x}{7}+\frac{9x}{7}$
54. $\frac{5x}{7}\cdot\frac{9x}{7}$
55. $\frac{x+3}{4}\div\frac{2x-1}{4}$
56. $\frac{x+3}{4}-\frac{2x-1}{4}$
57. $\frac{x^2}{x-6}-\frac{5x+6}{x-6}$
58. $\frac{x^2+5x}{x^2-25}\cdot\frac{3x-15}{x^2}$

59. $\dfrac{-2x}{x^3 - 8x} + \dfrac{3x}{x^3 - 8x}$

60. $\dfrac{-2x}{x^3 - 8x} \div \dfrac{3x}{x^3 - 8x}$

61. $\dfrac{12x - 6}{x^2 + 3x} \cdot \dfrac{4x^2 + 13x + 3}{4x^2 - 1}$

62. $\dfrac{x^3 + 7x^2}{3x^3 - x^2} \div \dfrac{5x^2 + 36x + 7}{9x^2 - 1}$

REVIEW AND PREVIEW

Perform each indicated operation. See Section 1.3.

63. $\dfrac{2}{3} + \dfrac{5}{7}$

64. $\dfrac{9}{10} - \dfrac{3}{5}$

65. $\dfrac{1}{6} - \dfrac{3}{4}$

66. $\dfrac{11}{15} + \dfrac{5}{9}$

67. $\dfrac{1}{12} + \dfrac{3}{20}$

68. $\dfrac{7}{30} + \dfrac{3}{18}$

CONCEPT EXTENSIONS

For Exercises 69 and 70, see the Concept Check in this section.

69. Choose the correct LCD of $\dfrac{11a^3}{4a - 20}$ and $\dfrac{15a^3}{(a - 5)^2}$.

a. $4a(a - 5)(a + 5)$ **b.** $a - 5$ **c.** $(a - 5)^2$
d. $4(a - 5)^2$ **e.** $(4a - 20)(a - 5)^2$

70. Choose the correct LCD of $\dfrac{5}{14x^2}$ and $\dfrac{y}{6x^3}$.

a. $84x^5$ **b.** $84x^3$
c. $42x^3$ **d.** $42x^5$

For Exercises 71 and 72, an algebra student approaches you with each incorrect solution. Find the error and correct the work shown below.

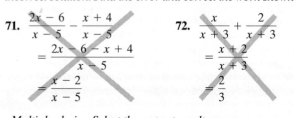

71. $\dfrac{2x - 6}{x - 5} - \dfrac{x + 4}{x - 5} = \dfrac{2x - 6 - x + 4}{x - 5} = \dfrac{x - 2}{x - 5}$

72. $\dfrac{x}{x + 3} + \dfrac{2}{x + 3} = \dfrac{x + 2}{x + 3} = \dfrac{2}{3}$

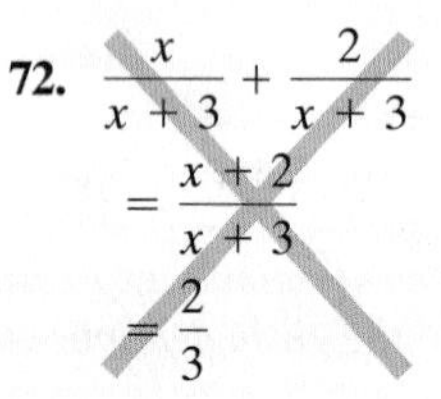

Multiple choice. Select the correct result.

73. $\dfrac{3}{x} + \dfrac{y}{x} =$

a. $\dfrac{3 + y}{x^2}$ **b.** $\dfrac{3 + y}{2x}$ **c.** $\dfrac{3 + y}{x}$

74. $\dfrac{3}{x} - \dfrac{y}{x} =$

a. $\dfrac{3 - y}{x^2}$ **b.** $\dfrac{3 - y}{2x}$ **c.** $\dfrac{3 - y}{x}$

75. $\dfrac{3}{x} \cdot \dfrac{y}{x} =$

a. $\dfrac{3y}{x}$ **b.** $\dfrac{3y}{x^2}$ **c.** $3y$

76. $\dfrac{3}{x} \div \dfrac{y}{x} =$

a. $\dfrac{3}{y}$ **b.** $\dfrac{y}{3}$ **c.** $\dfrac{3}{x^2y}$

Write each rational expression as an equivalent expression with a denominator of $x - 2$.

77. $\dfrac{5}{2 - x}$

78. $\dfrac{8y}{2 - x}$

79. $-\dfrac{7 + x}{2 - x}$

80. $\dfrac{x - 3}{-(x - 2)}$

△ **81.** A square has a side of length $\dfrac{5}{x - 2}$ meters. Express its perimeter as a rational expression.

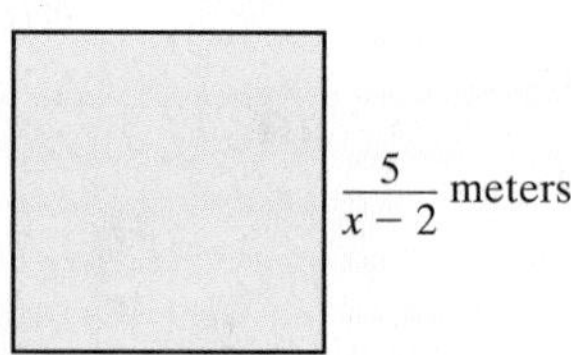

△ **82.** A trapezoid has sides of the indicated lengths. Find its perimeter.

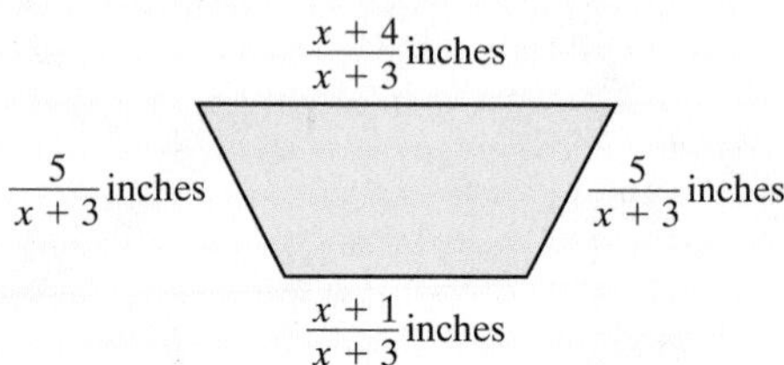

83. Write two rational expressions with the same denominator whose sum is $\dfrac{5}{3x - 1}$.

84. Write two rational expressions with the same denominator whose difference is $\dfrac{x - 7}{x^2 + 1}$.

85. The planet Mercury revolves around the sun in 88 Earth days. It takes Jupiter 4332 Earth days to make one revolution around the sun. (*Source:* National Space Science Data Center) If the two planets are aligned as shown in the figure, how long will it take for them to align again?

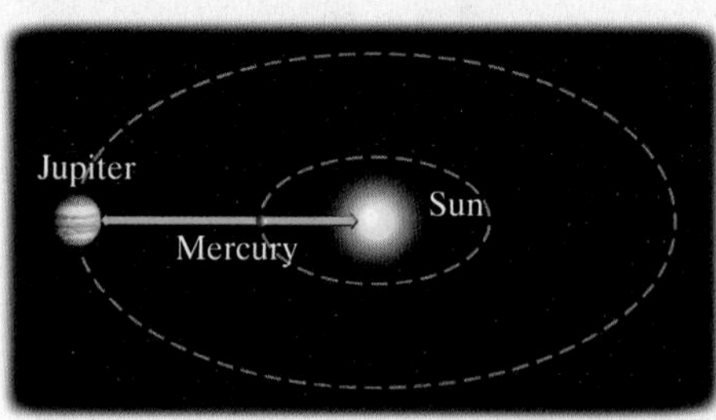

86. You are throwing a barbecue and you want to make sure that you purchase the same number of hot dogs as hot dog buns. Hot dogs come 8 to a package and hot dog buns come 12 to a package. What is the least number of each type of package you should buy?

87. Write some instructions to help a friend who is having difficulty finding the LCD of two rational expressions.

88. Explain why the LCD of the rational expressions $\dfrac{7}{x + 1}$ and $\dfrac{9x}{(x + 1)^2}$ is $(x + 1)^2$ and not $(x + 1)^3$.

89. In your own words, describe how to add or subtract two rational expressions with the same denominators.

90. Explain the similarities between subtracting $\dfrac{3}{8}$ from $\dfrac{7}{8}$ and subtracting $\dfrac{6}{x + 3}$ from $\dfrac{9}{x + 3}$.

7.4 Adding and Subtracting Rational Expressions with Unlike Denominators

OBJECTIVE

1 Add and Subtract Rational Expressions with Unlike Denominators.

OBJECTIVE

1 Adding and Subtracting Rational Expressions with Unlike Denominators

Let's add $\frac{3}{8}$ and $\frac{1}{6}$. From the previous section, the LCD of 8 and 6 is 24. Now let's write equivalent fractions with denominator 24 by multiplying by different forms of 1.

$$\frac{3}{8} = \frac{3}{8} \cdot 1 = \frac{3}{8} \cdot \frac{3}{3} = \frac{3 \cdot 3}{8 \cdot 3} = \frac{9}{24}$$

$$\frac{1}{6} = \frac{1}{6} \cdot 1 = \frac{1}{6} \cdot \frac{4}{4} = \frac{1 \cdot 4}{6 \cdot 4} = \frac{4}{24}$$

Now that the denominators are the same, we may add.

$$\frac{3}{8} + \frac{1}{6} = \frac{9}{24} + \frac{4}{24} = \frac{9 + 4}{24} = \frac{13}{24}$$

We add or subtract rational expressions the same way. You may want to use the steps below.

Adding or Subtracting Rational Expressions with Unlike Denominators

Step 1. Find the LCD of the rational expressions.

Step 2. Rewrite each rational expression as an equivalent expression whose denominator is the LCD found in Step 1.

Step 3. Add or subtract numerators and write the sum or difference over the common denominator.

Step 4. Simplify or write the rational expression in simplest form.

EXAMPLE 1 Perform each indicated operation.

a. $\frac{a}{4} - \frac{2a}{8}$ **b.** $\frac{3}{10x^2} + \frac{7}{25x}$

Solution

a. First, we must find the LCD. Since $4 = 2^2$ and $8 = 2^3$, the LCD $= 2^3 = 8$. Next we write each fraction as an equivalent fraction with the denominator 8, then we subtract.

$$\frac{a}{4} - \frac{2a}{8} = \frac{a(2)}{4(2)} - \frac{2a}{8} = \frac{2a}{8} - \frac{2a}{8} = \frac{2a - 2a}{8} = \frac{0}{8} = 0$$

Multiplying the numerator and denominator by 2 is the same as multiplying by $\frac{2}{2}$ or 1.

b. Since $10x^2 = 2 \cdot 5 \cdot x \cdot x$ and $25x = 5 \cdot 5 \cdot x$, the LCD $= 2 \cdot 5^2 \cdot x^2 = 50x^2$. We write each fraction as an equivalent fraction with a denominator of $50x^2$.

$$\frac{3}{10x^2} + \frac{7}{25x} = \frac{3(5)}{10x^2(5)} + \frac{7(2x)}{25x(2x)}$$

$$= \frac{15}{50x^2} + \frac{14x}{50x^2}$$

$$= \frac{15 + 14x}{50x^2}$$ Add numerators. Write the sum over the common denominator. □

PRACTICE 1 Perform each indicated operation.

a. $\frac{2x}{5} - \frac{6x}{15}$ **b.** $\frac{7}{8a} + \frac{5}{12a^2}$

EXAMPLE 2 Subtract: $\frac{6x}{x^2 - 4} - \frac{3}{x + 2}$

Solution Since $x^2 - 4 = (x + 2)(x - 2)$, the LCD $= (x - 2)(x + 2)$. We write equivalent expressions with the LCD as denominators.

$$\frac{6x}{x^2 - 4} - \frac{3}{x + 2} = \frac{6x}{(x - 2)(x + 2)} - \frac{3(x - 2)}{(x + 2)(x - 2)}$$

$$= \frac{6x - 3(x - 2)}{(x + 2)(x - 2)}$$ Subtract numerators. Write the difference over the common denominator.

$$= \frac{6x - 3x + 6}{(x + 2)(x - 2)}$$ Apply the distributive property in the numerator.

$$= \frac{3x + 6}{(x + 2)(x - 2)}$$ Combine like terms in the numerator.

Next we factor the numerator to see if this rational expression can be simplified.

$$= \frac{3(x + 2)}{(x + 2)(x - 2)}$$ Factor.

$$= \frac{3}{x - 2}$$ Divide out common factors to simplify. □

PRACTICE 2 Subtract: $\frac{12x}{x^2 - 25} - \frac{6}{x + 5}$

EXAMPLE 3 Add: $\frac{2}{3t} + \frac{5}{t + 1}$

Solution The LCD is $3t(t + 1)$. We write each rational expression as an equivalent rational expression with a denominator of $3t(t + 1)$.

$$\frac{2}{3t} + \frac{5}{t + 1} = \frac{2(t + 1)}{3t(t + 1)} + \frac{5(3t)}{(t + 1)(3t)}$$

$$= \frac{2(t + 1) + 5(3t)}{3t(t + 1)}$$ Add numerators. Write the sum over the common denominator.

$$= \frac{2t + 2 + 15t}{3t(t + 1)}$$ Apply the distributive property in the numerator.

$$= \frac{17t + 2}{3t(t + 1)}$$ Combine like terms in the numerator. □

PRACTICE 3 Add: $\frac{3}{5y} + \frac{2}{y + 1}$

EXAMPLE 4 Subtract: $\frac{7}{x - 3} - \frac{9}{3 - x}$

Solution To find a common denominator, we notice that $x - 3$ and $3 - x$ are opposites. That is, $3 - x = -(x - 3)$. We write the denominator $3 - x$ as $-(x - 3)$ and simplify.

$$\frac{7}{x-3} - \frac{9}{3-x} = \frac{7}{x-3} - \frac{9}{-(x-3)}$$

$$= \frac{7}{x-3} - \frac{-9}{x-3}$$ Apply $\frac{a}{-b} = \frac{-a}{b}$.

$$= \frac{7-(-9)}{x-3}$$ Subtract numerators. Write the difference over the common denominator.

$$= \frac{16}{x-3}$$ □

PRACTICE 4 Subtract: $\frac{6}{x-5} - \frac{7}{5-x}$

EXAMPLE 5 Add: $1 + \frac{m}{m+1}$

Solution Recall that 1 is the same as $\frac{1}{1}$. The LCD of $\frac{1}{1}$ and $\frac{m}{m+1}$ is $m + 1$.

$$1 + \frac{m}{m+1} = \frac{1}{1} + \frac{m}{m+1}$$ Write 1 as $\frac{1}{1}$.

$$= \frac{1(m+1)}{1(m+1)} + \frac{m}{m+1}$$ Multiply both the numerator and the denominator of $\frac{1}{1}$ by $m + 1$.

$$= \frac{m+1+m}{m+1}$$ Add numerators. Write the sum over the common denominator.

$$= \frac{2m+1}{m+1}$$ Combine like terms in the numerator. □

PRACTICE 5 Add: $2 + \frac{b}{b+3}$

EXAMPLE 6 Subtract: $\frac{3}{2x^2+x} - \frac{2x}{6x+3}$

Solution First, we factor the denominators.

$$\frac{3}{2x^2+x} - \frac{2x}{6x+3} = \frac{3}{x(2x+1)} - \frac{2x}{3(2x+1)}$$

The LCD is $3x(2x + 1)$. We write equivalent expressions with denominators of $3x(2x + 1)$.

$$= \frac{3(3)}{x(2x+1)(3)} - \frac{2x(x)}{3(2x+1)(x)}$$

$$= \frac{9-2x^2}{3x(2x+1)}$$ Subtract numerators. Write the difference over the common denominator. □

PRACTICE 6 Subtract: $\frac{5}{2x^2+3x} - \frac{3x}{4x+6}$

EXAMPLE 7 Add: $\frac{2x}{x^2 + 2x + 1} + \frac{x}{x^2 - 1}$

Solution First we factor the denominators.

$$\frac{2x}{x^2 + 2x + 1} + \frac{x}{x^2 - 1} = \frac{2x}{(x + 1)(x + 1)} + \frac{x}{(x + 1)(x - 1)}$$

Now we write the rational expressions as equivalent expressions with denominators of $(x + 1)(x + 1)(x - 1)$, the LCD.

$$= \frac{2x(x - 1)}{(x + 1)(x + 1)(x - 1)} + \frac{x(x + 1)}{(x + 1)(x - 1)(x + 1)}$$

$$= \frac{2x(x - 1) + x(x + 1)}{(x + 1)^2(x - 1)}$$ Add numerators. Write the sum over the common denominator.

$$= \frac{2x^2 - 2x + x^2 + x}{(x + 1)^2(x - 1)}$$ Apply the distributive property in the numerator.

$$= \frac{3x^2 - x}{(x + 1)^2(x - 1)} \quad \text{or} \quad \frac{x(3x - 1)}{(x + 1)^2(x - 1)}$$ □

PRACTICE 7 Add: $\frac{2x}{x^2 + 7x + 12} + \frac{3x}{x^2 - 9}$

The numerator was factored as a last step to see if the rational expression could be simplified further. Since there are no factors common to the numerator and the denominator, we can't simplify further.

Vocabulary, Readiness & Video Check

Match each exercise with the first step needed to perform the operation. Do not actually perform the operation.

1. $\frac{3}{4} - \frac{y}{4}$ **2.** $\frac{2}{a} \cdot \frac{3}{(a + 6)}$ **3.** $\frac{x + 1}{x} \div \frac{x - 1}{x}$ **4.** $\frac{9}{x - 2} - \frac{x}{x + 2}$

a. Multiply the first rational expression by the reciprocal of the second rational expression.
b. Find the LCD. Write each expression as an equivalent expression with the LCD as denominator.
c. Multiply numerators and multiply denominators.
d. Subtract numerators. Place the difference over a common denominator.

Martin-Gay Interactive Videos

See Video 7.4

Watch the section lecture video and answer the following question.

OBJECTIVE 1

5. What special case is shown in Example 2 and what's the purpose of presenting it?

7.4 Exercise Set

MyMathLab®

MIXED PRACTICE

Perform each indicated operation. Simplify if possible. See Examples 1 through 7.

1. $\frac{4}{2x} + \frac{9}{3x}$

2. $\frac{15}{7a} + \frac{8}{6a}$

3. $\frac{15a}{b} - \frac{6b}{5}$

4. $\frac{4c}{d} - \frac{8d}{5}$

5. $\frac{3}{x} + \frac{5}{2x^2}$

6. $\frac{14}{3x^2} + \frac{6}{x}$

7. $\frac{6}{x+1} + \frac{10}{2x+2}$

8. $\frac{8}{x+4} - \frac{3}{3x+12}$

9. $\frac{3}{x+2} - \frac{2x}{x^2-4}$

10. $\frac{5}{x-4} + \frac{4x}{x^2-16}$

11. $\frac{3}{4x} + \frac{8}{x-2}$

12. $\frac{5}{y^2} - \frac{y}{2y+1}$

13. $\frac{6}{x-3} + \frac{8}{3-x}$

14. $\frac{15}{y-4} + \frac{20}{4-y}$

15. $\frac{9}{x-3} + \frac{9}{3-x}$

16. $\frac{5}{a-7} + \frac{5}{7-a}$

17. $\frac{-8}{x^2-1} - \frac{7}{1-x^2}$

18. $\frac{-9}{25x^2-1} + \frac{7}{1-25x^2}$

19. $\frac{5}{x} + 2$

20. $\frac{7}{x^2} - 5x$

21. $\frac{5}{x-2} + 6$

22. $\frac{6y}{y+5} + 1$

23. $\frac{y+2}{y+3} - 2$

24. $\frac{7}{2x-3} - 3$

25. $\frac{-x+2}{x} - \frac{x-6}{4x}$

26. $\frac{-y+1}{y} - \frac{2y-5}{3y}$

27. $\frac{5x}{x+2} - \frac{3x-4}{x+2}$

28. $\frac{7x}{x-3} - \frac{4x+9}{x-3}$

29. $\frac{3x^4}{7} - \frac{4x^2}{21}$

30. $\frac{5x}{6} + \frac{11x^2}{2}$

31. $\frac{1}{x+3} - \frac{1}{(x+3)^2}$

32. $\frac{5x}{(x-2)^2} - \frac{3}{x-2}$

33. $\frac{4}{5b} + \frac{1}{b-1}$

34. $\frac{1}{y+5} + \frac{2}{3y}$

35. $\frac{2}{m} + 1$

36. $\frac{6}{x} - 1$

37. $\frac{2x}{x-7} - \frac{x}{x-2}$

38. $\frac{9x}{x-10} - \frac{x}{x-3}$

39. $\frac{6}{1-2x} - \frac{4}{2x-1}$

40. $\frac{10}{3n-4} - \frac{5}{4-3n}$

41. $\frac{7}{(x+1)(x-1)} + \frac{8}{(x+1)^2}$

42. $\frac{5}{(x+1)(x+5)} - \frac{2}{(x+5)^2}$

43. $\frac{x}{x^2-1} - \frac{2}{x^2-2x+1}$

44. $\frac{x}{x^2-4} - \frac{5}{x^2-4x+4}$

45. $\frac{3a}{2a+6} - \frac{a-1}{a+3}$

46. $\frac{1}{x+y} - \frac{y}{x^2-y^2}$

47. $\frac{y-1}{2y+3} + \frac{3}{(2y+3)^2}$

48. $\frac{x-6}{5x+1} + \frac{6}{(5x+1)^2}$

49. $\frac{5}{2-x} + \frac{x}{2x-4}$

50. $\frac{-1}{a-2} + \frac{4}{4-2a}$

51. $\frac{15}{x^2+6x+9} + \frac{2}{x+3}$

52. $\dfrac{2}{x^2 + 4x + 4} + \dfrac{1}{x + 2}$

53. $\dfrac{13}{x^2 - 5x + 6} - \dfrac{5}{x - 3}$

54. $\dfrac{-7}{y^2 - 3y + 2} - \dfrac{2}{y - 1}$

55. $\dfrac{70}{m^2 - 100} + \dfrac{7}{2(m + 10)}$

56. $\dfrac{27}{y^2 - 81} + \dfrac{3}{2(y + 9)}$

57. $\dfrac{x + 8}{x^2 - 5x - 6} + \dfrac{x + 1}{x^2 - 4x - 5}$

58. $\dfrac{x + 4}{x^2 + 12x + 20} + \dfrac{x + 1}{x^2 + 8x - 20}$

59. $\dfrac{5}{4n^2 - 12n + 8} - \dfrac{3}{3n^2 - 6n}$

60. $\dfrac{6}{5y^2 - 25y + 30} - \dfrac{2}{4y^2 - 8y}$

MIXED PRACTICE

Perform the indicated operations. Addition, subtraction, multiplication, and division of rational expressions are included here.

61. $\dfrac{15x}{x + 8} \cdot \dfrac{2x + 16}{3x}$

62. $\dfrac{9z + 5}{15} \cdot \dfrac{5z}{81z^2 - 25}$

63. $\dfrac{8x + 7}{3x + 5} - \dfrac{2x - 3}{3x + 5}$

64. $\dfrac{2z^2}{4z - 1} - \dfrac{z - 2z^2}{4z - 1}$

65. $\dfrac{5a + 10}{18} \div \dfrac{a^2 - 4}{10a}$

66. $\dfrac{9}{x^2 - 1} \div \dfrac{12}{3x + 3}$

67. $\dfrac{5}{x^2 - 3x + 2} + \dfrac{1}{x - 2}$

68. $\dfrac{4}{2x^2 + 5x - 3} + \dfrac{2}{x + 3}$

REVIEW AND PREVIEW

Solve the following linear and quadratic equations. See Sections 2.4 and 6.6.

69. $3x + 5 = 7$

70. $5x - 1 = 8$

71. $2x^2 - x - 1 = 0$

72. $4x^2 - 9 = 0$

73. $4(x + 6) + 3 = -3$

74. $2(3x + 1) + 15 = -7$

CONCEPT EXTENSIONS

Perform each indicated operation.

75. $\dfrac{3}{x} - \dfrac{2x}{x^2 - 1} + \dfrac{5}{x + 1}$

76. $\dfrac{5}{x - 2} + \dfrac{7x}{x^2 - 4} - \dfrac{11}{x}$

77. $\dfrac{5}{x^2 - 4} + \dfrac{2}{x^2 - 4x + 4} - \dfrac{3}{x^2 - x - 6}$

78. $\dfrac{8}{x^2 + 6x + 5} - \dfrac{3x}{x^2 + 4x - 5} + \dfrac{2}{x^2 - 1}$

79. $\dfrac{9}{x^2 + 9x + 14} - \dfrac{3x}{x^2 + 10x + 21} + \dfrac{x + 4}{x^2 + 5x + 6}$

80. $\dfrac{x + 10}{x^2 - 3x - 4} - \dfrac{8}{x^2 + 6x + 5} - \dfrac{9}{x^2 + x - 20}$

81. A board of length $\dfrac{3}{x + 4}$ inches was cut into two pieces. If one piece is $\dfrac{1}{x - 4}$ inches, express the length of the other piece as a rational expression.

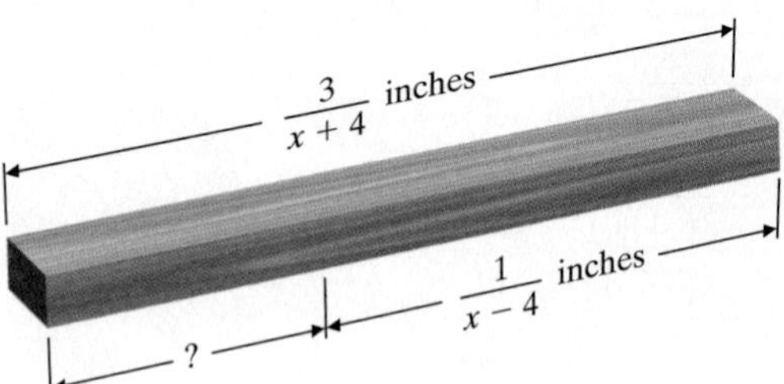

82. The length of a rectangle is $\dfrac{3}{y - 5}$ feet, while its width is $\dfrac{2}{y}$ feet. Find its perimeter and then find its area.

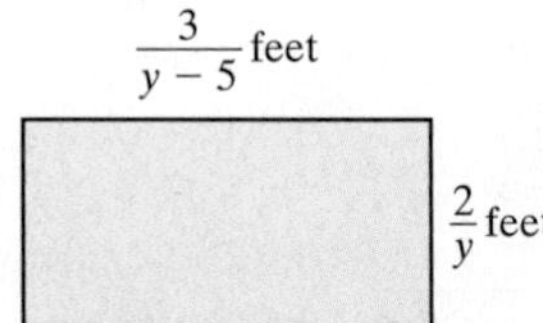

83. In ice hockey, penalty killing percentage is a statistic calculated as $1 - \dfrac{G}{P}$, where G = opponent's power play goals and P = opponent's power play opportunities. Simplify this expression.

84. The dose of medicine prescribed for a child depends on the child's age A in years and the adult dose D for the medication. Two expressions that give a child's dose are Young's Rule, $\dfrac{DA}{A + 12}$, and Cowling's Rule, $\dfrac{D(A + 1)}{24}$. Find an expression for the difference in the doses given by these expressions.

85. Explain when the LCD of the rational expressions in a sum is the product of the denominators.

86. Explain when the LCD is the same as one of the denominators of a rational expression to be added or subtracted.

△ **87.** Two angles are said to be complementary if the sum of their measures is 90°. If one angle measures $\frac{40}{x}$ degrees, find the measure of its complement.

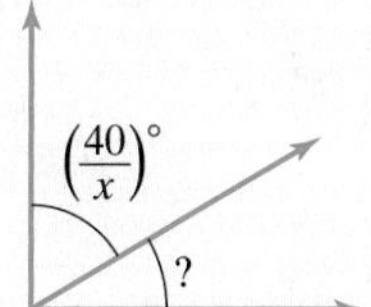

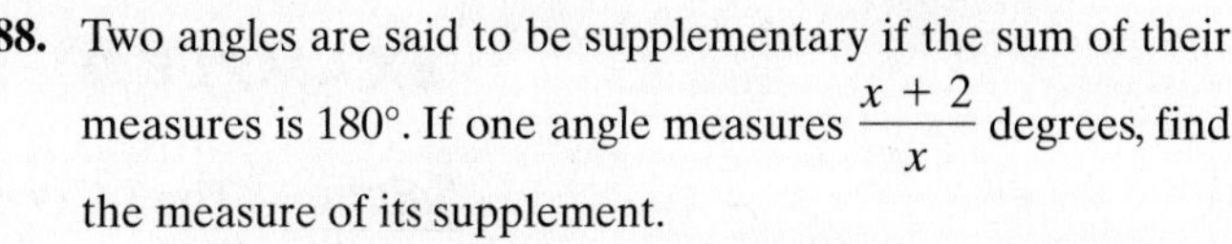

△ **88.** Two angles are said to be supplementary if the sum of their measures is 180°. If one angle measures $\frac{x+2}{x}$ degrees, find the measure of its supplement.

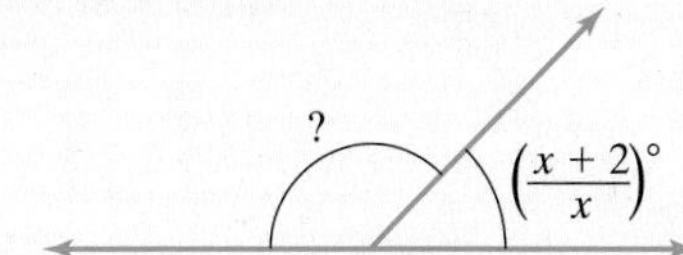

89. In your own words, explain how to add two rational expressions with different denominators.

90. In your own words, explain how to subtract two rational expressions with different denominators.

7.5 Solving Equations Containing Rational Expressions 

OBJECTIVES

1 Solve Equations Containing Rational Expressions.

2 Solve Equations Containing Rational Expressions for a Specified Variable.

OBJECTIVE

1 Solving Equations Containing Rational Expressions 

In Chapter 2, we solved equations containing fractions. In this section, we continue the work we began in Chapter 2 by solving equations containing rational expressions.

Examples of Equations Containing Rational Expressions

$$\frac{x}{2}+\frac{8}{3}=\frac{1}{6} \quad \text{and} \quad \frac{4x}{x^2+x-30}+\frac{2}{x-5}=\frac{1}{x+6}$$

To solve equations such as these, use the multiplication property of equality to clear the equation of fractions by multiplying both sides of the equation by the LCD.

EXAMPLE 1 Solve: $\frac{x}{2}+\frac{8}{3}=\frac{1}{6}$

Solution The LCD of denominators 2, 3, and 6 is 6, so we multiply both sides of the equation by 6.

$$6\left(\frac{x}{2}+\frac{8}{3}\right)=6\left(\frac{1}{6}\right)$$

$$6\left(\frac{x}{2}\right)+6\left(\frac{8}{3}\right)=6\left(\frac{1}{6}\right) \quad \text{Use the distributive property.}$$

$$3\cdot x+16=1 \quad \text{Multiply and simplify.}$$

$$3x=-15 \quad \text{Subtract 16 from both sides.}$$

$$x=-5 \quad \text{Divide both sides by 3.}$$

Helpful Hint
Make sure that *each* term is multiplied by the LCD, 6.

Check: To check, we replace x with -5 in the original equation.

$$\frac{x}{2}+\frac{8}{3}=\frac{1}{6}$$

$$\frac{-5}{2}+\frac{8}{3}\stackrel{?}{=}\frac{1}{6} \quad \text{Replace } x \text{ with } -5$$

$$\frac{1}{6}=\frac{1}{6} \quad \text{True}$$

This number checks, so the solution is -5. □

PRACTICE

1 Solve: $\frac{x}{3}+\frac{4}{5}=\frac{2}{15}$

EXAMPLE 2 Solve: $\frac{t-4}{2} - \frac{t-3}{9} = \frac{5}{18}$

Solution The LCD of denominators 2, 9, and 18 is 18, so we multiply both sides of the equation by 18.

$$18\left(\frac{t-4}{2} - \frac{t-3}{9}\right) = 18\left(\frac{5}{18}\right)$$

$$18\left(\frac{t-4}{2}\right) - 18\left(\frac{t-3}{9}\right) = 18\left(\frac{5}{18}\right) \quad \text{Use the distributive property.}$$

Helpful Hint
Multiply *each* term by 18.

$$9(t-4) - 2(t-3) = 5 \quad \text{Simplify.}$$

$$9t - 36 - 2t + 6 = 5 \quad \text{Use the distributive property.}$$

$$7t - 30 = 5 \quad \text{Combine like terms.}$$

$$7t = 35$$

$$t = 5 \quad \text{Solve for } t.$$

Check:

$$\frac{t-4}{2} - \frac{t-3}{9} = \frac{5}{18}$$

$$\frac{5-4}{2} - \frac{5-3}{9} \stackrel{?}{=} \frac{5}{18} \quad \text{Replace } t \text{ with 5.}$$

$$\frac{1}{2} - \frac{2}{9} \stackrel{?}{=} \frac{5}{18} \quad \text{Simplify.}$$

$$\frac{5}{18} = \frac{5}{18} \quad \text{True}$$

The solution is 5. □

PRACTICE
2 Solve: $\frac{x+4}{4} - \frac{x-3}{3} = \frac{11}{12}$

Recall from Section 7.1 that a rational expression is defined for all real numbers except those that make the denominator of the expression 0. This means that if an equation contains *rational expressions with variables in the denominator*, we must be certain that the proposed solution does not make the denominator 0. If replacing the variable with the proposed solution makes the denominator 0, the rational expression is undefined and this proposed solution must be rejected.

EXAMPLE 3 Solve: $3 - \frac{6}{x} = x + 8$

Solution In this equation, 0 cannot be a solution because if x is 0, the rational expression $\frac{6}{x}$ is undefined. The LCD is x, so we multiply both sides of the equation by x.

$$x\left(3 - \frac{6}{x}\right) = x(x+8)$$

$$x(3) - x\left(\frac{6}{x}\right) = x \cdot x + x \cdot 8 \quad \text{Use the distributive property.}$$

Helpful Hint
Multiply *each* term by x.

$$3x - 6 = x^2 + 8x \quad \text{Simplify.}$$

Now we write the quadratic equation in standard form and solve for x.

$$0 = x^2 + 5x + 6$$

$$0 = (x+3)(x+2) \quad \text{Factor.}$$

$$x + 3 = 0 \quad \text{or} \quad x + 2 = 0 \quad \text{Set each factor equal to 0 and solve.}$$

$$x = -3 \qquad\qquad x = -2$$

Notice that neither -3 nor -2 makes the denominator in the original equation equal to 0.

Check: To check these solutions, we replace x in the original equation by -3, and then by -2.

If $x = -3$:

$$3 - \frac{6}{x} = x + 8$$
$$3 - \frac{6}{-3} \stackrel{?}{=} -3 + 8$$
$$3 - (-2) \stackrel{?}{=} 5$$
$$5 = 5 \quad \text{True}$$

If $x = -2$:

$$3 - \frac{6}{x} = x + 8$$
$$3 - \frac{6}{-2} \stackrel{?}{=} -2 + 8$$
$$3 - (-3) \stackrel{?}{=} 6$$
$$6 = 6 \quad \text{True}$$

□

Both -3 and -2 are solutions.

PRACTICE 3 Solve: $8 + \frac{7}{x} = x + 2$

The following steps may be used to solve an equation containing rational expressions.

Solving an Equation Containing Rational Expressions

Step 1. Multiply both sides of the equation by the LCD of all rational expressions in the equation.

Step 2. Remove any grouping symbols and solve the resulting equation.

Step 3. Check the solution in the original equation.

EXAMPLE 4 Solve: $\frac{4x}{x^2 + x - 30} + \frac{2}{x - 5} = \frac{1}{x + 6}$

Solution The denominator $x^2 + x - 30$ factors as $(x + 6)(x - 5)$. The LCD is then $(x + 6)(x - 5)$, so we multiply both sides of the equation by this LCD.

$$(x + 6)(x - 5)\left(\frac{4x}{x^2 + x - 30} + \frac{2}{x - 5}\right) = (x + 6)(x - 5)\left(\frac{1}{x + 6}\right) \quad \text{Multiply by the LCD.}$$

$$(x + 6)(x - 5) \cdot \frac{4x}{x^2 + x - 30} + (x + 6)(x - 5) \cdot \frac{2}{x - 5} \quad \text{Apply the distributive property.}$$
$$= (x + 6)(x - 5) \cdot \frac{1}{x + 6}$$

$$4x + 2(x + 6) = x - 5 \quad \text{Simplify.}$$
$$4x + 2x + 12 = x - 5 \quad \text{Apply the distributive property.}$$
$$6x + 12 = x - 5 \quad \text{Combine like terms.}$$
$$5x = -17$$
$$x = -\frac{17}{5} \quad \text{Divide both sides by 5.}$$

Check: Check by replacing x with $-\frac{17}{5}$ in the original equation. The solution is $-\frac{17}{5}$.

□

PRACTICE 4 Solve: $\frac{6x}{x^2 - 5x - 14} - \frac{3}{x + 2} = \frac{1}{x - 7}$

EXAMPLE 5 Solve: $\frac{2x}{x-4} = \frac{8}{x-4} + 1$

Solution Multiply both sides by the LCD, $x - 4$.

$$(x-4)\left(\frac{2x}{x-4}\right) = (x-4)\left(\frac{8}{x-4} + 1\right)$$ Multiply by the LCD. Notice that 4 cannot be a solution.

$$(x-4)\cdot\frac{2x}{x-4} = (x-4)\cdot\frac{8}{x-4} + (x-4)\cdot 1$$ Use the distributive property.

$$2x = 8 + (x-4)$$ Simplify.

$$2x = 4 + x$$

$$x = 4$$

Notice that 4 makes the denominators 0 in the original equation. Therefore, 4 is *not* a solution.

This equation has *no solution*. □

PRACTICE 5 Solve: $\frac{7}{x-2} = \frac{3}{x-2} + 4$

Helpful Hint

As we can see from Example 5, it is important to check the proposed solution(s) in the *original* equation.

✓CONCEPT CHECK

When can we clear fractions by multiplying through by the LCD?

a. When adding or subtracting rational expressions

b. When solving an equation containing rational expressions

c. Both of these

d. Neither of these

EXAMPLE 6 Solve: $x + \frac{14}{x-2} = \frac{7x}{x-2} + 1$

Solution Notice the denominators in this equation. We can see that 2 can't be a solution. The LCD is $x - 2$, so we multiply both sides of the equation by $x - 2$.

$$(x-2)\left(x + \frac{14}{x-2}\right) = (x-2)\left(\frac{7x}{x-2} + 1\right)$$

$$(x-2)(x) + (x-2)\left(\frac{14}{x-2}\right) = (x-2)\left(\frac{7x}{x-2}\right) + (x-2)(1)$$

$$x^2 - 2x + 14 = 7x + x - 2$$ Simplify.

$$x^2 - 2x + 14 = 8x - 2$$ Combine like terms.

$$x^2 - 10x + 16 = 0$$ Write the quadratic equation in standard form.

$$(x-8)(x-2) = 0$$ Factor.

$$x - 8 = 0 \quad \text{or} \quad x - 2 = 0$$ Set each factor equal to 0.

$$x = 8 \qquad\qquad x = 2$$ Solve.

Answer to Concept Check: b

As we have already noted, 2 can't be a solution of the original equation. So we need only replace x with 8 in the original equation. We find that 8 is a solution; the only solution is 8. □

PRACTICE

6 Solve: $x + \frac{x}{x-5} = \frac{5}{x-5} - 7$

OBJECTIVE

2 Solving Equations for a Specified Variable 

The last example in this section is an equation containing several variables, and we are directed to solve for one of the variables. The steps used in the preceding examples can be applied to solve equations for a specified variable as well.

EXAMPLE 7 Solve $\frac{1}{a} + \frac{1}{b} = \frac{1}{x}$ for x.

Solution (This type of equation often models a work problem, as we shall see in Section 7.6.) The LCD is abx, so we multiply both sides by abx.

$$abx\left(\frac{1}{a} + \frac{1}{b}\right) = abx\left(\frac{1}{x}\right)$$

$$abx\left(\frac{1}{a}\right) + abx\left(\frac{1}{b}\right) = abx \cdot \frac{1}{x}$$

$$bx + ax = ab \quad \text{Simplify.}$$

$$x(b + a) = ab \quad \text{Factor out } x \text{ from each term on the left side.}$$

$$\frac{x(b + a)}{b + a} = \frac{ab}{b + a} \quad \text{Divide both sides by } b + a.$$

$$x = \frac{ab}{b + a} \quad \text{Simplify.}$$

This equation is now solved for x. □

PRACTICE

7 Solve: $\frac{1}{a} + \frac{1}{b} = \frac{1}{x}$ for b

Graphing Calculator Explorations

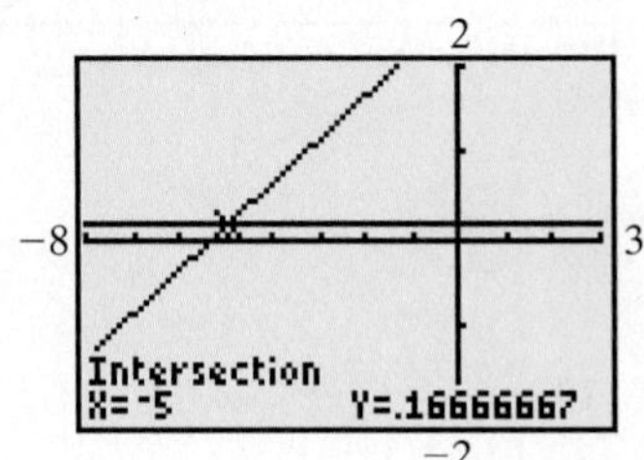

A graphing calculator may be used to check solutions of equations containing rational expressions. For example, to check the solution of Example 1, $\frac{x}{2} + \frac{8}{3} = \frac{1}{6}$, graph $y_1 = \frac{x}{2} + \frac{8}{3}$ and $y_2 = \frac{1}{6}$.

Use TRACE and ZOOM, or use INTERSECT, to find the point of intersection. The point of intersection has an x-value of -5, so the solution of the equation is -5.

Use a graphing calculator to check the examples of this section.

1. Example 2

2. Example 3

3. Example 5

4. Example 6

Vocabulary, Readiness & Video Check

Multiple choice. Choose the correct response.

1. Multiply both sides of the equation $\frac{3x}{2} + 5 = \frac{1}{4}$ by 4. The result is:

a. $3x + 5 = 1$ **b.** $6x + 5 = 1$ **c.** $6x + 20 = 1$ **d.** $6x + 9 = 1$

2. Multiply both sides of the equation $\frac{1}{x} - \frac{3}{5x} = 2$ by $5x$. The result is:

a. $1 - 3 = 10x$ **b.** $5 - 3 = 10x$ **c.** $1 - 3 = 7x$ **d.** $5 - 3 = 7x$

Choose the correct LCD for the fractions in each equation.

3. Equation: $\frac{9}{x} + \frac{3}{4} = \frac{1}{12}$; LCD: ______

a. $4x$ **b.** $12x$ **c.** $48x$ **d.** x

4. Equation: $\frac{8}{3x} - \frac{1}{x} = \frac{7}{9}$; LCD: ______

a. x **b.** $3x$ **c.** $27x$ **d.** $9x$

5. Equation: $\frac{9}{x-1} = \frac{7}{(x-1)^2}$; LCD: ______

a. $(x-1)^2$ **b.** $x - 1$ **c.** $(x-1)^3$ **d.** 63

6. Equation: $\frac{1}{x-2} - \frac{3}{x^2-4} = 8$; LCD: ______

a. $(x-2)$ **b.** $x + 2$ **c.** $x^2 - 4$ **d.** $(x-2)(x^2-4)$

Martin-Gay Interactive Videos

See Video 7.5 

Watch the section lecture video and answer the following questions.

OBJECTIVE 1 **7.** After multiplying through by the LCD and then simplifying, why is it important to take a moment and determine whether you have a linear or a quadratic equation before you finish solving the problem?

OBJECTIVE 1 **8.** From Examples 2–5, what extra step is needed when checking solutions to an equation containing rational expressions?

OBJECTIVE 2 **9.** The steps for solving Example 6 for a specified variable are the same as what other steps? How do you treat this specified variable?

7.5 Exercise Set

MyMathLab®  

Solve each equation and check each solution. See Examples 1 through 3.

1. $\frac{x}{5} + 3 = 9$

2. $\frac{x}{5} - 2 = 9$

3. $\frac{x}{2} + \frac{5x}{4} = \frac{x}{12}$

4. $\frac{x}{6} + \frac{4x}{3} = \frac{x}{18}$

5. $2 - \frac{8}{x} = 6$

6. $5 + \frac{4}{x} = 1$

7. $2 + \frac{10}{x} = x + 5$

8. $6 + \frac{5}{y} = y - \frac{2}{y}$

9. $\frac{a}{5} = \frac{a-3}{2}$

10. $\frac{b}{5} = \frac{b+2}{6}$

11. $\frac{x-3}{5} + \frac{x-2}{2} = \frac{1}{2}$

12. $\frac{a+5}{4} + \frac{a+5}{2} = \frac{a}{8}$

Solve each equation and check each proposed solution. See Examples 4 through 6.

13. $\frac{3}{2a-5} = -1$

14. $\frac{6}{4-3x} = -3$

15. $\frac{4y}{y-4} + 5 = \frac{5y}{y-4}$

16. $\frac{2a}{a+2} - 5 = \frac{7a}{a+2}$

17. $2 + \frac{3}{a-3} = \frac{a}{a-3}$

18. $\frac{2y}{y-2} - \frac{4}{y-2} = 4$

19. $\frac{1}{x+3} + \frac{6}{x^2-9} = 1$

20. $\frac{1}{x+2} + \frac{4}{x^2-4} = 1$

21. $\frac{2y}{y+4} + \frac{4}{y+4} = 3$

22. $\frac{5y}{y+1} - \frac{3}{y+1} = 4$

23. $\frac{2x}{x+2} - 2 = \frac{x-8}{x-2}$

24. $\frac{4y}{y-3} - 3 = \frac{3y-1}{y+3}$

MIXED PRACTICE

Solve each equation. See Examples 1 through 6.

25. $\frac{2}{y} + \frac{1}{2} = \frac{5}{2y}$

26. $\frac{6}{3y} + \frac{3}{y} = 1$

27. $\frac{a}{a-6} = \frac{-2}{a-1}$

28. $\frac{5}{x-6} = \frac{x}{x-2}$

29. $\frac{11}{2x} + \frac{2}{3} = \frac{7}{2x}$

30. $\frac{5}{3} - \frac{3}{2x} = \frac{3}{2}$

31. $\frac{2}{x-2} + 1 = \frac{x}{x+2}$

32. $1 + \frac{3}{x+1} = \frac{x}{x-1}$

33. $\frac{x+1}{3} - \frac{x-1}{6} = \frac{1}{6}$

34. $\frac{3x}{5} - \frac{x-6}{3} = -\frac{2}{5}$

35. $\frac{t}{t-4} = \frac{t+4}{6}$

36. $\frac{15}{x+4} = \frac{x-4}{x}$

37. $\frac{y}{2y+2} + \frac{2y-16}{4y+4} = \frac{2y-3}{y+1}$

38. $\frac{1}{x+2} = \frac{4}{x^2-4} - \frac{1}{x-2}$

39. $\frac{4r-4}{r^2+5r-14} + \frac{2}{r+7} = \frac{1}{r-2}$

40. $\frac{3}{x+3} = \frac{12x+19}{x^2+7x+12} - \frac{5}{x+4}$

41. $\frac{x+1}{x+3} = \frac{x^2-11x}{x^2+x-6} - \frac{x-3}{x-2}$

42. $\frac{2t+3}{t-1} - \frac{2}{t+3} = \frac{5-6t}{t^2+2t-3}$

Solve each equation for the indicated variable. See Example 7.

43. $R = \frac{E}{I}$ for I (Electronics: resistance of a circuit)

44. $T = \frac{V}{Q}$ for Q (Water purification: settling time)

45. $T = \frac{2U}{B+E}$ for B (Merchandising: stock turnover rate)

46. $i = \frac{A}{t+B}$ for t (Hydrology: rainfall intensity)

47. $B = \frac{705w}{h^2}$ for w (Health: body-mass index)

48. $\frac{A}{W} = L$ for W (Geometry: area of a rectangle)

49. $N = R + \frac{V}{G}$ for G (Urban forestry: tree plantings per year)

50. $C = \frac{D(A+1)}{24}$ for A (Medicine: Cowling's Rule for child's dose)

51. $\frac{C}{\pi r} = 2$ for r (Geometry: circumference of a circle)

52. $W = \frac{CE^2}{2}$ for C (Electronics: energy stored in a capacitor)

53. $\frac{1}{y} + \frac{1}{3} = \frac{1}{x}$ for x

54. $\frac{1}{5} + \frac{2}{y} = \frac{1}{x}$ for x

REVIEW AND PREVIEW

TRANSLATING

Write each phrase as an expression.

55. The reciprocal of x

56. The reciprocal of $x + 1$

57. The reciprocal of x, added to the reciprocal of 2

58. The reciprocal of x, subtracted from the reciprocal of 5

Answer each question.

59. If a tank is filled in 3 hours, what fractional part of the tank is filled in 1 hour?

60. If a strip of beach is cleaned in 4 hours, what fractional part of the beach is cleaned in 1 hour?

Identify the x- and y-intercepts. See Section 3.3.

61.

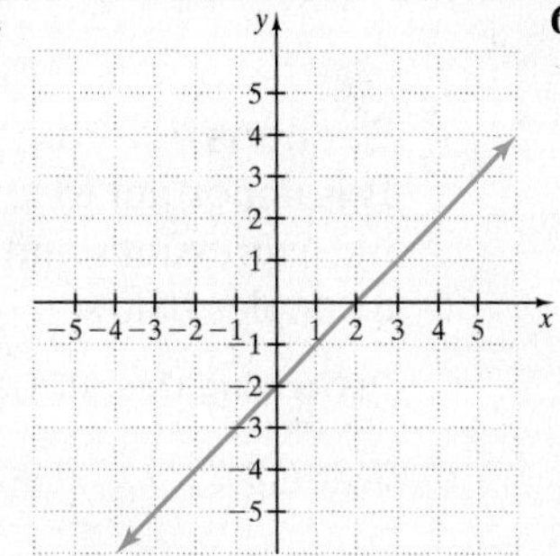

62.

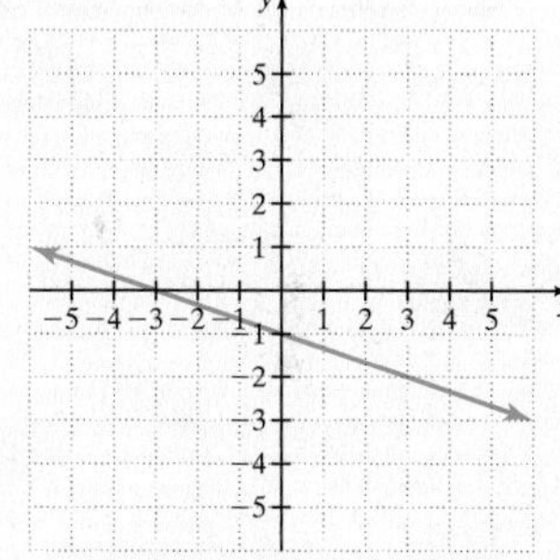

63.

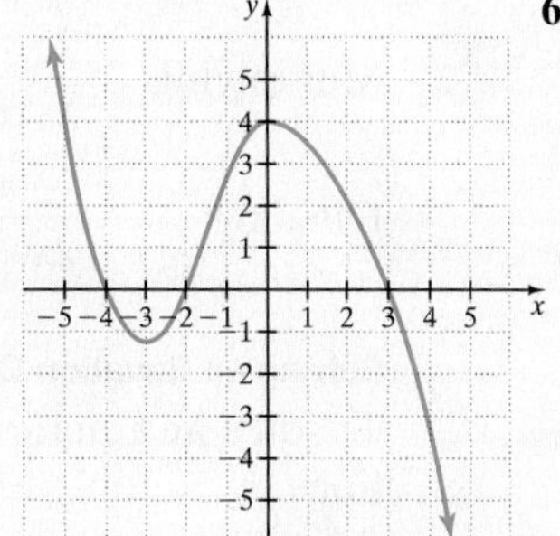

64.

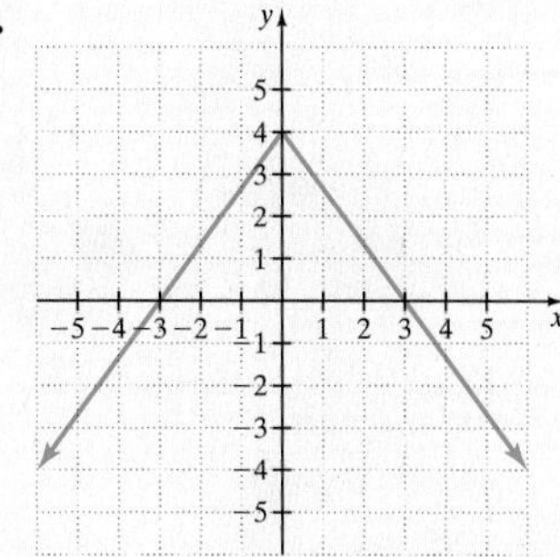

CONCEPT EXTENSIONS

65. Explain the difference between solving an equation such as $\frac{x}{2} + \frac{3}{4} = \frac{x}{4}$ for x and performing an operation such as adding $\frac{x}{2} + \frac{3}{4}$.

66. When solving an equation such as $\frac{y}{4} = \frac{y}{2} - \frac{1}{4}$, we may multiply all terms by 4. When subtracting two rational expressions such as $\frac{y}{2} - \frac{1}{4}$, we may not. Explain why.

Determine whether each of the following is an equation or an expression. If it is an equation, then solve it for its variable. If it is an expression, perform the indicated operation.

△ **67.** $\frac{1}{x} + \frac{5}{9}$

68. $\frac{1}{x} + \frac{5}{9} = \frac{2}{3}$

69. $\frac{5}{x-1} - \frac{2}{x} = \frac{5}{x(x-1)}$

70. $\frac{5}{x-1} - \frac{2}{x}$

Recall that two angles are supplementary if the sum of their measures is 180°. Find the measures of the following supplementary angles.

△ **71.** △ **72.**

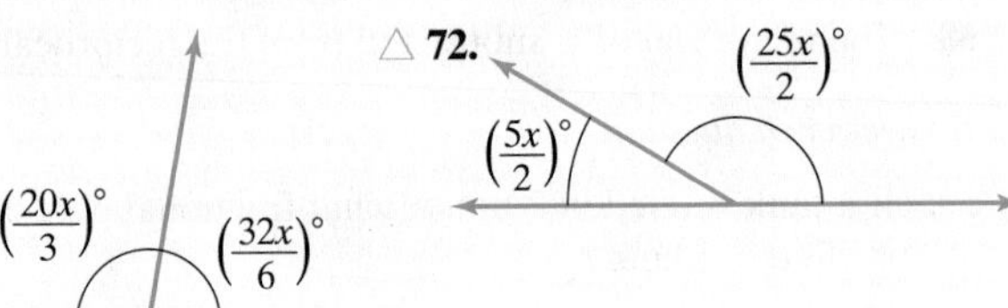

Recall that two angles are complementary if the sum of their measures is 90°. Find the measures of the following complementary angles.

△ **73.**

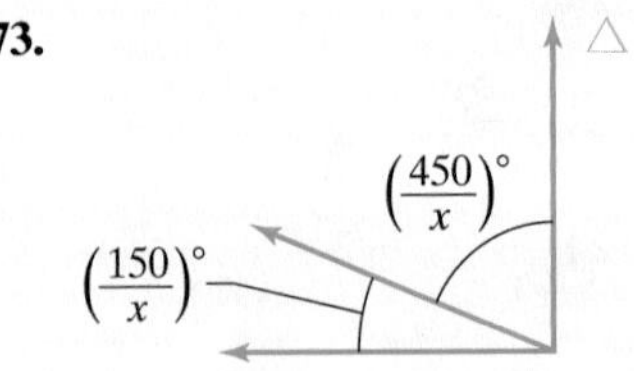

△ **74.**

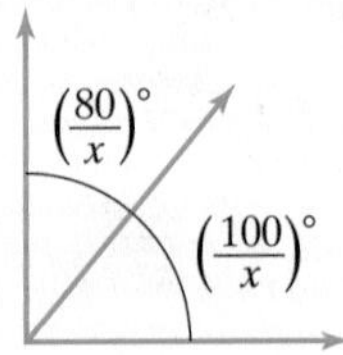

Solve each equation.

75. $\frac{5}{a^2 + 4a + 3} + \frac{2}{a^2 + a - 6} - \frac{3}{a^2 - a - 2} = 0$

76. $\frac{-2}{a^2 + 2a - 8} + \frac{1}{a^2 + 9a + 20} = \frac{-4}{a^2 + 3a - 10}$

Integrated Review SUMMARY ON RATIONAL EXPRESSIONS

Sections 7.1–7.5

It is important to know the difference between performing operations with rational expressions and solving an equation containing rational expressions. Study the examples below.

Performing Operations with Rational Expressions

Adding: $\frac{1}{x} + \frac{1}{x+5} = \frac{1 \cdot (x+5)}{x(x+5)} + \frac{1 \cdot x}{x(x+5)} = \frac{x+5+x}{x(x+5)} = \frac{2x+5}{x(x+5)}$

Subtracting: $\frac{3}{x} - \frac{5}{x^2y} = \frac{3 \cdot xy}{x \cdot xy} - \frac{5}{x^2y} = \frac{3xy - 5}{x^2y}$

Multiplying: $\frac{2}{x} \cdot \frac{5}{x-1} = \frac{2 \cdot 5}{x(x-1)} = \frac{10}{x(x-1)}$

Dividing: $\frac{4}{2x+1} \div \frac{x-3}{x} = \frac{4}{2x+1} \cdot \frac{x}{x-3} = \frac{4x}{(2x+1)(x-3)}$

Solving an Equation Containing Rational Expressions

To solve an equation containing rational expressions, we clear the equation of fractions by multiplying both sides by the LCD.

$$\frac{3}{x} - \frac{5}{x-1} = \frac{1}{x(x-1)} \quad \text{Note that } x \text{ can't be 0 or 1.}$$

$$x(x-1)\left(\frac{3}{x}\right) - x(x-1)\left(\frac{5}{x-1}\right) = x(x-1) \cdot \frac{1}{x(x-1)} \quad \text{Multiply both sides by the LCD.}$$

$$3(x-1) - 5x = 1 \quad \text{Simplify.}$$

$$3x - 3 - 5x = 1 \quad \text{Use the distributive property.}$$

$$-2x - 3 = 1 \quad \text{Combine like terms.}$$

$$-2x = 4 \quad \text{Add 3 to both sides.}$$

$$x = -2 \quad \text{Divide both sides by } -2.$$

Determine whether each of the following is an equation or an expression. If it is an equation, solve it for its variable. If it is an expression, perform the indicated operation.

1. $\frac{1}{x} + \frac{2}{3}$

2. $\frac{3}{a} + \frac{5}{6}$

3. $\frac{1}{x} + \frac{2}{3} = \frac{3}{x}$

4. $\frac{3}{a} + \frac{5}{6} = 1$

5. $\frac{2}{x - 1} - \frac{1}{x}$

6. $\frac{4}{x - 3} - \frac{1}{x}$

7. $\frac{2}{x + 1} - \frac{1}{x} = 1$

8. $\frac{4}{x - 3} - \frac{1}{x} = \frac{6}{x(x - 3)}$

9. $\frac{15x}{x + 8} \cdot \frac{2x + 16}{3x}$

10. $\frac{9z + 5}{15} \cdot \frac{5z}{81z^2 - 25}$

11. $\frac{2x + 1}{x - 3} + \frac{3x + 6}{x - 3}$

12. $\frac{4p - 3}{2p + 7} + \frac{3p + 8}{2p + 7}$

13. $\frac{x + 5}{7} = \frac{8}{2}$

14. $\frac{1}{2} = \frac{x - 1}{8}$

15. $\frac{5a + 10}{18} \div \frac{a^2 - 4}{10a}$

16. $\frac{9}{x^2 - 1} + \frac{12}{3x + 3}$

17. $\frac{x + 2}{3x - 1} + \frac{5}{(3x - 1)^2}$

18. $\frac{4}{(2x - 5)^2} + \frac{x + 1}{2x - 5}$

19. $\frac{x - 7}{x} - \frac{x + 2}{5x}$

20. $\frac{10x - 9}{x} - \frac{x - 4}{3x}$

21. $\frac{3}{x + 3} = \frac{5}{x^2 - 9} - \frac{2}{x - 3}$

22. $\frac{9}{x^2 - 4} + \frac{2}{x + 2} = \frac{-1}{x - 2}$

23. Explain the difference between solving an equation such as $\frac{x}{5} + \frac{3}{10} = \frac{x}{10}$, for x and performing an operation such as adding $\frac{x}{5} + \frac{3}{10}$.

24. When solving an equation such as $\frac{y}{10} = \frac{y}{5} - \frac{1}{10}$, we may multiply all terms by 4. When subtracting two rational expressions such as $\frac{y}{5} - \frac{1}{10}$, we may not. Explain why.

7.6 Proportion and Problem Solving with Rational Equations

OBJECTIVES

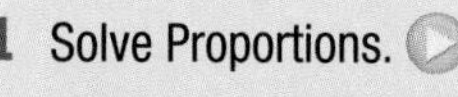

1 Solve Proportions.

2 Use Proportions to Solve Problems.

3 Solve Problems About Numbers.

4 Solve Problems About Work.

5 Solve Problems About Distance.

OBJECTIVE

1 Solving Proportions

A **ratio** is the quotient of two numbers or two quantities. For example, the ratio of 2 to 5 can be written as $\frac{2}{5}$, the quotient of 2 and 5.

If two ratios are equal, we say the ratios are **in proportion** to each other. A **proportion** is a mathematical statement that two ratios are equal.

For example, the equation $\frac{1}{2} = \frac{4}{8}$ is a proportion, as is $\frac{x}{5} = \frac{8}{10}$, because both sides of the equations are ratios. When we want to emphasize the equation as a proportion, we

read the proportion $\frac{1}{2} = \frac{4}{8}$ as "one is to two as four is to eight"

In a proportion, cross products are equal. To understand cross products, let's start with the proportion

$$\frac{a}{b} = \frac{c}{d}$$

and multiply both sides by the LCD, bd.

$$bd\left(\frac{a}{b}\right) = bd\left(\frac{c}{d}\right) \quad \text{Multiply both sides by the LCD, } bd.$$

$$\underbrace{ad}_{\text{Cross product}} = \underbrace{bc}_{\text{Cross product}} \quad \text{Simplify.}$$

Notice why ad and bc are called cross products.

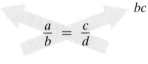

$$\frac{a}{b} = \frac{c}{d}$$

ad bc

Cross Products

If $\frac{a}{b} = \frac{c}{d}$, then $ad = bc$.

For example, if

$$\frac{1}{2} = \frac{4}{8}, \text{ then } 1 \cdot 8 = 2 \cdot 4 \text{ or}$$
$$8 = 8$$

Notice that a proportion contains four numbers (or expressions). If any three numbers are known, we can solve and find the fourth number.

EXAMPLE 1 Solve for x: $\frac{45}{x} = \frac{5}{7}$

Solution This is an equation with rational expressions, and also a proportion. Below are two ways to solve.

Since this is a rational equation, we can use the methods of the previous section.

$$\frac{45}{x} = \frac{5}{7}$$

$$7x \cdot \frac{45}{x} = 7x \cdot \frac{5}{7} \quad \text{Multiply both sides by the LCD } 7x.$$

$$7 \cdot 45 = x \cdot 5 \quad \text{Divide out common factors.}$$

$$315 = 5x \quad \text{Multiply.}$$

$$\frac{315}{5} = \frac{5x}{5} \quad \text{Divide both sides by 5.}$$

$$63 = x \quad \text{Simplify.}$$

Since this is also a proportion, we may set cross products equal.

$$\frac{45}{x} = \frac{5}{7}$$

$$45 \cdot 7 = x \cdot 5 \quad \text{Set cross products equal.}$$

$$315 = 5x \quad \text{Multiply.}$$

$$\frac{315}{5} = \frac{5x}{5} \quad \text{Divide both sides by 5.}$$

$$63 = x \quad \text{Simplify.}$$

Check: Both methods give us a solution of 63. To check, substitute 63 for x in the original proportion. The solution is 63. □

PRACTICE 1 Solve for x: $\frac{36}{x} = \frac{4}{11}$

In this section, if the rational equation is a proportion, we will use cross products to solve.

EXAMPLE 2 Solve for x: $\frac{x-5}{3} = \frac{x+2}{5}$

Solution

$$\frac{x-5}{3} = \frac{x+2}{5}$$

$$
\begin{aligned}
5(x-5) &= 3(x+2) && \text{Set cross products equal.}\\
5x - 25 &= 3x + 6 && \text{Multiply.}\\
5x &= 3x + 31 && \text{Add 25 to both sides.}\\
2x &= 31 && \text{Subtract } 3x \text{ from both sides.}\\
\frac{2x}{2} &= \frac{31}{2} && \text{Divide both sides by 2.}\\
x &= \frac{31}{2}
\end{aligned}
$$

Check: Verify that $\frac{31}{2}$ is the solution. □

PRACTICE

2 Solve for x: $\frac{3x+2}{9} = \frac{x-1}{2}$

OBJECTIVE

2 Using Proportions to Solve Problems 

Proportions can be used to model and solve many real-life problems. When using proportions in this way, it is important to judge whether the solution is reasonable. Doing so helps us decide if the proportion has been formed correctly. We use the same problem-solving steps that were introduced in Section 2.5.

EXAMPLE 3 **Calculating the Cost of Recordable Compact Discs**

Three boxes of CD-Rs (recordable compact discs) cost \$37.47. How much should 5 boxes cost?

Solution

1. UNDERSTAND. Read and reread the problem. We know that the cost of 5 boxes is more than the cost of 3 boxes, or \$37.47, and less than the cost of 6 boxes, which is double the cost of 3 boxes, or 2(\$37.47) = \$74.94. Let's suppose that 5 boxes cost \$60.00. To check, we see if 3 boxes is to 5 boxes as the *price* of 3 boxes is to the *price* of 5 boxes. In other words, we see if

$$\frac{3 \text{ boxes}}{5 \text{ boxes}} = \frac{\text{price of 3 boxes}}{\text{price of 5 boxes}}$$

or

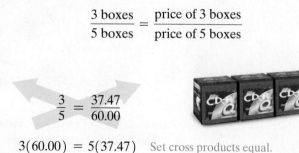

$$\frac{3}{5} = \frac{37.47}{60.00}$$

$$3(60.00) = 5(37.47) \quad \text{Set cross products equal.}$$

or

$$180.00 = 187.35 \quad \text{Not a true statement.}$$

(Continued on next page)

Thus, \$60 is not correct, but we now have a better understanding of the problem.

Let x = price of 5 boxes of CD-Rs.

2. TRANSLATE.

$$\frac{3 \text{ boxes}}{5 \text{ boxes}} = \frac{\text{price of 3 boxes}}{\text{price of 5 boxes}}$$

$$\frac{3}{5} = \frac{37.47}{x}$$

3. SOLVE.

$$\frac{3}{5} = \frac{37.47}{x}$$

$$3x = 5(37.47) \quad \text{Set cross products equal.}$$

$$3x = 187.35$$

$$x = 62.45 \quad \text{Divide both sides by } \mathbf{3}.$$

4. INTERPRET.

Check: Verify that 3 boxes is to 5 boxes as \$37.47 is to \$62.45. Also, notice that our solution is a reasonable one as discussed in Step 1.

State: Five boxes of CD-Rs cost \$62.45. □

PRACTICE

3 Four 2-liter bottles of Diet Pepsi cost \$5.16. How much will seven 2-liter bottles cost?

Helpful Hint

The proportion $\frac{5 \text{ boxes}}{3 \text{ boxes}} = \frac{\text{price of 5 boxes}}{\text{price of 3 boxes}}$ could also have been used to solve Example 3. Notice that the cross products are the same.

Similar triangles have the same shape but not necessarily the same size. In similar triangles, the measures of corresponding angles are equal, and corresponding sides are in proportion.

If triangle ABC and triangle XYZ shown are similar, then we know that the measure of angle A = the measure of angle X, the measure of angle B = the measure of angle Y, and the measure of angle C = the measure of angle Z. We also know that corresponding sides are in proportion: $\frac{a}{x} = \frac{b}{y} = \frac{c}{z}$.

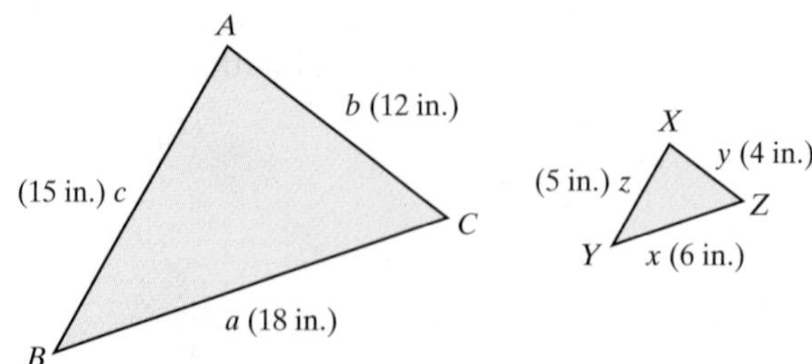

In this section, we will position similar triangles so that they have the same orientation.

To show that corresponding sides are in proportion for the triangles above, we write the ratios of the corresponding sides.

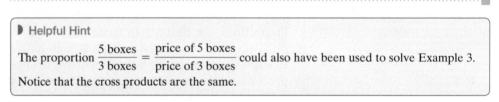

$$\frac{a}{x} = \frac{18}{6} = 3 \quad \frac{b}{y} = \frac{12}{4} = 3 \quad \frac{c}{z} = \frac{15}{5} = 3$$

EXAMPLE 4 **Finding the Length of a Side of a Triangle**

If the following two triangles are similar, find the missing length x.

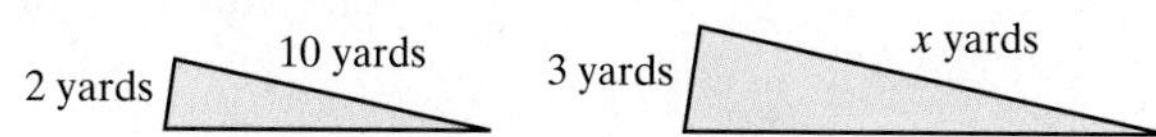

Solution

1. UNDERSTAND. Read the problem and study the figure.
2. TRANSLATE. Since the triangles are similar, their corresponding sides are in proportion and we have

$$\frac{2}{3} = \frac{10}{x}$$

3. SOLVE. To solve, we multiply both sides by the LCD, $3x$, or cross multiply.

$$2x = 30$$

$$x = 15 \quad \text{Divide both sides by 2.}$$

4. INTERPRET.

Check: To check, replace x with 15 in the original proportion and see that a true statement results.

State: The missing length is 15 yards. □

PRACTICE 4 If the following two triangles are similar, find x.

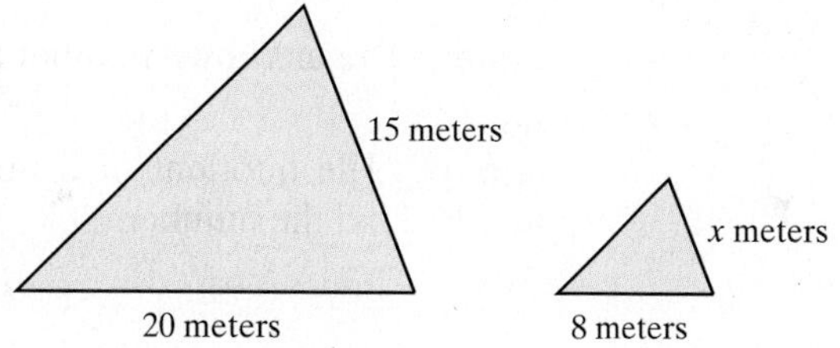

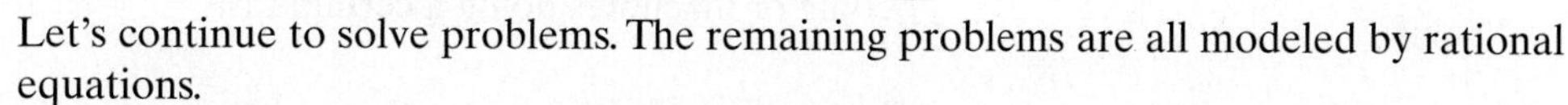

OBJECTIVE 3 Solving Problems About Numbers

Let's continue to solve problems. The remaining problems are all modeled by rational equations.

EXAMPLE 5 **Finding an Unknown Number**

The quotient of a number and 6, minus $\frac{5}{3}$, is the quotient of the number and 2. Find the number.

Solution

1. UNDERSTAND. Read and reread the problem. Suppose that the unknown number is 2, then we see if the quotient of 2 and 6, or $\frac{2}{6}$, minus $\frac{5}{3}$ is equal to the quotient of 2 and 2, or $\frac{2}{2}$.

$$\frac{2}{6} - \frac{5}{3} = \frac{1}{3} - \frac{5}{3} = -\frac{4}{3}, \text{ not } \frac{2}{2}$$

Don't forget that the purpose of a proposed solution is to better understand the problem.

Let $x =$ the unknown number.

2. TRANSLATE.

In words:	the quotient of x and 6	minus	$\frac{5}{3}$	is	the quotient of x and 2
	↓	↓	↓	↓	↓
Translate:	$\frac{x}{6}$	$-$	$\frac{5}{3}$	$=$	$\frac{x}{2}$

3. SOLVE. Here, we solve the equation $\frac{x}{6} - \frac{5}{3} = \frac{x}{2}$. We begin by multiplying both sides of the equation by the LCD, 6.

$$6\left(\frac{x}{6} - \frac{5}{3}\right) = 6\left(\frac{x}{2}\right)$$

$$6\left(\frac{x}{6}\right) - 6\left(\frac{5}{3}\right) = 6\left(\frac{x}{2}\right) \quad \text{Apply the distributive property.}$$

$$x - 10 = 3x \quad \text{Simplify.}$$

$$-10 = 2x \quad \text{Subtract } x \text{ from both sides.}$$

$$\frac{-10}{2} = \frac{2x}{2} \quad \text{Divide both sides by 2.}$$

$$-5 = x \quad \text{Simplify.}$$

4. INTERPRET.

Check: To check, we verify that "the quotient of -5 and 6 minus $\frac{5}{3}$ is the quotient of -5 and 2," or $-\frac{5}{6} - \frac{5}{3} = -\frac{5}{2}$.

State: The unknown number is -5. □

PRACTICE

5 The quotient of a number and 5, minus $\frac{3}{2}$, is the quotient of the number and 10. Find the number.

OBJECTIVE

4 Solving Problems About Work 

The next example is often called a work problem. Work problems usually involve people or machines doing a certain task.

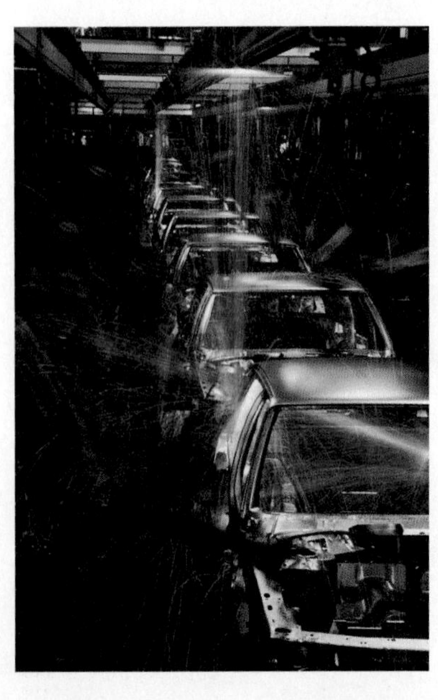

EXAMPLE 6 **Finding Work Rates**

Sam Waterton and Frank Schaffer work in a plant that manufactures automobiles. Sam can complete a quality control tour of the plant in 3 hours, while his assistant, Frank, needs 7 hours to complete the same job. The regional manager is coming to inspect the plant facilities, so both Sam and Frank are directed to complete a quality control tour together. How long will this take?

Solution

1. UNDERSTAND. Read and reread the problem. The key idea here is the relationship between the **time** (hours) it takes to complete the job and the **part of the job** completed in 1 unit of time (hour). For example, if the **time** it takes Sam to complete the job is 3 hours, the **part of the job** he can complete in 1 hour is $\frac{1}{3}$. Similarly, Frank can complete $\frac{1}{7}$ of the job in 1 hour.

Let $x =$ the **time** in hours it takes Sam and Frank to complete the job together. Then $\frac{1}{x} =$ the **part of the job** they complete in 1 hour.

	Hours to Complete Total Job	*Part of Job Completed in 1 Hour*
Sam	3	$\frac{1}{3}$
Frank	7	$\frac{1}{7}$
Together	x	$\frac{1}{x}$

2. TRANSLATE.

In words:	part of job Sam completes in 1 hour	added to	part of job Frank completes in 1 hour	is equal to	part of job they complete together in 1 hour
	↓	↓	↓	↓	↓
Translate:	$\frac{1}{3}$	$+$	$\frac{1}{7}$	$=$	$\frac{1}{x}$

3. SOLVE. Here, we solve the equation $\frac{1}{3} + \frac{1}{7} = \frac{1}{x}$. We begin by multiplying both sides of the equation by the LCD, $21x$.

$$21x\left(\frac{1}{3}\right) + 21x\left(\frac{1}{7}\right) = 21x\left(\frac{1}{x}\right)$$

$$7x + 3x = 21 \quad \text{Simplify.}$$

$$10x = 21$$

$$x = \frac{21}{10} \text{ or } 2\frac{1}{10} \text{ hours}$$

4. INTERPRET.

Check: Our proposed solution is $2\frac{1}{10}$ hours. This proposed solution is reasonable since $2\frac{1}{10}$ hours is more than half of Sam's time and less than half of Frank's time.

Check this solution in the originally *stated* problem.

State: Sam and Frank can complete the quality control tour in $2\frac{1}{10}$ hours. □

PRACTICE

6 Cindy Liu and Mary Beckwith own a landscaping company. Cindy can complete a certain garden planting in 3 hours, while Mary takes 4 hours to complete the same job. If both of them work together, how long will it take to plant the garden?

✓CONCEPT CHECK

Solve $E = mc^2$

a. for m. b. for c^2.

OBJECTIVE

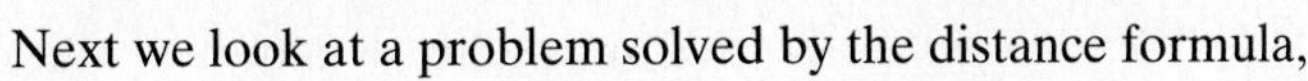

5 Solving Problems About Distance

Next we look at a problem solved by the distance formula,

$$d = r \cdot t$$

Answers to Concept Check:

a. $m = \frac{E}{c^2}$ **b.** $c^2 = \frac{E}{m}$

EXAMPLE 7 **Finding Speeds of Vehicles**

A car travels 180 miles in the same time that a truck travels 120 miles. If the car's speed is 20 miles per hour faster than the truck's, find the car's speed and the truck's speed.

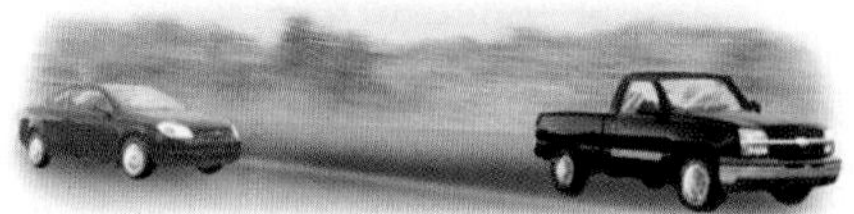

Solution

1. UNDERSTAND. Read and reread the problem. Suppose that the truck's speed is 45 miles per hour. Then the car's speed is 20 miles per hour more, or 65 miles per hour.

We are given that the car travels 180 miles in the same time that the truck travels 120 miles. To find the time it takes the car to travel 180 miles, remember that since $d = rt$, we know that $\frac{d}{r} = t$.

Car's Time

$$t = \frac{d}{r} = \frac{180}{65} = 2\frac{50}{65} = 2\frac{10}{13} \text{ hours}$$

Truck's Time

$$t = \frac{d}{r} = \frac{120}{45} = 2\frac{30}{45} = 2\frac{2}{3} \text{ hours}$$

Since the times are not the same, our proposed solution is not correct. But we have a better understanding of the problem.

Let $x =$ the speed of the truck.

Since the car's speed is 20 miles per hour faster than the truck's, then

$$x + 20 = \text{the speed of the car}$$

Use the formula $d = r \cdot t$ or **d**istance = **r**ate · **t**ime. Prepare a chart to organize the information in the problem.

> **Helpful Hint**
> If $d = r \cdot t$,
> then $t = \frac{d}{r}$
> or $time = \frac{distance}{rate}$.

	Distance	= ***Rate***	· ***Time***
Truck	120	x	$\frac{120}{x}$ ← distance / ← rate
Car	180	$x + 20$	$\frac{180}{x+20}$ ← distance / ← rate

2. TRANSLATE. Since the car and the truck travel the same amount of time, we have that

In words: car's time = truck's time

Translate: $\frac{180}{x + 20} = \frac{120}{x}$

3. SOLVE. We begin by multiplying both sides of the equation by the LCD, $x(x + 20)$, or cross multiplying.

$$\frac{180}{x + 20} = \frac{120}{x}$$

$$180x = 120(x + 20)$$

$$180x = 120x + 2400 \quad \text{Use the distributive property.}$$

$$60x = 2400 \quad \text{Subtract } 120x \text{ from both sides.}$$

$$x = 40 \quad \text{Divide both sides by 60.}$$

4. INTERPRET. The speed of the truck is 40 miles per hour. The speed of the car must then be $x + 20$ or 60 miles per hour.

Check: Find the time it takes the car to travel 180 miles and the time it takes the truck to travel 120 miles.

Car's Time	***Truck's Time***
$t = \frac{d}{r} = \frac{180}{60} = 3$ hours	$t = \frac{d}{r} = \frac{120}{40} = 3$ hours

Since both travel the same amount of time, the proposed solution is correct.

State: The car's speed is 60 miles per hour and the truck's speed is 40 miles per hour. □

PRACTICE

7 A bus travels 180 miles in the same time that a car travels 240 miles. If the car's speed is 15 mph faster than the speed of the bus, find the speed of the car and the speed of the bus.

Vocabulary, Readiness & Video Check

Without solving algebraically, select the best choice for each exercise.

1. One person can complete a job in 7 hours. A second person can complete the same job in 5 hours. How long will it take them to complete the job if they work together?

a. more than 7 hours
b. between 5 and 7 hours
c. less than 5 hours

2. One inlet pipe can fill a pond in 30 hours. A second inlet pipe can fill the same pond in 25 hours. How long before the pond is filled if both inlet pipes are on?

a. less than 25 hours
b. between 25 and 30 hours
c. more than 30 hours

TRANSLATING

Given the variable in the first column, use the phrase in the second column to translate to an expression and then continue to the phrase in the third column to translate to another expression.

3.	A number: x	The reciprocal of the number:	The reciprocal of the number, decreased by 3:
4.	A number: y	The reciprocal of the number:	The reciprocal of the number, increased by 2:
5.	A number: z	The sum of the number and 5:	The reciprocal of the sum of the number and 5:
6.	A number: x	The difference of the number and 1:	The reciprocal of the difference of the number and 1:
7.	A number: y	Twice the number:	Eleven divided by twice the number:
8.	A number: z	Triple the number:	Negative ten divided by triple the number:

Martin-Gay Interactive Videos

See Video 7.6

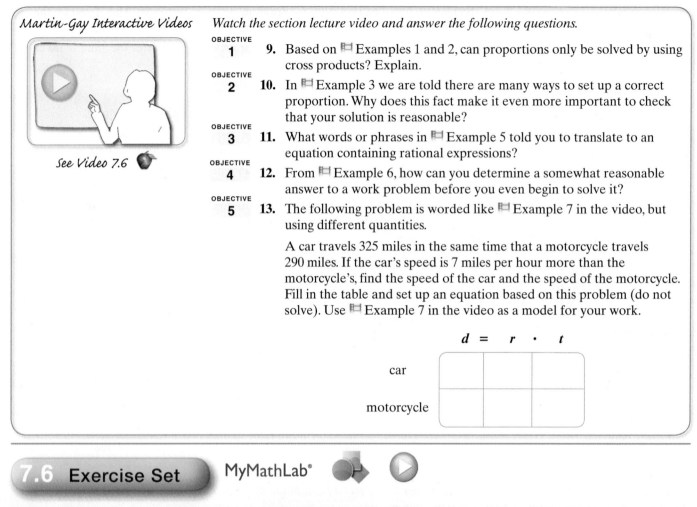

Watch the section lecture video and answer the following questions.

OBJECTIVE 1

9. Based on Examples 1 and 2, can proportions only be solved by using cross products? Explain.

OBJECTIVE 2

10. In Example 3 we are told there are many ways to set up a correct proportion. Why does this fact make it even more important to check that your solution is reasonable?

OBJECTIVE 3

11. What words or phrases in Example 5 told you to translate to an equation containing rational expressions?

OBJECTIVE 4

12. From Example 6, how can you determine a somewhat reasonable answer to a work problem before you even begin to solve it?

OBJECTIVE 5

13. The following problem is worded like Example 7 in the video, but using different quantities.

A car travels 325 miles in the same time that a motorcycle travels 290 miles. If the car's speed is 7 miles per hour more than the motorcycle's, find the speed of the car and the speed of the motorcycle. Fill in the table and set up an equation based on this problem (do not solve). Use Example 7 in the video as a model for your work.

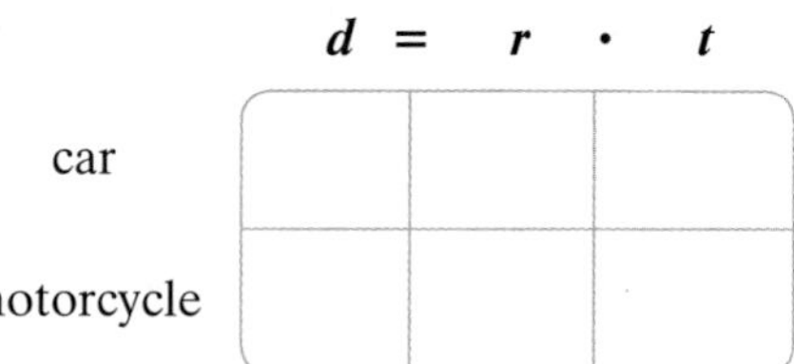

	d	$=$ r	$\cdot$ t
car			
motorcycle			

7.6 Exercise Set

MyMathLab®

Solve each proportion. See Examples 1 and 2. For additional exercises on proportion and proportion applications, see Appendix B.

1. $\frac{2}{3} = \frac{x}{6}$

2. $\frac{x}{2} = \frac{16}{6}$

3. $\frac{x}{10} = \frac{5}{9}$

4. $\frac{9}{4x} = \frac{6}{2}$

5. $\frac{x+1}{2x+3} = \frac{2}{3}$

6. $\frac{x+1}{x+2} = \frac{5}{3}$

7. $\frac{9}{5} = \frac{12}{3x+2}$

8. $\frac{6}{11} = \frac{27}{3x-2}$

Solve. See Example 3.

9. The ratio of the weight of an object on Earth to the weight of the same object on Pluto is 100 to 3. If an elephant weighs 4100 pounds on Earth, find the elephant's weight on Pluto.

10. If a 170-pound person weighs approximately 65 pounds on Mars, about how much does a 9000-pound satellite weigh? Round your answer to the nearest pound.

11. There are 110 calories per 28.8 grams of Frosted Flakes cereal. Find how many calories are in 43.2 grams of this cereal.

12. On an architect's blueprint, 1 inch corresponds to 4 feet. Find the length of a wall represented by a line that is $3\frac{7}{8}$ inches long on the blueprint.

Find the unknown length x or y in the following pairs of similar triangles. See Example 4.

△ 13.

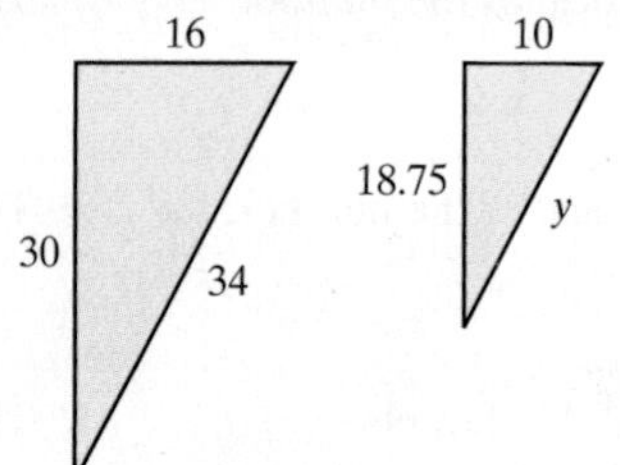

△ 14.

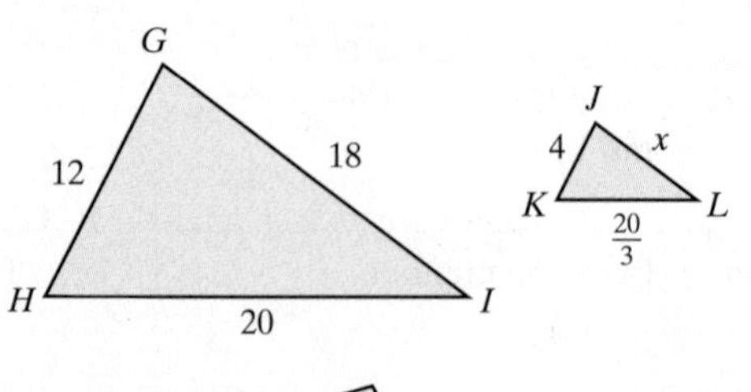

△ 15.

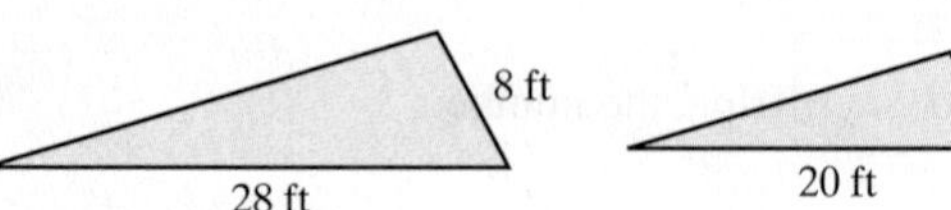

△ 16.

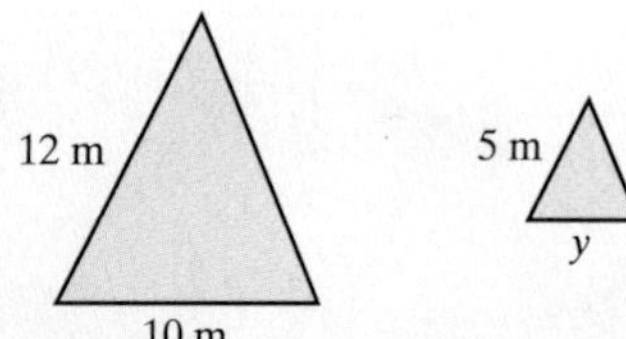

Solve the following. See Example 5.

17. Three times the reciprocal of a number equals 9 times the reciprocal of 6. Find the number.

18. Twelve divided by the sum of x and 2 equals the quotient of 4 and the difference of x and 2. Find x.

19. If twice a number added to 3 is divided by the number plus 1, the result is three halves. Find the number.

20. A number added to the product of 6 and the reciprocal of the number equals -5. Find the number.

See Example 6.

21. Smith Engineering found that an experienced surveyor surveys a roadbed in 4 hours. An apprentice surveyor needs 5 hours to survey the same stretch of road. If the two work together, find how long it takes them to complete the job.

22. An experienced bricklayer constructs a small wall in 3 hours. The apprentice completes the job in 6 hours. Find how long it takes if they work together.

23. In 2 minutes, a conveyor belt moves 300 pounds of recyclable aluminum from the delivery truck to a storage area. A smaller belt moves the same quantity of cans the same distance in 6 minutes. If both belts are used, find how long it takes to move the cans to the storage area.

24. Find how long it takes the conveyor belts described in Exercise 23 to move 1200 pounds of cans. (*Hint:* Think of 1200 pounds as four 300-pound jobs.)

See Example 7.

25. A jogger begins her workout by jogging to the park, a distance of 12 miles. She then jogs home at the same speed but along a different route. This return trip is 18 miles and her time is one hour longer. Find her jogging speed. Complete the accompanying chart and use it to find her jogging speed.

	Distance	=	*Rate*	·	*Time*
Trip to Park	12				
Return Trip	18				

26. A boat can travel 9 miles upstream in the same amount of time it takes to travel 11 miles downstream. If the current of the river is 3 miles per hour, complete the chart below and use it to find the speed of the boat in still water.

	Distance	=	*Rate*	·	*Time*
Upstream	9		$r - 3$		
Downstream	11		$r + 3$		

27. A cyclist rode the first 20-mile portion of his workout at a constant speed. For the 16-mile cooldown portion of his workout, he reduced his speed by 2 miles per hour. Each portion of the workout took the same time. Find the cyclist's speed during the first portion and find his speed during the cooldown portion.

28. A semi-truck travels 300 miles through the flatland in the same amount of time that it travels 180 miles through mountains. The rate of the truck is 20 miles per hour slower in the mountains than in the flatland. Find both the flatland rate and mountain rate.

MIXED PRACTICE

*Solve the following. See Examples 1 through 7. (*Note: *Some exercises can be modeled by equations without rational expressions.)*

29. A human factors expert recommends that there be at least 9 square feet of floor space in a college classroom for every student in the class. Find the minimum floor space that 40 students need.

30. Due to space problems at a local university, a 20-foot by 12-foot conference room is converted into a classroom. Find the maximum number of students the room can accommodate. (See Exercise 29.)

31. One-fourth equals the quotient of a number and 8. Find the number.

32. Four times a number added to 5 is divided by 6. The result is $\frac{7}{2}$. Find the number.

33. Marcus and Tony work for Lombardo's Pipe and Concrete. Mr. Lombardo is preparing an estimate for a customer. He knows that Marcus lays a slab of concrete in 6 hours. Tony lays the same size slab in 4 hours. If both work on the job and the cost of labor is \$45.00 per hour, decide what the labor estimate should be.

34. Mr. Dodson can paint his house by himself in 4 days. His son needs an additional day to complete the job if he works by himself. If they work together, find how long it takes to paint the house.

35. A pilot can travel 400 miles with the wind in the same amount of time as 336 miles against the wind. Find the speed of the wind if the pilot's speed in still air is 230 miles per hour.

36. A fisherman on Pearl River rows 9 miles downstream in the same amount of time he rows 3 miles upstream. If the current is 6 miles per hour, find how long it takes him to cover the 12 miles.

37. Find the unknown length y.

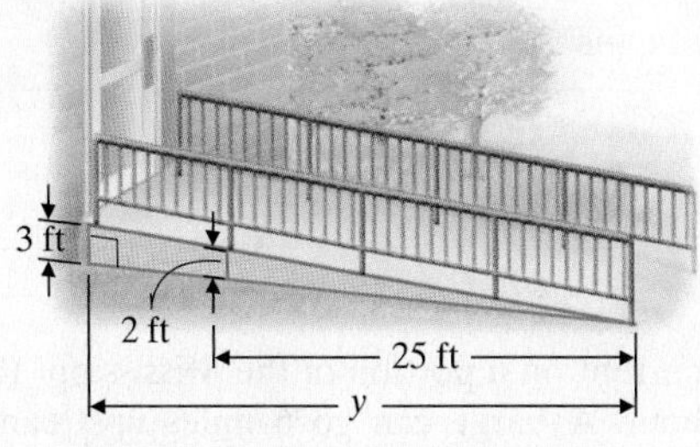

△ 38. Find the unknown length y.

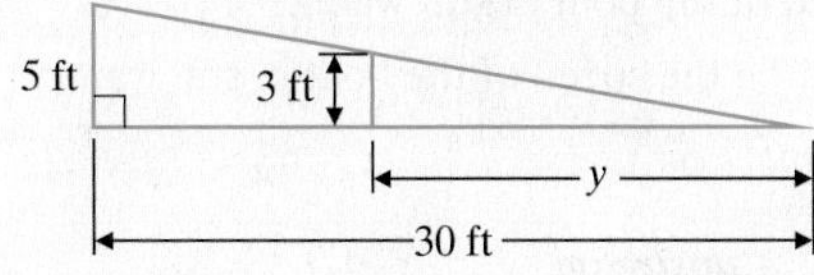

39. Suppose two trains leave Holbrook, Arizona, at the same time, traveling in opposite directions. One train travels 10 mph faster than the other. In 3.5 hours, the trains are 322 miles apart. Find the speed of each train.

40. Suppose two cars leave Brinkley, Arkansas, at the same time, traveling in opposite directions. One car travels 8 mph faster than the other car. In 2.5 hours, the cars are 280 miles apart. Find the speed of each car.

41. Two divided by the difference of a number and 3 minus 4 divided by the number plus 3, equals 8 times the reciprocal of the difference of the number squared and 9. What is the number?

42. If 15 times the reciprocal of a number is added to the ratio of 9 times the number minus 7 and the number plus 2, the result is 9. What is the number?

43. A pilot flies 630 miles with a tailwind of 35 miles per hour. Against the wind, he flies only 455 miles in the same amount of time. Find the rate of the plane in still air.

44. A marketing manager travels 1080 miles in a corporate jet and then an additional 240 miles by car. If the car ride takes one hour longer than the jet ride takes, and if the rate of the jet is 6 times the rate of the car, find the time the manager travels by jet and find the time the manager travels by car.

45. To mix weed killer with water correctly, it is necessary to mix 8 teaspoons of weed killer with 2 gallons of water. Find how many gallons of water are needed to mix with the entire box if it contains 36 teaspoons of weed killer.

46. The directions for a certain bug spray concentrate is to mix 3 ounces of concentrate with 2 gallons of water. How many ounces of concentrate are needed to mix with 5 gallons of water?

47. A boater travels 16 miles per hour on the water on a still day. During one particular windy day, he finds that he travels 48 miles with the wind behind him in the same amount of time that he travels 16 miles into the wind. Find the rate of the wind.

Let x be the rate of the wind.

	$r \times t = d$	
with wind	$16 + x$	48
into wind	$16 - x$	16

48. The current on a portion of the Mississippi River is 3 miles per hour. A barge can go 6 miles upstream in the same amount of time it takes to go 10 miles downstream. Find the speed of the boat in still water.

Let x be the speed of the boat in still water.

	$r \times t = d$	
upstream	$x - 3$	6
downstream	$x + 3$	10

49. Two hikers are 11 miles apart and walking toward each other. They meet in 2 hours. Find the rate of each hiker if one hiker walks 1.1 mph faster than the other.

50. On a 255-mile trip, Gary Alessandrini traveled at an average speed of 70 mph, got a speeding ticket, and then traveled at 60 mph for the remainder of the trip. If the entire trip took 4.5 hours and the speeding ticket stop took 30 minutes, how long did Gary speed before getting stopped?

51. One custodian cleans a suite of offices in 3 hours. When a second worker is asked to join the regular custodian, the job takes only $1\frac{1}{2}$ hours. How long does it take the second worker to do the same job alone?

52. One person proofreads a copy for a small newspaper in 4 hours. If a second proofreader is also employed, the job can be done in $2\frac{1}{2}$ hours. How long does it take for the second proofreader to do the same job alone?

△ **53.** An architect is completing the plans for a triangular deck. Use the diagram below to find the missing dimension.

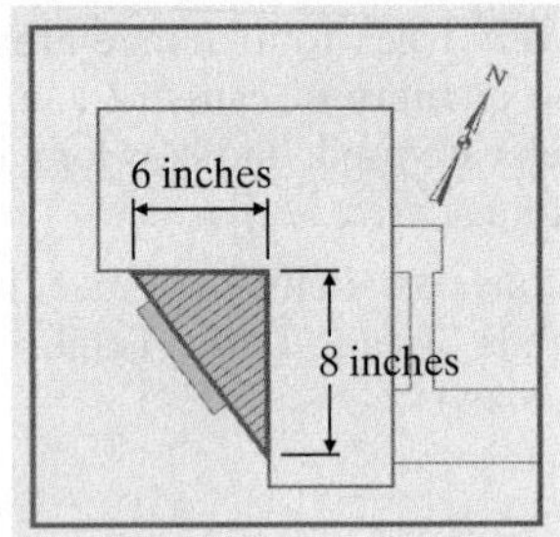

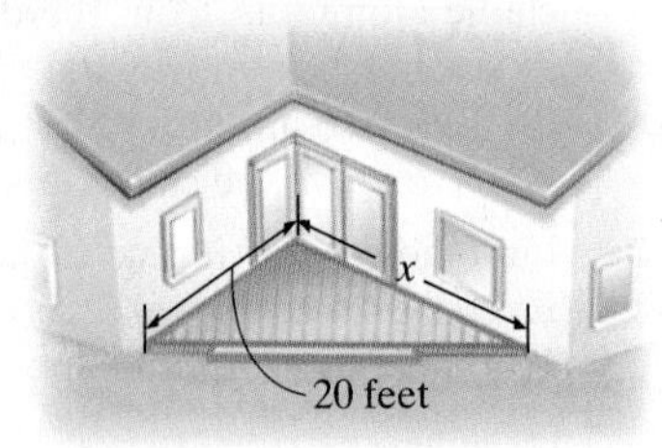

△ **54.** A student wishes to make a small model of a triangular mainsail to study the effects of wind on the sail. The smaller model will be the same shape as a regular-size sailboat's mainsail. Use the following diagram to find the missing dimensions.

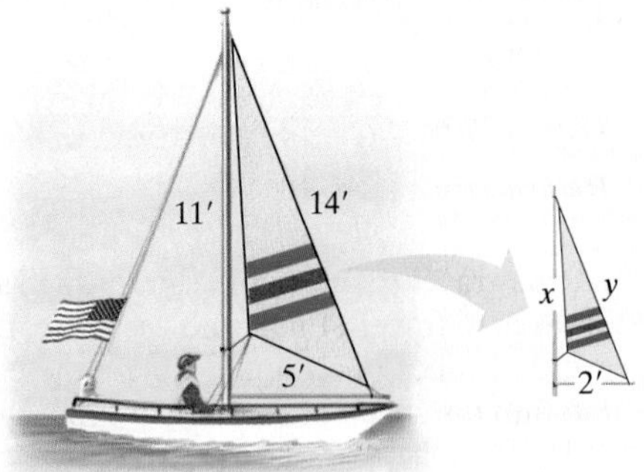

55. The manufacturers of cans of salted mixed nuts state that the ratio of peanuts to other nuts is 3 to 2. If 324 peanuts are in a can, find how many other nuts should also be in the can.

56. There are 1280 calories in a 14-ounce portion of Eagle Brand Milk. Find how many calories are in 2 ounces of Eagle Brand Milk.

57. A jet plane traveling at 500 mph overtakes a propeller plane traveling at 200 mph that had a 2-hour head start. How far from the starting point are the planes?

58. How long will it take a bus traveling at 60 miles per hour to overtake a car traveling at 40 mph if the car had a 1.5-hour head start?

59. One pipe fills a storage pool in 20 hours. A second pipe fills the same pool in 15 hours. When a third pipe is added and all three are used to fill the pool, it takes only 6 hours. Find how long it takes the third pipe to do the job.

60. One pump fills a tank in 9 hours. A second pump fills the same tank in 6 hours. When a third pump is added and all three are used to fill the tank, it takes only 3 hours. Find how long it takes the third pump to fill the tank.

61. A car travels 280 miles in the same time that a motorcycle travels 240 miles. If the car's speed is 10 miles per hour more than the motorcycle's, find the speed of the car and the speed of the motorcycle.

62. A bus traveled on a level road for 3 hours at an average speed 20 miles per hour faster than it traveled on a winding road. The time spent on the winding road was 4 hours. Find the average speed on the level road if the entire trip was 305 miles.

63. In 6 hours, an experienced cook prepares enough pies to supply a local restaurant's daily order. Another cook prepares the same number of pies in 7 hours. Together with a third cook, they prepare the pies in 2 hours. Find how long it takes the third cook to prepare the pies alone.

64. Mrs. Smith balances the company books in 8 hours. It takes her assistant 12 hours to do the same job. If they work together, find how long it takes them to balance the books.

65. The quotient of a number and 3, minus 1, equals $\frac{5}{3}$. Find the number.

66. The quotient of a number and 5, minus 1, equals $\frac{7}{5}$. Find the number.

67. Currently, the Toyota Corolla is the best-selling car in the world. Suppose that during a test drive of two Corollas, one car travels 224 miles in the same time that the second car travels 175 miles. If the speed of the first car is 14 miles per hour faster than the speed of the second car, find the speed of both cars. (*Source: Top Ten of Everything*)

68. The second best-selling car is the Volkswagen Golf. A driver of this car took a day trip around the California coastline driving at two speeds. He drove 70 miles at a slower speed and 300 miles at a speed 40 miles per hour faster. If the time spent driving the faster speed was twice that spent at the slower speed, find the two speeds during the trip. (*Source: Top Ten of Everything*)

69. A pilot can fly an MD-11 2160 miles with the wind in the same time she can fly 1920 miles against the wind. If the speed of the wind is 30 mph, find the speed of the plane in still air. (*Source*: Air Transport Association of America)

70. A pilot can fly a DC-10 1365 miles against the wind in the same time he can fly 1575 miles with the wind. If the speed of the plane in still air is 490 miles per hour, find the speed of the wind. (*Source*: Air Transport Association of America)

Given that the following pairs of triangles are similar, find each missing length.

71.

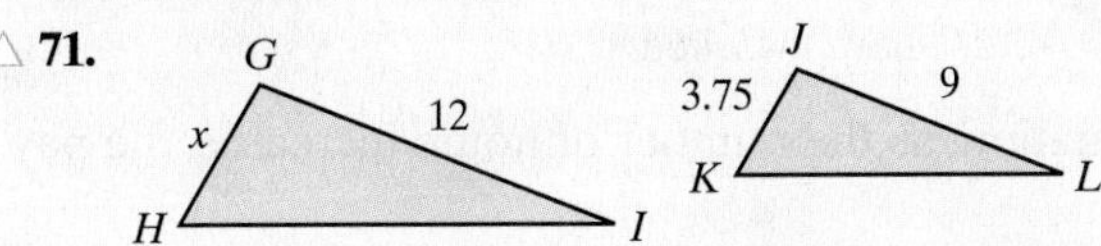

72. **73.** **74.**

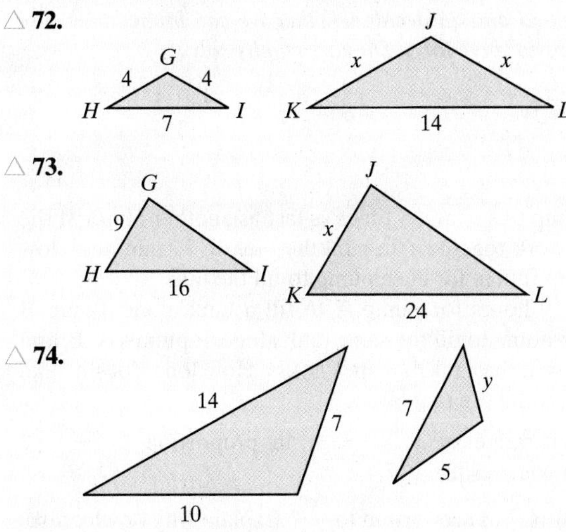

REVIEW AND PREVIEW

Find the slope of the line through each pair of points. Use the slope to determine whether the line is vertical, horizontal, or moves upward or downward from left to right. See Section 3.4.

75. $(-2, 5), (4, -3)$

76. $(0, 4), (2, 10)$

77. $(-3, -6), (1, 5)$

78. $(-2, 7), (3, -2)$

79. $(3, 7), (3, -2)$

80. $(0, -4), (2, -4)$

CONCEPT EXTENSIONS

The following bar graph shows the capacity of the United States to generate electricity from the wind in the years shown. Use this graph for Exercises 81 and 82.

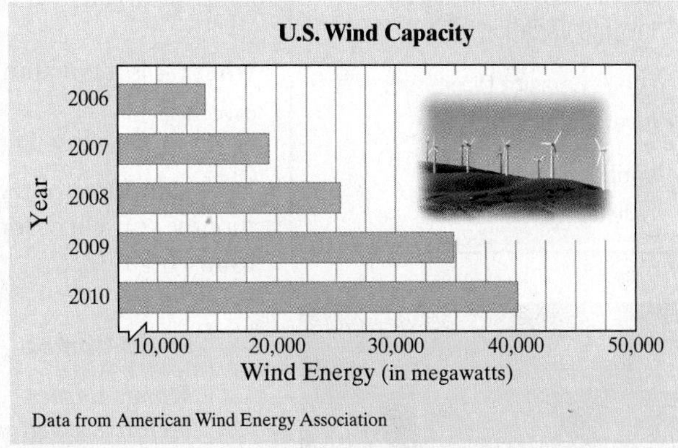

81. Find the approximate megawatt capacity in 2010.

82. Find the approximate megawatt capacity in 2009.

In general, 1000 megawatts will serve the average electricity needs of 560,000 people. Use this fact and the preceding graph to answer Exercises 83 and 84.

83. In 2010, the number of megawatts that were generated from wind would serve the electricity needs of how many people? (Round to the nearest ten-thousand.)

84. How many megawatts of electricity are needed to serve the city or town in which you live?

For Exercises 85 and 86 decide whether we can immediately use cross products to solve for x. Do not actually solve.

85. $\frac{2-x}{5} = \frac{1+x}{3}$

86. $\frac{2}{5} - x = \frac{1+x}{3}$

Solve.

87. One pump fills a tank 3 times as fast as another pump. If the pumps work together, they fill the tank in 21 minutes. How long does it take for each pump to fill the tank?

88. It takes 9 hours for pump A to fill a tank alone. Pump B takes 15 hours to fill the same tank alone. If pumps A, B, and C are used, the tank fills in 5 hours. How long does it take pump C to fill the tank alone?

89. For what value of x is $\frac{x}{x-1}$ in proportion to $\frac{x+1}{x}$? Explain your result.

90. If x is 10, is $\frac{2}{x}$ in proportion to $\frac{x}{50}$? Explain why or why not.

91. Person A can complete a job in 5 hours, and person B can complete the same job in 3 hours. Without solving algebraically, discuss reasonable and unreasonable answers for how long it would take them to complete the job together.

92. A hyena spots a giraffe 0.5 mile away and begins running toward it. The giraffe starts running away from the hyena just as the hyena begins running toward it. A hyena can run at a speed of 40 mph and a giraffe can run at 32 mph. How long will it take for the hyena to overtake the giraffe? (*Source: World Almanac and Book of Facts*)

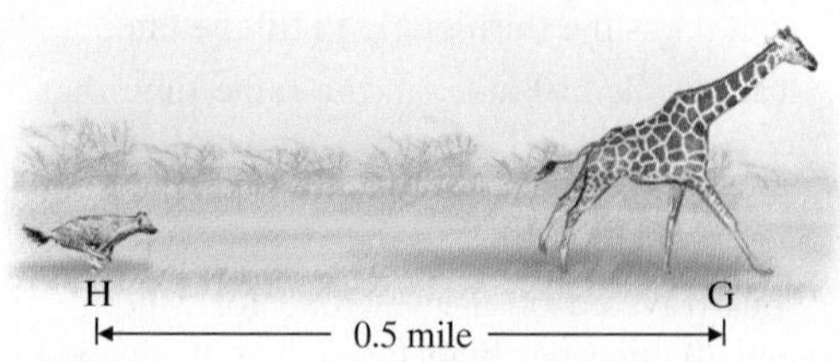

Solve. See the Concept Check in this section.

Solve $D = RT$

93. for R

94. for T

7.7 Variation and Problem Solving 

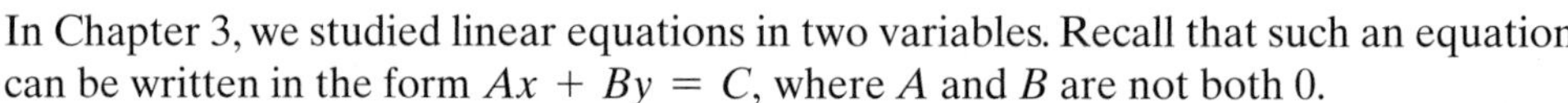

OBJECTIVES

1 Solve Problems Involving Direct Variation.

2 Solve Problems Involving Inverse Variation.

3 Other Types of Direct and Inverse Variation.

4 Variation and Problem Solving.

In Chapter 3, we studied linear equations in two variables. Recall that such an equation can be written in the form $Ax + By = C$, where A and B are not both 0.

Also recall that the graph of a linear equation in two variables is a line. In this section, we begin by looking at a particular family of linear equations—those that can be written in the form

$$y = kx,$$

where k is a constant. This family of equations is called *direct variation.*

OBJECTIVE

1 Solving Direct Variation Problems 

Let's suppose that you are earning \$7.25 per hour at a part-time job. The amount of money you earn depends on the number of hours you work. This is illustrated by the following table:

Hours Worked	0	1	2	3	4
Money Earned (before deductions)	0	7.25	14.50	21.75	29.00

and so on

In general, to calculate your earnings (before deductions) multiply the constant \$7.25 by the number of hours you work. If we let y represent the amount of money earned and x represent the number of hours worked, we get the direct variation equation

$$y = 7.25 \cdot x$$

earnings = \$7.25 · hours worked

Notice that in this direct variation equation, as the number of hours increases, the pay increases as well.

Direct Variation

y* varies directly as *x, or ***y* is directly proportional to *x***, if there is a nonzero constant k such that

$$y = kx$$

The number k is called the **constant of variation** or the **constant of proportionality.**

In our direct variation example $y = 7.25x$, the constant of variation is 7.25.

Let's use the previous table to graph $y = 7.25x$. We begin our graph at the ordered pair solution $(0, 0)$. Why? We assume that the least amount of hours worked is 0. If 0 hours are worked, then the pay is \$0.

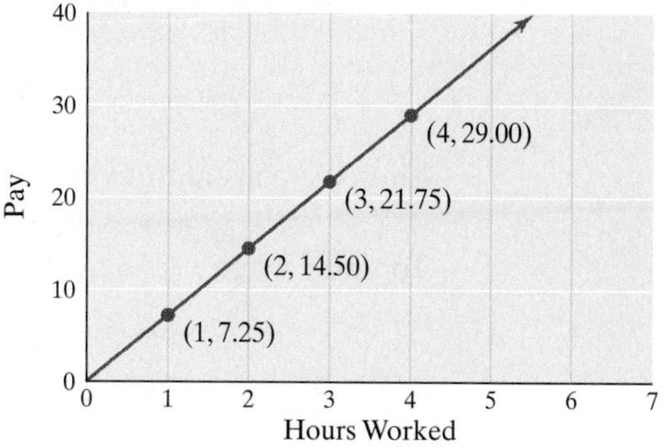

As illustrated in this graph, a direct variation equation $y = kx$ is linear. Also notice that $y = 7.25x$ is a function since its graph passes the vertical line test.

EXAMPLE 1 Write a direct variation equation of the form $y = kx$ that satisfies the ordered pairs in the table below.

x	2	9	1.5	−1
y	6	27	4.5	−3

Solution We are given that there is a direct variation relationship between x and y. This means that

$$y = kx$$

By studying the given values, you may be able to calculate k mentally. If not, to find k, we simply substitute one given ordered pair into this equation and solve for k. We'll use the given pair $(2, 6)$.

$$y = kx$$
$$6 = k \cdot 2$$
$$\frac{6}{2} = \frac{k \cdot 2}{2}$$
$$3 = k \quad \text{Solve for } k.$$

Since $k = 3$, we have the equation $y = 3x$.
To check, see that each given y is 3 times the given x. □

PRACTICE

1 Write a direct variation of the form $y = kx$ that satisfies the ordered pairs in the table below.

x	2	8	−4	1.3
y	10	40	−20	6.5

Let's try another type of direct variation example.

EXAMPLE 2 Suppose that y varies directly as x. If y is 17 when x is 34, find the constant of variation and the direct variation equation. Then find y when x is 12.

Solution Let's use the same method as in Example 1 to find x. Since we are told that y varies directly as x, we know the relationship is of the form

$$y = kx$$

Let $y = 17$ and $x = 34$ and solve for k.

$$17 = k \cdot 34$$

$$\frac{17}{34} = \frac{k \cdot 34}{34}$$

$$\frac{1}{2} = k \quad \text{Solve for } k.$$

Thus, the constant of variation is $\frac{1}{2}$ and the equation is $y = \frac{1}{2}x$.

To find y when $x = 12$, use $y = \frac{1}{2}x$ and replace x with 12.

$$y = \frac{1}{2}x$$

$$y = \frac{1}{2} \cdot 12 \quad \text{Replace } x \text{ with 12.}$$

$$y = 6$$

Thus, when x is 12, y is 6. □

PRACTICE

2 If y varies directly as x and y is 12 when x is 48, find the constant of variation and the direct variation equation. Then find y when x is 20.

Let's review a few facts about linear equations of the form $y = kx$.

Direct Variation: $y = kx$

- There is a direct variation relationship between x and y.
- The graph is a line.
- The line will always go through the origin $(0, 0)$. Why? Let $x = 0$. Then $y = k \cdot 0$ or $y = 0$.
- The slope of the graph of $y = kx$ is k, the constant of variation. Why? Remember that the slope of an equation of the form $y = mx + b$ is m, the coefficient of x.
- The equation $y = kx$ describes a function. Each x has a unique y and its graph passes the vertical line test.

EXAMPLE 3 The line is the graph of a direct variation equation. Find the constant of variation and the direct variation equation.

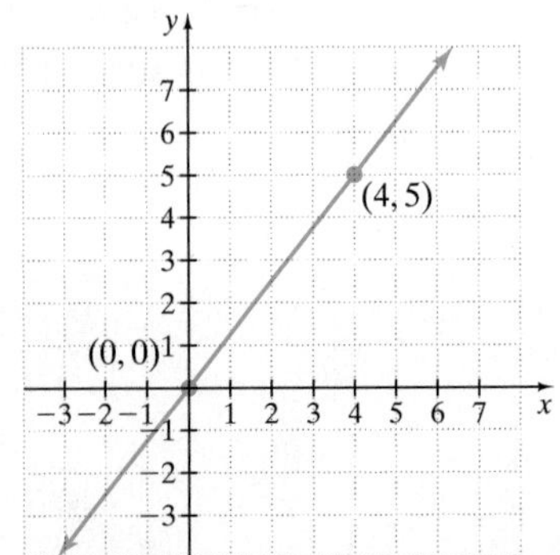

Solution Recall that k, the constant of variation, is the same as the slope of the line. Thus, to find k, we use the slope formula and find slope.

Using the given points $(0, 0)$, and $(4, 5)$, we have

$$\text{slope} = \frac{5 - 0}{4 - 0} = \frac{5}{4}.$$

Thus, $k = \frac{5}{4}$ and the variation equation is $y = \frac{5}{4}x$. □

PRACTICE 3 Find the constant of variation and the direct variation equation for the line below.

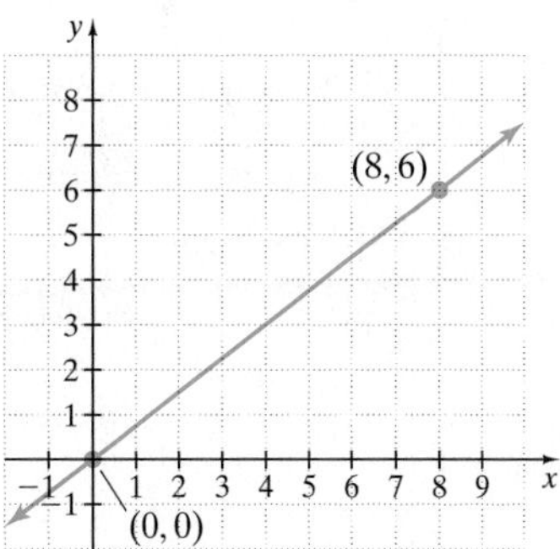

OBJECTIVE 2

Solving Inverse Variation Problems

In this section, we will introduce another type of variation, called inverse variation.

Let's suppose you need to drive a distance of 40 miles. You know that the faster you drive the distance, the sooner you arrive at your destination. Recall that there is a mathematical relationship between distance, rate, and time. It is $d = r \cdot t$. In our example, distance is a constant 40 miles, so we have $40 = r \cdot t$ or $t = \frac{40}{r}$.

For example, if you drive 10 mph, the time to drive the 40 miles is

$$t = \frac{40}{r} = \frac{40}{10} = 4 \text{ hours}$$

If you drive 20 mph, the time is

$$t = \frac{40}{r} = \frac{40}{20} = 2 \text{ hours}$$

Again, notice that as speed increases, time decreases. Below are some ordered pair solutions of $t = \frac{40}{r}$ and its graph.

Rate (mph)	r	5	10	20	40	60	80
Time (hr)	t	8	4	2	1	$\frac{2}{3}$	$\frac{1}{2}$

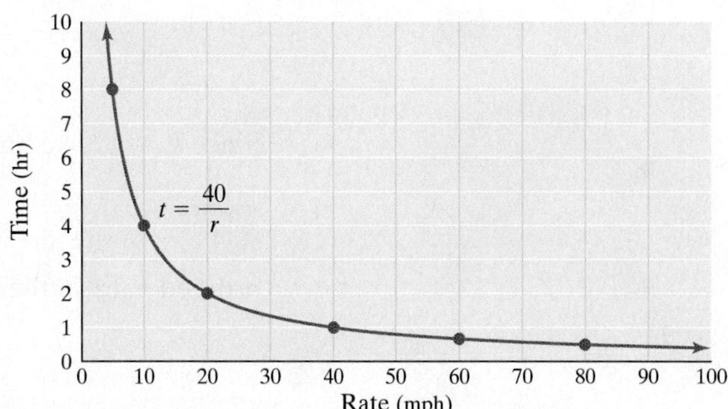

Notice that the graph of this variation is not a line, but it passes the vertical line test, so $t = \frac{40}{r}$ does describe a function. This is an example of inverse variation.

Inverse Variation

y* varies inversely as *x, or ***y* is inversely proportional to *x***, if there is a nonzero constant k such that

$$y = \frac{k}{x}$$

The number k is called the **constant of variation** or the **constant of proportionality.**

In our inverse variation example, $t = \frac{40}{r}$ or $y = \frac{40}{x}$, the constant of variation is 40.

We can immediately see differences and similarities in direct variation and inverse variation.

Direct variation	$y = kx$	linear equation	both functions
Inverse variation	$y = \frac{k}{x}$	rational equation	

Remember that $y = \frac{k}{x}$ is a rational equation and not a linear equation. Also notice that because x is in the denominator, x can be any value except 0.

We can still derive an inverse variation equation from a table of values.

EXAMPLE 4 Write an inverse variation equation of the form $y = \frac{k}{x}$ that satisfies the ordered pairs in the table below.

x	2	4	$\frac{1}{2}$
y	6	3	24

Solution Since there is an inverse variation relationship between x and y, we know that $y = \frac{k}{x}$. To find k, choose one given ordered pair and substitute the values into the equation. We'll use $(2, 6)$.

$$y = \frac{k}{x}$$

$$6 = \frac{k}{2}$$

$$2 \cdot 6 = 2 \cdot \frac{k}{2} \quad \text{Multiply both sides by 2.}$$

$$12 = k \quad \text{Solve for } k.$$

Since $k = 12$, we have the equation $y = \frac{12}{x}$. □

PRACTICE

4 Write an inverse variation equation of the form $y = \frac{k}{x}$ that satisfies the ordered pairs in the table below.

x	2	-1	$\frac{1}{3}$
y	4	-8	24

Helpful Hint

Multiply both sides of the inverse variation relationship equation $y = \frac{k}{x}$ by x (as long as x is not 0), and we have $xy = k$. This means that if y varies inversely as x, their product is always the constant of variation k. For an example of this, check the table from Example 4.

x	2	4	$\frac{1}{2}$
y	6	3	24

$2 \cdot 6 = 12 \quad 4 \cdot 3 = 12 \quad \frac{1}{2} \cdot 24 = 12$

EXAMPLE 5 Suppose that y varies inversely as x. If $y = 0.02$ when $x = 75$, find the constant of variation and the inverse variation equation. Then find y when x is 30.

Solution Since y varies inversely as x, the constant of variation may be found by simply finding the product of the given x and y.

$$k = xy = 75(0.02) = 1.5$$

To check, we will use the inverse variation equation

$$y = \frac{k}{x}.$$

Let $y = 0.02$ and $x = 75$ and solve for k.

$$0.02 = \frac{k}{75}$$

$$75(0.02) = 75 \cdot \frac{k}{75} \quad \text{Multiply both sides by 75.}$$

$$1.5 = k \quad \text{Solve for } k.$$

Thus, the constant of variation is 1.5 and the equation is $y = \frac{1.5}{x}$.

To find y when $x = 30$, use $y = \frac{1.5}{x}$ and replace x with 30.

$$y = \frac{1.5}{x}$$

$$y = \frac{1.5}{30} \quad \text{Replace } x \text{ with 30.}$$

$$y = 0.05$$

Thus, when x is 30, y is 0.05. □

PRACTICE

5 If y varies inversely as x and y is 0.05 when x is 42, find the constant of variation and the inverse variation equation. Then find y when x is 70.

OBJECTIVE

3 Solving Other Types of Direct and Inverse Variation Problems 

It is possible for y to vary directly or inversely as powers of x.

Direct and Inverse Variation as *n*th Powers of *x*

y* varies directly as a power of *x if there is a nonzero constant k and a natural number n such that

$$y = kx^n$$

y* varies inversely as a power of *x if there is a nonzero constant k and a natural number n such that

$$y = \frac{k}{x^n}$$

EXAMPLE 6 The surface area of a cube A varies directly as the square of a length of its side s. If A is 54 when s is 3, find A when $s = 4.2$.

Solution Since the surface area A varies directly as the square of side s, we have

$$A = ks^2.$$

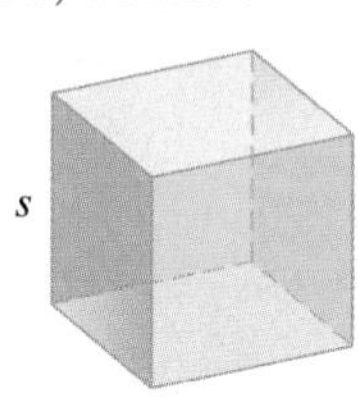

To find k, let $A = 54$ and $s = 3$.

$$A = k \cdot s^2$$

$$54 = k \cdot 3^2 \quad \text{Let } A = 54 \text{ and } s = 3.$$

$$54 = 9k \quad 3^2 = 9.$$

$$6 = k \quad \text{Divide by 9.}$$

The formula for surface area of a cube is then

$$A = 6s^2 \text{ where } s \text{ is the length of a side.}$$

To find the surface area when $s = 4.2$, substitute.

$$A = 6s^2$$

$$A = 6 \cdot (4.2)^2$$

$$A = 105.84$$

The surface area of a cube whose side measures 4.2 units is 105.84 square units. □

PRACTICE

6 The area of an isosceles right triangle A varies directly as the square of one of its legs x. If A is 32 when x is 8, find A when $x = 3.6$.

OBJECTIVE

4 Solving Applications of Variation 

There are many real-life applications of direct and inverse variation.

EXAMPLE 7 The weight of a body w varies inversely with the square of its distance from the center of Earth d. If a person weighs 160 pounds on the surface of Earth, what is the person's weight 200 miles above the surface? (Assume that the radius of Earth is 4000 miles.)

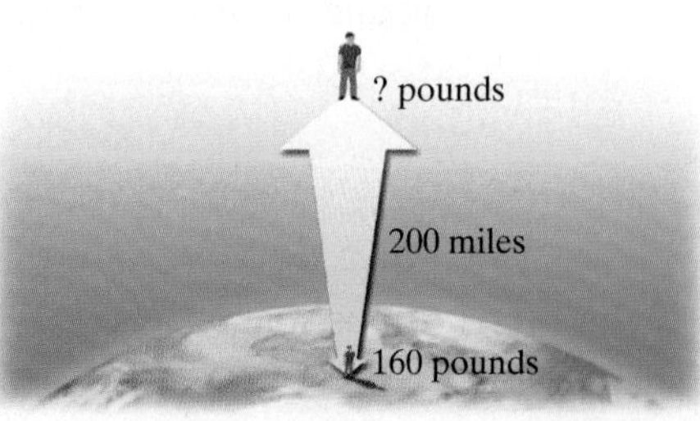

Solution

1. UNDERSTAND. Make sure you read and reread the problem.
2. TRANSLATE. Since we are told that weight w varies inversely with the square of its distance from the center of Earth, d, we have

$$w = \frac{k}{d^2}.$$

3. SOLVE. To solve the problem, we first find k. To do so, use the fact that the person weighs 160 pounds on Earth's surface, which is a distance of 4000 miles from Earth's center.

$$w = \frac{k}{d^2}$$

$$160 = \frac{k}{(4000)^2}$$

$$2{,}560{,}000{,}000 = k$$

Thus, we have $w = \dfrac{2{,}560{,}000{,}000}{d^2}$.

Since we want to know the person's weight 200 miles above Earth's surface, we let $d = 4200$ and find w.

$$w = \frac{2{,}560{,}000{,}000}{d^2}$$

$$w = \frac{2{,}560{,}000{,}000}{(4200)^2}$$ A person 200 miles above Earth's surface is 4200 miles from Earth's center.

$$w \approx 145$$ Simplify.

4. INTERPRET.

Check: Your answer is reasonable since the farther a person is from Earth, the less the person weighs.

State: Thus, 200 miles above the surface of Earth, a 160-pound person weighs approximately 145 pounds. □

PRACTICE

7 Robert Boyle investigated the relationship between volume of a gas and its pressure. He developed Boyle's law, which states that the volume of a gas varies inversely with pressure if the temperature is held constant. If 50 ml of oxygen is at a pressure of 20 atmospheres, what will the volume of the oxygen be at a pressure of 40 atmospheres?

Vocabulary, Readiness & Video Check

State whether each equation represents direct or inverse variation.

1. $y = \frac{k}{x}$, where k is a constant. ________
2. $y = kx$, where k is a constant. ________
3. $y = 5x$ ________
4. $y = \frac{5}{x}$ ________
5. $y = \frac{7}{x^2}$ ________
6. $y = 6.5x^4$ ________
7. $y = \frac{11}{x}$ ________
8. $y = 18x$ ________
9. $y = 12x^2$ ________
10. $y = \frac{20}{x^3}$ ________

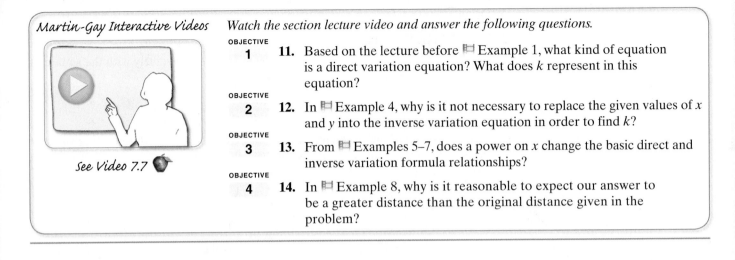

Martin-Gay Interactive Videos

See Video 7.7

Watch the section lecture video and answer the following questions.

OBJECTIVE 1

11. Based on the lecture before Example 1, what kind of equation is a direct variation equation? What does k represent in this equation?

OBJECTIVE 2

12. In Example 4, why is it not necessary to replace the given values of x and y into the inverse variation equation in order to find k?

OBJECTIVE 3

13. From Examples 5–7, does a power on x change the basic direct and inverse variation formula relationships?

OBJECTIVE 4

14. In Example 8, why is it reasonable to expect our answer to be a greater distance than the original distance given in the problem?

7.7 Exercise Set

MyMathLab®

Write a direct variation equation, $y = kx$, that satisfies the ordered pairs in each table. See Example 1.

1.

x	0	6	10
y	0	3	5

2.

x	0	2	−1	3
y	0	14	−7	21

3.

x	−2	2	4	5
y	−12	12	24	30

4.

x	3	9	−2	12
y	1	3	$-\frac{2}{3}$	4

Write a direct variation equation, $y = kx$, that describes each graph. See Example 3.

5.

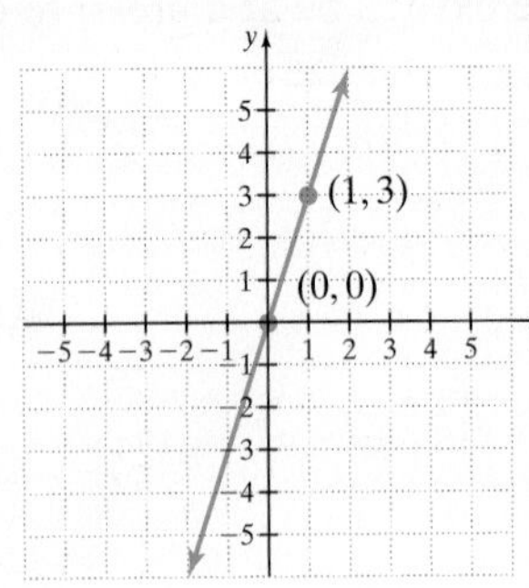

6.

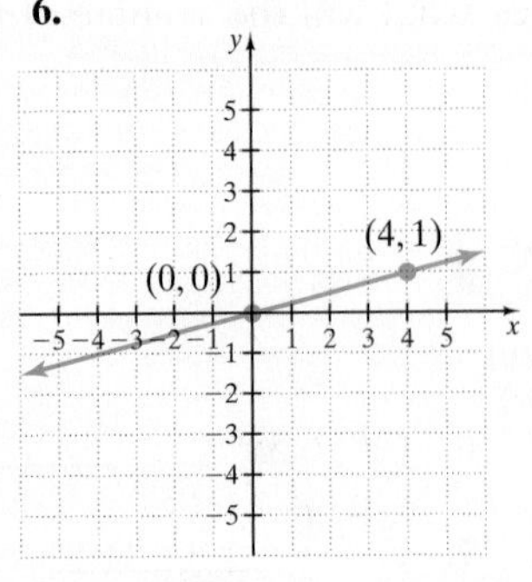

7.

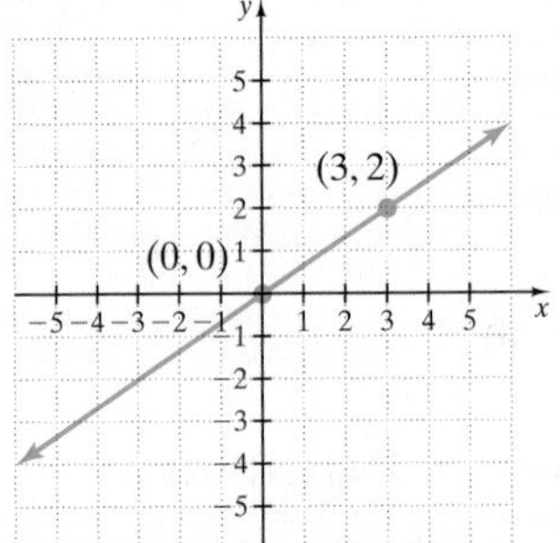

8.

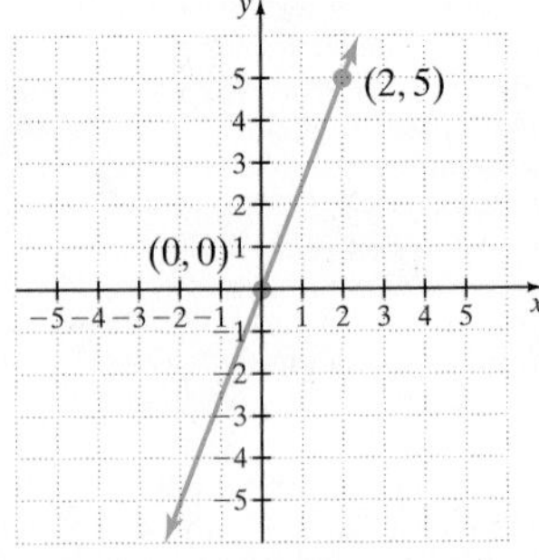

Write an inverse variation equation, $y = \frac{k}{x}$, that satisfies the ordered pairs in each table. See Example 4.

9.

x	1	−7	3.5	−2
y	7	−1	2	−3.5

10.

x	2	−11	4	−4
y	11	−2	5.5	−5.5

11.

x	10	$\frac{1}{2}$	$-\frac{3}{2}$
y	0.05	1	$-\frac{1}{3}$

12.

x	4	$\frac{1}{5}$	−8
y	0.1	2	−0.05

MIXED PRACTICE

Write an equation to describe each variation. Use k for the constant of proportionality. See Examples 1 through 6.

13. y varies directly as x

14. a varies directly as b

15. h varies inversely as t

16. s varies inversely as t

17. z varies directly as x^2

18. p varies inversely as x^2

19. y varies inversely as z^3

20. x varies directly as y^4

21. x varies inversely as $\sqrt{y}$

22. y varies directly as d^2

Solve. See Examples 2, 5, and 6.

23. y varies directly as x. If $y = 20$ when $x = 5$, find y when x is 10.

24. y varies directly as x. If $y = 27$ when $x = 3$, find y when x is 2.

25. y varies inversely as x. If $y = 5$ when $x = 60$, find y when x is 100.

26. y varies inversely as x. If $y = 200$ when $x = 5$, find y when x is 4.

27. z varies directly as x^2. If $z = 96$ when $x = 4$, find z when $x = 3$.

28. s varies directly as t^3. If $s = 270$ when $t = 3$, find s when $x = 1$.

29. a varies inversely as b^3. If $a = \frac{3}{2}$ when $b = 2$, find a when b is 3.

30. p varies inversely as q^2. If $p = \frac{5}{16}$ when $q = 8$, find p when $q = \frac{1}{2}$.

Solve. See Examples 1 through 7.

31. Your paycheck (before deductions) varies directly as the number of hours you work. If your paycheck is \$112.50 for 18 hours, find your pay for 10 hours.

32. If your paycheck (before deductions) is \$244.50 for 30 hours, find your pay for 34 hours. See Exercise 31.

33. The cost of manufacturing a certain type of headphone varies inversely as the number of headphones increases. If 5000 headphones can be manufactured for \$9.00 each, find the cost per headphone to manufacture 7500 headphones.

34. The cost of manufacturing a certain composition notebook varies inversely as the number of notebooks increases. If 10,000 notebooks can be manufactured for \$0.50 each, find the cost per notebook to manufacture 18,000 notebooks.

35. The distance a spring stretches varies directly with the weight attached to the spring. If a 60-pound weight stretches the spring 4 inches, find the distance that an 80-pound weight stretches the spring.

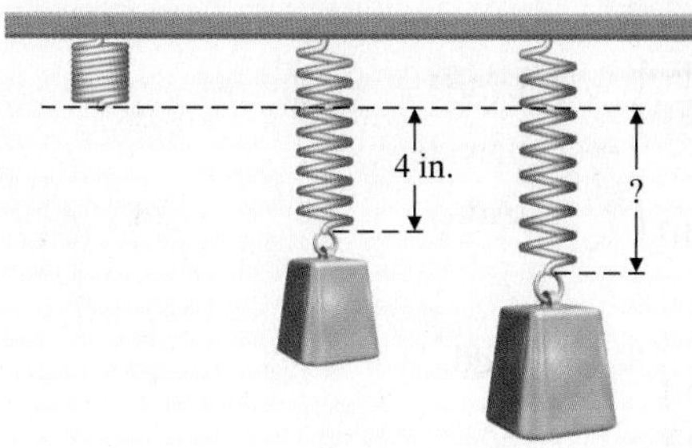

36. If a 30-pound weight stretches a spring 10 inches, find the distance a 20-pound weight stretches the spring. (See Exercise 35.)

37. The weight of an object varies inversely as the square of its distance from the *center* of Earth. If a person weighs 180 pounds on Earth's surface, what is his weight 10 miles above the surface of Earth? Round to the nearest tenth of a pound. (Assume that Earth's radius is 4000 miles.)

38. For a constant distance, the rate of travel varies inversely as the time traveled. If a family travels 55 mph and arrives at a destination in 4 hours, how long will the return trip take traveling at 60 mph?

39. The distance d that an object falls is directly proportional to the square of the time of the fall, t. A person who is parachuting for the first time is told to wait 10 seconds before opening the parachute. If the person falls 64 feet in 2 seconds, find how far he falls in 10 seconds.

40. The distance needed for a car to stop, d, is directly proportional to the square of its rate of travel, r. Under certain driving conditions, a car traveling 60 mph needs 300 feet to stop. With these same driving conditions, how long does it take a car to stop if the car is traveling 30 mph when the brakes are applied?

REVIEW AND PREVIEW

Simplify. Follow the circled steps in the order shown.

41. $$\frac{\frac{3}{4}+\frac{1}{4}}{\frac{3}{8}+\frac{13}{8}}$$ ① Add. (numerator) ← ③ Divide. ② Add. (denominator)

42. $\dfrac{\left.\dfrac{9}{5}+\dfrac{6}{5}\right\}}{\left.\dfrac{17}{6}+\dfrac{7}{6}\right\}}$ ① Add. ← ③ Divide. ② Add.

43. $\dfrac{\left.\dfrac{2}{5}+\dfrac{1}{5}\right\}}{\left.\dfrac{7}{10}+\dfrac{7}{10}\right\}}$ ① Add. ← ③ Divide. ② Add.

44. $\dfrac{\left.\dfrac{1}{4}+\dfrac{5}{4}\right\}}{\left.\dfrac{3}{8}+\dfrac{7}{8}\right\}}$ ① Add. ← ③ Divide. ② Add.

CONCEPT EXTENSIONS

45. Suppose that y varies directly as x. If x is tripled, what is the effect on y?

46. Suppose that y varies directly as x^2. If x is tripled, what is the effect on y?

47. The period, P, of a pendulum (the time of one complete back and forth swing) varies directly with the square root of its length, l. If the length of the pendulum is quadrupled, what is the effect on the period, P?

48. For a constant distance, the rate of travel r varies inversely with the time traveled, t. If a car traveling 100 mph completes a test track in 6 minutes, find the rate needed to complete the same test track in 4 minutes. (***Hint:*** Convert minutes to hours.)

7.8 Simplifying Complex Fractions

OBJECTIVES

1 Simplify Complex Fractions Using Method 1.

2 Simplify Complex Fractions Using Method 2.

A rational expression whose numerator or denominator or both numerator and denominator contain fractions is called a **complex rational expression** or a **complex fraction.** Some examples are

$$\frac{4}{2-\frac{1}{2}}, \quad \frac{\frac{3}{2}}{\frac{4}{7}-x}, \quad \frac{\frac{1}{x+2}}{x+2-\frac{1}{x}}$$

← Numerator of complex fraction
← Main fraction bar
← Denominator of complex fraction

Our goal in this section is to write complex fractions in simplest form. A complex fraction is in simplest form when it is in the form $\frac{P}{Q}$, where P and Q are polynomials that have no common factors.

OBJECTIVE

1 Simplifying Complex Fractions—Method 1

In this section, two methods of simplifying complex fractions are presented. The first method presented uses the fact that the main fraction bar indicates division.

Method 1: Simplifying a Complex Fraction

Step 1. Add or subtract fractions in the numerator or denominator so that the numerator is a single fraction and the denominator is a single fraction.

Step 2. Perform the indicated division by multiplying the numerator of the complex fraction by the reciprocal of the denominator of the complex fraction.

Step 3. Write the rational expression in simplest form.

EXAMPLE 1 Simplify the complex fraction $\dfrac{\frac{5}{8}}{\frac{2}{3}}$.

Solution Since the numerator and denominator of the complex fraction are already single fractions, we proceed to Step 2: Perform the indicated division by multiplying the numerator $\frac{5}{8}$ by the reciprocal of the denominator $\frac{2}{3}$.

$$\frac{\frac{5}{8}}{\frac{2}{3}} = \frac{5}{8} \div \frac{2}{3} = \frac{5}{8} \cdot \frac{3}{2} = \frac{15}{16}$$

The reciprocal of $\frac{2}{3}$ is $\frac{3}{2}$. □

PRACTICE

1 Simplify the complex fraction $\dfrac{\frac{3}{4}}{\frac{6}{11}}$.

EXAMPLE 2 Simplify: $\dfrac{\frac{2}{3} + \frac{1}{5}}{\frac{2}{3} - \frac{2}{9}}$

Solution Simplify above and below the main fraction bar separately. First, add $\frac{2}{3}$ and $\frac{1}{5}$ to obtain a single fraction in the numerator; then subtract $\frac{2}{9}$ from $\frac{2}{3}$ to obtain a single fraction in the denominator.

$$\frac{\frac{2}{3} + \frac{1}{5}}{\frac{2}{3} - \frac{2}{9}} = \frac{\frac{2(5)}{3(5)} + \frac{1(3)}{5(3)}}{\frac{2(3)}{3(3)} - \frac{2}{9}}$$ The LCD of the numerator's fractions is 15. The LCD of the denominator's fractions is 9.

$$= \frac{\frac{10}{15} + \frac{3}{15}}{\frac{6}{9} - \frac{2}{9}}$$ Simplify.

$$= \frac{\frac{13}{15}}{\frac{4}{9}}$$ Add the numerator's fractions. Subtract the denominator's fractions.

Next, perform the indicated division by multiplying the numerator of the complex fraction by the reciprocal of the denominator of the complex fraction.

$$\frac{\frac{13}{15}}{\frac{4}{9}} = \frac{13}{15} \cdot \frac{9}{4}$$ The reciprocal of $\frac{4}{9}$ is $\frac{9}{4}$.

$$= \frac{13 \cdot 3 \cdot 3}{3 \cdot 5 \cdot 4} = \frac{39}{20}$$ Simplify. □

PRACTICE

2 Simplify: $\dfrac{\frac{3}{4} + \frac{2}{3}}{\frac{3}{4} - \frac{1}{5}}$

EXAMPLE 3 Simplify: $\dfrac{\dfrac{1}{z} - \dfrac{1}{2}}{\dfrac{1}{3} - \dfrac{z}{6}}$

Solution Subtract to get a single fraction in the numerator and a single fraction in the denominator of the complex fraction.

$$\frac{\frac{1}{z} - \frac{1}{2}}{\frac{1}{3} - \frac{z}{6}} = \frac{\frac{2}{2z} - \frac{z}{2z}}{\frac{2}{6} - \frac{z}{6}}$$ The LCD of the numerator's fractions is $2z$. The LCD of the denominator's fractions is 6.

$$= \frac{\frac{2-z}{2z}}{\frac{2-z}{6}}$$

$$= \frac{2-z}{2z} \cdot \frac{6}{2-z}$$ Multiply by the reciprocal of $\frac{2-z}{6}$.

$$= \frac{2 \cdot 3 \cdot (2-z)}{2 \cdot z \cdot (2-z)}$$ Factor.

$$= \frac{3}{z}$$ Write in simplest form. □

PRACTICE 3 Simplify: $\dfrac{\dfrac{4}{x} - \dfrac{1}{2}}{\dfrac{1}{5} - \dfrac{x}{10}}$

OBJECTIVE 2

Simplifying Complex Fractions—Method 2 

Next we study a second method for simplifying complex fractions. In this method, we multiply the numerator and the denominator of the complex fraction by the LCD of all fractions in the complex fraction.

Method 2: Simplifying a Complex Fraction

Step 1. Find the LCD of all the fractions in the complex fraction.

Step 2. Multiply both the numerator and the denominator of the complex fraction by the LCD from Step 1.

Step 3. Perform the indicated operations and write the result in simplest form.

We use method 2 to rework Example 2.

EXAMPLE 4 Simplify: $\dfrac{\dfrac{2}{3} + \dfrac{1}{5}}{\dfrac{2}{3} - \dfrac{2}{9}}$

Solution The LCD of $\frac{2}{3}, \frac{1}{5}, \frac{2}{3}$, and $\frac{2}{9}$ is 45, so we multiply the numerator and the denominator of the complex fraction by 45. Then we perform the indicated operations and write in simplest form.

$$\frac{\frac{2}{3}+\frac{1}{5}}{\frac{2}{3}-\frac{2}{9}} = \frac{45\left(\frac{2}{3}+\frac{1}{5}\right)}{45\left(\frac{2}{3}-\frac{2}{9}\right)}$$

$$= \frac{45\left(\frac{2}{3}\right)+45\left(\frac{1}{5}\right)}{45\left(\frac{2}{3}\right)-45\left(\frac{2}{9}\right)}$$ Apply the distributive property.

$$= \frac{30+9}{30-10} = \frac{39}{20}$$ Simplify. □

PRACTICE 4 Simplify: $\dfrac{\frac{3}{4}+\frac{2}{3}}{\frac{3}{4}-\frac{1}{5}}$

> **Helpful Hint**
> The same complex fraction was simplified using different methods in Examples 2 and 4. Notice that the simplified results are the same.

EXAMPLE 5 Simplify: $\dfrac{\frac{x+1}{y}}{\frac{x}{y}+2}$

Solution The LCD of $\frac{x+1}{y}$, $\frac{x}{y}$, and $\frac{2}{1}$ is y, so we multiply the numerator and the denominator of the complex fraction by y.

$$\frac{\frac{x+1}{y}}{\frac{x}{y}+2} = \frac{y\left(\frac{x+1}{y}\right)}{y\left(\frac{x}{y}+2\right)}$$

$$= \frac{y\left(\frac{x+1}{y}\right)}{y\left(\frac{x}{y}\right)+y\cdot 2}$$ Apply the distributive property in the denominator.

$$= \frac{x+1}{x+2y}$$ Simplify. □

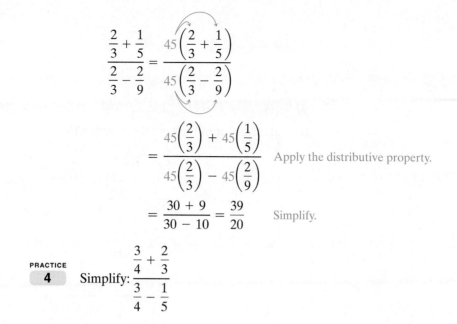

PRACTICE 5 Simplify: $\dfrac{\frac{a-b}{b}}{\frac{a}{b}+4}$

EXAMPLE 6 Simplify: $\dfrac{\dfrac{x}{y}+\dfrac{3}{2x}}{\dfrac{x}{2}+y}$

Solution The LCD of $\frac{x}{y}, \frac{3}{2x}, \frac{x}{2}$, and $\frac{y}{1}$ is $2xy$, so we multiply both the numerator and the denominator of the complex fraction by $2xy$.

$$\frac{\dfrac{x}{y}+\dfrac{3}{2x}}{\dfrac{x}{2}+y} = \frac{2xy\left(\dfrac{x}{y}+\dfrac{3}{2x}\right)}{2xy\left(\dfrac{x}{2}+y\right)}$$

$$= \frac{2xy\left(\dfrac{x}{y}\right)+2xy\left(\dfrac{3}{2x}\right)}{2xy\left(\dfrac{x}{2}\right)+2xy(y)} \quad \text{Apply the distributive property.}$$

$$= \frac{2x^2+3y}{x^2y+2xy^2}$$

$$\text{or } \frac{2x^2+3y}{xy(x+2y)}$$

□

PRACTICE 6 Simplify: $\dfrac{\dfrac{4}{3b}+\dfrac{b}{a}}{\dfrac{a}{3}-b}$

Vocabulary, Readiness & Video Check

Complete the steps by writing the simplified complex fraction.

1. $\dfrac{\dfrac{y}{2}}{\dfrac{5x}{2}} = \dfrac{2\left(\dfrac{y}{2}\right)}{2\left(\dfrac{5x}{2}\right)} = \dfrac{?}{?}$

2. $\dfrac{\dfrac{10}{x}}{\dfrac{z}{x}} = \dfrac{x\left(\dfrac{10}{x}\right)}{x\left(\dfrac{z}{x}\right)} = \dfrac{?}{?}$

3. $\dfrac{\dfrac{3}{x}}{\dfrac{5}{x^2}} = \dfrac{x^2\left(\dfrac{3}{x}\right)}{x^2\left(\dfrac{5}{x^2}\right)} = \dfrac{?}{?}$

4. $\dfrac{\dfrac{a}{10}}{\dfrac{b}{20}} = \dfrac{20\left(\dfrac{a}{10}\right)}{20\left(\dfrac{b}{20}\right)} = \dfrac{?}{?}$

Martin-Gay Interactive Videos

See Video 7.8

Watch the section lecture video and answer the following questions.

OBJECTIVE 1

5. From Example 2, before you can rewrite the complex fraction as a division problem, what must you make sure you have?

OBJECTIVE 2

6. How does finding an LCD in Method 2, as in Examples 4 and 5, differ from finding an LCD in Method 1? Mention the purpose of the LCD in each method.

7.8 Exercise Set

MyMathLab®

MIXED PRACTICE

Simplify each complex fraction. See Examples 1 through 6.

1. $\dfrac{\frac{1}{2}}{\frac{3}{4}}$

2. $\dfrac{\frac{1}{8}}{-\frac{5}{12}}$

3. $\dfrac{-\frac{4x}{9}}{-\frac{2x}{3}}$

4. $\dfrac{-\frac{6y}{11}}{\frac{4y}{9}}$

5. $\dfrac{\frac{1+x}{6}}{\frac{1+x}{3}}$

6. $\dfrac{\frac{6x-3}{5x^2}}{\frac{2x-1}{10x}}$

7. $\dfrac{\frac{1}{2}+\frac{2}{3}}{\frac{5}{9}-\frac{5}{6}}$

8. $\dfrac{\frac{3}{4}-\frac{1}{2}}{\frac{3}{8}+\frac{1}{6}}$

9. $\dfrac{2+\frac{7}{10}}{1+\frac{3}{5}}$

10. $\dfrac{4-\frac{11}{12}}{5+\frac{1}{4}}$

11. $\dfrac{\frac{1}{3}}{\frac{1}{2}-\frac{1}{4}}$

12. $\dfrac{\frac{7}{10}-\frac{3}{5}}{\frac{1}{2}}$

13. $\dfrac{-\frac{2}{9}}{-\frac{14}{3}}$

14. $\dfrac{\frac{3}{8}}{\frac{4}{15}}$

15. $\dfrac{-\frac{5}{12x^2}}{\frac{25}{16x^3}}$

16. $\dfrac{-\frac{7}{8y}}{\frac{21}{4y}}$

17. $\dfrac{\frac{m}{n}-1}{\frac{m}{n}+1}$

18. $\dfrac{\frac{x}{2}+2}{\frac{x}{2}-2}$

19. $\dfrac{\frac{1}{5}-\frac{1}{x}}{\frac{7}{10}+\frac{1}{x^2}}$

20. $\dfrac{\frac{1}{y^2}+\frac{2}{3}}{\frac{1}{y}-\frac{5}{6}}$

21. $\dfrac{1+\frac{1}{y-2}}{y+\frac{1}{y-2}}$

22. $\dfrac{x-\frac{1}{2x+1}}{1-\frac{x}{2x+1}}$

23. $\dfrac{\frac{4y-8}{16}}{\frac{6y-12}{4}}$

24. $\dfrac{\frac{7y+21}{3}}{\frac{3y+9}{8}}$

25. $\dfrac{\frac{x}{y}+1}{\frac{x}{y}-1}$

26. $\dfrac{\frac{3}{5y}+8}{\frac{3}{5y}-8}$

27. $\dfrac{1}{2+\frac{1}{3}}$

28. $\dfrac{3}{1-\frac{4}{3}}$

29. $\dfrac{\frac{ax+ab}{x^2-b^2}}{\frac{x+b}{x-b}}$

30. $\dfrac{\frac{m+2}{m-2}}{\frac{2m+4}{m^2-4}}$

31. $\dfrac{\frac{-3+y}{4}}{\frac{8+y}{28}}$

32. $\dfrac{\frac{-x+2}{18}}{\frac{8}{9}}$

33. $\dfrac{3+\frac{12}{x}}{1-\frac{16}{x^2}}$

34. $\dfrac{2+\frac{6}{x}}{1-\frac{9}{x^2}}$

35. $\dfrac{\frac{8}{x+4}+2}{\frac{12}{x+4}-2}$

36. $\dfrac{\frac{25}{x+5}+5}{\frac{3}{x+5}-5}$

37. $\dfrac{\frac{s}{r}+\frac{r}{s}}{\frac{s}{r}-\frac{r}{s}}$

38. $\dfrac{\frac{2}{x}+\frac{x}{2}}{\frac{2}{x}-\frac{x}{2}}$

39. $\dfrac{\frac{6}{x-5}+\frac{x}{x-2}}{\frac{3}{x-6}-\frac{2}{x-5}}$

40. $\dfrac{\frac{4}{x}+\frac{x}{x+1}}{\frac{1}{2x}+\frac{1}{x+6}}$

REVIEW AND PREVIEW

*Use the bar graph below to answer Exercises 41 through 44. See Section 1.2. **Note:** Some of these players are still competing; thus, their total prize money may increase.*

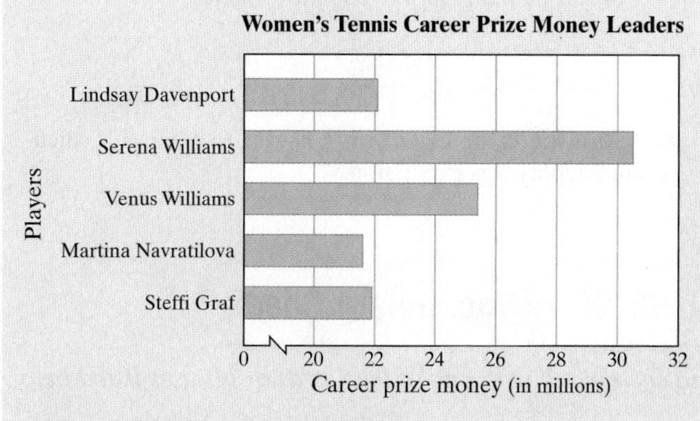

41. Which women's tennis player has earned the most prize money in her career?

42. Estimate how much more prize money Venus Williams has earned in her career than Steffi Graf.

43. What is the approximate spread in lifetime prize money between Lindsay Davenport and Serena Williams?

44. During her career, Martina Navratilova won 363 doubles and singles tournaments. Assuming her prize money was earned only for tournament titles, how much prize money did she earn, on average, per tournament title?

CONCEPT EXTENSIONS

45. Explain how to simplify a complex fraction using method 1.

46. Explain how to simplify a complex fraction using method 2.

To find the average of two numbers, we find their sum and divide by 2. For example, the average of 65 and 81 is found by simplifying $\frac{65+81}{2}$. *This simplifies to* $\frac{146}{2} = 73$. *Use this for Exercises 47 and 48.*

47. Find the average of $\frac{1}{3}$ and $\frac{3}{4}$.

48. Write the average of $\frac{3}{n}$ and $\frac{5}{n^2}$ as a simplified rational expression.

49. In electronics, when two resistors R_1 (read R sub 1) and R_2 (read R sub 2) are connected in parallel, the total resistance is given by the complex fraction

$$\frac{1}{\frac{1}{R_1}+\frac{1}{R_2}}.$$

Simplify this expression.

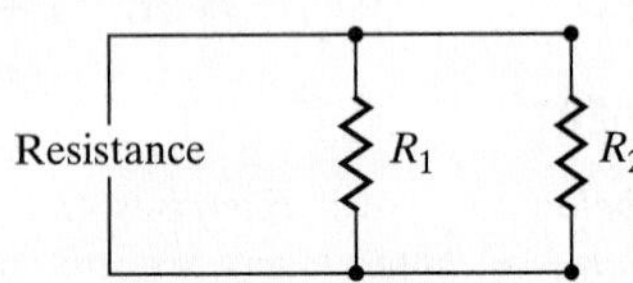

50. Astronomers occasionally need to know the day of the week a particular date fell on. The complex fraction

$$\frac{J+\frac{3}{2}}{7}$$

where J is the *Julian day number,* is used to make this calculation. Simplify this expression.

Simplify each expression in Exercises 51 through 54. First, write each expression with positive exponents. Then simplify the complex fraction. The first step has been completed for Exercise 51.

51. $\frac{x^{-1}+2^{-1}}{x^{-2}-4^{-1}} = \frac{\frac{1}{x}+\frac{1}{2}}{\frac{1}{x^2}-\frac{1}{4}}$

52. $\frac{3^{-1}-x^{-1}}{9^{-1}-x^{-2}}$

53. $\frac{y^{-2}}{1-y^{-2}}$

54. $\frac{4+x^{-1}}{3+x^{-1}}$

55. If the distance formula $d = r \cdot t$ is solved for t, then $t = \frac{d}{r}$. Use this formula to find t if distance d is $\frac{20x}{3}$ miles and rate r is $\frac{5x}{9}$ miles per hour. Write t in simplified form.

△ **56.** If the formula for area of a rectangle, $A = l \cdot w$, is solved for w, then $w = \frac{A}{l}$. Use this formula to find w if area A is $\frac{4x-2}{3}$ square meters and length l is $\frac{6x-3}{5}$ meters. Write w in simplified form.

Chapter 7 Vocabulary Check

Fill in each blank with one of the words or phrases listed below. Not all choices will be used.

least common denominator	simplifying	reciprocals	numerator	$\frac{-a}{b}$	$\frac{a}{-b}$
cross products	ratio	proportion	inverse variation		
rational expression	direct variation	complex fraction	denominator	$\frac{-a}{-b}$	

1. A(n) ________________ is an expression that can be written in the form $\frac{P}{Q}$, where P and Q are polynomials and Q is not 0.
2. In a(n) ________________, the numerator or denominator or both may contain fractions.
3. For a rational expression, $-\frac{a}{b} =$ ________________ $=$ ________________.
4. A rational expression is undefined when the ________________ is 0.
5. The process of writing a rational expression in lowest terms is called ________________.
6. The expressions $\frac{2x}{7}$ and $\frac{7}{2x}$ are called ________________.
7. The ________________ of a list of rational expressions is a polynomial of least degree whose factors include all factors of the denominators in the list.
8. A(n) ________________ is the quotient of two numbers.
9. $\frac{x}{2} = \frac{7}{16}$ is an example of a(n) ________________.

10. If $\frac{a}{b} = \frac{c}{d}$, then ad and bc are called ______________.

11. The equation $y = \frac{k}{x}$ is an example of ______________.

12. The equation $y = kx$ is an example of ______________.

Chapter 7 Highlights

DEFINITIONS AND CONCEPTS	EXAMPLES
Section 7.1 Simplifying Rational Expressions	
A **rational expression** is an expression that can be written in the form $\frac{P}{Q}$, where P and Q are polynomials and Q does not equal 0.	$\frac{7y^3}{4}, \frac{x^2 + 6x + 1}{x - 3}, \frac{-5}{s^3 + 8}$
To find values for which a rational expression is undefined, find values for which the denominator is 0.	Find any values for which the expression $\frac{5y}{y^2 - 4y + 3}$ is undefined. $y^2 - 4y + 3 = 0$ Set the denominator equal to 0. $(y - 3)(y - 1) = 0$ Factor. $y - 3 = 0$ or $y - 1 = 0$ Set each factor equal to 0. $y = 3$ $\quad y = 1$ Solve. The expression is undefined when y is 3 and when y is 1.
To Simplify a Rational Expression **Step 1.** Factor the numerator and denominator. **Step 2.** Divide out factors common to the numerator and denominator. (This is the same as removing a factor of 1.)	Simplify: $\frac{4x + 20}{x^2 - 25}$ $\frac{4x + 20}{x^2 - 25} = \frac{4(x + 5)}{(x + 5)(x - 5)} = \frac{4}{x - 5}$
Section 7.2 Multiplying and Dividing Rational Expressions	
To multiply rational expressions, **Step 1.** Factor numerators and denominators. **Step 2.** Multiply numerators and multiply denominators. **Step 3.** Write the product in simplest form. $\frac{P}{Q} \cdot \frac{R}{S} = \frac{PR}{QS}$	Multiply: $\frac{4x + 4}{2x - 3} \cdot \frac{2x^2 + x - 6}{x^2 - 1}$ $\frac{4x + 4}{2x - 3} \cdot \frac{2x^2 + x - 6}{x^2 - 1} = \frac{4(x + 1)}{2x - 3} \cdot \frac{(2x - 3)(x + 2)}{(x + 1)(x - 1)}$ $= \frac{4(x + 1)(2x - 3)(x + 2)}{(2x - 3)(x + 1)(x - 1)}$ $= \frac{4(x + 2)}{x - 1}$
To divide by a rational expression, multiply by the reciprocal. $\frac{P}{Q} \div \frac{R}{S} = \frac{P}{Q} \cdot \frac{S}{R} = \frac{PS}{QR}$	Divide: $\frac{15x + 5}{3x^2 - 14x - 5} \div \frac{15}{3x - 12}$ $\frac{15x + 5}{3x^2 - 14x - 5} \div \frac{15}{3x - 12} = \frac{5(3x + 1)}{(3x + 1)(x - 5)} \cdot \frac{3(x - 4)}{3 \cdot 5}$ $= \frac{x - 4}{x - 5}$

DEFINITIONS AND CONCEPTS	EXAMPLES

Section 7.3 Adding and Subtracting Rational Expressions with Common Denominators and Least Common Denominator

To add or subtract rational expressions with the same denominator, add or subtract numerators and place the sum or difference over a common denominator.

$$\frac{P}{R} + \frac{Q}{R} = \frac{P + Q}{R}$$

$$\frac{P}{R} - \frac{Q}{R} = \frac{P - Q}{R}$$

Perform indicated operations.

$$\frac{5}{x+1} + \frac{x}{x+1} = \frac{5+x}{x+1}$$

$$\frac{2y+7}{y^2-9} - \frac{y+4}{y^2-9} = \frac{(2y+7)-(y+4)}{y^2-9}$$

$$= \frac{2y+7-y-4}{y^2-9}$$

$$= \frac{y+3}{(y+3)(y-3)}$$

$$= \frac{1}{y-3}$$

To find the least common denominator (LCD),

Step 1. Factor the denominators.

Step 2. The LCD is the product of all unique factors, each raised to a power equal to the greatest number of times that it appears in any one factored denominator.

Find the LCD for

$$\frac{7x}{x^2+10x+25} \text{ and } \frac{11}{3x^2+15x}$$

$$x^2 + 10x + 25 = (x+5)(x+5)$$

$$3x^2 + 15x = 3x(x+5)$$

LCD is $3x(x+5)(x+5)$ or $3x(x+5)^2$

Section 7.4 Adding and Subtracting Rational Expressions with Unlike Denominators

To add or subtract rational expressions with unlike denominators,

Step 1. Find the LCD.

Step 2. Rewrite each rational expression as an equivalent expression whose denominator is the LCD.

Step 3. Add or subtract numerators and place the sum or difference over the common denominator.

Step 4. Write the result in simplest form.

Perform the indicated operation.

$$\frac{9x+3}{x^2-9} - \frac{5}{x-3}$$

$$= \frac{9x+3}{(x+3)(x-3)} - \frac{5}{x-3}$$

LCD is $(x+3)(x-3)$.

$$= \frac{9x+3}{(x+3)(x-3)} - \frac{5(x+3)}{(x-3)(x+3)}$$

$$= \frac{9x+3-5(x+3)}{(x+3)(x-3)}$$

$$= \frac{9x+3-5x-15}{(x+3)(x-3)}$$

$$= \frac{4x-12}{(x+3)(x-3)}$$

$$= \frac{4(x-3)}{(x+3)(x-3)} = \frac{4}{x+3}$$

Section 7.5 Solving Equations Containing Rational Expressions

To solve an equation containing rational expressions,

Step 1. Multiply both sides of the equation by the LCD of all rational expressions in the equation.

Solve: $\frac{5x}{x+2} + 3 = \frac{4x-6}{x+2}$

DEFINITIONS AND CONCEPTS	EXAMPLES
Section 7.5 Solving Equations Containing Rational Expressions (continued)	
Step 2. Remove any grouping symbols and solve the resulting equation. **Step 3.** Check the solution in the original equation.	$(x+2)\left(\frac{5x}{x+2}+3\right)=(x+2)\left(\frac{4x-6}{x+2}\right)$ $(x+2)\left(\frac{5x}{x+2}\right)+(x+2)(3)=(x+2)\left(\frac{4x-6}{x+2}\right)$ $5x+3x+6=4x-6$ $4x=-12$ $x=-3$ The solution checks and the solution is -3.
Section 7.6 Proportion and Problem Solving with Rational Equations	
A **ratio** is the quotient of two numbers or two quantities. A **proportion** is a mathematical statement that two ratios are equal.	***Proportions*** $\frac{2}{3}=\frac{8}{12}$ $\quad \frac{x}{7}=\frac{15}{35}$
Cross products: If $\frac{a}{b}=\frac{c}{d}$, then $ad=bc$.	***Cross Products*** $2 \cdot 12$ or 24 $\qquad \frac{2}{3}=\frac{8}{12}$ $\qquad 3 \cdot 8$ or 24 Solve: $\frac{3}{4}=\frac{x}{x-1}$ $\frac{3}{4}=\frac{x}{x-1}$ $3(x-1)=4x$ Set cross products equal. $3x-3=4x$ $-3=x$
Problem-Solving Steps **1.** UNDERSTAND. Read and reread the problem.	A small plane and a car leave Kansas City, Missouri, and head for Minneapolis, Minnesota, a distance of 450 miles. The speed of the plane is 3 times the speed of the car, and the plane arrives 6 hours ahead of the car. Find the speed of the car. Let $x =$ the speed of the car. Then $3x =$ the speed of the plane.

	Distance =	***Rate*** ·	***Time***
Car	450	x	$\frac{450}{x}\left(\frac{\text{distance}}{\text{rate}}\right)$
Plane	450	$3x$	$\frac{450}{3x}\left(\frac{\text{distance}}{\text{rate}}\right)$

2. TRANSLATE.

In words: plane's time + 6 hours = car's time

3. SOLVE.

Translate: $\frac{450}{3x} + 6 = \frac{450}{x}$

(continued)

DEFINITIONS AND CONCEPTS	EXAMPLES
Section 7.6 Proportion and Problem Solving with Rational Equations (continued)	
	$\frac{450}{3x} + 6 = \frac{450}{x}$ $3x\left(\frac{450}{3x}\right) + 3x(6) = 3x\left(\frac{450}{x}\right)$ $450 + 18x = 1350$ $18x = 900$ $x = 50$
4. INTERPRET.	**Check** this solution in the originally stated problem. **State** the conclusion: The speed of the car is 50 miles per hour.
Section 7.7 Variation and Problem Solving	
y **varies directly as** x, or y **is directly proportional to** x, if there is a nonzero constant k such that $y = kx$	The circumference of a circle C varies directly as its radius r. $C = \underbrace{2\pi}_{k} r$
y **varies inversely as** x, or y is **inversely proportional to** x, if there is a nonzero constant k such that $y = \frac{k}{x}$	Pressure P varies inversely with volume V. $P = \frac{k}{V}$
Section 7.8 Simplifying Complex Fractions	
Method 1: To Simplify a Complex Fraction **Step 1.** Add or subtract fractions in the numerator and the denominator of the complex fraction. **Step 2.** Perform the indicated division. **Step 3.** Write the result in lowest terms.	Simplify: $\frac{\frac{1}{x} + 2}{\frac{1}{x} - \frac{1}{y}} = \frac{\frac{1}{x} + \frac{2x}{x}}{\frac{y}{xy} - \frac{x}{xy}}$ $= \frac{\frac{1 + 2x}{x}}{\frac{y - x}{xy}}$ $= \frac{1 + 2x}{x} \cdot \frac{xy}{y - x}$ $= \frac{y(1 + 2x)}{y - x}$
Method 2: To Simplify a Complex Fraction **Step 1.** Find the LCD of all fractions in the complex fraction. **Step 2.** Multiply the numerator and the denominator of the complex fraction by the LCD. **Step 3.** Perform the indicated operations and write the result in lowest terms.	$\frac{\frac{1}{x} + 2}{\frac{1}{x} - \frac{1}{y}} = \frac{xy\left(\frac{1}{x} + 2\right)}{xy\left(\frac{1}{x} - \frac{1}{y}\right)}$ $= \frac{xy\left(\frac{1}{x}\right) + xy(2)}{xy\left(\frac{1}{x}\right) - xy\left(\frac{1}{y}\right)}$ $= \frac{y + 2xy}{y - x}$ or $\frac{y(1 + 2x)}{y - x}$

Chapter 7 Review

(7.1) *Find any real number for which each rational expression is undefined.*

1. $\dfrac{x+5}{x^2-4}$

2. $\dfrac{5x+9}{4x^2-4x-15}$

Find the value of each rational expression when $x = 5$, $y = 7$, and $z = -2$.

3. $\dfrac{2-z}{z+5}$

4. $\dfrac{x^2+xy-y^2}{x+y}$

Simplify each rational expression.

5. $\dfrac{2x+6}{x^2+3x}$

6. $\dfrac{3x-12}{x^2-4x}$

7. $\dfrac{x+2}{x^2-3x-10}$

8. $\dfrac{x+4}{x^2+5x+4}$

9. $\dfrac{x^3-4x}{x^2+3x+2}$

10. $\dfrac{5x^2-125}{x^2+2x-15}$

11. $\dfrac{x^2-x-6}{x^2-3x-10}$

12. $\dfrac{x^2-2x}{x^2+2x-8}$

Simplify each expression. This section contains four-term polynomials and sums and differences of two cubes.

13. $\dfrac{x^2+xa+xb+ab}{x^2-xc+bx-bc}$

14. $\dfrac{x^2+5x-2x-10}{x^2-3x-2x+6}$

15. $\dfrac{4-x}{x^3-64}$

16. $\dfrac{x^2-4}{x^3+8}$

(7.2) *Perform each indicated operation and simplify.*

17. $\dfrac{15x^3y^2}{z}\cdot\dfrac{z}{5xy^3}$

18. $\dfrac{-y^3}{8}\cdot\dfrac{9x^2}{y^3}$

19. $\dfrac{x^2-9}{x^2-4}\cdot\dfrac{x-2}{x+3}$

20. $\dfrac{2x+5}{x-6}\cdot\dfrac{2x}{-x+6}$

21. $\dfrac{x^2-5x-24}{x^2-x-12}\div\dfrac{x^2-10x+16}{x^2+x-6}$

22. $\dfrac{4x+4y}{xy^2}\div\dfrac{3x+3y}{x^2y}$

23. $\dfrac{x^2+x-42}{x-3}\cdot\dfrac{(x-3)^2}{x+7}$

24. $\dfrac{2a+2b}{3}\cdot\dfrac{a-b}{a^2-b^2}$

25. $\dfrac{2x^2-9x+9}{8x-12}\div\dfrac{x^2-3x}{2x}$

26. $\dfrac{x^2-y^2}{x^2+xy}\div\dfrac{3x^2-2xy-y^2}{3x^2+6x}$

27. $\dfrac{x-y}{4}\div\dfrac{y^2-2y-xy+2x}{16x+24}$

28. $\dfrac{5+x}{7}\div\dfrac{xy+5y-3x-15}{7y-35}$

(7.3) *Perform each indicated operation and simplify.*

29. $\dfrac{x}{x^2+9x+14}+\dfrac{7}{x^2+9x+14}$

30. $\dfrac{x}{x^2+2x-15}+\dfrac{5}{x^2+2x-15}$

31. $\dfrac{4x-5}{3x^2}-\dfrac{2x+5}{3x^2}$

32. $\dfrac{9x+7}{6x^2}-\dfrac{3x+4}{6x^2}$

Find the LCD of each pair of rational expressions.

33. $\dfrac{x+4}{2x}, \dfrac{3}{7x}$

34. $\dfrac{x-2}{x^2-5x-24}, \dfrac{3}{x^2+11x+24}$

Rewrite each rational expression as an equivalent expression whose denominator is the given polynomial.

35. $\dfrac{5}{7x}=\dfrac{\quad}{14x^3y}$

36. $\dfrac{9}{4y}=\dfrac{\quad}{16y^3x}$

37. $\dfrac{x+2}{x^2+11x+18}=\dfrac{\quad}{(x+2)(x-5)(x+9)}$

38. $\dfrac{3x-5}{x^2+4x+4}=\dfrac{\quad}{(x+2)^2(x+3)}$

(7.4) *Perform each indicated operation and simplify.*

39. $\dfrac{4}{5x^2}-\dfrac{6}{y}$

40. $\dfrac{2}{x-3}-\dfrac{4}{x-1}$

41. $\dfrac{4}{x+3}-2$

42. $\dfrac{3}{x^2+2x-8}+\dfrac{2}{x^2-3x+2}$

43. $\dfrac{2x-5}{6x+9}-\dfrac{4}{2x^2+3x}$

44. $\dfrac{x-1}{x^2-2x+1}-\dfrac{x+1}{x-1}$

Find the perimeter and the area of each figure.

△ **45.**

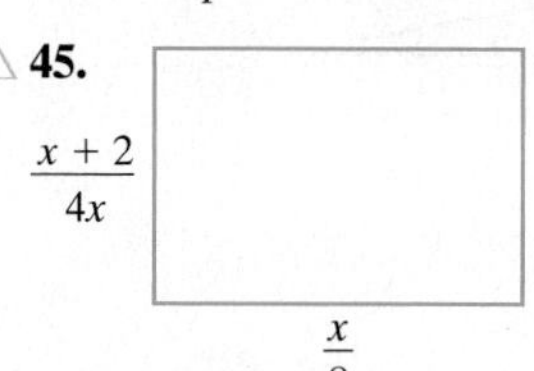

△ **46.**

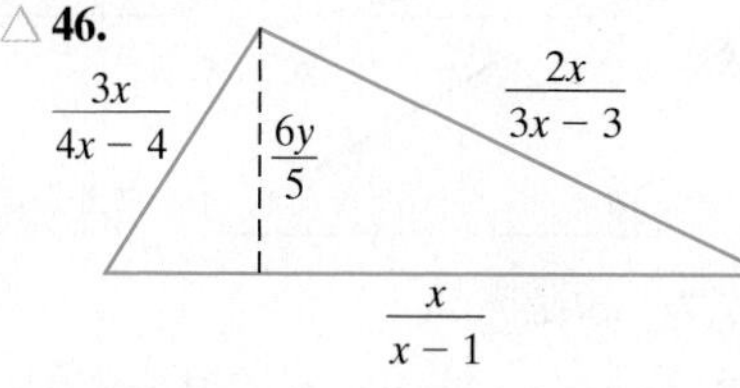

(7.5) *Solve each equation.*

47. $\dfrac{n}{10}=9-\dfrac{n}{5}$

48. $\dfrac{2}{x+1}-\dfrac{1}{x-2}=-\dfrac{1}{2}$

49. $\dfrac{y}{2y+2}+\dfrac{2y-16}{4y+4}=\dfrac{y-3}{y+1}$

50. $\frac{2}{x-3} - \frac{4}{x+3} = \frac{8}{x^2-9}$

51. $\frac{x-3}{x+1} - \frac{x-6}{x+5} = 0$

52. $x + 5 = \frac{6}{x}$

Solve the equation for the indicated variable.

53. $\frac{4A}{5b} = x^2$, for b

54. $\frac{x}{7} + \frac{y}{8} = 10$, for y

(7.6) Solve each proportion.

55. $\frac{x}{2} = \frac{12}{4}$

56. $\frac{20}{1} = \frac{x}{25}$

57. $\frac{2}{x-1} = \frac{3}{x+3}$

58. $\frac{4}{y-3} = \frac{2}{y-3}$

Solve.

59. A machine can process 300 parts in 20 minutes. Find how many parts can be processed in 45 minutes.

60. As his consulting fee, Mr. Visconti charges \$90.00 per day. Find how much he charges for 3 hours of consulting. Assume an 8-hour work day.

61. Five times the reciprocal of a number equals the sum of $\frac{3}{2}$ the reciprocal of the number and $\frac{7}{6}$. What is the number?

62. The reciprocal of a number equals the reciprocal of the difference of 4 and the number. Find the number.

63. A car travels 90 miles in the same time that a car traveling 10 miles per hour slower travels 60 miles. Find the speed of each car.

64. The current in a bayou near Lafayette, Louisiana, is 4 miles per hour. A paddle boat travels 48 miles upstream in the same amount of time it takes to travel 72 miles downstream. Find the speed of the boat in still water.

65. When Mark and Maria manicure Mr. Stergeon's lawn, it takes them 5 hours. If Mark works alone, it takes 7 hours. Find how long it takes Maria alone.

66. It takes pipe A 20 days to fill a fish pond. Pipe B takes 15 days. Find how long it takes both pipes together to fill the pond.

Given that the pairs of triangles are similar, find each missing length x.

△ **67.**

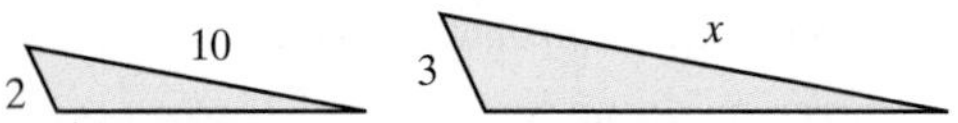

△ **68.**

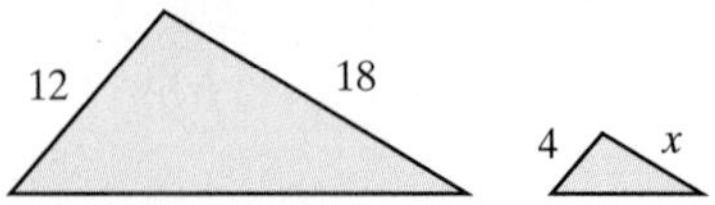

(7.7) Solve.

69. y varies directly as x. If $y = 40$ when $x = 4$, find y when x is 11.

70. y varies inversely as x. If $y = 4$ when $x = 6$, find y when x is 48.

71. y varies inversely as x^3. If $y = 12.5$ when $x = 2$, find y when x is 3.

72. y varies directly as x^2. If $y = 175$ when $x = 5$, find y when $x = 10$.

73. The cost of manufacturing a certain medicine varies inversely as the amount of medicine manufactured increases. If 3000 milliliters can be manufactured for \$6600, find the cost to manufacture 5000 milliliters.

74. The distance a spring stretches varies directly with the weight attached to the spring. If a 150-pound weight stretches the spring 8 inches, find the distance that a 90-pound weight stretches the spring.

(7.8) Simplify each complex fraction.

75. $\frac{\frac{5x}{27}}{-\frac{10xy}{21}}$

76. $\frac{\frac{3}{5}+\frac{2}{7}}{\frac{1}{5}+\frac{5}{6}}$

77. $\frac{3-\frac{1}{y}}{2-\frac{1}{y}}$

78. $\frac{\frac{6}{x+2}+4}{\frac{8}{x+2}-4}$

MIXED REVIEW

Simplify each rational expression.

79. $\frac{4x+12}{8x^2+24x}$

80. $\frac{x^3-6x^2+9x}{x^2+4x-21}$

Perform the indicated operations and simplify.

81. $\frac{x^2+9x+20}{x^2-25} \cdot \frac{x^2-9x+20}{x^2+8x+16}$

82. $\frac{x^2-x-72}{x^2-x-30} \div \frac{x^2+6x-27}{x^2-9x+18}$

83. $\frac{x}{x^2-36} + \frac{6}{x^2-36}$

84. $\frac{5x-1}{4x} - \frac{3x-2}{4x}$

85. $\frac{4}{3x^2+8x-3} + \frac{2}{3x^2-7x+2}$

86. $\frac{3x}{x^2+9x+14} - \frac{6x}{x^2+4x-21}$

Solve.

87. $\frac{4}{a-1} + 2 = \frac{3}{a-1}$

88. $\frac{x}{x+3} + 4 = \frac{x}{x+3}$

Solve.

89. The quotient of twice a number and three, minus one-sixth is the quotient of the number and two. Find the number.

90. Mr. Crocker can paint his house by himself in three days. His son will need an additional day to complete the job if he works alone. If they work together, find how long it takes to paint the house.

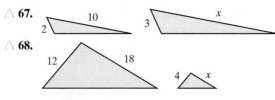

Given that the following pairs of triangles are similar, find each missing length.

91.

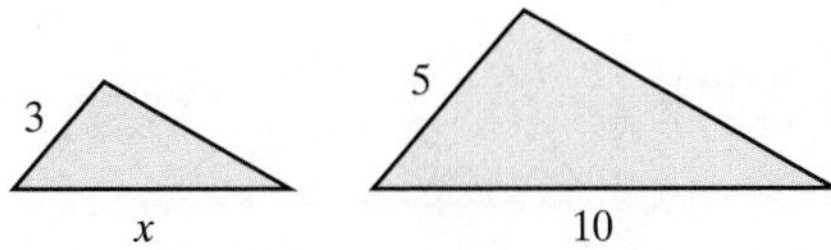

92.

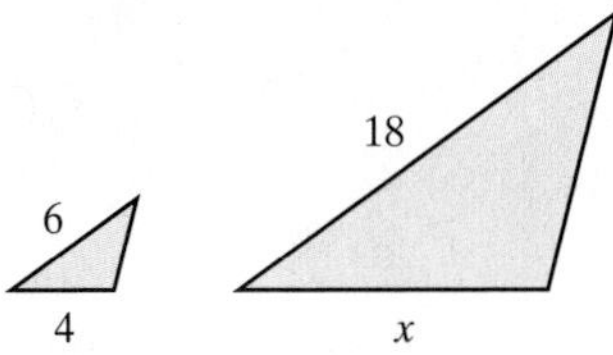

Simplify each complex fraction.

93. $\dfrac{\frac{1}{4}}{\frac{1}{3}+\frac{1}{2}}$

94. $\dfrac{4+\frac{2}{x}}{6+\frac{3}{x}}$

Convert as indicated.

95. 1.8 square yards = ______ square feet

96. 135 cubic feet = ______ cubic yards

Chapter 7 Test

YouTube

1. Find any real numbers for which the following expression is undefined.

$$\frac{x+5}{x^2+4x+3}$$

2. For a certain computer desk, the average cost C (in dollars) per desk manufactured is

$$C = \frac{100x + 3000}{x}$$

where x is the number of desks manufactured.

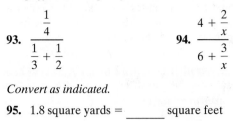

a. Find the average cost per desk when manufacturing 200 computer desks.

b. Find the average cost per desk when manufacturing 1000 computer desks.

Simplify each rational expression.

3. $\dfrac{3x-6}{5x-10}$

4. $\dfrac{x+6}{x^2+12x+36}$

5. $\dfrac{x+3}{x^3+27}$

6. $\dfrac{2m^3-2m^2-12m}{m^2-5m+6}$

7. $\dfrac{ay+3a+2y+6}{ay+3a+5y+15}$

8. $\dfrac{y-x}{x^2-y^2}$

Perform the indicated operation and simplify if possible.

9. $\dfrac{3}{x-1}\cdot(5x-5)$

10. $\dfrac{y^2-5y+6}{2y+4}\cdot\dfrac{y+2}{2y-6}$

11. $\dfrac{15x}{2x+5}-\dfrac{6-4x}{2x+5}$

12. $\dfrac{5a}{a^2-a-6}-\dfrac{2}{a-3}$

13. $\dfrac{6}{x^2-1}+\dfrac{3}{x+1}$

14. $\dfrac{x^2-9}{x^2-3x}\div\dfrac{xy+5x+3y+15}{2x+10}$

15. $\dfrac{x+2}{x^2+11x+18}+\dfrac{5}{x^2-3x-10}$

Solve each equation.

16. $\dfrac{4}{y}-\dfrac{5}{3}=\dfrac{-1}{5}$

17. $\dfrac{5}{y+1}=\dfrac{4}{y+2}$

18. $\dfrac{a}{a-3}=\dfrac{3}{a-3}-\dfrac{3}{2}$

19. $x-\dfrac{14}{x-1}=4-\dfrac{2x}{x-1}$

20. $\dfrac{10}{x^2-25}=\dfrac{3}{x+5}+\dfrac{1}{x-5}$

Simplify each complex fraction.

21. $\dfrac{\frac{5x^2}{yz^2}}{\frac{10x}{z^3}}$

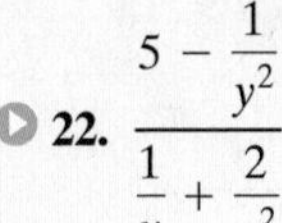

22. $\dfrac{5-\frac{1}{y^2}}{\frac{1}{y}+\frac{2}{y^2}}$

23. $\dfrac{\frac{b}{a}-\frac{a}{b}}{\frac{1}{b}+\frac{1}{a}}$

24. y varies directly as x. If $y = 10$ when $x = 15$, find y when x is 42.

25. y varies inversely as x^2. If $y = 8$ when $x = 5$, find y when x is 15.

26. In a sample of 85 fluorescent bulbs, 3 were found to be defective. At this rate, how many defective bulbs should be found in 510 bulbs?

27. One number plus five times its reciprocal is equal to six. Find the number.

28. A pleasure boat traveling down the Red River takes the same time to go 14 miles upstream as it takes to go 16 miles downstream. If the current of the river is 2 miles per hour, find the speed of the boat in still water.

29. An inlet pipe can fill a tank in 12 hours. A second pipe can fill the tank in 15 hours. If both pipes are used, find how long it takes to fill the tank.

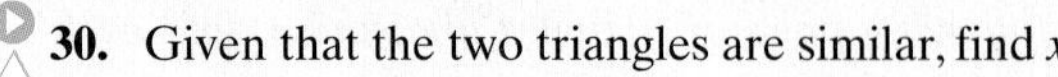

30. Given that the two triangles are similar, find x.

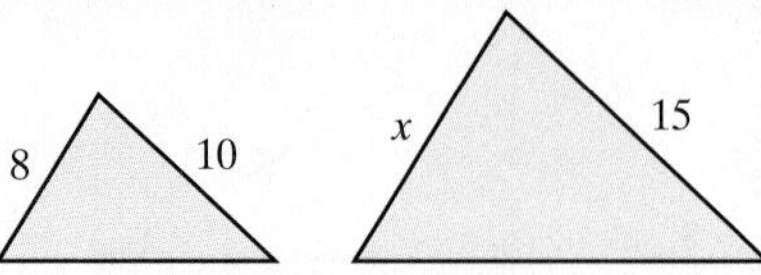

Chapter 7 Cumulative Review

TRANSLATING

1. Write each sentence as an equation or inequality. Let x represent the unknown number.
a. The quotient of 15 and a number is 4.
b. Three subtracted from 12 is a number.
c. Four times a number, added to 17, is not equal to 21.
d. Triple a number is less than 48.

2. Write each sentence as an equation. Let x represent the unknown number.
a. The difference of 12 and a number is -45.
b. The product of 12 and a number is -45.
c. A number less 10 is twice the number.

3. Rajiv Puri invested part of his $20,000 inheritance in a mutual funds account that pays 7% simple interest yearly and the rest in a certificate of deposit that pays 9% simple interest yearly. At the end of one year, Rajiv's investments earned $1550. Find the amount he invested at each rate.

4. The number of nonbusiness bankruptcies has increased over the years. In 2002, the number of nonbusiness bankruptcies was 80,000 less than twice the number in 1994. If the total of nonbusiness bankruptcies for these two years is 2,290,000, find the number of nonbusiness bankruptcies for each year. (*Source:* American Bankruptcy Institute)

5. Graph $x - 3y = 6$ by finding and plotting intercepts.

6. Find the slope of the line whose equation is $7x + 2y = 9$.

7. Use the product rule to simplify each expression.
a. $4^2 \cdot 4^5$ **b.** $x^4 \cdot x^6$
c. $y^3 \cdot y$ **d.** $y^3 \cdot y^2 \cdot y^7$
e. $(-5)^7 \cdot (-5)^8$ **f.** $a^2 \cdot b^2$

8. Simplify.
a. $\dfrac{x^9}{x^7}$ **b.** $\dfrac{x^{19}y^5}{xy}$
c. $(x^5y^2)^3$ **d.** $(-3a^2b)(5a^3b)$

9. Subtract $(5z - 7)$ from the sum of $(8z + 11)$ and $(9z - 2)$.

10. Subtract $(9x^2 - 6x + 2)$ from $(x + 1)$.

11. Multiply: $(3a + b)^3$

12. Multiply: $(2x + 1)(5x^2 - x + 2)$

13. Use a special product to square each binomial.
a. $(t + 2)^2$
b. $(p - q)^2$
c. $(2x + 5)^2$
d. $(x^2 - 7y)^2$

14. Multiply.
a. $(x + 9)^2$
b. $(2x + 1)(2x - 1)$
c. $8x(x^2 + 1)(x^2 - 1)$

15. Simplify each expression. Write results using positive exponents only.
a. $\dfrac{1}{x^{-3}}$ **b.** $\dfrac{1}{3^{-4}}$
c. $\dfrac{p^{-4}}{q^{-9}}$ **d.** $\dfrac{5^{-3}}{2^{-5}}$

16. Simplify. Write results with positive exponents.
a. 5^{-3} **b.** $\dfrac{9}{x^{-7}}$ **c.** $\dfrac{11^{-1}}{7^{-2}}$

17. Divide: $\dfrac{4x^2 + 7 + 8x^3}{2x + 3}$

18. Divide $(4x^3 - 9x + 2)$ by $(x - 4)$.

19. Find the GCF of each list of numbers.
a. 28 and 40
b. 55 and 21
c. 15, 18, and 66

20. Find the GCF of $9x^2$, $6x^3$, and $21x^5$.

Factor.

21. $-9a^5 + 18a^2 - 3a$

22. $7x^6 - 7x^5 + 7x^4$

23. $3m^2 - 24m - 60$

24. $-2a^2 + 10a + 12$

25. $3x^2 + 11x + 6$

26. $10m^2 - 7m + 1$

27. $x^2 + 12x + 36$

28. $4x^2 + 12x + 9$

29. $x^2 + 4$

30. $x^2 - 4$

31. $x^3 + 8$

32. $27y^3 - 1$

33. $2x^3 + 3x^2 - 2x - 3$

34. $3x^3 + 5x^2 - 12x - 20$

35. $12m^2 - 3n^2$

36. $x^5 - x$

37. Solve: $x(2x - 7) = 4$

38. Solve: $3x^2 + 5x = 2$

39. Find the x-intercepts of the graph of $y = x^2 - 5x + 4$.

40. Find the x-intercepts of the graph of $y = x^2 - x - 6$.

41. The height of a triangular sail is 2 meters less than twice the length of the base. If the sail has an area of 30 square meters, find the length of its base and the height.

42. The height of a parallelogram is 5 feet more than three times its base. If the area of the parallelogram is 182 square feet, find the length of its base and height.

43. Simplify: $\dfrac{5x - 5}{x^3 - x^2}$

44. Simplify: $\dfrac{2x^2 - 50}{4x^4 - 20x^3}$

45. Divide: $\dfrac{6x + 2}{x^2 - 1} \div \dfrac{3x^2 + x}{x - 1}$

46. Multiply: $\dfrac{6x^2 - 18x}{3x^2 - 2x} \cdot \dfrac{15x - 10}{x^2 - 9}$

47. Simplify: $\dfrac{\dfrac{x + 1}{y}}{\dfrac{x}{y} + 2}$

48. Simplify: $\dfrac{\dfrac{m}{3} + \dfrac{n}{6}}{\dfrac{m + n}{12}}$

CHAPTER 8

Roots and Radicals

Having spent the last chapter studying equations, we return now to algebraic expressions. We expand on our skills of operating on expressions—adding, subtracting, multiplying, dividing, and raising to powers—to include finding roots. Finding roots is defined by raising to powers. As we master finding roots, we will work with equations that contain roots and solve problems that can be modeled by such equations.

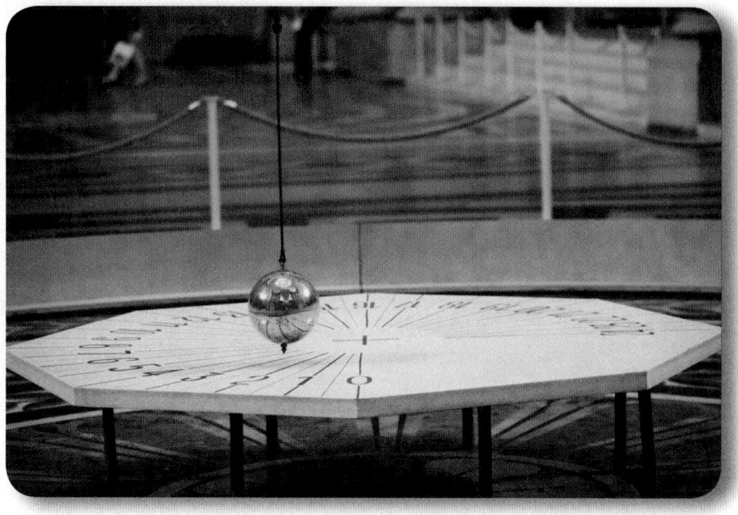

Did you know that pendulums can be used to demonstrate that the earth rotates on its axis?

In 1851, French physicist Léon Foucault developed a special pendulum in an experiment to demonstrate that the earth rotates on its axis. He connected his tall pendulum, capable of running for many hours, to the roof of the Paris Observatory. The pendulum's bob was able to swing back and forth in one plane but not to twist in other directions. So, when the pendulum bob appeared to move in a circle over time, he demonstrated that it was not the pendulum but the building that moved. And since the building was firmly attached to the earth, it must be the earth rotating that created the apparent circular motion of the bob. In Section 8.1, Exercise 93, roots are used to explore the time it takes Foucault's pendulum to complete one swing of its bob.

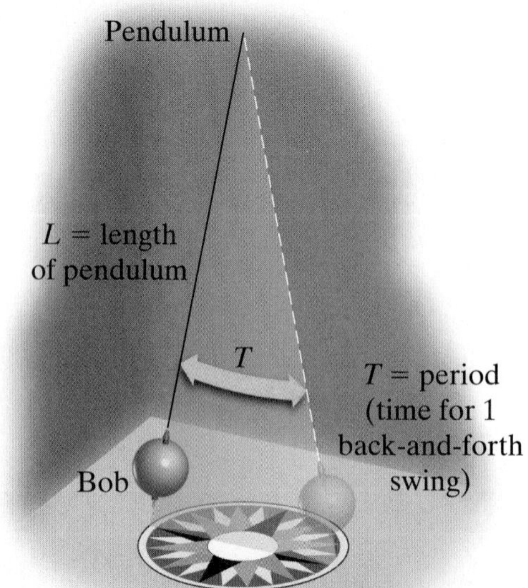

8.1 Introduction to Radicals

OBJECTIVES

1 Find Square Roots.

2 Find Cube Roots.

3 Find *n*th Roots.

4 Approximate Square Roots.

5 Simplify Radicals Containing Variables.

OBJECTIVE

1 Finding Square Roots

In this section, we define finding the **root** of a number by its reverse operation, raising a number to a power. We begin with squares and square roots.

The *square* of 5 is $5^2 = 25$.

The *square* of -5 is $(-5)^2 = 25$.

The *square* of $\frac{1}{2}$ is $\left(\frac{1}{2}\right)^2 = \frac{1}{4}$.

The reverse operation of squaring a number is finding the **square root** of a number. For example,

A *square root* of 25 is 5 because $5^2 = 25$.

A *square root* of 25 is also -5 because $(-5)^2 = 25$.

A *square root* of $\frac{1}{4}$ is $\frac{1}{2}$ because $\left(\frac{1}{2}\right)^2 = \frac{1}{4}$.

In general, a number *b* is a square root of a number *a* if $b^2 = a$.

Notice that both 5 and -5 are square roots of 25. The symbol $\sqrt{}$ denotes the **positive** or **principal square root** of a number. For example,

$$\sqrt{25} = 5 \text{ since } 5^2 = 25 \text{ and 5 is positive.}$$

The symbol $-\sqrt{}$ denotes the **negative square root.** For example,

$$-\sqrt{25} = -5$$

The symbol $\sqrt{}$ is called a **radical** or **radical sign.** The expression within or under a radical sign is called the **radicand.** An expression containing a radical is called a **radical expression.**

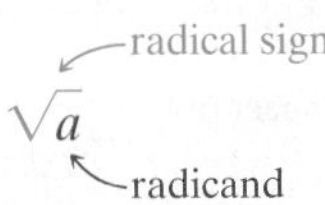

Square Root

If a is a positive number, then

$\sqrt{a}$ is the **positive square root** of a and

$-\sqrt{a}$ is the **negative square root** of a.

$$\sqrt{a} = b \quad \text{only if} \quad b^2 = a \text{ and } b > 0$$

Also, $\sqrt{0} = 0$.

EXAMPLE 1 Find each square root.

a. $\sqrt{36}$ **b.** $-\sqrt{16}$ **c.** $\sqrt{\frac{9}{100}}$ **d.** $\sqrt{0}$ **e.** $\sqrt{0.64}$

Solution

a. $\sqrt{36} = 6$ because $6^2 = 36$ and 6 is positive.

b. $-\sqrt{16} = -4$. The negative sign in front of the radical indicates the negative square root of 16.

c. $\sqrt{\frac{9}{100}} = \frac{3}{10}$ because $\left(\frac{3}{10}\right)^2 = \frac{9}{100}$ and $\frac{3}{10}$ is positive.

d. $\sqrt{0} = 0$ because $0^2 = 0$.

e. $\sqrt{0.64} = 0.8$ because $(0.8)^2 = 0.64$ and 0.8 is positive. □

PRACTICE

1 Find each square root.

a. $\sqrt{\frac{4}{81}}$ **b.** $-\sqrt{25}$ **c.** $\sqrt{144}$ **d.** $\sqrt{0.49}$ **e.** $-\sqrt{1}$

Is the square root of a negative number a real number? For example, is $\sqrt{-4}$ a real number? To answer this question, we ask ourselves, is there a real number whose square is -4? Since there is no real number whose square is -4, we say that $\sqrt{-4}$ is not a real number. In general,

A square root of a negative number is not a real number.

Study the following table to make sure you understand the differences discussed earlier.

Number	*Square Roots of Number*	$\sqrt{number}$	$-\sqrt{number}$
25	$-5, 5$	$\sqrt{25} = 5$ only	$-\sqrt{25} = -5$
$\frac{1}{4}$	$-\frac{1}{2}, \frac{1}{2}$	$\sqrt{\frac{1}{4}} = \frac{1}{2}$ only	$-\sqrt{\frac{1}{4}} = -\frac{1}{2}$
-4	No real square roots.	$\sqrt{-4}$ is not a real number.	

We will discuss numbers such as $\sqrt{-4}$ further in Chapter 9.

OBJECTIVE

2 Finding Cube Roots 

We can find roots other than square roots. For example, since $2^3 = 8$, we call 2 the **cube root** of 8. In symbols, we write

$\sqrt[3]{8} = 2$ The number 3 is called the **index.**

Also,

$\sqrt[3]{27} = 3$ Since $3^3 = 27$

$\sqrt[3]{-64} = -4$ Since $(-4)^3 = -64$

Notice that unlike the square root of a negative number, the cube root of a negative number is a real number. This is so because although we cannot find a real number whose **square** is negative, we **can** find a real number whose **cube** is negative. In fact, the cube of a negative number is a negative number. Therefore, the cube root of a negative number is a negative number.

EXAMPLE 2 Find each cube root.

a. $\sqrt[3]{1}$ **b.** $\sqrt[3]{-27}$ **c.** $\sqrt[3]{\frac{1}{125}}$

Solution

a. $\sqrt[3]{1} = 1$ because $1^3 = 1$.

b. $\sqrt[3]{-27} = -3$ because $(-3)^3 = -27$.

c. $\sqrt[3]{\frac{1}{125}} = \frac{1}{5}$ because $\left(\frac{1}{5}\right)^3 = \frac{1}{125}$. □

PRACTICE

2 Find each cube root.

a. $\sqrt[3]{0}$ **b.** $\sqrt[3]{-64}$ **c.** $\sqrt[3]{\frac{1}{8}}$

OBJECTIVE

3 Finding *n*th Roots 

Just as we can raise a real number to powers other than 2 or 3, we can find roots other than square roots and cube roots. In fact, we can take the *n*th root of a number where *n* is any natural number. An ***n*th root** of a number *a* is a number whose *n*th power is *a*. The natural number *n* is called the **index.**

In symbols, the *n*th root of *a* is written as $\sqrt[n]{a}$. The index 2 is usually omitted for square roots.

> **Helpful Hint**
> If the index is even, such as $\sqrt{\ }$, $\sqrt[4]{\ }$, $\sqrt[6]{\ }$, and so on, the radicand must be nonnegative for the root to be a real number. For example,
>
> $\sqrt[4]{16} = 2$ but $\sqrt[4]{-16}$ is not a real number
>
> $\sqrt[6]{64} = 2$ but $\sqrt[6]{-64}$ is not a real number

> ✓**CONCEPT CHECK**
> Which of the following is a real number?
>
> a. $\sqrt{-64}$ b. $\sqrt[4]{-64}$ c. $\sqrt[5]{-64}$ d. $\sqrt[6]{-64}$

EXAMPLE 3 Find each root.

a. $\sqrt[4]{16}$ **b.** $\sqrt[5]{-32}$ **c.** $-\sqrt[3]{8}$ **d.** $\sqrt[4]{-81}$

Solution

a. $\sqrt[4]{16} = 2$ because $2^4 = 16$ and 2 is positive.

b. $\sqrt[5]{-32} = -2$ because $(-2)^5 = -32$.

c. $-\sqrt[3]{8} = -2$ since $\sqrt[3]{8} = 2$.

d. $\sqrt[4]{-81}$ is not a real number since the index 4 is even and the radicand -81 is negative. □

PRACTICE

3 Find each root.

a. $\sqrt[4]{81}$ **b.** $\sqrt[5]{100{,}000}$ **c.** $\sqrt[6]{-64}$ **d.** $\sqrt[3]{-125}$

OBJECTIVE

4 Approximating Square Roots 

Recall that numbers such as 1, 4, 9, 25, and $\frac{4}{25}$ are called **perfect squares,** since $1^2 = 1, 2^2 = 4, 3^2 = 9, 5^2 = 25$, and $\left(\frac{2}{5}\right)^2 = \frac{4}{25}$. Square roots of perfect square

Answer to Concept Check: c

radicands simplify to rational numbers. What happens when we try to simplify a root such as $\sqrt{3}$? Since 3 is not a perfect square, $\sqrt{3}$ is not a rational number. It cannot be written as a quotient of integers. It is called an **irrational number** and we can find a decimal **approximation** of it. To find decimal approximations, use a calculator or an appendix. (For calculator help, see the box at the end of this section.)

EXAMPLE 4 Use a calculator or an appendix to approximate $\sqrt{3}$ to three decimal places.

Solution We may use an appendix or a calculator to approximate $\sqrt{3}$. To use a calculator, find the square root key $\boxed{\sqrt{\ }}$.

$$\sqrt{3} \approx 1.732050808$$

To three decimal places, $\sqrt{3} \approx 1.732$. □

PRACTICE

4 Use a calculator or an appendix to approximate $\sqrt{17}$ to three decimal places.

OBJECTIVE

5 Simplifying Radicals Containing Variables

Radicals can also contain variables. To simplify radicals containing variables, special care must be taken. To see how we simplify $\sqrt{x^2}$, let's look at a few examples in this form.

If $x = 3$, we have $\sqrt{3^2} = \sqrt{9} = 3$, or x.
If x is 5, we have $\sqrt{5^2} = \sqrt{25} = 5$, or x.

From these two examples, you may think that $\sqrt{x^2}$ simplifies to x. Let's now look at an example where x is a negative number. If $x = -3$, we have $\sqrt{(-3)^2} = \sqrt{9} = 3$, not -3, our original x. To make sure that $\sqrt{x^2}$ simplifies to a nonnegative number, we have the following.

> For any real number a,
>
> $$\sqrt{a^2} = |a|.$$

Thus,

$$\sqrt{x^2} = |x|,$$
$$\sqrt{(-8)^2} = |-8| = 8$$
$$\sqrt{(7y)^2} = |7y|, \quad \text{and so on.}$$

To avoid this, for the rest of the chapter we assume that **if a variable appears in the radicand of a radical expression, it represents positive numbers only.** Then

$\sqrt{x^2} = |x| = x$ since x is a positive number.

$\sqrt{y^2} = y$ Because $(y)^2 = y^2$

$\sqrt{x^8} = x^4$ Because $(x^4)^2 = x^8$

$\sqrt{9x^2} = 3x$ Because $(3x)^2 = 9x^2$

$\sqrt[3]{8z^{12}} = 2z^4$ Because $(2z^4)^3 = 8z^{12}$

EXAMPLE 5 Simplify each expression. Assume that all variables represent positive numbers.

a. $\sqrt{z^2}$ **b.** $\sqrt{x^6}$ **c.** $\sqrt[3]{27y^6}$ **d.** $\sqrt{16x^{16}}$ **e.** $\sqrt{\dfrac{x^4}{25}}$ **f.** $\sqrt[3]{-125a^{12}b^{15}}$

Solution

a. $\sqrt{z^2} = z$ because $(z)^2 = z^2$.

b. $\sqrt{x^6} = x^3$ because $(x^3)^2 = x^6$.

c. $\sqrt[3]{27y^6} = 3y^2$ because $(3y^2)^3 = 27y^6$.

d. $\sqrt{16x^{16}} = 4x^8$ because $(4x^8)^2 = 16x^{16}$.

e. $\sqrt{\frac{x^4}{25}} = \frac{x^2}{5}$ because $\left(\frac{x^2}{5}\right)^2 = \frac{x^4}{25}$.

f. $\sqrt[3]{-125a^{12}b^{15}} = -5a^4b^5$ because $(-5a^4b^5)^3 = -125a^{12}b^{15}$. □

PRACTICE 5 Simplify each expression. Assume that all variables represent positive numbers.

a. $\sqrt{x^{10}}$ **b.** $\sqrt{y^{14}}$ **c.** $\sqrt[3]{125z^9}$ **d.** $\sqrt{49x^2}$ **e.** $\sqrt{\frac{z^4}{36}}$ **f.** $\sqrt[3]{-8a^6b^{12}}$

Calculator Explorations

To simplify or approximate square roots using a calculator, locate the key marked [$\sqrt{\ }$]. To simplify $\sqrt{25}$ using a scientific calculator, press [25] [$\sqrt{\ }$]. The display should read [5]. To simplify $\sqrt{25}$ using a graphing calculator, press [$\sqrt{\ }$] [25] [ENTER].

To approximate $\sqrt{30}$, press [30] [$\sqrt{\ }$] (or [$\sqrt{\ }$] [30] [ENTER]). The display should read [5.4772256]. This is an approximation for $\sqrt{30}$. A three-decimal-place approximation is

$$\sqrt{30} \approx 5.477$$

Is this answer reasonable? Since 30 is between perfect squares 25 and 36, $\sqrt{30}$ is between $\sqrt{25} = 5$ and $\sqrt{36} = 6$. The calculator result is then reasonable since 5.4772256 is between 5 and 6.

Use a calculator to approximate each expression to three decimal places. Decide whether each result is reasonable.

1. $\sqrt{7}$ **2.** $\sqrt{14}$ **3.** $\sqrt{11}$

4. $\sqrt{200}$ **5.** $\sqrt{82}$ **6.** $\sqrt{46}$

Many scientific calculators have a key, such as [$\sqrt[x]{y}$], that can be used to approximate roots other than square roots. To approximate these roots using a graphing calculator, look under the [MATH] menu or consult your manual.

Use a calculator to approximate each expression to three decimal places. Decide whether each result is reasonable.

7. $\sqrt[3]{40}$ **8.** $\sqrt[3]{71}$ **9.** $\sqrt[4]{20}$

10. $\sqrt[4]{15}$ **11.** $\sqrt[5]{18}$ **12.** $\sqrt[6]{2}$

Vocabulary, Readiness & Video Check

Use the choices below to fill in each blank.

principal radical sign index radicand

1. In the expression $\sqrt[4]{16}$, the number 4 is called the ________, the number 16 is called the ________, and $\sqrt{\ }$ is called the ________.
2. The symbol $\sqrt{\ }$ is used to denote the positive, or ________, square root.

Answer each exercise true or false.

3. $\sqrt{-16}$ simplifies to a real number.
4. $\sqrt{64} = 8$, while $\sqrt[3]{64} = 4$.
5. The number 9 has two square roots.
6. $\sqrt{0} = 0$ and $\sqrt{1} = 1$.
7. If x is a positive number, $\sqrt{x^{10}} = x^5$.
8. If x is a positive number, $\sqrt{x^{16}} = x^4$.

Martin-Gay Interactive Videos

See Video 8.1

Watch the section lecture video and answer the following questions.

OBJECTIVE 1

9. Explain the differences between Examples 1 and 2, including how you know which one simplifies to a positive number and which one to a negative number.

OBJECTIVE 2

10. From Example 11, what is an important difference between the square root and the cube root of a negative number?

OBJECTIVE 3

11. From Examples 12–15, given a negative radicand, what kind of index must you have to have a real number?

OBJECTIVE 4

12. From Example 16, how do you determine if an approximate answer is reasonable?

OBJECTIVE 5

13. As explained in Example 19, when simplifying radicals containing variables, what is a shortcut you can use when dealing with exponents?

8.1 Exercise Set

MyMathLab®

Find each square root. See Example 1.

1. $\sqrt{16}$
2. $\sqrt{64}$
3. $\sqrt{\frac{1}{25}}$
4. $\sqrt{\frac{1}{64}}$
5. $-\sqrt{100}$
6. $-\sqrt{36}$
7. $\sqrt{-4}$
8. $\sqrt{-25}$
9. $-\sqrt{121}$
10. $-\sqrt{49}$
11. $\sqrt{\frac{9}{25}}$
12. $\sqrt{\frac{4}{81}}$
13. $\sqrt{900}$
14. $\sqrt{400}$
15. $\sqrt{144}$
16. $\sqrt{169}$
17. $\sqrt{\frac{1}{100}}$
18. $\sqrt{\frac{1}{121}}$
19. $\sqrt{0.25}$
20. $\sqrt{0.49}$

Find each cube root. See Example 2.

21. $\sqrt[3]{125}$
22. $\sqrt[3]{64}$
23. $\sqrt[3]{-64}$
24. $\sqrt[3]{-27}$
25. $-\sqrt[3]{8}$
26. $-\sqrt[3]{27}$
27. $\sqrt[3]{\frac{1}{8}}$
28. $\sqrt[3]{\frac{1}{64}}$
29. $\sqrt[3]{-125}$
30. $\sqrt[3]{-1}$

MIXED PRACTICE

Find each root. See Examples 1 through 3.

31. $\sqrt[5]{32}$
32. $\sqrt[4]{81}$
33. $\sqrt{81}$
34. $\sqrt{49}$
35. $\sqrt[4]{-16}$
36. $\sqrt{-9}$
37. $\sqrt[3]{-\frac{27}{64}}$
38. $\sqrt[3]{-\frac{8}{27}}$
39. $-\sqrt[4]{625}$
40. $-\sqrt[5]{32}$
41. $\sqrt[6]{1}$
42. $\sqrt[5]{1}$

Approximate each square root to three decimal places. See Example 4.

43. $\sqrt{7}$
44. $\sqrt{10}$
45. $\sqrt{37}$
46. $\sqrt{27}$
47. $\sqrt{136}$
48. $\sqrt{8}$

49. A standard baseball diamond is a square with 90-foot sides connecting the bases. The distance from home plate to second base is $90 \cdot \sqrt{2}$ feet. Approximate $\sqrt{2}$ to two decimal places and use your result to approximate the distance $90 \cdot \sqrt{2}$ feet.

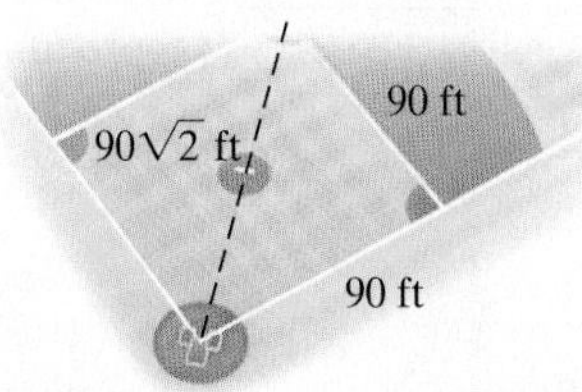

50. The roof of the warehouse shown needs to be shingled. The total area of the roof is exactly $240 \cdot \sqrt{41}$ square feet. Approximate $\sqrt{41}$ to two decimal places and use your result to approximate the area $240 \cdot \sqrt{41}$ square feet. Approximate this area to the nearest whole number.

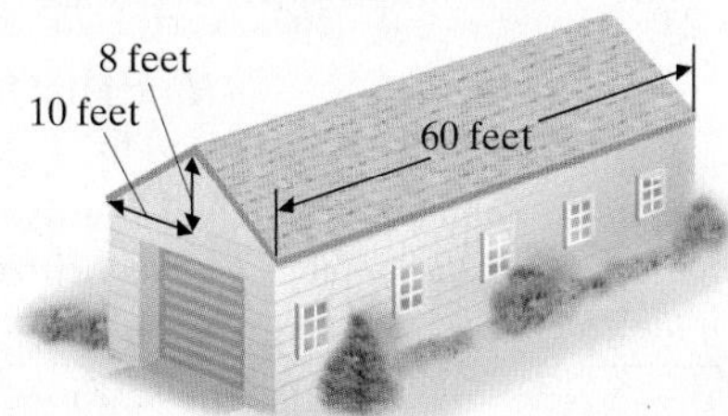

Find each root. Assume that all variables represent positive numbers. See Example 5.

51. $\sqrt{m^2}$
52. $\sqrt{y^{10}}$
53. $\sqrt{x^4}$
54. $\sqrt{y^{10}}$
55. $\sqrt{9x^8}$
56. $\sqrt{36x^{12}}$
57. $\sqrt{81x^2}$
58. $\sqrt{100z^4}$
59. $\sqrt{a^2b^4}$
60. $\sqrt{x^{12}y^{20}}$
61. $\sqrt{16a^6b^4}$
62. $\sqrt{4m^{14}n^2}$
63. $\sqrt[3]{a^6b^{18}}$
64. $\sqrt[3]{x^{12}y^{18}}$
65. $\sqrt[3]{-8x^3y^{27}}$
66. $\sqrt[3]{-27a^6b^{30}}$
67. $\sqrt{\frac{x^6}{36}}$
68. $\sqrt{\frac{y^8}{49}}$
69. $\sqrt{\frac{25y^2}{9}}$
70. $\sqrt{\frac{4x^2}{81}}$

REVIEW AND PREVIEW

Write each integer as a product of two integers such that one of the factors is a perfect square. For example, we can write $18 = 9 \cdot 2$, where 9 is a perfect square.

71. 50
72. 8
73. 32
74. 75
75. 28
76. 44
77. 27
78. 90

CONCEPT EXTENSIONS

Solve. See the Concept Check in this section.

79. Which of the following is a real number?
a. $\sqrt[7]{-1}$
b. $\sqrt[3]{-125}$
c. $\sqrt[6]{-128}$
d. $\sqrt[8]{-1}$

80. Which of the following is a real number?
a. $\sqrt{-1}$
b. $\sqrt[3]{-1}$
c. $\sqrt[4]{-1}$
d. $\sqrt[5]{-1}$

The length of a side of a square is given by the expression $\sqrt{A}$, where A is the square's area. Use this expression for Exercises 81 through 84. Be sure to attach the appropriate units.

△ 81. The area of a square is 49 square miles. Find the length of a side of the square.

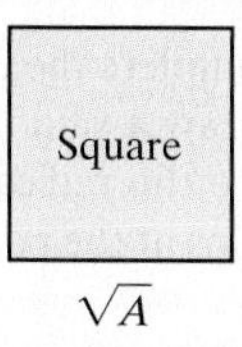

△ 82. The area of a square is $\frac{1}{81}$ square meter. Find the length of a side of the square.

△ 83. Sony currently makes the smallest portable mini disc player. It is approximately in the shape of a square with top area of 9.61 square inches. Find the length of a side. (*Source:* SONY)

△ 84. A parking lot is in the shape of a square with area 2500 square yards. Find the length of a side.

85. Simplify $\sqrt{\sqrt{81}}$.
86. Simplify $\sqrt[3]{\sqrt[3]{1}}$.
87. Simplify $\sqrt{\sqrt{10{,}000}}$.
88. Simplify $\sqrt{\sqrt{1{,}600{,}000{,}000}}$.

For each square root below, give two whole numbers that the square root lies between. For example, since $\sqrt{11}$ is between $\sqrt{9}$ and $\sqrt{16}$, $\sqrt{11}$ is between 3 and 4.

89. $\sqrt{18}$
90. $\sqrt{28}$
91. $\sqrt{80}$
92. $\sqrt{98}$

93. The formula for calculating the period (one back-and-forth swing) of a pendulum is $T = 2\pi\sqrt{\frac{L}{g}}$, where T is time of the period of the swing, L is the length of the pendulum, and g is the acceleration of gravity. Suppose there is a Foucault's pendulum with length = 30 ft and $g = 32 \text{ ft/sec}^2$. Using $\pi \approx 3.14$, find the period of this pendulum. (Round to the nearest tenth of a second.)

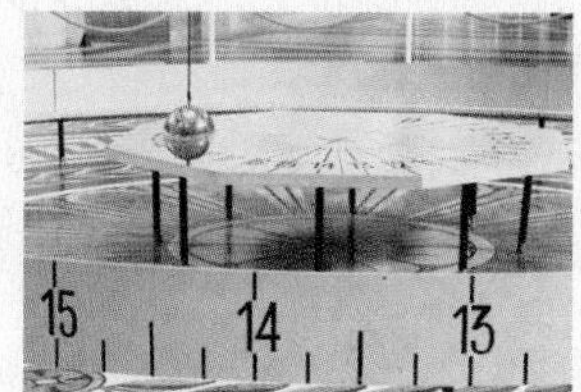

94. If the amount of gold discovered by humankind could be assembled in one place, it would be a cube with a volume of 195,112 cubic feet. Each side of the cube would be $\sqrt[3]{195{,}112}$ feet long. How long would one side of the cube be? (*Source: Reader's Digest*)

95. Explain why the square root of a negative number is not a real number.

96. Explain why the cube root of a negative number is a real number.

97. Graph $y = \sqrt{x}$. (Complete the table below, plot the ordered pair solutions, and draw a smooth curve through the points. Remember that since the radicand cannot be negative, this particular graph begins at the point with coordinates $(0, 0)$.)

x	y	
0	0	
1		
3		(approximate)
4		
9		

98. Graph $y = \sqrt[3]{x}$. (Complete the table below, plot the ordered pair solutions, and draw a smooth curve through the points.)

x	y	
−8		
−2		(approximate)
−1		
0		
1		
2		(approximate)
8		

Recall from this section that $\sqrt{a^2} = |a|$ for any real number a. Simplify the following given that x represents any real number.

99. $\sqrt{x^2}$

100. $\sqrt{4x^2}$

101. $\sqrt{(x + 2)^2}$

102. $\sqrt{x^2 + 6x + 9}$ (*Hint:* First factor $x^2 + 6x + 9$.)

Use a graphing calculator and graph each function. Observe the graph from left to right and give the ordered pair that corresponds to the "beginning" of the graph. Then tell why the graph starts at that point.

103. $y = \sqrt{x - 2}$

104. $y = \sqrt{x + 3}$

105. $y = \sqrt{x + 4}$

106. $y = \sqrt{x - 5}$

8.2 Simplifying Radicals

OBJECTIVES

1 Use the Product Rule to Simplify Square Roots.

2 Use the Quotient Rule to Simplify Square Roots.

3 Simplify Radicals Containing Variables.

4 Simplify Higher Roots.

OBJECTIVE

1 Simplifying Radicals Using the Product Rule

A square root is simplified when the radicand contains no perfect square factors (other than 1). For example, $\sqrt{20}$ is not simplified because $\sqrt{20} = \sqrt{4 \cdot 5}$ and 4 is a perfect square.

To begin simplifying square roots, we notice the following pattern.

$$\sqrt{9 \cdot 16} = \sqrt{144} = 12$$
$$\sqrt{9} \cdot \sqrt{16} = 3 \cdot 4 = 12$$

Since both expressions simplify to 12, we can write

$$\sqrt{9 \cdot 16} = \sqrt{9} \cdot \sqrt{16}$$

This suggests the following product rule for square roots.

Product Rule for Square Roots

If $\sqrt{a}$ and $\sqrt{b}$ are real numbers, then

$$\sqrt{a \cdot b} = \sqrt{a} \cdot \sqrt{b}$$

In other words, the square root of a product is equal to the product of the square roots.

To simplify $\sqrt{45}$, for example, we factor 45 so that one of its factors is a perfect square factor.

$$\begin{aligned} \sqrt{45} &= \sqrt{9 \cdot 5} && \text{Factor 45.} \\ &= \sqrt{9} \cdot \sqrt{5} && \text{Use the product rule.} \\ &= 3\sqrt{5} && \text{Write } \sqrt{9} \text{ as 3.} \end{aligned}$$

The notation $3\sqrt{5}$ means $3 \cdot \sqrt{5}$. Since the radicand 5 has no perfect square factor other than 1, the expression $3\sqrt{5}$ is in simplest form.

Helpful Hint

A radical expression in simplest form *does not mean* a decimal approximation. The simplest form of a radical expression is an exact form and may still contain a radical.

$$\sqrt{45} = \underbrace{3\sqrt{5}}_{\text{exact}} \qquad \sqrt{45} \approx \underbrace{6.71}_{\text{decimal approximation}}$$

EXAMPLE 1 Simplify.

a. $\sqrt{54}$ **b.** $\sqrt{12}$ **c.** $\sqrt{200}$ **d.** $\sqrt{35}$

Solution

a. Try to factor 54 so that at least one of the factors is a perfect square. Since 9 is a perfect square and $54 = 9 \cdot 6$,

$$\begin{aligned} \sqrt{54} &= \sqrt{9 \cdot 6} && \text{Factor 54 so that one factor is a perfect square.} \\ &= \sqrt{9} \cdot \sqrt{6} && \text{Apply the product rule.} \\ &= 3\sqrt{6} && \text{Write } \sqrt{9} \text{ as 3.} \end{aligned}$$

b.

$$\begin{aligned} \sqrt{12} &= \sqrt{4 \cdot 3} && \text{Factor 12 so that one factor is a perfect square.} \\ &= \sqrt{4} \cdot \sqrt{3} && \text{Apply the product rule.} \\ &= 2\sqrt{3} && \text{Write } \sqrt{4} \text{ as 2.} \end{aligned}$$

c.

$$\begin{aligned} \sqrt{200} &= \sqrt{100 \cdot 2} && \text{Factor 200 so that one factor is a perfect square.} \\ &= \sqrt{100} \cdot \sqrt{2} && \text{Apply the product rule.} \\ &= 10\sqrt{2} && \text{Write } \sqrt{100} \text{ as 10.} \end{aligned}$$

d. The radicand 35 contains no perfect square factors other than 1. Thus $\sqrt{35}$ is in simplest form. □

PRACTICE

1 Simplify.

a. $\sqrt{24}$ **b.** $\sqrt{60}$ **c.** $\sqrt{42}$ **d.** $\sqrt{300}$

In Example 1, part **(c)**, 100 is the largest perfect square factor of 200. What happens if we don't use the largest perfect square factor? Although using the largest perfect square factor saves time, the result is the same no matter what perfect square factor is used. For example, it is also true that $200 = 4 \cdot 50$. Then

$$\begin{aligned} \sqrt{200} &= \sqrt{4} \cdot \sqrt{50} \\ &= 2 \cdot \sqrt{50} \end{aligned}$$

Since $\sqrt{50}$ is not in simplest form, we continue.

$$\begin{aligned} \sqrt{200} &= 2 \cdot \sqrt{50} \\ &= 2 \cdot \sqrt{25 \cdot 2} \\ &= 2 \cdot \sqrt{25} \cdot \sqrt{2} \\ &= 2 \cdot 5 \cdot \sqrt{2} \\ &= 10\sqrt{2} \end{aligned}$$

EXAMPLE 2 Simplify $3\sqrt{8}$.

Solution Remember that $3\sqrt{8}$ means $3 \cdot \sqrt{8}$.

$$3 \cdot \sqrt{8} = 3 \cdot \sqrt{4 \cdot 2} \quad \text{Factor 8 so that one factor is a perfect square.}$$
$$= 3 \cdot \sqrt{4} \cdot \sqrt{2} \quad \text{Use the product rule.}$$
$$= 3 \cdot 2 \cdot \sqrt{2} \quad \text{Write } \sqrt{4} \text{ as 2.}$$
$$= 6 \cdot \sqrt{2} \text{ or } 6\sqrt{2} \quad \text{Write } 3 \cdot 2 \text{ as 6.}$$

□

PRACTICE 2 Simplify $5\sqrt{40}$.

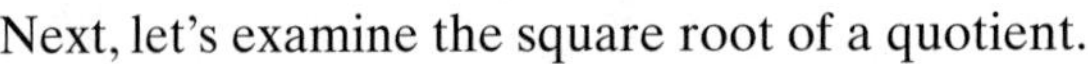

OBJECTIVE 2 Simplifying Radicals Using the Quotient Rule

Next, let's examine the square root of a quotient.

$$\sqrt{\frac{16}{4}} = \sqrt{4} = 2$$

Also,

$$\frac{\sqrt{16}}{\sqrt{4}} = \frac{4}{2} = 2$$

Since both expressions equal 2, we can write

$$\sqrt{\frac{16}{4}} = \frac{\sqrt{16}}{\sqrt{4}}$$

This suggests the following quotient rule.

Quotient Rule for Square Roots

If $\sqrt{a}$ and $\sqrt{b}$ are real numbers and $b \neq 0$, then

$$\sqrt{\frac{a}{b}} = \frac{\sqrt{a}}{\sqrt{b}}$$

In other words, the square root of a quotient is equal to the quotient of the square roots.

EXAMPLE 3 Simplify.

a. $\sqrt{\frac{25}{36}}$ **b.** $\sqrt{\frac{3}{64}}$ **c.** $\sqrt{\frac{40}{81}}$

Use the quotient rule.

Solution

a. $\sqrt{\frac{25}{36}} = \frac{\sqrt{25}}{\sqrt{36}} = \frac{5}{6}$

b. $\sqrt{\frac{3}{64}} = \frac{\sqrt{3}}{\sqrt{64}} = \frac{\sqrt{3}}{8}$

c. $\sqrt{\frac{40}{81}} = \frac{\sqrt{40}}{\sqrt{81}}$ Use the quotient rule.

$= \frac{\sqrt{4} \cdot \sqrt{10}}{9}$ Apply the product rule and write $\sqrt{81}$ as 9.

$= \frac{2\sqrt{10}}{9}$ Write $\sqrt{4}$ as 2.

□

PRACTICE 3 Simplify.

a. $\sqrt{\frac{5}{49}}$ **b.** $\sqrt{\frac{9}{100}}$ **c.** $\sqrt{\frac{18}{25}}$

OBJECTIVE 3 **Simplifying Radicals Containing Variables** 

Recall that $\sqrt{x^6} = x^3$ because $(x^3)^2 = x^6$. If an odd exponent occurs, we write the exponential expression so that one factor is the greatest even power contained in the expression. Then we use the product rule to simplify.

EXAMPLE 4 Simplify. Assume that all variables represent positive numbers.

a. $\sqrt{x^5}$ **b.** $\sqrt{8y^2}$ **c.** $\sqrt{\frac{45}{x^6}}$ **d.** $\sqrt{\frac{5p^3}{9}}$

Solution

a. $\sqrt{x^5} = \sqrt{x^4 \cdot x} = \sqrt{x^4} \cdot \sqrt{x} = x^2\sqrt{x}$

b. $\sqrt{8y^2} = \sqrt{4 \cdot 2 \cdot y^2} = \sqrt{4y^2 \cdot 2} = \sqrt{4y^2} \cdot \sqrt{2} = 2y\sqrt{2}$

c. $\sqrt{\frac{45}{x^6}} = \frac{\sqrt{45}}{\sqrt{x^6}} = \frac{\sqrt{9 \cdot 5}}{x^3} = \frac{\sqrt{9} \cdot \sqrt{5}}{x^3} = \frac{3\sqrt{5}}{x^3}$

d. $\sqrt{\frac{5p^3}{9}} = \frac{\sqrt{5p^3}}{\sqrt{9}} = \frac{\sqrt{p^2 \cdot 5p}}{3} = \frac{\sqrt{p^2} \cdot \sqrt{5p}}{3} = \frac{p\sqrt{5p}}{3}$

PRACTICE 4 Simplify. Assume that all variables represent positive numbers.

a. $\sqrt{x^7}$ **b.** $\sqrt{12a^4}$ **c.** $\sqrt{\frac{98}{z^8}}$ **d.** $\sqrt{\frac{11y^9}{49}}$

OBJECTIVE 4 **Simplifying Higher Roots** 

The product and quotient rules also apply to roots other than square roots. In general, we have the following product and quotient rules for radicals.

Product Rule for Radicals

If $\sqrt[n]{a}$ and $\sqrt[n]{b}$ are real numbers, then

$$\sqrt[n]{a \cdot b} = \sqrt[n]{a} \cdot \sqrt[n]{b}$$

Quotient Rule for Radicals

If $\sqrt[n]{a}$ and $\sqrt[n]{b}$ are real numbers and $b \neq 0$, then

$$\sqrt[n]{\frac{a}{b}} = \frac{\sqrt[n]{a}}{\sqrt[n]{b}}$$

To simplify cube roots, look for perfect cube factors of the radicand. For example, 8 is a perfect cube, since $2^3 = 8$.

To simplify $\sqrt[3]{48}$, factor 48 as $8 \cdot 6$.

$\sqrt[3]{48} = \sqrt[3]{8 \cdot 6}$ Factor 48.

$= \sqrt[3]{8} \cdot \sqrt[3]{6}$ Apply the product rule.

$= 2\sqrt[3]{6}$ Write $\sqrt[3]{8}$ as 2.

$2\sqrt[3]{6}$ is in simplest form since the radicand 6 contains no perfect cube factors other than 1.

EXAMPLE 5 Simplify.

a. $\sqrt[3]{54}$ **b.** $\sqrt[3]{18}$ **c.** $\sqrt[3]{\frac{7}{8}}$ **d.** $\sqrt[3]{\frac{40}{27}}$

Solution

a. $\sqrt[3]{54} = \sqrt[3]{27 \cdot 2} = \sqrt[3]{27} \cdot \sqrt[3]{2} = 3\sqrt[3]{2}$

b. The number 18 contains no perfect cube factors, so $\sqrt[3]{18}$ cannot be simplified further.

c. $\sqrt[3]{\frac{7}{8}} = \frac{\sqrt[3]{7}}{\sqrt[3]{8}} = \frac{\sqrt[3]{7}}{2}$

d. $\sqrt[3]{\frac{40}{27}} = \frac{\sqrt[3]{40}}{\sqrt[3]{27}} = \frac{\sqrt[3]{8 \cdot 5}}{3} = \frac{\sqrt[3]{8} \cdot \sqrt[3]{5}}{3} = \frac{2\sqrt[3]{5}}{3}$ □

PRACTICE

5 Simplify.

a. $\sqrt[3]{24}$ **b.** $\sqrt[3]{38}$ **c.** $\sqrt[3]{\frac{5}{27}}$ **d.** $\sqrt[3]{\frac{15}{64}}$

To simplify fourth roots, look for perfect fourth powers of the radicand. For example, 16 is a perfect fourth power since $2^4 = 16$.

To simplify $\sqrt[4]{32}$, factor 32 as $16 \cdot 2$.

$\sqrt[4]{32} = \sqrt[4]{16 \cdot 2}$ Factor 32.

$= \sqrt[4]{16} \cdot \sqrt[4]{2}$ Apply the product rule.

$= 2\sqrt[4]{2}$ Write $\sqrt[4]{16}$ as 2.

EXAMPLE 6 Simplify.

a. $\sqrt[4]{243}$ **b.** $\sqrt[4]{\frac{3}{16}}$ **c.** $\sqrt[5]{64}$

Solution

a. $\sqrt[4]{243} = \sqrt[4]{81 \cdot 3} = \sqrt[4]{81} \cdot \sqrt[4]{3} = 3\sqrt[4]{3}$ **b.** $\sqrt[4]{\frac{3}{16}} = \frac{\sqrt[4]{3}}{\sqrt[4]{16}} = \frac{\sqrt[4]{3}}{2}$

c. $\sqrt[5]{64} = \sqrt[5]{32 \cdot 2} = \sqrt[5]{32} \cdot \sqrt[5]{2} = 2\sqrt[5]{2}$ □

PRACTICE

6 Simplify.

a. $\sqrt[4]{80}$ **b.** $\sqrt[4]{\frac{5}{81}}$ **c.** $\sqrt[5]{96}$

Vocabulary, Readiness & Video Check

Use the choices below to fill in the blanks. Not all choices will be used.

$a \cdot b$ $\frac{a}{b}$ $\frac{\sqrt{a}}{\sqrt{b}}$ $\sqrt{a} \cdot \sqrt{b}$

1. If $\sqrt{a}$ and $\sqrt{b}$ are real numbers, then $\sqrt{a \cdot b} =$ ____________.

2. If $\sqrt{a}$ and $\sqrt{b}$ are real numbers, then $\sqrt{\frac{a}{b}} =$ ____________.

For Exercises 3 and 4, fill in the blanks using the example: $\sqrt{4 \cdot 9} = \sqrt{\underline{4}} \cdot \sqrt{\underline{9}} = \underline{2} \cdot \underline{3} = \underline{6}$.

3. $\sqrt{16 \cdot 25} = \sqrt{\_\_} \cdot \sqrt{\_\_} = \_\_ \cdot \_\_ = \_\_$

4. $\sqrt{36 \cdot 3} = \sqrt{\_\_} \cdot \sqrt{\_\_} = \_\_ \cdot \sqrt{\_\_} = \_\_\_\_$

Martin-Gay Interactive Videos

Watch the section lecture video and answer the following questions.

OBJECTIVE 1

5. From Example 3, if you have trouble finding a perfect square factor in the radicand, what is recommended?

OBJECTIVE 2

6. Based on the lecture before Example 5, complete the following statement. In words, the quotient rule for square roots says that the square root of a quotient is equal to the square root of the ______ over the square root of the ______.

OBJECTIVE 3

7. From Examples 6–8, we know that even powers of a variable are perfect square factors of the variable. Therefore, what must be true about the power of any variable left in the radicand of a simplified square root? Explain.

OBJECTIVE 4

8. From Example 9, how does factoring the radicand as a product of primes help simplify higher roots also?

8.2 Exercise Set

MyMathLab®

Use the product rule to simplify each radical. See Examples 1 and 2.

1. $\sqrt{20}$
2. $\sqrt{44}$
3. $\sqrt{50}$
4. $\sqrt{28}$
5. $\sqrt{33}$
6. $\sqrt{21}$
7. $\sqrt{98}$
8. $\sqrt{125}$
9. $\sqrt{60}$
10. $\sqrt{90}$
11. $\sqrt{180}$
12. $\sqrt{150}$
13. $\sqrt{52}$
14. $\sqrt{75}$
15. $3\sqrt{25}$
16. $9\sqrt{36}$
17. $7\sqrt{63}$
18. $11\sqrt{99}$
19. $-5\sqrt{27}$
20. $-6\sqrt{75}$

Use the quotient rule and the product rule to simplify each radical. See Example 3.

21. $\sqrt{\frac{8}{25}}$
22. $\sqrt{\frac{63}{16}}$
23. $\sqrt{\frac{27}{121}}$
24. $\sqrt{\frac{24}{169}}$
25. $\sqrt{\frac{9}{4}}$
26. $\sqrt{\frac{100}{49}}$
27. $\sqrt{\frac{125}{9}}$
28. $\sqrt{\frac{27}{100}}$
29. $\sqrt{\frac{11}{36}}$
30. $\sqrt{\frac{30}{49}}$
31. $-\sqrt{\frac{27}{144}}$
32. $-\sqrt{\frac{84}{121}}$

Simplify each radical. Assume that all variables represent positive numbers. See Example 4.

33. $\sqrt{x^7}$
34. $\sqrt{y^3}$
35. $\sqrt{x^{13}}$
36. $\sqrt{y^{17}}$
37. $\sqrt{36a^3}$
38. $\sqrt{81b^5}$
39. $\sqrt{96x^4}$
40. $\sqrt{40y^{10}}$
41. $\sqrt{\frac{12}{m^2}}$
42. $\sqrt{\frac{63}{p^2}}$
43. $\sqrt{\frac{9x}{y^{10}}}$
44. $\sqrt{\frac{6y^2}{z^{16}}}$
45. $\sqrt{\frac{88}{x^{12}}}$
46. $\sqrt{\frac{500}{y^{22}}}$

MIXED PRACTICE

Simplify each radical. See Examples 1 through 4.

47. $8\sqrt{4}$
48. $6\sqrt{49}$
49. $\sqrt{\frac{36}{121}}$
50. $\sqrt{\frac{25}{144}}$
51. $\sqrt{175}$
52. $\sqrt{700}$
53. $\sqrt{\frac{20}{9}}$
54. $\sqrt{\frac{45}{64}}$

55. $\sqrt{24m^7}$

56. $\sqrt{50n^{13}}$

57. $\sqrt{\frac{23y^3}{4x^6}}$

58. $\sqrt{\frac{41x^5}{9y^8}}$

Simplify each radical. See Example 5.

59. $\sqrt[3]{24}$

60. $\sqrt[3]{81}$

61. $\sqrt[3]{250}$

62. $\sqrt[3]{56}$

63. $\sqrt[3]{\frac{5}{64}}$

64. $\sqrt[3]{\frac{32}{125}}$

65. $\sqrt[3]{\frac{23}{8}}$

66. $\sqrt[3]{\frac{37}{27}}$

67. $\sqrt[3]{\frac{15}{64}}$

68. $\sqrt[3]{\frac{4}{27}}$

69. $\sqrt[3]{80}$

70. $\sqrt[3]{108}$

Simplify. See Example 6.

71. $\sqrt[4]{48}$

72. $\sqrt[4]{405}$

73. $\sqrt[4]{\frac{8}{81}}$

74. $\sqrt[4]{\frac{25}{256}}$

75. $\sqrt[5]{96}$

76. $\sqrt[5]{128}$

77. $\sqrt[5]{\frac{5}{32}}$

78. $\sqrt[5]{\frac{16}{243}}$

REVIEW AND PREVIEW

Perform the following operations. See Sections 5.2 and 5.3.

79. $6x + 8x$

80. $(6x)(8x)$

81. $(2x + 3)(x - 5)$

82. $(2x + 3) + (x - 5)$

83. $9y^2 - 9y^2$

84. $(9y^2)(-8y^2)$

CONCEPT EXTENSIONS

Simplify each radical. Assume that all variables represent positive numbers.

85. $\sqrt{x^6y^3}$

86. $\sqrt{a^{13}b^{14}}$

87. $\sqrt{98x^5y^4}$

88. $\sqrt{27x^8y^{11}}$

89. $\sqrt[3]{-8x^6}$

90. $\sqrt[3]{27x^{12}}$

Simplify.

△ **91.** If a cube is to have a volume of 80 cubic inches, then each side must be $\sqrt[3]{80}$ inches long. Simplify the radical representing the side length.

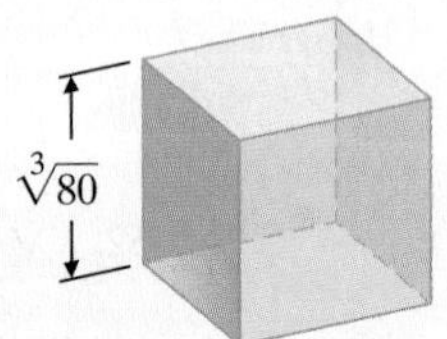

△ **92.** Jeannie Boswell is swimming across a 40-foot-wide river, trying to head straight across to the opposite shore. However, the current is strong enough to move her downstream 100 feet by the time she reaches land. (See the figure.) Because of the current, the actual distance she swam is $\sqrt{11{,}600}$ feet. Simplify this radical.

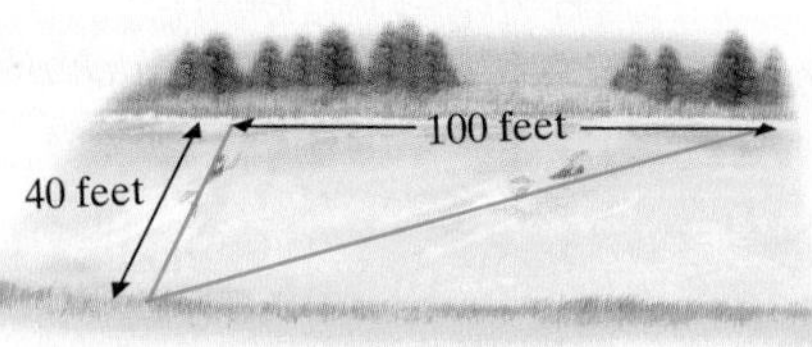

93. By using replacement values for a and b, show that $\sqrt{a^2 + b^2}$ does not equal $a + b$.

94. By using replacement values for a and b, show that $\sqrt{a + b}$ does not equal $\sqrt{a} + \sqrt{b}$.

The length of a side of a cube is given by the expression $\sqrt{\frac{A}{6}}$ units where A square units is the cube's surface area. Use this expression for Exercises 95 through 98. Be sure to attach the appropriate units.

△ **95.** The surface area of a cube is 120 square inches. Find the exact length of a side of the cube.

△ **96.** The surface area of a cube is 594 square feet. Find the exact length of a side of the cube.

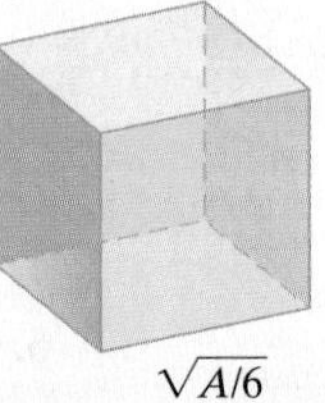

97. Rubik's cube, named after its inventor, Erno Rubik, was first imagined by him in 1974 and, by 1980, was a worldwide phenomenon. A standard Rubik's cube has a surface area of 30.375 square inches. Find the length of one side of a Rubik's cube. (A few world records are listed below. *Source: Guinness World Records*)

Fastest time to solve 1 Rubik's cube: 9.55 sec by Ron van Bruchem (Netherlands) in 2007.

Most Rubik's cubes solved in 1 hour: 185 by David Calvo (Spain) in 2008.

△ **98.** The Borg spaceship on *Star Trek: The Next Generation* is in the shape of a cube. Suppose a model of this ship has a surface area of 121 square inches. Find the length of a side of the ship.

The cost C in dollars per day to operate a small delivery service is given by $C = 100\sqrt[3]{n} + 700$, where n is the number of deliveries per day.

99. Find the cost if the number of deliveries is 1000.

100. Approximate the cost if the number of deliveries is 500.

The Mosteller formula for calculating body surface area is $B = \sqrt{\frac{hw}{3600}}$, where B is an individual's body surface area in square meters, h is the individual's height in centimeters, and w is the individual's weight in kilograms. Use this formula in Exercises 101 and 102. Round answers to the nearest tenth.

101. Find the body surface area of a person who is 169 cm tall and weighs 64 kilograms.

102. Approximate the body surface area of a person who is 183 cm tall and weighs 85 kilograms.

8.3 Adding and Subtracting Radicals 

OBJECTIVES

1. Add or Subtract Like Radicals.
2. Simplify Radical Expressions and Then Add or Subtract Any Like Radicals.

OBJECTIVE

1 Adding and Subtracting Like Radicals

To combine like terms, we use the distributive property.

$$5x + 3x = (5 + 3)x = 8x$$

The distributive property can also be applied to expressions containing radicals. For example,

$$5\sqrt{2} + 3\sqrt{2} = (5 + 3)\sqrt{2} = 8\sqrt{2}$$

Also,

$$9\sqrt{5} - 6\sqrt{5} = (9 - 6)\sqrt{5} = 3\sqrt{5}$$

Radical terms $5\sqrt{2}$ and $3\sqrt{2}$ are **like radicals,** as are $9\sqrt{5}$ and $6\sqrt{5}$.

Like Radicals

Like radicals are radical expressions that have the same index and the same radicand.

From the examples above, we can see that **only like radicals can be combined** in this way. For example, the expression $2\sqrt{3} + 3\sqrt{2}$ cannot be simplified further since the radicals are not like radicals. Also, the expression $4\sqrt{7} + 4\sqrt[3]{7}$ cannot be simplified further because the radicals are not like radicals since the indices are different.

EXAMPLE 1 Simplify by combining like radical terms.

a. $4\sqrt{5} + 3\sqrt{5}$ **b.** $\sqrt{10} - 6\sqrt{10}$ **c.** $\sqrt[3]{7} + \sqrt[3]{7} - 4\sqrt[3]{5}$ **d.** $2\sqrt{6} + 2\sqrt[3]{6}$

Solution

a. $4\sqrt{5} + 3\sqrt{5} = (4 + 3)\sqrt{5} = 7\sqrt{5}$

b. $\sqrt{10} - 6\sqrt{10} = 1\sqrt{10} - 6\sqrt{10} = (1 - 6)\sqrt{10} = -5\sqrt{10}$

c. $\sqrt[3]{7} + \sqrt[3]{7} - 4\sqrt[3]{5} = 1\sqrt[3]{7} + 1\sqrt[3]{7} - 4\sqrt[3]{5} = (1 + 1)\sqrt[3]{7} - 4\sqrt[3]{5} = 2\sqrt[3]{7} - 4\sqrt[3]{5}$
This expression cannot be simplified further since the radicands are not the same.

d. $2\sqrt{6} + 2\sqrt[3]{6}$ cannot be simplified further since the indices are not the same. □

PRACTICE

1 Simplify by combining like radical terms.

a. $3\sqrt{2} + 5\sqrt{2}$

b. $\sqrt{6} - 8\sqrt{6}$

c. $6\sqrt[4]{5} - 2\sqrt[4]{5} + 11\sqrt[4]{7}$

d. $4\sqrt{13} - 5\sqrt[3]{13}$

✓**CONCEPT CHECK**

Which is true?

a. $2 + 3\sqrt{5} = 5\sqrt{5}$ b. $2\sqrt{3} + 2\sqrt{7} = 2\sqrt{10}$ c. $\sqrt{3} + \sqrt{5} = \sqrt{8}$

d. $\sqrt{3} + \sqrt{3} = 3$ e. None of the above is true.

OBJECTIVE

2 Simplifying Radicals, Then Adding or Subtracting 

At first glance, it appears that the expression $\sqrt{50} + \sqrt{8}$ cannot be simplified further because the radicands are different. However, the product rule can be used to simplify each radical, and then further simplification might be possible.

EXAMPLE 2 Add or subtract by first simplifying each radical.

a. $\sqrt{50} + \sqrt{8}$ **b.** $7\sqrt{12} - \sqrt{75}$ **c.** $\sqrt{25} - \sqrt{27} - 2\sqrt{18} - \sqrt{16}$

Solution

a. First simplify each radical.

$\sqrt{50} + \sqrt{8} = \sqrt{25 \cdot 2} + \sqrt{4 \cdot 2}$ Factor radicands.

$= \sqrt{25} \cdot \sqrt{2} + \sqrt{4} \cdot \sqrt{2}$ Apply the product rule.

$= 5\sqrt{2} + 2\sqrt{2}$ Simplify $\sqrt{25}$ and $\sqrt{4}$.

$= 7\sqrt{2}$ Add like radicals.

b. $7\sqrt{12} - \sqrt{75} = 7\sqrt{4 \cdot 3} - \sqrt{25 \cdot 3}$ Factor radicands.

$= 7\sqrt{4} \cdot \sqrt{3} - \sqrt{25} \cdot \sqrt{3}$ Apply the product rule.

$= 7 \cdot 2\sqrt{3} - 5\sqrt{3}$ Simplify $\sqrt{4}$ and $\sqrt{25}$.

$= 14\sqrt{3} - 5\sqrt{3}$ Multiply.

$= 9\sqrt{3}$ Subtract like radicals.

c. $\sqrt{25} - \sqrt{27} - 2\sqrt{18} - \sqrt{16}$

$= 5 - \sqrt{9 \cdot 3} - 2\sqrt{9 \cdot 2} - 4$ Factor radicands.

$= 5 - \sqrt{9} \cdot \sqrt{3} - 2\sqrt{9} \cdot \sqrt{2} - 4$ Apply the product rule.

$= 5 - 3\sqrt{3} - 2 \cdot 3\sqrt{2} - 4$ Simplify.

$= 1 - 3\sqrt{3} - 6\sqrt{2}$ Write $5 - 4$ as 1 and $2 \cdot 3$ as 6. □

Answer to Concept Check:
e (a, b, and c are not true since each left side cannot be simplified further. For d, $\sqrt{3} + \sqrt{3} = 2\sqrt{3}$.)

PRACTICE 2 Add or subtract by first simplifying each radical.

a. $\sqrt{45} + \sqrt{20}$ **b.** $\sqrt{36} + 3\sqrt{24} - \sqrt{40} - \sqrt{150}$

c. $\sqrt{98} - 5\sqrt{8}$

If radical expressions contain variables, we proceed in a similar way. Simplify radicals using the product and quotient rules. Then add or subtract any like radicals.

EXAMPLE 3 Simplify $2\sqrt{x^2} - \sqrt{25x^5} + \sqrt{x^5}$. Assume variables represent positive numbers.

Solution $2\sqrt{x^2} - \sqrt{25x^5} + \sqrt{x^5}$

$= 2x - \sqrt{25x^4 \cdot x} + \sqrt{x^4 \cdot x}$ Factor radicands so that one factor is a perfect square. Simplify $\sqrt{x^2}$.

$= 2x - \sqrt{25x^4} \cdot \sqrt{x} + \sqrt{x^4} \cdot \sqrt{x}$ Use the product rule.

$= 2x - 5x^2\sqrt{x} + x^2\sqrt{x}$ Write $\sqrt{25x^4}$ as $5x^2$ and $\sqrt{x^4}$ as x^2.

$= 2x - 4x^2\sqrt{x}$ Add like radicals. □

PRACTICE 3 Simplify $\sqrt{x^3} - 8x\sqrt{x} + 3\sqrt{x^2}$. Assume variables represent positive numbers.

EXAMPLE 4 Simplify the radical expression: $5\sqrt[3]{16x^3} - \sqrt[3]{54x^3}$

Solution $5\sqrt[3]{16x^3} - \sqrt[3]{54x^3}$

$= 5\sqrt[3]{8x^3 \cdot 2} - \sqrt[3]{27x^3 \cdot 2}$ Factor radicands so that one factor is a perfect cube.

$= 5 \cdot \sqrt[3]{8x^3} \cdot \sqrt[3]{2} - \sqrt[3]{27x^3} \cdot \sqrt[3]{2}$ Use the product rule.

$= 5 \cdot 2x \cdot \sqrt[3]{2} - 3x \cdot \sqrt[3]{2}$ Write $\sqrt[3]{8x^3}$ as $2x$ and $\sqrt[3]{27x^3}$ as $3x$.

$= 10x\sqrt[3]{2} - 3x\sqrt[3]{2}$ Write $5 \cdot 2x$ as $10x$.

$= 7x\sqrt[3]{2}$ Subtract like radicands. □

PRACTICE 4 Simplify the radical expression: $4\sqrt[3]{81x^6} - \sqrt[3]{24x^6}$

Vocabulary, Readiness & Video Check

Fill in each blank.

1. Radicals that have the same index and same radicand are called ________.

2. The expressions $7\sqrt[3]{2x}$ and $-\sqrt[3]{2x}$ are called ________.

3. $11\sqrt{2} + 6\sqrt{2} =$ ________.

a. $66\sqrt{2}$ **b.** $17\sqrt{2}$ **c.** $17\sqrt{4}$

4. $\sqrt{5}$ is the same as ________.

a. $0\sqrt{5}$ **b.** $1\sqrt{5}$ **c.** $5\sqrt{5}$

5. $\sqrt{5} + \sqrt{5} =$ ________

a. $\sqrt{10}$ **b.** 5 **c.** $2\sqrt{5}$

6. $9\sqrt{7} - \sqrt{7} =$ ________

a. $8\sqrt{7}$ **b.** 9 **c.** 0

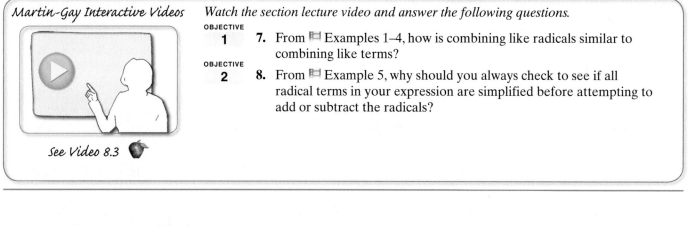

Martin-Gay Interactive Videos

See Video 8.3

Watch the section lecture video and answer the following questions.

OBJECTIVE 1

7. From Examples 1–4, how is combining like radicals similar to combining like terms?

OBJECTIVE 2

8. From Example 5, why should you always check to see if all radical terms in your expression are simplified before attempting to add or subtract the radicals?

8.3 Exercise Set

MyMathLab®

Simplify each expression by combining like radicals where possible. See Example 1.

1. $4\sqrt{3} - 8\sqrt{3}$
2. $2\sqrt{5} - 9\sqrt{5}$
3. $3\sqrt{6} + 8\sqrt{6} - 2\sqrt{6} - 5$
4. $12\sqrt{2} - 3\sqrt{2} + 8\sqrt{2} + 10$
5. $\sqrt{11} + \sqrt{11} + 11$
6. $\sqrt{13} + 13 + \sqrt{13}$
7. $6\sqrt{5} - 5\sqrt{5} + \sqrt{2}$
8. $4\sqrt{3} + \sqrt{5} - 3\sqrt{3}$
9. $\sqrt[3]{18} + \sqrt[3]{18} - 4\sqrt[3]{18}$
10. $\sqrt[3]{49} - 8\sqrt[3]{49} + \sqrt[3]{49}$
11. $2\sqrt[3]{3} + 5\sqrt[3]{3} - \sqrt{3}$
12. $8\sqrt[3]{5} + 2\sqrt[3]{5} + \sqrt{5}$
13. $2\sqrt[3]{2} - 7\sqrt[3]{2} - 6$
14. $5\sqrt[3]{9} + 2 - 11\sqrt[3]{9}$

MIXED PRACTICE

Add or subtract by first simplifying each radical and then combining any like radical terms. Assume that all variables represent positive real numbers. See Examples 2 and 3.

15. $\sqrt{12} + \sqrt{27}$
16. $\sqrt{50} + \sqrt{18}$
17. $\sqrt{45} + 3\sqrt{20}$
18. $\sqrt{28} + \sqrt{63}$
19. $2\sqrt{54} - \sqrt{20} + \sqrt{45} - \sqrt{24}$
20. $2\sqrt{8} - \sqrt{128} + \sqrt{48} + \sqrt{18}$
21. $4x - 3\sqrt{x^2} + \sqrt{x}$
22. $x - 6\sqrt{x^2} + 2\sqrt{x}$
23. $\sqrt{25x} + \sqrt{36x} - 11\sqrt{x}$
24. $3\sqrt{x^3} - x\sqrt{4x}$
25. $\sqrt{16x} - \sqrt{x^3}$
26. $\sqrt{8x^3} - \sqrt{x^2}$
27. $12\sqrt{5} - \sqrt{5} - 4\sqrt{5}$
28. $7\sqrt{3} + 2\sqrt{3} - 13\sqrt{3}$
29. $\sqrt{5} + \sqrt[3]{5}$
30. $\sqrt{5} + \sqrt{5}$
31. $4 + 8\sqrt{2} - 9$
32. $6 - 2\sqrt{3} - \sqrt{3}$
33. $8 - \sqrt{2} - 5\sqrt{2}$
34. $\sqrt{75} + \sqrt{48}$
35. $5\sqrt{32} - \sqrt{72}$
36. $2\sqrt{80} - \sqrt{45}$
37. $\sqrt{8} + \sqrt{9} + \sqrt{18} + \sqrt{81}$
38. $\sqrt{6} + \sqrt{16} + \sqrt{24} + \sqrt{25}$
39. $\sqrt{\frac{5}{9}} + \sqrt{\frac{5}{81}}$
40. $\sqrt{\frac{3}{64}} + \sqrt{\frac{3}{16}}$
41. $\sqrt{\frac{3}{4}} - \sqrt{\frac{3}{64}}$
42. $\sqrt{\frac{7}{25}} - \sqrt{\frac{7}{100}}$
43. $2\sqrt{45} - 2\sqrt{20}$
44. $5\sqrt{18} + 2\sqrt{32}$
45. $\sqrt{35} - \sqrt{140}$
46. $\sqrt{6} - \sqrt{600}$
47. $5\sqrt{2x} + \sqrt{98x}$
48. $3\sqrt{9x} + 2\sqrt{x}$
49. $5\sqrt{x} + 4\sqrt{4x} - 13\sqrt{x}$
50. $\sqrt{9x} + \sqrt{81x} - 11\sqrt{x}$
51. $\sqrt{3x^3} + 3x\sqrt{x}$
52. $x\sqrt{4x} + \sqrt{9x^3}$

Add or subtract by first simplifying each radical and then combining any like radical terms. Assume that all variables represent positive real numbers. See Example 4.

53. $\sqrt[3]{81} + \sqrt[3]{24}$

54. $\sqrt[3]{32} - \sqrt[3]{4}$

55. $4\sqrt[3]{9} - \sqrt[3]{243}$

56. $7\sqrt[3]{6} - \sqrt[3]{48}$

57. $\sqrt[3]{8} + \sqrt[3]{54} - 5$

58. $\sqrt[3]{64} + \sqrt[3]{14} - 9$

59. $\sqrt{32x^2} + \sqrt[3]{32} + \sqrt{4x^2}$

60. $\sqrt{18x^2} + \sqrt[3]{24} + \sqrt{2x^2}$

61. $2\sqrt[3]{8x^3} + 2\sqrt[3]{16x^3}$

62. $3\sqrt[3]{27z^3} + 3\sqrt[3]{81z^3}$

63. $12\sqrt[3]{y^7} - y^2\sqrt[3]{8y}$

64. $19\sqrt[3]{z^{11}} - z^3\sqrt[3]{125z^2}$

65. $\sqrt{40x} + \sqrt[3]{40} - 2\sqrt{10x} - \sqrt[3]{5}$

66. $\sqrt{72x^2} + \sqrt[3]{54} - x\sqrt{50} - 3\sqrt[3]{2}$

REVIEW AND PREVIEW

Square each binomial. See Section 5.4.

67. $(x + 6)^2$

68. $(3x + 2)^2$

69. $(2x - 1)^2$

70. $(x - 5)^2$

CONCEPT EXTENSIONS

71. In your own words, describe like radicals.

72. In the expression $\sqrt{5} + 2 - 3\sqrt{5}$, explain why 2 and -3 cannot be combined.

73. Find the perimeter of the rectangular picture frame.

74. Find the perimeter of the plot of land.

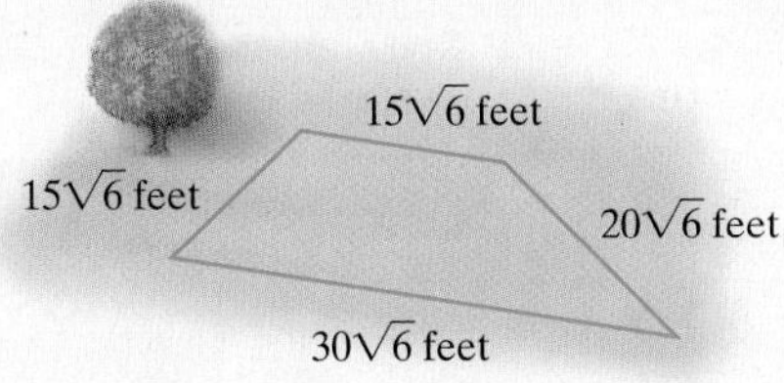

75. An 8-foot-long water trough is to be made of wood. Each of the two triangular end pieces has an area of $\frac{3\sqrt{27}}{4}$ square feet. The two side panels are both rectangular. In simplest radical form, find the total area of the wood needed.

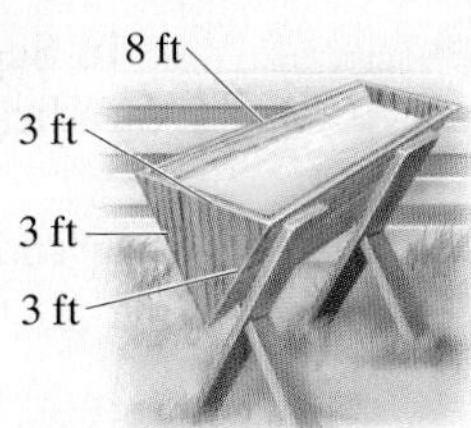

76. Eight wooden braces are to be attached along the diagonals of the vertical sides of a storage bin. Each of four of these diagonals has a length of $\sqrt{52}$ feet, while each of the other four has a length of $\sqrt{80}$ feet. In simplest radical form, find the total length of the wood needed for these braces.

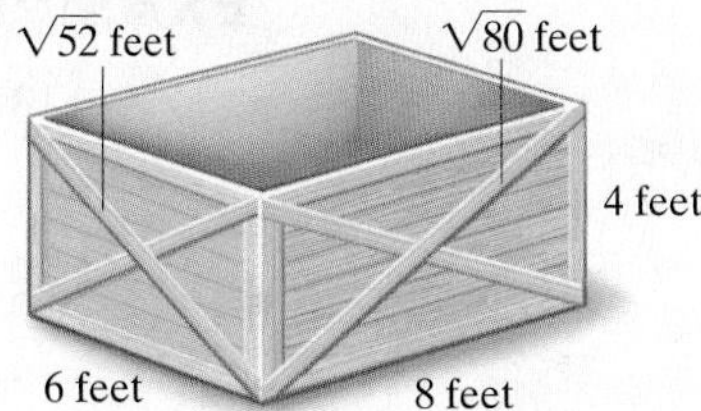

Determine whether each expression can be simplified. If yes, then simplify. See the Concept Check in this section.

77. $4\sqrt{2} + 3\sqrt{2}$

78. $3\sqrt{7} + 3\sqrt{6}$

79. $6 + 7\sqrt{6}$

80. $5x\sqrt{2} + 8x\sqrt{2}$

81. $\sqrt{7} + \sqrt{7} + \sqrt{7}$

82. $6\sqrt{5} - \sqrt{5}$

Simplify.

83. $\sqrt{\frac{x^3}{16}} - x\sqrt{\frac{9x}{25}} + \frac{\sqrt{81x^3}}{2}$

84. $7\sqrt{x^{11}y^7} - x^2y\sqrt{25x^7y^5} + \sqrt{8x^8y^2}$

8.4 Multiplying and Dividing Radicals 

OBJECTIVES

1 Multiply Radicals.

2 Divide Radicals.

3 Rationalize Denominators.

4 Rationalize Using Conjugates.

OBJECTIVE

1 Multiplying Radicals 

In Section 8.2, we used the product and quotient rules for radicals to help us simplify radicals. In this section, we use these rules to simplify products and quotients of radicals.

Product Rule for Radicals

If $\sqrt[n]{a}$ and $\sqrt[n]{b}$ are real numbers, then

$$\sqrt[n]{a} \cdot \sqrt[n]{b} = \sqrt[n]{a \cdot b}$$

This property says that the product of the nth roots of two numbers is the nth root of the product of the two numbers. For example,

$$\sqrt{3} \cdot \sqrt{2} = \sqrt{3 \cdot 2} = \sqrt{6} \quad \text{Also,} \quad \sqrt[3]{5} \cdot \sqrt[3]{7} = \sqrt[3]{5 \cdot 7} = \sqrt[3]{35}$$

EXAMPLE 1 Multiply. Then simplify if possible.

a. $\sqrt{7} \cdot \sqrt{3}$ **b.** $\sqrt{3} \cdot \sqrt{3}$ **c.** $\sqrt{3} \cdot \sqrt{15}$ **d.** $2\sqrt{3} \cdot 5\sqrt{2}$ **e.** $\sqrt{2x^3} \cdot \sqrt{6x}$

Solution

a. $\sqrt{7} \cdot \sqrt{3} = \sqrt{7 \cdot 3} = \sqrt{21}$

b. $\sqrt{3} \cdot \sqrt{3} = \sqrt{3 \cdot 3} = \sqrt{9} = 3$

c. $\sqrt{3} \cdot \sqrt{15} = \sqrt{45}$. Next, simplify $\sqrt{45}$.

$$\sqrt{45} = \sqrt{9 \cdot 5} = \sqrt{9} \cdot \sqrt{5} = 3\sqrt{5}$$

d. $2\sqrt{3} \cdot 5\sqrt{2} = 2 \cdot 5\sqrt{3 \cdot 2} = 10\sqrt{6}$

e. $\sqrt{2x^3} \cdot \sqrt{6x} = \sqrt{2x^3 \cdot 6x}$ Use the product rule.

$= \sqrt{12x^4}$ Multiply.

$= \sqrt{4x^4 \cdot 3}$ Write $12x^4$ so that one factor is a perfect square.

$= \sqrt{4x^4} \cdot \sqrt{3}$ Use the product rule.

$= 2x^2\sqrt{3}$ Simplify.

PRACTICE

1 Multiply. Then simplify if possible.

a. $\sqrt{11} \cdot \sqrt{7}$ **b.** $9\sqrt{10} \cdot 8\sqrt{3}$ **c.** $\sqrt{5} \cdot \sqrt{10}$

d. $\sqrt{17} \cdot \sqrt{17}$ **e.** $\sqrt{15y} \cdot \sqrt{5y^3}$

From Example 1b, we found that

$$\sqrt{3} \cdot \sqrt{3} = 3 \quad \text{or} \quad (\sqrt{3})^2 = 3$$

This is true in general.

If a is a positive number,

$$\sqrt{a} \cdot \sqrt{a} = a \quad \text{or} \quad (\sqrt{a})^2 = a$$

Answer to Concept Check:
a, c, d

✓CONCEPT CHECK

Identify the true statement(s).

a. $\sqrt{7} \cdot \sqrt{7} = 7$ **b.** $\sqrt{2} \cdot \sqrt{3} = 6$

c. $(\sqrt{131})^2 = 131$ **d.** $\sqrt{5x} \cdot \sqrt{5x} = 5x$ (Here, x is a positive number.)

EXAMPLE 2 Find $(3\sqrt{2})^2$.

Solution $(3\sqrt{2})^2 = 3^2 \cdot (\sqrt{2})^2 = 9 \cdot 2 = 18$ □

PRACTICE 2 Find $(2\sqrt{7})^2$.

EXAMPLE 3 Multiply $\sqrt[3]{4} \cdot \sqrt[3]{18}$. Then simplify if possible.

Solution $\sqrt[3]{4} \cdot \sqrt[3]{18} = \sqrt[3]{4 \cdot 18} = \sqrt[3]{4 \cdot 2 \cdot 9} = \sqrt[3]{8 \cdot 9} = \sqrt[3]{8} \cdot \sqrt[3]{9} = 2\sqrt[3]{9}$ □

PRACTICE 3 Multiply $\sqrt[3]{10} \cdot \sqrt[3]{50}$. Then simplify if possible.

When multiplying radical expressions containing more than one term, use the same techniques we use to multiply other algebraic expressions with more than one term.

EXAMPLE 4 Multiply. Then simplify if possible.

a. $\sqrt{5}(\sqrt{5} - \sqrt{2})$ **b.** $\sqrt{3x}(\sqrt{x} - 5\sqrt{3})$

c. $(\sqrt{x} + \sqrt{2})(\sqrt{3} - \sqrt{2})$

Solution

a. Using the distributive property, we have

$$\sqrt{5}(\sqrt{5} - \sqrt{2}) = \sqrt{5} \cdot \sqrt{5} - \sqrt{5} \cdot \sqrt{2}$$
$$= 5 - \sqrt{10}$$ Since $\sqrt{5} \cdot \sqrt{5} = 5$ and $\sqrt{5} \cdot \sqrt{2} = \sqrt{10}$

b. $\sqrt{3x}(\sqrt{x} - 5\sqrt{3}) = \sqrt{3x} \cdot \sqrt{x} - \sqrt{3x} \cdot 5\sqrt{3}$ Use the distributive property.

$= \sqrt{3x \cdot x} - 5\sqrt{3x \cdot 3}$ Use the product rule.

$= \sqrt{3 \cdot x^2} - 5\sqrt{9 \cdot x}$ Factor each radicand so that one factor is a perfect square.

$= \sqrt{3} \cdot \sqrt{x^2} - 5 \cdot \sqrt{9} \cdot \sqrt{x}$ Use the product rule.

$= x\sqrt{3} - 5 \cdot 3 \cdot \sqrt{x}$ Simplify.

$= x\sqrt{3} - 15\sqrt{x}$ Simplify.

c. Use the FOIL method of multiplication.

$$(\sqrt{x} + \sqrt{2})(\sqrt{3} - \sqrt{2}) = \overset{F}{\sqrt{x} \cdot \sqrt{3}} - \overset{O}{\sqrt{x} \cdot \sqrt{2}} + \overset{I}{\sqrt{2} \cdot \sqrt{3}} - \overset{L}{\sqrt{2} \cdot \sqrt{2}}$$

$= \sqrt{3x} - \sqrt{2x} + \sqrt{6} - \sqrt{4}$ Apply the product rule.

$= \sqrt{3x} - \sqrt{2x} + \sqrt{6} - 2$ Simplify. □

PRACTICE 4 Multiply. Then simplify if possible.

a. $\sqrt{3}(\sqrt{3} - \sqrt{5})$ **b.** $\sqrt{2z}(\sqrt{z} + 7\sqrt{2})$

c. $(\sqrt{x} - \sqrt{7})(\sqrt{x} + \sqrt{2})$

The special product formulas can be used to multiply expressions containing radicals.

EXAMPLE 5 Multiply. Then simplify if possible.

a. $(\sqrt{5} - 7)(\sqrt{5} + 7)$ **b.** $(\sqrt{7x} + 2)^2$

Solution

a. Recall from Chapter 5 that $(a - b)(a + b) = a^2 - b^2$. Then

$$(\sqrt{5} - 7)(\sqrt{5} + 7) = (\sqrt{5})^2 - 7^2$$
$$= 5 - 49$$
$$= -44$$

b. Recall that $(a + b)^2 = a^2 + 2ab + b^2$. Then

$$(\sqrt{7x} + 2)^2 = (\sqrt{7x})^2 + 2(\sqrt{7x})(2) + (2)^2$$
$$= 7x + 4\sqrt{7x} + 4$$

□

PRACTICE
5 Multiply. Then simplify if possible.

a. $(\sqrt{7} + 4)(\sqrt{7} - 4)$ **b.** $(\sqrt{3x} - 5)^2$

OBJECTIVE
2 Dividing Radicals 

To simplify quotients of radical expressions, we use the quotient rule.

Quotient Rule for Radicals

If $\sqrt[n]{a}$ and $\sqrt[n]{b}$ are real numbers and $b \neq 0$, then

$$\frac{\sqrt[n]{a}}{\sqrt[n]{b}} = \sqrt[n]{\frac{a}{b}}, \text{ providing } b \neq 0$$

EXAMPLE 6 Divide. Then simplify if possible.

a. $\frac{\sqrt{14}}{\sqrt{2}}$ **b.** $\frac{\sqrt{100}}{\sqrt{5}}$ **c.** $\frac{\sqrt{12x^3}}{\sqrt{3x}}$

Solution Use the quotient rule and then simplify the resulting radicand.

a. $\frac{\sqrt{14}}{\sqrt{2}} = \sqrt{\frac{14}{2}} = \sqrt{7}$

b. $\frac{\sqrt{100}}{\sqrt{5}} = \sqrt{\frac{100}{5}} = \sqrt{20} = \sqrt{4 \cdot 5} = \sqrt{4} \cdot \sqrt{5} = 2\sqrt{5}$

c. $\frac{\sqrt{12x^3}}{\sqrt{3x}} = \sqrt{\frac{12x^3}{3x}} = \sqrt{4x^2} = 2x$

□

PRACTICE
6 Divide. Then simplify if possible.

a. $\frac{\sqrt{21}}{\sqrt{7}}$ **b.** $\frac{\sqrt{48}}{\sqrt{6}}$ **c.** $\frac{\sqrt{45y^5}}{\sqrt{5y}}$

EXAMPLE 7 Divide $\frac{\sqrt[3]{32}}{\sqrt[3]{4}}$. Then simplify if possible.

Solution $\frac{\sqrt[3]{32}}{\sqrt[3]{4}} = \sqrt[3]{\frac{32}{4}} = \sqrt[3]{8} = 2$

□

PRACTICE
7 Divide $\frac{\sqrt[3]{625}}{\sqrt[3]{5}}$. Then simplify if possible.

OBJECTIVE

3 Rationalizing Denominators 

It is sometimes easier to work with radical expressions if the denominator does not contain a radical. To rewrite the expression so that the denominator does not contain a radical expression, we use the fact that we can multiply the numerator and the denominator of a fraction by the same nonzero number without changing the value of the expression. This is the same as multiplying the fraction by 1. For example, to get rid of the radical in the denominator of $\frac{\sqrt{5}}{\sqrt{2}}$, we multiply by 1 in the form of $\frac{\sqrt{2}}{\sqrt{2}}$. Then

$$\frac{\sqrt{5}}{\sqrt{2}} = \frac{\sqrt{5}}{\sqrt{2}} \cdot 1 = \frac{\sqrt{5}}{\sqrt{2}} \cdot \frac{\sqrt{2}}{\sqrt{2}} = \frac{\sqrt{5} \cdot \sqrt{2}}{\sqrt{2} \cdot \sqrt{2}} = \frac{\sqrt{10}}{2}$$

This process is called **rationalizing** the denominator.

EXAMPLE 8 Rationalize each denominator.

a. $\frac{2}{\sqrt{7}}$ **b.** $\frac{\sqrt{5}}{\sqrt{12}}$ **c.** $\sqrt{\frac{1}{18x}}$

Solution

a. To rewrite $\frac{2}{\sqrt{7}}$ so that there is no radical in the denominator, we multiply by 1 in the form of $\frac{\sqrt{7}}{\sqrt{7}}$.

$$\frac{2}{\sqrt{7}} = \frac{2}{\sqrt{7}} \cdot \frac{\sqrt{7}}{\sqrt{7}} = \frac{2 \cdot \sqrt{7}}{\sqrt{7} \cdot \sqrt{7}} = \frac{2\sqrt{7}}{7}$$

b. We can multiply by $\frac{\sqrt{12}}{\sqrt{12}}$, but see what happens if we simplify first.

$$\frac{\sqrt{5}}{\sqrt{12}} = \frac{\sqrt{5}}{\sqrt{4 \cdot 3}} = \frac{\sqrt{5}}{2\sqrt{3}}$$

To rationalize the denominator now, we multiply by $\frac{\sqrt{3}}{\sqrt{3}}$.

$$\frac{\sqrt{5}}{2\sqrt{3}} = \frac{\sqrt{5}}{2\sqrt{3}} \cdot \frac{\sqrt{3}}{\sqrt{3}} = \frac{\sqrt{5} \cdot \sqrt{3}}{2\sqrt{3} \cdot \sqrt{3}} = \frac{\sqrt{15}}{2 \cdot 3} = \frac{\sqrt{15}}{6}$$

c. First we simplify.

$$\sqrt{\frac{1}{18x}} = \frac{\sqrt{1}}{\sqrt{18x}} = \frac{1}{\sqrt{9} \cdot \sqrt{2x}} = \frac{1}{3\sqrt{2x}}$$

Now to rationalize the denominator, we multiply by $\frac{\sqrt{2x}}{\sqrt{2x}}$.

$$\frac{1}{3\sqrt{2x}} = \frac{1}{3\sqrt{2x}} \cdot \frac{\sqrt{2x}}{\sqrt{2x}} = \frac{1 \cdot \sqrt{2x}}{3\sqrt{2x} \cdot \sqrt{2x}} = \frac{\sqrt{2x}}{3 \cdot 2x} = \frac{\sqrt{2x}}{6x}$$ □

PRACTICE

8 Rationalize each denominator.

a. $\frac{4}{\sqrt{5}}$ **b.** $\frac{\sqrt{3}}{\sqrt{18}}$ **c.** $\sqrt{\frac{3}{14x}}$

As a general rule, simplify a radical expression first and then rationalize the denominator.

EXAMPLE 9 Rationalize each denominator.

a. $\dfrac{5}{\sqrt[3]{4}}$ **b.** $\dfrac{\sqrt[3]{7}}{\sqrt[3]{3}}$

Solution

a. Since the denominator contains a cube root, we multiply the numerator and the denominator by a factor that gives the **cube root of a perfect cube** in the denominator. Recall that $\sqrt[3]{8} = 2$ and that the denominator $\sqrt[3]{4}$ multiplied by $\sqrt[3]{2}$ is $\sqrt[3]{4 \cdot 2}$ or $\sqrt[3]{8}$.

$$\frac{5}{\sqrt[3]{4}} = \frac{5 \cdot \sqrt[3]{2}}{\sqrt[3]{4} \cdot \sqrt[3]{2}} = \frac{5\sqrt[3]{2}}{\sqrt[3]{8}} = \frac{5\sqrt[3]{2}}{2}$$

b. Recall that $\sqrt[3]{27} = 3$. Multiply the denominator $\sqrt[3]{3}$ by $\sqrt[3]{9}$ and the result is $\sqrt[3]{3 \cdot 9}$ or $\sqrt[3]{27}$.

$$\frac{\sqrt[3]{7}}{\sqrt[3]{3}} = \frac{\sqrt[3]{7} \cdot \sqrt[3]{9}}{\sqrt[3]{3} \cdot \sqrt[3]{9}} = \frac{\sqrt[3]{63}}{\sqrt[3]{27}} = \frac{\sqrt[3]{63}}{3}$$

□

PRACTICE

9 Rationalize each denominator.

a. $\dfrac{3}{\sqrt[3]{25}}$ **b.** $\dfrac{\sqrt[3]{6}}{\sqrt[3]{5}}$

OBJECTIVE

4 Rationalizing Denominators Using Conjugates

To rationalize a denominator that is a sum, such as the denominator in

$$\frac{2}{4 + \sqrt{3}}$$

we multiply the numerator and the denominator by $4 - \sqrt{3}$. The expressions $4 + \sqrt{3}$ and $4 - \sqrt{3}$ are called **conjugates** of each other. When a radical expression such as $4 + \sqrt{3}$ is multiplied by its conjugate, $4 - \sqrt{3}$, the product simplifies to an expression that contains no radicals.

$$(a + b)(a - b) = a^2 - b^2$$

$$(4 + \sqrt{3})(4 - \sqrt{3}) = 4^2 - (\sqrt{3})^2 = 16 - 3 = 13$$

Then

$$\frac{2}{4 + \sqrt{3}} = \frac{2(4 - \sqrt{3})}{(4 + \sqrt{3})(4 - \sqrt{3})} = \frac{2(4 - \sqrt{3})}{13}$$

EXAMPLE 10 Rationalize each denominator and simplify.

a. $\dfrac{2}{1 + \sqrt{3}}$ **b.** $\dfrac{\sqrt{5} + 4}{\sqrt{5} - 1}$ **c.** $\dfrac{3}{1 + \sqrt{x}}$

Solution

a. Multiply the numerator and the denominator of this fraction by the conjugate of $1 + \sqrt{3}$, that is, by $1 - \sqrt{3}$.

$$\frac{2}{1+\sqrt{3}} = \frac{2(1-\sqrt{3})}{(1+\sqrt{3})(1-\sqrt{3})}$$
$$= \frac{2(1-\sqrt{3})}{1^2 - (\sqrt{3})^2}$$
$$= \frac{2(1-\sqrt{3})}{1-3}$$
$$= \frac{2(1-\sqrt{3})}{-2}$$
$$= -\frac{2(1-\sqrt{3})}{2} \quad \frac{a}{-b} = -\frac{a}{b}$$
$$= -1(1-\sqrt{3}) \quad \text{Simplify.}$$
$$= -1 + \sqrt{3} \quad \text{Multiply.}$$

▶ Helpful Hint
Don't forget that $(\sqrt{3})^2 = 3$.

b. $$\frac{\sqrt{5}+4}{\sqrt{5}-1} = \frac{(\sqrt{5}+4)(\sqrt{5}+1)}{(\sqrt{5}-1)(\sqrt{5}+1)}$$ Multiply the numerator and denominator by $\sqrt{5}+1$, the conjugate of $\sqrt{5}-1$.

$$= \frac{5+\sqrt{5}+4\sqrt{5}+4}{5-1}$$ Multiply.

$$= \frac{9+5\sqrt{5}}{4}$$ Simplify.

c. $$\frac{3}{1+\sqrt{x}} = \frac{3(1-\sqrt{x})}{(1+\sqrt{x})(1-\sqrt{x})}$$ Multiply the numerator and denominator by $1-\sqrt{x}$, the conjugate of $1+\sqrt{x}$.

$$= \frac{3(1-\sqrt{x})}{1-x}$$ □

PRACTICE
10 Rationalize each denominator and simplify.

a. $\frac{4}{1+\sqrt{5}}$ **b.** $\frac{\sqrt{3}+2}{\sqrt{3}-1}$ **c.** $\frac{8}{5-\sqrt{x}}$

EXAMPLE 11 Simplify $\frac{12-\sqrt{18}}{9}$.

Solution First simplify $\sqrt{18}$.

$$\frac{12-\sqrt{18}}{9} = \frac{12-\sqrt{9\cdot 2}}{9} = \frac{12-3\sqrt{2}}{9}$$

Next, factor out a common factor of 3 from the terms in the numerator and the denominator and simplify.

$$\frac{12-3\sqrt{2}}{9} = \frac{3(4-\sqrt{2})}{3\cdot 3} = \frac{4-\sqrt{2}}{3}$$ □

PRACTICE
11 Simplify $\frac{14-\sqrt{28}}{6}$.

Vocabulary, Readiness & Video Check

Fill in each blank.

1. $\sqrt{7} \cdot \sqrt{3} =$ ________
2. $\sqrt{10} \cdot \sqrt{10} =$ ________
3. $\dfrac{\sqrt{15}}{\sqrt{3}} =$ ________
4. The process of eliminating the radical in the denominator of a radical expression is called ________________.
5. The conjugate of $2 + \sqrt{3}$ is ________.

Martin-Gay Interactive Videos

See Video 8.4

Watch the section lecture video and answer the following questions.

OBJECTIVE 1 — 6. In Examples 1 and 3, the product rule for radicals is applied twice, but in different ways. Explain.

OBJECTIVE 1 — 7. Starting with Example 2, what important reminder is made repeatedly about the square root of a positive number that is squared?

OBJECTIVE 2 — 8. From Examples 6 and 7, when you're looking at a quotient of two radicals, what would make you think to apply the quotient rule in order to simplify?

OBJECTIVE 3 — 9. From the lecture before Example 8, what is the goal of rationalizing a denominator?

OBJECTIVE 4 — 10. From Example 11, why will multiplying a denominator by its conjugate rationalize the denominator?

8.4 Exercise Set

MyMathLab®

Multiply and simplify. Assume that all variables represent positive real numbers. See Examples 1, 2, 4, and 5.

1. $\sqrt{8} \cdot \sqrt{2}$
2. $\sqrt{3} \cdot \sqrt{12}$
3. $\sqrt{10} \cdot \sqrt{5}$
4. $\sqrt{2} \cdot \sqrt{14}$
5. $(\sqrt{6})^2$
6. $(\sqrt{10})^2$
7. $\sqrt{2x} \cdot \sqrt{2x}$
8. $\sqrt{5y} \cdot \sqrt{5y}$
9. $(2\sqrt{5})^2$
10. $(3\sqrt{10})^2$
11. $(6\sqrt{x})^2$
12. $(8\sqrt{y})^2$
13. $\sqrt{3x^5} \cdot \sqrt{6x}$
14. $\sqrt{21y^7} \cdot \sqrt{3y}$
15. $\sqrt{2xy^2} \cdot \sqrt{8xy}$
16. $\sqrt{18x^2y^2} \cdot \sqrt{2x^2y}$
17. $\sqrt{6}(\sqrt{5} + \sqrt{7})$
18. $\sqrt{10}(\sqrt{3} - \sqrt{7})$
19. $\sqrt{10}(\sqrt{2} + \sqrt{5})$
20. $\sqrt{6}(\sqrt{3} + \sqrt{2})$
21. $\sqrt{7y}(\sqrt{y} - 2\sqrt{7})$
22. $\sqrt{5b}(2\sqrt{b} + \sqrt{5})$
23. $(\sqrt{3} + 6)(\sqrt{3} - 6)$
24. $(\sqrt{5} + 2)(\sqrt{5} - 2)$
25. $(\sqrt{3} + \sqrt{5})(\sqrt{2} - \sqrt{5})$
26. $(\sqrt{7} + \sqrt{5})(\sqrt{2} - \sqrt{5})$
27. $(2\sqrt{11} + 1)(\sqrt{11} - 6)$
28. $(5\sqrt{3} + 2)(\sqrt{3} - 1)$
29. $(\sqrt{x} + 6)(\sqrt{x} - 6)$
30. $(\sqrt{y} + 5)(\sqrt{y} - 5)$
31. $(\sqrt{x} - 7)^2$
32. $(\sqrt{x} + 4)^2$
33. $(\sqrt{6y} + 1)^2$
34. $(\sqrt{3y} - 2)^2$

Divide and simplify. Assume that all variables represent positive real numbers. See Example 6.

35. $\dfrac{\sqrt{32}}{\sqrt{2}}$
36. $\dfrac{\sqrt{40}}{\sqrt{10}}$
37. $\dfrac{\sqrt{21}}{\sqrt{3}}$
38. $\dfrac{\sqrt{55}}{\sqrt{5}}$
39. $\dfrac{\sqrt{90}}{\sqrt{5}}$
40. $\dfrac{\sqrt{96}}{\sqrt{8}}$
41. $\dfrac{\sqrt{75y^5}}{\sqrt{3y}}$
42. $\dfrac{\sqrt{24x^7}}{\sqrt{6x}}$
43. $\dfrac{\sqrt{150}}{\sqrt{2}}$
44. $\dfrac{\sqrt{120}}{\sqrt{3}}$
45. $\dfrac{\sqrt{72y^5}}{\sqrt{3y^3}}$
46. $\dfrac{\sqrt{54x^3}}{\sqrt{2x}}$
47. $\dfrac{\sqrt{24x^3y^4}}{\sqrt{2xy}}$
48. $\dfrac{\sqrt{96x^5y^3}}{\sqrt{3x^2y}}$

Rationalize each denominator and simplify. Assume that all variables represent positive real numbers. See Example 8.

49. $\frac{\sqrt{3}}{\sqrt{5}}$ 50. $\frac{\sqrt{2}}{\sqrt{3}}$ 51. $\frac{7}{\sqrt{2}}$

52. $\frac{8}{\sqrt{11}}$ 53. $\frac{1}{\sqrt{6y}}$ 54. $\frac{1}{\sqrt{10z}}$

55. $\sqrt{\frac{3}{x}}$ 56. $\sqrt{\frac{5}{x}}$ 57. $\sqrt{\frac{1}{8}}$

58. $\sqrt{\frac{1}{27}}$ 59. $\sqrt{\frac{2}{15}}$ 60. $\sqrt{\frac{11}{14}}$

61. $\frac{8y}{\sqrt{5}}$ 62. $\frac{7x}{\sqrt{2}}$ 63. $\sqrt{\frac{y}{12x}}$

64. $\sqrt{\frac{x}{20y}}$

Rationalize each denominator and simplify. Assume that all variables represent positive real numbers. See Example 10.

65. $\frac{3}{\sqrt{2}+1}$ 66. $\frac{6}{\sqrt{5}+2}$

67. $\frac{\sqrt{5}+1}{\sqrt{6}-\sqrt{5}}$ 68. $\frac{\sqrt{3}+1}{\sqrt{3}-\sqrt{2}}$

69. $\frac{3}{\sqrt{x}-4}$ 70. $\frac{4}{\sqrt{x}-1}$

MIXED PRACTICE

Rationalize each denominator and simplify.

71. $\sqrt{\frac{3}{20}}$ 72. $\sqrt{\frac{3}{50}}$

73. $\frac{4}{2-\sqrt{5}}$ 74. $\frac{2}{1-\sqrt{2}}$

75. $\frac{3x}{\sqrt{2x}}$ 76. $\frac{5y}{\sqrt{3y}}$

77. $\frac{5}{2+\sqrt{x}}$ 78. $\frac{9}{3+\sqrt{x}}$

Simplify the following. See Example 11.

79. $\frac{6+2\sqrt{3}}{2}$ 80. $\frac{9+6\sqrt{2}}{3}$

81. $\frac{18-12\sqrt{5}}{6}$ 82. $\frac{8-20\sqrt{3}}{4}$

83. $\frac{15\sqrt{3}+5}{5}$ 84. $\frac{8+16\sqrt{2}}{8}$

Multiply or divide as indicated. See Examples 3 and 7.

85. $\sqrt[3]{12}\cdot\sqrt[3]{4}$ 86. $\sqrt[3]{9}\cdot\sqrt[3]{6}$

87. $2\sqrt[3]{5}\cdot6\sqrt[3]{2}$ 88. $8\sqrt[3]{4}\cdot7\sqrt[3]{7}$

89. $\sqrt[3]{15}\cdot\sqrt[3]{25}$ 90. $\sqrt[3]{4}\cdot\sqrt[3]{4}$

91. $\frac{\sqrt[3]{54}}{\sqrt[3]{2}}$ 92. $\frac{\sqrt[3]{80}}{\sqrt[3]{10}}$

93. $\frac{\sqrt[3]{120}}{\sqrt[3]{5}}$ 94. $\frac{\sqrt[3]{270}}{\sqrt[3]{5}}$

Rationalize each denominator. See Example 9.

95. $\sqrt[3]{\frac{5}{4}}$ 96. $\sqrt[3]{\frac{7}{9}}$

97. $\frac{6}{\sqrt[3]{2}}$ 98. $\frac{3}{\sqrt[3]{5}}$

99. $\sqrt[3]{\frac{1}{9}}$ 100. $\sqrt[3]{\frac{8}{11}}$

101. $\sqrt[3]{\frac{2}{9}}$ 102. $\sqrt[3]{\frac{3}{4}}$

REVIEW AND PREVIEW

Solve each equation. See Sections 2.4 and 5.3.

103. $x + 5 = 7^2$

104. $2y - 1 = 3^2$

105. $4z^2 + 6z - 12 = (2z)^2$

106. $16x^2 + x + 9 = (4x)^2$

107. $9x^2 + 5x + 4 = (3x + 1)^2$

108. $x^2 + 3x + 4 = (x + 2)^2$

CONCEPT EXTENSIONS

△ 109. Find the area of a rectangle whose length is $13\sqrt{2}$ meters and width is $5\sqrt{6}$ meters.

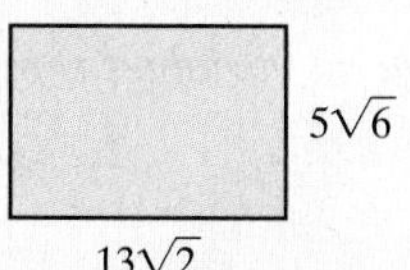

△ 110. Find the volume of a box whose length is $\sqrt{3}$ feet, width is $\sqrt{2}$ feet, and height is $\sqrt{2}$ feet.

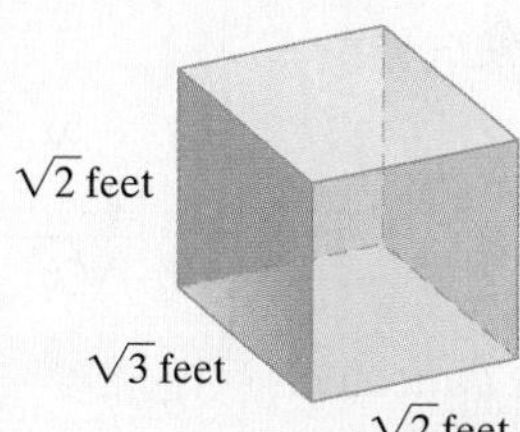

△ 111. If a circle has area A, then the formula for the radius r of the circle is

$$r = \sqrt{\frac{A}{\pi}}$$

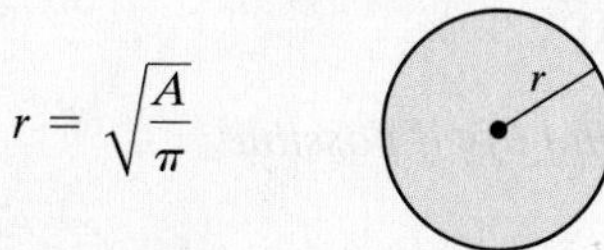

Simplify this expression by rationalizing the denominator.

△ 112. If a round ball has volume V, then the formula for the radius r of the ball is

$$r = \sqrt[3]{\frac{3V}{4\pi}}$$

Simplify this expression by rationalizing the denominator.

Identify each statement as true or false. See the Concept Check in this section.

113. $\sqrt{5} \cdot \sqrt{5} = 5$

114. $\sqrt{5} \cdot \sqrt{3} = 15$

115. $\sqrt{3x} \cdot \sqrt{3x} = 2\sqrt{3x}$

116. $\sqrt{3x} + \sqrt{3x} = 2\sqrt{3x}$

117. $\sqrt{11} + \sqrt{2} = \sqrt{13}$

118. $\sqrt{11} \cdot \sqrt{2} = \sqrt{22}$

119. When rationalizing the denominator of $\frac{\sqrt{2}}{\sqrt{3}}$, explain why both the numerator and the denominator must be multiplied by $\sqrt{3}$.

120. In your own words, explain why $\sqrt{6} + \sqrt{2}$ cannot be simplified further, but $\sqrt{6} \cdot \sqrt{2}$ can be.

121. When rationalizing the denominator of $\frac{\sqrt[3]{2}}{\sqrt[3]{3}}$, explain why both the numerator and the denominator must be multiplied by $\sqrt[3]{9}$.

122. When rationalizing the denominator of $\frac{5}{1+\sqrt{2}}$, explain why multiplying by $\frac{\sqrt{2}}{\sqrt{2}}$ will not accomplish this, but multiplying by $\frac{1-\sqrt{2}}{1-\sqrt{2}}$ will.

It is often more convenient to work with a radical expression whose numerator is rationalized. Rationalize the numerator of each expression by multiplying the numerator and denominator by the conjugate of the numerator.

123. $\frac{\sqrt{3}+1}{\sqrt{2}-1}$

124. $\frac{\sqrt{2}-2}{2-\sqrt{3}}$

Integrated Review SIMPLIFYING RADICALS

Sections 8.1–8.4

Simplify. Assume that all variables represent positive numbers.

1. $\sqrt{36}$ **2.** $\sqrt{48}$ **3.** $\sqrt{x^4}$ **4.** $\sqrt{y^7}$

5. $\sqrt{16x^2}$ **6.** $\sqrt{18x^{11}}$ **7.** $\sqrt[3]{8}$ **8.** $\sqrt[4]{81}$

9. $\sqrt[3]{-27}$ **10.** $\sqrt{-4}$ **11.** $\sqrt{\frac{11}{9}}$ **12.** $\sqrt[3]{\frac{7}{64}}$

13. $-\sqrt{16}$ **14.** $-\sqrt{25}$ **15.** $\sqrt{\frac{9}{49}}$ **16.** $\sqrt{\frac{1}{64}}$

17. $\sqrt{a^8b^2}$ **18.** $\sqrt{x^{10}y^{20}}$ **19.** $\sqrt{25m^6}$ **20.** $\sqrt{9n^{16}}$

Add or subtract as indicated.

21. $5\sqrt{7} + \sqrt{7}$ **22.** $\sqrt{50} - \sqrt{8}$

23. $5\sqrt{2} - 5\sqrt{3}$ **24.** $2\sqrt{x} + \sqrt{25x} - \sqrt{36x} + 3x$

Multiply and simplify if possible.

25. $\sqrt{2} \cdot \sqrt{15}$ **26.** $\sqrt{3} \cdot \sqrt{3}$ **27.** $(2\sqrt{7})^2$

28. $(3\sqrt{5})^2$ **29.** $\sqrt{3}(\sqrt{11} + 1)$ **30.** $\sqrt{6}(\sqrt{3} - 2)$

31. $\sqrt{8y} \cdot \sqrt{2y}$ **32.** $\sqrt{15x^2} \cdot \sqrt{3x^2}$ **33.** $(\sqrt{x} - 5)(\sqrt{x} + 2)$

34. $(3 + \sqrt{2})^2$

Divide and simplify if possible.

35. $\frac{\sqrt{8}}{\sqrt{2}}$ **36.** $\frac{\sqrt{45}}{\sqrt{15}}$ **37.** $\frac{\sqrt{24x^5}}{\sqrt{2x}}$ **38.** $\frac{\sqrt{75a^4b^5}}{\sqrt{5ab}}$

Rationalize each denominator.

39. $\sqrt{\frac{1}{6}}$ **40.** $\frac{x}{\sqrt{20}}$ **41.** $\frac{4}{\sqrt{6}+1}$ **42.** $\frac{\sqrt{2}+1}{\sqrt{x}-5}$

8.5 Solving Equations Containing Radicals 

OBJECTIVES

1. Solve Radical Equations by Using the Squaring Property of Equality Once.
2. Solve Radical Equations by Using the Squaring Property of Equality Twice.

OBJECTIVE

1 Using the Squaring Property Once 

In this section, we solve **radical equations** such as

$$\sqrt{x + 3} = 5 \quad \text{and} \quad \sqrt{2x + 1} = \sqrt{3x}$$

Radical equations contain variables in the radicand. To solve these equations, we rely on the following squaring property.

The Squaring Property of Equality

If $a = b$, then $a^2 = b^2$

Unfortunately, this squaring property does not guarantee that all solutions of the new equation are solutions of the original equation. For example, if we square both sides of the equation

$$x = 2$$

we have

$$x^2 = 4$$

This new equation has two solutions, 2 and -2, while the original equation $x = 2$ has only one solution. Thus, squaring both sides of the original equation resulted in an equation that has an **extraneous solution** that isn't a solution of the original equation. For this reason, we must **always check proposed solutions of radical equations in the original equation.** If a proposed solution does not work, we call that value an **extraneous solution.**

EXAMPLE 1 Solve: $\sqrt{x + 3} = 5$

Solution To solve this radical equation, we use the squaring property of equality and square both sides of the equation.

$$\begin{aligned} \sqrt{x + 3} &= 5 \\ (\sqrt{x + 3})^2 &= 5^2 && \text{Square both sides.} \\ x + 3 &= 25 && \text{Simplify.} \\ x &= 22 && \text{Subtract 3 from both sides.} \end{aligned}$$

Check: We replace x with 22 in the original equation.

$$\begin{aligned} \sqrt{x + 3} &= 5 && \text{Original equation} \\ \sqrt{22 + 3} &\stackrel{?}{=} 5 && \text{Let } x = 22. \\ \sqrt{25} &\stackrel{?}{=} 5 \\ 5 &= 5 && \text{True} \end{aligned}$$

Helpful Hint

Don't forget to check the proposed solutions of radical equations in the original equation.

Since a true statement results, 22 is the solution. □

PRACTICE

1 Solve: $\sqrt{x - 5} = 2$

EXAMPLE 2 Solve: $\sqrt{x} + 6 = 4$

Solution First we set the radical by itself on one side of the equation. Then we square both sides.

$$\begin{aligned} \sqrt{x} + 6 &= 4 \\ \sqrt{x} &= -2 && \text{Subtract 6 from both sides to get the radical by itself.} \end{aligned}$$

Recall that $\sqrt{x}$ is the principal or nonnegative square root of x so that $\sqrt{x}$ cannot equal -2 and thus this equation has no solution. We arrive at the same conclusion if we continue by applying the squaring property.

$$\begin{aligned} \sqrt{x} &= -2 \\ (\sqrt{x})^2 &= (-2)^2 && \text{Square both sides.} \\ x &= 4 && \text{Simplify.} \end{aligned}$$

Check: We replace x with 4 in the original equation.

$$\begin{aligned} \sqrt{x} + 6 &= 4 && \text{Original equation} \\ \sqrt{4} + 6 &\stackrel{?}{=} 4 && \text{Let } x = 4. \\ 2 + 6 &= 4 && \text{False} \end{aligned}$$

Since 4 *does not* satisfy the original equation, this equation has no solution. □

PRACTICE

2 Solve: $\sqrt{x} + 5 = 3$

Example 2 makes it very clear that we *must* check proposed solutions in the original equation to determine if they are truly solutions. Remember, if a proposed solution is not an actual solution, we say that the value is an **extraneous solution.**

The following steps can be used to solve radical equations containing square roots.

Solving a Radical Equation Containing Square Roots

Step 1. Arrange terms so that one radical is by itself on one side of the equation. That is, isolate a radical.

Step 2. Square both sides of the equation.

Step 3. Simplify both sides of the equation.

Step 4. If the equation still contains a radical term, repeat Steps 1 through 3.

Step 5. Solve the equation.

Step 6. Check all solutions in the original equation for extraneous solutions.

EXAMPLE 3 Solve: $\sqrt{x} = \sqrt{5x - 2}$.

Solution Each of the radicals is already isolated, since each is by itself on one side of the equation. So we begin solving by squaring both sides.

$$\begin{aligned} \sqrt{x} &= \sqrt{5x - 2} && \text{Original equation} \\ (\sqrt{x})^2 &= (\sqrt{5x - 2})^2 && \text{Square both sides.} \\ x &= 5x - 2 && \text{Simplify.} \\ -4x &= -2 && \text{Subtract } 5x \text{ from both sides.} \\ x &= \frac{-2}{-4} = \frac{1}{2} && \text{Divide both sides by } -4 \text{ and simplify.} \end{aligned}$$

Check: We replace x with $\frac{1}{2}$ in the original equation.

$$\begin{aligned} \sqrt{x} &= \sqrt{5x - 2} && \text{Original equation} \\ \sqrt{\frac{1}{2}} &\stackrel{?}{=} \sqrt{5 \cdot \frac{1}{2} - 2} && \text{Let } x = \frac{1}{2}. \\ \sqrt{\frac{1}{2}} &\stackrel{?}{=} \sqrt{\frac{5}{2} - 2} && \text{Multiply.} \end{aligned}$$

$$\sqrt{\frac{1}{2}} \stackrel{?}{=} \sqrt{\frac{5}{2} - \frac{4}{2}} \qquad \text{Write 2 as } \frac{4}{2}.$$

$$\sqrt{\frac{1}{2}} = \sqrt{\frac{1}{2}} \qquad \text{True}$$

This statement is true, so the solution is $\frac{1}{2}$. □

PRACTICE 3 Solve: $\sqrt{7x - 4} = \sqrt{x}$

EXAMPLE 4 Solve: $\sqrt{4y^2 + 5y - 15} = 2y$

Solution The radical is already isolated, so we start by squaring both sides.

$$\begin{aligned} \sqrt{4y^2 + 5y - 15} &= 2y \\ (\sqrt{4y^2 + 5y - 15})^2 &= (2y)^2 && \text{Square both sides.} \\ 4y^2 + 5y - 15 &= 4y^2 && \text{Simplify.} \\ 5y - 15 &= 0 && \text{Subtract } 4y^2 \text{ from both sides.} \\ 5y &= 15 && \text{Add 15 to both sides.} \\ y &= 3 && \text{Divide both sides by 5.} \end{aligned}$$

Check: We replace y with 3 in the original equation.

$$\begin{aligned} \sqrt{4y^2 + 5y - 15} &= 2y && \text{Original equation} \\ \sqrt{4 \cdot 3^2 + 5 \cdot 3 - 15} &\stackrel{?}{=} 2 \cdot 3 && \text{Let } y = 3. \\ \sqrt{4 \cdot 9 + 15 - 15} &\stackrel{?}{=} 6 && \text{Simplify.} \\ \sqrt{36} &\stackrel{?}{=} 6 \\ 6 &= 6 && \text{True} \end{aligned}$$

This statement is true, so the solution is 3. □

PRACTICE 4 Solve: $\sqrt{16y^2 + 4y - 28} = 4y$

EXAMPLE 5 Solve: $\sqrt{x + 3} - x = -3$

Solution First we isolate the radical by adding x to both sides. Then we square both sides.

$$\begin{aligned} \sqrt{x + 3} - x &= -3 \\ \sqrt{x + 3} &= x - 3 && \text{Add } x \text{ to both sides.} \\ (\sqrt{x + 3})^2 &= (x - 3)^2 && \text{Square both sides.} \\ x + 3 &= x^2 - 6x + 9 \end{aligned}$$

> **Helpful Hint**
> Don't forget that $(x - 3)^2 = (x - 3)(x - 3) = x^2 - 6x + 9$

To solve the resulting quadratic equation, we write the equation in standard form by subtracting x and 3 from both sides.

$$\begin{aligned} 3 &= x^2 - 7x + 9 && \text{Subtract } x \text{ from both sides.} \\ 0 &= x^2 - 7x + 6 && \text{Subtract 3 from both sides.} \\ 0 &= (x - 6)(x - 1) && \text{Factor.} \\ 0 = x - 6 \quad &\text{or} \quad 0 = x - 1 && \text{Set each factor equal to zero.} \\ 6 = x \quad & \qquad\;\; 1 = x && \text{Solve for } x. \end{aligned}$$

Check: We replace x with 6 and then x with 1 in the original equation.

Let $x = 6$.		Let $x = 1$.	
$\sqrt{x+3} - x = -3$		$\sqrt{x+3} - x = -3$	
$\sqrt{6+3} - 6 \stackrel{?}{=} -3$		$\sqrt{1+3} - 1 \stackrel{?}{=} -3$	
$\sqrt{9} - 6 \stackrel{?}{=} -3$		$\sqrt{4} - 1 \stackrel{?}{=} -3$	
$3 - 6 \stackrel{?}{=} -3$		$2 - 1 \stackrel{?}{=} -3$	
$-3 = -3$	True	$1 = -3$	False

Since replacing x with 1 resulted in a false statement, 1 is an extraneous solution. The only solution is 6. □

PRACTICE 5 Solve: $\sqrt{x+15} - x = -5$

OBJECTIVE 2

Using the Squaring Property Twice

If a radical equation contains two radicals, we may need to use the squaring property twice.

EXAMPLE 6 Solve: $\sqrt{x-4} = \sqrt{x} - 2$

Solution

$$\sqrt{x-4} = \sqrt{x} - 2$$

$$(\sqrt{x-4})^2 = (\sqrt{x} - 2)^2 \quad \text{Square both sides.}$$

$$x - 4 = x - 4\sqrt{x} + 4$$

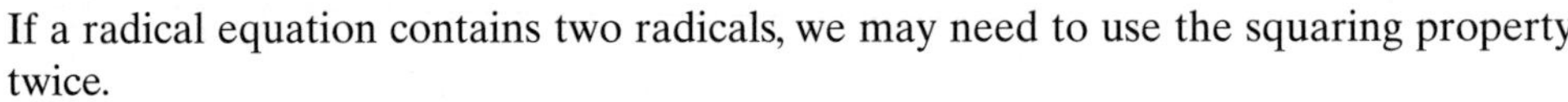

> **Helpful Hint**
>
> $(\sqrt{x} - 2)^2 = (\sqrt{x} - 2)(\sqrt{x} - 2)$
> $= \sqrt{x} \cdot \sqrt{x} - 2\sqrt{x} - 2\sqrt{x} + 4$
> $= x - 4\sqrt{x} + 4$

$$-8 = -4\sqrt{x}$$

$$2 = \sqrt{x} \quad \text{Divide both sides by } -4.$$

$$4 = x \quad \text{Square both sides again.}$$

Check the proposed solution in the original equation. The solution is 4. □

PRACTICE 6 Solve: $\sqrt{x} - 4 = \sqrt{x - 16}$

Vocabulary, Readiness & Video Check

Martin-Gay Interactive Videos

See Video 8.5

Watch the section lecture video and answer the following questions.

OBJECTIVE 1

1. From Examples 1 and 2, why must you be sure to check your proposed solution(s) in the original equation?

OBJECTIVE 2

2. Solving Example 5 requires using the squaring property twice. Is anything else done differently to solve these equations as compared to equations where the property is used only once?

8.5 Exercise Set

MyMathLab®

Solve each equation. See Examples 1 through 3.

1. $\sqrt{x} = 9$
2. $\sqrt{x} = 4$
3. $\sqrt{x + 5} = 2$
4. $\sqrt{x + 12} = 3$
5. $\sqrt{x} - 2 = 5$
6. $4\sqrt{x} - 7 = 5$
7. $3\sqrt{x} + 5 = 2$
8. $3\sqrt{x} + 8 = 5$
9. $\sqrt{x} = \sqrt{3x - 8}$
10. $\sqrt{x} = \sqrt{4x - 3}$
11. $\sqrt{4x - 3} = \sqrt{x + 3}$
12. $\sqrt{5x - 4} = \sqrt{x + 8}$

Solve each equation. See Examples 4 and 5.

13. $\sqrt{9x^2 + 2x - 4} = 3x$
14. $\sqrt{4x^2 + 3x - 9} = 2x$
15. $\sqrt{x} = x - 6$
16. $\sqrt{x} = x - 2$
17. $\sqrt{x + 7} = x + 5$
18. $\sqrt{x + 5} = x - 1$
19. $\sqrt{3x + 7} - x = 3$
20. $x = \sqrt{4x - 7} + 1$
21. $\sqrt{16x^2 + 2x + 2} = 4x$
22. $\sqrt{4x^2 + 3x + 2} = 2x$
23. $\sqrt{2x^2 + 6x + 9} = 3$
24. $\sqrt{3x^2 + 6x + 4} = 2$

Solve each equation. See Example 6.

25. $\sqrt{x - 7} = \sqrt{x} - 1$
26. $\sqrt{x - 8} = \sqrt{x} - 2$
27. $\sqrt{x} + 2 = \sqrt{x + 24}$
28. $\sqrt{x} + 5 = \sqrt{x + 55}$
29. $\sqrt{x + 8} = \sqrt{x} + 2$
30. $\sqrt{x} + 1 = \sqrt{x + 15}$

MIXED PRACTICE

Solve each equation. See Examples 1 through 6.

31. $\sqrt{2x + 6} = 4$
32. $\sqrt{3x + 7} = 5$
33. $\sqrt{x + 6} + 1 = 3$
34. $\sqrt{x + 5} + 2 = 5$
35. $\sqrt{x + 6} + 5 = 3$
36. $\sqrt{2x - 1} + 7 = 1$
37. $\sqrt{16x^2 - 3x + 6} = 4x$
38. $\sqrt{9x^2 - 2x + 8} = 3x$
39. $-\sqrt{x} = -6$
40. $-\sqrt{y} = -8$
41. $\sqrt{x + 9} = \sqrt{x} - 3$
42. $\sqrt{x} - 6 = \sqrt{x + 36}$
43. $\sqrt{2x + 1} + 3 = 5$
44. $\sqrt{3x - 1} + 1 = 4$
45. $\sqrt{x} + 3 = 7$
46. $\sqrt{x} + 5 = 10$
47. $\sqrt{4x} = \sqrt{2x + 6}$
48. $\sqrt{5x + 6} = \sqrt{8x}$
49. $\sqrt{2x + 1} = x - 7$
50. $\sqrt{2x + 5} = x - 5$
51. $x = \sqrt{2x - 2} + 1$
52. $\sqrt{2x - 4} + 2 = x$
53. $\sqrt{1 - 8x} - x = 4$
54. $\sqrt{2x + 5} - 1 = x$

REVIEW AND PREVIEW

TRANSLATING

Translate each sentence into an equation and then solve. See Section 2.5.

55. If 8 is subtracted from the product of 3 and x, the result is 19. Find x.

56. If 3 more than x is subtracted from twice x, the result is 11. Find x.

57. The length of a rectangle is twice the width. The perimeter is 24 inches. Find the length.

58. The length of a rectangle is 2 inches longer than the width. The perimeter is 24 inches. Find the length.

CONCEPT EXTENSIONS

Solve each equation.

59. $\sqrt{x - 3} + 3 = \sqrt{3x + 4}$

60. $\sqrt{2x + 3} = \sqrt{x - 2} + 2$

61. Explain why proposed solutions of radical equations must be checked in the original equation.

62. Is 8 a solution of the equation $\sqrt{x - 4} - 5 = \sqrt{x + 1}$? Explain why or why not.

63. The formula $b = \sqrt{\frac{V}{2}}$ can be used to determine the length b of a side of the base of a square-based pyramid with height 6 units and volume V cubic units.

a. Find the length of the side of the base that produces a pyramid with each volume. (Round to the nearest tenth of a unit.)

V	20	200	2000
b			

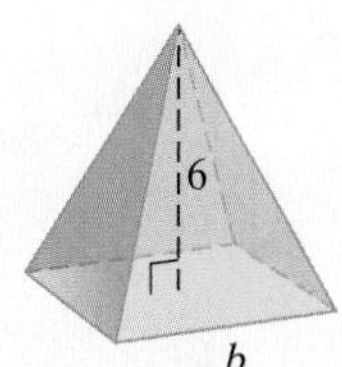

b. Notice in the table that volume V has been increased by a factor of 10 each time. Does the corresponding length b of a side increase by a factor of 10 each time also?

64. The formula $r = \sqrt{\frac{V}{2\pi}}$ can be used to determine the radius r of a cylinder with height 2 units and volume V cubic units.

a. Find the radius needed to manufacture a cylinder with each volume. (Round to the nearest tenth of a unit.)

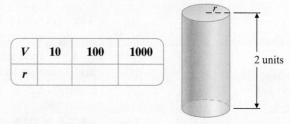

V	10	100	1000
r			

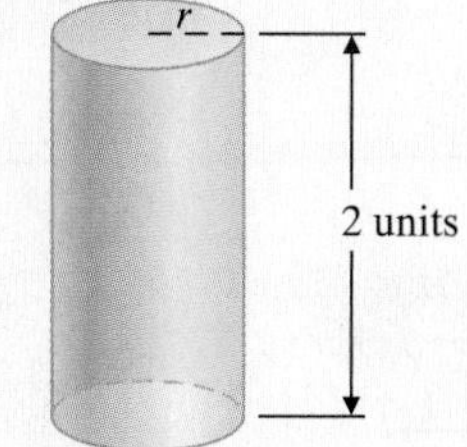

b. Notice in the table that volume V has been increased by a factor of 10 each time. Does the corresponding radius increase by a factor of 10 each time also?

Graphing calculators can be used to solve equations. To solve $\sqrt{x - 2} = x - 5$, for example, graph $y_1 = \sqrt{x - 2}$ and $y_2 = x - 5$ on the same set of axes. Use the Trace and Zoom features or an Intersect feature to find the point of intersection of the graphs. The x-value of the point is the solution of the equation. Use a graphing calculator to solve the equations below. Approximate solutions to the nearest hundredth.

65. $\sqrt{x - 2} = x - 5$
66. $\sqrt{x + 1} = 2x - 3$
67. $-\sqrt{x + 4} = 5x - 6$
68. $-\sqrt{x + 5} = -7x + 1$

8.6 Radical Equations and Problem Solving 

OBJECTIVES

1 Use the Pythagorean Formula to Solve Problems.

2 Use the Distance Formula. 

3 Solve Problems Using Formulas Containing Radicals.

OBJECTIVE

1 Using the Pythagorean Formula 

Applications of radicals can be found in geometry, finance, science, and other areas of technology. Our first application involves the Pythagorean theorem, giving a formula that relates the lengths of the three sides of a right triangle. We first studied the Pythagorean theorem in Chapter 6, and we review it here.

The Pythagorean Theorem

If a and b are lengths of the legs of a right triangle and c is the length of the hypotenuse, then $a^2 + b^2 = c^2$.

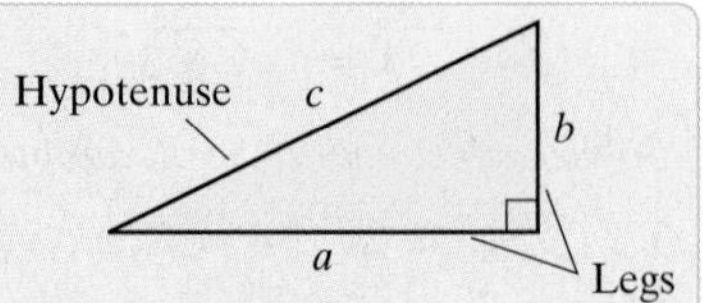

That is, the square of the length of the hypotenuse is equal to the sum of the squares of the lengths of the legs.

EXAMPLE 1 Find the length of the hypotenuse of a right triangle whose legs are 6 inches and 8 inches long.

Solution Because this is a right triangle, we use the Pythagorean theorem. We let $a = 6$ inches and $b = 8$ inches. Length c must be the length of the hypotenuse.

$$a^2 + b^2 = c^2 \quad \text{Use the Pythagorean theorem.}$$
$$6^2 + 8^2 = c^2 \quad \text{Substitute the lengths of the legs.}$$
$$36 + 64 = c^2 \quad \text{Simplify.}$$
$$100 = c^2$$

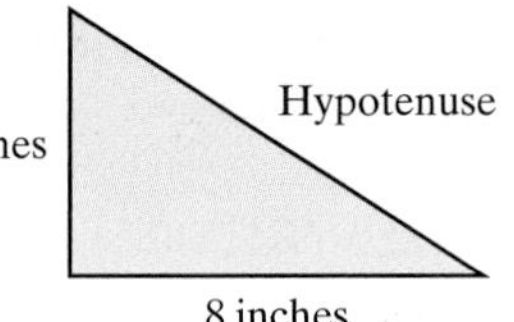

Since c represents a length, we know that c is positive and is the principal square root of 100.

$$100 = c^2$$
$$\sqrt{100} = c \quad \text{Use the definition of principal square root.}$$
$$10 = c \quad \text{Simplify.}$$

The hypotenuse has a length of 10 inches. □

PRACTICE

1 Find the length of the hypotenuse of a right triangle whose legs are 5 inches and 12 inches long.

EXAMPLE 2 Find the length of the leg of the right triangle shown. Give the exact length and a two-decimal-place approximation.

Solution We let $a = 2$ meters and b be the unknown length of the other leg. The hypotenuse is $c = 5$ meters.

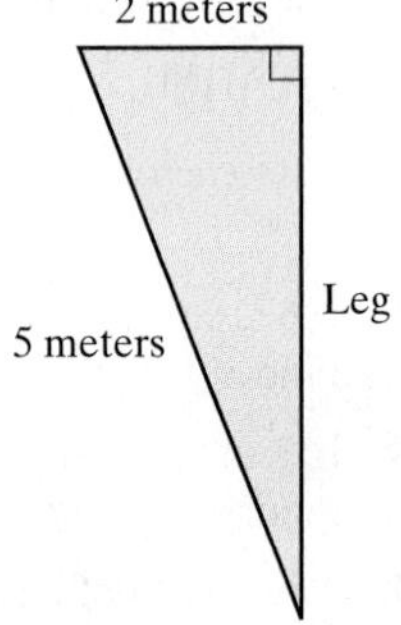

$$a^2 + b^2 = c^2 \quad \text{Use the Pythagorean theorem.}$$
$$2^2 + b^2 = 5^2 \quad \text{Let } a = 2 \text{ and } c = 5.$$
$$4 + b^2 = 25$$
$$b^2 = 21$$
$$b = \sqrt{21} \approx 4.58 \text{ meters}$$

The length of the leg is exactly $\sqrt{21}$ meters or approximately 4.58 meters. □

PRACTICE

2 Find the length of the leg of the right triangle shown. Give the exact length and a two-decimal-place approximation.

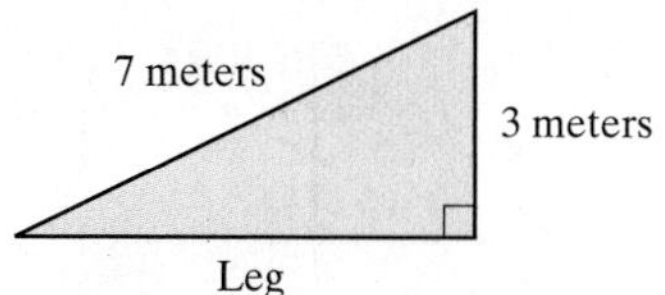

EXAMPLE 3 Finding a Distance

A surveyor must determine the distance across a lake at points P and Q as shown in the figure. To do this, she finds a third point R perpendicular to line PQ. If the length of $\overline{PR}$ is 320 feet and the length of $\overline{QR}$ is 240 feet, what is the distance across the lake? Approximate this distance to the nearest whole foot.

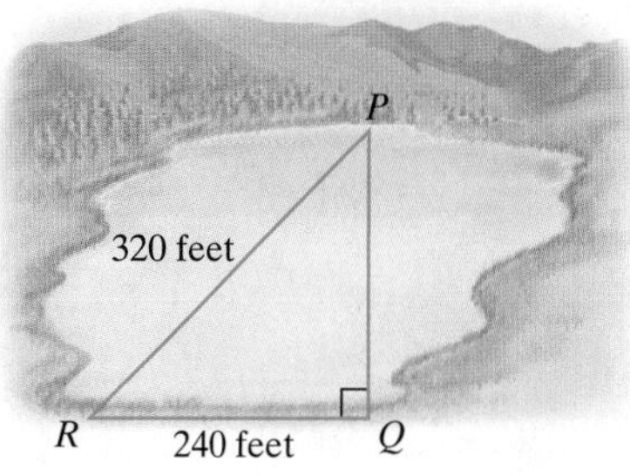

Solution

1. UNDERSTAND. Read and reread the problem. We will set up the problem using the Pythagorean theorem. By creating a line perpendicular to line PQ, the surveyor deliberately constructed a right triangle. The hypotenuse, $\overline{PR}$, has a length of 320 feet, so we let $c = 320$ in the Pythagorean theorem. The side $\overline{QR}$ is one of the legs, so we let $a = 240$ and $b =$ the unknown length.

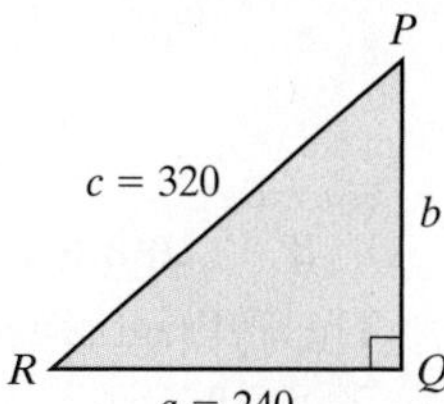

2. TRANSLATE.

$$a^2 + b^2 = c^2 \quad \text{Use the Pythagorean theorem.}$$
$$240^2 + b^2 = 320^2 \quad \text{Let } a = 240 \text{ and } c = 320.$$

3. SOLVE.

$$57{,}600 + b^2 = 102{,}400$$
$$b^2 = 44{,}800 \quad \text{Subtract 57,600 from both sides.}$$
$$b = \sqrt{44{,}800} \quad \text{Use the definition of principal square root.}$$

4. INTERPRET.

Check: See that $240^2 + (\sqrt{44{,}800})^2 = 320^2$.

State: The distance across the lake is **exactly** $\sqrt{44{,}800}$ feet. The surveyor can now use a calculator to find that $\sqrt{44{,}800}$ feet is **approximately** 211.6601 feet, so the distance across the lake is roughly 212 feet. □

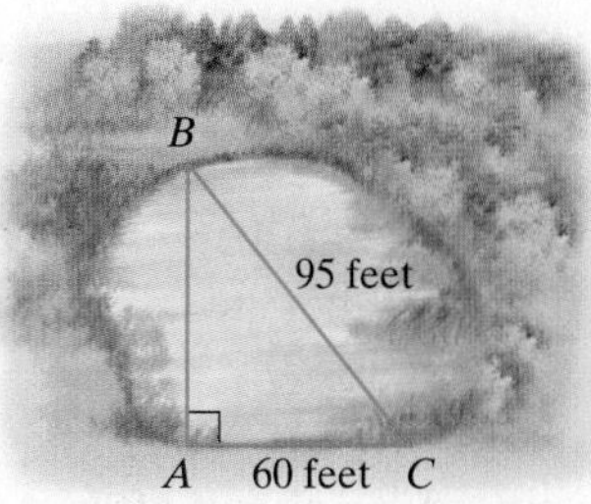

PRACTICE

3 Find the length of a bridge, to the nearest whole foot, to be constructed from point A to point B across Little Marsh. Use the distances shown in the diagram.

OBJECTIVE

2 Using the Distance Formula  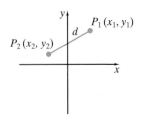

A second important application of radicals is in finding the distance between two points in the plane. By using the Pythagorean theorem, the following formula can be derived.

Distance Formula

The distance d between two points with coordinates (x_1, y_1) and (x_2, y_2) is given by

$$d = \sqrt{(x_2 - x_1)^2 + (y_2 - y_1)^2}$$

EXAMPLE 4 Find the distance between $(-1, 9)$ and $(-3, -5)$.

Solution Use the distance formula with $(x_1, y_1) = (-1, 9)$ and $(x_2, y_2) = (-3, -5)$.

$d = \sqrt{(x_2 - x_1)^2 + (y_2 - y_1)^2}$ The distance formula.

$= \sqrt{[-3 - (-1)]^2 + (-5 - 9)^2}$ Substitute known values.

$= \sqrt{(-2)^2 + (-14)^2}$ Simplify.

$= \sqrt{4 + 196}$

$= \sqrt{200} = 10\sqrt{2}$ Simplify the radical.

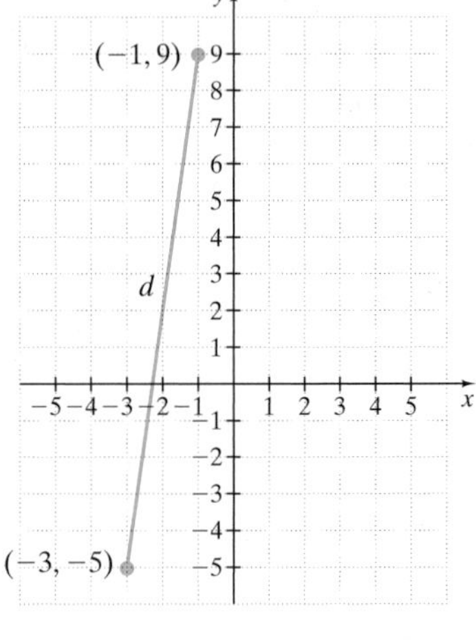

The distance is **exactly** $10\sqrt{2}$ units or **approximately** 14.1 units.

PRACTICE

4 Find the distance between $(-2, 5)$ and $(-4, -7)$.

OBJECTIVE

3 Using Formulas Containing Radicals 

The Pythagorean theorem is an extremely important result in mathematics and should be memorized. But there are other applications involving formulas containing radicals that are not quite as well known, such as the velocity formula used in the next example.

EXAMPLE 5 **Determining Velocity**

A formula used to determine the velocity v, in feet per second, of an object (neglecting air resistance) after it has fallen a certain height is $v = \sqrt{2gh}$, where g is the acceleration due to gravity, and h is the height the object has fallen. On Earth, the acceleration g due to gravity is approximately 32 feet per second per second. Find the velocity of a person after falling 5 feet.

Solution We are told that $g = 32$ feet per second per second. To find the velocity v when $h = 5$ feet, we use the velocity formula.

$v = \sqrt{2gh}$ Use the velocity formula.

$= \sqrt{2 \cdot 32 \cdot 5}$ Substitute known values.

$= \sqrt{320}$

$= 8\sqrt{5}$ Simplify the radicand.

The velocity of the person after falling 5 feet is **exactly** $8\sqrt{5}$ feet per second, or **approximately** 17.9 feet per second.

PRACTICE

5 Use the formula in Example 5 to find the velocity of an object after it has fallen 12 feet.

Vocabulary, Readiness & Video Check

Martin-Gay Interactive Videos

See Video 8.6

Watch the section lecture video and answer the following questions.

OBJECTIVE 1

1. From Examples 1 and 2, when solving exercises using the Pythagorean theorem, what two things must you keep in mind?

OBJECTIVE 1

2. What very important point is made as the final answer to Example 1 is being found?

OBJECTIVE 2

3. From Examples 3 and 4, this distance formula uses the coordinates of two points, similar to the slope formula. Therefore, how do you think you need to be careful when replacing values in the formula?

OBJECTIVE 3

4. In Examples 5 and 6, how do you know to give an estimated answer instead of an exact answer? In what form would the exact answer be given?

8.6 Exercise Set

MyMathLab®

Use the Pythagorean theorem to find the unknown side of each right triangle. See Examples 1 and 2.

△ **1.**
3
2

△ **2.**
3
5

△ **3.**

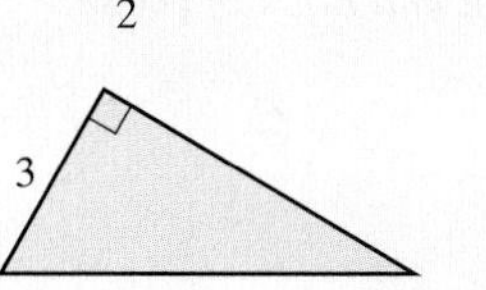

△ **4.**

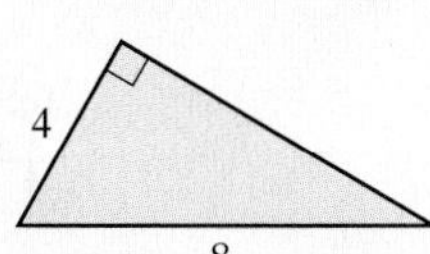

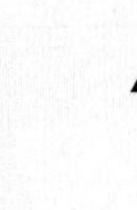

△ **5.**

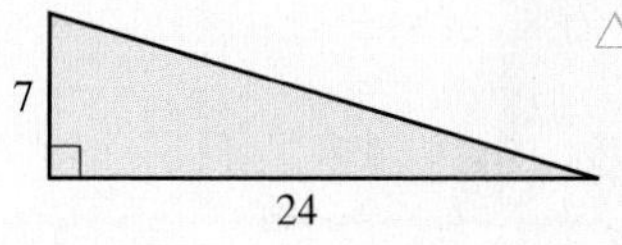

△ **6.**

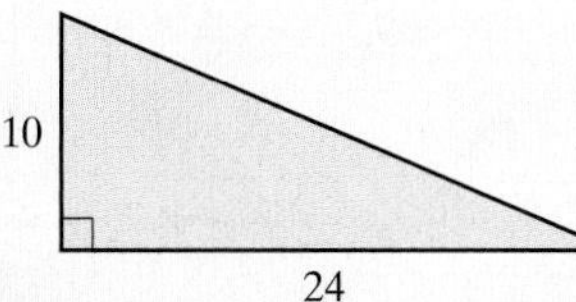

△ **7.**

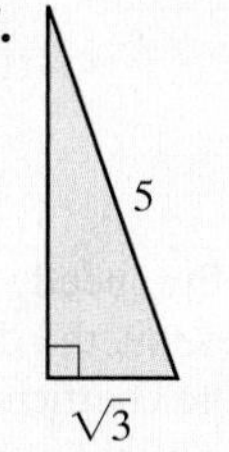

△ **8.**

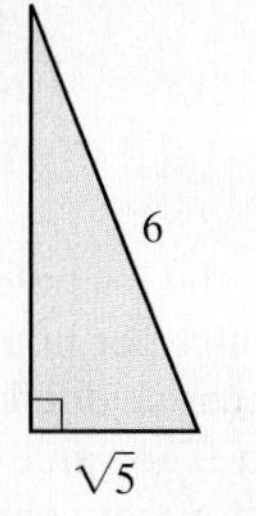

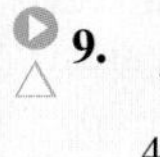

△ **9.**

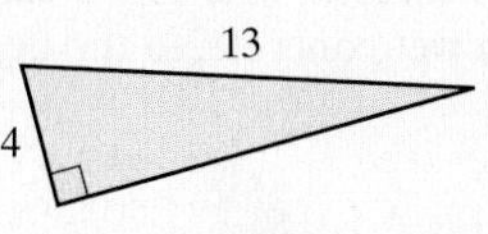

△ **10.**

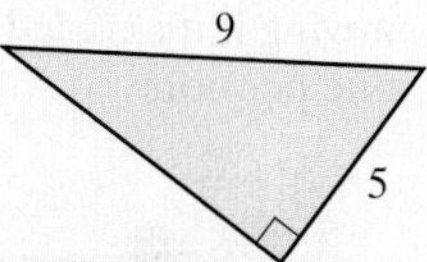

Find the length of the unknown side of each right triangle with sides a, b, and c, where c is the hypotenuse. See Examples 1 and 2.

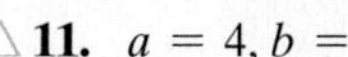

△ **11.** $a = 4, b = 5$

△ **12.** $a = 2, b = 7$

△ **13.** $b = 2, c = 6$

△ **14.** $b = 1, c = 5$

△ **15.** $a = \sqrt{10}, c = 10$

△ **16.** $a = \sqrt{7}, c = \sqrt{35}$

Solve. See Examples 1 through 5.

△ **17.** Evan and Noah Saacks want to determine the distance at certain points across a pond on their property. They are able to measure the distances shown on the following diagram. Find how wide the pond is to the nearest tenth of a foot.

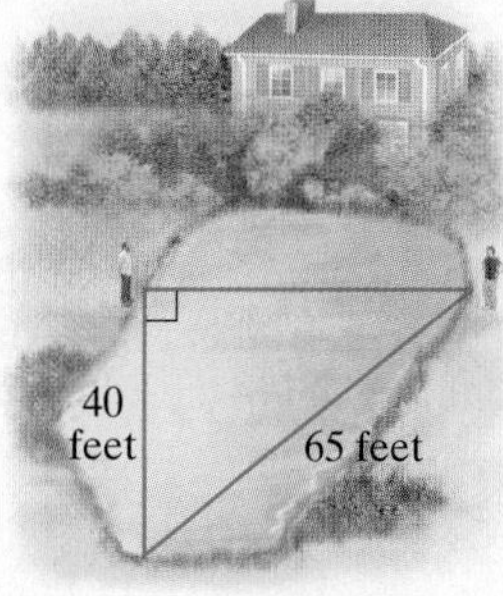

△ **18.** Use the formula from Example 5 and find the velocity of an object after it has fallen 20 feet.

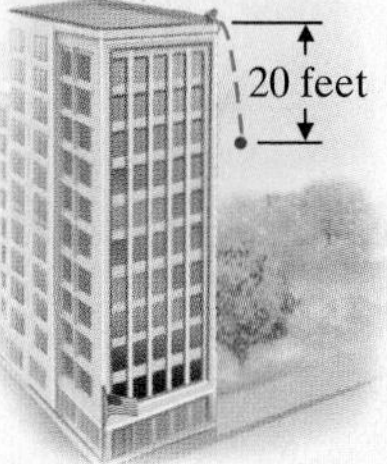

19. A wire is used to anchor a 20-foot-high pole. One end of the wire is attached to the top of the pole. The other end is fastened to a stake five feet away from the bottom of the pole. Find the length of the wire, to the nearest tenth of a foot.

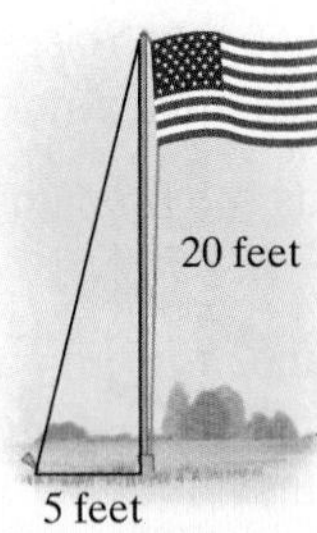

20. Jim Spivey needs to connect two underground pipelines, which are offset by 3 feet, as pictured in the diagram. Neglecting the joints needed to join the pipes, find the length of the shortest possible connecting pipe rounded to the nearest hundredth of a foot.

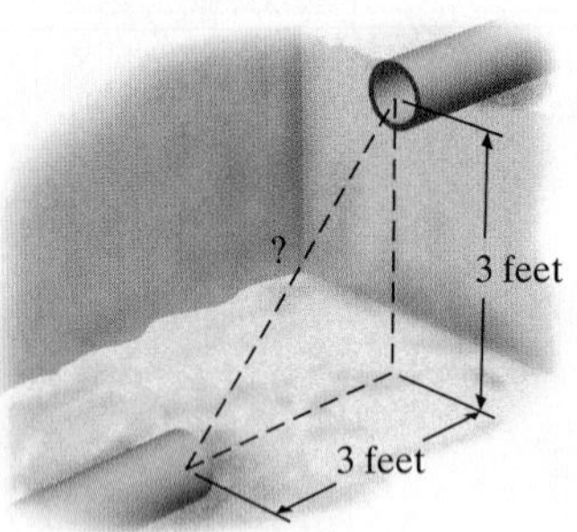

21. Robert Weisman needs to attach a diagonal brace to a rectangular frame to make it structurally sound. If the framework is 6 feet by 10 feet, find how long the brace needs to be to the nearest tenth of a foot.

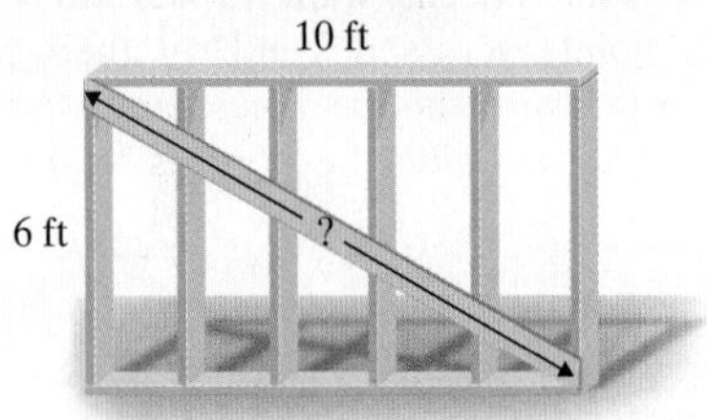

22. Elizabeth Kaster is flying a kite. She let out 80 feet of string and attached the string to a stake in the ground. The kite is now directly above her brother Mike, who is 32 feet away from Elizabeth. Find the height of the kite to the nearest foot.

Use the distance formula to find the distance between the points given. See Example 4.

23. $(3, 6), (5, 11)$
24. $(2, 3), (9, 7)$
25. $(-3, 1), (5, -2)$
26. $(-2, 6), (3, -2)$
27. $(3, -2), (1, -8)$
28. $(-5, 8), (-2, 2)$
29. $\left(\frac{1}{2}, 2\right), (2, -1)$
30. $\left(\frac{1}{3}, 1\right), (1, -1)$
31. $(3, -2), (5, 7)$
32. $(-2, -3), (-1, 4)$

Solve each problem. See Example 5.

33. For a square-based pyramid, the formula $b = \sqrt{\frac{3V}{h}}$ describes the relationship between the length b of one side of the base, the volume V, and the height h. Find the volume if each side of the base is 6 feet long, and the pyramid is 2 feet high.

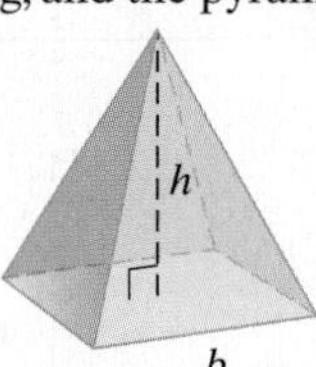

34. The formula $t = \frac{\sqrt{d}}{4}$ relates the distance d, in feet, that an object falls in t seconds, assuming that air resistance does not slow down the object. Find how long, to the nearest hundredth of a second, it takes an object to reach the ground from the top of the Willis Tower in Chicago, a distance of 1450 feet. (*Source: World Almanac and Book of Facts*)

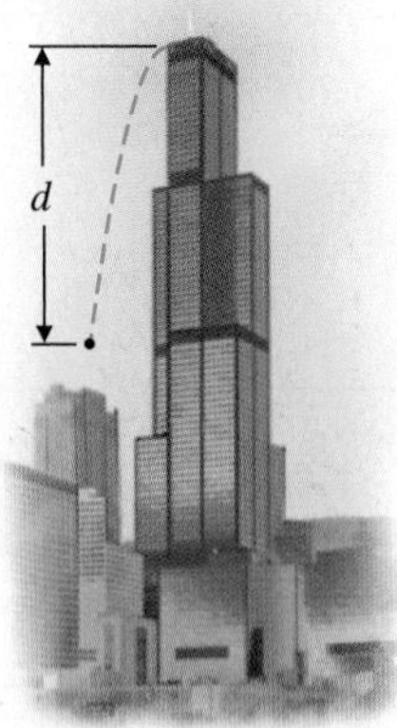

35. Police use the formula $s = \sqrt{30fd}$ to estimate the speed s of a car in miles per hour. In this formula, d represents the distance the car skidded in feet and f represents the coefficient of friction. The value of f depends on the type of road surface, and for wet concrete f is 0.35. Find how fast a car was moving if it skidded 280 feet on wet concrete, to the nearest mile per hour.

36. The coefficient of friction of a certain dry road is 0.95. Use the formula in Exercise 35 to find how far a car will skid on this dry road if it is traveling at a rate of 60 mph. Round the length to the nearest foot.

37. The formula $v = \sqrt{2.5r}$ can be used to estimate the maximum safe velocity, v, in miles per hour, at which a car can travel if it is driven along a curved road with a **radius of curvature,** r, in feet. To the nearest whole number, find the maximum safe speed if a cloverleaf exit on an expressway has a radius of curvature of 300 feet.

38. Use the formula from Exercise 37 to find the radius of curvature if the safe velocity is 30 mph.

39. The maximum distance d in kilometers that you can see from a height of h meters is given by $d = 3.5\sqrt{h}$. Find how far you can see from the top of the Bank One Tower in Indianapolis, a height of 285.4 meters. Round to the nearest tenth of a kilometer. (*Source: World Almanac and Book of Facts*)

40. Use the formula from Exercise 39 to find how far you can see from the top of the Chase Tower Building in Houston, Texas, a height of 305 meters. Round to the nearest tenth of a kilometer. (*Source:* Council on Tall Buildings and Urban Habitat)

41. Use the formula from Exercise 39 to find how far you can see from the top of the First Interstate Tower in Houston, Texas, a height of 295.7 meters. Round to the nearest tenth of a kilometer. (*Source:* Council on Tall Buildings and Urban Habitat)

42. Use the formula from Exercise 39 to find how far you can see from the top of the Gas Company Tower in Los Angeles, California, a height of 228.3 m. Round to the nearest tenth of a kilometer. (*Source:* Council on Tall Buildings and Urban Habitat)

REVIEW AND PREVIEW

Simplify using rules for exponents. See Section 5.1.

43. 2^5

44. $(-3)^3$

45. $\left(-\frac{1}{5}\right)^2$

46. $\left(\frac{2}{7}\right)^3$

47. $x^2 \cdot x^3$

48. $x^4 \cdot x^2$

49. $y^3 \cdot y$

50. $x \cdot x^7$

CONCEPT EXTENSIONS

For each triangle, find the length of y, then x.

51.

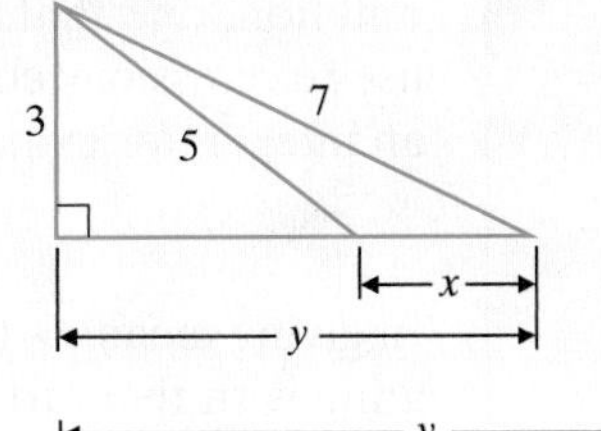

52.

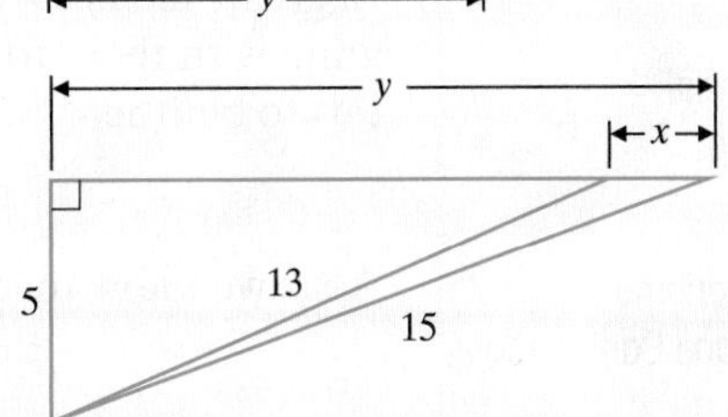

Solve.

53. Mike and Sandra Hallahan leave the seashore at the same time. Mike drives northward at a rate of 30 miles per hour while Sandra drives west at 60 mph. Find how far apart they are after 3 hours to the nearest mile.

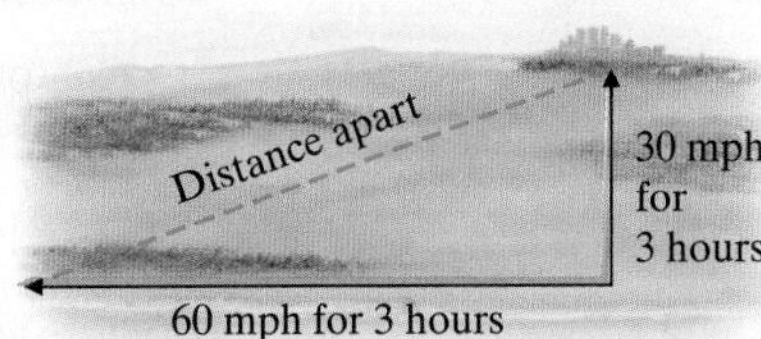

54. Railroad tracks are invariably made up of relatively short sections of rail connected by expansion joints. To see why this construction is necessary, consider a single rail 100 feet long (or 1200 inches). On an extremely hot day, suppose it expands 1 inch in the hot sun to a new length of 1201 inches. Theoretically, the track would bow upward as pictured.

Let us approximate the bulge in the railroad this way. Calculate the height h of the bulge.

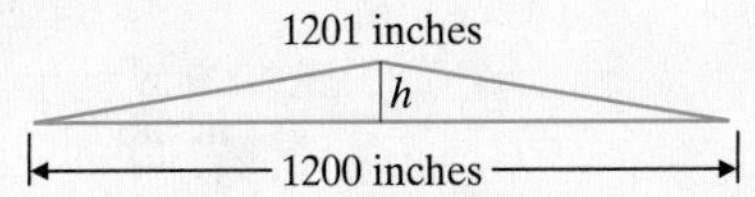

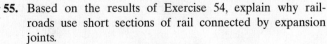

55. Based on the results of Exercise 54, explain why railroads use short sections of rail connected by expansion joints.

8.7 Rational Exponents 

OBJECTIVES

1 Evaluate Exponential Expressions of the Form $a^{1/n}$.

2 Evaluate Exponential Expressions of the Form $a^{m/n}$.

3 Evaluate Exponential Expressions of the Form $a^{-m/n}$.

4 Use Rules for Exponents to Simplify Expressions Containing Fractional Exponents.

OBJECTIVE

1 Evaluating $a^{1/n}$

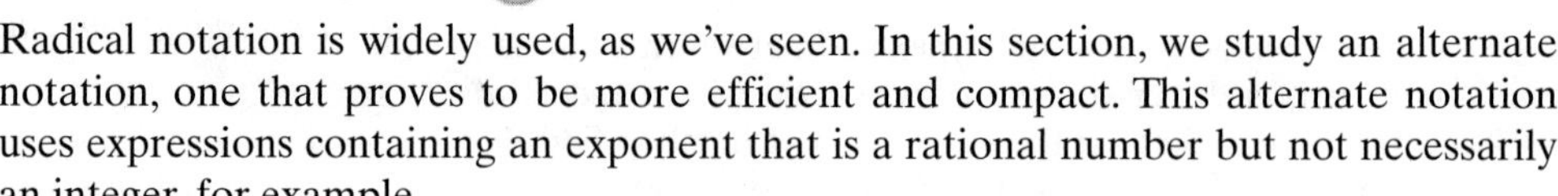

Radical notation is widely used, as we've seen. In this section, we study an alternate notation, one that proves to be more efficient and compact. This alternate notation uses expressions containing an exponent that is a rational number but not necessarily an integer, for example,

$$3^{1/2}, 2^{-3/4}, \text{ and } y^{5/6}$$

In giving meaning to rational exponents, keep in mind that we want the rules for operating with them to be the same as the rules for operating with integer exponents. For this to be true,

$$(3^{1/2})^2 = 3^{1/2 \cdot 2} = 3^1 = 3$$

Also, we know that

$$(\sqrt{3})^2 = 3$$

Since the square of both $3^{1/2}$ and $\sqrt{3}$ is 3, it would be reasonable to say that

$$3^{1/2} \text{ means } \sqrt{3}$$

In general, we have the following.

Definition of $a^{1/n}$

If n is a positive integer and $\sqrt[n]{a}$ is a real number, then

$$a^{1/n} = \sqrt[n]{a}$$

Notice that the denominator of the rational exponent is the same as the index of the corresponding radical.

EXAMPLE 1 Write in radical notation. Then simplify.

a. $25^{1/2}$ **b.** $8^{1/3}$ **c.** $-16^{1/4}$ **d.** $(-27)^{1/3}$ **e.** $\left(\frac{1}{9}\right)^{1/2}$

Solution

a. $25^{1/2} = \sqrt{25} = 5$

b. $8^{1/3} = \sqrt[3]{8} = 2$

c. In $-16^{1/4}$, the base of the exponent is 16. Thus **the negative sign is not affected by the exponent;** so $-16^{1/4} = -\sqrt[4]{16} = -2$.

d. The parentheses show that -27 is the base. $(-27)^{1/3} = \sqrt[3]{-27} = -3$.

e. $\left(\frac{1}{9}\right)^{1/2} = \sqrt{\frac{1}{9}} = \frac{1}{3}$ □

PRACTICE

1 Write in radical notation. Then simplify.

a. $36^{1/2}$ **b.** $125^{1/3}$ **c.** $-\left(\frac{1}{81}\right)^{1/4}$ **d.** $(-1000)^{1/3}$ **e.** $32^{1/5}$

OBJECTIVE

2 Evaluating $a^{m/n}$

In Example 1, each rational exponent has a numerator of 1. What happens if the numerator is some other positive integer? Consider $8^{2/3}$. Since $\frac{2}{3}$ is the same as $\frac{1}{3}\cdot 2$, we reason that

$$8^{2/3} = 8^{(1/3)2} = (8^{1/3})^2 = (\sqrt[3]{8})^2 = 2^2 = 4$$

The denominator 3 of the rational exponent is the same as the index of the radical. The numerator 2 of the fractional exponent indicates that the radical base is to be squared.

Definition of $a^{m/n}$

If m and n are integers with $n > 0$ and if a is a positive number, then

$$a^{m/n} = (a^{1/n})^m = (\sqrt[n]{a})^m$$

Also,

$$a^{m/n} = (a^m)^{1/n} = \sqrt[n]{a^m}$$

EXAMPLE 2 Simplify each expression.

a. $4^{3/2}$ **b.** $27^{2/3}$ **c.** $-16^{3/4}$

Solution

a. $4^{3/2} = (4^{1/2})^3 = (\sqrt{4})^3 = 2^3 = 8$

b. $27^{2/3} = (27^{1/3})^2 = (\sqrt[3]{27})^2 = 3^2 = 9$

c. The negative sign is **not** affected by the exponent since the base of the exponent is 16. Thus, $-16^{3/4} = -(16^{1/4})^3 = -(\sqrt[4]{16})^3 = -2^3 = -8$. □

PRACTICE

2 Simplify each expression.

a. $9^{3/2}$ **b.** $8^{5/3}$ **c.** $-625^{1/4}$

Helpful Hint

Recall that

$$-3^2 = -(3\cdot 3) = -9$$

and

$$(-3)^2 = (-3)(-3) = 9$$

In other words, without parentheses the exponent 2 applies to the base 3, **not** -3. The same is true of rational exponents. For example,

$$-16^{1/2} = -\sqrt{16} = -4$$

and

$$(-27)^{1/3} = \sqrt[3]{-27} = -3$$

OBJECTIVE

3 Evaluating $a^{-m/n}$

If the exponent is a negative rational number, use the following definition.

Definition of $a^{-m/n}$

If $a^{-m/n}$ is a nonzero real number, then

$$a^{-m/n} = \frac{1}{a^{m/n}}$$

EXAMPLE 3 Write each expression with a positive exponent and then simplify.

a. $36^{-1/2}$ **b.** $16^{-3/4}$ **c.** $-9^{1/2}$ **d.** $32^{-4/5}$

Solution

a. $36^{-1/2} = \frac{1}{36^{1/2}} = \frac{1}{\sqrt{36}} = \frac{1}{6}$

b. $16^{-3/4} = \frac{1}{16^{3/4}} = \frac{1}{(\sqrt[4]{16})^3} = \frac{1}{2^3} = \frac{1}{8}$

c. $-9^{1/2} = -\sqrt{9} = -3$

d. $32^{-4/5} = \frac{1}{32^{4/5}} = \frac{1}{(\sqrt[5]{32})^4} = \frac{1}{2^4} = \frac{1}{16}$ □

PRACTICE

3 Write each expression with a positive exponent and then simplify.

a. $25^{-1/2}$ **b.** $1000^{-2/3}$ **c.** $-49^{1/2}$ **d.** $1024^{-2/5}$

OBJECTIVE

4 Using Exponent Rules with Fractional Exponents 

It can be shown that the properties of integer exponents hold for rational exponents. By using these properties and definitions, we can now simplify products and quotients of expressions containing rational exponents.

EXAMPLE 4 Simplify each expression. Write results with positive exponents only. Assume that all variables represent positive numbers.

a. $3^{1/2} \cdot 3^{3/2}$ **b.** $\frac{5^{1/3}}{5^{2/3}}$ **c.** $(x^{1/4})^{12}$ **d.** $\frac{x^{1/5}}{x^{-4/5}}$ **e.** $\left(\frac{y^{3/5}}{z^{1/4}}\right)^2$

Solution

a. $3^{1/2} \cdot 3^{3/2} = 3^{(1/2)+(3/2)} = 3^{4/2} = 3^2 = 9$

b. $\frac{5^{1/3}}{5^{2/3}} = 5^{(1/3)-(2/3)} = 5^{-1/3} = \frac{1}{5^{1/3}}$

c. $(x^{1/4})^{12} = x^{(1/4)12} = x^3$

d. $\frac{x^{1/5}}{x^{-4/5}} = x^{(1/5)-(-4/5)} = x^{5/5} = x^1$ or x

e. $\left(\frac{y^{3/5}}{z^{1/4}}\right)^2 = \frac{y^{(3/5)2}}{z^{(1/4)2}} = \frac{y^{6/5}}{z^{1/2}}$ □

PRACTICE

4 Simplify each expression. Write results with positive exponents only. Assume that all variables represent positive numbers.

a. $6^{3/5} \cdot 6^{7/5}$ **b.** $\frac{7^{1/6}}{7^{3/6}}$ **c.** $(z^{3/8})^{16}$ **d.** $\frac{a^{3/7}}{a^{-4/7}}$ **e.** $\left(\frac{x^{5/8}}{y^{2/3}}\right)^{12}$

Vocabulary, Readiness & Video Check

Martin-Gay Interactive Videos

See Video 8.7

Watch the section lecture video and answer the following questions.

OBJECTIVE 1

1. From Examples 3 and 4, what do you think is the value of $(-16)^{1/4}$ and why?

OBJECTIVE 2

2. From Examples 6–8, in a fractional exponent, what do the numerator and denominator each represent in radical form?

OBJECTIVE 3

3. Based on Example 9, complete the following statement. A negative fractional exponent will move a base from the numerator to the _______ with the fractional exponent becoming _______.

OBJECTIVE 4

4. Based on Examples 11–13, complete the following statements. Assume you have an expression with fractional exponents. If applying the product rule of exponents, you would _______ the exponents. If applying the quotient rule of exponents, you would _______ the exponents. If applying the power rule of exponents, you would _______ the exponents.

8.7 Exercise Set

MyMathLab®  

Simplify each expression. See Examples 1 and 2.

1. $8^{1/3}$
2. $16^{1/4}$
3. $9^{1/2}$
4. $16^{1/2}$
5. $16^{3/4}$
6. $27^{4/3}$
7. $32^{2/5}$
8. $-64^{5/6}$

Simplify each expression. See Example 3.

9. $-16^{-1/4}$
10. $-8^{-1/3}$
11. $16^{-3/2}$
12. $27^{-4/3}$
13. $81^{-3/2}$
14. $32^{-2/5}$
15. $\left(\frac{4}{25}\right)^{-1/2}$
16. $\left(\frac{8}{27}\right)^{-1/3}$
17. Explain the meaning of the numbers 2, 3, and 4 in the exponential expression $4^{3/2}$.
18. Explain why $-4^{1/2}$ is a real number but $(-4)^{1/2}$ is not.

Simplify each expression. Write each answer with positive exponents. Assume that all variables represent positive numbers. See Example 4.

19. $2^{1/3} \cdot 2^{2/3}$
20. $4^{2/5} \cdot 4^{3/5}$
21. $\frac{4^{3/4}}{4^{1/4}}$
22. $\frac{9^{7/2}}{9^{3/2}}$
23. $\frac{x^{1/6}}{x^{5/6}}$
24. $\frac{x^{1/4}}{x^{3/4}}$
25. $(x^{1/2})^6$
26. $(x^{1/3})^6$
27. Explain how simplifying $x^{1/2} \cdot x^{1/3}$ is similar to simplifying $x^2 \cdot x^3$.
28. Explain how simplifying $(x^{1/2})^{1/3}$ is similar to simplifying $(x^2)^3$.

MIXED PRACTICE

Simplify each expression.

29. $81^{1/2}$
30. $(-32)^{6/5}$
31. $(-8)^{1/3}$
32. $36^{1/2}$
33. $-81^{1/4}$
34. $-64^{1/3}$
35. $\left(\frac{1}{81}\right)^{1/2}$
36. $\left(\frac{9}{16}\right)^{1/2}$
37. $\left(\frac{27}{64}\right)^{1/3}$
38. $\left(\frac{16}{81}\right)^{1/4}$
39. $9^{3/2}$
40. $16^{3/2}$
41. $64^{3/2}$
42. $64^{2/3}$
43. $-8^{2/3}$
44. $8^{2/3}$
45. $4^{5/2}$
46. $9^{4/2}$
47. $\left(\frac{4}{9}\right)^{3/2}$
48. $\left(\frac{8}{27}\right)^{2/3}$
49. $\left(\frac{1}{81}\right)^{3/4}$
50. $\left(\frac{1}{32}\right)^{3/5}$
51. $4^{-1/2}$
52. $9^{-1/2}$
53. $125^{-1/3}$
54. $216^{-1/3}$
55. $625^{-3/4}$
56. $256^{-5/8}$

Simplify each expression. Write each answer with positive exponents. Assume that all variables represent positive numbers.

57. $3^{4/3} \cdot 3^{2/3}$
58. $2^{5/4} \cdot 2^{3/4}$
59. $\frac{6^{2/3}}{6^{1/3}}$
60. $\frac{3^{3/5}}{3^{1/5}}$
61. $(x^{2/3})^9$
62. $(x^6)^{3/4}$

63. $\dfrac{6^{1/3}}{6^{-5/3}}$

64. $\dfrac{2^{-3/4}}{2^{5/4}}$

65. $\dfrac{3^{-3/5}}{3^{2/5}}$

66. $\dfrac{5^{1/4}}{5^{-3/4}}$

67. $\left(\dfrac{x^{1/3}}{y^{3/4}}\right)^2$

68. $\left(\dfrac{x^{1/2}}{y^{2/3}}\right)^6$

69. $\left(\dfrac{x^{2/5}}{y^{3/4}}\right)^8$

70. $\left(\dfrac{x^{3/4}}{y^{1/6}}\right)^3$

REVIEW AND PREVIEW

Solve each system of linear inequalities by graphing on a single coordinate system. See Section 4.5.

71. $\begin{cases} x + y < 6 \\ y \ge 2x \end{cases}$

72. $\begin{cases} 2x - y \ge 3 \\ x < 5 \end{cases}$

Solve each quadratic equation. See Section 6.6.

73. $x^2 - 4 = 3x$

74. $x^2 + 2x = 8$

75. $2x^2 - 5x - 3 = 0$

76. $3x^2 + x - 2 = 0$

CONCEPT EXTENSIONS

77. The formula $P = P_O(1.08)^N$ can be used to estimate the total population of a certain city P after N years have passed, assuming the original population is P_O. Find the population of this city after $1\frac{1}{2}$ years if the original population is 10,000 people.

78. Money grows in a certain savings account at a rate of 4% compounded annually. The amount of money A in the account at time t is given by the formula

$$A = P(1.04)^t$$

where P is the original amount deposited in the account. Find the amount of money in the account after $3\frac{3}{4}$ years if \$200 was initially deposited.

Use a calculator and approximate each to three decimal places.

79. $5^{3/4}$

80. $20^{1/8}$

81. $18^{3/5}$

82. $42^{3/10}$

Chapter 8 Vocabulary Check

Fill in each blank with one of the words or phrases listed below. Not all choices will be used.

index	radicand	like radicals
rationalizing the denominator	conjugate	leg
principal square root	radical	hypotenuse

1. The expressions $5\sqrt{x}$ and $7\sqrt{x}$ are examples of ____________.
2. In the expression $\sqrt[3]{45}$, the number 3 is the ____________, the number 45 is the ____________, and $\sqrt{}$ is called the ____________ sign.
3. The ____________ of $a + b$ is $a - b$.
4. The ____________ of 25 is 5.
5. The process of eliminating the radical in the denominator of a radical expression is called ____________.
6. The Pythagorean theorem states that for a right triangle, $(\text{leg})^2 + (\text{leg})^2 = ($____________$)^2$.

Chapter 8 Highlights

DEFINITIONS AND CONCEPTS	EXAMPLES
Section 8.1 Introduction to Radicals	
The **positive** or **principal square root** of a positive number a is written as $\sqrt{a}$. The **negative square root** of a is written as $-\sqrt{a}$. $\sqrt{a} = b$ only if $b^2 = a$ and $b > 0$.	$\sqrt{25} = 5$ $\quad \sqrt{100} = 10$ $-\sqrt{9} = -3$ $\quad \sqrt{\dfrac{4}{49}} = \dfrac{2}{7}$

DEFINITIONS AND CONCEPTS | **EXAMPLES**

Section 8.1 Introduction to Radicals (continued)

A square root of a negative number is not a real number.

$\sqrt{-4}$ is not a real number.

The **cube root** of a real number a is written as $\sqrt[3]{a}$ and $\sqrt[3]{a} = b$ only if $b^3 = a$.

$\sqrt[3]{64} = 4 \qquad \sqrt[3]{-8} = -2$

The ***n*th root** of a number a is written as $\sqrt[n]{a}$ and $\sqrt[n]{a} = b$ only if $b^n = a$.

$\sqrt[4]{81} = 3 \qquad \sqrt[5]{-32} = -2$

In $\sqrt[n]{a}$, the natural number n is called the **index,** the symbol $\sqrt{}$ is called a **radical,** and the expression within the radical is called the **radicand.**

index

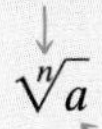

radicand

(*Note:* If the index is even, the radicand must be nonnegative for the root to be a real number.)

Section 8.2 Simplifying Radicals

Product rule for radicals

If $\sqrt[n]{a}$ and $\sqrt[n]{b}$ are real numbers, then $\sqrt[n]{a} \cdot \sqrt[n]{b} = \sqrt[n]{a \cdot b}$.

$\sqrt{2} \cdot \sqrt{3} = \sqrt{6}$

$\sqrt[3]{7} \cdot \sqrt[3]{2} = \sqrt[3]{14}$

A square root is in **simplified form** if the radicand contains no perfect square factors other than 1. To simplify a square root, factor the radicand so that one of its factors is a perfect square factor.

$$\begin{aligned}\sqrt{45} &= \sqrt{9 \cdot 5} \\ &= \sqrt{9} \cdot \sqrt{5} \\ &= 3\sqrt{5} \quad \text{in simplest form.}\end{aligned}$$

To simplify cube roots, factor the radicand so that one of its factors is a perfect cube.

$$\begin{aligned}\sqrt[3]{48} &= \sqrt[3]{8 \cdot 6} \\ &= \sqrt[3]{8} \cdot \sqrt[3]{6} \\ &= 2\sqrt[3]{6}\end{aligned}$$

Quotient rule for radicals

If $\sqrt[n]{a}$ and $\sqrt[n]{b}$ are real numbers and $b \neq 0$, then

$$\sqrt[n]{\frac{a}{b}} = \frac{\sqrt[n]{a}}{\sqrt[n]{b}}$$

$$\sqrt{\frac{18}{x^6}} = \frac{\sqrt{9 \cdot 2}}{\sqrt{x^6}} = \frac{\sqrt{9} \cdot \sqrt{2}}{x^3} = \frac{3\sqrt{2}}{x^3}$$

$$\sqrt[3]{\frac{18}{x^6}} = \frac{\sqrt[3]{18}}{\sqrt[3]{x^6}} = \frac{\sqrt[3]{18}}{x^2}$$

Section 8.3 Adding and Subtracting Radicals

Like radicals are radical expressions that have the same index and the same radicand.

Like Radicals

$5\sqrt{2}, -7\sqrt{2}, \sqrt{2}$ Also, $-\sqrt[3]{11}, 3\sqrt[3]{11}$

To combine like radicals, use the distributive property.

$2\sqrt{7} - 13\sqrt{7} = (2 - 13)\sqrt{7} = -11\sqrt{7}$

$$\begin{aligned}&\sqrt[3]{24} + \sqrt[3]{8} + \sqrt[3]{81} \\ &= \sqrt[3]{8 \cdot 3} + 2 + \sqrt[3]{27 \cdot 3} \\ &= \sqrt[3]{8} \cdot \sqrt[3]{3} + 2 + \sqrt[3]{27} \cdot \sqrt[3]{3} \\ &= 2\sqrt[3]{3} + 2 + 3\sqrt[3]{3} \\ &= (2 + 3)\sqrt[3]{3} + 2 \\ &= 5\sqrt[3]{3} + 2\end{aligned}$$

DEFINITIONS AND CONCEPTS	EXAMPLES
Section 8.4 Multiplying and Dividing Radicals	
The product and quotient rules for radicals may be used to simplify products and quotients of radicals.	Perform the indicated operations and simplify. $\sqrt{3}\cdot\sqrt{11}=\sqrt{33}$ $(2\sqrt{5})^2=2^2\cdot(\sqrt{5})^2=4\cdot5=20$ Multiply. $(\sqrt{3x}+1)(\sqrt{5}-\sqrt{3})$ $=\sqrt{15x}-\sqrt{9x}+\sqrt{5}-\sqrt{3}$ $=\sqrt{15x}-3\sqrt{x}+\sqrt{5}-\sqrt{3}$ $\frac{\sqrt[3]{56x^4}}{\sqrt[3]{7x}}=\sqrt[3]{\frac{56x^4}{7x}}=\sqrt[3]{8x^3}=2x$
The process of eliminating the radical in the denominator of a radical expression is called **rationalizing the denominator.**	Rationalize the denominator. $\frac{5}{\sqrt{11}}=\frac{5\cdot\sqrt{11}}{\sqrt{11}\cdot\sqrt{11}}=\frac{5\sqrt{11}}{11}$
The **conjugate** of $a+b$ is $a-b$. To rationalize a denominator that is a sum or difference of radicals, multiply the numerator and the denominator by the conjugate of the denominator.	The conjugate of $2+\sqrt{3}$ is $2-\sqrt{3}$. Rationalize the denominator. $\frac{5}{6-\sqrt{5}}=\frac{5(6+\sqrt{5})}{(6-\sqrt{5})(6+\sqrt{5})}$ $=\frac{5(6+\sqrt{5})}{36+6\sqrt{5}-6\sqrt{5}-5}$ $=\frac{5(6+\sqrt{5})}{31}$
Section 8.5 Solving Equations Containing Radicals	
To solve a radical equation containing square roots **Step 1.** Get one radical by itself on one side of the equation. **Step 2.** Square both sides of the equation. **Step 3.** Simplify both sides of the equation. **Step 4.** If the equation still contains a radical term, repeat Steps 1 through 3. **Step 5.** Solve the equation. **Step 6.** Check solutions in the original equation.	Solve $\sqrt{2x-1}-x=-2$. $\sqrt{2x-1}=x-2$ $(\sqrt{2x-1})^2=(x-2)^2$ Square both sides. $2x-1=x^2-4x+4$ $0=x^2-6x+5$ $0=(x-1)(x-5)$ Factor. $x-1=0$ or $x-5=0$ $x=1$ or $x=5$ Solve. Check both proposed solutions in the original equation. 5 checks but 1 does not. The only solution is 5.
Section 8.6 Radical Equations and Problem Solving	
Problem-solving steps **1.** UNDERSTAND. Read and reread the problem.	A gutter is mounted on the eaves of a house 15 feet above the ground. A garden is adjacent to the house so that the closest a ladder can be placed to the house is 6 feet. How long a ladder is needed for installing the gutter? Let $x=$ the length of the ladder. 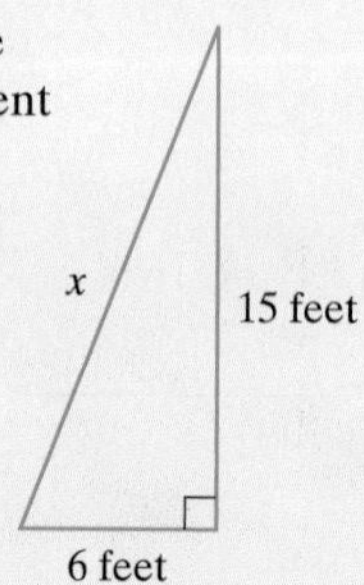

DEFINITIONS AND CONCEPTS	EXAMPLES
Section 8.6 Radical Equations and Problem Solving (continued)	
2. TRANSLATE.	Here, we use the Pythagorean theorem. The unknown length x is the hypotenuse. In words: $(\text{leg})^2 + (\text{leg})^2 = (\text{hypotenuse})^2$
3. SOLVE.	Translate: $6^2 + 15^2 = x^2$ $36 + 225 = x^2$ $261 = x^2$ $\sqrt{261} = x$ or $x = 3\sqrt{29}$
4. INTERPRET.	Check and state. The ladder needs to be $3\sqrt{29}$ feet or approximately 16.2 feet long.
Section 8.7 Rational Exponents	
If n is a positive integer and $\sqrt[n]{a}$ is a real number, then $a^{1/n} = \sqrt[n]{a}$.	$9^{1/2} = \sqrt{9} = 3$ $(-8)^{1/3} = \sqrt[3]{-8} = -2$
If m and n are integers with $n > 0$ and a is positive, then $a^{m/n} = (a^{1/n})^m = (\sqrt[n]{a})^m$ Also $a^{m/n} = (a^m)^{1/n} = \sqrt[n]{a^m}$	$25^{3/2} = (25^{1/2})^3 = (\sqrt{25})^3 = 5^3$ or 125
If $a^{m/n}$ is a nonzero real number, then $a^{-m/n} = \frac{1}{a^{m/n}}$	$81^{-3/4} = \frac{1}{81^{3/4}} = \frac{1}{(\sqrt[4]{81})^3} = \frac{1}{3^3} = \frac{1}{27}$
Properties for integer exponents hold for rational exponents also. $a^m \cdot a^n = a^{m+n}$ $(a^m)^n = a^{mn}$ $\frac{a^m}{a^n} = a^{m-n}$	$x^{1/2} \cdot x^{1/4} = x^{(1/2)+(1/4)} = x^{(2/4)+(1/4)} = x^{3/4}$ $(x^{2/3})^{1/5} = x^{(2/3)\cdot(1/5)} = x^{2/15}$ $\frac{x^{5/6}}{x^{1/6}} = x^{(5/6)-(1/6)} = x^{4/6} = x^{2/3}$

Chapter 8 Review

***(8.1)** Find the root.*

1. $\sqrt{81}$ **2.** $-\sqrt{49}$

3. $\sqrt[3]{27}$ **4.** $\sqrt[4]{16}$

5. $-\sqrt{\frac{9}{64}}$ **6.** $\sqrt{\frac{36}{81}}$

7. $\sqrt[4]{\frac{16}{81}}$ **8.** $\sqrt[3]{-\frac{27}{64}}$

9. Which radical(s) is not a real number?

a. $\sqrt{4}$ **b.** $-\sqrt{4}$ **c.** $\sqrt{-4}$ **d.** $\sqrt[3]{-4}$

10. Which radical(s) is not a real number?

a. $\sqrt{-5}$ **b.** $\sqrt[3]{-5}$ **c.** $\sqrt[4]{-5}$ **d.** $\sqrt[5]{-5}$

Find the following roots. Assume that variables represent positive numbers only.

11. $\sqrt{x^{12}}$ **12.** $\sqrt{x^8}$

13. $\sqrt{9x^6}$ **14.** $\sqrt{25x^4}$

15. $\sqrt{\frac{16}{y^{10}}}$ **16.** $\sqrt{\frac{y^{12}}{49}}$

***(8.2)** Simplify each expression using the product rule. Assume that variables represent nonnegative real numbers.*

17. $\sqrt{54}$ **18.** $\sqrt{88}$

19. $\sqrt{150x^3}$ **20.** $\sqrt{92y^5}$

21. $\sqrt[3]{54}$ **22.** $\sqrt[3]{88}$

23. $\sqrt[4]{48}$ **24.** $\sqrt[4]{162}$

Simplify each expression using the quotient rule. Assume that variables represent positive real numbers.

25. $\sqrt{\frac{18}{25}}$

26. $\sqrt{\frac{75}{64}}$

27. $\sqrt{\frac{45y^2}{4x^4}}$

28. $\sqrt{\frac{20x^5}{9x^2}}$

29. $\sqrt[4]{\frac{9}{16}}$

30. $\sqrt[3]{\frac{40}{27}}$

31. $\sqrt[3]{\frac{3}{8}}$

32. $\sqrt[4]{\frac{5}{81}}$

(8.3) *Add or subtract by combining like radicals.*

33. $3\sqrt[3]{2} + 2\sqrt[3]{3} - 4\sqrt[3]{2}$

34. $5\sqrt{2} + 2\sqrt[3]{2} - 8\sqrt{2}$

35. $\sqrt{6} + 2\sqrt[3]{6} - 4\sqrt[3]{6} + 5\sqrt{6}$

36. $3\sqrt{5} - \sqrt[3]{5} - 2\sqrt{5} + 3\sqrt[3]{5}$

Add or subtract by simplifying each radical and then combining like terms. Assume that variables represent nonnegative real numbers.

37. $\sqrt{28x} + \sqrt{63x} + \sqrt[3]{56}$

38. $\sqrt{75y} + \sqrt{48y} - \sqrt[4]{16}$

39. $\sqrt{\frac{5}{9}} - \sqrt{\frac{5}{36}}$

40. $\sqrt{\frac{11}{25}} + \sqrt{\frac{11}{16}}$

41. $2\sqrt[3]{125} - 5\sqrt[3]{8}$

42. $3\sqrt[3]{16} - 2\sqrt[3]{2}$

(8.4) *Find the product and simplify if possible.*

43. $3\sqrt{10} \cdot 2\sqrt{5}$

44. $2\sqrt[3]{4} \cdot 5\sqrt[3]{6}$

45. $\sqrt{3}(2\sqrt{6} - 3\sqrt{12})$

46. $4\sqrt{5}(2\sqrt{10} - 5\sqrt{5})$

47. $(\sqrt{3} + 2)(\sqrt{6} - 5)$

48. $(2\sqrt{5} + 1)(4\sqrt{5} - 3)$

49. $(\sqrt{x} - 2)^2$

50. $(\sqrt{y} + 4)^2$

Divide and simplify if possible. Assume that all variables represent positive numbers.

51. $\frac{\sqrt{27}}{\sqrt{3}}$

52. $\frac{\sqrt{20}}{\sqrt{5}}$

53. $\frac{\sqrt{160}}{\sqrt{8}}$

54. $\frac{\sqrt{96}}{\sqrt{3}}$

55. $\frac{\sqrt{30x^6}}{\sqrt{2x^3}}$

56. $\frac{\sqrt{54x^5y^2}}{\sqrt{3xy^2}}$

Rationalize each denominator and simplify.

57. $\frac{\sqrt{2}}{\sqrt{11}}$

58. $\frac{\sqrt{3}}{\sqrt{13}}$

59. $\sqrt{\frac{5}{6}}$

60. $\sqrt{\frac{7}{10}}$

61. $\frac{1}{\sqrt{5x}}$

62. $\frac{5}{\sqrt{3y}}$

63. $\sqrt{\frac{3}{x}}$

64. $\sqrt{\frac{6}{y}}$

65. $\frac{3}{\sqrt{5} - 2}$

66. $\frac{8}{\sqrt{10} - 3}$

67. $\frac{\sqrt{2} + 1}{\sqrt{3} - 1}$

68. $\frac{\sqrt{3} - 2}{\sqrt{5} + 2}$

69. $\frac{10}{\sqrt{x} + 5}$

70. $\frac{8}{\sqrt{x} - 1}$

71. $\sqrt[3]{\frac{7}{9}}$

72. $\sqrt[3]{\frac{3}{4}}$

73. $\sqrt[3]{\frac{3}{2}}$

74. $\sqrt[3]{\frac{5}{4}}$

(8.5) *Solve each radical equations.*

75. $\sqrt{2x} = 6$

76. $\sqrt{x + 3} = 4$

77. $\sqrt{x} + 3 = 8$

78. $\sqrt{x} + 8 = 3$

79. $\sqrt{2x + 1} = x - 7$

80. $\sqrt{3x + 1} = x - 1$

81. $\sqrt{x} + 3 = \sqrt{x + 15}$

82. $\sqrt{x - 5} = \sqrt{x} - 1$

(8.6) *Use the Pythagorean theorem to find the length of each unknown side. Give an exact answer and a two-decimal-place approximation.*

△ **83.**

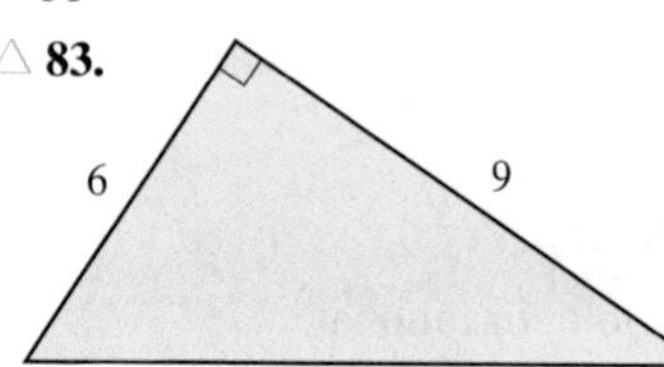

△ **84.**

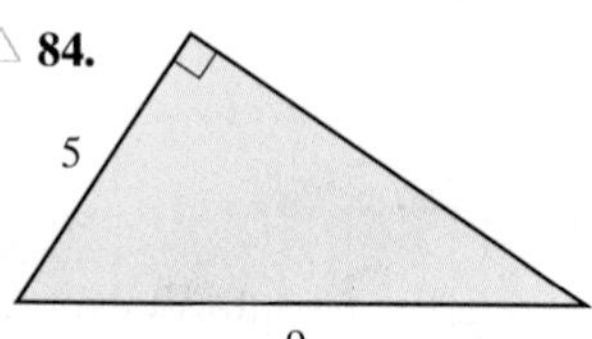

Solve. Give an exact answer and a two-decimal-place approximation.

△ **85.** Romeo is standing 20 feet away from the wall below Juliet's balcony during a school play. Juliet is on the balcony, 12 feet above the ground. Find how far apart Romeo and Juliet are.

△ **86.** The diagonal of a rectangle is 10 inches long. If the width of the rectangle is 5 inches, find the length of the rectangle.

Use the distance formula to find the distance between the points.

87. $(6, -2)$ and $(-3, 5)$

88. $(2, 8)$ and $(-6, 10)$

Use the formula $r = \sqrt{\frac{S}{4\pi}}$, where r = the radius of a sphere and S = the surface area of the sphere, for Exercises 89 and 90.

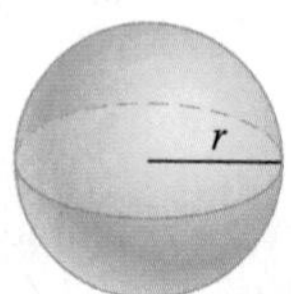

△ **89.** Find the radius of a sphere to the nearest tenth of an inch if the area is 72 square inches.

△ **90.** Find the exact surface area of a sphere if its radius is 6 inches. (Do not approximate π.)

(8.7) Write each of the following with fractional exponents and simplify if possible. Assume that variables represent nonnegative real numbers.

91. $\sqrt{a^5}$ **92.** $\sqrt[5]{a^3}$

93. $\sqrt[6]{x^{15}}$ **94.** $\sqrt[4]{x^{12}}$

Evaluate each of the following expressions.

95. $16^{1/2}$ **96.** $36^{1/2}$

97. $(-8)^{1/3}$ **98.** $(-32)^{1/5}$

99. $-64^{3/2}$ **100.** $-8^{2/3}$

101. $\left(\frac{16}{81}\right)^{3/4}$ **102.** $\left(\frac{9}{25}\right)^{3/2}$

Simplify each expression using positive exponents only. Assume that variables represent positive real numbers.

103. $8^{1/3} \cdot 8^{4/3}$ **104.** $4^{3/2} \cdot 4^{1/2}$

105. $\frac{3^{1/6}}{3^{5/6}}$ **106.** $\frac{2^{1/4}}{2^{-3/5}}$

107. $(x^{-1/3})^6$ **108.** $\left(\frac{x^{1/2}}{y^{1/3}}\right)^2$

MIXED REVIEW

Find each root. Assume all variables represent positive numbers.

109. $\sqrt{144}$ **110.** $-\sqrt[3]{64}$

111. $\sqrt{16x^{16}}$ **112.** $\sqrt{4x^{24}}$

Simplify each expression. Assume all variables represent positive numbers.

113. $\sqrt{18x^7}$ **114.** $\sqrt{48y^6}$

115. $25^{-1/2}$ **116.** $64^{-2/3}$

117. $\sqrt{\frac{y^4}{81}}$ **118.** $\sqrt{\frac{x^9}{9}}$

Add or subtract by simplifying and then combining like terms. Assume all variables represent positive numbers.

119. $\sqrt{12} + \sqrt{75}$ **120.** $\sqrt{63} + \sqrt{28} - \sqrt[3]{27}$

121. $\sqrt{\frac{3}{16}} - \sqrt{\frac{3}{4}}$ **122.** $\sqrt{45x^3} + x\sqrt{20x} - \sqrt{5x^3}$

Multiply and simplify if possible. Assume all variables represent positive numbers.

123. $\sqrt{7} \cdot \sqrt{14}$

124. $\sqrt{3}(\sqrt{9} - \sqrt{2})$

125. $(\sqrt{2} + 4)(\sqrt{5} - 1)$

126. $(\sqrt{x} + 3)^2$

Divide and simplify if possible. Assume all variables represent positive numbers.

127. $\frac{\sqrt{120}}{\sqrt{5}}$ **128.** $\frac{\sqrt{60x^9}}{\sqrt{15x^7}}$

Rationalize each denominator and simplify.

129. $\sqrt{\frac{2}{7}}$ **130.** $\frac{3}{\sqrt{2x}}$

131. $\frac{3}{\sqrt{x} - 6}$ **132.** $\frac{\sqrt{7} - 5}{\sqrt{5} + 3}$

Solve each radical equation.

133. $\sqrt{4x} = 2$ **134.** $\sqrt{x - 4} = 3$

135. $\sqrt{4x + 8} + 6 = x$ **136.** $\sqrt{x - 8} = \sqrt{x} - 2$

137. Use the Pythagorean theorem to find the length of the unknown side. Give an exact answer and a two-decimal-place approximation.

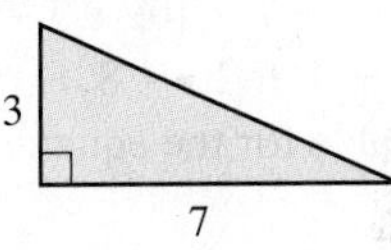

138. The diagonal of a rectangle is 6 inches long. If the width of the rectangle is 2 inches, find the length of the rectangle.

139. The Apple (computer) store on Fifth Avenue in New York City, which opened in 2006, features a distinctive glass cube. If the glass surface area of this cube (which includes only the four walls and the roof) is 5120 square feet, find the length of a side of the cube.

Chapter 8 Test

MyMathLab®

YouTube

Simplify the following. Indicate if the expression is not a real number.

1. $\sqrt{16}$ **2.** $\sqrt[3]{-125}$

3. $16^{3/4}$ **4.** $\left(\frac{9}{16}\right)^{1/2}$

5. $\sqrt[4]{-81}$ **6.** $27^{-2/3}$

Simplify each radical expression. Assume that variables represent positive numbers only.

7. $\sqrt{54}$ **8.** $\sqrt{92}$

9. $\sqrt{3x^6}$ **10.** $\sqrt{8x^4y^7}$

11. $\sqrt{9x^9}$ **12.** $\sqrt[3]{8}$

13. $\sqrt[3]{40}$ **14.** $\sqrt{x^{10}}$

15. $\sqrt{y^7}$ **16.** $\sqrt{\frac{5}{16}}$

17. $\sqrt{\frac{y^3}{25}}$ **18.** $\sqrt[3]{\frac{2}{27}}$

19. $3\sqrt{8x}$

Perform each indicated operation. Assume that all variables represent positive numbers.

20. $\sqrt{13} + \sqrt{13} - 4\sqrt{13}$

21. $\sqrt{12} - 2\sqrt{75}$

22. $\sqrt{2x^2} + \sqrt[3]{54} - x\sqrt{18}$

23. $\sqrt{\frac{3}{4}} + \sqrt{\frac{3}{25}}$

24. $\sqrt{7} \cdot \sqrt{14}$

25. $\sqrt{2}(\sqrt{6} - \sqrt{5})$

26. $(\sqrt{x} + 2)(\sqrt{x} - 3)$

27. $\frac{\sqrt{50}}{\sqrt{10}}$

28. $\frac{\sqrt{40x^4}}{\sqrt{2x}}$

Rationalize the denominator.

29. $\sqrt{\frac{2}{3}}$

30. $\sqrt[3]{\frac{5}{9}}$

31. $\sqrt{\frac{5}{12x^2}}$

32. $\frac{2\sqrt{3}}{\sqrt{3} - 3}$

Solve each of the following radical equations.

33. $\sqrt{x} + 8 = 11$

34. $\sqrt{3x - 6} = \sqrt{x + 4}$

35. $\sqrt{2x - 2} = x - 5$

36. Find the length of the unknown leg of a right triangle if the other leg is 8 inches long and the hypotenuse is 12 inches long.

37. Find the distance between $(-3, 6)$ and $(-2, 8)$.

Simplify each expression using positive exponents only.

38. $16^{-3/4} \cdot 16^{-1/4}$

39. $\left(\frac{x^{2/3}}{y^{2/5}}\right)^5$

Chapter 8 Cumulative Review

1. Simplify each expression.
   a. $\frac{(-12)(-3) + 3}{-7 - (-2)}$
   b. $\frac{2(-3)^2 - 20}{-5 + 4}$
2. Simplify each expression.
   a. $\frac{4(-3) - (-6)}{-8 + 4}$
   b. $\frac{3 + (-3)(-2)^3}{-1 - (-4)}$
3. Solve. $2x + 3x - 5 + 7 = 10x + 3 - 6x - 4$
4. Solve. $6y - 11 + 4 + 2y = 8 + 15y - 8y$
5. Complete the table for the equation $y = 3x$.

x	y
-1	
	0
	-9

6. Complete the table for the equation $2x + y = 6$.

x	y
0	
	-2
3	

7. Find an equation of the line with y-intercept $(0, -3)$ and slope of $\frac{1}{4}$.
8. Find an equation of a line with y-intercept $(0, 4)$ and slope of -2.
9. Find an equation of the line parallel to the line $y = 5$ and passing through $(-2, -3)$.
10. Find an equation of the line perpendicular to $y = 2x + 4$ and passing through $(1, 5)$.
11. Decide whether the equation describes a function.
   a. $y = x$
   b. $y = 2x + 1$
   c. $y = 5$
   d. $x = -1$
12. Decide whether the equation describes a function.
   a. $2x + 3 = y$
   b. $x + 4 = 0$
   c. $\frac{1}{2}y = 2x$
   d. $y = 0$
13. Determine whether $(12, 6)$ is a solution of the given system
$$\begin{cases} 2x - 3y = 6 \\ x = 2y \end{cases}$$
14. Determine whether the following ordered pairs are solutions of the given systems.
$$\begin{cases} 2x + y = 4 \\ x + y = 2 \end{cases}$$
   a. $(1, 1)$
   b. $(2, 0)$
15. Solve the system:
$$\begin{cases} 2x + y = 10 \\ x = y + 2 \end{cases}$$
16. Solve the system:
$$\begin{cases} 3y = x + 10 \\ 2x + 5y = 24 \end{cases}$$
17. Solve the system:
$$\begin{cases} -x - \frac{y}{2} = \frac{5}{2} \\ \frac{x}{6} - \frac{y}{2} = 0 \end{cases}$$
18. Solve the system:
$$\begin{cases} \frac{x}{2} + y = \frac{5}{6} \\ 2x - y = \frac{5}{6} \end{cases}$$

19. A chemistry teaching assistant needs 10 liters of a 20% saline solution (saltwater) for his 2 p.m. laboratory class. Unfortunately, the only mixtures on hand are a 5% saline solution and a 25% saline solution. How much of each solution should he mix to produce the 20% solution?

20. Two streetcars are 11 miles apart and traveling toward each other on parallel tracks. They meet in 12 minutes. Find the speed of each streetcar if one travels 15 miles per hour faster than the other.

21. Graph the solution of the system:
$$\begin{cases} 3x \geq y \\ x + 2y \leq 8 \end{cases}$$

22. Graph the solution of the system:
$$\begin{cases} x + y \leq 1 \\ 2x - y \geq 2 \end{cases}$$

23. Combine like terms to simplify. $-9x^2 + 3xy - 5y^2 + 7yx$

24. Combine like terms to simplify.
$4a^2 + 3a - 2a^2 + 7a - 5$

25. Factor: $x^2 + 7xy + 6y^2$

26. Factor: $3x^2 + 15x + 18$

27. Simplify: $\dfrac{4 - x^2}{3x^2 - 5x - 2}$

28. Simplify: $\dfrac{2x^2 + 7x + 3}{x^2 - 9}$

29. Divide: $\dfrac{3x^3y^7}{40} \div \dfrac{4x^3}{y^2}$

30. Divide: $\dfrac{12x^2y^3}{5} \div \dfrac{3y^3}{x}$

31. Subtract: $\dfrac{2y}{2y - 7} - \dfrac{7}{2y - 7}$

32. Subtract: $\dfrac{-4x^2}{x + 1} - \dfrac{4x}{x + 1}$

33. Add: $\dfrac{2x}{x^2 + 2x + 1} + \dfrac{x}{x^2 - 1}$

34. Add: $\dfrac{3x}{x^2 + 5x + 6} + \dfrac{1}{x^2 + 2x - 3}$.

35. Solve: $\dfrac{x}{2} + \dfrac{8}{3} = \dfrac{1}{6}$

36. Solve: $\dfrac{1}{21} + \dfrac{x}{7} = \dfrac{5}{3}$

37. If the following two triangles are similar, find the missing length x.

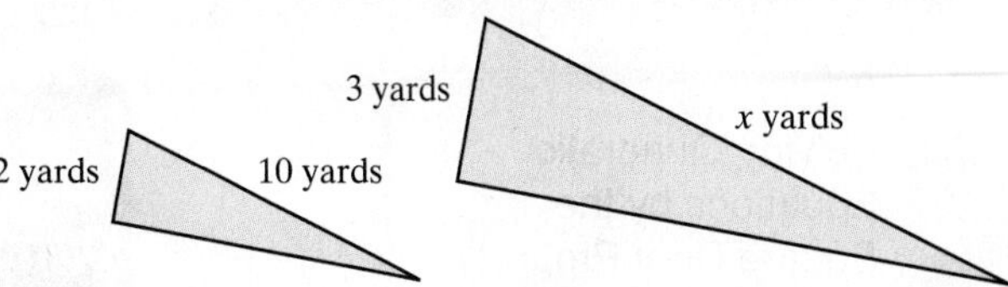

38. If the following two triangles are similar, find the missing length.

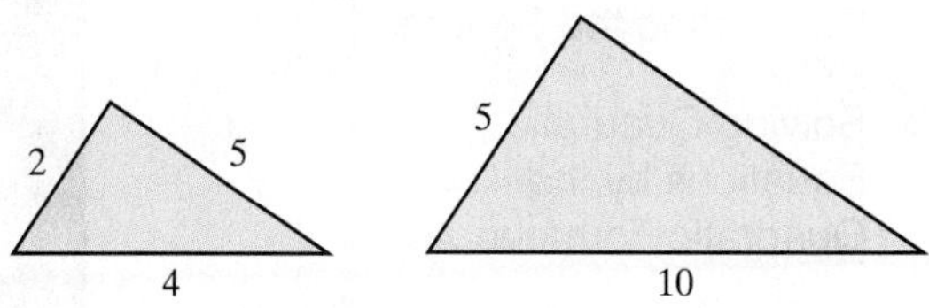

39. Simplify: $\dfrac{\frac{1}{z} - \frac{1}{2}}{\frac{1}{3} - \frac{z}{6}}$

40. Simplify: $\dfrac{x + 3}{\frac{1}{x} + \frac{1}{3}}$

41. Simplify.
a. $\sqrt{54}$ **b.** $\sqrt{12}$
c. $\sqrt{200}$ **d.** $\sqrt{35}$

42. Simplify.
a. $\sqrt{40}$ **b.** $\sqrt{500}$
c. $\sqrt{63}$ **d.** $\sqrt{169}$

43. Multiply. Then simplify if possible.
a. $(\sqrt{5} - 7)(\sqrt{5} + 7)$
b. $(\sqrt{7x} + 2)^2$

44. Multiply. Then simplify, if possible.
a. $(\sqrt{6} + 2)^2$
b. $(\sqrt{x} + 5)(\sqrt{x} - 5)$

45. Solve. $\sqrt{x} + 6 = 4$

46. Solve. $\sqrt{x + 4} = \sqrt{3x - 1}$

47. Find the length of the hypotenuse of a right triangle whose legs are 6 inches and 8 inches long.

48. Find the length of the unknown leg of a right triangle whose other leg is 9 and whose hypotenuse is 13.

49. Simplify each expression.
a. $4^{3/2}$ **b.** $27^{2/3}$ **c.** $-16^{3/4}$

50. Simplify each expression.
a. $9^{5/2}$ **b.** $-81^{1/4}$ **c.** $(-64)^{2/3}$

CHAPTER 9 Quadratic Equations

An important part of the study of algebra is learning to use methods for solving equations. In Chapter 2, we presented techniques for solving linear equations in one variable. In Chapter 6, we solved quadratic equations in one variable by factoring the quadratic expressions. We now present other methods for solving quadratic equations in one variable.

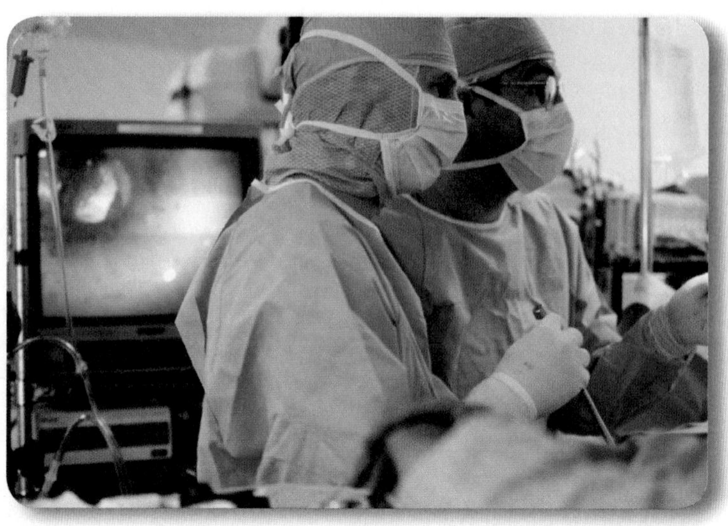

There are currently over 110,000 people living in the United States waiting for a transplant, and that number is increasing every year. The bars on the graph below show the yearly number of kidney transplants performed, and the broken line shows the number of patients waiting for a kidney transplant. Physicians use these data to help plan for the future. In Section 9.3, Exercise 91, we will use a formula called the quadratic formula to predict the number of patients on the waiting list at a future date.

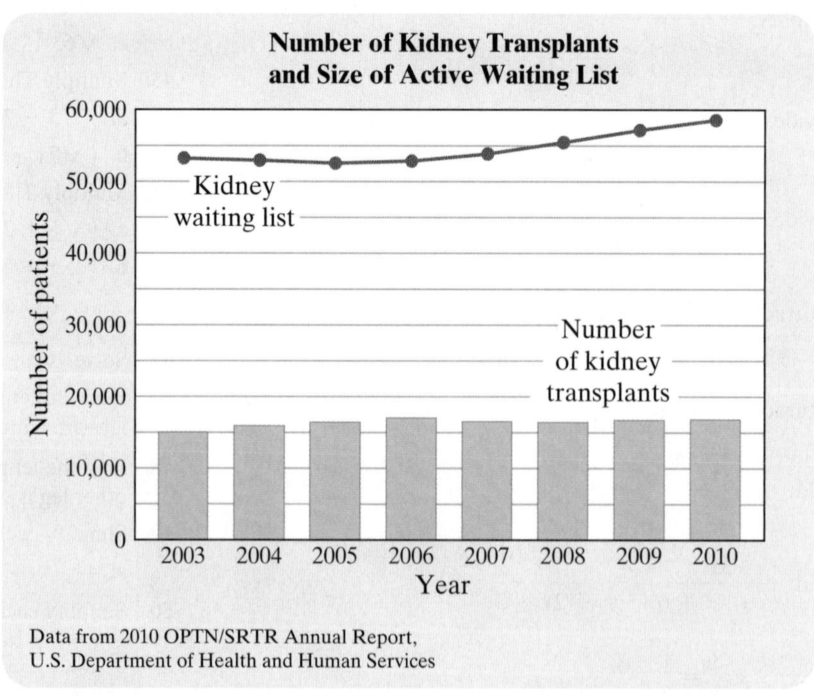

Data from 2010 OPTN/SRTR Annual Report, U.S. Department of Health and Human Services

9.1 Solving Quadratic Equations by the Square Root Property 

OBJECTIVES

1 Use the Square Root Property to Solve Quadratic Equations.

2 Solve Problems Modeled by Quadratic Equations.

OBJECTIVE

1 Using the Square Root Property 

Recall that a quadratic equation is an equation that can be written in the form

$$ax^2 + bx + c = 0$$

where a, b, and c are real numbers and $a \neq 0$.

To solve quadratic equations by factoring, use the **zero factor theorem:** If the product of two numbers is zero, then at least one of the two numbers is zero. For example, to solve $x^2 - 4 = 0$, we first factor the left side of the equation and then set each factor equal to 0.

$$x^2 - 4 = 0$$
$$(x + 2)(x - 2) = 0 \quad \text{Factor.}$$
$$x + 2 = 0 \quad \text{or} \quad x - 2 = 0 \quad \text{Apply the zero factor theorem.}$$
$$x = -2 \quad \text{or} \quad x = 2 \quad \text{Solve each equation.}$$

The solutions are -2 and 2.

Now let's solve $x^2 - 4 = 0$ another way. First, add 4 to both sides of the equation.

$$x^2 - 4 = 0$$
$$x^2 = 4 \quad \text{Add 4 to both sides.}$$

Now we see that the value for x must be a number whose square is 4. Therefore $x = \sqrt{4} = 2$ or $x = -\sqrt{4} = -2$. This reasoning is an example of the square root property.

Square Root Property

If $x^2 = a$ for $a \geq 0$, then

$$x = \sqrt{a} \quad \text{or} \quad x = -\sqrt{a}$$

EXAMPLE 1 Use the square root property to solve $x^2 - 9 = 0$.

Solution First we solve for x^2 by adding 9 to both sides.

$$x^2 - 9 = 0$$
$$x^2 = 9 \quad \text{Add 9 to both sides.}$$

Next we use the square root property.

$$x = \sqrt{9} \quad \text{or} \quad x = -\sqrt{9}$$
$$x = 3 \qquad\qquad x = -3$$

Check:

$$x^2 - 9 = 0 \quad \text{Original equation} \qquad\qquad x^2 - 9 = 0 \quad \text{Original equation}$$
$$3^2 - 9 \stackrel{?}{=} 0 \quad \text{Let } x = 3. \qquad\qquad (-3)^2 - 9 \stackrel{?}{=} 0 \quad \text{Let } x = -3.$$
$$0 = 0 \quad \text{True} \qquad\qquad 0 = 0 \quad \text{True}$$

The solutions are 3 and -3. □

PRACTICE

1 Use the square root property to solve $x^2 - 16 = 0$.

EXAMPLE 2 Use the square root property to solve $2x^2 = 7$.

Solution First we solve for x^2 by dividing both sides by 2. Then we use the square root property.

$$2x^2 = 7$$

$$x^2 = \frac{7}{2} \quad \text{Divide both sides by 2.}$$

$$x = \sqrt{\frac{7}{2}} \quad \text{or} \quad x = -\sqrt{\frac{7}{2}} \quad \text{Use the square root property.}$$

If the denominators are rationalized, we have

$$x = \frac{\sqrt{7}\cdot\sqrt{2}}{\sqrt{2}\cdot\sqrt{2}} \quad \text{or} \quad x = -\frac{\sqrt{7}\cdot\sqrt{2}}{\sqrt{2}\cdot\sqrt{2}} \quad \text{Rationalize the denominator.}$$

$$x = \frac{\sqrt{14}}{2} \qquad x = -\frac{\sqrt{14}}{2} \quad \text{Simplify.}$$

Remember to check both solutions in the original equation. The solutions are $\frac{\sqrt{14}}{2}$ and $-\frac{\sqrt{14}}{2}$. □

PRACTICE
2 Use the square root property to solve $5x^2 = 13$.

EXAMPLE 3 Use the square root property to solve $(x - 3)^2 = 16$.

Solution Instead of x^2, here we have $(x - 3)^2$. But the square root property can still be used.

$$(x - 3)^2 = 16$$

$$x - 3 = \sqrt{16} \quad \text{or} \quad x - 3 = -\sqrt{16} \quad \text{Use the square root property.}$$

$$x - 3 = 4 \qquad x - 3 = -4 \quad \text{Write } \sqrt{16} \text{ as 4 and } -\sqrt{16} \text{ as } -4.$$

$$x = 7 \qquad x = -1 \quad \text{Solve.}$$

Check:

$(x - 3)^2 = 16$	Original equation	$(x - 3)^2 = 16$	Original equation
$(7 - 3)^2 \stackrel{?}{=} 16$	Let $x = 7$.	$(-1 - 3)^2 \stackrel{?}{=} 16$	Let $x = -1$.
$4^2 \stackrel{?}{=} 16$	Simplify.	$(-4)^2 \stackrel{?}{=} 16$	Simplify.
$16 = 16$	True	$16 = 16$	True

Both 7 and -1 are solutions. □

PRACTICE
3 Use the square root property to solve $(x - 5)^2 = 36$.

EXAMPLE 4 Use the square root property to solve $(x + 1)^2 = 8$.

Solution $(x + 1)^2 = 8$

$$x + 1 = \sqrt{8} \quad \text{or} \quad x + 1 = -\sqrt{8} \quad \text{Use the square root property.}$$

$$x + 1 = 2\sqrt{2} \qquad x + 1 = -2\sqrt{2} \quad \text{Simplify the radical.}$$

$$x = -1 + 2\sqrt{2} \qquad x = -1 - 2\sqrt{2} \quad \text{Solve for } x.$$

> **Helpful Hint**
>
> read "plus or minus"
>
> The notation $-1 \pm \sqrt{5}$, for example, is just a shorthand notation for both $-1 + \sqrt{5}$ and $-1 - \sqrt{5}$.

Check both solutions in the original equation. The solutions are $-1 + 2\sqrt{2}$ and $-1 - 2\sqrt{2}$. This can be written compactly as $-1 \pm 2\sqrt{2}$. The notation $\pm$ is read as "plus or minus." □

PRACTICE

4 Use the square root property to solve $(x + 2)^2 = 12$.

EXAMPLE 5 Use the square root property to solve $(x - 1)^2 = -2$.

Solution This equation has no real solution because the square root of -2 is not a real number. □

PRACTICE

5 Use the square root property to solve $(x - 8)^2 = -5$.

EXAMPLE 6 Use the square root property to solve $(5x - 2)^2 = 10$.

Solution $(5x - 2)^2 = 10$

$5x - 2 = \sqrt{10}$ or $5x - 2 = -\sqrt{10}$ Use the square root property.

$5x = 2 + \sqrt{10}$ $\quad$ $5x = 2 - \sqrt{10}$ Add 2 to both sides.

$x = \dfrac{2 + \sqrt{10}}{5}$ $\quad$ $x = \dfrac{2 - \sqrt{10}}{5}$ Divide both sides by 5.

Check both solutions in the original equation. The solutions are $\dfrac{2 + \sqrt{10}}{5}$ and $\dfrac{2 - \sqrt{10}}{5}$, which can be written as $\dfrac{2 \pm \sqrt{10}}{5}$. □

PRACTICE

6 Use the square root property to solve $(3x - 5)^2 = 17$.

> **Helpful Hint**
>
> For some applications and graphing purposes, decimal approximations of exact solutions to quadratic equations may be desired.

Exact Solutions from Example 6		Decimal Approximations
$\dfrac{2 + \sqrt{10}}{5}$	$\approx$	1.032
$\dfrac{2 - \sqrt{10}}{5}$	$\approx$	-0.232

OBJECTIVE

2 Solving Problems Modeled by Quadratic Equations

Many real-world applications are modeled by quadratic equations.

EXAMPLE 7 **Finding the Length of Time of a Dive**

The record for the highest dive into a lake was made by Harry Froboess of Switzerland. In 1936, he dove 394 feet from the airship Hindenburg into Lake Constance. To the nearest tenth of a second, how long did his dive take? (*Source: The Guiness Book of Records*)

(Continued on next page)

Solution

1. UNDERSTAND. To approximate the time of the dive, we use the formula $h = 16t^2$* where t is time in seconds and h is the distance in feet traveled by a free-falling body or object. For example, to find the distance traveled in 1 second, or 3 seconds, we let $t = 1$ and then $t = 3$.

$$\text{If } t = 1, h = 16(1)^2 = 16 \cdot 1 = 16 \text{ feet}$$
$$\text{If } t = 3, h = 16(3)^2 = 16 \cdot 9 = 144 \text{ feet}$$

Since a body travels 144 feet in 3 seconds, we now know the dive of 394 feet lasted longer than 3 seconds.

2. TRANSLATE. Use the formula $h = 16t^2$, let the distance $h = 394$, and we have the equation $394 = 16t^2$.
3. SOLVE. To solve $394 = 16t^2$ for t, we will use the square root property.

$$394 = 16t^2$$
$$\frac{394}{16} = t^2 \quad \text{Divide both sides by 16.}$$
$$24.625 = t^2 \quad \text{Simplify.}$$
$$\sqrt{24.625} = t \quad \text{or} \quad -\sqrt{24.625} = t \quad \text{Use the square root property.}$$
$$5.0 \approx t \quad \text{or} \quad -5.0 \approx t \quad \text{Approximate.}$$

4. INTERPRET.

Check: We reject the solution -5.0 since the length of the dive is not a negative number.

State: The dive lasted approximately 5 seconds. □

PRACTICE

7 On August 16, 1960, as part of an Air Force research program, Air Force Captain Joseph W. Kittinger, Jr. stepped off a platform lifted to 102,800 feet above the New Mexico desert by a hot air balloon. As part of the research, Captain Kittinger free fell for 84,700 feet before opening his parachute. To the nearest tenth of a second, how long did Captain Kittinger free fall before he opened his chute? (*Source: PBS, Nova*)

*The formula $h = 16t^2$ does not take into account air resistance.

Vocabulary, Readiness & Video Check

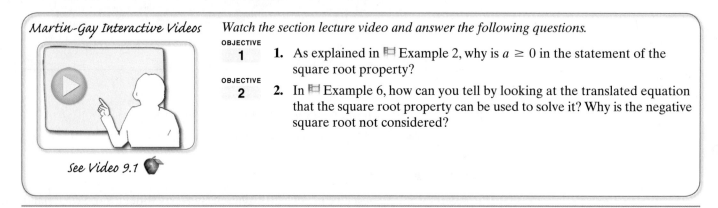

Martin-Gay Interactive Videos

See Video 9.1

Watch the section lecture video and answer the following questions.

OBJECTIVE 1

1. As explained in Example 2, why is $a \geq 0$ in the statement of the square root property?

OBJECTIVE 2

2. In Example 6, how can you tell by looking at the translated equation that the square root property can be used to solve it? Why is the negative square root not considered?

9.1 Exercise Set

MyMathLab®

Use the square root property to solve each quadratic equation. See Examples 1, 2, and 5.

1. $x^2 = 64$
2. $x^2 = 121$
3. $x^2 = 21$
4. $x^2 = 22$
5. $x^2 = \frac{1}{25}$
6. $x^2 = \frac{1}{16}$
7. $x^2 = -4$
8. $x^2 = -25$
9. $3x^2 = 13$
10. $5x^2 = 2$
11. $7x^2 = 4$
12. $2x^2 = 9$
13. $x^2 - 2 = 0$
14. $x^2 - 15 = 0$
15. $2x^2 - 10 = 0$
16. $7x^2 - 21 = 0$
17. $5x^2 + 15 = 0$
18. $9x^2 + 90 = 0$

Use the square root property to solve each quadratic equation. See Examples 3 through 6.

19. $(x - 5)^2 = 49$
20. $(x + 2)^2 = 25$
21. $(x + 2)^2 = 7$
22. $(x - 7)^2 = 2$
23. $\left(m - \frac{1}{2}\right)^2 = \frac{1}{4}$
24. $\left(m + \frac{1}{3}\right)^2 = \frac{1}{9}$
25. $(p + 2)^2 = 10$
26. $(p - 7)^2 = 13$
27. $(3y + 2)^2 = 100$
28. $(4y - 3)^2 = 81$
29. $(z - 4)^2 = -9$
30. $(z + 7)^2 = -20$
31. $(2x - 11)^2 = 50$
32. $(3x - 17)^2 = 28$
33. $(3x - 7)^2 = 32$
34. $(5x - 11)^2 = 54$
35. $(2p - 5)^2 = 121$
36. $(3p - 1)^2 = 4$

MIXED PRACTICE

Use the square root property to solve. See Examples 1 through 6.

37. $x^2 - 2 = 0$
38. $x^2 - 15 = 0$
39. $(x + 6)^2 = 24$
40. $(x + 5)^2 = 20$
41. $\frac{1}{2}n^2 = 5$
42. $\frac{1}{5}y^2 = 2$
43. $(4x - 1)^2 = 5$
44. $(7x - 2)^2 = 11$
45. $3z^2 = 36$
46. $3z^2 = 24$
47. $(8 - 3x)^2 - 45 = 0$
48. $(10 - 9x)^2 - 75 = 0$

The formula for area of a square is $A = s^2$ where s is the length of a side. Use this formula for Exercises 49 through 52. For each exercise, give an exact answer and a two-decimal-place approximation.

△ 49. If the area of a square is 20 square inches, find the length of a side.

△ 50. If the area of a square is 32 square meters, find the length of a side.

△ 51. The "Water Cube" National Swimming Center was constructed in Beijing for the 2008 Summer Olympics. Its square base has an area of 31,329 sq meters. Find the length of a side of this building. (*Source:* ARUP East Asia)

Note: The Beijing Water Cube was converted to an indoor Water Park and recently reopened.

△ 52. The Washington Monument has a square base whose area is approximately 3039 square feet. Find the length of a side. (*Source: The World Almanac*)

Solve. See Example 7. For Exercises 53 through 56, use the formula from $h = 16t^2$ and round answers to the nearest tenth of a second. (Recall that this formula does not take into account any air resistance.)

53. The highest regularly performed dives are made by professional divers from La Quebrada, in Acapulco, Mexico. The performer dives head first from a height of 115 feet (into 12 feet of water). Determine the time of a dive. (*Source: The Guinness Book of Records*)

54. In 1988, Eddie Turner saved Frank Fanan, who became unconscious after an injury while jumping out of an airplane. Fanan fell 11,136 feet before Turner pulled his ripcord. Determine the time of Fanan's unconscious free fall.

55. The Hualapai Indian Tribe allowed the Grand Canyon Skywalk to be built over the rim of the Grand Canyon on its tribal land. The skywalk extends 70 feet beyond the canyon's edge and is 4000 feet above the canyon floor. Determine the time, to the nearest tenth of a second, it would take an object, dropped off the skywalk, to land at the bottom of the Grand Canyon. (*Source: Boston Globe*)

56. If a sandblaster drops his goggles from a bridge 400 feet from the water below, find how long it takes for the goggles to hit the water.

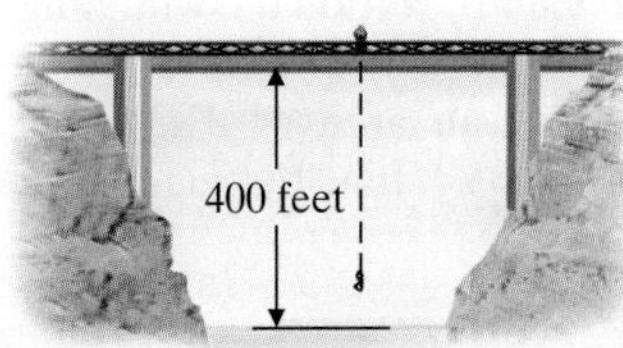

57. The area of a circle is found by the equation $A = \pi r^2$. If the area A of a certain circle is 36π square inches, find its radius r.

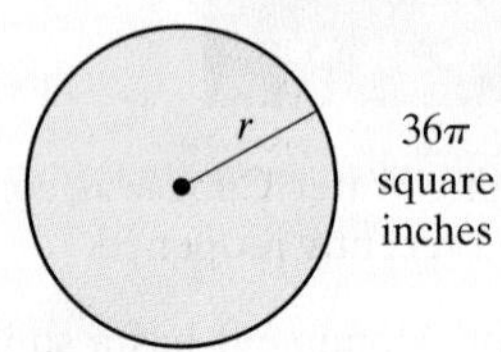

58. If the area of the circle below is 10π square units, find its exact radius. (See Exercise 57.)

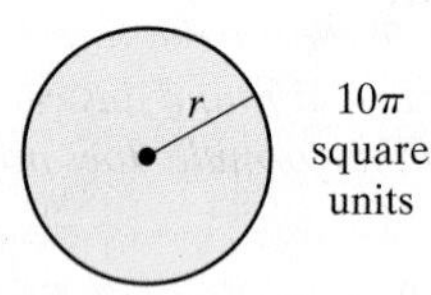

REVIEW AND PREVIEW

Factor each perfect square trinomial. See Section 6.3.

59. $x^2 + 6x + 9$

60. $y^2 + 10y + 25$

61. $x^2 - 4x + 4$

62. $x^2 - 20x + 100$

CONCEPT EXTENSIONS

63. Explain why the equation $x^2 = -9$ has no real solution.

64. Explain why the equation $x^2 = 9$ has two solutions.

Solve each quadratic equation by first factoring the perfect square trinomial on the left side. Then apply the square root property.

65. $x^2 + 4x + 4 = 16$

66. $z^2 - 6z + 9 = 25$

67. $x^2 + 14x + 49 = 31$

68. $y^2 - 10y + 25 = 11$

For Exercises 69 through 72, solve each quadratic equation by using the square root property. If necessary, use a calculator and round each solution to the nearest hundredth.

69. $x^2 = 1.78$

70. $(x - 1.37)^2 = 5.71$

71. The number of U.S. highway bridges for the years 2000 to 2007 can be modeled by the equation $y = 148(x + 1.5)^2 + 589{,}105$, where $x = 0$ represents the year 2000. Assume that this trend continued and find the year in which there were 605,000 highway bridges in the United States. (*Hint:* Replace y with 605,000 in the equation and solve for x. Round to the nearest year.) (*Source:* Based on data from the U.S. Department of Transportation, Bureau of Transportation Statistics)

72. The number y (in millions) of wireless subscribers in the United States from 2000 through mid-2008 can be estimated by $y = 0.6(x + 15)^2 - 37$, where $x = 0$ represents the year 2000. Assume that this trend continued and determine the first year in which the number of cellular subscribers surpassed 340 million. (*Hint:* Replace y with 340 in the equation and solve for x. Round to the nearest year.) (*Source:* Based on data from CTIA–The Wireless Association)

9.2 Solving Quadratic Equations by Completing the Square

OBJECTIVES

1 Solve Quadratic Equations of the Form $x^2 + bx + c = 0$ by Completing the Square. 

2 Solve Quadratic Equations of the Form $ax^2 + bx + c = 0$ by Completing the Square.

OBJECTIVE

1 Completing the Square to Solve $x^2 + bx + c = 0$ 

In the last section, we used the square root property to solve equations such as

$$(x + 1)^2 = 8 \quad \text{and} \quad (5x - 2)^2 = 3$$

Notice that one side of each equation is a quantity squared and that the other side is a constant.

To solve

$$x^2 + 2x = 4$$

notice that if we add 1 to both sides of the equation, the left side is a perfect square trinomial that can be factored.

$$x^2 + 2x + 1 = 4 + 1 \quad \text{Add 1 to both sides.}$$

$$(x + 1)^2 = 5 \quad \text{Factor.}$$

Now we can solve this equation as we did in the previous section by using the square root property.

$$x + 1 = \sqrt{5} \quad \text{or} \quad x + 1 = -\sqrt{5} \qquad \text{Use the square root property.}$$
$$x = -1 + \sqrt{5} \qquad x = -1 - \sqrt{5} \qquad \text{Solve.}$$

The solutions are $-1 \pm \sqrt{5}$.

Adding a number to $x^2 + 2x$ to form a perfect square trinomial is called **completing the square** on $x^2 + 2x$.

In general, we have the following.

Completing the Square

The number that completes the square on $x^2 + bx$ is $\left(\frac{b}{2}\right)^2$. To find $\left(\frac{b}{2}\right)^2$, **find half the coefficient of x, then square the result.**

By completing the square, a quadratic equation can be solved using the square root property.

EXAMPLE 1 Solve $x^2 + 6x + 3 = 0$ by completing the square.

Solution First we get the variable terms alone by subtracting 3 from both sides of the equation.

$$x^2 + 6x + 3 = 0$$
$$x^2 + 6x = -3 \qquad \text{Subtract 3 from both sides.}$$

Next we find half the coefficient of the x-term, then square it. Add this result to **both sides** of the equation. This will make the left side a perfect square trinomial. The coefficient of x is 6, and half of 6 is 3. So we add 3^2 or 9 to both sides.

$$x^2 + 6x + 9 = -3 + 9 \qquad \text{Complete the square.}$$
$$(x + 3)^2 = 6 \qquad \text{Factor the trinomial } x^2 + 6x + 9.$$
$$x + 3 = \sqrt{6} \quad \text{or} \quad x + 3 = -\sqrt{6} \qquad \text{Use the square root property.}$$
$$x = -3 + \sqrt{6} \qquad x = -3 - \sqrt{6} \qquad \text{Subtract 3 from both sides.}$$

Check by substituting $-3 + \sqrt{6}$ and $-3 - \sqrt{6}$ in the original equation. The solutions are $-3 \pm \sqrt{6}$. □

PRACTICE

1 Solve $x^2 + 2x - 5 = 0$ by completing the square.

Helpful Hint

Remember, when solving an equation by completing the square, add the number that completes the square to **both sides of the equation.**

EXAMPLE 2 Solve $x^2 - 10x = -14$ by completing the square.

Solution The variable terms are already alone on one side of the equation. The coefficient of x is -10. Half of -10 is -5, and $(-5)^2 = 25$. So we add 25 to both sides.

Helpful Hint

Add 25 to *both* sides of the equation.

$$x^2 - 10x = -14$$
$$x^2 - 10x + 25 = -14 + 25$$
$$(x - 5)^2 = 11 \qquad \text{Factor the trinomial and simplify } -14 + 25.$$
$$x - 5 = \sqrt{11} \quad \text{or} \quad x - 5 = -\sqrt{11} \qquad \text{Use the square root property.}$$
$$x = 5 + \sqrt{11} \qquad x = 5 - \sqrt{11} \qquad \text{Add 5 to both sides.}$$

The solutions are $5 \pm \sqrt{11}$. □

PRACTICE

2 Solve $x^2 - 8x = -8$ by completing the square.

OBJECTIVE

2 Completing the Square to Solve $ax^2 + bx + c = 0$

The method of completing the square can be used to solve *any* quadratic equation whether the coefficient of the squared variable is 1 or not. When the coefficient of the squared variable is not 1, we first divide both sides of the equation by the coefficient of the squared variable so that the new coefficient is 1. Then we complete the square.

EXAMPLE 3 Solve $4x^2 - 8x - 5 = 0$ by completing the square.

Solution $4x^2 - 8x - 5 = 0$

$$x^2 - 2x - \frac{5}{4} = 0 \quad \text{Divide both sides by 4.}$$

$$x^2 - 2x = \frac{5}{4} \quad \text{Get the variable terms alone on one side of the equation.}$$

The coefficient of x is -2. Half of -2 is -1, and $(-1)^2 = 1$. So we add 1 to both sides.

$$x^2 - 2x + 1 = \frac{5}{4} + 1$$

$$(x - 1)^2 = \frac{9}{4} \quad \text{Factor } x^2 - 2x + 1 \text{ and simplify } \frac{5}{4} + 1.$$

$$x - 1 = \sqrt{\frac{9}{4}} \quad \text{or} \quad x - 1 = -\sqrt{\frac{9}{4}} \quad \text{Use the square root property.}$$

$$x = 1 + \frac{3}{2} \qquad x = 1 - \frac{3}{2} \quad \text{Add 1 to both sides and simplify the radical.}$$

$$x = \frac{5}{2} \qquad x = -\frac{1}{2} \quad \text{Simplify.}$$

Both $\frac{5}{2}$ and $-\frac{1}{2}$ are solutions. □

PRACTICE

3 Solve $9x^2 - 36x - 13 = 0$ by completing the square.

The following steps may be used to solve a quadratic equation in x by completing the square.

Solving a Quadratic Equation in x by Completing the Square

Step 1. If the coefficient of x^2 is 1, go to Step 2. If not, divide both sides of the equation by the coefficient of x^2.

Step 2. Get all terms with variables on one side of the equation and constants on the other side.

Step 3. Find half the coefficient of x and then square the result. Add this number to both sides of the equation.

Step 4. Factor the resulting perfect square trinomial.

Step 5. Use the square root property to solve the equation.

EXAMPLE 4 Solve $2x^2 + 6x = -7$ by completing the square.

Solution The coefficient of x^2 is not 1. We divide both sides by 2, the coefficient of x^2.

$$2x^2 + 6x = -7$$

$$x^2 + 3x = -\frac{7}{2} \quad \text{Divide both sides by 2.}$$

$$x^2 + 3x + \frac{9}{4} = -\frac{7}{2} + \frac{9}{4} \quad \text{Add } \left(\frac{3}{2}\right)^2 \text{ or } \frac{9}{4} \text{ to both sides.}$$

$$\left(x + \frac{3}{2}\right)^2 = -\frac{5}{4} \quad \text{Factor the left side and simplify the right.}$$

There is no real solution to this equation since the square root of a negative number is not a real number. □

PRACTICE 4 Solve $2x^2 + 12x = -20$ by completing the square.

EXAMPLE 5 Solve $2x^2 = 10x + 1$ by completing the square.

Solution First we divide both sides of the equation by 2, the coefficient of x^2.

$$2x^2 = 10x + 1$$

$$x^2 = 5x + \frac{1}{2} \quad \text{Divide both sides by 2.}$$

Next we get the variable terms alone by subtracting $5x$ from both sides.

$$x^2 - 5x = \frac{1}{2}$$

$$x^2 - 5x + \frac{25}{4} = \frac{1}{2} + \frac{25}{4} \quad \text{Add } \left(-\frac{5}{2}\right)^2 \text{ or } \frac{25}{4} \text{ to both sides.}$$

$$\left(x - \frac{5}{2}\right)^2 = \frac{27}{4} \quad \text{Factor the left side and simplify the right side.}$$

$$x - \frac{5}{2} = \sqrt{\frac{27}{4}} \quad \text{or} \quad x - \frac{5}{2} = -\sqrt{\frac{27}{4}} \quad \text{Use the square root property.}$$

$$x - \frac{5}{2} = \frac{3\sqrt{3}}{2} \qquad x - \frac{5}{2} = -\frac{3\sqrt{3}}{2} \quad \text{Simplify.}$$

$$x = \frac{5}{2} + \frac{3\sqrt{3}}{2} \qquad x = \frac{5}{2} - \frac{3\sqrt{3}}{2}$$

The solutions are $\frac{5 \pm 3\sqrt{3}}{2}$. □

PRACTICE 5 Solve $2x^2 = 6x - 3$ by completing the square.

Vocabulary, Readiness & Video Check

Use the choices below to fill in each blank. Not all choices will be used, and these exercises come from Sections 9.1 and 9.2.

$\sqrt{a}$	linear equation	zero	$\left(\frac{b}{2}\right)^2$	$\frac{b}{2}$	6
$\pm\sqrt{a}$	quadratic equation	one	completing the square	9	3

1. By the zero factor property, if the product of two numbers is zero, then at least one of these two numbers must be ________________.
2. If a is a positive number, and if $x^2 = a$, then $x =$ ________________.
3. An equation that can be written in the form $ax^2 + bx + c = 0$ where a, b, and c are real numbers and a is not zero is called a ________________.
4. The process of solving a quadratic equation by writing it in the form $(x + a)^2 = c$ is called ________________.
5. To complete the square on $x^2 + 6x$, add ________________.
6. To complete the square on $x^2 + bx$, add ________________.

Martin-Gay Interactive Videos

See Video 9.2

Watch the section lecture video and answer the following questions.

OBJECTIVE 1

7. In  Examples 1 and 2, the constant to complete the square is added to both sides of the equation. Explain.

OBJECTIVE 2

8. In  Example 3, why is the equation first divided through by 2?

9.2 Exercise Set

MyMathLab®

Solve each quadratic equation by completing the square. See Examples 1 and 2.

1. $x^2 + 8x = -12$
2. $x^2 - 10x = -24$
3. $x^2 + 2x - 7 = 0$
4. $z^2 + 6z - 9 = 0$
5. $x^2 - 6x = 0$
6. $y^2 + 4y = 0$
7. $z^2 + 5z = 7$
8. $x^2 - 7x = 5$
9. $x^2 - 2x - 1 = 0$
10. $x^2 - 4x + 2 = 0$
11. $y^2 + 5y + 4 = 0$
12. $y^2 - 5y + 6 = 0$

Solve each quadratic equation by completing the square. See Examples 3 through 5.

13. $3x^2 - 6x = 24$
14. $2x^2 + 18x = -40$
15. $5x^2 + 10x + 6 = 0$
16. $3x^2 - 12x + 14 = 0$
17. $2x^2 = 6x + 5$
18. $4x^2 = -20x + 3$
19. $2y^2 + 8y + 5 = 0$
20. $4z^2 - 8z + 1 = 0$

MIXED PRACTICE

Solve each quadratic equation by completing the square. See Examples 1 through 5.

21. $x^2 + 6x - 25 = 0$
22. $x^2 - 6x + 7 = 0$
23. $x^2 - 3x - 3 = 0$
24. $x^2 - 9x + 3 = 0$
25. $2y^2 - 3y + 1 = 0$
26. $2y^2 - y - 1 = 0$
27. $x(x + 3) = 18$
28. $x(x - 3) = 18$
29. $3z^2 + 6z + 4 = 0$
30. $2y^2 + 8y + 9 = 0$
31. $4x^2 + 16x = 48$
32. $6x^2 - 30x = -36$

REVIEW AND PREVIEW

Simplify each expression. See Section 8.3.

33. $\frac{3}{4} - \sqrt{\frac{25}{16}}$
34. $\frac{3}{5} + \sqrt{\frac{16}{25}}$
35. $\frac{1}{2} - \sqrt{\frac{9}{4}}$
36. $\frac{9}{10} - \sqrt{\frac{49}{100}}$

Simplify each expression. See Section 8.4.

37. $\frac{6 + 4\sqrt{5}}{2}$
38. $\frac{10 - 20\sqrt{3}}{2}$
39. $\frac{3 - 9\sqrt{2}}{6}$
40. $\frac{12 - 8\sqrt{7}}{16}$

CONCEPT EXTENSIONS

41. In your own words, describe a perfect square trinomial.
42. Describe how to find the number to add to $x^2 - 7x$ to make a perfect square trinomial.
43. Write your own quadratic equation to be solved by completing the square. Write it in the form

perfect square trinomial = a number that is not a perfect square

$$\underbrace{x^2 + 6x + 9}_{} = 11$$

For example,

a. Solve $x^2 + 6x + 9 = 11$.

b. Write and solve your quadratic equation by completing the square.

44. Follow the directions of Exercise 43, except write your equation in the form

perfect square trinomial = negative number

Solve your quadratic equation by completing the square.

45. Find a value of k that will make $x^2 + kx + 16$ a perfect square trinomial.
46. Find a value of k that will make $x^2 + kx + 25$ a perfect square trinomial.
47. Retail sales y (in billions of dollars) from online shopping and mail order in the United States from 2002 through 2007 can be approximated by the equation $y = 2.5x^2 + 7.5x + 122$, where x is the number of years since 2002. Assume this trend continued and predict the year after 2002 in which the retail sales for online and mail-order shopping were \$392 billion. (Round to the nearest whole number.) (*Source:* Based on data from the U.S. Bureau of the Census, Retail Surveys Branch)

48. The average price y of gold (in dollars per ounce) from 2003 through 2007 is given by the equation $y = 12x^2 + 36x + 362$, where x is the number of years after 2003. Assume that this trend continues and find the year after 2003 in which the price of gold will be $2210 per ounce. (Round to the nearest whole number.) (*Source:* Based on data from the U.S. Geological Survey, Minerals Information)

Recall that a graphing calculator may be used to solve an equation. For example, to solve $x^2 + 8x = -12$ (Exercise 1), graph

$y_1 = x^2 + 8x$ *(left side of equation) and*

$y_2 = -12$ *(right side of equation)*

The x-coordinate of the point of intersection of the graphs is the solution. Use a graphing calculator and solve each equation. Round solutions to the nearest hundredth.

 **49.** Exercise 1  **50.** Exercise 2

51. Exercise 17 **52.** Exercise 18

9.3 Solving Quadratic Equations by the Quadratic Formula

OBJECTIVES

1. Use the Quadratic Formula to Solve Quadratic Equations.
2. Approximate Solutions to Quadratic Equations.
3. Determine the Number of Solutions of a Quadratic Equation by Using the Discriminant.

OBJECTIVE

1 Using the Quadratic Formula

We can use the technique of completing the square to develop a formula to find solutions of any quadratic equation. We develop and use the **quadratic formula** in this section.

Recall that a quadratic equation in **standard form** is

$$ax^2 + bx + c = 0, \quad a \neq 0$$

To develop the quadratic formula, let's complete the square for this quadratic equation in standard form.

First we divide both sides of the equation by the coefficient of x^2 and then get the variable terms alone on one side of the equation.

$$x^2 + \frac{b}{a}x + \frac{c}{a} = 0 \quad \text{Divide by } a\text{; recall that } a \text{ cannot be 0.}$$

$$x^2 + \frac{b}{a}x = -\frac{c}{a} \quad \text{Get the variable terms alone on one side of the equation.}$$

The coefficient of x is $\frac{b}{a}$. Half of $\frac{b}{a}$ is $\frac{b}{2a}$ and $\left(\frac{b}{2a}\right)^2 = \frac{b^2}{4a^2}$. So we add $\frac{b^2}{4a^2}$ to both sides of the equation.

$$x^2 + \frac{b}{a}x + \frac{b^2}{4a^2} = -\frac{c}{a} + \frac{b^2}{4a^2} \quad \text{Add } \frac{b^2}{4a^2} \text{ to both sides.}$$

$$\left(x + \frac{b}{2a}\right)^2 = -\frac{c}{a} + \frac{b^2}{4a^2} \quad \text{Factor the left side.}$$

$$\left(x + \frac{b}{2a}\right)^2 = -\frac{4ac}{4a^2} + \frac{b^2}{4a^2} \quad \text{Multiply } -\frac{c}{a} \text{ by } \frac{4a}{4a} \text{ so that both terms on the right side have a common denominator.}$$

$$\left(x + \frac{b}{2a}\right)^2 = \frac{b^2 - 4ac}{4a^2} \quad \text{Simplify the right side.}$$

Now we use the square root property.

$$x + \frac{b}{2a} = \sqrt{\frac{b^2 - 4ac}{4a^2}} \quad \text{or} \quad x + \frac{b}{2a} = -\sqrt{\frac{b^2 - 4ac}{4a^2}} \quad \text{Use the square root property.}$$

$$x + \frac{b}{2a} = \frac{\sqrt{b^2 - 4ac}}{2a} \qquad x + \frac{b}{2a} = -\frac{\sqrt{b^2 - 4ac}}{2a} \quad \text{Simplify the radical.}$$

$$x = -\frac{b}{2a} + \frac{\sqrt{b^2 - 4ac}}{2a} \qquad x = -\frac{b}{2a} - \frac{\sqrt{b^2 - 4ac}}{2a} \quad \text{Subtract } \frac{b}{2a} \text{ from both sides.}$$

$$x = \frac{-b + \sqrt{b^2 - 4ac}}{2a} \qquad x = \frac{-b - \sqrt{b^2 - 4ac}}{2a} \quad \text{Simplify.}$$

The solutions are $\frac{-b \pm \sqrt{b^2 - 4ac}}{2a}$. This final equation is called the **quadratic formula** and gives the solutions of any quadratic equation.

Quadratic Formula

If a, b, and c are real numbers and $a \neq 0$, a quadratic equation written in the form $ax^2 + bx + c = 0$ has solutions

$$x = \frac{-b \pm \sqrt{b^2 - 4ac}}{2a}$$

Helpful Hint

Don't forget that to identify a, b, and c correctly in the quadratic formula, you should write the equation in standard form.

Quadratic Equations in Standard Form

$5x^2 - 6x + 2 = 0$	$a = 5, b = -6, c = 2$
$4y^2 - 9 = 0$	$a = 4, b = 0, c = -9$
$x^2 + x = 0$	$a = 1, b = 1, c = 0$
$\sqrt{2}x^2 + \sqrt{5}x + \sqrt{3} = 0$	$a = \sqrt{2}, b = \sqrt{5}, c = \sqrt{3}$

EXAMPLE 1 Solve $3x^2 + x - 3 = 0$ using the quadratic formula.

Solution This equation is in standard form with $a = 3$, $b = 1$, and $c = -3$. By the quadratic formula, we have

$$x = \frac{-b \pm \sqrt{b^2 - 4ac}}{2a}$$

$$x = \frac{-1 \pm \sqrt{1^2 - 4 \cdot 3 \cdot (-3)}}{2 \cdot 3} \quad \text{Let } a = 3, b = 1, \text{ and } c = -3.$$

$$= \frac{-1 \pm \sqrt{1 + 36}}{6} \quad \text{Simplify.}$$

$$= \frac{-1 \pm \sqrt{37}}{6}$$

Check both solutions in the original equation. The solutions are $\frac{-1 + \sqrt{37}}{6}$ and $\frac{-1 - \sqrt{37}}{6}$. □

PRACTICE

1 Solve $5x^2 + x - 2 = 0$ using the quadratic formula.

EXAMPLE 2 Solve $2x^2 - 9x = 5$ using the quadratic formula.

Solution First we write the equation in standard form by subtracting 5 from both sides.

$$2x^2 - 9x = 5$$
$$2x^2 - 9x - 5 = 0$$

Next we note that $a = 2$, $b = -9$, and $c = -5$. We substitute these values into the quadratic formula.

Helpful Hint

Notice that the fraction bar is under the entire numerator of $-b \pm \sqrt{b^2 - 4ac}$.

$$x = \frac{-b \pm \sqrt{b^2 - 4ac}}{2a}$$

$$x = \frac{-(-9) \pm \sqrt{(-9)^2 - 4 \cdot 2 \cdot (-5)}}{2 \cdot 2} \quad \text{Substitute in the formula.}$$

$$= \frac{9 \pm \sqrt{81 + 40}}{4} \quad \text{Simplify.}$$

$$= \frac{9 \pm \sqrt{121}}{4} = \frac{9 \pm 11}{4}$$

Then,

$$x = \frac{9 - 11}{4} = -\frac{1}{2} \quad \text{or} \quad x = \frac{9 + 11}{4} = 5$$

Check $-\frac{1}{2}$ and 5 in the original equation. Both $-\frac{1}{2}$ and 5 are solutions. □

PRACTICE
2 Solve $3x^2 + 2x = 8$ using the quadratic formula.

The following steps may be useful when solving a quadratic equation by the quadratic formula.

Solving a Quadratic Equation by the Quadratic Formula

Step 1. Write the quadratic equation in standard form: $ax^2 + bx + c = 0$.

Step 2. If necessary, clear the equation of fractions to simplify calculations.

Step 3. Identify a, b, and c.

Step 4. Replace a, b, and c in the quadratic formula with the identified values and simplify.

✓CONCEPT CHECK

For the quadratic equation $2x^2 - 5 = 7x$, if $a = 2$ and $c = -5$ in the quadratic formula, the value of b is which of the following?

a. $\frac{7}{2}$ b. 7 c. -5 d. -7

EXAMPLE 3 Solve $7x^2 = 1$ using the quadratic formula.

Solution First we write the equation in standard form by subtracting 1 from both sides.

$$7x^2 = 1$$
$$7x^2 - 1 = 0$$

Helpful Hint
$7x^2 - 1 = 0$ can be written as $7x^2 + 0x - 1 = 0$. This form helps you see that $b = 0$.

Next we replace a, b, and c with the identified values: $a = 7$, $b = 0$ and $c = -1$.

$$x = \frac{0 \pm \sqrt{0^2 - 4 \cdot 7 \cdot (-1)}}{2 \cdot 7} \quad \text{Substitute in the formula.}$$

$$= \frac{\pm\sqrt{28}}{14} \quad \text{Simplify.}$$

$$= \frac{\pm 2\sqrt{7}}{14}$$

$$= \pm\frac{\sqrt{7}}{7}$$

The solutions are $\frac{\sqrt{7}}{7}$ and $-\frac{\sqrt{7}}{7}$. □

PRACTICE
3 Solve $3x^2 = 5$ using the quadratic formula.

Notice that the equation in Example 3, $7x^2 = 1$, could have been easily solved by dividing both sides by 7 and then using the square root property. We solved the equation by the quadratic formula to show that this formula can be used to solve any quadratic equation.

Answer to Concept Check: d

EXAMPLE 4 Solve $x^2 = -x - 1$ using the quadratic formula.

Solution First we write the equation in standard form.

$$x^2 + x + 1 = 0$$

Next we replace a, b, and c in the quadratic formula with $a = 1$, $b = 1$, and $c = 1$.

$$x = \frac{-1 \pm \sqrt{1^2 - 4 \cdot 1 \cdot 1}}{2 \cdot 1} \quad \text{Substitute in the formula.}$$

$$= \frac{-1 \pm \sqrt{-3}}{2} \quad \text{Simplify.}$$

There is no real number solution because $\sqrt{-3}$ is not a real number. □

PRACTICE 4 Solve $x^2 = 3x - 4$ using the quadratic formula.

EXAMPLE 5 Solve $\frac{1}{2}x^2 - x = 2$ by using the quadratic formula.

Solution We write the equation in standard form and then clear the equation of fractions by multiplying both sides by the LCD, 2.

$$\frac{1}{2}x^2 - x = 2$$

$$\frac{1}{2}x^2 - x - 2 = 0 \quad \text{Write in standard form.}$$

$$x^2 - 2x - 4 = 0 \quad \text{Multiply both sides by 2.}$$

Here, $a = 1$, $b = -2$, and $c = -4$, so we substitute these values into the quadratic formula.

$$x = \frac{-(-2) \pm \sqrt{(-2)^2 - 4 \cdot 1 \cdot (-4)}}{2 \cdot 1}$$

$$= \frac{2 \pm \sqrt{20}}{2} = \frac{2 \pm 2\sqrt{5}}{2} \quad \text{Simplify.}$$

$$= \frac{2(1 \pm \sqrt{5})}{2} = 1 \pm \sqrt{5} \quad \text{Factor and simplify.}$$

The solutions are $1 - \sqrt{5}$ and $1 + \sqrt{5}$. □

PRACTICE 5 Solve $\frac{1}{5}x^2 - x = 1$ using the quadratic formula.

Notice that in Example 5, although we cleared the equation of fractions, the coefficients $a = \frac{1}{2}$, $b = -1$, and $c = -2$ will give the same results.

Helpful Hint

When simplifying an expression such as

$$\frac{3 \pm 6\sqrt{2}}{6}$$

first factor out a common factor from the terms of the numerator and then simplify.

$$\frac{3 \pm 6\sqrt{2}}{6} = \frac{3(1 \pm 2\sqrt{2})}{2 \cdot 3} = \frac{1 \pm 2\sqrt{2}}{2}$$

OBJECTIVE

2 Approximate Solutions to Quadratic Equations 

Sometimes approximate solutions for quadratic equations are appropriate.

EXAMPLE 6 Approximate the exact solutions of the quadratic equation in Example 1. Round the approximations to the nearest tenth.

Solution From Example 1, we have exact solutions $\frac{-1 \pm \sqrt{37}}{6}$. Thus,

$$\frac{-1 + \sqrt{37}}{6} \approx 0.847127088 \approx 0.8 \text{ to the nearest tenth.}$$

$$\frac{-1 - \sqrt{37}}{6} \approx -1.180460422 \approx -1.2 \text{ to the nearest tenth.}$$

Thus approximate solutions to the quadratic equation in Example 1 are 0.8 and −1.2. □

PRACTICE

6 Approximate the exact solutions of the quadratic equation in Practice 1. Round the approximations to the nearest tenth.

OBJECTIVE

3 Using the Discriminant 

In the quadratic formula, $x = \frac{-b \pm \sqrt{b^2 - 4ac}}{2a}$, the radicand $b^2 - 4ac$ is called the **discriminant** because, by knowing its value, we can **discriminate** among the possible number and type of solutions of a quadratic equation. Possible values of the discriminant and their meanings are summarized next.

Discriminant

The following table corresponds the discriminant $b^2 - 4ac$ of a quadratic equation of the form $ax^2 + bx + c = 0$ with the number of solutions of the equation.

$b^2 - 4ac$	*Number of Solutions*
Positive	Two distinct real solutions
Zero	One real solution
Negative	No real solution*

*In this case, the quadratic equation will have two complex (but not real) solutions. See Section 9.4 for a discussion of complex numbers.

EXAMPLE 7 Use the discriminant to determine the number of solutions of $3x^2 + x - 3 = 0$.

Solution In $3x^2 + x - 3 = 0$, $a = 3$, $b = 1$, and $c = -3$. Then

$$b^2 - 4ac = (1)^2 - 4(3)(-3) = 1 + 36 = 37$$

Since the discriminant is 37, a positive number, this equation has two distinct real solutions.

We solved this equation in Example 1 of this section, and the solutions are $\frac{-1 + \sqrt{37}}{6}$ and $\frac{-1 - \sqrt{37}}{6}$, two distinct real solutions. □

PRACTICE

7 Use the discriminant to determine the number of solutions of $5x^2 + x - 2 = 0$.

EXAMPLE 8 Use the discriminant to determine the number of solutions of each quadratic equation.

a. $x^2 - 6x + 9 = 0$ **b.** $5x^2 + 4 = 0$

Solution

a. In $x^2 - 6x + 9 = 0$, $a = 1$, $b = -6$, and $c = 9$.

$$b^2 - 4ac = (-6)^2 - 4(1)(9) = 36 - 36 = 0$$

Since the discriminant is 0, this equation has one real solution.

b. In $5x^2 + 4 = 0$, $a = 5$, $b = 0$, and $c = 4$.

$$b^2 - 4ac = 0^2 - 4(5)(4) = 0 - 80 = -80$$

Since the discriminant is -80, a negative number, this equation has no real solution. □

PRACTICE

8 Use the discriminant to determine the number of solutions of each quadratic equation.

a. $x^2 - 10x + 35 = 0$ **b.** $5x^2 + 3x = 0$

Vocabulary, Readiness & Video Check

Fill in each blank.

1. The quadratic formula is ______________________.

Identify the values of a, b, and c in the quadratic equation.

2. $x^2 - 6 = 0$; $a =$ ________, $b =$ ________, $c =$ ________

Simplify the following.

3. $\dfrac{-1 \pm \sqrt{1^2 - 4(1)(-2)}}{2(1)}$

4. $\dfrac{-(-5) \pm \sqrt{(-5)^2 - 4(2)(3)}}{2(2)}$

5. $\dfrac{-5 \pm \sqrt{5^2 - 4(1)(2)}}{2(1)}$

6. $\dfrac{-7 \pm \sqrt{7^2 - 4(2)(1)}}{2(2)}$

Martin-Gay Interactive Videos

See Video 9.3

Watch the section lecture video and answer the following questions.

OBJECTIVE 1

7. Based on the lectures and Examples 1–3, answer the following.

a. Must a quadratic equation be written in standard form in order to use the quadratic formula? Why or why not?

b. Must fractions be cleared from the equation before using the quadratic formula? Why or why not?

OBJECTIVE 2

8. From Example 4, how are approximate solutions found?

OBJECTIVE 3

9. Based on Examples 5–7, complete the following statements. The discriminant is the ____________ portion of the quadratic formula and can be used to find out the number of ____________ solutions of a quadratic equation without solving the equation. To use the discriminant, the quadratic equation needs to be written in ____________ form.

9.3 Exercise Set

MyMathLab®

Use the quadratic formula to solve each quadratic equation. See Examples 1 through 4.

1. $x^2 - 3x + 2 = 0$

2. $x^2 - 5x - 6 = 0$

3. $3k^2 + 7k + 1 = 0$

4. $7k^2 + 3k - 1 = 0$

5. $49x^2 - 4 = 0$

6. $25x^2 - 15 = 0$

7. $5z^2 - 4z + 3 = 0$

8. $3z^2 + 2z + 1 = 0$

9. $y^2 = 7y + 30$

10. $y^2 = 5y + 36$

11. $2x^2 = 10$

12. $5x^2 = 15$

13. $m^2 - 12 = m$

14. $m^2 - 14 = 5m$

15. $3 - x^2 = 4x$

16. $10 - x^2 = 2x$

17. $2a^2 - 7a + 3 = 0$

18. $3a^2 - 7a + 2 = 0$

19. $x^2 - 5x - 2 = 0$

20. $x^2 - 2x - 5 = 0$

21. $3x^2 - x - 14 = 0$

22. $5x^2 - 13x - 6 = 0$

23. $6x^2 + 9x = 2$

24. $3x^2 - 9x = 8$

25. $7p^2 + 2 = 8p$

26. $11p^2 + 2 = 10p$

27. $a^2 - 6a + 2 = 0$

28. $a^2 - 10a + 19 = 0$

29. $2x^2 - 6x + 3 = 0$

30. $5x^2 - 8x + 2 = 0$

31. $3x^2 = 1 - 2x$

32. $5y^2 = 4 - y$

33. $20y^2 = 3 - 11y$

34. $2z^2 = z + 3$

35. $x^2 + x + 1 = 0$

36. $k^2 + 2k + 5 = 0$

37. $4y^2 = 6y + 1$

38. $6z^2 = 2 - 3z$

Use the quadratic formula to solve each quadratic equation. See Example 5.

39. $\frac{m^2}{2} = m + \frac{1}{2}$

40. $\frac{m^2}{2} = 3m - 1$

41. $3p^2 - \frac{2}{3}p + 1 = 0$

42. $\frac{5}{2}p^2 - p + \frac{1}{2} = 0$

43. $4p^2 + \frac{3}{2} = -5p$

44. $4p^2 + \frac{3}{2} = 5p$

45. $5x^2 = \frac{7}{2}x + 1$

46. $2x^2 = \frac{5}{2}x + \frac{7}{2}$

47. $x^2 - \frac{11}{2}x - \frac{1}{2} = 0$

48. $\frac{2}{3}x^2 - 2x - \frac{2}{3} = 0$

49. $5z^2 - 2z = \frac{1}{5}$

50. $9z^2 + 12z = -1$

MIXED PRACTICE

Use the quadratic formula to solve each quadratic equation. Find the exact solutions; then approximate these solutions to the nearest tenth. See Examples 1 through 6.

51. $3x^2 = 21$

52. $2x^2 = 26$

53. $x^2 + 6x + 1 = 0$

54. $x^2 + 4x + 2 = 0$

55. $x^2 = 9x + 4$

56. $x^2 = 7x + 5$

57. $3x^2 - 2x - 2 = 0$

58. $5x^2 - 3x - 1 = 0$

Use the discriminant to determine the number of solutions of each quadratic equation. See Examples 7 and 8.

59. $x^2 + 3x - 1 = 0$

60. $x^2 - 5x - 3 = 0$

61. $3x^2 + x + 5 = 0$

62. $2x^2 + x + 4 = 0$

63. $4x^2 + 4x = -1$

64. $x^2 + 10x = -25$

65. $9x^2 + 2x = 0$

66. $7x^2 - x = 0$

67. $5x^2 + 1 = 0$

68. $10x^2 + 2 = 0$

69. $x^2 + 36 = -12x$

70. $4x^2 + 9 = 12x$

REVIEW AND PREVIEW

Simplify each radical. See Section 8.2.

71. $\sqrt{48}$

72. $\sqrt{104}$

73. $\sqrt{50}$

74. $\sqrt{80}$

Solve the following. See Section 2.6.

△ **75.** The height of a triangle is 4 times the length of the base. The area of the triangle is 18 square feet. Find the height and base of the triangle.

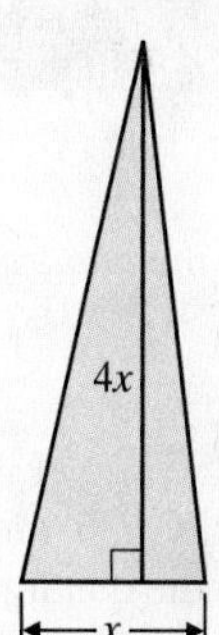

△ **76.** The length of a rectangle is 6 inches more than its width. The area of the rectangle is 391 square inches. Find the dimensions of the rectangle.

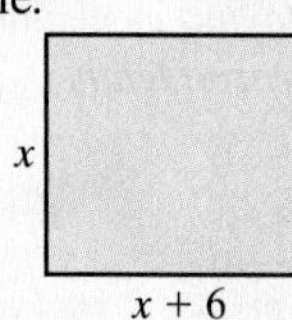

CONCEPT EXTENSIONS

Solve. See the Concept Check in this section. For the quadratic equation $5x^2 + 2 = x$, if $a = 5$ in the quadratic formula,

77. What is the value of b?

a. $\frac{1}{5}$ **b.** 0 **c.** -1 **d.** 1

78. What is the value of c?

a. 5 **b.** x **c.** -2 **d.** 2

For the quadratic equation $7y^2 = 3y$, if $b = 3$, in the quadratic formula,

79. What is the value of a?

a. 7 **b.** −7 **c.** 0 **d.** 1

80. What is the value of c?

a. 7 **b.** 3 **c.** 0 **d.** 1

△ **81.** In a recent year, Nestle created a chocolate bar that the company claimed weighed more than 2 tons. The rectangular bar had a base area of approximately 34.65 square feet, and its length was 0.6 foot shorter than three times its width. Find the length and width of the bar. (*Source:* Nestle)

△ **82.** The area of a rectangular conference room table is 95 square feet. If its length is six feet longer than its width, find the dimensions of the table. Round each dimension to the nearest tenth.

Solve each equation using the quadratic formula.

83. $x^2 + 3\sqrt{2}x - 5 = 0$

84. $y^2 - 2\sqrt{5}y - 1 = 0$

85. Explain how to identify a, b, and c correctly when solving a quadratic equation by the quadratic formula.

86. Explain how the quadratic formula is developed and why it is useful.

Use the quadratic formula and a calculator to solve each equation. Round solutions to the nearest tenth.

87. $7.3z^2 + 5.4z - 1.1 = 0$

88. $1.2x^2 - 5.2x - 3.9 = 0$

A rocket is launched from the top of an 80-foot cliff with an initial velocity of 120 feet per second. The height, h, of the rocket after t seconds is given by the equation $h = -16t^2 + 120t + 80$. Use this for Exercises 89 and 90.

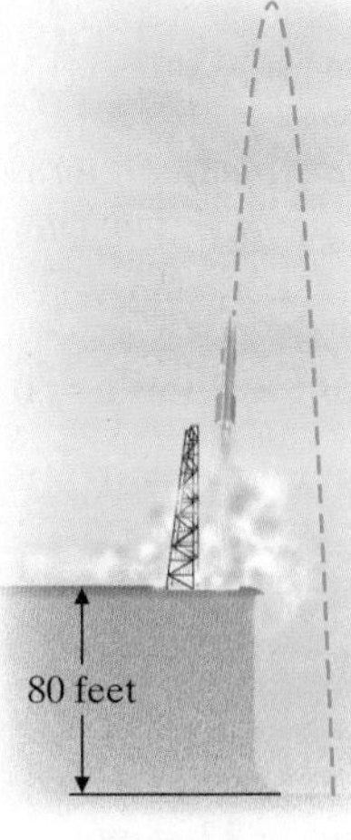

89. How long after the rocket is launched will it be 30 feet from the ground? Round to the nearest tenth of a second.

90. How long after the rocket is launched will it strike the ground? Round to the nearest tenth of a second. (*Hint:* The rocket will strike the ground when its height $h = 0$.)

91. The number of patients on the kidney waiting list from 2003 to 2010 can be modeled by the equation $y = 209x^2 - 1917x + 57{,}131$, where $x = 0$ represents the year 2000. Assume that this trend continues and predict the year in which 71,257 patients will be on the waiting list.

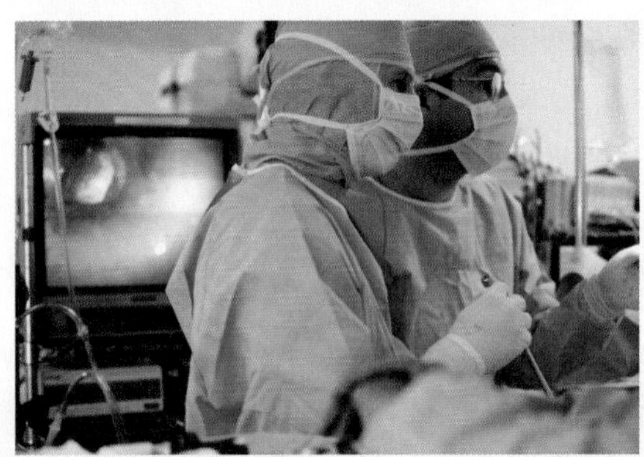

92. The revenues from product sales y (in thousands of dollars) of ABIOMED, Inc., maker of the AbioCor artificial heart, from 2004 through 2008 can be modeled by the equation $y = -587x^2 + 10{,}243x + 26{,}047$, where $x = 0$ represents the year 2004. Assume that this trend continues and predict the first year in which ABIOMED's revenues from product sales will be \$70,423 thousand. (*Source:* Based on data from ABIOMED Corporation)

Integrated Review SUMMARY ON SOLVING QUADRATIC EQUATIONS

Sections 9.1–9.3

An important skill in mathematics is learning when to use one technique in favor of another. We now practice this by deciding which method to use when solving quadratic equations. Although both the quadratic formula and completing the square can be used to solve any quadratic equation, the quadratic formula is usually less tedious and thus preferred. The following steps may be used to solve a quadratic equation.

To Solve a Quadratic Equation

Step 1. If the equation is in the form $ax^2 = c$ or $(ax + b)^2 = c$, use the square root property and solve. If not, go to Step 2.

Step 2. Write the equation in standard form: $ax^2 + bx + c = 0$.

Step 3. Try to solve the equation by the factoring method. If not possible, go to Step 4.

Step 4. Solve the equation by the quadratic formula.

Study the examples below to help you review these steps.

EXAMPLE 1 Solve $m^2 - 2m - 7 = 0$.

Solution The equation is in standard form, but the quadratic expression $m^2 - 2m - 7$ is not factorable, so use the quadratic formula with $a = 1$, $b = -2$, and $c = -7$.

$$m^2 - 2m - 7 = 0$$

$$m = \frac{-(-2) \pm \sqrt{(-2)^2 - 4 \cdot 1 \cdot (-7)}}{2 \cdot 1} = \frac{2 \pm \sqrt{32}}{2}$$

$$m = \frac{2 \pm 4\sqrt{2}}{2} = \frac{2(1 \pm 2\sqrt{2})}{2} = 1 \pm 2\sqrt{2}$$

The solutions are $1 - 2\sqrt{2}$ and $1 + 2\sqrt{2}$. □

PRACTICE

1 Solve $y^2 - 3y - 4 = 0$.

EXAMPLE 2 Solve $(3x + 1)^2 = 20$.

Solution This equation is in a form that makes the square root property easy to apply.

$$(3x + 1)^2 = 20$$

$$3x + 1 = \pm\sqrt{20} \quad \text{Apply the square root property.}$$

$$3x + 1 = \pm 2\sqrt{5} \quad \text{Simplify } \sqrt{20}.$$

$$3x = -1 \pm 2\sqrt{5}$$

$$x = \frac{-1 \pm 2\sqrt{5}}{3}$$

The solutions are $\frac{-1 - 2\sqrt{5}}{3}$ and $\frac{-1 + 2\sqrt{5}}{3}$. □

PRACTICE

2 Solve $(2x + 5)^2 = 45$.

EXAMPLE 3 Solve $x^2 - \frac{11}{2}x = -\frac{5}{2}$.

Solution The fractions make factoring more difficult and complicate the calculations for using the quadratic formula. Clear the equation of fractions by multiplying both sides of the equation by the LCD 2.

$$x^2 - \frac{11}{2}x = -\frac{5}{2}$$

$$x^2 - \frac{11}{2}x + \frac{5}{2} = 0 \quad \text{Write in standard form.}$$

$$2x^2 - 11x + 5 = 0 \quad \text{Multiply both sides by 2.}$$

$$(2x - 1)(x - 5) = 0 \quad \text{Factor.}$$

$$2x - 1 = 0 \quad \text{or} \quad x - 5 = 0 \quad \text{Apply the zero factor theorem.}$$

$$2x = 1 \qquad x = 5$$

$$x = \frac{1}{2} \qquad x = 5$$

The solutions are $\frac{1}{2}$ and 5. □

PRACTICE 3 Solve $x^2 - \frac{5}{2}x = -\frac{3}{2}$.

Choose and use a method to solve each equation.

1. $5x^2 - 11x + 2 = 0$
2. $5x^2 + 13x - 6 = 0$
3. $x^2 - 1 = 2x$
4. $x^2 + 7 = 6x$
5. $a^2 = 20$
6. $a^2 = 72$
7. $x^2 - x + 4 = 0$
8. $x^2 - 2x + 7 = 0$
9. $3x^2 - 12x + 12 = 0$
10. $5x^2 - 30x + 45 = 0$
11. $9 - 6p + p^2 = 0$
12. $49 - 28p + 4p^2 = 0$
13. $4y^2 - 16 = 0$
14. $3y^2 - 27 = 0$
15. $x^4 - 3x^3 + 2x^2 = 0$
16. $x^3 + 7x^2 + 12x = 0$
17. $(2z + 5)^2 = 25$
18. $(3z - 4)^2 = 16$
19. $30x = 25x^2 + 2$
20. $12x = 4x^2 + 4$
21. $\frac{2}{3}m^2 - \frac{1}{3}m - 1 = 0$
22. $\frac{5}{8}m^2 + m - \frac{1}{2} = 0$
23. $x^2 - \frac{1}{2}x - \frac{1}{5} = 0$
24. $x^2 + \frac{1}{2}x - \frac{1}{8} = 0$
25. $4x^2 - 27x + 35 = 0$
26. $9x^2 - 16x + 7 = 0$
27. $(7 - 5x)^2 = 18$
28. $(5 - 4x)^2 = 75$
29. $3z^2 - 7z = 12$
30. $6z^2 + 7z = 6$
31. $x = x^2 - 110$
32. $x = 56 - x^2$
33. $\frac{3}{4}x^2 - \frac{5}{2}x - 2 = 0$
34. $x^2 - \frac{6}{5}x - \frac{8}{5} = 0$
35. $x^2 - 0.6x + 0.05 = 0$
36. $x^2 - 0.1x - 0.06 = 0$
37. $10x^2 - 11x + 2 = 0$
38. $20x^2 - 11x + 1 = 0$
39. $\frac{1}{2}z^2 - 2z + \frac{3}{4} = 0$
40. $\frac{1}{5}z^2 - \frac{1}{2}z - 2 = 0$
41. Explain how you will decide which method to use when solving quadratic equations.

9.4 Complex Solutions of Quadratic Equations 

OBJECTIVES

1. Write Complex Numbers Using *i* Notation.
2. Add and Subtract Complex Numbers.
3. Multiply Complex Numbers.
4. Divide Complex Numbers.
5. Solve Quadratic Equations That Have Complex Solutions.

In Chapter 8, we learned that $\sqrt{-4}$, for example, is not a real number because there is no real number whose square is -4. However, our real number system can be extended to include numbers like $\sqrt{-4}$. This extended number system is called the **complex number** system. The complex number system includes the **imaginary unit** i, which is defined next.

Imaginary Unit *i*

The imaginary unit, written i, is the number whose square is -1. That is,

$$i^2 = -1 \quad \text{and} \quad i = \sqrt{-1}$$

OBJECTIVE

1 Writing Complex Numbers Using *i* Notation 

We use i to write numbers like $\sqrt{-6}$ as the product of a real number and i. Since $i = \sqrt{-1}$, we have

$$\sqrt{-6} = \sqrt{-1 \cdot 6} = \sqrt{-1} \cdot \sqrt{6} = i\sqrt{6}$$

EXAMPLE 1 Write each radical as the product of a real number and i.

a. $\sqrt{-4}$ **b.** $\sqrt{-11}$ **c.** $\sqrt{-20}$

Solution Write each negative radicand as a product of a positive number and -1. Then write $\sqrt{-1}$ as i.

a. $\sqrt{-4} = \sqrt{-1 \cdot 4} = \sqrt{-1} \cdot \sqrt{4} = i \cdot 2 = 2i$

b. $\sqrt{-11} = \sqrt{-1 \cdot 11} = \sqrt{-1} \cdot \sqrt{11} = i\sqrt{11}$

c. $\sqrt{-20} = \sqrt{-1 \cdot 20} = \sqrt{-1} \cdot \sqrt{20} = i \cdot 2\sqrt{5} = 2i\sqrt{5}$ □

PRACTICE

1 Write each radical as the product of a real number and i.

a. $\sqrt{-36}$ **b.** $\sqrt{-15}$ **c.** $\sqrt{-48}$

The numbers $2i$, $i\sqrt{11}$, and $2i\sqrt{5}$ are called **pure imaginary numbers.** Both real numbers and pure imaginary numbers are complex numbers.

Complex Numbers and Pure Imaginary Numbers

A complex number is a number that can be written in the form

$$a + bi$$

where a and b are real numbers. A complex number that can be written in the form

$$0 + bi$$

$b \neq 0$, is also called a pure imaginary number.

A complex number written in the form $a + bi$ is in **standard form.** We call a the real part and bi the imaginary part of the complex number $a + bi$.

EXAMPLE 2 Identify each number as a complex number by writing it in standard form $a + bi$. (Solution on next page)

a. 7 **b.** 0 **c.** $\sqrt{20}$ **d.** $\sqrt{-27}$ **e.** $2 + \sqrt{-4}$

Solution

a. 7 is a complex number since $7 = 7 + 0i$.

b. 0 is a complex number since $0 = 0 + 0i$.

c. $\sqrt{20}$ is a complex number since $\sqrt{20} = 2\sqrt{5} = 2\sqrt{5} + 0i$.

d. $\sqrt{-27}$ is a complex number since $\sqrt{-27} = i \cdot 3\sqrt{3} = 0 + 3i\sqrt{3}$.

e. $2 + \sqrt{-4}$ is a complex number since $2 + \sqrt{-4} = 2 + 2i$. □

PRACTICE 2 Identify each number as a complex number by writing it in standard form $a + bi$.

a. 6 **b.** 0 **c.** $\sqrt{24}$

d. $\sqrt{-1}$ **e.** $5 + \sqrt{-9}$

OBJECTIVE 2

2 Adding and Subtracting Complex Numbers 

We now present arithmetic operations—addition, subtraction, multiplication, and division—for the complex number system. Complex numbers are added and subtracted in the same way as we add and subtract polynomials.

EXAMPLE 3 Simplify the sum or difference. Write the result in standard form.

a. $(2 + 3i) + (-6 - i)$ **b.** $-i + (3 + 7i)$ **c.** $(5 - i) - 4$

Solution Add the real parts and then add the imaginary parts.

a. $(2 + 3i) + (-6 - i) = [2 + (-6)] + (3i - i) = -4 + 2i$

b. $-i + (3 + 7i) = 3 + (-i + 7i) = 3 + 6i$

c. $(5 - i) - 4 = (5 - 4) - i = 1 - i$ □

PRACTICE 3 Simplify the sum or difference. Write the result in standard form.

a. $(4 + 3i) + (-8 - 2i)$ **b.** $5i + (6 - 9i)$

c. $(3 - 2i) - 4$

EXAMPLE 4 Subtract $(11 - i)$ from $(1 + i)$.

Solution $(1 + i) - (11 - i) = 1 + i - 11 + i = (1 - 11) + (i + i) = -10 + 2i$ □

PRACTICE 4 Subtract $(13 - 5i)$ from $3i$.

OBJECTIVE 3

3 Multiplying Complex Numbers 

Use the distributive property and the FOIL method to multiply complex numbers.

EXAMPLE 5 Find the following products and write in standard form.

a. $5i(2 - i)$ **b.** $(7 - 3i)(4 + 2i)$ **c.** $(2 + 3i)(2 - 3i)$

Solution

a. By the distributive property, we have

$$\begin{aligned} 5i(2 - i) &= 5i \cdot 2 - 5i \cdot i && \text{Apply the distributive property.} \\ &= 10i - 5i^2 \\ &= 10i - 5(-1) && \text{Write } i^2 \text{ as } -1. \\ &= 10i + 5 \\ &= 5 + 10i && \text{Write in standard form.} \end{aligned}$$

b. $(7 - 3i)(4 + 2i) = \overset{F}{28} + \overset{O}{14i} - \overset{I}{12i} - \overset{L}{6i^2}$

$= 28 + 2i - 6(-1)$ Write i^2 as -1.

$= 28 + 2i + 6$

$= 34 + 2i$

c. $(2 + 3i)(2 - 3i) = 4 - 6i + 6i - 9i^2$

$= 4 - 9(-1)$ Write i^2 as -1.

$= 13 + 0i$ Write in standard form. □

PRACTICE
5 Find the following products and write in standard form.

a. $2i(3 - 4i)$ **b.** $(3 + i)(2 - 3i)$ **c.** $(5 - 2i)(5 + 2i)$

The product in part (c) is the real number 13. Notice that one factor is the sum of 2 and $3i$, and the other factor is the difference of 2 and $3i$. When complex number factors are related as these two are, their product is a real number. In general,

$$\underset{\text{sum}}{(a + bi)}\underset{\text{difference}}{(a - bi)} = \underset{\text{real number}}{a^2 + b^2}$$

The complex numbers $a + bi$ and $a - bi$ are called **complex conjugates** of each other. For example, $2 - 3i$ is the conjugate of $2 + 3i$, and $2 + 3i$ is the conjugate of $2 - 3i$. Also,

The conjugate of $3 - 10i$ is $3 + 10i$.

The conjugate of 5 is 5. (Note that $5 = 5 + 0i$ and its conjugate is $5 - 0i = 5$.)

The conjugate of $4i$ is $-4i$. ($0 - 4i$ is the conjugate of $0 + 4i$.)

OBJECTIVE
4 Dividing Complex Numbers

The fact that the product of a complex number and its conjugate is a real number provides a method for dividing by a complex number and for simplifying fractions whose denominators are complex numbers.

EXAMPLE 6 Write $\dfrac{4 + i}{3 - 4i}$ in standard form.

Solution To write this quotient as a complex number in the standard form $a + bi$, we need to find an equivalent fraction whose denominator is a real number. By multiplying both numerator and denominator by the denominator's conjugate, we obtain a new fraction that is an equivalent fraction with a real number denominator.

$\dfrac{4 + i}{3 - 4i} = \dfrac{(4 + i)}{(3 - 4i)} \cdot \dfrac{(3 + 4i)}{(3 + 4i)}$ Multiply numerator and denominator by $3 + 4i$.

$= \dfrac{12 + 16i + 3i + 4i^2}{9 - 16i^2}$

$= \dfrac{12 + 19i + 4(-1)}{9 - 16(-1)}$ Recall that $i^2 = -1$.

$= \dfrac{12 + 19i - 4}{9 + 16} = \dfrac{8 + 19i}{25}$

$= \dfrac{8}{25} + \dfrac{19}{25}i$ Write in standard form.

Note that our last step was to write $\dfrac{4 + i}{3 - 4i}$ in standard form $a + bi$, where a and b are real numbers. □

PRACTICE
6 Write $\dfrac{3 - i}{2 + 5i}$ in standard form.

OBJECTIVE

5 Solving Quadratic Equations with Complex Solutions 

Some quadratic equations have complex solutions.

EXAMPLE 7 Solve $(x + 2)^2 = -25$.

Solution Begin by applying the square root property.

$$(x + 2)^2 = -25$$

$$x + 2 = \pm\sqrt{-25} \quad \text{Apply the square root property.}$$

$$x + 2 = \pm 5i \quad \text{Write } \sqrt{-25} \text{ as } 5i.$$

$$x = -2 \pm 5i$$

The solutions are $-2 + 5i$ and $-2 - 5i$. □

PRACTICE

7 Solve $(x - 3)^2 = -16$.

EXAMPLE 8 Solve $m^2 = 4m - 5$.

Solution Write the equation in standard form and use the quadratic formula to solve.

$$m^2 = 4m - 5$$

$$m^2 - 4m + 5 = 0 \quad \text{Write the equation in standard form.}$$

Apply the quadratic formula with $a = 1, b = -4$, and $c = 5$.

$$m = \frac{4 \pm \sqrt{16 - 4 \cdot 1 \cdot 5}}{2 \cdot 1}$$

$$= \frac{4 \pm \sqrt{-4}}{2}$$

$$= \frac{4 \pm 2i}{2} \quad \text{Write } \sqrt{-4} \text{ as } 2i.$$

$$= \frac{2(2 \pm i)}{2} = 2 \pm i$$

The solutions are $2 - i$ and $2 + i$. □

PRACTICE

8 Solve $y^2 = 3y - 5$.

EXAMPLE 9 Solve $x^2 + x = -1$.

Solution

$$x^2 + x = -1$$

$$x^2 + x + 1 = 0 \quad \text{Write in standard form.}$$

$$x = \frac{-1 \pm \sqrt{1 - 4 \cdot 1 \cdot 1}}{2 \cdot 1} \quad \text{Apply the quadratic formula with } a = 1, b = 1, \text{ and } c = 1.$$

$$= \frac{-1 \pm \sqrt{-3}}{2}$$

$$= \frac{-1 \pm i\sqrt{3}}{2}$$

The solutions are $\dfrac{-1 - i\sqrt{3}}{2}$ and $\dfrac{-1 + i\sqrt{3}}{2}$. □

PRACTICE

9 Solve $x^2 + x = -4$.

Vocabulary, Readiness & Video Check

Use the choices below to fill in each blank.

real	complex	standard
pure imaginary	conjugate	

1. A number that can be written in the form $a + bi$ is called a ____________ number.
2. A complex number that can be written in the form $0 + bi$ is also called a ____________ number.
3. A complex number that can be written in the form $a + 0i$ is also called a ____________ number.
4. The ____________ of $a + bi$ is $a - bi$.
5. The form $a + bi$ is called ____________ form.

Martin-Gay Interactive Videos

See Video 9.4

Watch the section lecture video and answer the following questions.

OBJECTIVE 1
6. Besides factoring out a $\sqrt{-1}$, which is i, what other differences are there in simplifying single square roots with negative radicands?

OBJECTIVE 2
7. In Examples 3 and 4, what is the process of adding and subtracting complex numbers compared to? What important reminder is given about i?

OBJECTIVE 3
8. In Examples 5 and 6, what part of the definition of the imaginary unit i may be useful during the multiplication of complex numbers to help simplify products?

OBJECTIVE 4
9. Multiplying a complex number by its conjugate results in a real number. Why do you think we use this to get a real number in the denominator of Example 7? Use the definition of i in your answer.

OBJECTIVE 5
10. The discriminant of the quadratic formula tells us how many real solutions a quadratic equation has. From Example 9, how does the discriminant tell us when we'll have complex solutions?

9.4 Exercise Set

MyMathLab®

Write each expression in i notation. See Example 1.

1. $\sqrt{-9}$
2. $\sqrt{-64}$
3. $\sqrt{-100}$
4. $\sqrt{-16}$
5. $\sqrt{-50}$
6. $\sqrt{-98}$
7. $\sqrt{-63}$
8. $\sqrt{-44}$

Add or subtract as indicated. See Examples 2 through 4.

9. $(2 - i) + (-5 + 10i)$
10. $(-7 + 2i) + (5 - 3i)$
11. $(-11 + 3i) - (1 - 3i)$
12. $(1 + i) - (-6 - 8i)$
13. $(3 - 4i) - (2 - i)$
14. $(-6 + i) - (3 + i)$
15. $(16 + 2i) + (-7 - 6i)$
16. $(-3 - i) + (-4 - i)$

Multiply. See Example 5.

17. $4i(3 - 2i)$
18. $9i(4 - i)$
19. $-5i(10 + i)$
20. $-2i(5 + 4i)$
21. $(6 - 2i)(4 + i)$
22. $(6 + 2i)(4 - i)$
23. $(3 + 8i)(3 - 8i)$
24. $(-9 + 2i)(-9 - 2i)$

Divide. Write each answer in standard form. See Example 6.

25. $\dfrac{8 - 12i}{4}$
26. $\dfrac{14 + 28i}{-7}$
27. $\dfrac{7 - i}{4 - 3i}$
28. $\dfrac{4 - 3i}{7 - i}$

Solve the following quadratic equations for complex solutions. See Example 7.

29. $(x + 1)^2 = -9$
30. $(y - 2)^2 = -25$
31. $(2z - 3)^2 = -12$
32. $(3p + 5)^2 = -18$

Solve the following quadratic equations for complex solutions. See Examples 8 and 9.

33. $y^2 + 6y + 13 = 0$
34. $y^2 - 2y + 5 = 0$

35. $4x^2 + 7x + 4 = 0$

36. $8x^2 - 7x + 2 = 0$

37. $2m^2 - 4m + 5 = 0$

38. $5m^2 - 6m + 7 = 0$

MIXED PRACTICE

Perform the indicated operations. Write results in standard form.

39. $3 + (12 - 7i)$
40. $(-14 + 5i) + 3i$
41. $-9i(5i - 7)$
42. $10i(4i - 1)$
43. $(2 - i) - (3 - 4i)$
44. $(3 + i) - (-6 + i)$
45. $\dfrac{15 + 10i}{5i}$
46. $\dfrac{-18 + 12i}{-6i}$
47. Subtract $(2 + 3i)$ from $(-5 + i)$.
48. Subtract $(-8 - i)$ from $(7 - 4i)$.
49. $(4 - 3i)(4 + 3i)$
50. $(12 - 5i)(12 + 5i)$
51. $\dfrac{4 - i}{1 + 2i}$
52. $\dfrac{9 - 2i}{-3 + i}$
53. $(5 + 2i)^2$
54. $(9 - 7i)^2$

Solve the following quadratic equations for complex solutions.

55. $(y - 4)^2 = -64$
56. $(x + 7)^2 = -1$
57. $4x^2 = -100$
58. $7x^2 = -28$
59. $z^2 + 6z + 10 = 0$
60. $z^2 + 4z + 13 = 0$
61. $2a^2 - 5a + 9 = 0$
62. $4a^2 + 3a + 2 = 0$
63. $(2x + 8)^2 = -20$
64. $(6z - 4)^2 = -24$
65. $3m^2 + 108 = 0$
66. $5m^2 + 80 = 0$
67. $x^2 + 14x + 50 = 0$
68. $x^2 + 8x + 25 = 0$

REVIEW AND PREVIEW

Graph the following linear equations in two variables. See Section 3.2.

69. $y = -3$
70. $x = 4$
71. $y = 3x - 2$
72. $y = 2x + 3$

CONCEPT EXTENSIONS

Answer the following true or false.

73. Every real number is a complex number.
74. Every complex number is a real number.
75. If a complex number such as $2 + 3i$ is a solution of a quadratic equation, then its conjugate $2 - 3i$ is also a solution.
76. Some pure imaginary numbers are real numbers.
77. Earlier in this text, we learned that $\sqrt{-4}$ is not a real number. Explain what that means and explain what type of number $\sqrt{-4}$ is.
78. Describe how to find the conjugate of a complex number.

9.5 Graphing Quadratic Equations

OBJECTIVES

1 Graph Quadratic Equations of the Form $y = ax^2$.

2 Graph Quadratic Equations of the Form $y = ax^2 + bx + c$.

3 Use the Vertex Formula to Determine the Vertex of a Parabola.

OBJECTIVE

1 Graphing $y = ax^2$

Recall from Section 3.2 that the graph of a linear equation in two variables $Ax + By = C$ is a straight line. Also recall from Section 6.6 that the graph of a quadratic equation in two variables $y = ax^2 + bx + c$ is a parabola. In this section, we further investigate the graph of a quadratic equation.

To graph the quadratic equation $y = x^2$, select a few values for x and find the corresponding y-values. Make a table of values to keep track. Then plot the points corresponding to these solutions.

If $x = -3$, then $y = (-3)^2$, or 9.

If $x = -2$, then $y = (-2)^2$, or 4.

If $x = -1$, then $y = (-1)^2$, or 1.

If $x = 0$, then $y = 0^2$, or 0.

If $x = 1$, then $y = 1^2$, or 1.

If $x = 2$, then $y = 2^2$, or 4.

If $x = 3$, then $y = 3^2$, or 9.

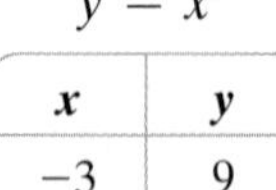
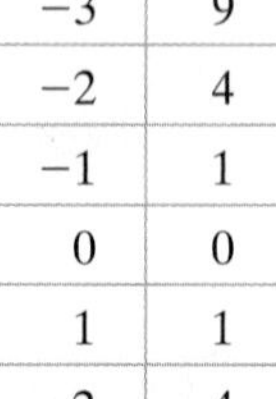
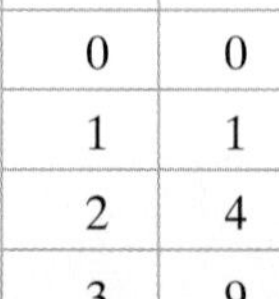

$y = x^2$

x	y
−3	9
−2	4
−1	1
0	0
1	1
2	4
3	9

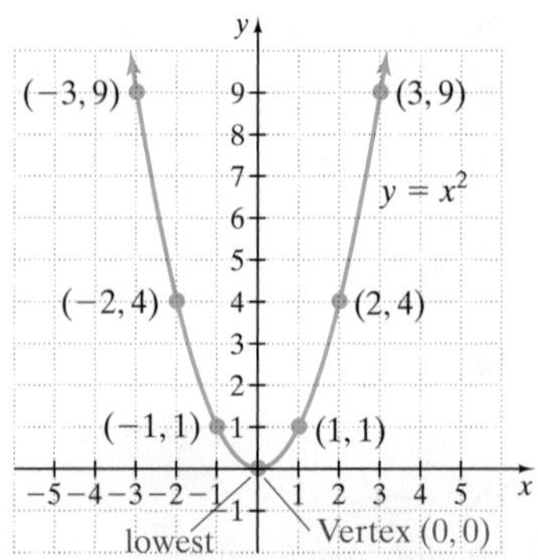

Clearly, these points are not on one straight line. As we saw in Chapter 6, the graph of $y = x^2$ is a smooth curve through the plotted points. This curve is called a **parabola.** The lowest point on a parabola opening upward is called the **vertex.** The vertex is (0, 0) for the parabola $y = x^2$. If we fold the graph paper along the y-axis, the two pieces of the parabola match perfectly. For this reason, we say the graph is **symmetric about the y-axis,** and we call the y-axis the **axis of symmetry.**

Notice that the parabola that corresponds to the equation $y = x^2$ opens upward. This happens when the coefficient of x^2 is positive. In the equation $y = x^2$, the coefficient of x^2 is 1. Example 1 shows the graph of a quadratic equation whose coefficient of x^2 is negative.

EXAMPLE 1 Graph $y = -2x^2$.

Solution Select x-values and calculate the corresponding y-values. Plot the ordered pairs found. Then draw a smooth curve through those points. When the coefficient of x^2 is negative, the corresponding parabola opens downward. When a parabola opens downward, the vertex is the highest point of the parabola. The vertex of this parabola is (0, 0) and the axis of symmetry is again the y-axis.

$y = -2x^2$

x	y
0	0
1	−2
2	−8
3	−18
−1	−2
−2	−8
−3	−18

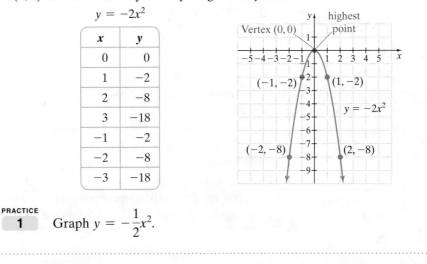

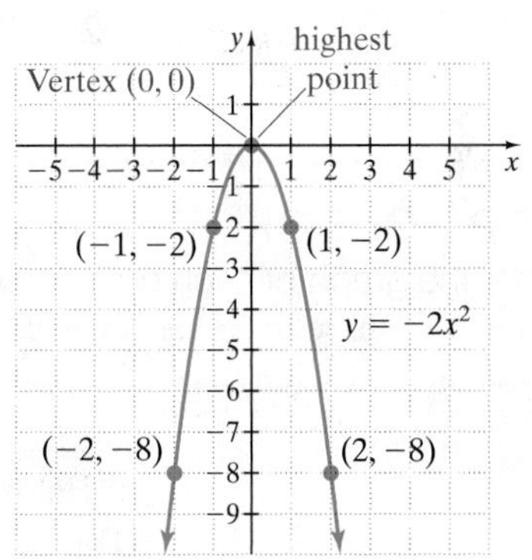

□

PRACTICE 1 Graph $y = -\frac{1}{2}x^2$.

OBJECTIVE 2

Graphing $y = ax^2 + bx + c$

Just as for linear equations, we can use x- and y-intercepts to help graph quadratic equations. Recall from Chapter 3 that an x-intercept is the point where the graph intersects the x-axis. A y-intercept is the point where the graph intersects the y-axis. We find intercepts just as we did in Chapter 3.

> **Helpful Hint**
> Recall that:
> To find x-intercepts, let $y = 0$ and solve for x.
> To find y-intercepts, let $x = 0$ and solve for y.

EXAMPLE 2 Graph $y = x^2 - 4$.

Solution First, find intercepts. To find the y-intercept, let $x = 0$. Then

$$y = 0^2 - 4 = -4$$

To find x-intercepts, we let $y = 0$.

$$0 = x^2 - 4$$
$$0 = (x - 2)(x + 2)$$
$$x - 2 = 0 \quad \text{or} \quad x + 2 = 0$$
$$x = 2 \qquad\qquad x = -2$$

Thus far, we have the y-intercept $(0, -4)$ and the x-intercepts $(2, 0)$ and $(-2, 0)$. Now we can select additional x-values, find the corresponding y-values, plot the points, and draw a smooth curve through the points.

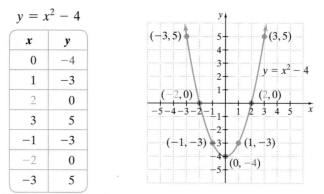

$y = x^2 - 4$

x	y
0	−4
1	−3
2	0
3	5
−1	−3
−2	0
−3	5

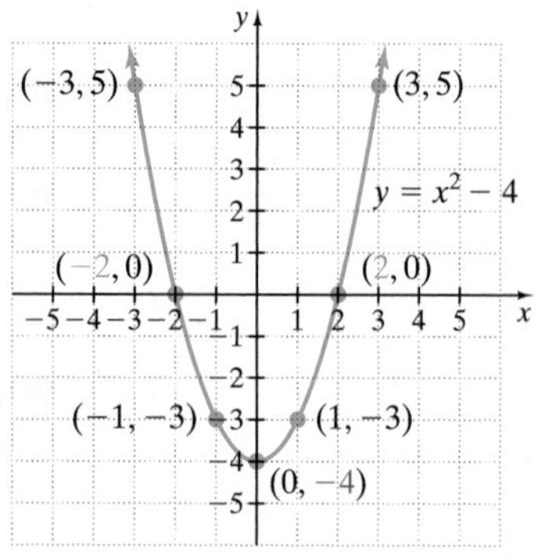

Notice that the vertex of this parabola is $(0, -4)$. □

PRACTICE

2 Graph $y = x^2 + 1$.

✓CONCEPT CHECK

Tell whether the graph of each equation opens upward or downward.

a. $y = 2x^2$ b. $y = 3x^2 + 4x - 5$ c. $y = -5x^2 + 2$

Helpful Hint

For the graph of $y = ax^2 + bx + c$,
If a is positive, the parabola opens upward.
If a is negative, the parabola opens downward.

✓CONCEPT CHECK

For which of the following graphs of $y = ax^2 + bx + c$ would the value of a be negative?

a.

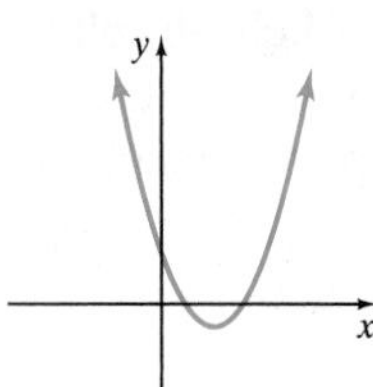

b.

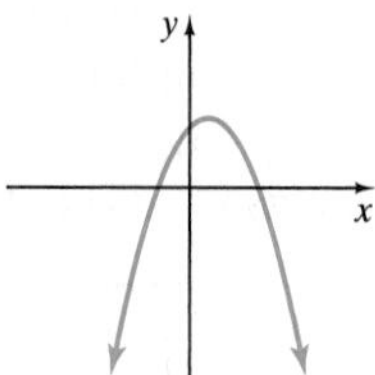

OBJECTIVE

3 Using the Vertex Formula

Thus far, we have accidentally stumbled upon the vertex of each parabola that we have graphed. However, our choice of values for x may not yield an ordered pair for the vertex of the parabola. It would be helpful if we could first find the vertex of a parabola, next determine whether the parabola opens upward or downward, and finally calculate additional points such as x- and y-intercepts as needed. In fact, there is a formula that may be used to find the vertex of a parabola.

One way to develop this formula is to notice that the x-value of the vertex of the parabolas that we are considering lies halfway between its x-intercepts. We can use this fact to find a formula for the vertex.

Answers to Concept Checks:
a. upward **b.** upward
c. downward; b

Recall that the x-intercepts of a parabola may be found by solving the equation $0 = ax^2 + bx + c$. These solutions, by the quadratic formula, are

$$x = \frac{-b - \sqrt{b^2 - 4ac}}{2a}, \quad x = \frac{-b + \sqrt{b^2 - 4ac}}{2a}$$

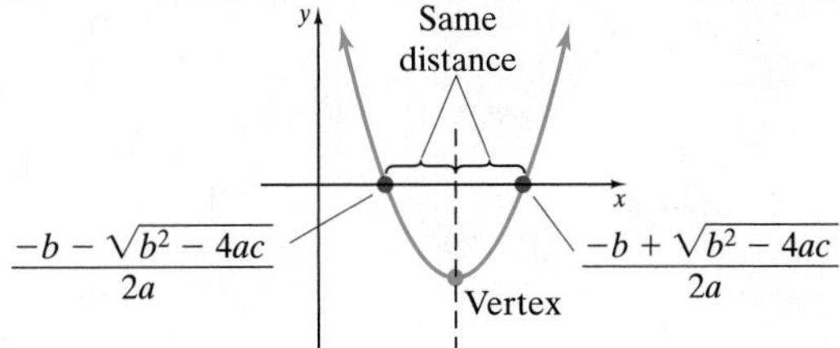

The x-coordinate of the vertex of a parabola is halfway between its x-intercepts, so the x-value of the vertex may be found by computing the average, or $\frac{1}{2}$ of the sum of the intercepts.

$$\begin{aligned} x &= \frac{1}{2}\left(\frac{-b - \sqrt{b^2 - 4ac}}{2a} + \frac{-b + \sqrt{b^2 - 4ac}}{2a}\right) \\ &= \frac{1}{2}\left(\frac{-b - \sqrt{b^2 - 4ac} - b + \sqrt{b^2 - 4ac}}{2a}\right) \\ &= \frac{1}{2}\left(\frac{-2b}{2a}\right) \\ &= \frac{-b}{2a} \end{aligned}$$

Vertex Formula

The vertex of the parabola $y = ax^2 + bx + c$ has x-coordinate

$$\frac{-b}{2a}$$

The corresponding y-coordinate of the vertex is found by substituting the x-coordinate into the equation and evaluating y.

EXAMPLE 3 Graph $y = x^2 - 6x + 8$.

Solution In the equation $y = x^2 - 6x + 8$, $a = 1$ and $b = -6$. The x-coordinate of the vertex is

$$x = \frac{-b}{2a} = \frac{-(-6)}{2 \cdot 1} = 3 \quad \text{Use the vertex formula, } x = \frac{-b}{2a}.$$

To find the corresponding y-coordinate, we let $x = 3$ in the original equation.

$$y = x^2 - 6x + 8 = 3^2 - 6 \cdot 3 + 8 = -1$$

The vertex is $(3, -1)$ and the parabola opens upward since a is positive. We now find and plot the intercepts.

To find the x-intercepts, we let $y = 0$.

$$0 = x^2 - 6x + 8$$

We factor the expression $x^2 - 6x + 8$ to find $(x - 4)(x - 2) = 0$. The x-intercepts are $(4, 0)$ and $(2, 0)$.

If we let $x = 0$ in the original equation, then $y = 8$ and the y-intercept is $(0, 8)$. Now we plot the vertex $(3, -1)$ and the intercepts $(4, 0)$, $(2, 0)$, and $(0, 8)$. Then we can sketch the parabola. These and two additional points are shown in the table.

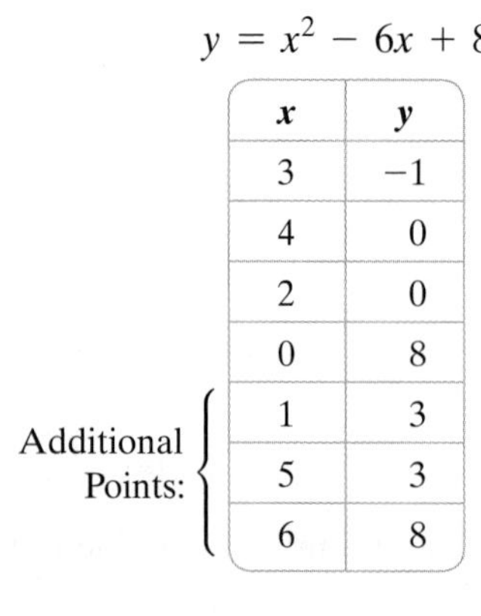

$y = x^2 - 6x + 8$

x	y
3	−1
4	0
2	0
0	8
1	3
5	3
6	8

Additional Points: (1, 3), (5, 3), (6, 8)

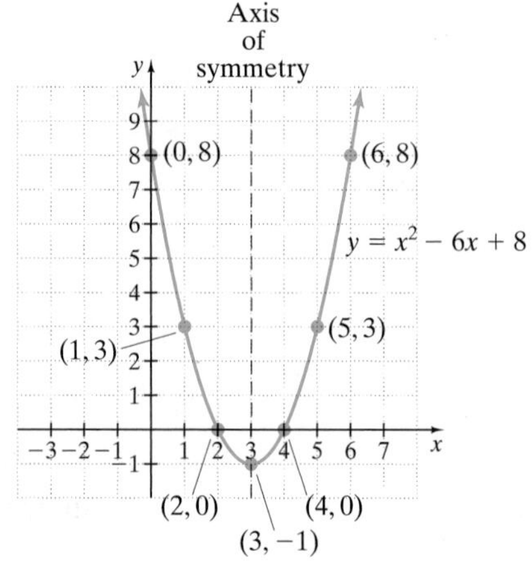

PRACTICE 3 Graph $y = x^2 - 3x - 4$.

Study Example 3 and let's use it to write down a general procedure for graphing quadratic equations.

Graphing Parabolas Defined by $y = ax^2 + bx + c$

1. **Find the vertex by using the formula $x = \frac{-b}{2a}$.** Don't forget to find the y-value of the vertex.
2. **Find the intercepts.**
   - Let $x = 0$ and solve for y to find the y-intercept. There will be only one.
   - Let $y = 0$ and solve for x to find any x-intercepts. There may be 0, 1, or 2.
3. **Plot the vertex and the intercepts.**
4. **Find and plot additional points on the graph.** Then draw a smooth curve through the plotted points. Keep in mind if $a > 0$, the parabola opens up and, if $a < 0$, the parabola opens down.

EXAMPLE 4 Graph $y = x^2 + 2x - 5$.

Solution In the equation $y = x^2 + 2x - 5$, $a = 1$ and $b = 2$. Using the vertex formula, we find that the x-coordinate of the vertex is

$$x = \frac{-b}{2a} = \frac{-2}{2 \cdot 1} = -1$$

The y-coordinate of the vertex is

$$y = (-1)^2 + 2(-1) - 5 = -6$$

Thus the vertex is $(-1, -6)$.

To find the x-intercepts, we let $y = 0$.

$$0 = x^2 + 2x - 5$$

This cannot be solved by factoring, so we use the quadratic formula.

$$x = \frac{-2 \pm \sqrt{2^2 - 4(1)(-5)}}{2 \cdot 1} \quad \text{Let } a = 1, b = 2, \text{ and } c = -5.$$

$$x = \frac{-2 \pm \sqrt{24}}{2}$$

$$x = \frac{-2 \pm 2\sqrt{6}}{2} \quad \text{Simplify the radical.}$$

$$x = \frac{2(-1 \pm \sqrt{6})}{2} = -1 \pm \sqrt{6}$$

The x-intercepts are $(-1 + \sqrt{6}, 0)$ and $(-1 - \sqrt{6}, 0)$. We use a calculator to approximate these so that we can easily graph these intercepts.

$$-1 + \sqrt{6} \approx 1.4 \quad \text{and} \quad -1 - \sqrt{6} \approx -3.4$$

To find the y-intercept, we let $x = 0$ in the original equation and find that $y = -5$. Thus the y-intercept is $(0, -5)$.

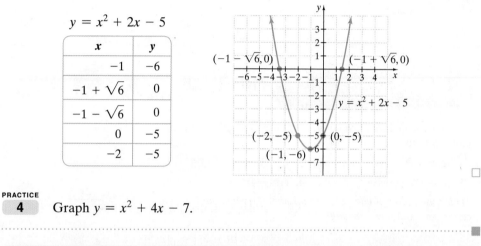

$y = x^2 + 2x - 5$

x	y
-1	-6
$-1 + \sqrt{6}$	0
$-1 - \sqrt{6}$	0
0	-5
-2	-5

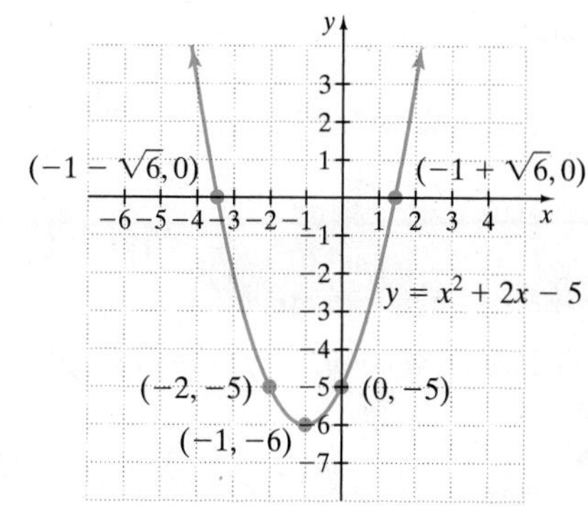

PRACTICE

4 Graph $y = x^2 + 4x - 7$.

Helpful Hint

Notice that the number of x-intercepts of the graph of the parabola $y = ax^2 + bx + c$ is the same as the number of real solutions of $0 = ax^2 + bx + c$.

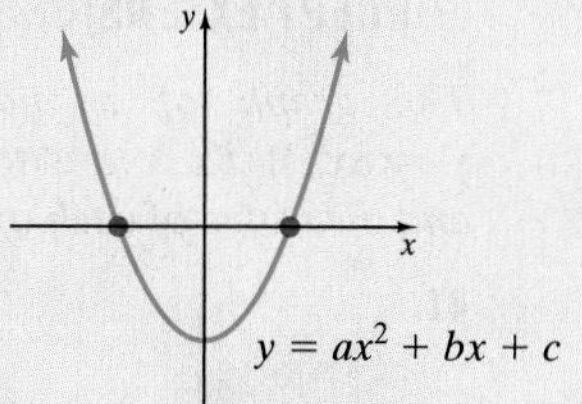

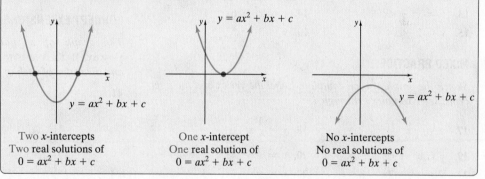

Two x-intercepts
Two real solutions of
$0 = ax^2 + bx + c$

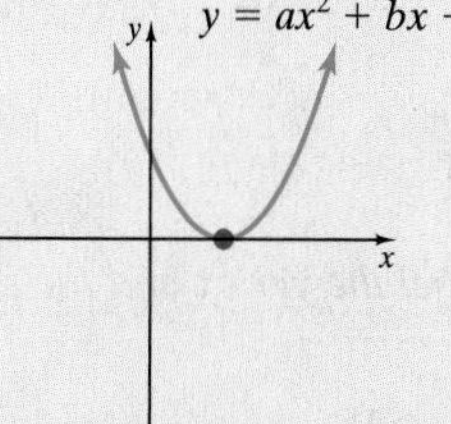

One x-intercept
One real solution of
$0 = ax^2 + bx + c$

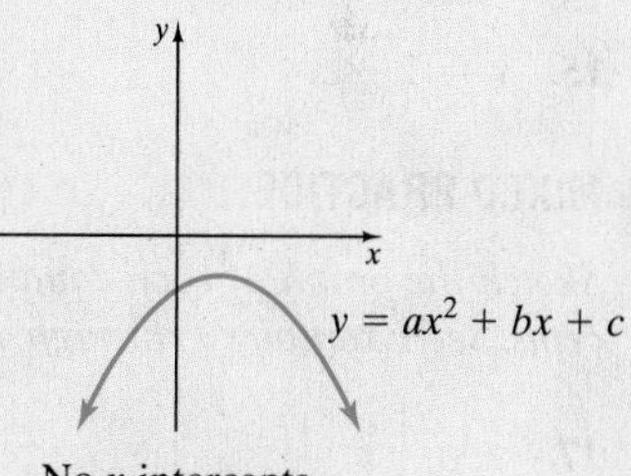

No x-intercepts
No real solutions of
$0 = ax^2 + bx + c$

Graphing Calculator Explorations

Recall that a graphing calculator may be used to solve quadratic equations. The x-intercepts of the graph of $y = ax^2 + bx + c$ are solutions of $0 = ax^2 + bx + c$. To solve $x^2 - 7x - 3 = 0$, for example, graph $y_1 = x^2 - 7x - 3$. The x-intercepts of the graph are the solutions of the equation.

Use a graphing calculator to solve each quadratic equation. Round solutions to two decimal places.

1. $x^2 - 7x - 3 = 0$

2. $2x^2 - 11x - 1 = 0$

3. $-1.7x^2 + 5.6x - 3.7 = 0$

4. $-5.8x^2 + 2.3x - 3.9 = 0$

5. $5.8x^2 - 2.6x - 1.9 = 0$

6. $7.5x^2 - 3.7x - 1.1 = 0$

Vocabulary, Readiness & Video Check

Martin-Gay Interactive Videos

See Video 9.5

Watch the section lecture video and answer the following questions.

OBJECTIVE 1

1. In Example 1, how are the vertex and axis of symmetry of a parabola explained?

OBJECTIVE 2

2. In Example 2, what important point was accidentally found? Why would it be useful to have an algebraic way to find this point?

OBJECTIVE 3

3. From Example 3, how can finding the vertex and noting whether the parabola opens up or down possibly help save you time and work? Explain.

9.5 Exercise Set

MyMathLab®

Graph each quadratic equation by finding and plotting ordered pair solutions. See Example 1.

1. $y = 2x^2$
2. $y = 3x^2$
3. $y = -x^2$
4. $y = -4x^2$

Sketch the graph of each equation. Label the vertex and the intercepts. See Examples 2 through 4.

5. $y = x^2 - 1$
6. $y = x^2 - 16$
7. $y = x^2 + 4$
8. $y = x^2 + 9$
9. $y = -x^2 + 4x - 4$
10. $y = -x^2 - 2x - 1$
11. $y = x^2 + 5x + 4$
12. $y = x^2 + 7x + 10$
13. $y = x^2 - 4x + 5$
14. $y = x^2 - 6x + 10$
15. $y = 2 - x^2$
16. $y = 3 - x^2$

MIXED PRACTICE

Sketch the graph of each equation. Label the vertex and the intercepts. See Examples 1 through 4.

17. $y = \frac{1}{3}x^2$
18. $y = \frac{1}{2}x^2$
19. $y = x^2 + 6x$
20. $y = x^2 - 4x$
21. $y = x^2 + 2x - 8$
22. $y = x^2 - 2x - 3$
23. $y = -\frac{1}{2}x^2$
24. $y = -\frac{1}{3}x^2$
25. $y = 2x^2 - 11x + 5$
26. $y = 2x^2 + x - 3$
27. $y = -x^2 + 4x - 3$
28. $y = -x^2 + 6x - 8$
29. $y = x^2 + 2x - 2$
30. $y = x^2 - 4x - 3$
31. $y = x^2 - 3x + 1$
32. $y = x^2 - 2x - 5$

REVIEW AND PREVIEW

Simplify the following complex fractions. See Section 7.8.

33. $\dfrac{\frac{1}{7}}{\frac{2}{5}}$
34. $\dfrac{\frac{3}{8}}{\frac{1}{7}}$

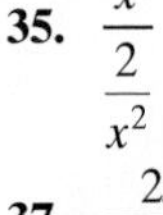

35. $\dfrac{\frac{1}{x}}{\frac{2}{x^2}}$
36. $\dfrac{\frac{x}{5}}{\frac{2}{x}}$
37. $\dfrac{2x}{1 - \frac{1}{x}}$
38. $\dfrac{x}{x - \frac{1}{x}}$
39. $\dfrac{\frac{a - b}{2b}}{\frac{b - a}{8b^2}}$
40. $\dfrac{\frac{2a^2}{a - 3}}{\frac{a}{3 - a}}$

CONCEPT EXTENSIONS

The graph of a quadratic equation that takes the form $y = ax^2 + bx + c$ is the graph of a function. Write the domain and the range of each of the functions graphed.

41.

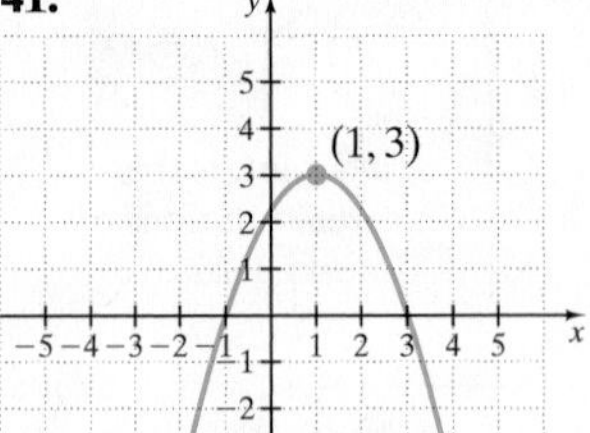

42.

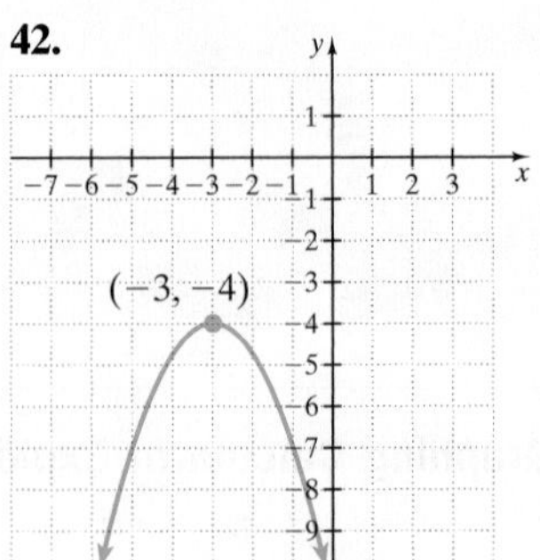

43.

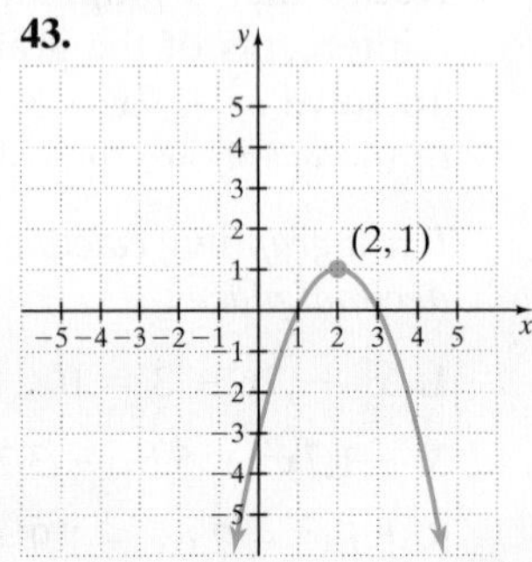

44.

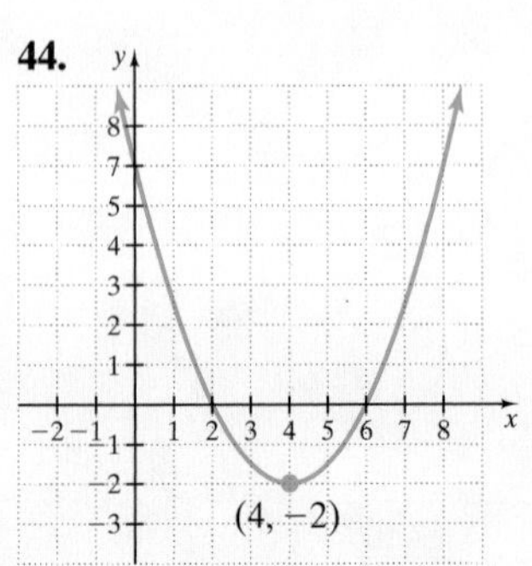

45. The height h of a fireball launched from a Roman candle with an initial velocity of 128 feet per second is given by the equation

$$h = -16t^2 + 128t$$

where t is time in seconds after launch.
Use the graph of this function to answer the questions.

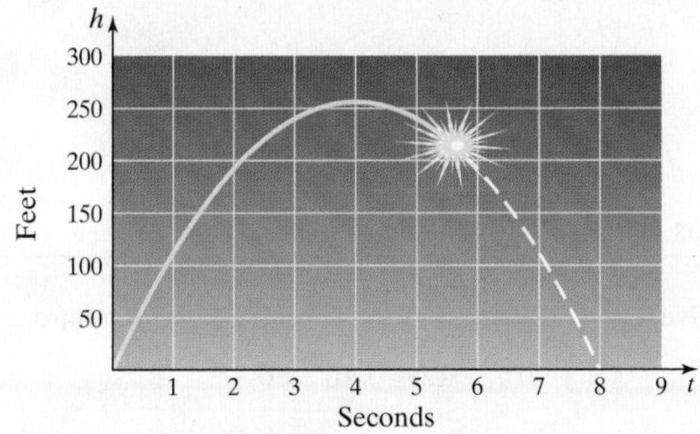

a. Estimate the maximum height of the fireball.

b. Estimate the time when the fireball is at its maximum height.

c. Estimate the time when the fireball returns to the ground.

46. Determine the maximum number and the minimum number of x-intercepts for a parabola. Explain your answer.

Match the values given with the correct graph of each quadratic equation of the form $y = a(x - h)^2 + k$.

47. $a > 0, h > 0, k > 0$

48. $a < 0, h > 0, k > 0$

49. $a > 0, h > 0, k < 0$

50. $a < 0, h > 0, k < 0$

A

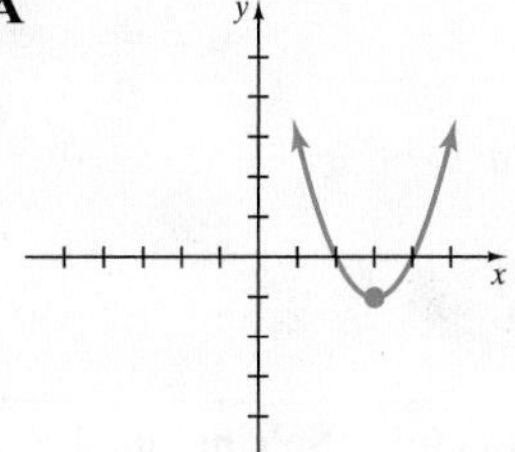

B

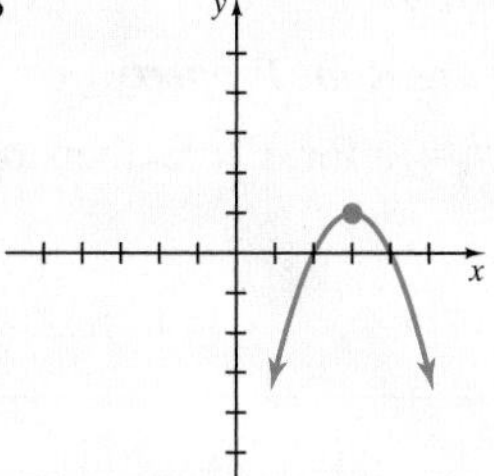

C

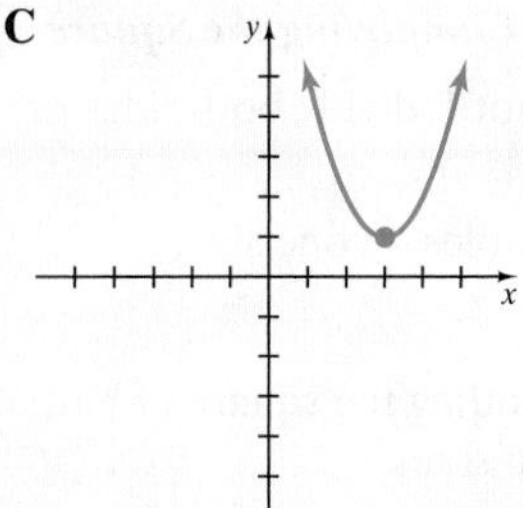

D

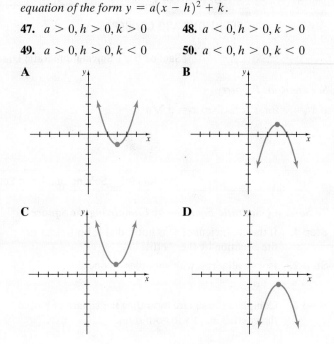

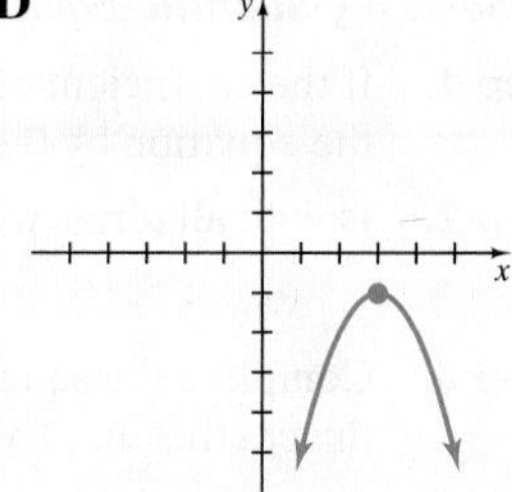

Chapter 9 Vocabulary Check

Fill in each blank with one of the words listed below.

square root	complex	pure imaginary	i
completing the square	quadratic	conjugate	vertex

1. If $x^2 = a$, then $x = \sqrt{a}$ or $x = -\sqrt{a}$. This property is called the ________________ property.

2. A number that can be written in the form $a + bi$ is called a ________________ number.

3. The formula $\frac{-b}{2a}$ where $y = ax^2 + bx + c$ is called the ________________ formula.

4. A complex number that can be written in the form $0 + bi$ is also called a ________________ number.

5. The ________________ of $2 + 3i$ is $2 - 3i$.

6. $\sqrt{-1} =$ ________________.

7. The process of solving a quadratic equation by writing it in the form $(x + a)^2 = c$ is called ________________.

8. The formula $x = \dfrac{-b \pm \sqrt{b^2 - 4ac}}{2a}$ is called the ________________ formula.

Chapter 9 Highlights

DEFINITIONS AND CONCEPTS	EXAMPLES
Section 9.1 Solving Quadratic Equations by the Square Root Property	
Square Root Property If $x^2 = a$ for $a \geq 0$, then $x = \pm\sqrt{a}$	***Solve the equation.*** $(x-1)^2 = 15$ $x - 1 = \pm\sqrt{15}$ $x = 1 \pm \sqrt{15}$
Section 9.2 Solving Quadratic Equations by Completing the Square	
To Solve a Quadratic Equation by Completing the Square	Solve $2x^2 + 12x - 10 = 0$ by completing the square.
Step 1. If the coefficient of x^2 is not 1, divide both sides of the equation by the coefficient.	$\frac{2x^2}{2} + \frac{12x}{2} - \frac{10}{2} = \frac{0}{2}$ Divide by 2.
Step 2. Isolate all terms with variables on one side.	$x^2 + 6x - 5 = 0$ Simplify. $x^2 + 6x = 5$ Add 5.
Step 3. Complete the square by adding the square of half of the coefficient of x to both sides.	The coefficient of x is 6. Half of 6 is 3 and $3^2 = 9$. Add 9 to both sides. $x^2 + 6x + 9 = 5 + 9$
Step 4. Factor the perfect square trinomial.	$(x+3)^2 = 14$ Factor.
Step 5. Apply the square root property to solve.	$x + 3 = \pm\sqrt{14}$ $x = -3 \pm \sqrt{14}$
Section 9.3 Solving Quadratic Equations by the Quadratic Formula	
Quadratic Formula If a, b, and c are real numbers and $a \neq 0$, the quadratic equation $ax^2 + bx + c = 0$ has solutions $x = \frac{-b \pm \sqrt{b^2 - 4ac}}{2a}$	Identify a, b, and c in the quadratic equation $4x^2 - 6x = 5$ First, subtract 5 from both sides. $4x^2 - 6x - 5 = 0$ $a = 4, b = -6$, and $c = -5$
To Solve a Quadratic Equation by the Quadratic Formula **Step 1.** Write the equation in standard form: $ax^2 + bx + c = 0$. **Step 2.** If necessary, clear the equation of fractions. **Step 3.** Identify a, b, and c. **Step 4.** Replace a, b, and c in the quadratic formula by known values and simplify.	Solve $3x^2 - 2x - 2 = 0$. In this equation, $a = 3, b = -2$, and $c = -2$. $x = \frac{-(-2) \pm \sqrt{(-2)^2 - 4(3)(-2)}}{2 \cdot 3}$ $= \frac{2 \pm \sqrt{4-(-24)}}{6}$ $= \frac{2 \pm \sqrt{28}}{6} = \frac{2 \pm \sqrt{4 \cdot 7}}{6} = \frac{2 \pm 2\sqrt{7}}{6}$ $= \frac{2(1 \pm \sqrt{7})}{2 \cdot 3} = \frac{1 \pm \sqrt{7}}{3}$

DEFINITIONS AND CONCEPTS	EXAMPLES
Section 9.3 Solving Quadratic Equations by the Quadratic Formula (continued)	
To Solve a Quadratic Equation **Step 1.** If the equation is in the form $(ax + b)^2 = c$, use the square root property and solve. If not, go to Step 2. **Step 2.** Write the equation in standard form: $ax^2 + bx + c = 0$. **Step 3.** Try to solve by factoring. If not, go to Step 4. **Step 4.** Solve by the quadratic formula.	Solve $(3x - 1)^2 = 10$. $3x - 1 = \pm\sqrt{10}$ Square root property. $3x = 1 \pm \sqrt{10}$ Add 1. $x = \frac{1 \pm \sqrt{10}}{3}$ Divide by 3. Solve $x(2x + 9) = 5$. $2x^2 + 9x - 5 = 0$ $(2x - 1)(x + 5) = 0$ $2x - 1 = 0$ or $x + 5 = 0$ $2x = 1$ $x = -5$ $x = \frac{1}{2}$
Section 9.4 Complex Solutions of Quadratic Equations	
The **imaginary unit,** written i, is the number whose square is -1. That is, $i^2 = -1$ and $i = \sqrt{-1}$	Write $\sqrt{-10}$ as the product of a real number and i. $\sqrt{-10} = \sqrt{-1 \cdot 10} = \sqrt{-1} \cdot \sqrt{10} = i\sqrt{10}$
A **complex number** is a number that can be written in the form $a + bi$ where a and b are real numbers. A complex number that can be written in the form $0 + bi$, $b \neq 0$, is also called a **pure imaginary number.**	Identify each number as a complex number by writing it in **standard form** $a + bi$. $7 = 7 + 0i$ $\sqrt{-5} = 0 + i\sqrt{5}$ Also a pure imaginary number $1 - \sqrt{-9} = 1 - 3i$
Complex numbers are added and subtracted in the same way as polynomials are added and subtracted.	Simplify the sum or difference. $(2 + 3i) - (1 - 6i) = 2 + 3i - 1 + 6i$ $= 1 + 9i$ $2i + (5 - 3i) = 2i + 5 - 3i$ $= 5 - i$
Use the distributive property to multiply complex numbers.	Multiply: $(4 - i)(-2 + 3i)$ $= -8 + 12i + 2i - 3i^2$ $= -8 + 14i + 3$ $= -5 + 14i$
The complex numbers $a + bi$ and $a - bi$ are called **complex conjugates.**	The conjugate of $5 - 6i$ is $5 + 6i$.
To write a quotient of complex numbers in standard form $a + bi$, multiply numerator and denominator by the denominator's conjugate.	Write $\frac{3 - 2i}{2 + i}$ in standard form. $\frac{3 - 2i}{2 + i} \cdot \frac{(2 - i)}{(2 - i)} = \frac{6 - 7i + 2i^2}{4 - i^2}$ $= \frac{6 - 7i + 2(-1)}{4 - (-1)}$ $= \frac{4 - 7i}{5}$ or $\frac{4}{5} - \frac{7}{5}i$

(continued)

DEFINITIONS AND CONCEPTS | **EXAMPLES**

Section 9.4 Complex Solutions of Quadratic Equations (continued)

Some quadratic equations have complex solutions.

Solve $x^2 - 3x = -3$.

$$x^2 - 3x + 3 = 0$$
$$a = 1, b = -3, c = 3$$
$$x = \frac{-(-3) \pm \sqrt{(-3)^2 - 4(1)(3)}}{2(1)}$$
$$= \frac{3 \pm \sqrt{-3}}{2} = \frac{3 \pm i\sqrt{3}}{2}$$

Section 9.5 Graphing Quadratic Equations

The graph of a quadratic equation $y = ax^2 + bx + c, a \neq 0$, is called a **parabola.** The lowest point on a parabola opening upward or the highest point on a parabola opening downward is called the **vertex.** The vertical line through the vertex is the **axis of symmetry.**

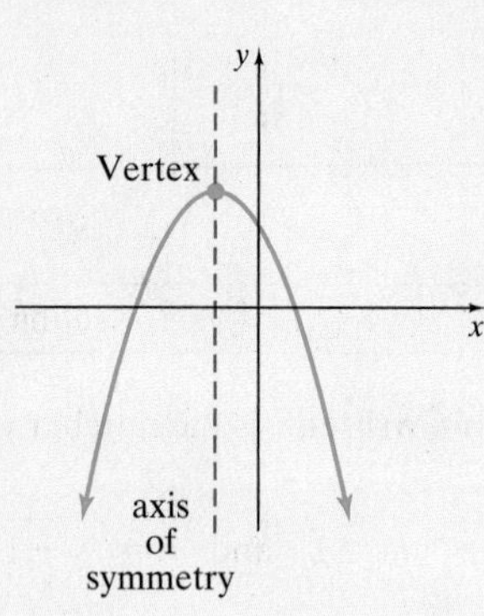

The vertex of the parabola $y = ax^2 + bx + c$ has x-value $\frac{-b}{2a}$.

Graph $y = 2x^2 - 6x + 4$.

The x-value of the vertex is

$$x = \frac{-b}{2a} = \frac{-(-6)}{2(2)} = \frac{6}{4} = \frac{3}{2}$$

The y-value is

$$y = 2\left(\frac{3}{2}\right)^2 - 6\left(\frac{3}{2}\right) + 4 = -\frac{1}{2}$$

The vertex is $\left(\frac{3}{2}, -\frac{1}{2}\right)$.

The y-intercept is

$$y = 2 \cdot 0^2 - 6 \cdot 0 + 4 = 4$$

The x-intercepts are found by solving

$$0 = 2x^2 - 6x + 4$$
$$0 = 2(x^2 - 3x + 2)$$
$$0 = 2(x - 2)(x - 1)$$
$$x - 2 = 0 \quad \text{or} \quad x - 1 = 0$$
$$x = 2 \quad \text{or} \quad x = 1$$

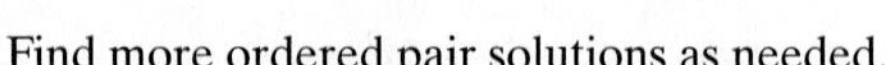
Find more ordered pair solutions as needed.

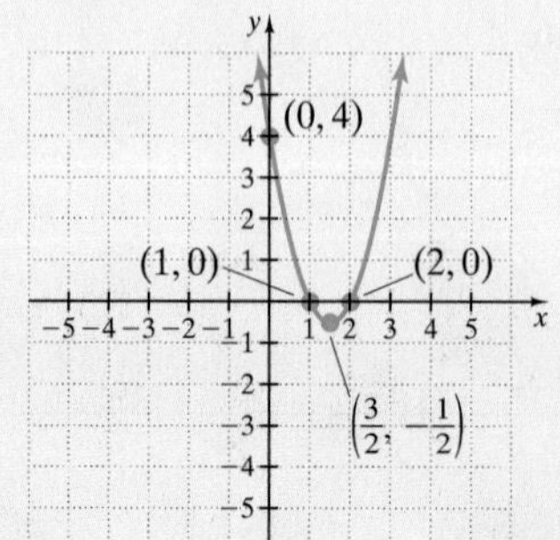

Chapter 9 Review

(9.1) Use the square root property to solve each quadratic equation.

1. $x^2 = 36$

2. $x^2 = 81$

3. $k^2 = 50$

4. $k^2 = 45$

5. $(x - 11)^2 = 49$

6. $(x + 3)^2 = 100$

7. $(4p + 5)^2 = 41$

8. $(3p + 7)^2 = 37$

Solve. For Exercises 9 and 10, use the formula $h = 16t^2$, where h is the height in feet at time t seconds.

9. If Kara Washington dives from a height of 100 feet, how long before she hits the water?

10. How long does a 5-mile free fall take? Round your result to the nearest tenth of a second. (*Hint:* 1 mi = 5280 ft)

Solve each quadratic equation by completing the square.

11. $x^2 - 9x = -8$

12. $x^2 + 8x = 20$

13. $x^2 + 4x = 1$

14. $x^2 - 8x = 3$

15. $x^2 - 6x + 7 = 0$

16. $x^2 + 6x + 7 = 0$

17. $2y^2 + y - 1 = 0$

18. $y^2 + 3y - 1 = 0$

(9.3) Use the quadratic formula to solve each quadratic equation.

19. $9x^2 + 30x + 25 = 0$

20. $16x^2 - 72x + 81 = 0$

21. $7x^2 = 35$

22. $11x^2 = 33$

23. $x^2 - 10x + 7 = 0$

24. $x^2 + 4x - 7 = 0$

25. $3x^2 + x - 1 = 0$

26. $x^2 + 3x - 1 = 0$

27. $2x^2 + x + 5 = 0$

28. $7x^2 - 3x + 1 = 0$

For the Exercise numbers given, approximate the exact solutions to the nearest tenth.

29. Exercise 25

30. Exercise 26

Use the discriminant to determine the number of solutions of each quadratic equation.

31. $x^2 - 7x - 1 = 0$

32. $x^2 + 6x = 5$

33. $x^2 + x + 5 = 0$

34. $5x^2 + 4 = 0$

35. $9x^2 + 1 = 6x$

36. $x^2 + 25 = 10x$

37. The average price of platinum (in dollars per ounce) from 2003 to 2008 is modeled by the equation $y = 8x^2 + 109x + 700$. In this equation, x is the number of years since 2003. Assume that this trend continues and find the year after 2003 in which the price of platinum will be \$2590 per ounce. (*Source:* U.S. Geological Survey, Minerals Information)

38. The average price of silver (in cents per ounce) from 2003 to 2007 is modeled by the equation $y = 26x^2 + 115x + 493$. In this equation, x is the number of years since 2003. Assume that this trend continued and find the year after 2003 in which the price of silver was 3077 cents per ounce. (*Source:* U.S. Geological Survey, Minerals Information)

(9.4) Perform the indicated operations. Write the resulting complex number in standard form.

39. $\sqrt{-144}$

40. $\sqrt{-36}$

41. $\sqrt{-108}$

42. $\sqrt{-500}$

43. $2i(3 - 5i)$

44. $i(-7 - i)$

45. $(7 - i) + (14 - 9i)$

46. $(10 - 4i) + (9 - 21i)$

47. $3 - (11 + 2i)$

48. $(-4 - 3i) + 5i$

49. $(2 - 3i)(3 - 2i)$

50. $(2 + 5i)(5 - i)$

51. $(3 - 4i)(3 + 4i)$

52. $(7 - 2i)(7 - 2i)$

53. $\frac{2 - 6i}{4i}$

54. $\frac{5 - i}{2i}$

55. $\frac{4 - i}{1 + 2i}$

56. $\frac{1 + 3i}{2 - 7i}$

Solve each quadratic equation.

57. $3x^2 = -48$

58. $5x^2 = -125$

59. $x^2 - 4x + 13 = 0$

60. $x^2 + 4x + 11 = 0$

(9.5) Graph each quadratic equation and find and plot any intercept points.

61. $y = 5x^2$

62. $y = -\frac{1}{2}x^2$

Graph each quadratic equation. Label the vertex and the intercept points with their coordinates.

63. $y = x^2 - 25$

64. $y = x^2 - 36$

65. $y = x^2 + 3$

66. $y = x^2 + 8$

67. $y = -4x^2 + 8$

68. $y = -3x^2 + 9$

69. $y = x^2 + 3x - 10$

70. $y = x^2 + 3x - 4$

71. $y = -x^2 - 5x - 6$

72. $y = 3x^2 - x - 2$

73. $y = 2x^2 - 11x - 6$

74. $y = -x^2 + 4x + 8$

Match each quadratic equation with its graph.

75. $y = 2x^2$

76. $y = -x^2$

77. $y = x^2 + 4x + 4$

78. $y = x^2 + 5x + 4$

A

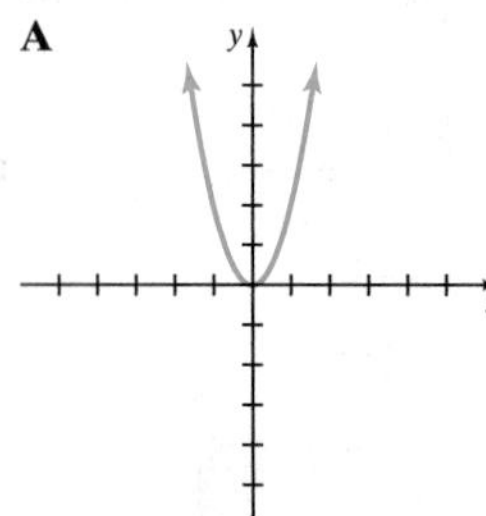

B

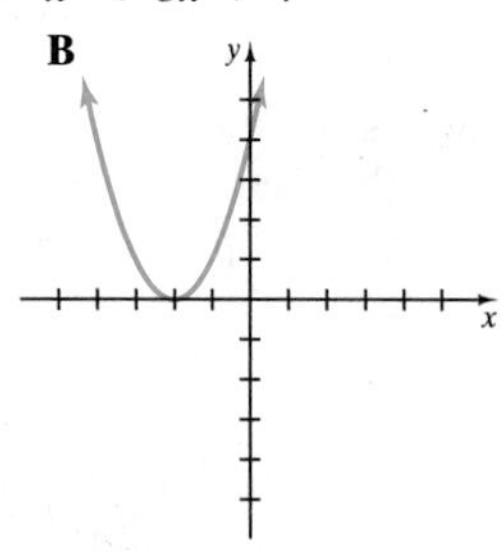

C

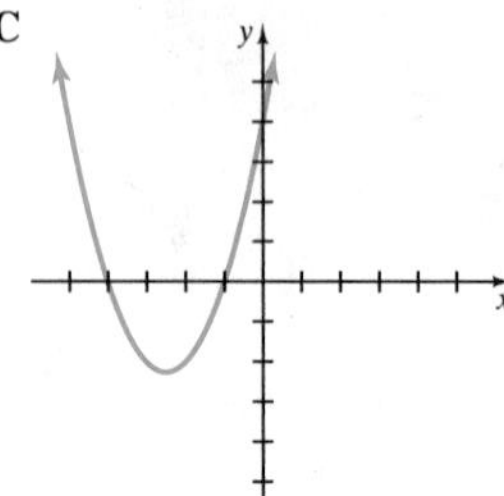

D

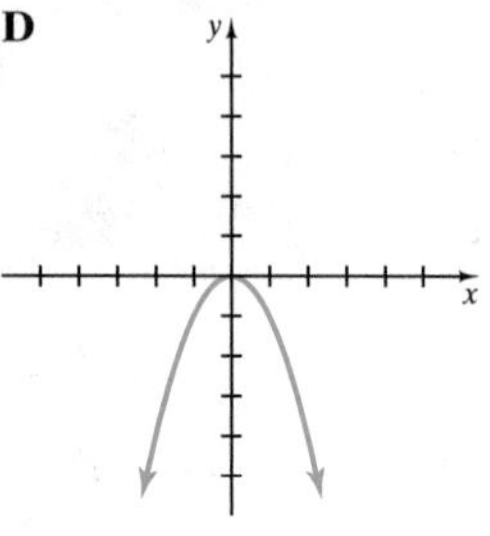

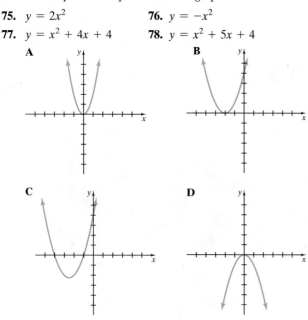

Quadratic equations in the form $y = ax^2 + bx + c$ are graphed below and in the next column. Determine the number of real solutions for the related equation $0 = ax^2 + bx + c$ from each graph.

79.

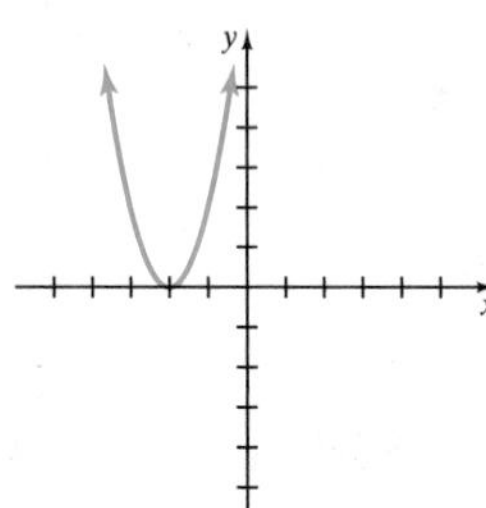

80.

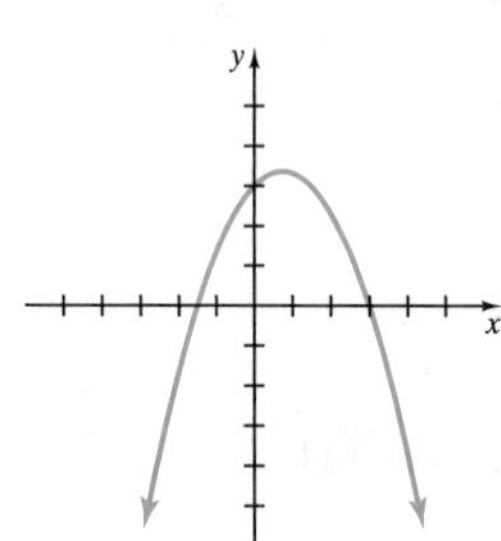

81.

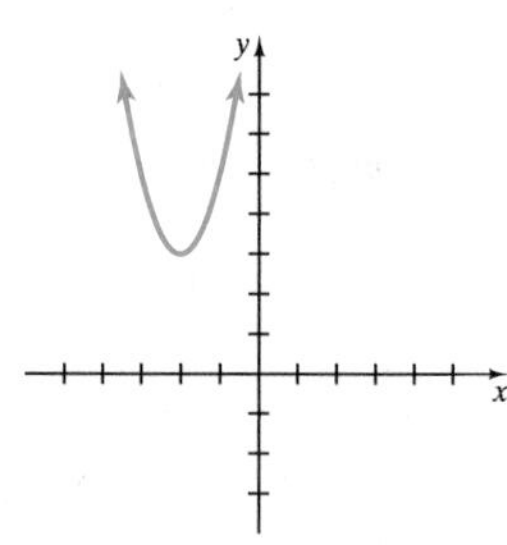

82.

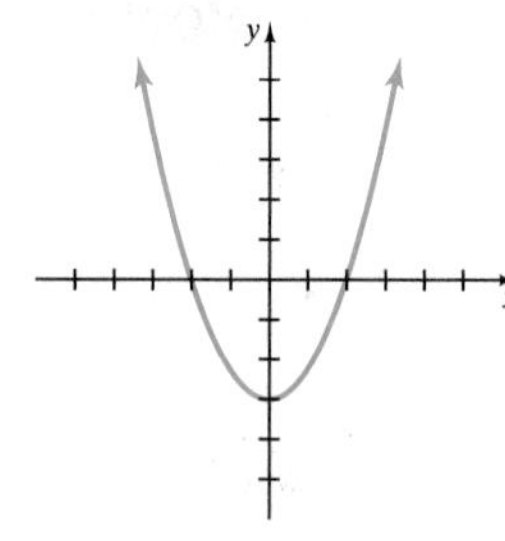

MIXED REVIEW

Use the square root property to solve each quadratic equation.

83. $x^2 = 49$

84. $y^2 = 75$

85. $(x - 7)^2 = 64$

Solve each quadratic equation by completing the square.

86. $x^2 + 4x = 6$

87. $3x^2 + x = 2$

88. $4x^2 - x - 2 = 0$

Use the quadratic formula to solve each quadratic equation.

89. $4x^2 - 3x - 2 = 0$

90. $5x^2 + x - 2 = 0$

91. $4x^2 + 12x + 9 = 0$

92. $2x^2 + x + 4 = 0$

Graph each quadratic equation. Label the vertex and the intercept points with their coordinates.

93. $y = 4 - x^2$

94. $y = x^2 + 4$

95. $y = x^2 + 6x + 8$

96. $y = x^2 - 2x - 4$

Chapter 9 Test

MyMathLab®

YouTube

Solve using the square root property.

1. $5k^2 = 80$

2. $(3m - 5)^2 = 8$

Solve by completing the square.

3. $x^2 - 26x + 160 = 0$

4. $3x^2 + 12x - 4 = 0$

Solve using the quadratic formula.

5. $x^2 - 3x - 10 = 0$

6. $p^2 - \frac{5}{3}p - \frac{1}{3} = 0$

Solve by the most appropriate method.

7. $(3x - 5)(x + 2) = -6$

8. $(3x - 1)^2 = 16$

9. $3x^2 - 7x - 2 = 0$

10. $x^2 - 4x + 5 = 0$

11. $3x^2 - 7x + 2 = 0$

12. $2x^2 - 6x + 1 = 0$

13. $9x^3 = x$

Perform the indicated operations. Write the resulting complex number in standard form.

14. $\sqrt{-25}$

15. $\sqrt{-200}$

16. $(3 + 2i) + (5 - i)$

17. $(3 + 2i) - (3 - 2i)$

18. $(3 + 2i)(3 - 2i)$

19. $\frac{3 - i}{1 + 2i}$

Graph each quadratic equation. Label the vertex and the intercept points with their coordinates.

20. $y = -5x^2$

21. $y = x^2 - 4$

22. $y = x^2 - 7x + 10$

23. $y = 2x^2 + 4x - 1$

Solve.

24. The height of a triangle is 4 times the length of the base. The area of the triangle is 18 square feet. Find the height and base of the triangle.

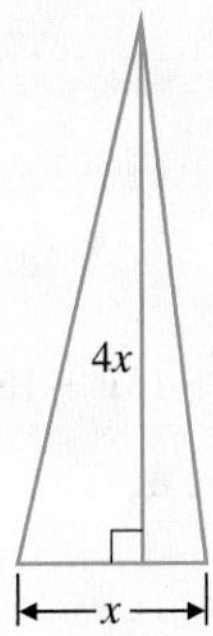

25. The number of diagonals d that a polygon with n sides has is given by the formula

$$d = \frac{n^2 - 3n}{2}$$

Find the number of sides of a polygon if it has 9 diagonals.

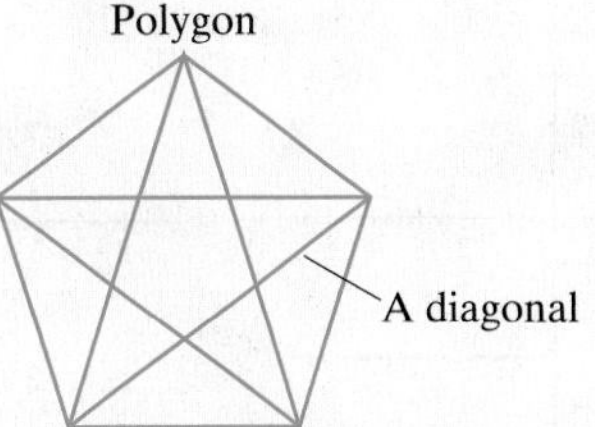

Solve.

26. The highest dive from a diving board by a woman was made by Lucy Wardle of the United States. She dove from a height of 120.75 feet at Ocean Park, Hong Kong, in 1985. To the nearest tenth of a second, how long did the dive take? Use the formula $h = 16t^2$.

Chapter 9 Cumulative Review

1. Find the value of each expression when $x = 2$ and $y = -5$.

   a. $\dfrac{x - y}{12 + x}$ b. $x^2 - 3y$

2. Find the value of each expression when $x = -4$ and $y = 7$.

   a. $\dfrac{x - y}{7 - x}$ b. $x^2 + 2y$

3. Simplify each expression by combining like terms.

   a. $2x + 3x + 5 + 2$

   b. $-5a - 3 + a + 2$

   c. $4y - 3y^2$

   d. $2.3x + 5x - 6$

   e. $-\dfrac{1}{2}b + b$

4. Simplify each expression by combining like terms.

   a. $4x - 3 + 7 - 5x$

   b. $-6y + 3y - 8 + 8y$

   c. $2 + 8.1a + a - 6$

   d. $2x^2 - 2x$

5. Identify the x- and y-intercepts.

   a.

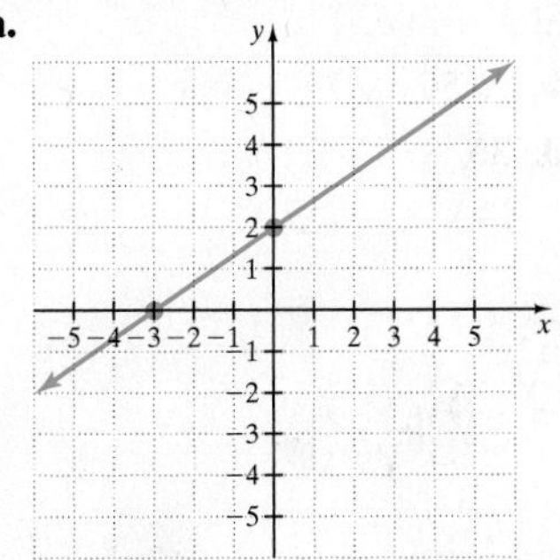

   b.

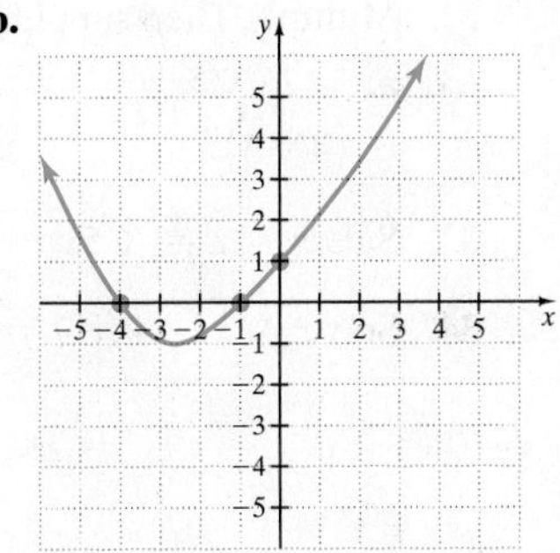

   c.

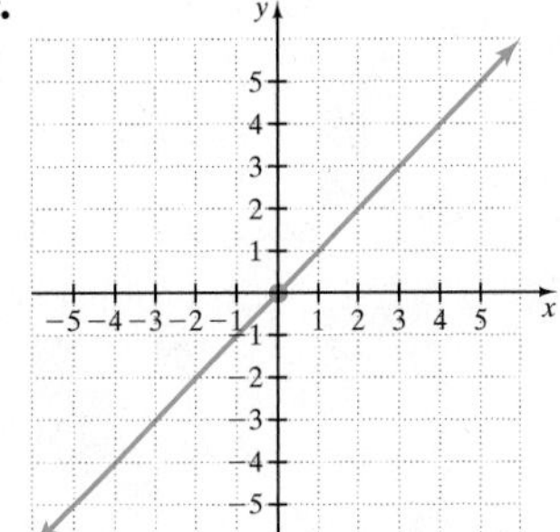

   d.

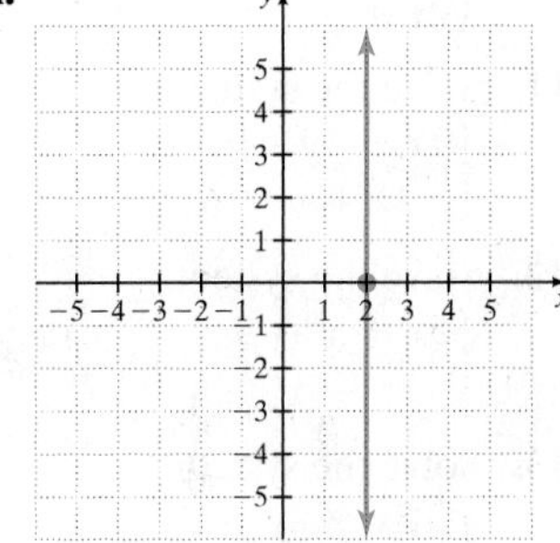

   e.

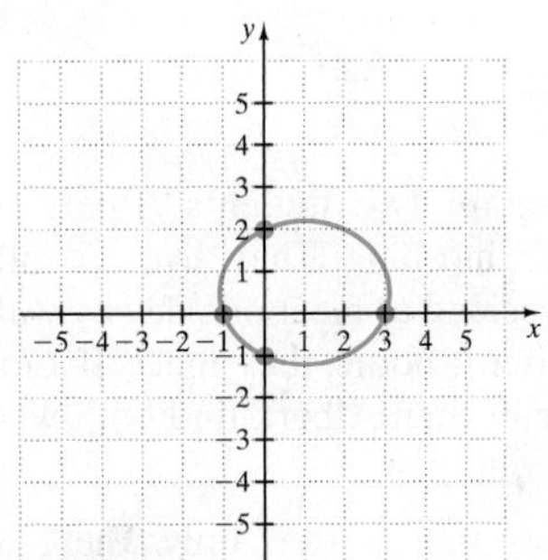

6. Identify the x- and y-intercepts.

   a.

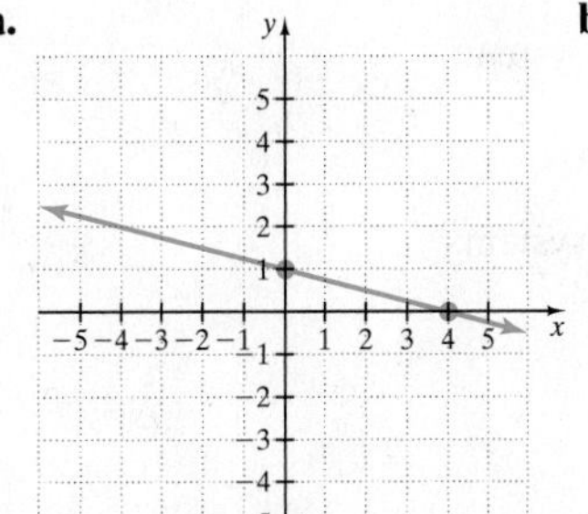

   b.

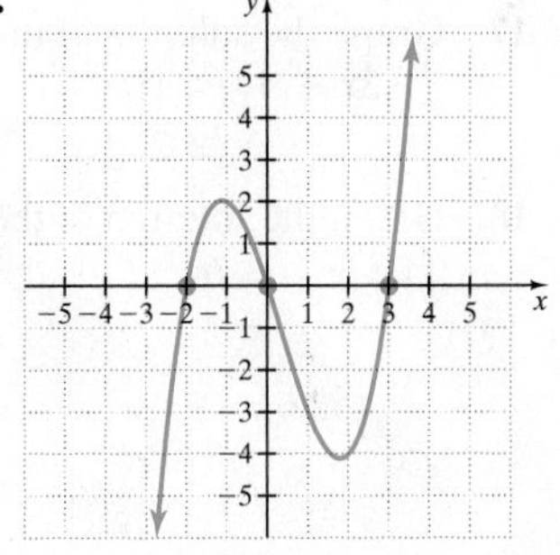

c.

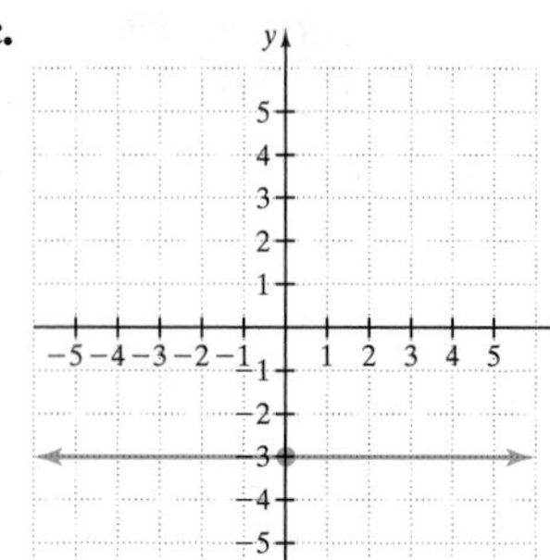

d.

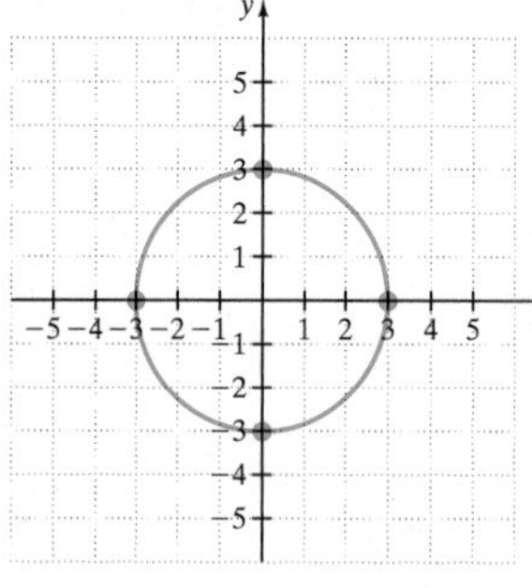

7. Determine whether the graphs of $y = -\frac{1}{5}x + 1$ and $2x + 10y = 3$ are parallel lines, perpendicular lines, or neither.

8. Determine whether the graphs of $y = 3x + 7$ and $x + 3y = -15$ are parallel lines, perpendicular lines, or neither.

9. Solve the following system of equations by graphing.
$$\begin{cases} 2x + y = 7 \\ 2y = -4x \end{cases}$$

10. Solve the following system by graphing.
$$\begin{cases} y = x + 2 \\ 2x + y = 5 \end{cases}$$

11. Solve the system.
$$\begin{cases} 7x - 3y = -14 \\ -3x + y = 6 \end{cases}$$

12. Solve the system.
$$\begin{cases} 5x + y = 3 \\ y = -5x \end{cases}$$

13. Solve the system.
$$\begin{cases} 3x - 2y = 2 \\ -9x + 6y = -6 \end{cases}$$

14. Solve the system.
$$\begin{cases} -2x + y = 7 \\ 6x - 3y = -21 \end{cases}$$

15. As part of an exercise program, two students, Albert and Louis, started walking each morning. They live 15 miles away from each other and decided to meet one day by walking toward one another. After 2 hours, they meet. If Louis walks one mile per hour faster than Albert, find both walking speeds.

16. A coin purse contains dimes and quarters only. There are 15 coins totalling \$2.85. How many dimes and how many quarters are in the purse?

17. Graph the solution of the system:
$$\begin{cases} -3x + 4y < 12 \\ x \geq 2 \end{cases}$$

18. Graph the solution of the system:
$$\begin{cases} 2x - y \leq 6 \\ y \geq 2 \end{cases}$$

19. Simplify each expression.

a. $x^7 \cdot x^4$ **b.** $\left(\frac{t}{2}\right)^4$ **c.** $(9y^5)^2$

20. Simplify.

a. $\left(\frac{-6x}{y^3}\right)^3$ **b.** $\frac{a^2b^7}{(2b^2)^5}$

c. $\frac{(3y)^2}{y^2}$ **d.** $\frac{(x^2y^4)^2}{xy^3}$

21. Solve $(5x - 1)(2x^2 + 15x + 18) = 0$.

22. Solve $(x + 1)(2x^2 - 3x - 5) = 0$.

23. Solve for x: $\frac{45}{x} = \frac{5}{7}$

24. Solve for x: $\frac{2x + 7}{3} = \frac{x - 6}{2}$

25. Find each root.

a. $\sqrt[4]{16}$ **b.** $\sqrt[5]{-32}$ **c.** $-\sqrt[3]{8}$

d. $\sqrt[4]{-81}$

26. Find each root.

a. $\sqrt[3]{27}$ **b.** $-\sqrt[4]{256}$ **c.** $\sqrt[5]{-125}$

d. $\sqrt[5]{1}$

27. Simplify the following expressions.

a. $\sqrt{\frac{25}{36}}$ **b.** $\sqrt{\frac{3}{64}}$ **c.** $\sqrt{\frac{40}{81}}$

28. Simplify the following expressions.

a. $\sqrt{\frac{4}{25}}$ **b.** $\sqrt{\frac{16}{121}}$ **c.** $\sqrt{\frac{2}{49}}$

29. Add or subtract by first simplifying each radical.

a. $\sqrt{50} + \sqrt{8}$

b. $7\sqrt{12} - \sqrt{75}$

c. $\sqrt{25} - \sqrt{27} - 2\sqrt{18} - \sqrt{16}$

30. Add or subtract by first simplifying each radical.

a. $\sqrt{80} + \sqrt{20}$

b. $2\sqrt{98} - 2\sqrt{18}$

c. $\sqrt{32} + \sqrt{121} - \sqrt{12}$

31. Multiply. Then simplify if possible.

a. $\sqrt{7} \cdot \sqrt{3}$ **b.** $\sqrt{3} \cdot \sqrt{3}$

c. $\sqrt{3} \cdot \sqrt{15}$ **d.** $2\sqrt{3} \cdot 5\sqrt{2}$

e. $\sqrt{2x^3} \cdot \sqrt{6x}$

32. Multiply. Then simplify if possible.

a. $\sqrt{2} \cdot \sqrt{5}$ **b.** $\sqrt{56} \cdot \sqrt{7}$

c. $(4\sqrt{3})^2$ **d.** $3\sqrt{8} \cdot 7\sqrt{2}$

33. Solve: $\sqrt{x} = \sqrt{5x - 2}$

34. Solve: $\sqrt{x - 4} + 7 = 2$

35. A surveyor must determine the distance across a lake at points P and Q. To do this, she finds a third point R perpendicular to line PQ. If the length of $\overline{PR}$ is 320 feet and the length of $\overline{QR}$ is 240 feet, what is the distance across the lake? Approximate this distance to the nearest whole foot.

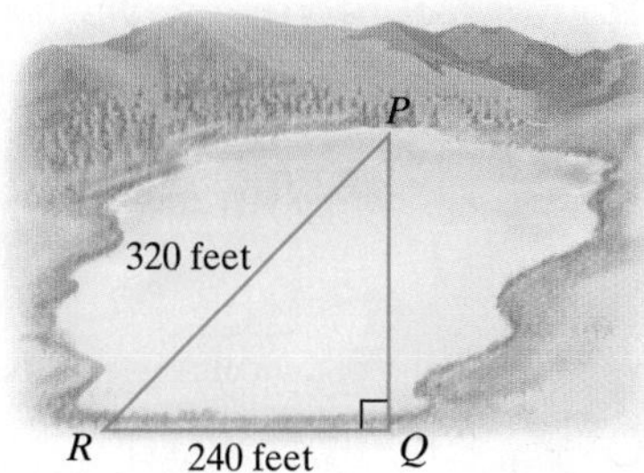

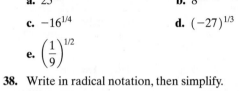

36. Find the distance between $(-7, 4)$ and $(2, 5)$.

37. Write in radical notation. Then simplify.

a. $25^{1/2}$ **b.** $8^{1/3}$

c. $-16^{1/4}$ **d.** $(-27)^{1/3}$

e. $\left(\frac{1}{9}\right)^{1/2}$

38. Write in radical notation, then simplify.

a. $-49^{1/2}$ **b.** $256^{1/4}$

c. $(-64)^{1/3}$ **d.** $\left(\frac{25}{36}\right)^{1/2}$

e. $(32)^{1/5}$

39. Use the square root property to solve $2x^2 = 7$.

40. Use the square root property to solve $(x - 4)^2 = 3$.

41. Solve $x^2 - 10x = -14$ by completing the square.

42. Solve $x^2 + 4x = 8$ by completing the square.

43. Solve $2x^2 - 9x = 5$ using the quadratic formula.

44. Solve $2x^2 + 5x = 7$ using the quadratic formula.

45. Write each radical as the product of a real number and i.

a. $\sqrt{-4}$

b. $\sqrt{-11}$

c. $\sqrt{-20}$

46. Write each radical as the product of a real number and i.

a. $\sqrt{-7}$

b. $\sqrt{-16}$

c. $\sqrt{-27}$

47. Graph $y = x^2 - 4$.

48. Graph $y = x^2 + 2x + 3$.

Appendix A

Geometry

A.1 Geometric Formulas

Rectangle

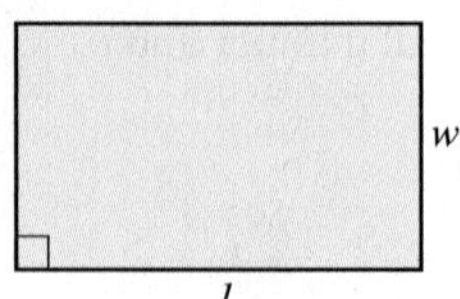

Perimeter: $P = 2l + 2w$
Area: $A = lw$

Square

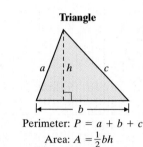

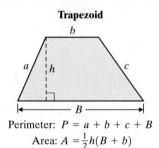

Perimeter: $P = 4s$
Area: $A = s^2$

Triangle

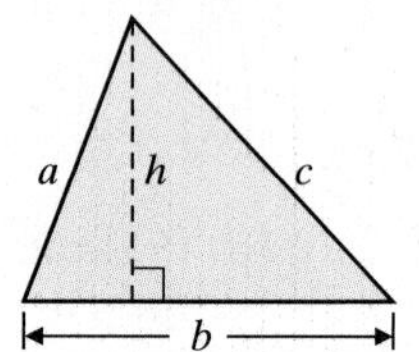

Perimeter: $P = a + b + c$
Area: $A = \frac{1}{2}bh$

Sum of Angles of Triangle

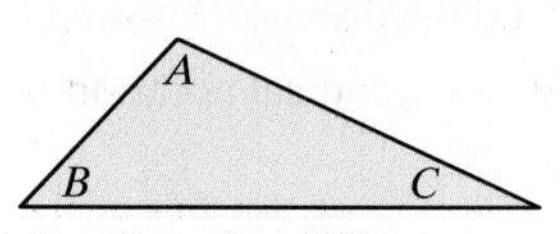

$A + B + C = 180°$
The sum of the measures of the three angles is 180°.

Right Triangles

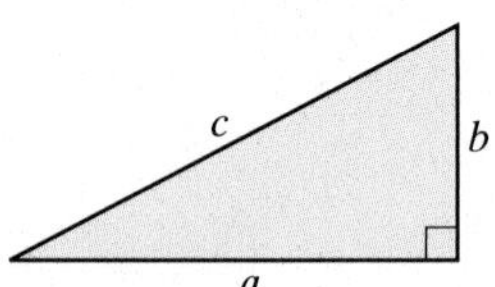

Perimeter: $P = a + b + c$
Area: $A = \frac{1}{2}ab$
One 90° (right) angle

Pythagorean Theorem (for right triangles)

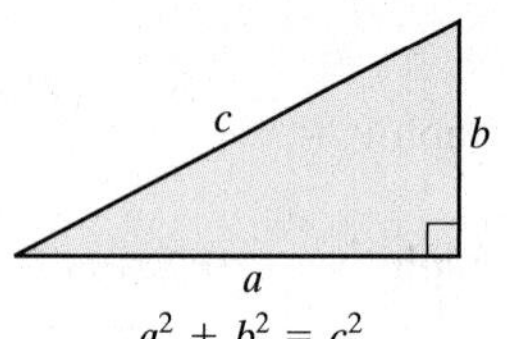

$a^2 + b^2 = c^2$

Isosceles Triangle

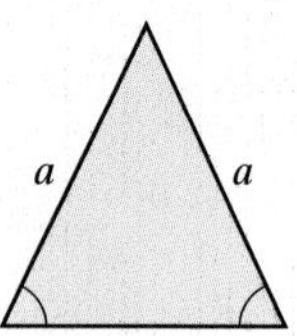

Triangle has:
two equal sides and
two equal angles.

Equilateral Triangle

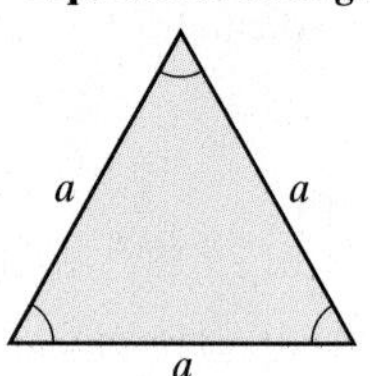

Triangle has:
three equal sides and
three equal angles.
Measure of each angle is 60°.

Trapezoid

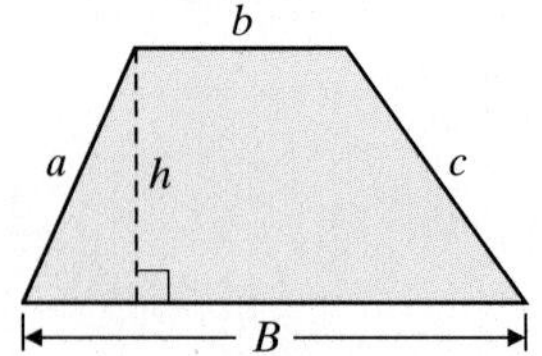

Perimeter: $P = a + b + c + B$
Area: $A = \frac{1}{2}h(B + b)$

Parallelogram

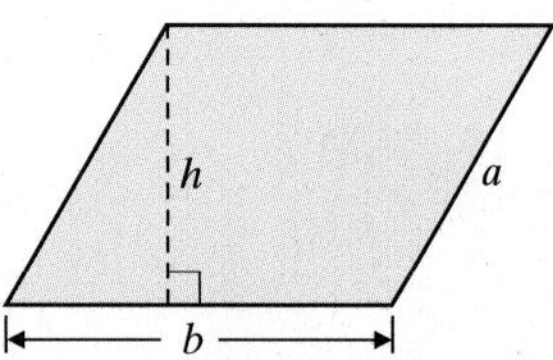

Perimeter: $P = 2a + 2b$
Area: $A = bh$

Circle

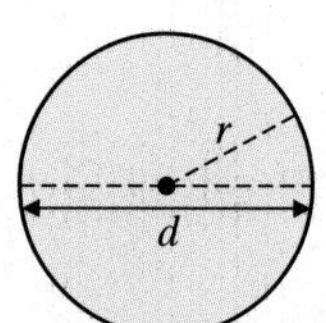

Circumference: $C = \pi d$
$C = 2\pi r$
Area: $A = \pi r^2$

Rectangular Solid

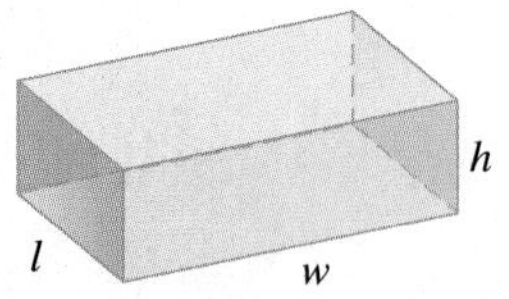

Volume: $V = lwh$
Surface Area:
$S = 2lw + 2hl + 2hw$

Cube

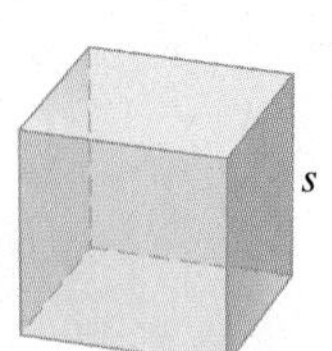

Volume: $V = s^3$
Surface Area: $S = 6s^2$

Cone

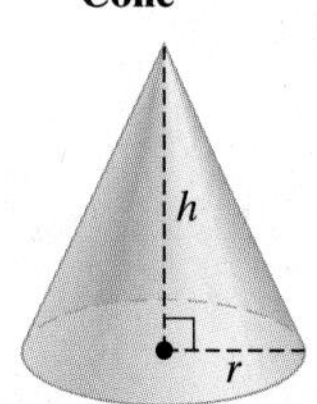

Volume: $V = \frac{1}{3}\pi r^2 h$
Lateral Surface Area:
$S = \pi r\sqrt{r^2 + h^2}$

Right Circular Cylinder

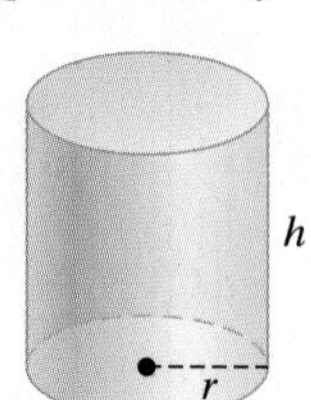

Volume: $V = \pi r^2 h$
Surface Area: $S = 2\pi r^2 + 2\pi rh$

Sphere

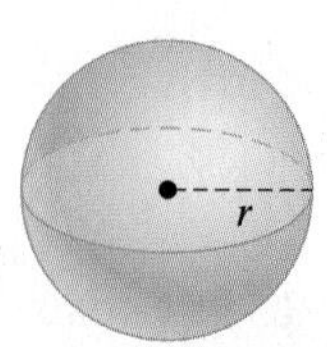

Volume: $V = \frac{4}{3}\pi r^3$
Surface Area: $S = 4\pi r^2$

Other Formulas
Distance: $d = rt$ (r = rate, t = time)
Percent: $p = br$ (p = percentage, b = base, r = rate)

Temperature: $F = \frac{9}{5}C + 32 \quad C = \frac{5}{9}(F - 32)$
Simple Interest: $I = Prt$
(P = principal, r = annual interest rate, t = time in years)

A.2 Review of Geometric Figures

Plane figures have length and width but no thickness or depth		
Name	***Description***	***Figure***
Polygon	Union of three or more coplanar line segments that intersect with each other only at each end point, with each end point shared by two segments.	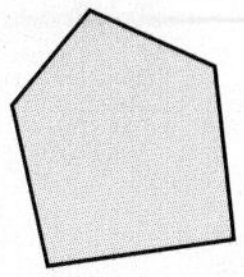
Triangle	Polygon with three sides (sum of measures of three angles is 180°).	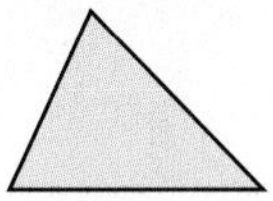
Scalene Triangle	Triangle with no sides of equal length.	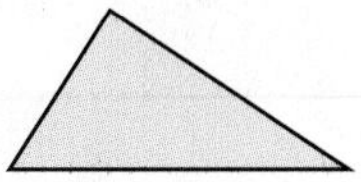
Isosceles Triangle	Triangle with two sides of equal length.	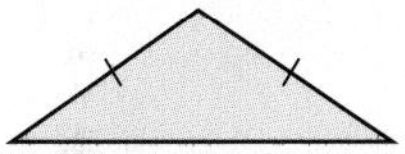
Equilateral Triangle	Triangle with all sides of equal length.	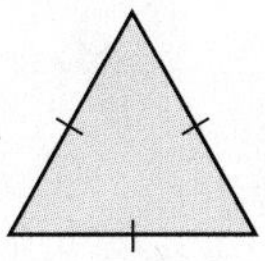
Right Triangle	Triangle that contains a right angle.	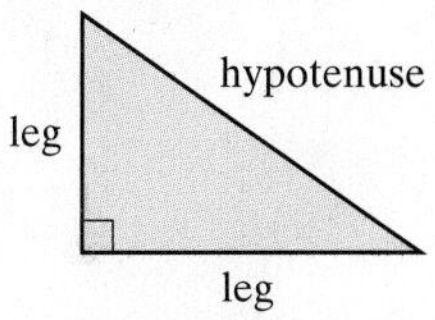 hypotenuse leg leg
Quadrilateral	Polygon with four sides (sum of measures of four angles is 360°).	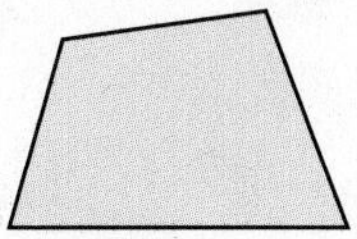
Trapezoid	Quadrilateral with exactly one pair of opposite sides parallel.	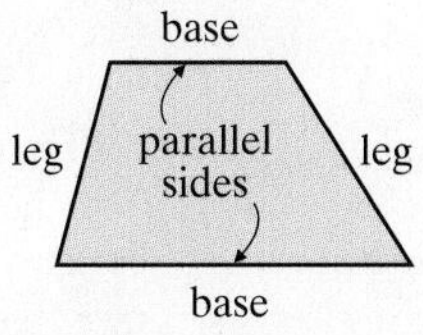 base leg parallel sides leg base
Isosceles Trapezoid	Trapezoid with legs of equal length.	
Parallelogram	Quadrilateral with both pairs of opposite sides parallel and equal in length.	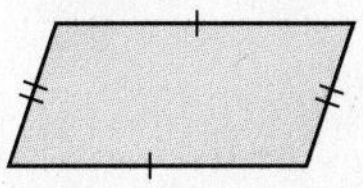

Plane figures have length and width but no thickness or depth		
Name	***Description***	***Figure***
Rhombus	Parallelogram with all sides of equal length.	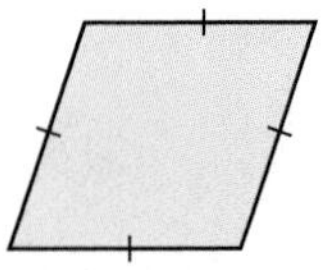
Rectangle	Parallelogram with four right angles.	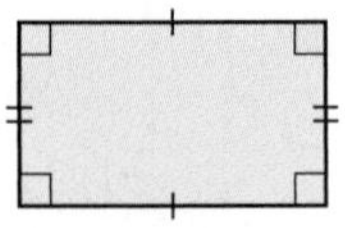
Square	Rectangle with all sides of equal length.	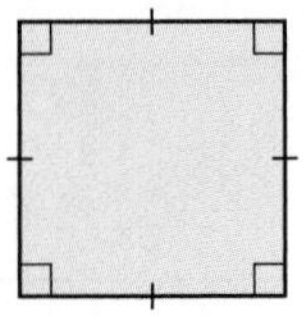
Circle	All points in a plane the same distance from a fixed point called the **center.**	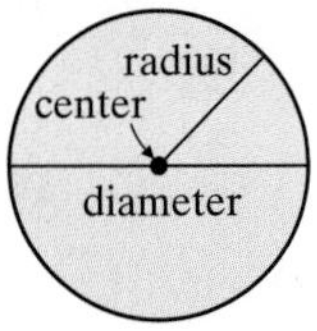 radius center diameter
Rectangular Solid	A solid with six sides, all of which are rectangles.	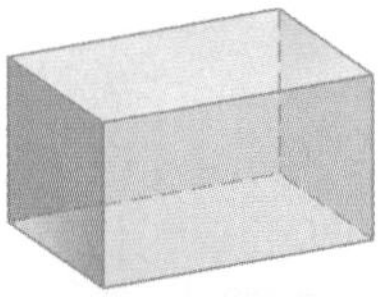
Cube	A rectangular solid whose six sides are squares.	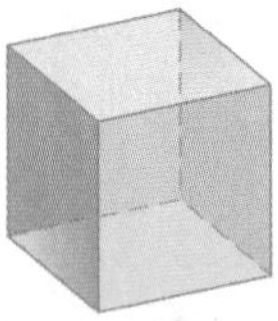
Sphere	All points the same distance from a fixed point, called the center.	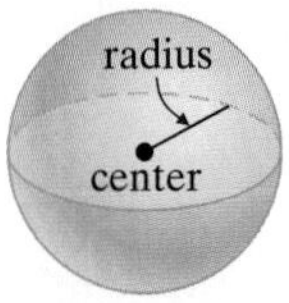 radius center
Right Circular Cylinder	A cylinder consisting of two circular bases that are perpendicular to its altitude.	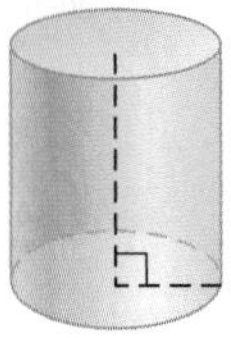
Right Circular Cone	A cone with a circular base that is perpendicular to its altitude.	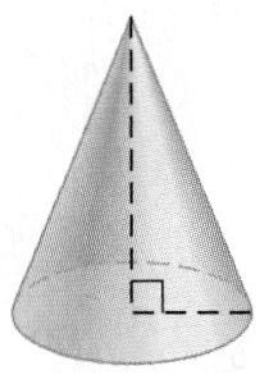

A.3 Review of Angles, Lines, and Special Triangles

The word **geometry** is formed from the Greek words, **geo,** meaning earth, and **metron,** meaning measure. Geometry literally means to measure the earth.

This section contains a review of some basic geometric ideas. It will be assumed that fundamental ideas of geometry such as point, line, ray, and angle are known. In this appendix, the notation $\angle 1$ is read "angle 1" and the notation $m\angle 1$ is read "the measure of angle 1."

We first review types of angles.

Angles

A **right angle** is an angle whose measure is 90°. A right angle can be indicated by a square drawn at the vertex of the angle, as shown below.

An angle whose measure is more than 0° but less than 90° is called an **acute angle.**

An angle whose measure is greater than 90° but less than 180° is called an **obtuse angle.**

An angle whose measure is 180° is called a **straight angle.**

Two angles are said to be **complementary** if the sum of their measures is 90°. Each angle is called the **complement** of the other.

Two angles are said to be **supplementary** if the sum of their measures is 180°. Each angle is called the **supplement** of the other.

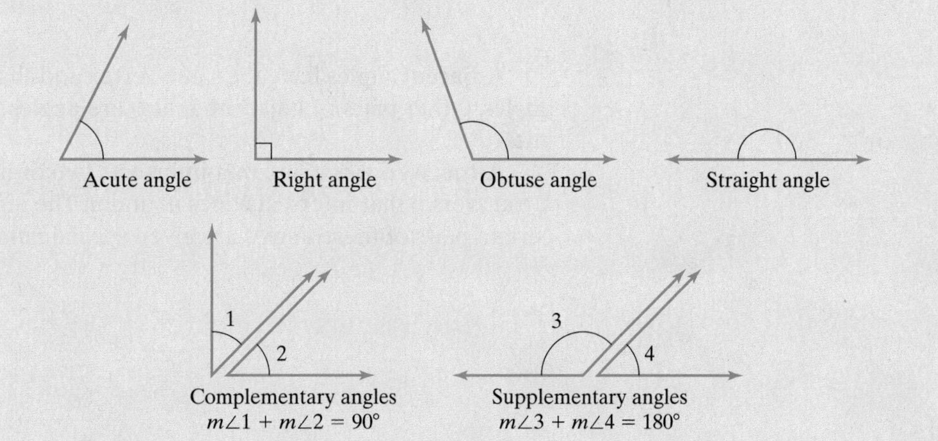

EXAMPLE 1 If an angle measures 28°, find its complement.

Solution Two angles are complementary if the sum of their measures is 90°. The complement of a 28° angle is an angle whose measure is $90° - 28° = 62°$. To check, notice that $28° + 62° = 90°$. □

Plane is an undefined term that we will describe. A plane can be thought of as a flat surface with infinite length and width, but no thickness. A plane is two dimensional. The arrows in the following diagram indicate that a plane extends indefinitely and has no boundaries.

Figures that lie on a plane are called **plane figures.** (See the descriptions of common plane figures in Appendix A.2.) Lines that lie in the same plane are called **coplanar**.

Lines

Two lines are **parallel** if they lie in the same plane but never meet.

Intersecting lines meet or cross in one point.

Two lines that form right angles when they intersect are said to be **perpendicular.**

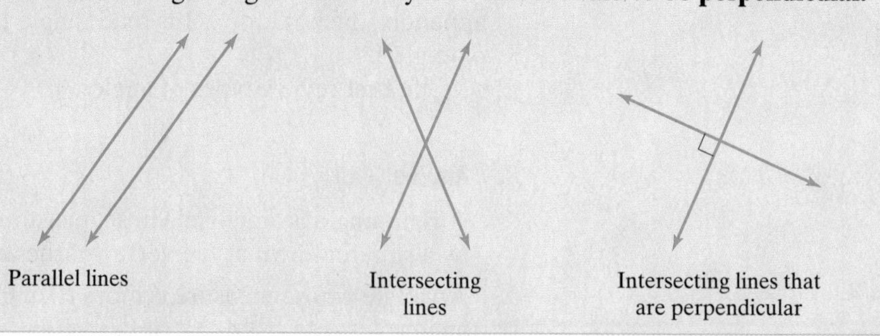

Two intersecting lines form **vertical angles**. Angles 1 and 3 are vertical angles. Also angles 2 and 4 are vertical angles. It can be shown that **vertical angles have equal measures**.

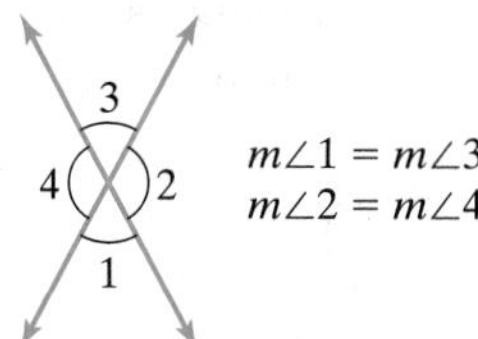

Adjacent angles have the same vertex and share a side. Angles 1 and 2 are adjacent angles. Other pairs of adjacent angles are angles 2 and 3, angles 3 and 4, and angles 4 and 1.

A **transversal** is a line that intersects two or more lines in the same plane. Line l is a transversal that intersects lines m and n. The eight angles formed are numbered and certain pairs of these angles are given special names.

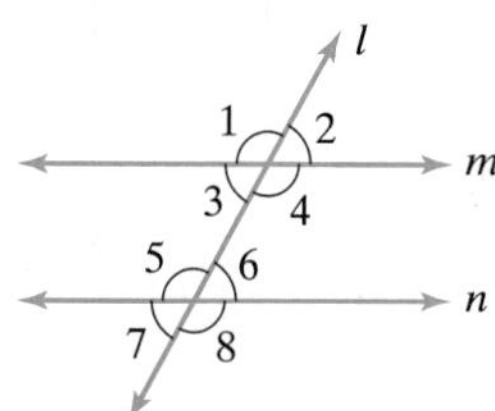

Corresponding angles: $\angle 1$ and $\angle 5$, $\angle 3$ and $\angle 7$, $\angle 2$ and $\angle 6$, and $\angle 4$ and $\angle 8$.

Exterior angles: $\angle 1$, $\angle 2$, $\angle 7$, and $\angle 8$.

Interior angles: $\angle 3$, $\angle 4$, $\angle 5$, and $\angle 6$.

Alternate interior angles: $\angle 3$ and $\angle 6$, $\angle 4$ and $\angle 5$.

These angles and parallel lines are related in the following manner.

Parallel Lines Cut by a Transversal

1. If two parallel lines are cut by a transversal, then
   a. corresponding angles are equal and
   b. alternate interior angles are equal.
2. If corresponding angles formed by two lines and a transversal are equal, then the lines are parallel.
3. If alternate interior angles formed by two lines and a transversal are equal, then the lines are parallel.

EXAMPLE 2 Given that lines m and n are parallel and that the measure of angle 1 is 100°, find the measures of angles 2, 3, and 4.

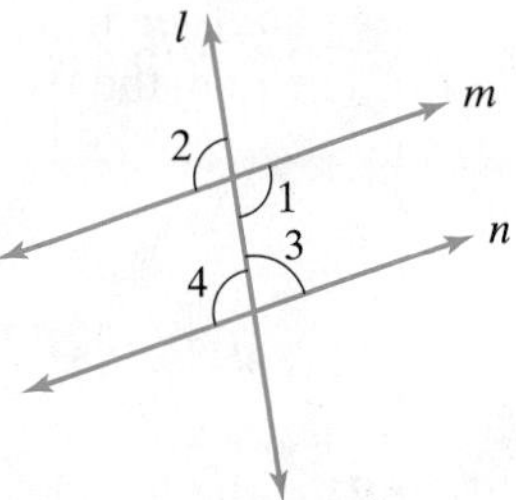

Solution $m\angle 2 = 100°$, since angles 1 and 2 are vertical angles.

$m\angle 4 = 100°$, since angles 1 and 4 are alternate interior angles.

$m\angle 3 = 180° - 100° = 80°$, since angles 4 and 3 are supplementary angles.

□

A **polygon** is the union of three or more coplanar line segments that intersect each other only at each end point, with each end point shared by exactly two segments.

A **triangle** is a polygon with three sides. The sum of the measures of the three angles of a triangle is 180°. In the following figure, $m\angle 1 + m\angle 2 + m\angle 3 = 180°$.

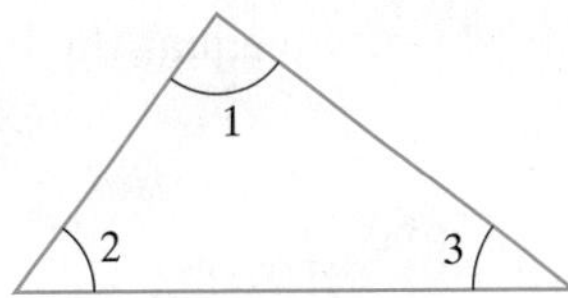

EXAMPLE 3 Find the measure of the third angle of the triangle shown.

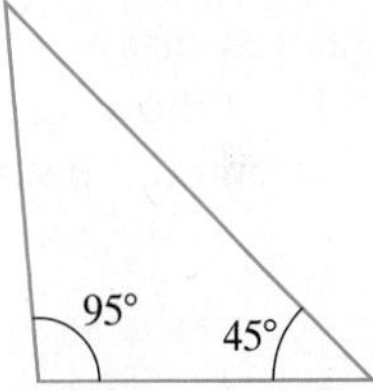

Solution The sum of the measures of the angles of a triangle is 180°. Since one angle measures 45° and the other angle measures 95°, the third angle measures $180° - 45° - 95° = 40°$. □

Two triangles are **congruent** if they have the same size and the same shape. In congruent triangles, the measures of corresponding angles are equal and the lengths of corresponding sides are equal. The following triangles are congruent.

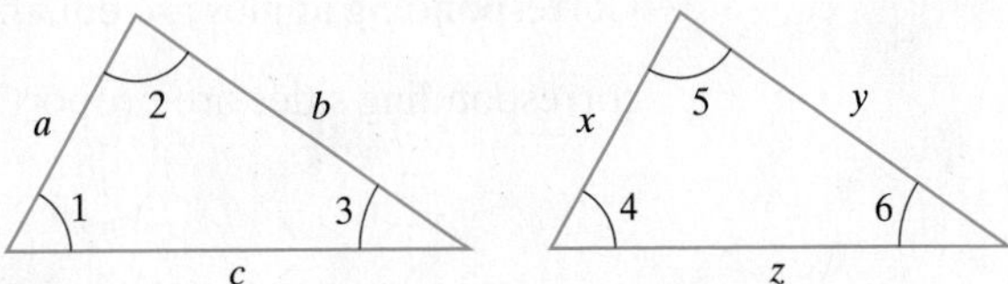

Corresponding angles are equal: $m\angle 1 = m\angle 4, m\angle 2 = m\angle 5$, and $m\angle 3 = m\angle 6$. Also, lengths of corresponding sides are equal: $a = x$, $b = y$, and $c = z$.

Any one of the following may be used to determine whether two triangles are congruent.

Congruent Triangles

1. If the measures of two angles of a triangle equal the measures of two angles of another triangle and the lengths of the sides between each pair of angles are equal, the triangles are congruent.

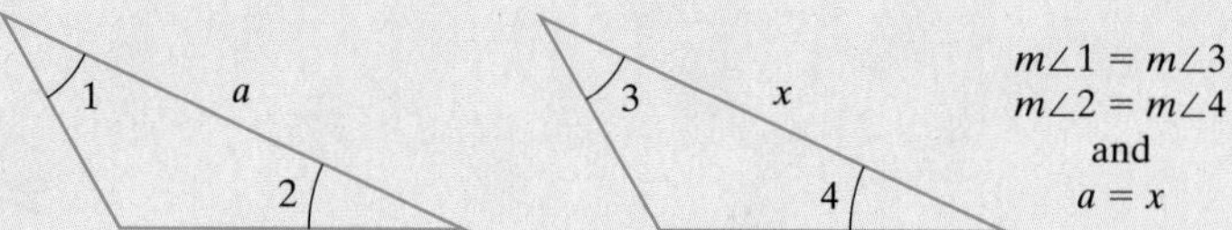

2. If the lengths of the three sides of a triangle equal the lengths of corresponding sides of another triangle, the triangles are congruent.

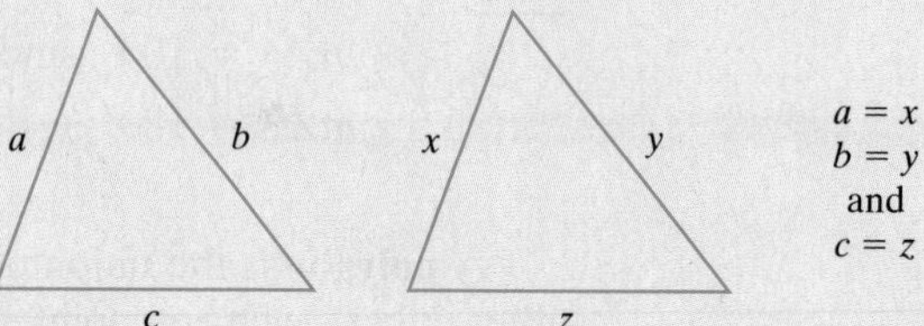

3. If the lengths of two sides of a triangle equal the lengths of corresponding sides of another triangle, and the measures of the angles between each pair of sides are equal, the triangles are congruent.

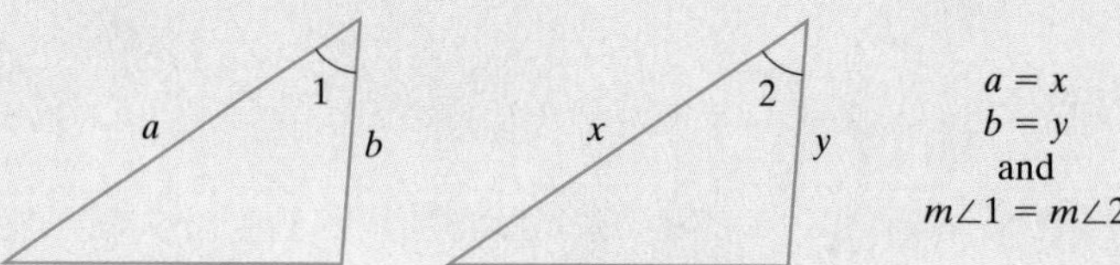

Two triangles are **similar** if they have the same shape. In similar triangles, the measures of corresponding angles are equal and corresponding sides are in proportion. The following triangles are similar. (All similar triangles drawn in this appendix will be oriented the same.)

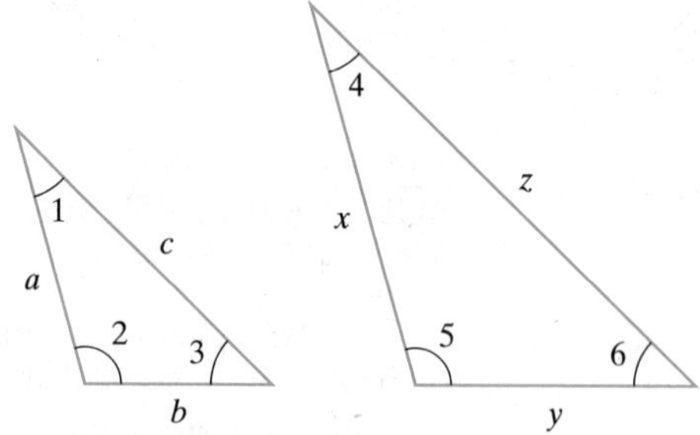

Corresponding angles are equal: $m\angle 1 = m\angle 4$, $m\angle 2 = m\angle 5$, and $m\angle 3 = m\angle 6$. Also, corresponding sides are proportional: $\frac{a}{x} = \frac{b}{y} = \frac{c}{z}$.

Any one of the following may be used to determine whether two triangles are similar.

Similar Triangles

1. If the measures of two angles of a triangle equal the measures of two angles of another triangle, the triangles are similar.

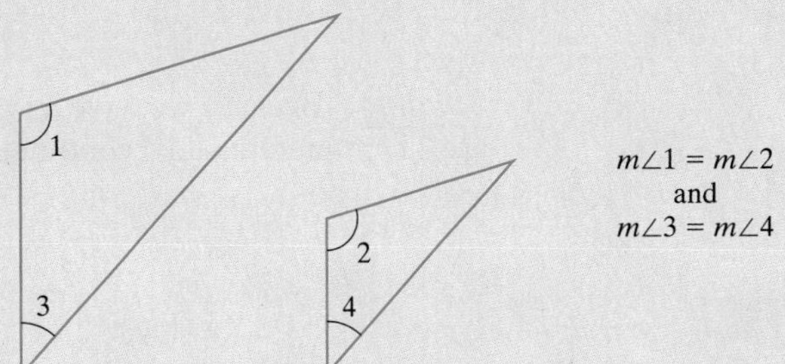

2. If three sides of one triangle are proportional to three sides of another triangle, the triangles are similar.

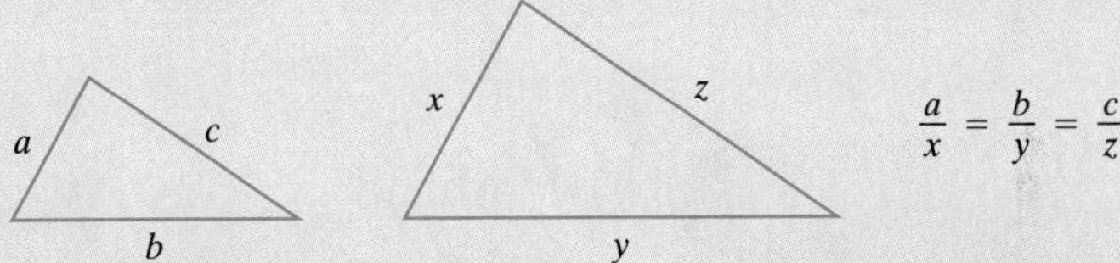

3. If two sides of a triangle are proportional to two sides of another triangle and the measures of the included angles are equal, the triangles are similar.

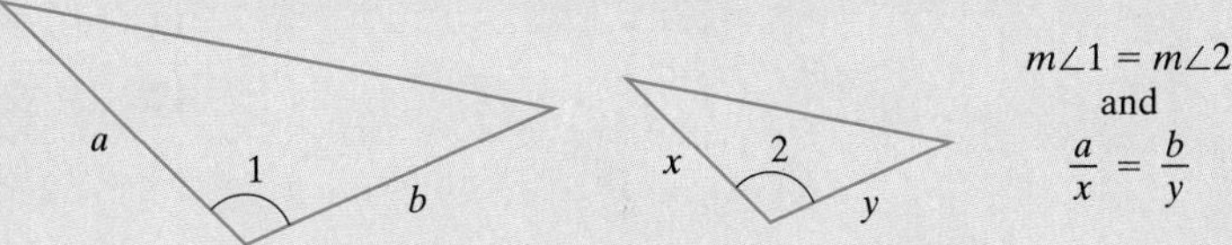

EXAMPLE 4 Given that the following triangles are similar, find the missing length x.

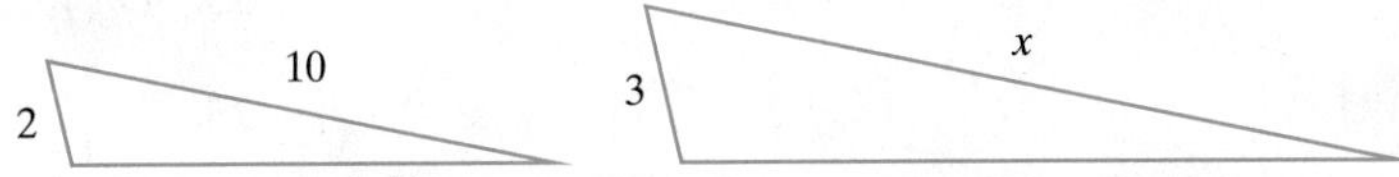

Solution Since the triangles are similar, corresponding sides are in proportion. Thus, $\frac{2}{3} = \frac{10}{x}$. To solve this equation for x, we multiply both sides by the LCD, $3x$.

$$3x\left(\frac{2}{3}\right) = 3x\left(\frac{10}{x}\right)$$
$$2x = 30$$
$$x = 15$$

The missing length is 15 units. □

A **right triangle** contains a right angle. The side opposite the right angle is called the **hypotenuse,** and the other two sides are called the **legs.** The **Pythagorean theorem** gives a formula that relates the lengths of the three sides of a right triangle.

The Pythagorean Theorem

If a and b are the lengths of the legs of a right triangle, and c is the length of the hypotenuse, then $a^2 + b^2 = c^2$.

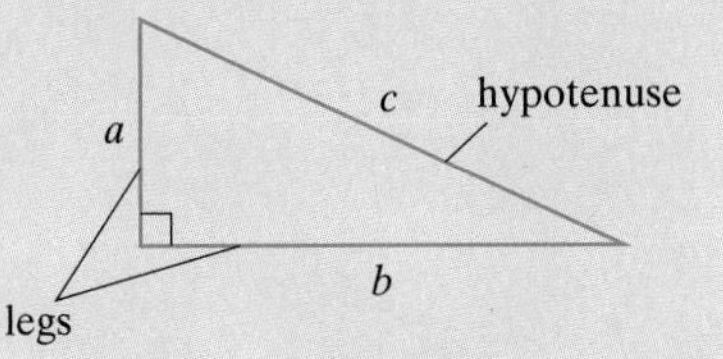

EXAMPLE 5 Find the length of the hypotenuse of a right triangle whose legs have lengths of 3 centimeters and 4 centimeters.

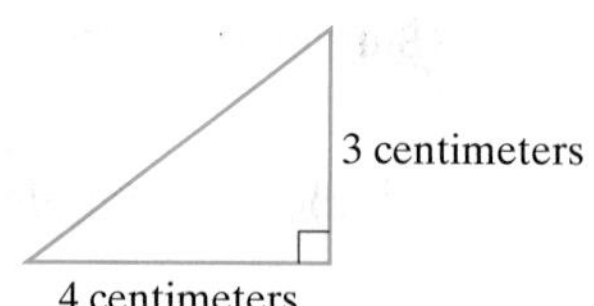

Solution Because we have a right triangle, we use the Pythagorean theorem. The legs are 3 centimeters and 4 centimeters, so let $a = 3$ and $b = 4$ in the formula.

$$a^2 + b^2 = c^2$$
$$3^2 + 4^2 = c^2$$
$$9 + 16 = c^2$$
$$25 = c^2$$

Since c represents a length, we assume that c is positive. Thus, if c^2 is 25, c must be 5. The hypotenuse has a length of 5 centimeters. □

A.3 Exercise Set

MyMathLab®

Find the complement of each angle. See Example 1.

1. 19°
2. 65°
3. 70.8°
4. $45\frac{2}{3}°$
5. $11\frac{1}{4}°$
6. 19.6°

Find the supplement of each angle.

7. 150°
8. 90°
9. 30.2°
10. 81.9°
11. $79\frac{1}{2}°$
12. $165\frac{8}{9}°$

13. If lines m and n are parallel, find the measures of angles 1 through 7. See Example 2.

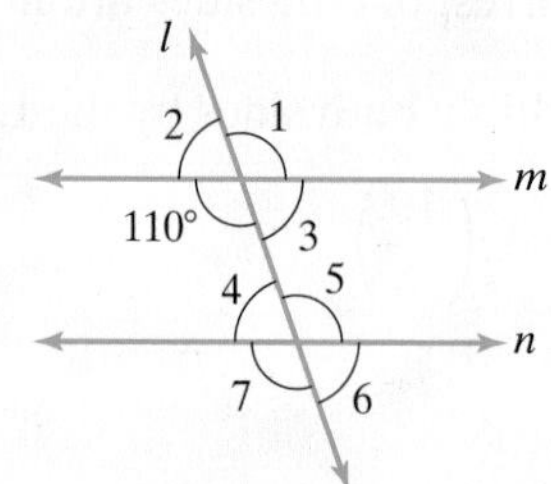

14. If lines m and n are parallel, find the measures of angles 1 through 5. See Example 2.

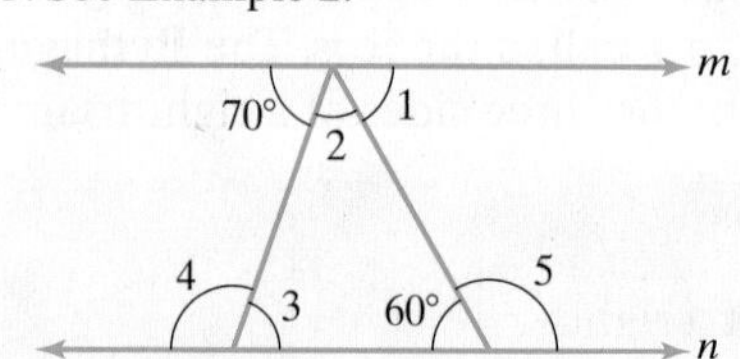

In each of the following, the measures of two angles of a triangle are given. Find the measure of the third angle. See Example 3.

15. 11°, 79°
16. 8°, 102°
17. 25°, 65°
18. 44°, 19°
19. 30°, 60°
20. 67°, 23°

In each of the following, the measure of one angle of a right triangle is given. Find the measures of the other two angles.

21. 45°
22. 60°
23. 17°
24. 30°
25. $39\frac{3}{4}°$
26. 72.6°

Given that each of the following pairs of triangles is similar, find the missing lengths. See Example 4.

27.

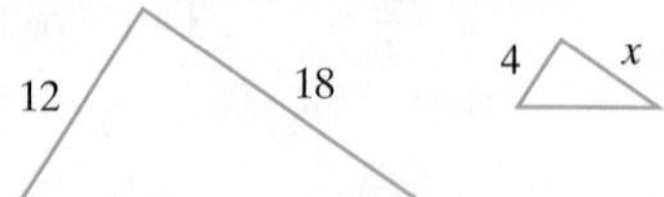

28.

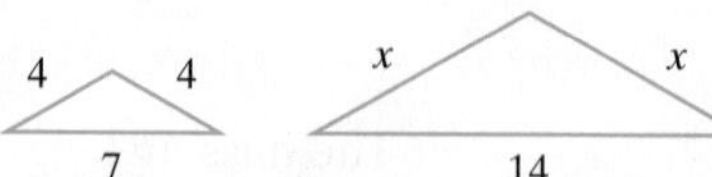

29.

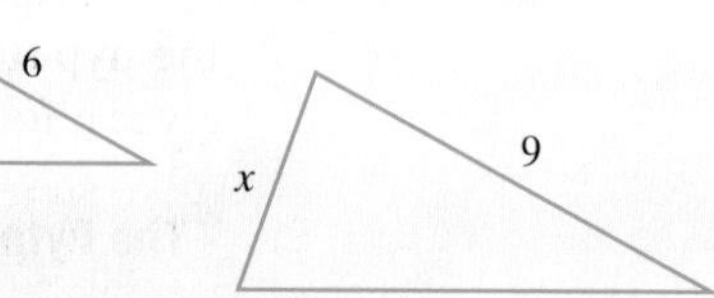

30.

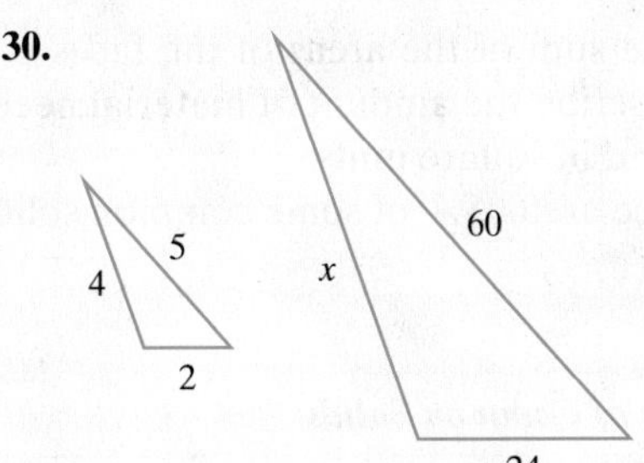

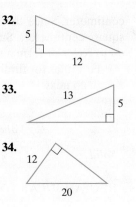

Use the Pythagorean theorem to find the missing lengths in the right triangles. See Example 5.

31.

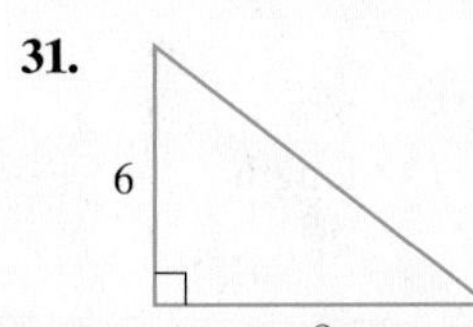

32.

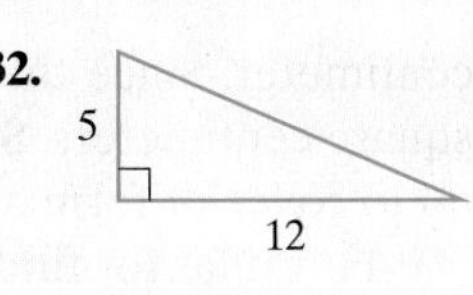

33.

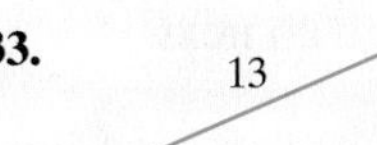

34.

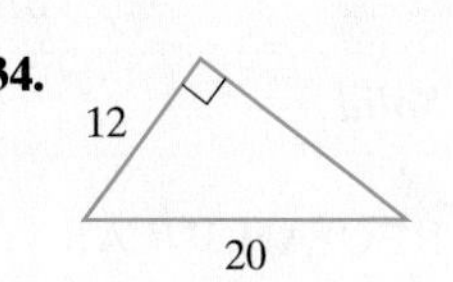

A.4 Review of Volume and Surface Area

A **convex solid** is a set of points, S, not all in one plane, such that for any two points A and B in S, all points between A and B are also in S. In this appendix, we will find the volume and surface area of special types of solids called polyhedrons. A solid formed by the intersection of a finite number of planes is called a **polyhedron.** The box below is an example of a polyhedron.

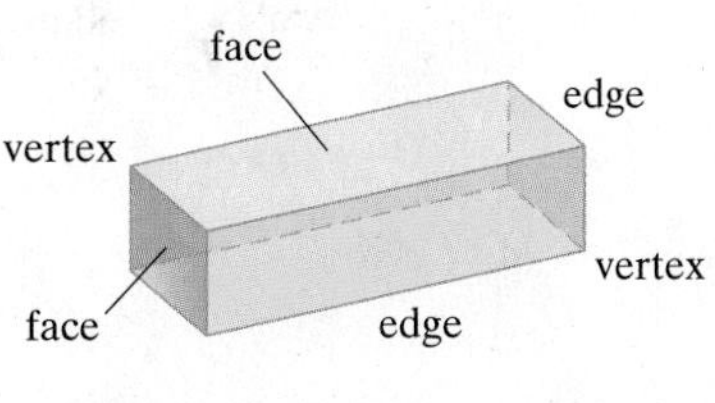

Polyhedron

Each of the plane regions of the polyhedron is called a **face** of the polyhedron. If the intersection of two faces is a line segment, this line segment is an **edge** of the polyhedron. The intersections of the edges are the **vertices** of the polyhedron.

Volume is a measure of the space of a solid. The volume of a box or can, for example, is the amount of space inside. Volume can be used to describe the amount of juice in a pitcher or the amount of concrete needed to pour a foundation for a house.

The volume of a solid is the number of **cubic units** in the solid. A cubic centimeter and a cubic inch are illustrated.

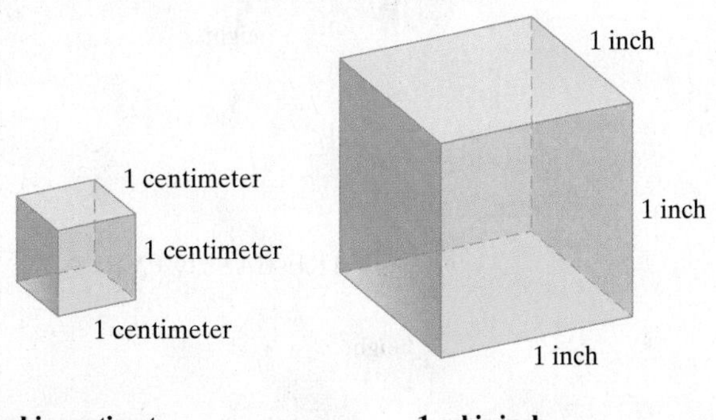

1 cubic centimeter **1 cubic inch**

The **surface area** of a polyhedron is the sum of the areas of the faces of the polyhedron. For example, each face of the cube to the left above has an area of 1 square centimeter. Since there are 6 faces of the cube, the sum of the areas of the faces is 6 square centimeters. Surface area can be used to describe the amount of material needed to cover or form a solid. Surface area is measured in square units.

Formulas for finding the volumes, V, and surface areas, SA, of some common solids are given next.

Volume and Surface Area Formulas of Common Solids

Solid	***Formulas***
RECTANGULAR SOLID 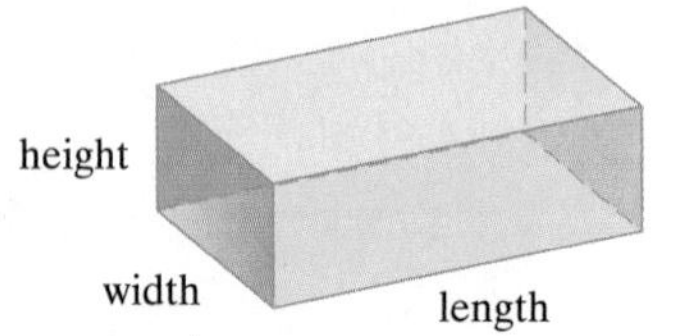	$V = lwh$ $SA = 2lh + 2wh + 2lw$ where h = height, w = width, l = length
CUBE 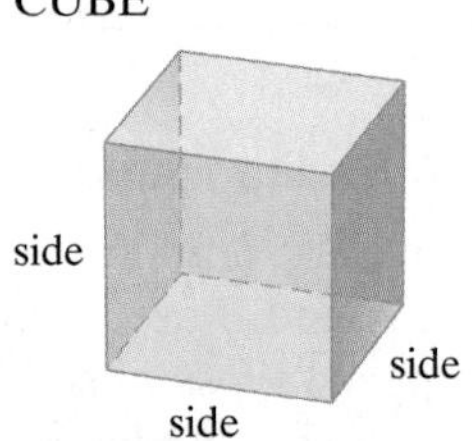	$V = s^3$ $SA = 6s^2$ where s = side
SPHERE 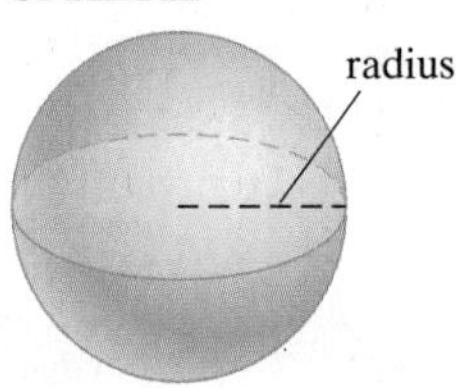	$V = \frac{4}{3}\pi r^3$ $SA = 4\pi r^2$ where r = radius
CIRCULAR CYLINDER 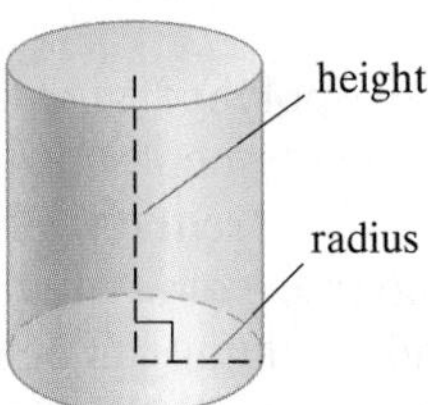	$V = \pi r^2 h$ $SA = 2\pi rh + 2\pi r^2$ where h = height, r = radius
CONE 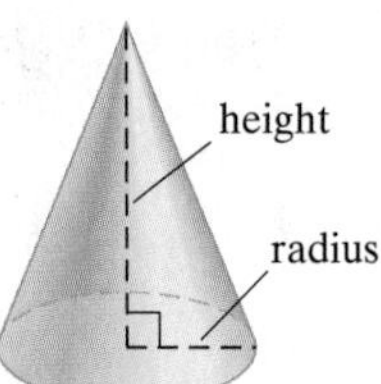	$V = \frac{1}{3}\pi r^2 h$ $SA = \pi r\sqrt{r^2 + h^2} + \pi r^2$ where h = height, r = radius
SQUARE-BASED PYRAMID 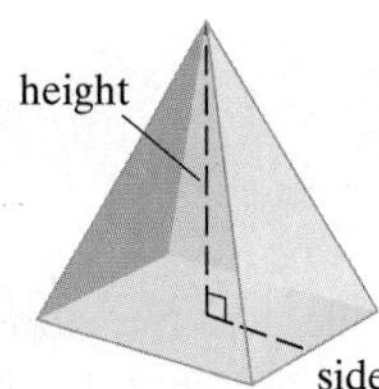	$V = \frac{1}{3}s^2 h$ $SA = B + \frac{1}{2}pl$ where B = area of base; p = perimeter of base, h = height, s = side, l = slant height

> **Helpful Hint**
> Volume is measured in cubic units. Surface area is measured in square units.

EXAMPLE 1 Find the volume and surface area of a rectangular box that is 12 inches long, 6 inches wide, and 3 inches high.

Solution Let $h = 3$ in., $l = 12$ in., and $w = 6$ in.

$$V = lwh$$
$$V = 12 \text{ inches} \cdot 6 \text{ inches} \cdot 3 \text{ inches} = 216 \text{ cubic inches}$$

The volume of the rectangular box is 216 cubic inches.

$$\begin{aligned} SA &= 2lh + 2wh + 2lw \\ &= 2(12 \text{ in.})(3 \text{ in.}) + 2(6 \text{ in.})(3 \text{ in.}) + 2(12 \text{ in.})(6 \text{ in.}) \\ &= 72 \text{ sq in.} + 36 \text{ sq in.} + 144 \text{ sq in.} \\ &= 252 \text{ sq in.} \end{aligned}$$

The surface area of rectangular box is 252 square inches. □

EXAMPLE 2 Find the volume and surface area of a ball of radius 2 inches. Give the exact volume and surface area and then use the approximation $\frac{22}{7}$ for π.

Solution

$V = \frac{4}{3}\pi r^3$ Formula for volume of a sphere

$V = \frac{4}{3} \cdot \pi(2 \text{ in.})^3$ Let $r = 2$ inches.

$= \frac{32}{3}\pi$ cu in. Simplify.

$\approx \frac{32}{3} \cdot \frac{22}{7}$ cu in. Approximate π with $\frac{22}{7}$.

$= \frac{704}{21}$ or $33\frac{11}{21}$ cu in.

The volume of the sphere is exactly $\frac{32}{3}\pi$ cubic inches or approximately $33\frac{11}{21}$ cubic inches.

$SA = 4\pi r^2$ Formula for surface area

$SA = 4 \cdot \pi(2 \text{ in.})^2$ Let $r = 2$ inches.

$= 16\pi$ sq in. Simplify.

$\approx 16 \cdot \frac{22}{7}$ sq in. Approximate π with $\frac{22}{7}$.

$= \frac{352}{7}$ or $50\frac{2}{7}$ sq in.

The surface area of the sphere is exactly 16π square inches or approximately $50\frac{2}{7}$ square inches. □

A.4 Exercise Set

MyMathLab®

Find the volume and surface area of each solid. See Examples 1 and 2. For formulas that contain π, give an exact answer and then approximate using $\frac{22}{7}$ for π.

1.

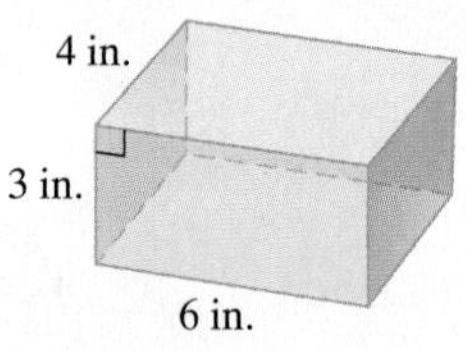

2.

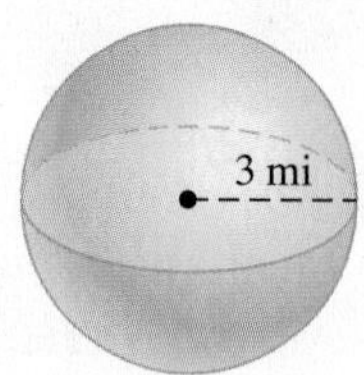

3.

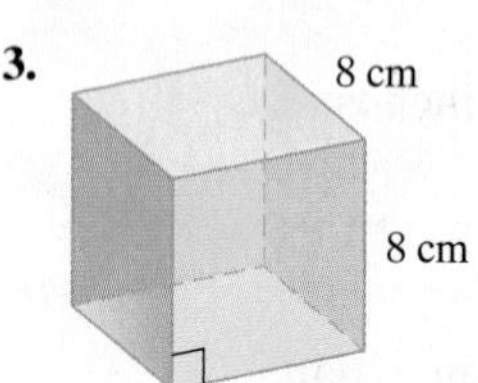

4.

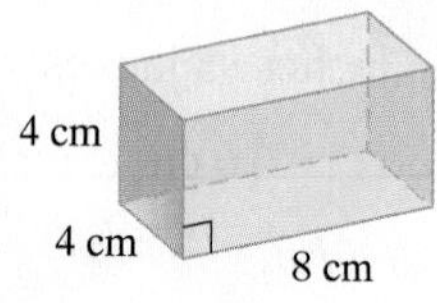

5. (For surface area, use 3.14 for π and round to two decimal places.)

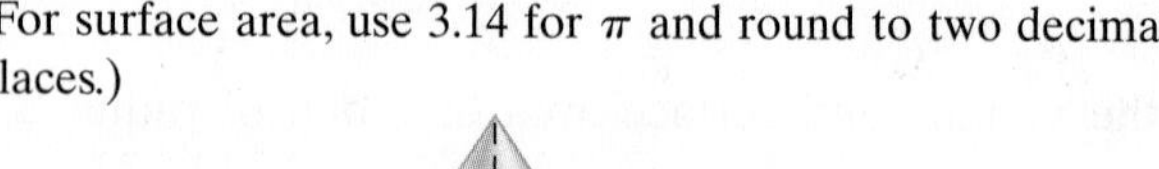

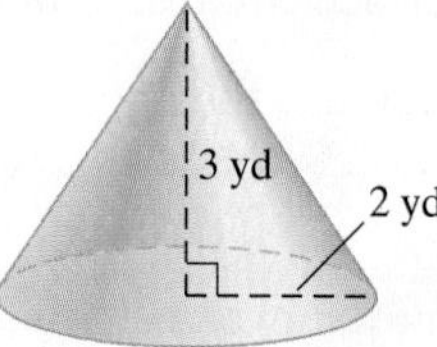

6.

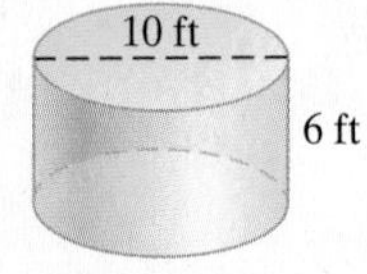

7.

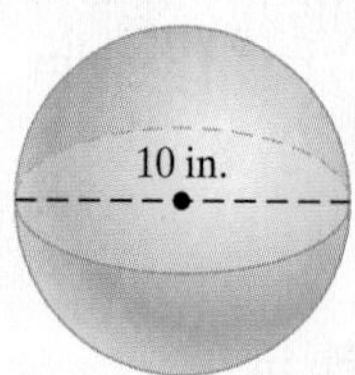

8. Find the volume only.

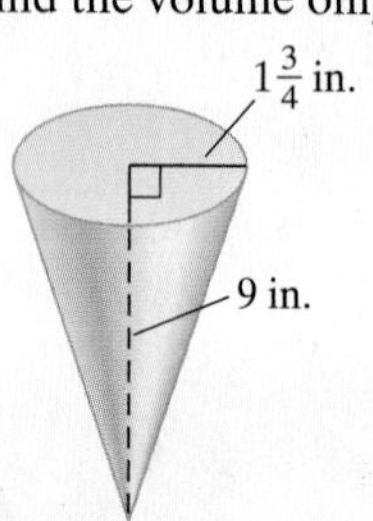

9.

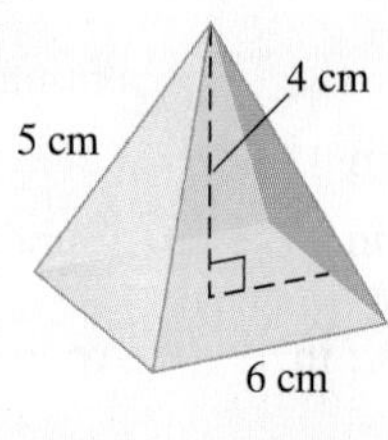

10.

Solve.

11. Find the volume of a cube with edges of $1\frac{1}{3}$ inches.

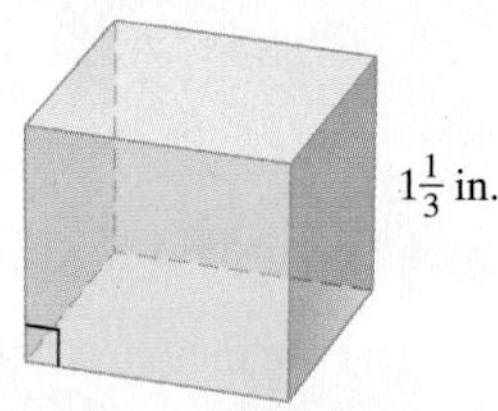

12. A water storage tank is in the shape of a cone with the pointed end down. If the radius is 14 ft and the depth of the tank is 15 ft, approximate the volume of the tank in cubic feet. Use $\frac{22}{7}$ for π.

13. Find the surface area of a rectangular box 2 ft by 1.4 ft by 3 ft.

14. Find the surface area of a box in the shape of a cube that is 5 ft on each side.

15. Find the volume of a pyramid with a square base 5 in. on a side and a height of $1\frac{3}{10}$ in.

16. Approximate to the nearest hundredth the volume of a sphere with a radius of 2 cm. Use 3.14 for π.

17. A paperweight is in the shape of a square-based pyramid 20 cm tall. If an edge of the base is 12 cm, find the volume of the paperweight.

18. A bird bath is made in the shape of a hemisphere (half-sphere). If its radius is 10 in., approximate the volume. Use $\frac{22}{7}$ for π.

19. Find the exact surface area of a sphere with a radius of 7 in.

20. A tank is in the shape of a cylinder 8 ft tall and 3 ft in radius. Find the exact surface area of the tank.

21. Find the volume of a rectangular block of ice 2 ft by $2\frac{1}{2}$ ft by $1\frac{1}{2}$ ft.

22. Find the capacity (volume in cubic feet) of a rectangular ice chest with inside measurements of 3 ft by $1\frac{1}{2}$ ft by $1\frac{3}{4}$ ft.

23. An ice cream cone with a 4-cm diameter and 3-cm depth is filled exactly level with the top of the cone. Approximate how much ice cream (in cubic centimeters) is in the cone. Use $\frac{22}{7}$ for π.

24. A child's toy is in the shape of a square-based pyramid 10 in. tall. If an edge of the base is 7 in., find the volume of the toy.

Appendix B

Additional Exercises on Proportion and Proportion Applications

1. $\frac{30}{10} = \frac{15}{y}$

2. $\frac{16}{20} = \frac{z}{35}$

3. $\frac{8}{15} = \frac{z}{6}$

4. $\frac{12}{10} = \frac{z}{16}$

5. $\frac{-3.5}{12.5} = \frac{-7}{n}$

6. $\frac{-0.2}{0.7} = \frac{-8}{n}$

7. $\frac{n}{0.6} = \frac{0.05}{12}$

8. $\frac{7.8}{13} = \frac{n}{2.6}$

9. $\frac{8}{\frac{2}{3}} = \frac{24}{n}$

10. $\frac{12}{\frac{3}{4}} = \frac{48}{n}$

11. $\frac{7}{9} = \frac{35}{3x}$

12. $\frac{5}{6} = \frac{40}{3x}$

13. $\frac{7x}{18} = \frac{5}{3}$

14. $\frac{6x}{5} = \frac{19}{10}$

15. $\frac{11}{7} = \frac{4}{x + 1}$

16. $\frac{12}{5} = \frac{7}{x - 1}$

17. $\frac{x - 3}{2x + 1} = \frac{4}{9}$

18. $\frac{x + 1}{3x - 4} = \frac{3}{8}$

19. $\frac{2x + 1}{4} = \frac{6x - 1}{5}$

20. $\frac{11x + 2}{3} = \frac{10x - 5}{2}$

Solve. A self-tanning lotion advertises that a 3-oz bottle will provide four applications.

21. Jen Haddad found a great deal on a 14-oz bottle of the self-tanning lotion she had been using. Based on the advertising claims, how many applications of the self-tanner should Jen expect? Round down to the whole number.

22. The Community College thespians need fake tans for a play they are doing. If the play has a cast of 35, how many ounces of self-tanning lotion should the cast purchase? Round up to the next whole number onces.

The school's computer lab goes through 5 reams of printer paper every 3 weeks.

23. Find out how long a case of printer paper is likely to last (a case of paper holds 8 reams of paper). Round to the nearest week.

24. How many cases of printer paper should be purchased to last the entire semester of 15 weeks? Round to the next case.

A recipe for pancakes calls for 2 cups flour and $1\frac{1}{2}$ cups milk to make a serving for four people.

25. Ming has plenty of flour, but only 4 cups milk. How many servings can he make?

26. The swim team has a weekly breakfast after early practice. How much flour will it take to make pancakes for 18 swimmers?

Solve.

27. A 16-oz grande Tazo Black Iced Tea at Starbucks has 80 calories. How many calories are there in a 24-oz venti Tazo Black Iced Tea? (*Source:* Starbucks Coffee Company)

28. A 16-oz nonfat Caramel Macchiato at Starbucks has 220 calories. How many calories are in a 12-oz nonfat Caramel Macchiato? (*Source:* Starbucks Coffee Company)

29. Mosquitos are annoying insects. To eliminate mosquito larvae, a certain granular substance can be applied to standing water in a ratio of 1 tsp per 25 sq ft of standing water.
    a. At this rate, find how many teaspoons of granules must be used for 450 square feet.
    b. If 3 tsp = 1 tbsp, how many tablespoons of granules must be used?

30. Another type of mosquito control is liquid, where 3 oz of pesticide is mixed with 100 oz of water. This mixture is sprayed on roadsides to control mosquito breeding grounds hidden by tall grass.
    a. If one mixture of water with this pesticide can treat 150 feet of roadway, how many ounces of pesticide are needed to treat one mile? (*Hint:* 1. mile = 5280 feet)
    b. If 8 liquid ounces equals one cup, write your answer to part **a** in cups. Round to the nearest cup.

31. The daily supply of oxygen for one person is provided by 625 square feet of lawn. A total of 3750 square feet of lawn would provide the daily supply of oxygen for how many people? (*Source:* Professional Lawn Care Association of America)

32. In the United States, approximately 71 million of the 200 million cars and light trucks in service have driver-side air bags. In a parking lot containing 800 cars and light trucks, how many would be expected to have driver-side air bags? (*Source:* Insurance Institute for Highway Safety)

33. A student would like to estimate the height of the Statue of Liberty in New York City's harbor. The length of the Statue of Liberty's right arm is 42 feet. The student's right arm is 2 feet long and her height is $5\frac{1}{3}$ feet. Use this information to estimate the height of the Statue of Liberty. How close is your estimate to the statue's actual height of 111 feet, 1 inch from heel to top of head? (*Source:* National Park Service)

34. The length of the Statue of Liberty's index finger is 8 feet while the height to the top of the head is about 111 feet. Suppose your measurements are proportionally the same as this statue and your height is 5 feet.

a. Use this information to find the proposed length of your index finger. Give an exact measurement and then a decimal rounded to the nearest hundredth.

b. Measure your index finger and write it as a decimal in feet rounded to the nearest hundredth. How close is the length of your index finger to the answer to **a**? Explain why.

35. There are 72 milligrams of cholesterol in a 3.5 ounce serving of lobster. How much cholesterol is in 5 ounces of lobster? Round to the nearest tenth of a milligram. (*Source:* The National Institute of Health)

36. There are 76 milligrams of cholesterol in a 3-ounce serving of skinless chicken. How much cholesterol is in 8 ounces of chicken? (*Source:* USDA)

37. Medication is prescribed in 7 out of every 10 hospital emergency room visits that involve an injury. If a large urban hospital had 620 emergency room visits involving an injury in the past month, how many of these visits would you expect included a prescription for medication? (*Source:* National Center for Health Statistics)

38. Currently in the American population of people aged 65 years old and older, there are 145 women for every 100 men. In a nursing home with 280 male residents over the age of 65, how many female residents over the age of 65 would be expected? (*Source:* U.S. Bureau of the Census)

39. One out of three American adults has worked in the restaurant industry at some point during his or her life. In an office of 84 workers, how many of these people would you expect to have worked in the restaurant industry at some point? (*Source:* National Restaurant Association)

40. One pound of firmly packed brown sugar yields $2\frac{1}{4}$ cups. How many pounds of brown sugar will be required in a recipe that calls for 6 cups of firmly packed brown sugar? (*Source:* Based on data from *Family Circle* magazine)

Appendix C

Operations on Decimals

To **add** or **subtract** decimals, write the numbers vertically with decimal points lined up. Add or subtract as with whole numbers and place the decimal point in the answer directly below the decimal points in the problem.

EXAMPLE 1 Add $5.87 + 23.279 + 0.003$.

Solution

$$\begin{array}{r} 5.87 \\ 23.279 \\ +0.003 \\ \hline 29.152 \end{array}$$

□

EXAMPLE 2 Subtract $32.15 - 11.237$.

Solution

$$\begin{array}{rrcrrr} & 3 & \overset{1}{\not{2}} & . & \overset{11}{\not{1}} & \overset{4}{\not{5}} & \overset{10}{\not{0}} \\ - & 1 & 1 & . & 2 & 3 & 7 \\ \hline & 2 & 0 & . & 9 & 1 & 3 \end{array}$$

□

To **multiply** decimals, multiply the numbers as if they were whole numbers. The decimal point in the product is placed so that the number of decimal places in the product is the same as the sum of the number of decimal places in the factors.

EXAMPLE 3 Multiply 0.072×3.5.

Solution

$$\begin{array}{rl} 0.072 & \text{3 decimal places} \\ \times \quad 3.5 & \text{1 decimal place} \\ \hline 360 & \\ 216 & \\ \hline 0.2520 & \text{4 decimal places} \end{array}$$

□

To **divide** decimals, move the decimal point in the divisor to the right of the last digit. Move the decimal point in the dividend the same number of places that the decimal point in the divisor was moved. The decimal point in the quotient lies directly above the decimal point in the dividend.

EXAMPLE 4 Divide $9.46 \div 0.04$.

Solution

$$\begin{array}{r} 236.5 \\ 04.\overline{)946.0} \\ -8 \\ \hline 14 \\ -12 \\ \hline 26 \\ -24 \\ \hline 20 \\ -20 \\ \hline \end{array}$$

□

C Exercise Set

MyMathLab®

Perform the indicated operations.

1. $9.076 + 8.004$

2. $\begin{array}{r} 6.3 \\ \times\ 0.05 \\ \hline \end{array}$

3. $\begin{array}{r} 27.004 \\ -14.2 \\ \hline \end{array}$

4. $\begin{array}{r} 0.0036 \\ 7.12 \\ 32.502 \\ +0.05 \\ \hline \end{array}$

5. $\begin{array}{r} 107.92 \\ +3.04 \\ \hline \end{array}$

6. $7.2 \div 4$

7. $10 - 7.6$

8. $40 \div 0.25$

9. $126.32 - 97.89$

10. $\begin{array}{r} 3.62 \\ 7.11 \\ 12.36 \\ 4.15 \\ +2.29 \\ \hline \end{array}$

11. $\begin{array}{r} 3.25 \\ \times\ 70 \\ \hline \end{array}$

12. $\begin{array}{r} 26.014 \\ -\ 7.8 \\ \hline \end{array}$

13. $8.1 \div 3$

14. $\begin{array}{r} 1.2366 \\ 0.005 \\ 15.17 \\ +\ 0.97 \\ \hline \end{array}$

15. $55.405 - 6.1711$

16. $8.09 + 0.22$

17. $60 \div 0.75$

18. $20 - 12.29$

19. $7.612 \div 100$

20. $\begin{array}{r} 8.72 \\ 1.12 \\ 14.86 \\ 3.98 \\ +\ 1.99 \\ \hline \end{array}$

21. $12.312 \div 2.7$

22. $0.443 \div 100$

23. $\begin{array}{r} 569.2 \\ 71.25 \\ +\ \ 8.01 \\ \hline \end{array}$

24. $3.706 - 2.91$

25. $768 - 0.17$

26. $63 \div 0.28$

27. $12 + 0.062$

28. $0.42 + 18$

29. $76 - 14.52$

30. $1.1092 \div 0.47$

31. $3.311 \div 0.43$

32. $7.61 + 0.0004$

33. $\begin{array}{r} 762.12 \\ 89.7 \\ +\ 11.55 \\ \hline \end{array}$

34. $444 \div 0.6$

35. $23.4 - 0.821$

36. $3.7 + 5.6$

37. $476.12 - 112.97$

38. $19.872 \div 0.54$

39. $0.007 + 7$

40. $\begin{array}{r} 51.77 \\ +\ 3.6 \\ \hline \end{array}$

Appendix D

Mean, Median, and Mode

It is sometimes desirable to be able to describe a set of data, or a set of numbers, by a single "middle" number. Three such **measures of central tendency** are the mean, the median, and the mode.

The most common measure of central tendency is the mean (sometimes called the arithmetic mean or the average). The **mean** of a set of data items, denoted by $\bar{x}$, is the sum of the items divided by the number of items.

EXAMPLE 1 Seven students in a psychology class conducted an experiment on mazes. Each student was given a pencil and asked to successfully complete the same maze. The timed results are below.

Student	Ann	Thanh	Carlos	Jesse	Melinda	Ramzi	Dayni
Time (Seconds)	13.2	11.8	10.7	16.2	15.9	13.8	18.5

a. Who completed the maze in the shortest time? Who completed the maze in the longest time?

b. Find the mean.

c. How many students took longer than the mean time? How many students took shorter than the mean time?

Solution

a. Carlos completed the maze in 10.7 seconds, the shortest time. Dayni completed the maze in 18.5 seconds, the longest time.

b. To find the mean, $\bar{x}$, find the sum of the data items and divide by 7, the number of items.

$$\bar{x} = \frac{13.2 + 11.8 + 10.7 + 16.2 + 15.9 + 13.8 + 18.5}{7} = \frac{100.1}{7} = 14.3$$

c. Three students, Jesse, Melinda, and Dayni, had times longer than the mean time. Four students, Ann, Thanh, Carlos, and Ramzi, had times shorter than the mean time. □

Two other measures of central tendency are the median and the mode.

The **median** of an ordered set of numbers is the middle number. If the number of items is even, the median is the mean of the two middle numbers. The **mode** of a set of numbers is the number that occurs most often. It is possible for a data set to have no mode or more than one mode.

EXAMPLE 2 Find the median and the mode of the following list of numbers. These numbers were high temperatures for fourteen consecutive days in a city in Montana.

76, 80, 85, 86, 89, 87, 82, 77, 76, 79, 82, 89, 89, 92

Solution

First, write the numbers in order.

$$76, 76, 77, 79, 80, 82, \underbrace{82, 85}_{\text{two middle numbers}}, 86, 87, \underbrace{89, 89, 89}_{\text{mode}}, 92$$

Since there are an even number of items, the median is the mean of the two middle numbers.

$$\text{median} = \frac{82 + 85}{2} = 83.5$$

The mode is 89, since 89 occurs most often. □

D Exercise Set

MyMathLab®

For each of the following data sets, find the mean, the median, and the mode. If necessary, round the mean to one decimal place.

1. 21, 28, 16, 42, 38

2. 42, 35, 36, 40, 50

3. 7.6, 8.2, 8.2, 9.6, 5.7, 9.1

4. 4.9, 7.1, 6.8, 6.8, 5.3, 4.9

5. 0.2, 0.3, 0.5, 0.6, 0.6, 0.9, 0.2, 0.7, 1.1

6. 0.6, 0.6, 0.8, 0.4, 0.5, 0.3, 0.7, 0.8, 0.1

7. 231, 543, 601, 293, 588, 109, 334, 268

8. 451, 356, 478, 776, 892, 500, 467, 780

Eight of the tallest buildings in the United States are listed below. Use this table for Exercises 9 through 12.

Building	*Height (feet)*
Willis Tower, Chicago, IL	1454
Empire State, New York, NY	1250
Aon Center, Chicago, IL	1136
John Hancock Center, Chicago, IL	1127
U.S. Bank Tower, Los Angeles, CA	1107
Chrysler, New York, NY	1046
Bank of America Plaza, Atlanta, GA	1023
JPMorgan Chase Tower, Houston, TX	1002

9. Find the mean height for the five tallest buildings.

10. Find the median height for the five tallest buildings.

11. Find the median height for the eight tallest buildings.

12. Find the mean height for the eight tallest buildings.

During an experiment, the following times (in seconds) were recorded: 7.8, 6.9, 7.5, 4.7, 6.9, 7.0.

13. Find the mean.

14. Find the median.

15. Find the mode.

In a mathematics class, the following test scores were recorded for a student: 86, 95, 91, 74, 77, 85.

16. Find the mean. Round to the nearest hundredth.

17. Find the median.

18. Find the mode.

The following pulse rates were recorded for a group of fifteen students: 78, 80, 66, 68, 71, 64, 82, 71, 70, 65, 70, 75, 77, 86, 72.

19. Find the mean.

20. Find the median.

21. Find the mode.

22. How many rates were higher than the mean?

23. How many rates were lower than the mean?

24. Have each student in your algebra class take his/her pulse rate. Record the data and find the mean, the median, and the mode.

Find the missing numbers in each list of numbers. (These numbers are not necessarily in numerical order.)

25. __, __, 16, 18, __
The mode is 21. The mean is 20.

26. __, __, __, __, 40
The mode is 35. The median is 37. The mean is 38.

Appendix E

Tables

E.1 Table of Squares and Square Roots

n	n^2	$\sqrt{n}$	n	n^2	$\sqrt{n}$
1	1	1.000	51	2601	7.141
2	4	1.414	52	2704	7.211
3	9	1.732	53	2809	7.280
4	16	2.000	54	2916	7.348
5	25	2.236	55	3025	7.416
6	36	2.449	56	3136	7.483
7	49	2.646	57	3249	7.550
8	64	2.828	58	3364	7.616
9	81	3.000	59	3481	7.681
10	100	3.162	60	3600	7.746
11	121	3.317	61	3721	7.810
12	144	3.464	62	3844	7.874
13	169	3.606	63	3969	7.937
14	196	3.742	64	4096	8.000
15	225	3.873	65	4225	8.062
16	256	4.000	66	4356	8.124
17	289	4.123	67	4489	8.185
18	324	4.243	68	4624	8.246
19	361	4.359	69	4761	8.307
20	400	4.472	70	4900	8.367
21	441	4.583	71	5041	8.426
22	484	4.690	72	5184	8.485
23	529	4.796	73	5329	8.544
24	576	4.899	74	5476	8.602
25	625	5.000	75	5625	8.660
26	676	5.099	76	5776	8.718
27	729	5.196	77	5929	8.775
28	784	5.292	78	6084	8.832
29	841	5.385	79	6241	8.888
30	900	5.477	80	6400	8.944
31	961	5.568	81	6561	9.000
32	1024	5.657	82	6724	9.055
33	1089	5.745	83	6889	9.110
34	1156	5.831	84	7056	9.165
35	1225	5.916	85	7225	9.220
36	1296	6.000	86	7396	9.274
37	1369	6.083	87	7569	9.327
38	1444	6.164	88	7744	9.381
39	1521	6.245	89	7921	9.434
40	1600	6.325	90	8100	9.487
41	1681	6.403	91	8281	9.539
42	1764	6.481	92	8464	9.592
43	1849	6.557	93	8649	9.644
44	1936	6.633	94	8836	9.695
45	2025	6.708	95	9025	9.747
46	2116	6.782	96	9216	9.798
47	2209	6.856	97	9409	9.849
48	2304	6.928	98	9604	9.899
49	2401	7.000	99	9801	9.950
50	2500	7.071	100	10,000	10.000

E.2 Table of Percent, Decimal, and Fraction Equivalents

Percent, Decimal, and Fraction Equivalents		
Percent	***Decimal***	***Fraction***
1%	0.01	$\frac{1}{100}$
5%	0.05	$\frac{1}{20}$
10%	0.1	$\frac{1}{10}$
12.5% or $12\frac{1}{2}$%	0.125	$\frac{1}{8}$
$16.\overline{6}$% or $16\frac{2}{3}$%	$0.1\overline{6}$	$\frac{1}{6}$
20%	0.2	$\frac{1}{5}$
25%	0.25	$\frac{1}{4}$
30%	0.3	$\frac{3}{10}$
$33.\overline{3}$% or $33\frac{1}{3}$%	$0.\overline{3}$	$\frac{1}{3}$
37.5% or $37\frac{1}{2}$%	0.375	$\frac{3}{8}$
40%	0.4	$\frac{2}{5}$
50%	0.5	$\frac{1}{2}$
60%	0.6	$\frac{3}{5}$
62.5% or $62\frac{1}{2}$%	0.625	$\frac{5}{8}$
$66.\overline{6}$% or $66\frac{2}{3}$%	$0.\overline{6}$	$\frac{2}{3}$
70%	0.7	$\frac{7}{10}$
75%	0.75	$\frac{3}{4}$
80%	0.8	$\frac{4}{5}$
$83.\overline{3}$% or $83\frac{1}{3}$%	$0.8\overline{3}$	$\frac{5}{6}$
87.5% or $87\frac{1}{2}$%	0.875	$\frac{7}{8}$
90%	0.9	$\frac{9}{10}$
100%	1.0	1
110%	1.1	$1\frac{1}{10}$
125%	1.25	$1\frac{1}{4}$
$133.\overline{3}$% or $133\frac{1}{3}$%	$1.\overline{3}$	$1\frac{1}{3}$
150%	1.5	$1\frac{1}{2}$
$166.\overline{6}$% or $166\frac{2}{3}$%	$1.\overline{6}$	$1\frac{2}{3}$
175%	1.75	$1\frac{3}{4}$
200%	2.0	2

Contents of Student Resources

Study Skills Builders

Student Resources

Study Skills Builders

Attitude and Study Tips

Study Skills Builder 1

Have You Decided to Complete This Course Successfully?

Ask yourself if one of your current goals is to complete this course successfully.

If it is not a goal of yours, ask yourself why. One common reason is fear of failure. Amazingly enough, fear of failure alone can be strong enough to keep many of us from doing our best in any endeavor.

Another common reason is that you simply haven't taken the time to think about or write down your goals for this course. To help accomplish this, answer the questions below.

Exercises

1. Write down your goal(s) for this course.
2. Now list steps you will take to make sure your goal(s) in Exercise 1 are accomplished.
3. Rate your commitment to this course with a number between 1 and 5. Use the diagram below to help.

High Commitment		Average Commitment		Not Commited at All
5	4	3	2	1

4. If you have rated your personal commitment level (from the exercise above) as a 1, 2, or 3, list the reasons why this is so. Then determine whether it is possible to increase your commitment level to a 4 or 5.

Good luck, and don't forget that a positive attitude will make a big difference.

Study Skills Builder 2

Tips for Studying for an Exam

To prepare for an exam, try the following study techniques:

- Start the study process days before your exam.
- Make sure that you are up-to-date on your assignments.
- If there is a topic that you are unsure of, use one of the many resources that are available to you. For example,
  - See your instructor.
  - View a lecture video on the topic.
  - Visit a learning resource center on campus.
  - Read the textbook material and examples on the topic.
- Reread your notes and carefully review the Chapter Highlights at the end of any chapter.
- Work the review exercises at the end of the chapter.
- Find a quiet place to take the Chapter Test found at the end of the chapter. Do not use any resources when taking this sample test. This way, you will have a clear indication of how prepared you are for your exam. Check your answers and use the Chapter Test Prep Videos to make sure that you correct any missed exercises.

Good luck, and keep a positive attitude.

Exercises

Let's see how you did on your last exam.

1. How many days before your last exam did you start studying for that exam?
2. Were you up-to-date on your assignments at that time or did you need to catch up on assignments?
3. List the most helpful text supplement (if you used one).
4. List the most helpful campus supplement (if you used one).
5. List your process for preparing for a mathematics test.
6. Was this process helpful? In other words, were you satisfied with your performance on your exam?
7. If not, what changes can you make in your process that will make it more helpful to you?

Study Skills Builder 3

What to Do the Day of an Exam

Your first exam may be soon. On the day of an exam, don't forget to try the following:

- Allow yourself plenty of time to arrive.
- Read the directions on the test carefully.
- Read each problem carefully as you take your test. Make sure that you answer the question asked.
- Watch your time and pace yourself so that you may attempt each problem on your test.
- Check your work and answers.
- ***Do not turn your test in early.*** If you have extra time, spend it double-checking your work.

Good luck!

Exercises

Answer the following questions based on your most recent mathematics exam, whenever that was.

1. How soon before class did you arrive?

2. Did you read the directions on the test carefully?

3. Did you make sure you answered the question asked for each problem on the exam?

4. Were you able to attempt each problem on your exam?

5. If your answer to Exercise 4 is no, list reasons why.

6. Did you have extra time on your exam?

7. If your answer to Exercise 6 is yes, describe how you spent that extra time.

Study Skills Builder 4

Are You Satisfied with Your Performance on a Particular Quiz or Exam?

If not, don't forget to analyze your quiz or exam and look for common errors. Were most of your errors a result of:

- *Carelessness?* Did you turn in your quiz or exam before the allotted time expired? If so, resolve to use any extra time to check your work.
- *Running out of time?* Answer the questions you are sure of first. Then attempt the questions you are unsure of, and delay checking your work until all questions have been answered.
- *Not understanding a concept?* If so, review that concept and correct your work so that you make sure you understand it before the next quiz or the final exam.
- *Test conditions?* When studying for a quiz or exam, make sure you place yourself in conditions similar to test conditions. For example, before your next quiz or exam, take a sample test without the aid of your notes or text.

(For a sample test, see your instructor or use the Chapter Test at the end of each chapter.)

Exercises

1. Have you corrected all your previous quizzes and exams?

2. List any errors you have found common to two or more of your graded papers.

3. Is one of your common errors not understanding a concept? If so, are you making sure you understand all the concepts for the next quiz or exam?

4. Is one of your common errors making careless mistakes? If so, are you now taking all the time allotted to check over your work so that you can minimize the number of careless mistakes?

5. Are you satisfied with your grades thus far on quizzes and tests?

6. If your answer to Exercise 5 is no, are there any more suggestions you can make to your instructor or yourself to help? If so, list them here and share these with your instructor.

Study Skills Builder 5

How Are You Doing?

If you haven't done so yet, take a few moments and think about how you are doing in this course. Are you working toward your goal of successfully completing this course? Is your performance on homework, quizzes, and tests satisfactory? If not, you might want to see your instructor to see if he/she has any suggestions on how you can improve your performance. Reread Section 1.1 for ideas on places to get help with your mathematics course.

Exercises

Answer the following.

1. List any textbook supplements you are using to help you through this course.

2. List any campus resources you are using to help you through this course.

3. Write a short paragraph describing how you are doing in your mathematics course.

4. If improvement is needed, list ways that you can work toward improving your situation as described in Exercise 3.

Study Skills Builder 6

Are You Preparing for Your Final Exam?

To prepare for your final exam, try the following study techniques:

- Review the material that you will be responsible for on your exam. This includes material from your textbook, your notebook, and any handouts from your instructor.
- Review any formulas that you may need to memorize.
- Check to see if your instructor or mathematics department will be conducting a final exam review.
- Check with your instructor to see whether final exams from previous semesters/quarters are available to students for review.
- Use your previously taken exams as a practice final exam. To do so, rewrite the test questions in mixed order on blank sheets of paper. This will help you prepare for exam conditions.
- If you are unsure of a few concepts, see your instructor or visit a learning lab for assistance. Also, view the video segment of any troublesome sections.
- If you need further exercises to work, try the Cumulative Reviews at the end of the chapters.

Once again, good luck! I hope you are enjoying this textbook and your mathematics course.

Organizing Your Work

Study Skills Builder 7

Learning New Terms

Many of the terms used in this text may be new to you. It will be helpful to make a list of new mathematical terms and symbols as you encounter them and to review them frequently. Placing these new terms (including page references) on 3×5 index cards might help you later when you're preparing for a quiz.

Exercises

1. Name one way you might place a word and its definition on a 3×5 card.

2. How do new terms stand out in this text so that they can be found?

Study Skills Builder 8

Are You Organized?

Have you ever had trouble finding a completed assignment? When it's time to study for a test, are your notes neat and organized? Have you ever had trouble reading your own mathematics handwriting? (Be honest—I have.)

When any of these things happens, it's time to get organized. Here are a few suggestions:

- Write your notes and complete your homework assignments in a notebook with pockets (spiral or ring binder).
- Take class notes in this notebook, and then follow the notes with your completed homework assignment.
- When you receive graded papers or handouts, place them in the notebook pocket so that you will not lose them.
- Mark (possibly with an exclamation point) any note(s) that seem extra important to you.
- Mark (possibly with a question mark) any notes or homework that you are having trouble with.
- See your instructor or a math tutor to help you with the concepts or exercises that you are having trouble understanding.
- If you are having trouble reading your own handwriting, *slow down* and write your mathematics work clearly!

Exercises

1. Have you been completing your assignments on time?
2. Have you been correcting any exercises you may be having difficulty with?
3. If you are having trouble with a mathematical concept or correcting any homework exercises, have you visited your instructor, a tutor, or your campus math lab?
4. Are you taking lecture notes in your mathematics course? (By the way, these notes should include worked-out examples solved by your instructor.)
5. Is your mathematics course material (handouts, graded papers, lecture notes) organized?
6. If your answer to Exercise 5 is no, take a moment and review your course material. List at least two ways that you might better organize it.

Study Skills Builder 9

Organizing a Notebook

It's never too late to get organized. If you need ideas about organizing a notebook for your mathematics course, try some of these:

- Use a spiral or ring binder notebook with pockets and use it for mathematics only.
- Start each page by writing the book's section number you are working on at the top.
- When your instructor is lecturing, take notes. *Always* include any examples your instructor works for you.
- Place your worked-out homework exercises in your notebook immediately after the lecture notes from that section. This way, a section's worth of material is together.
- Homework exercises: Attempt and check all assigned homework.
- Place graded quizzes in the pockets of your notebook or a special section of your binder.

Exercises

Check your notebook organization by answering the following questions.

1. Do you have a spiral or ring binder notebook for your mathematics course only?
2. Have you ever had to flip through several sheets of notes and work in your mathematics notebook to determine what section's work you are in?
3. Are you now writing the textbook's section number at the top of each notebook page?
4. Have you ever lost or had trouble finding a graded quiz or test?
5. Are you now placing all your graded work in a dedicated place in your notebook?
6. Are you attempting all of your homework and placing all of your work in your notebook?
7. Are you checking and correcting your homework in your notebook? If not, why not?
8. Are you writing in your notebook the examples your instructor works for you in class?

Study Skills Builder 10

How Are Your Homework Assignments Going?

It is very important in mathematics to keep up with homework. Why? Many concepts build on each other. Often your understanding of a day's concepts depends on an understanding of the previous day's material.

Remember that completing your homework assignment involves a lot more than attempting a few of the problems assigned.

To complete a homework assignment, remember these four things:

- Attempt all of it.
- Check it.
- Correct it.
- If needed, ask questions about it.

Exercises

Take a moment and review your completed homework assignments. Answer the questions below based on this review.

1. Approximate the fraction of your homework you have attempted.

2. Approximate the fraction of your homework you have checked (if possible).

3. If you are able to check your homework, have you corrected it when errors have been found?

4. When working homework, if you do not understand a concept, what do you do?

MyMathLab and MathXL

Study Skills Builder 11

Tips for Turning In Your Homework on Time

It is very important to keep up with your mathematics homework assignments. Why? Many concepts in mathematics build upon each other.

Remember these four tips to help ensure your work is completed on time:

- Know the assignments and due dates set by your instructor.
- Do not wait until the last minute to submit your homework.
- Set a goal to submit your homework 6–8 hours before the scheduled due date in case you have unexpected technology trouble.
- Schedule enough time to complete each assignment.

Following these tips will also help you avoid potentially losing points for late or missed assignments.

Exercises

Take a moment to consider your work on your homework assignments to date and answer the following questions:

1. What percentage of your assignments have you turned in on time?

2. Why might it be a good idea to submit your homework 6–8 hours before the scheduled deadline?

3. If you have missed submitting any homework by the due date, list some of the reasons why this occurred.

4. What steps do you plan to take in the future to ensure your homework is submitted on time?

Study Skills Builder 12

Tips for Doing Your Homework Online

Practice is one of the main keys to success in any mathematics course. Did you know that MyMathLab/MathXL provides you with **immediate feedback** for each exercise? If you are incorrect, you are given hints to work the exercise correctly. You have **unlimited practice opportunities** and can rework any exercises you have trouble with until you master them, and submit homework assignments unlimited times before the deadline.

Remember these success tips when doing your homework online:

- Attempt all assigned exercises.
- Write down (neatly) your step-by-step work for each exercise before entering your answer.
- Use the immediate feedback provided by the program to help you check and correct your work for each exercise.
- Rework any exercises you have trouble with until you master them.
- Work through your homework assignment as many times as necessary until you are satisfied.

Exercises

Take a moment to think about your homework assignments to date and answer the following:

1. Have you attempted all assigned exercises?

2. Of the exercises attempted, have you also written out your work before entering your answer—so that you can check it?

3. Are you familiar with how to enter answers using the MathXL player so that you avoid answer entry type errors?

4. List some ways the immediate feedback and practice supports have helped you with your homework. If you have not used these supports, how do you plan to use them with the success tips above on your next assignment?

Study Skills Builder 13

Organizing Your Work

Have you ever used any readily available paper (such as the back of a flyer, another course assignment, post-its, etc.) to work out homework exercises before entering the answer in MathXL? To save time, have you ever entered answers directly into MathXL without working the exercises on paper? When it's time to study, have you ever been unable to find your completed work or read and follow your own mathematics handwriting?

When any of these things happen, it's time to get organized. Here are some suggestions:

- Write your step-by-step work for each homework exercise, (neatly) on lined, loose-leaf paper and keep this in a 3-ring binder.
- Refer to your step-by-step work when you receive feedback that your answer is incorrect in MathXL. Double-check against the steps and hints provided by the program and correct your work accordingly.
- Keep your written homework with your class notes for that section.
- Identify any exercises you are having trouble with and ask questions about them.
- Keep all graded quizzes and tests in this binder as well to study later.

If you follow the suggestions above, you and your instructor or tutor will be able to follow your steps and correct any mistakes. You will have a written copy of your work to refer to later to ask questions and study for tests.

Exercises

1. Why is it important that you write out your step-by-step work on homework exercises and keep a hard copy of all work submitted online?

2. If you have gotten an incorrect answer, are you able to follow your steps and find your error?

3. If you were asked today to review your previous homework assignments and first test, could you find them? If not, list some ways you might better organize your work.

Study Skills Builder 14

Getting Help with Your Homework Assignments

There are many helpful resources available to you through MathXL to help you work through any homework exercises you may have trouble with. It is important that you know what these resources are and know when and how to use them.

Let's review these features found in the homework exercises:

- **Help Me Solve This**—provides step-by-step help for the exercise you are working. You must work an additional exercise of the same type (without this help) before you can get credit for having worked it correctly.
- **View an Example**—allows you to view a correctly worked exercise similar to the one you are having trouble with. You can go back to your original exercise and work it on your own.
- **E-Book**—allows you to read examples from your text and find similar exercises.
- **Video**—your text author, Elayn Martin-Gay, works an exercise similar to the one you need help with. **Not all exercises have an accompanying video clip.
- **Ask My Instructor**—allows you to email your instructor for help with an exercise.

Exercises

1. How does the "Help Me Solve This" feature work?
2. If the "View an Example" feature is used, is it necessary to work an additional problem before continuing the assignment?
3. When might be a good time to use the "Video" feature? Do all exercises have an accompanying video clip?
4. Which of the features above have you used? List those you found the most helpful to you.
5. If you haven't used the features discussed, list those you plan to try on your next homework assignment.

Study Skills Builder 15

Tips for Preparing for an Exam

Did you know that you can rework your previous homework assignments in MyMathLab and MathXL? This is a great way to prepare for tests. To do this, open a previous homework assignment and click "similar exercise." This will generate new exercises similar to the homework you have submitted. You can then rework the exercises and assignments until you feel confident that you understand them.

To prepare for an exam, follow these tips:

- Review your written work for your previous homework assignments along with your class notes.
- Identify any exercises or topics that you have questions on or have difficulty understanding.
- Rework your previous assignments in MyMathLab and MathXL until you fully understand them and can do them without help.
- Get help for any topics you feel unsure of or for which you have questions.

Exercises

1. Are your current homework assignments up to date and is your written work for them organized in a binder or notebook? If the answer is no, it's time to get organized. For tips on this, see Study Skills Builder 13—Organizing Your Work.
2. How many days in advance of an exam do you usually start studying?
3. List some ways you think that practicing previous homework assignments can help you prepare for your test.
4. List two or three resources you can use to get help for any topics you are unsure of or have questions on.

Good luck!

Study Skills Builder 16

How Well Do You Know the Resources Available to You in MyMathLab?

There are many helpful resources available to you in MyMathLab. Let's take a moment to locate and explore a few of them now. Go into your MyMathLab course, and visit the multimedia library, tools for success, and E-book.

Let's see what you found.

Exercises

1. List the resources available to you in the Multimedia Library.
2. List the resources available to you in the Tools for Success folder.
3. Where did you find the English/Spanish Audio Glossary?
4. Can you view videos from the E-Book?
5. Did you find any resources you did not know about? If so, which ones?
6. Which resources have you used most often or found most helpful?

Additional Help Inside and Outside Your Textbook

Study Skills Builder 17

How Well Do You Know Your Textbook?

The questions below will help determine whether you are familiar with your textbook. For additional information, see Section 1.1 in this text.

1. What does the ▶ icon mean?
2. What does the ╲ icon mean?
3. What does the △ icon mean?
4. Where can you find a review for each chapter? What answers to this review can be found in the back of your text?
5. Each chapter contains an overview of the chapter along with examples. What is this feature called?
6. Each chapter contains a review of vocabulary. What is this feature called?
7. There are practice exercises that are contained in this text. What are they and how can they be used?
8. This text contains a student section in the back entitled Student Resources. List the contents of this section and how they might be helpful.
9. What exercise answers are available in this text? Where are they located?

Study Skills Builder 18

Are You Familiar with Your Textbook Supplements?

Below is a review of some of the student supplements available for additional study. Check to see if you are using the ones most helpful to you.

- Chapter Test Prep Videos. These videos provide video clip solutions to the Chapter Test exercises in this text. You will find this extremely useful when studying for tests or exams.
- Interactive DVD Lecture Series. These are keyed to each section of the text. The material is presented by me, Elayn Martin-Gay, and I have placed a ▶ by the exercises in the text that I have worked on the video.
- The *Student Solutions Manual*. This contains worked-out solutions to odd-numbered exercises as well as Practice Exercises and every exercise in the Integrated Reviews, Chapter Reviews, Chapter Tests, and Cumulative Reviews.
- Pearson Tutor Center. Mathematics questions may be phoned, faxed, or e-mailed to this center.
- MyMathLab is a text-specific online course. MathXL is an online homework, tutorial, and assessment system. Take a moment and determine whether these are available to you.

As usual, your instructor is your best source of information.

Exercises

Let's see how you are doing with textbook supplements.

1. Name one way the Lecture Videos can be helpful to you.
2. Name one way the Chapter Test Prep Video can help you prepare for a chapter test.
3. List any textbook supplements that you have found useful.
4. Have you located and visited a learning resource lab located on your campus?
5. List the textbook supplements that are currently housed in your campus's learning resource lab.

Study Skills Builder 19

Are You Getting All the Mathematics Help That You Need?

Remember that, in addition to your instructor, there are many places to get help with your mathematics course. For example:

- This text has an accompanying video lesson for every section, and the Chapter Test Prep videos provide worked-out solutions to every Chapter Test exercise.
- The back of the book contains answers to odd-numbered exercises.
- A *Student Solutions Manual* is available that contains worked-out solutions to odd-numbered exercises as well as solutions to Practice Exercises and every exercise in the Integrated Reviews, Chapter Reviews, Chapter Tests, and Cumulative Reviews.
- Don't forget to check with your instructor for other local resources available to you, such as a tutoring center.

Exercises

1. List items you find helpful in the text and all student supplements to this text.
2. List all the campus help that is available to you for this course.
3. List any help (besides the textbook) from Exercises 1 and 2 above that you are using.
4. List any help (besides the textbook) that you feel you should try.
5. Write a goal for yourself that includes trying everything you listed in Exercise 4 during the next week.

Bigger Picture—Study Guide Outline

I. Simplifying Expressions

A. Real Numbers

1. Add: (Sec. 1.5)

$-1.7 + (-0.21) = -1.91$ Adding like signs.
Add absolute values. Attach common sign.

$-7 + 3 = -4$ Adding different signs.
Subtract absolute values. Attach the sign of the number with the larger absolute value.

2. Subtract: Add the first number to the opposite of the second number. (Sec. 1.6)

$$17 - 25 = 17 + (-25) = -8$$

3. Multiply or divide: Multiply or divide the two numbers as usual. If the signs are the same, the answer is positive. If the signs are different, the answer is negative. (Sec. 1.7)

$$-10 \cdot 3 = -30, \qquad -81 \div (-3) = 27$$

B. Exponents (Sec. 5.1)

$$x^7 \cdot x^5 = x^{12};\ (x^7)^5 = x^{35};\ \frac{x^7}{x^5} = x^2;\ x^0 = 1;\ 8^{-2} = \frac{1}{8^2} = \frac{1}{64}$$

C. Polynomials

1. Add: Combine like terms. (Sec. 5.2)

$$\begin{aligned}(3y^2 + 6y + 7) + (9y^2 - 11y - 15) &= 3y^2 + 6y + 7 + 9y^2 - 11y - 15\\ &= 12y^2 - 5y - 8\end{aligned}$$

2. Subtract: Change the sign of the terms of the polynomial being subtracted, then add. (Sec. 5.2)

$$\begin{aligned}(3y^2 + 6y + 7) - (9y^2 - 11y - 15) &= 3y^2 + 6y + 7 - 9y^2 + 11y + 15\\ &= -6y^2 + 17y + 22\end{aligned}$$

3. Multiply: Multiply each term of one polynomial by each term of the other polynomial. (Sec. 5.3)

$$\begin{aligned}(x + 5)(2x^2 - 3x + 4) &= x(2x^2 - 3x + 4) + 5(2x^2 - 3x + 4)\\ &= 2x^3 - 3x^2 + 4x + 10x^2 - 15x + 20\\ &= 2x^3 + 7x^2 - 11x + 20\end{aligned}$$

4. Divide: (Sec. 5.6)

a. To divide by a monomial, divide each term of the polynomial by the monomial.

$$\frac{8x^2 + 2x - 6}{2x} = \frac{8x^2}{2x} + \frac{2x}{2x} - \frac{6}{2x} = 4x + 1 - \frac{3}{x}$$

b. To divide by a polynomial other than a monomial, use long division.

$$\begin{array}{r} x - 6 + \dfrac{40}{2x + 5} \\ 2x + 5\overline{)2x^2 - 7x + 10} \\ \underline{\overset{-}{2x^2} \overset{-}{\not+} 5x} \\ -12x + 10 \\ \underline{\overset{+}{\not-}12x \overset{+}{\not-} 30} \\ 40 \end{array}$$

D. Factoring Polynomials

See the Chapter 6 Integrated Review for steps.

$$3x^4 - 78x^2 + 75 = 3(x^4 - 26x^2 + 25) \quad \text{Factor out GCF—always first step.}$$
$$= 3(x^2 - 25)(x^2 - 1) \quad \text{Factor trinomial.}$$
$$= 3(x + 5)(x - 5)(x + 1)(x - 1) \quad \text{Factor further—each difference of squares.}$$

E. Rational Expressions

1. **Simplify:** Factor the numerator and denominator. Then divide out factors of 1 by dividing out common factors in the numerator and denominator. (Sec. 7.1)

$$\frac{x^2 - 9}{7x^2 - 21x} = \frac{(x + 3)(x - 3)}{7x(x - 3)} = \frac{x + 3}{7x}$$

2. **Multiply:** Multiply numerators, then multiply denominators. (Sec. 7.2)

$$\frac{5z}{2z^2 - 9z - 18} \cdot \frac{22z + 33}{10z} = \frac{5 \cdot z}{(2z + 3)(z - 6)} \cdot \frac{11(2z + 3)}{2 \cdot 5 \cdot z} = \frac{11}{2(z - 6)}$$

3. **Divide:** First fraction times the reciprocal of the second fraction. (Sec. 7.2)

$$\frac{14}{x + 5} \div \frac{x + 1}{2} = \frac{14}{x + 5} \cdot \frac{2}{x + 1} = \frac{28}{(x + 5)(x + 1)}$$

4. **Add or subtract:** Must have same denominator. If not, find the LCD and write each fraction as an equivalent fraction with the LCD as denominator. (Sec. 7.4)

$$\frac{9}{10} - \frac{x + 1}{x + 5} = \frac{9(x + 5)}{10(x + 5)} - \frac{10(x + 1)}{10(x + 5)}$$
$$= \frac{9x + 45 - 10x - 10}{10(x + 5)} = \frac{-x + 35}{10(x + 5)}$$

F. Radicals

1. **Simplify square roots:** If possible, factor the radicand so that one factor is a perfect square. Then use the product rule and simplify. (Sec. 8.2)

$$\sqrt{75} = \sqrt{25 \cdot 3} = \sqrt{25} \cdot \sqrt{3} = 5\sqrt{3}$$

2. **Add or subtract:** Only like radicals (same index and radicand) can be added or subtracted. (Sec. 8.3)

$$8\sqrt{10} - \sqrt{40} + \sqrt{5} = 8\sqrt{10} - 2\sqrt{10} + \sqrt{5} = 6\sqrt{10} + \sqrt{5}$$

3. **Multiply or divide:** $\sqrt{a} \cdot \sqrt{b} = \sqrt{ab}$; $\dfrac{\sqrt{a}}{\sqrt{b}} = \sqrt{\dfrac{a}{b}}$. (Sec. 8.4)

$$\sqrt{11} \cdot \sqrt{3} = \sqrt{33}; \frac{\sqrt{140}}{\sqrt{7}} = \sqrt{\frac{140}{7}} = \sqrt{20} = \sqrt{4 \cdot 5} = 2\sqrt{5}$$

4. **Rationalizing the denominator:** (Sec. 8.4)

   a. If denominator is one term,

$$\frac{5}{\sqrt{11}} = \frac{5 \cdot \sqrt{11}}{\sqrt{11} \cdot \sqrt{11}} = \frac{5\sqrt{11}}{11}$$

   b. If denominator is two terms, multiply by 1 in the form of $\dfrac{\text{conjugate of denominator}}{\text{conjugate of denominator}}$.

$$\frac{13}{3 + \sqrt{2}} = \frac{13}{3 + \sqrt{2}} \cdot \frac{3 - \sqrt{2}}{3 - \sqrt{2}} = \frac{13(3 - \sqrt{2})}{9 - 2} = \frac{39 - 13\sqrt{2}}{7}$$

II. Solving Equations and Inequalities

A. Linear Equations: Power on variable is 1 and there are no variables in denominator. (Sec. 2.4)

$7(x-3) = 4x+6$	Linear equation. (If fractions, multiply by LCD.)
$7x - 21 = 4x + 6$	Use the distributive property.
$7x = 4x + 27$	Add 21 to both sides.
$3x = 27$	Subtract $4x$ from both sides.
$x = 9$	Divide both sides by 3.

B. Linear Inequalities: Same as linear equation except if you multiply or divide by a negative number, then reverse direction of inequality. (Sec. 2.9)

$-4x + 11 \le -1$	Linear inequality.
$-4x \le -12$	Subtract 11 from both sides.
$\frac{-4x}{-4} \ge \frac{-12}{-4}$	Divide both sides by -4 and reverse the direction of the inequality symbol.
$x \ge 3$	Simplify.

C. Quadratic and Higher Degree Equations: Solve: first write the equation in standard form (one side is 0).

1. If the polynomial on one side factors, solve by factoring. (Sec. 6.6)

2. If the polynomial does not factor, solve by the quadratic formula. (Sec. 9.3)

By factoring:

$$x^2 + x = 6$$
$$x^2 + x - 6 = 0$$
$$(x-2)(x+3) = 0$$
$$x - 2 = 0 \text{ or } x + 3 = 0$$
$$x = 2 \quad \text{or} \quad x = -3$$

By quadratic formula:

$$x^2 + x = 5$$
$$x^2 + x - 5 = 0$$
$$a = 1, b = 1, c = -5$$
$$x = \frac{-1 \pm \sqrt{1^2 - 4(1)(-5)}}{2 \cdot 1}$$
$$x = \frac{-1 \pm \sqrt{21}}{2}$$

D. Equations with Rational Expressions: Make sure the proposed solution does not make any denominator 0. (Sec. 7.5)

$\frac{3}{x} - \frac{1}{x-1} = \frac{4}{x-1}$	Equation with rational expressions.
$x(x-1) \cdot \frac{3}{x} - x(x-1) \cdot \frac{1}{x-1} = x(x-1) \cdot \frac{4}{x-1}$	Multiply through by $x(x-1)$.
$3(x-1) - x \cdot 1 = x \cdot 4$	Simplify.
$3x - 3 - x = 4x$	Use the distributive property.
$-3 = 2x$	Simplify and move variable terms to right side.
$-\frac{3}{2} = x$	Divide both sides by 2.

E. Proportions: An equation with two ratios equal. Set cross products equal, then solve. Make sure the proposed solution does not make any denominator 0. (Sec. 7.6)

$$\frac{5}{x} = \frac{9}{2x - 3}$$

$$
\begin{aligned}
5(2x - 3) &= 9 \cdot x && \text{Set cross products equal.} \\
10x - 15 &= 9x && \text{Multiply.} \\
x &= 15 && \text{Write equation with variable terms on one side and constants on the other.}
\end{aligned}
$$

F. Equations with Radicals: To solve, isolate a radical, then square both sides. You may have to repeat this. Check possible solution in the original equation. (Sec. 8.5)

$$
\begin{aligned}
\sqrt{x + 49} + 7 &= x \\
\sqrt{x + 49} &= x - 7 && \text{Subtract 7 from both sides.} \\
x + 49 &= x^2 - 14x + 49 && \text{Square both sides.} \\
0 &= x^2 - 15x && \text{Set terms equal to 0.} \\
0 &= x(x - 15) && \text{Factor.} \\
\cancel{x = 0} \text{ or } x &= 15 && \text{Set each factor equal to 0 and solve.}
\end{aligned}
$$

PRACTICE FINAL EXAM

Preparing for your Final Exam? Take this Practice Final and watch the full video solutions to any of the exercises you want to review. You will find the Practice Final video in the Video Lecture Series. The video also provides you with an overview to help you approach different problem types just as you will need to do on a Final Exam. To build your own study guide, use the Bigger Picture—Study Guide Outline in this text previous to this Practice Final Exam.

Evaluate.

1. -3^4

2. 4^{-3}

3. $6[5 + 2(3 - 8) - 3]$

Perform the indicated operations and simplify if possible.

4. $(5x^3 + x^2 + 5x - 2) - (8x^3 - 4x^2 + x - 7)$

5. $(4x - 2)^2$

6. $(3x + 7)(x^2 + 5x + 2)$

Factor.

7. $y^2 - 8y - 48$

8. $9x^3 + 39x^2 + 12x$

9. $180 - 5x^2$

10. $3a^2 + 3ab - 7a - 7b$

11. $8y^3 - 64$

Simplify. Write answers with positive exponents only.

12. $\left(\dfrac{x^2y^3}{x^3y^{-4}}\right)^2$

13. $\dfrac{5 - \dfrac{1}{y^2}}{\dfrac{1}{y} + \dfrac{2}{y^2}}$

Perform the indicated operations and simplify if possible.

14. $\dfrac{x^2 - 9}{x^2 - 3x} \div \dfrac{xy + 5x + 3y + 15}{2x + 10}$

15. $\dfrac{5a}{a^2 - a - 6} - \dfrac{2}{a - 3}$

Solve each equation or inequality.

16. $4(n - 5) = -(4 - 2n)$

17. $x(x + 6) = 7$

18. $3x - 5 \geq 7x + 3$

19. $2x^2 - 6x + 1 = 0$

20. $\dfrac{4}{y} - \dfrac{5}{3} = -\dfrac{1}{5}$

21. $\dfrac{5}{y + 1} = \dfrac{4}{y + 2}$

22. $\dfrac{a}{a - 3} = \dfrac{3}{a - 3} - \dfrac{3}{2}$

23. $\sqrt{2x - 2} = x - 5$

Graph the following.

24. $5x - 7y = 10$

25. $x - 3 = 0$

26. $y > -4x$

Find the slope of each line.

27. through $(6, -5)$ and $(-1, 2)$

28. $-3x + y = 5$

Write equations of the following lines. Write each equation in standard form.

29. through $(2, -5)$ and $(1, 3)$

30. through $(-5, -1)$ and parallel to $x = 7$

Solve each system of equations.

31. $\begin{cases} 3x - 2y = -14 \\ y = x + 5 \end{cases}$

32. $\begin{cases} 4x - 6y = 7 \\ -2x + 3y = 0 \end{cases}$

Answer the questions about functions.

33. If $h(x) = x^3 - x$, find

a. $h(-1)$ **b.** $h(0)$ **c.** $h(4)$

34. **a.** Identify the x- and y-intercepts.
b. Find the domain and range of the function graphed.

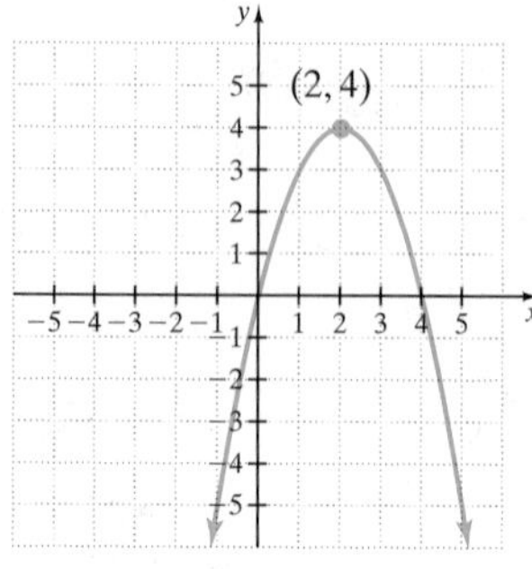

Evaluate.

35. $\sqrt{16}$

36. $27^{-2/3}$

37. $\left(\dfrac{9}{16}\right)^{1/2}$

Simplify.

38. $\sqrt{54}$

39. $\sqrt{9x^9}$

Perform the indicated operations and simplify if possible.

40. $\sqrt{12} - 2\sqrt{75}$

41. $\dfrac{\sqrt{40x^4}}{\sqrt{2x}}$

42. $\sqrt{2}(\sqrt{6} - \sqrt{5})$

Rationalize each denominator.

43. $\sqrt{\dfrac{5}{12x^2}}$

44. $\dfrac{2\sqrt{3}}{\sqrt{3} - 3}$

Solve each application.

45. One number plus five times its reciprocal is equal to six. Find the number.

46. Some states have a single area code for the entire state. Two such states have area codes where one is double the other. If the sum of these integers is 1203, find the two area codes.

47. Two trains leave Los Angeles simultaneously traveling on the same track in opposite directions at speeds of 50 and 64 mph. How long will it take before they are 285 miles apart?

48. Find the amount of a 12% saline solution a lab assistant should add to 80 cc (cubic centimeters) of a 22% saline solution in order to have a 16% solution.

Answers to Selected Exercises

CHAPTER 1 REVIEW OF REAL NUMBERS

Section 1.2

Practice Exercises

1. a. $<$ **b.** $>$ **c.** $<$ **2. a.** True **b.** False **c.** True **d.** True **3. a.** $3 < 8$ **b.** $15 \geq 9$ **c.** $6 \neq 7$ **4.** -10 **5. a.** 25 **b.** 25 **c.** $25, -15, -99$ **d.** $25, \frac{7}{3}, -15, -\frac{3}{4}, -3.7, 8.8, -99$ **e.** $\sqrt{5}$ **f.** $25, \frac{7}{3}, -15, -\frac{3}{4}, \sqrt{5}, -3.7, 8.8, -99$ **6. a.** $<$ **b.** $>$ **c.** $=$ **7. a.** 8 **b.** 9 **c.** 2.5 **d.** $\frac{5}{11}$ **e.** $\sqrt{3}$ **8. a.** $=$ **b.** $>$ **c.** $<$ **d.** $>$ **e.** $<$

Vocabulary, Readiness & Video Check 1.2

1. whole **3.** inequality **5.** real **7.** irrational **9.** To form a true statement: $0 < 7$. **11.** 0 belongs to the whole numbers, the integers, the rational numbers, and the real numbers; because 0 is a rational number, it cannot also be an irrational number

Exercise Set 1.2

1. $>$ **3.** $=$ **5.** $<$ **7.** $<$ **9.** $32 < 212$ **11.** $30 \leq 45$ **13.** true **15.** false **17.** false **19.** true **21.** false **23.** $8 < 12$ **25.** $5 \geq 4$ **27.** $15 \neq -2$ **29.** 14,494; -282 **31.** $-28{,}000$ **33.** 350; -126 **35.** whole, integers, rational, real **37.** integers, rational, real **39.** natural, whole, integers, rational, real **41.** rational, real **43.** irrational, real **45.** false **47.** true **49.** true **51.** true **53.** false **55.** $>$ **57.** $>$ **59.** $<$ **61.** $<$ **63.** $>$ **65.** $=$ **67.** $<$ **69.** $<$ **71.** 2009 **73.** 2009, 2010 **75.** 280 million < 281 million **77.** 49 million > 16 million **79.** 38 million pounds less, or -38 million **81.** $-0.04 > -26.7$ **83.** sun **85.** sun **87.** $20 \leq 25$ **89.** $6 > 0$ **91.** $-12 < -10$ **93.** answers may vary

Section 1.3

Practice Exercises

1. a. $2 \cdot 2 \cdot 3 \cdot 3$ **b.** $2 \cdot 2 \cdot 2 \cdot 5 \cdot 5$ **2. a.** $\frac{7}{8}$ **b.** $\frac{16}{3}$ **c.** $\frac{7}{25}$ **3.** $\frac{7}{24}$ **4. a.** $\frac{27}{16}$ **b.** $\frac{1}{36}$ **c.** $\frac{5}{2}$ **5. a.** 1 **b.** $\frac{6}{5}$ **c.** $\frac{4}{5}$ **d.** $\frac{1}{2}$ **6.** $\frac{14}{21}$ **7. a.** $\frac{46}{77}$ **b.** $\frac{1}{14}$ **c.** $\frac{1}{2}$ **8.** $22\frac{11}{15}$ **9.** $40\frac{5}{6}$

Vocabulary, Readiness & Video Check 1.3

1. fraction **3.** product **5.** factors, product **7.** equivalent **9.** The division operation changes to multiplication and the second fraction $\frac{1}{20}$ changes to its reciprocal $\frac{20}{1}$. **11.** The number $4\frac{7}{6}$ is not in proper mixed number form as the fraction part, $\frac{7}{6}$, should not be an improper fraction.

Exercise Set 1.3

1. $\frac{3}{8}$ **3.** $\frac{5}{7}$ **5.** $3 \cdot 11$ **7.** $2 \cdot 7 \cdot 7$ **9.** $2 \cdot 2 \cdot 5$ **11.** $3 \cdot 5 \cdot 5$ **13.** $3 \cdot 3 \cdot 5$ **15.** $\frac{1}{2}$ **17.** $\frac{2}{3}$ **19.** $\frac{3}{7}$ **21.** $\frac{3}{5}$ **23.** $\frac{30}{61}$ **25.** $\frac{3}{8}$ **27.** $\frac{1}{2}$ **29.** $\frac{6}{7}$ **31.** 15 **33.** $\frac{1}{6}$ **35.** $\frac{25}{27}$ **37.** $\frac{11}{20}$ sq mi **39.** $\frac{7}{36}$ sq ft **41.** $\frac{3}{5}$ **43.** 1 **45.** $\frac{1}{3}$ **47.** $\frac{9}{35}$ **49.** $\frac{21}{30}$ **51.** $\frac{4}{18}$ **53.** $\frac{16}{20}$ **55.** $\frac{23}{21}$ **57.** $\frac{11}{60}$ **59.** $\frac{5}{66}$ **61.** $\frac{7}{5}$ **63.** $\frac{1}{5}$ **65.** $\frac{3}{8}$ **67.** $\frac{1}{9}$ **69.** $18\frac{20}{27}$ **71.** $2\frac{28}{29}$ **73.** $48\frac{1}{15}$ **75.** $7\frac{1}{12}$ **77.** $\frac{5}{7}$ **79.** $\frac{65}{21}$ **81.** $\frac{2}{5}$ **83.** $\frac{10}{9}$ **85.** $\frac{17}{3}$ **87.** 37 **89.** $\frac{5}{66}$ **91.** $\frac{1}{5}$ **93.** $5\frac{1}{6}$ **95.** $\frac{17}{18}$ **97.** $55\frac{1}{4}$ ft **99.** answers may vary **101.** $3\frac{3}{8}$ mi **103.** $\frac{21}{100}$ **105.** multiplexes **107.** incorrect; $\frac{12}{24} = \frac{2 \cdot 2 \cdot 3}{2 \cdot 2 \cdot 2 \cdot 3} = \frac{1}{2}$ **109.** incorrect; $\frac{2}{7} + \frac{9}{7} = \frac{11}{7}$

Section 1.4

Practice Exercises

1. a. 1 **b.** 25 **c.** $\frac{1}{100}$ **d.** 9 **e.** $\frac{8}{125}$ **2. a.** 33 **b.** 11 **c.** $\frac{32}{9}$ or $3\frac{5}{9}$ **d.** 36 **e.** $\frac{3}{16}$ **3.** $\frac{31}{11}$ **4.** 4 **5.** $\frac{9}{22}$ **6. a.** 9 **b.** $\frac{8}{15}$ **c.** $\frac{19}{10}$ **d.** 33 **7.** No **8. a.** $6x$ **b.** $x - 8$ **c.** $x \cdot 9$ or $9x$ **d.** $2x + 3$ **e.** $7 + x$ **9. a.** $x + 7 = 13$ **b.** $x - 2 = 11$ **c.** $2x + 9 \neq 25$ **d.** $5(11) \geq x$

Calculator Explorations 1.4

1. 625 **3.** 59,049 **5.** 30 **7.** 9857 **9.** 2376

Vocabulary, Readiness & Video Check 1.4

1. base; exponent **3.** variable **5.** equation **7.** solving **9.** The replacement value for z is not used because it's not needed–there is no variable z in the given algebraic expression. **11.** We translate phrases to mathematical expressions and sentences to mathematical equations.

Exercise Set 1.4

1. 243 **3.** 27 **5.** 1 **7.** 5 **9.** 49 **11.** $\frac{16}{81}$ **13.** $\frac{1}{125}$ **15.** 1.44 **17.** 0.000064 **19.** 17 **21.** 20 **23.** 10 **25.** 21 **27.** 45 **29.** 0 **31.** 30 **33.** 2 **35.** $\frac{7}{18}$ **37.** $\frac{27}{10}$ **39.** $\frac{7}{5}$ **41.** 32 **43.** $\frac{23}{27}$ **45. a.** 64 **b.** 43 **c.** 19 **d.** 22 **47.** 9 **49.** 1 **51.** 1 **53.** 11 **55.** 8 **57.** 45 **59.** 27 **61.** 132 **63.** $\frac{37}{18}$ **65.** 16; 64; 144; 256 **67.** yes **69.** no **71.** no **73.** yes **75.** no **77.** $x + 15$ **79.** $x - 5$ **81.** $\frac{x}{4}$ **83.** $3x + 22$ **85.** $1 + 2 = 9 \div 3$ **87.** $3 \neq 4 \div 2$ **89.** $5 + x = 20$ **91.** $7.6x = 17$ **93.** $13 - 3x = 13$ **95.** multiply **97.** subtract **99.** no; answers may vary **101.** 14 in., 12 sq in. **103.** 14 in., 9.01 sq in. **105.** Rectangles with the same perimeter can have different areas. **107.** $(20 - 4) \cdot 4 \div 2$ **109. a.** expression **b.** equation **c.** equation **d.** expression **e.** expression **111.** answers may vary **113.** answers may vary, for example, $-2(5) - 1$ **115.** 12,000 sq ft **117.** 51 mph

Section 1.5

Practice Exercises

1.

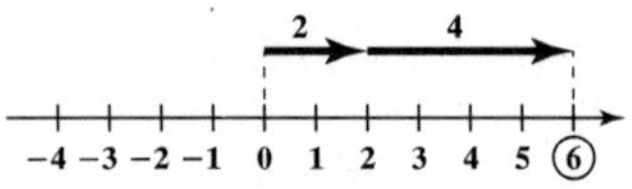

2.

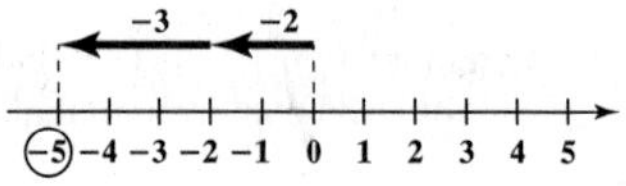

3. a. −13 **b.** −32 **4.**

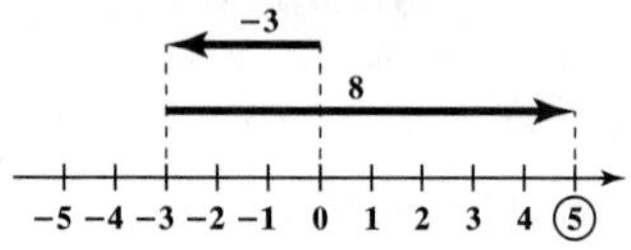

5. a. −3 **b.** 1 **c.** −0.2 **6. a.** −1 **b.** −6 **c.** 0.5 **d.** $\frac{1}{70}$ **7. a.** −6 **b.** −6 **8.** 4°F **9. a.** $\frac{5}{9}$ **b.** −8 **c.** −6.2 **d.** 3 **10. a.** −15 **b.** $\frac{3}{5}$ **c.** $5y$ **d.** 8

Vocabulary, Readiness & Video Check 1.5

1. opposites **3.** n **5.** absolute values **7.** Negative temperatures; the high temperature for the day was −6°F

Exercise Set 1.5

1. 9 **3.** −14 **5.** 1 **7.** −12 **9.** −5 **11.** −12 **13.** −4 **15.** 7 **17.** −2 **19.** 0 **21.** −19 **23.** 31 **25.** −47 **27.** −2.1 **29.** −8 **31.** 38 **33.** −13.1 **35.** $\frac{2}{8} = \frac{1}{4}$ **37.** $-\frac{3}{16}$ **39.** $-\frac{13}{10}$ **41.** −8 **43.** −59 **45.** −9 **47.** 5 **49.** 11 **51.** −18 **53.** 19 **55.** −0.7 **57.** −6° **59.** 146 ft **61.** −$6.9 million **63.** −16 **65.** −6 **67.** 2 **69.** 0 **71.** −6 **73.** −2 **75.** 0 **77.** $-\frac{2}{3}$ **79.** yes **81.** no **83.** July **85.** October **87.** 4.7°F **89.** −3 **91.** −22 **93.** negative **95.** positive **97.** true **99.** false **101.** answers may vary **103.** answers may vary

Section 1.6

Practice Exercises

1. a. −13 **b.** −7 **c.** 12 **d.** −2 **2. a.** 10.9 **b.** $-\frac{1}{2}$ **c.** $-\frac{19}{20}$ **3.** −7 **4. a.** −6 **b.** 6.1 **5. a.** −20 **b.** 13 **6. a.** 2 **b.** 13 **7.** $357 **8. a.** 28° **b.** 137°

Vocabulary, Readiness & Video Check 1.6

1. $7 - x$ **3.** $x - 7$ **5.** $7 - x$ **7.** $-10 - (-14)$; d **9.** addition; opposite **11.** There's a minus sign in the numerator and the replacement value is negative (notice parentheses are used around the replacement value), and it's always good to be careful when working with negative signs. **13.** In Example 9, you have two supplementary angles and know the measure of one of them. From the definition, you know that two supplementary angles must sum to 180°. Therefore you can subtract the known angle measure from 180° to get the measure of the other angle.

Exercise Set 1.6

1. −10 **3.** −5 **5.** 19 **7.** $\frac{1}{6}$ **9.** 2 **11.** −11 **13.** 11 **15.** 5 **17.** 37 **19.** −6.4 **21.** −71 **23.** 0 **25.** 4.1 **27.** $\frac{2}{11}$ **29.** $-\frac{11}{12}$ **31.** 8.92 **33.** 13 **35.** −5 **37.** −1 **39.** −23 **41.** −26 **43.** −24 **45.** 3 **47.** −45 **49.** −4 **51.** 13 **53.** 6 **55.** 9 **57.** −9 **59.** −7 **61.** $\frac{7}{5}$ **63.** 21 **65.** $\frac{1}{4}$ **67.** 100° **69.** 265°F **71.** 35,653 ft **73.** −308 ft **75.** 19,852 ft **77.** 130° **79.** 30° **81.** no **83.** no **85.** yes **87.** $-5 + x$ **89.** $-20 - x$ **91.** −4.4°; 2.6°; 12°; 23.5°; 15.3° **93.** May **95.** answers may vary **97.** 16 **99.** −20 **101.** true; answers may vary **103.** false; answers may vary **105.** negative, −30,387

Integrated Review

1. negative **2.** negative **3.** positive **4.** 0 **5.** positive **6.** 0 **7.** positive **8.** positive **9.** $-\frac{1}{7}; \frac{1}{7}$ **10.** $\frac{12}{5}; \frac{12}{5}$ **11.** 3; 3 **12.** $-\frac{9}{11}; \frac{9}{11}$ **13.** −42 **14.** 10 **15.** 2 **16.** −18 **17.** −7 **18.** −39 **19.** −2 **20.** −9 **21.** −3.4 **22.** −9.8 **23.** $-\frac{25}{28}$ **24.** $-\frac{5}{24}$ **25.** −4 **26.** −24 **27.** 6 **28.** 20 **29.** 6 **30.** 61 **31.** −6 **32.** −16 **33.** −19 **34.** −13 **35.** −4 **36.** −1 **37.** $\frac{13}{20}$ **38.** $-\frac{29}{40}$ **39.** 4 **40.** 9 **41.** −1 **42.** −3 **43.** 8 **44.** 10 **45.** 47 **46.** $\frac{2}{3}$

Section 1.7

Practice Exercises

1. a. -40 **b.** 12 **c.** -54 **2. a.** -30 **b.** 24 **c.** 0 **d.** 26 **3. a.** -0.046 **b.** $-\frac{4}{15}$ **c.** 14 **4. a.** 36 **b.** -36 **c.** -64 **d.** -64 **5. a.** $\frac{3}{8}$ **b.** $\frac{1}{15}$ **c.** $-\frac{7}{2}$ **d.** $-\frac{1}{5}$ **6. a.** -8 **b.** -4 **c.** 5 **7. a.** 3 **b.** -16 **c.** $-\frac{6}{5}$ **d.** $-\frac{1}{18}$ **8. a.** 0 **b.** undefined **c.** undefined **9. a.** $-\frac{84}{5}$ **b.** 11 **10. a.** -9 **b.** 33 **c.** $\frac{5}{3}$ **11.** -52

Calculator Explorations 1.7

1. 38 **3.** -441 **5.** $163.\overline{3}$ **7.** 54,499 **9.** 15,625

Vocabulary, Readiness & Video Check 1.7

1. 0; 0 **3.** positive **5.** negative **7.** positive **9.** The parentheses, or lack of them, determine the base of the expression. In Example 6, $(-2)^4$, the base is -2 and all of -2 is raised to 4. In Example 7, -2^4, the base is 2 and only 2 is raised to 4. **11.** Yes; because division of real numbers is defined in terms of multiplication. **13.** The football team lost 4 yards on each play and a loss of yardage is represented by a negative number.

Exercise Set 1.7

1. -24 **3.** -2 **5.** 50 **7.** -12 **9.** 0 **11.** -18 **13.** $\frac{3}{10}$ **15.** $\frac{2}{3}$ **17.** -7 **19.** 0.14 **21.** -800 **23.** -28 **25.** 25 **27.** $-\frac{8}{27}$ **29.** -121 **31.** $-\frac{1}{4}$ **33.** -30 **35.** 23 **37.** -7 **39.** true **41.** false **43.** 16 **45.** -1 **47.** 25 **49.** -49 **51.** $\frac{1}{9}$ **53.** $\frac{3}{2}$ **55.** $-\frac{1}{14}$ **57.** $-\frac{11}{3}$ **59.** $\frac{1}{0.2}$ **61.** -6.3 **63.** -9 **65.** 4 **67.** -4 **69.** 0 **71.** -5 **73.** undefined **75.** 3 **77.** -15 **79.** $-\frac{18}{7}$ **81.** $\frac{20}{27}$ **83.** -1 **85.** $-\frac{9}{2}$ **87.** -4 **89.** 16 **91.** -3 **93.** $-\frac{16}{7}$ **95.** 2 **97.** $\frac{6}{5}$ **99.** -5 **101.** $\frac{3}{2}$ **103.** -21 **105.** 41 **107.** -134 **109.** 3 **111.** 0 **113.** $-71 \cdot x$ or $-71x$ **115.** $-16 - x$ **117.** $-29 + x$ **119.** $\frac{x}{-33}$ or $x \div (-33)$ **121.** $3 \cdot (-4) = -12$; a loss of 12 yd **123.** $5(-20) = -100$; a depth of 100 ft **125.** yes **127.** no **129.** yes **131.** -162°F **133.** answers may vary **135.** 1, -1 **137.** positive **139.** not possible **141.** negative **143.** $-2 + \frac{-15}{3}$; -7 **145.** $2[-5 + (-3)]$; -16

Section 1.8

Practice Exercises

1. a. $8 \cdot x$ **b.** $17 + x$ **2. a.** $2 + (9 + 7)$ **b.** $(-4 \cdot 2) \cdot 7$ **3. a.** $x + 14$ **b.** $-30x$ **4. a.** $5x - 5y$ **b.** $-24 - 12t$ **c.** $6x - 8y - 2z$ **d.** $-3 + y$ **e.** $-x + 7 - 2s$ **f.** $x + 11$ **5. a.** $5(w + 3)$ **b.** $9(w + z)$ **6. a.** commutative property of multiplication **b.** associative property of addition **c.** identity element for addition **d.** multiplicative inverse property **e.** commutative property of addition **f.** additive inverse property **g.** commutative and associative properties of multiplication

Vocabulary, Readiness & Video Check 1.8

1. commutative property of addition **3.** distributive property **5.** associative property of addition **7.** opposites or additive inverses **9.** 2 is outside the parentheses, so the point is made that you should only distribute the -9 to the terms within the parentheses and not also to the 2.

Exercise Set 1.8

1. $16 + x$ **3.** $y \cdot (-4)$ **5.** yx **7.** $13 + 2x$ **9.** $x \cdot (yz)$ **11.** $(2 + a) + b$ **13.** $(4a) \cdot b$ **15.** $a + (b + c)$ **17.** $17 + b$ **19.** $24y$ **21.** y **23.** $26 + a$ **25.** $-72x$ **27.** s **29.** $2 + x$ **31.** $4x + 4y$ **33.** $9x - 54$ **35.** $6x + 10$ **37.** $28x - 21$ **39.** $18 + 3x$ **41.** $-2y + 2z$ **43.** $-21y - 35$ **45.** $5x + 20m + 10$ **47.** $-4 + 8m - 4n$ **49.** $-5x - 2$ **51.** $-r + 3 + 7p$ **53.** $3x + 4$ **55.** $-x + 3y$ **57.** $6r + 8$ **59.** $-36x - 70$ **61.** $-16x - 25$ **63.** $4(1 + y)$ **65.** $11(x + y)$ **67.** $-1(5 + x)$ **69.** $30(a + b)$ **71.** commutative property of multiplication **73.** associative property of addition **75.** distributive property **77.** associative property of multiplication **79.** identity element of addition **81.** distributive property **83.** commutative and associative properties of multiplication **85.** -8; $\frac{1}{8}$ **87.** $-x$; $\frac{1}{x}$ **89.** $2x$; $-2x$ **91.** false **93.** no **95.** yes **97.** yes **99.** no **101. a.** commutative property of addition **b.** commutative property of addition **c.** associative property of addition **103.** answers may vary **105.** answers may vary

Chapter 1 Vocabulary Check

1. inequality symbols **2.** equation **3.** absolute value **4.** variable **5.** opposites **6.** numerator **7.** solution **8.** reciprocals **9.** base; exponent **10.** denominator **11.** grouping symbols **12.** set

Chapter 1 Review

1. $<$ **3.** $>$ **5.** $<$ **7.** $=$ **9.** $>$ **11.** $4 \geq -3$ **13.** $0.03 < 0.3$ **15. a.** 1, 3 **b.** 0, 1, 3 **c.** $-6, 0, 1, 3$ **d.** $-6, 0, 1, 1\frac{1}{2}, 3, 9.62$ **e.** π **f.** $-6, 0, 1, 1\frac{1}{2}, 3, \pi, 9.62$ **17.** Friday **19.** $2 \cdot 2 \cdot 3 \cdot 3$ **21.** $\frac{12}{25}$ **23.** $\frac{13}{10}$ **25.** $9\frac{3}{8}$ **27.** 15 **29.** $\frac{7}{12}$ **31.** $A = 1\frac{1}{6}$ sq m; $P = 4\frac{5}{12}$ m **33.** $14\frac{1}{8}$ lb **35.** $18\frac{7}{16}$ lb **37.** Baby E **39.** c **41.** $\frac{4}{49}$ **43.** 37 **45.** $\frac{18}{7}$ **47.** $20 - 12 = 2 \cdot 4$ **49.** 18 **51.** 5 **53.** 63° **55.** yes **57.** 9 **59.** -2 **61.** -11 **63.** $-\frac{3}{16}$ **65.** -13.9 **67.** -14

69. 5 **71.** −19 **73.** a **75.** \$51 **77.** $-\frac{1}{6}$ **79.** −48 **81.** 3 **83.** undefined **85.** undefined **87.** −12 **89.** 9
91. $-7 \cdot x$ or $-7x$ **93.** $-20 - x$ **95.** commutative property of addition **97.** distributive property **99.** associative property of addition **101.** distributive property **103.** multiplicative inverse property **105.** $5y - 10$ **107.** $-7 + x - 4z$ **109.** $-12z - 27$ **111.** $<$
113. −15.3 **115.** −80 **117.** $-\frac{1}{4}$ **119.** 16 **121.** −5 **123.** $-\frac{5}{6}$ **125.** $1\frac{3}{8}$ ft

Chapter 1 Test

1. $|-7| > 5$ **2.** $9 + 5 \geq 4$ **3.** −5 **4.** −11 **5.** −3 **6.** −39 **7.** 12 **8.** −2 **9.** undefined **10.** −8 **11.** $-\frac{1}{3}$ **12.** $4\frac{5}{8}$
13. 1.275 **14.** −32 **15.** −48 **16.** 3 **17.** 0 **18.** $>$ **19.** $>$ **20.** $<$ **21.** $=$ **22.** $2221 < 10{,}993$ **23. a.** $1, 7$ **b.** $0, 1, 7$
c. $-5, -1, 0, 1, 7$ **d.** $-5, -1, 0, \frac{1}{4}, 1, 7, 11.6$ **e.** $\sqrt{7}, 3\pi$ **f.** $-5, -1, 0, \frac{1}{4}, 1, 7, 11.6, \sqrt{7}, 3\pi$ **24.** 40 **25.** 12 **26.** 22 **27.** −1
28. associative property of addition **29.** commutative property of multiplication **30.** distributive property **31.** multiplicative inverse property
32. 9 **33.** −3 **34.** second down **35.** yes **36.** 17° **37.** \$650 million **38.** \$420

CHAPTER 2 EQUATIONS, INEQUALITIES, AND PROBLEM SOLVING

Section 2.1

Practice Exercises

1. a. 1 **b.** −7 **c.** $-\frac{1}{5}$ **d.** 43 **e.** −1 **2. a.** like terms **b.** unlike terms **c.** like terms **d.** like terms **3. a.** $8y$ **b.** $5x^2$
c. $5x + 5x^2$ **d.** $21y^2$ **4. a.** $11y - 5$ **b.** $5x - 6$ **c.** $-\frac{1}{4}t$ **d.** $12.2y + 13$ **e.** $5z - 3z^4$ **5. a.** $6x - 21$ **b.** $-5x + 2.5z + 25$
c. $-2x + y - z + 2$ **6. a.** $36x + 10$ **b.** $-11x + 1$ **c.** $-30x - 17$ **7.** $-5x + 4$ **8. a.** $2x + 3$ **b.** $x - 1$ **c.** $2x + 10$ **d.** $\frac{13}{2}x$

Vocabulary, Readiness & Video Check 2.1

1. expression; term **3.** numerical coefficient **5.** numerical coefficient **7.** Although these terms have exactly the same variables, the exponents on each are not exactly the same—the exponents on x differ in each term. **9.** −1

Exercise Set 2.1

1. −7 **3.** 1 **5.** 17 **7.** like **9.** unlike **11.** like **13.** $15y$ **15.** $13w$ **17.** $-7b - 9$ **19.** $-m - 6$ **21.** −8 **23.** $7.2x - 5.2$
25. $4x - 3$ **27.** $5x^2$ **29.** $1.3x + 3.5$ **31.** $5y - 20$ **33.** $-2x - 4$ **35.** $7d - 11$ **37.** $-10x + 15y - 30$ **39.** $-3x + 2y - 1$
41. $2x + 14$ **43.** $10x - 3$ **45.** $-4x - 9$ **47.** $-4m - 3$ **49.** $k - 6$ **51.** $-15x + 18$ **53.** 16 **55.** $x + 5$ **57.** $x + 2$
59. $2k + 10$ **61.** $-3x + 5$ **63.** −11 **65.** $3y + \frac{5}{6}$ **67.** $-22 + 24x$ **69.** $0.9m + 1$ **71.** $10 - 6x - 9y$ **73.** $-x - 38$ **75.** $5x - 7$
77. $2x - 4$ **79.** $2x + 7$ **81.** $\frac{3}{4}x + 12$ **83.** $-2 + 12x$ **85.** $8(x + 6)$ or $8x + 48$ **87.** $x - 10$ **89.** $7x - 7$ **91.** 2 **93.** −23
95. −25 **97.** $(18x - 2)$ ft **99.** balanced **101.** balanced **103.** answers may vary **105.** $(15x + 23)$ in. **107.** answers may vary
109. $5b^2c^3 + b^3c^2$ **111.** $5x^2 + 9x$ **113.** $-7x^2y$

Section 2.2

Practice Exercises

1. −8 **2.** −1.8 **3.** $\frac{1}{10}$ **4.** 10 **5.** −10 **6.** 18 **7.** −8 **8. a.** 7 **b.** $9 - x$ **c.** $(9 - x)$ ft **9.** $(s - 3.8)$ mph

Vocabulary, Readiness & Video Check 2.2

1. equation; expression **3.** solution **5.** addition **7.** both sides **9.** $\frac{1}{7}x$

Exercise Set 2.2

1. 3 **3.** −2 **5.** 3 **7.** 0.5 **9.** $\frac{5}{12}$ **11.** −0.7 **13.** 3 **15.** $-\frac{14}{25}$ **17.** −3 **19.** −9 **21.** −10 **23.** 2 **25.** −7 **27.** −1
29. −9 **31.** −12 **33.** $-\frac{1}{2}$ **35.** 11 **37.** 21 **39.** 25 **41.** −14 **43.** $\frac{1}{4}$ **45.** 11 **47.** 13 **49.** −30 **51.** −0.4 **53.** −7
55. $-\frac{1}{3}$ **57.** −17.9 **59.** $-\frac{3}{4}$ **61.** 0 **63.** 1.83 **65.** $20 - p$ **67.** $(10 - x)$ ft **69.** $(180 - x)°$ **71.** $n - 28{,}000$ **73.** $7x$ sq mi
75. $\frac{8}{5}$ **77.** $\frac{1}{2}$ **79.** −9 **81.** x **83.** y **85.** x **87.** $(173 - 3x)°$ **89.** answers may vary **91.** answers may vary **93.** 250 ml
95. answers may vary **97.** 4 **99.** answers may vary **101.** solution **103.** not a solution

Section 2.3

Practice Exercises

1. 20 **2.** −12 **3.** 65 **4.** 1.5 **5.** $-\frac{12}{35}$ **6.** 11 **7.** −3 **8.** 6 **9.** $\frac{11}{12}$ **10.** $3x + 6$

Vocabulary, Readiness & Video Check 2.3

1. multiplication **3.** false **5.** same **7.** $(x + 1) + (x + 3) = 2x + 4$

Exercise Set 2.3

1. 4 **3.** 0 **5.** 12 **7.** −12 **9.** 3 **11.** 2 **13.** 0 **15.** 6.3 **17.** 10 **19.** −20 **21.** 0 **23.** −9 **25.** −3 **27.** −30 **29.** 3 **31.** $\frac{10}{9}$ **33.** −1 **35.** −4 **37.** $-\frac{1}{2}$ **39.** 0 **41.** 4 **43.** $-\frac{1}{14}$ **45.** 0.21 **47.** 5 **49.** 6 **51.** −5.5 **53.** −5 **55.** 0 **57.** −3 **59.** $-\frac{9}{28}$ **61.** $\frac{14}{3}$ **63.** −9 **65.** −2 **67.** $\frac{11}{2}$ **69.** $-\frac{1}{4}$ **71.** $\frac{9}{10}$ **73.** $-\frac{17}{20}$ **75.** −16 **77.** $2x + 2$ **79.** $2x + 2$ **81.** $5x + 20$ **83.** $7x - 12$ **85.** $12z + 44$ **87.** $-a - 3$ **89.** > **91.** = **93.** −48 **95.** answers may vary **97.** answers may vary **99.** $\frac{700}{3}$ mg **101.** −2.95 **103.** 0.02

Section 2.4

Practice Exercises

1. 3 **2.** $\frac{21}{13}$ **3.** −15 **4.** 3 **5.** 0 **6.** no solution **7.** all real numbers

Calculator Explorations 2.4

1. solution **3.** not a solution **5.** solution

Vocabulary, Readiness & Video Check 2.4

1. equation **3.** expression **5.** expression **7.** equation **9.** 3; distributive property, addition property of equality, multiplication property of equality **11.** The number of decimal places in each number helps you determine what power of 10 you can multiply through by so you are no longer dealing with decimals.

Exercise Set 2.4

1. −6 **3.** 3 **5.** 1 **7.** $\frac{3}{2}$ **9.** 0 **11.** −1 **13.** 4 **15.** −4 **17.** −3 **19.** 2 **21.** 50 **23.** 1 **25.** $\frac{7}{3}$ **27.** 0.2 **29.** all real numbers **31.** no solution **33.** no solution **35.** all real numbers **37.** 18 **39.** $\frac{19}{9}$ **41.** $\frac{14}{3}$ **43.** 13 **45.** 4 **47.** all real numbers **49.** $-\frac{3}{5}$ **51.** −5 **53.** 10 **55.** no solution **57.** 3 **59.** −17 **61.** −4 **63.** 3 **65.** all real numbers **67.** $-8 - x$ **69.** $-3 + 2x$ **71.** $9(x + 20)$ **73.** $(6x - 8)$ m **75. a.** all real numbers **b.** answers may vary **c.** answers may vary **77.** A **79.** B **81.** C **83.** answers may vary **85. a.** $x + x + x + 2x + 2x = 28$ **b.** $x = 4$ **c.** $x = 4$ cm; $2x = 8$ cm **87.** answers may vary **89.** 15.3 **91.** −0.2 **93.** $-\frac{7}{8}$ **95.** no solution

Integrated Review

1. 6 **2.** −17 **3.** 12 **4.** −26 **5.** −3 **6.** −1 **7.** $\frac{27}{2}$ **8.** $\frac{25}{2}$ **9.** 8 **10.** −64 **11.** 2 **12.** −3 **13.** no solution **14.** no solution **15.** −2 **16.** −2 **17.** $-\frac{5}{6}$ **18.** $\frac{1}{6}$ **19.** 1 **20.** 6 **21.** 4 **22.** 1 **23.** $\frac{9}{5}$ **24.** $-\frac{6}{5}$ **25.** all real numbers **26.** all real numbers **27.** 0 **28.** −1.6 **29.** $\frac{4}{19}$ **30.** $-\frac{5}{19}$ **31.** $\frac{7}{2}$ **32.** $-\frac{1}{4}$ **33.** 2 **34.** 2 **35.** no solution **36.** no solution **37.** $\frac{7}{6}$ **38.** $\frac{1}{15}$

Section 2.5

Practice Exercises

1. 9 **2.** 2 **3.** 9 in. and 36 in. **4.** 29 Republican and 20 Democratic governors **5.** 25°, 75°, 80° **6.** 46, 48, 50

Vocabulary, Readiness & Video Check 2.5

1. $2x; 2x - 31$ **3.** $x + 5; 2(x + 5)$ **5.** $20 - y; \frac{20 - y}{3}$ or $(20 - y) \div 3$ **7.** in the statement of the application **9.** That the 3 angle measures are consecutive even integers and that they sum to 180°.

Exercise Set 2.5

1. $6x + 1 = 5x; -1$ **3.** $3x - 6 = 2x + 8; 14$ **5.** $2(x - 8) = 3(x + 3); -25$ **7.** $2(-2 + x) = x - \frac{1}{2}; \frac{7}{2}$ **9.** 3 in.; 6 in.; 16 in. **11.** 1st piece: 5 in.; 2nd piece: 10 in.; 3rd piece: 25 in. **13.** 7837 screens; 31,710; In 2010, 7837 screens were 3D. **15.** 1st angle: 37.5°; 2nd angle: 37.5°; 3rd angle: 105° **17.** $3x + 3$ **19.** $x + 2;\ x + 4;\ 2x + 4$ **21.** $x + 1; x + 2; x + 3; 4x + 6$ **23.** $x + 2; x + 4; 2x + 6$ **25.** 234, 235 **27.** Belgium: 32; France: 33; Spain: 34 **29.** Sahara: 3,500,000 sq mi; Gobi: 500,000 sq mi **31.** 5 ft, 12 ft **33.** $\frac{5}{4}$ **35.** Botswana: 32,000,000 carats; Angola: 8,000,000 carats **37.** 58°, 60°, 62° **39.** South Korea: 14; Russia: 15; Austria: 16 **41.** −16 **43.** 43°, 137° **45.** 1 **47.** $\frac{3}{2}$ **49.** Maglev: 361 mph; TGV: 357.2 mph **51.** $\frac{5}{2}$ **53.** California: 58; Montana: 56 **55.** 111° **57.** 1st piece: 3 ft; 2nd piece: 12 ft; 3rd piece: 15 ft **59.** Eagles: *Their Greatest Hits, 1971–1975* **61.** *Thriller*: \$27 million; *The Wall*: \$23 million **63.** answers may vary

65. 34 **67.** 225π **69.** 15 ft by 24 ft **71.** 5400 chirps per hour; 129,600 chirps per day; 47,304,000 chirps per year **73.** answers may vary **75.** answers may vary **77.** c

Section 2.6

Practice Exercises

1. 116 sec or 1 min 56 sec **2.** 9 ft **3.** 46.4°F **4.** length: 28 in.; width: 5 in. **5.** $r = \frac{I}{Pt}$ **6.** $s = \frac{H - 10a}{5a}$ **7.** $d = \frac{N - F}{n - 1}$ **8.** $B = \frac{2A - ab}{a}$

Vocabulary, Readiness & Video Check 2.6

1. relationships **3.** To show that the process of solving this equation for x—dividing both sides by 5, the coefficient of x—is the same process used to solve a formula for a specific variable. Treat whatever is multiplied by that specific variable as the coefficient—the coefficient is all the factors except that specific variable.

Exercise Set 2.6

1. $h = 3$ **3.** $h = 3$ **5.** $h = 20$ **7.** $c = 12$ **9.** $r \approx 2.5$ **11.** $T = 3$ **13.** $h \approx 15$ **15.** $h = \frac{f}{5g}$ **17.** $w = \frac{V}{lh}$ **19.** $y = 7 - 3x$ **21.** $R = \frac{A - P}{PT}$ **23.** $A = \frac{3V}{h}$ **25.** $a = P - b - c$ **27.** $h = \frac{S - 2\pi r^2}{2\pi r}$ **29.** 120 ft **31. a.** area: 103.5 sq ft; perimeter: 41 ft **b.** baseboard: perimeter; carpet: area **33. a.** area: 480 sq in.; perimeter: 120 in. **b.** frame: perimeter; glass: area **35.** −10°C **37.** length: 78 ft; width: 52 ft **39.** 18 ft, 36 ft, 48 ft **41.** 55.2 mph **43.** 96 piranhas **45.** 61.5°F **47.** 60 chirps per minute **49.** increases **51.** 2 bags **53.** one 16-in. pizza **55.** $x = 6$ m, $2.5x = 15$ m **57.** 22 hr **59.** 13 in. **61.** 2.25 hr **63.** 12,090 ft **65.** 50°C **67.** 332.6°F **69.** 449 cu in. **71.** 0.32 **73.** 2.00 or 2 **75.** 17% **77.** 720% **79.** multiplies the volume by 8; answers may vary **81.** $53\frac{1}{3}$ **83.** $V = G(N - R)$ **85.** ● = (■ − ▲)/▮ **87.** 500 sec or $8\frac{1}{3}$ min **89.** 608.33 ft **91.** 4.42 min **93.** $35\frac{11}{17}$ mph

Section 2.7

Practice Exercises

1. 62.5% **2.** 360 **3. a.** 42% **b.** 98% **c.** 79.3 million dogs **4.** discount: \$408; new price: \$72 **5.** 42.6% **6.** 3554 screens **7.** 2 liters of 5% eyewash; 4 liters of 2% eyewash

Vocabulary, Readiness & Video Check 2.7

1. no **3.** yes **5. a.** equals; = **b.** multiplication; • **c.** Drop the percent symbol and move the decimal point two places to the left. **7.** You must first find the actual amount of increase in price by subtracting the original price from the new price.

Exercise Set 2.7

1. 11.2 **3.** 55% **5.** 180 **7.** 4% **9.** 9990 **11.** discount: \$1480; new price: \$17,020 **13.** \$46.58 **15.** 9.8% **17.** 30% **19.** \$104 **21.** \$42,500 **23.** 2 gal **25.** 7 lb **27.** 4.6 **29.** 50 **31.** 30% **33.** 23% **35.** 90,405 **37.** 59%; 5%; 26%; 2%; 99% due to rounding **39.** decrease: \$64; sale price: \$192 **41.** 27.2% **43.** 239 million **45.** 300% **47.** 400 oz **49.** 66.7% **51.** 120 employees **53.** 361 college students **55.** 400 oz **57.** 854 thousand Scoville units **59.** > **61.** = **63.** > **65.** no; answers may vary **67.** no; answers may vary **69.** 9.6% **71.** 26.9%; yes **73.** 17.1%

Section 2.8

Practice Exercises

1. 2.2 hr **2.** eastbound: 62 mph; westbound: 52 mph **3.** 106 \$5 bills; 59 \$20 bills **4.** \$18,000 at 11.5%; \$12,000 at 6%

Vocabulary, Readiness & Video Check 2.8

1.

	r ·	t =	d
bus	55	x	$55x$
car	50	$x + 3$	$50(x + 3)$

; $55x = 50(x + 3)$

3.

P ·	R ·	T =	I
x	0.06	1	$0.06x$
$36{,}000 - x$	0.04	1	$0.04(36{,}000 - x)$

; $0.06x = 0.04(36{,}000 - x)$

Exercise Set 2.8

1. $666\frac{2}{3}$ mi **3.** 55 mph **5.** $0.10y$ **7.** $0.05(x + 7)$ **9.** $20(4y)$ or $80y$ **11.** $50(35 - x)$ **13.** 12 \$10 bills; 32 \$5 bills **15.** \$11,500 at 8%; \$13,500 at 9% **17.** \$7000 at 11% profit; \$3000 at 4% loss **19.** 187 adult tickets; 313 child tickets **21.** 2 hr **23.** \$30,000 at 8%; \$24,000 at 10% **25.** 2 hr $37\frac{1}{2}$ min **27.** 36 mph; 46 mph **29.** 483 dimes; 161 nickels **31.** 4 hr **33.** 2.5 hr **35.** \$4500 **37.** 2.2 mph; 3.3 mph **39.** 27.5 mi **41.** −4 **43.** $\frac{9}{16}$ **45.** −4 **47.** 25 \$100 bills; 71 \$50 bills; 175 \$20 bills **49.** 25 skateboards **51.** 800 books **53.** answers may vary

Section 2.9

Practice Exercises

1. 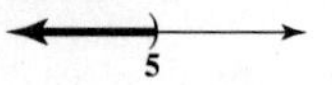

$(-\infty, 5)$ **2.** 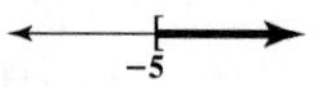

$[-5, \infty)$ **3.** 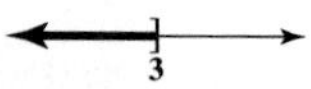

$(-\infty, 3]$ **4.** 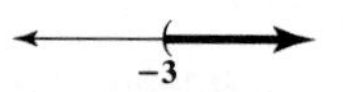

$(-3, \infty)$

5. 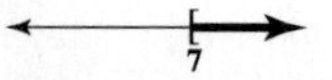

$[7, \infty)$ **6.** 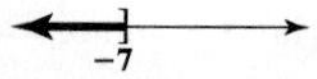

$(-\infty, -7]$ **7.** 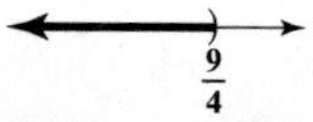

$\left(-\infty, \frac{9}{4}\right)$ **8.** 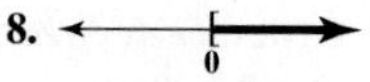

$[0, \infty)$

9. $[-3, 1)$ **10.** $(-2, 2]$ **11.** $\left(-\frac{16}{3}, \frac{4}{3}\right)$ **12.** all numbers less than 10

13. Kasonga can afford at most 3 classes.

Vocabulary, Readiness & Video Check 2.9

1. expression **3.** inequality **5.** -5 **7.** The graph of Example 1 is shaded from $-\infty$ to and including -1, as indicated by a bracket. To write interval notation, you write down what is shaded for the inequality from left to right. A parenthesis is always used with $-\infty$, so from the graph, the interval notation is $(-\infty, -1]$. **9.** You would divide the left, middle, and right by -3 instead of 3, which would reverse the directions of both inequality signs.

Exercise Set 2.9

1. $x \geq 2$ **3.** $x < -5$ **5.** $(-\infty, -1]$ **7.** $\left(-\infty, \frac{1}{2}\right)$
9. $[5, \infty)$ **11.** $x < -3$, $(-\infty, -3)$ **13.** $x \geq -5$, $[-5, \infty)$
15. $x \geq -2$ $[2, \infty)$ **17.** $x > -3$, $(-3, \infty)$ **19.** $x \leq 1$, $(-\infty, 1]$
21. $x > -5$, $(-5, \infty)$ **23.** $x \leq -2$, $(-\infty, -2]$ **25.** $x \leq -8$, $(-\infty, -8]$
27. $x > 4$, $(4, \infty)$ **29.** $x \geq 20$, $[20, \infty)$ **31.** $x > 16$, $(16, \infty)$
33. $x > -3$, $(-3, \infty)$ **35.** $x \leq -\frac{2}{3}$, $\left(-\infty, -\frac{2}{3}\right]$ **37.** $x > \frac{8}{3}$, $\left(\frac{8}{3}, \infty\right)$
39. $x > -13$, $(-13, \infty)$ **41.** $x < 0$, $(-\infty, 0)$ **43.** $x \leq 0$, $(-\infty, 0]$
45. $x > 3$, $(3, \infty)$ **47.** $x > \frac{8}{3}$ $\left(\frac{8}{3}, \infty\right)$ **49.** $(-1, 3)$ **51.** $[0, 2)$
53. $(-1, 2)$ **55.** $[4, 5]$ **57.** $(1, 5]$ **59.** $(1, 4)$
61. $\left(0, \frac{14}{3}\right]$ **63.** all numbers greater than -10 **65.** 35 cm **67.** at least 193 **69.** 86 people **71.** at least 35 min
73. $-3 < x < 3$ **75.** 8 **77.** 1 **79.** $\frac{16}{49}$ **81.** $>$ **83.** $\geq$ **85.** when multiplying or dividing by a negative number
87. final exam score ≥ 78.5 **89.** answers may vary **91.** answers may vary **93.** $0.924 \leq d \leq 0.987$ **95.** $(1, \infty)$
97. $\left(-\infty, \frac{5}{8}\right)$

Chapter 2 Vocabulary Check

1. like terms **2.** unlike terms **3.** linear equation in one variable **4.** linear inequality in one variable **5.** compound inequalities **6.** formula **7.** numerical coefficient **8.** equivalent equations **9.** all real numbers **10.** no solution **11.** the same **12.** reversed

Chapter 2 Review

1. $6x$ **3.** $4x - 2$ **5.** $3n - 18$ **7.** $-6x + 7$ **9.** $3x - 7$ **11.** 4 **13.** 6 **15.** 0 **17.** -23 **19.** 5; 5 **21.** b **23.** b **25.** -12
27. 0 **29.** 0.75 **31.** -6 **33.** -1 **35.** $-\frac{1}{5}$ **37.** $3x + 3$ **39.** -4 **41.** 2 **43.** no solution **45.** $\frac{3}{4}$ **47.** 20 **49.** $\frac{23}{7}$
51. 102 **53.** 6665.5 in. **55.** Kellogg: 35 plants; Keebler: 18 plants **57.** 3 **59.** $w = 9$ **61.** $m = \frac{y - b}{x}$ **63.** $x = \frac{2y - 7}{5}$ **65.** $\pi = \frac{C}{D}$
67. 15 m **69.** 1 hr 20 min **71.** 20% **73.** 110 **75.** mark-up: \$209; new price: \$2109 **77.** 40% solution: 10 gal; 10% solution: 20 gal
79. 18% **81.** 966 customers **83.** 32% **85.** 50 km **87.** 80 nickels **89.** $(0, \infty)$ **91.** $[0.5, 1.5)$
93. $(-\infty, -4)$ **95.** $(-\infty, 4]$ **97.** $\left(-\frac{1}{2}, \frac{3}{4}\right)$ **99.** $\left(-\infty, \frac{19}{3}\right]$
101. \$2500 **103.** 4 **105.** $-\frac{3}{2}$ **107.** all real numbers **109.** -13 **111.** $h = \frac{3V}{A}$ **113.** 160 **115.** $(9, \infty)$
117. $(-\infty, 0]$

Chapter 2 Test

1. $y - 10$ **2.** $5.9x + 1.2$ **3.** $-2x + 10$ **4.** $10y + 1$ **5.** -5 **6.** 8 **7.** $\frac{7}{10}$ **8.** 0 **9.** 27 **10.** 3 **11.** 0.25 **12.** $\frac{25}{7}$
13. no solution **14.** $x = 6$ **15.** $h = \frac{V}{\pi r^2}$ **16.** $y = \frac{3x - 10}{4}$ **17.** $(-\infty, -2]$ **18.** $(-\infty, 4)$
19. $\left(-1, \frac{7}{3}\right)$ **20.** $\left(\frac{2}{5}, \infty\right)$ **21.** 21 **22.** 62 ft by 64 ft **23.** 401, 802 **24.** \$8500 at 10%; \$17,000 at 12%
25. $2\frac{1}{2}$ hr **26.** 552 **27.** 40% **28.** 16%

Chapter 2 Cumulative Review

1. a. 11, 112 **b.** 0, 11, 112 **c.** $-3, -2, 0, 11, 112$ **d.** $-3, -2, -1.5, 0, \frac{1}{4}, 11, 112$ **e.** $\sqrt{2}$ **f.** all numbers in the given set; Sec. 1.2, Ex. 5 **3. a.** 4 **b.** 5 **c.** 0 **d.** $\frac{1}{2}$ **e.** 5.6; Sec. 1.2, Ex. 7 **5. a.** $2 \cdot 2 \cdot 2 \cdot 5$ **b.** $3 \cdot 3 \cdot 7$; Sec. 1.3, Ex. 1 **7.** $\frac{8}{20}$; Sec. 1.3, Ex. 6 **9.** 66; Sec. 1.4, Ex. 4 **11.** 2 is a solution; Sec. 1.4, Ex. 7 **13.** -3; Sec. 1.5, Ex. 2 **15.** 2; Sec. 1.5, Ex. 4 **17. a.** 10 **b.** $\frac{1}{2}$ **c.** $2x$ **d.** -6; Sec. 1.5, Ex. 10 **19. a.** 9.9 **b.** $-\frac{4}{5}$ **c.** $\frac{2}{15}$; Sec. 1.6, Ex. 2 **21. a.** 52° **b.** 118°; Sec. 1.6, Ex. 8 **23. a.** -0.06 **b.** $-\frac{7}{15}$ **c.** 16; Sec. 1.7, Ex. 3 **25. a.** 6 **b.** -12 **c.** $-\frac{8}{15}$ **d.** $-\frac{1}{6}$; Sec. 1.7, Ex. 7 **27. a.** $5 + x$ **b.** $x \cdot 3$; Sec. 1.8, Ex. 1 **29. a.** $8(2 + x)$ **b.** $7(s + t)$; Sec. 1.8, Ex. 5 **31.** $-2x - 1$; Sec. 2.1, Ex. 7 **33.** $\frac{5}{4}$; Sec. 2.2, Ex. 3 **35.** 19; Sec. 2.2, Ex. 6 **37.** 140; Sec. 2.3, Ex. 3 **39.** 2; Sec. 2.4, Ex. 1 **41.** 10; Sec. 2.5, Ex. 2 **43.** $\frac{V}{wh} = l$; Sec. 2.6, Ex. 5 **45.** $(-\infty, -10]$ 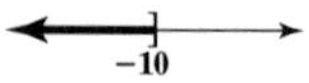

; Sec. 2.9, Ex. 2

CHAPTER 3 GRAPHING

Section 3.1

Practice Exercises

1. a. Africa/Middle East region, 145 million Internet users **b.** 640 million more Internet users **2. a.** 70 beats per minute **b.** 60 beats per minute **c.** 5 minutes after lighting **3.**

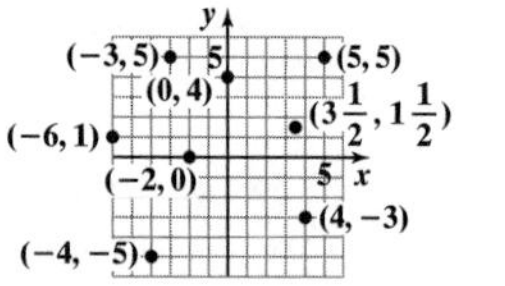

4. a. (2004, 65), (2005, 67), (2006, 96), (2007, 86), (2008, 79), (2009, 79), (2010, 72)

b.

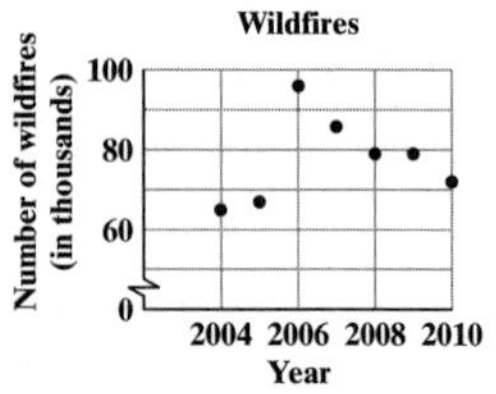

5. a. yes **b.** yes **c.** no **6. a.** $(0, -8)$ **b.** $(6, 4)$ **c.** $(-3, -14)$

7.

	x	y
a.	-2	8
b.	3	-12
c.	0	0

8.

	x	y
a.	-10	-4
b.	0	-2
c.	10	0

9.

x	0	1	2	3	4
y	12,000	10,200	8400	6600	4800

Vocabulary, Readiness & Video Check 3.1

1. x-axis; y-axis **3.** quadrants; four **5.** one **7.** horizontal: top tourist countries; vertical: number of arrivals (in millions) to these countries **9.** Data occurring in pairs of numbers can be written as ordered pairs, called paired data, and then graphed on a coordinate system. **11.** a linear equation in one variable

Exercise Set 3.1

1. France **3.** France, U.S., Spain, Italy, and China **5.** 43 million **7.** 71,000 **9.** 2011; 103,000 **11.** 15.9 **13.** from 1998 to 2000 **15.** 2014 **17.**

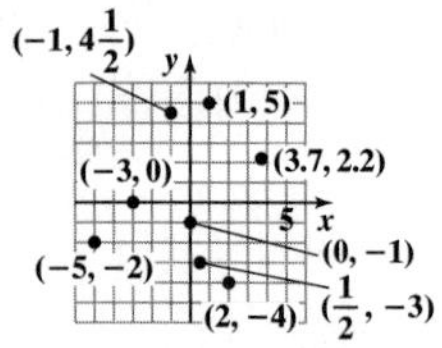

(1, 5) and (3.7, 2.2) are in quadrant I, $\left(-1, 4\frac{1}{2}\right)$ is in quadrant II, $(-5, -2)$ is in quadrant III, $(2, -4)$ and $\left(\frac{1}{2}, -3\right)$ are in quadrant IV, $(-3, 0)$ lies on the x-axis, $(0, -1)$ lies on the y-axis **19.** (0, 0) **21.** (3, 2) **23.** $(-2, -2)$ **25.** $(2, -1)$ **27.** $(0, -3)$ **29.** (1, 3) **31.** $(-3, -1)$

33. a. (2006, 25.5), (2007, 26.3), (2008, 27.7), (2009, 29.4), (2010, 31.8) **b.** In the year 2010, the worldwide box office was \$31.8 billion. **c.**

d. The worldwide box office increased every year. **35. a.** (0.50, 10), (0.75, 12), (1.00, 15), (1.25, 16), (1.50, 18), (1.50, 19), (1.75, 19), (2.00, 20) **b.** When Minh studied 1.25 hours, her quiz score was 16. **c.**

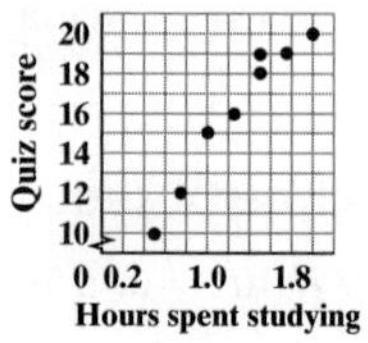

d. answers may very

37. a. (2313, 2), (2085, 1), (2711, 21), (2869, 39), (2920, 42), (4038, 99), (1783, 0), (2493, 9) **b.**

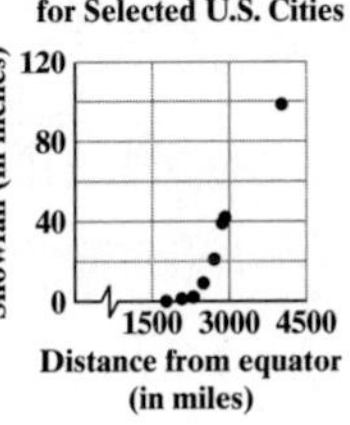

c. The farther from the equator, the more snowfall. **39.** yes; no; yes **41.** yes; yes **43.** no; yes; yes **45.** $(-4, -2), (4, 0)$ **47.** $(-8, -5), (16, 1)$ **49.** $0; 7; -\frac{2}{7}$ **51.** 2; 2; 5 **53.** 0; −3; 2 **55.** 2; 6; 3 **57.** −12; 5; −6 **59.** $\frac{5}{7}; \frac{5}{2}; -1$ **61.** 0; −5; −2

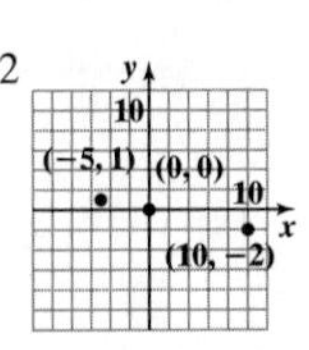

63. 2; 1; −6 **65. a.** 13,000; 21,000; 29,000 **b.** 45 desks **67. a.** 5.52; 6.00; 6.48 **b.** year 9; 2009 **c.** 2016 **d.** In 2005, the average cinema admission price was $6.48. **69.** In 2010, there were 3755 Walmart stores in the U.S. **71.** year 8: 100 stores; year 9: 105 stores; year 10: 100 stores **73.** The y-values are all 0. **75.** $y = 5 - x$ **77.** $y = -\frac{1}{2}x + \frac{5}{4}$ **79.** $y = -2x$ **81.** $y = \frac{1}{3}x - 2$ **83.** false **85.** true **87.** negative; negative **89.** positive; negative **91.** 0; 0 **93.** y **95.** no; answers may vary **97.** answers may vary **99.** $(4, -7)$ **101.** 26 units **103. a.** $(-2, 6)$ **b.** 28 units **c.** 45 sq units

Section 3.2

Practice Exercises

1. a. yes **b.** no **c.** yes **d.** yes **2.** **3.** **4.** **5.** **6.** **7.** The graph of $y = -2x + 3$ is the same as the graph of $y = -2x$ except that the graph of $y = -2x + 3$ is moved 3 units upward. **8. a.** **b.** We predict 725 thousand computer software application engineers in the year 2020.

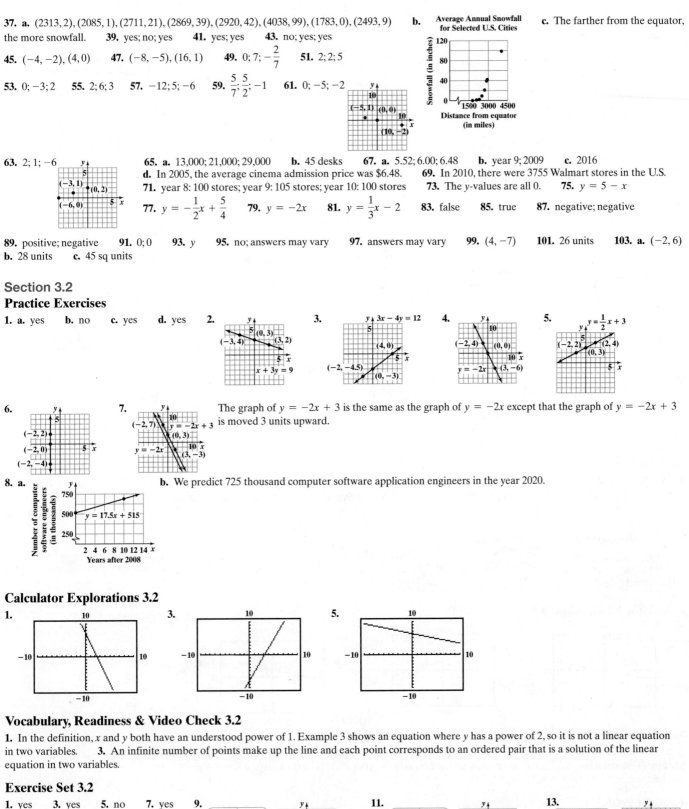

Calculator Explorations 3.2

1.

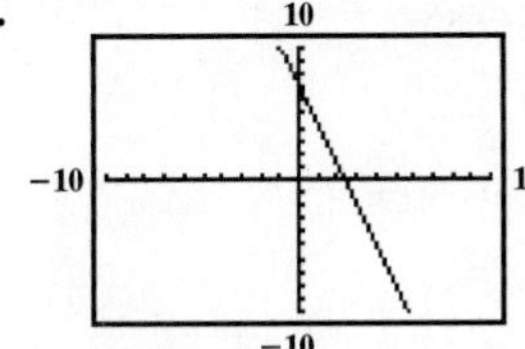

3.

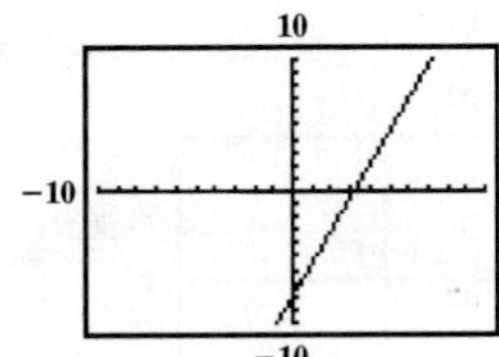

5.

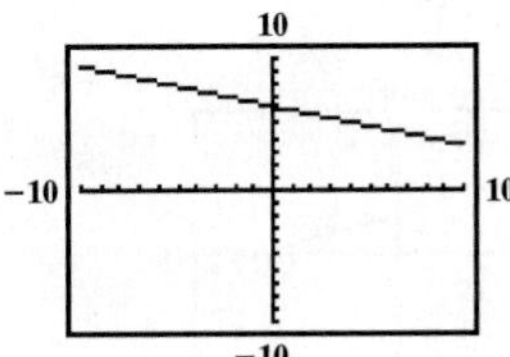

Vocabulary, Readiness & Video Check 3.2

1. In the definition, x and y both have an understood power of 1. Example 3 shows an equation where y has a power of 2, so it is not a linear equation in two variables. **3.** An infinite number of points make up the line and each point corresponds to an ordered pair that is a solution of the linear equation in two variables.

Exercise Set 3.2

1. yes **3.** yes **5.** no **7.** yes **9.**

x	y
6	0
4	−2
5	−1

11.

x	y
1	−4
0	0
−1	4

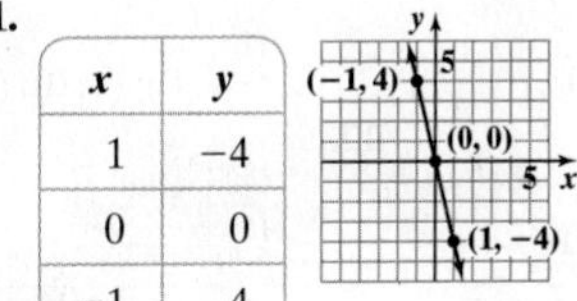

13.

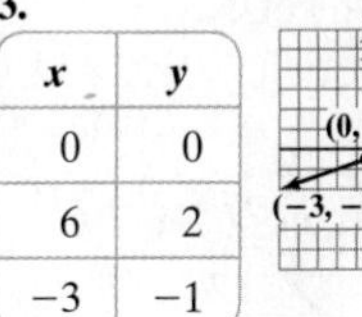

x	y
0	0
6	2
−3	−1

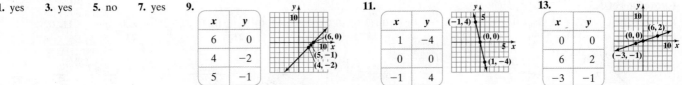

15. 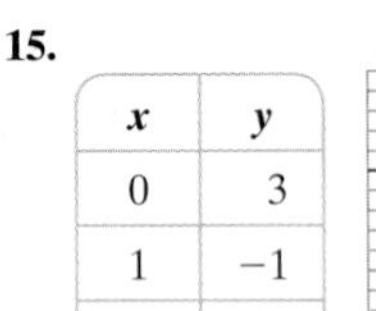

x	y
0	3
1	−1
2	−5

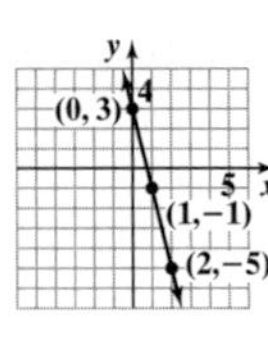

17. 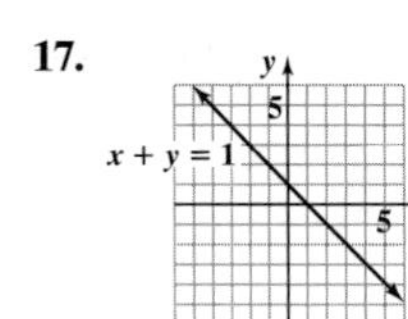

19. 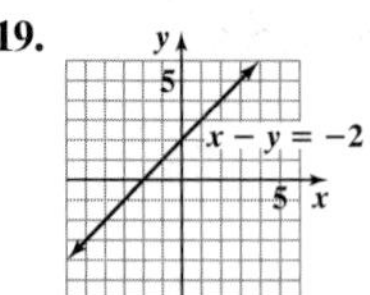

21. 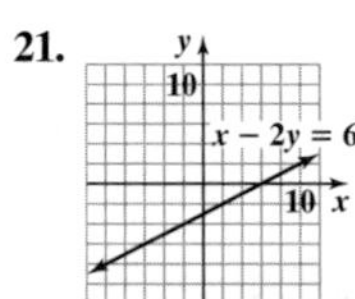

23. 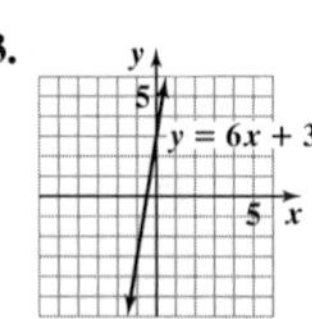

25. 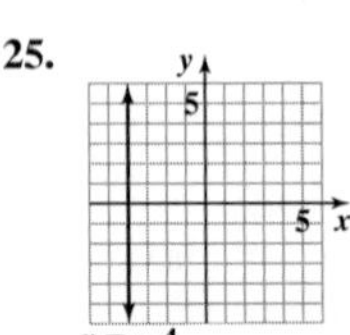

27. 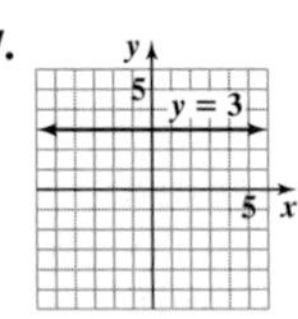

29. 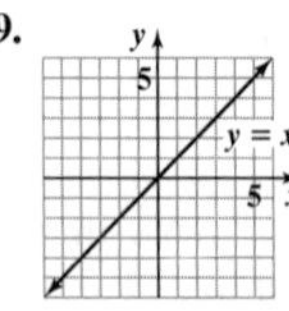

31. 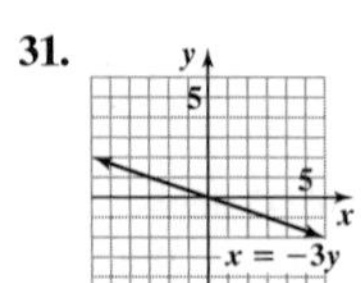

33. 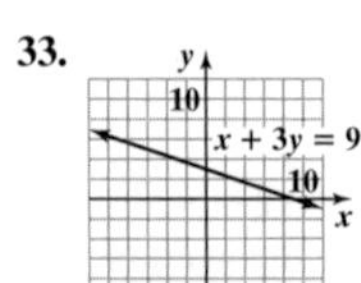

35. 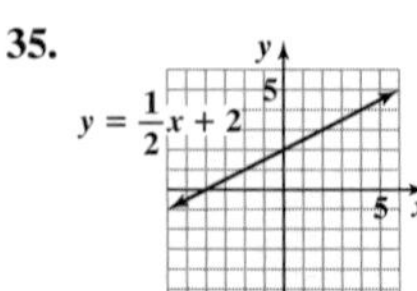

37. 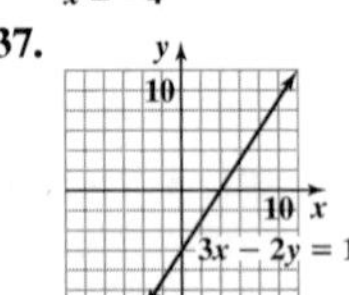

39. 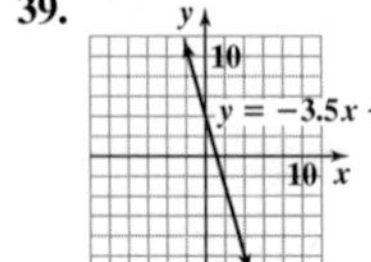

41. 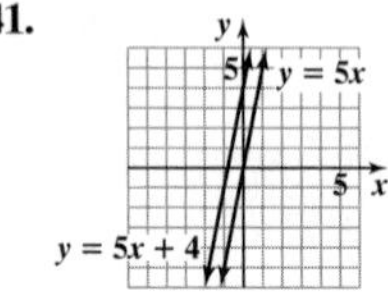

43. 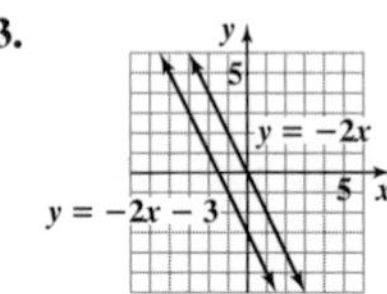

45. 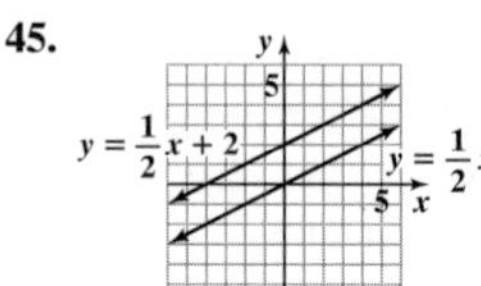

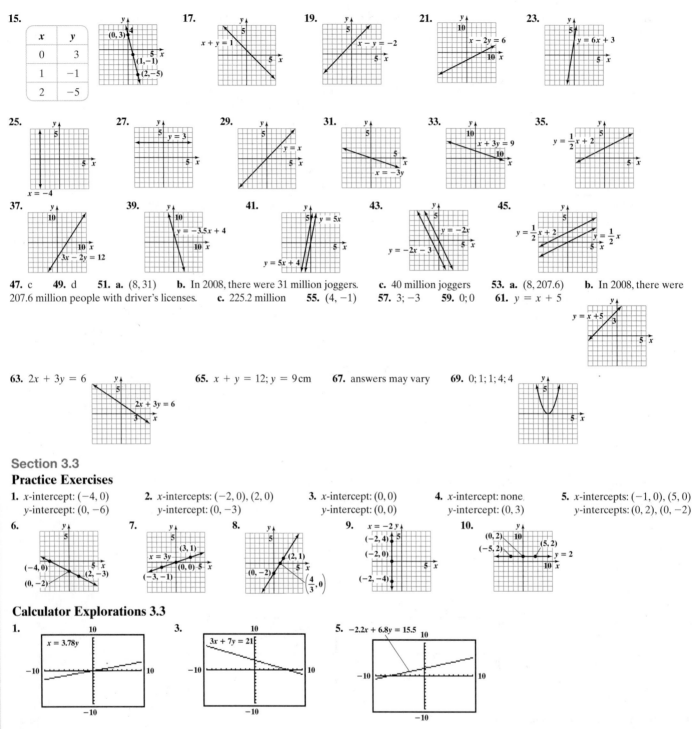

47. c **49.** d **51. a.** (8, 31) **b.** In 2008, there were 31 million joggers. **c.** 40 million joggers **53. a.** (8, 207.6) **b.** In 2008, there were 207.6 million people with driver's licenses. **c.** 225.2 million **55.** $(4, -1)$ **57.** $3; -3$ **59.** $0; 0$ **61.** $y = x + 5$

63. $2x + 3y = 6$ 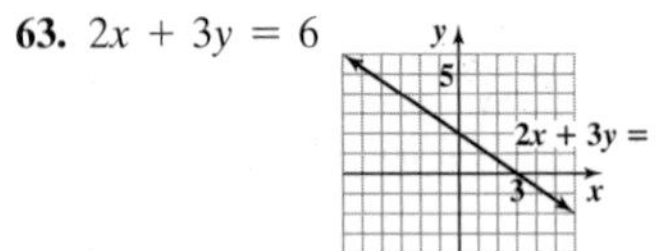

65. $x + y = 12$; $y = 9$ cm **67.** answers may vary **69.** 0; 1; 1; 4; 4 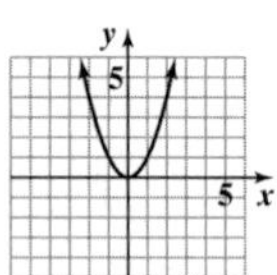

Section 3.3

Practice Exercises

1. x-intercept: $(-4, 0)$; y-intercept: $(0, -6)$ **2.** x-intercepts: $(-2, 0)$, $(2, 0)$; y-intercept: $(0, -3)$ **3.** x-intercept: $(0, 0)$; y-intercept: $(0, 0)$ **4.** x-intercept: none; y-intercept: $(0, 3)$ **5.** x-intercepts: $(-1, 0)$, $(5, 0)$; y-intercepts: $(0, 2)$, $(0, -2)$

6. 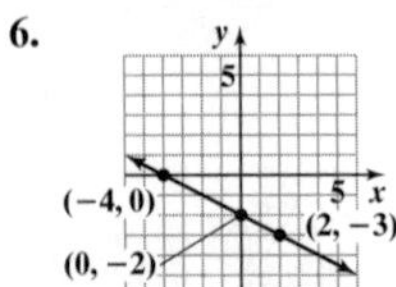

7. 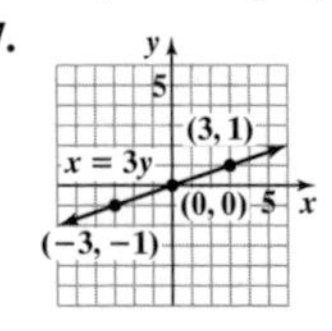

8. 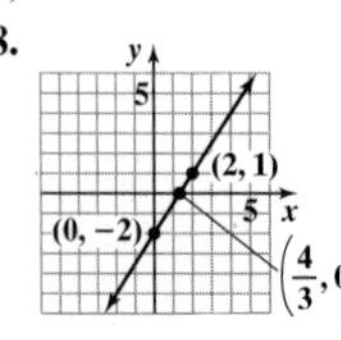

9. 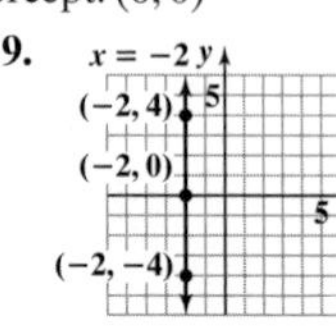

10. 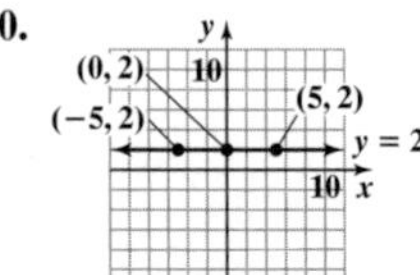

Calculator Explorations 3.3

1. 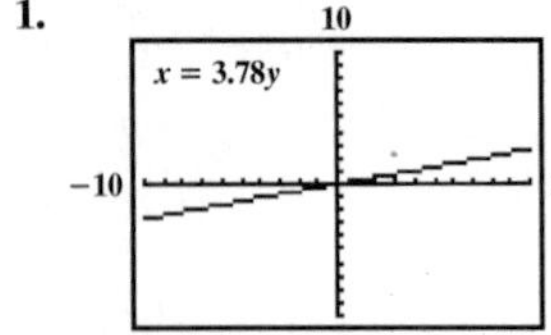

3. 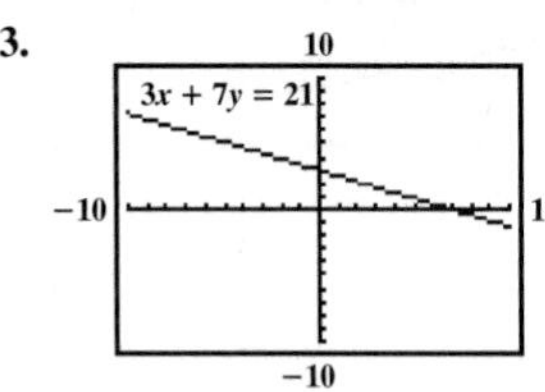

5. 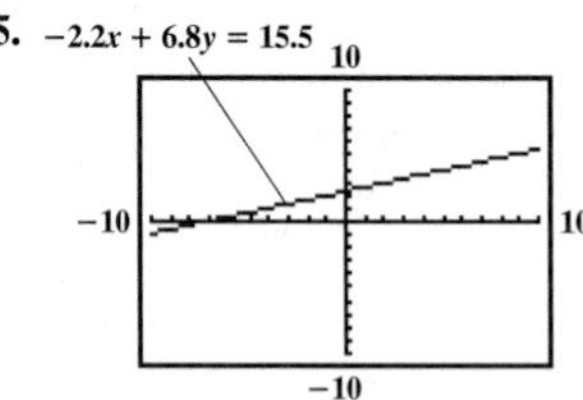

Vocabulary, Readiness & Video Check 3.3

1. linear **3.** horizontal **5.** y-intercept **7.** y; x **9.** Because x-intercepts lie on the x-axis; because y-intercepts lie on the y-axis. **11.** For a horizontal line, the coefficient of x will be 0 and the coefficient of y will be 1; for a vertical line, the coefficient of y will be 0 and the coefficient of x will be 1.

Exercise Set 3.3

1. $(-1, 0)$; $(0, 1)$ **3.** $(-2, 0)$; $(2, 0)$; $(0, -2)$ **5.** $(-2, 0)$; $(1, 0)$; $(3, 0)$; $(0, 3)$ **7.** $(-1, 0)$; $(1, 0)$; $(0, 1)$; $(0, -2)$ **9.** infinite **11.** 0

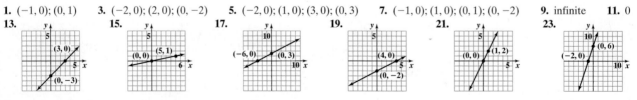

13. (3, 0), (0, −3) **15.** (0, 0), (5, 1) **17.** (−6, 0), (0, 3) **19.** (4, 0), (0, −2) **21.** (0, 0), (1, 2) **23.** (−2, 0), (0, 6)

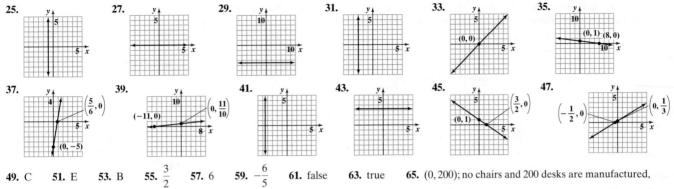

25. **27.** 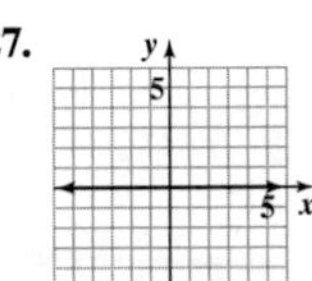 **29.** **31.** 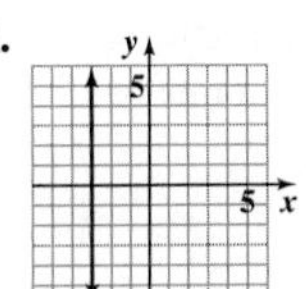 **33.** 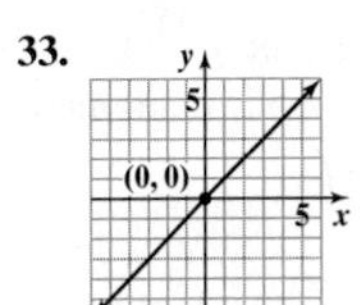

35. 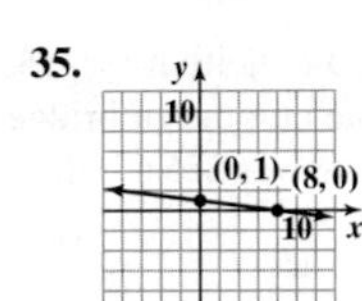

37. 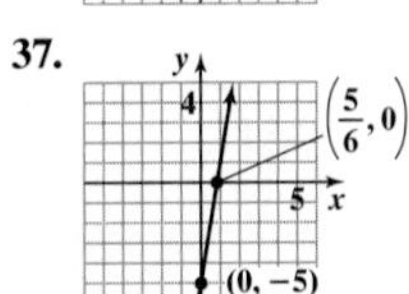

39. 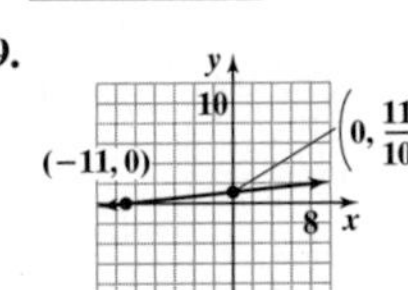

41. 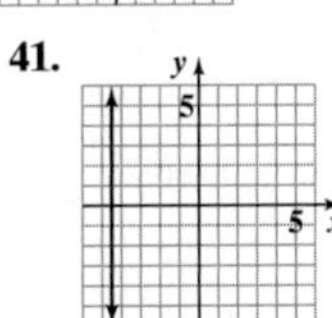 **43.** 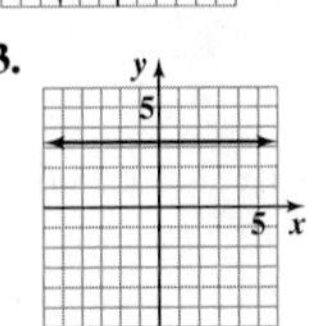 **45.** 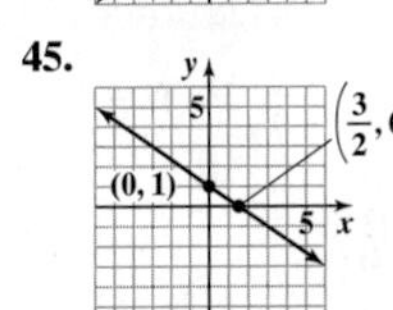

47. 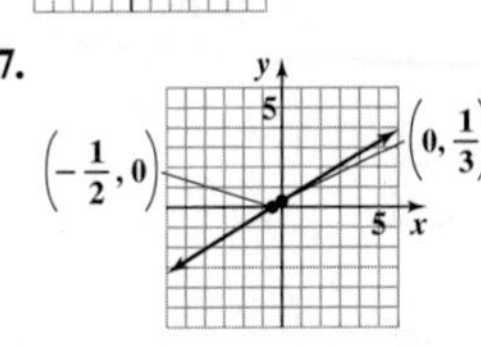

49. C **51.** E **53.** B **55.** $\frac{3}{2}$ **57.** 6 **59.** $-\frac{6}{5}$ **61.** false **63.** true **65.** (0, 200); no chairs and 200 desks are manufactured, **67.** 300 chairs **69. a.** (62, 0) **b.** 62 years after 2002, 0 people will attend the movies at the theater. **c.** answers may very **71.** $x = 1$ **73.** answers may vary **75.** answers may vary

Section 3.4

Practice Exercises

1. −1 **2.** $\frac{1}{3}$ **3.** $m = \frac{2}{3}$; y-intercept: (0, −2) **4.** $m = 6$; y-intercept: (0, −5) **5.** $m = -\frac{5}{2}$; y-intercept: (0, 4) **6.** $m = 0$ **7.** slope is undefined **8. a.** perpendicular **b.** neither **c.** parallel **9.** 25% **10.** $m = \frac{0.75 \text{ dollar}}{1 \text{ pound}}$; The Wash-n-Fold charges $0.75 per pound of laundry.

Calculator Explorations 3.4

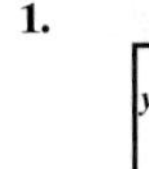
1. 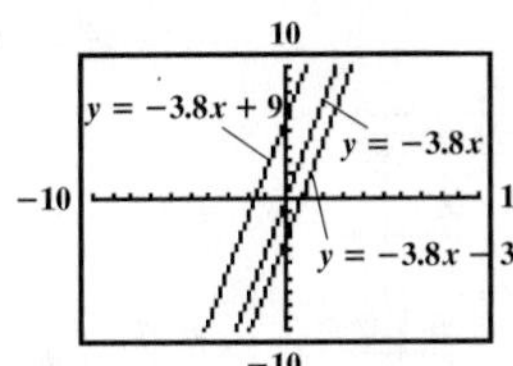

3. 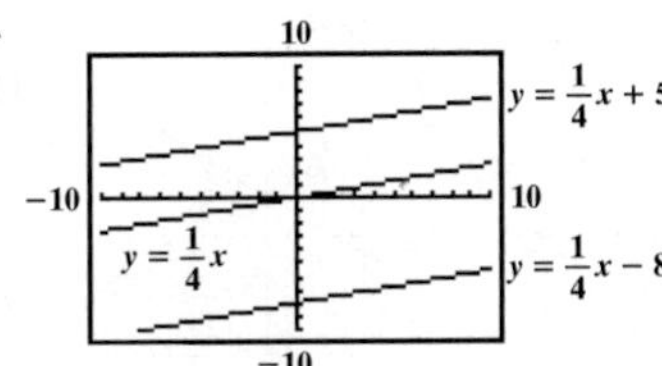

Vocabulary, Readiness & Video Check 3.4

1. slope **3.** 0 **5.** positive **7.** *y*; *x* **9.** solve the equation for *y*; the slope is the coefficient of *x*. **11.** slope-intercept form; this form makes the slope easy to see, and you need to compare slopes to determine if two lines are parallel or perpendicular.

Exercise Set 3.4

1. −1 **3.** undefined **5.** $-\frac{2}{3}$ **7.** 0 **9.** $m = -\frac{4}{3}$ **11.** undefined slope **13.** $m = \frac{5}{2}$ **15.** negative **17.** undefined **19.** upward **21.** horizontal **23.** line 1 **25.** line 2 **27.** D **29.** B **31.** E **33.** undefined slope **35.** $m = 0$ **37.** undefined slope **39.** $m = 0$ **41.** $m = 5$ **43.** $m = -0.3$ **45.** $m = -2$ **47.** $m = \frac{2}{3}$ **49.** undefined slope **51.** $m = \frac{1}{2}$ **53.** $m = 0$ **55.** $m = -\frac{3}{4}$ **57.** $m = 4$ **59. a.** 1 **b.** −1 **61. a.** $\frac{9}{11}$ **b.** $-\frac{11}{9}$ **63.** neither **65.** neither **67.** parallel **69.** perpendicular **71.** $\frac{3}{5}$ **73.** 12.5% **75.** 40% **77.** 37%; 35% **79.** $m = \frac{1}{1}$ or 1; Every 1 year, there are 1 million more U.S. households with televisions. **81.** $m = 0.47$; It costs $0.47 per 1 mile to own and operate a compact car. **83.** $y = 2x - 14$ **85.** $y = -6x - 11$ **87.** $m = \frac{1}{2}$ **89.** answers may vary **91.** 2005 to 2006 **93.** 2000; 28.5 mi per gallon **95.** from 2008 to 2009 **97.** $x = 6$ **99. a.** (2007, 2207); (2010, 2333) **b.** 42 **c.** For the years 2007 through 2010, the number of heart transplants increased at a rate of 42 per year. **101.** The slope through (−3, 0) and (1, 1) is $\frac{1}{4}$. The slope through (−3, 0) and (−4, 4) is −4. The product of the slopes is −1, so the sides are perpendicular. **103.** −0.25 **105.** 0.875 **107.** The line becomes steeper.

Integrated Review

1. $m = 2$ **2.** $m = 0$ **3.** $m = -\frac{2}{3}$ **4.** undefined slope **5.** 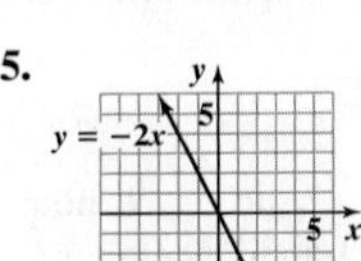

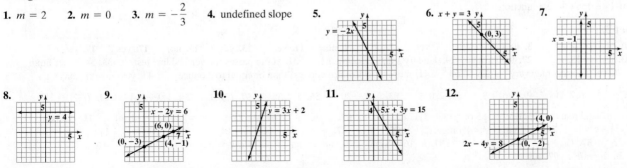

6. $x + y = 3$ 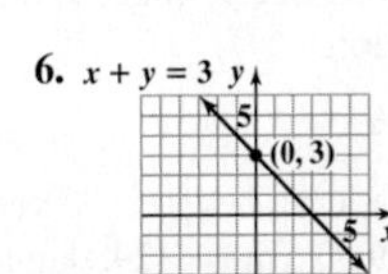

7. 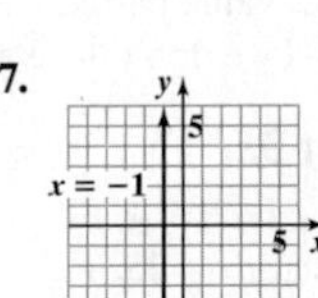

8. 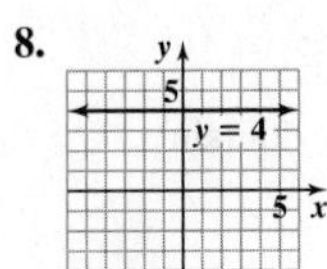

9. 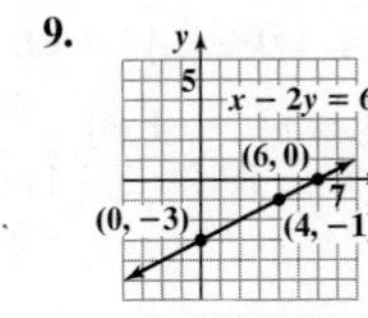

10. 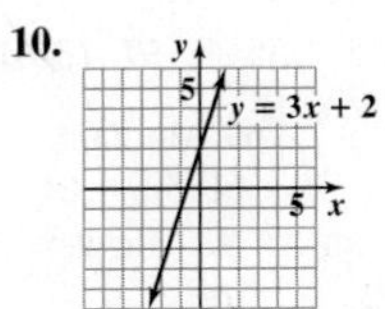

11. 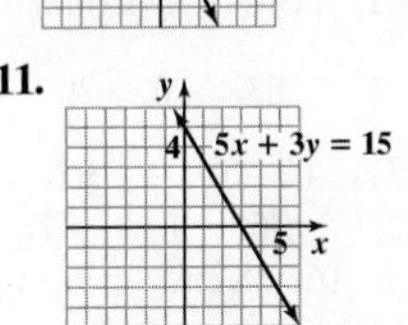

12. 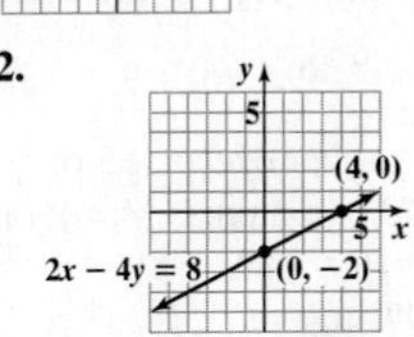

13. parallel **14.** neither **15. a.** (0, 587) **b.** In 2000, there were 587 thousand bridges on public roads. **c.** 1.7 **d.** For the years 2000 through 2009, the number of bridges on public roads increased at a rate of 1.7 thousand per year. **16. a.** (9, 26.6) **b.** In 2009, the revenue for online advertising was $26.6 billion.

Section 3.5

Practice Exercises

1. 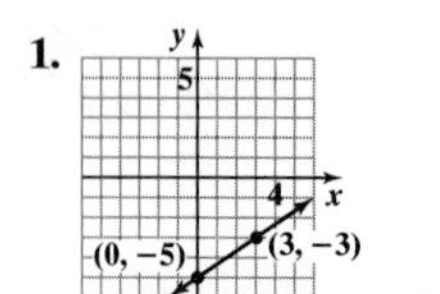

2. 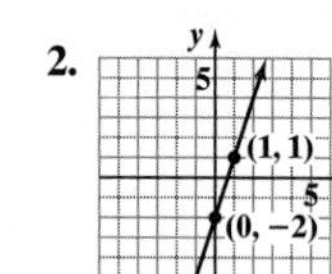

3. $y = \frac{1}{2}x + 7$ **4.** $4x - y = 5$ **5.** $5x + 4y = 19$ **6.** $x = 3$ **7.** $y = 3$ **8. a.** $y = -1500x + 195{,}000$ **b.** $105,000

Calculator Explorations 3.5

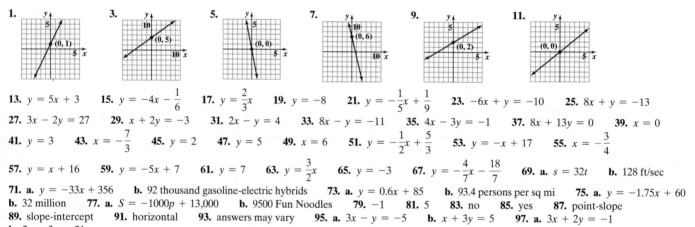

1. 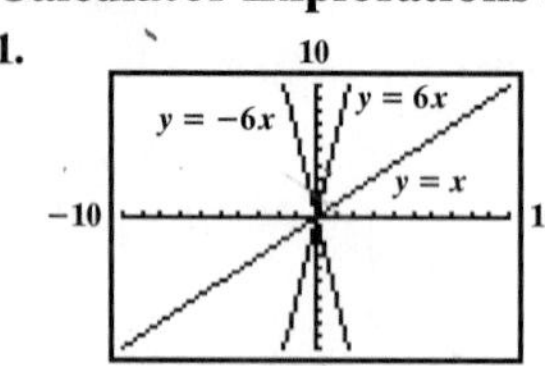

3. 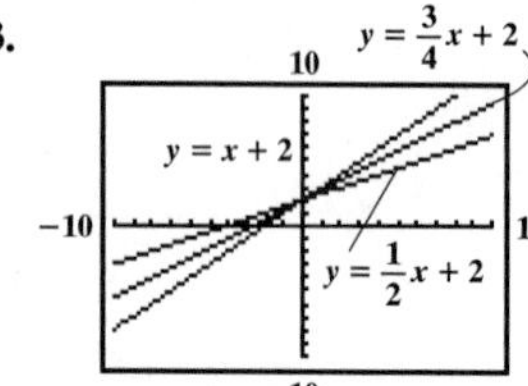

5. 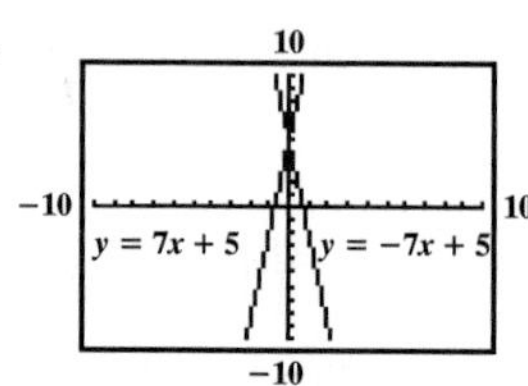

Vocabulary, Readiness & Video Check 3.5

1. slope-intercept; m; b **3.** y-intercept; fraction **5.** Write the equation with x- and y- terms on one side of the equal sign and a constant on the other side. **7.** Example 6: $y = -3$; Example 7: $x = -2$

Exercise Set 3.5

1. 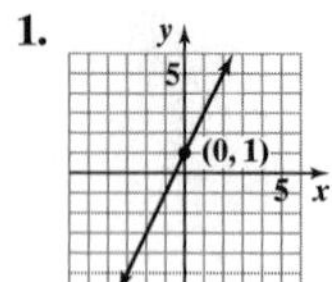

3. 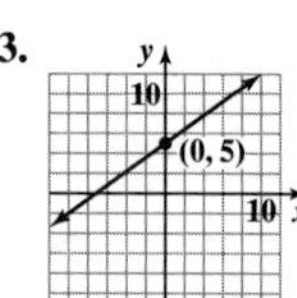

5. 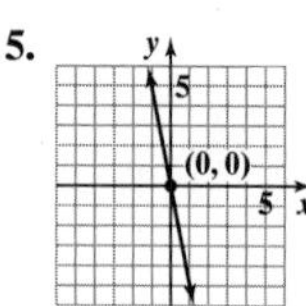

7. 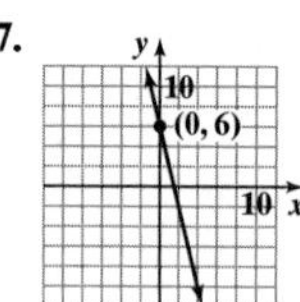

9. 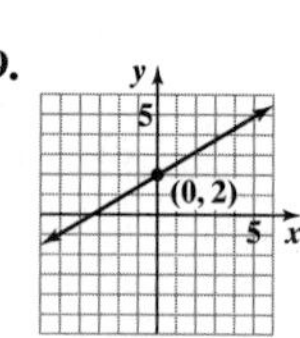

11. 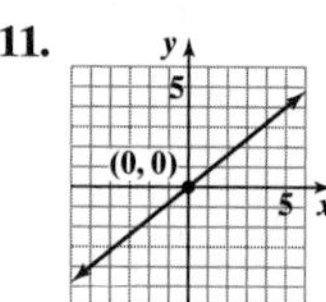

13. $y = 5x + 3$ **15.** $y = -4x - \frac{1}{6}$ **17.** $y = \frac{2}{3}x$ **19.** $y = -8$ **21.** $y = -\frac{1}{5}x + \frac{1}{9}$ **23.** $-6x + y = -10$ **25.** $8x + y = -13$ **27.** $3x - 2y = 27$ **29.** $x + 2y = -3$ **31.** $2x - y = 4$ **33.** $8x - y = -11$ **35.** $4x - 3y = -1$ **37.** $8x + 13y = 0$ **39.** $x = 0$ **41.** $y = 3$ **43.** $x = -\frac{7}{3}$ **45.** $y = 2$ **47.** $y = 5$ **49.** $x = 6$ **51.** $y = -\frac{1}{2}x + \frac{5}{3}$ **53.** $y = -x + 17$ **55.** $x = -\frac{3}{4}$ **57.** $y = x + 16$ **59.** $y = -5x + 7$ **61.** $y = 7$ **63.** $y = \frac{3}{2}x$ **65.** $y = -3$ **67.** $y = -\frac{4}{7}x - \frac{18}{7}$ **69. a.** $s = 32t$ **b.** 128 ft/sec **71. a.** $y = -33x + 356$ **b.** 92 thousand gasoline-electric hybrids **73. a.** $y = 0.6x + 85$ **b.** 93.4 persons per sq mi **75. a.** $y = -1.75x + 60$ **b.** 32 million **77. a.** $S = -1000p + 13{,}000$ **b.** 9500 Fun Noodles **79.** −1 **81.** 5 **83.** no **85.** yes **87.** point-slope **89.** slope-intercept **91.** horizontal **93.** answers may vary **95. a.** $3x - y = -5$ **b.** $x + 3y = 5$ **97. a.** $3x + 2y = -1$ **b.** $2x - 3y = 21$

Section 3.6

Practice Exercises

1. Domain: {0, 1, 5}: Range: {−2, 0, 3, 4} **2. a.** function **b.** not a function **3. a.** not a function **b.** function **4. a.** function **b.** function **c.** function **d.** not a function **5. a.** function **b.** function **c.** function **d.** not a function **6. a.** 69°F **b.** November **c.** yes **7. a.** $h(2) = 9$; (2, 9) **b.** $h(-5) = 30$; (−5, 30) **c.** $h(0) = 5$; (0, 5) **8. a.** domain: $(-\infty, \infty)$ **b.** domain: $(-\infty, 0) \cup (0, \infty)$ **9. a.** domain: $[-4, 6]$; range: $[-2, 3]$ **b.** domain: $(-\infty, \infty)$; range: $(-\infty, 3]$

Vocabulary, Readiness & Video Check 3.6

1. relation **3.** range **5.** vertical **7.** A relation is a set of ordered pairs and an equation in two variables defines a set of ordered pairs. Therefore, an equation in two variables can also define a relation. **9.** A vertical line represents one x-value paired with many y-values. A function only allows an x-value paired with exactly one y-value, so if a vertical line intersects a graph more than once, there's an x-value paired with more than one y-value, and we don't have a function.

Exercise Set 3.6

1. {−7, 0, 2, 10}; {−7, 0, 4, 10} **3.** {0, 1, 5}; {−2} **5.** yes **7.** no **9.** no **11.** yes **13.** yes **15.** no **17.** yes **19.** yes **21.** yes **23.** no **25.** no **27.** 9:30 p.m. **29.** January 1 and December 1 **31.** yes; it passes the vertical line test **33.** $4.25 per hour **35.** 2009 **37.** yes; answers may vary **39.** $1.50 **41.** more than 1 ounce and less than or equal to 2 ounces **43.** yes; answers may vary **45.** −9, −5, 1 **47.** 6, 2, 11 **49.** −6, 0, 9 **51.** 2, 0, 3 **53.** 5, 0, −20 **55.** 5, 3, 35 **57.** (3, 6) **59.** $\left(0, -\frac{1}{2}\right)$ **61.** (−2, 9) **63.** $(-\infty, \infty)$ **65.** all real number except −5 or $(-\infty, -5) \cup (-5, \infty)$ **67.** $(-\infty, \infty)$ **69.** domain: $(-\infty, \infty)$; range: $[-4, \infty)$ **71.** domain: $(-\infty, \infty)$; range: $(-\infty, \infty)$ **73.** domain: $(-\infty, \infty)$; range: {2} **75.** −1 **77.** −1 **79.** −1, 5 **81.** (−2, 1) **83.** (−3, −1) **85.** $f(-5) = 12$ **87.** (3, −4) **89.** $f(5) = 0$ **91. a.** 166.38 cm **b.** 148.25 cm **93.** answers may vary **95.** $f(x) = x + 7$ **97. a.** 11 **b.** $2a + 7$ **99. a.** 16 **b.** $a^2 + 7$

Chapter 3 Vocabulary Check

1. solution **2.** y-axis **3.** linear **4.** x-intercept **5.** standard **6.** y-intercept **7.** slope-intercept **8.** point-slope **9.** y **10.** x-axis **11.** x **12.** slope **13.** function **14.** domain **15.** range **16.** relation

Chapter 3 Review

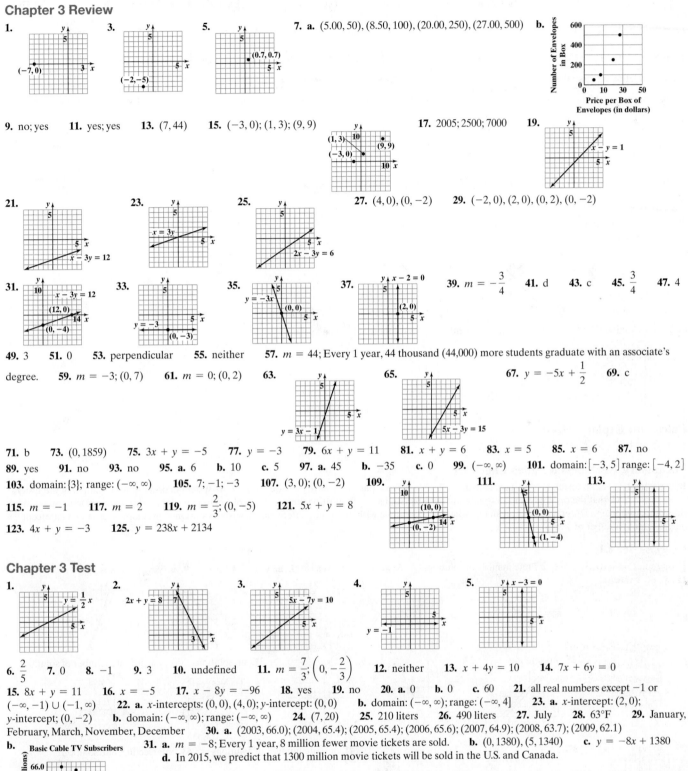

1. **3.** **5.** **7. a.** (5.00, 50), (8.50, 100), (20.00, 250), (27.00, 500) **b.**

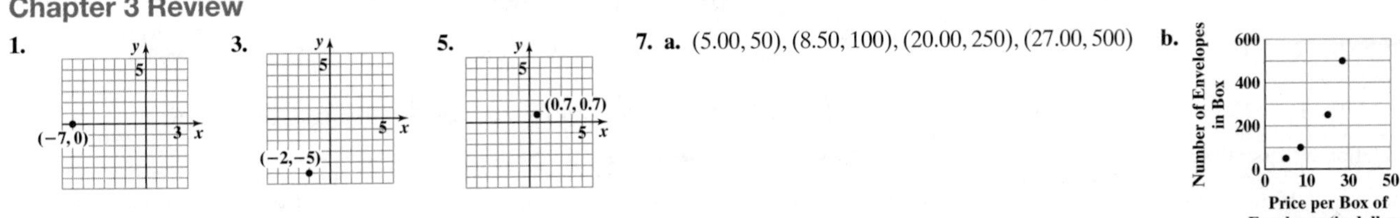

9. no; yes **11.** yes; yes **13.** (7, 44) **15.** (−3, 0); (1, 3); (9, 9) **17.** 2005; 2500; 7000 **19.**

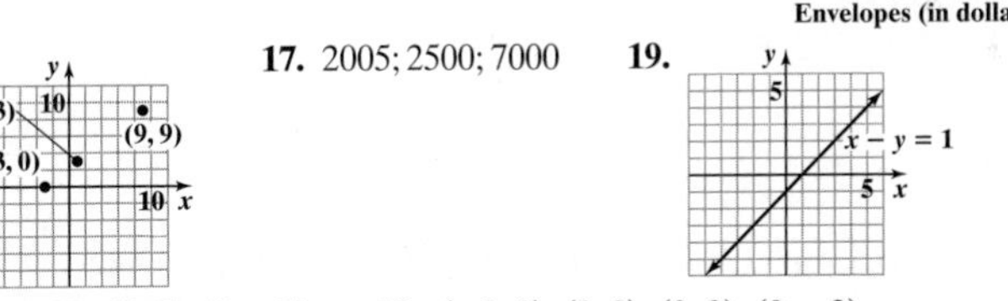

21. **23.** **25.** **27.** (4, 0), (0, −2) **29.** (−2, 0), (2, 0), (0, 2), (0, −2)

31. **33.** **35.** **37.** **39.** $m = -\frac{3}{4}$ **41.** d **43.** c **45.** $\frac{3}{4}$ **47.** 4

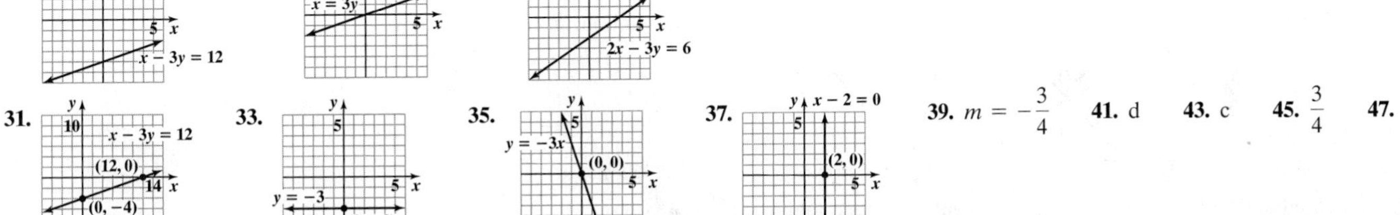

49. 3 **51.** 0 **53.** perpendicular **55.** neither **57.** $m = 44$; Every 1 year, 44 thousand (44,000) more students graduate with an associate's degree. **59.** $m = -3$; (0, 7) **61.** $m = 0$; (0, 2) **63.** **65.** **67.** $y = -5x + \frac{1}{2}$ **69.** c

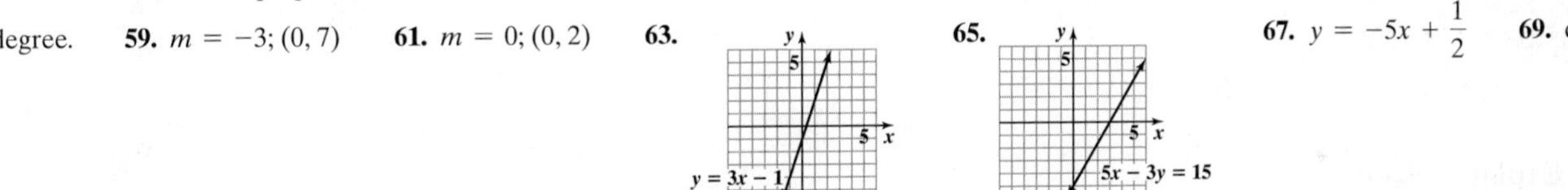

71. b **73.** (0, 1859) **75.** $3x + y = -5$ **77.** $y = -3$ **79.** $6x + y = 11$ **81.** $x + y = 6$ **83.** $x = 5$ **85.** $x = 6$ **87.** no **89.** yes **91.** no **93.** no **95. a.** 6 **b.** 10 **c.** 5 **97. a.** 45 **b.** −35 **c.** 0 **99.** $(-\infty, \infty)$ **101.** domain: $[-3, 5]$ range: $[-4, 2]$ **103.** domain: {3}; range: $(-\infty, \infty)$ **105.** 7; −1; −3 **107.** (3, 0); (0, −2) **109.** **111.** **113.** **115.** $m = -1$ **117.** $m = 2$ **119.** $m = \frac{2}{3}$; (0, −5) **121.** $5x + y = 8$ **123.** $4x + y = -3$ **125.** $y = 238x + 2134$

Chapter 3 Test

1. **2.** **3.** **4.** **5.**

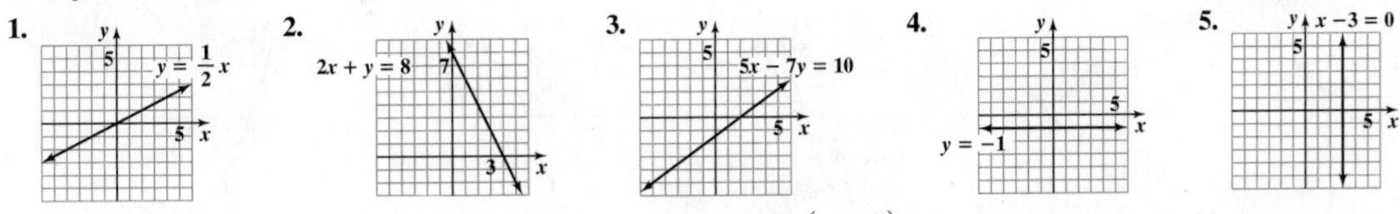

6. $\frac{2}{5}$ **7.** 0 **8.** −1 **9.** 3 **10.** undefined **11.** $m = \frac{7}{3}; \left(0, -\frac{2}{3}\right)$ **12.** neither **13.** $x + 4y = 10$ **14.** $7x + 6y = 0$ **15.** $8x + y = 11$ **16.** $x = -5$ **17.** $x - 8y = -96$ **18.** yes **19.** no **20. a.** 0 **b.** 0 **c.** 60 **21.** all real numbers except −1 or $(-\infty, -1) \cup (-1, \infty)$ **22. a.** x-intercepts: (0, 0), (4, 0); y-intercept: (0, 0) **b.** domain: $(-\infty, \infty)$; range: $(-\infty, 4]$ **23. a.** x-intercept: (2, 0); y-intercept; (0, −2) **b.** domain: $(-\infty, \infty)$; range: $(-\infty, \infty)$ **24.** (7, 20) **25.** 210 liters **26.** 490 liters **27.** July **28.** 63°F **29.** January, February, March, November, December **30. a.** (2003, 66.0); (2004, 65.4); (2005, 65.4); (2006, 65.6); (2007, 64.9); (2008, 63.7); (2009, 62.1) **b.**

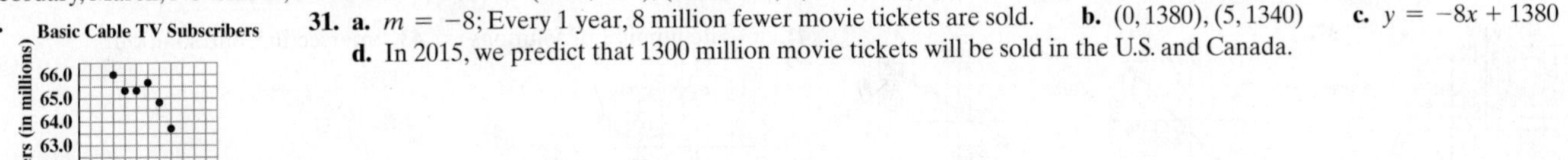

31. a. $m = -8$; Every 1 year, 8 million fewer movie tickets are sold. **b.** (0, 1380), (5, 1340) **c.** $y = -8x + 1380$ **d.** In 2015, we predict that 1300 million movie tickets will be sold in the U.S. and Canada.

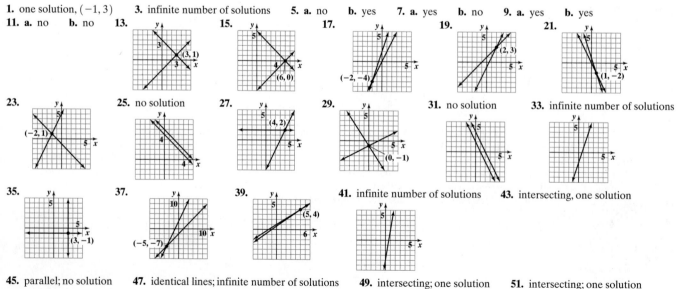

Chapter 3 Cumulative Review

1. a. $<$ **b.** $>$ **c.** $>$; Sec. 1.2, Ex. 1 **3.** $\frac{2}{39}$; Sec. 1.3, Ex. 3 **5.** $\frac{8}{3}$; Sec. 1.4, Ex. 3 **7. a.** -19 **b.** 30 **c.** -0.5 **d.** $-\frac{4}{5}$ **e.** 6.7 **f.** $\frac{1}{40}$; Sec. 1.5, Ex. 6 **9. a.** -6; **b.** 6.3; Sec. 1.6, Ex. 4 **11. a.** -6 **b.** 0 **c.** $\frac{3}{4}$; Sec. 1.7, Ex. 10 **13. a.** $22 + x$ **b.** $-21x$; Sec. 1.8, Ex. 3 **15. a.** -3 **b.** 22 **c.** 1 **d.** -1 **e.** $\frac{1}{7}$; Sec. 2.1, Ex. 1 **17.** -1.6; Sec. 2.2, Ex. 2 **19.** $\frac{15}{4}$; Sec. 2.3, Ex. 5 **21.** $3x + 3$; Sec. 2.3, Ex. 10 **23.** 0; Sec. 2.4, Ex. 4 **25.** 242 Republicans, 193 Democrats; Sec. 2.5, Ex. 4 **27.** 40 ft; Sec. 2,6, Ex. 2 **29.** $\frac{y - b}{m} = x$; Sec. 2.6, Ex. 6 **31.** 40% solution: 8 liters; 70% solution: 4 liters; Sec. 2.7, Ex. 7 **33.** 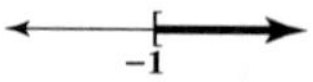 $[-1, \infty)$; Sec. 2.9, Ex. 1 **35.**  $[1, 4)$; Sec. 2.9, Ex. 10 **37. a.** solution **b.** not a solution **c.** solution; Sec. 3.1, Ex. 5 **39. a.** yes **b.** yes **c.** no **d.** yes; Sec. 3.2, Ex. 1 **41.** 0; Sec. 3.4, Ex. 6 **43.** $y = \frac{1}{4}x - 3$; Sec. 3.5, Ex. 3

CHAPTER 4 SOLVING SYSTEMS OF LINEAR EQUATIONS AND INEQUALITIES

Section 4.1

Practice Exercises

1. no **2.** yes **3.** $(8, 5)$

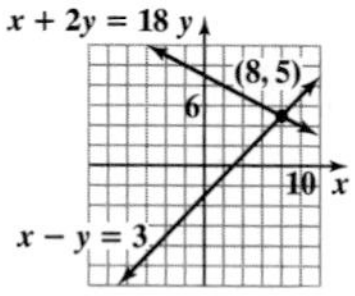

4. $(-3, -5)$

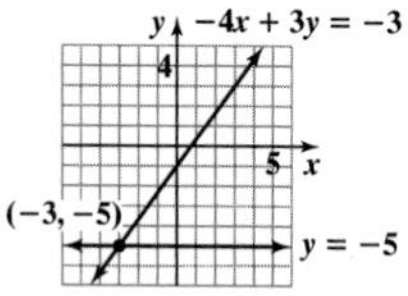

5. no solution; inconsistent system

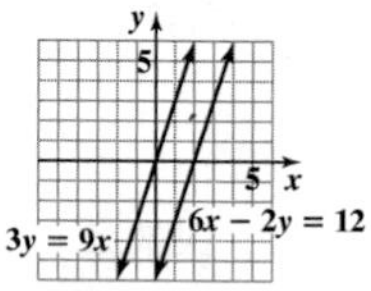

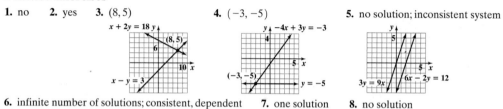

6. infinite number of solutions; consistent, dependent

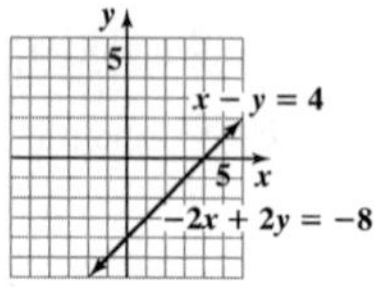

7. one solution **8.** no solution

Calculator Explorations 4.1

1. $(0.37, 0.23)$ **3.** $(0.03, -1.89)$

Vocabulary, Readiness & Video Check 4.1

1. dependent **3.** consistent **5.** inconsistent **7.** The ordered pair must satisfy all equations of the system in order to be a solution of the system, so we must check that the ordered pair is a solution of both equations. **9.** Writing the equations of a system in slope-intercept form lets you see their slope and y-intercept. Different slopes mean one solution; same slope with different y-intercepts means no solution; same slope with same y-intercept means infinite number of solutions.

Exercise Set 4.1

1. one solution, $(-1, 3)$ **3.** infinite number of solutions **5. a.** no **b.** yes **7. a.** yes **b.** no **9. a.** yes **b.** yes **11. a.** no **b.** no

13. 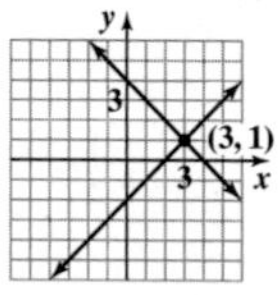

15. 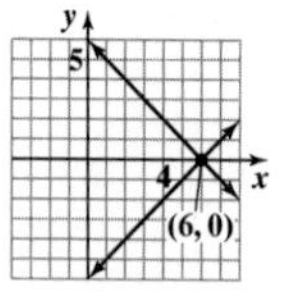

17. 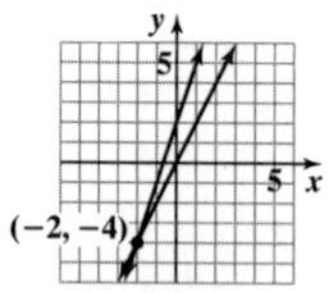

19. 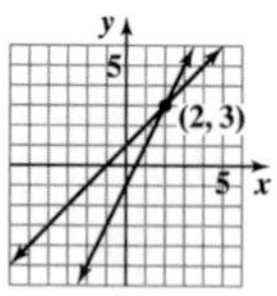

21. 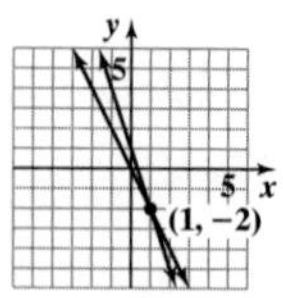

23. 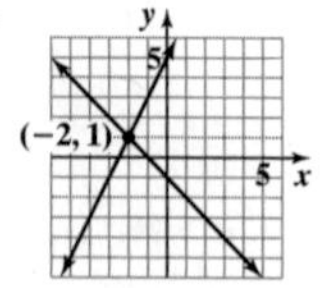

25. no solution 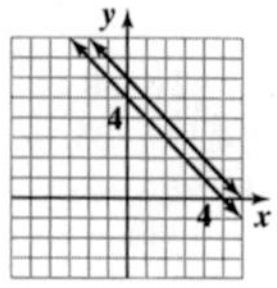

27. 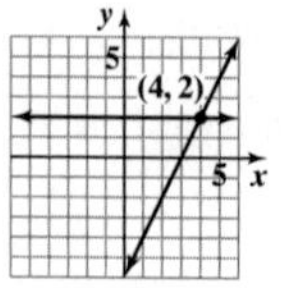

29. 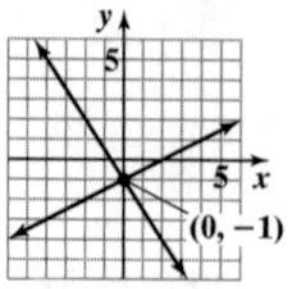

31. no solution 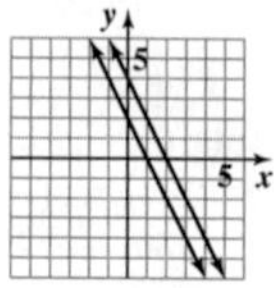

33. infinite number of solutions 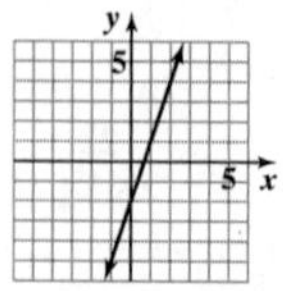

35. 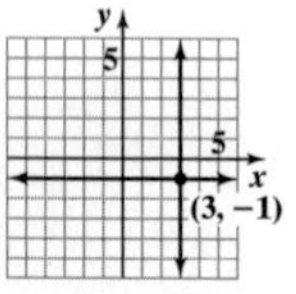

37. 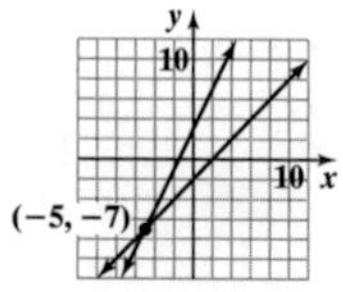

39. 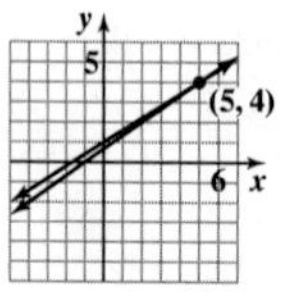

41. infinite number of solutions 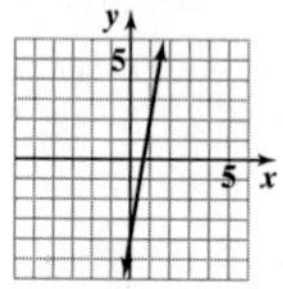

43. intersecting, one solution **45.** parallel; no solution **47.** identical lines; infinite number of solutions **49.** intersecting; one solution **51.** intersecting; one solution

53. identical lines; infinite number of solutions **55.** parallel; no solution **57.** 2 **59.** $-\frac{2}{5}$ **61.** 2

63. answers may vary; possible answer

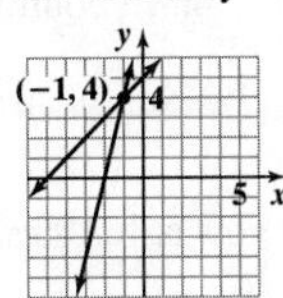

65. answers may vary; possible answer

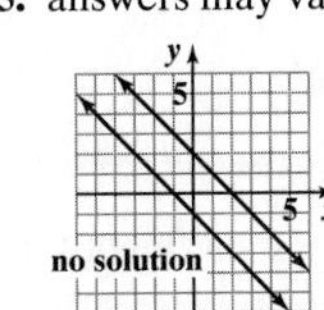

67. answers may vary **69.** 1988–1989; 2001–2002

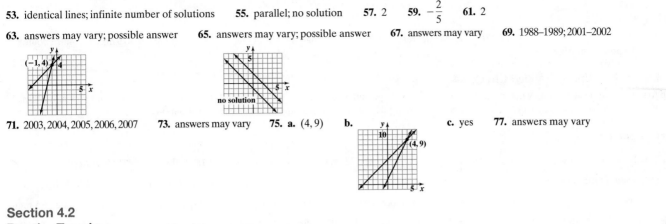

71. 2003, 2004, 2005, 2006, 2007 **73.** answers may vary **75. a.** $(4, 9)$ **b.**

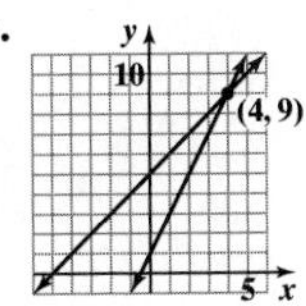

c. yes **77.** answers may vary

Section 4.2

Practice Exercises

1. $(8, 7)$ **2.** $(-3, -6)$ **3.** $\left(4, \frac{2}{3}\right)$ **4.** $(-3, 2)$ **5.** infinite number of solutions **6.** no solution

Vocabulary, Readiness & Video Check 4.2

1. $(1, 4)$ **3.** infinite number of solutions **5.** $(0, 0)$ **7.** You solved one equation for a variable. Now be sure to substitute this expression for the variable into the *other* equation.

Exercise Set 4.2

1. $(2, 1)$ **3.** $(-3, 9)$ **5.** $(2, 7)$ **7.** $\left(-\frac{1}{5}, \frac{43}{5}\right)$ **9.** $(2, -1)$ **11.** $(-2, 4)$ **13.** $(4, 2)$ **15.** $(-2, -1)$ **17.** no solution **19.** $(3, -1)$ **21.** $(3, 5)$ **23.** $\left(\frac{2}{3}, -\frac{1}{3}\right)$ **25.** $(-1, -4)$ **27.** $(-6, 2)$ **29.** $(2, 1)$ **31.** no solution **33.** infinite number of solutions **35.** $\left(\frac{1}{2}, 2\right)$ **37.** $(1, -3)$ **39.** $-6x - 4y = -12$ **41.** $-12x + 3y = 9$ **43.** $5n$ **45.** $-15b$ **47.** answers may vary **49.** no; answers may vary **51. c;** answers may vary **53. a.** $(13, 492)$ **b.** In $1970 + 13 = 1983$, the number of men and women receiving bachelor's degrees was the same. **c.** answers may vary

y = 14.2x + 314
y = 3.9x + 443

55. $(-2.6, 1.3)$ **57.** $(3.28, 2.11)$

Section 4.3

Practice Exercises

1. $(5, 3)$ **2.** $(3, -4)$ **3.** no solution **4.** infinite number of solutions **5.** $(2, 2)$ **6.** $\left(-\frac{8}{5}, \frac{6}{5}\right)$

Vocabulary, Readiness & Video Check 4.3

1. false **3.** true **5.** The multiplication property of equality; be sure to multiply *both* sides of the equation by the number chosen.

Exercise Set 4.3

1. $(1, 2)$ **3.** $(2, -3)$ **5.** $(-2, -5)$ **7.** $(5, -2)$ **9.** $(-7, 5)$ **11.** $(6, 0)$ **13.** no solution **15.** infinite number of solutions **17.** $\left(2, -\frac{1}{2}\right)$ **19.** $(-2, 0)$ **21.** $(1, -1)$ **23.** infinite number of solutions **25.** $\left(\frac{12}{11}, -\frac{4}{11}\right)$ **27.** $\left(\frac{3}{2}, 3\right)$ **29.** infinite number of solutions **31.** $(1, 6)$ **33.** $\left(-\frac{1}{2}, -2\right)$ **35.** infinite number of solutions **37.** $\left(-\frac{2}{3}, \frac{2}{5}\right)$ **39.** $(2, 4)$ **41.** $(-0.5, 2.5)$ **43.** $(2, 5)$ **45.** $(-3, 2)$ **47.** $(0, 3)$ **49.** $(5, 7)$ **51.** $\left(\frac{1}{3}, 1\right)$ **53.** infinite number of solutions **55.** $(-8.9, 10.6)$ **57.** $2x + 6 = x - 3$ **59.** $20 - 3x = 2$ **61.** $4(n + 6) = 2n$ **63.** 2; $6x - 2y = -24$ **65. b;** answers may vary **67.** answers may vary **69. a.** $b = 15$ **b.** any real number except 15 **71.** $(-4.2, 9.6)$ **73. a.** $(7, 22)$ **b.** In 2015 $(2008 + 7)$, the percent of workers age 25–34 and the percent of workers age 55 and older will be the same. **c.** 22% of the workforce for each of these age groups.

Integrated Review

1. $(2, 5)$ **2.** $(4, 2)$ **3.** $(5, -2)$ **4.** $(6, -14)$ **5.** $(-3, 2)$ **6.** $(-4, 3)$ **7.** $(0, 3)$ **8.** $(-2, 4)$ **9.** $(5, 7)$ **10.** $(-3, -23)$ **11.** $\left(\frac{1}{3}, 1\right)$ **12.** $\left(-\frac{1}{4}, 2\right)$ **13.** no solution **14.** infinite number of solutions **15.** $(0.5, 3.5)$ **16.** $(-0.75, 1.25)$ **17.** infinite number of solutions **18.** no solution **19.** $(7, -3)$ **20.** $(-1, -3)$ **21.** answers may vary **22.** answers may vary

Section 4.4

Practice Exercises

1. 18, 12 **2. a.** Adult: $19 **b.** Child: $6 **c.** No, the regular rates are less than the group rate. **3.** 1.75 mph and 3.75 mph **4.** 15 pounds of Kona and 5 pounds of Blue Mountain

Vocabulary, Readiness & Video Check 4.4

1. Up to now we've been working with one variable/unknown and one equation. Because systems involve two equations with two unknowns, for these applications, you need to choose two variables to represent two unknowns and translate the problem into two equations.

Exercise Set 4.4

1. c **3.** b **5.** a **7.** $\begin{cases} x + y = 15 \\ x - y = 7 \end{cases}$ **9.** $\begin{cases} x + y = 6500 \\ x = y + 800 \end{cases}$ **11.** 33 and 50 **13.** 14 and −3 **15.** Cabrera: 126; Rodriguez: 125 **17.** child's ticket: $18; adult's ticket: $29 **19.** quarters: 53; nickels: 27 **21.** McDonald's: $73.50; The Ohio Art Company: $3.50 **23.** daily fee: $32; mileage charge: $0.25 per mi **25.** distance downstream = distance upstream = 18 mi; time downstream: 2 hr; time upstream: $4\frac{1}{2}$ hr; still water: 6.5 mph; current: 2.5 mph **27.** still air: 455 mph; wind: 65 mph **29.** $4\frac{1}{2}$ hr **31.** 12% solution: $7\frac{1}{2}$ liters; 4% solution: $4\frac{1}{2}$ liters **33.** $4.95 beans: 113 lb; $2.65 beans: 87 lb **35.** 60°, 30° **37.** 20°, 70° **39.** number sold at $9.50: 23; number sold at $7.50: 67 **41.** $2\frac{1}{4}$ mph and $2\frac{3}{4}$ mph **43.** 30%: 50 gal; 60%: 100 gal **45.** length: 42 in.; width: 30 in. **47.** $(3, \infty)$ **49.** $\left[\frac{1}{2}, \infty\right)$ **51.** a **53.** width: 9 ft; length: 15 ft **55. a.** (27.1, 39.5) **b.** For viewers 27.1 years over 18 (or 45.1 years old) the percent who watch cable news and network news is the same, 39.5%. **c.** answers may vary

Section 4.5

Practice Exercises

1. a. no **b.** yes **2.**

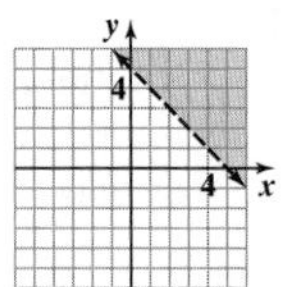

3.

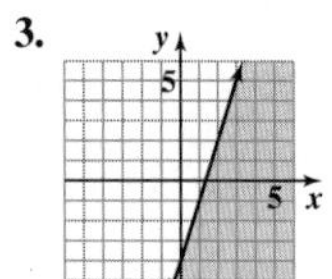

4.

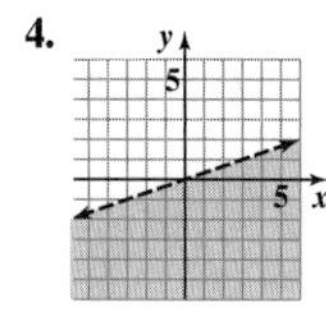

5.

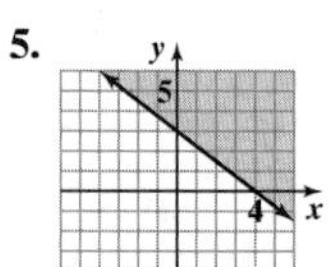

6.

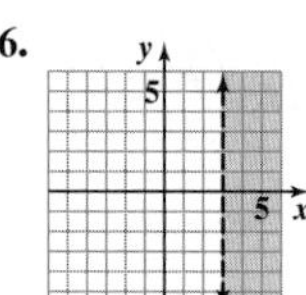

7.

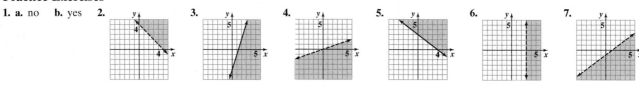

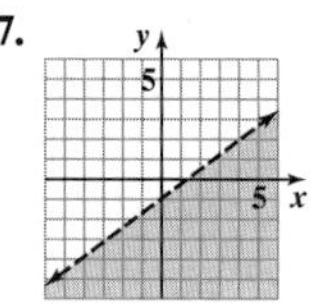

Vocabulary, Readiness & Video Check 4.5

1. linear inequality in two variables **3.** false **5.** true **7.** An ordered pair is a solution of an inequality if replacing the variables with the coordinates of the ordered pair results in a true statement.

Exercise Set 4.5

1. no; yes **3.** no; no **5.** no; yes **7.**

9. **11.** **13.**

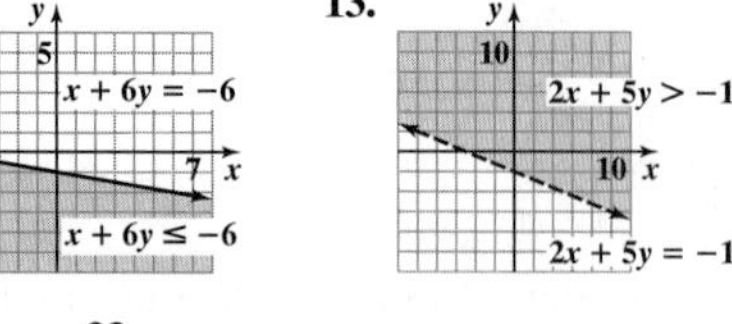

15.

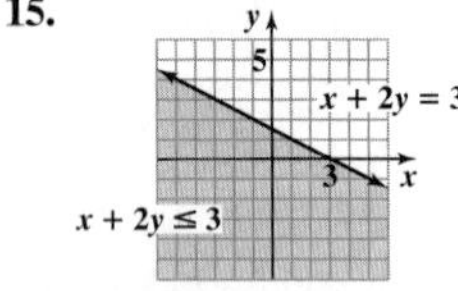

17.

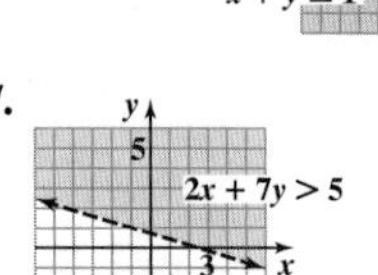

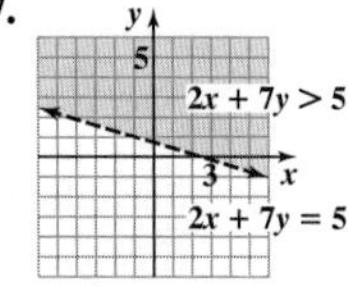

19.

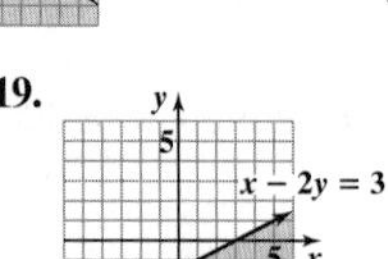

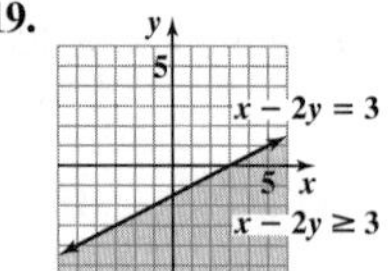

21.

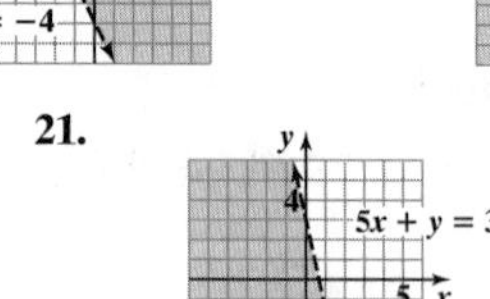

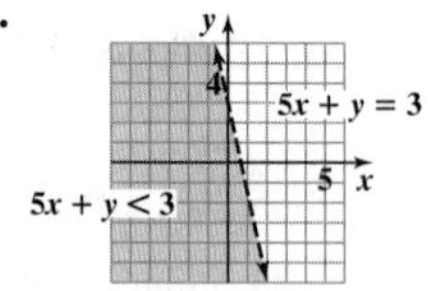

23.

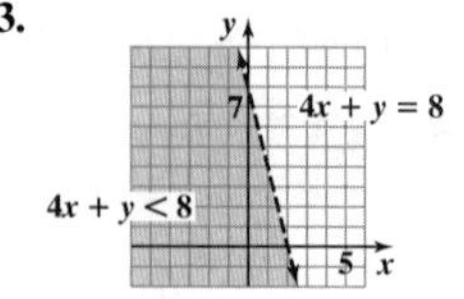

25.

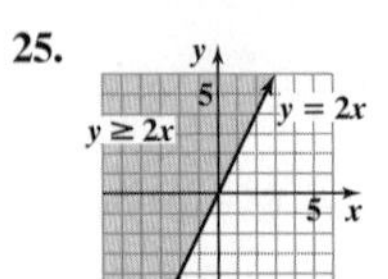

27.

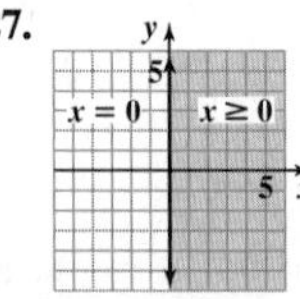

29.

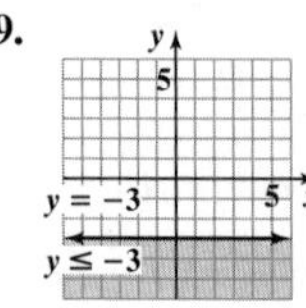

31.

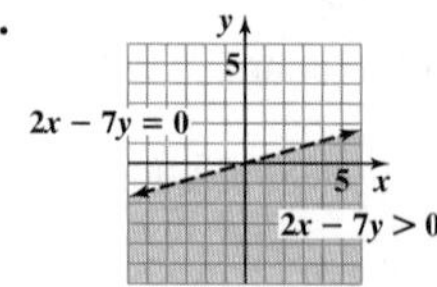

33.

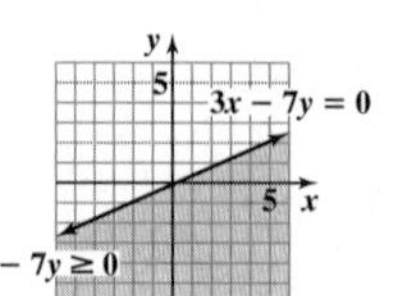

35.

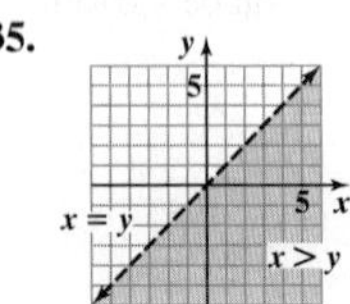

37.

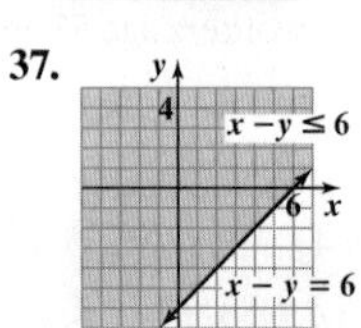

39.

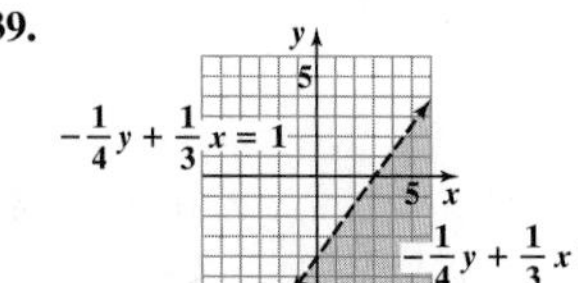

41.

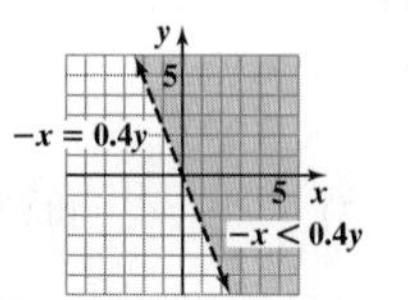

43. e **45.** c **47.** f

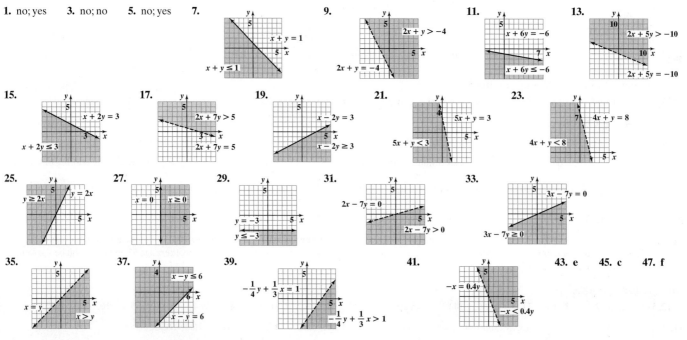

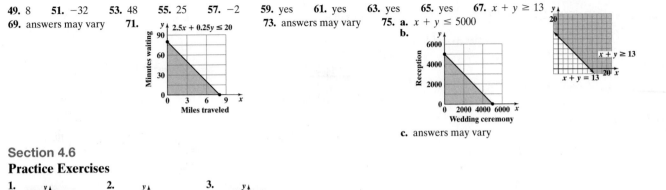

49. 8 **51.** −32 **53.** 48 **55.** 25 **57.** −2 **59.** yes **61.** yes **63.** yes **65.** yes **67.** $x + y \geq 13$

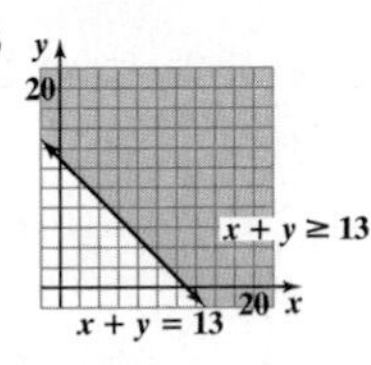

69. answers may vary **71.** $2.5x + 0.25y \leq 20$ (Minutes waiting vs. Miles traveled) **73.** answers may vary **75. a.** $x + y \leq 5000$ **b.**

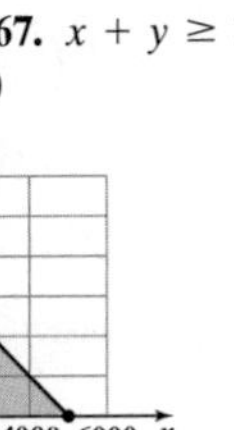

c. answers may vary

Section 4.6

Practice Exercises

1.

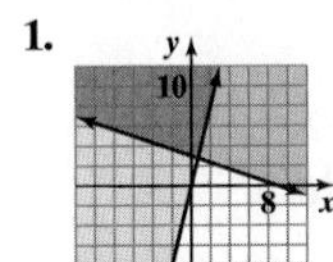

2.

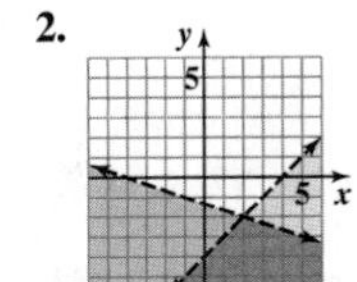

3.

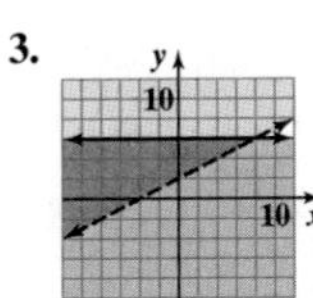

Vocabulary, Readiness & Video Check 4.6

1. No; we can choose any test point except a point on the second inequality's own boundary line.

Exercise Set 4.6

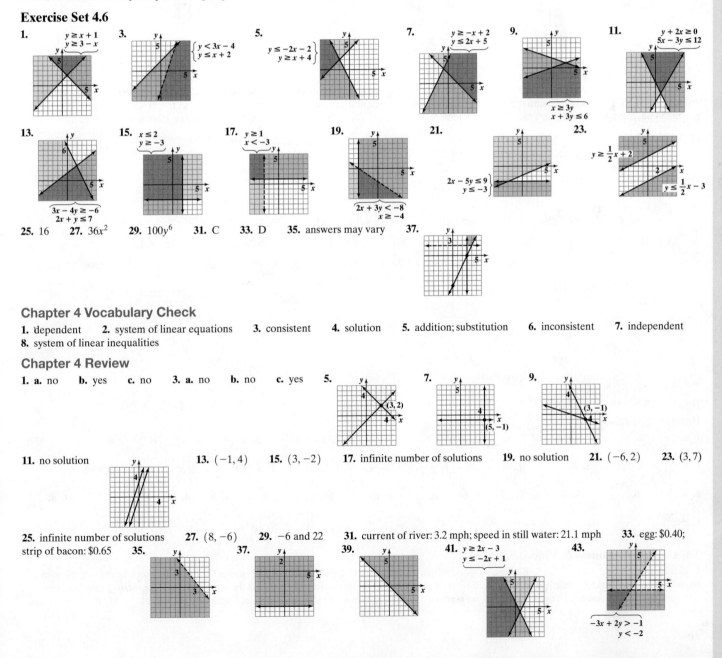

1. $y \geq x + 1$, $y \geq 3 - x$ **3.**

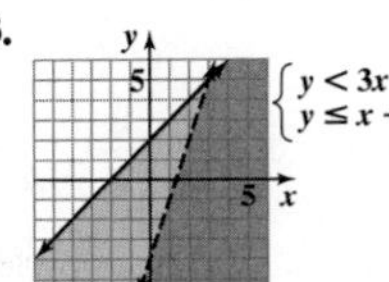

5.

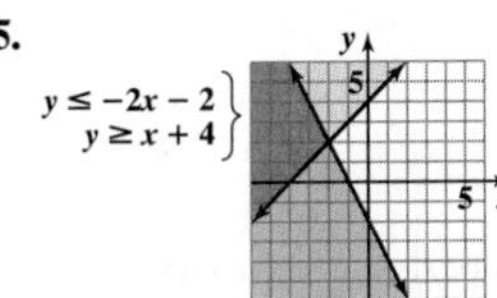

7.

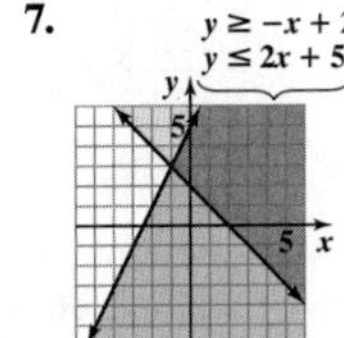

9.

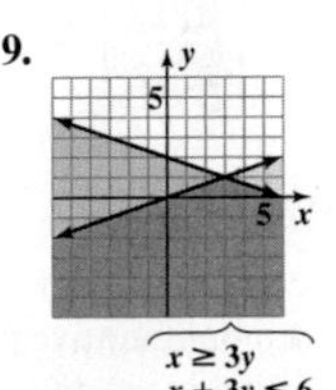

11.

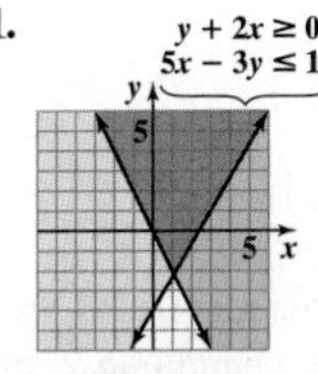

13.

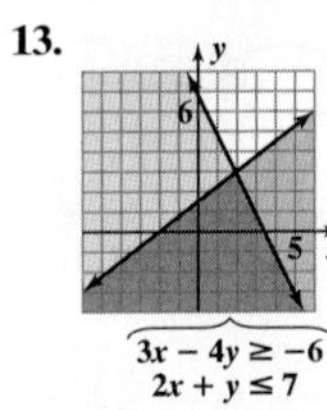

15.

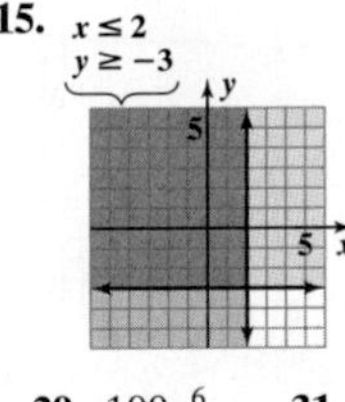

17.

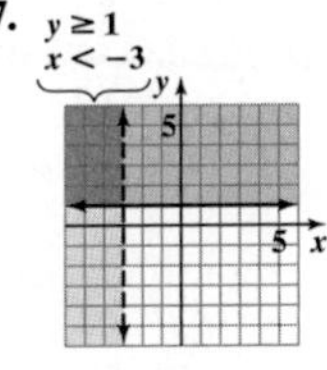

19.

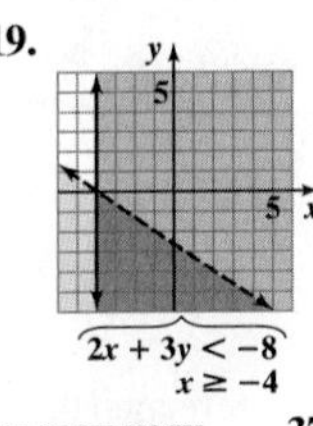

21.

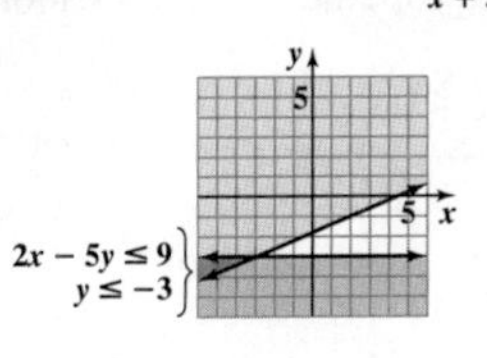

23.

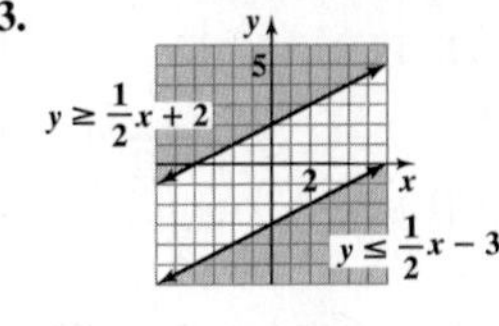

25. 16 **27.** $36x^2$ **29.** $100y^6$ **31.** C **33.** D **35.** answers may vary **37.**

Chapter 4 Vocabulary Check

1. dependent **2.** system of linear equations **3.** consistent **4.** solution **5.** addition; substitution **6.** inconsistent **7.** independent **8.** system of linear inequalities

Chapter 4 Review

1. a. no **b.** yes **c.** no **3. a.** no **b.** no **c.** yes **5.** (3, 2) **7.**

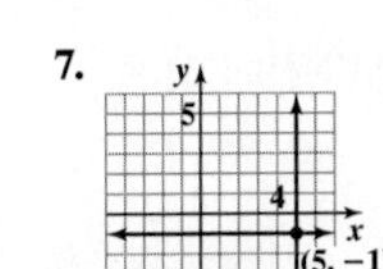

9.

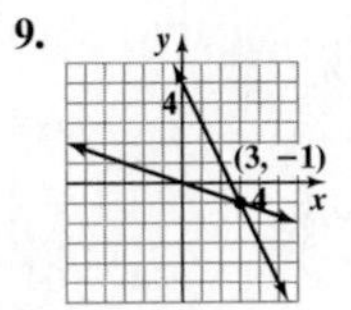

11. no solution

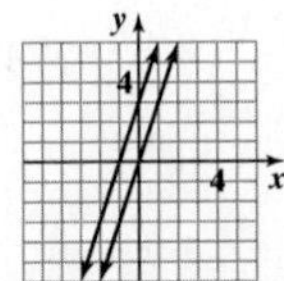

13. $(-1, 4)$ **15.** $(3, -2)$ **17.** infinite number of solutions **19.** no solution **21.** $(-6, 2)$ **23.** $(3, 7)$

25. infinite number of solutions **27.** $(8, -6)$ **29.** −6 and 22 **31.** current of river: 3.2 mph; speed in still water: 21.1 mph **33.** egg: \$0.40; strip of bacon: \$0.65 **35.**

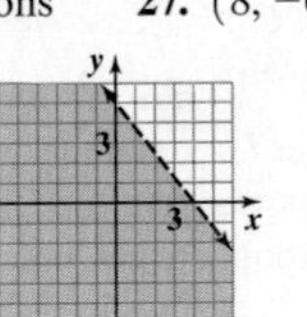

37. **39.**

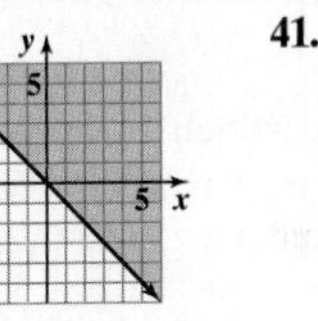

41. $y \geq 2x - 3$, $y \leq -2x + 1$

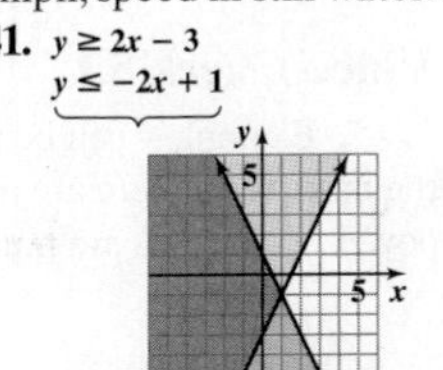

43.

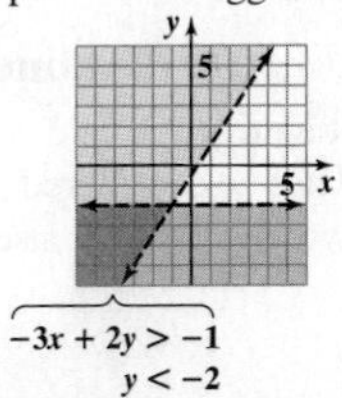

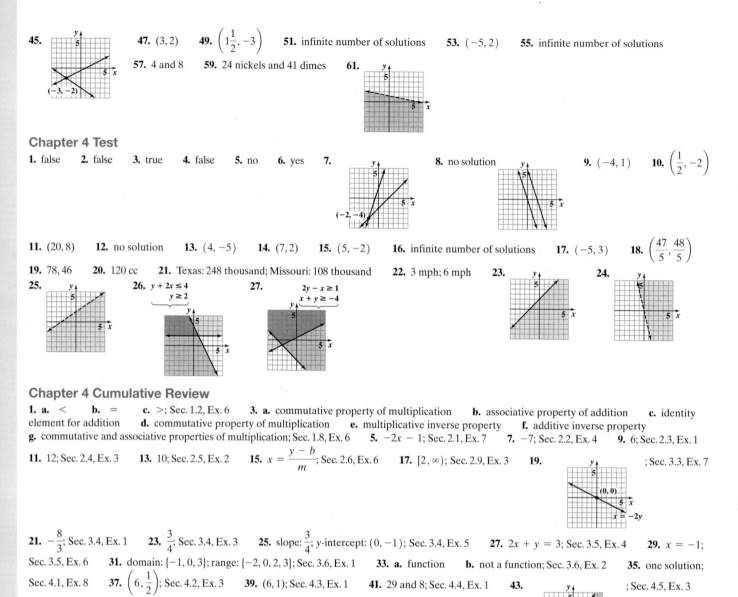

45. **47.** (3, 2) **49.** $\left(1\frac{1}{2}, -3\right)$ **51.** infinite number of solutions **53.** $(-5, 2)$ **55.** infinite number of solutions **57.** 4 and 8 **59.** 24 nickels and 41 dimes **61.**

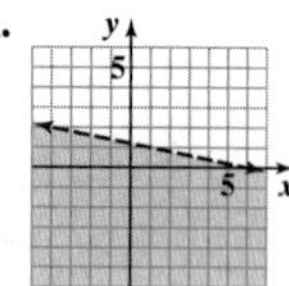

Chapter 4 Test

1. false **2.** false **3.** true **4.** false **5.** no **6.** yes **7.** **8.** no solution

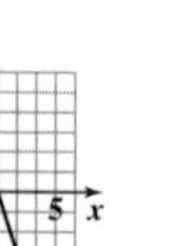

9. $(-4, 1)$ **10.** $\left(\frac{1}{2}, -2\right)$

11. $(20, 8)$ **12.** no solution **13.** $(4, -5)$ **14.** $(7, 2)$ **15.** $(5, -2)$ **16.** infinite number of solutions **17.** $(-5, 3)$ **18.** $\left(\frac{47}{5}, \frac{48}{5}\right)$

19. 78, 46 **20.** 120 cc **21.** Texas: 248 thousand; Missouri: 108 thousand **22.** 3 mph; 6 mph **23.**

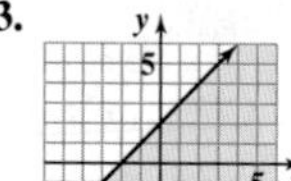

24.

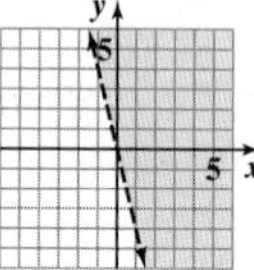

25. **26.** $y + 2x \le 4$, $y \ge 2$ **27.** $2y - x \ge 1$, $x + y \ge -4$

Chapter 4 Cumulative Review

1. a. < **b.** = **c.** >; Sec. 1.2, Ex. 6 **3. a.** commutative property of multiplication **b.** associative property of addition **c.** identity element for addition **d.** commutative property of multiplication **e.** multiplicative inverse property **f.** additive inverse property **g.** commutative and associative properties of multiplication; Sec. 1.8, Ex. 6 **5.** $-2x - 1$; Sec. 2.1, Ex. 7 **7.** -7; Sec. 2.2, Ex. 4 **9.** 6; Sec. 2.3, Ex. 1 **11.** 12; Sec. 2.4, Ex. 3 **13.** 10; Sec. 2.5, Ex. 2 **15.** $x = \frac{y - b}{m}$; Sec. 2.6, Ex. 6 **17.** $[2, \infty)$; Sec. 2.9, Ex. 3 **19.** ; Sec. 3.3, Ex. 7

21. $-\frac{8}{3}$; Sec. 3.4, Ex. 1 **23.** $\frac{3}{4}$; Sec. 3.4, Ex. 3 **25.** slope: $\frac{3}{4}$; y-intercept: $(0, -1)$; Sec. 3.4, Ex. 5 **27.** $2x + y = 3$; Sec. 3.5, Ex. 4 **29.** $x = -1$; Sec. 3.5, Ex. 6 **31.** domain: $\{-1, 0, 3\}$; range: $\{-2, 0, 2, 3\}$; Sec. 3.6, Ex. 1 **33. a.** function **b.** not a function; Sec. 3.6, Ex. 2 **35.** one solution; Sec. 4.1, Ex. 8 **37.** $\left(6, \frac{1}{2}\right)$; Sec. 4.2, Ex. 3 **39.** $(6, 1)$; Sec. 4.3, Ex. 1 **41.** 29 and 8; Sec. 4.4, Ex. 1 **43.** ; Sec. 4.5, Ex. 3

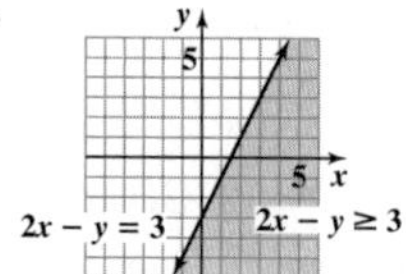

CHAPTER 5 EXPONENTS AND POLYNOMIALS

Section 5.1

Practice Exercises

1. a. 27 **b.** 4 **c.** 64 **d.** -64 **e.** $\frac{27}{64}$ **f.** 0.0081 **g.** 75 **2. a.** 243 **b.** $\frac{3}{8}$ **3. a.** 3^{10} **b.** y^5 **c.** z^5 **d.** x^{11} **e.** $(-2)^8$ **f.** $b^3 \cdot t^5$ **4.** $15y^7$ **5. a.** $y^{12}z^4$ **b.** $-7m^5n^{14}$ **6. a.** z^{21} **b.** 4^{18} **c.** $(-2)^{15}$ **7. a.** p^5r^5 **b.** $36b^2$ **c.** $\frac{1}{64}x^6y^3$ **d.** $81a^{12}b^{16}c^4$ **8. a.** $\frac{x^5}{y^{10}}$ **b.** $\frac{32a^{20}}{b^{15}}$ **9. a.** z^4 **b.** 25 **c.** 64 **d.** $\frac{q^5}{t^2}$ **e.** $6x^2y^2$ **10. a.** -1 **b.** 1 **c.** 1 **d.** 1 **e.** 1 **f.** 7 **11. a.** $\frac{z^2}{144}$ **b.** $64x^{18}$ **c.** y^{13} **12. a.** 63 **b.** 2 **c.** $\frac{x^6}{9y^8}$ **d.** $-z^{30}$

Vocabulary, Readiness & Video Check 5.1

1. exponent **3.** add **5.** 1 **7.** Example 4 can be written as $-4^2 = -1 \cdot 4^2$, which is similar to Example 7, $4 \cdot 3^2$, and shows why the negative sign should not be considered part of the base when there are no parentheses. **9.** Be careful not to confuse the power rule with the product rule. The power rule involves a power raised to a power (exponents are multiplied), and the product rule involves a product (exponents are added). **11.** the quotient rule

Exercise Set 5.1

1. exponent: 2; base: 3 **3.** exponent: 2; base: 4 **5.** exponent: 2; base: x **7.** 49 **9.** -5 **11.** -16 **13.** 16 **15.** 0.00001 **17.** $\frac{1}{81}$ **19.** 224 **21.** -250 **23.** 4 **25.** 135 **27.** 150 **29.** $\frac{32}{5}$ **31.** x^7 **33.** $(-3)^{12}$ **35.** $15y^5$ **37.** $x^{19}y^6$ **39.** $-72m^3n^8$ **41.** $-24z^{20}$ **43.** $20x^5$ sq ft **45.** x^{36} **47.** p^8q^8 **49.** $8a^{15}$ **51.** $x^{10}y^{15}$ **53.** $49a^4b^{10}c^2$ **55.** $\frac{r^9}{s^9}$ **57.** $\frac{m^5p^5}{n^5}$ **59.** $\frac{4x^2z^2}{y^{10}}$ **61.** $64z^{10}$ sq dm **63.** $27y^{12}$ cu ft **65.** x^2 **67.** -64 **69.** p^6q^5 **71.** $\frac{y^3}{2}$ **73.** 1 **75.** 1 **77.** -7 **79.** 2 **81.** -81 **83.** $\frac{1}{64}$ **85.** b^6 **87.** a^9 **89.** $-16x^7$ **91.** $a^{11}b^{20}$ **93.** $26m^9n^7$ **95.** z^{40} **97.** $64a^3b^3$ **99.** $36x^2y^2z^6$ **101.** $3x$ **103.** $81x^2y^2$ **105.** 9 **107.** $\frac{y^{15}}{8x^{12}}$ **109.** $2x^2y$ **111.** 2 **113.** $\frac{x^{18}}{4y^{22}}$ **115.** $-b^5$ **117.** $2y - 10$ **119.** $-x - 4$ **121.** $-x + 5$ **123.** c **125.** e **127.** answers may vary **129.** answers may vary **131.** 343 cu m **133.** volume **135.** answers may vary **137.** answers may vary **139.** x^{9a} **141.** a^{5b} **143.** x^{5a} **145.** \$1045.85

Section 5.2

Practice Exercises

1. a. degree 3 **b.** degree 8 **c.** degree 0 **2. a.** trinomial, degree 2 **b.** binomial, degree 1 **c.** none of these, degree 3

3.

Term	*Numerical Coefficient*	*Degree of Term*
$-3x^3y^2$	-3	5
$4xy^2$	4	3
$-y^2$	-1	2
$3x$	3	1
-2	-2	0

4. a. 31 **b.** 36 **5.** 114 ft; 66 ft **6. a.** $-2y$ **b.** $z + 5z^3$ **c.** $14x^3$ **d.** $4a^2 - 12$ **e.** $\frac{1}{3}x^4 + \frac{11}{24}x^3 - x^2$ **7.** $-3x^2 + 5xy + 5y^2$ **8.** $x^2 + 2x + 4 + 5x + 3x^2$; $4x^2 + 7x + 4$ **9.** $-x^2 - 6x + 9$ **10.** $-3x^3 + 10x^2 - 6x - 4$ **11.** $5z^3 + 8z^2 + 8z$ **12.** $5x - 1$ **13.** $2x^3 - 4x^2 + 4x - 6$ **14.** $-8z^2 - 11z + 12$ **15.** $7x - 11$ **16. a.** $-5a^2 - ab + 6b^2$ **b.** $3x^2y^2 - 10xy - 4xy^2 - 6y^2 + 5$

Vocabulary, Readiness & Video Check 5.2

1. binomial **3.** trinomial **5.** constant **7.** The degree of the polynomial, which is the greatest degree of any of its terms, so we need to find the degree of each term first. **9.** simplifying it

Exercise Set 5.2

1. 1; binomial **3.** 3; none of these **5.** 6; trinomial **7.** 8; binomial **9.** 3 **11.** 2 **13. a.** -6 **b.** -11 **15. a.** -2 **b.** 4 **17. a.** 1 **b.** 21 **19.** 1134 ft **21.** 1006 ft **23.** 164 thousand **25.** $-11x$ **27.** $23x^2$ **29.** $12x^2 - y$ **31.** $7s$ **33.** $-1.1y^2 + 4.8$ **35.** $\frac{5}{6}x^4 - 7x^3 - 19$ **37.** $-\frac{7}{12}x^3 + \frac{7}{5}x^2 + 6$ **39.** $5a^2 - 9ab + 16b^2$ **41.** $-x + 14$ **43.** $-x^2 + 14$ **45.** $-3x^2 + 10$ **47.** $-2x + 9$ **49.** $2x^2 + 7x - 16$ **51.** $8t^2 - 4$ **53.** $-2z^2 - 16z + 6$ **55.** $2x^3 - 2x^2 + 7x + 2$ **57.** $62x^2 + 5$ **59.** $12x + 2$ **61.** $7y - 10$ **63.** $-2x^2 + 8x - 1$ **65.** $14x + 18$ **67.** $3a^2 - 6a + 11$ **69.** $3x - 3$ **71.** $7x^2 - 4x + 2$ **73.** $7x^2 - 2x + 2$ **75.** $4y^2 + 12y + 19$ **77.** $4x^2 + 7x + x^2 + 5x$; $5x^2 + 12x$ **79.** $-2a - b + 1$ **81.** $3x^2 + 5$ **83.** $6x^2 - 2xy + 19y^2$ **85.** $8r^2s + 16rs - 8 + 7r^2s^2$ **87.** $-5.42x^2 + 7.75x - 19.61$ **89.** $3.7y^4 - 0.7y^3 + 2.2y - 4$ **91.** $6x^2$ **93.** $-12x^8$ **95.** $200x^3y^2$ **97.** $18x + 44$ **99.** $(x^2 + 7x + 4)$ ft **101.** $(3y^2 + 4y + 11)$ m **103.** 7.33 million **105.** answers may vary **107.** answers may vary **109.** b **111.** e **113. a.** $4z$ **b.** $3z^2$ **c.** $-4z$ **d.** $3z^2$; answers may vary **115. a.** m^3 **b.** $3m$ **c.** $-m^3$ **d.** $-3m$; answers may vary **117.** 2; 2 **119.** 4; 3; 3; 4 **121.** $2x^2 + 4xy$

Section 5.3

Practice Exercises

1. $10y^2$ **2.** $-2z^8$ **3.** $\frac{7}{72}b^9$ **4. a.** $27x^6 + 33x$ **b.** $-12x^5 + 54x^4 - 12x^3$ **5.** $10x^2 + 11x - 6$ **6.** $25x^2 - 30xy + 9y^2$ **7.** $2y^3 + 5y^2 - 7y + 20$ **8.** $s^3 + 6s^2t + 12st^2 + 8t^3$ **9.** $5x^3 - 23x^2 + 17x - 20$ **10.** $x^5 - 2x^4 + 2x^3 - 3x^2 + 2$ **11.** $5x^4 - 3x^3 + 11x^2 + 8x - 6$

Vocabulary, Readiness & Video Check 5.3

1. distributive **3.** $(5y - 1)(5y - 1)$ **5.** No. The monomials are unlike terms. **7.** Yes. The parentheses have been removed for the vertical format, but every term in the first polynomial is still distributed to every term in the second polynomial.

Exercise Set 5.3

1. $-28n^{10}$ **3.** $-12.4x^{12}$ **5.** $-\frac{2}{15}y^3$ **7.** $-24x^8$ **9.** $6x^2 + 15x$ **11.** $-2a^2 - 8a$ **13.** $6x^3 - 9x^2 + 12x$ **15.** $-6a^4 + 4a^3 - 6a^2$ **17.** $-4x^3y + 7x^2y^2 - xy^3 - 3y^4$ **19.** $4x^4 - 3x^3 + \frac{1}{2}x^2$ **21.** $x^2 + 7x + 12$ **23.** $a^2 + 5a - 14$ **25.** $x^2 + \frac{1}{3}x - \frac{2}{9}$ **27.** $12x^4 + 25x^2 + 7$ **29.** $4y^2 - 16y + 16$ **31.** $12x^2 - 29x + 15$ **33.** $9x^4 + 6x^2 + 1$ **35. a.** $-4y^4$ **b.** $3y^2$ **c.** answers may vary **37.** $x^3 - 5x^2 + 13x - 14$ **39.** $x^4 + 5x^3 - 3x^2 - 11x + 20$ **41.** $10a^3 - 27a^2 + 26a - 12$ **43.** $x^3 + 6x^2 + 12x + 8$ **45.** $8y^3 - 36y^2 + 54y - 27$ **47.** $12x^2 - 64x - 11$ **49.** $10x^3 + 22x^2 - x - 1$ **51.** $2x^4 + 3x^3 - 58x^2 + 4x + 63$ **53.** $8.4y^7$ **55.** $-3x^3 - 6x^2 + 24x$ **57.** $2x^2 + 39x + 19$ **59.** $x^2 - \frac{2}{7}x - \frac{3}{49}$ **61.** $9y^2 + 30y + 25$ **63.** $a^3 - 2a^2 - 18a + 24$ **65.** $8x^3 - 60x^2 + 150x - 125$

67. $32x^3 + 48x^2 - 6x - 20$ **69.** $6x^4 - 8x^3 - 7x^2 + 22x - 12$ **71.** $(4x^2 - 25)$ sq yd **73.** $(6x^2 - 4x)$ sq in.
75. $5a + 15a = 20a; 5a - 15a = -10a; 5a \cdot 15a = 75a^2; \frac{5a}{15a} = \frac{1}{3}$ **77.** $-3y^5 + 9y^4$, cannot be simplified; $-3y^5 - 9y^4$, cannot be simplified; $-3y^5 \cdot 9y^4 = -27y^9; \frac{-3y^5}{9y^4} = -\frac{y}{3}$ **79. a.** $6x + 12$ **b.** $9x^2 + 36x + 35$; answers may vary **81.** $13x - 7$ **83.** $30x^2 - 28x + 6$
85. $-7x + 5$ **87.** $x^2 + 3x$ **89.** $x^2 + 5x + 6$ **91.** $11a$ **93.** $25x^2 + 4y^2$ **95. a.** $a^2 - b^2$ **b.** $4x^2 - 9y^2$ **c.** $16x^2 - 49$
d. answers may vary **97.** $(x^2 + 6x + 5)$ sq units

Section 5.4

Practice Exercises

1. $x^2 - 3x - 10$ **2.** $4x^2 - 13x + 9$ **3.** $9x^2 + 42x - 15$ **4.** $16x^2 - 8x + 1$ **5. a.** $b^2 + 6b + 9$ **b.** $x^2 - 2xy + y^2$
c. $9y^2 + 12y + 4$ **d.** $a^4 - 10a^2b + 25b^2$ **6. a.** $3x^2 - 75$ **b.** $16b^2 - 9$ **c.** $x^2 - \frac{4}{9}$ **d.** $25s^2 - t^2$ **e.** $4y^2 - 9z^4$
7. a. $4x^2 - 21x - 18$ **b.** $49b^2 - 28b + 4$ **c.** $x^2 - 0.16$ **d.** $3x^6 - \frac{9}{7}x^4 + \frac{2}{7}x^2 - \frac{6}{49}$ **e.** $x^3 + 6x^2 + 3x - 2$

Vocabulary, Readiness & Video Check 5.4

1. false **3.** false **5.** a binomial times a binomial **7.** Multiplying gives you four terms, and the two like terms will always subtract out.

Exercise Set 5.4

1. $x^2 + 7x + 12$ **3.** $x^2 + 5x - 50$ **5.** $5x^2 + 4x - 12$ **7.** $20y^2 - 125y + 30$ **9.** $6x^2 + 13x - 5$ **11.** $x^2 + \frac{1}{3}x - \frac{2}{9}$ **13.** $x^2 + 4x + 4$
15. $4x^2 - 4x + 1$ **17.** $9a^2 - 30a + 25$ **19.** $25x^2 + 90x + 81$ **21.** $a^2 - 49$ **23.** $9x^2 - 1$ **25.** $9x^2 - \frac{1}{4}$ **27.** $81x^2 - y^2$
29. $4x^2 - 0.01$ **31.** $a^2 + 9a + 20$ **33.** $a^2 + 14a + 49$ **35.** $12a^2 - a - 1$ **37.** $x^2 - 4$ **39.** $9a^2 + 6a + 1$ **41.** $4x^3 - x^2y^4 + 4xy - y^5$
43. $x^3 - 3x^2 - 17x + 3$ **45.** $4a^2 - 12a + 9$ **47.** $25x^2 - 36z^2$ **49.** $x^{10} - 8x^5 + 15$ **51.** $x^2 - 0.64$ **53.** $a^7 - 3a^3 + 11a^4 - 33$
55. $3x^2 - 12x + 12$ **57.** $6b^2 - b - 35$ **59.** $49p^2 - 64$ **61.** $\frac{1}{9}a^4 - 49$ **63.** $15x^4 - 5x^3 + 10x^2$ **65.** $4r^2 - 9s^2$
67. $9x^2 - 42xy + 49y^2$ **69.** $16x^2 - 25$ **71.** $64x^2 + 64x + 16$ **73.** $a^2 - \frac{1}{4}y^2$ **75.** $\frac{1}{25}x^2 - y^2$ **77.** $3a^3 + 2a^2 + 1$
79. $(2x + 1)(2x + 1)$ sq ft or $(4x^2 + 4x + 1)$ sq ft **81.** $\frac{5b^5}{7}$ **83.** $-2a^{10}b^5$ **85.** $\frac{2y^8}{3}$ **87.** $\frac{1}{3}$ **89.** 1 **91.** c **93.** d **95.** 2; 2
97. $(x^4 - 3x^2 + 1)$ sq m **99.** $(24x^2 - 32x + 8)$ sq m **101.** $(x^2 + 10x + 25)$ sq units **103.** answers may vary **105.** answers may vary
107. answers may vary **109.** $x^2 + 2xy + y^2 - 9$ **111.** $a^2 - 6a + 9 - b^2$

Integrated Review

1. $35x^5$ **2.** $32y^9$ **3.** -16 **4.** 16 **5.** $2x^2 - 9x - 5$ **6.** $3x^2 + 13x - 10$ **7.** $3x - 4$ **8.** $4x + 3$ **9.** $7x^6y^2$ **10.** $\frac{10b^6}{7}$
11. $144m^{14}n^{12}$ **12.** $64y^{27}z^{30}$ **13.** $48y^2 - 27$ **14.** $98x^2 - 2$ **15.** $x^{63}y^{45}$ **16.** $27x^{27}$ **17.** $2x^2 - 2x - 6$ **18.** $6x^2 + 13x - 11$
19. $2.5y^2 - 6y - 0.2$ **20.** $8.4x^2 - 6.8x - 5.7$ **21.** $x^2 + 8xy + 16y^2$ **22.** $y^2 - 18yz + 81z^2$ **23.** $2x + 8y$ **24.** $2y - 18z$
25. $7x^2 - 10xy + 4y^2$ **26.** $-a^2 - 3ab + 6b^2$ **27.** $x^3 + 2x^2 - 16x + 3$ **28.** $x^3 - 2x^2 - 5x - 2$ **29.** $6x^5 + 20x^3 - 21x^2 - 70$
30. $20x^7 + 25x^3 - 4x^4 - 5$ **31.** $2x^3 - 19x^2 + 44x - 7$ **32.** $5x^3 + 9x^2 - 17x + 3$ **33.** cannot simplify **34.** $25x^3y^3$ **35.** $125x^9$
36. $\frac{x^3}{y^3}$ **37.** $2x$ **38.** x^2

Section 5.5

Practice Exercises

1. a. $\frac{1}{125}$ **b.** $\frac{3}{y^4}$ **c.** $\frac{5}{6}$ **d.** $\frac{1}{25}$ **e.** $\frac{1}{x^5}$ **2. a.** s^5 **b.** 8 **c.** $\frac{y^5}{x^7}$ **d.** $\frac{9}{64}$ **3. a.** $\frac{1}{x^5}$ **b.** $5y^7$ **c.** z^5 **d.** $\frac{81}{25}$ **4. a.** $\frac{b^{15}}{a^{20}}$
b. x^{10} **c.** $\frac{q^2}{25p^{16}}$ **d.** $\frac{6y^2}{x^7}$ **e.** $-27x^6y^9$ **5. a.** 7×10^{-6} **b.** 2.07×10^7 **c.** 4.3×10^{-3} **d.** 8.12×10^8 **6. a.** 0.000367
b. 8,954,000 **c.** 0.00002009 **d.** 4054 **7. a.** 4000 **b.** 20,000,000,000

Calculator Explorations 5.5

1. 5.31 EE 3 **3.** 6.6 EE −9 **5.** 1.5×10^{13} **7.** 8.15×10^{19}

Vocabulary, Readiness & Video Check 5.5

1. $\frac{1}{x^3}$ **3.** scientific notation **5.** A negative exponent has nothing to do with the sign of the simplified result. **7.** When you move the decimal point to the left, the sign of the exponent will be positive; when you move the decimal point to the right, the sign of the exponent will be negative.
9. the quotient rule

Exercise Set 5.5

1. $\frac{1}{64}$ **3.** $\frac{1}{81}$ **5.** $\frac{7}{x^3}$ **7.** 32 **9.** -64 **11.** $\frac{8}{15}$ **13.** p^3 **15.** $\frac{q^4}{p^5}$ **17.** $\frac{1}{x^3}$ **19.** z^3 **21.** $\frac{4}{9}$ **23.** $-p^4$ **25.** -2 **27.** x^4
29. p^4 **31.** m^{11} **33.** r^6 **35.** $\frac{1}{x^{15}y^9}$ **37.** $\frac{1}{x^4}$ **39.** $\frac{1}{a^2}$ **41.** $4k^3$ **43.** $3m$ **45.** $-\frac{4a^5}{b}$ **47.** $-\frac{6x^2}{y^3}$ **49.** $\frac{a^{30}}{b^{12}}$ **51.** $\frac{1}{x^{10}y^6}$

53. $\frac{z^2}{4}$ **55.** $\frac{1}{32x^5}$ **57.** $\frac{49a^4}{b^6}$ **59.** $a^{24}b^8$ **61.** x^9y^{19} **63.** $-\frac{y^8}{8x^2}$ **65.** $-\frac{6x}{7y^2}$ **67.** $\frac{25b^{33}}{a^{16}}$ **69.** 7.8×10^4 **71.** 1.67×10^{-6} **73.** 6.35×10^{-3} **75.** 1.16×10^6 **77.** 2×10^9 **79.** 2.4×10^3 **81.** 0.0000000008673 **83.** 0.033 **85.** 20,320 **87.** 700,000,000 **89.** 9,460,000,000,000 **91.** 1.84×10^{11} **93.** 155,000,000,000 **95.** 35,000 **97.** 0.000036 **99.** 0.0000000000000000028 **101.** 0.0000005 **103.** 200,000 **105.** $\frac{5x^3}{3}$ **107.** $\frac{5z^3y^2}{7}$ **109.** $5y - 6 + \frac{5}{y}$ **111.** $\frac{27}{x^6z^3}$ cu in. **113.** $9a^{13}$ **115.** −5 **117.** answers may vary **119. a.** 1.3×10^1 **b.** 4.4×10^7 **c.** 6.1×10^{-2} **121.** answers may vary **123.** a^m **125.** $27y^{6z}$ **127.** −394.5 **129.** 1.3 sec

Section 5.6

Practice Exercises

1. $2t + 1$ **2.** $4x^4 + 5x - \frac{3}{x}$ **3.** $3x^3y^3 - 2 + \frac{1}{5x}$ **4.** $x + 3$ **5.** $2x + 3 + \frac{-10}{2x + 1}$ or $2x + 3 - \frac{10}{2x + 1}$ **6.** $3x^2 - 2x + 5 + \frac{-13}{3x + 2}$ or $3x^2 - 2x + 5 - \frac{13}{3x - 2}$ **7.** $3x^2 - 2x - 9 + \frac{5x + 22}{x^2 + 2}$ **8.** $x^2 - 3x + 9$

Vocabulary, Readiness & Video Check 5.6

1. dividend, quotient, divisor **3.** a^2 **5.** y **7.** the common denominator

Exercise Set 5.6

1. $12x^3 + 3x$ **3.** $4x^3 - 6x^2 + x + 1$ **5.** $5p^2 + 6p$ **7.** $-\frac{3}{2x} + 3$ **9.** $-3x^2 + x - \frac{4}{x^3}$ **11.** $-1 + \frac{3}{2x} - \frac{7}{4x^4}$ **13.** $x + 1$ **15.** $2x + 3$ **17.** $2x + 1 + \frac{7}{x - 4}$ **19.** $3a^2 - 3a + 1 + \frac{2}{3a + 2}$ **21.** $4x + 3 - \frac{2}{2x + 1}$ **23.** $2x^2 + 6x - 5 - \frac{2}{x - 2}$ **25.** $x + 6$ **27.** $x^2 + 3x + 9$ **29.** $-3x + 6 - \frac{11}{x + 2}$ **31.** $2b - 1 - \frac{6}{2b - 1}$ **33.** $ab - b^2$ **35.** $4x + 9$ **37.** $x + 4xy - \frac{y}{2}$ **39.** $2b^2 + b + 2 - \frac{12}{b + 4}$ **41.** $5x - 2 + \frac{2}{x + 6}$ **43.** $x^2 - \frac{12x}{5} - 1$ **45.** $6x - 1 - \frac{1}{x + 3}$ **47.** $6x - 1$ **49.** $-x^3 + 3x^2 - \frac{4}{x}$ **51.** $x^2 + 3x + 9$ **53.** $y^2 + 5y + 10 + \frac{24}{y - 2}$ **55.** $-6x - 12 - \frac{19}{x - 2}$ **57.** $x^3 - x^2 + x$ **59.** $2a^3 + 2a$ **61.** $2x^3 + 14x^2 - 10x$ **63.** $-3x^2y^3 - 21x^3y^2 - 24xy$ **65.** $9a^2b^3c + 36ab^2c - 72ab$ **67.** $(3x^3 + x - 4)$ ft **69.** c **71.** answers may vary **73.** $(2x + 5)$ m **75.** $9x^{7a} - 6x^{5a} + 7x^{2a} - 1$

Chapter 5 Vocabulary Check

1. term **2.** FOIL **3.** trinomial **4.** degree of a polynomial **5.** binomial **6.** coefficient **7.** degree of a term **8.** monomial **9.** polynomials **10.** distributive

Chapter 5 Review

1. base: 7; exponent: 9 **3.** base: 5; exponent: 4 **5.** 512 **7.** −36 **9.** 1 **11.** y^9 **13.** $-6x^{11}$ **15.** x^8 **17.** $81y^{24}$ **19.** x^5 **21.** a^4b^3 **23.** $\frac{4}{x^3y^4}$ **25.** $40a^{19}$ **27.** 3 **29.** b **31.** 7 **33.** 8 **35.** 5 **37.** 5 **39.** 4000 ft; 3984 ft; 3856 ft; 3600 ft **41.** $2a^2$ **43.** $15a^2 + 4a$ **45.** $-6a^2b - 3b^2 - q^2$ **47.** $8x^2 + 3x + 6$ **49.** $-7y^2 - 1$ **51.** $-5x^2 + 5x + 1$ **53.** $4x - 13y$ **55.** $8a + 28$ **57.** $-7x^3 - 35x$ **59.** $-6a^4 + 8a^2 - 2a$ **61.** $2x^2 - 12x - 14$ **63.** $x^2 - 18x + 81$ **65.** $4a^2 + 27a - 7$ **67.** $25x^2 + 20x + 4$ **69.** $x^4 + 7x^3 + 4x^2 + 23x - 35$ **71.** $x^4 + 4x^3 + 4x^2 - 16$ **73.** $x^3 + 21x^2 + 147x + 343$ **75.** $x^2 + 14x + 49$ **77.** $9x^2 - 42x + 49$ **79.** $25x^2 - 90x + 81$ **81.** $49x^2 - 16$ **83.** $4x^2 - 36$ **85.** $(9x^2 - 6x + 1)$ sq m **87.** $\frac{1}{49}$ **89.** $\frac{2}{x^4}$ **91.** 125 **93.** $\frac{17}{16}$ **95.** x^8 **97.** r **99.** c^4 **101.** $\frac{1}{x^6y^{13}}$ **103.** a^{11m} **105.** $27x^3y^{6z}$ **107.** 2.7×10^{-4} **109.** 8.08×10^7 **111.** 9.1×10^7 **113.** 867,000 **115.** 0.00086 **117.** 1,431,280,000,000,000 **119.** 0.016 **121.** $\frac{1}{7} + \frac{3}{x} + \frac{7}{x^2}$ **123.** $a + 1 + \frac{6}{a - 2}$ **125.** $a^2 + 3a + 8 + \frac{22}{a - 2}$ **127.** $2x^3 - x^2 + 2 - \frac{1}{2x - 1}$ **129.** $\left(5x - 1 + \frac{20}{x^2}\right)$ ft **131.** $-\frac{1}{8}$ **133.** $\frac{2x^6}{3}$ **135.** $\frac{x^{16}}{16y^{12}}$ **137.** $11x - 5$ **139.** $5y^2 - 3y - 1$ **141.** $28x^3 + 12x$ **143.** $x^3 + x^2 - 18x + 18$ **145.** $25x^2 + 40x + 16$ **147.** $4a - 1 + \frac{2}{a^2} - \frac{5}{2a^3}$ **149.** $2x^2 + 7x + 5 + \frac{19}{2x - 3}$

Chapter 5 Test

1. 32 **2.** 81 **3.** −81 **4.** $\frac{1}{64}$ **5.** $-15x^{11}$ **6.** y^5 **7.** $\frac{1}{r^5}$ **8.** $\frac{y^{14}}{x^2}$ **9.** $\frac{1}{6xy^8}$ **10.** 5.63×10^5 **11.** 8.63×10^{-5} **12.** 0.0015 **13.** 62,300 **14.** 0.036 **15. a.** 4, 3; 7, 3; 1, 4; −2, 0 **b.** 4 **16.** $-2x^2 + 12xy + 11$ **17.** $16x^3 + 7x^2 - 3x - 13$ **18.** $-3x^3 + 5x^2 + 4x + 5$ **19.** $x^3 + 8x^2 + 3x - 5$ **20.** $3x^3 + 22x^2 + 41x + 14$ **21.** $6x^4 - 9x^3 + 21x^2$ **22.** $3x^2 + 16x - 35$ **23.** $9x^2 - \frac{1}{25}$ **24.** $16x^2 - 16x + 4$ **25.** $64x^2 + 48x + 9$ **26.** $x^4 - 81b^2$ **27.** 1001 ft; 985 ft; 857 ft; 601 ft **28.** $(4x^2 - 9)$ sq in. **29.** $\frac{x}{2y} + 3 - \frac{7}{8y}$ **30.** $x + 2$ **31.** $9x^2 - 6x + 4 - \frac{16}{3x + 2}$

Chapter 5 Cumulative Review

1. a. true **b.** true **c.** false **d.** true; Sec. 1.2, Ex. 2 **3. a.** $\frac{64}{25}$ **b.** $\frac{1}{20}$ **c.** $\frac{5}{4}$; Sec. 1.3, Ex. 4 **5. a.** 9 **b.** 125 **c.** 16 **d.** 7 **e.** $\frac{9}{49}$; Sec. 1.4, Ex. 1 **7. a.** −10 **b.** −21 **c.** −12; Sec. 1.5, Ex. 3 **9.** −12; Sec. 1.6, Ex. 3 **11. a.** $\frac{1}{22}$ **b.** $\frac{16}{3}$ **c.** $-\frac{1}{10}$

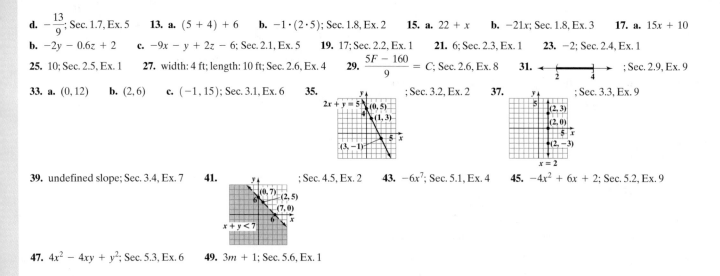

d. $-\frac{13}{9}$; Sec. 1.7, Ex. 5 **13. a.** $(5 + 4) + 6$ **b.** $-1 \cdot (2 \cdot 5)$; Sec. 1.8, Ex. 2 **15. a.** $22 + x$ **b.** $-21x$; Sec. 1.8, Ex. 3 **17. a.** $15x + 10$ **b.** $-2y - 0.6z + 2$ **c.** $-9x - y + 2z - 6$; Sec. 2.1, Ex. 5 **19.** 17; Sec. 2.2, Ex. 1 **21.** 6; Sec. 2.3, Ex. 1 **23.** -2; Sec. 2.4, Ex. 1 **25.** 10; Sec. 2.5, Ex. 1 **27.** width: 4 ft; length: 10 ft; Sec. 2.6, Ex. 4 **29.** $\frac{5F - 160}{9} = C$; Sec. 2.6, Ex. 8 **31.** (number line: 2, 4); Sec. 2.9, Ex. 9 **33. a.** $(0, 12)$ **b.** $(2, 6)$ **c.** $(-1, 15)$; Sec. 3.1, Ex. 6 **35.** ; Sec. 3.2, Ex. 2 **37.** ; Sec. 3.3, Ex. 9

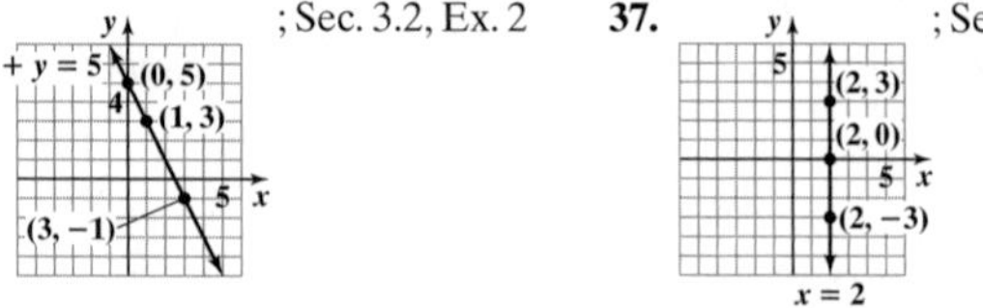

39. undefined slope; Sec. 3.4, Ex. 7 **41.** ; Sec. 4.5, Ex. 2 **43.** $-6x^7$; Sec. 5.1, Ex. 4 **45.** $-4x^2 + 6x + 2$; Sec. 5.2, Ex. 9

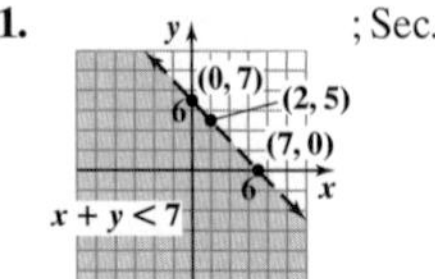

47. $4x^2 - 4xy + y^2$; Sec. 5.3, Ex. 6 **49.** $3m + 1$; Sec. 5.6, Ex. 1

CHAPTER 6 FACTORING POLYNOMIALS

Section 6.1

Practice Exercises

1. a. 6 **b.** 1 **c.** 4 **2. a.** y^4 **b.** x **3. a.** $5y^2$ **b.** x^2 **c.** a^2b^2 **4. a.** $4(t + 3)$ **b.** $y^4(y^4 + 1)$ **5.** $8b^2(-b^4 + 2b^2 - 1)$ or $-8b^2(b^4 - 2b^2 + 1)$ **6.** $5x(x^3 - 4)$ **7.** $\frac{1}{9}z^3(5z^2 + z - 2)$ **8.** $4ab^3(2ab - 5a^2 + 3)$ **9.** $(y - 2)(8 + x)$ **10.** $(p + q)(7xy^3 - 1)$ **11.** $(x + 3)(y + 4)$ **12.** $(5x - 3)(8x^2 + 3)$ **13.** $(2x + 3y)(y - 1)$ **14.** $(7a + 5)(a^2 + 1)$ **15.** $(y - 3)(4x - 5)$ **16.** cannot be factored by grouping **17.** $3(x - a)(y - 2a)$

Vocabulary, Readiness & Video Check 6.1

1. factors **3.** least **5.** false **7.** The GCF of a list of numbers is the largest number that is a factor of all numbers in the list. **9.** When factoring out a GCF, the number of terms in the other factor should have the same number of terms as your original polynomial.

Exercise Set 6.1

1. 4 **3.** 6 **5.** 1 **7.** y^2 **9.** z^7 **11.** xy^2 **13.** 7 **15.** $4y^3$ **17.** $5x^2$ **19.** $3x^3$ **21.** $9x^2y$ **23.** $10a^6b$ **25.** $3(a + 2)$ **27.** $15(2x - 1)$ **29.** $x^2(x + 5)$ **31.** $2y^3(3y + 1)$ **33.** $4(x - 2y + 1)$ **35.** $3x(2x^2 - 3x + 4)$ **37.** $a^2b^2(a^5b^4 - a + b^3 - 1)$ **39.** $4(2x^5 + 4x^4 - 5x^3 + 3)$ **41.** $\frac{1}{3}x(x^3 + 2x^2 - 4x^4 + 1)$ **43.** $(x^2 + 2)(y + 3)$ **45.** $(y + 4)(z - 3)$ **47.** $(z^2 - 6)(r + 1)$ **49.** $-2(x + 7)$ **51.** $-x^5(2 - x^2)$ **53.** $-3a^2(2a^2 - 3a + 1)$ **55.** $(x + 2)(x^2 + 5)$ **57.** $(x + 3)(5 + y)$ **59.** $(3x - 2)(2x^2 + 5)$ **61.** $(5m^2 + 6n)(m + 1)$ **63.** $(y - 4)(2 + x)$ **65.** $(2x - 1)(x^2 + 4)$ **67.** not factorable by grouping **69.** $(x - 2y)(4x - 3)$ **71.** $(5q - 4p)(q - 1)$ **73.** $x(x^2 + 1)(2x + 5)$ **75.** $2(2y - 7)(3x^2 - 1)$ **77.** $2x(16y - 9x)$ **79.** $(x + 2)(y - 3)$ **81.** $7xy(2x^2 + x - 1)$ **83.** $(4x - 1)(7x^2 + 3)$ **85.** $-8x^8y^5(5y + 2x)$ **87.** $3(2a + 3b^2)(a + b)$ **89.** $x^2 + 7x + 10$ **91.** $b^2 - 3b - 4$ **93.** 2, 6 **95.** $-1, -8$ **97.** $-2, 5$ **99.** b **101.** factored **103.** not factored **105.** answers may vary **107.** answers may vary **109. a.** 1200 million **b.** 1120 million **c.** $-20(x^2 - 15x - 6)$ **111.** $12x^3 - 2x$; $2x(6x^2 - 1)$ **113.** $(n^3 - 6)$ units **115.** $(x^n + 2)(x^n + 3)$ **117.** $(3x^n - 5)(x^n + 7)$

Section 6.2

Practice Exercises

1. $(x + 2)(x + 3)$ **2.** $(x - 10)(x - 7)$ **3.** $(x + 7)(x - 2)$ **4.** $(p - 9)(p + 7)$ **5.** prime polynomial **6.** $(x + 3y)(x + 4y)$ **7.** $(x^2 + 12)(x^2 + 1)$ **8.** $(x - 6)(x - 8)$ **9.** $4(x - 3)(x - 3)$ **10.** $3y^2(y - 7)(y + 1)$

Vocabulary, Readiness & Video Check 6.2

1. true **3.** false **5.** +5 **7.** -3 **9.** +2 **11.** 15 is positive, so its factors would have to either be both positive or both negative. Since the factors need to sum to -8, both factors must be negative.

Exercise Set 6.2

1. $(x + 6)(x + 1)$ **3.** $(y - 9)(y - 1)$ **5.** $(x - 3)(x - 3)$ or $(x - 3)^2$ **7.** $(x - 6)(x + 3)$ **9.** $(x + 10)(x - 7)$ **11.** prime **13.** $(x + 5y)(x + 3y)$ **15.** $(a^2 - 5)(a^2 + 3)$ **17.** $(m + 13)(m + 1)$ **19.** $(t - 2)(t + 12)$ **21.** $(a - 2b)(a - 8b)$ **23.** $2(z + 8)(z + 2)$ **25.** $2x(x - 5)(x - 4)$ **27.** $(x - 4y)(x + y)$ **29.** $(x + 12)(x + 3)$ **31.** $(x - 2)(x + 1)$ **33.** $(r - 12)(r - 4)$ **35.** $(x + 2y)(x - y)$ **37.** $3(x + 5)(x - 2)$ **39.** $3(x - 18)(x - 2)$ **41.** $(x - 24)(x + 6)$ **43.** prime **45.** $(x - 5)(x - 3)$ **47.** $6x(x + 4)(x + 5)$ **49.** $4y(x^2 + x - 3)$ **51.** $(x - 7)(x + 3)$ **53.** $(x + 5y)(x + 2y)$ **55.** $2(t + 8)(t + 4)$ **57.** $x(x - 6)(x + 4)$ **59.** $2t^3(t - 4)(t - 3)$ **61.** $5xy(x - 8y)(x + 3y)$ **63.** $3(m - 9)(m - 6)$ **65.** $-1(x - 11)(x - 1)$ **67.** $\frac{1}{2}(y - 11)(y + 2)$ **69.** $x(xy - 4)(xy + 5)$ **71.** $2x^2 + 11x + 5$ **73.** $15y^2 - 17y + 4$ **75.** $9a^2 + 23ab - 12b^2$ **77.** $x^2 + 5x - 24$ **79.** answers may vary

81. $2x^2 + 28x + 66; 2(x + 3)(x + 11)$ **83.** $-16(t - 5)(t + 1)$ **85.** $\left(x + \frac{1}{4}\right)\left(x + \frac{1}{4}\right)$ or $\left(x + \frac{1}{4}\right)^2$ **87.** $(x + 1)(z - 10)(z + 7)$ **89.** $(x^n + 10)(x^n - 2)$ **91.** 5; 8; 9 **93.** 3; 4 **95.** 8; 16 **97.** 6; 26

Section 6.3

Practice Exercises

1. $(2x + 5)(x + 3)$ **2.** $(5x - 4)(3x - 2)$ **3.** $(4x - 1)(x + 3)$ **4.** $(7x - y)(3x + 2y)$ **5.** $(2x^2 - 7)(x^2 + 1)$ **6.** $x(3x + 2)(x + 5)$ **7.** $-1(4x - 3)(2x + 1)$ **8.** $(x + 7)^2$ **9.** $(2x + 9y)(2x + y)$ **10.** $(6n^2 - 1)^2$ **11.** $3x(2x - 7)^2$

Vocabulary, Readiness & Video Check 6.3

1. perfect square trinomial **3.** perfect square trinomial **5.** d **7.** Consider the factors of the first and last terms and the signs of the trinomial. Continue to check by multiplying until you get the middle term of the trinomial. **9.** The first and last terms are squares, a^2 and b^2, and the middle term is $2 \cdot a \cdot b$ or $-2 \cdot a \cdot b$.

Exercise Set 6.3

1. $x + 4$ **3.** $10x - 1$ **5.** $5x - 2$ **7.** $(2x + 3)(x + 5)$ **9.** $(y - 1)(8y - 9)$ **11.** $(2x + 1)(x - 5)$ **13.** $(4r - 1)(5r + 8)$ **15.** $(10x + 1)(x + 3)$ **17.** prime **19.** $(3x - 5y)(2x - y)$ **21.** $(3m - 5)(5m + 3)$ **23.** $x(3x + 2)(4x + 1)$ **25.** $3(7b + 5)(b - 3)$ **27.** $(3z + 4)(4z - 3)$ **29.** $2y^2(3x - 10)(x + 3)$ **31.** $(2x - 7)(2x + 3)$ **33.** $-1(x - 6)(x + 4)$ **35.** $x(4x + 3)(x - 3)$ **37.** $(4x - 9)(6x - 1)$ **39.** $(x + 11)^2$ **41.** $(x - 8)^2$ **43.** $(4a - 3)^2$ **45.** $(x^2 + 2)^2$ **47.** $2(n - 7)^2$ **49.** $(4y + 5)^2$ **51.** $(2x + 11)(x - 9)$ **53.** $(8x + 3)(3x + 4)$ **55.** $(3a + b)(a + 3b)$ **57.** $(x - 4)(x - 5)$ **59.** $(p + 6q)^2$ **61.** $(xy - 5)^2$ **63.** $b(8a - 3)(5a + 3)$ **65.** $2x(3x + 2)(5x + 3)$ **67.** $2y(3y + 5)(y - 3)$ **69.** $5x^2(2x - y)(x + 3y)$ **71.** $-1(2x - 5)(7x - 2)$ **73.** $p^2(4p - 5)^2$ **75.** $(3x - 2)(x + 1)$ **77.** $(4x + 9y)(2x - 3y)$ **79.** prime **81.** $(3x - 4y)^2$ **83.** $(6x - 7)(3x + 2)$ **85.** $(7t + 1)(t - 4)$ **87.** $(7p + 1)(7p - 2)$ **89.** $m(m + 9)^2$ **91.** prime **93.** $a(6a^2 + b^2)(a^2 + 6b^2)$ **95.** $x^2 - 4$ **97.** $a^3 + 27$ **99.** 25–34 **101.** answers may vary **103.** no **105.** answers may vary **107.** $4x^2 + 21x + 5; (4x + 1)(x + 5)$ **109.** $\left(2x + \frac{1}{2}\right)^2$ **111.** $(y - 1)^2(4x^2 + 10x + 25)$ **113.** 8 **115.** $a^2 + 2ab + b^2$ **117.** 2; 14 **119.** 2 **121.** $-3xy^2(4x - 5)(x + 1)$ **123.** $(y - 1)^2(2x + 5)^2$ **125.** $(3x^n + 2)(x^n + 5)$ **127.** answers may vary

Section 6.4

Practice Exercises

1. $(5x + 1)(x + 12)$ **2.** $(4x - 5)(3x - 1)$ **3.** $2(5x + 1)(3x - 2)$ **4.** $5m^2(8m - 7)(m + 1)$ **5.** $(4x + 3)^2$

Vocabulary, Readiness & Video Check 6.4

1. a **3.** b **5.** This gives us a four-term polynomial, which may be factored by grouping.

Exercise Set 6.4

1. $(x + 3)(x + 2)$ **3.** $(y + 8)(y - 2)$ **5.** $(8x - 5)(x - 3)$ **7.** $(5x^2 - 3)(x^2 + 5)$ **9. a.** 9, 2 **b.** $9x + 2x$ **c.** $(3x + 1)(2x + 3)$ **11. a.** $-20, -3$ **b.** $-20x - 3x$ **c.** $(3x - 4)(5x - 1)$ **13.** $(3y + 2)(7y + 1)$ **15.** $(7x - 11)(x + 1)$ **17.** $(5x - 2)(2x - 1)$ **19.** $(2x - 5)(x - 1)$ **21.** $(2x + 3)^2$ **23.** $(2x + 3)(2x - 7)$ **25.** $(5x - 4)(2x - 3)$ **27.** $x(2x + 3)(x + 5)$ **29.** $2(8y - 9)(y - 1)$ **31.** $(2x - 3)(3x - 2)$ **33.** $3(3a + 2)(6a - 5)$ **35.** $a(4a + 1)(5a + 8)$ **37.** $3x(4x + 3)(x - 3)$ **39.** $y(3x + y)(x + y)$ **41.** prime **43.** $5(x + 5y)^2$ **45.** $6(a + b)(4a - 5b)$ **47.** $p^2(15p + q)(p + 2q)$ **49.** $2(9a^2 - 2)^2$ **51.** $(7 + x)(5 + x)$ or $(x + 7)(x + 5)$ **53.** $(6 - 5x)(1 - x)$ or $(5x - 6)(x - 1)$ **55.** $x^2 - 4$ **57.** $y^2 + 8y + 16$ **59.** $81z^2 - 25$ **61.** $x^3 - 27$ **63.** $10x^2 + 45x + 45; 5(2x + 3)(x + 3)$ **65.** $(x^n + 2)(x^n + 3)$ **67.** $(3x^n - 5)(x^n + 7)$ **69.** answers may vary

Section 6.5

Practice Exercises

1. $(x + 9)(x - 9)$ **2. a.** $(3x - 1)(3x + 1)$ **b.** $(6a - 7b)(6a + 7b)$ **c.** $\left(p + \frac{5}{6}\right)\left(p - \frac{5}{6}\right)$ **3.** $(p^2 - q^5)(p^2 + q^5)$ **4. a.** $(z^2 + 9)(z + 3)(z - 3)$ **b.** prime polynomial **5.** $y(6y + 5)(6y - 5)$ **6.** $5(4y^2 + 1)(2y + 1)(2y - 1)$ **7.** $-1(3x + 10)(3x - 10)$ or $(10 + 3x)(10 - 3x)$ **8.** $(x + 4)(x^2 - 4x + 16)$ **9.** $(x - 5)(x^2 + 5x + 25)$ **10.** $(3y + 1)(9y^2 - 3y + 1)$ **11.** $4(2x - 5y)(4x^2 + 10xy + 25y^2)$

Calculator Explorations 6.5

	$x^2 - 2x + 1$	$x^2 - 2x - 1$	$(x - 1)^2$
$x = 5$	16	14	16
$x = -3$	16	14	16
$x = 2.7$	2.89	0.89	2.89
$x = -12.1$	171.61	169.61	171.61
$x = 0$	1	-1	1

Vocabulary, Readiness & Video Check 6.5

1. difference of two cubes **3.** sum of two cubes **5.** $(7x)^2$ **7.** $(2y)^3$ **9.** In order to recognize the binomial as a difference of squares and also to identify the terms to use in the special factoring formula. **11.** First rewrite the original binomial with terms writtten as cubes. Answers will then vary depending on your interpretation.

Exercise Set 6.5

1. $(x + 2)(x - 2)$ **3.** $(9p + 1)(9p - 1)$ **5.** $(5y - 3)(5y + 3)$ **7.** $(11m + 10n)(11m - 10n)$ **9.** $(xy - 1)(xy + 1)$ **11.** $\left(x - \frac{1}{2}\right)\left(x + \frac{1}{2}\right)$ **13.** $-1(2r + 1)(2r - 1)$ **15.** prime **17.** $(-6 + x)(6 + x)$ or $-1(6 + x)(6 - x)$ **19.** $(m^2 + 1)(m + 1)(m - 1)$ **21.** $(m^2 + n^9)(m^2 - n^9)$ **23.** $(x + 5)(x^2 - 5x + 25)$ **25.** $(2a - 1)(4a^2 + 2a + 1)$ **27.** $(m + 3n)(m^2 - 3mn + 9n^2)$ **29.** $5(k + 2)(k^2 - 2k + 4)$ **31.** $(xy - 4)(x^2y^2 + 4xy + 16)$ **33.** $2(5r - 4t)(25r^2 + 20rt + 16t^2)$ **35.** $(r + 8)(r - 8)$ **37.** $(x + 13y)(x - 13y)$ **39.** $(3 - t)(9 + 3t + t^2)$ **41.** $2(3r + 2)(3r - 2)$ **43.** $x(3y + 2)(3y - 2)$ **45.** $8(m + 2)(m^2 - 2m + 4)$ **47.** $xy(y - 3z)(y + 3z)$ **49.** $4(3x - 4y)(3x + 4y)$ **51.** $9(4 - 3x)(4 + 3x)$ **53.** $(xy - z^2)(x^2y^2 + xyz^2 + z^4)$ **55.** $\left(7 - \frac{3}{5}m\right)\left(7 + \frac{3}{5}m\right)$ **57.** $(t + 7)(t^2 - 7t + 49)$ **59.** $n(n^2 + 49)$ **61.** $x^2(x^2 + 9)(x + 3)(x - 3)$ **63.** $pq(8p + 9q)(8p - 9q)$ **65.** $xy^2(27xy + 1)$ **67.** $a(5a - 4b)(25a^2 + 20ab + 16b^2)$ **69.** $16x^2(x + 2)(x - 2)$ **71.** 6 **73.** -2 **75.** $\frac{1}{5}$ **77.** $(x + 2 + y)(x + 2 - y)$ **79.** $(a + 4)(a - 4)(b - 4)$ **81.** $(x + 3 + 2y)(x + 3 - 2y)$ **83.** $(x^n + 10)(x^n - 10)$ **85.** $(x + 6)$ **87.** answers may vary **89. a.** 2560 ft **b.** 1920 ft **c.** 13 sec **d.** $16(13 - t)(13 + t)$ **91. a.** 1456 ft **b.** 816 ft **c.** 10 sec **d.** $16(10 + t)(10 - t)$

Integrated Review

Practice Exercises

1. $(3x - 1)(2x - 3)$ **2.** $(3x + 1)(x - 2)(x + 2)$ **3.** $3(3x - y)(3x + y)$ **4.** $(2a + b)(4a^2 - 2ab + b^2)$ **5.** $6xy^2(5x + 2)(2x - 3)$

Exercise Set

1. $(x + y)^2$ **2.** $(x - y)^2$ **3.** $(a + 12)(a - 1)$ **4.** $(a - 10)(a - 1)$ **5.** $(a + 2)(a - 3)$ **6.** $(a - 1)^2$ **7.** $(x + 1)^2$ **8.** $(x + 2)(x - 1)$ **9.** $(x + 1)(x + 3)$ **10.** $(x + 3)(x - 2)$ **11.** $(x + 3)(x + 4)$ **12.** $(x + 4)(x - 3)$ **13.** $(x + 4)(x - 1)$ **14.** $(x - 5)(x - 2)$ **15.** $(x + 5)(x - 3)$ **16.** $(x + 6)(x + 5)$ **17.** $(x - 6)(x + 5)$ **18.** $(x + 8)(x + 3)$ **19.** $2(x + 7)(x - 7)$ **20.** $3(x + 5)(x - 5)$ **21.** $(x + 3)(x + y)$ **22.** $(y - 7)(3 + x)$ **23.** $(x + 8)(x - 2)$ **24.** $(x - 7)(x + 4)$ **25.** $4x(x + 7)(x - 2)$ **26.** $6x(x - 5)(x + 4)$ **27.** $2(3x + 4)(2x + 3)$ **28.** $(2a - b)(4a + 5b)$ **29.** $(2a + b)(2a - b)$ **30.** $(4 - 3x)(7 + 2x)$ **31.** $(5 - 2x)(4 + x)$ **32.** prime **33.** prime **34.** $(3y + 5)(2y - 3)$ **35.** $(4x - 5)(x + 1)$ **36.** $y(x + y)(x - y)$ **37.** $4(t^2 + 9)$ **38.** $(x + 1)(x + y)$ **39.** $(x + 1)(a + 2)$ **40.** $9x(2x^2 - 7x + 1)$ **41.** $4a(3a^2 - 6a + 1)$ **42.** $(x + 16)(x - 2)$ **43.** prime **44.** $(4a - 7b)^2$ **45.** $(5p - 7q)^2$ **46.** $(7x + 3y)(x + 3y)$ **47.** $(5 - 2y)(25 + 10y + 4y^2)$ **48.** $(4x + 3)(16x^2 - 12x + 9)$ **49.** $-(x - 5)(x + 6)$ **50.** $-(x - 2)(x - 4)$ **51.** $(7 - x)(2 + x)$ **52.** $(3 + x)(1 - x)$ **53.** $3x^2y(x + 6)(x - 4)$ **54.** $2xy(x + 5y)(x - y)$ **55.** $5xy^2(x - 7y)(x - y)$ **56.** $4x^2y(x - 5)(x + 3)$ **57.** $3xy(4x^2 + 81)$ **58.** $2xy^2(3x^2 + 4)$ **59.** $(2 + x)(2 - x)$ **60.** $(3 + y)(3 - y)$ **61.** $(s + 4)(3r - 1)$ **62.** $(x - 2)(x^2 + 3)$ **63.** $(4x - 3)(x - 2y)$ **64.** $(2x - y)(2x + 7z)$ **65.** $6(x + 2y)(x + y)$ **66.** $2(x + 4y)(6x - y)$ **67.** $(x + 3)(y + 2)(y - 2)$ **68.** $(y + 3)(y - 3)(x^2 + 3)$ **69.** $(5 + x)(x + y)$ **70.** $(x - y)(7 + y)$ **71.** $(7t - 1)(2t - 1)$ **72.** prime **73.** $(3x + 5)(x - 1)$ **74.** $(7x - 2)(x + 3)$ **75.** $(x + 12y)(x - 3y)$ **76.** $(3x - 2y)(x + 4y)$ **77.** $(1 - 10ab)(1 + 2ab)$ **78.** $(1 + 5ab)(1 - 12ab)$ **79.** $(3 + x)(3 - x)(1 + x)(1 - x)$ **80.** $(3 + x)(3 - x)(2 + x)(2 - x)$ **81.** $(x + 4)(x - 4)(x^2 + 2)$ **82.** $(x + 5)(x - 5)(x^2 + 3)$ **83.** $(x - 15)(x - 8)$ **84.** $(y + 16)(y + 6)$ **85.** $2x(3x - 2)(x - 4)$ **86.** $2y(3y + 5)(y - 3)$ **87.** $(3x - 5y)(9x^2 + 15xy + 25y^2)$ **88.** $(6y - z)(36y^2 + 6yz + z^2)$ **89.** $(xy + 2z)(x^2y^2 - 2xyz + 4z^2)$ **90.** $(3ab + 2)(9a^2b^2 - 6ab + 4)$ **91.** $2xy(1 + 6x)(1 - 6x)$ **92.** $2x(x + 3)(x - 3)$ **93.** $(x + 2)(x - 2)(x + 6)$ **94.** $(x - 2)(x + 6)(x - 6)$ **95.** $2a^2(3a + 5)$ **96.** $2n(2n - 3)$ **97.** $(a^2 + 2)(a + 2)$ **98.** $(a - b)(1 + x)$ **99.** $(x + 2)(x - 2)(x + 7)$ **100.** $(a + 3)(a - 3)(a + 5)$ **101.** $(x - y + z)(x - y - z)$ **102.** $(x + 2y + 3)(x + 2y - 3)$ **103.** $(9 + 5x + 1)(9 - 5x - 1)$ **104.** $(b + 4a + c)(b - 4a - c)$ **105.** answers may vary **106.** yes; $9(x^2 + 9y^2)$ **107.** a, c **108.** b, c

Section 6.6

Practice Exercises

1. $-4, 5$ **2.** $-\frac{3}{4}, 12$ **3.** $0, \frac{6}{7}$ **4.** $-4, 12$ **5.** $\frac{4}{3}$ **6.** $-3, \frac{2}{3}$ **7.** $-6, 4$ **8.** $-3, 0, 3$ **9.** $\frac{2}{3}, \frac{3}{2}, 5$ **10.** $-3, 0, 2$ **11.** The x-intercepts are $(2, 0)$ and $(4, 0)$.

Calculator Explorations 6.6

1. $-0.9, 2.2$ **3.** no real solution **5.** $-1.8, 2.8$

Vocabulary, Readiness & Video Check 6.6

1. quadratic **3.** $3, -5$ **5.** One side of the equation must be a factored polynomial and the other side must be zero. **7.** To find the x-intercepts of any graph in two variables, we let $y = 0$. Doing this with our quadratic equation gives us an equation = 0, which we can solve by factoring.

Exercise Set 6.6

1. 6, 7 **3.** $2, -1$ **5.** $-9, -17$ **7.** $0, -6$ **9.** 0, 8 **11.** $-\frac{3}{2}, \frac{5}{4}$ **13.** $\frac{7}{2}, -\frac{2}{7}$ **15.** $\frac{1}{2}, -\frac{1}{3}$ **17.** $-0.2, -1.5$ **19.** 9, 4 **21.** $-4, 2$ **23.** 0, 7 **25.** $8, -4$ **27.** $4, -4$ **29.** $-3, 12$ **31.** $\frac{7}{3}, -2$ **33.** $-5, 5$ **35.** $-2, \frac{1}{6}$ **37.** 0, 4, 8 **39.** $\frac{3}{4}$ **41.** $-\frac{1}{2}, 0, \frac{1}{2}$ **43.** $-\frac{3}{8}, 0, \frac{1}{2}$

45. $-3, 2$ **47.** $-20, 0$ **49.** $\frac{17}{2}$ **51.** $-\frac{1}{2}, \frac{1}{2}$ **53.** $-\frac{3}{2}, -\frac{1}{2}, 3$ **55.** $-5, 3$ **57.** $-\frac{5}{6}, \frac{6}{5}$ **59.** $2, -\frac{4}{5}$ **61.** $-\frac{4}{3}, 5$ **63.** $-4, 3$ **65.** $\frac{8}{3}, -9, 0$ **67.** -7 **69.** $0, \frac{3}{2}$ **71.** $0, 1, -1$ **73.** $-6, \frac{4}{3}$ **75.** $\frac{6}{7}, 1$ **77.** $\left(-\frac{4}{3}, 0\right), (1, 0)$ **79.** $(-2, 0), (5, 0)$ **81.** $(-6, 0), \left(\frac{1}{2}, 0\right)$ **83.** e **85.** b **87.** c **89.** $\frac{47}{45}$ **91.** $\frac{17}{60}$ **93.** $\frac{15}{8}$ **95.** $\frac{7}{10}$ **97.** didn't write equation in standard form; should be $x = 4$ or $x = -2$ **99.** answers may vary; for example $(x - 6)(x + 1) = 0$ **101.** answers may vary; for example, $x^2 - 12x + 35 = 0$ **103. a.** 300; 304; 276; 216; 124; 0; -156 **b.** 5 sec **c.** 304 ft **d.**

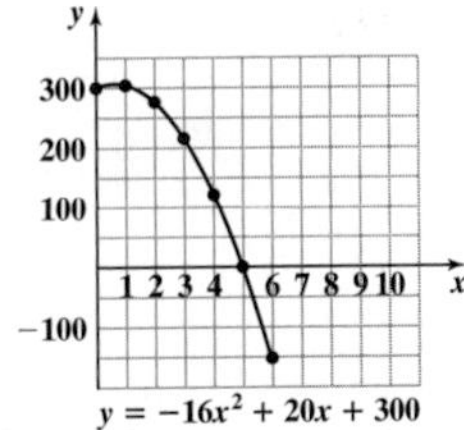

105. $0, \frac{1}{2}$ **107.** $0, -15$

Section 6.7

Practice Exercises

1. 2 sec **2.** There are two numbers. They are -4 and 12. **3.** base: 35 ft; height: 12 ft **4.** 7 and 8 or -6 and -5 **5.** leg: 8 units; leg: 15 units; hypotenuse: 17 units

Vocabulary, Readiness & Video Check 6.7

1. In applications, the context of the problem needs to be considered. Each exercise resulted in both a positive and a negative solution, and a negative solution is not appropriate for any of the problems.

Exercise Set 6.7

1. width $= x$; length $= x + 4$ **3.** x and $x + 2$ if x is an odd integer **5.** base $= x$; height $= 4x + 1$ **7.** 11 units **9.** 15 cm, 13 cm, 70 cm, 22 cm **11.** base $=$ 16 mi; height $=$ 6 mi **13.** 5 sec **15.** width $=$ 5 cm; length $=$ 6 cm **17.** 54 diagonals **19.** 10 sides **21.** -12 or 11 **23.** 14, 15 **25.** 13 feet **27.** 5 in. **29.** 12 mm, 16 mm, 20 mm **31.** 10 km **33.** 36 ft **35.** 9.5 sec **37.** 20% **39.** length: 15 mi; width: 8 mi **41.** 105 units **43.** 2 million **45.** 1.9 million **47.** 2003 **49.** answers may vary **51.** $\frac{4}{7}$ **53.** $\frac{3}{2}$ **55.** $\frac{1}{3}$ **57.** slow boat: 8 mph; fast boat: 15 mph **59.** 13 and 7 **61.** width: 29 m; length: 35 m **63.** answers may vary

Chapter 6 Vocabulary Check

1. quadratic equation **2.** factoring **3.** greatest common factor **4.** perfect square trinomial **5.** difference of two squares **6.** difference of two cubes **7.** sum of two cubes **8.** 0 **9.** hypotenuse **10.** leg **11.** hypotenuse

Chapter 6 Review

1. $2x - 5$ **3.** $4x(5x + 3)$ **5.** $(2x + 3)(3x - 5)$ **7.** $(x - 1)(3x + 2)$ **9.** $(2a + b)(5a + 7b)$ **11.** $(x + 4)(x + 2)$ **13.** prime **15.** $(x + 6y)(x - 2y)$ **17.** $2(3 - x)(12 + x)$ **19.** $10a(a - 1)(a - 10)$ **21.** $-48, 2$ **23.** $(2x + 1)(x + 6)$ **25.** $(3x + 4y)(2x - y)$ **27.** $5y(2y - 3)(y + 4)$ **29.** $2(3x - 5)^2$ **31.** $(2x + 3)(2x - 3)$ **33.** prime **35.** $(2x + 3)(4x^2 - 6x + 9)$ **37.** $2(3 - xy)(9 + 3xy + x^2y^2)$ **39.** $(4x^2 + 1)(2x + 1)(2x - 1)$ **41.** $-6, 2$ **43.** $-\frac{1}{5}, -3$ **45.** $-4, 6$ **47.** $2, 8$ **49.** $-\frac{2}{7}, \frac{3}{8}$ **51.** $-\frac{2}{5}$ **53.** 3 **55.** $0, -\frac{7}{4}, 3$ **57.** c **59.** 9 units **61.** width: 20 in.; length: 25 in. **63.** 19 and 20 **65. a.** 17.5 sec and 10 sec; The rocket reaches a height of 2800 ft on its way up and on its way back down. **b.** 27.5 sec **67.** $7(x - 9)$ **69.** $\left(m + \frac{2}{5}\right)\left(m - \frac{2}{5}\right)$ **71.** $(y + 2)(x - 1)$ **73.** $3x(x - 9)(x - 1)$ **75.** $2(x + 3)(x - 3)$ **77.** $5(x + 2)^2$ **79.** $2xy(2x - 3y)$ **81.** $3(8x^2 - x - 6)$ **83.** $(x + 3)(x + 2)(x - 2)$ **85.** $5x^2 - 9x - 2$; $(5x + 1)(x - 2)$ **87.** $-\frac{7}{2}, 4$ **89.** $0, -7, -4$ **91.** $0, 16$ **93.** length: 6 in.; width: 2 in. **95.** $28x^2 - \pi x^2$; $x^2(28 - \pi)$

Chapter 6 Test

1. $(x + 7)(x + 4)$ **2.** $(7 - m)(7 + m)$ **3.** $(y + 11)^2$ **4.** $(a + 3)(4 - y)$ **5.** prime **6.** $(y - 12)(y + 4)$ **7.** prime **8.** $3x(3x + 1)(x + 4)$ **9.** $(3a - 7)(a + b)$ **10.** $(3x - 2)(x - 1)$ **11.** $(x + 12y)(x + 2y)$ **12.** $5(6 + x)(6 - x)$ **13.** $(6t + 5)(t - 1)$ **14.** $(y + 2)(y - 2)(x - 7)$ **15.** $x(1 + x^2)(1 + x)(1 - x)$ **16.** $-xy(y^2 + x^2)$ **17.** $(4x - 1)(16x^2 + 4x + 1)$ **18.** $8(y - 2)(y^2 + 2y + 4)$ **19.** $-9, 3$ **20.** $-7, 2$ **21.** $-7, 1$ **22.** $0, \frac{3}{2}, -\frac{4}{3}$ **23.** $0, 3, -3$ **24.** $-3, 5$ **25.** $0, \frac{5}{2}$ **26.** 17 ft **27.** 8 and 9 **28.** 7 sec **29.** hypotenuse: 25 cm; legs: 15 cm, 20 cm

Chapter 6 Cumulative Review

1. a. $9 \le 11$ **b.** $8 > 1$ **c.** $3 \ne 4$; Sec. 1.2, Ex. 3 **3. a.** $\frac{6}{7}$ **b.** $\frac{11}{27}$ **c.** $\frac{22}{5}$; Sec. 1.3, Ex. 2 **5.** $\frac{14}{3}$; Sec. 1.4, Ex. 5 **7. a.** -12 **b.** -1; Sec. 1.5, Ex. 7 **9. a.** -32 **b.** -14 **c.** 90; Sec. 1.7, Ex. 1 **11. a.** $4x$ **b.** $11y^2$ **c.** $8x^2 - x$; **d.** $5n^2$; Sec. 2.1, Ex. 3 **13.** -4; Sec. 2.2, Ex. 7 **15.** -11; Sec. 2.3, Ex. 2 **17.** $\frac{16}{3}$; Sec. 2.4, Ex. 2 **19.** shorter: 12 in.; longer: 36 in.; Sec. 2.5, Ex. 3

21. 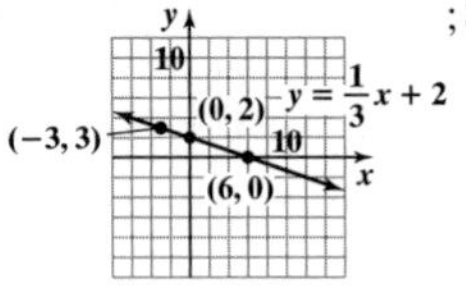

; Sec. 3.2, Ex. 5 **23.** $m = \frac{3}{4}$; y-intercept: $(0, -1)$; Sec. 3.4, Ex. 5 **25. a.** 250 **b.** 1; Sec. 5.1, Ex. 2 **27. a.** 2 **b.** 5 **c.** 0; Sec. 5.2, Ex. 1 **29.** $9x^2 - 6x - 1$; Sec. 5.2, Ex. 13 **31.** $6x^2 - 11x - 10$; Sec. 5.3, Ex. 5 **33.** $9y^2 + 6y + 1$; Sec. 5.4, Ex. 4 **35. a.** $\frac{1}{9}$ **b.** $\frac{2}{x^3}$ **c.** $\frac{3}{4}$ **d.** $\frac{1}{16}$ **e.** y^4 Sec. 5.5, Ex. 1 **37. a.** 3.67×10^8 **b.** 3.0×10^{-6} **c.** 2.052×10^{10} **d.** 8.5×10^{-4}; Sec. 5.5, Ex. 5 **39.** $x + 4$; Sec. 5.6, Ex. 4 **41. a.** x^3 **b.** y; Sec. 6.1, Ex. 2 **43.** $(x + 3)(x + 4)$; Sec. 6.2, Ex. 1 **45.** $(4x - 1)(2x - 5)$; Sec. 6.3, Ex. 2 **47.** $(5a + 3b)(5a - 3b)$; Sec. 6.5, Ex. 2b **49.** 3, −1; Sec. 6.6, Ex. 1

CHAPTER 7 RATIONAL EXPRESSIONS

Section 7.1

Practice Exercises

1. a. $\frac{9}{7}$ **b.** $-\frac{3}{11}$ **2. a.** $x = -6$ **b.** none **c.** $x = -2$ or $x = -4$ **d.** none **3.** $\frac{x^5}{6}$ **4.** $\frac{x + 1}{x - 2}$ **5.** $\frac{x^2}{x + 9}$ **6.** $\frac{1}{x + 7}$ **7. a.** −1 **b.** 1 **8.** $-\frac{2x + 3}{x + 4}$ or $\frac{-2x - 3}{x + 4}$ **9.** $\frac{-(x + 3)}{6x - 11}; \frac{-x - 3}{6x - 11}; \frac{x + 3}{-(6x - 11)}; \frac{x + 3}{-6x + 11}; \frac{x + 3}{11 - 6x}$

Vocabulary, Readiness & Video Check 7.1

1. rational expression **3.** −1 **5.** 2 **7.** $\frac{-a}{b}; \frac{a}{-b}$ **9.** Rational expressions are fractions and are therefore undefined if the denominator is zero; if a denominator contains variables, set it equal to zero and solve. **11.** You would need to write parentheses around the numerator or denominator if it had more than one term because the negative sign needs to apply to the entire numerator or denominator.

Exercise Set 7.1

1. $\frac{7}{4}$ **3.** 3 **5.** $-\frac{8}{3}$ **7.** $-\frac{11}{2}$ **9.** $x = 0$ **11.** $x = -2$ **13.** $x = \frac{5}{2}$ **15.** none **17.** $x = 6, x = -1$ **19.** none **21.** $x = -2, x = -\frac{7}{3}$ **23.** $\frac{-(x - 10)}{x + 8}; \frac{-x + 10}{x + 8}; \frac{x - 10}{-(x + 8)}; \frac{x - 10}{-x - 8}$ **25.** $\frac{-(5y - 3)}{y - 12}; \frac{-5y + 3}{y - 12}; \frac{5y - 3}{-(y - 12)}; \frac{5y - 3}{-y + 12}$ **27.** 1 **29.** −1 **31.** $\frac{1}{4(x + 2)}$ **33.** $\frac{1}{x + 2}$ **35.** can't simplify **37.** −5 **39.** $\frac{7}{x}$ **41.** $\frac{1}{x - 9}$ **43.** $5x + 1$ **45.** $\frac{x^2}{x - 2}$ **47.** $7x$ **49.** $\frac{x + 5}{x - 5}$ **51.** $\frac{x + 2}{x + 4}$ **53.** $\frac{x + 2}{2}$ **55.** $-(x + 2)$ or $-x - 2$ **57.** $\frac{x + 1}{x - 1}$ **59.** $\frac{m - 3}{m + 2}$ **61.** $\frac{11x}{6}$ **63.** $x + y$ **65.** $\frac{5 - y}{2}$ **67.** $x^2 - 2x + 4$ **69.** $-x^2 - x - 1$ **71.** $\frac{2y + 5}{3y + 4}$ **73.** correct **75.** correct **77.** $\frac{3}{11}$ **79.** $\frac{4}{3}$ **81.** $\frac{117}{40}$ **83.** correct **85.** incorrect; $\frac{1 + 2}{1 + 3} = \frac{3}{4}$ **87.** no **89.** yes **91.** answers may vary **93.** answers may vary **95. a.** \$37.5 million **b.** ≈ \$85.7 million **c.** ≈ \$48.2 million **97.** 400 mg **99.** $C = 78.125$; medium **101.** 59.6% **103.** 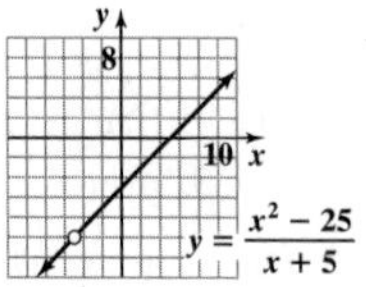

105. 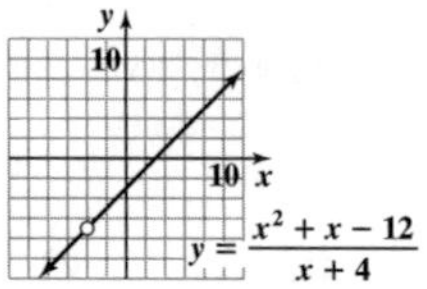

Section 7.2

Practice Exercises

1. a. $\frac{12a}{5b^2}$ **b.** $-\frac{2q}{3}$ **2.** $\frac{3}{x + 1}$ **3.** $-\frac{3x - 5}{2x(x + 2)}$ **4.** $\frac{b^2}{8a^2}$ **5.** $\frac{3(x - 5)}{4}$ **6.** $\frac{2}{x(x - 3)}$ **7.** 1 **8. a.** $\frac{(y + 9)^2}{16x^2}$ **b.** $\frac{1}{4x}$ **c.** $-\frac{7(x - 2)}{x + 4}$ **9.** 2 sq ft **10.** 504 sq in. **11.** 549,000 sq ft **12.** 70.0 miles per hour

Vocabulary, Readiness & Video Check 7.2

1. reciprocals **3.** $\frac{a \cdot d}{b \cdot c}$ or $\frac{ad}{bc}$ **5.** $\frac{6}{7}$ **7.** fractions; reciprocal **9.** The units in the unit fraction consist of $\frac{\text{units converting to}}{\text{original units}}$.

Exercise Set 7.2

1. $\frac{21}{4y}$ **3.** x^4 **5.** $-\frac{b^2}{6}$ **7.** $\frac{x^2}{10}$ **9.** $\frac{1}{3}$ **11.** $\frac{m + n}{m - n}$ **13.** $\frac{x + 5}{x}$ **15.** $\frac{(x + 2)(x - 3)}{(x - 4)(x + 4)}$ **17.** $\frac{2x^4}{3}$ **19.** $\frac{12}{y^6}$ **21.** $x(x + 4)$ **23.** $\frac{3(x + 1)}{x^3(x - 1)}$ **25.** $m^2 - n^2$ **27.** $-\frac{x + 2}{x - 3}$ **29.** $\frac{x + 2}{x - 3}$ **31.** $\frac{5}{6}$ **33.** $\frac{3x}{8}$ **35.** $\frac{3}{2}$ **37.** $\frac{3x + 4y}{2(x + 2y)}$ **39.** $\frac{2(x + 2)}{x - 2}$ **41.** $-\frac{y(x + 2)}{4}$ **43.** $\frac{(a + 5)(a + 3)}{(a + 2)(a + 1)}$ **45.** $\frac{5}{x}$ **47.** $\frac{2(n - 8)}{3n - 1}$ **49.** $4x^3(x - 3)$ **51.** $\frac{(a + b)^2}{a - b}$ **53.** $\frac{3x + 5}{x^2 + 4}$ **55.** $\frac{4}{x - 2}$

57. $\frac{a-b}{6(a^2+ab+b^2)}$ **59.** 1440 **61.** 5 **63.** 81 **65.** 73 **67.** 56.7 **69.** 1,201,500 sq ft **71.** 244.9 miles/hour **73.** 1 **75.** $-\frac{10}{9}$ **77.** $-\frac{1}{5}$ **79.**

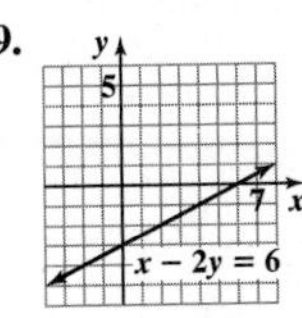

81. true **83.** false; $\frac{x^2+3x}{20}$ **85.** $\frac{2}{9(x-5)}$ sq ft **87.** $\frac{x}{2}$ **89.** $\frac{5a(2a+b)(3a-2b)}{b^2(a-b)(a+2b)}$ **91.** answers may vary

Section 7.3

Practice Exercises

1. $\frac{2a}{b}$ **2.** 1 **3.** $4x-5$ **4. a.** 42 **b.** $45y^3$ **5. a.** $(y-5)(y-4)$ **b.** $a(a+2)$ **6.** $3(2x-1)^2$ **7.** $(x+4)(x-4)(x+1)$ **8.** $3-x$ or $x-3$ **9. a.** $\frac{21x^2y}{35xy^2}$ **b.** $\frac{18x}{8x+14}$ **10.** $\frac{3x-6}{(x-2)(x+3)(x-5)}$

Vocabulary, Readiness & Video Check 7.3

1. $\frac{9}{11}$ **3.** $\frac{a+c}{b}$ **5.** $\frac{5-(6+x)}{x}$ **7.** We factor denominators into the smallest factors—including coefficients—so we can determine the most number of times each unique factor occurs in any one denominator for the LCD.

Exercise Set 7.3

1. $\frac{a+9}{13}$ **3.** $\frac{3m}{n}$ **5.** 4 **7.** $\frac{y+10}{3+y}$ **9.** $5x+3$ **11.** $\frac{4}{a+5}$ **13.** $\frac{1}{x-6}$ **15.** $\frac{5x+7}{x-3}$ **17.** $x+5$ **19.** 3 **21.** $4x^3$ **23.** $8x(x+2)$ **25.** $(x+3)(x-2)$ **27.** $3(x+6)$ **29.** $5(x-6)^2$ **31.** $6(x+1)^2$ **33.** $x-8$ or $8-x$ **35.** $(x-1)(x+4)(x+3)$ **37.** $(3x+1)(x+1)(x-1)(2x+1)$ **39.** $2x^2(x+4)(x-4)$ **41.** $\frac{6x}{4x^2}$ **43.** $\frac{24b^2}{12ab^2}$ **45.** $\frac{9y}{2y(x+3)}$ **47.** $\frac{9ab+2b}{5b(a+2)}$ **49.** $\frac{x^2+x}{x(x+4)(x+2)(x+1)}$ **51.** $\frac{18y-2}{30x^2-60}$ **53.** $2x$ **55.** $\frac{x+3}{2x-1}$ **57.** $x+1$ **59.** $\frac{1}{x^2-8}$ **61.** $\frac{6(4x+1)}{x(2x+1)}$ **63.** $\frac{29}{21}$ **65.** $-\frac{7}{12}$ **67.** $\frac{7}{30}$ **69.** d **71.** answers may vary **73.** c **75.** b **77.** $-\frac{5}{x-2}$ **79.** $\frac{7+x}{x-2}$ **81.** $\frac{20}{x-2}$ m **83.** answers may vary **85.** 95,304 Earth days **87.** answers may vary **89.** answers may vary

Section 7.4

Practice Exercises

1. a. 0 **b.** $\frac{21a+10}{24a^2}$ **2.** $\frac{6}{x-5}$ **3.** $\frac{13y+3}{5y(y+1)}$ **4.** $\frac{13}{x-5}$ **5.** $\frac{3b+6}{b+3}$ or $\frac{3(b+2)}{b+3}$ **6.** $\frac{10-3x^2}{2x(2x+3)}$ **7.** $\frac{x(5x+6)}{(x+4)(x+3)(x-3)}$

Vocabulary, Readiness & Video Check 7.4

1. d **3.** a **5.** The problem adds two rational expressions with denominators that are opposites of each other. Recognizing this special case can save you time and effort. If you recognize that one denominator is −1 times the other denominator, you may save time.

Exercise Set 7.4

1. $\frac{5}{x}$ **3.** $\frac{75a-6b^2}{5b}$ **5.** $\frac{6x+5}{2x^2}$ **7.** $\frac{11}{x+1}$ **9.** $\frac{x-6}{(x-2)(x+2)}$ **11.** $\frac{35x-6}{4x(x-2)}$ **13.** $-\frac{2}{x-3}$ **15.** 0 **17.** $-\frac{1}{x^2-1}$ **19.** $\frac{5+2x}{x}$ **21.** $\frac{6x-7}{x-2}$ **23.** $-\frac{y+4}{y+3}$ **25.** $\frac{-5x+14}{4x}$ or $-\frac{5x-14}{4x}$ **27.** 2 **29.** $\frac{9x^4-4x^2}{21}$ **31.** $\frac{x+2}{(x+3)^2}$ **33.** $\frac{9b-4}{5b(b-1)}$ **35.** $\frac{2+m}{m}$ **37.** $\frac{x^2+3x}{(x-7)(x-2)}$ or $\frac{x(x+3)}{(x-7)(x-2)}$ **39.** $\frac{10}{1-2x}$ **41.** $\frac{15x-1}{(x+1)^2(x-1)}$ **43.** $\frac{x^2-3x-2}{(x-1)^2(x+1)}$ **45.** $\frac{a+2}{2(a+3)}$ **47.** $\frac{y(2y+1)}{(2y+3)^2}$ **49.** $\frac{x-10}{2(x-2)}$ **51.** $\frac{2x+21}{(x+3)^2}$ **53.** $\frac{-5x+23}{(x-2)(x-3)}$ **55.** $\frac{7}{2(m-10)}$ **57.** $\frac{2x^2-2x-46}{(x+1)(x-6)(x-5)}$ or $\frac{2(x^2-x-23)}{(x+1)(x-6)(x-5)}$ **59.** $\frac{n+4}{4n(n-1)(n-2)}$ **61.** 10 **63.** 2 **65.** $\frac{25a}{9(a-2)}$ **67.** $\frac{x+4}{(x-2)(x-1)}$ **69.** $x=\frac{2}{3}$ **71.** $x=-\frac{1}{2}, x=1$ **73.** $x=-\frac{15}{2}$ **75.** $\frac{6x^2-5x-3}{x(x+1)(x-1)}$ **77.** $\frac{4x^2-15x+6}{(x-2)^2(x+2)(x-3)}$ **79.** $\frac{-2x^2+14x+55}{(x+2)(x+7)(x+3)}$ **81.** $\frac{2x-16}{(x+4)(x-4)}$ in. **83.** $\frac{P-G}{P}$ **85.** answers may vary **87.** $\left(\frac{90x-40}{x}\right)^\circ$ **89.** answers may vary

Section 7.5

Practice Exercises

1. −2 **2.** 13 **3.** −1, 7 **4.** $-\frac{19}{2}$ **5.** 3 **6.** −8 **7.** $b=\frac{ax}{a-x}$

Graphing Calculator Explorations 7.5

1.

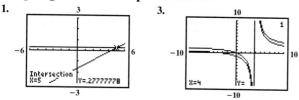

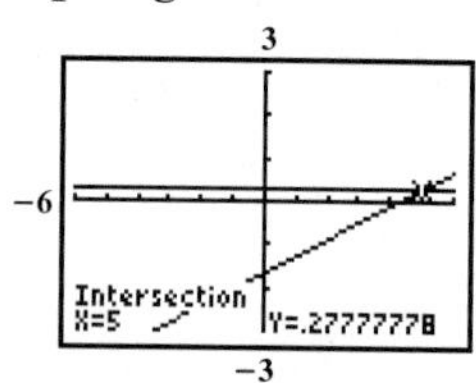

3.

Vocabulary, Readiness & Video Check 7.5

1. c **3.** b **5.** a **7.** These equations are solved in very different ways, so you need to determine the next correct move to make. For a linear equation, you first "move" variable terms on one side and numbers on the other; for a quadratic equation, you first set the equation equal to 0. **9.** the steps for solving an equation containing rational expressions; as if it's the only variable in the equation

Exercise Set 7.5

1. 30 **3.** 0 **5.** -2 **7.** $-5, 2$ **9.** 5 **11.** 3 **13.** 1 **15.** 5 **17.** no solution **19.** 4 **21.** -8 **23.** $6, -4$ **25.** 1 **27.** $3, -4$ **29.** -3 **31.** 0 **33.** -2 **35.** $8, -2$ **37.** no solution **39.** 3 **41.** $-11, 1$ **43.** $I = \frac{E}{R}$ **45.** $B = \frac{2U - TE}{T}$ **47.** $w = \frac{Bh^2}{705}$ **49.** $G = \frac{V}{N - R}$ **51.** $r = \frac{C}{2\pi}$ **53.** $x = \frac{3y}{3 + y}$ **55.** $\frac{1}{x}$ **57.** $\frac{1}{x} + \frac{1}{2}$ **59.** $\frac{1}{3}$ **61.** $(2, 0), (0, -2)$ **63.** $(-4, 0), (-2, 0), (3, 0), (0, 4)$ **65.** answers may vary **67.** $\frac{5x + 9}{9x}$ **69.** no solution **71.** 100°, 80° **73.** 22.5°, 67.5° **75.** $\frac{17}{4}$

Integrated Review

1. expression; $\frac{3 + 2x}{3x}$ **2.** expression; $\frac{18 + 5a}{6a}$ **3.** equation; 3 **4.** equation; 18 **5.** expression; $\frac{x + 1}{x(x - 1)}$ **6.** expression; $\frac{3(x + 1)}{x(x - 3)}$ **7.** equation; no solution **8.** equation; 1 **9.** expression; 10 **10.** expression; $\frac{z}{3(9z - 5)}$ **11.** expression; $\frac{5x + 7}{x - 3}$ **12.** expression; $\frac{7p + 5}{2p + 7}$ **13.** equation; 23 **14.** equation; 5 **15.** expression; $\frac{25a}{9(a - 2)}$ **16.** expression; $\frac{4x + 5}{(x + 1)(x - 1)}$ **17.** expression; $\frac{3x^2 + 5x + 3}{(3x - 1)^2}$ **18.** expression; $\frac{2x^2 - 3x - 1}{(2x - 5)^2}$ **19.** expression; $\frac{4x - 37}{5x}$ **20.** expression; $\frac{29x - 23}{3x}$ **21.** equation; $\frac{8}{5}$ **22.** equation; $-\frac{7}{3}$ **23.** answers may vary **24.** answers may vary

Section 7.6
Practice Exercises

1. 99 **2.** $\frac{13}{3}$ **3.** \$9.03 **4.** 6 **5.** 15 **6.** $1\frac{5}{7}$ hr **7.** bus: 45 mph; car: 60 mph

Vocabulary, Readiness & Video Check 7.6

1. c **3.** $\frac{1}{x}; \frac{1}{x} - 3$ **5.** $z + 5; \frac{1}{z + 5}$ **7.** $2y; \frac{11}{2y}$ **9.** No. Proportions are actually equations containing rational expressions, so they can also be solved by using the steps to solve those equations. **11.** divided by, quotient **13.** $\frac{325}{x + 7} = \frac{290}{x}$

Exercise Set 7.6

1. 4 **3.** $\frac{50}{9}$ **5.** -3 **7.** $\frac{14}{9}$ **9.** 123 lb **11.** 165 cal **13.** $y = 21.25$ **15.** $y = 5\frac{5}{7}$ ft **17.** 2 **19.** -3 **21.** $2\frac{2}{9}$ hr **23.** $1\frac{1}{2}$ min **25.** trip to park rate: r; to park time: $\frac{12}{r}$; return trip rate: r; return time: $\frac{18}{r} = \frac{12}{r} + 1$; $r = 6$ mph **27.** 1st portion: 10 mph; cooldown: 8 mph **29.** 360 sq ft **31.** 2 **33.** \$108.00 **35.** 20 mph **37.** $y = 37\frac{1}{2}$ ft **39.** 41 mph; 51 mph **41.** 5 **43.** 217 mph **45.** 9 gal **47.** 8 mph **49.** 2.2 mph; 3.3 mph **51.** 3 hr **53.** $26\frac{2}{3}$ ft **55.** 216 nuts **57.** $666\frac{2}{3}$ mi **59.** 20 hr **61.** car: 70 mph; motorcycle: 60 mph **63.** $5\frac{1}{4}$ hr **65.** 8 **67.** first car: 64 mph; second car: 50 mph **69.** 510 mph **71.** $x = 5$ **73.** $x = 13.5$ **75.** $-\frac{4}{3}$; downward **77.** $\frac{11}{4}$; upward **79.** undefined slope; vertical **81.** 40,200 megawatts **83.** 22,510,000 people **85.** yes **87.** first pump: 28 min; second pump: 84 min **89.** none; answers may vary **91.** answers may vary **93.** $R = \frac{D}{T}$

Section 7.7
Practice Exercises

1. $y = 5x$ **2.** $k = \frac{1}{4}; y = \frac{1}{4}x$; when x is 20, $y = 5$. **3.** $k = \frac{3}{4}; y = \frac{3}{4}x$ **4.** $y = \frac{8}{x}$ **5.** $k = 2.1; y = \frac{2.1}{x}$; when x is 70, $y = 0.03$. **6.** 6.48 sq units **7.** 25 ml

Vocabulary, Readiness & Video Check 7.7

1. inverse **3.** direct **5.** inverse **7.** inverse **9.** direct **11.** linear; slope **13.** No. The direct relationship is the power of x times a constant, and the inverse relationship is the reciprocal of the power of x times a constant.

Exercise Set 7.7

1. $y = \frac{1}{2}x$ **3.** $y = 6x$ **5.** $y = 3x$ **7.** $y = \frac{2}{3}x$ **9.** $y = \frac{7}{x}$ **11.** $y = \frac{0.5}{x}$ **13.** $y = kx$ **15.** $h = \frac{k}{t}$ **17.** $z = kx^2$ **19.** $y = \frac{k}{z^3}$ **21.** $x = \frac{k}{\sqrt{y}}$ **23.** $y = 40$ **25.** $y = 3$ **27.** $z = 54$ **29.** $a = \frac{4}{9}$ **31.** \$62.50 **33.** \$6 **35.** $5\frac{1}{3}$ in. **37.** 179.1 lb **39.** 1600 ft **41.** $\frac{1}{2}$ **43.** $\frac{3}{7}$ **45.** multiplied by 3 **47.** it is doubled

Section 7.8

Practice Exercises

1. $\frac{11}{8}$ **2.** $\frac{85}{33}$ **3.** $\frac{5(8-x)}{x(2-x)}$ **4.** $\frac{85}{33}$ **5.** $\frac{a-b}{a+4b}$ **6.** $\frac{4a+3b^2}{ab(a-3b)}$

Vocabulary, Readiness & Video Check 7.8

1. $\frac{y}{5x}$ **3.** $\frac{3x}{5}$ **5.** a single fraction in the numerator and in the denominator

Exercise Set 7.8

1. $\frac{2}{3}$ **3.** $\frac{2}{3}$ **5.** $\frac{1}{2}$ **7.** $-\frac{21}{5}$ **9.** $\frac{27}{16}$ **11.** $\frac{4}{3}$ **13.** $\frac{1}{21}$ **15.** $-\frac{4x}{15}$ **17.** $\frac{m-n}{m+n}$ **19.** $\frac{2x(x-5)}{7x^2+10}$ **21.** $\frac{1}{y-1}$ **23.** $\frac{1}{6}$ **25.** $\frac{x+y}{x-y}$ **27.** $\frac{3}{7}$ **29.** $\frac{a}{x+b}$ **31.** $\frac{7(y-3)}{8+y}$ **33.** $\frac{3x}{x-4}$ **35.** $-\frac{x+8}{x-2}$ **37.** $\frac{s^2+r^2}{s^2-r^2}$ **39.** $\frac{(x-6)(x+4)}{x-2}$ **41.** Serena Williams **43.** about \$8.4 million **45.** answers may vary **47.** $\frac{13}{24}$ **49.** $\frac{R_1R_2}{R_2+R_1}$ **51.** $\frac{2x}{2-x}$ **53.** $\frac{1}{y^2-1}$ **55.** 12 hr

Chapter 7 Vocabulary Check

1. rational expression **2.** complex fraction **3.** $\frac{-a}{b}; \frac{a}{-b}$ **4.** denominator **5.** simplifying **6.** reciprocals **7.** least common denominator **8.** ratio **9.** proportion **10.** cross products **11.** inverse variation **12.** direct variation

Chapter 7 Review

1. $x = 2, x = -2$ **3.** $\frac{4}{3}$ **5.** $\frac{2}{x}$ **7.** $\frac{1}{x-5}$ **9.** $\frac{x(x-2)}{x+1}$ **11.** $\frac{x-3}{x-5}$ **13.** $\frac{x+a}{x-c}$ **15.** $-\frac{1}{x^2+4x+16}$ **17.** $\frac{3x^2}{y}$ **19.** $\frac{x-3}{x+2}$ **21.** $\frac{x+3}{x-4}$ **23.** $(x-6)(x-3)$ **25.** $\frac{1}{2}$ **27.** $-\frac{2(2x+3)}{y-2}$ **29.** $\frac{1}{x+2}$ **31.** $\frac{2x-10}{3x^2}$ **33.** $14x$ **35.** $\frac{10x^2y}{14x^3y}$ **37.** $\frac{x^2-3x-10}{(x+2)(x-5)(x+9)}$ **39.** $\frac{4y-30x^2}{5x^2y}$ **41.** $\frac{-2x-2}{x+3}$ **43.** $\frac{x-4}{3x}$ **45.** $\frac{x^2+2x+4}{4x}; \frac{x+2}{32}$ **47.** 30 **49.** no solution **51.** $\frac{9}{7}$ **53.** $b = \frac{4A}{5x^2}$ **55.** $x = 6$ **57.** $x = 9$ **59.** 675 parts **61.** 3 **63.** fast car speed: 30 mph; slow car speed: 20 mph **65.** $17\frac{1}{2}$ hr **67.** $x = 15$ **69.** $y = 110$ **71.** $y = \frac{100}{27}$ **73.** \$3960 **75.** $-\frac{7}{18y}$ **77.** $\frac{3y-1}{2y-1}$ **79.** $\frac{1}{2x}$ **81.** $\frac{x-4}{x+4}$ **83.** $\frac{1}{x-6}$ **85.** $\frac{2}{(x+3)(x-2)}$ **87.** $\frac{1}{2}$ **89.** 1 **91.** $x = 6$ **93.** $\frac{3}{10}$ **95.** 16.2

Chapter 7 Test

1. $x = -1, x = -3$ **2. a.** \$115 **b.** \$103 **3.** $\frac{3}{5}$ **4.** $\frac{1}{x+6}$ **5.** $\frac{1}{x^2-3x+9}$ **6.** $\frac{2m(m+2)}{m-2}$ **7.** $\frac{a+2}{a+5}$ **8.** $-\frac{1}{x+y}$ **9.** 15 **10.** $\frac{y-2}{4}$ **11.** $\frac{19x-6}{2x+5}$ **12.** $\frac{3a-4}{(a-3)(a+2)}$ **13.** $\frac{3}{x-1}$ **14.** $\frac{2(x+5)}{x(y+5)}$ **15.** $\frac{x^2+2x+35}{(x+9)(x+2)(x-5)}$ **16.** $\frac{30}{11}$ **17.** -6 **18.** no solution **19.** $-2, 5$ **20.** no solution **21.** $\frac{xz}{2y}$ **22.** $\frac{5y^2-1}{y+2}$ **23.** $b - a$ **24.** 28 **25.** $\frac{8}{9}$ **26.** 18 bulbs **27.** 5 or 1 **28.** 30 mph **29.** $6\frac{2}{3}$ hr **30.** $x = 12$

Chapter 7 Cumulative Review

1. a. $\frac{15}{x} = 4$ **b.** $12 - 3 = x$ **c.** $4x + 17 \neq 21$ **d.** $3x < 48$; Sec. 1.4, Ex. 9 **3.** amount at 7%: \$12,500; amount at 9%: \$7500; Sec. 2.8, Ex. 4 **5.**

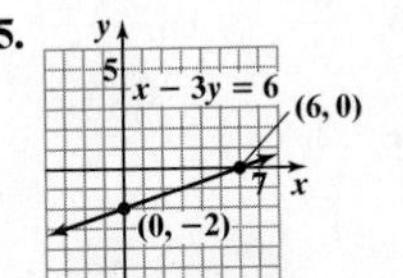

; Sec. 3.3, Ex. 6 **7. a.** 4^7 **b.** x^{10} **c.** y^4 **d.** y^{12} **e.** $(-5)^{15}$ **f.** a^2b^2; Sec. 5.1, Ex. 3 **9.** $12z + 16$; Sec. 5.2, Ex. 15

11. $27a^3 + 27a^2b + 9ab^2 + b^3$; Sec. 5.3, Ex. 8 **13. a.** $t^2 + 4t + 4$ **b.** $p^2 - 2pq + q^2$ **c.** $4x^2 + 20x + 25$ **d.** $x^4 - 14x^2y + 49y^2$; Sec. 5.4, Ex. 5 **15. a.** x^3 **b.** 81 **c.** $\frac{q^9}{p^4}$ **d.** $\frac{32}{125}$; Sec. 5.5, Ex. 2 **17.** $4x^2 - 4x + 6 + \frac{-11}{2x+3}$; Sec. 5.6, Ex. 6 **19. a.** 4 **b.** 1 **c.** 3; Sec. 6.1, Ex. 1 **21.** $-3a(3a^4 - 6a + 1)$; Sec. 6.1, Ex. 5 **23.** $3(m+2)(m-10)$; Sec. 6.2, Ex. 9 **25.** $(3x+2)(x+3)$; Sec. 6.3, Ex. 1 **27.** $(x+6)^2$; Sec. 6.3, Ex. 8 **29.** prime polynomial; Sec. 6.5, Ex. 4b **31.** $(x+2)(x^2 - 2x + 4)$; Sec. 6.5, Ex. 8 **33.** $(2x+3)(x+1)(x-1)$; Ch. 6 Int. Rev., Ex. 2 **35.** $3(2m+n)(2m-n)$; Ch. 6 Int. Rev., Ex. 3 **37.** $-\frac{1}{2}, 4$; Sec. 6.6, Ex. 6 **39.** $(1, 0), (4, 0)$; Sec. 6.6, Ex. 11 **41.** base: 6 m; height: 10 m; Sec. 6.7, Ex. 3 **43.** $\frac{5}{x^2}$; Sec. 7.1, Ex. 3 **45.** $\frac{2}{x(x+1)}$; Sec. 7.2, Ex. 6 **47.** $\frac{x+1}{x+2y}$; Sec. 7.8, Ex. 5

CHAPTER 8 ROOTS AND RADICALS

Section 8.1

Practice Exercises

1. a. $\frac{2}{9}$ **b.** -5 **c.** 12 **d.** 0.7 **e.** -1 **2. a.** 0 **b.** -4 **c.** $\frac{1}{2}$ **3. a.** 3 **b.** 10 **c.** not a real number **d.** -5 **4.** 4.123 **5. a.** x^5 **b.** y^7 **c.** $5z^3$ **d.** $7x$ **e.** $\frac{z^2}{6}$ **f.** $-2a^2b^4$

Calculator Explorations 8.1

1. 2.646; yes **3.** 3.317; yes **5.** 9.055; yes **7.** 3.420; yes **9.** 2.115; yes **11.** 1.783; yes

Vocabulary, Readiness & Video Check 8.1

1. index; radicand; radical sign **3.** false **5.** true **7.** true **9.** The radical sign, $\sqrt{\ }$, indicates a positive square root only. A negative sign before the radical sign, $-\sqrt{\ }$, indicates a negative square root. **11.** an odd-numbered index **13.** Divide the index into each exponent in the radicand—but still check by squaring your answer.

Exercise Set 8.1

1. 4 **3.** $\frac{1}{5}$ **5.** -10 **7.** not a real number **9.** -11 **11.** $\frac{3}{5}$ **13.** 30 **15.** 12 **17.** $\frac{1}{10}$ **19.** 0.5 **21.** 5 **23.** -4 **25.** -2 **27.** $\frac{1}{2}$ **29.** -5 **31.** 2 **33.** 9 **35.** not a real number **37.** $-\frac{3}{4}$ **39.** -5 **41.** 1 **43.** 2.646 **45.** 6.083 **47.** 11.662 **49.** $\sqrt{2} \approx 1.41$; 126.90 ft **51.** m **53.** x^2 **55.** $3x^4$ **57.** $9x$ **59.** ab^2 **61.** $4a^3b^2$ **63.** a^2b^6 **65.** $-2xy^9$ **67.** $\frac{x^3}{6}$ **69.** $\frac{5y}{3}$ **71.** $25 \cdot 2$ **73.** $16 \cdot 2$ or $4 \cdot 8$ **75.** $4 \cdot 7$ **77.** $9 \cdot 3$ **79.** a, b **81.** 7 mi **83.** 3.1 in. **85.** 3 **87.** 10 **89.** 4, 5 **91.** 8, 9 **93.** 6.1 sec **95.** answers may vary **97.** 1; 1.7; 2; 3

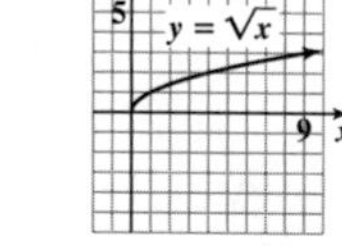

99. $|x|$ **101.** $|x+2|$ **103.** $(2, 0)$; answers may vary **105.** $(-4, 0)$; answers may vary

Section 8.2

Practice Exercises

1. a. $2\sqrt{6}$ **b.** $2\sqrt{15}$ **c.** $\sqrt{42}$ **d.** $10\sqrt{3}$ **2.** $10\sqrt{10}$ **3. a.** $\frac{\sqrt{5}}{7}$ **b.** $\frac{3}{10}$ **c.** $\frac{3\sqrt{2}}{5}$ **4. a.** $x^3\sqrt{x}$ **b.** $2a^2\sqrt{3}$ **c.** $\frac{7\sqrt{2}}{z^4}$ **d.** $\frac{y^4\sqrt{11y}}{7}$ **5. a.** $2\sqrt[3]{3}$ **b.** $\sqrt[3]{38}$ **c.** $\frac{\sqrt[3]{5}}{3}$ **d.** $\frac{\sqrt[3]{15}}{4}$ **6. a.** $2\sqrt[4]{5}$ **b.** $\frac{\sqrt[4]{5}}{3}$ **c.** $2\sqrt[5]{3}$

Vocabulary, Readiness & Video Check 8.2

1. $\sqrt{a} \cdot \sqrt{b}$ **3.** 16; 25; 4; 5; 20 **5.** Factor until you have a product of primes. A repeated prime factor means a perfect square—if more than one factor is repeated you can multiply all the repeated factors together to get one larger perfect square factor. **7.** The power must be 1. Any even power is a perfect square and can be simplified; any higher odd power is the product of an even power times the variable with a power of 1.

Exercise Set 8.2

1. $2\sqrt{5}$ **3.** $5\sqrt{2}$ **5.** $\sqrt{33}$ **7.** $7\sqrt{2}$ **9.** $2\sqrt{15}$ **11.** $6\sqrt{5}$ **13.** $2\sqrt{13}$ **15.** 15 **17.** $21\sqrt{7}$ **19.** $-15\sqrt{3}$ **21.** $\frac{2\sqrt{2}}{5}$ **23.** $\frac{3\sqrt{3}}{11}$ **25.** $\frac{3}{2}$ **27.** $\frac{5\sqrt{5}}{3}$ **29.** $\frac{\sqrt{11}}{6}$ **31.** $-\frac{\sqrt{3}}{4}$ **33.** $x^3\sqrt{x}$ **35.** $x^6\sqrt{x}$ **37.** $6a\sqrt{a}$ **39.** $4x^2\sqrt{6}$ **41.** $\frac{2\sqrt{3}}{m}$ **43.** $\frac{3\sqrt{x}}{y^5}$ **45.** $\frac{2\sqrt{22}}{x^6}$ **47.** 16 **49.** $\frac{6}{11}$ **51.** $5\sqrt{7}$ **53.** $\frac{2\sqrt{5}}{3}$ **55.** $2m^3\sqrt{6m}$ **57.** $\frac{y\sqrt{23y}}{2x^3}$ **59.** $2\sqrt[3]{3}$ **61.** $5\sqrt[3]{2}$ **63.** $\frac{\sqrt[3]{5}}{4}$ **65.** $\frac{\sqrt[3]{23}}{2}$ **67.** $\frac{\sqrt[3]{15}}{4}$ **69.** $2\sqrt[3]{10}$ **71.** $2\sqrt[4]{3}$ **73.** $\frac{\sqrt[4]{8}}{3}$ **75.** $2\sqrt[5]{3}$ **77.** $\frac{\sqrt[5]{5}}{2}$ **79.** $14x$ **81.** $2x^2 - 7x - 15$ **83.** 0 **85.** $x^3y\sqrt{y}$ **87.** $7x^2y^2\sqrt{2x}$ **89.** $-2x^2$ **91.** $2\sqrt[3]{10}$ **93.** answers may vary **95.** $2\sqrt{5}$ in. **97.** 2.25 in. **99.** $1700 **101.** 1.7 sq m

Section 8.3

Practice Exercises

1. a. $8\sqrt{2}$ **b.** $-7\sqrt{6}$ **c.** $4\sqrt[4]{5} + 11\sqrt[4]{7}$ **d.** $4\sqrt{13} - 5\sqrt[3]{13}$ **2. a.** $5\sqrt{5}$ **b.** $6 - 2\sqrt{10} + \sqrt{6}$ **c.** $-3\sqrt{2}$ **3.** $-7x\sqrt{x} + 3x$ **4.** $10x^2\sqrt[3]{3}$

Vocabulary, Readiness & Video Check 8.3

1. like radicals **3.** $17\sqrt{2}$ **5.** $2\sqrt{5}$ **7.** Both like terms and like radicals are combined using the distributive property; also only like (vs. unlike) terms can be combined, as with like radicals (same index and same radical).

Exercise Set 8.3

1. $-4\sqrt{3}$ **3.** $9\sqrt{6} - 5$ **5.** $2\sqrt{11} + 11$ **7.** $\sqrt{5} + \sqrt{2}$ **9.** $-2\sqrt[3]{18}$ **11.** $7\sqrt[3]{3} - \sqrt{3}$ **13.** $-5\sqrt[3]{2} - 6$ **15.** $5\sqrt{3}$ **17.** $9\sqrt{5}$ **19.** $4\sqrt{6} + \sqrt{5}$ **21.** $x + \sqrt{x}$ **23.** 0 **25.** $4\sqrt{x} - x\sqrt{x}$ **27.** $7\sqrt{5}$ **29.** $\sqrt{5} + \sqrt[3]{5}$ **31.** $-5 + 8\sqrt{2}$ **33.** $8 - 6\sqrt{2}$ **35.** $14\sqrt{2}$ **37.** $5\sqrt{2} + 12$ **39.** $\frac{4\sqrt{5}}{9}$ **41.** $\frac{3\sqrt{3}}{8}$ **43.** $2\sqrt{5}$ **45.** $-\sqrt{35}$ **47.** $12\sqrt{2x}$ **49.** 0 **51.** $x\sqrt{3x} + 3x\sqrt{x}$ **53.** $5\sqrt[3]{3}$ **55.** $\sqrt[3]{9}$ **57.** $-3 + 3\sqrt[3]{2}$ **59.** $4x\sqrt{2} + 2\sqrt[3]{4} + 2x$ **61.** $4x + 4x\sqrt[3]{2}$ **63.** $10y^2\sqrt[3]{y}$ **65.** $\sqrt[3]{5}$ **67.** $x^2 + 12x + 36$ **69.** $4x^2 - 4x + 1$ **71.** answers may vary **73.** $8\sqrt{5}$ in. **75.** $\left(48 + \frac{9\sqrt{3}}{2}\right)$ sq ft **77.** yes; $7\sqrt{2}$ **79.** no **81.** yes; $3\sqrt{7}$ **83.** $\frac{83x\sqrt{x}}{20}$

Section 8.4

Practice Exercises

1. a. $\sqrt{77}$ **b.** $72\sqrt{30}$ **c.** $5\sqrt{2}$ **d.** 17 **e.** $5y^2\sqrt{3}$ **2.** 28 **3.** $5\sqrt[3]{4}$ **4. a.** $3 - \sqrt{15}$ **b.** $z\sqrt{2} + 14\sqrt{z}$ **c.** $x + \sqrt{2x} - \sqrt{7x} - \sqrt{14}$ **5. a.** -9 **b.** $3x - 10\sqrt{3x} + 25$ **6. a.** $\sqrt{3}$ **b.** $2\sqrt{2}$ **c.** $3y^2$ **7.** 5 **8. a.** $\frac{4\sqrt{5}}{5}$ **b.** $\frac{\sqrt{6}}{6}$ **c.** $\frac{\sqrt{42x}}{14x}$ **9. a.** $\frac{3\sqrt[3]{5}}{5}$ **b.** $\frac{\sqrt[3]{150}}{5}$ **10. a.** $-1 + \sqrt{5}$ **b.** $\frac{5 + 3\sqrt{3}}{2}$ **c.** $\frac{8(5 + \sqrt{x})}{25 - x}$ **11.** $\frac{7 - \sqrt{7}}{3}$

Vocabulary, Readiness & Video Check 8.4

1. $\sqrt{21}$ **3.** $\sqrt{\frac{15}{3}}$ or $\sqrt{5}$ **5.** $2 - \sqrt{3}$ **7.** The square root of a positive number times the square root of the same positive number (or the square root of a positive number squared) is that positive number. **9.** To write an equivalent expression without a radical in the denominator.

Exercise Set 8.4

1. 4 **3.** $5\sqrt{2}$ **5.** 6 **7.** $2x$ **9.** 20 **11.** $36x$ **13.** $3x^3\sqrt{2}$ **15.** $4xy\sqrt{y}$ **17.** $\sqrt{30} + \sqrt{42}$ **19.** $2\sqrt{5} + 5\sqrt{2}$ **21.** $y\sqrt{7} - 14\sqrt{y}$ **23.** -33 **25.** $\sqrt{6} - \sqrt{15} + \sqrt{10} - 5$ **27.** $16 - 11\sqrt{11}$ **29.** $x - 36$ **31.** $x - 14\sqrt{x} + 49$ **33.** $6y + 2\sqrt{6y} + 1$ **35.** 4 **37.** $\sqrt{7}$ **39.** $3\sqrt{2}$ **41.** $5y^2$ **43.** $5\sqrt{3}$ **45.** $2y\sqrt{6}$ **47.** $2xy\sqrt{3y}$ **49.** $\frac{\sqrt{15}}{5}$ **51.** $\frac{7\sqrt{2}}{2}$ **53.** $\frac{\sqrt{6y}}{6y}$ **55.** $\frac{\sqrt{3x}}{x}$ **57.** $\frac{\sqrt{2}}{4}$ **59.** $\frac{\sqrt{30}}{15}$ **61.** $\frac{8y\sqrt{5}}{5}$ **63.** $\frac{\sqrt{3xy}}{6x}$ **65.** $3\sqrt{2} - 3$ **67.** $5 + \sqrt{30} + \sqrt{6} + \sqrt{5}$ **69.** $\frac{3\sqrt{x} + 12}{x - 16}$ **71.** $\frac{\sqrt{15}}{10}$ **73.** $-8 - 4\sqrt{5}$ **75.** $\frac{3\sqrt{2x}}{2}$ **77.** $\frac{10 - 5\sqrt{x}}{4 - x}$ **79.** $3 + \sqrt{3}$ **81.** $3 - 2\sqrt{5}$ **83.** $3\sqrt{3} + 1$ **85.** $2\sqrt[3]{6}$ **87.** $12\sqrt[3]{10}$ **89.** $5\sqrt[3]{3}$ **91.** 3 **93.** $2\sqrt[3]{3}$ **95.** $\frac{\sqrt[3]{10}}{2}$ **97.** $3\sqrt[3]{4}$ **99.** $\frac{\sqrt[3]{3}}{3}$ **101.** $\frac{\sqrt[3]{6}}{3}$ **103.** 44 **105.** 2 **107.** 3 **109.** $130\sqrt{3}$ sq m **111.** $\frac{\sqrt{A\pi}}{\pi}$ **113.** true **115.** false **117.** false **119.** answer may vary **121.** answer may vary **123.** $\frac{2}{\sqrt{6} - \sqrt{2} - \sqrt{3} + 1}$

Integrated Review

1. 6 **2.** $4\sqrt{3}$ **3.** x^2 **4.** $y^3\sqrt{y}$ **5.** $4x$ **6.** $3x^5\sqrt{2x}$ **7.** 2 **8.** 3 **9.** -3 **10.** not a real number **11.** $\frac{\sqrt{11}}{3}$ **12.** $\frac{\sqrt[3]{7}}{4}$ **13.** -4 **14.** -5 **15.** $\frac{3}{7}$ **16.** $\frac{1}{8}$ **17.** a^4b **18.** x^5y^{10} **19.** $5m^3$ **20.** $3n^8$ **21.** $6\sqrt{7}$ **22.** $3\sqrt{2}$ **23.** cannot be simplified **24.** $\sqrt{x} + 3x$ **25.** $\sqrt{30}$ **26.** 3 **27.** 28 **28.** 45 **29.** $\sqrt{33} + \sqrt{3}$ **30.** $3\sqrt{2} - 2\sqrt{6}$ **31.** $4y$ **32.** $3x^2\sqrt{5}$ **33.** $x - 3\sqrt{x} - 10$ **34.** $11 + 6\sqrt{2}$ **35.** 2 **36.** $\sqrt{3}$ **37.** $2x^2\sqrt{3}$ **38.** $ab^2\sqrt{15a}$ **39.** $\frac{\sqrt{6}}{6}$ **40.** $\frac{x\sqrt{5}}{10}$ **41.** $\frac{4\sqrt{6} - 4}{5}$ **42.** $\frac{\sqrt{2x} + 5\sqrt{2} + \sqrt{x} + 5}{x - 25}$

Section 8.5

Practice Exercises

1. 9 **2.** no solution **3.** $\frac{2}{3}$ **4.** 7 **5.** 10 **6.** 16

Vocabulary, Readiness & Video Check 8.5

1. The squaring property can cause extraneous solutions, so you need to check your solutions in the original equation—before the squaring property was applied—to make sure they are actually solutions.

Exercise Set 8.5

1. 81 **3.** -1 **5.** 49 **7.** no solution **9.** 4 **11.** 2 **13.** 2 **15.** 9 **17.** -3 **19.** $-1, -2$ **21.** no solution **23.** $0, -3$ **25.** 16 **27.** 25 **29.** 1 **31.** 5 **33.** -2 **35.** no solution **37.** 2 **39.** 36 **41.** no solution **43.** $\frac{3}{2}$ **45.** 16 **47.** 3 **49.** 12 **51.** 3, 1 **53.** -1 **55.** $3x - 8 = 19; x = 9$ **57.** $2(2x + x) = 24$; length $= 8$ in. **59.** 4, 7 **61.** answers may vary **63. a.** 3.2, 10, 31.6 **b.** no **65.** 7.30 **67.** 0.76

Section 8.6

Practice Exercises

1. 13 in. **2.** $2\sqrt{10}$ m ≈ 6.32 m **3.** $\sqrt{5425}$ ft ≈ 74 ft **4.** $2\sqrt{37}$ units **5.** $16\sqrt{3}$ ft/sec ≈ 27.71 ft/sec

Vocabulary, Readiness & Video Check 8.6

1. The Pythagorean theorem applies to right triangles only, and in the formula $a^2 + b^2 = c^2$, c is the length of the hypotenuse. **3.** Be careful of signs because you are subtracting.

Exercise Set 8.6

1. $\sqrt{13}$ **3.** $3\sqrt{3}$ **5.** 25 **7.** $\sqrt{22}$ **9.** $3\sqrt{17}$ **11.** $\sqrt{41}$ **13.** $4\sqrt{2}$ **15.** $3\sqrt{10}$ **17.** 51.2 ft **19.** 20.6 ft **21.** 11.7 ft **23.** $\sqrt{29}$ **25.** $\sqrt{73}$ **27.** $2\sqrt{10}$ **29.** $\frac{3\sqrt{5}}{2}$ **31.** $\sqrt{85}$ **33.** 24 cu ft **35.** 54 mph **37.** 27 mph **39.** 59.1 km **41.** 60.2 km **43.** 32 **45.** $\frac{1}{25}$ **47.** x^5 **49.** y^4 **51.** $y = 2\sqrt{10}, x = 2\sqrt{10} - 4$ **53.** 201 mi **55.** answers may vary

Section 8.7

Practice Exercises

1. a. $\sqrt{36} = 6$ **b.** $\sqrt[3]{125} = 5$ **c.** $-\sqrt[4]{\frac{1}{81}} = -\frac{1}{3}$ **d.** $\sqrt[3]{-1000} = -10$ **e.** $\sqrt[5]{32} = 2$ **2. a.** 27 **b.** 32 **c.** -5 **3. a.** $\frac{1}{25^{1/2}} = \frac{1}{5}$ **b.** $\frac{1}{1000^{2/3}} = \frac{1}{100}$ **c.** $-\sqrt{49} = -7$ **d.** $\frac{1}{1024^{2/5}} = \frac{1}{16}$ **4. a.** 36 **b.** $\frac{1}{7^{1/3}}$ **c.** z^6 **d.** a **e.** $\frac{x^{15/2}}{y^8}$

Vocabulary, Readiness & Video Check 8.7

1. $(-16)^{1/4}$ is not a real number because it is equivalent to $\sqrt[4]{-16}$, which is not a real number since it has an even index and a negative radicand. **3.** denominator; positive

Exercise Set 8.7

1. 2 **3.** 3 **5.** 8 **7.** 4 **9.** $-\frac{1}{2}$ **11.** $\frac{1}{64}$ **13.** $\frac{1}{729}$ **15.** $\frac{5}{2}$ **17.** answers may vary **19.** 2 **21.** 2 **23.** $\frac{1}{x^{2/3}}$ **25.** x^3 **27.** answers may vary **29.** 9 **31.** -2 **33.** -3 **35.** $\frac{1}{9}$ **37.** $\frac{3}{4}$ **39.** 27 **41.** 512 **43.** -4 **45.** 32 **47.** $\frac{8}{27}$ **49.** $\frac{1}{27}$ **51.** $\frac{1}{2}$ **53.** $\frac{1}{5}$ **55.** $\frac{1}{125}$ **57.** 9 **59.** $6^{1/3}$ **61.** x^6 **63.** 36 **65.** $\frac{1}{3}$ **67.** $\frac{x^{2/3}}{y^{3/2}}$ **69.** $\frac{x^{16/5}}{y^6}$ **71.** $\begin{cases} x + y < 6 \\ y \ge 2x \end{cases}$ **73.** $-1, 4$ **75.** $-\frac{1}{2}, 3$ **77.** 11,224 people **79.** 3.344 **81.** 5.665

Chapter 8 Vocabulary Check

1. like radicals **2.** index; radicand; radical **3.** conjugate **4.** principal square root **5.** rationalizing the denominator **6.** hypotenuse

Chapter 8 Review

1. 9 **3.** 3 **5.** $-\frac{3}{8}$ **7.** $\frac{2}{3}$ **9.** c **11.** x^6 **13.** $3x^3$ **15.** $\frac{4}{y^5}$ **17.** $3\sqrt{6}$ **19.** $5x\sqrt{6x}$ **21.** $3\sqrt[3]{2}$ **23.** $2\sqrt[4]{3}$ **25.** $\frac{3\sqrt{2}}{5}$ **27.** $\frac{3y\sqrt{5}}{2x^2}$ **29.** $\frac{\sqrt[4]{9}}{2}$ **31.** $\frac{\sqrt[3]{3}}{2}$ **33.** $2\sqrt[3]{3} - \sqrt[3]{2}$ **35.** $6\sqrt{6} - 2\sqrt[3]{6}$ **37.** $5\sqrt{7x} + 2\sqrt[3]{7}$ **39.** $\frac{\sqrt{5}}{6}$ **41.** 0 **43.** $30\sqrt{2}$ **45.** $6\sqrt{2} - 18$ **47.** $3\sqrt{2} - 5\sqrt{3} + 2\sqrt{6} - 10$ **49.** $x - 4\sqrt{x} + 4$ **51.** 3 **53.** $2\sqrt{5}$ **55.** $x\sqrt{15x}$ **57.** $\frac{\sqrt{22}}{11}$ **59.** $\frac{\sqrt{30}}{6}$ **61.** $\frac{\sqrt{5x}}{5x}$ **63.** $\frac{\sqrt{3x}}{x}$ **65.** $3\sqrt{5} + 6$ **67.** $\frac{\sqrt{6} + \sqrt{2} + \sqrt{3} + 1}{2}$ **69.** $\frac{10\sqrt{x} - 50}{x - 25}$ **71.** $\frac{\sqrt[3]{21}}{3}$ **73.** $\frac{\sqrt[3]{12}}{2}$ **75.** 18 **77.** 25 **79.** 12 **81.** 1 **83.** $3\sqrt{13}$; 10.82 **85.** $4\sqrt{34}$ ft; 23.32 ft **87.** $\sqrt{130}$ **89.** 2.4 in. **91.** $a^{5/2}$ **93.** $x^{5/2}$ **95.** 4 **97.** -2

99. -512 **101.** $\frac{8}{27}$ **103.** 32 **105.** $\frac{1}{3^{2/3}}$ **107.** $\frac{1}{x^2}$ **109.** 12 **111.** $4x^8$ **113.** $3x^3\sqrt{2x}$ **115.** $\frac{1}{5}$ **117.** $\frac{y^2}{9}$ **119.** $7\sqrt{3}$ **121.** $-\frac{\sqrt{3}}{4}$ **123.** $7\sqrt{2}$ **125.** $\sqrt{10}-\sqrt{2}+4\sqrt{5}-4$ **127.** $2\sqrt{6}$ **129.** $\frac{\sqrt{14}}{7}$ **131.** $\frac{3\sqrt{x}+18}{x-36}$ **133.** 1 **135.** 14 **137.** $\sqrt{58}$; 7.62 **139.** 32 ft

Chapter 8 Test

1. 4 **2.** -5 **3.** 8 **4.** $\frac{3}{4}$ **5.** not a real number **6.** $\frac{1}{9}$ **7.** $3\sqrt{6}$ **8.** $2\sqrt{23}$ **9.** $x^3\sqrt{3}$ **10.** $2x^2y^3\sqrt{2y}$ **11.** $3x^4\sqrt{x}$ **12.** 2 **13.** $2\sqrt[3]{5}$ **14.** x^5 **15.** $y^3\sqrt{y}$ **16.** $\frac{\sqrt{5}}{4}$ **17.** $\frac{y\sqrt{y}}{5}$ **18.** $\frac{\sqrt[3]{2}}{3}$ **19.** $6\sqrt{2x}$ **20.** $-2\sqrt{13}$ **21.** $-8\sqrt{3}$ **22.** $3\sqrt[3]{2}-2x\sqrt{2}$ **23.** $\frac{7\sqrt{3}}{10}$ **24.** $7\sqrt{2}$ **25.** $2\sqrt{3}-\sqrt{10}$ **26.** $x-\sqrt{x}-6$ **27.** $\sqrt{5}$ **28.** $2x\sqrt{5x}$ **29.** $\frac{\sqrt{6}}{3}$ **30.** $\frac{\sqrt[3]{15}}{3}$ **31.** $\frac{\sqrt{15}}{6x}$ **32.** $-1-\sqrt{3}$ **33.** 9 **34.** 5 **35.** 9 **36.** $4\sqrt{5}$ in. **37.** $\sqrt{5}$ **38.** $\frac{1}{16}$ **39.** $\frac{x^{10/3}}{y^2}$

Chapter 8 Cumulative Review

1. a. $-\frac{39}{5}$ **b.** 2; Sec. 1.7, Ex. 9 **3.** -3; Sec. 2.2, Ex. 5

5.

x	y
-1	-3
0	0
-3	-9; Sec. 3.1, Ex. 7

7. $y=\frac{1}{4}x-3$; Sec. 3.5, Ex. 3 **9.** $y=-3$; Sec. 3.5, Ex. 7 **11. a.** function **b.** function **c.** function **d.** not a function; Sec. 3.6, Ex. 5 **13.** yes; Sec. 4.1, Ex. 1 **15.** (4, 2); Sec. 4.2, Ex. 1 **17.** $\left(-\frac{15}{7}, -\frac{5}{7}\right)$; Sec. 4.3, Ex. 6 **19.** 5% saline solution: 2.5 L; 25% saline solution: 7.5L; Sec. 4.4, Ex. 4

21. 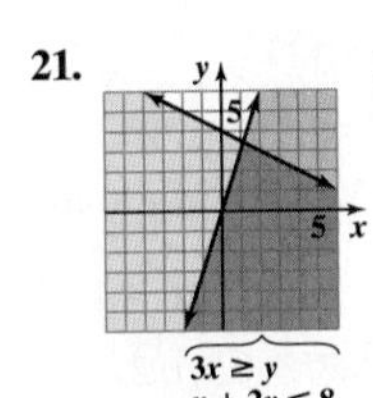

; Sec. 4.6, Ex. 1 **23.** $-9x^2+10xy-5y^2$; Sec. 5.2, Ex. 7 **25.** $(x+6y)(x+y)$; Sec. 6.2, Ex. 6 **27.** $\frac{-2-x}{3x+1}$; Sec. 7.1, Ex. 8 **29.** $\frac{3y^9}{160}$; Sec. 7.2, Ex. 4 **31.** 1; Sec. 7.3, Ex. 2 **33.** $\frac{x(3x-1)}{(x+1)^2(x-1)}$; Sec. 7.4, Ex. 7 **35.** -5; Sec. 7.5, Ex. 1 **37.** 15 yd; Sec. 7.6, Ex. 4 **39.** $\frac{3}{z}$; Sec. 7.8, Ex. 3 **41. a.** $3\sqrt{6}$ **b.** $2\sqrt{3}$ **c.** $10\sqrt{2}$ **d.** $\sqrt{35}$; Sec. 8.2, Ex. 1 **43. a.** -44 **b.** $7x+4\sqrt{7x}+4$; Sec. 8.4, Ex. 5 **45.** no solution; Sec. 8.5, Ex. 2 **47.** 10 in.; Sec. 8.6, Ex. 1 **49. a.** 8 **b.** 9 **c.** -8; Sec. 8.7, Ex. 2

CHAPTER 9 QUADRATIC EQUATIONS

Section 9.1

Practice Exercises

1. $-4, 4$ **2.** $-\frac{\sqrt{65}}{5}, \frac{\sqrt{65}}{5}$ **3.** $-1, 11$ **4.** $-2 \pm 2\sqrt{3}$ **5.** no real solution **6.** $\frac{5 \pm \sqrt{17}}{3}$ **7.** 72.8 sec

Vocabulary, Readiness & Video Check 9.1

1. To solve, *a* becomes the radicand and the square root of a negative number is not a real number.

Exercise Set 9.1

1. ± 8 **3.** $\pm\sqrt{21}$ **5.** $\pm\frac{1}{5}$ **7.** no real solution **9.** $\pm\frac{\sqrt{39}}{3}$ **11.** $\pm\frac{2\sqrt{7}}{7}$ **13.** $\pm\sqrt{2}$ **15.** $\pm\sqrt{5}$ **17.** no real solution **19.** $12, -2$ **21.** $-2 \pm \sqrt{7}$ **23.** 1, 0 **25.** $-2 \pm \sqrt{10}$ **27.** $\frac{8}{3}, -4$ **29.** no real solution **31.** $\frac{11 \pm 5\sqrt{2}}{2}$ **33.** $\frac{7 \pm 4\sqrt{2}}{3}$ **35.** $8, -3$ **37.** $\pm\sqrt{2}$ **39.** $-6 \pm 2\sqrt{6}$ **41.** $\pm\sqrt{10}$ **43.** $\frac{1 \pm \sqrt{5}}{4}$ **45.** $\pm 2\sqrt{3}$ **47.** $\frac{-8 \pm 3\sqrt{5}}{-3}$ or $\frac{8 \pm 3\sqrt{5}}{3}$ **49.** $2\sqrt{5}$ in. $\approx$ 4.47 in. **51.** 177 m **53.** 2.7 sec **55.** 15.8 sec **57.** 6 in. **59.** $(x+3)^2$ **61.** $(x-2)^2$ **63.** answers may vary **65.** $2, -6$ **67.** $-7 \pm \sqrt{31}$ **69.** ± 1.33 **71.** 2009

Section 9.2

Practice Exercises

1. $-1 \pm \sqrt{6}$ **2.** $4 \pm 2\sqrt{2}$ **3.** $\frac{13}{3}, -\frac{1}{3}$ **4.** no real solution **5.** $\frac{3 \pm \sqrt{3}}{2}$

Vocabulary, Readiness & Video Check 9.2

1. zero **3.** quadratic equation **5.** 9 **7.** In these examples we are working with equations and whatever is added to one side must also be added to the other to keep equality.

Exercise Set 9.2

1. $-6, -2$ **3.** $-1 \pm 2\sqrt{2}$ **5.** $0, 6$ **7.** $\frac{-5 \pm \sqrt{53}}{2}$ **9.** $1 \pm \sqrt{2}$ **11.** $-1, -4$ **13.** $-2, 4$ **15.** no real solution **17.** $\frac{3 \pm \sqrt{19}}{2}$ **19.** $-2 \pm \frac{\sqrt{6}}{2}$ **21.** $-3 \pm \sqrt{34}$ **23.** $\frac{3 \pm \sqrt{21}}{2}$ **25.** $\frac{1}{2}, 1$ **27.** $-6, 3$ **29.** no real solution **31.** $2, -6$ **33.** $-\frac{1}{2}$ **35.** -1 **37.** $3 + 2\sqrt{5}$ **39.** $\frac{1 - 3\sqrt{2}}{2}$ **41.** answers may vary **43. a.** $-3 \pm \sqrt{11}$ **b.** answers may vary **45.** $k = 8$ or $k = -8$ **47.** 2011 **49.** $-6, -2$ **51.** $\approx -0.68, 3.68$

Section 9.3

Practice Exercises

1. $\frac{-1 \pm \sqrt{41}}{10}$ **2.** $\frac{4}{3}, -2$ **3.** $\frac{\pm\sqrt{15}}{3}$ **4.** no real solution **5.** $\frac{5 \pm 3\sqrt{5}}{2}$ **6.** $\frac{-1 + \sqrt{41}}{10} \approx 0.5$; $\frac{-1 - \sqrt{41}}{10} \approx -0.7$ **7.** two distinct real solutions **8. a.** no real solutions **b.** two distinct real solutions

Vocabulary, Readiness & Video Check 9.3

1. $x = \frac{-b \pm \sqrt{b^2 - 4ac}}{2a}$ **3.** $-2, 1$ **5.** $\frac{-5 \pm \sqrt{17}}{2}$ **7. a.** Yes, in order to make sure you have correct values for a, b, and c. **b.** No; it simplifies calculations, but you would still get a correct answer using fraction values in the formula. **9.** radicand; real; standard

Exercise Set 9.3

1. $1, 2$ **3.** $\frac{-7 \pm \sqrt{37}}{6}$ **5.** $\pm\frac{2}{7}$ **7.** no real solution **9.** $-3, 10$ **11.** $\pm\sqrt{5}$ **13.** $-3, 4$ **15.** $-2 \pm \sqrt{7}$ **17.** $\frac{1}{2}, 3$ **19.** $\frac{5 \pm \sqrt{33}}{2}$ **21.** $-2, \frac{7}{3}$ **23.** $\frac{-9 \pm \sqrt{129}}{12}$ **25.** $\frac{4 \pm \sqrt{2}}{7}$ **27.** $3 \pm \sqrt{7}$ **29.** $\frac{3 \pm \sqrt{3}}{2}$ **31.** $-1, \frac{1}{3}$ **33.** $-\frac{3}{4}, \frac{1}{5}$ **35.** no real solution **37.** $\frac{3 \pm \sqrt{13}}{4}$ **39.** $1 \pm \sqrt{2}$ **41.** no real solution **43.** $-\frac{3}{4}, -\frac{1}{2}$ **45.** $\frac{7 \pm \sqrt{129}}{20}$ **47.** $\frac{11 \pm \sqrt{129}}{4}$ **49.** $\frac{1 \pm \sqrt{2}}{5}$ **51.** $\pm\sqrt{7}; -2.6, 2.6$ **53.** $-3 \pm 2\sqrt{2}; -5.8, -0.2$ **55.** $\frac{9 \pm \sqrt{97}}{2}; 9.4, -0.4$ **57.** $\frac{1 \pm \sqrt{7}}{3}; 1.2, -0.5$ **59.** 2 real solutions **61.** no real solution **63.** 1 real solution **65.** 2 real solutions **67.** no real solution **69.** 1 real solution **71.** $4\sqrt{3}$ **73.** $5\sqrt{2}$ **75.** base: 3 ft; height: 12 ft **77.** c **79.** b **81.** width: 3.5 ft; length: 9.9 ft **83.** $\frac{-3\sqrt{2} \pm \sqrt{38}}{2}$ **85.** answers may vary **87.** $-0.9, 0.2$ **89.** 7.9 sec **91.** 2014

Integrated Review

Practice Exercises

1. $-1, 4$ **2.** $\frac{-5 \pm 3\sqrt{5}}{2}$ **3.** $\frac{3}{2}, 1$

Exercise Set

1. $2, \frac{1}{5}$ **2.** $\frac{2}{5}, -3$ **3.** $1 \pm \sqrt{2}$ **4.** $3 \pm \sqrt{2}$ **5.** $\pm 2\sqrt{5}$ **6.** $\pm 6\sqrt{2}$ **7.** no real solution **8.** no real solution **9.** 2 **10.** 3 **11.** 3 **12.** $\frac{7}{2}$ **13.** ± 2 **14.** ± 3 **15.** $0, 1, 2$ **16.** $0, -3, -4$ **17.** $0, -5$ **18.** $\frac{8}{3}, 0$ **19.** $\frac{3 \pm \sqrt{7}}{5}$ **20.** $\frac{3 \pm \sqrt{5}}{2}$ **21.** $\frac{3}{2}, -1$ **22.** $\frac{2}{5}, -2$ **23.** $\frac{5 \pm \sqrt{105}}{20}$ **24.** $\frac{-1 \pm \sqrt{3}}{4}$ **25.** $5, \frac{7}{4}$ **26.** $1, \frac{7}{9}$ **27.** $\frac{7 \pm 3\sqrt{2}}{5}$ **28.** $\frac{5 \pm 5\sqrt{3}}{4}$ **29.** $\frac{7 \pm \sqrt{193}}{6}$ **30.** $\frac{-7 \pm \sqrt{193}}{12}$ **31.** $11, -10$ **32.** $7, -8$ **33.** $4, -\frac{2}{3}$ **34.** $2, -\frac{4}{5}$ **35.** $0.5, 0.1$ **36.** $0.3, -0.2$ **37.** $\frac{11 \pm \sqrt{41}}{20}$ **38.** $\frac{11 \pm \sqrt{41}}{40}$ **39.** $\frac{4 \pm \sqrt{10}}{2}$ **40.** $\frac{5 \pm \sqrt{185}}{4}$ **41.** answers may vary

Section 9.4

Practice Exercises

1. a. $6i$ **b.** $i\sqrt{15}$ **c.** $4i\sqrt{3}$ **2. a.** $6 + 0i$ **b.** $0 + 0i$ **c.** $2\sqrt{6} + 0i$ **d.** $0 + i$ **e.** $5 + 3i$ **3. a.** $-4 + i$ **b.** $6 - 4i$ **c.** $-1 - 2i$ **4.** $-13 + 8i$ **5. a.** $8 + 6i$ **b.** $9 - 7i$ **c.** $29 + 0i$ **6.** $\frac{1}{29} - \frac{17}{29}i$ **7.** $3 \pm 4i$ **8.** $\frac{3 \pm i\sqrt{11}}{2}$ **9.** $\frac{-1 \pm i\sqrt{15}}{2}$

Vocabulary, Readiness & Video Check 9.4

1. complex **3.** real **5.** standard **7.** combining like terms; i is *not* a variable, but a constant, $\sqrt{-1}$ **9.** To write a quotient of complex numbers in standard form $a + bi$, there should be no i's, but only real numbers in the denominator.

Exercise Set 9.4

1. $3i$ **3.** $10i$ **5.** $5i\sqrt{2}$ **7.** $3i\sqrt{7}$ **9.** $-3 + 9i$ **11.** $-12 + 6i$ **13.** $1 - 3i$ **15.** $9 - 4i$ **17.** $8 + 12i$ **19.** $5 - 50i$ **21.** $26 - 2i$ **23.** 73 **25.** $2 - 3i$ **27.** $\frac{31}{25} + \frac{17}{25}i$ **29.** $-1 \pm 3i$ **31.** $\frac{3 \pm 2i\sqrt{3}}{2}$ **33.** $-3 \pm 2i$ **35.** $\frac{-7 \pm i\sqrt{15}}{8}$

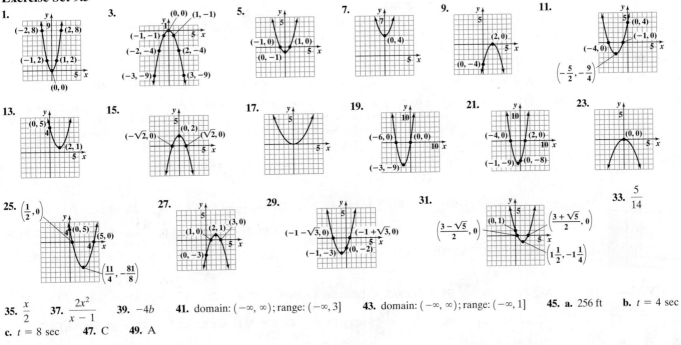

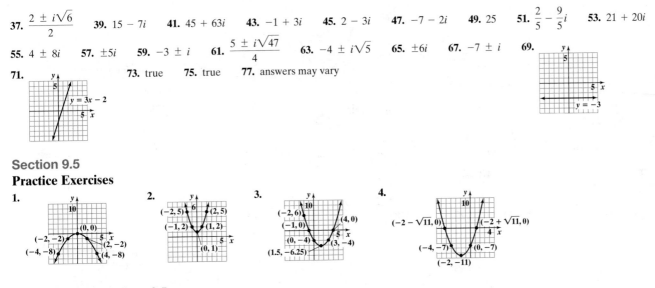

37. $\frac{2 \pm i\sqrt{6}}{2}$ **39.** $15 - 7i$ **41.** $45 + 63i$ **43.** $-1 + 3i$ **45.** $2 - 3i$ **47.** $-7 - 2i$ **49.** 25 **51.** $\frac{2}{5} - \frac{9}{5}i$ **53.** $21 + 20i$ **55.** $4 \pm 8i$ **57.** $\pm 5i$ **59.** $-3 \pm i$ **61.** $\frac{5 \pm i\sqrt{47}}{4}$ **63.** $-4 \pm i\sqrt{5}$ **65.** $\pm 6i$ **67.** $-7 \pm i$ **69.**

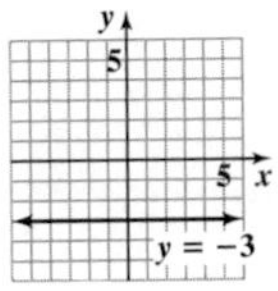

71.

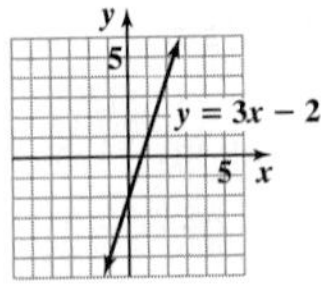

73. true **75.** true **77.** answers may vary

Section 9.5

Practice Exercises

1.

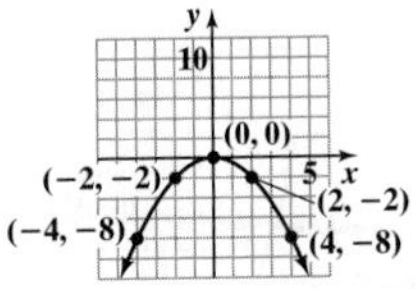

2.

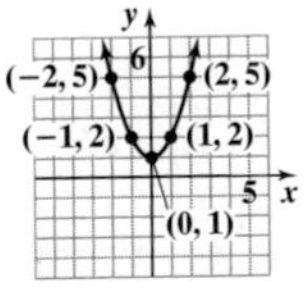

3.

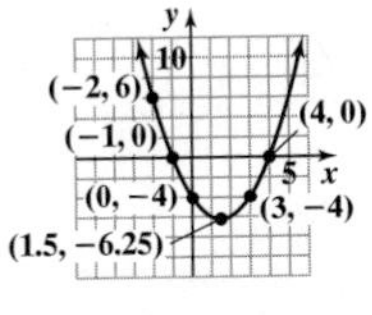

4.

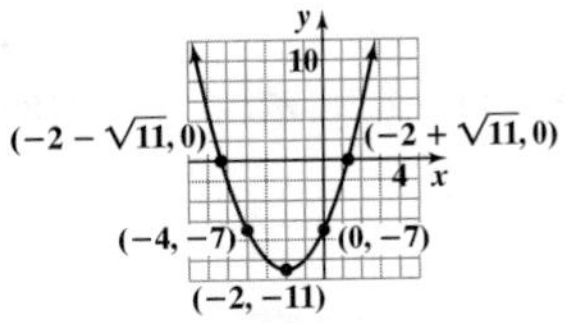

Calculator Explorations 9.5

1. $x = -0.41, 7.41$ **3.** $x = 0.91, 2.38$ **5.** $x = -0.39, 0.84$

Vocabulary, Readiness & Video Check 9.5

1. If a parabola opens upward, the lowest point is called the vertex; if a parabola opens downward, the highest point is called the vertex. If a graph can be folded along a line such that the two sides coincide or form mirror images of each other, we say the graph is symmetric with respect to that line and that line is the axis of symmetry. **3.** For example, if the vertex is in quadrant III or IV and the parabola opens downward, then there won't be any x-intercepts, and there's no need to let $y = 0$ and solve the equation for x.

Exercise Set 9.5

1.

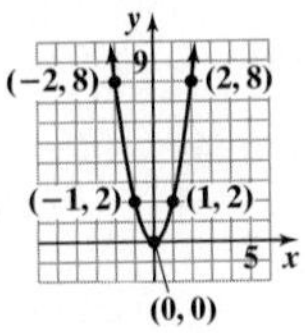

3.

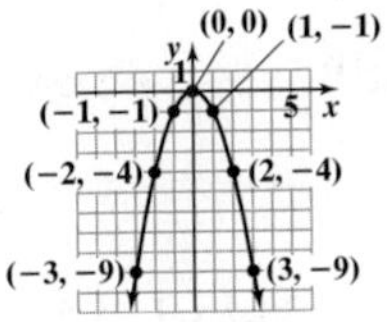

5.

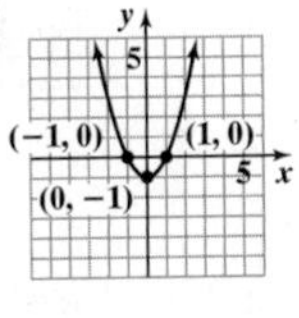

7.

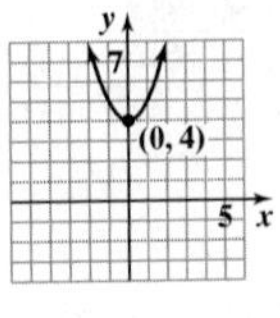

9.

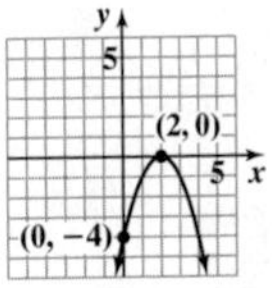

11.

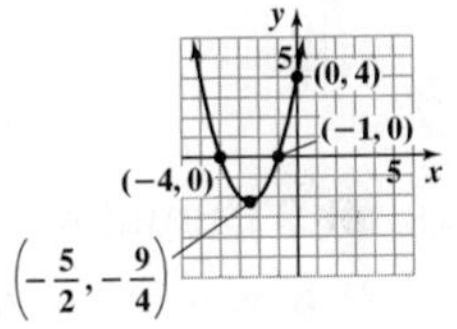

13.

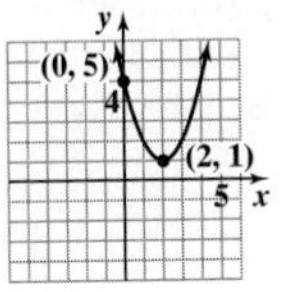

15.

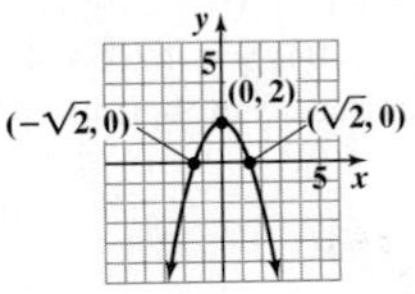

17.

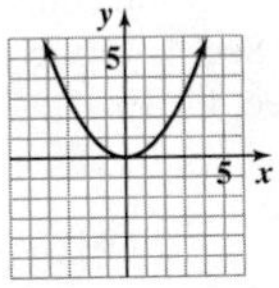

19.

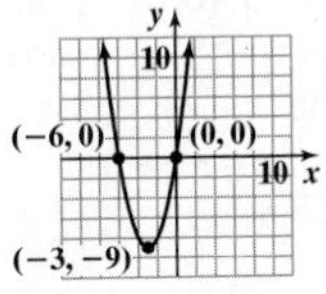

21.

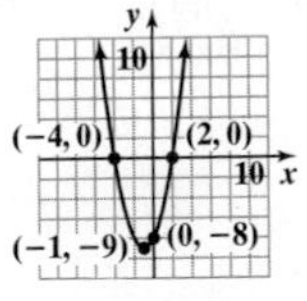

23.

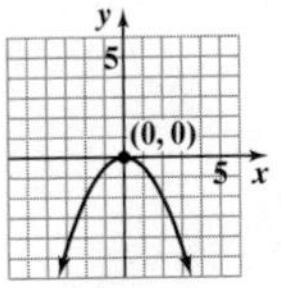

25.

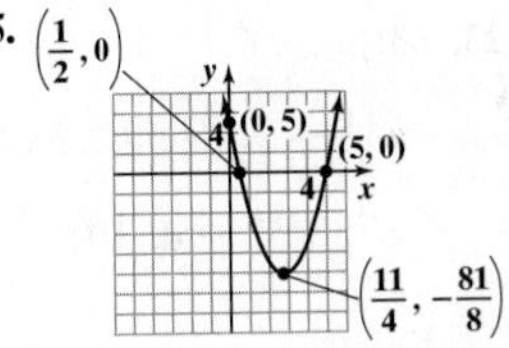

27.

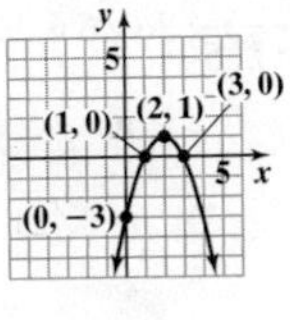

29.

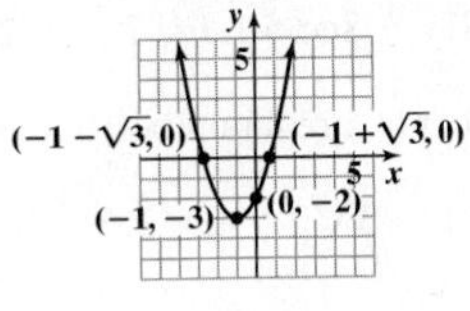

31.

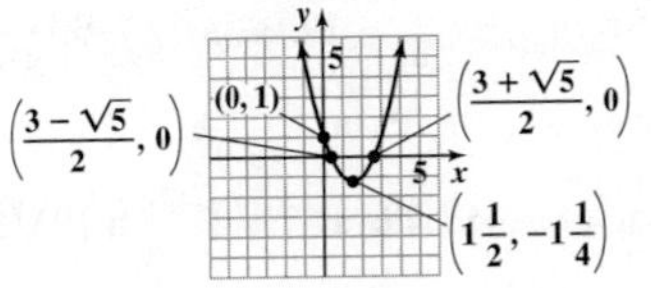

33. $\frac{5}{14}$

35. $\frac{x}{2}$ **37.** $\frac{2x^2}{x - 1}$ **39.** $-4b$ **41.** domain: $(-\infty, \infty)$; range: $(-\infty, 3]$ **43.** domain: $(-\infty, \infty)$; range: $(-\infty, 1]$ **45. a.** 256 ft **b.** $t = 4$ sec **c.** $t = 8$ sec **47.** C **49.** A

Chapter 9 Vocabulary Check

1. square root **2.** complex **3.** vertex **4.** pure imaginary **5.** conjugate **6.** i **7.** completing the square **8.** quadratic

Chapter 9 Review

1. ± 6 **3.** $\pm 5\sqrt{2}$ **5.** 4, 18 **7.** $\frac{-5 \pm \sqrt{41}}{4}$ **9.** 2.5 sec **11.** 1, 8 **13.** $-2 \pm \sqrt{5}$ **15.** $3 \pm \sqrt{2}$ **17.** $\frac{1}{2}, -1$ **19.** $-\frac{5}{3}$ **21.** $\pm\sqrt{5}$ **23.** $5 \pm 3\sqrt{2}$ **25.** $\frac{-1 \pm \sqrt{13}}{6}$ **27.** no real solution **29.** 0.4, −0.8 **31.** 2 real solutions **33.** no real solution **35.** 1 real solution **37.** 2013 **39.** $12i$ **41.** $6i\sqrt{3}$ **43.** $10 + 6i$ **45.** $21 - 10i$ **47.** $-8 - 2i$ **49.** $-13i$ **51.** 25 **53.** $-\frac{3}{2} - \frac{1}{2}i$

55. $\frac{2}{5} - \frac{9}{5}i$ **57.** $\pm 4i$ **59.** $2 \pm 3i$ **61.**

63. 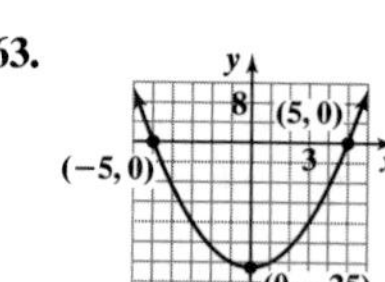

65. 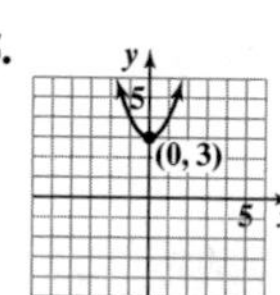

67. 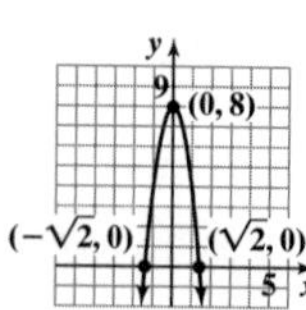

69. 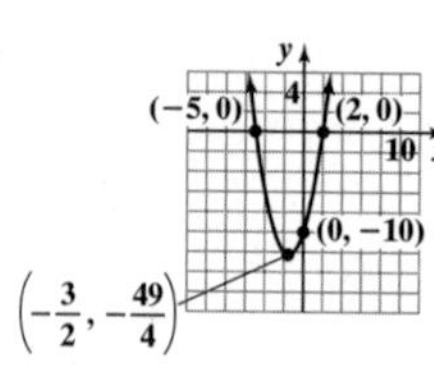

71. 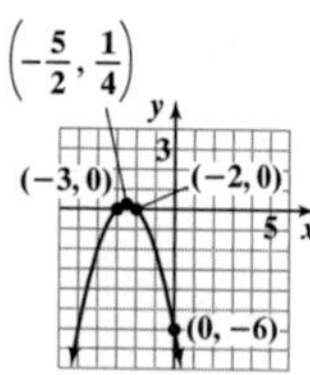

73. 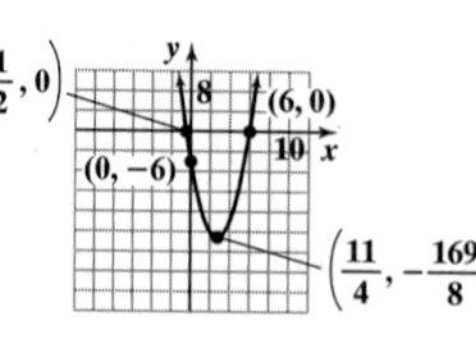

75. A **77.** B **79.** one real solution

81. no real solution **83.** ± 7 **85.** 15, −1 **87.** $\frac{2}{3}, -1$ **89.** $\frac{3 \pm \sqrt{41}}{8}$ **91.** $-\frac{3}{2}$

93. 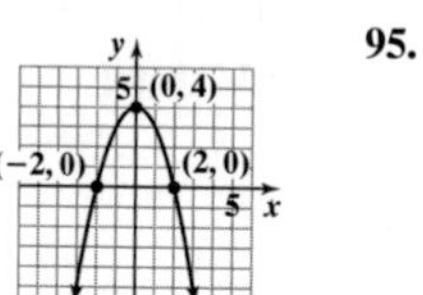

95.

Chapter 9 Test

1. ± 4 **2.** $\frac{5 \pm 2\sqrt{2}}{3}$ **3.** 10, 16 **4.** $\frac{-6 \pm 4\sqrt{3}}{3}$ **5.** −2, 5 **6.** $\frac{5 \pm \sqrt{37}}{6}$ **7.** $-\frac{4}{3}, 1$ **8.** $-1, \frac{5}{3}$ **9.** $\frac{7 \pm \sqrt{73}}{6}$ **10.** $2 \pm i$ **11.** $\frac{1}{3}, 2$ **12.** $\frac{3 \pm \sqrt{7}}{2}$ **13.** $0, \pm\frac{1}{3}$ **14.** $5i$ **15.** $10i\sqrt{2}$ **16.** $8 + i$ **17.** $0 + 4i$ **18.** $13 + 0i$ **19.** $\frac{1}{5} - \frac{7}{5}i$

20. 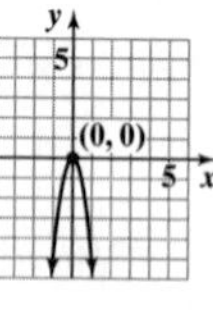

21. 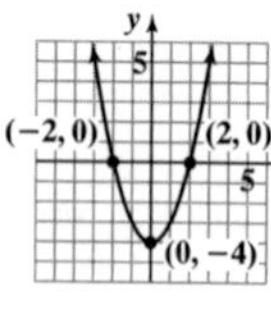

22. 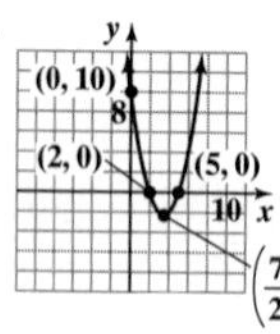

23. 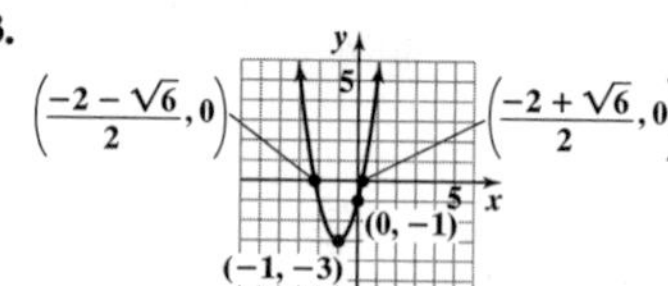

24. base: 3 ft; height: 12 ft **25.** 6 sides **26.** 2.7 sec

Chapter 9 Cumulative Review

1. a. $\frac{1}{2}$ **b.** 19; Sec. 1.6, Ex. 6 **3. a.** $5x + 7$ **b.** $-4a - 1$ **c.** $4y - 3y^2$ **d.** $7.3x - 6$ **e.** $\frac{1}{2}b$; Sec. 2.1, Ex. 4 **5. a.** x-int: $(-3, 0)$; y-int: $(0, 2)$ **b.** x-int: $(-4, 0), (-1, 0)$; y-int: $(0, 1)$ **c.** x-int and y-int: $(0, 0)$ **d.** x-int: $(2, 0)$; y-int: none **e.** x-int: $(-1, 0), (3, 0)$; y-int: $(0, -1), (0, 2)$; Sec. 3.3, Ex. 1–5 **7.** parallel; Sec. 3.4 Ex. 8a **9.** no solution; Sec. 4.1, Ex. 5 **11.** $(-2, 0)$; Sec. 4.2, Ex. 4 **13.** infinite number of solutions; Sec. 4.3, Ex. 4 **15.** Albert: 3.25 mph; Louis 4.25 mph; Sec. 4.4, Ex. 3

17. ; Sec. 4.6, Ex. 3 **19. a.** x^{11} **b.** $\frac{t^4}{16}$ **c.** $81y^{10}$; Sec. 5.1, Ex. 11 **21.** $-6, -\frac{3}{2}, \frac{1}{5}$; Sec. 6.6, Ex. 9 **23.** 63; Sec. 7.6, Ex. 1 **25. a.** 2 **b.** −2 **c.** −2 **d.** not a real number; Sec. 8.1, Ex. 3 **27. a.** $\frac{5}{6}$ **b.** $\frac{\sqrt{3}}{8}$ **c.** $\frac{2\sqrt{10}}{9}$; Sec. 8.2, Ex. 3 **29. a.** $7\sqrt{2}$ **b.** $9\sqrt{3}$ **c.** $1 - 3\sqrt{3} - 6\sqrt{2}$; Sec. 8.3, Ex. 2 **31. a.** $\sqrt{21}$ **b.** 3 **c.** $3\sqrt{5}$ **d.** $10\sqrt{6}$ **e.** $2x^2\sqrt{3}$; Sec. 8.4, Ex. 1 **33.** $\frac{1}{2}$; Sec. 8.5, Ex. 3 **35.** $\sqrt{44{,}800}$ ft $\approx$ 212 ft; Sec. 8.6, Ex. 3 **37. a.** $\sqrt{25} = 5$ **b.** $\sqrt[3]{8} = 2$ **c.** $-\sqrt[4]{16} = -2$ **d.** $\sqrt[3]{-27} = -3$ **e.** $\sqrt{\frac{1}{9}} = \frac{1}{3}$; Sec. 8.7, Ex. 1 **39.** $\frac{\sqrt{14}}{2}, -\frac{\sqrt{14}}{2}$; Sec. 9.1, Ex. 2 **41.** $5 \pm \sqrt{11}$; Sec. 9.2, Ex. 2 **43.** $-\frac{1}{2}, 5$; Sec. 9.3, Ex. 2 **45. a.** $2i$ **b.** $i\sqrt{11}$ **c.** $2i\sqrt{5}$; Sec. 9.4, Ex. 1

47. 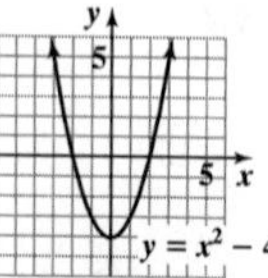

; Sec. 9.5, Ex. 2

APPENDIX A GEOMETRY

A.3 Review of Angles, Lines, and Special Triangles

1. $71°$ **3.** $19.2°$ **5.** $78\frac{3}{4}°$ **7.** $30°$ **9.** $149.8°$ **11.** $100\frac{1}{2}°$ **13.** $m\angle 1 = m\angle 5 = m\angle 7 = 110°$; $m\angle 2 = m\angle 3 = m\angle 4 = m\angle 6 = 70°$ **15.** $90°$ **17.** $90°$ **19.** $90°$ **21.** $45°, 90°$ **23.** $73°, 90°$ **25.** $50\frac{1}{4}°, 90°$ **27.** $x = 6$ **29.** $x = 4.5$ **31.** 10 **33.** 12

A.4 Review of Volume and Surface Area

1. $V = 72$ cu in.; $SA = 108$ sq in. **3.** $V = 512$ cu cm; $SA = 384$ sq cm **5.** $V = 4\pi$ cu yd $\approx 12\frac{4}{7}$ cu yd.; $SA = (2\pi\sqrt{13} + 4\pi)$ sq yd ≈ 35.20 sq yd **7.** $V = \frac{500}{3}\pi$ cu in. $\approx 523\frac{17}{21}$ cu in.; $SA = 100\pi$ sq in. $\approx 314\frac{2}{7}$ sq in. **9.** $V = 48$ cu cm; $SA = 96$ sq cm **11.** $2\frac{10}{27}$ cu in. **13.** 26 sq ft **15.** $10\frac{5}{6}$ cu in. **17.** 960 cu cm **19.** 196π sq in. **21.** $7\frac{1}{2}$ cu ft **23.** $12\frac{4}{7}$ cu cm

APPENDIX B ADDITIONAL EXERCISES ON PROPORTION AND PROPORTION APPLICATIONS

1. 5 **3.** 3.2 or $\frac{16}{5}$ **5.** 25 **7.** 0.0025 **9.** 2 **11.** 15 **13.** $\frac{30}{7}$ **15.** $\frac{17}{11}$ **17.** 31 **19.** $\frac{9}{14}$ **21.** 18 applications **23.** 5 weeks **25.** $10\frac{2}{3}$ servings **27.** 120 calories **29. a.** 18 tsp **b.** 6 tbsp **31.** 6 people **33.** 112 ft; 11-in. difference **35.** 102.9 mg **37.** 434 emergency room visits **39.** 28 workers

APPENDIX C OPERATIONS ON DECIMALS

1. 17.08 **3.** 12.804 **5.** 110.96 **7.** 2.4 **9.** 28.43 **11.** 227.5 **13.** 2.7 **15.** 49.2339 **17.** 80 **19.** 0.07612 **21.** 4.56 **23.** 648.46 **25.** 767.83 **27.** 12.062 **29.** 61.48 **31.** 7.7 **33.** 863.37 **35.** 22.579 **37.** 363.15 **39.** 7.007

APPENDIX D MEAN, MEDIAN, AND MODE

1. mean: 29, median: 28, no mode **3.** mean: 8.1, median: 8.2, mode: 8.2 **5.** mean: 0.6, median: 0.6, mode: 0.2 and 0.6 **7.** mean: 370.9, median: 313.5, no mode **9.** 1214.8 ft **11.** 1117 ft **13.** 6.8 **15.** 6.9 **17.** 85.5 **19.** 73 **21.** 70 and 71 **23.** 9 **25.** 21, 21, 24

PRACTICE FINAL EXAM

1. -81 **2.** $\frac{1}{64}$ **3.** -48 **4.** $-3x^3 + 5x^2 + 4x + 5$ **5.** $16x^2 - 16x + 4$ **6.** $3x^3 + 22x^2 + 41x + 14$ **7.** $(y - 12)(y + 4)$ **8.** $3x(3x + 1)(x + 4)$ **9.** $5(6 + x)(6 - x)$ **10.** $(3a - 7)(a + b)$ **11.** $8(y - 2)(y^2 + 2y + 4)$ **12.** $\frac{y^{14}}{x^2}$ **13.** $\frac{5y^2 - 1}{y + 2}$ **14.** $\frac{2(x + 5)}{x(y + 5)}$ **15.** $\frac{3a - 4}{(a - 3)(a + 2)}$ **16.** 8 **17.** $-7, 1$ **18.** $(-\infty, -2]$ **19.** $\frac{3 \pm \sqrt{7}}{2}$ **20.** $\frac{30}{11}$ **21.** -6 **22.** no solution **23.** 9 **24.** **25.** 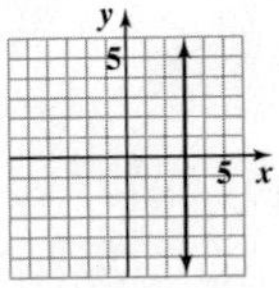 **26.** 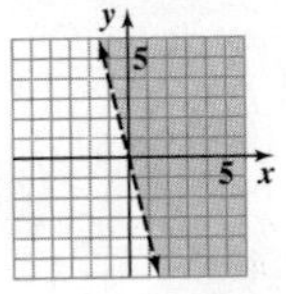 **27.** $m = -1$ **28.** $m = 3$ **29.** $8x + y = 11$ **30.** $x = -5$ **31.** $(-4, 1)$ **32.** no solution **33. a.** 0 **b.** 0 **c.** 60 **34. a.** x-intercepts: $(0, 0), (4, 0)$; y-intercept: $(0, 0)$ **b.** Domain: $(-\infty, \infty)$; Range: $(-\infty, 4]$ **35.** 4 **36.** $\frac{1}{9}$ **37.** $\frac{3}{4}$ **38.** $3\sqrt{6}$ **39.** $3x^4\sqrt{x}$ **40.** $-8\sqrt{3}$ **41.** $2x\sqrt{5x}$ **42.** $2\sqrt{3} - \sqrt{10}$ **43.** $\frac{\sqrt{15}}{6x}$ **44.** $-1 - \sqrt{3}$ **45.** 5 or 1 **46.** Rhode Island: 401; Vermont: 802 **47.** $2\frac{1}{2}$ hr **48.** 120 cc

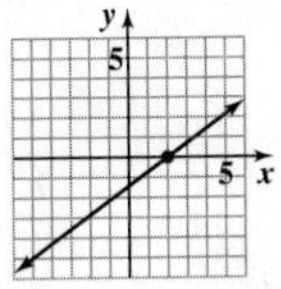

Index

Photo Credits

Chapter 1

Page 5 Stylephotographs/Dreamstime
Page 9 Martin Fischer/Shutterstock
Page 15 Thinkstock
Page 28 Digital Vision/Thinkstock
Page 39 Dorling Kindersley, Inc.
Page 41 Endless Traveller/Shutterstock
Page 41 ALCE/Fotolia
Page 49 Peter Zaharov/Shutterstock
Page 49 Rjlerich/Dreamstime
Page 57 Lunamarina/Fotolia

Chapter 2

Page 89 Georgiy/Fotolia
Page 89 Fitter85/Dreamstime
Page 91 Dmitryp/Dreamstime
Page 91 Dmitryp/Dreamstime
Page 92 Jupiterimages/Thinkstock
Page 100 Wavebreakmedia, Ltd./Shutterstock
Page 111 Editorial Image, LLC/Alamy
Page 112 Kristoffer Tripplaar/Alamy
Page 113 Bpperry/Dreamstime
Page 114 Yuri Arcurs/Fotolia
Page 118 Eastimages/Shutterstock
Page 118 Evgeny Prokofyev/Shutterstock
Page 119 Nmaverick/Fotolia
Page 121 Diana Jo Currier/Shutterstock
Page 128 Michael Dwyer/Alamy
Page 129 Bruce MacQueen/Shutterstock
Page 130 Guy Cali/Glow Images
Page 131 WaterFrame/Alamy
Page 131 Gracious Tiger/Shutterstock
Page 135 Digital Vision/Thinkstock
Page 136 300dpi/Shutterstock
Page 136 Deklofenak/Dreamstime
Page 141 IlFede/Shutterstock
Page 141 Kuzma/Shutterstock
Page 158 Deepfrog17/Dreamstime
Page 168 Amalia Ferreira-Espinoza/Shutterstock

Chapter 3

Page 173 Interlight/Shutterstock
Page 174 Stephen Coburn/Shutterstock

Page 182 Pcruciatti/Dreamstime
Page 185 Franckgrondin/Dreamstime
Page 185 Einstein/Shutterstock
Page 194 Alexander Raths/Shutterstock
Page 197 Stockbyte/Thinkstock
Page 197 Kenneystudios/Dreamstime
Page 197 Petdcat/Dreamstime
Page 206 Corbis/Fotolia
Page 215 Georgeburba/Dreamstime
Page 221 Still Representation/Fotolia
Page 221 Uzi Tzur/Fotolia
Page 222 Ralukatudor/Dreamstime
Page 226 GoodMood Photo/Fotolia
Page 230 Fabinus08/Dreamstime
Page 230 Fotovampir/Dreamstime
Page 247 Cpbackpacker/Shutterstock

Chapter 4

Page 254 Wavebreakmedia/Shutterstock
Page 270 Comstock/Thinkstock
Page 276 Wavebreakmedia/Shutterstock
Page 277 Kzenon/Fotolia
Page 279 Paha_I/Dreamstime
Page 284 Smontgom65/Dreamstime
Page 285 Olaf Loose/iStockphoto
Page 285 Ekaterinasemenova/Dreamstime
Page 304 DCW Creations/Shutterstock

Chapter 5

Page 349 Astrofireball/Dreamstime
Page 349 Imredesiuk/Dreamstime
Page 361 Daivaluk/Dreamstime
Page 362 Ablestock/Thinkstock

Chapter 6

Page 366 CandyBoxPhoto/Shutterstock
Page 401 Tondafoto/Dreamstime
Page 402 Chin-Cheng Liu/Shutterstock
Page 402 Paha_l/Dreamstime
Page 415 Zensu/Dreamstime

Chapter 7

Page 432 Jerry Sharp/Shutterstock
Page 432 Aspenphoto/Dreamstime
Page 441 Wavebreakmedia, Ltd./Shutterstock
Page 447 Christopher Parypa/Shutterstock

Page 448 Testing/Shutterstock
Page 478 Digital Vision/Thinkstock
Page 486 James Steidl/Shutterstock
Page 495 Sashkin/Shutterstock
Page 495 Budda/Dreamstime
Page 509 James Woodson/Thinkstock

Chapter 8

Page 512 Jose Ignacio Soto/Fotolia
Page 519 Marco Cannizzaro/Shutterstock
Page 526 Cozyta/Dreamstime
Page 527 Pictorial Press Ltd/Alamy
Page 527 Stephen Coburn/Shutterstock
Page 548 Forgiss/Dreamstime

Chapter 9

Page 564 Keith Brofsky/Thinkstock
Page 568 Jgroup/Dreamstime
Page 569 Danilo Tee/Shutterstock
Page 569 Danilo Tee/Shutterstock
Page 569 Daivaluk/Dreamstime
Page 582 Keith Brofsky/Thinkstock
Page 582 Jewish Hospital/ZUMA Press/Newscom
Page 601 Gary Whitton/Fotolia